PENGUIN REFERENCE

The Penguin Dictionary of

Michael Thain was born on Hampstead Heath during Halloween of 1946 and was fortunate enough to be sent to University College School, where access to the microscope cabinet during lunch breaks enabled him to spend many formative hours examining pond life. This led to life-long interests in natural history and biology. Major influences at this time, and subsequently, were Alister Hardy's *The Open Sea* and *The Invertebrata* (Borradaile, et. al.). Membership of the Royal Society for the Protection of Birds provided him access to bird conferences and bird-ringing courses, while discovery of Ernst Mayr's *Animal Species and Evolution* in a laboratory cupboard gave life new meaning. Oxford University's undergraduate Zoology course encouraged more detailed study of genetics and evolution, deepened his interest in birds, and introduced him to the biology of plants. His fourth year at Keble was spent largely in the Department of Human Anatomy (and the Cross Keys at South Hinksey), studying human cell biology and evolution, which have also remained enduring interests.

He returned to London in 1969, where the able and avuncular Head Master of Harrow School, Robert (Jimmy) James, offered him a teaching post. Remaining on the Harrow teaching staff until 2003, he developed interests in philosophy (initiated at Oxford) and in the histories of science and technology, leading to degrees from Birkbeck and Imperial Colleges, London. A memorable meeting in 1985 at Penguin Books in London brought a commission to co-author the 8th edition of the *Dictionary of Biology*, the 12th edition of which is in preparation. A commission to write the *Dictionary of Human Biology* came in 1999. Since leaving Harrow he has taught privately and at Davies, Laing & Dick College, Queen's Gate School for Girls, and Winchester College.

The Penguin Dictionary of

HUMAN BIOLOGY

Michael Thain

PENGUIN BOOKS

For my mother Jeanne, and in memory of my father, Jack

PENGUIN BOOKS

Published by the Penguin Group
Penguin Books Ltd, 80 Strand, London WC2R 0RL, England
Penguin Group (USA) Inc., 375 Hudson Street, New York, New York 10014, USA
Penguin Group (Canada), 90 Eglinton Avenue East, Suite 700, Toronto, Ontario, Canada M4P 2Y3
(a division of Pearson Penguin Canada Inc.)
Penguin Ireland, 25 St Stephen's Green, Dublin 2, Ireland (a division of Penguin Books Ltd)
Penguin Group (Australia), 250 Camberwell Road, Camberwell, Victoria 3124, Australia
(a division of Pearson Australia Group Pty Ltd)
Penguin Books India Pvt Ltd, 11 Community Centre, Panchsheel Park, New Delhi – 110 017, India
Penguin Group (NZ), 67 Apollo Drive, Rosedale, North Shore 0632, New Zealand
(a division of Pearson New Zealand Ltd)
Penguin Books (South Africa) (Pty) Ltd, 24 Sturdee Avenue, Rosebank, Johannesburg 2196, South Africa

Penguin Books Ltd, Registered Offices: 80 Strand, London WC2R 0RL, England

www.penguin.com

First published 2009
1

Set in ITC Stone Sans and ITC Stone Serif
Typeset by Data Standards Ltd, Frome, Somerset
Printed in England by Clays Ltd, St Ives plc

ISBN: 978-0-140-51482-7

www.greenpenguin.co.uk

Contents

Preface

In preparing this work I had to decide how many of the headwords included in the *Dictionary of Biology* I would also include here. I decided that the two dictionaries would read better as a pair, and have therefore tended to include entries present in *Biology* only if there was something that made them especially biologically important and interesting in the human context. The MITOCHONDRION entries in the two dictionaries, for instance, are doing rather different jobs. I hope that in this and similar cases the pair of entries act synergistically. In most entries, I have assumed I am writing about the eukaryote *Homo sapiens*, and have tried to indicate where a different context is intended. I am only too aware how very eclectic I have been in my choice of headwords. The result is akin to an anthology of entries and does not attempt to be exhaustive. Rather, as with *Biology*, the attempt is to say something interesting about the headwords that have been included. The result is for entries to resemble more those in an encyclopaedia than those in a typical dictionary.

Ideally, more entries would have been included from the fields of primatology, and physical and social anthropology. It is hoped that a subsequent edition will rectify this. Fields such as GENOMICS and PROTEOMICS are advancing so rapidly that it is difficult to do justice to them; but I hope that what is included at least sets their stalls out helpfully. Indeed, my hope is that readers may turn to this dictionary as a first port of call when seeking the definition and biological context of any term relating to the human body. It is a pleasure to acknowledge the help and support of many individuals during the preparation of this book. Those with, or formerly with, Penguin Books include Martin Toseland, Kristen Harrison, Ruth Stimson and Adam Freudenheim. To my copy editor, Pat Croucher, I owe an enormous debt of gratitude for unstinting professionalism. The untimely death of the first of my two designers, David Graham was a tragedy for his wife and family, to whom I formally register my condolences. David's work was admirably sustained, at unavoidably short notice, by Jeff Edwards. Two friends, Drs Matt Dryden and Bill Richmond, took on the task of checking the manuscript for errors. Any that remain are entirely my fault. It goes without saying that I should be most grateful to readers for any suggestions that would improve the dictionary.

MT
Northwood, 2008

THE PENGUIN DICTIONARY OF HUMAN BIOLOGY

α For these entries, see under the headword prefixed by α-.

ABC transporter family (ATP-binding cassette family) The largest family of transmembrane ION PUMPS. They are multidrug transporters, each containing two conserved ATP-binding cassette domains on their cytosolic surface. Found in all cells of all organisms, they have two transmembrane domains with six transmembrane helices. In humans, they are involved mostly in active transport of a variety of substrates, including inorganic ions, sugars, amino acids, peptides, vitamins, steroids, and even polysaccharides and proteins, the transporter binding and hydrolysing ATP, conformational changes then 'flipping' the substrate across the membrane. Some can 'flip' lipids from one layer of the membrane to the other while others transport hydrophobic drugs out of the cytosol, leading to resistance to a wide variety of cytotoxic drugs, including ANTIBIOTICS (see BLOOD–BRAIN BARRIER, CANCER, MALARIA, MULTIDRUG RESISTANCE). The CFTR protein is a family member (see CYSTIC FIBROSIS and ION CHANNELS). ABC transporters have a key role in delivering CHOLESTEROL and phospholipids to apolipoproteins (see LIPOPROTEINS).

abdomen The region lying between the diaphragm and the pelvis. The abdominal cavity (see BODY CAVITIES, Fig. 23) is the superior portion of the abdominopelvic cavity and houses the stomach, spleen, liver, gall bladder, pancreas, small intestine and most of the large intestine.

abdominal cavity See ABDOMEN.

abducens nerve The sixth CRANIAL NERVE. Its nucleus lies in the pons. A mixed nerve, it conveys proprioceptive information from the lateral rectus muscle (an extrinsic eyeball muscle) to the pons, while its motor origins lie in the pons and innervate the same muscle.

abduction A movement away from the midline, or axis, of the body.

A beta (Aβ) fibres Thick myelinated fibres (see AXON) with cell bodies in the segmental dorsal root ganglia, conveying mechanosensory (tactile) information; branches terminate in the deep dorsal horn of the grey matter of the spinal cord. See A DELTA FIBRES, C FIBRES.

***abl* gene** A proto-oncogene originally discovered through its presence in the Abelson murine leukaemia virus, a rapidly tumorigenic retrovirus. The gene maps to human chromosome 9q34 but is found fused to sequences clustered in a narrow region of chromosome 22q11 known as the breakpoint cluster region, *bcr*. The resulting fusion of Abl with Bcr amino acid sequences deregulates the normally well-controlled Abl protein and causes it to emit growth-promoting signals in a strong and deregulated manner. Imatinib mesylate (Gleevec) is a potent BCR-ABL tyrosine kinase inhibitor. See LEUKAEMIAS.

abiotic environment The non-biological component of an organism's external environment, including climate, aspect, photoperiod, atmosphere, inorganic nutrients, etc.

ABO blood group system The four major

BLOOD GROUPS in this system, A, B, AB and O, are determined by a locus on chromosome 9q. Three alleles, I^A, I^B and *i*, determine the presence of the ABO blood group antigens on the surfaces of an individual's red blood cells, the *i* allele resulting in no antigenic protein of this type and having a recessive expression to both the other alleles, whose expressions are codominant. People with an antigen of the ABO system on their red blood cells automatically lack the antibody to it in their plasma, whereas lack of an ABO antigen is coupled to presence of the antibody to it in the plasma. They are important in blood transfusion (see GRAFTS). See SECRETOR.

absolute refractory period The time taken for inactivated voltage-gated sodium channels to return to their closed conformation in an axon or muscle fibre membrane. They cannot be activated again until the membrane potential goes sufficiently negative for the channels to be de-inactivated. It is responsible for the unidirectionality of propagated ACTION POTENTIALS, and in part for the upper limit to the frequency of impulses a membrane can support. See REFRACTORY PERIOD, CARDIAC MUSCLE; compare RELATIVE REFRACTORY PERIOD.

abstinence syndrome See WITHDRAWAL SYNDROME.

abstraction In general, the subordination of the particular to the general, and a critical step in the acquisition of knowledge. It has been suggested that without this ability, our brains would need to recall every detail, a memory demand it cannot satisfy. The neural correlate for abstraction seems to be CONVERGENCE of cell connections, as found in the RETINA. Light-sensitive receptor cells feed information to bipolar cells, which feed ganglion cells, which are then surveyed by cells in the LATERAL GENICULATE NUCLEUS, a group of which converge onto a stellate cell in layer 4Cβ of the PRIMARY VISUAL CORTEX, a group of which converges onto a simple cell, a group of which in turn converges onto a complex cell. Inevitably, some information is lost by this processing. It is of interest that there appear to be some cells in the visual cortex which are able to *recognize* objects in a view-invariant manner. Sensation is an abstraction of the real world. The sensory systems and cerebral cortex construct an internal representation of external physical events, its component parts being first separately analysed and then reassembled according to 'rules' implicit in the properties of the nervous system (see NEURAL CODING). Somehow, a unified conscious perception is the result. Recent work on mice and rats on the manner in which the neural-clique-based hierarchical organization of the HIPPOCAMPUS is involved in encoding memory also suggests how the brain might achieve abstraction and generalization. See BINDING PROBLEM, BISTABILITY, CATEGORIZATION, COGNITION, GENERALIZATION, PARALLEL PROCESSING.

α-bungarotoxin A snake venom toxin which, like cobrotoxin and tubocurarine (an active component of curare), blocks nicotinic acetylcholine receptors at neuromuscular junctions and/or prevents their ion channels from opening.

abzymes ANTIBODIES with enzyme-like actions.

acapnia Condition in which lower-than-normal levels of carbon dioxide are present in the blood and tissues.

ACAT See ACYL-COA-CHOLESTEROL ACYL TRANSFERASE.

accessory ligaments Extracapsular ligaments and intracapsular ligaments lying outside and inside the articular capsule of a SYNOVIAL JOINT in addition to the ligaments of the fibrous capsule itself.

accessory molecules Leukocyte surface antigens, of which the CD system of nomenclature (see CD ANTIGENS) recognizes the following major kinds: CD4 characterizes T_H1 and T_H2 lymphocytes (but also occurs on dendritic cells and macrophages), binds to MHC class II molecules on the antigen-presenting cell (APC) and acts as a co-receptor to enhance the signal received through the T-CELL RECEPTOR; CD8 characterizes cytotoxic T cells (T_C cells), binds to MHC class I molecules on the APC and acts as a co-receptor to enhance the signal received at the T-cell receptor; CD40, produced by B cells, among others, interacts with the CD40 ligand (CD40L) on activated cells, a requirement if T cells are to activate B cells and initiate their isotype switching.

accessory muscle (accessorius m.) A muscle aiding the action of another.

accessory nerve Cranial nerve XI. A mixed nerve (mainly motor). Its sensory input derives from proprioceptors in muscles of the pharynx, larynx and soft palate. Its motor output leaves via two routes: cranially, from nuclei originating in the medulla and innervating muscles of the pharynx, larynx and soft palate and involved in the control of swallowing; and spinally, originating in the anterior grey horn of the first five cervical segments and innervating muscles (including the trapezius) involved in raising the shoulders and turning the head.

accessory organs Organs whose embryological origins distinguish them from the organ system which they serve. Thus, the salivary glands, liver and pancreas are accessory organs of the gastrointestinal tract, although they are usually included with them in the digestive system.

accessory proteins Any proteins that are necessary for the assembly, spatial distribution, or maintenance, of a functional protein-based system. Within sarcomeres, α-actinin which binds to the plus ends of actin filaments and forms the Z-disc, and nebulin which acts as a 'molecular ruler' in determining actin filament length are accessory proteins, as are the numerous cytoskeletal proteins, including motor proteins, which control actin's assembly and functioning.

accessory receptors T-cell-surface proteins that help stabilize the interaction between a T-cell receptor and MHC molecule. If they play a direct role in activating the T cell they are termed CO-RECEPTORS.

acclimatization Naturally induced physiological compensation to changes in external environment, e.g., increased ALTITUDE, leading to TOLERANCE.

accommodation **1**. The process of focusing resulting from deformation of the lens. Although the CORNEA achieves most of the refraction of light entering the EYES (see Fig. 73), the lens contributes a further dozen or so dioptres. On viewing a close object, the ciliary muscles contract, pulling the ciliary process and choroid forward toward the lens, tension is reduced in the suspensory ligament attaching the lens to the ciliary body and the lens becomes more biconvex through natural elasticity, decreasing its focal length and focusing the image on the fovea. On viewing more distant objects, the ciliary muscles relax and the focal length of the lens increases, refocusing the image. The ability to accommodate changes with age, and artificial lenses can compensate for this. Refraction abnormalities include nearsightedness, farsightedness and astigmatism, most of which can be corrected by eyeglasses or contact lenses. **2**. A time-dependent decrease in excitability of a sensory neuron, and of excitable membranes in general, causing the threshold potential to increase leading to conduction of action potentials at a slower rate despite an unchanged level of receptor stimulus. If stimulated by a series of subthreshold depolarizations, then the slower the rate of increase in the intensity of the stimulating current, the greater the rise in threshold potential. See PHASIC RECEPTORS.

acentric chromosome A chromosome lacking a centromere, as may arise when the wrong broken ends of chromosomes are joined together, e.g., during TRANSLOCATIONS. They are not stable during nuclear division.

acetabulum The rounded cavity on the external surface of the hip bone, receiving the head of the femur. See PELVIC GIRDLE, Fig. 132.

acetaldehyde (ethanal) See ALCOHOL DEHYDROGENASE.

acetoacetate (acetoacetic acid) The product of condensation of two acetyl CoA molecules, which liberates the CoA molecules. It is produced by liver cells (hepatocytes) as a part of normal fatty acid metabolism, other cells taking it up and using it to form two acetyl CoA molecules for entry into the KREBS CYCLE. See KETONE BODIES.

acetone See KETONE BODIES.

acetylcholine (ACh) A NEUROTRANSMITTER of many interneural, all neuromuscular, and other cholinergic effector synapses (e.g., those of the vagus nerve) where it relays electrical signals in chemical form across the synaptic cleft (see Fig. 1). Formed in mitochondria, it initiates depolarization of postsynaptic membrane receptors to which it binds and is degraded by ACETYLCHOLINESTERASE.

ACh receptors (AChRs) are Cys-loop receptor ION CHANNELS that seem to have evolved with (animal) multicellularity, when cell–cell interactions such as those in the nervous system became important. Their neurotransmitter binding site is located on the N terminus and in some cases may have been co-opted from elsewhere. AChRs permit flux of a variety of cations, but exclude anions. There are two kinds of ACh receptor: nicotinic and muscarinic receptors.

Nicotinic receptors for ACh are IONOTROPIC and comprise two identical α and single β, γ and δ subunits, each having four transmembrane helical segments. They are located on neuron cell bodies, notably on postganglionic neurons of both the parasympathetic and sympathetic divisions of the autonomic nervous system, at neuromuscular junctions, and on macrophages (see NICOTINE). Drugs that bind to and block nicotinic receptors are termed ganglionic blocking agents and have greater effect on the sympathetic ganglions. Trimethaphan camsylate is used to treat high BLOOD PRESSURE by blocking sympathetic stimulation of blood vessels, causing blood vessels to dilate (e.g., see PENIS). An anaesthetist can relax skeletal muscles by

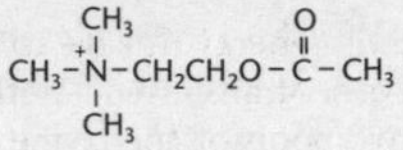

$$CH_3-\overset{+}{N}(CH_3)_2-CH_2CH_2O-\overset{O}{\overset{\|}{C}}-CH_3$$

Acetylcholine (ACh)

FIG 1. The structural formula of ACETYLCHOLINE.

administering small quantities of curare, leaving heart muscle unaffected. See α-BUNGAROTOXIN.

Muscarinic receptors are METABOTROPIC and occur on cardiac muscle fibres, in the brain and on cells of the ENDOTHELIUM, being blocked and antagonized by atropine. Such drugs dilate the pupil of the eye and are used during eye examinations to reveal the retina. They are also used during surgery to reduce salivary secretion and prevent choking during anaesthesia. See MYASTHENIA GRAVIS.

acetylcholinesterase (AChE) An extremely fast-acting enzyme which degrades acetylcholine to acetic acid and choline, as in the cleft of NEUROMUSCULAR JUNCTIONS. Secreted into the synaptic cleft by presynaptic membranes of cholinergic (and some other) neurons. Much of the resulting choline is taken up by the presynaptic axon and reused. AChE is the target of many nerve gases and insecticides that therefore prevent ACh breakdown at cholinergic synapses in skeletal and cardiac muscle, making muscle relaxation impossible.

acetyl-CoA carboxylase The enzyme converting ACETYL COENZYME A to malonyl-CoA, the building block for long-chain fatty acid synthesis by liver cells (see FATTY ACIDS, Fig. 75).

acetyl coenzyme A, acetyl-CoA A thioester product of fatty acid β-oxidation (see FATTY ACID METABOLISM), catabolism of ten amino acids entering the KREBS CYCLE directly, and the decarboxylation of pyruvate in mitochondria (where it is a positive allosteric modulator of PYRUVATE CARBOXYLASE). It is used to add two carbon units in the biosynthesis of larger molecules, e.g., FATTY ACIDS, and in the production of KETONE BODIES by the liver. The acetyl group is carried in high-energy linkage to CoA and can readily be transferred. Within cells there are two pools of coenzyme A (CoA), one in the cytosol and another in the mitochondrial matrix. The pool in the matrix is mainly used in oxidative breakdown of pyruvate, fatty acids and certain amino acids; that in the cytosol is required in the synthesis of fatty acids. See Appendix I for structural formula.

acetylglucosamine See O-LINKED B-N-ACETYLGLUCOSAMINE.

acetyltransferases See HISTONE ACETYLTRANSFERASES, N-ACETYLTRANSFERASES, O-LINKED B-N-ACETYLGLUCOSAMINE.

Acheulian culture One of the stone tool industries and traditions of the Lower Palaeolithic (Lower and Middle Pleistocene), characterized by a large proportion of bifacial tools (flaked on both sides) and cleavers. Commonly found in Africa, Southwest Asia and Western Europe, but nearly absent elsewhere, the earliest material comes from Ethiopia, ~1.4 Mya. The earliest known HOMO ERGASTER fossils appear ~2 Mya and either later populations of this species invented this technology after employing the simpler OLDOWAN CULTURE, or the material may have been an innovation of HOMO ERECTUS. Its bifacial handaxes are symmetrical with a tear-drop shape, sharpened all round, and made of flint or quartz. It is not certain whether some of the material was used for throwing, cutting or scraping. As a technological innovation it endured ~1 Myr without being superseded and without the appearance of spearheads, arrow points, daggers or needles. It is almost impossible to improve on the bifacial handaxe as a tool for butchering large game in the absence of steel, being perfected by the careful use of 'soft hammers' of bone – apparently used once or a few times then discarded. See MOUSTERIAN CULTURE.

Achilles' tendon (calcaneal tendon) Tendon connecting the soleus, gastrocnemius and plantaris muscles to the back of the heel. The most important mechanical spring involved in running, comparatively longer than in chimpanzees and other modern apes (see Appendices VIII, Xa and Xb, and GAIT). Evidence from early australopithecines (Hadar specimens) suggests that the transverse groove into which the tendon inserts on the posterior surface of the calcaneus is chimpanzee-like in size, contrasting with the wider and taller area of *H. sapiens*, suggesting that a developed Achilles tendon was absent in *Australopithecus* and probably originated in the genus *Homo*.

achondroplasia (ACH) The most common form of dwarfism (1:26,000 live births), a DYSPLASIA primarily affecting the long bones although other skeletal defects include a large skull, small mid-face, short fingers and accentuated spinal curvature. It is due to a dominant mutation in the *FGFR3* gene (chromosome 4p16; see FIBROBLAST GROWTH FACTORS), seems to arise in the male germ line (80% spontaneously), and is an example of an age-dependent 'hot-spot' mutation (see GENETIC DISEASES AND DISORDERS). The rare instances of homozygosity do not usually survive to term. See ACROCEPHALOSYNDACTYLY.

achromatopsia A rare clinical disorder in which there is partial or complete loss of colour vision despite the presence of normally functioning cone cells. Afflicted persons describe the world as consisting only of shades of grey and often also have difficulties with the perception of form. It is associated with damage to the occipital and temporal lobes (some would say to a human area V4) if there is no damage to area V1, the lateral geniculate nuclei or to the retina, and indicates the existence of a specialized colour-processing pathway in the ventral visual stream (see COLOUR PERCEPTION, EXTRASTRIATE CORTEX).

acid–base balance Buffers almost immediately resist changes in the pH of body fluids, in addition to which the pH of the body fluids is regulated homeostatically by the lungs and the kidneys. There are three default buffering systems: (i) Acidic and basic AMINO ACID R-groups of proteins can be reversibly protonated and deprotonated in response to changing proton availability, the *protein buffer system* consisting of a large pool of both intracellular and plasma proteins, which because of their high concentrations provide approximately three-quarters of the body's buffer capacity. Important intracellular buffering molecules include haemoglobin of red blood cells and histone proteins of nucleated cells, while extracellular protein buffers include plasma albumin. (ii) The *carbonic acid/hydrogen carbonate buffer system* (CHBS), an important extracellular buffering mechanism, is of limited capacity and depends upon the equilibrium quickly established between H_2CO_3 and H^+ and HCO_3^-. It can respond quickly to addition of CO_2 and lactate produced during EXERCISE, and to the increased fatty acid and ketone body production released by fatty acid metabolism. It can also respond to addition of basic substances, as after consumption of the antacid $NaHCO_3$. (iii) Intracellular molecules such as DNA, RNA and ATP, and ions such as HPO_4^{2-} also act as buffers, comprising the *phosphate buffer system*. Phosphate ions such as HPO_4^{2-} act as weak acids and can bind H^+ in acid conditions, forming $H_2PO_4^-$ and can release H^+ again as pH increases.

The lack of protein buffer in cerebrospinal fluid results in its pH changing more readily than that of tissue fluid. The lungs regulate acid–base balance by influencing the CHBS. The reaction between CO_2 and H_2O is catalysed by the enzyme carbonic anhydrase located in erythrocytes and on the surfaces of capillary endothelial cells. This enzyme speeds up the rate of attainment of the equilibrium position of the reaction:

$$H^+ + HCO_3^- \leftrightharpoons H_2CO_3 \leftrightharpoons H_2O + CO_2$$

Decrease in pH of the body fluids, regardless of cause, stimulates the RESPIRATORY CENTRE of the medulla oblongata, increasing the rate and depth of breathing. The CHBS responds to the reduced CO_2 concentration of body fluids by favouring the reaction to the right (above), so that the H^+ concentration falls accordingly, CO_2 leaving at the alveoli. Increase in pH of body fluids decreases the rate and depth of breathing, the rising CO_2 concentration then favours the reaction to the left (above), causing the dissociation of H^+ and restoration of pH towards its normal range.

Carbonic anhydrase is also present in cells of the distal convoluted tubules and collecting ducts of the kidney nephrons, and these cells regulate acid–base balance homeostatically by varying their rate of H^+ secretion into the filtrate and their rate of HCO_3^- reabsorption (see KIDNEYS for further details).

acid hydrolases Hydrolytic enzymes that require an acidic environment for optimal activity. They include the cathepsin acid proteases. Several occur in ENDOSOMES and LYSOSOMES (see these entries for some details).

acidosis See KETOSIS.

acid phosphatases See ACID HYDROLASES.

acid proteases See ACID HYDROLASES.

acid rain Rain with a pH below the normal value of 5.6 (which it has as a result of dissolving carbon dioxide). Acidic gases such as oxides of sulphur and nitrogen arising naturally through fires and anthropogenically by combustion of fossil fuels are major contributors to lowering the pH.

acidosis A lowering of the pH of the blood and body fluids (acidaemia in the case of the blood) as can occur if KETONE BODIES accumulate beyond the ability of other tissues to oxidize them. In extreme cases, it can cause coma and death.

acinus (pl. acini) Flask-like (rather than tubular) secretory portion of multicellular exocrine glands (known in consequence as acinar glands).

acoustic nerve See VESTIBULOCOCHLEAR NERVE.

acquired immune deficiency syndrome (AIDS) A newly emerging infectious DISEASE, first recognized in the early 1980s, caused by the human immunodeficiency virus (see HIV) and commonly referred to as HIV/AIDS. The number of cases has risen exponentially, and it might never have emerged had it not been for disruptions in the economic and social infrastructure in post-colonial sub-Saharan Africa. Increased travel, movement of rural populations to large cities, urban poverty and weakening of family structure all promoted sexual practices, such as promiscuity and prostitution, which favour HIV transmission. Increasing decline in CD4 cell numbers and increasing viraemia herald the end of the latent phase and before becoming truly symptomatic, many patients develop generalized lymphoadenopathy (lumpy lymph nodes). Opportunistic infections that are common in immunosuppressed individuals tend to occur at this time, including herpes zoster (see SHINGLES), and bacterial infections including those by pneumococci and *Salmonella*. As CD4 cell numbers decline further, malignancies with an infectious aetiology such as Kaposi's sarcoma, lymphomas or invasive carcinoma of the cervix represent a defect of immune control of latent, potentially oncogenic, infections.

acquired immunity (specific immunity) See IMMUNITY.

acrocentric chromosome A CHROMOSOME whose centromere is close to one end: chromosomes 13–15 and 21–22 (see Appendix III).

acrocephalosyndactyly A dominant skeletal malformation; see ACHONDROPLASIA.

acromegaly Condition arising in adulthood from hypersecretion of human GROWTH HORMONE, hGH. Compare GIANTISM.

acromion Flattened, expanded process of the spine of the scapula (felt as the high point of the shoulder), articulating with the acromial extremity of the clavicle. See PECTORAL GIRDLE, Fig. 131.

acrosome (acrosomal vesicle) A Golgi apparatus-derived vesicle, anterior to the nucleus of a sperm cell, and with it comprising the head of the SPERM. It contains hydrolases that digest proteins and complex sugars, helping the sperm tunnel through the zona pellucida after first binding it during FERTILIZATION (see Fig. 76). The sperm then undergoes the *acrosome reaction*, in which the acrosome contents are released by exocytosis, an increase in cytosolic calcium ions being both necessary and sufficient. The trigger for the acrosome reaction is the species-specific ZP3 glycoprotein in the zona; but the acrosome reaction also exposes other proteins on the sperm surface that mediate binding and fusion of the sperm plasma membrane with that of the ovum, causing the activation reaction. See CUMULUS CELLS, FERTILIZATION.

acrosome reaction See ACROSOME.

ACTH (adrenocorticotropic hormone) See CORTICOTROPIN.

actin One of the most abundant structural proteins in eukaryotic cells and one of the three major proteins of the CYTOSKELETON. It interacts with myosin motor proteins during contraction of muscle fibres, and in adhesion belts (see ADHERENS JUNCTIONS). It has been highly conserved during evolution, each of the six tissue-specific actin isoforms sharing 100% amino acid sequence identity across species from birds to humans, only four out of 373 amino acids differing between the β and γ isoforms (encoded by an IMMEDIATE EARLY GENE). The G-actin subunits are single globular polypeptides that diffuse readily within the cytoplasm so that they are available for assembly into filaments; disassembly releases them again. Each G-actin polypeptide bears one molecule of ATP or ADP within a deep cleft inside it and can hydrolyse ATP. Free subunits are usually in the ATP-form (T-subunits), but shortly after addition to a filament the ATP is hydrolysed to ADP (forming D-subunits), much of the free energy release being stored in the polymer lattice. Filaments (F-actin) consist of two parallel protofilaments, held together by lateral contacts, twisting around each other in a helical structure (termed microfilaments in muscle). The critical phase of assembly of G-actin monomers into F-actin polymers is *nucleation*, when a small monomer cluster is orientated to initiate rapid polymerization. Unlike microtubule nucleation, which occurs near the nucleus, actin filament nucleation usually occurs at the plasma membrane. It is usually regulated by actin-related proteins (ARPs), each with ~45% similarity to actin. The Arp2/3 complex nucleates actin filament growth from the less dynamic ('minus') end, allowing faster growth at the 'plus' end. The 'plus' end is that at which growth and shrinkage are rapid. Actin filaments are organized into *bundles* and gel-like *networks*, different cross-linking proteins being involved (see SHIGELLA). Fimbrin and α-actinin are actin-bundling proteins; spectrin and filamin are two web-forming proteins (see RED BLOOD CELLS). Different forms of actin filament aggregation are found in stress fibres (antiparallel, contractile and exerting tension), in the cell cortex which underlies the plasma membrane (gel-like network), and in the spike-like filopodia which are projections of the plasma membrane enabling a cell to explore its environment (tightly parallel bundles). In the core of a microvillus (see MICROVILLI), a bundle of actin filaments is cross-linked by the actin-binding proteins villin and fimbrin, while lateral side-arms (made of myosin I and calmodulin) connect the lateral sides of the bundle to the overlying plasma membrane. Actin can assist fibroblasts in forming the extracellular matrix (see FIBRONECTINS). Numerous ACCESSORY PROTEINS, such as integrins, bind to actin, as they do to TUBULIN.

action potentials (APs) Brief and dramatic alterations in an axon or muscle membrane potential caused by rapid opening and closing of voltage-gated ION CHANNELS which, when propagated, constitutes a NERVE IMPULSE. Depolarization of the membrane during passage of an AP in skeletal muscle and in axons is due to influx of sodium ions, and repolarization is caused by efflux of potassium ions. Depolarization of pacemaker cells of the heart is due mainly to a calcium current (see SA NODE) while in other cardiac muscle fibres it is due to a combination of a sodium and a later calcium influx. In neurons, APs are initiated at the axon hillock, where voltage-gated Na^+ channels, three kinds of K^+ channel, and one kind of Ca^{2+} channel are plentiful. The delayed K^+ channel is voltage-gated, but because of its slow kinetics opens only during the falling, repolarization, phase of an action potential, driving the membrane back to the K^+ membrane equilibrium value. This is so negative, however, that Na^+ channels rapidly recover from their inactivated state, closing the delayed K^+ channels again and effectively limiting the rate of action potential generation. In order to reflect the intensity of stimulation, the axon hillock must be able to adapt. Early K^+ channels enable this to happen. Their voltage-gating specificity and kinetics enables them to reduce the rate of AP firing at levels of stimulation only just above the threshold for firing resulting in firing rates proportional to the strength of the depolarizing stimulus over a very broad range. The two further channels, voltage-gated Ca^{2+} channels and Ca^{2+}-activated K^+ channels, operate together so as to decrease the response to unchanging, prolonged, stimulation (termed *adaptation*). The latter opens in response to the raised intracellular $[Ca^{2+}]$ produced by a period of intense stimulation and the increased K^+-permeability makes the membrane harder to depolarize, increasing the delay between one AP and the next (see SENSORY ADAPTATION). See AXON HILLOCK, MYELIN SHEATH, NODES OF RANVIER, SPIKE, SYNAPSES.

action selection The task of resolving conflicts between competing behavioural alternatives; i.e., the processes involved in deciding what to do next. A central goal of behavioural science is to provide the proximate causes as to why an observed transition, or sequence of such transitions, occurs in a given context, and possibly to context such an account within the framework of biological utility, or fitness. Relevant behavioural switching might be a global property of the nervous system, and of its embedding in a body and environment, making attributions of these events to specific subcomponents of brain architecture inappropriate. However, there is evidence that biological control systems may include specialized action-selection components. There is a growing body of evidence that the BASAL GANGLIA meet the criteria that make it a candidate action-selection mechanism. These criteria are: (i) that such a candidate should have inputs carrying information about internal and external cues relevant to decision-making; (ii) that some internal mechanism should be present that calculates the 'urgency' or 'salience' that should be attached to each alternative action; (iii) that there should be mechanisms permitting resolution of conflicts between competing actions based on their relative salience; and (iv) that the outputs of the system should be configured to enable the expression of 'winning' actions while preventing the expression of 'losers'.

activating enzymes Any enzyme that, by acting on a substrate molecule, promotes the substrate's activity and thereby its functioning. Kinases are a large class of such enzymes. The term is also used for the amino-acyl-tRNA synthetases, which bind the correct amino acid to the correct transfer RNA molecule, without which the genetic code could not be translated (see TRANSLATION).

activation reaction See FERTILIZATION.

activator proteins For gene activator proteins, see GENE EXPRESSION, TRANSCRIPTION FACTORS.

active immunity Any immune response in an individual which involves its *de novo* production of antibodies during ADAPTIVE IMMUNITY. Contrast PASSIVE IMMUNITY. See IMMUNITY.

active site A fold or groove in an enzyme's conformation to which a substrate molecule binds in order to form the enzyme–substrate complex prior to the catalysed reaction.

active transport Movement of an ion or molecule across a membrane, or other barrier, driven by energy (often ATP hydrolysis) other than that stored in the chemical gradient of the particle transported (see ION CHANNELS). *Secondary active transport* involves facilitated diffusion across a membrane, where the concentration gradient is generated by active transport elsewhere in the cell (see, e.g., ENTEROCYTES, PROXIMAL CONVOLUTED TUBULE).

activins Members of the same TRANSFORMING GROWTH FACTOR-B SUPERFAMILY of glycoproteins as INHIBINS, being homodimers or heterodimers of the β_A/β_B inhibin subunits. They appear to have autocrine or paracrine functions in the many adult tissues (e.g., anterior pituitary gland and gonads) and cell types in which they are produced, since these cells also have receptors to them. In females, activin promotes granulosa cell production (see GRAAFIAN FOLLICLE) and up-regulates their FSH receptor expression during folliculogenesis. It also modulates steroidogenesis by both granulosa and theca cells. Pituitary activin is an antagonist of inhibin and interacts with the negative feedback regulation of inhibin B, stimulating FSH synthesis (see GONADOTROPIN-RELEASING HORMONE, TESTES). Activins have been shown to modulate the activity of IGF-1 (see INSULIN-LIKE GROWTH FACTORS).

acuity See VISUAL ACUITY, AUDIBILITY.

acupuncture Some of the effects of acupuncture may be due to opioid peptides that act peripherally and reduce release of SUBSTANCE P, since these effects can be blocked by the opiate antagonist naloxone.

acute disease Abrupt occurrence of disease symptoms (*contra* chronic disease). In acute RENAL FAILURE, the main feature is suppression of urine flow caused by low blood volume (e.g., due to haemorrhage), decreased cardiac output, damaged renal tubules, kidney stones, and some antibiotics. In acute LEUKAEMIA, immature leukocytes accumulate in the bloodstream and may be treated with X-rays and antileukaemic drugs (e.g., vincristine from *Catharanthus*).

acute hypoxia See HYPOXIA.

acute leukaemias See LEUKAEMIAS.

acute pain See PAIN.

acute phase proteins Antibody-like proteins whose synthesis is induced by a stimulus that triggers phagocyte release of the cytokines TNF-α, IL-1 and IL-6. They are involved in the ACUTE PHASE RESPONSE that occurs during the early stages of INFECTION. Unlike antibodies, they have no structural diversity. Instead, they have broad specificity for the molecular patterns of pathogen surfaces and products. Examples include opsonins such as C-REACTIVE PROTEIN and MANNOSE-BINDING LECTIN, which bind to bacterial and fungal cell-wall LIPOPOLYSACCHARIDES. Other opsonins among acute phase proteins include the pulmonary surfactant proteins SP-A and SP-D whose globular lectin domains are attached to a collagen-like stalk. Found with macrophages in the alveolar fluid of the lungs, these are important in promoting phagocytosis of respiratory pathogens such as *Pneumocystis carinii*, a major cause of pneumonia in patients with AIDS. FERRITIN is also an example.

acute phase response One of the most important INNATE IMMUNITY effects of the cytokines TNF-α, IL-1 and IL-6 released by activated MACROPHAGES (see Fig. 119) at a site of infection is the liver's response in which it shifts its synthesis and release of plasma proteins in favour of ACUTE PHASE PROTEINS, which include mannose-binding lectin (see COMPLEMENT) and LIPOPOLYSACCHARIDE-binding protein, LBP. Within a few days, this provides the individual with several proteins with the functional properties of antibodies but with the ability to bind a broader range of pathogens.

acylation-stimulating protein Adipokine whose paracrine signalling increases efficiency of triglyceride synthesis in adipocytes, leading to more rapid lipid clearance after meals.

acyl-CoA-cholesterol acyl transferase (ACAT) An enzyme found, e.g., in liver cell and macrophage endoplasmic reticula, forming cholesteryl esters through transfer of a fatty acid from CoA to the hydroxyl group of CHOLESTEROL, making cholesterol more hydrophobic. Cholesteryl esters are transported in secreted LIPOPROTEIN particles to other tissues using cholesterol, or are stored in the liver. See ATHEROSCLEROSIS, LECITHIN-CHOLESTEROL ACYL TRANSFERASE.

acyl-CoA synthetases Enzymes (in several isozyme forms) present in the outer mitochondrial membrane, catalysing the general reaction:

$$\text{Fattyacid} + \text{CoA} + \text{ATP} \rightleftharpoons \text{fatty-acyl} - \text{CoA} + \text{AMP} + \text{PP}_i$$

ADA deficiency See ADENOSINE DEAMINASE DEFICIENCY.

adaptation **1**. Evolutionary adaptation. **2**. A decrease in a receptor potential with time, even though the level of stimulus is kept constant; or the decrease in rate of impulses travelling past a point on a sensory neuron, despite a maintained stimulus. Different systems are slow-adapting (e.g., some joint capsule and muscle spindle receptors), moderately rapidly adapting (e.g., hair receptors) or very rapidly adapting (e.g., PACINIAN CORPUSCLE). These differences are due to the properties of different combination of ion channels at the neuron's axon hillock (see ACTION POTENTIALS for details). See MICROSACCADES, SENSORY ADAPTATION.

adaptive immunity (acquired immunity) The response of antigen-specific lymphocytes to antigen, initiated in LYMPH NODES and involving their clonal selection and cooperation (see Appendix XI) and production of ANTIBODIES. Unlike responses of INNATE IMMUNITY (which are a prerequisite for adaptive immune responses), adaptive immunity involves lymphocyte 'learning' and 'memory', enabling it to track and adapt to changes in pathogen strategy. This adaptability, analogous to behavioural adaptability, relies on genetically programmed modifiability. Adaptive immunity can be treated in two phases: the primary adaptive response, in which a foreign antigen is encountered for the first time, and the secondary adaptive response, in which it is re-encountered.

DENDRITIC CELLS initiate adaptive immune responses. In a *primary immune response*, once these antigen-presenting cells (and to a lesser extent MACROPHAGES) are stimulated by foreign antigen to express cell-surface co-stimulatory molecules (e.g., CD80 and CD86), they are targeted to enter lymphoid tissues (e.g., SPLEEN, LYMPH NODES), where they encounter naïve itinerant blood-borne T CELLS. Although capable of presenting antigen to CD4 T cells, B CELLS in the lymph nodes are relatively inefficient at initiating adaptive immune responses. However, those T-CELL RECEPTORS that recognize (i.e., are selected by) and bind to antigen presented on the dendritic cell and macrophage cell surfaces then activate the T cells to divide clonally and mature into effector T cells that re-enter the blood circulation. Where an infection produces local INFLAMMATION, these effector cells will creep out of the capillaries by DIAPEDESIS and migrate to the infected region.

When antigen-specific lymphocytes are activated through their antigen receptors, they initially undergo 'blast transformation' and start to proliferate exponentially by cell division, a clonal expansion that can last for 7–8 days leading to their predominance in the population. In a response to certain viruses, 50% of the cytotoxic T cells may be specific for a single virus-derived class 1 MHC:peptide complex. These T cells then undergo final differentiation into effector cells, which removes the pathogenic stimulus, after which most of them undergo Fas-mediated apoptosis.

When a B cell with the appropriate antigen:class II MHC molecule complex is recognized by an armed T helper cell with the appropriate receptor specificity, it is activated to divide and differentiate into a mature plasma cell, during which the T_H cell's secretion of IL-4 is an important signal to the B cell. The B cell then secretes antibodies with the same antigen specificity as its own receptor (see B CELLS, Fig. 19 and entry for details), helping to clear the antigen from the circulation.

Immunological memory is the ability of the immune system to respond more rapidly and efficiently to pathogens that have been previously encountered, the events involved constituting the *secondary immune response*. It depends upon the existence of pre-existing populations of antigen-specific memory B cells and memory T cells which can respond to reoccurrence of an antigen by rapid clonal expansion. This is the principle underlying vaccination programmes. Apart from the greater speed and concentration of antibody production during secondary and subsequent responses, the antibodies produced also differ qualitatively from those in the primary response owing to their increased affinity for the antigen (see SOMATIC HYPERMUTATION). Further details may be found in the B CELLS and T CELLS entries. See VACCINES.

adaptive radiation A characteristic evolutionary process, subsequent to the origin of some evolutionary novelty (e.g., bipedalism in hominins), in which the originating species quickly gives rise to descendant species, each with its own adaptive variant of the novelty.

adaptor proteins Proteins linking receptor activation to downstream members of signalling pathways, all bearing at least one SH2 domain.

addiction Formerly defined as 'DRUG DEPENDENCE as evidenced by craving, increased TOLERANCE and WITHDRAWAL', it is nowadays accepted that substance dependence requires that an individual display a loss of autonomy characterized by an inability to control his/her drug-seeking behaviour, despite adverse consequences, that is often observed in spite of a stated desire on the part of the addict to stop. HEROIN, NICOTINE and COCAINE all act on brain circuitry that motivates drug-seeking behaviour, although cocaine is highly addictive but causes little withdrawal, and a person addicted to morphine may stop taking the drug without developing an obsession with it (see HABIT, REINFORCEMENT AND REWARD). A craving-generating system would receive sensory cues (e.g., sights, smells) and compare them with memories of rewarding objects (e.g., food) and produce craving to motivate and direct drug-seeking and -use APPETITIVE BEHAVIOURS. Many human fMRI studies (see MAGNETIC RESONANCE IMAGING) provide evidence that cue-induced craving for nicotine, alcohol, cocaine, opiates and chocolate increase metabolic activity in the anterior CINGULATE GYRUS and other frontal lobe regions. A craving-inhibition system would signal satisfaction, inhibiting appetitive behaviour when appropriate.

Excitatory brain synapses are strengthened or weakened in response to specific patterns of synaptic activation, changes in SYNAPTIC STRENGTH which are thought to underlie persistent pathologies such as addiction. A key site of action for cocaine is the NUCLEUS ACCUMBENS (NA), a major target of the ascending dopaminergic axons in the forebrain. There is evidence that even a single dose of cocaine can make neurons of the ventral tegmental area (VTA; see BRAIN, Fig. 29) more sensitive to the drug. Chronic overstimulation of this 'reward' pathway leads to down-regulation and drug tolerance. Heroin and nicotine also act on the VTA, a component of the mesolimbic system where dopaminergic neurons bearing opiate and nicotinic receptors project axons through the lateral hypothalamus to the forebrain. Nicotine enhances the synthesis and release of dopamine by acting on nicotinic cholinergic receptors on the dopamine cell bodies in the VTA, while cannabinoids act on excitatory cannabinoid receptors on dopamine terminals to increase dopamine release. Mice lacking the β2 subunit of the nicotinic ACETYLCHOLINE receptor (nAChR) do not self-administer nicotine, but those in which this receptor is reintroduced into the VTA do (see DOPAMINE). Long-term potentiation in $GABA_A$-mediated synaptic transmission (LTP_{GABA}) onto dopamine neurons in the rat VTA has recently been discovered. This LTP is heterosynaptic, requiring postsynaptic NMDA RECEPTOR activation at glutamate receptors, but results from increased GABA release at neighbouring inhibitory nerve terminals. The dopamine neurons release NITRIC OXIDE (NO), which initiates LTP_{GABA} by activating guanylate cyclase in GABAergic neurons.

With opiates, the situation is more complex. Opiates activate the mesolimbic system by hyperpolarizing VTA GABAergic interneurons, preventing LTP_{GABA} both *in vivo* (interrupting signalling from NO to guanylate cyclase) and *in vitro* (by inhibiting presynaptic glutamate release). These neuroadaptations to opioid drugs may contribute to early stages of addiction, and indicate possible therapy involving drugs targeting $GABA_A$ receptors. Tolerance arises not because the *number* of μ opioid receptors changes but because they couple less well with G_i proteins and fail to inhibit adenylyl cyclase as effectively.

Ethanol (see ALCOHOL), benzodiazepines and barbiturates are all positively reinforcing, in part by reducing anxiety. They increase GABAergic inhibition by binding $GABA_A$ receptors, enhancing the Cl^- ion influx, and $GABA_A$ antagonists injected into the rat amygdala decreases ethanol self-stimulation.

There is increasing interest in research into whether brain activity and biochemistry are affected in the same way in so-called 'behavioural addictions' (e.g., gambling, food) as they are by substance abuse, in which surges in dopamine and other brain neurotransmitters signalling pleasure and reward occur. Addiction seems to rely on some of the same neurobiological mechanisms that underlie LEARNING and MEMORY, and cravings are triggered by memories and situations associated with drug use. Imaging studies of addicts consistently reveal abnormalities in blood flow in the ORBITOFRONTAL CORTEX, alcohol and cocaine addicts displaying reductions in baseline measurements of OFC activation during acute withdrawal, and even after long periods of abstinence. A growing number of animal studies indicate that prolonged exposure to addictive drugs, especially psychostimulants, results in relatively long-lasting brain and behavioural changes.

Addison's disease An AUTOIMMUNE DISEASE caused by undersecretion of adrenocortical hormones, notably aldosterone, resulting from autoimmune destruction of the gland. A common feature, in addition to lethargy, weight loss and hypotension, is excessive melanocyte activity and hyperpigmentation of the skin due to excessive CORTICOTROPIN secretion. Autoantibodies and the enzyme 21-hydroxylase are highly predictive in children.

additive alleles See POLYGENIC INHERITANCE.

A delta (Aδ) fibres Thin, lightly myelinated afferent fibres (see AXON) of certain skin nociceptors, whose cell bodies lie in the segmental dorsal root ganglia, the fibres branching and running in the zone of Lissauer in the spinal cord and synapsing on cells in the outer part of the dorsal horn in the substantia gelatinosa and responsible for the sharp 'first pain' of skin damage. See A BETA FIBRES, C FIBRES, PAIN.

adenine A purine base, synthesized *de novo* from 5-phosphoribosyl 1-pyrophosphate (PRPP) by several steps including incorporation of amino acid moieties (glutaminyl, glycyl, aspartyl) via inosinate (IMP) to yield adenylate (AMP). Adenine nucleotides include ADENOSINE TRIPHOSPHATE and CYCLIC AMP, the former (in its ribosyl or deoxyribosyl form) being a substrate of RNA POLYMERASES and DNA POLYMERASES, respectively. Deamination of adenine, producing hypoxanthine, occurs very much more slowly than does that of the pyrimidine CYTOSINE to URACIL (see DNA REPAIR MECHANISMS). Degradation of purines produces URIC ACID.

adeno- Prefix indicating 'glandular'.

adenohypophysis The anterior PITUITARY GLAND.

adenoids Mucosal-associated lymphoid tissues located in the nasopharynx (see PHARYNX). See GUT-ASSOCIATED LYMPHOID TISSUES, TONSILS.

adenomas A general term for any CANCER arising in epithelial gland tissue. Also termed polyps, adenomatous polyps, papillomas and (in the skin) warts, these are often large growths. See COLORECTAL CANCERS, ONCOSUPPRESSORS, PAPILLOMAVIRUSES.

adenomatous polyposis coli (APC) protein A key protein in the WNT SIGNALLING PATHWAY (see CANCER CELLS, Fig. 36) which is mutated in about 70% of COLORECTAL CANCERS (see COLON). The *APC* gene can be inherited, causing FAMILIAL ADENOMATOUS POLYPOSIS, but like other CANCERS affecting tumour suppressor genes it occurs more commonly in the non-heritable sporadic form due to somatic mutation. The normal protein has highly conserved nuclear export signals, although these are lost in the mutant protein because the protein is truncated. The ability of APC protein to leave the nucleus seems to be important to its tumour-suppressing function. The increased amounts of β-catenin in cells with the *APC* mutation inappropriately activates transcription of the intestinal *Tcf* family member and transcriptional co-repressor *TCF4*. Recognition of nuclear β-catenin/Tcf4 complexes as the key effectors in cancer initiation was confirmed by rare mutations in other Wnt signalling components. Malignant transformation in intestinal epithelial cells is thus caused by inappropriate, constitutive, expression of *Tcf* target genes. See APC GENE, TUMOUR SUPPRESSOR GENES.

adenosine A PURINERGIC breakdown product of the ATP released by axon termini during impulse conduction, stimulates OLIGODENDROCYTES in the brain to mature and form myelin. It stimulates A_1-adenosine receptors and opens acetylcholine-sensitive K^+ channels, hyperpolarizing the cell membrane in the atrioventricular node by inhibiting Ca^{2+} channels, and is used in treatment of supraventricular tachycardia (see ARRHYTHMIAS). It also reduces neuronal activity, being a hypothetical inducer of SLEEP.

adenosine 3′,5′-monophosphate (cyclic AMP) See ADENOSINE MONOPHOSPHATE.

adenosine deaminase deficiency (ADA deficiency) An autosomal recessive enzyme deficiency leading to accumulation of toxic purine nucleotides and nucleosides, resulting in death of most developing lymphocytes within the THYMUS and causing severe combined immunodeficiency (see SCID). Affected children present in the first year of life with recurrent viral and bacterial infections. The diagnosis is confirmed by deficient red blood cell adenosine deaminase activity. Bone marrow transplantation, even antenatally, has been known to correct the immunodeficiency and trials are under way using a combination of mild chemotherapy to kill diseased cells and a novel retroviral vector containing a functioning ADA gene to infect bone marrow cells *ex vivo* prior to reintroduction to the patient. Some success has been reported for this method although not, apparently, in the absence of simultaneous treatment with an enzyme preparation (making the efficacy of the transgene approach uncertain). See LESCH-NYHAN SYNDROME, RNA EDITING.

adenosine diphosphate (ADP) With inorganic phosphate, the product of ATP hydrolysis and a substrate for its resynthesis. It is a positive modulator of phosphofructokinase-1, the key regulatory enzyme of glycolysis, and activates AMP-ACTIVATED PROTEIN KINASE.

adenosine monophosphate (AMP) A product of ADENOSINE TRIPHOSPHATE hydrolysis during free energy transfer reactions, often becoming covalently attached to a substrate in a metabolic pathway, or to an amino acid of an enzyme. An important modulator of GLYCOLYSIS and of GLUCONEOGENESIS, it is a major second messenger in its cyclic form (see CYCLIC AMP). See AMP-ACTIVATED PROTEIN KINASE.

adenosine triphosphate (ATP) A nucleoside triphosphate (see Fig. 2), and the 'common energy currency' of all cells. It is generated mainly by F-type ATPases (ATP synthases) in MITOCHONDRIA during oxidative phosphorylation, and outside mitochondria by glycolysis, by whose energy group transfers (not merely hydrolysis) the energy for most cellular activity is supplied. Gibbs energy, G, is defined by the equation

$$G = H - TS$$

where H is the enthalpy, T is the thermodynamic temperature and S is the entropy. Under constant pressure and temperature, the direction of a reaction's progress is in the direction of lower Gibbs energy. At constant temperature, the Gibbs energy change, ΔG, is related to the enthalpy change ΔH and the entropy change ΔS by

$$\Delta G = \Delta H - T\Delta S$$

ATP hydrolysis *in vitro* usually releases energy as heat, which cannot drive a chemical reaction in an isothermal system. However, *in vivo* the process is usually a controlled two-step process, the first of which involves the covalent transfer of part of the molecule (a phosphoryl or pyrophosphoryl group, or the adenyl moiety (AMP)) to a substrate molecule or to an amino acid in a protein, thereby raising its free-energy content. The second step involves displacement of the phosphate-containing moiety from the recipient, generating P_i, PP_i (inorganic pyrophosphate) or AMP. Thus, in enzyme-catalysed reactions to which it contributes free energy, ATP contributes *covalently*. This is how, e.g., ATP is used in active transport mechanisms (e.g., the P-type Na^+K^+-ATPase ion pump; see the ion pumps part of ION CHANNELS) and autophosphorylation (see PROTEIN KINASES). However, *non-covalent* binding of ATP (or GTP) followed by its direct hydrolysis to ADP (if GTP, to GDP) does provide the energy to cycle some MOTOR PROTEINS between two conformations, producing mechanical motion, as in muscle contraction, the movement of enzymes along DNA, and of ribosomes along mRNA. ATP is the source of CYCLIC AMP (see ADENYLYL CYCLASE), a negative modulator of phosphofructokinase-1, the key regulatory enzyme in GLYCOLYSIS with numerous crucial roles as a SECOND MESSENGER. Cells normally contain ~10 times as much ATP as ADP and AMP; but when they are metabolically active, the ATP/(ADP + AMP) ratio falls, resulting in acceleration of glycolysis and oxidative phosphorylation. Conversely, GLUCONEOGENESIS is favoured by a high ATP/(ADP + AMP) ratio.

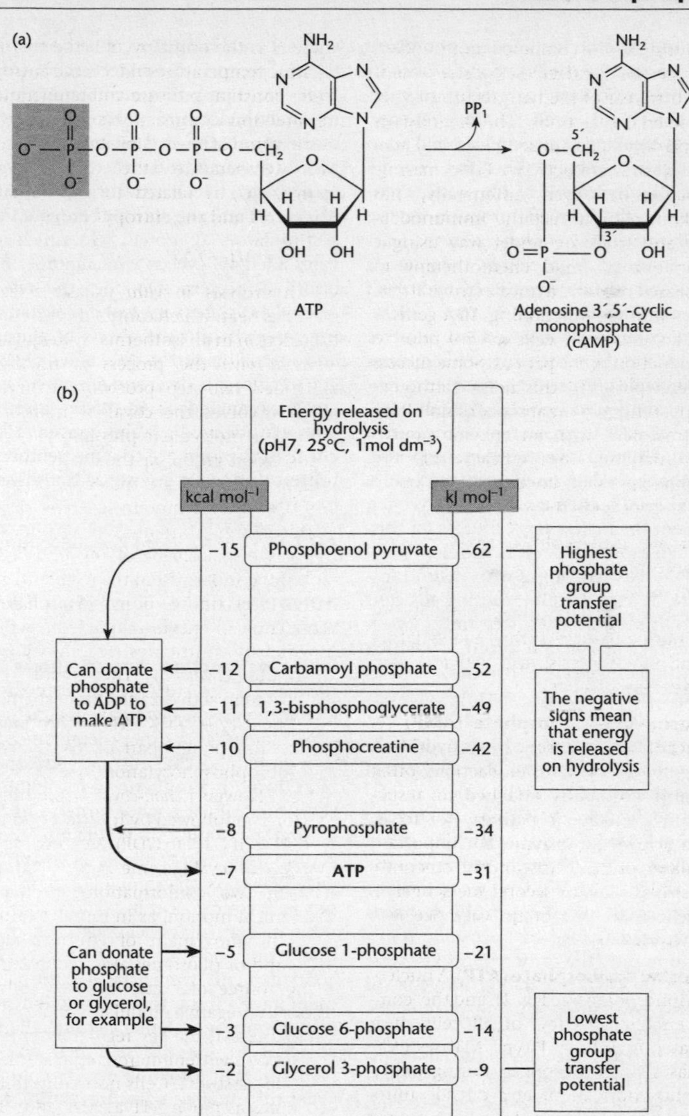

FIG 2. (a) The structural formulae of ADENOSINE TRIPHOSPHATE and of the cyclic AMP formed from it by adenylyl cyclase activity. Removal of the shaded terminal phosphoryl group of ATP is highly exergonic and is a reaction coupled to many endergonic reactions in the cell. (b) Diagram indicating the flow of phosphoryl groups from substances with higher transfer potential (indicated by a more negative free energy release on hydrolysis) to those with lower transfer potential, assuming molar concentrations of reactants and products.

Use of the term 'high-energy phosphate bond' to describe the P–O bond broken in hydrolysis reactions incorrectly suggests that the bond itself *contains* energy, whereas breaking of a chemical bond actually *requires* energy. The free energy released by hydrolysis of phosphate compounds results from the products having smaller free-energy content than the reactants. So-called 'high energy' compounds have standard free energies of hydrolysis more negative than -25 kJ mol^{-1}, 'low-energy' compounds having a less negative value. ATP is therefore a high-energy compound, while glucose-6-phosphate is a low energy compound (see Fig. 2). Only ~80–100 g of ATP is present at a time in a human, although even a sedentary person resynthesizes a daily amount equal to ~75% of their total body mass. Fat and glycogen represent the main energy sources for this, although PHOSPHOCREATINE is another.

ATP is released, along with neurotransmitters, at axon termini, causing adjacent glial cells such as ASTROCYTES and SCHWANN CELLS to take up calcium ions (Ca^{2+}) which causes them to release ATP in turn. See AMP-ACTIVATED PROTEIN KINASE, MULTIPLE SCLEROSIS, PAIN.

adenovirus A family of VIRUSES with double-stranded DNA as the genetic material (genome size ~36–38 kb), lacking an envelope. They cause lower respiratory tract infections, through either a lytic cycle which releases infectious virions, or through latent infection (usually in tonsils or lymphoid tissue). In oncogenic transformation only the early replicative stages occur, viral DNA integrating into that of the host without production of infective virions. They encode a protein that binds to class I MHC molecules in the endoplasmic reticulum, preventing their transport to the cell surface and avoiding detection by CD8 cytotoxic T CELLS. They have been used, sadly unsuccessfully, as vectors of the gene for the *CFTR* gene as a therapy in cases of CYSTIC FIBROSIS.

adenylyl cyclase (adenyl cyclase, adenylate cyclase, AC) A membrane-bound enzyme catalysing the conversion of ATP to the second messenger CYCLIC AMP. A large multi-pass transmembrane protein, its catalytic domain, on the cytosolic side of the membrane, is regulated by G PROTEINS and CALCIUM ions, and is therefore activated by pathways which activate PHOSPHOLIPASE C. All receptors acting via cAMP are coupled to a stimulatory G protein (G_s) which activates adenylyl cyclase while another inhibitory G protein (G_i) does the reverse. Adenylyl cyclases were first discovered during research into the effects of glucagon and adrenaline on the breakdown of glycogen to glucose by liver and muscle cells; but enzymes analogous to these hormone-sensitive adenylyl cyclases convert ATP to cAMP in the presence of some neurotransmitters during slow synaptic transmission (see SYNAPSE).

ADH See **1**. VASOPRESSIN, **2**. ALCOHOL DEHYDROGENASE.

ADHD See ATTENTION DEFICIT HYPERACTIVITY DISORDER.

adherens junction An INTERCELLULAR JUNCTION (see Fig. 106) in which the cytoplasmic face of the plasma membrane is attached to ACTIN filaments. The transmembrane adhesion protein is E-cadherin and the extracellular ligand is a cadherin in the adjoining cell. Intracellular anchor proteins include α- and β-catenins and vinculin and α-actinin (see WNT SIGNALLING PATHWAYS, Fig. 181). The classical examples occur in *adhesion belts* which occur just below TIGHT JUNCTIONS in epithelia, encircling each of the interacting cells in the cell sheet. The actin filaments, lying parallel to the plasma membrane, form (in conjunction with myosin motor proteins) a contractile ring adjacent to the adhesion belt and are responsible for folding of an epithelial sheet into tubular structures during morphogenesis. Adherens junction formation appears to be a prerequisite for tight junction formation.

adhesion belt (zonula adherens) See ADHERENS JUNCTION.

adhesion molecules Proteins mediating

binding of one cell to another (CELL ADHESION MOLECULES) or of cells to proteins of the EXTRACELLULAR MATRIX. They include integrins, selectins, connexins and beta-catenin (see CANCER CELLS, Fig. 35) and members of the immunoglobulin superfamily (e.g., ICAM-1 and VCAM-1 of the activated ENDOTHELIUM). See INTERCELLULAR JUNCTIONS, Fig. 106.

adipocytes (fat cells) Connective tissue cells specialized for storage of triglycerides, they and their precursor cells having different powers of increase, developmental attributes and responses to hormone signals depending upon their body location (see ADIPOSE TISSUE).

Adipocytes have membrane receptors for VLDLs (see LIPOPROTEINS) from which they obtain fatty acids through LIPOPROTEIN LIPASE activity for the resynthesis of triglycerides. Accumulation and coalescence of the resulting fat droplets requires the cell to be greatly distensible (~1000-fold), the nucleus and cytoplasm being pushed to the rim in 'fat' adipocytes. They have a crucial role in ENERGY BALANCE and glucose homeostasis and secrete an array of ADIPOKINES (see entry). Those bloated with lipids secrete different adipokines (e.g., RBP4, IL-6) from 'lean adipocytes' (which secrete ADIPONECTIN). RBP4 is believed to promote INSULIN RESISTANCE and to trigger INFLAMMATION (see OBESITY). Adipocytes contain lipases and respond to lipolytic stimuli by releasing fatty acids into the circulation.

The central engine of adipocyte differentiation is peroxisome-activated receptor-γ (PPAR-γ) which, when activated by an agonist in fibroblast cells, initiates morphological changes, lipid accumulation and gene expression. The nuclear coactivator PGC-1α (see PGC-1) has been shown to regulate adaptive thermogenesis in brown adipocytes, including stimulation of fuel intake, mitochondrial fatty-acid oxidation and heat production through expression of mitochondrial uncoupling protein-1 (*UCP1*; see UCPS) and its expression is strongly induced by cold exposure (see TEMPERATURE REGULATION). Ectopic expression of PGC-1α is sufficient but not necessary to promote differentiation towards the brown adipocyte lineage and expression of *UCP1*. PGC-1β is also expressed during brown adipocyte differentiation (see MITOCHONDRION). GROWTH HORMONE inhibits adipocyte differentiation, reduces triglyceride accumulation and stimulates lipolysis and β-oxidation of FATTY ACIDS. These effects are mediated by a reduction in the activity of LIPOPROTEIN LIPASE.

Both hydrogen peroxide and nicotine suppress adipokine expression in adipocytes and smoking may directly regulate adiponectin concentration through lipolysis (see CIGARETTE SMOKING). In normal individuals, binding of insulin to its adipocyte cell membrane receptor stimulates glucose uptake through the glucose transporter GLUT-4. Overall, GROWTH HORMONE counteracts these effects of insulin. The expression of GLUT-4 is greatly reduced in adipocytes of rodents and humans suffering from INSULIN RESISTANCE and OBESITY, but not in their muscle fibres, and is accompanied by increased expression and secretion of adipocyte-derived retinol binding protein-4 (RBP4). This operates in part by blocking the action of insulin in muscle and liver. Obese and type 2 diabetic patients have higher RBP4 levels in their blood than healthy controls. See STEM CELLS.

adipogenesis The formation of ADIPOSE TISSUE.

adipokines Proteins acting in an endocrine or paracrine manner, many released as cytokines, that may be produced as developing pre-adipocytes differentiate into ADIPOCYTES. They include adipocyte differentiation factor, ADIPONECTIN, ACYLATION-STIMULATING PROTEIN, ANGIOTENSINOGEN, COMPLEMENT D, LEPTIN and RESISTIN, IL-6, lipoprotein lipase, RETINOL BINDING PROTEIN-4 (RBP4) and TNF-α. ADIPONECTIN, TNF-α and resistin, signal changes in the mass of adipose tissue and energy status to other organs controlling fuel usage (See ENERGY BALANCE). The term is sometimes restricted to hormones and pro-inflammatory cytokines derived from adipose tissue with effects on feeding behaviour and energy expenditure, so affecting body weight (see DIABETES, INSULIN RESISTANCE, OBESITY).

adiponectin (AdipoQ) Protein ADIPOKINE produced by 'lean' ADIPOCYTES in negative correlation to levels of plasma triglycerides, but produced decreasingly so in obese individuals. It has been regarded as a 'starvation signal', signalling that an adipocyte can accept additional triglyceride for storage, promoting glucose uptake and its accumulation in tissues as triglyceride. It also increases insulin sensitivity and tissue fat oxidation, reducing levels of circulating fatty acid levels and muscle and liver triglycerides. CIGARETTE SMOKING causes a dose-dependent increase in plasma ICAM-1 levels and a decrease in adiponectin levels. See ENERGY BALANCE, INSULIN RESISTANCE.

adipose tissue Loose connective tissue whose cells (ADIPOCYTES) derive from fibroblasts and are specialized for storage of triglycerides and releasing non-esterified FATTY ACIDS (NEFA) into the circulation. Now considered to be an active endocrine organ, it is found wherever areolar tissue is located and, being a good thermal insulator, it can reduce heat loss through the skin.

Fat deposits are located throughout the body. Although some, such as the *fat pads* of the heels, fingers, toes and periorbital fat supporting the eyes, are thought to provide mechanical support but to contribute relatively little to energy balance, other adipocytes exist in loose association with the skin (subcutaneous fat), are the cause of 'cellulite' and the target of cosmetic liposuction. There are also several distinct fat depots within the body cavity, termed *visceral fat*. Surrounding the heart and other organs and associated with the intestinal mesentery and retroperitoneum, many draining directly into the portal circulation, these have been linked to morbidities associated with OBESITY. An independent curvilinear association exists between visceral adiposity and mortality, and the presence of large visceral adipose tissue (VAT) depots is associated with increased risk of HYPERTENSION, CARDIOVASCULAR DISEASE (CVD) and type 2 DIABETES MELLITUS. The waist-to-hip ratio (WHR), an indicator of VAT, reflects abdominal fat in predicting type 2 diabetes, stroke, myocardial infarction and cardiovascular mortality in middle-aged individuals (see BODY MASS INDEX). Women with the highest WHR have an elevated relative risk of death from CVD. The level of adiposity in males is directly related to the expression of AROMATASE, and therefore with the conversion of testosterone to oestradiol.

Visceral fat accumulation is associated not only with quantitative changes in serum lipids, but also with qualitative changes in lipoproteins (e.g., small dense LDL) and it conveys greater insulin resistance than does subcutaneous fat. Its accumulation is associated with impaired suppression of adipocyte lipolysis and elevated NEFA levels, potentiating vascular ENDOTHELIUM dysfunction.

Adipose tissue contains most of the energy stores in a healthy individual and, in adults, most adipose tissue is *white fat* composed of white adipocytes forming the depots described. Brown adipocytes differ in storing less lipid and in having more mitochondria, which express uncoupling protein-1 (UCP-1; see UCPS) so that each MITOCHONDRION can dissipate the proton gradient across the inner mitochondrial membrane as heat rather than generating ATP. In humans, *brown fat* (brown adipose tissue, BAT) surrounds the heart and great vessels in the foetus and infancy (see TEMPERATURE REGULATION) but tends to disappear over time so that only scattered cells eventually remain within the white fat pads. Its 'brown' colour results from its rich blood supply and densely packed mitochondria (whose cytochrome pigments are coloured).

Two hormones released in proportion to *body fat* are INSULIN and LEPTIN (an ADIPOKINE). Both are transported to the brain, where they modulate the expression of hypothalamic neuropeptides that regulate feeding and body weight resulting in reduction in food intake and increase in energy expenditure (see ENERGY BALANCE). Secretion of hormones and cytokines by adipose tissue is known to be dysregulated by both excess and deficiency of body fat. Estimation of *body fat* by using skinfold calipers (calibrated for correct jaw tension and gap width) to measure skinfold thickness involves taking readings from 3 to 9 standard anatomical sites, reliability varying with the skill and experience of the tester. It is best to use the sum of several sites and compare body fat measures over time. Usually, the right side only is measured, the tester pinching the skin to raise a double layer of skin and the underlying connective tissue, but not the muscle. Calipers are then applied 1 cm below and at right angles to the pinch, a reading being taken 2 seconds later. The mean of two readings is found, a third reading being required if the first two differ widely and the median then found. Standard site terms include triceps, biceps, subscapular, iliac crest, supraspinale, abdominal, frontal thigh, medial calf, chest and axilla. In each case, it is important to locate the correct pinch site by following the standard anatomical descriptions for each landmark. These measurements are not a good predictor of percentage body fat (% BF), although equations exist for calculating this. See BODY COMPOSITION, Fig. 24; CACHEXIA.

adiposity negative feedback model See ENERGY BALANCE.

adjuvant Any substance enhancing an immune response to an antigen with which it is mixed; often components of non-replicating VACCINES.

adolescence A universally recognized life stage, the period from puberty until about three or four years later, characterized by the ADOLESCENT GROWTH SPURT in skeletal and muscular dimensions, closely related to the rapid development of the reproductive system. Other than for a brief postnatal period, the hypothalamic-pituitary-gonadal (HPG) axis is quiescent until PUBERTY, circulating gonadal steroid levels being low. Many of the sex differences in adult body size and shape are the result of differential growth patterns at adolescence. Brain grey matter thickens in childhood, but then thins in a wave beginning at the rear of the brain and reaching the front by early adulthood. Probably reflecting a selective 'synaptic pruning', some find it supports the 'use-it-or-lose-it' view that the more environmental input there is to guide the pruning, the better able a brain will be able to react to complex situations in later life (by no means proven). Completed earlier in girls than in boys, the thinning lends support to the long-held wisdom that adolescence sees the prefrontal cortical regions handling executive functions 'wake up', and that this maturation occurs earlier in girls than in boys.

People aged 10–19 now constitute a 'demographic bulge', being the largest age group globally (20% of the 6.5 billion world population estimate in 2005, 85% of whom live in developing countries). Adolescence has been described as 'demographically dense', a period in life when a large percentage of people experience a large percentage of key life-course events, such as leaving or completing school, bearing a child, and becoming economically productive. It has been said that the full glory of human intelligence requires the extensive, culturally supported developmental learning process that takes place over the first decades of life, and it is relevant that the PREFRONTAL CORTEX does not fully mature until completion of adolescence. Neurophysiological and brain imaging studies indicate brain reorganization during adolescence coincident with the onset of puberty, which could make adolescents more sensitive to experiences affecting judgement. Younger people find negative information more attention-grabbing and memorable than positive information whereas there seems to be a shift to stronger focus on positive information in old age.

adolescent growth spurt (pubertial growth spurt) An increase in height velocity during PUBERTY, peaking in boys between 14 and 15 years (~10 cm per year) and in girls between 12 and 13 years (~9 cm per year). Virtually every aspect of muscular and skeletal growth is affected, underlying hormonal mechanisms involving cooperation between pituitary GROWTH HORMONE and gonadal steroids. Testosterone-derived OESTROGEN is the critical sex hormone involved. It stimulates chondrogenesis in the EPIPHYSEAL GROWTH PLATE (e.p.g), increasing pubertial linear growth. At puberty, oestrogen stimulates skeletal maturation and the gradual, progressive closure of the e.p.g and termination of chondrogenesis. Although a constant phenomenon, puberty's intensity and duration vary. Before it, boys and girls differ only by some 2% in height, but after it by ~8%. This is due partly to the later occurrence of the male spurt, allowing an extra period of growth, and partly because of its greater intensity. In absolute terms, the adult sex difference is ~13 cm. Muscles appear to have their spurt ~3 months after the height peak, and the weight peak occurs ~6 months after the height peak. Although the height growth spurt is unique to humans, the weight-velocity curve is shared by monkeys and non-human hominids. See OESTROGENS.

adoptive immunity Immunity conferred on a naïve or irradiated person by transfer of lymphoid cells from an actively immunized donor.

ADP See ADENOSINE DIPHOSPHATE.

adrenal cortex See ADRENAL GLANDS.

adrenal glands Paired endocrine glands lying either side of the mid-line, one atop each kidney. Each is a composite of an outer cortex derived from coelomic mesoderm, making up the bulk of the gland, and an inner medulla derived from neural crest cells of the ectoderm. They play important roles in adaptive STRESS RESPONSES, in the maintenance of body water and salt balance, and in the control of blood pressure.

The cortex produces the mineralocorticoid ALDOSTERONE (from the zona glomerulosa), the GLUCOCORTICOIDS cortisol and corticosterone (from the zona fasciculata) and the androgen dehydroepiandrosterone (from the zona reticularis, postnatally). The medulla produces the catecholamines NORADRENALINE and ADRENALINE and, to a lesser extent, DOPAMINE.

The adrenal medulla is richly vascularized. The neural crest cells that found it by migrating away from the neural folds after closure of the neural tube are effectively sympathetic postganglionic neurons that lack axons. Staining yellow with chrome salts, these large cells are termed *chromaffin* (chromaphil) *cells* and are widely distributed during embryonic life but restricted to the adrenal medulla in the adult, almost entirely surrounded by the cortex. The more numerous of these cells are ADRENALINE-secreting, smaller numbers secreting NORADRENALINE. Release of these hormones (and of certain peptides of unknown function with them) is by preganglionic release of ACh from sympathetic nervous stimulation and binding of ACh to nicotinic receptors on the chromaffin cells. The medulla also secretes β-endorphin, along with adrenaline and noradrenaline, as part of the stress response.

adrenal hyperplasia See CONGENITAL ADRENAL HYPERPLASIA.

adrenal medulla See ADRENAL GLANDS.

adrenaline (epinephrine) A hormone derivative of the AMINO ACID tyrosine produced by the chromaffin (chromafil) cells of the adrenal medulla (see ADRENAL GLANDS), its synthetic pathway being via L-dopa, dopamine and noradrenaline (see CATECHOLAMINES). It is released on sympathetic activation of the adrenal medulla as part of the stress response and typically has diverse short-term effects mediated by its binding to METABOTROPIC receptors (see also ADRENERGIC), with greatest affinity for α-adrenergic receptors, stimulating release of glucagon and suppressing release of insulin, so causing an increase in the glucagon:insulin ratio and an overall increase in BLOOD GLUCOSE REGULATION. It is a bronchodilator, and employed in therapy for anaphylactic reactions, where it stimulates reformation of tight junctions between endothelial cells. See Appendix I for its structural formula, and STRESS RESPONSE entries for some details of its effects, which include raising STROKE VOLUME and HEART RATE.

adrenergic (sympathomimetic) Describing (i) postganglionic nerve endings of the sympathetic division of the ANS that secrete noradrenaline, and (ii) those G PROTEIN-COUPLED RECEPTORS (adrenoceptors) that bind the catecholamines ADRENALINE or NORADRENALINE. Adrenergic receptors are METABOTROPIC and components of very specific coupled signal transduction pathways, and occur in numerous subtly different subtypes. They can undergo up-regulation by increased transcription of the appropriate genes; but chronic exposure to receptor agonists down-regulates their expression.

α-adrenergic receptors (of which there are three subtypes) have greater affinity for ADRENALINE, and are predominantly stimulatory (although α-stimulation of gut muscle is inhibitory). Those of the α_1 subtype usually activate PHOSPHOLIPASE Cα (increasing intracellular Ca^{2+} concentration) or PHOSPHOLIPASE A_2. Those of the α_2 subtype (of which there are also three subtypes) operate via G_i proteins and may decrease the activity of adenylyl cyclase (opposing the effects of β-adrenergic receptors), activate K^+ channels, or inhibit Ca^{2+} channels and activate phospholipase C_β or phospholipase A_2 (similarly to the α_1-adrenergic receptors). The effects of α-adrenergic stimulation include vasoconstriction, iris dilation, intestinal relaxation, pilomotor contraction, bladder sphincter contraction, bronchoconstriction, uterine smooth muscle constriction, increased cardiac contractility and hepatic glucose production. As a consequence, α_2-adrenergic receptors are implicated in several cardiovascular and central nervous system disorders and their agonists (e.g., phenylephrine, which binds smooth muscle in blood vessel walls and causes vasoconstriction) are used in the treatment of hypotension, glaucoma and attention deficit disorder, in suppression of opiate withdrawal, and as adjuncts to general anaesthesia. α-adrenergic-receptor antagonists (*α-blockers*) have limited clinical application, although are sometimes used in treatment of hypertension.

β-adrenergic receptors are all 'serpentine receptors' (having seven transmembrane helices), and have greater affinity for the synthetic agonist isoproterenol than for adrenaline or noradrenaline. They are linked to the $G\alpha_s$ G protein and activate adenylyl cyclase, being predominantly inhibitory (although β-stimulation of the heart is excitatory) and found in membranes of muscle, liver and adipose tissue. There are three subtypes, all of which activate adenylyl cyclase. β_1-adrenergic receptors play an important part in regulating contraction, relaxation, and strength of CARDIAC MUSCLE fibres and their antagonists (*beta-blockers*, e.g., propranolol) are used in treatment of HYPERTENSION, some cardiac ARRHYTHMIAS, coronary heart disease and chronic heart failure. They vary in their lipid solubility and cardioselectivity but are effective in reducing blood pressure and preventing angina. The more lipid-soluble ones are more likely to enter the brain and cause such central effects as bad dreams. β_2-adrenergic receptors mediate vasodilation (especially in skeletal muscle), bronchial and smooth muscle relaxation, and lipolysis, in various tissues, their agonists causing bronchial dilatation and of clinical use in ASTHMA therapy; β_3-adrenergic receptors play an important role in mediating catecholamine-stimulated thermogenesis and lipolysis. The effects of β-adrenergic stimulation include vasodilation, increased heart rate, increased myocardial strength, intestinal and bladder wall relaxation, uterus relaxation, bronchodilation, glycogenolysis and lipolysis.

adrenoceptor An ADRENERGIC receptor.

adrenocorticotropic hormone (ACTH) See CORTICOTROPIN.

adult stem cells STEM CELLS located in adult tissues and normally involved in homeostatic self-renewal processes, but also capable of recruitment and differentiation in the repair of injured tissues. Their possible future use in therapeutic cloning would obviate the need to use EMBRYONIC STEM CELLS; but both their source and potential are matters of dispute. They are comparatively scarce in tissues – fewer than $1:10^4$ cells; nor are they generally predictably located. They are also very slow to culture (a labour-intensive procedure). Many claims have been made, but few have been rigorously established. We now realize, for instance, that bone marrow stem cells can fuse to cells in the heart, liver and brain – sometimes explaining their alleged transdifferentiation. See MULTIPOTENT STEM CELLS.

aerobic capacity See MAXIMAL OXYGEN CAPACITY.

aerobic respiration In humans, those metabolic pathways occurring within mitochondria (e.g, the KREBS CYCLE and oxidative phosphorylation; see MITOCHONDRION) by which pyruvate, fatty acids (see FATTY ACID METABOLISM), ketones and certain amino acids, are *completely* oxidized to carbon dioxide and water. See EXERCISE, OXYGEN DEBT.

aerobic training See EXERCISE.

aerosol A dispersion of a solid or a liquid in the atmosphere, including smoke, smog, fog and droplet INFECTION.

affective disorders Those disorders characterized by a disturbance of EMOTION or MOOD that are health-threatening, having a tendency to be associated with alterations in behaviour, energy balance, appetite, sleep and body mass (e.g., OBESITY). They include (depending upon phenotypic classification) mania (extremes of mood change, from intense elation to severe depressive states), hypomania, schizoaffective disorder, major DEPRESSION, cycles of manic and depressive episodes, and anxiety disorders (see ANTIDEPRESSANTS, BIPOLAR DISORDER, FEAR AND ANXIETY; (see also PHOBIAS, SCHIZOPHRENIA). PARKINSON'S DISEASE (PD) is often accompanied by depression, and the distinction between disorders of neurology such as PD and Alzheimer's disease and disorders of psychiatry, such as affective disorders, may turn out to be increasingly subtle.

Major affective disorder, affecting 20% or more of the population, is often subdivided into *unipolar affective disorder* (major depressive disorder), in which depressed mood, lack of interest and pleasure (including sexual), insomnia, weight loss, fatigue and feelings of worthlessness are common; and *bipolar affective disorder* (mania), in which mood is abnormally and persistently elevated and expansive, often taking an irritable form, often involving inflated self-esteem (grandiosity), reduced need for sleep, tendency to be over-talkative, experience of thoughts racing, and easy distractibility (see BIPOLAR DISORDER). Psychotic symptoms (see PSYCHOSIS), auditory hallucinations and delusions may be present in affective disorders, whose understanding is made difficult by the interplay of genetic and non-genetic factors (i.e., they are COMPLEX DISORDERS), and by variable penetrance of genes involved. In twin studies, up to 72% of identical twins were concordant for the disorder while 14% of non-identical twins were found to be. Decreased docosahexaenoic acid (DHA) and brain-derived neurotrophic factor (BDNF) have been implicated in bipolar disease. Deprivation of n-3 polyunsaturated fatty acids (PUFAs) for 15 weeks in rats has been shown to decrease frontal cortex DHA level and BDNF expression level as well as CREB activity and p38 MAPK activity. Elevated CREB activity in the NUCLEUS ACCUMBENS, as during exposure to drugs of abuse, produces depressive-like effects, or stress, in rodents (contrast effects of CREB in the HIPPOCAMPUS). DOPAMINE antagonists reduce mania, while dopamine agonists can precipitate it.

Depression and anxiety disorders often coexist and have GENETIC PREDISPOSITIONS, although the particular genes that contribute to the pathology are largely unknown (but see SEROTONIN, VASOPRESSIN). Recently discovered links between the muscarinic receptor *CHRM2*, alcoholism (see ALCOHOL) and depression may provide avenues for pharmacological treatment. Another candidate gene is that encoding BDNF, because of its known roles in neuronal survival, differentiation and synaptic plasticity. A common polymorphism in humans is a SINGLE NUCLEOTIDE POLYMORPHISM in the *BDNF* gene (Val66Met), with an allele frequency of 20–30% in Caucasians. Those heterozygous for the allele have smaller hippocampus volumes and perform poorly on hippocam-

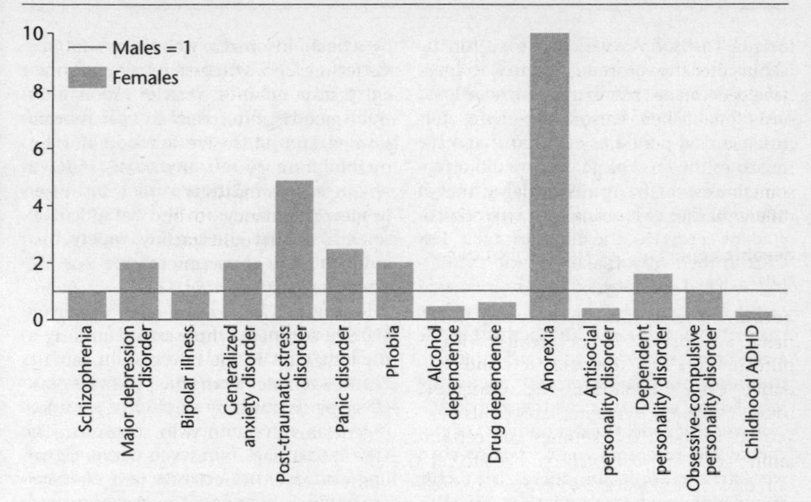

FIG 3. Bar chart showing the ratios of occurrence in the two sexes for selected AFFECTIVE DISORDERS, with male occurrence in each case being represented as unity.

pal-dependent memory tasks, notably fear conditioning, in both MOUSE MODELS and humans. Although mice have been shown to display increased anxiety-related behaviour when homozygous for the mouse *BDNF*^{met} allele, this phenotype has not yet been established in humans. As expected from their wide range of symptoms, affective disorders often arise through dysfunction in more than one part of the brain.

A remarkably high number of patients with major depression report appalling personal histories of child abuse, physical and sexual, as well as severe emotional neglect and early loss. Major depressive disorder is associated with cognitive deficits and dysregulation of the hypothalamic-pituitary-adrenal axis (HPA), a part of the neuroendocrine system controlling the STRESS RESPONSE. Epidemiological studies indicate that women are more vulnerable than men to most affective disorders, including major depression and many anxiety-related disorders (generalized anxiety disorder, panic disorder, post-traumatic stress disorder, and phobias; see Fig. 3). Female depression rates start to rise during puberty: apparently just the right amount of oestrogen is required for emotional balance, and too much oestrogen can overactivate the HPA and result in depression. Testosterone, on the other hand, may protect against stress and depression by damping HPA reactivity through blunting the cortisol response to CRH (see SEX DIFFERENCES). There is postmortem evidence of marked CRH hypersecretion and increased CRH gene expression in depression. Injection of CRH into the brains of animals produces behavioural effects reminiscent of major depression: insomnia, decreased appetite, decreased sexual interest and increased behavioural expression of anxiety. Stressful life events often precipitate depressive episodes in patients with major depression, and adrenal gland enlargement is common.

According to the *diathesis-stress hypothesis*, the HPA axis is the main site where genetic and environmental influences converge to cause mood disorders, and CRH is now thought to be responsible for the endocrine pathology of depression and its major behavioural effects. Early traumatic experiences in humans seem to be associated with persistent hyperactivity of CRH neurons, due largely to disruption of the negative feedback inhibition of the HPA axis by cortisol. A molecular basis for the diminished hippocampal response to cortisol is a decreased number of glucocorticoid receptors. In rats, sensory experience during a critical period of early postnatal life regulates the level of glucocorticoid receptor gene expression, pups receiving a lot of maternal care expressing more glucocorticoid receptors in the hippocampus, less CRH in their hypothalamus, and exhibiting reduced anxiety as adults. Increased tactile experience by the pups can replace the maternal influence, this activating the ascending sensory serotonergic inputs to the hippocampus, serotonin increasing long-lasting expression of the glucocorticoid receptor gene. Greater numbers of glucocorticoid receptors enable rats to cope with stress as adults, but increasing tactile stimulation in adult rats does not have this effect.

Neuroimaging studies of humans with major depressive disorder have tended to implicate prefrontal regions, especially in the midline subgenual anterior cingulate cortex (area 25), a region enriched with the serotonin transporter and which appears abnormal on structural and functional scans. Those inheriting a risk allele within the promoter of the gene for this transporter have reduced volume of area 25 and reduced functional coupling of this region to the amygdala. Early indications are that DEEP BRAIN STIMULATION adjacent to area 25 relieves symptoms of major depressive (unipolar affective) disorder. Rodent studies using voltage-sensitive dyes and optical imaging implicate the hippocampus in depression-like symptoms whereas the midline prefrontal cortex is implicated in humans (in whom such dyes cannot be used). It may well turn out that understanding the neurobiology of abnormal mood regulation in humans involves a functional network including both the ventral hippocampus and midline prefrontal cortex.

In the 1960s, the drug reserpine was introduced to control high blood pressure, but caused psychotic depression in 20% of patients. Reserpine depletes levels of the catecholamine and serotonin neurotransmitters at CNS synapses by slowing their entry into synaptic vesicles. Soon afterwards another drug, used to treat tuberculosis, was found to elevate mood. It did so by inhibiting MONOAMINE OXIDASE (MAO) at synapses, causing these same transmitters to remain for longer in the synaptic cleft, something that the already widely used antidepressant drug imipramine was also discovered to do.

During the decade up to 2005, a theory evolved that many drugs used currently in the long term to treat depression, such as Prozac (an SSRI), exert their ANTIDEPRESSANT effect by inhibiting reuptake of SEROTONIN (5-hydroxytryptamine) by neurons. The view is that serotonin levels trigger signalling cascades that activate GENE EXPRESSION programmes to enhance neuronal survival and connectivity. Such drugs also augment neurogenesis in experimental animals – taking about the same time that it takes to elevate mood – tallying with the view that depression is in part due to decreased neurogenesis in the HIPPOCAMPUS, which does shrink in chronically depressed patients. Serotonin signalling pathways may also become defective if one of its receptor subtypes, or a G PROTEIN normally associated with a serotonin-bound receptor, malfunctions. In one mouse model of depression, genetic under-expression of one such protein, p11, that associates with receptor subtype 5-HT_{1B}, increased depression-like behaviour while its over-expression caused the opposite behaviour. There is some evidence in humans for reduced 5-HT_{1B} receptor function in depressed patients, and MAO inhibitors (which increase levels of noradrenaline and serotonin by suppressing their catabolism), SSRIs and 5-HT_{1A} receptor agonists (which down-regulate presynaptic AUTORECEPTORS) appear to work by decreasing sensitivity of 5-HT_{1A} and 5-HT_{1B} inhibitory autoreceptors.

afferent Leading, or carrying, towards a structure; used particularly of nerves, axons (see AFFERENT NEURONS), and blood and lymph vessels. Contrast EFFERENT.

afferent neurons In a functional classification, those neurons (or their axons) which synapse with sensory receptors, or which are themselves capable of responding directly to physiological stimuli. Their axons typically project into the central nervous system (see SOMATIC SENSORY SYSTEM).

affinity The strength of binding of one molecule to another at a single site.

affinity maturation The selection for survival of B cells with high affinity for a specific antigen, resulting in increased affinity of the antibodies produced during an adaptive immune response (see SOMATIC HYPERMUTATION). It is especially prevalent in secondary and subsequent immunizations. See VACCINES.

aflatoxin A mycotoxin produced by *Aspergillus flavus* growing on improperly stored food, such as grain. It is a highly toxic guanine-binding MUTAGEN (see DNA REPAIR MECHANISMS) inducing tumours in some animals.

afterbirth The PLACENTA, once it has detached from the uterus after birth of the baby.

after-images Visual sensations which outlast the stimulus, reflecting the bleaching of photopigment. Negative after-images are seen as complementary (opponent) colours (dark for a bright image, red for a green image, etc.). After a single bright flash followed by observing a white or pale grey background, one experiences at first a positive after-image due to bleaching and then a negative after-image due to neural mechanisms, fatigue and surround inhibition.

agammaglobulinaemia See X-LINKED AGAMMAGLOBULINAEMIA.

ageing (senescence) Progressive deterioration in the body's structure and function, the most evident external signs of which are wrinkling of the skin, loss and greying of hair, alopecia, stooping posture and weight loss. Internal, often imperceptible, changes include: osteoporosis, anaemia, decline in hearing (presbycusis) and vision (e.g., presbyopia, and AGE-RELATED MACULA DEGENERATION), in the ability of the immune system to respond to antigens, in memory and reaction response times, cardiomyopathy leading to decreases in maximum heart rate and cardiac output, kidney function (as the number of functional nephrons decreases), muscle strength (as myofibril numbers decrease), lung capacity and gonadal atrophy (see AGE-RELATED DISEASES). The gait tends to be slow, often on a widened basis, and shuffling is common. Standing posture is less erect due to loss of muscle power, and ability to maintain posture is reduced, especially when the eyes are closed. These are due to degenerative changes in the vestibular apparatus, cerebellum, skeletal muscles and proprioceptors. Some mouse models implicate activation of the WNT SIGNALLING PATHWAY in the age-related conversion of myogenic to fibrogenic pathways in progenitor cells, while recent work indicates that ageing and CANCER are complex biological tapestries that are often, although not always, woven by similar molecular threads.

GC Williams has argued that decline in old age could be caused by harmful pleiotropic effects of genes selected for advantages they offered during youth. In humans, it is likely that ageing, and particularly dementia, is accompanied by a global disturbance of the circadian metabolic order. A progressive advance in timing and loss of precision of the SLEEP–WAKE CYCLE tends to accompany ageing, being especially pronounced in ALZHEIMER'S DISEASE (see NEURAL DEGENERATION). Premature TELOMERE shortening has been invoked as a further biological ageing mechanism, although any involvement is not clear-cut.

The *disposable soma theory of ageing* posits that ageing results from accumulated damage. Accumulation of DNA damage leading to adult STEM CELL exhaustion has been suggested as a principal contributor to ageing. Multiple types of cellular and molecular damage (e.g., GENOMIC INSTABILITY) accumulate with age, probably resulting in increased cell death, anti-oxidant defences, or cell SENESCENCE (see GENOTOXIC STRESS). Although this suppresses carcinogenesis, it enhances ageing. It is not known how unrepaired DNA damage accelerates ageing, or how it is relevant to normal ageing; but there seems to be a link between genome maintenance and the somatotroph axis, leading to reduced growth hormone, IGF1/insulin, and other growth-promoting influences, resembling the effects of caloric restriction (see LIFE EXPECTANCY). There is also evidence that increased p53 activity can accelerate ageing (see P53 GENE). The main tenet of one 'disposable soma' approach, the *free radical theory of ageing*, is that ageing is promoted by progressive accrual of damage inflicted by reactive oxygen species, including nuclear and mitochondrial mutations and mitochondrial dysfunction. These have long been implicated in ageing, PARKINSON DISEASE and other age-related neurodegenerative diseases (see MTDNA). Recent studies implicate an accumulation of mtDNA mutations in the degeneration of dopaminergic neurons of the substantia nigra, and accumulation of mtDNA mutations in mice correlates negatively with mitochondrial function, notably in respiratory activity and ATP production. Permeation of the inner mitochondrial membrane and regulation of the autophagy of these damaged organelles (see MITOCHONDRION) by lysosomes may be a further piece in the jigsaw of ageing research. The serine/threonine kinase mTOR (see MTOR) is a critical inhibitor of such autophagy. There is a growing list of inherited 'progeroid' syndromes that resemble, but do not equate with, an accelerated process of normal ageing (see PROGERIAS).

As for physiological changes with likely ageing effects, the circulating levels of GROWTH HORMONE and sex steroids decline with age, with a consequent fall in IGF-I (see IGFS). In men, ageing is associated with decreased testosterone:oestradiol ratio, decreased LH pulse frequency, loss of diurnal rhythm of testosterone release and reduced accumulation of 5α-reduced steroids in reproductive tissues (collectively termed *andropause*). Circulating levels of DHEA and androstenedione also decline progressively in the adult male. For women, MENOPAUSE is responsible for decreased oestrogen output, with consequent general metabolic effects, as well as reproductive ones. In addition to biological processes that contribute to ageing, LIFESTYLE, nutrition, socio-economic status and (especially in older persons) a sense of being part of a community, and/or of not feeling socially isolated, are also factors promoting or delaying ageing.

age-related diseases/disorders Other than the general effects of AGEING, specifically age-related diseases include OSTEOPOROSIS, ARTHRITIS, ATHEROSCLEROSIS, ALZHEIMER'S DISEASE, CANCER, DEAFNESS, HAEMOPHILIA B, AGE-RELATED MACULA DEGENERATION and RHEUMATOID ARTHRITIS. Some are known to be caused by age-dependent 'hot-spot' mutations (see ACHONDROPLASIA, APERT'S SYNDROME), and declining levels of certain hormones certainly contribute to bone disorders (see AGEING). In human oocytes, the fidelity of chromosome segregation declines markedly as women age (see NONDISJUNCTION), although the underlying molecular defects are largely unknown. This is not apparent in men; however, germ cell point mutations in certain genes are age-related in men (see MUTATIONS, SPERMATOGENESIS). New evidence suggests that damage to the mitochondrial genome (see MTDNA) accumulates with age, specifically in the dopaminergic neurons of the substantia nigra implicated in PARKINSON DISEASE. Also, many autosomal dominant disorders have delayed age of onset (see SCREENING).

age-related macula degeneration (AMD) The leading cause of blindness in the elderly in the developed world. A chronic multifactorial disorder in which yellow extracellular deposits of proteins and other materials (drusen) form in and around the MACULA DENSA of the retina, between the pigment epithelium and the choroid, causing breakdown of the photoreceptor cells. In its more severe 'wet' form, AMD involves abnormal growth and breakdown of blood vessels in the macula. Its molecular pathogenesis is often characterized by a sequence-specific intronic variant in the gene (chromosome 1q32) encoding COMPLEMENT factor H involved in inflammation. The Y402H variant of the complement factor H gene *CFH* is a strong susceptibility factor for AMD (and for myocardial infarction). Variants of the closely linked *BF* (*CFB*) and *C2* genes, also major regulators in the alternative complement pathway, also contribute significantly to risk. Recent work implicates a SINGLE NUCLEOTIDE POLYMORPHISM on chromosome 10 (10q26) in the regulatory sequence of the *HTRA1* gene whose product is a protease known to interact with TGF-β, which is important in the final stages of VASCULOGENESIS. There is twice the risk of developing this condition in chronic cigarette smokers. Recent studies indicated that more protective than susceptibility haplotypes for AMD were found, that susceptibility and protective haplotypes conferred effects of comparable magnitude but in opposite directions (i.e., resistance versus susceptibility), and that protective haplotypes occurred at an appreciable frequency in the general population. The strong effects of protective alleles might be exploited pharmacologically to modulate inherited diseases, such as AMD. See GENE THERAPY.

agglutination Clumping together of large particles, e.g., bacteria, by the antigen-binding regions of antibodies. See HAEMAGGLUTINATION.

aggressive behaviour See ANGER AND AGGRESSION.

agnosia The inability to recognize objects visually, despite having normal simple sensory skills. It often involves lesions of the INFEROTEMPORAL CORTEX. One form, astereognosia, is the inability to recognize objects by touch, which may accompany similar difficulties in object recognition by sight or sound. It can result from lesions in the POSTERIOR PARIETAL CORTEX, often to the contralateral side.

agonist A substance binding to and activating a receptor, thereby mimicking the effect of a natural neurotransmitter or hormone. Thus phenylephrine activates α_1 ADRENERGIC receptors, constricting blood vessels in the nasal mucosa, reducing mucus and relieving symptoms of the common cold. Compare ANTAGONIST.

agouti-related protein (AgRP) An orexigenic neuropeptide released by some neurons of the ARCUATE NUCLEUS that stimulates APPETITE by blocking α-MSH action.

agraphia Impaired ability to write that can result from damage to the CORPUS CALLOSUM. Compare ALEXIA.

AgRP See AGOUTI-RELATED PROTEIN.

AHR (aryl hydrocarbon receptor, AhR) A cytosolic ligand-activated BASIC-HELIX-LOOP-HELIX MOTIF transcription factor increasingly recognized as functioning in CELL CYCLE control. It mediates the toxic effects of many environmental contaminants including halogenated aromatic hydrocarbons and non-halogenated polycyclic aromatic hydrocarbons (PAHs, e.g., dioxin; see TOXINS). Residing in the cytoplasm in complex with Hsp90 (see HEAT-SHOCK PROTEINS), binding of ligand leads to conformational changes resulting in exchange of Hsp90 for the AhR nuclear translocator ARNT. The AhR/ARNT complex is often referred to as the *xenobiotic response element*, XRE, and binds specific sequence in the promoter region of AhR-responsive genes. AhR also regulates T_{reg} and T_H17 cell differentiation (see T CELLS) in a ligand-specific manner, affording a unique target for therapeutic immunomodulation. Inhibition of AhR signalling is a potential chemoprotective mechanism, and the green tea component polyphenol epigallocatechin gallate (EGCG) is capable of antagonizing AhR-mediated gene transcription. EGCG appears to bind the AhR chaperone protein Hsp90, preventing DNA-binding.

AIDS See ACQUIRED IMMUNE DEFICIENCY SYNDROME, HIV.

akinesia Difficulty in initiating movements. See INTENTIONAL MOVEMENTS, PARKINSON DISEASE.

akinetopsia Loss of ability to perceive motion. Lesions in area V5 of the MEDIAL TEMPORAL CORTEX are involved. See VISUAL MOTION CENTRE.

Akt (protein kinase B) Alternative acronym and name for protein kinase B (see PKB).

alactacid oxygen debt See OXYGEN DEBT.

alactic aerobic system See AEROBIC RESPIRATION.

alar plates Distinct bilateral sensory areas in the rhombencephalon and mesencephalon of the embryonic spinal cord, whose neuroblasts develop into association (relay) neurons. Each plate contains three groups of sensory relay nuclei: the somatic afferent group, receiving impulses from the ear and head surface via cranial nerves V and VIII; the intermediate group, receiving impulses from the palate and taste buds of the tongue; and the medial group, receiving impulses from the gastrointestinal tract and heart. Compare BASAL PLATES.

alarm response See STRESS RESPONSE.

albinism Two major, partially overlapping, albino phenotypes occur. In *oculocutaneous albinism* (OCA), melanin is absent or partially deficient in the eyes, skin and hair, accompanied by poor vision, squinting and nystagmus (rapid, jerky involuntary eyeball movements). Mutations in several loci are responsible, producing at least ten heterogeneous phenotypes ranging from absent to nearly normal skin pigmentation. In its classic form, which is recessive (homozygous oculocutaneous albinism occurs in ~1:1000 births), tyrosinase enzyme lesions prevent melanocytes from converting tyrosine to DOPA and dopaquinone, preventing formation of MELANINS. This produces very pale hair and skin, blue or pink irides, red pupils and photophobia. It is relatively common in the Hopi (Arizona) and Kuna San Blas (Panama) Indians. The much less frequent *ocular albinism* (OA), of which four forms have been described, may be either X-linked or autosomal, defects being limited to the eyes, skin and hair being normal or only slightly underpigmented. See Albinism Database: http://www.cbc.umn.edu/tad/genes.htm

albumins The major group of plasma proteins (~54% of the total plasma protein mass), generally being small proteins and almost entirely produced by the liver, with proportionate roles in blood colloid osmotic pressure (see TISSUE FLUID), buffering (see ACID–BASE BALANCE), and binding of FATTY ACIDS (each molecule binding 10 fatty acid molecules), oestrogen and progesterone, and bilirubin. Albumins also bind ~35% of plasma calcium, in a pH-sensitive manner. The albumin gene cluster is on chromosome 4q12.

alcohol (ethanol, ethyl alcohol) Used socially to provide a sense of well-being, a general depressant drug with effects resembling those of general ANAESTHETICS. It inhibits presynaptic Ca^{2+} entry (and hence release of transmitter) and potentiates GABA-mediated inhibition. Alcohol dehydrogenase (ADH) produced by liver (80%) and stomach (20%) converts ethanol (C_2H_5OH) to acetaldehyde (CH_3HCO) and methanol (CH_3OH) to formaldehyde (H_2CO) (damaging to many tissues). It has several forms, but the *ADH1* and *ADH4* genes in particular enhance the risk of alcoholism. The high NADH levels generated by the cells involved inhibit glycolysis and the KREBS CYCLE. In consequence, sugars and amino acids are not broken down but are converted into fats that accumulate in the liver, leading to cirrhosis. The enzyme aldehyde dehydrogenase breaks down acetaldehyde, but slight variations in its sequence lead to differences in its rate of activity. The slow-acting enzyme encoded by the *ALDH1* allele is common in Asian populations, but rare in people of European descent; it causes increased plasma acetaldehyde concentrations leading to warmth of the skin, palpitations and weakness. Chronic alcohol abuse (*alcoholism*), a COMPLEX DISORDER, also leads to physical dependence, often to psychosis, and to symptoms of thiamine deficiency (see BERIBERI). EEG patterns differ in characteristic ways between alcoholics and non-alcoholics, with excitation exceeding inhibition in the former. This disinhibition can also be observed in the children of alcoholics, and may point to vulnerability: an inherited predisposition to alcoholism. Markers appearing most often in families exhibiting alcoholism-associated phenotypes have been detected on CHROMOSOMES 1, 2, 4 and 7 and include *ADH4* and *GABRA2* (a gene for one of the $GABA_A$ receptor subunits) on hsa4 and *CHRM2* (encoding the M2 muscarinic ACh receptor) on hsa7. As with *GABRA2*, the *CHRM2* variants appear to alter the rate of production of the receptor, not its structure.

The considerable tolerance to alcohol is poorly understood, but presynaptic Ca^{2+} channels may increase in number. As a result, when alcohol is withdrawn transmitter release is abnormally high and may lead to the withdrawal syndrome, which may need diazepam (Valium) to prevent seizures (see BENZODIAZEPINES). The clumsiness accompanying ethanol abuse results from depression of circuits in the CEREBELLUM. Ethanol enhances the function of the $GABA_A$ receptor, its effects depending on the receptor's composition of α, β and γ subunits. It also has complex effects on NMDA, glycine, nicotinic ACh and serotonin receptors. Alcohol is a teratogen if taken in high doses by a pregnant mother during the third week of embryonic development as it kills cells in the anterior midline of the germ disc, causing HOLOPROSENCEPHALY. See FOETAL ALCOHOL SYNDROME, REINFORCEMENT AND REWARD, SIRTUINS.

alcohol dehydrogenase (ADH) See ALCOHOL.

alcoholism See ALCOHOL, DRUG DEPENDENCE.

aldosterone The major human MINERALOCORTICOID, synthesized from CHOLESTEROL (see Fig. 48) by the cortical cells of the adrenal glands. The most important mechanism controlling its secretion is the renin-angiotensin pathway (see ANGIOTENSINOGENS). Aldosterone release by the adrenal gland can be stimulated independently by either blood volume contraction or hyperkalaemia (a serum potassium concentration greater than 5.5 mmol L^{-1} (mEq L^{-1}); a concentration above 6.5 is considered critical). In volume contraction, stimulation is mainly mediated by angiotensin II. Acting on fewer cell types than do glucocorticoids, its main function is to regulate renal sodium reabsorption and so maintain water balance. It binds mineralocorticoid receptors on the cells of DISTAL CONVOLUTED TUBULES (DCTs) and COLLECTING DUCTS, but also diffuses through cell membranes and binds a nuclear hormone receptor (type I glucocorticoid receptor) in the cytosol before entering the nucleus as a transcription factor and promoting selective gene transcription. It up-regulates synthesis, in PRINCIPAL CELLS and INTERCALATED CELLS appropriately, of Na^+ channels and of the H^+-ATPase in the apical membranes, and of Na^+/K^+-ATPase and the Cl^-/HCO_3^- exchanger in the basolateral membranes. This increases the rate of Na^+ transport out of the tubular fluid back into the blood, causing water to follow osmotically where it can. Reduced aldosterone secretion decreases the rate of Na^+ transport, reducing the capacity for water to move out of the collecting ducts and DCTs, so that urine volume and Na^+ concentration increase.

WNK1 and WNK4, members of a small family of kinases, are expressed in the distal nephron, specifically in the DCTs and collecting ducts. They seem to interact with the Na^+-Cl^- co-transporter (NCC) located in the DCT. NCC is the target of thiazide diuretics widely used in treatment of HYPERTENSION. In mouse models, mice made homozygous for the wild type human transgene *Wnk4* had reduced blood pressures and a tendency for hypokalaemia (low plasma potassium); but mice bearing a transgene engineered to replicate a missense Wnk4 mutation from humans with pseudohypoaldosteronism type II (PHAII) developed hypertension, hyperkalaemia and hypercalciuria, all features of PHAII. Through its control of NCC, especially in the more distal portions of the DCT, WNK4 can alter the balance between electrogenic and electroneutral sodium reabsorption and therefore determine whether uptake of sodium is accompanied by chloride reabsorption or potassium secretion and may therefore directly shape the renal response to aldosterone. Undersecretion of aldosterone results in rising blood K^+ levels and cardiac ARRHYTHMIAS, even cardiac arrest.

Plasma concentrations of glucocorticoids are much higher than those of aldosterone, yet this does not lead to their maximal occupancy of the aldosterone receptor despite high affinity for it (see GLUCOCORTICOIDS). Aldosterone-sensitive cells contain the enzyme 11β-hydroxysteroid dehydrogenase type 2, which converts glucocorticoids to inactive forms with less affinity for the mineralocorticoid receptor, and aldosterone dissociates from the receptor five times more slowly than do glucocorticoids, so that it is less easily displaced than cortisol. See ESSENTIAL HYPERTENSION.

alexia Inability to understand written material. Can accompany left FRONTAL LOBE lesions. Compare AGRAPHIA.

alimentary canal (digestive tract, gut) Developing from the endoderm-lined yolk sac cavity, the alimentary canal is already differentiated anteriorly into foregut (which gives rise to the lung bud; see LUNGS) and, in the tail region, into hindgut. The midgut is originally continuous with the yolk sac, but this vitelline duct narrows and becomes much longer so that by the end of the 1st month it is becoming part of the UMBILICAL CORD. Much later, the vitelline duct is obliterated and the midgut becomes free in the abdominal cavity. During cephalocaudal folding of the embryo, endoderm forms the epithelial lining of what will become the adult digestive tract, whose muscle, connective tissue and peritoneal contributions derive from splanchnic mesoderm, its differentiation resulting from reciprocal interactions between the mesoderm and endoderm (see Fig. 4).

Portions are suspended from the dorsal and ventral body walls by mesenteries (e.g., the greater OMENTUM). Its component organs are: the ORAL CAVITY, pharynx, oesophagus, STOMACH, SMALL INTESTINE and LARGE INTESTINE. Primates eating high-quality meat-rich diets tend to have smaller digestive tracts and larger BRAIN SIZES than do those eating high carbohydrate diets. For digestive processes, see the entries on the gut regions named above; and for absorption

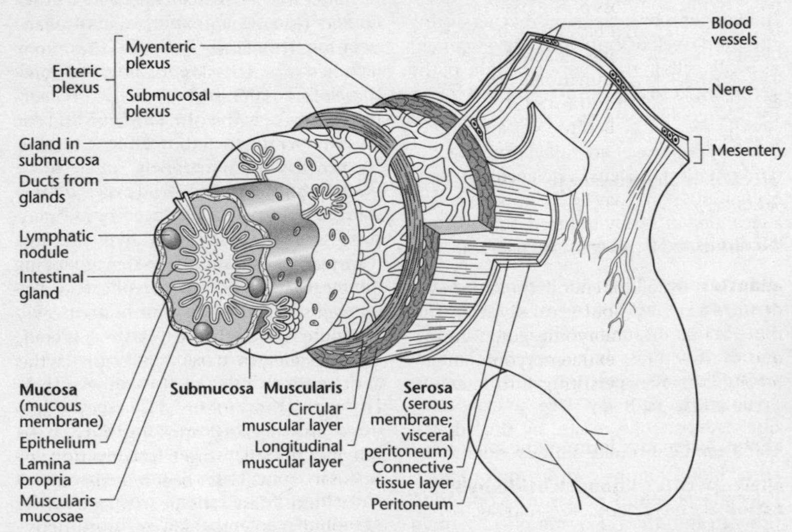

FIG 4. The basic histology of the ALIMENTARY CANAL. Regions differ in their detailed anatomies; e.g., glands in the submucosa are typical of the duodenum but not elsewhere.

of digestion products, see ENTEROCYTES. See ENTERIC NERVOUS SYSTEM, SWALLOWING.

alkaloids A large group of chemically unrelated plant compounds, many ultimately synthesized from a few common amino acids, whose three principal features are that they (i) are soluble in water, (ii) possess at least one nitrogen atom (which may be basic – hence the name) and (iii) are biologically highly active. Noted mainly for their pharmacological properties, most are heterocyclic although a few (e.g., mescaline, colchicine) are aliphatic. They include vinblastine and vincristine (from the Madagascar periwinkle), cocaine (from leaves of the coca plant), codeine (from the opium poppy), nicotine (tobacco), caffeine (coffee beans and tea leaves). Many have antibiotic properties and some are toxic in high concentrations. They tend to have an extremely bitter TASTE.

alkylating agents Among the most potent mutagens, these chemical species are capable of attaching alkyl (methyl) groups covalently to DNA bases. This may destabilize its covalent bond to deoxyribose, resulting in loss of the base from the DNA. The alkylated bases may be misread by the DNA polymerase machinery during DNA REPLICATION. Bifunctional alkylating agents can cause intra-strand or inter-strand cross-links. Because certain cancer therapeutic drugs are alkylating agents (and potent mutagens) a delayed outcome of chemotherapy may be the appearance of a secondary, therapy-induced, tumour. See DNA METHYLATION, DNA REPAIR MECHANISMS.

allantois A stalk of endoderm and mesoderm which grows out ventrally from the posterior of the embryonic gut, forming one of the three extraembryonic membranes, part of it persisting into later life as the urinary BLADDER.

alleles Alternative forms of the same GENE.

allele-specific oligonucleotide hybridization See DNA SEQUENCING, SINGLE NUCLEOTIDE POLYMORPHISMS.

Allen's rule Published in 1877, this states that populations of a geographically widespread species living in warm regions will have longer extremities than those living in cold regions. See BODY MASS, SIZE AND SHAPE. Compare BERGMANN'S RULE.

allergen Antigens eliciting hypersensitivity or allergic reactions. See ALLERGIES.

allergies Diseases following immune responses to otherwise innocuous substances. One of a class of immune responses termed HYPERSENSITIVITY responses, allergic reactions occur when individuals who have produced IgE antibodies in response to an innocuous antigen, and who subsequently encounter the same allergen, produce a series of responses triggered by activation of IgE-binding cells (including MAST CELLS and BASOPHILS). Allergic responses can be regarded as having immediate and late-phase components. The *immediate reaction* is due to activities of histamine, prostaglandins and other preformed or rapidly synthesized mediators causing increased vascular permeability (INFLAMMATION) and smooth muscle contraction. The *late-phase reaction* is due to induced synthesis and release of prostaglandins, leukotrienes, chemokines and cytokines from activated mast cells and includes a second phase of smooth muscle contraction mediated by T CELLS, with lasting oedema and tissue remodelling including smooth muscle hypertrophy and hyperplasia.

The more sustained effects of the late-phase reaction depend upon the anatomy of the site where the immediate reaction occurred, and the clinical syndrome of an allergic reaction depends on the amount of allergen-specific IgE present on mast cells, the route by which the allergen entered, and the allergen dose. Inhalation is the most common route of allergen (e.g., pollen) entry, inducing mild allergies manifested as sneezing and a runny nose (allergic rhinitis) through activation of mucosal mast cells beneath the nasal epithelium. Mast cells degranulate when they bind an allergen. Allergic conjunctivitis is a similar response to allergens on the conjunctiva of the eye. If introduced directly into the bloodstream, or rapidly absorbed across the gut, an allergen will activate connective tissue mast cells associated with all blood vessels, leading to systemic anaphylaxis. Such disseminated mast cell activation can lead to catastrophic loss of blood pressure, airways constriction and potentially suffocating swelling of the epiglottis (see ANAPHYLACTIC SHOCK).

Atopy is an exaggerated tendency to mount IGE responses to a wide variety of environmental allergens, atopic individuals having a higher total level of circulating IgE and higher eosinophil counts than normal and being more susceptible to allergic diseases such as hay fever (rhinitis) and allergic ASTHMA. Atopy and its associated susceptibility to asthma, hay fever and eczema, can be determined by different genes in different populations, affecting responses to drug treatment (see COMPLEX DISEASES). Regions on chromosomes 11q and 5q appear to be important in determining atopy, the candidate gene on chromosome 11 encoding the β subunit of the high affinity IgE receptor. Chromosome 5q is more complicated because 5q23-25 contains a plethora of possible candidate genes for atopy. Some of these are tightly linked genes for cytokines that promote T_H2 (see T CELL) responses by enhancing IgE isotype switching, EOSINOPHIL survival and mast-cell proliferation. Genetic variation in the promoter region of one of the genes, that for IL-4, has been associated with raised IgE levels in atopic individuals. A second set of genes on 5q encodes T-cell surface proteins, inherited variation in which has been associated with varying susceptibility to allergic inflammation of the airways in mice. A third 5q region encodes the p40 subunit of IL-12, a cytokine promoting CD4 T_H1 responses. Variation in its expression causing reduced IL-12 production is associated with more severe forms of asthma. The β-ADRENERGIC receptor is also encoded in this region, variation in which could be associated in smooth muscle responsiveness to endogenous and pharmacological ligands. Again, many studies show that variation in IgE production in response to particular allergens is associated with having certain MHC class II alleles and that particular MHC:peptide combinations may favour a strong T_H2 response. Allergies to common drugs such as penicillin show no such association, or with the presence or absence of atopy.

Inflammatory bowel diseases (IBDs) such as Crohn's disease (CD) and ulcerative colitis (UC) are complex diseases believed to be caused by inappropriate immune responses to commensal intestinal bacteria, or possibly defective mucosal barrier function or bacterial clearance. Increasing evidence indicates that these disorders have a genetic component (see COLON). Genetic predisposition to CD is suggested by a sibling recurrence risk, λ_s, of 17–35 and twin studies that contrast monozygotic phenotypic concordance rates of 50%, with only 10% in dizygotic pairs. Crohn's disease is characterized by aggregation of macrophages that frequently cause non-caseating granulomas, involvement of the ileum being most common and the earliest mucosal lesions to appear occurring over Peyer's patches. Unlike UC, it may be patchy and segmental and seems to result from an exaggerated T_H1 response in which large amounts of IL-12, IL-18 and interferon-γ are produced (see INTERFERONS). Reliable evidence supports involvement of the gene *NOD2* (aka *CARD15* and *IBD1*; see NOD PROTEINS); but several additional susceptibility genes have been implicated in IBDs, including *IBD5*, *IL23R* and *ATG16L1*. A number of genes associated with Crohn's disease are also associated with other autoimmune disorders. Early linkage analysis indications that the MHC region contributes to IBD susceptibility have been difficult to confirm with precision (unlike the situation in rheumatoid arthritis and multiple sclerosis).

The corticosteroids used in treating allergic disease are often given in combination with other drugs in an attempt to keep their toxic effects to a minimum. Monoclonal antibodies against TNF-α (see TUMOUR NECROSIS FACTORS), and TNF-binding proteins, have had some therapeutic success. Animals deficient in IL-10 develop a chronic IBD and are susceptible to a more severe form of collagen-induced arthritis. See PHOSPHATIDYLINOSITOL 3-KINASE.

allodynia A form of hypersensitivity in which PAIN is experienced in response to normally innocuous stimuli.

allogeneic Two individuals of the same species that are genetically different (e.g., in their MHC loci). Rejected GRAFT tissue from unrelated donors frequently results from T-cell responses to allogeneic MHC molecules expressed by the grafted tissues (see GRAFTS). Contrast SYNGENEIC, XENOGENEIC.

allometric growth The faster rate of growth of one part of the body relative to the rest; or, size-dependent alterations in organic form and process – the effects of scale. In humans, limb size and shape have an allometric relationship with BODY SIZE and are strongly influenced by such factors as climate.

allopatric speciation Speciation in which one adapted gene pool gives rise to two by a process that involves geographical, and consequent reproductive, isolation of the two populations.

allozyme genetic markers See GENETIC MARKERS.

alpha activity See ATTENTION; ELECTROENCEPHALOGRAM; SLEEP, Fig. 158.

alpha blocker See ADRENERGIC.

alpha cells (α cells) See PANCREAS.

alpha-defensins See DEFENSINS.

alpha rhythms See ELECTROENCEPHALOGRAM.

alpha satellite repeat Large, highly repetitive stretch of (A + T)-rich DNA sequences in the human genome that are usually not transcribed. See SATELLITE DNA.

ALS See AMYOTROPHIC LATERAL SCLEROSIS.

alternative splicing (AS) See EXON SHUFFLING, RNA PROCESSING.

altitude The decreased ambient partial pressure of oxygen (PO_2) as air density decreases is the major challenge of altitude, changing from its sea level value of 100 mm Hg to 78 mm Hg at 1980 m. Only a small change occurs in the percentage saturation of HAEMOGLOBIN with oxygen until ~3050 m, remaining 90% saturated at 1980 m. Arrival at altitudes higher than 2300 m initiates HYPERVENTILATION and increased resting systemic blood pressure through raised submaximal heart rate and cardiac output (stroke volume initially remaining constant). Body fluids become more alkaline on account of the reduction in CO_2 (H_2CO_3) with hyperventilation. Additional longer-term adjustments to altitude include greater renal HCO_3^- excretion, submaximal cardiac output falling to below sea-level values, decreases in stroke volume and maximal cardiac output, decreased blood volume; increased haematocrit, haemoglobin concentration and red cell count (see ERYTHROPOIETIN); increased red-blood-cell 2,3-DPG; increased mitochondrial density and aerobic enzymes in skeletal muscle, and loss of body weight and lean body mass. While such relatively short-term acclimatization contributes to the ability of high-altitude populations to maintain themselves, it appears that either developmental change or genetic adaptation is necessary for people to become fully functional at the higher elevations. Most research does not support endurance training at altitude as improving subsequent sea-level EXERCISE performance.

altruism For Charles Darwin, altruism in the form of sterile worker castes of bees that appeared to 'sacrifice' themselves for the good of the colony was a 'special difficulty' for evolutionary theory. The phenomenon of kin altruism has a biological explanation in HAMILTON'S RULE. But additional theories of cooperation are required for a complete explanation of the complex interactions between human individuals. These invoke reward and punishment (including reciprocity), fairness, reputation, emotion, the sense of self, empathy, and other aspects of morality. See KIN DETECTION, KIN SELECTION.

***Alu* insertion polymorphism** A polymorphism defined by the insertion of an *Alu* element into one allele. See ALU REPEAT FAMILY.

***Alu* repeat family** A primate-specific family of SINES (see DNA REPEATS), these 300–400 bp short interspersed elements, so-called because they contain a target site for the *Alu* restriction enzyme, originated from copies of 7S RNA. There are over one million *Alu* sequences in the HUMAN GENOME, or 10.6% of the draft human genome, so that virtually all human genes are closely linked to one or more *Alu* sequences. Some of these elements are human-specific, ~2000 of them being absent from chimpanzee and gorilla DNA, a subset of these still being active in transcription (see TRANSPOSABLE ELEMENTS). *Alu* repeats appeared in primate genomes over 60 Myr ago, expanding rapidly 30–50 Myr ago. Placed in several subfamilies on the basis of sequence identity, some *Alu* elements may inactivate gene function. Although *Alu* repeats are not long enough to encode reverse transcriptase, they sometimes make use of it from elsewhere and get successfully 'copied' from DNA to RNA and then to double-stranded DNA and become inserted at other sites, modifying function (e.g., in some cases of HAEMOPHILIA and Duchenne muscular dystrophy). They also carry an internal promoter for RNA polymerase III, although this is insufficient for active transcription because it lacks appropriate flanking sequences needed for activation. However, an *Alu* repeat may integrate in a region enabling its promoter to become active. The taxonomic grouping of rodents, primates and tree shrews is supported by their being the only eutherian mammals known to have *Alu*-like SINEs derived from 7SL RNA (see RNA). See MICROSATELLITES.

alveolus (pl. alveoli) **1**. Minute air-filled sac (see Fig. 5) at the end of each alveolar duct and grouped together as alveolar sacs to form the termini of bronchioles in the lungs, with a total surface area of 140 m^2. Averaging ~250 μm in diameter, there are ~150 million per lung, their thin walls being composed of a single layer of squamous epithelium (type I alveolar cells), on the inner surface of which are type II alveolar cells which secrete SURFACTANT, and macrophages which remove bacteria and dust. The ACUTE PHASE PROTEINS SP-A and SP-D, which opsonize pathogens, are also found along with macrophages in the alveolar fluid, as are β-DEFENSINS. A rich capillary network in contact with the alveolar surface provides the surface for the exchange of respiratory gases (a total of ~100 m^2, or ~40 × the body's external surface). Each red blood cell spends ~0.5–1.0 seconds in the exchange capillaries per pulmonary circuit, traversing an average of 2–3 alveoli before entering a venule. The large alveolar surface area:volume ratio increases the probability that newly inhaled oxygen molecules will make contact with the alveolar epithelium which, along with the capillary endothelial wall and their basement membranes, comprises the *respiratory membrane* across which gases diffuse. With an average thickness of just 0.3 μm this allows steep concentration gradients for oxygen and carbon dioxide to be generated and offers little resistance to their diffusion. Elastic fibres in the connective tissue surrounding the alveoli are important for their recoil during exhalation, as are stretch receptors in the control of VENTILATION (see also for *alveolar pressures*). *Minute alveolar ventilation* is the total volume of new air entering the alveoli and adjacent gas exchange areas per minute and equals the product of the respiratory rate and the volume of new air entering these areas per minute. With a normal tidal volume of 500 mL, a normal dead space of 150 mL and 12 breaths per minute, alveolar ventilation is 4200 mL min^{-1}.

Quiet breathing around the FUNCTIONAL RESIDUAL CAPACITY does not alter breath-by-breath alveolar surface very much and as a result the tightly compressed surface film of surfactant molecules is slowly squeezed away from the surface. The progressive lowering of surfactant concentration which results raises the surface tension, giving the lung greater elastic resistance. A sigh then reflexly occurs, the large pulmonary inflation which results stretches and unfolds the alveolar surface area, spreading out the surfactant molecules and permitting recruitment of further molecules, returning the alveolar surface tension to its usual ~12 dyn cm^{-1}. **2**. Expanded sac of secretory epithelium forming internal termini of ducts of many glands, e.g., MAMMARY GLANDS. **3**. Bony sockets into which teeth fit in the mandible and maxilla, lying in the *alveolar processes* of the jaws.

Alzheimer('s) disease (AD) A genetically complex and heterogeneous disorder following an age-related dichotomy. Rare autosomal dominant mutations cause early-onset familial AD, whereas risk for late-onset familial AD is probably modulated by genetic variants with fairly low penetrance but high prevalence. AD has been compared with the destruction of a computer's hard drive, beginning with

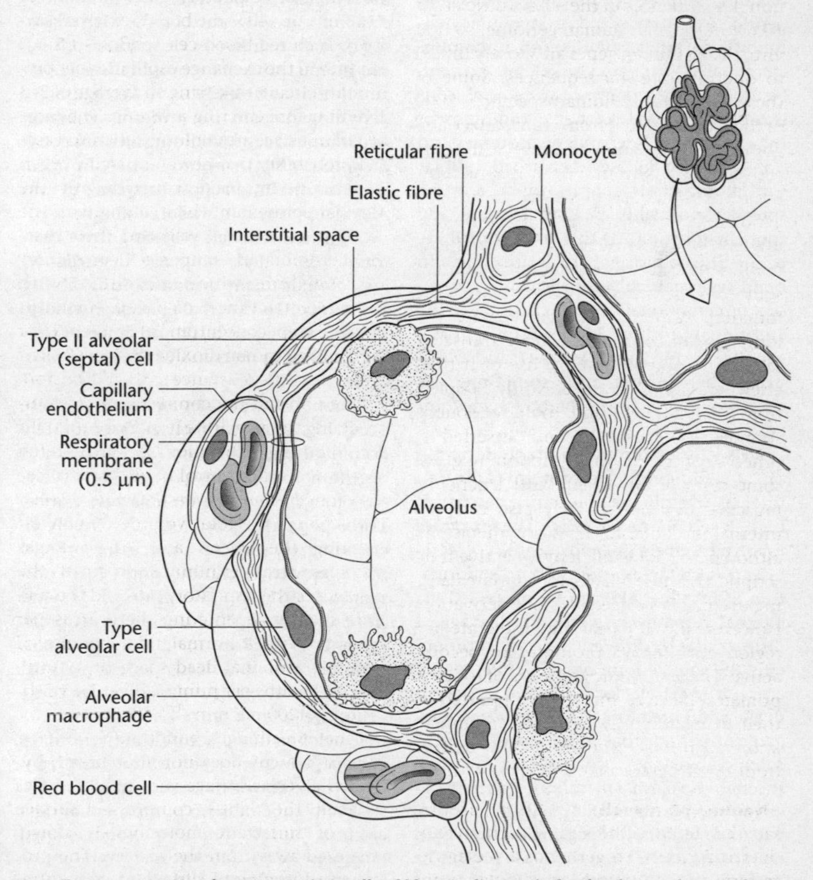

FIG 5. Transverse section of an ALVEOLUS (normally ~200 μm in diameter), the alveolar membrane is shown enlarged relative to the rest of the structure.

the most recent files and working backwards. However, reloading of such 'brain files' is impossible because AD destroys the neurons that store this information (see MEMORY). Drugs that prevent acetylcholine breakdown at cortical synapses, and those that block excessive glutamate activity, stave off memory loss, but cannot halt the neuron loss. The phenomenon of EXCITOTOXICITY has sometimes been implicated.

AD is characterized clinically by progressive cognitive decline and pathologically by the presence in the brain of senile plaques composed primarily of β-amyloid peptide (Aβ, A beta) and intracellular neurofibrillary tangles (NFTs) of hyperphosphorylated TAU, a cytoskeletal protein. Neurofibrillary tangle-bearing neurons exhibit high levels of Ca^{2+}, and signs of hyperactive Ca^{2+}-dependent proteases and Ca^{2+}-activated kinases. Perturbed cellular Ca^{2+} homeostasis appears to be a widespread abnormality in both familial and sporadic forms of AD that may contribute to the disease process. The lesions form in brain regions involved in learning, memory and emotional behaviour, notably the entorhinal cortex, hippocampus, basal ganglia, the nucleus of Meinert, and amygdala. The disease may sometimes be detected by PET scans before symptoms appear.

Aβ is released through proteolytic action on amyloid precursor protein (APP), an integral membrane protein present at high levels in nerve cells. The enzymes involved are beta (β-) and gamma (γ-)secretases (β-secretase is essential for the formation of myelin). APP can be processed via alternative pathways through cleavage at distinct sites by different secretases, producing the soluble form of APP, sAPP, which acts as a growth factor for neural stem cells in the adult brain or, in AD, the amyloidogenic Aβ peptide that goes to form the neurotoxic Aβ plaques. Molecules of APP stick through the plasma membrane, with intracellular and extracellular portions exposed, with a hydrophobic portion in the lipid membrane itself. The γ-secretase inhibitors developed for AD therapy may also turn out to be helpful as regulators of ANGIOGENESIS. Mutations in the *APP* gene on chromosome 21 (see DOWN SYNDROME) cause some hereditary Alzheimer's cases. In mouse studies, over-expression of APP led to blockage of protein import in the outer mitochondrial membrane leading to impaired energy metabolism. In 2002, a radioactive compound (PIB) was discovered that reveals β-amyloid in living brains, and the latest fluorine-containing PIB should enable PIB PET scans to monitor the effectiveness of candidate drugs in reducing β-amyloid build-up. There is much evidence for the involvement of dysfunctional mitochondria in the pathogenesis of AD, with resulting oxidative stress (see MITOCHONDRION). Aβ can interact with mitochondria (see Fig. 6) and inhibit complex IV and α-ketoglutarate dehydrogenase in the inner membrane, and it can bind to Aβ-binding alcohol dehydrogenase (ABAD) in the matrix. In mice, blocking the interaction between Aβ and ABAD suppressed Aβ-induced apoptosis.

Cognitive disorders resulting from neuronal loss include language disturbance, loss of judgement and reasoning ability and difficulties in verbal episodic memory. There is also a progressive advance in timing and loss of precision of the sleep/wake cycle.

About 5–10% of cases are familial, these occurring in an early-onset, autosomally dominant manner, most people with it beginning to express it by 60 years, although these are of variable penetrance. Three separate gene loci are involved, encoding (i) amyloid-β precursor protein (APP), located on chromosome 21; (ii) presenilin-1 (PS1), on chromosome 14; and (iii) presenilin-2 (PS2), located on chromosome 1. Presenilins are mainly found in neurons, and to a lesser extent in glia. The presenilin proteins encoded by these genes are part of the γ-secretase enzyme that helps to manufacture the harmful Aβ peptide, which may perturb calcium regulation by inducing oxidative stress, impairing membrane calcium pumps, enhancing calcium uptake through voltage-gated channels and ionotropic glutamate receptors. But presenilin-based mutations are very rare causes of AD. APP is present

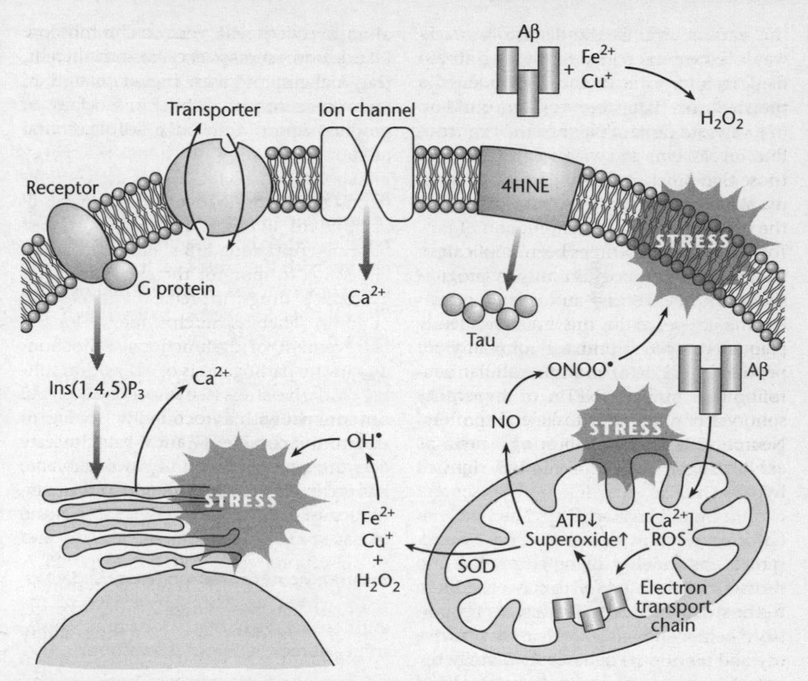

FIG 6. Diagram indicating the hypothetical neurotoxic effect of Aβ peptide in ALZHEIMER'S DISEASE. Aβ is believed to interact with Fe^{2+} and Cu^{+} when it aggregates at the cell membrane, resulting in production of hydrogen peroxide (H_2O_2) and lipid peroxidation, with resulting 4-hydroxynonenal (4HNE), a neurotoxic aldehyde covalently modifying cysteine, lysine and histidine side chains. Proteins so modified include membrane glutamate and glucose transporters, voltage-dependent chloride and NMDA receptor channels, and G proteins. Oxidative modifications of tau protein by 4HNE and other reactive oxygen species can promote tau aggregation, inducing neurofibrillary tangle formation. Mitochondrial oxidative stress and dysregulation of calcium (Ca^{2+}) homeostasis by Aβ can impede the electron transport chain (see MITOCHONDRION), while superoxide conversion to H_2O_2 by superoxide dismutases (SOD) and superoxide also interact with NITRIC OXIDE (NO) to generate peroxynitrite ($ONOO^{\cdot}$), while H_2O_2 interacts with Fe^{2+} or Cu^{+} to generate the hydroxyl radical ($OH^{\cdot}$), a potent contributor to dysfunction of the ENDOPLASMIC RETICULUM.

in glia and neurons – especially in synapses, where it may affect neuronal communication by loss of SYNAPTIC PLASTICITY. Associated GENETIC RISK factor mutations include those for apolipoprotein E4 (apoE4, see APOE) which promotes Aβ aggregation and decreases its clearance, and interleukin-1. Individuals who inherit *APOE4* have an increased risk of late-onset AD; those homozygous for the allele have a 16-fold increased risk of developing AD, those who do so having a mean onset of disease of just under 70 years; but those homozygous for *APOE3* have a mean onset in excess of 90 years.

Genetic linkage studies implicate the long arm of chromosome 10, which may contain one or more loci associated with late onset AD. Recent work suggests that it is absence of the protein encoded by *SORLA* (aka *LR11*) that results in APP being trafficked off to compartments such as the late endosome that contain PS1, a misdirected protein transport that may contribute to the late onset form of AD.

A vaccine against the disease is a long way off; but mice genetically engineered to develop a form of the disease show a positive correlation between the build-up of β amyloid protein plaques and cognitive decline. Vaccination with Aβ protects mice with Alzheimer-like pathology from both plaque build-up and loss of memory and learning ability.

Low dietary FOLATE increases the risk of AD (probably through increased levels of HOMOCYSTEINE). Dietary lipids and metals such as copper and iron are also positively correlated. One promising approach for preventing and treating AD involves stimulating the immune system to remove Aβ from the brain, an initial clinical trial suggesting that immunization with anti-Aβ antibodies proved promising but required refinement.

Although increased CREB function can enhance memory under certain circumstances, its widely ranging effects can also disrupt cognitive performance – the brain regions in which it has these effects possibly being different, so that its putative therapeutic use in diseases involving cognitive decline, such as AD, may depend on improved understanding of the molecular elements in the different pathways by which CREB influences different parts of the brain.

amacrine cells Very diverse small interneurons of the inner plexiform layer of the RETINA (see Fig. 149) with short processes (hence, no axons), serving to relay information from cones in the surround of a ganglion cell's receptive field to the ganglion cell via bipolar cells. Their extensive neurites share properties of both axons and dendrites, and there are ~30 types, using at least eight different neurotransmitters. They are implicated in signalling by rod cells, surround inhibition and motion detection. Some relay information from distant bipolar cells to ganglion cells. Dopaminergic amacrine cells (~1%) have long interconnecting dendrites, forming a network that is possibly held together by gap junctions. These cells get input from cone bipolar cells, and are able to signal average illumination, used to produce surround inhibition. Like photoreceptors and horizontal cells, they can respond with graded changes in membrane potential, but some types do produce action potentials. See HORIZONTAL CELLS.

amblyopia ('lazy eye') The loss of vision in an otherwise normal eye that, through muscle imbalance, cannot focus in synchrony with the other eye. It is the leading cause of vision loss in one eye among 20 to 70 year olds. See MICROSACCADES, NEURAL PLASTICITY, OCULAR DOMINANCE.

AMD See AGE-RELATED MACULA DEGENERATION.

amenorrhoea Severe obesity, malnutrition and strenuous physical exercise may prevent MENSTRUATION; but a woman who has not had a menstrual period by the completion of puberty or by age eighteen is said to have *primary amenorrhoea*.

aminergic neuronal systems Those systems which act through release of the amines acetylcholine, dopamine, histamine, noradrenaline and serotonin.

amino acid-gated ion channels ION CHANNELS mediating most of the fast synaptic transmission in the central nervous system. They include GLUTAMATE-, GABA- and GLYCINE-gated channels.

amino acids (AAs) The 20 naturally occurring AAs encoded by the genetic code are α-amino acids, in which a carboxyl group and an amino group are bonded to the same carbon atom (the α carbon) (see Fig. 8). Non-coded AAs are formed by modification of standard residues. The rare AA, SELENOCYSTEINE, resembles cysteine except for a selenium atom in place of sulphur. Derived from serine, it is an important component of deiodinase enzymes. Some 300 additional AAs have been found in cells, but are not constituents of proteins, ornithine and citrulline being important because of their roles in the synthesis of arginine and in the UREA cycle, creatine because of its role in the ATP:PCR system (see PHOSPHOCREATINE).

Only proline, a cyclic AA (strictly, an

imino acid), has a structure fundamentally different from that shown in Fig. 8. The R group (side chain) differs for each AA and enables it to be placed into one of four 'families' (acidic, basic, uncharged polar and nonpolar) befitting the R group's properties. The sequence of these R groups in a protein has enormous implications for its tertiary structure, and function. Acidic and basic R groups are capable of ionization and therefore of carrying a unit charge at appropriate pH. Nonpolar R groups tend to be present within lipid interiors of transmembrane proteins and are important in hydrophobic interactions and associations of a protein. In the following, the AAs underlined are *essential amino acids* for humans and cannot be synthesized by our cells, and must be present in the diet. Conventional three-letter and single-letter abbreviations follow each AA. See Appendix II for structural formulae.

Acidic side chains: aspartic acid (Asp, D), glutamic acid (Glu, E).

Basic side chains: lysine (Lys, K), arginine (Arg, R), histidine (His, H; see Appendix I).

Uncharged polar side chains: asparagine (Asn, N), glutamine (Gln, Q), serine (Ser, S), threonine (Thr, T), tyrosine (Tyr, Y).

Nonpolar side chains: alanine (Ala, A), valine (Val, V), leucine (Leu, L), isoleucine (Ile, I), proline (Pro, P), phenylalanine (Phe, F), methionine (Met, M), tryptophan (Try, W), glycine (Gly, G), cysteine (Cys, C).

Amino acids are the monomers of peptides, polypeptides and proteins. AA sequences of proteins are encoded by the exons within the gene(s) for the protein, all encoded amino acids being L strereoisomers (rotating the plane of polarized light anticlockwise, D-amino acids having been found in only a few small peptides), this seeming to be a requisite if the stable, repeating, structures of proteins are to be possible. Serine, threonine and tyrosine are especially important because they are targets for kinases in intracellular signalling cascades (see SERINE/THREONINE KINASES, TYROSINE KINASES). Tryptophan is a metabolite of SEROTONIN, while tyrosine is a metabolite of MELANINS, DOPAMINE, ADRENALINE and NORADRENALINE. After DEAMINATION or transamination of their amino groups, AAs (especially the branched-chain leucine, isoleucine, valine, glutamine and aspartate) form important respiratory substrates for most tissues. Some are also glucogenic (see GLUCONEOGENESIS) and some are NEUROTRANSMITTERS. If not reused for synthesis of other AAs or nitrogenous products, the removed amino groups are channelled into UREA synthesis. See AMMONIA, GLUCONEOGENESIS, TRANSFER RNA.

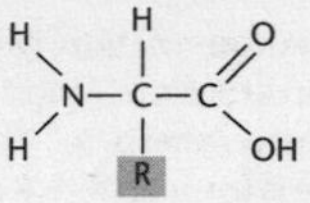

H N C C O OH H H R

FIG 7. The generalized formula of an AMINO ACID. See Appendix II for further details.

aminopeptidase An enzyme of the intestinal epithelial brush borders (microvilli) hydrolysing amino-terminal residues from short peptides, yielding a mixture of free amino acids that can be taken up by epithelial cells of the small intestine. See dipeptidase; compare CARBOXYPEPTIDASES.

aminotransferases See DEAMINATION.

Ammon's horns See HIPPOCAMPUS and Fig. 91.

ammonia Ammonia is quite toxic in low concentrations, which it can reach quickly because of its high solubility in water. It is released by pyrimidine nucleotide degradation, especially in the brain (see LESCH–NYHAN SYNDROME). Combined with glutamate by glutamine synthase, it is converted to the amide nitrogen of the non-toxic glutamine and transported to the liver for UREA formation, and to the kidneys for deamination by glutaminase, NH_4^+. The NH_4^+ is either excreted directly in the urine, or in combination with metabolic acids (counteracting acidosis; see KETONE BODIES). Hepatocytes can transfer the amino groups from most amino acids to α-ketoglutarate to form glutamate. This enters the mitochondria, when its amino group is either removed by *glutamate dehydrogenase* during oxidative DEAMINATION to form NH_4^+, α-ketoglutarate and reduced NAD or NADP or undergoes transamination with oxaloacetate to form aspartate, another nitrogen donor in urea synthesis.

Glutamine is formed from excess NH_4^+ in most other tissues, passing to the liver and into the mitochondria; but alanine transports ammonia from tissues (e.g., skeletal muscles) that use amino acids for fuel to the liver. The amino groups are first collected by glutamate by transamination, glutamate either being converted to glutamine for liver transport, or to pyruvate to form alanine by transamination of its α-amino group by alanine aminotransferase. Alanine then passes in the blood to the liver, where alanine aminotransferase in the cytosol moves the amino group to α-ketoglutarate to form pyruvate and glutamate, the latter entering the mitochondria as above.

ammonification The release of ammonium ions from organic substrates during decomposition of dead organic matter by (ammonifying) bacteria.

amnesia See MEMORY.

amniocentesis Withdrawal of amniotic fluid, with its sloughed embryonic cells, in order to test for chromosomal and other genetic disorders (between 14 and 16 weeks of gestation) such as Down syndrome, spina bifida, haemophilia, Tay–Sachs disorder, sickle cell disease and some muscular dystrophies. Also employed (after the 35th week of gestation) to test for foetal well-being and expected time of delivery. Compare CHORIONIC VILLUS SAMPLING.

amnion A foetal membrane formed by proliferation on the cells of the inner cytotrophoblast (see BLASTOCYST, Fig. 20) and separated from the developing ectoderm by the AMNIOTIC CAVITY. The amnion surrounds the embryo as it develops, so that by the end of the 3rd month it has come into contact with the chorion, obliterating the chorionic cavity.

amniotic cavity Cavity forming within the epiblast of the embryoblast (see Fig. 8) and causing part of the inner cell mass to separate as a flat disc of tissue, the EMBRYONIC DISC. The amniotic cavity enlarges rapidly at the expense of the chorionic cavity as the amnion begins to envelop the connecting and yolk sac stalks, giving rise to the primitive umbilical cord. The amniotic cavity is filled with a clear, watery AMNIOTIC FLUID (removed in Fig. 8b to reveal the embryonic disc).

amniotic fluid Most amniotic fluid is derived initially from cells of the AMNION, but later mainly by filtration from the maternal blood. Its volume increases from ~30 mL at 10 weeks to 450 mL at 20 weeks and ~800–1000 mL at 37 weeks (its final volume varying greatly, from a few millilitres to several litres). The foetus contributes to it daily by excreting urine into it, and similarly a certain amount of absorption occurs via the foetal gastrointestinal tract and lungs, and amniotic membranes, enabling its replacement roughly every 3 hours. It serves as a 'shock absorber' for the foetus and helps regulate foetal temperature and prevents adhesion of the foetus to the surrounding tissues. Oligohydramniotic foetuses have insufficient amniotic fluid (see BIRTH DEFECTS). Cells sloughed off into the fluid by the embryo can be examined for genetic disorders by amniocentesis.

AMP See ADENOSINE MONOPHOSPHATE.

AMPA (α-amino-3-hydroxy-5-methyl-4-isoxazolepropionic acid) A GLUTAMATE analogue used to distinguish between two classes of glutamate receptor (AMPA receptors, or AMPARS) and NMDA RECEPTORS. See SYNAPTIC PLASTICITY.

AMP-activated protein kinase (AMPK) A sensor of nutrient insufficiency. Intracellular rise in the AMP/ATP ratio signals a critical drop in fuel availability, whereupon AMPK activation by AMP increases substrate oxidation to replenish depleted ATP levels. Increased activation of AMPK leads to oxidation of long chain fatty acyl-CoA, a possible cause of its orexigenic effects. It phosphorylates and inhibits ACETYL-COA CARBOXYLASE, inhibiting fatty acid synthesis, and by increasing malonyl-CoA levels

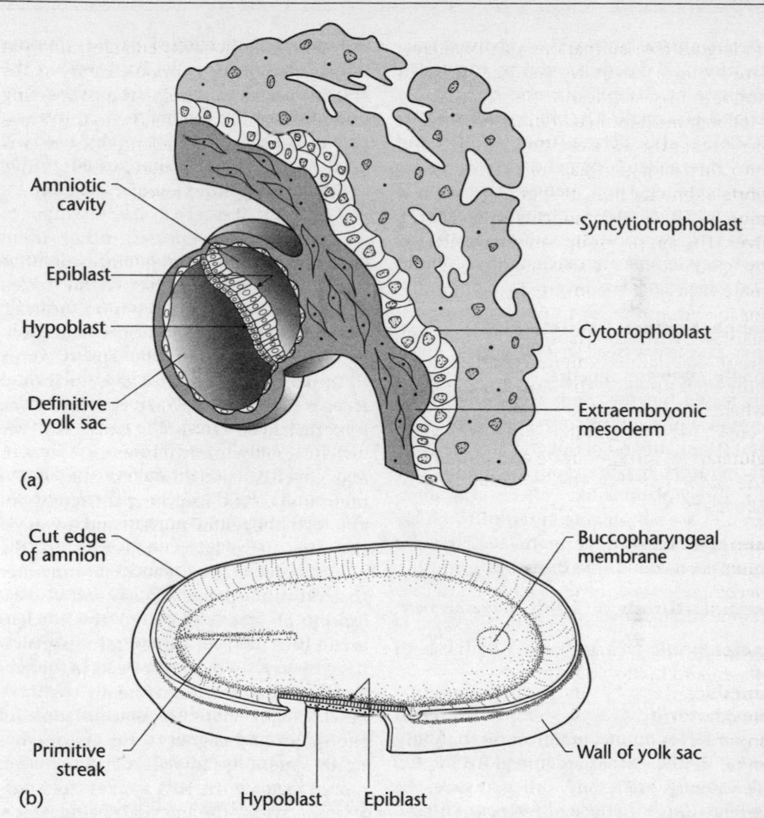

FIG 8. (a) The site of implantation at the end of the 2nd week, showing the AMNIOTIC CAVITY and section through the EMBRYONIC DISC. (b) With the AMNIOTIC CAVITY opened, a dorsal view of the 2nd week epiblast of the embryonic disc, the primitive streak forming a shallow groove in the caudal region.

promotes fatty acid oxidation. It also decreases CHOLESTEROL synthesis in the liver by inhibiting HMG-CoA reductase. AMPK also activates the PGC-1 promoter (see PGC-1 GENES) and its activation will therefore increase metabolism in muscles. It is a prime target for treatment of 'metabolic syndrome', one leading drug (metformin) being thought to work by activating this kinase. See ENERGY BALANCE.

AMPARs (AMPA receptors) GLUTAMATE receptor ionotropic ION CHANNELS mediating fast excitatory synaptic transmission in the brain. Most are formed from GluR2 subunits dimerized to either GluR1 or GluR3 subunits. The GluR2 subunit controls the transport of AMPARs to the postsynaptic membrane and, by binding an unidentified protein, stimulates growth of dendritic spines on cortical pyramidal cells. Permeable to both Na^+ and K^+, but not usually to Ca^+, their opening produces a net Na^+ entry to the cell (but see CREB, Fig. 60). They deactivate and desensitize rapidly, which determines the time course of synaptic transmission. The membrane protein stargazin (one of several transmembrane AMPA receptor regulatory proteins, or TARPs) mediates trafficking of AMPARs to the membrane but also slows channel reactivation and desensitization by increasing the rate of channel opening once it has bound glutamate. On binding glutamate, AMPARs also detach from TARPs and are internalized by endocytosis in a glutamate-regulated manner and recycle in a way that may contribute to SYNAPTIC PLASTICITY. By contrast, NMDA RECEPTORS do not recycle and are much more stable in the membrane. See AMPA.

amphetamines (amfetamines) PSYCHOACTIVE DRUGS (see Fig. 144) which, given orally, decrease appetite, give a sense of increased energy and well-being, and enhance physical performance. Their central stimulant and peripheral actions (e.g., tachycardia, hypertension) are mainly due to catecholamine-like effects (amphetamines are DOPAMINE agonists) or to blockage of catecholamine reuptake. They are resistant to MONOAMINE OXIDASES.

amphiarthrosis A slightly movable JOINT.

amphiphilic Of a substance which is both water- and lipid-soluble.

amphoterin A heparin-binding protein essential for normal neurite growth. Abundant in the extracellular regions of the developing brain and other organs. Its interactions with the membranes of neuronal surfaces seem to be required for extension of neuronal processes. Its cell surface receptor is the protein RAGE.

AMPK See AMP-ACTIVATED PROTEIN KINASE.

ampliconic, amplicons Euchromatic regions of the Y CHROMOSOME exhibiting marked similarity (up to 99.9%) to other sequences in the male-specific (non-recombining) region (MSY). These MSY-repeats are scattered in seven segments over the euchromatic long arm and proximal short arm, and are known as *amplicons*.

ampulla 1. The widest part of each uterine tube, close to the ovary, where FERTILIZATION usually occurs. **2**. Enlargements of each of the three semicircular canals (see VESTIBULAR APPARATUS).

amygdala A subcortical, paired, almond-shaped cluster of nuclei lying within the amygdaloid body within the anterior temporal lobe of the cerebral cortex, forming part of the LIMBIC SYSTEM (see Fig. 114, and HIPPOCAMPUS, Fig. 99), implicated in the regulation of EMOTIONS and responsible for fear learning. It receives olfactory input to its corticomedial nucleus; other input (sensory, arousal state and cognitive input) is relayed to its basolateral nuclei. Other sensory modalities (vision, hearing, somatosensory) provide inputs via respective thalamic nuclei and the sensory cortex to the basolateral nuclei. These nuclei also receive inputs concerning viscera via the hypothalamus, arousal status via the locus coeruleus and nucleus of Meynert, and cognitive processing via the orbital prefrontal cortex. Its principal efferent originates in the central nucleus and leaves via the stria terminalis, projecting to the HYPOTHALAMUS and to brainstem structures that organize motor, arousal and visceral aspects of emotions, or to the nucleus accumbens (NA) via the ventral amygdaloidal pathway. NA output projects to the substantia nigra, and therefore to cognitive basal ganglia loops, generating some of the motor and cognitive aspects of emotional states. The amygdala and the HIPPOCAMPUS regulate the HPA system (see STRESS RESPONSE, STRESS), the amygdala being largely concerned with innate and learned fear responses, figuring prominently in current theories about anxiety disorders (see FEAR AND ANXIETY, PAPEZ CIRCUIT).

The amygdala evaluates a stimulus before there is conscious cognitive appraisal of the situation. Lesions of the amygdala prevent acquisition of new conditioned fear responses, or the expression of pre-existing ones, but do not affect autonomic responses to aversive stimuli (organized by the hypothalamus). Its role in implicit fear learning entails that the fear responses cannot be consciously generated, although explicit aspects of the fearful situation may be consciously recalled. But later presentation of the appropriate stimulus does generate fear and escape responses. During infancy, the amygdala grows more rapidly than the hippocampus, and fearful memories acquired during this time cannot be consciously accounted for – which may account for specific phobias. *Emotional memories* are probably not stored in the amygdala, but in the associated cerebral cortex. In *post-traumatic stress disorder*, exposure to inescapable stress, an image or sound triggers inappropriate fear responses, often accompanied by recall of some past horrific experience. It is probably due to cortical activation of the amygdala. An experimental treatment for this condition is propranolol, a β-adrenoceptor antagonist that inhibits the cAMP-PKA pathway and disrupts memory (see CREB and ENDOCANNABINOIDS). Enhancement of CREB function within the rodent amygdala produces anxiety-like effects (contrast the NUCLEUS ACCUMBENS).

Although the amygdala is not thought to be a primary location for memory storage, it is selectively activated during REM SLEEP and does seem to give emotional content to memories. It is necessary for recognition of facial expressions of emotion. Neurons in the amygdala appear to 'learn' to respond to stimuli associated with pain so that these stimuli then evoke a fearful response, lesions to the amygdala eliminating learned visceral responses to painful stimuli, such as increased heart rate and blood pressure. fMRI images indicate that feared visual and aversive sound stimuli activate the amygdala and the cerebral cortex, including the cingulate and insular cortices; and one current model of social cognition holds that amygdala function – and with it fear response – is regulated by both the orbitofrontal cortex and the medial prefrontal cortex (see WILLIAMS–BEUREN SYNDROME).

In children with AUTISM between 18 and 35 months, the amygdala is abnormally large. This persists through early childhood (the period of sex-differential amygdala growth in normal boys), but by adolescence the enlargement has disappeared and by early adulthood the amygdala in autistic individuals is abnormally small.

Since recognition that removal of the amygdala reduces aggression in animal models, clinical amygdalectomy in humans has been reported to reduce aggressive asocial behaviour, increasing powers of concentration, decreasing hyperactivity and reducing seizures. See ALZHEIMER'S DISEASE, REINFORCEMENT AND REWARD.

amylases Enzymes digesting α-1,4–glycosidic bonds in starch and glycogen to maltose and maltotriose. Secreted by SALIVARY GLANDS and by the PANCREAS (pancreatic amylase can digest uncooked starch, unlike salivary amylase). Both have pH optima of ~6.9 and are examples of GLYCOSIDE HYDROLASES.

amyloid, amyloid precursor protein (APP) See ALZHEIMER'S DISEASE.

amyloidoses Collective term for diseases resulting from abnormal deposits of protein. This tendency of proteins to aggregate may be increased by mutations in their encoding genes or induced by an external factor such as a prion. The former include ALZHEIMER'S DISEASE and HUNTINGTON'S DISEASE; the latter include mad cow disease and kuru. See NEURAL DEGENERATION.

amyotrophic lateral sclerosis (ALS, Lou Gehrig's disease) A neurological disorder involving dysfunction and degeneration of spinal MOTOR NEURONS and those in the cerebral cortex and brainstem. Muscle weakness, atrophy and spasticity progress rapidly and death results in a few years. About 90% of cases are sporadic and 10% familial, of which ~20% are caused by mutations in Cu/Zn-superoxide dismutase (*SOD1*), of outer and inner mitochondrial membranes and the matrix, impairing mitochondrial electron-transport-chain activities and decreasing mitochondrial calcium-loading (*SOD1* is located on chromosome 21). These mutations also promote production of aberrant mitochondrial reactive oxygen species (ROS), thereby causing cytochrome *c* release and APOPTOSIS. Although therapy involving neuronal replacement is probably a long way away, a nearer clinical target may involve use of stem cells to release neurotrophic factors that prevent motor neurons from dying.

The phenomenon of EXCITOTOXICITY has been implicated. See NEURAL DEGENERATION, NEUROROBOTICS.

anabolic Term used for any factor promoting ANABOLISM.

anabolic steroids Androgens and other steroids, such as nandrolone and stanozolol, used to promote protein synthesis after surgery, in treatment of osteoporosis and muscle-wasting diseases. Their adverse effects may include masculinization in women and prepubertial children, and suppression of FSH and LH. Their illegal use was first highlighted in 1955 when the US weight-lifting team was discovered to have used the synthetic testosterone methandrostenolone in training. Estimates based on interviews are that up to 30% of professional athletes currently use androgens, often in association with stimulants and diuretics, in the belief that they augment training effectiveness.

anabolism Synthetic, and hence energy-demanding, metabolic pathways and processes leading to production of polymers, energy stores, and cell components from simpler substrates.

anaemia Reduction in the concentration of circulating haemoglobin, arising through decreased production, or increased destruction or loss, of red blood cells. Low production arises from iron deficiency (women routinely experience high iron losses through menstrual bleeding, pregnancy and lactation); low ERYTHROPOIETIN levels; deficiency of vitamin B_{12} (see VITAMIN B COMPLEX for symptoms of *pernicious anaemia*), or folate, or both; and defective globin synthesis (e.g., THALASSAEMIAS and haemoglobinopathies). Increased destruction may be caused by intracellular defects (e.g., of the membrane, haemoglobin or enzymes) or extracellular defects (e.g., immune and autoimmune disease; or trauma, drugs or bacterial infection). Bleeding, both acute and chronic, may also result in anaemia. See HAEMOGLOBINOPATHIES, HAEMOLYTIC ANAEMIA.

anaerobic Term indicating a lack of molecular oxygen.

anaerobic respiration A term which generally refers to the process of glycolysis, in which ATP is synthesized anaerobically. Strictly, however, this is a form of fermentation and not a form of RESPIRATION, which by definition requires electron transport. The term 'anaerobic metabolism' is perhaps more appropriate.

anaerobic threshold The work rate above which the concentration of LACTIC ACID in the blood builds up rapidly (see LACTATE THRESHOLD). See EXERCISE.

anaerobic training See EXERCISE.

anaesthesia (i) Local anaesthetics prevent pain by causing a reversible block of nerve impulse conduction. Most are weak bases that, in a protonated form, penetrate the nerve in a non-ionized lipid-soluble form; but inside the axon form ionized molecules that block the sodium channels. Small diameter fibres tend to be more sensitive than large diameter ones. (ii) General anaesthetics generate loss of consciousness while depressing all excitable tissues – central neurons, cardiac muscle, smooth and skeletal muscle. The mechanisms of action are unclear, although potency correlates with lipid solubility and probably involves alteration of membrane fluidity and conductances to sodium and/or potassium ions. High pressure reverses general anaesthesia, probably by reordering of membranes. The α_2-ADRENERGIC receptors are implicated in several cardiovascular and central nervous system disorders and their agonists are used as adjuncts to general anaesthesia.

anaesthetics General anaesthesia is the absence of sensation, associated with a reversible loss of CONSCIOUSNESS. Anaesthetic potency correlates well with lipid solubility, these including historically ether, chloroform and trichloroethylene. Thiopental is among the most commonly used, intravenously injected. All anaesthetics depress excitable tissues, including central neurons, cardiac muscle, smooth and skeletal muscle, tissues differing in their sensitivities. The RETICULAR FORMATION is among the most sensitive, so that it is possible to administer anaesthetics at concentrations producing unconsciousness without undue effects on the cardiovascular and respiratory centres, or the myocardium, although the margin of safety is small. Local anaesthetics (e.g., lidocaine) are drugs used to prevent PAIN by reversible block of conduction along axons. Most are weak bases existing in a protonated form which, on entering the neuron, becomes ionized and blocks Na^+ channels. Although not yet certain, it is likely that ligand-gated ION CHANNELS are the main site of anaesthetic action. The GABA and NMDA receptors, in particular, are affected by concentrations of anaesthetic agents similar to those used clinically, and anaesthetics are thought to increase the function of $GABA_A$ receptors by binding cavities or attaching to specific amino acids, prolonging channel opening and extending their inhibitory effect. Small diameter fibres are more sensitive to anaesthetics than larger ones. Compare ANALGESICS.

anagenesis The process of gradual change within an evolutionary lineage, in the absence of lineage branching (cladogenesis; see PHYLOGENETIC TREES). It may lead in time to sufficient morphological change for the lineage to be recognized as a different species, perhaps usually by passing through a small demographic bottleneck.

analgesia, analgesics Analgesia is the absence of normal sensations of PAIN; analgesics are drugs that achieve this effect. The 'endogenous analgesia system' of the brain and spinal cord has three components: (i) the periaqueductal grey and periventricular areas of the mesencephalon and the upper pons surrounding the aqueduct of Sylvius and parts of the 3rd and 4th brain ventricles, whose neurons secrete ENKEPHALINS and project to (ii) the raphe magnus nucleus, located in the lower pons and upper medulla, from where serotonin-releasing neurons project down the dorsolateral columns of the SPINAL CORD to (iii) a pain inhibitory complex in the dorsal horns of the spinal cord, at which point analgesia signals can block pain before it is relayed to the brain. Serotonin release in the dorsal horns of the spinal cord causes enkephalin release there, which is thought to cause presynaptic and postsynaptic inhibition of incoming type C and type Aδ pain fibres (see PAIN). CAPSAICIN, when applied in large quantities, can cause analgesia as it depletes substance P from nerve terminal surroundings. Complete inability to sense pain in an otherwise healthy individual is a very rare congenital phenotypic trait, usually categorized as 'congenital indifference to pain', or 'autosomal recessive congenital analgesia'. Recent work has mapped the condition as an autosomal-recessive trait to chromosome 2q24.3, a region containing the *SCN9A* gene encoding the α-subunit of the voltage-gated sodium channel $Na_v1.7$, strongly expressed in nociceptive neurons. Compare ANAESTHETICS; see ASPIRIN, BRADYKININ, NON-STEROIDAL ANTI-INFLAMMATORY DRUGS, OPIOIDS.

analogous Structures with similar biological functions, and often appearances, but having different evolutionary and/or embryological origins. They are homoplasies. Contrast HOMOLOGY.

anaphase **1**. The period during mitosis of the CELL CYCLE and of the 2nd division of meiosis when the centromeres split and chromatids move to opposite poles of the spindle. It begins with the sudden disruption of sister-chromatid cohesion, initiated by a signalling cascade involving the ANAPHASE-PROMOTING COMPLEX and M-Cdk (see G2). **2**. The phase during the 1st meiotic division, when the two chromosomes of a homologous pair (bivalent) separate and move to opposite poles of the spindle.

anaphase-promoting complex (APC) A highly regulated ubiquitin kinase, a major role of which is ubiquitylation of M-cyclin in late mitosis, on binding the activating subunit Cdc20. See CYCLINS.

anaphylactic shock A general circulatory collapse generated when the small

COMPLEMENT fragments C3a, C4a and C5a are produced in large amounts, or injected systemically. It is similar in kind to the systemic allergic reaction involving IgE antibodies. Although C5a is the most potent of the three, they all produce smooth muscle contraction and increase vascular permeability. C5a and C3a also stimulate endothelial cells to produce adhesion molecules, while activating mast cells to release histamine and TNF-α from submucosal tissues. Resulting tracheal swelling may result in suffocation. Damage to endothelia can be moderated by treatment with adrenaline, which stimulates reformation of tight junctions. See ALLERGIES, IMMUNOSUPPRESSION.

anaplerosis (adj. anaplerotic) The replenishment of KREBS CYCLE intermediates as they are removed for biosynthetic reactions, in most tissues involving the conversion of pyruvate or phosphoenolpyruvate to oxaloacetate or malate (rather than acetyl-CoA). In mammalian liver and kidney, the most important anaplerotic reaction is the reversible carboxylation of pyruvate by CO_2 to form oxaloacetate, when this or any of the other cycle's intermediates is in short supply. See PYRUVATE CARBOXYLASE.

anastomosis Any branching and rejoining, e.g., of vessels or muscle fibres (notably cardiac muscle). Arterial anastomoses often provide alternative routes for blood supply to a particular region, e.g., arteries in the myocardium, and those forming the CIRCLE OF WILLIS.

anatomic dead space See DEAD SPACE.

anatomical planes of section Universally employed planes of reference and terms of relative anatomical position when the body is erect, facing the observer, the upper limbs at the sides with the palms facing forward and the feet flat on the ground (see Fig. 9).

anatomically modern human (AMH) See HOMO SAPIENS.

anatomy The structure, or study of structure, of the body and the relation of its parts to one another.

ancient DNA DNA that has been recovered from the bones of archaeological specimens. It tends to be of short lengths but can be amplified by PCR. It is thought that most DNA degrades within 10^5 yr. See DIVERGENCE, HOMO NEANDERTHALENSIS, mtDNA, POLYMERASE CHAIN REACTION.

androgen insensitivity syndrome (formerly, testicular feminization) An X-linked recessive trait (a form of PSEUDOHERMAPHRODITISM) occurring in about 1:65,000 individuals who are chromosomally male (46,XY) and have the external appearance of normal females. The testes are intact within the inguinal or labial regions, releasing plenty of androgen, although spermatogenesis does not occur and 33% of these individuals develop testicular malignancies prior to age 50. Because MÜLLERIAN INHIBITING SUBSTANCE is produced, no Wolffian ducts develop. Androgen receptors on target cells are absent or defective so that the individuals develop as phenotypic females who are sterile, lacking oviducts, uterus and upper vagina (which is short and ends as a blind pouch). The mutated gene is located on the long arm of the X chromosome, near the centromere.

androgen receptor See ANDROGENS.

androgens Steroid hormones, derivatives of CHOLESTEROL (see Fig. 49), produced mainly by LEYDIG CELLS of the post-pubertial testes, but also by the adrenal medulla (in women, this amounts to less than 5% of the total androgen potency of males). The testes produce TESTOSTERONE (much the most abundant androgen), the more potent dihydrotestosterone (DHT) and androstenedione. In the serum, only 0.5–4% of androgens (or other steroid hormones) are free and not complexed with serum proteins. Most testosterone and DHT (~60%) is bound to albumin, most of the remainder being complexed with SEX HORMONE-BINDING GLOBULIN (SHBG). The *androgen receptor* (AR) is a member of the NUCLEAR RECEPTOR SUPERFAMILY. Existing in the cytosol as an oligomer

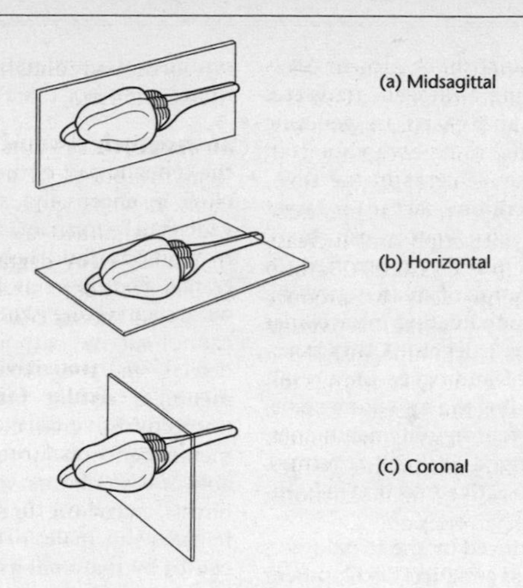

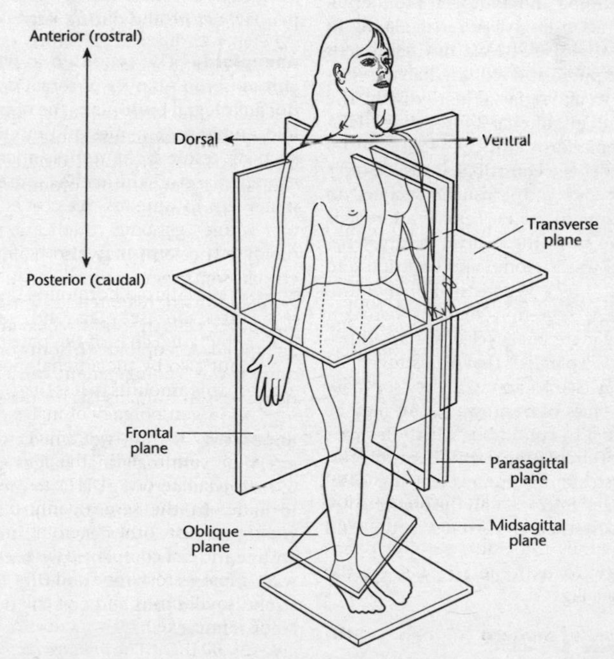

FIG 9. Diagrams illustrating some major ANATOMICAL PLANES OF SECTION, their relationships, and some relevant terminology.

complexed with heat-shock protein AR is released on binding androgen, translates to the nucleus, and exerts its genomic effects by binding as a homodimer to androgen response elements in the DNA, acting as a transcription factor for target genes to which coactivators can bind. Androgens have also been reported to induce rapid activation of kinase-signalling cascades, thereby modulating intracellular calcium ion (Ca^{2+}) levels. Other non-genomic effects of androgens may result through their binding the SEX HORMONE-BINDING GLOBULIN RECEPTOR on cell membranes, possibly activating a distinct G PROTEIN-COUPLED RECEPTOR to activate second messenger systems.

Androgens produced by the foetal testes usually permit, and enhance, development of the Wolffian ducts into male-type structures, and Leydig cells are then numerous; but these cells almost disappear during male childhood and androgen production decreases accordingly. Adrenal glands in both sexes release at least five androgens and cause pubic and axillary hair production and, in males, facial hair growth. The most abundant adrenal androgen is DHEA (dehydroepiandrosterone), whose secretion is under the control of ADRENOCORTICOTROPIC HORMONE and is usually secreted in low amounts in males. Female androgens also derive from the ovaries (androstenedione and testosterone) and contribute to libido (sex drive). Whereas adrenal androgen (DHEA) secretion remains constant during the menstrual cycle, ovarian androgen secretion parallels that of oestrogen. By conversion (see AROMATASE), they are additional sources of oestrogens, and may be oversecreted in conditions which decrease the level of circulating CORTISOL. Androgens, such as testosterone produced by the foetal and neonatal testes, act on the brain to produce SEX DIFFERENCES in neural structure and function.

See SECONDARY SEXUAL CHARACTERISTICS; compare OESTROGENS.

andropause See AGEING.

androstenedione Intermediate in the synthesis of testosterone and 17β-oestradiol from CHOLESTEROL (see Fig. 49).

anencephaly A NEURAL TUBE defect in which the cephalic part of the neural tube fails to close (exencephaly) so that the cranial vaults fail to form and brain tissue exposed to amniotic fluid becomes necrotic. The cranial neuropore is left open. Children with such severe skull and brain defects cannot survive, although relatively small defects in the SKULL, through which meninges and/or brain tissue herniate (respectively, cranial meningocoele and meningoencephalocoele), may be treatable. It is a common abnormality (1:1,500 births), occurring four times as often in female as in male foetuses. Hyperthermia caused by maternal infection, or by sauna baths during pregnancy, may cause SPINA BIFIDA (see entry) and exencephaly. As with spina bifida, up to 70% of cases can be avoided by a woman taking 400 μg of FOLATE per day, before and during PREGNANCY.

aneuploidy The condition in which the chromosome number of somatic nuclei is not an integral multiple of the normal haploid number. Trisomies and monosomies, such as result from nondisjunction, are examples and are the leading cause of pregnancy loss in humans. See GENE DUPLICATION for whole genome duplication (polyploidy). The term may also be applied to chromosome segments (see SEGMENTAL DUPLICATIONS), and in this extended sense the karyotypes of cells in solid TUMOURS are always aneuploid. Aneuploidy (trisomy or monosomy) is the leading genetic cause of pregnancy loss in humans (see MISCARRIAGE).

aneurysm A thin, weakened, section of artery or vein wall that bulges outwards to form a balloon-like sac. Common causes include atherosclerosis, syphilis and congenital blood vessel defects. If untreated, an aneurysm will eventually burst as the wall becomes thinner, resulting in shock, stroke, severe pain and possibly death. See BLOOD PRESSURE.

Angelman syndrome A rare BIRTH DEFECT involving maternal microdeletion in the q11-13 region of chromosome 15 and GENOMIC IMPRINTING. Affected children tend to be hyperactive and attention-seeking, moving jerkily with repetitive puppet-like sequences. Although lacking speech, and having severe mental and motor retardation, individuals appear happy and exhibit excessive laughter. They have abnormal brain waves and suffer from seizures, and there is a high incidence of AUTISM among them. If the deletion is inherited from the father, resulting in a maternal bias, PRADER–WILLI SYNDROME results. Higher than expected frequencies arise though use of ASSISTED REPRODUCTIVE TECHNOLOGY. See COPY NUMBER, SELFISH GENE THEORY.

anger and aggression Aggressive behaviour has traditionally been treated as a symptom of certain disorders rather than a disorder in its own right. Current research suggests that violent behaviour, brought out or reinforced by the social environment, is the end result of multiple risk factors which may include 'biological vulnerability' (either genetic or created by the prenatal environment). See AFFECTIVE DISORDERS, AMYGDALA, EMOTIONS, HYPOTHALAMUS, MONOAMINE OXIDASES.

angina pectoris Severe chest pain that usually accompanies exertion, as when the heart requires more oxygen, but especially when this is reduced by myocardial ischaemia. See ATHEROSCLEROSIS, CORONARY HEART DISEASE, REFERRED PAIN.

angiogenesis The formation of new blood vessels by sprouting from existing ones. Dissolved oxygen has an effective diffusion distance of 0.2 mm in the body, and cells further than this from an oxygen-bearing blood capillary will suffer moderate or severe HYPOXIA (see entry for relevant details). Such tissues are in danger of becoming necrotic. More than a dozen factors, mostly small peptides, have been shown to promote angiogenesis in tissues with inadequate blood supplies, the best characterized being vascular endothelial growth factor (VEGF-A), fibroblast growth factor (FGF) and angiogenin. In sprouting angiogenesis, specialized endothelial tip cells lead the outgrowth of blood-vessel sprouts towards gradients of VEGF-A. First, the basement membrane around the existing endothelium dissolves at the point of sprouting; then rapid division of endothelial cells occurs whose migration causes them to stream out along the vessel wall in extended cords towards the source of the angiogenic factor, in a manner resembling AXONAL GUIDANCE. These cells then fold up into a tube, which connects (anastomoses) with an adjacent tube budding from an adjacent donor vessel, to form a capillary loop through which blood starts to flow. If flow is sufficient, smooth muscle invades the wall (see VASCULOGENESIS). Phosphoinositide 3-kinases (see PHOSPHATIDYLINOSITOL 3-KINASE) have been implicated in angiogenesis, and of many isoforms of the p100 catalytic subunit it appears that only the p110α isoform is essential for vascular development. Its activity is especially high in endothelial cells, and is preferentially induced by tyrosine kinase ligands such as VEGF-A prior to endothelial cell migration. Naturally occurring inhibitors of angiogenesis also hinder growth of experimental tumours, and synthetic inhibitors of VEGF inhibit tumour progression in patients with cancer. In mice, inhibition of Notch signalling using γ-secretase inhibitors (see ALZHEIMER'S DISEASE), genetic inactivation of one allele of the endothelial Notch ligand Dll4, or endothelial-specific deletion of *Notch1*, all promote increased numbers of endothelial tip cells and vessel branches. Further, inactivation of the tumour-suppressor gene VHL supports renal cell tumour growth by favouring tumour angiogenesis. Membrane METALLOPROTEINASES are thought to play a major role in angiogenesis. Angiogenesis must be shut down once adequately achieved, which is often achieved by suppression of HIF-1 transcription factor, normally produced by anoxic conditions but whose assembly is suppressed under normal oxygen concentrations. See ENDOTHELIUM.

angioplasty Procedure used especially in treatment of coronary artery disease. A balloon catheter (plastic tube) is inserted into an arm or leg artery and guided into the CORONARY ARTERY (balloon angioplasty). Dye is released and X-rays (angiograms) taken to locate plaques, whereupon the catheter is advanced to the obstruction and inflated with air to squash the plaque against the vessel wall. A stent (stainless steel spring coil-like device) is inserted to keep the artery open, although this too can become blocked. See BYPASS SURGERY.

angiotensinogens, angiotensins The walls of afferent arterioles entering Bowman's capsule are stretched to a greater or lesser degree by blood volume and BLOOD PRESSURE changes, and baroreceptors of the juxtaglomerular apparatus sense this, causing the juxtaglomerular cells to secrete the enzyme renin when less stretching occurs (see MECHANORECEPTORS). Renin removes a 10-amino acid peptide (angiotensin I) from the liver-produced plasma protein angiotensinogen, and angiotensin-converting enzyme in endothelial cells, especially of the lung, converts this to angiotensin II, which is the active hormone. It binds to cell-surface G-PROTEIN COUPLED RECEPTORS (AT_1 and AT_2 receptors), producing direct arteriolar vasoconstriction via AT_1; but it also binds AT_2 receptors in adrenocortical cells of the zona glomerulosa, stimulating PHOSPHOLIPASE C (see Fig. 135), causing raised intracellular Ca^{2+} and consequent synthesis and release of ALDOSTERONE. Increased Na^+ and water retention by the kidneys follows, causing increased blood volume. Drugs such as captopril are used in treatment of HYPERTENSION, reducing angiotensin and aldosterone levels by blocking angiotensin-converting enzyme. See MACULA DENSA.

angular gyrus (area 39) Gyrus at the junction of the occipital, parietal and temporal lobes (see BRAIN, Fig. 30) providing an important link between auditory and visual association areas and involved in naming, reading, writing and calculation. See COGNITION; LANGUAGE AREAS, Fig. 109.

aniridia Absence of the IRIS. See PAX GENES.

anoikis That form of non-p53-dependent APOPTOSIS resulting from detachment of a cell from its neighbours, or from the basal stroma. See INTEGRINS.

anomia The inability to name familiar objects. See APHASIA.

***Anopheles* mosquito** Genus of dipteran insects and primary host and vector of *Plasmodium*, the cause of MALARIA. It requires standing water to complete its life cycle.

anorexia nervosa An eating disorder, mostly of young women, whose defining character is the voluntary maintenance of body weight at an abnormally low level. It is often accompanied by DEPRESSION, and a distorted perception of body image (regarding themselves as overweight, when they are not), or denial of sexuality by retaining or acquiring a pre-pubertial physique. Difficulties arise in deciding whether neurochemical changes are the cause or the effect of the starvation. For anorexics, eating is stressful, and stress is thought to reduce appetite by exciting the neurons in the paraventricular nucleus of the HYPOTHALAMUS releasing CORTICOTROPIN RELEASING HORMONE (CRH). Anorexics have heightened levels of CRH and of glucocorticoids (see STRESS RESPONSE), which may be reinforced by the normal STRESS RESPONSE to STARVATION CONDITIONS. It is not clear why this is not overridden by the low LEPTIN concentration of untreated anorexics, which for some reason does not lead to increases in NEUROPEPTIDE Y. Leptin concentrations do rise in weight-recovering anorexics, to levels higher than in normal women with the same BODY MASS INDEX. See FEEDING CENTRE, GHRELIN, OBESITY, SEX DIFFERENCES.

anoxia Deficiency or absence of free (gaseous or dissolved) oxygen. It is a stimulus for ANGIOGENESIS (see HYPOXIA).

ANS See AUTONOMIC NERVOUS SYSTEM.

antagonist 1. A substance which binds to, and blocks, a receptor, and prevents a natural neurotransmitter or hormone from exerting its effect. Thus, atropine blocks muscarinic ACETYLCHOLINE receptors; and propranolol is a non-selective BETA-BLOCKER.

Contrast AGONIST. **2**. One of a pair of ANTAGONISTIC MUSCLES.

antagonistic muscles Muscles, usually in pairs (e.g., FLEXORS and EXTENSORS at a joint), whose contractions bring about opposing effects. See MUSCLE SPINDLES.

anterior (rostral) (adj.) Directionally, towards the head end of the body (an anatomical reference; see ANATOMICAL PLANES OF SECTION, Fig. 9), determined at the EMBRYONIC DISC stage. Contrast POSTERIOR (caudal).

anterior commissure Association fibres running in front of the fornix and linking the anterior temporal cortices: the inferior and middle temporal gyri and olfactory regions of the two cerebral hemispheres. It plays an important role in unifying the emotional responses of the two sides of the brain. See CORPUS CALLOSUM.

anterior visceral endoderm See EMBRYONIC DISC.

anterograde amnesia Inability to form new memories, following trauma.

anterograde transport Movement of material within vesicles along microtubules from the cell body of an axon in the direction of the terminal. Employed, as with RETROGRADE TRANSPORT, in tracing neural connections in the brain.

anterolateral system, anterolateral pathway See SOMATIC SENSORY SYSTEM.

anthrax A disease of cattle, and occasionally of humans, caused by the bacterium *Bacillus anthracis*, whose spores can remain potent in soil for decades. The cutaneous form of the disease is not as dangerous as the pulmonary form (caused by inhaling spores). The toxin produced by the bacterium has three components: edema factor (EF) prevents macrophages from engulfing the bacteria; lethal factor (LF) kills macrophages, and eventually the host; protective antigen (PA, a major component of current anthrax vaccines) assists the other two components to enter the cell. Once seven PA molecules have formed a ring-like complex on the cell surface, they bind an EF/LF complex and become engulfed by the membrane. Once inside, PA molecules form a pore in the membrane of an endosome, pierce it, and allow EF and LF into the cell proper to cause damage to and death of the macrophage.

Anthropoidea The extant Anthropoidea can be clearly resolved anatomically into three groups: new world monkeys (Ceboidea), old world monkeys, here abbreviated as 'OWM' (Cercopithecoidea, defined in dental terms by the emergence of bilophodonty) and humans and apes (Hominoidea), of which the OWM and hominoids share a more recent common ancestor.

antibiotic resistance See ANTIBIOTICS.

antibiotics Secondary metabolites produced by microorganisms which kill competing microorganisms and now produced commercially, normally being ingested by mouth. Once in the blood they kill bacteria and fungi, but are ineffective against viruses. Of the many natural antibiotics discovered, <1% have proved of practical medical value, and even these are often chemically modified as *semisynthetic* antibiotics for improved effectiveness. Some are TOXINS. Antibiotics and other chemotherapeutic agents (e.g., the sulphur drugs, and other analogues of substances required by bacteria in their medium) target cell wall synthesis, protein and nucleic acid synthesis. GRAM-POSITIVE BACTERIA are usually more sensitive to them than GRAM-NEGATIVE BACTERIA, although some target only the latter. Antibiotics acting on both bacterial types are *broad-spectrum antibiotics*. Antibiotics may cause prominent side effects, e.g., the observed increase of pseudomembranous colitis (PMC) when clindamycin was introduced in the 1970s. Nearly all antibiotics are capable of inducing antibiotic-associated diarrhoea or its fatal form, PMC.

One of the most important antibiotic groups is the β-lactam group, including penicillins, cephalosporins and cephamycins. Penicillin G, the first antibiotic discovered, was active primarily against Gram-positive bacteria, although some newer forms are quite effective against

Gram-negative bacteria. Fleming, Florey and Chain all contributed to the discovery and mass production of penicillin G (produced by the fungus *Penicillium notatum*) to which some individuals are allergic. The β-lactam group are potent inhibitors of cell wall synthesis, preventing transpeptidation required for cross-linking of two glycan-linked peptide chains and causing bacteria to break open. However, an important group of bacteria, the mycoplasmas, do not have cell walls and are therefore resistant to penicillins (though not to all antibiotics). A bacterium may be resistant because its coat or wall is impermeable to the antibiotic, as Gram-negative bacteria are to penicillin G. Further, a bacterium may be able to alter the antibiotic to an inactive form, e.g., many staphylococci contain β-lactamases, enzymes that cleave part of the penicillin molecule; and other bacteria pump the antibiotic out as it enters. Cephalosporins generally have a broader spectrum of activity than penicillins and are more resistant to β-lactamases. Some are very useful against NEISSERIA gonorrhoeae. Prokaryotic antibiotics include the aminoglycosides (e.g., streptomycin, kanamycin, gentamicin and neomycin), macrolides (e.g., erythromycin) and tetracyclines (e.g., chlortetracycline, oxytetracycline). Aminoglycosides (e.g., kanamycin), which inhibit protein synthesis at the 30S ribosomal subunit, are now 'reserve antibiotics' used primarily when others fail and streptomycin has been supplanted in the treatment of TUBERCULOSIS because of side effects and bacterial resistance to it. Erythromycin, very effective against *Legionella pneumophila*, prevents protein synthesis by targeting the 50S prokaryotic ribosomal subunit, as does chloramphenicol. Tetracyclines, some of the earliest broad-spectrum antibiotics, interfere with the 30S ribosomal subunit and, with the β-lactam group, are the most important antibiotics in the medical field. Whereas streptomycin, tetracyclines, kanamycin and gentamicin inhibit the initiation phase of translation, puromycin, chloramphenicol, cycloheximide and tetracycline inhibit elongation.

Some genetic resistance to antibiotics is due to random mutation in the bacterial chromosome; but resistance commonly arises through genetic elements on bacterial plasmids (R plasmids). These usually encode enzymes that inactivate the drug, prevent its uptake, alter its molecular target or actively pump it out. Such plasmids often confer multiple *antibiotic resistance* since they may contain a number of genes encoding different antibiotic inactivating enzymes (see MULTIDRUG RESISTANCE). Vancomycin is a 'drug of last resort', preventing cross-linking of peptidoglycan units in the bacterial cell wall and used to treat hospital-acquired bacterial infections that are already resistant to all other antibiotics. Very effective resistance to it is due to a transposon encoding seven genes which collectively detect vancomycin presence, shut down the normal synthesis of the cell wall and generate a new cell wall using subunits that do not bind vancomycin. The transposon can now be transmitted to many other pathogenic bacterial species.

Antibiotic resistance is an increasing problem in hospitals (see MRSA, CLOSTRIDIUM DIFFICILE), probably coming about by transfer of genetic elements from the antibiotic-producing organisms themselves – selected because they enable these bacteria to survive in an environment containing their own antibiotics. It now seems that resistance in pneumococci is spreading not because diverse strains are acquiring plasmids, but through clonal expansion of a few 'superbugs', whose low natural variability makes them good targets for vaccines. Unless patients complete their course of antibiotics, premature fall in antibiotic levels may allow the bacterial population to rise, increasing the numerical potential for newly mutating resistant strains to emerge and crowd out the wild-type strains (the equivalent process may occur in anti-retroviral treatments, which do not however involve antibiotics: see HIV).

A strain of *E.coli* frozen since 1946 has been shown to contain resistance plasmids to the antibiotics tetracycline and streptomycin – several years before these drugs were used clinically – and R plasmids conferring antibiotic resistance have been discovered in non-pathogenic Gram-negative soil bacteria (presumably because the bacterium *Streptomyces* and the fungus *Penicillium* are soil organisms). Unfortunately, the widespread use of antibiotics in clinical and veterinary practice has been paralleled by the emergence of bacteria resistant to them – and high levels of antibiotic use result in high levels of resistance. When non-pathogenic bacteria of the GUT FLORA are killed by antibiotics, pathogenic micro-organisms often replace them and cause disease. If use of an antibiotic is stopped, resistance to it appears to reverse over time, so that through prudent long-term monitoring and use some antibiotics may then become useful again. Resistance could be minimized if antibiotics were used only for serious diseases, not at all when they would be useless in any case, and given in sufficiently high doses that bacterial populations are reduced before resistant mutants begin to appear. The overuse in developing countries of antibiotics in livestock as a pre-emptive disease-reducing and growth-promoting measure has undoubtedly increased the spread of antibiotic resistance.

antibodies Immunoglobulin glycoproteins secreted by mature B CELLS during ADAPTIVE IMMUNITY and structurally identical to the secreting B cell's receptor except for a small portion of the carboxy-terminal part of the heavy-chain C region (which is hydrophobic in the receptor and hydrophilic in the antibody). Secreted antibodies are Y-shaped molecules, whose two arms bind to pathogens and their toxic products in the extracellular spaces and clump them together (termed agglutination when the particles bound are large, e.g., bacteria; see Appendix XI). They are constructed from paired heavy and light polypeptide chains, each light chain containing two folded polypeptide domains 110 amino acids in length (immunoglobulin domains) while each heavy chain contains four. The specificity of an antibody response is determined by the *antigen-binding site*, which consists of the two variable V domains, V_H and V_L (see Fig. 10, and ANTIBODY DIVERSITY), is structurally very similar to the region giving a T-CELL RECEPTOR its specificity, and in both cells is responsible for RECEPTOR CLUSTERING on their respective membranes when antigen is bound. Antigen-binding sites tend to contain many AMINO ACIDS with aromatic side chains, participating in van der Waals and hydrophobic interactions, and occasionally in hydrogen bonds. The hydrophobic and van der Waals forces operate over very short ranges, serving to pull the interacting surfaces of the antigen and antibody together.

Antibodies may contain either of two types of *light chain*, the lambda (λ) or kappa (κ) chains, and any given antibody has either λ or κ chains, but not both. Although either type of *light* chain can be found in antibodies of each of the five major antibody isotypes (IgM, IgD, IgG, IgA and IgE), the particular isotype to which an antibody belongs, and therefore its effector function, is determined by particular C region sequences that are joined to the VDJ region to give the eventual F_C portion of the molecule *heavy chain* (see F_C DOMAIN). There are five main C regions that can be joined to VDJ to give the five major antibody isotypes: the Cμ exons in IgM; the Cδ exons in IgG; the Cγ exons in IgG; the Cε exon in IgE and the Cα exon in IgA. The IgG isotype has several subtypes because there are a variety of Cγ exons that can be recombined with VDJ (see ISOTYPE SWITCHING).

Antibodies participate in adaptive immunity in three main ways (see B CELL, Fig. 19). (i) They can bind to and neutralize a bacterial toxin, preventing it from contacting host cells and causing pathology. The antibody:toxin complex will eventually be scavenged and digested by macrophages. (ii) They can coat an antigen and make it recognizable by macrophages and neutrophils (opsonization) so that these ingest and degrade it. (iii) By coating plasma-borne bacteria, they can form a receptor for the first protein of the complement system, so activating the system. This may kill the bacterium directly, or more

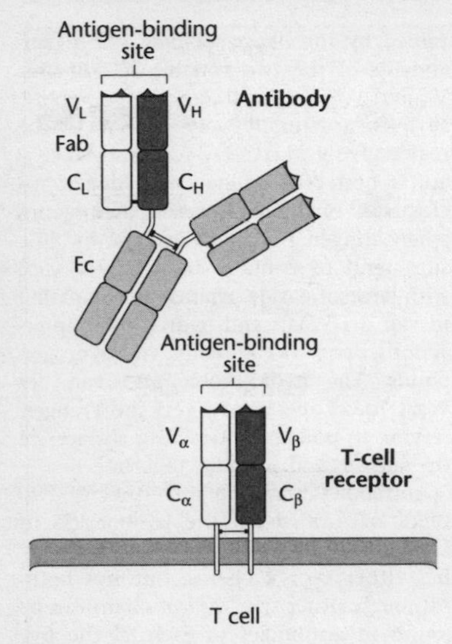

FIG 10. The structure of ANTIBODIES includes the Fab fragment, a disulphide-linked heterodimer, each of the two chains containing one C domain and one V domain (shaded). Each light chain is composed of two immunoglobulin domains, each heavy chain containing four. The two V domain termini create the antigen-binding site while their F_C DOMAINS determine their effector function. A hinge region links the F_C and Fab portions.

The structure of T CELL RECEPTORS (see Fig. 171) is also that of a disulphide-linked heterodimer, each of the two chains containing an immunoglobulin C-like domain and an immunoglobulin V-like domain, the juxtaposition of the two V-like domains similarly forming the antigen-binding site.

usually encourage uptake and destruction by phagocytes.

The major class of plasma antibody is IgG, produced in large amounts during a secondary immune response. The four subclasses of IgG in rodents and primates have evolved specialized effector responses, such as cytotoxicity, phagocytosis and release of inflammatory mediators, their expression being influenced by the prevailing cytokine environment (see ISOTYPE SWITCHING) and by the nature of the stimulating antigen. Protein antigens elicit a thymus-dependent response generally dominated by IgG1, 2a and 2b, whereas carbohydrate antigens can induce 'thymus-independent' responses resulting in IgG3 antibody expression. IgG2a and 2b are the most potent subclasses for activating effector responses and dominate antiviral immunity and autoimmune diseases. In addition to activating COMPLEMENT, each IgG molecule can bind at its tail region (the F_C region) to specific F_C receptors on the surfaces of macrophages and neutrophils, which can then bind IgG-coated microorganisms, phagocytose, and destroy them. F_C receptors are also located on mast cells, monocytes and NATURAL KILLER CELLS (see F_C DOMAIN). The destruction of antibody-coated target cells by these cells (antibody-dependent cell-mediated cytotoxicity) is triggered when antibody bound to a foreign cell interacts with the F_C receptors. In the case of NK cells, the mechanism of attack is analogous to that of cytotoxic T cells, involving release of granules containing perforin and granzymes. IgG molecules also contain a Fab component (the fragment antigen-binding regions), joined by the flexible *hinge region* to the F_C region of the intact IgG molecule. One of three fragments produced by papain digestion, it consists of the light chain and the amino-terminal half of the heavy chain held together by an interchain disulphide bond. The hinge region allows independent movement of the two Fab arms. Another protease, pepsin, cuts the antibody on the carboxy-terminal side of the disulphide bonds to release an $F(ab')_2$ fragment with a few more amino acids.

The normal antibody response to an antigen is *polyclonal*: many B cells are stimulated to produce antibodies to a complex antigen. EXON SHUFFLING at the DNA level has been exploited in *antibody engineering*, where antibodies with novel combinations of protein domains can be artificially constructed. Such work produced humanized antibodies in which a lesser or greater part of the rodent MONOCLONAL ANTIBODY is replaced by the human equivalent. In recent work, hybridoma construction has been replaced by phage display technology.

antibody diversity The wide range of

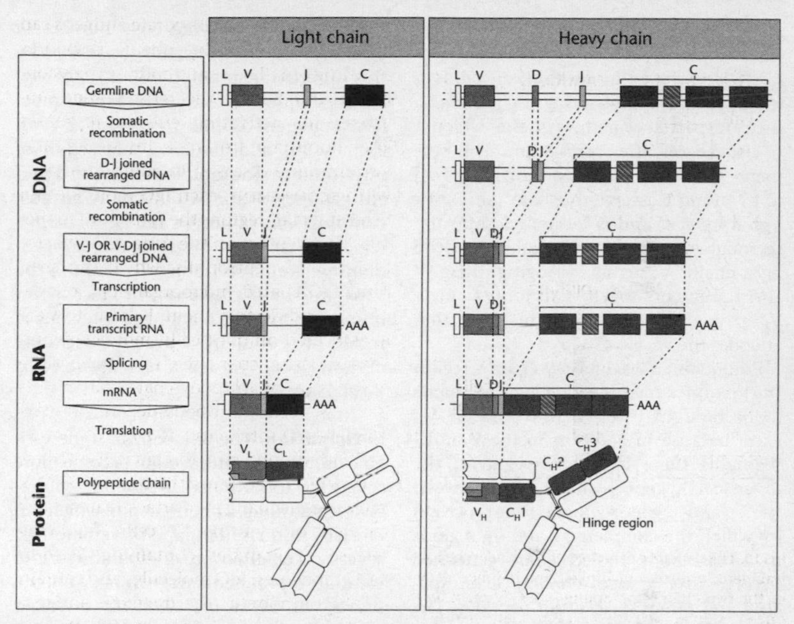

FIG 11. A comparison of the gene rearrangements involved in immunoglobulin light-chain and heavy-chain production (see ANTIBODY DIVERSITY). *Light-chain V-region sequences* are constructed from two gene segments in the genomic DNA: a variable (V) and a joining (J) gene segment. The light-chain C region is encoded by a separate exon and joined to the V-region exon by splicing the light-chain RNA to remove the L-to-V and the J-to-C introns. *Heavy-chain V-region sequences* are constructed from three gene segments: the diversity (D) and J gene segments are first spliced together followed by joining of this DJ sequence to the V gene segment, forming a complete V_H exon. The heavy-chain C region is encoded by several exons and these, along with the leader sequence (L), are spliced to the V-domain sequence during RNA processing of the heavy-chain transcript. After translation, the leader sequence is removed and disulphide bonds linking the heavy and light chains are formed. © From *Immunobiology* (6th ed), by C.A. Janeway, P. Travers, M. Walport and M.J. Shlomchik. Reproduced by permission of Garland Science/ Taylor Francis LLC.

antigen specificities found in the antigen-binding region of ANTIBODIES and B-CELL receptors is due to variation in amino acid sequence of the variable (V) regions of both its light and heavy chains. The variability of these V regions is particularly located in three hypervariable regions (HVs), the six combined HVs forming hypervariable loops, also termed complementarity-determining regions (CDRs) which are brought together in the antibody or IMMUNOGLOBULIN (see Fig. 99) to create a single hypervariable site, the *antigen-binding site*. The centre of the antigen-binding site of an antibody is formed by the combination of loops formed by the highly variable CDR3s of the heavy and light chains.

In each chain, the V region is linked to an invariant constant (C) region. Each V region is encoded by a separate block of DNA termed a *gene segment* and in order to generate the repertoire of immunoglobulins and secreted antibodies of which we are capable, the immunoglobulin genes are rearranged in immune cells (see Fig. 11).

There is more than one gene segment encoding the V domain (V region) of an immunoglobulin or antibody heavy or light chain. The first 95–101 amino acids (most of the V region) are encoded by a *V gene segment*. Immunoglobulin chains are extracellular proteins and the V gene segment is preceded by an exon encoding a leader peptide (L) directing the protein to the cell's secretory pathways, after which it is cleaved off after translation. This segment, plus the small remaining part of up to 13 amino acids, is termed the 'joining' or *J gene segment* and is located close to the segment encoding the C region. For the light-chain V region, bringing these V and J segments together to form a single V_L exon creates a continuous exon that encodes the whole of V_L.

Production of each heavy-chain V region (V_H) requires *three* separate gene segments to be brought together into a single V_H exon because in addition to the V and J segments there is a third segment, the diversity (D_H gene) segment, lying between the V_H and J_H gene segments. The processes by which the complete V_L and V_H regions of an antibody molecule are generated involve SOMATIC RECOMBINATION, exon splicing and removal of leader peptides. The disulphide bonds linking the polypeptide chains (see ANTIBODIES, Fig. 10) are also formed post-translationally.

The light chains may be produced from either the λ or κ loci, located on chromosomes 22 and 2 respectively. Each λ and κ locus has numerous alternative V and J gene segments and the λ light-chain locus has numerous alternative C regions, and it is a random affair as to precisely which combinations of V, J and C are spliced together to produce a complete V_LC_L Fab component.

The heavy-chain locus is on chromosome 14 and has about 40 functional V_H gene segments and a cluster of about 25 D_H segments lying between the V_H and six J_H gene segments. It also contains a large cluster of C_H genes downstream of these, each corresponding to a different isotype (see ISOTYPE SWITCHING). The antibody class depends upon which isotype is formed by alternative mRNA splicing during exon splicing (see RNA PROCESSING, EXON SHUFFLING).

The diversity of the antibody and immunoglobulin repertoire is therefore generated by (i) the varying copy number of each type of gene segment, as well as by the different combinations of gene segments, employed during the chromosomal rearrangement events in particular cells; and by (ii) addition and subtraction of nucleotides at the junctions between different gene segments; (iii) the many possible combinations of heavy- and light-chain V regions that pair to form the antigen-binding site of the molecule, and (iv) the SOMATIC HYPERMUTATION that introduces point mutations into the rearranged V-regions of B cells once they are activated, generating diversity even among a clonally selected B-cell class.

anticipation The tendency of some variable dominant conditions to become more severe (or to have earlier onset) in successive generations, a phenomenon which can be easily mimicked by random variations in severity. The discovery of expanding trinucleotide repeat sequences in FRAGILE-X SYNDROME, in myotonic MUSCULAR DYSTROPHY and HUNTINGTON'S DISEASE gave the needed molecular explanation. Severity of onset in these disorders correlates with repeat length, which grows as the sequence is passed from one generation to the next. A claim for anticipation needs careful statistical backing rather than a clinical impression. See GENETIC COUNSELLING.

anticoagulants Factors preventing blood clotting in the normal vasculature include the smoothness of the endothelial wall (see ENDOTHELIUM); a layer of *glycocalyx* on the endothelium repelling clotting factors; a protein, *thrombomodulin*, bound to the endothelial surface which binds thrombin, removing it from the plasma and slowing clotting but also forming a thrombomodulin:thrombin complex that activates a plasma protein, *protein C*, inactivating Factors V and VIII. Factors in the blood itself that remove plasma thrombin include an α-globulin (*antithrombin III*), while the *fibrin fibres* of the clot also sequester thrombin. HEPARIN is another powerful anticoagulant, binding antithrombin III and activating it. Prostacyclin, a product of endothelia and

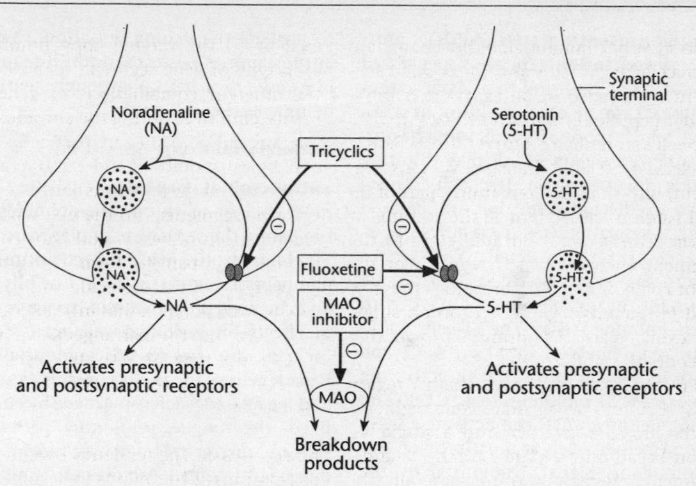

FIG 12. Some of the events in the secretion, breakdown and recycling of noradrenaline and serotonin and the effects on these of ANTIDEPRESSANTS such as monoamine oxidase (MAO) inhibitors, tricyclics and SSRIs. MAO inhibitors enhance NA and 5-HT actions by preventing their enzymatic destruction, tricyclics do so by blocking their reuptake and SSRIs (such as fluoxetine) do the same, but selectively for serotonin. © 2001 from *Neuroscience* (2nd edition) by M.R. Bear, B.W. Connors and M.A. Paradiso, published by Lippincott, Williams & Wilkins. Reproduced by kind permission of the publisher.

some leukocytes, is a powerful inhibitor of platelet adhesion and release. Warfarin is an anticoagulant of clinical use.

anticodon The triplet nucleotide sequence of a transfer RNA (tRNA) molecule which base-pairs with a codon on mRNA during the TRANSLATION phase of protein synthesis.

antidepressants Most DRUGS used in treatment of DEPRESSION inhibit the reuptake of noradrenaline and/or serotonin (see Fig. 12). They also increase hippocampal neurogenesis and synaptogenesis in rodents. The tricyclics (e.g., imipramine, amitriptyline) are older drugs with proven efficacy but often have sedative and autonomic side effects. They are also the most dangerous in overdose. The selective SEROTONIN reuptake inhibitors (SSRIS), such as fluoxetine, are newer drugs with a wider margin of safety and different (e.g., gastrointestinal) side effects. Monoamine oxidase inhibitors (MAOIs) are less often used, because of dangerous interactions with some foods and other drugs. Benefits of antidepressants do not become apparent for 2–3 weeks, possibly due to gradual alterations in sensitivity of serotonin receptors and adrenoceptors.

The mania (much less common than depression) of bipolar disorder responds to lithium, although it is not clear why. Valium has an effect on second messenger systems, including the phosphoinositide and adenylyl cyclase systems, any combination of which might be antidepressant. In laboratory animals at least, CREB activity in the HIPPOCAMPUS appears to be a crucial mediator of antidepressant effects, a wide variety of standard antidepressant treatments (noradrenaline-reuptake inhibitors, SSRIs and electroconvulsive seizures) increasing CREB activity within the region. Many treatments with antidepressant effects in humans also increase expression of CREB-regulated neural growth factors, such as BDNF (see NEURAL PLASTICITY).

antidiuretic hormone (ADH) See VASOPRESSIN.

antigen-presenting cells (APCs), antigen presentation Highly specialized cells (DENDRITIC CELLS and to a lesser extent MACROPHAGES and B CELLS), residing in LYMPH NODES, which process foreign antigens during ADAPTIVE IMMUNITY and display resulting peptide fragments on the cell surface bound to MHC proteins that are required for T CELL activation (see MAJOR HISTOCOMPATIBILITY COMPLEX for some details). Cell-surface INTERCELLULAR ADHESION MOLECULES (ICAMs) are important in establishing contact with other immune cells. The T-cell CD40 ligand, e.g., binds to CD40 on APCs and transmits signals to the T cell, activating it to express B7 molecules (e.g., CD80 and CD86). Before an APC can activate a cytotoxic or helper T cell, it must provide this co-stimulatory signal, recognized by the co-receptor protein CD28 on the T cell surface. CD40 also activates B CELLS. Drugs such as chloroquine that raise the pH of ENDOSOMES, making them less acidic, inhibit the presentation of antigens.

antigen receptors See ANTIBODIES, ANTIGEN-PRESENTING CELLS, B CELLS, T-CELL RECEPTORS.

antigenic determinants See EPITOPES.

antigenic drift, antigenic shift See INFLUENZA.

antigens *Foreign antigens* are molecules, typically glycoproteins or LIPOPOLYSACCHARIDES, on the surfaces of pathogens or other foreign cells that are capable of initiating innate and adaptive immune responses (see Appendix XI). *Self-antigens* are molecules identifying an individual's own cells to T cells and should therefore not initiate an immune response. See ADAPTIVE IMMUNITY, B CELLS, INFECTION, INNATE IMMUNITY.

antigravity muscles Muscles involved in posture, including extensors of the back and legs, and arm flexors. See POSTURAL REFLEX PATHWAYS.

antihistamine Substances blocking HISTAMINE receptors. Relevant H_1 receptors include those on the endothelium of capillaries that cause increased permeability, and those on unmyelinated axons thought to mediate the itching sensation. Certain antihistamines possess serotonin-reuptake blocking properties, although without being selective (see SSRIS).

antimalarial drugs See MALARIA.

antimicrobial peptides Possibly the most ancient of INNATE IMMUNITY defences, pre-dating the separation of plant and animal lineages, the many related antimicrobial peptides exhibit a variety of physical and chemical properties and have a variety of effects on microbial pathogens. They include the α-defensins produced by Paneth cells of the small intestinal mucosa and the related β-defensins made by epithelia of the tongue, respiratory pathways, skin and urogenital tract, bacteriocins produced by vaginal lactobacilli, and the CATHELICIDIN LL37. Activated MACROPHAGES also produce them. Antimicrobial proteins include lysozyme of tears and saliva and surfactant proteins SP-A and SP-D of the epithelial surfaces of the lung (see ALVEOLUS).

antimüllerian hormone (AMH) See MÜLLERIAN INHIBITING HORMONE.

antioxidants See COENZYME Q, FLAVONOIDS, FREE RADICALS, VITAMIN C.

antiporters Those ion pumps using the free energy change caused by collapse of one ion's concentration gradient across a membrane to drive another ion or molecule in the opposite direction. One such includes the ATP-independent Na^+-Ca^{2+} exchanger pumping calcium out of cells and driven by the Na^+ electrochemical gradient. See the ion pumps part of ION CHANNELS, and SYMPORTERS.

antipsychotic drugs See PSYCHOSIS.

antiretroviral drugs See HIV.

antisaccade Eye movements away from a target. Contrast PROSACCADE.

antisense RNA RNA that is transcribed from the DNA strand of duplex DNA complementary to that from which mRNA is transcribed. See GENOMIC IMPRINTING.

antiserum The fluid component from

clotted blood of an immune individual, containing antibodies against the antigen used for immunization. Antisera contain a mixture of antibodies that bind this antigen, each with its own structure, epitope on the antigen and its own set of cross-reactions so that each antiserum is unique. However, they can be produced in only limited volumes and even when purified by affinity chromatography will still contain minor populations of antibodies giving unexpected cross-reactions, problems reduced by use of MONOCLONAL ANTIBODIES. See PASSIVE IMMUNITY.

α_1 anti-trypsin (P_i system) An anticlotting factor, synthesized by the liver and inhibiting factor IX, trypsin and some other enzymes. Its physiological role may be to inhibit proteolytic enzymes released from macrophages at sites of inflammation. Those homozygous for the P_i^s allele show marked deficiency of anti-tryptic activity. P_i^z homozygotes are unusually susceptible to CHRONIC OBSTRUCTIVE PULMONARY DISEASE.

antiviral drugs See under appropriate viral diseases.

anus The exterior opening of the hindgut, developing from the proctodaeum at the end of the anorectal canal through regression of the embryonic cloaca, the cloacal membrane giving rise to the anal membrane while the associated ectoderm and mesoderm give rise to the PERINEUM, the anal triangle of which contains the anus. The last 2–3 cm of the digestive tract is the *anal canal*, beginning at the inferior end of the RECTUM and ending at the anus. This region has thicker smooth muscle than the rectum and forms the internal anal sphincter at the canal's inferior end.

anxiety See FEAR AND ANXIETY.

anxiolytics Anxiety-relieving drugs. See FEAR AND ANXIETY.

aorta The elastic ARTERY forming the main systemic trunk of the arterial system, emerging from the left ventricle. The *arch of the aorta* (not an AORTIC ARCH, see Fig. 14), is supplied with baroreceptors (the AORTIC BODIES), and gives off the brachiocephalic trunk, left common carotid and left subclavian arteries, while the main trunk gives off arteries to the viscera, skeletal and dermal structures, branching into the two common iliac arteries.

aortic arches During development, the aortic arches are arteries supplying the pharyngeal arches (see PHARYNGEAL POUCHES, Fig. 133) and are embedded in mesenchyme, terminating in the left and right dorsal aortae (see Fig. 13). The transformation from the embryonic to the adult arterial system involves the following modifications. By day 30, most of the 1st arch and the entire 2nd arch have disappeared. The 3rd arch forms the common carotid artery and the first part of the internal carotid artery, the external carotid artery being a branch of the third arch; the 4th aortic arch persists on both sides, on the left forming part of the arch of the aorta between the left common carotid and the left subclavian arteries while on the right it forms the most proximal segment of the right subclavian artery, whose distal part is formed by a portion of the right dorsal aorta and the 7th intersegmental artery. The 5th arch never forms, or does so incompletely before regressing. Most of the distal part of the 6th arch, or pulmonary arch, degenerates; but proximally it gives off an important branch that grows towards the developing lung bud. On the right side, this becomes the proximal segment of the right pulmonary artery, while on the left side the distal portion persists during intrauterine life as the DUCTUS ARTERIOSUS. See CARDIOVASCULAR SYSTEM, Fig. 38.

aortic bodies Small bilateral structures attached to a small branch of the aorta, near its arch. It contains peripheral chemoreceptors responding primarily to decreases in blood oxygen (pO_2) but less sensitive to blood pH and pCO_2. They also contain baroreceptors that help monitor systolic blood pressure. Compare CAROTID BODIES, CAROTID SINUSES. See CARDIOVASCULAR CENTRE, RESPIRATORY CENTRE.

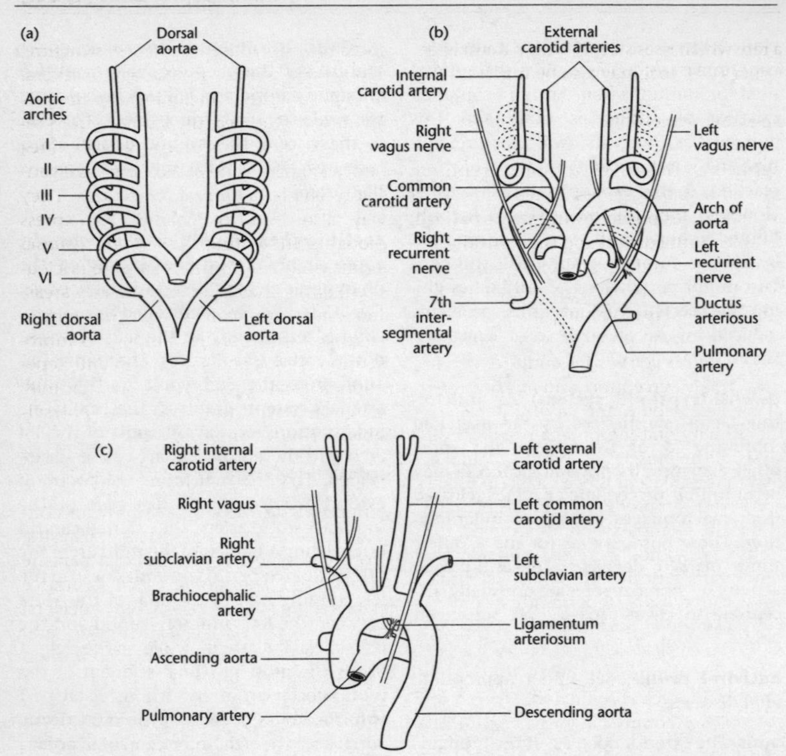

FIG 13. (a) The arterial condition of the embryo. (b) The arterial system after transformation to the adult condition (broken lines indicating obliterated vessels). (c) The great arteries of the adult after cephalic folding, growth of the forebrain and elongation of the neck have pushed the heart into the thoracic cavity. This caudal shift results in elongation of the carotid and brachiocephalic arteries, the left subclavian artery shifting its point of origin from the level of the 7th intersegmental artery to a higher position closer to the origin of the left common carotid artery. Redrawn from *Langman's Medical Embryology*, T.W. Sadler. © 2004 Lippincott, Williams & Wilkins, with the permission of Wolters Kluwer.

aortic reflex See CARDIOVASCULAR CENTRE, CARDIAC CYCLE.

Apaf (apoptotic protease-activating factor) A pro-apoptotic complex which binds to CYTOCHROME C (see entry for some details). *Apaf-1* expression is enhanced by E2F.

APC gene A tumour suppressor gene originally discovered as the cause, when mutated and inherited, of FAMILIAL ADENOMATOUS POLYPOSIS (see entry), and when originating by somatic mutation to ADENOMATOUS POLYPOSIS COLI. See CANCER CELLS, Fig. 35; COLORECTAL CANCER.

Apert's syndrome A craniosynostosis (abnormal development and premature fusion of the flat bones of the SKULL), leading to an oversized brain housed within a misshapen skull. It also involves malformations of hands, feet and internal organs. It is clinically important. A dominant 'hot-spot' mutation in gene for FIBROBLAST GROWTH FACTOR receptor 3 (*FGFR3*) on chromosome 4p, occurring almost exclusively in male parent germ line and with a rate which rises steeply with paternal age. Other FGFR loci may also be involved.

apes See HOMINOIDEA.

aphasia The loss, or impairment, of spoken LANGUAGE, generally resulting from focal brain lesions following STROKE or head trauma. Both cortical and subcortical (e.g., thalamus, caudate nucleus) may be involved. The perisylvian region of the left hemisphere contains, in addition to LANGUAGE AREAS (BROCA'S AREA, WERNICKE'S AREA), cortical centres for auditory perception, tactile perception, and motor control of the face, mouth and larynx. Damage to the primary oral-motor area produces paralysis of the muscles of sound production; that to the adjacent Broca's area (in Broca's aphasiacs) can produce problems in generating the complex sound sequences of words, but without paralysis, and may prevent grammatical and syntactical usage, but without loss of understanding content. Brain damage to the PRIMARY AUDITORY CORTEX results in cortical deafness, without affecting either comprehension of the written word or speech production. Wernicke's aphasiacs often speak fluently, but unintelligibly.

Global aphasia, the inability to comprehend or produce intelligible speech, results from damage to both Broca's and Wernicke's areas. *Anomia*, the inability to name familiar objects, often very specifically, results from damage to a diverse range of brain regions, usually overlapping the major language areas. *Transcortical sensory (receptive) aphasias* involve deficit in auditory comprehension and confusion of word meaning without loss of ability to form correct word sounds, and are often associated with damage to Wernicke's area and regions adjacent to it. The person not only fails to understand others' speech, but cannot monitor their own, which may become unintelligible, though fluent. Phonemes are frequently substituted incorrectly. Reading is impaired if the angular gyrus is included in the damage. *Transcortical motor (expressive) aphasias* do not disrupt comprehension but lead to laboured speech with reduced sentence structure, and are associated with damage to several frontal structures, including the supplementary motor area, the left posterior prefrontal cortex, including Broca's area (Broca's aphasia) and parts of the basal ganglia receiving prefrontal projections. They may also lead to inability to express thoughts in writing. *Conduction aphasia* results from lesions in the arcuate fasciculus, linking Broca's and Wernicke's areas, and results in difficulty in repeating written or spoken words orally, although comprehension is good and speech is fluent. Repetition is usually best with nouns, but may fail entirely with function words, polysyllabic words or nonsense words.

apical ectodermal ridge See LIMBS.

Apicomplexa A protoctist phylum containing more than 5000 species of parasitic protozoa, including the malarial parasite *Plasmodium*, *Toxoplasma gondii* (a source of congenital neurological birth defects) and *Cryptosporidium*, an opportunistic parasite when the immune system is compromised, as in AIDS.

apo- Prefix designating a protein in a form free from any organic or inorganic cofactors or prosthetic groups, as in apoenzyme, apolipoprotein, apoproetin.

apo (a) See APOLIPOPROTEIN (a).

apo A See APOLIPOPROTEIN A.

apo B See APOLIPOPROTEIN B, RNA INTERFERENCE.

apo C See APOLIPOPROTEINS C.

apocrine glands Glands accumulating their secretory product at the apical surface of the secreting cell. This portion then pinches off, the remaining part of the cell repeating the process. Examples include MAMMARY GLANDS and certain SWEAT GLANDS.

apo E See APOLIPOPROTEIN E.

apolipoprotein A (apo A) The major protein in HDL; some particles contain only apo A-I, others both apo A-I and apo A-II. Apo A-I is an acceptor of cholesterol efflux from cells delivered by ABCAI transporters (see ABC TRANSPORTER FAMILY) and an activator of the enzyme LCAT in recipient HDL particles in plasma. Apo A-I therefore plays a key role in promoting cholesterol transport from the peripheral tissues to the liver for disposal (see LCAT and LIPOPROTEINS). Apo A-I and Apo A-II are also found in low concentrations in chylomicron and VLDL particles, and apo A-IV is a minor component of HDL and CHYLOMICRONS.

apolipoprotein (a) Synthesized in hepatocytes, apo (a) forms an atherogenic LDL-like lipoprotein particle, Lp(a), by covalent binding to apo B-100. It has been demonstrated that apo (a) has high homology with plasminogen, and may inhibit fibrinolysis by competing with plasminogen in binding to fibrin, thus conferring on Lp(a) thrombogenic as well as atherogenic potential.

apolipoprotein B (apo B) Apo B-100 is the binding protein for LDL (or B,E) receptors. It is synthesized in the liver and occurs in VLDL, IDL and LDL particles in plasma (see LIPOPROTEINS), each particle containing one molecule of this apoprotein. In most circumstances, more than 90% of plasma apo B-100 is associated with the LDL fraction. Plasma apo B-100 concentrations have therefore been proposed as a good measure of the number of LDL particles in blood for the purposes of CHD risk assessment, or monitoring response to statin therapy.

Apo B-48 is found only in CHYLOMICRONS and their remnants. It is synthesized in the intestine using the apo B-100 gene by insertion of a stop codon into mRNA (see RNA EDITING), resulting in a molecule with only 48% of the molecular mass of apo B-100 and lacking the LDL receptor ligand.

apolipoprotein C (apo C) Apo C-I, C-II and C-III occur as minor constituents of HDL and major constituents of CHYLOMICRONS and VLDL, the triglyceride-rich LIPOPROTEINS. Apo C-II is an activator of lipoprotein lipase whereas apo C-III appears to be an inhibitor of this enzyme.

apolipoprotein E (apo E) A ubiquitous apoprotein modulating transport of CHOLESTEROL in the central nervous system, and playing a complex role in lipoprotein metabolism. Apo E exhibits genetic polymorphism, there being three major ISOFORMS, E2, E3 and E4, whose distributions are highly variable in different populations. Apo E is the binding ligand for hepatic receptors clearing chylomicron remnants, VLDL remnants and IDL from the blood.

It is thought that ApoE may also have a role in neural plasticity by stabilizing the cytoskeletal structure of neurons, ApoE2 and ApoE3 (but not ApoE4) binding a number of proteins associated with their microtubules. It participates in the transport of plasma lipids and in the redistribution of lipids among cells, modulating susceptibility to the most common diseases involving NEURAL DEGENERATION, a role for apoE being implicated in regeneration of synaptic circuitry after neural injury (see NEURAL REGENERATION). The *ApoE4* allele is a major risk factor for late-onset familial and sporadic ALZHEIMER'S DISEASE (AD) and is associated with a poor outcome after brain injury. ApoE isoforms are suggested to have differential effects on neuronal repair mechanisms, and *in vitro* work has demonstrated the neurotrophic properties of ApoE3 on neurite outgrowth. Three common variants of the gene encoding ApoE occur in human populations, modulating susceptibility to Alzheimer's disease and PARKINSON'S DISEASE. The *ApoE4* allele increases risk for AD and has a frequency of ~15% in most populations, although its frequency increases in sub-Saharan peoples; *ApoE3* is neutral with respect to the disease and is the commonest allele at ~78%; *ApoE2* is protective and has a frequency of ~7%. Variation in the *ApoE* gene is also known to modulate susceptibility to CORONARY HEART DISEASE (see MODIFIERS).

apolipoproteins Lipid-binding proteins ('apo' designating the protein in its lipid-free form) possessing amphipathic helical regions in their structure, one surface of which is hydrophobic while the other is hydrophilic. Hydrophobic bonding of apoproteins with the fatty acyl chains of phospholipids and hydrophobic interaction with their polar head groups enables the assembly of spherical, micellar-like, particles in which the hydrophilic moieties of the apoprotein, phospholipid and the polar alcohol group of the unesterified cholesterol form the surface of the particle and interact with its aqueous environment, while the non-polar components project into the core of the particle which accommodates non-polar cholesterol esters and triglycerides, thereby enabling their mobilization from tissues and their transport in the plasma.

In addition to this structural role, a number of apoproteins play functional roles in lipoprotein metabolism as enzyme modulators and cell surface receptor ligands. Many minor apolipoproteins have been isolated and characterized, although their functions are not yet elucidated. See LIPOPROTEINS and named apolipoproteins.

apomorphy In phylogenetics, any character derived as a novelty from a pre-existing (plesiomorphous) character, the two forming a homologous pair of characters. Synapomorphous characters (termed homologous characters in many writings on evolution) are characters which are shared homologues of two or more taxa and which are believed to have originated in their closest common ancestor and not in an earlier one. See PLESIOMORPHY.

apoptosis (programmed cell death, type I programmed cell death) From the Greek 'falling away', this is the best understood form of programmed cell death. The apoptotic programme is available in all human cells and can be triggered by an absence of appropriate SURVIVAL SIGNALS as well as by external stimuli (e.g., FAS). Cells also produce a set of proteins that can inhibit the programme. Apoptosis is a normal cellular response that is crucial to tissue remodelling during morphogenesis, as when gaps are introduced between fingers and toes at ~45 days of gestation as interdigital cells in the hand and foot plates die, and when neurons with unproductive connections are pruned out so that neuronal circuitry is progressively refined (this can approach 80% of neurons in the retina). It is also the manner by which cytotoxic T CELLS kill their targets and the preferred solution to genotoxic stress. Under the microscope, cells are seen to roll up, form blebs, undergo zeigosis (appearing to 'boil' through rapid bleb formation), chromatin condensation, nuclear fragmentation and the shedding of membrane-bound vesicles: apoptotic bodies. Much of this is the result of activation of the 'cell-suicide' cysteine proteases, the CASPASES. A hallmark of this type of cell death is the fragmentation of nuclear DNA into 200-bp pieces through the action of endogenous nucleases that cleave the DNA between nucleosomes. Cells undergoing apoptosis are rapidly ingested by phagocytic cells, without induction of co-stimulatory molecules, on recognition of phosphatidylserine (normally only found in the inner membrane leaflet) as it replaces phosphatidylcholine as the major outer leaflet phospholipid.

The *extrinsic pathway* to apoptosis may be initiated by activation of pro-apoptotic transmembrane cell surface receptors ('death receptors'). Ligands of these death receptors include TNF-α and FasL (see FAS) and, on binding, activate the receptor's death domain which in turn activates FADD in the cytoplasm. The resulting protein complex, DISC, attracts the inactive forms of caspases 8 and 10, triggering their self-cleavage. These initiator CASPASES (see entry for details) then trigger the caspase cascade which converges on the signalling pathway through which the intrinsic apoptotic programme operates. Alternatively, a cell may undergo ANOIKIS.

The *intrinsic pathway* to apoptosis is p53-dependent (see P53 GENE). Once p53 is activated and in an apoptosis-inducing conformation, it can drive expression of genes encoding proteins such as Bax, which open mitochondrial channels (see CYTOCHROME C), initiating the activities of caspases – which is where the extrinsic and intrinsic pathways converge. Once in the cytosol, cytochrome *c* molecules associate with the Apaf-1 protein to form an 'apoptosome', which activates procaspase 9, converting it to the active caspase 9. This then cleaves procaspase 3, the final caspases of the cascade cleaving 'death substrates' (lamin, ICAD, actin, vimentin, etc.) resulting in the apoptotic phenotype. ICAD is a caspase-activated DNase, whose liberation fragments chromosomal DNA. Cleavage of nuclear envelope and cytoskeletal proteins leads to cytoskeletal collapse, resulting in nuclear fragmentation and bleb formation. Other proteins released by damaged mitochondria inactivate IAPs (inhibitors of apoptosis) which normally block caspases. See CRYSTALLINS.

The protein p53 has a crucial role in bringing about apoptosis in those cells with damage to DNA that is not repairable. Its stabilization by phosphorylation increases p53-dependent transcription of *PUMA* and *PIG3*, genes encoding Bcl-2-related proteins that control apoptosis. The presence of free 'DNA ends' in a cell that retains the capacity for DNA repair would lead to activation of the enzyme poly(ADP-ribose) polymerase (PARP), which adds poly-A tails to the ends, and would exhaust the cell of its energy supplies. Clusters of resulting dead cells would also impair the cell–cell and cell–matrix interactions upon which tissue modelling depends. Apoptosis deletes such cells rapidly by tagging them for phagocytosis, delaying energy exhaustion by uncoupling (through caspase activation) the catalytic and DNA-binding domains of PARP. Failure to initiate apoptosis in response to DNA damage is associated with appearance of cells with twice the normal mutation number. See SENESCENCE, SURVIVAL FACTORS.

apoptosome See APOPTOSIS.

APP (amyloid precursor protein) See ALZHEIMER'S DISEASE.

appendicitis Inflammation of the APPENDIX.

appendicular skeleton One of the two divisions of the SKELETAL SYSTEM. It includes the pectoral and pelvic girdles which append the AXIAL SKELETON, and the bones of the FORE LIMBS and HIND LIMBS.

appendix (vermiform appendix) Attached to the posteromedial surface of the caecum, a small blind finger-sized pouch containing lymphoid tissue, but with no essential function. See GUT-ASSOCIATED LYMPHOID TISSUES.

appetite A major factor in the regulation of energy intake, and consequently of ENERGY BALANCE, in which hypothalamic nuclei play a major role (see Fig. 14). Primary neurons in the hypothalamic arcuate nucleus receive peripheral humoral signals across a relaxed BLOOD–BRAIN BARRIER from adipocyte and gut peptides, and second order neurons integrate these with vagal inputs from the gastrointestinal tract via the NUCLEUS TRACTUS SOLITARIUS and the neighbouring circumventricular organ, the AREA POSTREMA of the medulla oblongata. Over short time intervals, the medulla regulates food intake on the basis of inhibitory satiety signals which may be both humoral and neural. Inputs from taste buds and the olfactory epithelium, stomach distension, CHOLECYSTOKININ, and neural inputs from the liver all inhibit feeding. Positive inputs from the sight and memory of food and their social contexts are also integrated, with resulting effects on meal size and frequency, the handling of ingested food by the gut, and energy expenditure. Food ingestion and digestion raises plasma osmolarity and normally stimulates drinking; but if water is not accessible the decreased water potential is detected by osmoreceptors in the hepatic portal vein. The resulting modulated output from the hypothalamus results in expression of genes involved in regulating satiety (reducing the motivation to eat) and is implicated in the development of OBESITY.

Food intake regulation by the CNS involves two parallel neural circuits originating in the ARCUATE NUCLEUS of the HYPOTHALAMUS (see Fig. 98). Here, LEPTIN-sensitive neurons project to two lateral hypothalamic neuron types, one group of which bears MC4 receptors that bind both

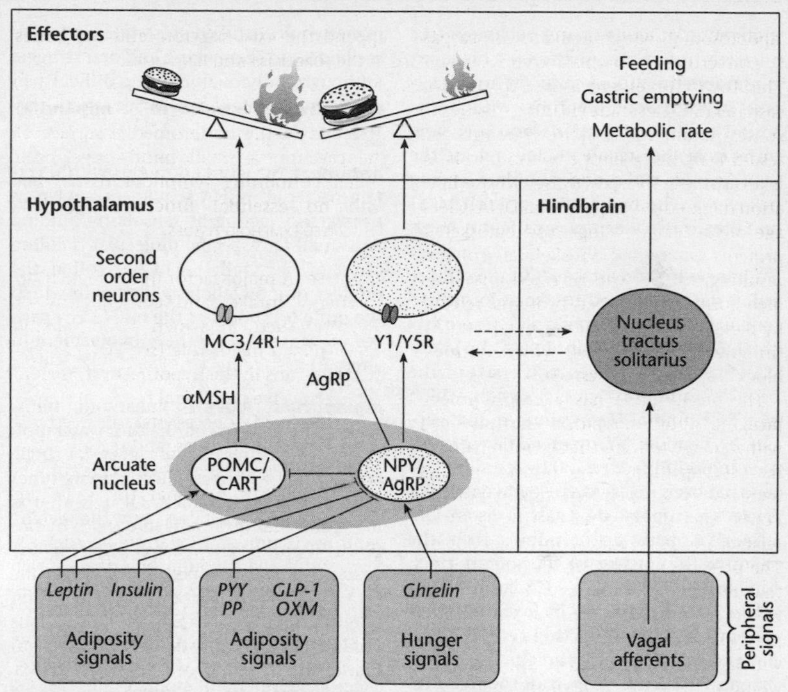

FIG 14. A simplified diagram of the potential actions of gut and adipocyte peptides on the hypothalamus (see APPETITE for details). Appetite-inhibiting pathways are shown stippled; appetite-stimulating pathways are unstippled. Directly stimulatory pathways are indicated by undotted arrows, directly inhibitory pathways by T-junctions, and indirect pathways by dotted arrows. The satiety signals are: PP = pancreatic polypeptide; OXM = oxyntomodulin; GLP-1 = glucagon-like peptide-1. From *Science* Vol. 307 No, 5717 (25 March 2005), p. 1911. Reprinted with permission from AAAS.

α-MSH (anorexigenic, catabolic, and a byproduct of POMC) and agouti-related peptide (AgRP; orexigenic, anabolic) neurotransmitters also released by the arcuate nucleus. α-MSH activates MC4 and inhibits feeding, while AgRP (an antagonist of MC3/4 receptor activity) blocks MC4 activation, inhibits CORTICOTROPIN and TSH release, and stimulates feeding. Another anorexigenic peptide, CART, is released in response to leptin by other arcuate nucleus projections to the lateral hypothalamus and inhibits feeding when leptin levels rise. A further neuropeptide, NEUROPEPTIDE Y (NPY), is released by the same neurons of the arcuate nucleus that produce AgRP, in response to leptin, insulin and cortisol (see also GHRELIN). It binds to NPY5 receptors on neurons of the paraventricular nucleus and lateral hypothalamus, inhibits TSH and corticotropin release and activates the parasympathetic division of the ANS to stimulate feeding. Leptin receptors are also present on neurons of the ventromedial and dorsomedial nuclei, regions regulating feeding behaviour (see FEEDING CENTRE). On binding its receptors in the hypothalamus, insulin inhibits feeding.

The other group of arcuate nucleus neurons releases the orexigenic cyclic neuropeptide melanin-concentrating hormone (MCH) and projects widely in the brain, including the cerebral cortex, where MCH may contribute significantly in the motivation of food-seeking behaviour (see DOPAMINE, DARPP-32). SEROTONIN, however, causes appetite suppression. Yet further lateral hypothalamic neurons with widespread cortical projections and receiving input from the arcuate nucleus release the neuropeptide OREXIN and stimulate feeding behaviour. The brain levels of both MCH and orexin rise when plasma leptin levels fall.

Changes in food intake affect the cellular redox state, and food intake is known to modulate the circadian clock (e.g., see GLUCOSE). Over short time intervals, the brainstem (medulla) regulates food intake on the basis of inhibitory satiety signals which may be humoral (e.g., glucagon-like peptide-1, peptide YY, pancreatic peptide, oxyntomodulin) or neural (when sufficient food has been ingested, the paraventricular nucleus is stimulated by gastric distension, through a pathway of impulses via the vagus to the nucleus of the solitary tract, inhibiting feeding). Cholecystokinin inhibits feeding partly by exciting these afferents but also by acting centrally as a neurotransmitter. Psychological factors may also influence or override these centres (see ANOREXIA NERVOSA, BULIMIA NERVOSA and OBESITY).

appetitive behaviours Motivated, goal-seeking behaviours such as food-seeking (see TASTE), and sexual behaviour, which are motivated by 'drives' which can be sated, and lead to the consummatory act (feeding, copulation, etc.). The motivations for many human goal-orientated behaviours (e.g., listening to music) are still unexplained. The HYPOTHALAMUS is involved in many such behaviours.

apraxia A selective inability to perform complex motor acts (e.g., buttoning clothing). It may involve lesions in the supplementary motor area (SMA) of the PREMOTOR CORTEX. Another form involves motor speech disorders resulting from damage to BROCA'S AREA, resulting in impaired capacity to programme the positioning of muscles employed in speech execution despite normal muscle strength, which will be expressed as disturbed prosody, inappropriate phoneme sequencing and oral struggle (although individuals have no difficulty in producing rote sequences). See PARIETAL CORTEX.

aptamers Structural domains within certain bacterial mRNA sequences known as RIBOSWITCHES, acting as sensors by binding to a small and specific molecular building block (a metabolite) which is often the product of a metabolic pathway in which the protein product of the mRNA is a part. See RNA PROCESSING for their involvement in alternative splicing.

aquaporins (AQPs) Eukaryotic transmembrane protein channels for water molecules, consisting of multiples of a single polypeptide whose α-helices pass six times through the membrane. AQP-1 is a tetrameric complex internally lined with hydrophilic R-group side chains allowing the passage of water molecules in single file in the direction dictated by the water potential gradient. Aquaporins of various kinds occur especially densely in the membranes of red blood cells, PROXIMAL CONVOLUTED TUBULE cell membranes, and (variably) in those of COLLECTING DUCTS (compare PORINS), where they are involved in osmoregulation.

aqueous humour Watery fluid filling the anterior cavity of the EYE (see Fig. 73), continually secreted by capillaries in the ciliary body nourishing the lens and cornea. Drains into the scleral venous sinus, being replaced about every 90 minutes. Largely responsible for creating the intraocular pressure (approx. 16 mm Hg), maintaining the shape of the eyeball and preventing the eyeball from collapsing. Excess pressure of the aqueous humour is the major cause of GLAUCOMA.

arachidonic acid A 20-carbon polyunsaturated fatty acid elongation product of linoleic acid metabolism (see ESSENTIAL FATTY ACIDS) and a product of DIACYLGLYCEROL cleavage. It is an essential precursor of regulatory lipids, the EICOSANOIDS, to which it is converted in the smooth endoplasmic reticulum. Its modification by CYCLOOXYGENASES gives rise to prostaglandins, thromboxanes and prostacyclin, while lipoxygenases convert it to leukotrienes and lipoxins. It is released on cell damage at the periphery.

arachnoid (arachnoid mater) The middle of the three MENINGES which, like the dura mater, loosely envelops the brain. It is separated from the dura mater by a narrow subdural space, through which veins pass on their way to the venous sinuses; and from the pia mater by the subarachnoid space. Arachnoid villi are finger-like extensions of the arachnoid that project into the dural venous sinuses, notably the superior sagittal sinus.

archaecortex See PALAEOCORTEX.

'archaic *Homo sapiens*' See HOMO HEIDELBERGENSIS.

arcuate fasciculus A fascicle linking Broca's area with Wernicke's area. See LANGUAGE AREAS.

arcuate nucleus The most ventral hypothalamic nucleus (see HYPOTHALAMUS, Fig. 98), important in integrating peripheral circulating ENERGY BALANCE signals. It is the origin of two parallel pathways: an anabolic orexigenic pathway promoting feeding, and a catabolic anorexigenic pathway inhibiting feeding (see APPETITE). The hypothalamic PARAVENTRICULAR NUCLEUS is a critical target for its neuronal projections, those using NEUROPEPTIDE Y and AGOUTI-RELATED PROTEIN (AgRP) as co-transmitters being the first-order neurons of the orexigenic pathway, those using melanocortin (derived from proopiomelanocortin) being the first-order neurons of the anorectic pathway. Other first-order cells in this pathway express CART, promoting negative energy balance. LEPTIN receptors are expressed on all these first-order neurons; but leptin inhibits the orexigenic and stimulates the anorectic pathway.

Circulating free fatty acids exert insulin-like effects upon it, possibly by favouring its intracellular accumulation of long-chain fatty acyl-CoA (in rats, selective reduction of this in the arcuate nucleus leads to obesity).

Ardipithecus ramidus Discovered in 1994 from the Middle Awash, Ethiopia, and dated to ~4.4 Myr (earlier than the then oldest known hominin, *Australopithecus afarensis*) this putative hominin seems to possess rather few hominin traits, and some teams regard these fossils as on the chimpanzee lineage. In 2001, hominid fossils from the Middle Awash dated 5.2–5.8 Mya were assigned to *A. r. kadabba*. Further work in November 2002 produced more material. An upper canine (diamond-shaped in cross section, like those of later hominids, rather than V-shaped like those of non-human hominids) and a lower third premolar were among six new teeth recovered from Asa Koma Locality 3 (ASK-VP-3), with an estimated age of 5.6–5.8 Mya. The new fossils illuminate hominid dental evolution near the divergence of apes and humans, and they require revision of the taxonomic status of *A. r. kadabba*. A full description of a partial skeleton, expected in 2007, is still awaited (2008). See BIPEDALISM, ORRORNIN TUGENENSIS, SAHELANTHROPUS TCHADENSIS.

area IT Regions of the monkey INFERIOR TEMPORAL CORTEX lying beyond AREA V4 in the ventral visual processing stream (see EXTRASTRIATE CORTEX, Fig. 72) and containing neurons with complex spatial receptive fields responding to a variety of colours and abstract shapes. While a small proportion of these neurons in monkeys respond especially strongly to pictures of faces, and to some faces more than others, they also respond less strongly to non-facial stimuli. See FACIAL RECOGNITION.

area MT (V5) An area of neocortex at the junction of the parietal and temporal lobes receiving input from the primary visual cortex. It seems to be involved in detection of stimulus movement. See VISUAL CORTEX, Fig. 184.

area postrema The most caudal aspect of the floor of the fourth ventricle of the brain (the rostral medulla), where the blood-brain barrier is absent, and the central site of action of substances which cause VOMITING. It acts with the neighbouring NUCLEUS TRACTUS SOLITARIUS, integrating input from the periphery: taste buds, pharynx and stomach distension. Neurons project from it to the hypothalamus, relaying neural and hormonal signals from the gastrointestinal tract (see APPETITE).

areas V1, V2, V3 See VISUAL CORTEX, Fig. 178.

area V4 An area of neocortex anterior to the primary VISUAL CORTEX (see Fig. 184) in some monkeys, and controversially in humans, forming part of the ventral visual processing stream (see EXTRASTRIATE CORTEX, Fig. 72) involved in the perception of shape and colour (see COLOUR PERCEPTION). It projects to AREA IT.

***ARF* gene (*CDKN2a* gene, *INK4a* gene)** A gene encoding cyclin-dependent kinase inhibitor 4a. It encodes two separate proteins: p16INK4a, which regulates RB1 (see RETINOBLASTOMA PROTEIN) activity by directly inhibiting cyclin-dependent kinases; and p19ARF, which regulates the function of P53 (see P53 GENE).

ARF ($p14^{ARF}$) Protein encoded by the $p14^{ARF}$ gene, a gene which carries an E2F recognition sequence in its promoter. High ARF levels block Hdm2 action, allowing P53 to accumulate and trigger APOPTOSIS.

Arf proteins A family of small GTPases, operating in conjunction with RAB PROTEINS and acting as molecular switches controlling the assembly of multiprotein complexes onto membranes and the consequent regulation of protein and membrane trafficking (see VESICLES) throughout the cell, as well as membrane-associated signalling events (e.g., see INSULIN). They are inactive when bound to GDP but become activated on binding GTP, in which they are assisted by guanine nucleotide exchange factors, GEFs (see GUANOSINE TRIPHOSPHATE), which also help determine membrane-attachment specificity by limiting Arfs to plasma membranes and the GOLGI APPARATUS. As with Rab proteins, the GDP-bound form is cytosolic whereas the GTP-bound form is membrane-associated. Also like Rabs, Arf GTPases usually carry a lipid moiety, in this case an amino-terminal myristoyl group, which mediates rapidly reversible membrane-binding. At plasma membranes, Arf GEFs are recruited in response to external signals; thus GEFs of the cytohesin/ARNO family contain PLECKSTRIN HOMOLOGY DOMAINS that mediate binding of PHOSPHOINOSITIDES generated by signal transduction pathways. Arfs are not to be confused with ARF. See CELL MEMBRANES.

arginine vasopressin (AVP) See VASOPRESSIN.

Argonaute proteins (Ago) Endonuclease components of RISC, involved in RNA INTERFERENCE. MicroRNAs bind to Ago1; conventional siRNAs bind to Ago2; and piRNAs bind to Piwi, another member of the Argonaute family.

arms See FORE LIMB.

aromatase Enzyme converting testosterone to 17β-oestradiol and expressed in granulosa cells of the GRAAFIAN FOLLICLE in LEYDIG CELLS of the testis, in adipose tissue and the placenta. Roughly 20% of circulating oestrogens in males is due to this activity. In males, aromatase expression is directly related to the degree of adiposity and dependent upon cytokine stimulation in the presence of glucocorticoids.

arousal **1**. It is generally recognized that *arousal from* SLEEP is controlled by the ascending RETICULAR FORMATION, operating in close association with sensory input. High frequency electrical stimulation of the midbrain of intact cats causes arousal and desynchronized electroencephalograms. **2**. During *sexual arousal*, the sensory and motor neural responses controlling sexual responses are the same for both sexes. Sensory impulses are conducted from the genitals to the sacral region of the spinal cord, where integration of sexual reflex responses occurs. Sensory impulses are then conducted along the anterolateral pathway of the SOMATIC SENSORY SYSTEM to the thalamus and cortex, cerebral influences moderating the sacral reflexes. Motor impulses reach the reproductive organs through both divisions of the autonomic nervous system, and to skeletal muscles by the SOMATIC MOTOR SYSTEM. Psychological factors play a major role in sexual excitement during which, in females, erectile tissue within the CLITORIS and around the opening of the vagina swell through parasympathetic stimulation, and the nipples of the breast often become erect. Mucous glands in the vestibule of the vagina secrete small amounts of mucus, and larger amounts are secreted by the vagina. During the height of arousal, often during coitus, the vaginal, uterine and perineal muscles contract rhythmically, and muscle tension increases more widely. For males, events are described in the entries for EJACULATION and PENIS. See STRESS RESPONSE.

ARPs (actin-related proteins) Proteins catalysing the nucleation of ACTIN at the plasma membrane (the ARP complex; aka Arp2/3 complex) and mediating attachment of DYNEIN to membranous organelles.

arrector pili muscle See HAIR FOLLICLE.

arrestins A protein family whose members bind to the phosphorylated carboxy-terminal portions of serpentine receptors, preventing their interaction with G PROTEINS and so terminating the signal through those receptors.

arrhythmias Irregular heart rhythms (see CARDIAC CYCLE, ELECTROCARDIOGRAM). More than 400,000 people in the United States die annually from sudden cardiac death and lethal arrhythmias. No effective therapy is available, although implantable defibrillators have had limited clinical success. As the cells of the electrical conduction system arise from cardiac muscle precursors, lesions in the pathways guiding or maintaining the differentiation between ventricular myocyte and Purkinje cell lineages may cause susceptibility to cardiac-associated sudden death. Research on mice has revealed a role for miR-1 (see MICRORNAS) in development of arrhythmia, miR-1 being crucial for normal cardiac conduction. Its targets seem to be a gap junction protein in CARDIAC MUSCLE fibres and a cardiac transcription factor, Irx5, regulating formation and maturation of the heart's electrical conduction system.

In adults, *tachycardia* (heart rate >100 bpm) may be caused by elevated body temperature, excessive sympathetic stimulation or toxicity; *paroxysmal atrial tachycardia* (sudden increase in heart rate to 95–150 bpm for a few seconds or hours) is caused by excessive sympathetic stimulation or abnormally elevated permeability of slow (voltage-gated Ca^{2+}) channels; *atrial flutter* (300 P waves min^{-1}, 125 QRS complexes min^{-1}) and *atrial fibrillation* (no P waves, normal QRS complexes, ventricles constantly stimulated by atria) are due to ectopic action potentials in the atria; *ventricular fibrillation* (no QRS complexes or rhythmic contraction of the myocardium, with many patches of asynchronously contracting ventricular muscle) is caused by ectopic action potentials in the ventricles; *bradycardia* (heart rate <60 bpm) results from elevated stroke volume in athletes, excessive vagal stimulation and carotid sinus syndrome; *SA node block* (no P wave, new low heart rate due to AV node acting as pacemaker, normal QRS complex and T wave) is due to ischaemia and tissue damage due to infarction; *AV node block* (PR interval >0.2 sec in first-degree cases, caused by inflammation of the AV bundle; 0.25–0.45 sec with some P waves triggering QRS complexes while others do not (second-degree, caused by excessive vagal stimulation)); P wave dissociated from QRS complex, atrial rhythm ~100 bpm, ventricular rhythm <40 bpm (complete heart block) caused by ischaemia of AV nodal fibres or compression of AV bundle; *premature atrial contractions* (occasional shortened intervals between one contraction and the next, frequent in healthy individuals) caused by excessive smoking, lack of sleep, too much caffeine or alcoholism; *premature ventricular contractions* (prolonged QRS complex, exaggerated voltage since only one ventricle may depolarize, inverted T wave and increased probability of fibrillation) caused by ectopic foci in ventricles, lack of sleep, too much caffeine, irritability and sometimes occurs with coronary thrombosis. AV node block at the level of the node occurs because of increased AV node refractory period, which does not permit conduction at a rate of 300 bpm although allowing it at 150 bpm. This is caused by idiopathic fibrosis and sclerosis of the conduction system in ~50% of patients and by ischaemic heart disease in ~40%, the remaining cases being due largely to medications involving BETA-BLOCKERS and calcium channel blockers (e.g., digoxin), increased vagal tone, CONGENITAL HEART DEFECTS, genetic or other disorders. See DISTAL CONVOLUTED TUBULE, SA NODE (for artificial pacemakers).

artemether An antimalarial drug (see MALARIA).

arteries Major blood vessels carrying blood away from the heart and to the organs. Their walls comprise a *tunica interna* (endothelium resting on an outer basement membrane), *tunica media* (usually the thickest layer, containing smooth muscle within an elastic lamina) and *tunica externa* (composed largely of elastic and collagen fibres). Large arteries branch into smaller ones which eventually branch into arterioles supplying capillary beds. *Elastic arteries* (see ELASTIN) have the largest diameters (e.g., the aorta, brachiocephalic, common carotid, subclavian, pulmonary, common iliac and vertebral arteries); *muscular arteries* are medium sized and their tunica media contains more smooth muscle and fewer elastic fibres than does that of an elastic artery, and an external elastic lamina composed of elastic tissue separates the tunica externa from the tunica media. The normal elasticity of an artery wall is required to prevent overproliferation of its smooth muscle cells (see ATHEROSCLEROSIS). Compare VEINS; see ARTERIOLES, BLOOD PRESSURE, VASCULOGENESIS.

arterioles Small arteries delivering blood to CAPILLARIES. Those near arteries from which they branch have an artery-like tunica interna, a tunica media of smooth muscle fibres and few elastic fibres, and a tunica externa composed mainly of elastic and collagen fibres. Those closest to a capillary bed, and the metarterioles to which they give rise, consist of little else but an ENDOTHELIUM, and scattered smooth muscle fibres for vasoconstriction (see VASODILATION AND VASOCONSTRICTION).

arteriosclerosis A group of diseases characterized by thickening, and loss of elasticity, of the arterial walls. One form is ATHEROSCLEROSIS. See CARDIOVASCULAR DISEASE.

arthritis A form of rheumatism (painful disorders of the supporting structures of the body) in which the JOINTS have become inflamed (see INFLAMMATION). Three forms occur: (i) diffuse connective tissue diseases, e.g., rheumatoid arthritis; (ii) degenerative joint disease, e.g., osteoarthritis; and (iii) metabolic and endocrine disorders with associated arthritis, such as gouty arthritis (see GOUT). *Rheumatoid arthritis* is an AUTOIMMUNE DISEASE, typically occurring bilaterally, in the smaller joints first, and involving inflammation of the synovial membrane. If untreated, this thickens and fluid accumulates causing PAIN. The membrane produces abnormal granulation tissue, *pannus*, which adheres to the articular cartilage and sometimes erodes it completely, after which fibrous tissue joins the exposed bone ends. This ossifies and fuses the joint, having a crippling effect. Pain relief can be achieved with NSAIDS, including COX-2 inhibitors. Autoantibodies to citrulline, a component of many modified proteins, have been found to appear as long as 10 years prior to symptoms. *Osteoarthritis* is apparently a degenerative disease accompanied by AGEING and irritation and general wear of the synovial joints, striking weight-bearing ones (knees and hips) first. Articular cartilage deteriorates, spurring new osseous tissue to be deposited on the spurs that are formed. The synovial membrane may become inflamed late in the disease.

articular cartilage The hyaline cartilage covering the bone surfaces in SYNOVIAL JOINTS.

articular surface A smooth surface of a bone (e.g., a FACET) involved in a JOINT.

artificial chromosomes See HUMAN ARTIFICIAL CHROMOSOMES.

artificial insemination See INTRA-UTERINE INSEMINATION.

artificial intelligence (AI) Its inception dating to the 1950s, AI is the attempt to examine cognitive processes by means of computer programs and other information-processing devices. It is unlikely that we will discover or be able to describe programs in the brain analogous to those of computers because, unlike a computer, the brain does not seem to use a linear series of steps in processing information or executing operations. A more promising approach models cognitive processes by simulating the actual mechanisms used in the brain, where computations are distributed through richly interconnected elements of artificial neural networks. In these parallel-distributed processing (PDP) models, each element is influenced in positive or negative ways by other elements of the network, one element's activity on another being the product of its output level and connection strength. An element sums up the effects of its various inputs and produces an output that is a linear or non-linear function of those inputs, and although the elements in PDP models may be thought of as analogues of neurons, they might also be higher order elements, such as words, percepts and ideas. An artificial network can learn to approximate an input–output function by being given examples of the in- and out-values of such a function through any of a variety of weight-adjustment procedures, or 'learning algorithms', and in the case of face recognition success can be achieved when it would be difficult for a programmer to work out a function *a priori* to distinguish between facial patterns. In what is probably a major advance, chess-playing computer algorithms can now be improved by allowing 'self-selection' of those artificial neural networks that exhibit success in beating artificial opponents, new batches being derived from successful networks until some good playing strategies have evolved, and even beating human experts. In real brains, where neurons connect at synapses, learning seems to take place through alteration in the 'weighting' (see SYNAPTIC STRENGTH) required to cause a neuron to fire action potentials along its axon. But evolutionary methods can develop machine intelligence without being programmed.

artificial pacemaker See SA NODE.

ascending arousal system See AROUSAL, RETICULAR FORMATION.

ascending reticular activating system (ARAS) A key structure of the RETICULAR FORMATION controlling the state of consciousness by influencing the excitability of cortical neurons. It plays a part in the SLEEP–WAKE CYCLE. See AROUSAL.

ascending sensory pathways See DESCENDING MOTOR PATHWAYS.

asexual reproduction The only occurrence of asexual reproduction in humans is monozygotic twinning. See TWINS.

aspartame (L-**aspartyl**-L-**phenylalanine methyl ester**) A commercially synthesized artificial sweetener.

aspartate See MALATE-ASPARTATE SHUTTLE.

aspirin The first of the NON-STEROIDAL ANTI-INFLAMMATORY DRUGS (NSAIDs) to be discovered, whose pK_a of 3.5 causes it to remain largely un-ionized in the stomach. It becomes ionized and therefore unable to pass back into the stomach lumen on passing through the mucosal barrier into the intracellular compartment whose pH is closer to neutral. Aspirin and prostaglandin A1, whose anti-inflammatory effects are mediated through their modulation of the NF-κB pathway, prevent activation of NF-κB-responsive genes. Aspirin also has important antiplatelet activity, and is an inhibitor of prostaglandin synthesis (like ibuprofen) through its inhibition of CYCLOOXYGENASE in the smooth ENDOPLASMIC

RETICULUM. Low doses of aspirin, taken regularly, inhibit vasoconstriction and platelet aggregation by blocking synthesis of thromboxane A2, reducing the risk of heart attacks, transient ischaemic attacks and STROKES.

***ASPM* (abnormal spindle-like microcephaly associated) gene** A gene spanning 62 kb of chromosome 1p31 and encoding a huge protein of 3477 amino acids, nonsense mutations in which lead to primary MICROCEPHALY. Just prior to cell division in neuroepithelial cells, the gene product ASPM concentrates at opposite poles of the cell and helps to organize the microtubules involved in chromosome separation. Reduction in the quantity of ASPM in a dividing neuroepithelial cell causes it to divide asymmetrically, so that instead of forming two daughter stem cells one of the daughter cells enters the specialization pathway and becomes a neuron, which short-circuits expansion of the cortex. It has been suggested that human *ASPM* underwent an episode of accelerated sequence evolution by positive selection (see NATURAL SELECTION) after the split of hominin and chimpanzee lineages but before the separation of modern non-Africans from Africans, and that it may be a major genetic component underlying the evolution of the human brain. But other evidence indicates that the ASPM gene sequence shows accelerated evolution in the African hominoid clade, preceding hominid brain expansion by several million years. Gorilla and human lineages show particularly accelerated evolution in the IQ domain of ASPM, containing multiple IQ repeats (calmodulin-binding motifs).

assisted reproductive technology (ART) See IN VITRO FERTILIZATION for some of the issues. See also KLINEFELTER SYNDROME.

association cortex (association areas) Large expanses of the cortex, often adjacent to the primary sensory centres, which cannot be simply allocated functionally to sensory or motor roles but which are often involved in recognition. They include the PREFRONTAL CORTEX, the POSTERIOR PARIETAL CORTEX and the INFEROTEMPORAL CORTEX associated with basic language skills and spatial attention.

association fibres Nerve fibres connecting parts of the cerebral cortex within the same hemisphere. Thus, the inferior longitudinal fasciculus links the occipital and temporal lobes, contributing to visual recognition. Compare COMMISSURAL FIBRES.

associative learning Forms of LEARNING dependent upon the form of neural plasticity termed HEBBIAN MODIFICATION. See also CLASSICAL CONDITIONING, INSTRUMENTAL CONDITIONING.

aster The microtubule array growing out from each centriole-based microtubule organizing centre during nuclear division (mitosis or meiosis). See CENTROSOME.

astereognosis Inability to identify the shape and nature of objects by touch alone. It is a symptom of MULTIPLE SCLEROSIS.

asthma Asthma is now recognized as an epidemic in the developed world, characterized by chronic inflammation of the LUNGS resulting from an allergic reaction (see ALLERGIES) to the same allergens as cause allergic rhinitis and conjunctivitis. It is marked by elevated IgE levels, lower airway submucosal mast cells and eosinophils, and airway-constricting T_H2 cytokines such as IL-5, IL-9 and IL-13. During an asthma attack, inflammation, mucus secretions and the shortening of smooth muscle fibres around the bronchial walls all cause narrowing of the airways, obstructing airflow, causing uneven ventilation, impairing gas exchange, and making breathing difficult by trapping inhaled air in the lungs. Most breathing difficulty seems to arise from severe constriction of the terminal bronchioles; but PET SCANNING reveals patchiness in the bronchoconstriction, giving large clusters of poorly ventilated lung units amid open units with good ventilation. In time, asthmatic lungs show permanent changes, including goblet-cell metaplasia, oversecretion of mucus, hyperplasia and hypertrophy of smooth muscle fibres, and deposition of collagen just beneath the lining of the epithelial surface.

The current understanding of allergic asthma is centred on the role of bronchoconstrictor substances and inflammatory mediators. The airways of most asthmatics are hyper-responsive, leading to goblet-cell metaplasia and secretion of mucus, and remodelling of the airways leading to permanent narrowing. The cholinergic agonist methacholine causes bronchospasm in normal individuals, but a much greater response in the asthmatic lung (used in diagnosis). Therapy typically involves use of inhaled BRONCHODILATORS (β_2-ADRENERGIC receptor agonists) and anticholinergic drugs (corticosteroids and leukotriene antagonists). But endogenous airway relaxants also exist, one of them being nitric oxide, and the enzymes synthesizing NO (nitric oxide synthases, NOSs) are constitutively expressed in lung epithelium. Some evidence suggests that insufficient endogenous *S*-nitrosothiols (SNOs; see NITRIC OXIDE) promote airway hyper-responsiveness.

astigmatism A refraction disorder of the eye, in which either the cornea or lens has an irregular curvature. Parts of the image are out of focus as a result. See ACCOMMODATION.

astrocytes Star-shaped glial cells of the central nervous system whose processes extend into the surrounding network of unmyelinated neurons (neuropil). Some appear to be stem cells in their own right; and by secreting a variety of growth factors and cytokines (e.g., BMPs, PDGF, CNTF, FGF-2) they can determine whether a local niche is neurogenic or gliogenic. They act as a structural support ('scaffolding') for neurons, absorb extracellular potassium ions in active parts of the brain and can pass these to one another through gap junctions. They have other neuroprotective roles, up-regulating expression of the anti-inflammatory crystallin gene, *CRYAB*, in MULTIPLE SCLEROSIS. They also remove neurotransmitters such as glutamate after their release. Their morphology varies widely although they often appear stellate in histological preparations. They ensheath synaptic junctions, associate with nodes of Ranvier, and respond to disease and infection by clearing cell debris, secreting trophic factors and forming scars. They release two substances that stimulate synapse formation – the fatty complex apoE/cholesterol and the protein thrombospondin. Firing of neurons induces astrocytes to release ATP, which binds receptors on nearby astrocytes and prompts Ca^{2+} ION CHANNELS to open, triggering release of further ATP. They clear neurotransmitter from synaptic clefts, but may also release transmitter themselves (e.g., GLUTAMATE; see SYNAPTIC STRENGTH), implying considerable involvement of astrocytes in information processing by the brain. Culture studies indicate that astrocytes play a powerful role in regulating the formation and maintenance of synaptic connections between neurons in development. Some provide scaffolding for neuron migration, even spanning the entire radial width of the brain – from the ventricles to the pial surface; or stretch from blood capillaries to neurons and, by transporting ions and other substances, sustain neurons and maintain the extracellular environment. Processes of adjacent astrocytes now appear not to overlap greatly and astrocytes may effectively divide the brain into separate compartments, each the sole domain of a single astrocyte. The HIPPOCAMPUS, where memories are formed, may be such a region. Compare OLIGODENDROCYTES; see NEUROGENESIS.

asymmetry Some aspects of early development are inherently asymmetrical, such as the point of entry of the sperm during fertilization, the attachment of the embryo to the wall of the uterus during implantation, and the location of cells with respect to their neighbours. Direct cell-to-cell signalling can provide a means of identifying cell position and triggering selective GENE EXPRESSION. Despite the bilateral symmetry of the SOMITES and the body's appearance of external symmetry, internal organs are in an asymmetrical, yet reproducible, arrangement. The heart is on the left and the lung next to it is smaller than is that on the right; within the abdominal cavity, the intestines form a clockwise loop; the stomach, pancreas and spleen are on the left while the liver and gall bladder are on the right. These differences are rooted in a cascade of molecular asymmetries established in development. See SITUS INVERSUS.

ataxias Uncoordinated (dyssynergic) and inaccurate (dysmetric) movements, often beginning with unsteady gait, caused by cerebellar lesions, resembling those caused by ALCOHOL abuse.

Selective ablation of Golgi cells in the cerebellar cortex causes severe acute motor disorders such as ataxia, although partial recovery may occur. See REELIN, REPETITIVE DNA.

ataxia telangiectasia mutated protein (ATM) An extremely large (~350 kilodaltons) protein kinase encoded by the *ATM* gene. One of a group of PI(3)K-related kinases (PIKKs) and crucial in detecting double-stranded DNA breaks (see DNA REPAIR MECHANISMS), ATM- and ATR- (ATM- and Rad3-related) activation are two of the first steps in the activation of signal transduction pathways inhibiting cell cycle progression after DNA damage (see CELL CYCLE, Fig. 40). It has a DNA-binding domain and a PI(3)K catalytic domain. Like DNA-PK, it can bind to free DNA ends (e.g., of damaged DNA) and initiate cascades which transmit damage signals to checkpoints and repair proteins. Deficiency leads to the disease *ataxia telangiectasia* (AT; lit. ataxia with skin redness; a PROGERIA in which redness is due to dilations of capillaries, esp. on eyes, ears and neck) and to hypersensitivity of cells to X-IRRADIATION. Its activity is minimal or low in unstressed cells and engages to help with cellular stresses affecting DNA or chromatin structure, but phosphorylation by ATM kinase activates it following DNA damage. Its substrates include p53, E2F-1 and c-Abl. About 15% of patients with AT (frequency ~1:40,000 children) develop malignancies, especially LEUKAEMIAS and lymphomas, and immunological deficiencies are common, contributing to a high frequency of lung infections, a major cause of death, which usually occurs in young childhood. Germline mutations in the *ATM* gene are associated with Li–Fraumeni syndrome, AT and common forms of familial breast and ovarian cancer, strengthening the link between the maintenance of genomic integrity (see GENOMIC INSTABILITY) and cancer susceptibility. See COCKAYNE SYNDROME, TELOMERES.

atheroma See ATHEROSCLEROSIS.

atherosclerosis A disease of the large arteries, and the major human disease associated with cholesterol and lipid metabolism. A chronic, progressive, COMPLEX DISEASE, it begins in teenage years and is associated with lifelong high plasma cholesterol levels (see LOW-DENSITY LIPOPROTEIN). Atherogenesis is first detectable by build-up of plasma LIPOPROTEINS in the subendothelium (the intima) of focal areas of the arterial tree, their accumulation resulting from proteoglycan binding and lipoprotein aggregation on the EXTRACELLULAR MATRIX. This is accelerated in those with INSULIN RESISTANCE (see entry), where an increased plasma concentration of small, dense LDL particles, which are also more prone to oxidation and glycation, enables their easier movement through endothelia into the subendothelial space, favouring inflammation and transformation into plaque. Macrophages ingest the oxidized and retained lipoproteins, and by acquiring lipoprotein-derived cholesterol turn into lipid-laden *foam cells*. These adhere to the artery's smooth muscle to form a fatty streak which may become artery-blocking plaque (*atheroma*), producing cytokines and growth factors which attract further monocytes and T cells, induce migration of smooth muscle progenitors from bone marrow to the plaque, and instruct smooth muscle fibres of the tunica media to hypertrophy, narrowing the lumen. The most dramatic changes occur in the large elastic arteries, involving thickening of the intima and loss of elasticity of the tunica media. This loss of elasticity seems to be a factor in

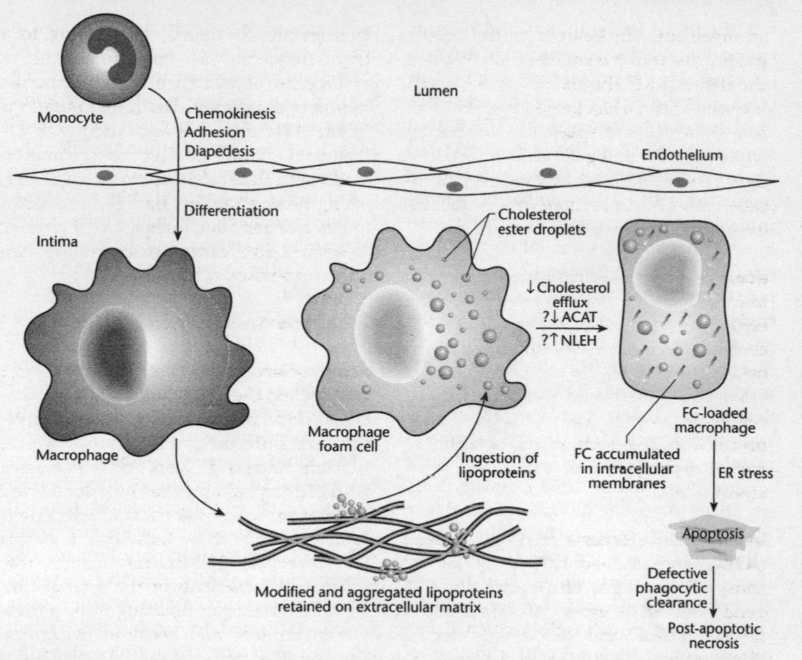

FIG 15. Diagram indicating some of the events influencing the entry and loading of cholesterol into macrophages during the formation of atherosclerotic plaques (lesions). See the ATHEROSCLEROSIS entry for clarification. From F.R. Maxfield and I. Tadas, *Nature* © 2005, with permission from Macmillian Publishers Ltd.

permitting the overproliferation of smooth muscle cells, which is a feature of the condition. The narrowed arterial lumen leads to raised BLOOD PRESSURE and further damage to the endothelium, with consequent risk to THROMBUS formation.

Current opinion is that INFLAMMATION and immune responses contribute to atherogenesis (the formation of atheroma). Reactive oxygen species (ROS) production by endothelial cells, which may be the initial signal for inflammation, may be reduced by UCP2 (see UCPS). One endothelial-leukocyte adhesion molecule that appears to be involved in the early adhesion of mononuclear leukocytes to an arterial endothelium at sites of atheroma formation is vascular cell adhesion molueule-1 (VCAM-1), particularly attractive to the cell membranes of monocytes and T CELLS. The mechanism of VCAM-1 induction soon after initiating an atherogenic diet probably depends on inflammation initiated by ROS-oxidized LIPOPROTEIN particles accumulating in the intima (see MITOCHONDRIA). The ENDOTHELIUM (see entry for relevant details) is then triggered to express monocyte-attracting chemokines and pro-inflammatory cytokines (IL-1β, TNF-α). The monocytes cross the endothelium and differentiate into MACROPHAGES (see Fig. 14), their scavenger receptors having greater affinity for the modified LDL than do normal LDL receptors. When low doses of the CANNABINOID THC were added to the diet of $ApoE^{-/-}$ mice, progression of atherosclerosis was markedly slowed as fewer plaque-infiltrating macrophages were recruited (although this does not mean that smoking cannabis is beneficial to the heart).

It might be thought that antioxidants would counter this tendency, but trials in humans have *not* shown that vitamin E or other antioxidants decrease the incidence of atherosclerotic heart disease. This has led some to doubt about the role of oxidized lipoprotein as a major factor in foam-cell production. Instead, a number of 'atherogenic' lipoproteins may induce this state in intimal macrophages, including the 'remnant lipoproteins' formed by partial catabolism and later cholesterol enrichment of triglyceride-rich lipoproteins formed by gut and liver cells. Internalized by macrophages in culture, this leads to ACAT-mediated re-esterification of lipoprotein-derived CHOLESTEROL (see Fig. 48) in the endoplasmic reticula of the intimal macrophages and formation of cytoplasmic membranous neutral lipid droplets, leading them to acquire a 'foamy' appearance (hence 'foam cells'). These foam cells adhere to the artery's smooth muscle to form a fatty streak which may become artery-blocking plaque (*atheroma*), produce cytokines and growth factors which attract further monocytes and T cells, and instruct smooth muscle fibres of the tunica media to hypertrophy, narrowing the lumen. In advanced lesions, unesterified 'free' cholesterol (FC) accumulates, leading to ER stress (see ENDOPLASMIC RETICULUM), macrophage apoptosis and necrosis.

Both human and animal studies indicate that lipoproteins containing apolipoprotein B (apoB), such as LDLs, are required for atherosclerotic development. The amount of functional LDL receptor present on hepatocytes is one of the most important factors influencing plasma LDL concentration, and the development of STATINS has provided a highly effective therapy. Plasma concentrations of atherogenic lipoproteins, such as LDLs, are also influenced by hepatic production rates of very-low-density lipoproteins (VLDLs), which are the metabolic precursors of LDLs. Overproduction of VLDLs is a common feature of those with INSULIN RESISTANCE.

Until the late 1990s, the notion was that fatty streaks inevitably develop progressively into complicated atheroma through proliferation of smooth muscle cells in arterial walls as they accumulate in the plaque and lay down an abundant extracellular matrix. Recent serial angiographic work suggests that many coronary arterial lesions develop stenoses discontinuously. In any event, by becoming bulkier the lesion causes the arterial lumen to narrow, leading to unstable angina pectoris and/or myocardial infarction, and desquamations of the endothelium which form the nidus of a PLATELET thrombus (accounting for ~25% of fatal coronary thromboses). Like those forming in the diabetic retina, microvascular channels that develop in the atheroma as a result of neoangiogenesis may be particularly fragile and prone to micro-haemorrhage.

The renin-angiotensin-aldosterone system (see ANGIOTENSINS), recognized to be a crucial regulator of BLOOD PRESSURE, has consistently been shown to have a prominent role in atherogenesis in both humans and experimental animals. Effective therapies for atherosclerotic cardiovascular disease which attempted to inhibit this system led to research suggesting that the proatherogenic effects of activating this system derived not solely through resulting increases in blood pressure.

Atherosclerotic patients are advised to take non-atherogenic diets, to increase carefully their level of EXERCISE, and are often given vasodilating drugs such as angiotensin-converting enzyme- (ACE-) inhibitors (see HYPERTENSION), in addition to lipid-lowering drugs. In recent work employing an atherosclerosis MOUSE MODEL, chemically modified short interfering RNAs (siRNAs) were used to silence the endogenous gene for apolipoprotein B in normal mice and those made hypercholesterolaemic transgenically, suggesting that siRNAs may eventually have therapeutic value. See CARDIOVASCULAR DISEASE.

athletes See EXERCISE.

ATM See ATAXIA TELANGIECTASIA MUTATED PROTEIN.

atopy See ALLERGIES.

ATP See ADENOSINE TRIPHOSPHATE.

ATP:PCr system See PHOSPHOCREATINE.

ATR (ATM- and Rad3-related protein) A very large protein kinase (a proximal checkpoint kinase), phosphorylating numerous substrates in processes critical to cellular responses to the arrest of DNA replication forks (see CELL CYCLE, Fig. 40; DNA REPLICATION). DNA damage activates ATAXIA TELANGIECTASIA MUTATED PROTEIN, leading to replication fork arrest and activation of cellular ATR's kinase activity. This relocates to single-stranded DNA in the nucleus and can phosphorylate critical substrates, such as RAD17 and CHK1. See DNA REPAIR MECHANISMS.

atrial diastole Relaxation of the atria during the CARDIAC CYCLE.

atrial natriuretic peptide (ANP) Hormone secreted by cardiac muscle fibres of the atria when atria are stretched by high blood pressure. It increases Na^+ excretion by the kidneys by suppressing reabsorption of Na^+ and water in the proximal convoluted tubules and collecting ducts and causes smooth muscle cells in blood vessel walls to relax. Both of these effects reduce blood pressure. It is a diuretic, decreasing blood volume. See NATRIURETIC PEPTIDES.

atrial systole Atrial contraction during the CARDIAC CYCLE.

atrioventricular node See AV NODE.

atrioventricular valves Valves between the atria and ventricles of the HEART (see Fig. 90). The tricuspid valve is on the right side, the bicuspid (mitral) valve is on the left side.

atrium One of the two receiving chambers of the HEART (see Fig. 90). See CARDIAC CYCLE.

atropine Derived from *Belladonna* plants, an antagonist of ACETYLCHOLINE at muscarinic receptors. Related compounds are used by ophthalmologists to dilate pupils.

attention Humans are visual organisms and spend much of their waking hours engaged in visual search behaviour, attempting to make the current object of interest into the current object of visual attention and motor action (see ELECTROENCEPHALOGRAM). The act of differential processing of simultaneous information sources (selective attention) may be adaptive simply because, even if the brain could process all such information, there might be a selective advantage (improved performance) in focusing on one source of information at a time. Behavioural and physiological research over the past 30 years and more have emphasized one of two types of mechanisms involved in visual search tasks: PARALLEL PROCESSING (in which all or many objects are analysed at once), and serial processing (in which one, or a few, objects are selected for detailed analysis). On being asked to search for a blue diamond among a variety of objects on a card, such as differently coloured and oriented squares (those on an apex being diamond-like), and complex objects composed of differently coloured vertical, horizontal and oblique bars, all blue items may seem to make themselves available simultaneously. But if asked to search for a yellow square, the blue items may seem to recede into the background as yellow ones become more apparent. Colour and orientation features seem to present almost immediately, but binding of a colour to an orientation seems to require more. It also seems possible to select individual items for further scrutiny even if it is not fixated. These two processes appear to interact to produce an effective visual search, and some evidence indicates that, in macaques at least, both sorts of processing occur in area V4 of the visual cortex (see EXTRASTRIATE CORTEX). Covert attention shifts to an object before it is fixated by the eyes, the act of attentional fixation preceding serial fixation and also enhancing the response of a V4 neuron sensitive to stimulation in one specific region of the retina (it may also have preferred sensitivity to colour and/or shape). It appears that parallel feature processes may guide serial selection of plausible targets for further scrutiny.

Most of the THALAMUS is concerned with controlling which sensory information is sent to the cortex, a major factor in attention. Increased vigilance is associated with β brain rhythms (see ELECTROENCEPHALOGRAM), the STRESS RESPONSE and, notably, with activation of the noradrenergic diffuse modulatory system of the BRAIN (see Fig. 30) arising from the LOCUS COERULEUS and other hindbrain nuclei. It is known that noradrenaline augments the effects of excitability (glutamatergic) or inhibitory (GABAergic) inputs relative to the basal firing rates of neurons, making cortical neurons more likely to respond to visual stimulation (an improvement in signal-to-noise ratio). The locus coeruleus may participate, via noradrenaline, in a general arousal of the brain during new, unexpected and non-painful external sensory stimuli, and GLUCOCORTICOIDS and CRH similarly enhance attention to biologically significant events, including those that are novel or threatening. This may initiate reflexive eye movements (saccades) and head-turning, involving the SUPERIOR COLLICULUS; and eye movements are preceded by a shift of attention to the new location (saccade goal). When we know where an important stimulus is more likely to appear, we move our attention to it and process the information with greater sensitivity and speed (see MICROSACCADES). Only cells with large enough RECEPTIVE FIELDS (RFs) to respond to several objects are involved: generally those in late-processing visual areas (INFEROTEMPORAL and MEDIAL TEMPORAL CORTEX). The RFs appear to shrink and target only the target stimulus (an increase in signal-to-noise ratio). Functional neuroimaging studies and PET scanning have revealed selective activation of areas of the PARIETAL CORTEX during shifts of visual attention (see LATERAL INTRAPARIETAL REGION, NEGLECT SYNDROME). Different areas of the cortex have higher activities when different attributes of a stimulus are being discriminated. Whereas the ventromedial occipital cortex is affected by attention involved in colour and shape discrimination tasks, it was unaffected by speed discrimination tasks; and whereas the parietal cortex is involved in attention to visual motion, it is not affected by the other tasks mentioned. In visual perception, only part of the image is selected for attention, the rest becoming background. The part of the image that is the central focus of attention is its boundaries and edges (see ORIENTATION-SELECTIVE CELLS). It is possible that those neurons from the visual cortex and reticular formation making back-projections to the lateral geniculate nucleus play a role in visual attention, modifying geniculostriate neuron responses in such a way that only a selected portion of retinal input is passed to the visual cortex. Such sensory filtering is a feature of the somatic sensory system (see PRIMARY SOMATOSENSORY CORTEX).

There is no clear resolution of the question of how attention is directed, although the pulvinar nucleus of the thalamus has several properties that make it an interesting candidate. It has reciprocal connections with most visual cortical regions of the occipital, parietal and temporal lobes, and people with pulvinar lesions have abnormally slow responses to contralateral stimuli, especially when stimuli competing for attention are presented to the ipsilateral side. A GABA antagonist (muscinol) suppresses pulvinar activity on injection in macaques, accompanied by difficulty in shifting attention. However, unilateral deactivation of the superior colliculus or posterior parietal cortex has similar effects, and bilateral deactivation of the pulvinar nucleus have no greater behavioural effect than do unilateral deactivations. Attention probably involves all these, and other, brain regions (see ATTENTION DEFICIT HYPERACTIVITY DISORDER). Attention is a key feature in stabilizing the PKA-dependent machinery involved in the protein synthesis-dependent phase of MEMORY storage (see also NEURAL PLASTICITY). Mice that do not attend to the spatial surroundings they walk through are able to form a spatial plan lasting only a few hours, whereas mice forced to attend to them retain a map lasting for days (see HIPPOCAMPUS).

attention deficit hyperactivity disorder (ADHD) A prevalent condition in children (more so than in adults) in which the prefrontal cortex is not properly functioning and behaviour becomes less consistent and coherent over time. Estimates are that 5–10% of schoolchildren worldwide exhibit ADHD, some of whom continue to exhibit symptoms as adults. Genetic and non-genetic (e.g., brain injury, premature birth) factors are implicated. Genes implicated include the D2 DOPAMINE receptor gene and the dopamine transporter gene. The α_2-ADRENERGIC receptors are implicated in several central nervous system disorders and their agonists are used in the treatment of attention deficit disorder. Ritalin (methylphenidate, which binds the dopamine transporter) is also employed, increasing the postsynaptic effect of dopamine. See TOURETTE SYNDROME.

attenuation reflex The reflex caused by the onset of a loud sound, reducing sound conduction by the ear ossicles to the inner ear. The reflex, which is probably activated as we speak, has a 50–100 msec delay but causes the muscles anchoring the malleus and the stapes to the bone in the middle ear cavity to contract, making the ossicles more rigid. The delay reduces the protection afforded the cochlea from very sudden loud sounds, such as a loud explosion (or Walkman). The reflex suppresses low frequencies more than high frequencies, tending to make high-frequency sounds easier to discriminate in an environment rich in low-frequency sound, enabling us to understand speech more easily in a noisy environment.

audibility The audibility of a tone depends upon both its sound pressure and its frequency. Adults can hear sounds whose frequencies range from 20 to 16 kHz and between 4 and 130 phon (if a reference sound is set at 70 $N m^{-2}$, or 70 dB SPL, it has a loudness of 70 phon). Those who are 'hard of hearing' have raised auditory thresholds. See HEARING.

auditory association cortex Lying immediately posterior to the PRIMARY AUDITORY CORTEX, processing and interpreting auditory information. In the dominant hemisphere, this is Wernicke's area. See HEARING, LANGUAGE AREAS.

auditory tubes See EUSTACHIAN TUBES.

Auerbach's plexus See ENTERIC NERVOUS SYSTEM.

auricle (pinna) See EARS, HEARING.

Aurignacian culture (Mode 4/5) The Upper Palaeolithic blade-based culture of modern humans, with use of bone, ivory and antler not only to make points but also ornamental beads, with sculpturing and engraving appearing for the first time. Tool kits emphasizing blades are labelled Mode 4 technologies, while those emphasizing spears, axes, etc., are labelled Mode 5 technologies. The sites are larger than those of the Middle Palaeolithic, open-air as opposed to rock-shelter or cave, with artifacts including simple bone flutes and indicating long-distance contact, and even trade. Early sites appear to represent a dispersal of modern humans from Africa into Europe (see MIGRATION (2)).

Australopithecines Hominins (Tribe Hominini) lacking the expanded brain, the reduced molar teeth and skeletal features of the *Homo* lineage, of which at least five, possibly six, species antedate or overlap the earliest representatives of *Homo*. Their relationship to *Sahelanthropus tchadensis* is not yet resolved. They are conventionally split into two groups: gracile (lightly built), and robust (megadontic, more heavily built), forms. Conservative taxonomy places all these forms in the genus *Australopithecus*; but it is usual nowadays to place the robust forms in the genus *Paranthropus*. As yet, it is not established whether *Ardipithecus ramidus* is a member.

With their huge molar teeth and massive jaw muscles, robust australopithecines (see PARANTHROPUS entries) are thought to have been dietary specialists that fed mainly on small, tough, fibrous plant foods, and their extinction between 1.0 and 1.4 Mya is often attributed to their low-nutrient diets. But recent studies of their tooth anatomy and dental microwear using very nearly non-destructive laser ablation of their tooth enamel suggest that their diets may have been less specialized than this and more similar to those of their ancestors and hominin competitors. The ratios of carbon isotopes in the teeth are thought to reflect differences in diet. Tropical grasses fix carbon using C_4 photosynthesis and have high $^{13}C/^{12}C$ ratios whereas most broad-leaved plants use the C_3 pathway and have low $^{13}C/^{12}C$ ratios. Browsing herbivores eat mainly broad-leaved plants while grazers tend to eat grasses. Oxygen isotope ratios can also shed light on climate and diet. The $^{18}O/^{16}O$ ratio of surface water increases with temperature and evaporation as well as with low humidity, an enrichment amplified in 'leaf water', which is often dietarily adequate in browsing herbivores. The gradual deposition of tooth enamel preserves an excellent record of seasonal chemical and isotopic variations. Recent studies of *Paranthropus* teeth reveal an isotopic composition characteristic of mixed feeders that could survive in environments ranging from cool and humid to warm and dry. The powerful teeth and jaws of *Paranthropus* may have been crucial to survival only when they resorted, like modern gorillas, to tough foods in the leanest season. This could have mitigated competition with *Homo*, for *Homo* and robust australopithecines appeared ~2.5 Mya and their divergence and coexistence may have been achieved by different strategies of adaptation to the increased seasonality and predominance of cooler, drier climates that began at that time. Such arguments are in need of evidence, currently lacking, from a comprehensive modern comparative database of a wide range of species (including baboons) from different climates and environments. Curators of fossils will probably need to be convinced of the value of laser ablation.

Australopithecus afarensis Dating from between ~3.0 and 3.9 Mya, fossils of this hominin have been discovered in several east African sites, from Ethiopia to Tanzania. Held by many to have been responsible for the Laetoli footprints preserved in volcanic ash dated to 3.5 Myr which demonstrate BIPEDALISM, the animal is estimated to have been 1–1.5 m tall and, with a large measure of sexual dimorphism, weighing between 25 and 50 kg. The specimen discovered at the Hadar site (Ethiopia) in 1974 and known as 'Lucy' (Afar Locality (AL) 288-1) represented almost 40% of an entire skeleton. The following year, dozens of bones representing ~13 individuals and including 4 infants were found scattered along a Hadar hillside. The species is morphologically archaic, its brain case, jaws and limbs being much more ape-like than those of later taxa included in the genus *Homo*. Upper body features (e.g., long curved fingers) indicate that it spent some time in trees, and the semicircular canals appear similar to those of *A. africanus* and African apes, while the lower body had bipedal adaptations. Brain size was ~400–500 cm^3 which, when adjusted for its body size, is not much larger than that of a chimpanzee. Although it has lost the large canines distinguishing apes from hominins, its large chewing teeth are still primitive. A remarkably complete 3.3 Myr skeleton of a three-year-old individual ('Selam') found in 2000 in Dikika, Ethiopia, already has features distinguishing it from *A. africanus*, including a characteristic rounded area above the upper teeth, a separation of the bone covering the roots of the upper canines from the edge of the nasal opening, and hourglass-shaped nasal bones fitting into a recess in the frontal bone. Its cranial endocast suggests that its rate of brain growth was slower than that of chimpanzees. It also has preserved scapulae, which the discoverers say resemble those of a gorilla; others say it is more human-like, particularly in the relative proportions of the depressed areas (fossae) for muscle attachment on either side of the spine dividing the blade. Its hyoid (only the second hominin hyoid found) suggests that *A. afarensis* had a chimp-like voice box. The fossils from Chad ascribed to *Australopithecus bahrelghazali* are morphologically similar and may be a regional variant of this species.

Australopithecus africanus The first member of the genus to be described, fossils have been found in numerous sites – all from the south of the African continent and dating mostly to the period 2.4–2.9 Mya. Ape-like in appearance, with a prognathous face and small brain, it had 'un-ape-like' dentition, including small canines and large, flat molars.

Australopithecus anamensis Dating from around 3.9–4.2 Mya, this is the earliest hominin species currently placed in the genus *Australopithecus*. The context of its discovery at Lake Turkana ('anam' means 'lake') in Kenya, and subsequently at other sites in the same region of east Africa, suggests that its habitat was relatively well wooded, possibly even forested. The fossils comprise upper and lower jaws, cranial fragments and the upper and lower parts of a tibia. The tibia suggests that at 46–55 kg this was a larger animal, with greater sexual dimorphism, than *Ardipithecus ramidus* or *Australopithecus afarensis*. Much is uncertain about the phylogeny of these early hominins, but it is possible that *A. ramidus* was ancestral to *A. anamensis*, and that this in turn was ancestral to *A. afarensis*.

Australopithecus bahrelghazali The species name given to fossils from Chad and dated to 3.0–3.5 Mya, which are more gracile than other contemporary hominins and may represent a regional variant of *A. afarensis*.

Australopithecus garhi Fossils, discovered between 1990 and 1998, and including craniodental and postcranial elements, from the Bouri formation of the Middle Awash of Ethiopia (see ARDIPITHECUS RAMIDUS) and ascribed to this species date from ~2.5 Mya and have small brains and large teeth, especially the postcanine teeth. The face is prognathous. The limb proportions are unusual, combining long fore limbs (as in *A. afarensis*) and quite long hind limbs (as in *Homo*). A foot phalanx is similar to remains of *A. afarensis* in size, length and curvature.

Australopithecus rudolfensis See HOMO RUDOLFENSIS.

autacoids So-called 'local hormones', locally released and believed to alter local blood flow through their role in processes such as inflammation and blood clotting. They include prostaglandins, leukotrienes and platelet-derived growth factor, histamine and KININS such as bradykinin.

autism A postnatal neurodevelopmental disorder. Classical autism, first defined in the 1950s, is now regarded as part of a spectrum of at least six related conditions: *autism spectrum disorder* (ASD), which affects roughly 6 in every 1000 children. Classic autism and Asperger syndrome (AS) are the two clearest subgroups on the autistic spectrum of conditions. Those with classic autism have empathy deficits – degrees of 'mind-blindness' – one proposal being that they are late in developing a 'theory of mind' (see LANGUAGE) in childhood, and joint attention in infancy. Brain imaging techniques show that those with autism process visual information about faces in the same part of the brain that they process information about non-facial objects (see FACIAL RECOGNITION), which contrasts with the situation in most children. Children with autism have been shown to have characteristic abnormalities of the cerebellum, although this is unlikely to be the sole cause and may even be a side-effect of pleiotropic genes that are real causes of the disorder. Young children with autism tend to have larger-than-average heads, containing larger-than-average brains. This is due largely to white matter hyperplasia, particularly in short-distance tracts, the CORPUS CALLOSUM and INTERNAL CAPSULE tending to be proportionately reduced. One theory of autism adopts the empathizing-systemizing theory of psychological SEX DIFFERENCES and extends it by proposing that individuals on the autistic spectrum display psychometric deficiencies in empathizing combined with intact or even superior ability in systemizing. This 'extreme male brain' (EMB) theory holds that autism represents an extreme of the male pattern: relatively impaired empathizing compared to females, combined with relatively enhanced systemizing ability. Those with AS typically show many of the social impairments found in autism, but have less severe deficits in language and cognitive skills.

The chief diagnostic signs are social isolation, lack of eye contact, poor language capacity, and inability to empathize. Other regular features include a tendency to interpret metaphors literally, to have difficulty miming (imitating) others' actions, and to be preoccupied with trifles while ignoring important aspects of the environment. Trusting behaviour also tends to be diminished. Autism and BIPOLAR DISORDER are two common complex traits which show high frequencies of phenotypic discordance in monozygotic twins. In autism, using strict criteria, only 60% concordance has been reported, with differences in cerebellar grey and white matter volumes. Some studies have shown parent-of-origin-specific linkage, with excess transmission of paternal genes in cases of autism, suggestive of GENOMIC IMPRINTING. Although causative genes remain largely unknown (but see MECP2) AUTOIMMUNE DISEASES, karyotype analyses show chromosomal rearrangements in 3–6% of individuals, the most common being deletions and duplications on chromosomes 15q, 22q and 7q (see COGNITION for the 22q13.3 deletion). The protein SHANK3 can bind to the cell adhesion molecules neuroligins, and genes (*NLGN3*, *NLGN4*) encoding these proteins have been found to be mutated in individuals with autism and Asperger syndrome (see DENDRITES). One estimate is that mutations in these, and in neurexin genes, probably explain only 5% at most of ASD cases.

Although no single cause of the condition has been identified, a new study is being undertaken to examine the biological and behavioural markers specific to its different subtypes. The effectiveness of treatments that remove heavy metals from the blood is also under examination.

Children with autism tend to have ring (2nd) fingers as long as or longer than their index fingers, as do their family members, an attribute that has been linked to high levels of testosterone in the womb. Studies involving minocycline are examining the possibility that this antibiotic might help treat autism. See RETT SYNDROME, SAVANT SYNDROME.

autoantibodies Antibodies specific for self antigens. See AUTOIMMUNE DISEASE.

autocrine signalling A signal loop in which a cell manufactures its own signal molecule (e.g., a mitogen). Compare ENDOCRINE GLANDS, PARACRINE SIGNALLING.

autoimmune diseases (ADs) Improperly controlled immune responses to self tissues, major causes of worldwide morbidity and mortality, resulting from the binding of B-cell and T-cell receptors to self-antigen molecules (autoantigens). More than 40 autoimmune conditions have been identified, constituting the third leading cause of ill-health and death after heart disease and cancer. Two siblings affected by the same autoimmune disease are far more likely than not to share the same MHC haplotypes. Many ADs are commoner in women than men. Broadly categorized as either organ-specific (e.g., insulin-dependent DIABETES MELLITUS (IDDM), GRAVES' DISEASE, Hashimoto disease and autoimmune MYASTHENIA GRAVIS) or systemic (e.g., SYSTEMIC LUPUS ERYTHEMATOSUS (SLE), MULTIPLE SCLEROSIS), these syndromes result from disorders of immune homeostasis, in which the proper regulation of the proliferation and apoptosis of activated lymphocytes, central tolerance (see IMMUNE TOLERANCE), has partially failed (see T CELLS). In primary biliary cirrhosis, self-reactive lymphocytes slip through the thymus' 'immunity education programme' and injure bile ducts, causing toxic bile acid accumulation. Injured liver cells fail to clear hormones, leading to pressures on the liver circulation and even to rupture of its venous system; under-production of clotting proteins may cause kidneys and brain to fail on account of blood loss after trauma.

No known ADs are attributable to a complete failure of self-tolerance, which would be incompatible with life. Some ADs, such as rheumatic fever, may be triggered by infectious agents. In some, such as SLE, autoantibodies are involved, and early screening for these may one day enable prediction that disease is under way (see ADDISON'S DISEASE, RHEUMATOID ARTHRITIS). In others, such as IDDM, T cells are thought to have a major role, although autoantibodies are present at diagnosis in 70–90% of those with IDDM. Identification of which is causally responsible is experimentally testable by transferring either autoantibodies or self-reactive T cells from a diseased individual to a healthy animal recipient. Autoantibody-induced INFLAMMATION of target tissues can lead to their release of intracellular autoantigens, promoting further activation of autoreactive B cells, as in SLE. Diseases of the foetus during pregnancy are sometimes the result of transmission of IgG autoantibodies across the placenta and, although usually not long-lasting, permanent damage can ensue as is the case with damage to the foetal heart when the mother is suffering from SLE. It is, in any case, better to regard ADs as the result of responses of the integrated immune system rather than of just one component of it, humoral or cell-mediated.

The immune system has evolved multiple mechanisms to prevent damage to self tissues. A cell bearing a forbidden receptor can 'edit' it by further V(D)J recombination or somatic hypermutation to display a non-self-reactive receptor. Intrinsic biochemical and gene-expression modifications (clonal anergy) can reduce the ability of a cell to be triggered by its own self-reactive receptor. And potentially autoreactive cells that have broken these safeguards (i.e., are *immunologically ignorant*) are still subject to extrinsic controls. Thus, B cells that display self-reactive receptors in the bone marrow do not go on to express the homing receptors needed to enter the lymph nodes. Other potentially autoreactive cells can have the supply of essential growth factor, costimuli and pro-inflammatory mediators limited, or be actively suppressed by regulatory T (T_R) cells (see T CELLS). Nevertheless, ignorance can be overcome if the stimulus is strong enough, as may occur during infection if an activated dendritic cell is displaying a sufficiently high density of co-stimulatory molecules, or if the autoantigens for self-reactive B cells are normally intracellular. The corticosteroids used in treating inflammatory autoimmune diseases are often given in combination with other drugs in an attempt to keep their toxic effects to a minimum, and often in combination with cytotoxic drugs in an attempt to provide IMMUNOSUPPRESSION (see this entry for dangers). Mice and humans with a mutant variant of Fas develop a lymphoproliferative disease associated with severe autoimmunity, while mutations in *Foxp3* (whose product Foxp3 is indispensable for T_R cell development and function) result in fatal multiorgan inflammatory disease in both humans and mice, highlighting its essential role in preventing lethal autoimmunity. See PHOSPHATIDYLINOSITOL 3-KINASE.

autonomic nervous system (ANS) Together with the somatic motor system, the ANS constitutes the total neural output of the CNS. Both systems have upper motor neurons in the brain carrying commands in series to lower motor neurons innervating the peripheral target organs; but whereas cell bodies of all somatic motor neurons lie within the CNS, those of all ANS lower motor neurons lie *outside* the CNS within cell clusters termed *autonomic ganglia*. The *postganglionic neurons* originating in these ganglia are driven by *preganglionic neurons* whose cell bodies lie in the SPINAL CORD and brain stem (see Fig. 16). The ANS therefore controls its targets by disynaptic pathways, contrasting with the monosynaptic pathways of the somatic motor system. The main input to the ANS is provided by autonomic sensory neurons, most associated with interoceptors, such as chemoreceptors monitoring blood CO_2 level and mechanoreceptors detecting stretch in the walls of blood vessels and other organs. Some input to the ANS also comes from somatic sensory and special sensory neurons (e.g., those inducing PAIN). Autonomic reflexes include, among many others, those involved in the control of blood pressure, heart rate, temperature, and adrenal, gastric and pancreatic secretion.

The two divisions of the ANS operate in

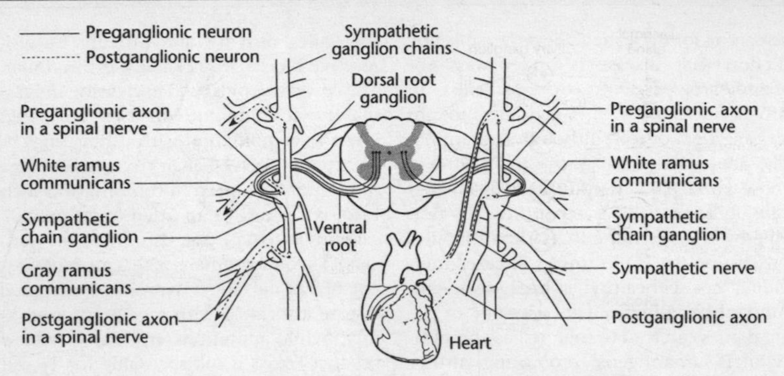

FIG 16. Diagram illustrating the arrangement of neuronal connections between the sympathetic ganglion chain, the spinal cord, and spinal nerves in the thoracic and the first two lumbar segments and described in the AUTONOMIC NERVOUS SYSTEM entry. Redrawn from *Anatomy & Physiology*, R.R. Seeley, T.D. Stephens and P. Tate © 2006 McGraw-Hill, with the permission of The McGraw-Hill Companies.

parallel, but use distinct pathways. Its actions are often integrated by the HYPOTHALAMUS. The preganglionic fibres of the *sympathetic division* emerge only from the grey matter in the lateral horns (intermediate zone) of the thoracic and first two lumbar segments, sending their axons through the ventral roots of spinal nerves (see Fig. 18). These myelinated axons then enter a short pathway (a *white ramus communicans*) before passing to the nearest sympathetic trunk ganglion on that side (one of a chain of sympathetic trunk ganglia running parallel to, anterior and lateral to, the spinal column). A preganglionic fibre may then synapse with as many as 20 postganglionic neurons, either in the first ganglion reached after ascending or descending to a higher or lower ganglion, or after continuing through a trunk ganglion to synapse in a *prevertebral ganglion* (or *collateral ganglion*) such as occur in the SOLAR PLEXUS. This pattern helps to explain why many sympathetic responses affect almost the entire body simultaneously. The *grey rami communicantes* are structures containing unmyelinated postganglionic fibres connecting the ganglia of the sympathetic trunk to spinal nerves.

Preganglionic fibres of the *parasympathetic* division emerge only from the brain stem and from the lowest (sacral) body segments, their axons travelling within several CRANIAL NERVES and in nerves of the sacral spinal cord. They synapse with postganglionic neurons in *terminal ganglia* located close to, or actually within, the wall of the visceral organ they innervate, and are therefore longer than the axons of most sympathetic preganglionic neurons.

All preganglionic sympathetic neurons are CHOLINERGIC, releasing acetylcholine, as are all postganglionic parasympathetic neurons and postganglionic sympathetic neurons innervating most sweat glands. Most sympathetic postganglionic neurons are ADRENERGIC, releasing noradrenaline.

Most organs receive dual innervation, the physiological influences of the two divisions generally opposing one another, and neural circuits in the CNS prevent both from being activated simultaneously. The sympathetic division tends to be most active during a crisis, real or perceived (see STRESS RESPONSE), and during sexual arousal; the parasympathetic division mediates such processes as digestion, growth, immune responses and energy storage.

autonomic reflex A reflex arc involving the AUTONOMIC NERVOUS SYSTEM, the effector typically being cardiac or smooth muscle, or a gland.

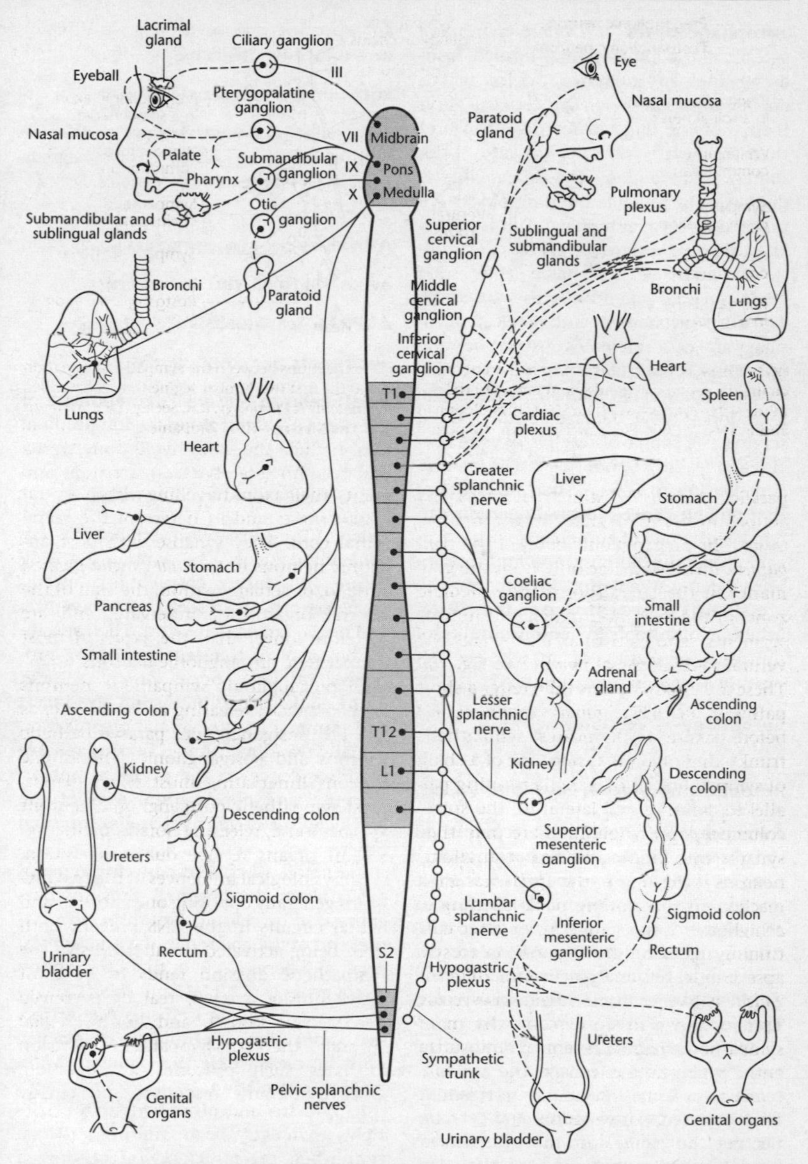

FIG 17. A simplified diagram of the arrangement of the AUTONOMIC NERVOUS SYSTEM, parasympathetic components shown only on the left side, sympathetic components only on the right (although both occur on both sides of the body). Cranial nerve components have roman numerals. Thoracic and lumbar segments are indicated by T1–T12, L1, etc., and sacral segments by S2, etc. Preganglionic fibres are shown by solid lines, postganglionic fibres by broken lines. The coeliac ganglion and mesenteric ganglia are examples of collateral ganglia (see text).

autophagosomes Double-membrane cytoplasmic compartments, formed from membranes of unknown origin, which enclose organelles (e.g., MITOCHONDRIA) and their proteins targeted for digestion and then fuse with LYSOSOMES (or late endosomes), whereupon the organelle is digested – to reusable building blocks – a process known as autophagy. This is a regulated process, occurs under basal conditions, and is stimulated by such environmental conditions as STARVATION. The exact mechanistic link between autophagy and cancer remains obscure, but one possibility is that interfering with the normal degradation of organelles leads to the retention of older, damaged mitochondria serving as a source of reactive oxygen species and leading to GENOMIC INSTABILITY, also contributing to AGEING. There is evidence that proteins linked to tumorigenesis can regulate the rate of autophagy, with oncogenes generally blocking, and tumour suppressor genes stimulating, the process. The protein mTOR (see entry), e.g., is a critical inhibitor of autophagy, and a number of upstream regulators of mTOR are frequently altered in human tumours.

autophagy See AUTOPHAGOSOMES, LYSOSOMES.

autophosphorylation See PROTEIN KINASES.

autoreceptors Presynaptic receptors, all METABOTROPIC, binding the neurotransmitter released at the same presynaptic membrane and usually (but not invariably) decreasing the release of the transmitter homeostatically by reducing calcium influx into the presynaptic terminal (neuromodulation). This negative feedback either avoids excess excitation (e.g., see EXCITOTOXICITY) or reduces postsynaptic receptor desensitization, which would reduce the synaptic sensitivity. Some increase neurotransmitter release (e.g., of serotonin and catecholamines), and some dopaminergic neurons have D_3 receptors for DOPAMINE on their dendrites, regulating neuron firing rates.

autosomal Referring to a gene locus located on a non-sex chromosome, or to an aspect of phenotype (e.g., a disease) determined by the locus.

autosome A non-sex CHROMOSOME.

AV node (atrioventricular node) A 'mini-pacemaker' which relays signals from the SA node to the ventricles. See CARDIAC CYCLE.

AV valves See ATRIOVENTRICULAR VALVES.

avian 'flu (bird 'flu) See INFLUENZA.

AVPR1a See VASOPRESSIN.

awareness See CONSCIOUSNESS.

axial skeleton The medial components of the skeletal system, along the longitudinal axis. Include the skull, hyoid bone, vertebral column (see VERTEBRAE), sternum and RIBS. Contrast APPENDICULAR SKELETON.

axis (1) The 2nd cervical vertebra. (2) Very generalized plane(s) of symmetry arising before and during GASTRULATION: antero-posterior (craniocaudal), dorso-ventral and medio-lateral (left-right). Their specification involves transient organizing processes, in the oocyte or at a later multicellular stage, producing the initial and minimal level of spatial organization (usually two-dimensional) bearing reliable correspondence to the eventual antero-posterior and dorso-ventral axes of the body, normally at right-angles to each other.

Induction of antero-posterior (craniocaudal) and dorso-ventral axes involves expression by cells at the prospective cranial end of the embryo in the anterior visceral endoderm (AVE) of the transcription factor genes *OTX2*, *LIM1* and *HESX1* and of the secreted protein Cerberus (see PATTERN FORMATION). These contribute to head development, establishing the cephalic region prior to gastrulation.

In humans and other vertebrates, left–right asymmetry begins during gastrulation when the anticlockwise beating of cilia in the midline PRIMITIVE NODE wafts morphogens (including FGF-8) specifically to the left side of the embryo. This induces expression of *Nodal* (a member of the

TGF-β gene family) on the left side, near the node. Later, during neural plate induction, FGF-8 induces expression of *Nodal* and *Lefty-1* on the left side of the ventral aspect of the neural tube. *Lefty-1* is an important inhibitor of side-biasing information flow. Subsequent expressions of *Nodal* and *Lefty-2* in the lateral plate mesoderm regulate expression of the transcription factor *PITX2* which, through its induction of further downstream effectors, establishes left-sidedness.

As gastrulation proceeds, BMP-4 secretion by the embryonic disc interacts with FGF to ventralize mesoderm into intermediate mesoderm and lateral plate structures. Primitive node production of the transcription factor Goosecoid activates *chordin* expression, whose gene product, in association with Noggin and Follistatin, antagonizes the ventralizing activity of BMP-4 to dorsalize mesoderm into notochord and paraxial mesoderm for the head region. Subsequent expression and secretion of *Brachyury (T)* by the notochord, at least in mice, antagonizes BMP-4 to dorsalize mesoderm in caudal embryonic regions. See HOX GENES.

axo-axonic synapse A synapse in which one axon terminal makes synaptic contact with the presynaptic membrane of another neuron. See SYNAPSES.

axon The long, slender process which grows from the cell body of a NEURON and conducts impulses to the cell which it targets. Those with a MYELIN SHEATH are *myelinated*, those without *unmyelinated*. Axon diameter usually varies from 0.1–20 μm and contains mitochondria, microtubules, microfilaments and synaptic vesicles. During its growth, signals involved in AXONAL GUIDANCE are detected by a filopodia-bearing terminal GROWTH CONE. Axons are usually longer than dendrites, and usually branch at their ends to form axon terminals which synapse with other cells. Primary afferent axons have widely varying diameters, depending on the sensory receptor to which they are attached. Axons from PRIMARY AFFERENTS, from receptors, are in order of decreasing size (with their speeds of conduction), usually designated Aα (13–20 μm; 80–120 m sec^{-1}; proprioceptors); Aβ (6–12 μm; 35–75 sec^{-1}; mechanoreceptors of the skin), Aδ (1–5 μm; 5–30 sec^{-1}; pain and temperature receptors) and C (0.2–1.5 μm; 0.5–2 sec^{-1}; pain, temperature and itch). Axons of similar size innervating tendons and muscles are called I (not Aα), II (not Aβ), III (not Aδ) and IV (not C).

axon hillock The *initial segment* of an AXON, adjacent to the cell body, whose membrane has a higher density of voltage-gated sodium channels (see ION CHANNELS) than that of the cell body and also requires a lower depolarization increase (10 mV) before reaching the threshold required (–55 mV) for initiation of ACTION POTENTIALS (see entry), so that these are usually initiated here. See SYNAPSES.

axon targeting See AXONAL GUIDANCE.

axonal guidance Santiago Ramon y Cajal was the first to suggest that axons might grow along a chemoattractant gradient, David Ferrier developing the concept. At least six major families of proteins (EPHRINS, INTEGRINS, NETRINS, ROBO, SEMAPHORINS and SLIT PROTEINS) play roles in human axonal guidance, being detected at the filopodia of the *growth cones* of their terminal arbors. These are all multifunctional proteins, acting as both chemoattractants and repellents at different times. Receptors on a growth cone change during axon growth enabling a former chemoattractant to become a repellent.

Once their earlier developmental roles are accomplished, the *Hedgehog*, *Wnt* and *BMP* families of morphogens become involved in axonal growth, ensuring that guidance is overspecified. Many axons make very long journeys, and it is now clear that the cues received differ depending on the segment of its pathway that an axon is engaged in. Some neurological disorders (e.g., HGPPS) result from failure of axonal guidance mechanisms. Axon branching is also highly directed, both positively and negatively, the EGF-like Slit proteins being branching factors for sensory neurons but repellents for other axon classes. Electrical activity in developing networks also influences the branching of the arbors, apparently more so in early development, less active axons losing out in *neuronal competition* with their more active peers.

Chemoattractants secreted by neurons in the ventral floor plate of the spinal cord include Netrin, which attracts axons of dorsal horn neurons that will cross the midline (see DECUSSATION) to form the spinothalamic tract. Netrin receptors on their membranes bind the molecule and cause the axon to grow along the netrin concentration gradient towards the midline; but crossing the midline requires them to pick up another molecular signal which boosts their production of another cell-surface receptor – ROBO (roundabout), which is the receptor to another chemoattractant, Slit. This signal repels the axons from the midline so that they grow away from it. Cues for axonal guidance are often very localized, only a few hundred microns (μ) in extent. Making contact with one molecular signal triggers the production of a membrane receptor for the next, so that by 'joining the dots' axons find their way to their final destination.

When axons grow in specific relational positions to one another, e.g., in the production of NEURAL MAPS, an important difference between them can be the relative concentrations of a specific receptor which is expressed on their growth cone surfaces. Roughly speaking, every retinal ganglion cell carries the same molecular marker, the ephrin receptor (Eph). But their individual growth cones employ the information about how many of these they and others express to guide them to their appropriate relational position in the SUPERIOR COLLICULUS in the midbrain. See DENDRITES, NEURAL PLASTICITY, NEURAL REGENERATION.

axoplasm The cytoplasm of an AXON.

B

β For these entries, see under the headword with the β-prefix.

B7 proteins Co-stimulatory proteins on an ANTIGEN-PRESENTING CELL which act with its MHC–peptide complex to activate a T CELL.

bacteria Prokaryotic cells of enormous ecological and medical importance. See ANTIBIOTICS, GRAM-NEGATIVE BACTERIA, GRAM-POSITIVE BACTERIA, and Gram-related entries, INFECTION, and entries for appropriate species and diseases.

bacterial artificial chromosomes A cloning vector that can accommodate pieces of DNA up to ~1 million bp. See GENOME MAPPING, HUMAN ARTIFICIAL CHROMOSOME.

bacterial inhibition assay (Guthrie test) Technique used in NEWBORN SCREENING to test for level of a specific amino acid in the blood. The dried blood samples used are suitable for tandem mass spectrograph or more sophisticated DNA-BASED DIAGNOSTIC TESTING.

Bacteroides thetaiotaomicron See GUT FLORA.

Bad See BAX, BAD.

Bainbridge reflex The increase in HEART RATE mediated by the sympathetic system through stimulation of atrial stretch receptors, helping to regulate venous pressure to the heart. During EXERCISE, blood returns via the venae cavae rich in CARBON DIOXIDE, thereby relaxing the muscles of the veins and allowing a greater volume of blood to return per minute. Stretching of the wall of the right atrium, and stimulation of either right or left atrial receptors, results in a reflex increase in heart rate – the sensory (afferent) pathway to the medullary cardio-acceleratory centre being via the vagus. The motor (efferent) pathway of this reflex lies in the sympathetic cardio-acceleratory nerves, especially those fibres innervating the sinoatrial node, heart rate and stroke volume increasing as a result. See CARDIAC CYCLE, SA NODE.

balance The static and dynamic equilibrium of the body, involving maintenance of bodily position, mainly of the head, relative to the force of gravity. It is maintained by involuntary reflexes organized in the brainstem (see POSTURAL REFLEX PATHWAYS) and, although the conscious analysis of spatial sense is possible, it is normally overshadowed, except when head acceleration is high, by visual and proprioceptive signals from head position and motion. The main organs of balance lie in the otolith organs (utricle and saccule) and semicircular canals of the VESTIBULAR APPARATUS of the inner ear. In addition to this sensory input, the vestibular nucleus in the medulla receives proprioceptive input from neurons throughout the body, notably the legs, and from the visual system. Axons of vestibular neurons project to the cerebellum, influencing postural muscles, to motor nuclei (oculomotor, trochlear and abducens) controlling extrinsic eye muscles, and to the posterior ventral nucleus of the thalamus which projects to the vestibular area of the cerebral cortex. See CEREBELLUM, GAIT.

balanced diet See HEALTHY DIET.

balancing selection A form of positive selection (see NATURAL SELECTION) that maintains POLYMORPHISM in the population. One example is HETEROZYGOUS ADVANTAGE; another is FREQUENCY-DEPENDENT SELECTION.

ball-and-socket joint (spheroid joint) A SYNOVIAL JOINT where the rounded surface of articular cartilage on one bone moves within a cup-shaped depression (fossa) formed by the cartilage on another bone; e.g., shoulder joint, hip joint.

balloon angioplasty See ANGIOPLASTY.

barbiturates Anxiolytic and hypnotic drugs, effective in the presence of GABA. Each, e.g., phenobarbital, and other sedatives and anticonvulsants, binds its own specific site on the outside face of the $GABA_A$ receptor, increasing duration of chloride channel opening (see ADDICTION).

baroreceptors Nerve endings responding to stretch, monitoring arterial blood pressure, air or other fluid pressure and involved in the regulation of BLOOD PRESSURE, heart rate (see CARDIOVASCULAR CENTRE), ventilation (see RESPIRATORY CENTRE). See AORTIC BODIES, CAROTID SINUSES, STRETCH RECEPTORS.

Barr body See X CHROMOSOME INACTIVATION.

basal ganglia (basal nuclei, cerebral nuclei) Bilaterally paired groups of nuclei and interconnected parallel and closed-loop circuits deep in the cerebral cortex and thought to be important in regulating its function and possibly also in ACTION SELECTION. Once thought to be the brain regions involved in motor control, they are probably more concerned with the learning and execution of new motor skills (e.g., the conditioned reflexes involved in piano playing), and with cognitive and affective aspects of motivation and reward (see DOPAMINE). They are: the dorsal striatum, globus pallidus, substantia nigra and subthalamus. Their circuitry is responsible for the execution of appropriate preprogrammed motor sequences during intentional movement. They constitute part of the extrapyramidal system, in that lesions here produce quite different symptoms from those in the CORTICOSPINAL TRACTS. The largest nucleus is the *corpus striatum*, comprising the *caudate nucleus* and *lentiform nucleus*, each lentiform nucleus being subdivided into a lateral *putamen* and a medial *globus pallidus* (see Fig. 18), the latter having a pars interna (GPi) and pars externa (GPe). The ventral striatum (nucleus accumbens) and ventral pallidum (ventral globus pallidus) are major components of the LIMBIC SYSTEM.

All regions of the CEREBRAL CORTEX project to the basal ganglia, which in turn relay impulses back along pallidothalamic fibres to target thalamic nuclei (ventral anterior and ventral lateral nuclei) which in turn project to the motor regions of the frontal lobe of the cortex, a loop which may select for promotion of specific patterns of cortical activity. A smaller group of medial pallidal efferents passes caudally to terminate in the pedunculopontine nucleus.

The *dorsal striatum* (aka NEOSTRIATUM), comprising the caudate nucleus and the putamen, is one of two primary input ports to the basal ganglia, the other being the *subthalamic nucleus* (STN), especially well-developed in primates. The STN axons use glutamate as transmitter, are excitatory, and send axons mainly to the GPi and substantia nigra. The neostriatum and subthalamic nucleus receive excitatory afferents directly from the cerebral cortex, limbic system and THALAMUS, and modulatory inputs from the midbrain substantia nigra (dopaminergic) and the dorsal raphe nucleus (serotonergic). Signals are then relayed, both directly and indirectly, to the principal output nuclei: the medial segment of the globus pallidus and the pars reticulata of the substantia nigra (SNpr; see Fig. 18). These nuclei project directly to the thalamus, midbrain and medulla, and indirectly via the thalamus to the target cortical and limbic regions whence the basal ganglia input originated: the so-called '*motor loop*' through the basal ganglia. The protein DARPP-32 is located only in the cells of these output nuclei, neurons which must integrate information entering the striatum from other brain regions.

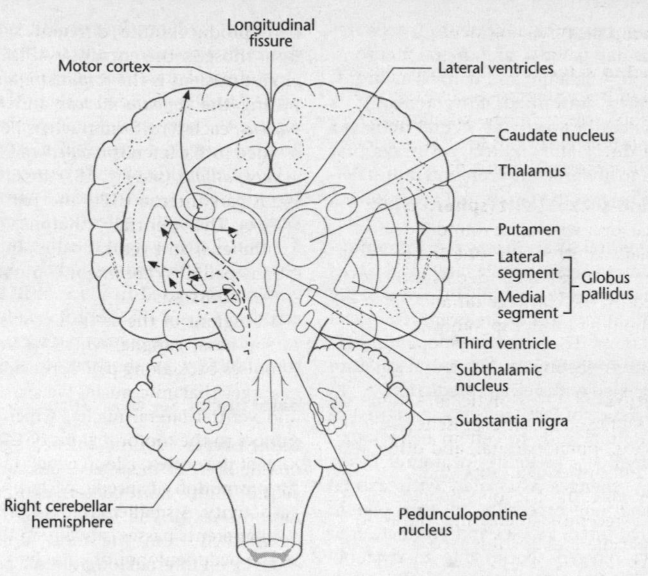

FIG 18. Oblique frontal section of the medial temporal region of the brain, showing the BASAL GANGLIA (labels in bold type) and the principal connections of the neostriatum, globus pallidus and thalamus. The afferents from the intralaminar thalamic nuclei and raphe nuclei are omitted, all efferents from the basal ganglia are shown as originating in the medial globus pallidus, and those from the pars reticulata of the substantia nigra are omitted for simplicity. Only those efferents from the putamen are shown, not those from the caudate nucleus. Glutamatergic pathways are shown by solid line; GABAergic pathways by dashed line; dopaminergic pathways by dashed and dotted line. From A.R. Crossman and D. Neary © 1995 *Neuroanatomy, An Illustrated Colour Text*; redrawn with permission from Elsevier Ltd.

Although neurons of the neostriatum appear randomly scattered compared with those in the layered cortex, this hides a complexity we are only beginning to appreciate. The group of cells in the cortex and spinal cord that represents a given movement sequence are normally inhibited by discharge from the GPe and SNpr, so that damage to the striatum can result in uncontrolled body movements. The striatum can control output neurons of the medial segment of the globus pallidus and the SNpr through two routes (see INTENTIONAL MOVEMENTS). In the first of these (the '*direct pathway*'), stimulation of the putamen by glutamatergic fibres of the frontal and parietal cortices inhibits the spontaneously inhibitory GABAergic neurons of the globus pallidus. Striatopallidal (from striatum to globus pallidus) and striatonigral (from striatum to substantia nigra) neurons directly inhibit medial pallidal and nigral (pars reticulata) neurons. But since medial pallidal and pars reticulata output is inhibitory in a resting subject, the result is disinhibition of target neurons, including those of the ventrolateral nuclei (VL) of the thalamus. The resulting increase in thalamic activity leads to more excitatory output to the supplementary motor areas (SMAs) of the PREMOTOR CORTEX, boosting them beyond some threshold and thereby enabling either new movements, or support of ongoing movements. The second route (the so-called '*indirect pathway*') involves the subthalamic nuclei. Efferents from the putamen terminate in the lateral globus pallidus, inducing inhibition of its neurons. The main efferent projection of the lateral pallidum is to the ipsilateral subthalamic nucleus, which becomes disinhibited as a result. Consequent increasing discharge of subthalamic neurons activates the medial pallidal and nigral neurons, inhibiting thalamic and cortical neurons, preventing unwanted movements. It is imbalances between the neural activities in the direct and indirect pathways that appear to underlie the profound motor deficits in HUNTINGTON DISEASE and PARKINSON DISEASE.

The ventral striatum may process reinforcing stimuli serving basic biological needs and may be more generally involved in the evaluation of stimuli relevant to social interactions. The activity of dopaminergic neurons projecting to the ventral striatum is associated with reward prediction.

Deficiency of PGC-1α (see PGC-1 GENES) in the striatum leads to certain behavioural abnormalities, including profound *hyperactivity*, changes associated with axonal degeneration, especially in this region. Impaired ENERGY BALANCE and production of reactive oxygen species due to dysfunctional oxidative metabolism by MITOCHONDRIA are likely causes.

basal lamina The EXTRACELLULAR MATRIX, ~40–120 nm thick, between epithelial cells and the underlying stroma of the connective tissue, composed mainly of type IV collagen, laminin and a heparin sulphate proteoglycan. Synthesized largely by the cells resting on it, it is a major component of the BASEMENT MEMBRANE and plays a major role in epithelial cell adhesion (see CELL ADHESION MOLECULES), in determining cell polarity, in promoting cell survival, proliferation and differentiation, and in providing cues for cell migration.

basal metabolic rate (BMR, basal energy expenditure) The rate of metabolism of an awake individual lying in conditions of physical and mental rest in a thermoneutral environment at least 12 hours after the previous meal and intake of tea, coffee or inhalation of nicotine. If any of these conditions is not met, the energy expenditure is termed *resting metabolic rate* (RMR). It is measured indirectly by oxygen consumption, values usually ranging from 160–290 cm^3 min^{-1} (0.8–1.43 kcal min^{-1}), depending on age, sex, body size and composition, and nutritional and physiological state (see BODY MASS, SIZE AND SHAPE). Usually recorded early in the morning, it is the largest single component of 24-hour energy expenditure in most individuals, contributing to body TEMPERATURE REGULATION (see LOWER CRITICAL TEMPERATURE). Resting metabolic rate (RMR), measured 3–4 hours after a light meal without prior physical activity, is only slightly higher and is sometimes used in place of BMR. RMR accounts for ~60–75% of total daily energy expenditure (see ENERGY BALANCE) and is stimulated by DEHYDROEPIANDROSTERONE.

basal nuclei See BASAL GANGLIA.

basal plates Ventral thickenings of the mantle layer of the NEURAL TUBE containing ventral motor horn cells and forming the motor areas of the SPINAL CORD. They are demarcated from the more dorsal ALAR PLATES by an internal longitudinal and medial groove on either side of the neural tube, the sulcus limitans.

base Any of the purines and pyrimidines present in nucleosides, nucleotides and nucleic acids.

base pairing, base pairs The hydrogen-bonding that permits adenine:thymine, adenine:uracil and guanine:cytosine pairing in the activities of DNA polymerases and RNA polymerases, and in complementary codon:anticodon complexes during the TRANSLATION of mRNA on ribosomes.

basement membrane Term used to describe the composite of the BASAL LAMINA and the anchoring type VII COLLAGEN molecules that tether it to the underlying connective tissue. It plays a major role in reducing the protein contents of the glomerular filtrate and INTERSTITIAL FLUID.

basic helix–loop–helix motif The motif of a class of DNA-binding proteins which dimerize at an interface comprising two helices joined by a loop. The bHLH proteins function as potent transcriptional activators of tissue-specific genes by forming heterodimers between ubiquitous and cell-restricted family members. In addition, cell-restricted bHLH members play an important role in specifying cell fate (e.g., MyoD for muscle). See CIRCADIAN CLOCKS, E-BOX, MITF GENE.

basicranium The SKULL as viewed from the underside (posterior aspect), the lower jaw being absent.

basic reproductive number (R_0) The expected number of secondary cases caused by the first infectious individual in a wholly susceptible population. This acts as a threshold criterion because disease invasion can succeed only if $R_0 > 1$. A pathogen with $R_0 < 1$ could not sustain transmission, so any epidemic it caused would inevitably die out. See EPIDEMIOLOGY.

basilar membrane Membrane separating the cochlear duct from the scala tympani on which the organ of Corti rests. See COCHLEA.

basket cells GABAergic inhibitory interneurons with widespread axonal collaterals. In the cerebellar cortex, they intersect the parallel fibres of granule cells, which excite them, each synapsing with several Purkinje cells; while in the visual cortex they are critical for mediating ocular dominance plasticity through synapses containing alpha-GABAergic receptors on pyramidal cells.

basophils Leukocytes containing granules staining with basic dyes normally present in very low numbers in the circulation. They are thought to have roles similar to those of mast cells and likewise release HISTAMINE from their granules. They share a common stem-cell precursor with EOSINOPHILS and like them are recruited to the sites of allergic reactions. Eosinophil degranulation causes degranulation of basophils and mast cells. See ALLERGIES; STEM CELLS.

Bax, Bad Cell-death-promoting genes whose protein products act as dimers but can interact with the BCL-2 and BCL-X gene products in determining whether a cell survives or dies. See CELL DEATH.

B-cell co-receptor A complex of CD19, CD81 and CR2 proteins which when co-ligated with the B-cell antigen receptor, increases antigen responsiveness by about a hundred-fold.

B-cell receptors (BCRs) See B CELLS.

B cells (B lymphocytes) B lymphocytes comprise one of the two major lymphocyte classes, originating in bone marrow but, unlike T CELLS, also maturing there. B cells are instrumental in the primary adaptive immune response, maturing and differentiating into either antibody-secreting plasma cells, or into B memory cells. The latter are responsible for the rapidity and specificity of the secondary adaptive immune response.

The key molecular determinant of the B-versus T-cell-fate decision is signalling by the membrane protein Notch (see NOTCH SIGNALLING PATHWAY), probably of stromal cell origin. Notch signalling in haematopoietic progenitor cells drives T-cell development at the expense of B-cell development, the factor required to block Notch being encoded by the *LRF* (leukaemia/lymphoma-related factor) gene, aka *Pokemon*. The PAX5 gene product is an essential factor in B-lineage commitment and maintenance, acting downstream of the *E2A* gene and *EBF1* to activate expression of other B-lineage specific genes and repress expression of alternative lineage genes. Pro-B cells are progenitor cells committed to B-cell lineages and in which rearrangement of immunoglobulin genes begins (see ANTIBODY DIVERSITY, Fig. 11). Primary B-cell development in bone marrow also requires contact with factors on stromal cell surfaces (see STROMA). One such, stem cell factor (SCF, aka Kit ligand), interacts with the cell-surface receptor tyrosine kinase, Kit, on B-cell precursors, which activates the kinase and induces proliferation. Other adhesion molecules, growth factors, and chemokines secreted by stromal cells are involved in these early stages, defined by the sequential rearrangement and expression of heavy- and light-chain immunoglobulin genes to produce a cell-surface IMMUNOGLOBULIN (see Fig. 19) that

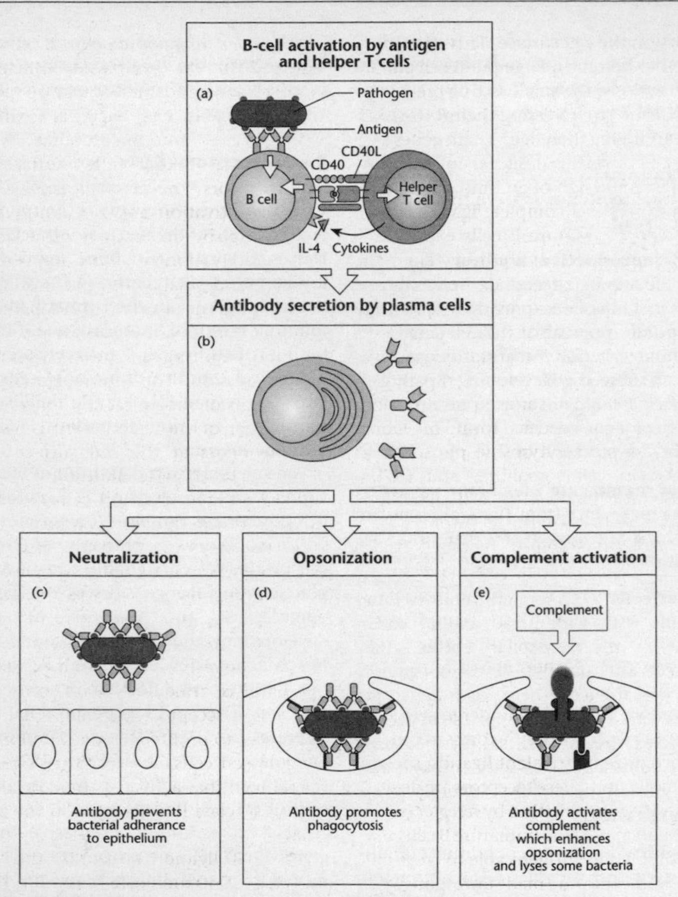

FIG 19. Diagram illustrating some canonical events in the initiation of a primary immune response (see B CELLS, ADAPTIVE IMMUNITY). Antigen on a pathogen surface binds the antigen binding site of a B-cell receptor, signalling the B cell and being internalized and processed into peptides that are expressed by the B cell in association with MHC Class II molecules (a, centre). Armed T helper cells with the appropriate receptor specificity then activate the B cell through cell adhesion and by secreted cytokines (e.g., IL-4), inducing the B cell to differentiate into a mature antibody-secreting plasma cell (b). The antibodies can inhibit toxic effects if there is infection through neutralization (c), opsonization (d) or complement activation (e). See ANTIBODIES. © 2005 From *Immunobiology* (6th edition), by C.A. Janeway Jr, P. Travers, M. Walport and M.J. Shlomchick. Reproduced by permission of Garland Science/Taylor & Francies LLC.

resembles the final secreted antibody. Once they have completed their maturation, immature B cells enter the blood stream and home to the peripheral lymphoid organs.

The heavy-chain (H-chain) locus rearranges first, a D gene segment being linked to a J_H gene segment, followed by V_H-to-DJ_H joining. If successful, this leads to cytoplasmic expression of a complete immunoglobulin heavy μ chain as part of the pre-B-cell receptor, some of which appears on the cell surface. This stimulates the cell to become a large pre-B cell and to divide actively, the small resting pre-B cells which result re-expressing their *RAG* genes and rearranging their light-chain genes (see SOMATIC RECOMBINATION), successful assembly of which is the sign of an immature B cell, which expresses a complete IgM molecule on its surface. Mature B cells express a δ heavy chain as well as a μ heavy chain by means of alternative mRNA splicing (see RNA PROCESSING) and express the additional IgD immunoglobulin at the cell surface.

Throughout their maturation in the bone marrow, the developing B cell has expressed other cell-surface proteins, notably the B-cell-specific form of CD45 (CD45R), a protein tyrosine phosphatase involved in cell signalling, and CD19, which is also involved in cell signalling. Receptors for IL-7 and the receptor tyrosine kinase Kit which binds stem-cell factor (SCF) on the stromal cell are expressed in pro-B-cell development. Kit activation is required for proliferation of pro-B cells. After Kit is turned off at the large pre-B-cell stage CD25, the high affinity IL-2 receptor is expressed.

Immature B cells now undergo selection for *self-tolerance*. If they express receptors that recognize multivalent ligands, such as MHC molecules, these receptors are deleted from the repertoire either by *receptor editing* or through apoptosis. Immature B cells that bind soluble self antigens able to cross-link the B-cell receptor are made unresponsive to antigen (made anergic) and bear little IgM. Immature B cells that bind with low affinity to monovalent antigens or soluble self antigens do not cross-link their receptors (are clonally ignorant), receive no signal as a result, and mature normally, expressing IgM and IgD, but are potentially self-reactive (see AUTOIMMUNE DISEASE). Newly formed B cells enter the spleen via the blood and if they survive this enter LYMPH NODES (see Fig. 121) from the blood via high endothelial vessels located within T-cell zones and then leave the blood coming to rest in the lymphoid follicles. The B-cell receptor is associated non-covalently on the cell surface with invariant Igα and Igβ chains.

In a *primary immune response*, B cells concentrated in the lymphoid follicles of lymph nodes are adapted to take up specific soluble antigens (e.g., bacterial toxins) at their surface immunoglobulin which results in cross-linking of these surface antigen receptors (*receptor clustering*). The resulting activation and phosphorylation of Igβ and Igα by Src-family tyrosine kinases leads to activation of PHOSPHOLIPASE C (see Fig. 135) and small G PROTEINS, leading to signalling via the NF-κB and MAPK pathways (see MAPK (MITOGEN-ACTIVATED PROTEIN KINASE) PATHWAY), resulting in new patterns of gene expression. The internalization and degradation of the antigen is followed by presentation of the degraded antigen's peptide fragments at the cell surface complexed to class II MHC molecules.

If such antigen-specific B cells have been activated by the signalling pathways mentioned to express CO-STIMULATORY MOLECULES, they can then activate naïve CD4 helper T cells, although because only those with the appropriate receptor specificity can present a particular antigen, they are inefficient at *initiating* the responses of ADAPTIVE IMMUNITY. Rather, it is armed helper T cells with appropriate receptor specificity that activate B cells to proliferate and differentiate, when those T cells recognize the same peptide antigen that activated them displayed with MHC class II molecules on the B-cell surface.

The CD40 ligand (CD40L) on the T cell binds the CD40 molecule of the B-cell surface and drives the resting B cell into the cell cycle, leading to further expression of B7 co-stimulatory molecules sustaining T-cell growth (see Fig. 19). The T_H cell secretes IL-4 and other cytokines that lead to B-cell proliferation, after several rounds of which the B cell can differentiate into an antibody-secreting plasma cell or a memory B cell. Some B cells activated at the T-cell/B-cell border in a lymph node migrate to form a *germinal centre* within a primary follicle, thereafter called a *secondary follicle*. These are sites for rapid B-cell proliferation and differentiation during which SOMATIC HYPERMUTATION alters the V regions of immunoglobulin genes, and AFFINITY MATURATION and ISOTYPE SWITCHING occur, under the influence of cytokines released by helper T cells. This enhances the later part of the primary immune response. Some germinal centre cells differentiate first into plasmablasts and then into plasma cells, migrating to the bone marrow and living for a long period. Other germinal centre B cells differentiate into *memory B cells*, divide very slowly (if at all) and express immunoglobulin on their surface, but secrete antibody either very slowly or not at all. They inherit the genetic changes that their predecessor underwent in the germinal centre, including isotype specificity.

Although some microbial antigens can activate B cells to produce antibodies without T-cell help, the antibodies so induced are less variable and less versatile than those induced by T-cell help.

During such a thymus-dependent B-cell response, the V regions of the B cell's assembled immunoglobulin genes undergo a high rate of point mutation (SOMATIC HYPERMUTATION) during clonal expansion. This creates further diversity within the expanding B-cell clone responding to the particular antigen. Furthermore, by a further somatic recombination event, the heavy-chain V regions become associated with different heavy-chain C regions, enabling different functional C regions to be represented among antibodies with the same antigen specificity.

When terminally differentiated, a mature B cell (plasma cell) secretes antibodies with the same specificity as its antigen receptor, and this constitutes its effector function. When activated, a B cell oxidizes fatty acids in preference to glucose to fuel movement and protein synthesis and build its extensive lamellipodia.

Two B-cell classes are recognized: B-1 cells (CD5 B cells) are self-renewing and found mainly in the adult peritoneal and pleural cavities. B-2 cells are generated in the bone marrow throughout life and populate the blood and lymphoid tissues. The first antibodies B cells produce remain bound to the cell surface membrane and serve as receptors for antigen. The receptors, of which there are ~10^5 per cell, are bound to proteins which signal to the cell interior when an antigen is bound. Each B cell produces one type of antibody, with a unique antigen-binding site, and on activation by antigen each naïve B cell, or memory B cell, proliferates and matures into an antibody-secreting effector cell, the antibodies now being soluble and not membrane-bound. See BURKITT'S LYMPHOMA (and EPSTEIN–BARR VIRUS), STEM CELLS.

B lymphocytes See B CELLS.

BCG (Bacille–Calmette–Guerin) vaccine A live attenuated strain derived from *Mycobacterium bovis*, this vaccine stimulates a protective immunity to *M. tuberculosis*, *M. bovis* and probably to other members of the human TUBERCULOSIS group. It also protects against the development of lepromatous leprosy. Given as a single intradermal dose, a local reaction develops at the immunization site within 2–6 weeks, increasing in size and often ulcerating. It should not be used for patients with immunosuppression, including asymptomatic HIV-positive individuals. Estimates of protection by BCG have varied, from zero (in one trial in India) to 90%; many UK studies indicate a protection rate of ~70%, providing better protection in young children than in the adult population. Where child infection is common, it is routinely administered at birth. Elsewhere, it may only be offered to those at greatest risk.

Bcl-2, Bcl-x Members of a small family of closely-related genes, some of which inhibit cell death by APOPTOSIS and some of which promote it. Among the possible roles of these Bcl proteins is in blocking release of CYTOCHROME C (see entry) from MITOCHONDRIA and binding to Apaf. *Bcl-2* (B-cell leukaemia/lymphoma 2) and *Bcl-x* are death-inhibiting genes (*contra* BAX and *Bad*), encoding proteins that act as homodimers that are activated by the EGF signalling pathway. The Bcl-2 oncoprotein encoded by the oncogene *Bcl-2* inhibits apoptosis when up-regulated, but because Bcl-2 and Bax proteins can form heterodimers, the more abundant protein determines whether the cell lives or dies (see

FUSION GENES for enhanced expression of *Bcl-2* in B-cell lymphomas). Bcl-2 is a target of the microphthalmia transcription factor, MITF, and is required for MELANOCYTE viability.

BCR 1. A B CELL antigen receptor. **2**. A gene known as the *breakage cluster region* which normally encodes a serine kinase but which is interrupted by a reciprocal translocation to generate the fusion gene *BCR-ABL* common in chronic myeloid LEUKAEMIA. See ABL GENE.

BDNF (brain-derived neurotrophic factor) A gene activated by CREB, whose decreased production has been implicated in bipolar disorder and whose expression in the HIPPOCAMPUS increases during many antidepressant treatments (see AFFECTIVE DISORDERS, NEURAL PLASTICITY).

Becker muscular dystrophy See MUSCULAR DYSTROPHIES.

Beckwith–Wiedemann syndrome A rare disorder, in which 85% of cases are sporadic and 15% familial. Characterized by prenatal overgrowth, a midline abdominal wall, other malformations (e.g., a large, protruding, tongue) and sometimes a predisposition to childhood cancers (particularly WILMS' TUMOUR). Associated with a cluster of genes at p15.5 on chromosome 11, the two main candidate genes are closely linked and differently expressed on account of GENOMIC IMPRINTING. Disrupted imprinting of either or both of two neighbouring imprinted subdomains on 11p15 reveals clustering of imprinted genes. One is the *H19/IGF2* (imprinted, maternally expressed, untranslated mRNA/insulin-like growth factor 2) subdomain, regulated by a differentially methylated region (DMR) that is methylated on the paternal but not the maternal allele. The other includes $p57^{KIP2}$ (a cyclin-dependent kinase inhibitor), with a subdomain regulated by a second DMR that is normally methylated on the maternal but not the paternal allele. Some individuals with the syndrome show loss of imprinting of *IGF2* (see FOLATE) leading to a double dose of this autocrine factor that results in tissue overgrowth and increased cancer risk. The cause is aberrant methylation of the maternal *H19* DMR. Higher than expected frequencies arise from IN VITRO FERTILIZATION, hypomethylation of *LIT1* being found in the great majority of these (13 out of 14 reported cases), despite the presence of this abnormality in roughly one-third of patients in general. In humans, Beckwith–Wiedemann syndrome is expressed when both the paternal and maternal copies of *IGF2* are expressed; but if both copies are silenced undergrowth of the foetus occurs, as featured in Silver–Russell syndrome.

behavioural sensitization See SENSITIZATION.

Bence Jones proteins A monoclonal globulin protein found in the blood or urine and described by the English physician Henry Bence Jones in 1847, whose isolated finding is today known as 'monoclonal gammopathy of uncertain significance'. Finding this protein in the context of end-organ manifestations such as renal failure, lytic bone disease or anaemia, or large numbers of plasma cells in the bone marrow of patients can be diagnostic of multiple MYELOMA. They do not react with reagents normally utilized in urinalysis dipsticks and are detectable by specific heating-and-precipitation tests, or by electrophoresis. Various rarer conditions which can produce Bence Jones proteins, such as Waldenström's macroglobulinaemia and other malignancies.

benign tumour A non-invasive TUMOUR.

benzodiazepines (BDZs) General depressant drugs dominating the treatment of sleep disorders (hypnotics) and acute anxiety states (anxiolytics), of which Valium (diazepam) is probably the best known. In general, they induce sleep when given in high doses at night and will provide sedation and reduce anxiety when given in low, separated, doses during the day. These effects, and their muscle relaxing and anticonvulsant effects, are thought to be caused mainly by their enhancement of

GABA-mediated inhibition in the central nervous system. Diazepam is often administered in ALCOHOL withdrawal. Temazepam is a popular drug of abuse, especially with opiate addicts who use it to tide themselves over withdrawals. See BARBITURATES, FEAR AND ANXIETY.

Bergmann's rule Published in 1847, this states that in a geographically widespread species, populations in warmer parts of the range will be smaller-bodied than those in cooler parts. See BODY MASS, SIZE AND SHAPE. Compare ALLEN'S RULE.

beriberi A vitamin B_1 (thiamine) deficiency disease, occurring either through dietary deficiency, poor absorption or increased metabolic demand (see VITAMIN B COMPLEX, vitamin B_1). Wet beriberi follows several years of progressive weakness, symptoms being dyspnoea and oedema, wide pulse pressure variation, tachycardia, enlarged heart and other signs of high-output cardiac failure. Dry beriberi is usually superimposed on wet beriberi, including severe muscle weakness and progressive polyneuropathy. Severe thiamine deficiency also occurs in chronic alcohol abuse and anorexia. See MALNUTRITION.

beta-blockers Drugs that are antagonists of β_1-ADRENERGIC receptors. Propranolol (Inderal) is used in treatment of patients with hypertension, decreasing heart rate and force of contraction. See ALPHA BLOCKER, CIRCADIAN RHYTHMS, MARFAN SYNDROME.

beta-catenin See FAMILIAL ADENOMATOUS POLYPOSIS, WNT SIGNALLING PATHWAY.

beta cell (β cell) Endocrine cell in the islets of Langerhans of the PANCREAS, secreting INSULIN into blood capillaries.

beta-globulins See β-GLOBULINS.

beta-oxidation See β-OXIDATION.

beta-pleated sheet (β–pleated sheet, beta-sheet) One of the two commoner forms of protein/polypeptide secondary structure.

beta receptors For β-adrenergic receptors, see ADRENERGIC.

Betz cells Giant pyramidal cells (~30,000) in Area 4 of the MOTOR CORTEX, with large myelinated axons within the CORTICOSPINAL TRACTS conducting at ~60–120 m s^{-1}.

bicuspid tooth A tooth with two CUSPS. In humans, the premolar teeth. See DENTITION, SECTORIAL.

bicuspid valve The MITRAL VALVE of the heart. The atrioventricular valve on the left side of the heart.

bilateral symmetry A property of most metazoans and of all vertebrates, humans included, in which a single plane separates the body into two roughly mirror-image halves. Usually the plane separating the left and right halves of the embryo, it is located at the interface of the antero-posterior and dorso-ventral axes (see AXIS). The left–right axis originates during GASTRULATION as a result of a directional extracellular flow of morphogens to the left side of the embryo. This flow is generated by dynein-dependent rotation of monocilia on the ventral surface of the embryonic PRIMITIVE NODE.

bile An aqueous secretion of the LIVER, containing BILE SALTS (bile acids), BILE PIGMENTS, CHOLESTEROL, lecithin, fat-soluble hormones and several ions, notably hydrogen carbonate (bile's pH of 7.8–8 helping to neutralize chyme from the stomach). It contains no digestive enzymes. The main factor regulating active secretion of bile salts by hepatocytes is the return of bile salts to them via the enterohepatic circulation (the bile-acid-dependent fraction of bile secretion). The bile-acid-independent fraction of bile secretion covers the secretion of water and electrolytes by hepatocytes and epithelial cells of the bile duct. Na^+ is actively transported into the bile canaliculi, followed by passive movement of Cl^- and water, while active transport of HCO_3^- is followed by passive movement of Na^+ and water. Although not under hormonal control, secretion of this bile-acid-independent fraction is stimulated by SECRETIN release from the duodenum, and to a lesser extent by glucagon and gastrin.

Bile is stored in the gall bladder (where cholesterol crystals may form gallstones) and released under CHOLECYSTOKININ control. Excess cholesterol that cannot be dispersed into micelles may crystallize and form gallstones in the hepatic ducts or gall bladder by acting as nucleating agents for deposition of calcium and phosphate salts. Roughly 600 cm^3–1 dm^3 is secreted daily, insufficient release resulting in inadequate fat digestion and absorption of lipids (e.g., VITAMIN K).

bile acids See BILE SALTS.

bile canaliculi Narrow intercellular canals by which bile, secreted by hepatocytes, enters bile ductules and then bile ducts at the periphery of liver lobules. See LIVER, Fig. 116.

bile duct Originating in the 3rd week as a narrowing of the connection between the liver diverticulum and the foregut, the functional bile duct forms by merging of bile ductules (see bile canaliculi) leaving the corners of liver lobules (see LIVER, Fig. 116) and merging to form the right and left hepatic ducts which unite to exit the liver as the common hepatic duct. This joins the cystic duct from the gall bladder to form the *common bile duct* which enters the duodenum. See AUTOIMMUNE DISEASES.

bile pigments Excreted breakdown products of the haem group of haemoglobin. Initially released from ageing and damaged red blood cells by macrophages in the spleen after removal of the iron, biliverdin (green) is formed, and then bilirubin (yellow-orange). Bilirubin binds to serum albumin and travels to the liver, where hepatocytes conjugate it to glucuronic acid, forming the water-soluble bilirubin diglucuronide. It is then excreted in the BILE, leaving via the large intestine where some is converted to urobilinogen by colon bacteria and a little is absorbed into the blood and converted by the kidney to urobilin and excreted in the urine. See PLACENTA for removal of foetal bilirubin.

bile salts Amphipathic compounds, serving as biological detergents by converting dietary fats into mixed micelles of bile salts and triglycerides (see SREBPS). During synthesis in liver cells, excess CHOLESTEROL is converted first into cholic and chenoleoxycholic acids (oxysterols). These combine with glycine and taurine to form conjugated glycol- and tauro-cholic *bile acids*, whose salts (mainly sodium salts) are secreted into the BILE. Their detergent, emulsifying, effects on fat particles in the food increase the total surface area of fat accessible to pancreatic LIPASE digestion. But more importantly, they assist absorption of fatty acids, monoglycerides, cholesterol, and other lipids from the ileum. This they do by forming MICELLES, 3–6 nm in diameter, which are soluble in the chyme because of the charges on the bile salts, and ferry the lipids to the microvilli of the mucosal epithelium membranes, where they diffuse out of the micelles and across the microvilli, leaving the micelles to be recycled again and again. See ENTEROCYTES; compare CHYLOMICRONS.

bilharzia See SCHISTOSOMIASIS.

bilirubin The major BILE PIGMENT. See PLACENTA.

biliverdin A precursor of bilirubin (see BILE PIGMENTS).

bilophodont Of cheek teeth having four cusps linked in pairs by transverse ridges to form shearing crests. Characteristic of Old World monkeys (cercopithecoids), they confer an advantage in enabling the shearing of tough, mature leaves. See DENTITION.

binaural neurons Neurons whose firing is influenced by sound at both ears. See HEARING.

binding problem A central problem in neural cognitive science is how the brain, e.g., scanning the visual field, separately but simultaneously analyses the form of objects, their movement and colour, before constructing an image according to its own rules. How this reconstitution of neurally encoded information occurs is termed 'the binding problem'. It is known that neuronal oscillations, revealed by the ELECTROEN-

CEPHALOGRAM, facilitate perceptual binding. See OBJECT RECOGNITION, PARALLEL PROCESSING.

binding proteins (BPs) **1**. Proteins, or glycoproteins, binding to particular water-insoluble molecules and carrying them in the body fluids (e.g., α2-globulin for cortisol; IGFBPs for IGFs), often acting as storage sites of the transported molecule, preventing its degradation and prolonging its half-life; or, **2**. proteins, often serving as transcription factors, that bind particular response element sequences in the genome (e.g., CREBP; SEE CREB).

binocular cells, b. neurons Cells in the PRIMARY VISUAL CORTEX, particularly in layers IVB, II and III, which have binocular properties in that they can be driven by either eye. See BINOCULAR VISION, OCULAR DOMINANCE.

binocular vision The slightly different perspectives provided by the two eyes (see BINOCULAR ZONE) yield two retinal images whose features have different relative positions. The principal benefit of this is *stereopsis*, the perception of relative depth stimulated by horizontal retinal image disparity. This does not require form, movement or colour. The eyes are ~6.3 cm apart, so that the image of a nearby object falls onto different horizontal positions on the two retinas (termed *retinal*, or *binocular, disparity*). When the eyes converge on a nearby point, the images are formed at the fovea of each retina and are perceived as fused into just one point. All other points in the visual fields appear fused because their images lie at corresponding positions on the two retinas. The position of each feature in the image is described on the retina by directions relative to the eye's visual axis, and these directions are compared by the brain's system of binocular correspondence which encodes as disparity any differences in direction relative to the visual axes of similar features in the two eyes. The brain can compute depth from disparity by comparing where the same pattern lies on the left and right retinas (see HYPERCOLUMNS). BINOCULAR NEURONS compare the amount of light falling in their receptive fields in the two eyes, the effect of parallax being detected by a group of such neurons. A shift in the image of only a few microns (μm), the smallest parallax humans can detect, alters the intensity of light reaching a cone cell by about 5%, sufficient for detection by single cells (see VISUAL FIELD, Fig. 179). CREB activity within the ocular cortex is involved with the loss, but not the recovery, of binocular vision. See OCULAR DOMINANCE, VISUAL PERCEPTION.

binocular zone The area of the VISUAL FIELD (see Fig. 179) occupied by light entering both eyes. See OPTIC CHIASMA, Fig. 130.

biochemical pathway Any enzyme-mediated pathway in which the product of one reaction is the substrate for the next. GLYCOLYSIS is a linear pathway, the KREBS CYCLE is a cyclical one.

biofeedback A form of INSTRUMENTAL CONDITIONING in which a patient learns to enhance or suppress responses of which they are ordinarily unaware, such as a slight muscle contraction after a stroke. The patient can be made aware of them, however, by providing an immediate auditory or visual cue that signals that the response has occurred, after which the patient can increase or decrease the strength of the response as desired, the feedback cue being used as a reinforcer. See MIRROR VISUAL FEEDBACK, PAIN, STRESS.

biolistics See DNA VACCINES.

biological clock See CIRCADIAN CLOCKS.

biomarkers Characteristics that can be objectively measured and evaluated as indicators of normal biological processes, pathogenic processes, or pharmacologic responses to therapeutic intervention. They include GENETIC MARKERS, neuroimages and clinical parameters. They are most useful when they enable early detection of disorders before classical disease symptoms present themselves. See CANCER, SCREENING.

biorhythms See CENTRAL PATTERN GENERATORS, CIRCADIAN CLOCKS, ELECTROENCEPHALOGRAM (for cortical rhythms).

biotin See VITAMIN B COMPLEX.

bipedalism Evidence for a bipedal mode of locomotion appears amongst the earliest hominins (contrast BRAIN SIZE) and has been a contributory factor in their adaptive radiation. Postural bipedalism is the ability to stand, and perhaps shuffle short distances, using just the hind legs for support. Locomotory bipedalism involves using just the hind legs for support during LOCOMOTION. Hand-assisted bipedalism, prominent in the most arboreal great ape (the orangutan), permits walking to be sustained on flexible supports that are otherwise too small to access. Until quite recently, locomotory bipedalism has been a defining feature of hominins, and was thought to have evolved just once among primates – in the most recent common ancestor of chimpanzees (which exhibit postural bipedalism) and humans; but discoveries from 1997 onwards of fossils in Kenya and Ethiopia, apparently from high-altitude woodland, are forcing us to rethink this. Dated at 5.8–5.2 Myr BP, an archaic and controversial Ethiopian subspecies of *Ardipithecus*, *A. ramidus kadabba*, and a still earlier Kenyan fossil, *Orrorin tugenensis*, dated to 6.0 Myr BP, have both been described by their discoverers as being bipedal. Bipedal locomotion might have conferred substantial advantage to arboreal apes, including crown hominoids and protohominins, since their long prehensile toes can grip multiple small branches while freeing one or both hands for balance and weight transfer. The discovery in the 1990s of the 4.4 Myr BP fossil *A. ramidus*, and its description as bipedal, was at that time the earliest hominid evidence; but these new primitive fossils raise the possibility that bipedalism may have evolved after forest-dwelling apes moved out into open savannas, and may not even define hominids (if *Orrorin* was not a hominid, but on an already separated 'chimpanzee lineage'). They also raise the possibility that bipedalism may have evolved more than once in hominoids, and that the knuckle-walking locomotion of chimpanzees may be a derived feature from an ancestor that had locomotory (as opposed to postural) bipedalism: i.e. that bipedalism was once more widespread than its current humans-only distribution. The position of the lunate sulcus on endocasts of australopithecines has been a debating point as it seems to mark the border between the primary and secondary visual areas. A relatively posterior position on the endocranium would suggest functional brain changes towards a human pattern. The adoption of a more upright stance during bipedal evolution lowered the position of the cerebellum with respect to the occipital lobes, displacing some of the visual cortex on the lateral surface into the midline, producing a more posterior position of the lunate sulcus.

The suite of anatomical features underlying hominin bipedalism includes: a curved lower spine, a shorter, broader, pelvic girdle and an angled femur (with associated reorganized musculature; see VALGUS ANGLE), lengthened lower limbs with enlarged joint surfaces, an extensible knee joint, a platform foot with the enlarged great toe brought in line with the other toes; and a repositioning of the foramen magnum towards the centre of the basicranium (see SAHELANTHROPUS TCHADENSIS, and Fig. 154). For the many biomechanical contrasts between walking and running, see GAIT, and Fig. 79.

bipolar cells Bipolar interneurons of the RETINA (see also Figs 150 and 151) relaying information from photoreceptors either directly to ganglion cells (vertical pathway) or indirectly via horizontal and amacrine cells (lateral pathway). Two types can be distinguished: invaginating bipolar cells, synapsing deep in the photoreceptor terminal and depolarizing when light strikes the photoreceptor; and flat bipolar cells, forming superficial basal synapses with photoreceptors and hyperpolarizing in response to light striking the photoreceptor. For *invaginating bipolar cells*, tonic release of GLUTAMATE from photoreceptors in the dark is inhibitory; but in the light, hyperpolarization of the photoreceptor suppresses glutamate release and the inhibition is lifted, depolarizing the bipolar cell. *Flat bipolar cells* respond to glutamate release with excitatory postsynaptic potentials so that light, reducing the excitation, hyperpolarizes the bipolar cell. This is because the two types of bipolar cell have different glutamate receptors at the synapse with the photoreceptor, gating different ION CHANNELS in their membranes.

Like GANGLION CELLS, bipolar cells have RECEPTIVE FIELDS with antagonistic centre-surround organization and, as with ganglion cells, there are on-centre or off-centre types. When cones in the centre of the receptive field are active, an on-centre bipolar cell depolarizes while an off-centre bipolar cell hyperpolarizes (using G PROTEIN-COUPLED RECEPTORS); but when cones in the surround are active, an off-centre bipolar cell depolarizes while an on-centre bipolar cell hyperpolarizes. The names 'on' and 'off' refer to whether these cells depolarize in response to light off (more glutamate released) or to light on (less glutamate released).

CONE CELLS form synapses with either invaginating or flat *midget bipolar cells,* these synapsing with ganglion cells which respond to light in the same manner as their bipolar cells: e.g., on-centre bipolar cells depolarized by light depolarize on-centre ganglion cells. In this way 'on-channels' are composed of cone-depolarizing bipolar cell (on ganglion cell) and 'off-channels' by cone-hyperpolarizing bipolar cell (off ganglion cell). On-channels respond with increasing firing to light levels greater than the local average, whereas off-channels increase their firing in response to dark regions, where light levels are less than the local average. The combined effect is to enhance the boundaries between regions reflecting different amounts of light, enabling the retina to respond preferentially to stimulus change rather than to steady-state illumination (see RETINA, Fig. 150; HORIZONTAL CELLS). See BIPOLAR NEURON.

bipolar disorder (b. disease, b. affective disorder, manic depressive illness) An episodic recurrent pathological disturbance of mood (see AFFECTIVE DISORDERS) ranging from extreme elation (mania) to severe depression; usually accompanied by disturbances in thinking and behaviour, commonly with psychotic features (delusions and hallucinations). There is evidence of a substantial genetic component to risk (estimated sibling recurrence risk (l_s) is 7–10, and heritability 80–90%). Bipolar disorder and AUTISM are two common complex traits which show high frequencies of phenotypic discordance in monozygotic twins. In bipolar disorder, 30% of monozygotic twins are discordant. Some studies have shown parent-of-origin-specific linkage, with excess transmission of maternal genes in cases of bipolar disorder, suggestive of GENOMIC IMPRINTING. Increasing evidence suggests an overlap in genetic susceptibility with SCHIZOPHRENIA (e.g., D-amino acid oxidase activator *DAOA*, and disrupted in schizophrenia 1, *DISC1*), including the role of dysfunctional OLIGODENDROCYTES.

bipolar neuron A neuron with two neurites (one axon and one dendrite) extending from the cell body.

bird 'flu See INFLUENZA.

birth An infant is delivered ~38 weeks after conception (~40 weeks after the last menstruation), parturition being the expulsion of the foetus as the cervix dilates and the myometrium contracts efficiently (labour). The mechanisms that result in the precise timing of birth are still open to speculation, but it is thought that both physical and endocrine factors are involved. The former include stretching of the myometrium as the foetus grows (becoming more excitable as it becomes thinner), and placental insufficiency: the decreasing ability of the PLACENTA to supply foetal nutritional needs. During labour, spontaneous myometrial contractions tend to squeeze the foetus towards the cervix, small areas of myometrium acting as pacemakers initiating action potentials that spread through the increased number of gap junctions between the syncytium-like smooth muscle fibres. The pregnant uterus becomes increasingly sensitive to oxytocin during labour on account of an increased density of oxytocin receptors on these fibres. Maturation of the foetal adrenal cortex may ultimately trigger this, foetal paraventricular nuclei increasing their CRH and vasopressin content as corticotropin (ACTH) levels rise towards term. Foetal cortisol certainly plays an important part in onset of parturition in sheep and some other mammals.

During most of human gestation, placental progesterone secretion exceeds that of oestrogens. But increasing foetal cortisol output would promote conversion of progesterone to oestrogens by the placenta during the final days of pregnancy. Since oestrogens are myometrial stimulants whereas progesterone is an inhibitor, this would remove the 'progesterone block' that has so far dominated pregnancy, favouring myometrial excitability. Declining progesterone levels release hypothalamic oxytocin-releasing neurons from its inhibitory effect, leading to increased oxytocin output. Rising oestrogens also stimulate placental prostaglandin $F_{2\alpha}$ ($PGF_{2\alpha}$) production, which may also improve the rhythmicity of uterine contractions. The contractions also stimulate oxytocin release by the posterior pituitary, in turn increasing myometrial excitability during labour. The stages of labour (parturition) are: (i) the start of regular uterine contractions, ~every 20–30 minutes, increasing to every 2–3 minutes. These dilate the cervix to ~10 cm, allowing the baby to move into the vagina (see RELAXIN). Full cervical dilation may take 15 hours for a first delivery. The amniotic sac may rupture at any time in this period, often at the onset of labour; (ii) delivery of the baby, in which intra-abdominal pressure is raised and the mother pushes the baby (normally head first) through the cervix and vagina, taking from minutes to hours. Forceps or suction may be needed; (iii) ~30 minutes after birth, further uterine contractions expel the 'afterbirth' (placenta, foetal membranes and remaining amniotic fluid) from the uterus, helping to seal blood vessels ruptured as the placenta is separated from the uterus. Oxytocin is thought to be involved in mother–child bonding after childbirth. See LACTATION.

birth defects (congenital abnormalities) Structural abnormalities that can occur in an embryo, foetus or newborn infant. Approximately 7% of major birth defects are due to chromosomal abnormalities and include those due to large DELETIONS (e.g., cri-du-chat syndrome) and microdeletions (e.g., ANGELMAN SYNDROME and PRADER–WILLI SYNDROME), while ~8% result from gene mutations, which may be dominant or recessive. Trisomies and monosomies resulting from NONDISJUNCTION arise during mitosis or meiosis. *Single abnormalities* may be either genetically or environmentally caused (multiple abnormalities are generally due to chromosomal mutations). Malformations are primary structural defects in an organ or organ part due to developmental arrestation or misdirection, usually multifactorial, and include congenital heart abnormalities, cleft lip or palate, and neural tube defects such as anencephaly and lumbo-sacral myelomeningocoele (spina bifida). Multiple malformations are generally due to chromosomal abnormalities. Disruptions are abnormalities in the structure of an organ or tissue due to developmental disturbance by external factors, such as ischaemia (as when a strand of amnion becomes entwined around the forearm or digits), infection or trauma. Deformations are defects resulting from abnormal force that distorts an otherwise normal structure and include hip dislocation and mild talipes ('club foot'), both caused by lack of AMNIOTIC FLUID or to crowding *in utero*. Other single abnormalities include DYSPLASIAS.

Multiple abnormalities, of which several thousand are known, often involve a sequence of cascading events initiated by a primary factor (e.g., RETINOIC ACID in congenital scoliosis). The term 'SYNDROME' is often applied loosely to such multiple abnormalities at birth, their study being dysmorphology. Despite the improvements in computer databases, many dysmorphic children are born for whom no diagnosis can as yet be given, with very little information as to likely prognosis or recurrence risk in siblings. See HEDGEHOG GENE.

Recent work on starfish eggs indicates that successful chromosome capture by microtubules requires prior formation of a filamentous actin mesh in the nuclear region, whose contraction delivers chromosomes to the spindle. This mechanism, essential for preventing chromosome loss and egg aneuploidy, may be of relevance to humans, where this is a leading cause of birth defects and pregnancy loss. See CONGENITAL HEART DEFECTS.

bisexual See SEXUAL ORIENTATION.

bistability The ability of certain systems (e.g., neurons, GENE EXPRESSION) to switch between two stable states, 'on' and 'off', much as bits in a computer. Switching between these states requires a gating mechanism – in the case of neurons, gated ION CHANNELS requiring specific threshold levels of depolarization for activation – analogous to the role of a transistor in a digital computer. Such bistability has been detected in neurons of the PREFRONTAL CORTEX (PFC), whereas the rest of the cortex seems to operate on a fundamentally analog system, operating on graded, distributed information. Such digital PFC dynamics may also support more abstract forms of reasoning in a form describable by rule-like representations enabling encoding of categories and other task-relevant information. Encoding of analog (graded) information using bistable neurons requires distributed binary representations operating somewhat like the binary encoding of a floating-point number on a computer in which many neurons (bits) cooperate so that the combined pattern of activity represents different values. This is less efficient than a direct analog representation (possible with just a single neuron), but is more robust to background 'noise', as are digital computers compared with analog ones.

bivalent A pair of homologous chromosomes while they are paired up during synapsis of the first meiotic prophase.

Black Death See BUBONIC PLAGUE.

blackwater fever See MALARIA.

bladder (urinary bladder) A hollow, muscular container for urine, lying in the pelvic cavity just posterior to the pubic symphysis. In males, it is just anterior to the rectum; in females, it is just anterior to the vagina and inferior and anterior to the uterus. During weeks 4–7 of development, the cloaca divides into the urogenital sinus rostrally and the anal canal caudally. The largest, and uppermost, portion of the urogenital sinus gives rise to the urinary bladder, which is initially continuous with the allantois; but when the lumen of the allantois disappears, the bladder

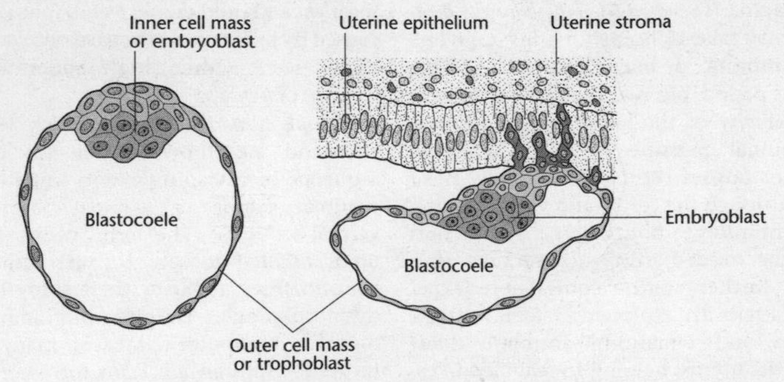

FIG 20. Sections through (left) a ~4-day BLASTOCYST, showing the inner cell mass (embryoblast) and cells of the trophoblast, and (right) a ~7-day blastocyst, with trophoblast cells penetrating the uterine mucosa. Redrawn from *Langman's Medical Embryology*, T.W. Sadler. © 2004 Lippincott, Williams & Wilkins, with permission of Wolters Kluwer.

remains attached to the umbilicus. Hindgut endoderm forms its internal epithelial lining, which is covered by a lamin propria, smooth muscle (detrusor muscle) and this is covered by connective tissue. See MICTURITION.

blastocoel, blastocoele The cavity within the BLASTOCYST (see Fig. 20).

blastocyst Developmental stage arising from the MORULA at about the 32-cell stage, ~5 days after fertilization, emerging from the zona pellucida. The morula, after entering the uterine cavity, swells with fluid that has crossed the zona pellucida, filling the blastocoele cavity within the blastocyst. The blastocyst is surrounded for the most part by a one-cell thick layer, the TROPHOBLAST. But where the inner nonpolar cells of the embryoblast (the INNER CELL MASS) are congregated in the upper hemisphere, the wall may be four cells thick (see Fig. 20). Only ~15% of blastocyst cells contribute to the inner cell mass, most going on to form supportive tissues, such as those of the placenta. By day 11–12 of development (see Fig. 21), the blastocyst is completely embedded in the endometrial stroma of the uterus and lacunae in its SYNCYTIOTROPHOBLAST layer fill with maternal blood. Although the DNA METHYLATION present in the gametes is removed in the zygote and early morula (the total genome methylation reaches an all-time low at the blastocyst stage), it is believed that most GENE SILENCING of imprinted genes by non-coding RNA and histone modification have been newly established by the blastocyst stage of development (see GENOMIC IMPRINTING). During the differentiation of EMBRYONIC STEM CELLS and other pluripotent cells, pluripotency-associated genes are repressed, possibly permanently, as a result of DNA methylation, while developmental genes start to be expressed and histone H3K4 methylation takes place. See SOMATIC CELL NUCLEAR TRANSFER.

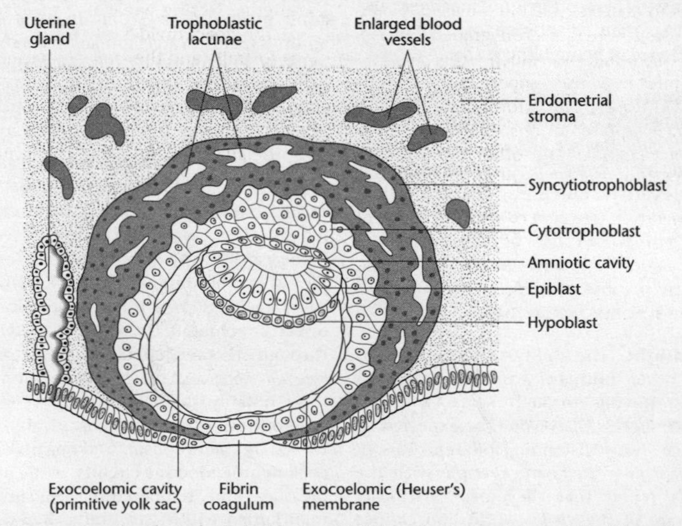

FIG 21. Transverse section through a 9-day blastocyst (~3 days after implantation). Cells of the epiblast are columnar, those of the hypoblast cuboidal. The original implantation site is closed with a fibrin coagulum. Redrawn from *Langman's Medical Embryology*, T.W. Sadler. © 2004 Lippincott, Williams & Wilkins, with permission of Wolters Kluwer.

blastomere One of the cells resulting from early cleavage of the fertilized egg. See FERTILIZATION.

blastula In humans and other mammals, referred to as a BLASTOCYST.

bleaching For bleaching of photopigments, see RHODOPSIN.

blind spot The optic disc contains no photoreceptors and, in both eyes, is located closer to the midline than is the fovea. But since light entering the eye from any single point in the binocular zone never strikes both optic discs simultaneously, we are normally unaware of this. But its existence can be demonstrated by drawing a black cross on white paper, closing one eye, and looking at the cross with the open eye while slowly moving the head further from the paper. When the paper is ~45 cm from the eye the image of the cross falls on the optic disc and disappears from view. This illustrates why damage to large parts of the peripheral retina can go undetected since the missing perception is not blackness, but nothing. This is presumably what the experience of being blind is like.

blindness Trachoma (see CHLAMYDIA TRACHOMATIS) is the leading cause of blindness world wide, but see also AGE-RELATED MACULA DEGENERATION, CATARACT, GLAUCOMA and ONCHOCERCIASIS (river blindness). Diabetic retinal degeneration (see DIABETES MELLITUS) is one of the commonest causes of blindness in the USA. Lesions anywhere in the RETINOFUGAL PROJECTION cause blindness. See AMBLYOPIA, COLOUR BLINDNESS, XEROPHTHALMIA.

blindsight The ability of some humans and other primates to avoid obstacles while moving through space with far greater success than would be expected by chance, notwithstanding bilateral loss of the PRIMARY VISUAL CORTEX. People with the ability report that they are consciously unaware of the visual world and cannot offer any explanation of their ability. It is mediated by a pathway leading from the magnocellular lateral geniculate nucleus to the thick stripe of V2 in the EXTRASTRIATE CORTEX, providing input into the 'where' system.

blobs Regions staining heavily for cytochrome oxidase, each about 0.2 mm diameter, and first detected on the surface of the PRIMARY VISUAL CORTEX (V1) (see Fig. 142) where they form a polka dot-like pattern, separated by paler interblob regions. The blobs extend as vertical functional columns through the six anatomical layers of the cortex, forming one component of each HYPERCOLUMN. The cells involved are part of the parvocellular-blob pathway and are involved in COLOUR PERCEPTION. See PARVOCELLULAR-BLOB/INTERBLOB PATHWAYS.

α-blockers, β-blockers See ADRENERGIC.

blood The fluid component of the cardiovascular system, being a mixed connective tissue comprising liquid plasma, and a cellular and cell fragment component (the 'formed elements') deriving from BONE MARROW. Blood constitutes ~8% of total body mass, occupying ~5–6 dm^3 in an average adult male and ~4–5 dm^3 in an average adult female. Water intake through the digestive tract closely matches loss through the kidneys, lungs, digestive tract and skin, so that plasma volume remains relatively constant (see VASOPRESSIN). *Blood plasma* is a colloid, most suspended substances being PLASMA PROTEINS. Water-soluble ions, nutrients and waste products are all carried in solution, water-insoluble molecules being carried within lipoproteins or complexed to binding proteins. Respiratory gases are carried in physical solution, by direct attachment to HAEMOGLOBIN (O_2) or through other red blood cells' participation (see CARBONIC ANHYDRASE). Plasma composition, notably its glucose and lipid content, is monitored and homeostatically controlled by chemosensors and negative feedback neuroendocrine circuits, while its pH, pressure and temperature are similarly maintained within set limits. Besides RED BLOOD CELLS, its formed elements include PLATELETS and various WHITE BLOOD CELLS (see cell types, STEM CELLS and HAEMATOPOIESIS) which play major roles in defence (see

IMMUNITY) and haemostasis (e.g., BLOOD CLOTTING).

blood–brain barrier (BBB) A barrier provided by the brain's blood capillaries and ependymal cells of the ventricles, governing what is allowed to cross into the brain extracellular fluid from the blood. Capillary endothelial cells are coupled by tight junctions far more tightly than in other capillaries so that even small ions cannot pass between them. The covering of the capillaries by astrocytes is implicated in the promotion of these junctions, and the relative lack of pinocytosis and receptor-mediated endocytosis by the endothelium restrict transport of bulk fluid and lipoprotein at the choroid plexus and at capillary membranes in most of the brain parenchyma (other than some parts of the hypothalamus, pineal gland and area postrema) where large molecular substances hardly pass from the blood into the interstitial fluids of the brain, despite doing so into the usual interstitial fluids of the body (a similar barrier occurs between the blood and cerebrospinal fluid). It is highly permeable to water, CO_2, O_2 and most lipid-soluble substances (e.g., alcohol, most anaesthetics), slightly permeable to electrolytes (e.g., Na^+, K^+, Cl^- and H^+), urea and creatinine, but almost entirely impermeable to plasma proteins and most non-lipid-soluble organic molecules. This makes delivery of some therapeutic drugs (e.g., antibodies) difficult. The BBB can actively exclude many potentially neurotoxic lipophilic compounds, many of them ingested in a normal diet. The transport protein P-glycoprotein (a member of the ABC TRANSPORTER FAMILY), expressed in high levels in the plasma membrane of endothelial cells, pumps lipophilic toxins that diffuse into the cells back into the blood. Because many brain tumours also express P-glycoprotein, lipophilic toxins and other chemically unrelated drugs of potential therapeutic value are unable to reach their targets (see MULTIDRUG RESISTANCE). However, mechanisms do exist for transporting insulin and leptin across the BBB and insulin receptors are expressed in appetite-controlling areas of the brain and the 'perivascular end-feet' of astrocytes, where they make contact with capillaries, may play a part. Otherwise, the tight junctions between the endothelial cells of the capillaries and the surrounding basement membrane are probably the major influences on permeability restriction, although PERICYTES are involved in its stability.

blood buffering See ACID–BASE BALANCE.

blood–cerebrospinal fluid barrier Physiological barrier between the blood and the cerebrospinal fluid, generated by the CHOROID PLEXUSES. See BLOOD–BRAIN BARRIER.

blood clot (thrombus) See BLOOD CLOTTING.

blood clotting (blood coagulation) A major haemostatic process, involving more than 50 procoagulants and anticoagulants in blood and tissues, the latter normally predominating in plasma so that coagulation does not occur while blood is circulating. ANTICOAGULANTS include the plasma protein antithrombin (from the liver, slowly inactivating thrombin), heparin (from basophils and endothelial cells) and prostacyclin (an endothelial prostaglandin counteracting thrombin by causing vasodilation and inhibiting release of coagulation factors from platelets). As with the COMPLEMENT system, blood coagulation involves enzyme cascades.

When a blood vessel is ruptured, procoagulants in the vicinity become activated and override the anticoagulants so that a clot develops. Inflammatory mediators stimulate endothelial cells to express proteins that trigger blood clotting in small local vessels, blocking them and cutting off blood flow – helping prevent the spread of a pathogen via the blood. The ESSENTIAL FATTY ACID EPA is a precursor for prostaglandin synthesis, which decreases blood clotting.

Two pathways, the extrinsic and the intrinsic pathways (see Appendix IX), involve a series of plasma proteins called clotting factors (see entries for individual Factors) whose enzymatic cascadings converge on the production of thrombin from prothrombin, after which the common pathway ensues. The *extrinsic pathway* begins with trauma to the vessel wall and to surrounding tissues, several factors, termed 'thromboplastin' being released from damaged cells. These include phospholipids from the cell membranes, plus a lipoprotein complex acting as a proteolytic enzyme. This complexes with Factor VII and, in the presence of Ca^{2+}, acts on Factor X to form activated Factor X. This combines with tissue phospholipids or with those from PLATELETS, and with Factor V, to form a complex termed prothrombin activator which, again in the presence of Ca^{2+}, splits PROTHROMBIN to form THROMBIN (see entries). The resulting polymerization of *fibrin* to form long fibrin fibres, and their cross-linking, provides the reticulum of the clot which is stabilized by fibrin-stabilizing factor (Factor XIII) released by platelets trapped in the clot. This causes covalent bonding between the fibrin monomers and adjacent fibres. The blood clot is composed of a meshwork of fibrin fibres that entraps blood cells, platelets and plasma. The fibrin also attaches to damaged blood vessel surfaces so that the clot adheres to any vascular opening and prevents blood loss.

The *intrinsic pathway* begins in the blood itself, by trauma to the vessel endothelium, leading to platelet damage and release of their phospholipids (platelet factor III), and/or by exposure of Factor XII to collagen within a damaged vessel wall. In the presence of platelet factor III and HMW (high molecular weight) kininogen, activated Factor XII acts on Factor XI, activating it in turn. Activated Factor XI activates Factor IX. In conjunction with activated Factor VIII and platelet factor III, activated Factor IX then activates Factor X. From this point onwards, the intrinsic and extrinsic pathways share the common pathway to fibrin clot formation outlined above.

A blood clot will normally trap a large amount of *plasminogen*, a plasma protein which is converted to PLASMIN when activated, once the clot has stopped the bleeding, by slow release of TISSUE PLASMINOGEN ACTIVATOR (tPA) from the vascular endothelium and damaged tissues. Plasmin becomes part of the clot and is activated by thrombin, Factor XII, tPA, urokinase and lysosomal enzymes released by damaged tissues. Plasmin then hydrolyses the fibrin in the remaining blood clot, many small blood vessels blocked by clots after tissue damage being reopened (see BYPASS SURGERY).

Factor VIII therapy has reduced the selective disadvantage of HAEMOPHILIA A mutations in recent years and will lead to increased incidence and prevalence of this condition unless the mutation is reduced or unless natural selection is replaced by artificial selection through genetic counselling and prenatal diagnosis. See CARDIOVASCULAR DISEASES, THROMBUS.

blood fluke See SCHISTOSOMIASIS.

blood glucose regulation Despite intermittent ingestion of dietary carbohydrates, plasma glucose concentration remains relatively steady throughout the day, within a range of 3.9–6.1 mM (70–110 mg dL^{-1}), being homeostatically regulated and maintained by interactions between hormonal, neural and hepatic autoregulatory mechanisms (see ENERGY BALANCE). Essentially, this involves the effects of pancreatic hormones on liver GLUCONEOGENESIS. Decrease in plasma levels within and below this range (*hypoglycaemia*) triggers activation of a counterregulatory neuroendocrine response. Both the pancreas and the hypothalamus/vagus are involved in this, the pancreatic hormones INSULIN and GLUCAGON playing central roles in short-term regulation. Following a meal (the postprandial state), insulin release causes glucose absorption by muscle and adipose tissue, while liver glucose output is suppressed as its GLYCOGEN synthesis is increased. In the post-absorptive state, release of glucose from the liver through glycogenolysis and gluconeogenesis is the key regulated process. Following an overnight fast, glycogenolysis provides ~50% of overall liver glucose output, and as glycogen stores are reduced during prolonged fasting (~60 hours) the contribution of glycogenolysis to hepatic glucose output is negligible and liver gluconeogenesis predominates. ADRENALINE stimulates hepatic glycogenolysis and hepatic and renal gluconeogenesis and also limits glucose absorption by insulin-sensitive tissues; but its role in hepatic glucose output is only critical when glucagon release is deficient. Both glucagon and adrenaline act within minutes to raise plasma glucose concentrations. Acute insulin-induced hypoglycaemia increases neural activity in the NUCLEUS TRACTUS SOLITARIUS and lateral hypothalamic glucosensors, resulting in increased sympathetic activity and raising liver glucose output. Release of the counterregulatory hormones glucagon, adrenaline, GROWTH HORMONE and CORTISOL also contribute to this, the resulting raised glucose levels decreasing firing of peripheral glucose receptors in the hepatic portal vein, small intestine and liver, afferent signals relayed to the hypothalamus and nucleus solitarius via the vagus nerve so that the central nervous system can integrate these signals into an appropriate response involving inhibition of hepatic and adrenal nerve activity which removes inhibition of pancreatic insulin release. The kidney is not a net producer of blood glucose, although adrenaline-stimulated renal gluconeogenesis by proximal convoluted tubule cells replaces the glucose used in the medulla so that the kidney's activity is not a drain on plasma glucose. All aspects of glucose metabolism seem to be enhanced by THYROID HORMONES, including glycogenolysis, gluconeogenesis and absorption of glucose across the alimentary canal. Elevated blood glucose concentration (*hyperglycaemia*) is a feature of DIABETES MELLITUS and INSULIN RESISTANCE. Several studies indicate that healthy individuals with fasting plasma glucose (FPG) levels at the high end of the normal range have an increased risk of mortality from CORONARY HEART DISEASE. One genome-wide association study (2008) suggests that a SINGLE NUCLEOTIDE POLYMORPHISM in the *G6PC2* gene encoding the GLUCOSE-6-PHOSPHATASE catalytic subunit-related protein (aka IGRP) is associated with FPG and with pancreatic β cell function (see PANCREAS), although not with type 2 diabetes. See EXERCISE.

blood group system A person's BLOOD GROUPS are genetically determined, each person belonging to several and most being determined by autosomal loci. Alleles at these loci determine self-antigens on the red blood cell membrane, and a blood group system refers to the range of red cell antigens determined by the alleles at one such locus, or sometimes a group of closely-linked loci. Some non-human primates share certain human blood group systems. In humans, 14 systems occur: ABO, RHESUS, MNS, P, Kell, Lewis, Lutheran, DUFFY, Kidd, Diego, Yt, I, Dombrock and Xg (only the last is sex-linked; but see KELL BLOOD GROUP SYSTEM). Host and donor ABO, Rhesus and Kell systems need to be matched in donor blood transfusions, the other groups only requiring to be matched if a recipient has developed antibodies to them through repeated transfusions. Apart from Lewis and Lutheran systems, no linkage disequilibrium is apparent between the loci.

blood groups Either a group of people bearing the same antigen(s) on their red blood cells within a particular blood group system, or the blood characteristic used to distinguish groups of individuals within a blood group system. The biological significance of blood groups, and of the loci involved, remains elusive. Their main clinical significance lies in blood transfusion and Rhesus incompatibility between mother and foetus. Without due matching of donor and recipient blood, death or severe illness of a recipient may result from a single transfusion (in the ABO BLOOD GROUP SYSTEM), or after repeated blood transfusion (with the RHESUS BLOOD GROUP SYSTEM). Danger arises when donor antigens encounter antibodies to them in recipient plasma and elsewhere, resulting in a response due to INNATE IMMUNITY. *Plasmodium*, cause of MALARIA, enters red blood cells by adhering to blood group determinants. Until the advent of DNA PROFILING, blood groups were used in linkage studies and in paternity testing.

blood island Isolated mass and cord of mesenchyme in the mesoderm from which blood vessels develop.

blood plasma See BLOOD, EXTRACELLULAR FLUID.

blood platelets See PLATELETS.

blood poisoning See SEPTIC SHOCK.

blood pressure (BP) The hydrostatic pressure exerted by blood on the walls of a blood vessel. Mean arterial blood pressure (MABP) is age-dependent, highest in the aorta (~120 mm Hg) and major arteries during ventricular systole, dropping most markedly in the arterioles, further still in the capillary beds, and remaining very low in venules and veins. Pressures in the venae cavae are lowest near the right atrium (see Fig. 22 and CARDIAC CYCLE). MABP is given by the formula:

$$MABP = \text{diastolic blood pressure} + \tfrac{1}{3}(\text{systolic BP} - \text{diastolic BP})$$

so that a person whose BP is 120/80 mm Hg has a MABP of ~93 mm Hg.

Resistance (R), the opposition to blood flow, arises mainly from friction between blood and vessel walls, being inversely proportional to the 4th power of the blood vessel radius, $R \propto 1/r^4$, and directly proportional to blood viscosity and total vessel length. Most of the *total peripheral resistance* (TPR) is contributed by the systemic arterioles, capillaries and venules, and arteriole walls need only to dilate or constrict slightly to have a large effect on TPR. Laplace's Law predicts that as the diameter of a vessel increases, the force applied to its wall increases even if the pressure remains constant. If part of an artery's wall becomes weakened so that a bulge (ANEURYSM) develops in it, the force applied to this weakened part is greater than it is along the undeformed regions, so that it bulges even more and is put under still greater force. Ruptured aneurysms of brain blood vessels or aorta are often fatal. Vessels with a large VASCULAR COMPLIANCE exhibit a large increase in volume when blood pressure increases only slightly, in contrast to those with smaller compliance values. Venous compliance is ~24-fold greater than is that of arteries.

The principal centre for regulating TPR (aka systemic vascular resistance) is the vasomotor centre of the medulla (see CARDIOVASCULAR CENTRE, OBESITY). If cardiac output rises due to increased STROKE VOLUME or HEART RATE, mean arterial blood pressure rises as long as resistance remains steady. Increased TPR also increases MABP. Decrease in blood pressure (*hypotension*) or volume, as through haemorrhage or during SEPTIC SHOCK, are detected by stretch receptors in the atria of the heart, aorta and carotid sinuses and relayed to the medulla oblongata (the NUCLEUS TRACTUS SOLITARIUS, NST) by neurons of the glossopharyngeal and vagus nerves. The NST activates noradrenergic neurons in the ventrolateral medulla which project to the paraventricular and supraoptic nuclei of the hypothalamus and bring about release of VASOPRESSIN, decreases in blood pressure greater than 10% causing its release. Heart atrial muscle fibres secrete ATRIAL NATRIURETIC PEPTIDE when blood pressure rises, its diuretic and vasodilatory effects tending to oppose the rise. The signalling molecule NITRIC OXIDE and, in mice at least, hydrogen sulphide, also regulate blood vessel dilation.

Detection by baroreceptors of increases in arterial and venous pressure cause decreased output by sympathetic axons to the afferent arterioles of the KIDNEYS, causing them to dilate and increase glomerular filtrate production and urine output. Reduced blood pressure and resultant reduced baroreceptor firing rate increases sympathetic stimulation of the afferent arterioles, causing vasoconstriction of arterioles, reduced filtrate production and reduced urine output. The JUXTAGLOMERULAR APPARATUS of the kidneys detect decrease in blood pressure and, through renin secretion, cause an increase in circulating angiotensin II which sensitizes the osmoreceptors and leads to increased VASOPRESSIN release (see the entries ANGIOTENSINOGENS and ALDOSTERONE). Chronically elevated

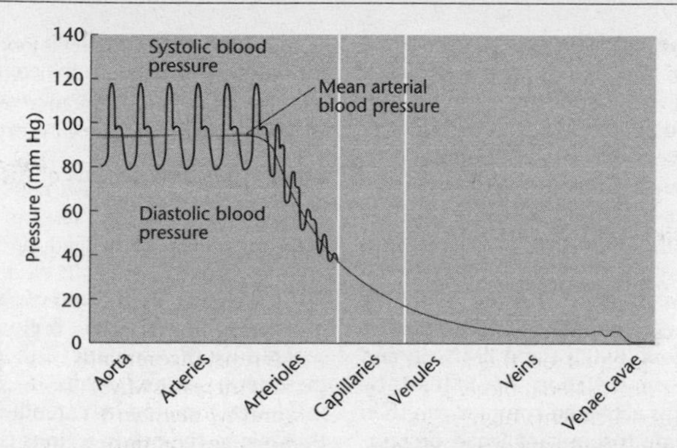

FIG 22. The approximate values and distributions of BLOOD PRESSURE within the blood vessels. © 1990 John Wiley & Sons. *Principles of Anatomy & Physiology*, by G.J. Tortora and N.P. Anagostakos. Reprinted with permission of the publisher.

blood pressure (see HYPERTENSION) results in progressive heart enlargement. For control of blood pressure, see CARDIOVASCULAR CENTRE (and Fig. 37).

blood proteins See PLASMA PROTEINS.

blood reservoir Systemic veins containing large volumes of blood that can be moved quickly to parts of the body in need of blood. See VEINS.

blood serum Fluid expressed from clotted blood, or from clotted blood plasma. It contains far lower levels of fibrinogen and other BLOOD CLOTTING proteins.

blood sugar See BLOOD GLUCOSE.

blood system See CIRCULATORY SYSTEM.

blood–testis barrier An immunologically privileged barrier formed by the SERTOLI CELLS in the seminiferous tubules of the TESTES, preventing blood cells from reaching the spermatocytes and mounting an immune response to their antigens. Since testosterone is a lipid, it crosses the blood–testis barrier by passive diffusion, where it is instrumental in development of spermatozoa.

blood vessels Formed by ANGIOGENESIS and VASCULOGENESIS, these include ARTERIES, ARTERIOLES, CAPILLARIES, venules and VEINS.

blood volume For control of blood volume, see ANGIOTENSINOGENS, CARDIOVASCULAR CENTRE, and pertinent references in these entries.

Bloom syndrome An autosomal recessive disorder, mapping to chromosome 15q26. It is characterized by UV- and chemotherapeutic agent-sensitivity, short stature, disfiguring facial rashes made worse by sunlight, reduced immunoglobulin (IgA and IgM) levels, predisposition to cancers at several sites, and GENOMIC INSTABILITY (UV-induced) chromosome breaks caused by defective DNA helicase and consequent impaired DNA REPLICATION. When homozygous, DNA repair is faulty and the rate of RECOMBINATION between sister chromatids in mitosis increases dramatically. Mutations in the *BLM* gene are responsible, BLM being a part of the BRCA1-associated genome surveillance complex, BASC (see DNA REPAIR MECHANISMS).

blue baby syndrome Disorder resulting from persistence of the FORAMEN OVALE into neonatal life so that blood ejected by the right ventricle often increases and admixture of oxygenated and deoxygenated blood occurs. Surgical intervention is required to correct the defect.

BMI See BODY MASS INDEX.

BMPs (bone morphogenetic proteins) Members of the TGF-β superfamily (see entry) of ligands, with important roles in myriad biological activities, including roles as morphogens in AXIS formation and in AXONAL GUIDANCE. Growth factors, originally identified as molecules inducing cartilage and bone formation, they now rival Wnts (see WNT SIGNALLING PATHWAYS) as general regulators of development. They act by binding to complexes of two type I and two type II BMP receptors (BMPR-I and BMPR-II) and induce phosphorylation of BMPR-I by BMPR-II. This activated complex can then phosphorylate a subset of Smad proteins (Smads1, 5 and 8), the receptor-activated Smads then forming heteromeric complexes with Smad4, translocating into the nucleus and regulating transcription of a specific target gene set. BMP4 and BMP7 are secreted by ectoderm overlying the NEURAL TUBE (see entry). BMP signalling is involved in the differentiation pathways of cells in the intestinal villi (see SMALL INTESTINE) and, with FGF2, in the embryonic differentiation of liver. See JUVENILE POLYPOSIS SYNDROME, NOTCH SIGNALLING PATHWAY.

BNST The basal nucleus of the STRIA TERMINALIS.

body axes See AXIS, HOX GENES.

body cavities By the end of the 3rd week of development, the coelomic cavity separating the somatic and splanchnic mesoderm layers forms a continuous intraembryonic cavity extending from the thoracic to the pelvic regions (see Fig. 23). With the expansion of the lungs, the body wall mesoderm is separated by the pleuropericardial membrane (the adult fibrous pericardium) into pleural and pericardial cavities. Of these, the pleural cavities remain confluent with the abdominal (peritoneal) cavity, but by the 7th week pleuroperitoneal folds extend medially

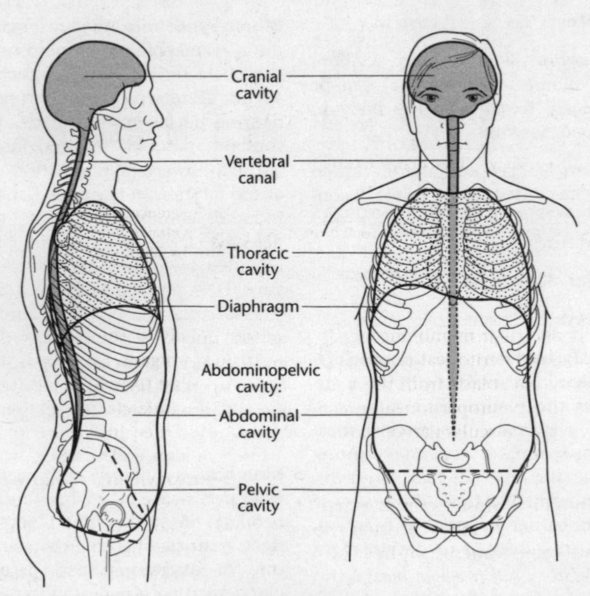

FIG 23. The major BODY CAVITIES. The dorsal body cavity is shaded grey; the ventral body cavity (thoracic and abdominopelvic cavities) is shown stippled (thoracic cavity) and in open outline. The dashed lines indicate the border between the abdominal and pelvic cavities. © 1990 John Wiley & Sons. *Principles of Anatomy & Physiology*, by G.J. Tortora and N.P. Anagostakos. Reprinted with permission of the publisher.

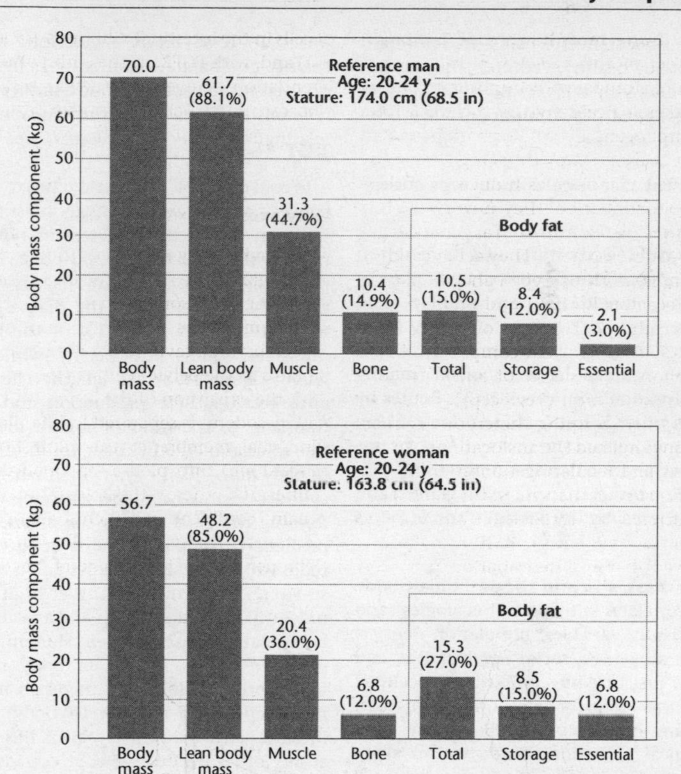

FIG 24. Bar charts showing the BODY COMPOSITIONS for Behnke's reference man and reference woman. Figures in brackets indicate percentage of total body mass. © 2001 Redrawn from *Exercise Physiology* (5th edition), by W.D. McArdle, F.I. Katch and V.L. Katch, published by Lippincott, Williams & Wilkins. Reproduced with the kind permission of the publisher.

and ventrally and their membranes separate the pleural and peritoneal portions of the body cavity. Myoblasts from the body wall invade the pleuroperitoneal membranes to form the muscular part of the diaphragm.

body composition Most methods of representing human body mass partition the body into two compartments: fat-free body mass, and fat mass. *Fat-free body mass* is the body mass devoid of *all* extractable fat, whereas *lean body mass* (a theoretical entity) includes the small percentage (~3%) of non-sex-specific essential fat located chiefly in the CNS, bone marrow and internal organs. Fat mass is subdivided into storage fat (primarily fat in ADIPOSE TISSUE) and essential fat components, these being represented in Fig. 24 for Behnke's 'reference' man and woman on the basis of large-scale civilian and military anthropometric surveys combined with data from laboratory studies of tissue composition. Adult females have roughly twice as much body fat as males, and a smaller percentage mass of skeletal muscle, changes in composition which begin at PUBERTY. Sex-specific essential fat includes, in women, additional fat in the breasts and as lower body subcutaneous fat. In normally hydrated, healthy, adults, the fat-free mass and lean body mass differ only in the essential fat component. See BODY MASS, SIZE AND SHAPE.

body fat See ADIPOSE TISSUE, BODY COMPOSITION.

body fluid Body water and its dissolved solutes, comprising ~60% of body mass in healthy individuals, extracellular fluid comprising ~20% of total body mass. Sources of body water are by ingestion (~90%) and cellular metabolism (mainly condensation reactions, ~10%). Routes by which water is lost include urine (~61%); evaporation from the skin (~35%) by sensible perspiration (sweating), insensible perspiration (other than via sweat glands) and from the respiratory passages; and in faeces (~4%).

body mass, size and shape Primate body size correlates with several ecological and life history variables: population density, home range size, social organization, and age at first breeding. Robusticity declined with the origin of anatomically modern humans, one likely cause being that technological inventions made use of sheer strength less important. Eventually, food preparation by cooking probably made large tooth size, heavy jaws and large masseter and temporalis musculature less important. In Australian populations during the five millennia since the ice age, the following changes were found: tooth size reduction, 4.5%; facial size reduction, 6–12%; brain size reduction, 9.5%; and stature reduction, 7%. In 1990, based on 149 geographically varied human populations, average body mass is 57 kg (male average 61 kg, female average 53 kg). Hominin body mass estimates are given in the HOMININI entry.

Different anthropometric methodologies have quantified physique status. Visual appraisal is often used to describe individual *somatotypes* as thin (ectomorphic), muscular (mesomorphic) or overweight (endomorphic). Body shape is closely linked to temperature regulation, water balance and habitat. Humans rely heavily on SWEATING to cool their body surface, the relationship between anatomy and climate relating to TEMPERATURE REGULATION (see BODY FLUID).

In cold climates, a low ratio tends to exist between surface area and body mass (e.g., Inuits in the Arctic), whereas in hot climates a high ratio tends to exist (e.g., Nilotics at the equator). Recent analysis views the body as a cylinder, the diameter of which represents the width of the body (more especially, the pelvis) while its length represents trunk length. The balance between heat production and loss translates to the ratio of surface area to the volume of the cylinder (proportional to body mass). This ratio is high when the cylinder is narrow, and low when it is wide, forming the basis of BERGMANN'S RULE. The prediction that people living at low latitudes will have narrow bodies and a linear structure, while those living at high latitudes will have wide bodies and a relatively bulky stature, bears up well. Human BASAL METABOLIC RATE is related to body mass, from birth to adult, and increases much more rapidly in relation to body mass increase during the first year of life than at any other period. See BODY COMPOSITION for relative masses of different body components, and BODY MASS INDEX. See also BRAIN SIZE, SEXUAL DIMORPHISM.

Body Mass Index (BMI) Derived from body mass and stature (height expressed in metric units), the BMI is used to assess 'normalcy' for a person's body weight and is calculated as follows:

$$\text{BMI} = \text{Bodymass (kg)} \div \text{Stature (m}^2\text{)}$$

Thus, a person weighing 65 kg and 1.73 m in height would have a BMI value (= QI value) of:

$$65 \div 1.73^2 = 21.7$$

which, using the often more convenient chart (see Fig. 25) showing the 20, 25, 30 and 40 boundaries of Quetelet's Index (QI), is in the desirable range. However, the limi-

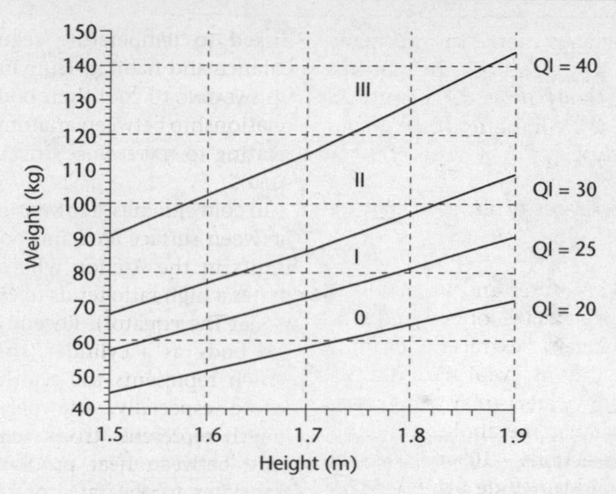

FIG 25. Graphs showing adult height–weight variation, line 0 indicating the desirable range of weight for height (Quetelet's Index value 20). Values below this line are underweight, three grades of overweight being given by lines I, II and III, representing Quetelet's Index values of 25, 30 and 40, respectively. See BODY MASS INDEX for more information. From J.S. Garrow & W.P.T. James © 1993 *Human Nutrition and Dietetics*, with permission from Elsevier Ltd.

tations of height–weight tables include that they employ unvalidated estimates of body frame size, are developed from data derived mainly from white populations, focus specifically on mortality data that may not reflect obesity-related comorbidities, and provide no assessment of BODY COMPOSITION.

This index exhibits a curvilinear relationship with the all-cause mortality ratio, and as BMI increases throughout the range of moderate and severe overweight, so does the risk increase for cardiovascular complications (e.g., hypertension and stroke), certain cancers, type 2 diabetes, gallstones, osteoarthritis and renal disease. A recent recommendation that waist circumference rather than BMI should be measured recognizes the important part played by abdominal OBESITY in the above-mentioned morbidities, often collectively described as 'metabolic syndrome'.

body size and shape See BODY MASS, SIZE AND SHAPE.

body temperature See TEMPERATURE REGULATION.

Bohr effect See HAEMOGLOBIN.

bolus A soft, rounded mass (usually of food) that is swallowed.

bone Connective tissue laid down by specialized mesodermal cells, OSTEOCYTES, lying in lacunae within a calcified matrix which they secrete, containing ~65% by weight of inorganic salts (mainly hydroxyapatite, providing hardness), the remainder being largely organic and comprising mostly collagen fibres which provide tensile strength. It contains ~85% of total body phosphorus. Differentiation of bone depends on FGFs (see DYSPLASIAS). The KIDNEYS control bone health through their influence on PARATHYROID HORMONE and VITAMIN D, two hormones critical for calcium and phosphorus balance. 'In return', the osteoblast lineage is the source of the FIBROBLAST GROWTH FACTOR FGF23, which acts as a hormone and regulates kidney function. Bone shape is partly genetically determined, although its architecture is modified to withstand load-bearing stresses and muscle contraction. Physical activity contributes to high peak bone mass, load-bearing EXERCISE being more effective than other forms. Peak bone mass is attained in the 3rd decade and maintained until the 5th decade, when bone removal exceeds its rate of

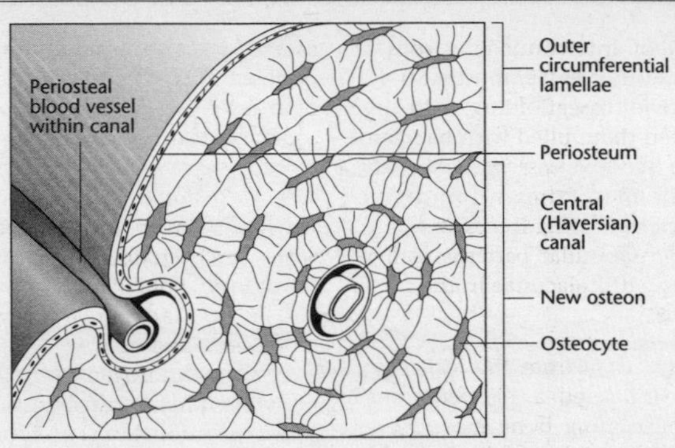

FIG 26. Diagram of newly forming osteon as osteoblasts in the periosteum build new outer circumferential lamellae and periosteal ridges fold over a blood vessel which will lie within a new Haversian canal. See BONE. From *Principles of Anatomy and Physiology*, G.J. Tortora & S.R. Grabowski, © 2000. Reproduced with permission of John Wiley & Sons, Inc.

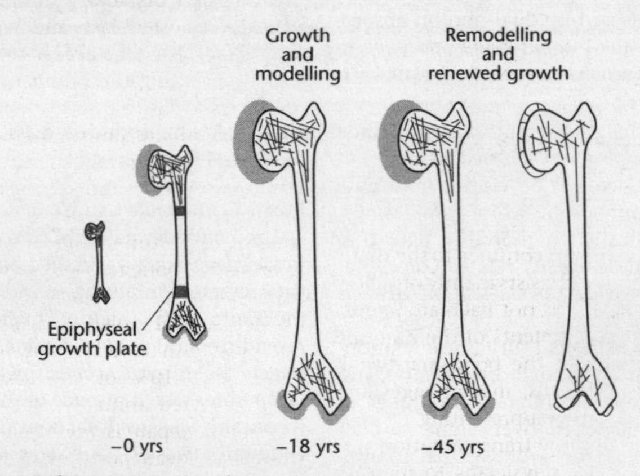

FIG 27. Diagram indicating stages in a human long BONE during a lifetime. Cartilage is indicated by shading, bone by thick black borders. Trabeculae are shown as thin lines. When the growth-plate cartilage degenerates at epiphyseal fusion, longitudinal growth ceases and the only cartilage remaining is at the articular surfaces. From *Encyclopedia of Human Biology* (2nd edition), J. Gallagher & J. Dillon © 1997. Reproduced with permission from Harcourt, Inc.

replacement and age-related bone loss begins in both sexes. This is particularly severe in post-menopausal women, leading to OSTEOPOROSIS. Ageing is also associated with a loss of articular cartilage (see Fig. 27).

Compact bone forms the outer cylinder of the shafts (*diaphyses*) of long bones of limbs and is typified by a HAVERSIAN SYSTEM of organization (see Fig. 26). Very much a living tissue, bone is supplied with nerves and blood vessels, its constitution changing under hormonal influence (see BONE REMODELLING). Besides its skeletal role in providing levers for movement and the support of soft parts of the body, it indirectly protects many delicate tissues and organs. See ADOLESCENT GROWTH SPURT, CARTILAGE, HAEMOPOIESIS, JOINTS, OSSIFICATION.

Spongy (cancellous) bone is located in the interiors of bones of the skull, vertebrae, sternum and pelvis, and in the ends (*epiphyses*) of the long bones. It has a lattice-like network of interconnecting rods or plates of bone (trabeculae, mostly 50–400 μm thick) providing scaffolding, with large spaces between them filled with haematopoietic tissue (see BONE MARROW). Trabeculae align along the lines of stress within a bone, and if the weight-bearing forces alter (e.g., post-fracture), trabecular patterns realign. Osteocytes lie within lacunae in the trabeculae. See BMPS.

bone density Data from the MIR space station crew indicated a 1% per month loss in weight-bearing bone mass during missions, and coincided with significant decreases in muscle mass and strength. Muscle strength under standard tests is a good predictor of bone mineral density (BMD), measured by dual-photon absorptiometry in the lumbar spine and neck of the femur. See BONE REMODELLING, OSTEOMALACIA, OSTEOPOROSIS.

bone marrow The soft connective tissue located within the marrow cavities of BONE: *red marrow* is confined to the cancellous bone and it contains HAEMATOPOIETIC STEM CELLS; *yellow marrow* is confined to the shafts of the long bones, consists mainly of yellow adipose tissue, and is not haematopoietic. Most cellular components of the immune system originate in the bone marrow: B cells, T cells, NK cells, macrophages and specialized antigen-presenting cells (APCs). Bone marrow transplantation was invented to allow physicians to increase chemotherapy and radiotherapy to myeloblastive doses to enable elimination of endogenous cancer cells. The first successful transplants were between identical twins, with no host-versus-donor histocompatibility barrier, and no immune-based response against the host. Allogeneic haematopoietic grafts contain donor T cells, which respond to host antigens in all tissues, leading to graft-versus-host syndrome (see GRAFTS). See OSSIFICATION.

bone morphogenetic proteins See BMPS.

bone remodelling Process of bone degradation and reconstruction, continuous throughout life, during which several OSTEOCLASTS create small tunnels in old bone as collagen and other organic compounds are digested (see COLLAGENASES) and inorganic salts dissolved by acid; then osteoblasts rebuild bone afresh (see BONE, see Figs 26 and 27). There is a homeostatic balance between the two opposing processes, orchestrated by several circulating hormones, notably by GROWTH HORMONE, and INSULIN-LIKE GROWTH FACTORS. Sex steroids (androgens and oestrogens) decrease bone resorption. Androgen receptors are found predominantly in osteoblasts at the site of bone formation, but also in osteoclasts and osteocytes. Osteoclast regulation by androgen is controlled mainly by 17β–oestradiol and the oestrogen receptor (see ADOLESCENT GROWTH SPURT). OESTROGENS stimulate the proliferation of osteoblasts and expressions of type I collagen and alkaline phosphatase while decreasing the number and activity of osteoclasts (increasing their apoptosis), and of cytokines promoting bone resorption. They also up-regulate the expression of receptors to vitamin D, growth hormone and progesterone. Reduction in ovarian steroids in post-menopausal women leads to increased bone resorption (see OSTEOPOROSIS).

After skeletal growth is complete, remodelling of both cortical and trabecular bone continues, ~10% of the adult skeleton being recycled annually. The final phase of fracture repair is bone remodelling of the callus, dead portions of the original broken bone being resorbed gradually by osteoclasts. Sex steroids slow resorption of old bone and promote new bone formation (thus oestrogens promote apoptosis of osteoclasts). PARATHYROID HORMONE and activated vitamin D both influence blood calcium levels. Calcitonin accelerates uptake of calcium and phosphate into the bone matrix.

bony labyrinth A series of cavities within the temporal bone forming the VESTIBULE, COCHLEA and SEMICIRCULAR CANALS of the inner ear.

bootstrapping A method for assessing how well supported by available data are the individual clades within a phylogeny. Once a putative phylogeny has been arrived at based on DNA sequence data, a subset of these data is first removed and then replaced by an equivalent data set of randomly chosen sites from the original data set. The new synthetic data, containing this replacement pseudosequence, are then fed through the appropriate software. If a data set strongly supports a particular statistical result (e.g., a gene tree) then the insertion of randomly chosen subsets into the data should continue to support the same result. Each of a large number of such synthetic data sets (~1,000) is used to reconstruct a phylogeny, and each time a node in the original phylogeny is precisely replicated, it is noted. Bootstrap values in the original phylogeny are usually represented as percentage values at each node so that a value of 80 means that the same node was reconstructed from 80% of the synthetic data sets. There is no definitive bootstrap value that confers unequivocal support for a phylogeny although values of 95–100 indicate a high degree of support for a node. Values below this do not imply incorrectness of a node; rather that the available data do not provide such convincing support. Values of 70 are often considered to be reasonably reliable. See PHYLOGENETIC TREES.

Bordetella pertussis The causative agent of WHOOPING COUGH.

bottleneck (genetic bottleneck) Loss of genetic variation owing to fall in population numbers. Even if population numbers recover rapidly, the variation (see GENETIC DIVERSITY) will remain low until supplemented by mutation, gene flow, etc. See SELECTIVE SWEEP for distinctions.

botulism The obligately anaerobic soil bacterium *Clostridium botulinum* rarely grows directly in the human body, but does so in improperly preserved foods, where it produces seven potent exotoxins (BoNTs). These bind to SNAP proteins (one class of SNARE) on presynaptic membranes, preventing Ca^{2+}-dependent fusion of synaptic vesicles carrying acetylcholine at neuromuscular junctions. This causes long-lasting paralysis, gradually descending from the head downwards. In rare severe cases, respiratory failure can result in death. Botulinum neurotoxin type A is routinely used in clinical neurology (and in beauty clinics) as BOTOX. Compare TETANUS; see VESICLES.

bouton Swelling at an axon tip from which neurotransmitter is released at a SYNAPSE.

bovine spongiform encephalopathy (BSE, mad cow disease) An emerging PRION zoonosis, which recently became epizootic (underwent a large-scale animal outbreak), primarily in Great Britain. Although its recent expansion is suspected to have originated through the now-abandoned practice of supplementing animal feed with pulverized meat and bones of previously slaughtered cattle, it is thought to have emerged earlier through use of cattle feed containing the agent of sheep scrapie. See CREUTZFELDT–JAKOB DISEASE.

Bowman's capsule (glomerular capsule) A two-layered capsule surrounding the glomerulus at the proximal end of each KIDNEY nephron. Its outer, parietal, cell layer is continuous with the proximal convoluted tubule, while its inner, visceral, cell layer consists of podocytes resting on the basement membrane of the glomerular endothelium. Filtration slits between the podocytes allow glomerular filtrate to pass into the capsule.

Boxgrove man See HOMO HEIDELBERGENSIS.

brachial plexus The nerve plexus formed from the anterior (ventral) rami of SPINAL NERVES C5–C8 and T1 and providing the entire nerve supply to the shoulders and FORELIMBS.

brachiation Movement through trees by hanging from branches, and swinging alternate arms from branch to branch, as used especially by gibbons (Superfamily Hominoidea, Family Hylobatidae) although all apes use the method at times. True brachiation includes a free flight phase. See GAIT.

brachiocephalic trunk (b. artery) The arterial trunk leaving the arch of the AORTA and giving rise to the right common carotid artery and the right subclavian artery. See AORTIC ARCHES, Fig. 13.

bradycardia Decrease in heart rate. See CARDIOVASCULAR CENTRE, ARRHYTHMIAS. Contrast TACHYCARDIA.

bradykinesia The condition in which any INTENTIONAL MOVEMENTS that are made are slow. See PARKINSON'S DISEASE.

bradykinin Peptide formed from the extracellular peptide *kininogen*on release of proteases by cell and tissue damage, and a potent excitant of NOCICEPTORS. Competitive antagonists of bradykinin (B_2) receptors are analgesic and anti-inflammatory in laboratory studies. See KININS, PAIN.

***Braf* gene** See RAS SIGNALLING PATHWAYS.

brain The brain is at first a simple tube produced by NEURULATION, although as it develops its walls do not thicken to the same extent throughout, and various flexures and outpushings are generated as a result. At its rostral end, the central canal of the spinal cord widens to form the primary BRAIN VESICLES (see Fig. 30b). Neuro-

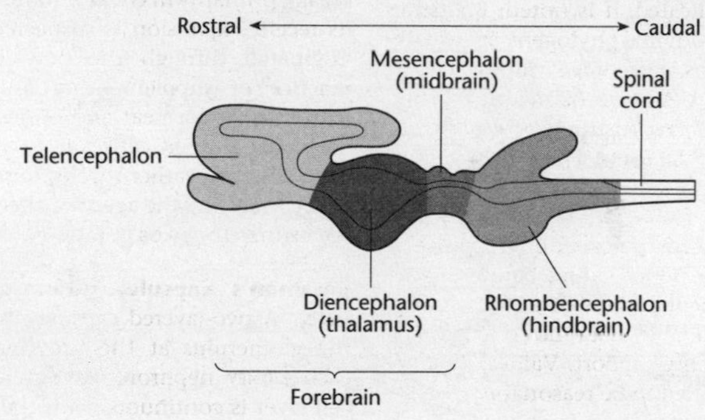

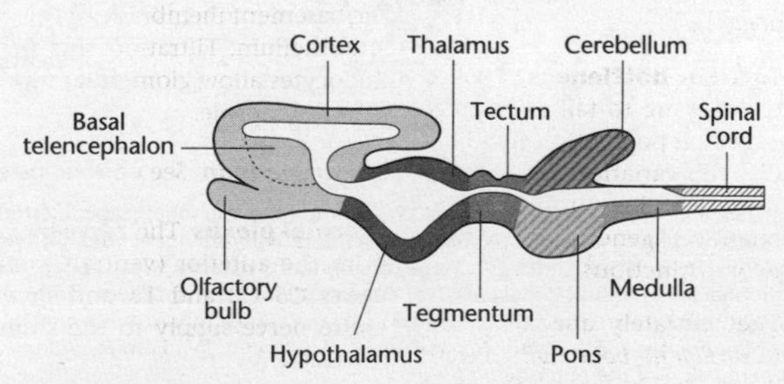

FIG 28. Diagrammatic representations of the primary vesicles of the early BRAIN, after the prosencephalon has differentiated into telencephalon (olfactory bulbs, hippocampus and the two cerebral hemispheres of the adult) and diencephalon (retina, pineal, thalamus and hypothalamus of the adult). The adult midbrain comprises fibre tracts linking the rostral and caudal brain, as well as the tectum and tegmentum. The rostral components of the adult hindbrain (metencephalon) comprise the cerebellum and pons, while the caudal region (myelencephalon) becomes the medulla oblongata. © 2001 from *Neuroscience* (2nd edition) by M.F. Bear, B.W. Connors and M.A. Paradiso, published by Lippincott, Williams & Wilkins. Reproduced with the kind permission of the publisher.

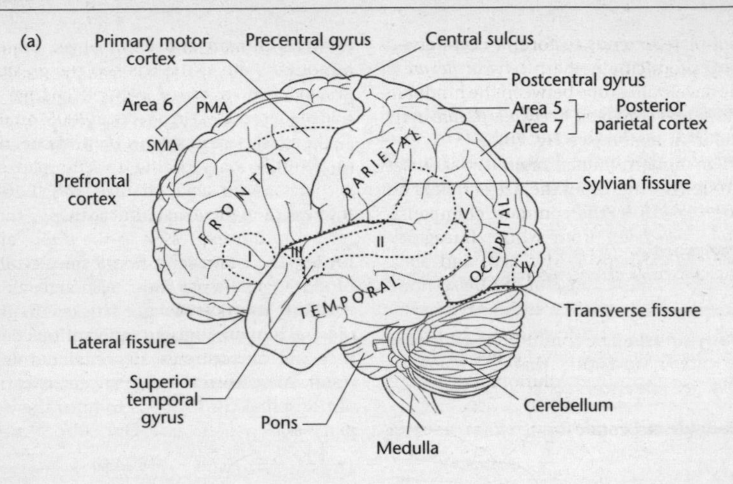

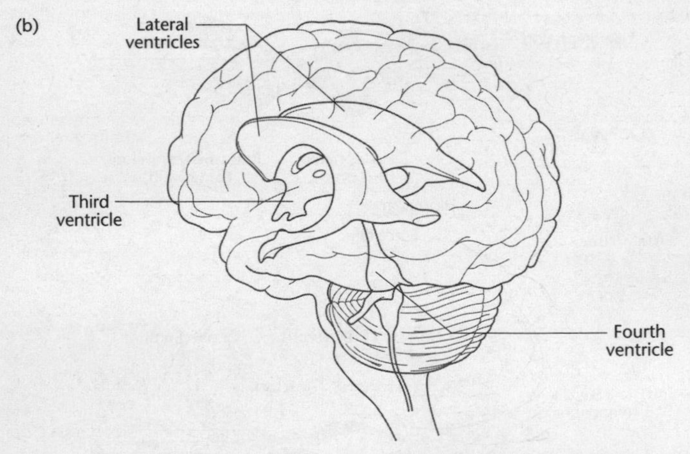

FIG 29. (a) Lateral view of the left surface of the BRAIN, showing the four lobes (capital lettering) and the approximate locations of Broca's area (I), Wernicke's area (II), primary auditory cortex (III), primary visual cortex (IV) and visual association cortex (V). Some other prominent surface features are also indicated, with some additional labels of functional areas. (b) The locations of the ventricles within the brain, the diagram being slightly rotated to the right compared with (a). © 2001 from *Neuroscience* (2nd edition) by M.F. Bear, B.W. Connors and M.A. Paradiso, published by Lippincott, Williams & Wilkins. Reproduced with the kind permission of the publisher.

epithelial cells of the neural tube give rise to primitive nerve cells, neuroblasts, forming the mantle layer which will later form the grey matter of the spinal cord. The outer layer of the spinal cord, the marginal layer, contains neurons whose axons emerge from the mantle layer. These become myelinated and form the white matter of the spinal cord. The two sides of the mantle layer fold outwards as the roof-plate extends to form a non-nervous roof, producing a sharp *cervical flexure* of the developing tube between the hindbrain and spinal cord, and a *cephalic flexure* in the midbrain region (see Fig. 28).

The modern human brain (see BRAIN SIZE) is roughly three times the size of that of our closest relative, the common chimpanzee (*Pan troglodytes*) from whose lineage we diverged perhaps 7 Myr ago, and about twice the size of those of pre-*Homo* hominins living as recently as 2.5 Myr ago. At present, there is little evidence to suggest that the addition of novel genes has been a major factor in the evolution of disparities between human and chimpanzee brains, although brain tissue has had the greatest acceleration in GENE EXPRESSION change of five tissues studied in a recent (2005) analysis. For genes expressed in brain tissue, the ratio of human-specific to chimpanzee-specific amino acid changes was 1.40 – higher, though not significantly so, than for genes not expressed in the brain, and higher than for genes expressed in any other single tissue. This is in agreement with other work showing a faster evolution on the human lineage for a set of genes involved in brain function and development, and there is need to establish the

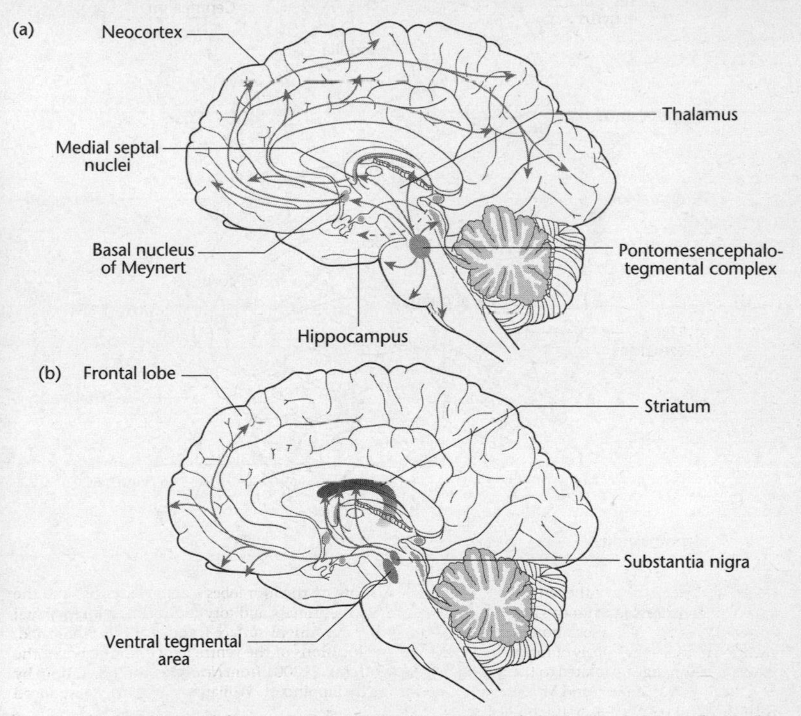

FIG 30. Representations of four DIFFUSE MODULATORY SYSTEMS of the BRAIN. Arrows indicate the approximate projections of axons releasing the relevant neurotransmitter (see relevant entries, and RETICULAR FORMATION, Fig. 152). (a) The cholinergic system, in which the medial septal nuclei and basal nucleus of Meynert project widely into the cerebral cortex, including the hippocampus (seen as a bulge on the right medial temporal lobe), while the pontomesencephalotegmental complex projects to the thalamus and some forebrain regions. (b) The dopaminergic systems arising from the substantia nigra and ventral tegmental area and projecting to the striatum, and limbic and frontal cortex, respectively. ... (*continued overleaf*)

full list of genes imprinted in the human brain (See GENOMIC IMPRINTING), discover their effects, and relate their expressions to, e.g., autistic and psychotic tendencies. Further work determining the phenotypic effect on the brain of genetic changes on the human lineage is needed to establish this. Recent studies of NON-CODING DNA suggest that *cis*-regulatory and other non-coding changes may have contributed to the modifications in brain development and function that are presumed to have given rise to uniquely human cognitive traits.

There are about ten times more glial cells than neurons in the brain, occupying about half its volume. Brain volume quadruples between birth and adulthood, not because of increase in cell number (except for a few brain regions which add neurons throughout life, the bulk of the brain's neurons are present by the seventh month of gestation) but through huge postnatal growth of syn-

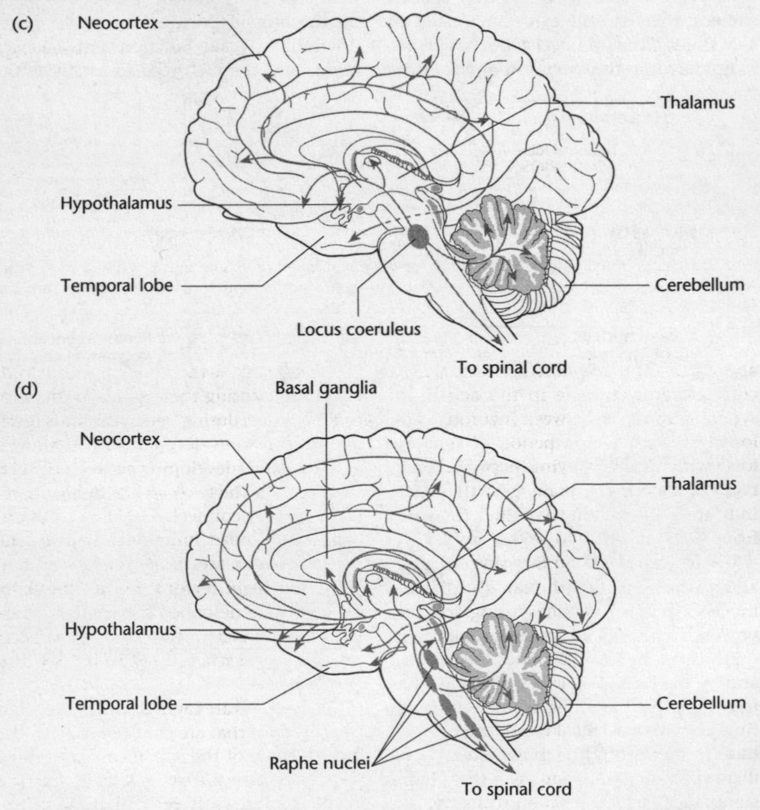

FIG 30. (*continued*)
(c) The noradrenergic system, arising from a small cluster of neurons in the locus coeruleus but innervating huge areas of the CNS, including the spinal cord, cerebellum, thalamus and cerebral cortex. (d) The serotonergic system, arising from the raphe nuclei clustered along the midline of the brain stem and projecting widely to the CNS. © 2001 from *Neuroscience* (2nd edition) by M.F. Bear, B.W. Connors and M.A. Paradiso, published by Lippincott, Williams & Wilkins. Reproduced with the kind permission of the publisher.

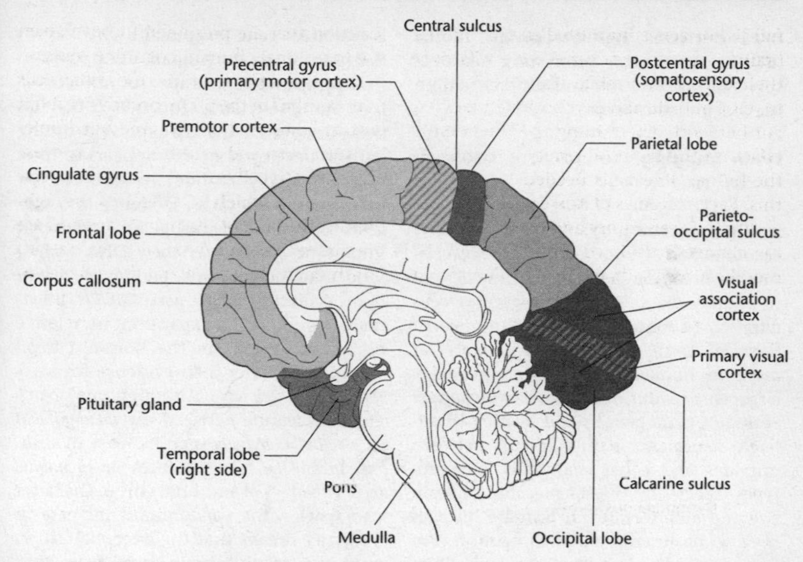

FIG 31. Sagittal section of the BRAIN, showing some of the principal gyri, sulci and functional areas. From A.R. Crossman and D. Neary © 1995 *Neuroanatomy, An Illustrated Colour Text*, redrawn with permission from Elsevier Ltd.

apses, dendrites and fibre bundles, with a corresponding increase in the density of synaptic contacts between neurons. Following this there is a period of synaptic loss, with timing varying between brain regions. Percentage brain growth (100% indicating mean adult volume) increases from 30% at birth to ~60% by 1st yr, ~70% by 2nd year and 80% by 3rd year, and reaches ~90% by 6th year. The massive increase in brain volume during the first years has a bearing on SPEECH development.

Protected by cranial bones (see SKULL), and by the cranial MENINGES, the brain has four principal parts: the brain stem (continuous with the spinal cord); the cerebellum (posterior to the brain stem); the diencephalon (superior to the brain stem), in large measure comprising the THALAMUS and HYPOTHALAMUS, but also including the epithalamus and subthalamus; and the paired CEREBRAL HEMISPHERES lying over the diencephalon and occupying most of the cranium. The superficial grey matter of the cerebral hemispheres constitutes the CEREBRAL CORTEX, which rolls up and folds upon itself during the expansion phase of brain volume, during which the folds (gyri) and deep fissures develop, with shallower grooves (sulci) developing between the gyri (see Fig. 31). The two cerebral hemispheres are internally connected by the CORPUS CALLOSUM. The broad functional areas of the cerebral cortex are shown in Figs 29 and 30. Other brain regions, not all mutually exclusive, but described elsewhere, include the BASAL GANGLIA, HIPPOCAMPUS, MEDULLA OBLONGATA and RETICULAR FORMATION. See also CRANIAL NERVES.

The *ventricles* are cavities within the brain (see Fig. 29b) that are continuous with the central canal of the spinal cord and filled with CEREBROSPINAL FLUID. A lateral ventricle occurs within each cerebral hemisphere; the narrow third ventricle lies between the right and left halves of the thalamus, while the fourth ventricle lies between the brain stem and the cerebellum.

brain asymmetry See CEREBRAL HEMISPHERES.

brain-derived neurotrophic factor (BDNF) A molecule promoting neuronal survival, proliferation and differentiation. See NEUROTROPHIC FACTORS.

brain imaging See MAGNETIC RESONANCE IMAGING, PET.

brain size The human BRAIN (see entry) accounts for 20% of bodily energy consumption although it forms only 5% of total body mass. Cortical progenitor cells undergo 11 rounds of division in mice, at least 28 in macaques, and probably far more in humans, creating progressively larger cortices during a progressively longer period of neurogenesis and leading to different neuronal 'circuit diagrams' between primates and other mammals. Cerebrotypes (species-by-species measure of brain size and architecture) in primate lineages need to be interpreted with caution. But there are indications that the cerebellum has occupied a constant fraction of total brain volume during the evolution of diverse mammals while the telencephalon in primates has grown considerably at the expense of the medulla, mesencephalon and diencephalon – more so in humans and other hominoids than in lower primates. In hominoids, the neocortex component of the telencephalon shows the greatest expansion, whereas the hippocampus, septum, schizocortex, piriform cortex and olfactory bulb components have all decreased. Brain mass in humans is 250% greater than in chimpanzees although the body as a whole has only 20% more mass. Relative to other primates, humans show greater intrahemispheric connectivity via cerebral white matter, but lower interhemispheric connectivity via the corpus callosum and anterior commissure. This may simply reflect the larger brain volume of humans (~1300 cm^3) compared with those of chimpanzees (~340 cm^3), gorillas (~380 cm^3) and rhesus monkeys (~80 cm^3). See MICROCEPHALY, SEX DIFFERENCES.

There is some evidence that human brain evolution is associated with changes in gene expression specifically restricted to the brain, some studies indicating more accelerated gene expression changes in the brain along the human, as opposed to the chimpanzee's, lineage (see NON-CODING DNA). An RNA-coding sequence (*HAR_1*) has been discovered that is expressed during cortical development and appears to have evolved rapidly in humans (see CEREBRAL HEMISPHERES, HAR1F GENE). Evidence also suggests that some neural genes underwent important changes in their DNA coding sequences during this period of recent brain evolution (FOXP2, ASPM, *MCPH1*). Evidence for a rapid expansion in relative brain size appears in the hominin fossil record much later than evidence for BIPEDALISM, at ~2–2.5 Mya. Although one distinguishing feature between early *Homo* and *H. erectus/H. ergaster* is an increase in absolute brain size, from ~550 cm^3 in *H. habilis* to between ~850 and 1100 cm^3 in the latter two species, the concomitant increase in body size means that the encephalization quotient (relative brain size) may have increased little. One 1999 estimate of relative brain sizes, derived by dividing the cube root of the mean cranial capacity of the species by the square root of the mean species' orbital area and multiplying by 10, gives relative brain size values for *erectus* of 2.87 and for *ergaster* of 2.76, compared with 3.08 for *H. sapiens*. Relative brain size among hominins is a proxy for neocortex size; but although there are substantial differences in mean brain size of australopithecines and *Homo*, some of these have dubious relevance when differences in the body size proxy are considered. At ~1.6 Mya, archaeological sites indicate that *H. erectus* was moving carcasses to campsites for further butchering/sharing, coincident with a reduction in tooth size. One suggestion is that, if *erectus* was cooking its food (there is no clear evidence of this), they would have spent less energy chewing and digesting it, and more energy might have been available for brain growth. An older view is that more extensive meat-eating enabled shrinkage of the 'expensive tissue' of the digestive tract (it is only 60% of the size expected for a primate of similar size), allowing more brain growth (see ALIMENTARY CANAL). A relationship between primate brain size and life history strategy has been detected, indicating that 28 species have slowed their metabolic rates, prolonged the juvenile years, postponed the age of first reproduction, and live longer. This allows the brain to grow larger, and more complex, before adulthood (see 'LESS-IS-MORE' HYPOTHESIS).

brain stem (brainstem) The MIDBRAIN, PONS and MEDULLA OBLONGATA. Concerned with vital life support functions (see CARDIO-VASCULAR CENTRE, VENTILATION), it contains nuclei for most of the cranial nerves. A core of highly interconnected brain stem nuclei constitutes the RETICULAR FORMATION. It receives ascending impulses from the spinal cord, and a special input from the viscera; it adds inputs from the eyes, ears and face, mouth and head and sends outputs to eye muscles, face and jaw. The caudal brain stem (medulla and posterior part of the pons) contains neurons which actively generate SLEEP, while an alleged MESENCEPHALIC LOCOMOTOR REGION is also located here.

brain vesicles The three primary brain vesicles are, in a rostrocaudal direction (see BRAIN, Fig. 29), the *prosencephalon* (forebrain), *mesencephalon* (midbrain) and *rhombencephalon* (hindbrain), developing by ballooning of the cephalic part of the neural tube. From the primary vesicles, four further secondary vesicles develop, two (*telencephalon* and *diencephalon*) from the prosencephalon and two (*metencephalon* and *myelencephalon*) from the rhombencephalon.

brain waves See ELECTROENCEPHALOGRAM.

branchial arches See PHARYNGEAL ARCHES.

BRCA1, BRCA2 The *BRCA1* gene has been extensively studied because of its role in hereditary breast, ovarian and other cancers (see BREAST CANCER), although it is not clear why the BRCA1 protein suppresses tumour development specifically in ovarian-hormone-sensitive tissues. It is thought to participate in the response to DNA damage, centrosome amplification regulation, and mitotic spindle assembly. It may also function in neuronal development (in particular, neurite outgrowth), NEURAL REGENERATION, and the pathology of neurological conditions.

In some estimates, 70–80% of all familial ovarian cancers are due to mutant germline alleles of these two genes. Rather than being tumour suppressor genes, they appear to be 'caretaker genes' involved in maintenance of genomic integrity. BRCA1 interacts with the *RAD50* product RAD50 to form the massive multiprotein machine known as the BRCA1-associated genome surveillance complex (BASC), involved in homologous chromosome recombination and DNA REPAIR MECHANISMS (see entry). In female mice, X chromosome structure and expression of some of its genes are abnormal when the cells contain one mutated *BRCA1* gene (i.e., in carriers) which appears to prevent the normal completion of X chromosome inactivation. The *BRCA2* gene is located at 13q12.3. Its product, BRCA2, has no structural similarity to BRCA1 but interacts with RAD51 in repairing radiation-damaged DNA. CDK-dependent phosphorylation of *BRCA2* is required as a regulatory mechanism for recombinational repair.

Mouse cells with mutant *BRCA2* fail to divide properly, chromosomes in offspring having structural defects (as observed in FANCONI ANAEMIA). Mammary glands of nulliparous *Brca1/p53*-deficient mice develop mammary tissue that normally occurs only in pregnancy and progesterone (but not oestrogen) receptors are over-expressed in the mutant mammary epithelial cells due to a defect in their PROTEASOME pathway. Treatment of these mice with the progesterone antagonist mifepristone (RU 486; see CONTRACEPTION) prevented mammary tumorigenesis, suggesting that antiprogesterone treatment may be useful for breast cancer prevention in *BRCA1*-mutant women. See GENOMIC INSTABILITY.

breast cancer A CARCINOMA of the MAMMARY GLAND, which is almost twice as common in sisters and mothers of affected individuals than it is in the general population. Known susceptibility genes account for <25% of the familial risk of breast cancer, i.e., inactivating mutations in *BRCA1*, *BRCA2* and *TP53* (see P53 GENE) confer a high risk of developing breast cancer (10–20-fold by the age of 60) whereas inactivating mutations in *CHEK2* and *ATM* are associated with more modest risks (~twofold). Most familial breast cancer remains unexplained, very likely due to variants causing more moderate risk leading to the proposal that breast cancer susceptibility is largely polygenic.

In 1987, amplification of the *erbB*-related gene (aka *ERBB2*, *neu*, *HER2*) was reported in many breast cancers, and increases in gene COPY NUMBER of more than five were found to correlate with a decreased survival of patients. It is not a tumour-specific antigen but an EGF receptor-related protein, often displayed at many times its normal level by breast cancer cells. The monoclonal antibody Herceptin reacts strongly with HER2 protein and has resulted in extension of life span of breast cancer patients in combination with other chemotherapeutic agents. Herceptin may block proliferation of residual cancer cells left behind after removal of primary tumours: in its absence, cells might be responsive to the wound-healing mitogens following surgery. Amplification of the c-*ERBB2* and c-*MYC* ONCOGENES and of cyclin D1 (see ONCOGENE, CYCLINS) is a feature of 20% of breast cancers, possibly correlating with such prognostic factors as lymph node status, OESTROGEN and PROGESTERONE receptor status and tumour size. Recently, SNPs in five novel independent loci have been found to exhibit strong consistent association with breast cancer, four of them in plausible causative genes: intron 2 of *FGFR2*, *TNRC9* (possibly a transcription factor gene), *MAP3K1* and *LSP1*, and apparently none of them in DNA repair pathways.

Lifetime risk of breast cancer in women, according to one epidemiological estimate, decreases by 20% for each year that menarche is delayed during adolescence. Women who stop menstruating before age 45 have only ~½ the risk of breast cancer of those who continue to menstruate to age 55 or beyond, almost certainly associated with the number of cycles of proliferation and regression of cells forming the mammary gland ducts. Removal of the ovaries, the prime source of oestrogen in women, causes breast cancer risk to decline dramatically. Women entering menopause before the age of 36, as a result of side effects of chemotherapy for Hodgkin's lymphoma, have a 90% reduced risk of subsequently developing breast cancer; but postmenopausal women who contract breast cancer have on average a 15% higher level of circulating oestrogen than do unaffected women. Primary tumours can be detected by mammography, which has been proven to reduce mortality from breast cancer. Primary tumours progress to metastatic tumours (see METASTASIS). miR-10b (see MICRORNAS) is highly expressed in metastatic breast cancer cells and positively regulates cell migration and invasion. See CANCER, CANCER CELLS, LIFESTYLE.

breasts See MAMMARY GLANDS.

breathing See VENTILATION.

brittle bone disease (osteogenesis imperfecta) A group of crippling genetic disorders producing very brittle bones which fracture easily because of insufficient collagen in the bone matrix. Some forms are inherited in an autosomal dominant manner, others as an autosomal recessive; some are mild in expression, others lethal. COLLAGEN genes on chromosomes 7 and 17 are implicated. Contrast OSTEOPOROSIS.

Broca's aphasia Condition in which speech is produced with extreme difficulty, and includes brief utterances with word errors (paraphasia). Word repetition is impaired although powers of comprehension are little affected. It may result from left FRONTAL LOBE lesions. See APHASIAS.

Broca's area (premotor speech area; areas 44 and 45) Located far laterally in the left inferior frontal lobe and almost always dominant on the left side, this SPEECH area causes the formation of words and is also required for writing. It receives input from the INSULA and simultaneously excites the laryngeal muscles, respiratory and mouth muscles. It is also involved in moving the hands and fingers (see LANGUAGE) and is reciprocally interconnected by the arcuate fasciculus to WERNICKE'S AREA. Its loss causes motor APHASIA. The sulcal markings associated with Broca's area have been identified on endocasts of *Homo habilis* and *H. erectus*, although this does not equate with a functional language role. See APRAXIA, DYSLEXIA.

bromodomains Domains, ~100 amino acids in length, on the general transcription factor TFIID, and on enzymes associated with CHROMATIN REMODELLING, which bind acetylated lysine side chains on HISTONES.

bronchial-associated lymphoid tissue (BALT) Lymphoid cells and tissues located in the respiratory tract, and very important in immune responses to inhaled antigens. See LYMPHOID TISSUES.

bronchial tree The trachea, bronchi and their branches up to and including the bronchioles.

bronchiole A branch of a tertiary BRONCHUS, which divides further into respiratory bronchioles which are distributed to alveolar sacs. The plates of cartilage gradually disappear as does the ciliated epithelium of the bronchi, being replaced in the terminal bronchioles by non-ciliated cuboidal epithelium.

bronchitis See CHRONIC BRONCHITIS.

bronchodilators Inhaled substances (e.g., ADRENALINE) that act on β-ADRENERGIC receptors to relax constricted muscle during ASTHMA attacks.

bronchus (pl. bronchi) The TRACHEA divides into right and left *primary bronchi* (with incomplete rings of cartilage and columnar ciliated epithelium), one to each LUNG, and on entering the lung each divides to form smaller *secondary bronchi*, one for each lobe. These continue to branch, forming *tertiary bronchi* which in turn divide to form BRONCHIOLES. The amount of cartilage gradually decreases, being replaced by plates of cartilage that gradually disappear. Correspondingly, the amount of smooth muscle increases, so that spasms (see ASTHMA, BRONCHODILATORS) can close off the airways.

brown fat ADIPOSE TISSUE composed of thermogenic brown ADIPOCYTES.

Brunner's gland (duodenal gland) Mucus-secreting gland in the submucosa of the DUODENUM (see ALIMENTARY CANAL, Fig. 4). Its alkaline product helps protect the duodenal lining from the acidity of the chyme as it leaves the stomach.

brush border See MICROVILLI.

BSE See BOVINE SPONGIFORM ENCEPHALOPATHY.

bubonic plague A flea-borne disease of rodents existing globally in many parts of rural and wooded areas, humans being increasingly bitten as animal hosts die. Urban foci of transmission also occur, where feral animals and humans share the environment. Most human cases today arise from flea-bite transmission or by handling infected rodents. Inhalation via the inhalation route is very efficient, causing a severe and highly infectious pneumonitis. It made a dramatic entry into historical records in 542 AD, when it swept through the Roman Empire. In 1347 it created the Black Death, killing up to a third of all Europeans. Caused by the Gram-negative rod bacterium *Yersinia pestis*, whose DNA analysis indicates a fairly recent molecular history, evolving from the related *Y. pseudotuberculosis* (a bacterium shed in rodent faeces) between 1.5 and 20.5 Kyr BP. See INFECTION.

buccal cavity See MOUTH.

buffers See ACID–BASE BALANCE.

bulbourethral gland See COWPER'S GLAND.

bulimia nervosa An eating disorder, characterized by frequent eating binges, often compensated for by forced vomiting. It is commonly accompanied by DEPRESSION, and linked to reduced brain serotonin levels (which reduce satiety). Antidepressant drugs (e.g., fluoxetine, Prozac) that elevate brain serotonin levels are also effective treatments for the majority of bulimics.

Recent evidence links abnormally low secretion of CHOLECYSTOKININ to bulimia. Epidemiological studies have shown that in families of women with bulimia, the men often have alcoholism and other addictions. See ANOREXIA NERVOSA, FEEDING CENTRE, OBESITY.

bundle A collection of axons running together but not necessarily having the same origin or destination. Contrast TRACT.

bundle of His See CARDIAC CYCLE.

Burkitt's lymphoma A childhood malignancy of B CELLS, especially in bones of the jaws and in abdominal organs, occurring commonly in parts of Central Africa (~1:10,000 births per year) but rarely elsewhere. The malignancy is caused by insertion of EPSTEIN–BARR VIRUS, probably involving failure of T-cell surveillance. With an average age of onset at 7 years, it is characterized by specific gene amplifications and reciprocal chromosome translocations between chromosome 8 (q24.13) and any of the three chromosomes carrying genes for the immunoglobulin heavy chain (IgH, an antigen receptor), the IgL λ light chain or the IgK κ light chain. This places expression of the c-*myc* proto-oncogene gene on chromosome 8 under the control of transcription-controlling enhancer sequences of an immunoglobulin gene. In all cases (see Fig. 32), one chromosome breakpoint occurs on chromosome 8 at the c-*myc* proto-oncogene, the other occurring on either chromosome 14, 22 or 2. In the commonest translocation, involving chromosomes 8 and 14, the break on chromosome 14 occurs at q32.33, near the so-called joining (*J*) segment of the transcriptionally active immunoglobin IgH gene, leading to overexpression of the *Myc* gene (see CANCER, MYC).

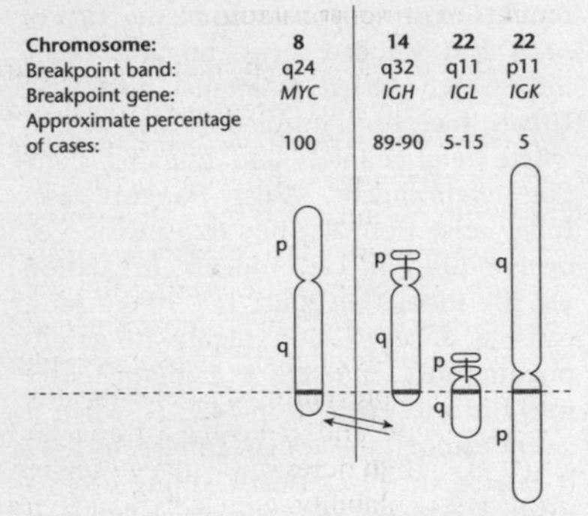

FIG 32. Diagram illustrating the chromosome breakpoints and genes involved in the reciprocal TRANSLOCATIONS producing BURKITT'S LYMPHOMA prior to the chromosomal rearrangements (see text for details). Arrows indicate the regions below the breakpoint bands on the horizontal dotted line that are juxtaposed by the translocation, regions above this line remaining as they are. From *Basic Human Genetics* (2nd edition), A.P. Mange and E.J. Mange © 1999 Sinaur Associates, Inc. Redrawn with Permission of the publisher.

bursa Sac or pouch of synovial fluid located at friction points, e.g., about JOINTS.

bypass surgery Procedure to relieve the effects of obstruction to one or more of the CORONARY ARTERIES, in which healthy parts of veins are removed from other parts of the body and inserted so as to bypass the obstructions. Blood clots, commonly the cause of heart attacks, are broken down using streptokinase, TISSUE PLASMINOGEN ACTIVATOR or urokinase, all of which activate plasminogen (see BLOOD CLOTTING).

C

CA1, CA3 Two of the four divisions of Ammon's horns in the HIPPOCAMPUS (see Fig. 91).

Ca^{2+}/calmodulin See CALMODULIN.

Ca^{2+}/calmodulin-dependent kinases (CaM-kinases) See CALCIUM/CALMODULIN-DEPENDENT PROTEIN KINASES.

cachexia A physical wasting syndrome, involving loss of BODY FAT and lean body mass which can occur in disease, e.g., cancer or inability to absorb nutrients. It often occurs with anorexia (loss of APPETITE). See OBESITY.

cadherins The major CELL ADHESION MOLECULES, most of them single-pass transmembrane glycoproteins, responsible for calcium-dependent cell–cell adhesion in tissues and linked to the actin cytoskeleton by catenins to form ADHERENS JUNCTIONS (see WNT SIGNALLING PATHWAYS). Among *classical cadherins*, E-cadherins are found on epithelial cells (see *Listeria monocytogenes*), N-cadherins on nerve, muscle and lens cells, and P-cadherins on cells of the placenta and epidermis. Vascular endothelial cadherin (VE-cadherin) is required for endothelial cell survival, mediating adhesion between endothelial cells (see ENDOTHELIUM). *Non-classical cadherins* (>50 expressed in the brain alone) include desmosomal cadherins and a diverse group of protocadherins in the brain. See CANCER CELLS, Fig. 35; NON-CODING DNA.

caecum A blind-ending tube, ~7 cm in length, leading from the ileocaecal valve to the COLON. Important in cellulose digestion in herbivores, but with no significant digestive role in humans. See GUT FLORA.

café-au-lait spots Lightly pigmented areas of the skin, the colour of coffee or milk. See NEUROFIBROMATOSIS.

caffeine A powerful alkaloid stimulant of the central nervous system, used socially to provide a sense of well-being and inducing increased alertness, clearer flow of thought, wakefulness and restlessness. It seems to owe its stimulant effects to blockage of adenosine (A_2) receptors, increasing the state of phosphorylation of a dopamine- and cyclic AMP-regulated phosphoprotein (DARPP-32), preventing the sedating effect of adenosine and reducing asthma and migraine. High doses cause *caffeinism*: nervousness, irritability, muscle hyperactivity and twitching.

Cajal–Retzius cells A transient population of neurons of the human embryonic (weeks 7–19 of gestation) marginal zone of the neocortex, an important period for cortical neuron specification and migration (see CEREBRAL CORTEX, HAR1F GENE, NATURAL SELECTION). A salient feature is their radial ascending processes which contact the pial surface, and their horizontal axon plexus located deep in the marginal zone. In mice, their migration takes them along the meninges. There seem to be homologous elements in the non-primate neocortex (initially described by Cajal in 1891), although their morphology is simpler. They appear to be critical in lamination of the cerebral cortex and at least some produce reelin, a known regulator of cortical development (i.e., are reelin-positive).

They seem to be generated in at least three focal sites at the borders of the developing pallium and distributed by tangential migration. There is a possible relation between reelin-expression-related defects in the brain and SCHIZOPHRENIA.

calcaneus The heel of the FOOT.

calcarine fissure (c. sulcus) Horizontal fissure in the occipital lobe of the cerebral cortex whose surrounding grey matter contains cells of the PRIMARY VISUAL CORTEX. See CEREBRAL HEMISPHERES, Fig. 46.

calciferol See VITAMIN D.

calcification (mineralization) Deposition of mineral salts, especially hydroxyapatite (calcium phosphate), in a collagen fibre framework, so that the tissue hardens. Under the control of OSTEOBLASTS. See bone, ossification.

calcineurin (CaN) A Ca^{2+}-dependent protein phosphatase with a high affinity for CALCIUM ions (Ca^{2+}). When activated, it can activate other phosphatases and is therefore a key player in control of Ca^{2+}-dependent signalling cascades. Its activity, e.g., favours a form of SYNAPTIC PLASTICITY known as LONG-TERM DEPRESSION, while inhibition of its activity favours LONG-TERM POTENTIATION. The immunosuppressive drug cyclosporin A inhibits its activity. See CALCIUM/CALMODULIN-DEPENDENT PROTEIN KINASES, IMMUNOSUPPRESSION, MEMORY.

calcitonin A 32-amino acid peptide hormone synthesized and secreted by PARAFOLLICULAR CELLS (C cells) of the thyroid gland, although also produced by bronchial cells of the lung, and cells of the prostate and brain. The calcitonin gene transcript also encodes a potent vasodilator, calcitonin gene-related peptide, released by NOCICEPTORS. Its release is regulated by plasma calcium levels (positively by Ca^{2+} levels >9 mg/dL) and its main effect is to reduce these levels, largely by inhibiting bone resorption. It directly inhibits OSTEOCLAST differentiation, motility and secretory activity. Its cellular effects are mediated by G PROTEIN-COUPLED RECEPTORS, leading to activation of PKA. It does not appear to be critical in the regulation of CALCIUM homeostasis.

calcitriol The hormonally active form of VITAMIN D, to which it is converted in the liver and kidneys. Its absence is a contributory factor in the occurrence of OSTEOMALACIA. See PHOSPHATE.

calcium Dietary intake of calcium averages 1000 mg per day, although only 30% of this is absorbed across the digestive tract. Of the ~1.1 kg of calcium in the human body, 99% is deposited in BONES (see BONE REMODELLING) and teeth. In addition to skeletal roles, calcium is involved in synaptic transmission and muscle contraction, having numerous effects as a second messenger. A pool of ~550 mg Ca^{2+} is interchanged between bone and extracellular fluid each day, and a normal plasma calcium concentration of between 85 and 105 $mg.dm^{-3}$ (~1 mM) is maintained largely by the actions of PARATHYROID HORMONE on the DISTAL CONVOLUTED TUBULES of kidney nephrons and by VITAMIN D and CALCITONIN on bones, kidneys and the small intestine. Plasma calcium exists as Ca^{2+} ions (50%), in protein-bound form (40%) and complexed to citrate and phosphate as soluble complexes (10%). Deficiency symptoms (of hypocalcaemia) include spontaneous action potential generation in neurons, and tetany.

The extracellular Ca^{2+} concentration, $[Ca^{2+}]$, is four orders of magnitude higher than its cytosolic concentration (~100 nM), largely as a result of CALCIUM PUMPS in the plasma membrane. However, it is sequestered in intracellular compartments, notably the mitochondria and endoplasmic reticulum, and the intracellular $[Ca^{2+}]$ may rise to 1 μM by release of Ca^{2+} from these stores or by uptake from extracellular sources in response to cell activation (e.g., the CORTICAL REACTION, the arrival of nerve impulses at presynaptic membranes, and during the depolarization phase of cells of the SA NODE and cardiac muscle fibres in general), or other pacemaking Ca^{2+} current (see SUBSTANTIA NIGRA). Calcium can enter neurons through many pathways, although the period of entry is usually brief.

Calcium is a major SECOND MESSENGER, a cofactor for several enzymes, involved in BLOOD CLOTTING, ribosome complexing, maintenance of skeletal integrity and the molecular events of muscle contraction, and required in many CELL ADHESION MOLECULES. The numerous hormone- and transmitter-initiated receptor-mediated signalling pathways in which it acts as a second messenger involve the G protein/PHOSPHOLIPASE C axis (see Fig. 139) leading to release of Ca^{2+} from the endoplasmic reticulum (ER) and a 10–20-fold rise in cytosolic [Ca^{2+}]. This ER Ca^{2+}-depletion causes opening of plasma membrane Ca^{2+} channels, sustaining the Ca^{2+} signal. Most of the isoforms of ADENYLYL CYCLASE are regulated by Ca^{2+}, a major effect of the ion. Raised free cytosolic [Ca^{2+}] activates PKC; but it also leads to activation of the Ca^{2+}-binding proteins CALCINEURIN and CALMODULIN, which in turn regulate activities of several other cytosolic proteins and enzymes involved in pathways that converge on the nucleus.

Calmodulin functions as a multi-purpose intracellular Ca^{2+} receptor, two ions being required to activate it so that it acts like a 'switch', producing a 50-fold increase in activity in response to the 10–20-fold rise in [Ca^{2+}] – the calcium signal – noted above (e.g., see SYNAPTIC PLASTICITY). Ca^{2+}/calmodulin promotes removal of PIP_2 from the plasma membrane (see MARCKS), freeing it to interact with other important proteins (see PHOSPHATIDYLINOSITOL SIGNALLING SYSTEM). Calmodulin's effects are also mediated by the CALCIUM/CALMODULIN-DEPENDENT PROTEIN KINASES and Ca^{2+}/calmodulin is a cofactor for NITRIC OXIDE synthase. Also, by binding and activating the plasma membrane Ca^{2+} pumps that pump Ca^{2+} out of the cell, cytosolic Ca^{2+} levels are brought back to normal.

Calcium channels mediating Ca^{2+} signalling include: (i) voltage-gated Ca^{2+} channels in the plasma membrane that open in response to membrane depolarization (see ION CHANNELS), notably those involved in synaptic plasticity and retinal photoreceptor activation (see ROD CELLS); (ii) IP_3-gated Ca^{2+} release channels, allowing Ca^{2+} to escape from the ER; (iii) synaptically activated NMDA-receptor channels (see GLUTAMATE); (iv) ryanodine receptors that react to a change in plasma membrane potential to release Ca^{2+} from the sarcoplasmic reticulum of striated muscle fibres during muscle contraction, but also present in the ER of many non-muscle cells, e.g., neurons. If the extracellular signal is particularly strong and prolonged, the initial Ca^{2+} influx can propagate through the cytosol rather like an action potential. Such a *Ca^{2+} spike* may be followed by further spikes, each lasting seconds, for as long as cell surface receptors are activated. The Ca^{2+} waves and oscillations depend on positive and negative feedback of Ca^{2+} on both ryanodine receptors and IP_3-gated Ca^{2+} release channels. The endocrine cells of the anterior PITUITARY generate oscillatory Ca^{2+} waves in response to extracellular signals, and the frequency-dependent Ca^{2+} response may be non-oscillatory, one frequency of Ca^{2+} spikes activating one set of nuclear genes and a higher frequency activating a different set (see GENE EXPRESSION).

calcium/calmodulin-dependent protein kinases (CaM-kinases) A family of serine/threonine kinases, members of which (as with PKA and PKC) phosphorylate available target proteins within cells. Narrow-specificity members include *myosin light-chain kinase*, which activates smooth muscle fibre contraction, and *phosphorylase kinase*, which activates glycogenolysis. The most abundant and widespread broad-specificity member is *CaM-kinase II* (CaMKII) found in all human cells and making up to 2% of total protein mass in some brain regions. CaMKII has four subunits. Activated by Ca^{2+}/calmodulin (see CALCIUM) it can remain so for long periods, even in the absence of the original Ca^{2+} signal, because it autophosphorylates itself and is only inactivated when phosphorylases eventually overwhelm it. It can therefore serve as a 'memory trace' of a cell's prior Ca^{2+} pulse, a property favouring its involvement in spatial memory and learning (see CREB, LONG-TERM POTENTIATION, SYNAPTIC PLASTICITY). See CALCINEURIN, MARCKS.

calcium/calmodulin-dependent protein kinase II (CaM-kinase II) See CALCIUM/CALMODULIN-DEPENDENT PROTEIN KINASES.

calcium pumps (Ca^{2+} pumps) All eukaryotic cells have a Ca^{2+} pump in the plasma membrane, using ATP hydrolysis to pump Ca^{2+} out of the cytosol (see ADENOSINE TRIPHOSPHATE and the ion pumps part of ION CHANNELS). Muscle and nerve cells have an additional Ca^{2+} pump that couples its efflux to the influx of Na^+ (itself removed by a Na^+K^+-ATPase). A Ca^{2+} pump in the endoplasmic reticulum membrane maintains cytosolic $[Ca^{2+}]$ low, and a low affinity, high capacity, Ca^{2+} pump in the inner mitochondrial membrane is involved in returning cytosolic $[Ca^{2+}]$ to normal after a Ca^{2+} signal (it uses the electrochemical gradient across the membrane generated by oxidative phosphorylation to take up Ca^{2+} from the cytosol). See CALCIUM, ENDOPLASMIC RETICULUM, MITOCHONDRION.

calcium signalling See CALCIUM and its references.

callus Growth of new bone tissue in and around a fracture, ultimately replaced by mature bone. See BONE REMODELLING, OSSIFICATION.

calmodulins Small, multiply-allosteric calcium-binding proteins, making up to 1% of the total protein mass of a cell. They are required for the calcium-dependent activities of many cellular enzymes, especially those that are membrane-bound (e.g., see CONTRACTILE RING). See CALCIUM, CALCIUM/CALMODULIN-DEPENDENT PROTEIN KINASES.

calvarium (pl. calvaria) The dome-like roof of the skull; or a skull (e.g., fossil) lacking lower jaw.

calyx of Held A large glutamatergic nerve terminal in the auditory pathway.

CaM-kinases See CALCIUM/CALMODULIN-DEPENDENT PROTEIN KINASES.

cAMP CYCLIC AMP. See also ADENOSINE MONOPHOSPHATE, PKA.

cAMP-dependent protein kinase See PKA.

Campylobacter Responsible for as many infections as SALMONELLA in most parts of the world, campylobacters are microaerophilic bacteria, producing less morbidity in high-income groups because they rarely cause metastatic or bacteraemic disease. Most cases of infection follow ingestion of undercooked meat (esp. poultry), unpasteurized milk, untreated water, and from domestic pets with diarrhoea. *C. jejuni*, which is thermophilic and produces enterotoxin and cytotoxin, is normally killed in the stomach, the risk of infection increasing when consumed with milk (protecting it from gastric acid). The enterotoxin causes fluid accumulation in ileal loops, with a mode of action similar to cholera toxin. Rarely passed from person to person, large food- and water-borne outbreaks of the bacterium have occurred. Periodontal *Campylobacter* spp. include *C. concisus* and *C. gracilis*. Campylobacters are distinguished by molecular techniques from HELICOBACTER, campylobacters possess a cell wall lipid A moiety and are divisible into three groups.

cancellous bone Spongy BONE.

cancer The term 'cancer' covers a large variety of conditions resulting in unscheduled and uncontrolled (i.e., deregulated) cell growth and division (oncogenic transformation) leading to a mass of CANCER CELLS (see entry) known as a tumour. More than half of a cancer mass can be made of supporting cells such as fibroblasts, tissue macrophages and endothelial cells, and progression to life-threatening metastatic lesions cannot occur without them. Any genetic predisposition or spontaneous mutation responsible for the deregulation will stand a chance of recruiting further mutations to the cancer cause through the increased number of DNA replications resulting from deregulation, so that cancer is favourably selected (see GENOMIC INSTABILITY). Cancer is both a genetic and an epigenetic disease, with lesions in both systems leading to altered GENE EXPRESSION. There are numerous subtypes, whose causes include GENETIC PREDISPOSITION, environmental factors (e.g., those favouring somatic mutation), infectious agents and AGEING. As cancer-related deaths worldwide increase with increased life expectancy, research into its causes and control has revealed details of cell and molecular biology of the utmost significance (see CANCER CELLS), not least in the fields of APOPTOSIS, DNA metabolism (e.g., DNA REPLICATION, DNA REPAIR MECHANISMS and DNA METHYLATION), and SIGNAL TRANSDUCTION (see DISEASE, Fig. 64; ONCOGENES, WNT SIGNALLING PATHWAY). Recent work indicates that cancer and AGEING are complex biological tapestries that are often, although not always, woven by similar molecular threads.

Some oncologists and surgeons restrict the term 'cancer' to malignant growths (see METASTASIS); but within contemporary research it is generally used more loosely to include all forms of abnormal growth. A person's LIFESTYLE can affect the risk of their suffering from cancer (see CARCINOGENS), and epidemiology indicates that one physiological effect of the increasing number of menstrual cycles during a woman's lifetime is a higher risk of her contracting BREAST CANCER. A recent search among half the known human genes using breast and colorectal cancer cells revealed a complex situation. Not only did these cells differ in their mutated genes, but each tumour had a different pattern of mutations, the number of affected genes suggesting that there may be more steps to cancer than was previously thought. The development of a cancer is progressive, giving rise to a broad spectrum of tissue architecture – from the apparently almost normal (hyperplastic), through metaplastic, dysplastic, premalignant and increasing degrees of aggressiveness and malignancy. It has been common to regard cancer as resulting from disorders of the CELL CYCLE, and most cancer-causing mutations have been discovered by searching among the genes controlling it. But it is now at least as fashionable to perceive it as a failure of apoptosis. Cancer cells appear more susceptible to effects of PROTEASOME inhibitors than do normal cells, and some cancers result from insufficient apoptotic turnover rather than from unrestrained cell proliferation. They also often display gross changes to the organization of the nuclear matrix likely to affect DNA metabolism and subnuclear organization. Programmed cell death protein 4 (PDCD4) normally blocks translation of mRNA on ribosomes and suppresses CELL GROWTH, so that loss of PDCD4 function is expected to result in a growth advantage to cells and ultimately to lead to cancer. It is therefore interesting that PDCD4 is over-expressed in cell cycle-arrested cells but under-expressed in cancer cells and that re-expression of PDCD4 in cancer cells induces their apoptosis and inhibits tumour growth. Not surprisingly, therefore, cancer has often been regarded as a failure of apoptosis.

Most human organs and cell types can develop cancer, producing a huge array of clinical symptoms. However, most cancers arise in tissues (e.g., bone marrow, skin, gut, breast, prostate, muscle and brain) where there exists a developmental cell hierarchy in which a small population of stem cells gives rise to progenitor cells that regenerate mature tissue cells. Most, if not all, tumours contain 'cancer stem cells' (see CANCER CELLS), just as they also involve multiple mutations affecting known *cancer genes*, such as those involved in the RETINOBLASTOMA PROTEIN, p53 (see *p53* GENE) and MYC pathways. Indeed, stem and progenitor cells are maintained in a state favouring their self-renewal (also a feature of cancer cells) by the Wnt and Hedgehog signalling pathways. Human papillomaviruses subvert the p53 system (see VIRUSES).

One outcome of the HUMAN GENOME PROJECT has been the identification of genetic alterations in cancers in unprecedented detail. Some laboratories see large-scale sequencing of cancer genomes as unfocused 'fishing' experiments. But the Sanger Centre of the Wellcome Trust lists over 350 genes, located on every chromosome except the Y, that have been causally implicated in cancer because they have appeared in somatically mutated form (amplified, deleted, translocated or damaged by missense, nonsense or frameshift mutations) in one or more cancer types. In addition, germline mutations associated with cancers have been detected in 66 genes, making them candidates for assessment of GENETIC RISKS of certain cancers. One 2006 study sequenced 13,000 genes in 11 breast and 11 colon cancer samples and found that each tumour contains, on average, the rather large number of 90 mutant genes and that, in the breast tumours, the DNA letter sequence CG was swapped to GC at a high frequency. One 2007 study found almost 120 mutant kinase enzymes that may have roles in human cancers.

A normal cellular gene, such as one involved in responses to GROWTH FACTORS (a proto-oncogene) can be converted into a GAIN-OF-FUNCTION MUTANT (an ONCOGENE) in several ways (see Fig. 33). Point mutation, gene duplication (e.g., see MYC) and chromosomal rearrangements that bring the gene under the control of a strong promoter (e.g., see BURKITT'S LYMPHOMA), may each be responsible. In many types of cancer, tumour cells acquire the ability to express a growth factor not normally expressed by cells of their lineage, although they do express the cognate receptor for this ligand. This creates an autostimulatory autocrine signalling loop as occurs in some invasive breast cancers. Thus, the ErbB oncoprotein releases growth-and-proliferation signals inside the cell that mimic those emitted by the normal ligand-activated EGF receptor, which is probably why cancer cells have a greatly reduced dependence on growth factors for growth and survival. An oncogene-driven cell division cycle would usually trigger DNA replication-associated DNA damage that would normally raise a barrier to sustained proliferation. However, if such barrier mechanisms are mutationally inactivated, full-blown cancer can result. 'Oncogene addiction', the notion that tumour cells become reliant on the continued function of activated oncogenes, is attractive to developers of therapeutic drugs (see CHEMOTHERAPY).

However, newly acquired mutations during the complex ontogeny of a tumour may reduce the dependency on early oncogenic events such as Ras activation (see CANCER CELLS). Although drugs targeting Ras itself have not been developed, there is much interest in developing drugs inhibiting downstream signalling cascades controlled by Ras, such as the RAF-mitogen-activated protein kinase kinase (MEK) pathway (see RAS SIGNALLING PATHWAY, Fig. 145) and the PHOSPHOINOSITOL-3-KINASE (PI3K)-Akt (PKB) pathway (see PROTEIN KINASE B). The pathological association of PKB with the plasma membrane is a common thread that connects it to cancer. Somatic mutations in one TUMOUR SUPPRESSOR GENE, *PI3KCA*, which

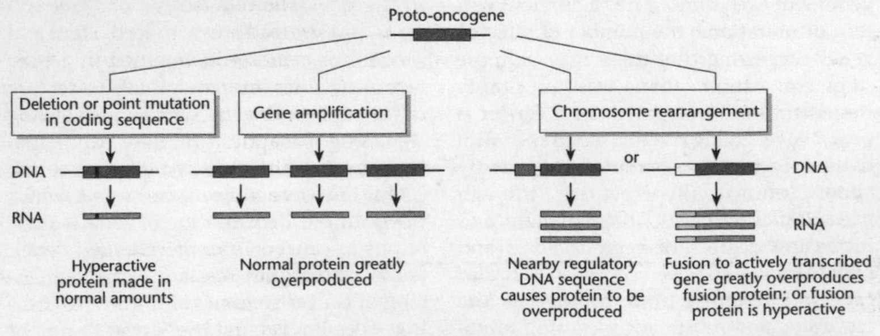

FIG 33. Three ways in which a proto-oncogene can be made overactive, converting it into an oncogene. See CANCER for more details. © 2002 From *Molecular Biology of the Cell* (4th edition) by B. Alberts, A. Johnson, J. Lewis, M. Raff, K. Roberts and P. Walter. Reproduced by permission of Garland Science/Taylor & Francis LLC.

encodes the p110α subunit of PI3K, have been identified in a wide range of cancers. Deletion mutations in another tumour suppressor gene, the *PTEN* (phosphatase and tensin homologue) gene, on chromosome 10, also occur in a wide range of human cancers (see DISEASE, Fig. 64). The tumour suppressor gene *PTEN* encodes a phosphatase that dephosphorylates and reverses the action of PI3K and several companies have drugs inhibiting PI3K in late pre-clinical trials but the therapeutic value of inhibiting the Ras-PI3K-PKB(Akt) signalling pathway is still uncertain. In human breast cancers, loss of responsiveness to anti-HER2 antibody occurs once *PTEN* is mutated, while loss of dependence on the c-*Myc* oncogene in mouse breast tumours results when mutation occurs in *Ras*. Such a 'causal hierarchy' of mutations in different types of tumour places a premium on selecting the most appropriate molecular targets for future multi-agent cancer therapies, because more than one hundred mutated genes occur repeatedly in human cancers and many more await discovery (mutations in the *Ras* oncogenes play a causal role in more than a quarter of human cancers). Such 'cancer-critical genes' contributing causally to cancer are either (i) GAIN-OF-FUNCTION mutations (cellular oncogenes) in previously normally functioning (wild-type) genes (proto-oncogenes); or (ii) loss-of-function mutations in TUMOUR SUPPRESSOR GENES whose function when unmutated (wild-type) is to regulate the normal CELL CYCLE.

Both oncogene and tumour-suppressor gene mutations deregulate the cell cycle, and all operate similarly at the physiological level, driving the neoplastic process by increasing tumour cell number by stimulating cell birth, the inhibition of cell death, or cell cycle arrest. But oncogenes are usually dominant in effect so that only one copy of the gene is needed to drive a cell towards cancer. They can transform cells in culture by TRANSFECTION. By contrast, cancer-causing tumour-suppressor mutations are usually recessive, so that both of a cell's normal copies need to be removed or mutated for cancer to result, which may be achieved in a variety of different ways (e.g., see Fig. 34, and GENE SILENCING). This made the identification of tumour suppressor genes particularly diffi-

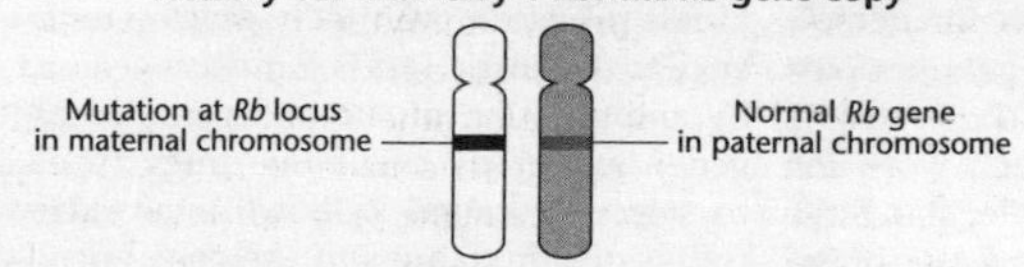

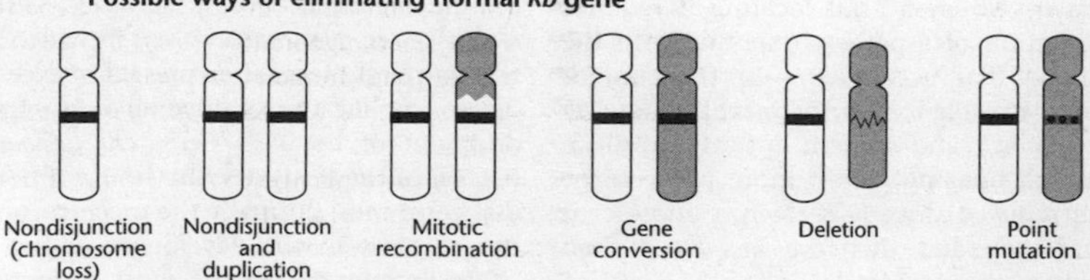

FIG 34. Diagram indicating six ways in which the remaining good copy of a tumour suppressor gene can be lost. A cell that is defective in only one of its two copies of such a gene (e.g., the *Rb* gene; see RETINOBLASTOMA PROTEIN) is normally unaffected, but loss of function of the second allele can drive the cell towards CANCER (see entry). A tumour suppressor gene may also be silenced by epigenetic change, without alteration of its DNA sequence (see CANCER CELLS). © 2002 From *Molecular Biology of the Cell* (4th edition) by B. Alberts, A. Johnson, J. Lewis, M. Raff, K. Roberts and P. Walter. Reproduced by permission of Garland Science/Taylor & Francis LLC.

cult, a major clue to their role being their repeated loss in many independent cases of a particular cancer. In non-heritable cancers, it is possible to compare levels of heterozygosity at common sites of SINGLE NUCLEOTIDE POLYMORPHISMS in cancerous and non-cancerous cells of the same patient – simultaneous loss of such heterozygosity at several linked sites in the cancer cells pointing to the location of a deletion.

But, when mutated, a third class of cancer genes promotes tumorigenesis differently. These stability genes, or caretakers, include the mismatch repair (MMR), nucleotide excision repair (NER) and base-excision repair (BER) genes responsible for DNA REPAIR MECHANISMS. Other stability genes (e.g., BRCA1, *BLM* and ATM) control processes involving larger sections of chromosomes, such as mitotic recombination and chromosome segregation.

So-called *cancer biomarkers* are measures of a wide range of homeostatic variables that indicate 'normality' of a system, and against which abnormal values can be compared. They include patient physiological performance status, mammograms (for breast cancer), specific molecules (e.g., prostate-specific antigen, PSA), gene mutations, gene or protein expression profiles (serum protein electrophoresis for detection of monoclonal immunogobulinopathies) and cell-based markers (e.g., circulating tumour cells). Much work in progress aims to discover molecular biomarkers in the blood betraying the presence of a tumour at the earliest stages of its development; but locating it requires injection of a probe compound into the blood that accumulates in the tumour and is visible to imaging machines (see PET SCANNING), and modern optical and ultrasound imaging permit more probe to be introduced since it is often a fluorescent antibody and therefore less toxic than radioactive probes.

Early precursor lesions (but not normal tissues) of urinary bladder, breast, lung and colon commonly express markers of an activated DNA-damage response (DDR) including the phosphorylated kinases ATM and CHK2, phosphorylated histone H2AX and p53 (see CELL CYCLE). Recently, single nucleotide polymorphisms, genomic profiling, transcriptome and proteome analyses have enlarged the range of biomarkers, leading us away from a single event-based view of cancer pathogenesis to a more system-based approach.

In the USA in 2005, cancer became the leading cause of death for those under 85, and survival rates for patients with the most common cancers remain low: $<10\%$ of patients with metastatic colon cancer and ~5% of those with pancreatic cancer survive 5 years or more (see LIFESTYLE). Use of DNA microarray-based gene expression signatures may enable more rapid definition of cancer subtypes, monitoring of disease recurrence and of response to specific therapies. It may offer the chance of providing a better basis for guiding the use of pathway-specific therapeutic drugs (see RAS SIGNALLING PATHWAY, Fig. 145).

Cancer-causing mutant alleles of certain RECEPTOR TYROSINE KINASES (RTKs) can be transmitted in the human germ line, explaining the origins of a number of familial cancer syndromes in which family members show greatly increased risks of contracting certain cancers. Unfortunately, the majority of patients responding to receptor tyrosine kinase inhibitors eventually develop resistance to the drugs. This is sometimes caused by amplification of the oncogenic kinase gene; but in any event the drugs appear to select for cancer cells with secondary (resistance) mutations in the gene encoding the targeted kinase. Possibly up to 40% of human cancers develop MULTIDRUG RESISTANCE (see ABC TRANSPORTER FAMILY) in which a transporter protein (often present in excess due to amplification of its gene) pumps the drug out of the cell. *Germ cell tumours* (GCTs) of the testis develop from primordial germ cells, normally the precursors of the gametes in the developing embryo. GCTs in humans are the most common malignancy among young adult males between the ages of 20 and 34 years.

Genome sequence analysis in different tumour types has identified specific activating mutations in other tyrosine kinase genes that drive tumour growth, including

BRAF mutations in melanoma and mutations in the Flt3 RTK in one third of acute myeloid leukaemias. The clustering of mutations in the *PI3KCA* gene could make this an excellent marker for early detection of cancers or for monitoring tumour progression. Future research on molecularly targeted therapies will involve the identification of new drugs and drug targets, improved selection of tumours sensitive to them, and the rational design of combination therapies to reduce the occurrence of drug resistance. Although the precise mechanisms are unclear, injection of the BCG vaccine into bladders of patients suffering from early-stage bladder carcinomas often halts or delays the progression of these tumours and functions through its ability to recruit a variety of immunocytes ($CD4^+$ T_H and $CD8^+$ T_C cells, macrophages and NK cells) to the bladder, where they create local inflammatory responses.

The three commonly used immunosuppressant drugs, azathioprine, cyclophosphamide and mycophenolate, were developed originally in cancer treatment. Cancer cells often depend on hyperactive growth factor signalling to generate intracellular anti-apoptotic signals sustaining them so many cancer therapies under development are directed towards activating pro-apoptotic signals within cancer cells. Even in the most aggressive tumours, important components of the apoptotic circuitry remain intact and could be potential targets. Effective therapies might be devised that interfere with the Akt/PKB pathway at one or another upstream signalling cascade.

Ever since the 1990s, tumour immunologists have tried to boost the numbers of tumour-specific T cells in melanoma patients. NK cells and γδ and αβ TCR-bearing T CELLS are both major players in immune surveillance and antitumour immunity – so linking the innate and adaptive immune systems in the arena of cancer immunology. Approaches to cancer treatment using GENE THERAPY and immunotherapy (60% of all approved gene therapy trials) are many and varied. They include gene supplementation to restore tumour suppressor function, gene inactivation to prevent expression of an activated oncogene, genetic manipulation of tumour cells to trigger apoptosis, *ex vivo* modification of tumour cells by introduction of expression vectors to make them express GM-CSF and attract and activate DENDRITIC CELLS (DCs) when reintroduced to the patient, *ex vivo* manipulation of dendritic precursor cells by culturing them with a cocktail of tumour antigen fragments, GM-CSF, TNF and IL-4 prior to reintroduction in the hope of increasing a tumour-specific immune response, transfection of tumour cell mRNA into DCs, use of oncolytic viruses engineered to kill tumour cells selectively, and genetic modification of tumour cells so that they convert a non-toxic prodrug into a toxic compound that selectively kills them and not the surrounding non-tumour cells. One or other of these approaches has increased the lifespans of advanced cancer patients, notably those with advanced melanomas, and it is probable that protocols that use mature rather than immature DCs will be more effective still.

Increasing the number of tumour-targeting T cells can be achieved by vaccination with a tumour-specific antigen or adoptive transfer of tumour-specific lymphocytes that have been grown in culture. Because SELF-TOLERANCE limits the immune response to tumour-associated autoantigens, researchers try, using MOUSE MODELS initially, to introduce genes encoding tumour-specific T-cell receptors into adoptively transferred primary T cells. It is hoped that the vectors used to transfer transgenes will be improved over time to prevent silencing of the transgene. Several malignant cancers occurring in humans also occur in domestic dogs: lymphomas resembling HODGKIN'S DISEASE, osteosarcomas resembling those in adolescents, bladder cancer, melanoma and mouth cancer. They also often metastasize preferentially to the same tissues: canine prostate cancers tend to spread to the skeleton, as in humans. For these and other reasons, dogs are probably more suitable animal models than are mice in advancing several cancer therapies in time for the challenge posed a few years ago by the director of the National Cancer Institute in the USA to 'eliminate the suffering and death caused by cancer by 2015'. See BREAST CANCER, CANCER CELLS, COLORECTAL CANCER, HEALTHY DIET.

cancer cells Cells which have undergone neoplastic TRANSFORMATION. Cells tend to become cancerous only after they have suffered mutations in several genes, including those controlling the mechanisms whereby cells multiply, die and migrate. Although different combinations of mutations are found in different patients, producing cancers that respond differently to treatment, there appears to be a step-by-step mutational progression involved, with mutations inactivating the APC GENE apparently among the first to occur (e.g., see METASTASIS, Fig. 124). The resulting increase in cell proliferation (without changing differentiation pattern) may predispose the cells to *Ras* gene mutation (see RAS SIGNALLING PATHWAY), associated in culture with aspects of transformed phenotype such as loss of anchorage to the substratum and found in ~30% of human tumours. Subsequent mutations in the *DCC* and *p53* genes are associated with malignant transformation, loss of p53 function probably allowing cells to accumulate further mutations by failed DNA REPAIR MECHANISMS so that they now progress through the cell cycle despite being unfit to do so (see CELL CYCLE). The amplification of the human Hdm2 gene in some lung cancer cells gives them an advantage (see P53 GENE). Many types of cancer cell have deregulated the R point decision-making machinery (see G1), or inactivated one or more of their checkpoint controls, and as a result undergo complex genetic recombinations, often involving three or more chromosomes. Because of this GENOMIC INSTABILITY (increased mutability), there are commonly several deletions accumulated in tumour cells. This has been seen as a method by which incipient cancer cells 'try out' various combinations of mutant alleles, accelerating the rate at which advantageous combinations of alleles are acquired and hastening the overall pace of tumour progression. One source of the gross chromosomal derangements seen in tumour cells may be depletion of TELOMERES to the point of crisis (see SENESCENCE (1)), although 85–90% of full-blown metastatic cancers have their telomerase activity upregulated and such cells always acquire the ability to maintain their telomeres and to replicate indefinitely. MICROSATELLITE instability is also a feature of certain colorectal cancers.

It is increasingly clear that cancer development not only depends upon genetic changes but also on abnormal epigenetic mechanisms, or 'cellular memory', including changes in both DNA METHYLATION and HISTONE modification of CpG island-associated PROMOTERS of tumour suppressor genes (e.g., hypermethylation of *RB*). Pioneering work in the 1980s showed that normal DNA methylation patterns that are required for maintaining gene expression states and chromosome stability are severely disrupted in cancer. An early observation in cancer epigenomics was that silencing of a gene can occur via aberrant tumour-specific transcriptionally repressive chromatin or methylation of an adjacent CpG island (see CPG DINUCLEOTIDES). Many cancer/testis (C/T) genes that are normally expressed in the healthy testis are activated in other cells by hypomethylation in cancer. It appears that 10% of the coding genes on the X chromosome are CT antigen genes of one sort or another, indicating a possible advantage to the male sex of these genes. Recently, oestrogen- and tamoxifen-induced activation of *PAX2* and endometrial proliferation was found to be cancer-specific because of *PAX2* hypomethylation in the tumours.

Cancer cells also often display gross changes in the organization of the nuclear matrix. These are likely to affect DNA metabolism and subnuclear organization, and are probably the result of global changes in CHROMATIN packaging. An emerging theme is that an understanding of STEM CELL chromatin control of GENE EXPRESSION may provide clues to the origins of cancer epigenetic changes. One current view is that acquisition through PRC activity (see

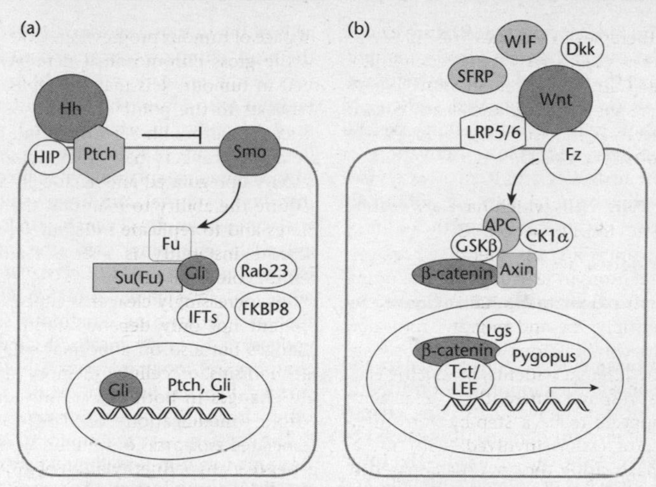

FIG 35. Simplified diagrams of (a) the hedgehog (see HEDGEHOG GENE) and (b) the WNT SIGNALLING PATHWAY (see Fig. 187), emphasizing components implicated in the origin of CANCER CELLS (see entry). Pathway components with primarily positive influences (dark stippling) and negative influences (pale stippling) in pathway activation are indicated, all stippled figures having been causally implicated in tumorigenesis.

Activation of the Hh pathway is initiated by binding of Hh ligand to Ptch, which lifts suppression of Smo and activates a cascade leading to nuclear entry of Gli and activation of target genes. The membrane protein HIP antagonizes pathway activity by binding Hh ligands. The proteins active downstream of Ptch and Smo, regulating Gli, include Fu, Su(Fu), the GTPase Rab23, FKBP8 and the intraflagellar transport proteins, IFTs.

Activation of the Wnt signalling pathway involves binding of Wnt ligands to their receptors Frizzled (Fz) and LRP, leading to release of β-catenin from the degradation complex, its entry into the nucleus, and activation of its target genes by association with TcF/LEF, Legless (Lgs) and Pygopus. SFRP, WIF and Dkk are secreted antagonists of Wnt signalling. From P.A. Beachey, S.S. Karhadkar and D.M. Berman, *Nature* © 2004, with permission from Macmillan Publishers Ltd.

POLYCOMB GROUP) of promoter DNA methylation of repressed developmental genes may lock stem cells into a permanent pluripotent state and initiate abnormal clonal expansion, predisposing to cancer. Tumour heterogeneity, and shared features of normal stem cells and cancer cells, has given rise to the concept of *cancer stem cells*. Such cells result from transformation of a PROGENITOR CELL, leading to activation of its self-renewal. The predisposition of embryonic stem cell PRC2 targets to cancer-specific DNA hypermethylation indicates crosstalk between PRC2 and *de novo* DNA methyltransferases in an early precursor cell with a PRC2 distribution similar to that of EMBRYONIC STEM CELLS. Such crosstalk may occur early in oncogenesis, when the PRC2 distribution resembles that of a stem cell.

Tumorigenesis is a complex and step-by-step process producing a pathologic 'organ' in which host and cancer cells cooperate in the production of an entity which grows, invades locally, and then spreads (metastasizes) and ultimately kills its host (see METASTASIS). Evidence indicates that the mutated *myc* ONCOGENE is required for maintaining the malignant phenotype, and that being turned off leads to apoptosis of malignant osteocytes. The roles of HEDGEHOG (Hh) and Wnt pathways in stem cell SELF-RENEWAL have prompted research on these pathways' roles in initiation and growth of a significant fraction of lethal cancers (see Fig. 35 and DISEASE, Fig. 64). Approximately one-third of total cancer deaths are caused by those cancer types in which malfunctioning of the hedgehog (Hh) or WNT SIGNALLING PATHWAY is implicated, often in progenitor cells of brain, skin, skeletal muscle, liver and colon. Activation of both the hedgehog and ICF pathways (see INSULIN-LIKE GROWTH FACTORS) are potentially supportive of cancer-cell growth.

cancer genes See CANCER.

cancer-testis antigens (CT antigens) See CANCER CELLS.

canine fossa A depression external to, and a little above, the prominence on the surface of the superior maxillary bone, characteristic of modern humans. It is caused by the socket of the canine tooth. The infraorbital surface is orientated coronally and slopes downward and backwards.

canine teeth Male primates often have enlarged canine teeth compared with females. In monkeys and (non-human) apes the upper canine occludes with the lower first premolar (which is sectorial), the lower premolar and upper and lower canines fitting together to form a complex that can hone, or sharpen. A DIASTEMA between the lateral incisor and the upper canine for the occlusion of the lower canine is present in all but prosimians. Modern humans have relatively small canines, with very little SEXUAL DIMORPHISM. See DENTITION.

cannabinoids For centuries, perhaps millennia, marijuana (hashish) has been used to treat epileptic convulsions. Mechoulam identified Δ^9-tetrahydrocannabinol (THC) as the compound responsible for most of the pharmacological activity of marijuana. This, and the other plant-derived cannabinoid receptor agonist, cannabinol, relax hepatic or mesenteric arteries *in vitro* by activating capsaicin-sensitive, CGRP-containing perivascular sensory nerve endings innervating the smooth muscle (see TRP RECEPTORS). The cannabinoid receptors are CB_1 (or, CB1R) within the brain and spinal cord, and CB_2 (or, CB2R) outside the nervous system, and particularly on cells of the immune system. CB_1 is one of the most abundant G-PROTEIN-COUPLED RECEPTORS in the brain, located only on those neurons that release the transmitter GABA, most densely in the cerebral cortex (explaining its psychoactive properties), hippocampus (explaining its effects on memory), hypothalamus (involved in APPETITE), cerebellum, basal ganglia, brain stem, amygdala and spinal cord (the last two explaining its analgesic effects). The challenge to developing cannabinoid-based therapies for treating conditions such as epilepsy is to target localized circuits on the principal forebrain neurons, blocking further hyperexcitability, while sparing other circuits. See FEAR AND ANXIETY.

The capacity of cannabinoids to regulate immune function is well established, suppressing the production of protective cytokines and increasing the production of immunosuppressive cytokines. Work on mice indicates that adhesion of immune cells to the endothelium is reduced by THC (see ATHEROSCLEROSIS). The plant-derived cannabinoid receptor agonists THC and cannabinol relax hepatic or mesenteric arteries *in vitro* by activating capsaicin-sensitive, CGRP-containing perivascular sensory nerve endings innervating the smooth muscle (see TRP RECEPTORS). Cannabidiol, a non-psychotropic constituent of cannabis, is effective in treating seizures in animals, while the ENDOCANNABINOIDS anandamide and 2-AG, and the synthetic cannabinoid HU-211, are neuroprotective.

Aside from the immune effects, smoking marijuana seems to pose health risks similar to those of smoking tobacco (see CIGARETTE SMOKING), and although it is not dangerously addictive mild DEPENDENCE may occur. It has both hallucinatory and depressive actions, producing feelings of euphoria, well-being and relaxation. However, it also impairs short-term MEMORY and cognition, adversely affecting motor coordination – all reversible once the drug is purged from the body. There is also evidence linking heavy cannabis use with symptoms of SCHIZOPHRENIA in predisposed individuals. Through binding CB_1 receptors in the brain it also increases pulse rate, producing an acute rise in blood pressure but sudden falls in blood pressure upon standing, lowering the exercise threshold for angina and is thus a further risk factor in heart attack and stroke.

Cannon–Bard theory See EMOTIONS.

cap, capping See RNA CAPPING.

capacitation An approximately 5-hour period of conditioning, triggered by hydrogen carbonate (HCO_3^-) ions in the vagina, which SPERMATOZOA must undergo within the female reproductive tract (mainly in the oviduct) before they can pass through the corona cells and undergo the acrosome reaction required for FERTILIZATION. HCO_3^- ions enter the sperm and directly activate a soluble adenylyl cyclase. The rising cAMP level activates PKA and results in a series of phosphorylations leading to the removal of the glycoprotein coat and seminal plasma proteins from the plasma membrane overlying the acrosomal region of the spermatozoa, increasing sperm motility and decreasing its membrane potential. Other reactions are triggered by progesterone and other factors which bind to receptor tyrosine kinases. See ACROSOME, CUMULUS CELLS.

capillaries (1) Minute blood vessels, usually connecting arterioles to venules and usually occurring within a few cells of any body cell (other than in the lens and cornea of the eye). The only *exchange vessels* of the blood system, producing TISSUE FLUID, they are especially abundant in metabolically active tissues. Venous blood capillaries are more numerous and more permeable than are arterial capillaries. Produced by ANGIOGENESIS, true capillaries emerge from ARTERIOLES or metarterioles at a precapillary sphincter, these vessels supplying a group of 10–100 capillaries constituting a *capillary bed*. *Continuous capillaries* (7–9 μm in diameter) are continuous tubes of endothelium resting on a covering BASEMENT MEMBRANE, with tight junctions and no gaps between the cells. They are therefore less permeable to large molecules than are other types, occurring in muscle, nerve tissue, and in capillaries of the BLOOD–BRAIN BARRIER, where the tight junctions are especially dense. In *fenestrated capillaries*, pores (70–100 nm in diameter) resembling those in the nuclear envelope occur in the endothelial cells where only a thin porous diaphragm exists, thinner than the normal plasma membrane or absent altogether. Fenestrated capillaries, which are therefore highly permeable, occur in intestinal villi, the ciliary process, CHOROID PLEXUSES and KIDNEY glomeruli. *Sinusoidal capillaries* are larger than the foregoing capillaries, with a less prominent, or absent, basement membrane. Their fenestrae are larger than in fenestrated capillaries, and gaps may occur between the endothelial cells. Highly permeable, they occur in endocrine tissues, while those in the secondary plexus provide blood supply to anterior pituitary. *Sinusoids* are larger in diameter than sinusoidal capillaries, and their basement membrane is also sparse or absent. They are abundant in the liver (where KUPFFER CELLS are closely associated with them) and in BONE MARROW. Capillary modification is a requirement during inflammatory responses (see INFLAMMATION). (2) Capillaries are also a component of the LYMPHATIC SYSTEM.

capsaicin Active ingredient of peppers (capsicum), causing release of substance P from some nociceptors, including some in the mouth (see NOCICEPTION). When applied in large amounts, it can deplete substance P and reduce PAIN (e.g., in shingles).

capsule **1**. A collection of axons (e.g., the INTERNAL CAPSULE) connecting the cerebrum with the brain stem. **2**. An outer covering, as of the LENS of the eye.

carbo-loading diets See EXERCISE.

carbon dioxide (CO_2) A small hydrophobic molecule, and therefore membrane-permeable, produced by decarboxylation of respiratory substrates during the KREBS CYCLE and converted by CARBONIC ANHYDRASE to hydrogen carbonate (see ACID–BASE BALANCE). Its blood concentration is monitored by the CAROTID BODIES and AORTIC BODIES and impinges on the control of both ventilation and blood pressure (see CARDIOVASCULAR CENTRE, RESPIRATORY CENTRE). Compare CARBON MONOXIDE; see HAEMOGLOBIN for the Bohr effect.

carbon monoxide (CO) A small hydrophobic molecule used as an intercellular signal and acting in the same manner as NITRIC OXIDE (both products of CIGARETTE SMOKING) in stimulating guanylyl cyclase (see ENDOTHELIUM). It is produced by haem oxygenase 2, an enzyme activated by calcium ions, and can activate guanylyl cyclase (see OLFACTORY RECEPTOR NEURONS); but it binds porphyrin groups irreversibly, so is a poison of cytochromes and excludes molecular oxygen from haemoglobin and myoglobin. See CARBON DIOXIDE.

carbonic acid/hydrogen carbonate buffer system See ACID–BASE BALANCE.

carbonic anhydrase An enzyme catalysing the reaction between carbon dioxide and water to form carbonic acid. It is found, *inter alia*, in RED BLOOD CELLS, the DISTAL CONVOLUTED TUBULE and COLLECTING DUCT, on the surfaces of capillary endothelial cells, and in PARIETAL CELLS of the stomach. See ACID–BASE BALANCE.

carboxypeptidases Zinc-containing pancreatic enzymes (A and B), released as zymogens (pro-enzymes, procarboxypeptidases) under the influence of CHOLECYSTOKININ and converted to the active enzymes by trypsin and hydrolysing peptides to free amino acids from the C-terminal end. Compare AMINOPEPTIDASE.

carcinogens An agent contributing to TUMOUR formation (see Table 1). Non-mutagenic (non-genotoxic) carcinogens, or 'tumour promoters', were revealed in the 1940s through work on induction of skin cancer in mice, involving exposure of mouse skin to carcinogenic tar constituents such as the polycyclic aromatic hydrocarbons benzo(*a*)pyrene (BP), 7,12-dimethylbenz(*a*)anthracene (DMBA), dibenz(*a*,*h*)anthracene and 3-methylchloranthene (3MC) – all of which arise through

Cancer	*Exposure*
Scrotal carcinoma	Chimney smoke condensates
Liver angiosarcoma	Vinyl chloride
Acute leukaemias	Benzene
Nasal adenocarcinoma	Hardwood dust
Osteosarcoma	Radium
Skin carcinoma	Arsenic
Mesothelioma	Asbestos
Vaginal carcinoma	Diethylstilbestrol
Oral carcinoma	Snuff

TABLE 1 *Forms of cancer associated with exposure to certain* CARCINOGENS.

incomplete combustion of organic compounds. BP and 3MC, in particular, were subsequently found in the condensates from CIGARETTE SMOKING (see CYTOCHROMES for P450-detoxification).

In 1927, Muller found he could induce mutations in fruit flies by exposing them to X-IRRADIATION. By the late 1940s, other chemicals, many of which were ALKYLATING AGENTS of the type used in World War I mustard gas, were found to be mutagenic for fruit flies and also carcinogenic for laboratory animals, prompting the suggestion that such carcinogenic agents as X-rays and some chemicals induced cancer through their ability to mutate genes. Despite the ozone layer, ULTRAVIOLET IRRADIATION from the sun is a far more common source of environmental radiation than are X-rays and the pyrimidine dimers formed are carcinogenic (see DNA REPAIR MECHANISMS).

carcinomas Tumours arising from epithelial tissue and considered benign as long as they remain on the epithelial side of the basement membrane. If they breach this membrane and invade the nearby stroma (see METASTASIS), the resulting neoplastic mass is termed malignant. Malignant carcinomas are responsible for >80% of the cancer-related deaths in the Western world. They include cancers of epithelia in the gastrointestinal tract, skin, mammary gland, pancreas, lung, liver, ovary, gall bladder and urinary bladder. Basal-cell skin carcinomas are the commonest form of cancer in humans and typically

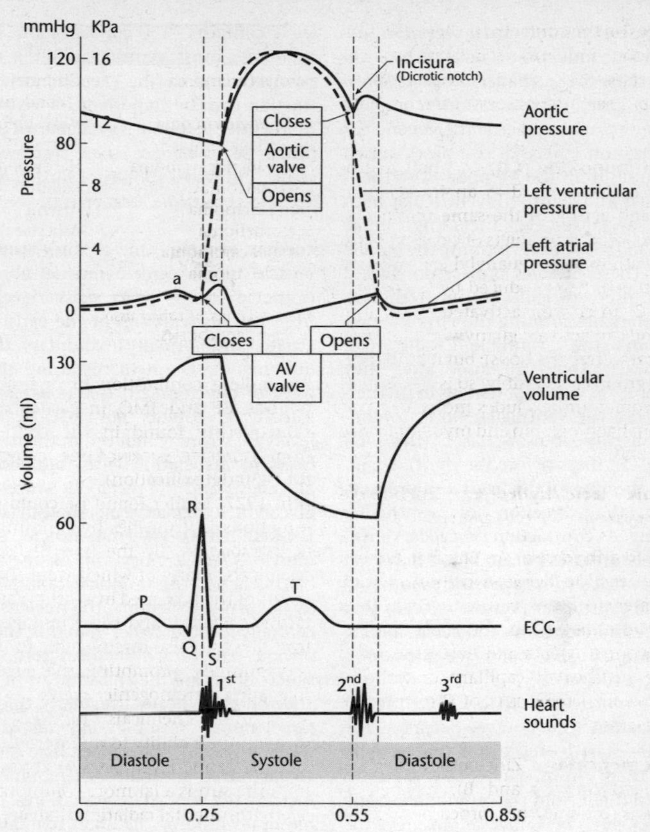

FIG 36. The major mechanical and electrical events during one CARDIAC CYCLE (see entry for explanation), the pressure changes being for the left side of the heart. Accompanying heart sounds and ELECTROCARDIOGRAM tracings are also shown. © 2006, Oxford University Press. Redrawn from *Human Physiology, The Basis of Medicine*, G. Pollock & C.D. Richards, with permission of the publisher.

occur in fair-skinned people over the age of 40. See CANCER; compare SARCOMAS.

cardiac accelerator centre See CARDIOVASCULAR CENTRES.

cardiac arrhythmias See ARRHYTHMIAS, CONGENITAL HEART DEFECTS.

cardiac centres See CARDIOVASCULAR CENTRES.

cardiac cycle One heart cycle, including alternating diastole and systole of both atria and the associated systole and diastole of both ventricles (see Fig. 36). *Systole* is the contraction phase of a heart chamber; *diastole* its relaxation phase. An isolated heart will continue to beat rhythmically if perfused with an appropriate fluid. As an impulse is generated by the sinoatrial node (SAN, or 'pacemaker'; see HEART, Fig. 90; SA NODE), it immediately spreads as a P wave of depolarization (see ELECTROCARDIOGRAM) in all directions through the gap junctions in the intercalated discs of the atrial fibres, reaching the AV NODE situated at the base of the interatrial region in ~50 msec. Here conduction is delayed by ~100 msec because of the smaller diameter fibres, allowing atrial depolarization (complete atrial emptying) to precede the ventricular depolarization (the QRS complex) caused by impulses from the AV node (AVN) activating left and right atrioventricular bundles of specialized cells including cardiac muscle and nerve fibres forming the bundle of His. These bundles project downward into the interventricular septum, terminating in branches forming the Purkinje network in the ventricle walls (see ARRHYTHMIAS). About 50 msec after they leave the AVN, action potentials in this network cause the ventricular fibres to contract slightly out-of-phase (the left starting ~50 msec before the right), beginning at the apex of the heart and proceeding in the direction of ventricular emptying. As contraction proceeds, ventricular pressure rises but no blood flows out of the ventricles (*isovolumetric contraction*) until the ventricular pressure exceeds that in the pulmonary trunk and aorta, forcing the semilunar valves open (see CARDIAC OUTPUT, STROKE VOLUME).

After ventricular systole, ventricular repolarization (the T wave) occurs as the ventricles start to relax and the pressure within them drops. Blood now starts to flow into them from the pulmonary trunk and aorta, but is prevented by closure of the semilunar valves as rebound of blood off the closed cusps produces the *dichrotic wave* on the aortic pressure curve. Soon after this, the atrioventricular (AV) and semilunar valves are all closed and ventricular volume does not change (*isovolumetric relaxation*), although as ventricular pressure drops below atrial pressure the AV valves open and ventricular filling begins as blood that has been flowing into the atria rushes into them through the AV valves. As it does so, their internal pressure rises, stretching the myocardial fibres and placing them under a degree of tension, termed *preload*. In the final third of ventricular filling, the P wave initiates atrial contraction, forcing the final 20–25 cm^3 of blood into the ventricles so that each ventricle contains ~130 cm^3 of blood. At rest therefore, most ventricular filling occurs passively; but during exercise, atrial contraction is more important for ventricular filling since less time is available for passive filling. See BAINBRIDGE REFLEX, CARDIOVASCULAR CENTRE and STROKE VOLUME for further details. See also CONGENITAL HEART DEFECTS.

cardiac muscle One of three vertebrate muscle types, cardiac muscle fibres are restricted to the heart walls, developing from the myocardium of the early heart. Striated and normally involuntary, the cardiac myoblasts originate during the 3rd week of development (see HEART) and are initially glycolytic; but PGC-1α mRNA levels are strongly induced in the neonatal heart in association with mitochondrial biogenesis and the metabolic switch from glycolytic to oxidative metabolism (see PGC-1 GENES). Unlike skeletal muscle, cardiac muscle is not a syncytium of fused myoblasts. Each fibre is uninucleate and limited by its own sarcolemma, the nucleus being centrally located. Other than this, the most obvious structural differences from SKELETAL MUSCLE fibres are the anastomosis of fibres, and the periodically irregularly thickened sarcolemma, forming *intercalated discs* where adjacent fibres meet. The discs are characterized by folded membranes that increase the contact between fibres, by desmosomes holding the fibres together, and by GAP JUNCTIONS (see ARRHYTHMIAS) that serve as areas of low electrical resistance, allowing action potentials to pass between the fibres so that they act as a single unit, essential for the coordinated contractions of the CARDIAC CYCLE.

Cardiac muscle action potentials are between 150 and 300 ms in duration, far longer than those of neurons and skeletal muscle. Atrial and ventricular fibres are the most rapidly activating; the myocytes of the SA NODE (pacemaker), Purkinje fibres and atrioventricular node (AVN) are far more slowly activating. The resting potentials of atrial and ventricular fibres are determined mainly by their permeability to potassium (K^+) ions, the depolarization upstroke being due to a rapid increase in permeability to sodium ions (Na^+), as in neurons and skeletal muscle. Repolarization is initially due to inactivation of these channels and transient opening of K^+-channels, while the long 'plateau phase' of repolarization is due to slowly activated calcium- (Ca^{2+}) channels, increasing membrane permeability to calcium ions. These channels progressively inactivate, repolarizing the membrane to a value approximating to the potassium equilibrium potential.

Cardiac muscle tissue, because of its extended plateau phase of repolarization, has a longer REFRACTORY PERIOD than skeletal muscle and consequently does not fatigue (see EXERCISE). This also ensures that, after contraction, relaxation is nearly complete before another action potential can be initiated, so preventing tetanic contractions. Unlike skeletal muscle, cardiac muscle is not stretched to the point at which it contracts with maximal force, so an increased preload causes cardiac muscle fibres to contract with greater force, increasing STROKE VOLUME. Cardiac muscle fibres hypertrophy during chronic HYPERTENSION (see MICRORNAS) but the reduced efficiency of diffusion of oxygen to the fibres from capillaries will eventually lead to left ventricular failure.

cardiac output (CO) The volume of blood leaving the left (or right) ventricle into the aorta (or pulmonary arteries) per minute, being the product of HEART RATE and STROKE VOLUME. It is also given by the mean arterial BLOOD PRESSURE (MABP) divided by the resistance (R). Sympathetic stimulation can increase heart rate to 250 or, occasionally, 300 bpm, while the increased force of contraction causes a lower end-systolic volume, emptying the heart to a greater extent. This can increase CO by an extra 50–100% of resting values, whereas parasympathetic stimulation can only cause a 10–20% decrease. If CO rises due to increased stroke volume or heart rate, mean arterial blood pressure rises as long as resistance remains steady. See EXERCISE.

cardiac veins The great cardiac vein is the major vessel draining the left side of the heart, the small cardiac vein draining the right margin. They converge toward the posterior part of the coronary sulcus and empty into the coronary sinus, a large venous cavity, which empties into the right atrium. Smaller veins empty into the cardiac veins, the coronary sinus, or directly into the right atrium. See CORONARY ARTERIES.

cardiovascular centre Brain regions involved in control of heart rate and blood pressure, extending from the cerebral cortex to the spinal cord, including the dorsal and ventral medulla oblongata, the hypothalamus, cerebellum and limbic system – although often shown in texts as located in the medulla. The *cardioregulatory centre* in the medulla (see NUCLEUS TRACTUS SOLITARIUS) receives inputs from the baroreceptors in the AORTIC BODIES of the aortic arch (via the vagus nerve) and carotid sinuses (via the glossopharyngeal nerve) indicating the level of systolic blood pressure, and from chemoreceptors in the carotid bodies (via the glossopharyngeal) and aortic bodies (via the vagus), although the chemoreceptors do not normally respond strongly unless blood oxygen tension decreases markedly. It also receives inputs from PROPRIOCEPTORS monitoring joint movements. The cardioregulatory centre has a cardioacceleratory component and a cardioinhibitory component, output from the former leaving along sympathetic fibres from the rostral vasopressor area via the spinal cord, output from the latter leaving along parasympathetic fibres via the dorsal motor nucleus of the vagus and the nucleus ambiguus, decreasing heart rate (see Fig. 37). The *vasomotor centre* in the lower pons and upper medulla is tonically active, sending a continuous low frequency of action potentials through the sympathetic vasoconstrictive fibres and causing the continual partial constriction of peripheral blood vessels known as vasomotor tone. This tone can be increased or decreased by input from the cerebral cortex. A part of the vasomotor centre inhibits vasomotor tone, increasing vasodilation, as when the hypothalamic temperature receptors detect increasing body temperature (see VEINS).

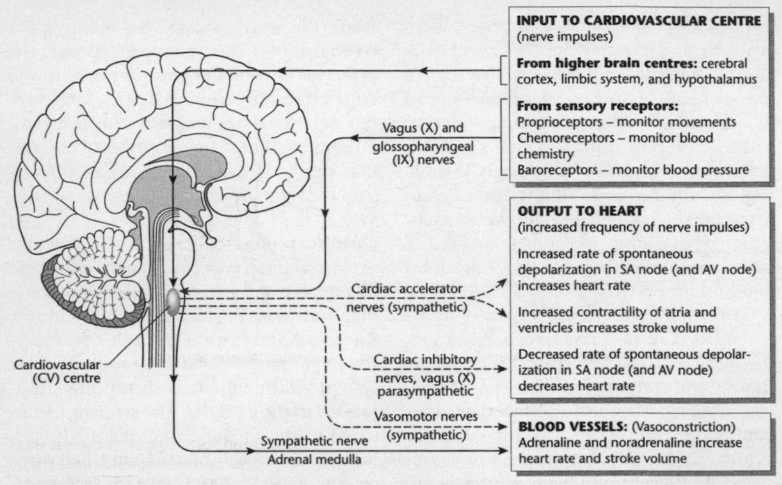

FIG 37. Diagram showing the major role of the CARDIOVASCULAR CENTRE in the medulla oblongata in the control of heart rate and vasomotor tone. Reproduced with modifications from *Principles of Anatomy and Physiology*, G.J. Tortora & S.R. Grabowski, © 2000. Reproduced with permission of John Wiley & Sons, Inc.

In the *baroreceptor reflex*, comprising the *carotid sinus reflex* and *aortic body reflex*, increasing or decreasing systolic BLOOD PRESSURE respectively raise and lower the output of baroreceptors in the carotid sinuses and aortic arch. An increased firing rate on account of increased blood pressure causes the vasomotor centre to decrease its sympathetic stimulation of blood vessels, leading to vasodilation and a fall in blood pressure, while the cardioinhibitory portion of the cardioregulatory centre responds by increasing its parasympathetic stimulation of the heart, leading to decrease in HEART RATE. Conversely, a sudden decrease in blood pressure and reduction in vagal and glossopharyngeal input from baroreceptors to the medulla leads to increased output from its cardioacceleratory centre, resulting in increased sympathetic vasomotor tone and heart rate. Any pressure on the carotid sinus may slow heart rate and cause *carotid sinus syncope*: fainting due to inappropriate stimulation of the carotid sinus baroreceptors.

In the *chemoreceptor reflex* control of blood pressure, a large decrease in blood oxygen concentration (pO_2), increasing blood CO_2, and decreasing pH all lead to increased firing by chemoreceptors in the carotid and aortic bodies and in the medulla itself. The resulting input to the cardioregulatory centre decreases its parasympathetic and increases its sympathetic stimulation of the heart, increasing heart rate and stroke volume, while the vasomotor centre increases peripheral vasoconstriction, thereby increasing peripheral resistance. All of these raise blood pressure. If blood pH increases, as through decrease in pCO_2, the vasomotor centre decreases its sympathetic stimulation of blood vessels, so decreasing peripheral resistance, while the cardioregulatory centre increases its parasympathetic and decreases its sympathetic stimulation of the heart, decreasing CARDIAC OUTPUT. The effect is for blood pH to decrease through raised blood pCO_2, since decreased blood pressure leads to decreased blood flow to the lungs and reduced exhalation of CO_2 (see RESPIRATORY CENTRE).

cardiovascular disease (CVD) Currently the leading cause of death and illness in

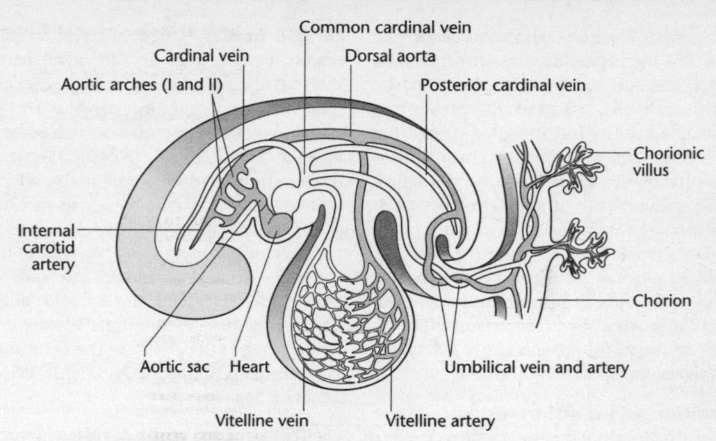

FIG 38. Diagram of the major intra- and extraembryonic arteries (shaded) and veins (open) on the left side of a 4-mm embryo (at the end of the 4th week of DEVELOPMENT). Many of the named structures have their own entries. Redrawn from *Langman's Medical Embryology*, T.W. Sadler. © 2004 Lippincott, Williams & Wilkins, with permission of Wolters Kluwer.

developed countries, CVD will soon become the pre-eminent health problem worldwide, and the single most important contributor to the burden is ATHEROSCLEROSIS (see this entry for causes and therapy, and OBESITY). Its commonest form is CORONARY HEART DISEASE, which is a COMPLEX DISEASE. Many forms of heart disease start with an ischaemic event in which the heart receives inadequate blood supply leading to accumulation of toxic metabolites causing irreparable tissue damage. Women develop heart diseases later in life than do men. Probably the most important factors in preventing clotting in the normal vascular system are (i) smoothness of the endothelial surface, which prevents contact activation of the intrinsic clotting system; (ii) a layer of endothelial glycocalyx which repels clotting factors and platelets; and (iii) the protein thrombomodulin bound to the endothelial membrane which binds thrombin, not only slowing the clotting process by removing thrombin but also activating protein C, a plasma protein that acts as an anticoagulant by inactivating activated Factors V and VIII (see THROMBOSIS). Circadian incidence of CVD is caused by variation in haematogenous factors. The tendency of platelets to aggregate is increased after waking; thrombosis is more likely in the morning, and the efficacy of TISSUE PLASMINOGEN ACTIVATOR and other thrombolytic agents in breaking down clots is lowest in the morning (see CIRCADIAN RHYTHMS). One study published in 2005 revealed that laboratory rats produced by 11 generations of selective breeding for high and low capacity for aerobic EXERCISE, as judged by ability to run on a treadmill (see MAXIMAL OXYGEN CONSUMPTION), had a 350% difference in their running abilities, and that the least able ranked high on the CVD risk factor scale. Compared to high-capacity runners, they were more obese, had higher blood pressure, higher blood lipid levels and increased INSULIN RESISTANCE. The obesity accounted for no more than 20% of the decreased aerobic capacity (see MITOCHONDRION). See CONGENITAL HEART DEFECTS, HYPERTENSION, MYOCARDIAL INFARCTION.

cardiovascular system The HEART and BLOOD VESSELS, appearing in the middle of the third week of development, when diffusion is no longer able to supply the embryo's nutritional needs (see Fig. 38). See CONGENITAL HEART DEFECTS.

caretaker genes Genes helping to maintain genetic integrity, whose mutation can lead to microsatellite or chromosomal instability. Among the 'caretaker', or 'stability', proteins, essential for preventing GENOMIC INSTABILITY and affording protection from oncogenic mutations or chromosome rearrangements, are the non-homologous-end-joining (NHEJ) proteins, notably RAD51 (see DNA REPAIR MECHANISMS, GENOTOXIC STRESS). These repair both environmentally induced random breaks as well as specific breaks introduced during lymphocyte differentiation (e.g., by *V(D)J* recombination). Lack of the NHEJ protein Ku is associated with genomic instability in mice.

carnitine acyl-transferases Enzymes in the mitochondrial membranes involved in transfer of fatty acids into the mitochondrial matrix. L-carnitine is a nitrogenous short-chain fatty acid synthesized in the liver and kidneys from methionine and lysine, ~95% of it located in SKELETAL MUSCLE cells. Dietary carnitine supplements do not appear to increase fatty acid oxidation. See FATTY ACID METABOLISM.

β-carotene A yellow-orange tetraterpene pigment, the principal source of VITAMIN A for humans and other primates and required for synthesis of retinal pigments. It is abundant in carrots, sweet potato, watercress and peas. As with other carotenoids, it has antioxidant properties. It is located in the stratum corneum and fatty areas of the DERMIS and subcutaneous layer.

carotid arteries A pair of arteries developing from the 3rd and 4th AORTIC ARCHES (see Fig. 13). In the late embryonic and adult circulation, the right and left common carotid arteries branch respectively from the brachiocephalic artery and aorta, each dividing into internal and external carotid arteries. The internal carotids supply structures internal to the skull (e.g., most of the cerebrum, the eyeball, pituitary gland, external nose and ear) while the external carotids branch into facial, temporal, occipital and maxillary arteries supplying the skull and meninges and structures external to the skull. See CAROTID BODIES, CAROTID SINUSES, CIRCLE OF WILLIS.

carotid bodies Small bilateral vascular organs in the walls at the bases of the internal carotid arteries containing chemoreceptors; communicating with the RESPIRATORY CENTRE and CARDIOVASCULAR CENTRE via the IXth cranial (glossopharyngeal) nerve; stimulated by a fall in blood pO_2 and to a lesser extent by a rise in blood pCO_2 or a fall in its pH. However, if blood pressure or barometric air pressure is low, the pO_2 in arterial blood can fall low enough to stimulate the carotid bodies and AORTIC BODIES, increasing alveolar ventilation, so that the ability of the respiratory system to eliminate CO_2 is not greatly affected. See EMPHYSEMA.

carotid sinuses, and c.s. reflex Normal systolic blood pressure in the carotid sinus partially stretches the arterial wall and causes baroreceptors in the carotid sinuses to generate a low background frequency of action potentials along the glossopharyngeal nerves to the cardioregulatory and vasomotor centres on the medulla oblongata (see CARDIOVASCULAR CENTRES for *carotid sinus reflex*). See CARDIAC CYCLE.

carpal bones Eight small bones of the WRIST (see Fig. 182).

carrier **1**. In genetics, an individual heterozygous for a recessive condition (see SCREENING). **2**. In disease, an infected individual with no obvious signs of clinical disease who may be a potential source of infection for others. *Acute carriers* are those in whom the carrier state lasts only a short time; *chronic carriers* are individuals who had a clinical disease and recovered, or who have a subclinical infection that has remained inapparent.

CART (cocaine- and amphetamine-regulated transcript) Peptide neurotransmitter released by neurons of the arcuate nucleus of the hypothalamus in response to leptin. See APPETITE.

cartilage With bone, the most important skeletal connective tissue. Cells (chondroblasts) derive from mesenchyme, becoming chondrocytes when surrounded within lacunae by the ground substance they secrete in which the huge proteoglycan aggrecan is a major component. This amorphous matrix (chondrin) contains glycoproteins, basophilic chondroitin and fine collagen fibres, whose varying proportions determine whether the cartilage is: *hyaline*, with a shiny bluish-white ground substance and fine collagen fibres (e.g., ends of long bones, anterior ends of ribs, parts of the larynx, trachea, etc.); *elastic*, containing threadlike network of elastic fibres (e.g., epiglottis, external ear); or *fibrocartilage*, containing bundles of collagen fibres (e.g., pubic symphysis, intervertebral discs). Differentiation of cartilage depends on FIBROBLAST GROWTH FACTORS (see DYSPLASIAS).

cartilage bones See OSSIFICATION.

cascades, cascade reactions biochemical pathways involving a series of enzymes which activate one another in turn, rapidly amplifying the original signal. Examples occur in BLOOD CLOTTING, COMPLEMENT pathways and SIGNAL TRANSDUCTION pathways.

casein The major protein in MILK. It has many phosphoserine groups that bind calcium.

caspases Cysteine-containing aspartate-specific proteases, each rich in cysteines, 14 of which have been identified in humans. When activated, they cleave target proteins ('death substrates') at a specific aspartate, these then initiating fragmentation of DNA (by cleaving an endonuclease-binding protein, ICAD), disruption of organelles, actin filament destruction, and other events characterizing 'cell suicide', or APOPTOSIS. (See also INFLAMMATION). In normal cells, each caspase is present in its inactive zymogen form, two classes being recognized: *initiators* (caspases 8 and 10, activated in response to signals from other proteins) and *executioners* (caspases 3, 6 and 7, cleaved by activators and cleaving other executioner caspases until they are all active). Caspase 1 is involved in processing inflammatory cytokines. Caspase 3 can cleave and activate a Bcl-2-related protein (Bid) in the cytosol, this migrating to the mitochondrion, where it opens the outer membrane channel. Caspase 9 is part of the apoptosome and is activated by release of CYTOCHROME C from mitochondria. It is phosphorylated and inactivated by PKB. See FAS.

CAT scan See CT SCAN.

catabolism, catabolite Catabolism refers to all degradative metabolic reactions, and a catabolite is any substance produced by such reactions.

catalase The enzyme catalysing breakdown of hydrogen peroxide generated during amino acid and fatty acid breakdown within PEROXISOMES.

cataplexy A common symptom of narcolepsy, involving a sudden loss of muscle tone most often triggered by an emotional change, resulting in tendency to collapse on the floor. Sometimes, at least, it is due to deficiency of OREXINS, and treatable with the antidepressant clomipramine.

cataract The leading cause of blindness worldwide, involving increasing lens cloudiness (opacity), and ability to delay this process would reduce cataract-related blindness significantly. However, the molecular basis of the process is as yet poorly understood. Most hereditary cases follow an autosomal dominant pattern of inheritance, although autosomal recessiveness has also been documented and is probably underrecorded. Mutations in connexin genes (See CONNEXINS) have been implicated in some cateracts.

Catarrhini (catarrhines) The infraorder of Primates encompassing the Old World monkeys, apes and humans (catarrhines). In contrast to the PLATYRRHINI, they have nostrils typically facing outward and separated by a narrow nasal septum.

catecholamines Tyrosine-derived hormones (adrenaline, noradrenaline) and neurotransmitters (dopamine, noradrenaline), all containing the catechol group and released as part of the STRESS RESPONSE to a physical or psychological insult, e.g., severe haemorrhage, decrease in blood glucose level, fearful experience, or in anticipation of a competitive or otherwise stressful situation. Adrenaline and noradrenaline are released as a direct response to cholinergic stimulation of the adrenal medulla by sympathetic nerve terminals, depolarization of the chromaffin cells leading to voltage-gated Ca^{2+} channel activation and influx of Ca^{2+}, leading to exocytosis of secretory granules into the interstitial fluid and hence the blood.

Their systemic effects are mediated by binding to cell membrane G PROTEIN-COUPLED RECEPTORS, with effects depending upon the subtype of G protein and the SIGNAL TRANSDUCTION pathway to which it is linked.

As part of the 'flight-or-fight' response, their physiological effects include arousal, increased attention, mood change, sweating, tachycardia, inhibition of gastrointestinal smooth muscle activity, constriction of sphincters and relaxation of uterine muscles. They ensure substrate mobilization from the liver, muscle and adipose tissue by promoting breakdown of glycogen and fat but do not readily cross the BLOOD-BRAIN BARRIER.

Catecholamines are degraded by catechol-*O*-methyltransferase (COMT) at their target cells and can undergo reuptake by extraneuronal sites and degradation by COMT or monoamine oxidase, especially in the liver, producing metabolites metanephrine and normetanephrine. These two enzymes acting on adrenaline and noradrenaline produce the metabolite vanillylmandelic acid, while their action on dopamine produces homovanillic acid. The half-life of catecholamine hormones in the circulation is estimated to range from ~10 to 100 sec, ~60% of them circulating bound with low affinity to albumin. Excreted in the urine, they are important in the clinical detection of tumours that produce them in excess. See PSYCHOACTIVE DRUGS, Fig. 144).

catechol-*O*-methyltransferase (COMT) Enzyme of most cells, including neurons, but especially of liver and kidney, degrading biogenic amines. The concentration of its metabolites in body fluids is a diagnostic clue to the efficiency of drugs affecting synthesis and degradation of biogenic amines (e.g., DOPAMINE, SEROTONIN, NORADRENALINE) in nervous tissue. It is thought to modulate the μ-opioid system that helps control PAIN. See CATECHOLAMINES.

categorization The process by which the brain assigns meaning to sensory stimuli, a task requiring the concerted action of many cortical neurons. Visual categorization for instance, a requirement for visual perception, is much less well understood than are the processes involved in the neural representation of simple visual stimulus features, such as orientation, direction and colour (e.g., see RETINA and its references). Relatively little is known about how we and our brains learn, and encode, the meanings of stimuli. Studies on macaque monkeys, trained to classify 360° of visual motion directions into two discrete categories, implicated the lateral intraparietal (LIP) region in encoding directions of movements according to their category membership whereas the middle temporal (MT) region, although strongly direction-sensitive, carried little if any explicit category information. This suggests that the LIP may be an important nexus for transforming visual direction selectivity into more abstract representations encoding the behavioural relevance of stimuli. See BISTABILITY; EXTRASTRIATE CORTEX, Fig. 75; VISUAL MOTION CENTRE.

γ-catenin, catenins See WNT SIGNALLING PATHWAY.

Catharanthus A genus of flowering plant including the rosy periwinkle, *C. roseus*, native of Madagascar, which is the natural source of two highly effective anti-cancer drugs: vinblastine (used in treatment of Hodgkin's disease) and vincristine (used in cases of acute leukaemia).

cathelicidins A large family of ANTIMICROBIAL PEPTIDES of mammals and other vertebrates, expressed in neutrophils and epithelial cells of many tissues. Studies in mice have demonstrated their importance in innate immune responses to bacterial infections. Additional to their antimicrobial activity, they coordinate such actions as migration of leukocytes and wound-healing. Humans have only one known member, LL37, which contributes to efficient antibacterial defence in psoriasis.

cathepsins Acid proteases involved in antigen processing by ANTIGEN-PRESENTING CELLS. They include the cysteine proteases cathepsins B, D, S and L, the last-named being the most active of those mentioned. Cathepsins S and L may be the most important proteases in the processing of vesicular antigens by antigen-presenting cells.

cation A positively charged ion; e.g., most proteins are cations at physiological pH.

caudal POSTERIOR in position; e.g. applicable to any tail-like structure. Contrast ROSTRAL.

caudal dysgenesis (sirenomelia) Condition in which insufficient mesoderm forms in the most caudal region of the embryo, possibly through genetic abnormality or toxic insult (i.e., from a teratogen). Whereas this mesoderm normally contributes to the formation of the lower limbs, urogenital system (intermediate mesoderm) and lumbosacral vertebrae, in caudal dysgenesis these structures develop abnormally. The condition is associated with maternal diabetes.

caudate nucleus Paired component of the BASAL GANGLIA, lying deep in the brain between the thalamus and cerebral cortex of each CEREBRAL HEMISPHERE and forming, with the putamen, the NEOSTRIATUM (dorsal striatum).

C-banding See HETEROCHROMATIN.

CBP The canonical CREB coactivator.

CCK, CCK-PZ See CHOLECYSTOKININ.

CCL2, CCL3, etc. CC chemokine ligands. See CHEMOKINES.

CCRs CC CHEMOKINE receptors. They include CCR1 and CCR5, present on dendritic cell surfaces. See CXCRS.

CCR5 A receptor for β-cytokines that is a co-receptor with CD4 for entry of HIV into macrophages and their precursors, monocytes. It binds CCL3, CCL4 and CCL5. See HIV; VIRUS, Fig. 177.

CD2, CD4, CD8, etc. (cluster of differentiation) proteins See ACCESSORY MOLECULES.

CD antigens (cluster of differentiation antigens) Cell-surface molecules, including some cytokine receptors, of leukocytes (e.g., see B CELLS, T CELLS). Originally identified by monoclonal antibodies, each is designated by a following number, e.g., CD2, CD4, CD8, etc. See ACCESSORY MOLECULES, DENDRITIC CELLS.

***cdc* genes** See CELL-DIVISION-CYCLE GENES.

cdk A gene encoding a CYCLIN-DEPENDENT KINASE (Cdk, CDK). See CELL CYCLE, CYCLINS.

Cdks, CDKs (cyclin-dependent kinases) See CELL CYCLE, CYCLINS.

cDNA See COMPLEMENTARY DNA.

CDR1, CDR2, CDR3 The three complementarity-determining regions in each V domain of IMMUNOGLOBULINS and T-CELL RECEPTORS which are the most variable part of the molecule and together determine its antigen-binding specificity.

celiac disease See COELIAC DISEASE.

cell adhesion molecules (CAMs) Proteins involved in holding cells together and maintaining tissue integrity at INTERCELLULAR JUNCTIONS (see Fig. 103) and at contacts between cells and the EXTRACELLULAR MATRIX (see BASAL LAMINA). Most, especially those on leukocytes, were named after the effects of specific monoclonal antibodies directed against them. Only later were these characterized by gene cloning, so their names bear no relation to their structure. Those whose activity is Ca^{2+}- (or sometimes Mg^{2+}-) dependent include CADHERINS, SELECTINS and INTEGRINS. Members of the immunoglobulin superfamily of proteins involved in cell adhesion (e.g., ICAMs) do not depend on these cations, and include the neural cell adhesion molecules, N-CAMs. See INTERCELLULAR ADHESION MOLECULES, NON-CODING DNA.

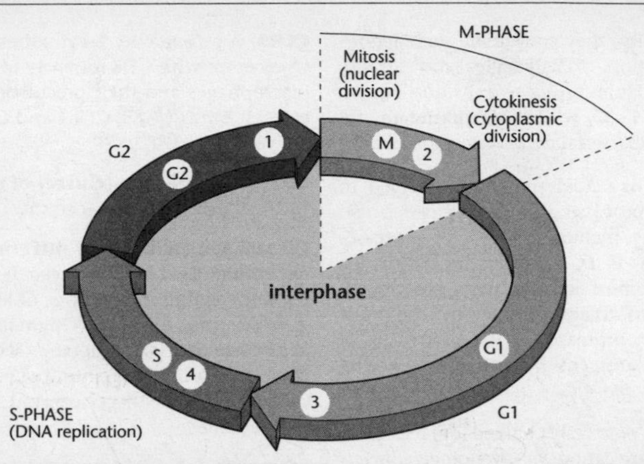

FIG 39. The phases of the CELL CYCLE (G1, S, G2 and M) and the positions of checkpoints within it (numbered). Entrance into M is blocked (1) if DNA replication is not completed; anaphase is blocked (2) if chromatid assembly on the mitotic spindle is not successful; entrance into S is blocked (3) if the genome is damaged, and DNA replication is halted (4) if subsequent DNA damage occurs. See Fig. 41 for molecular details of (3) → (1). © 2002 From *Molecular Biology of the Cell* (4th edition) by B. Alberts, A. Johnson, J. Lewis, M. Raff, K. Roberts and P. Walter. Reproduced by permission of Garland Science/ Taylor & Francis LLC.

cell cycle (cell-division cycle) The complex cycle of growth, content duplication and division which all eukaryotic cells undergo if the two daughter cells are to receive a copy of the entire genome. A gene controlling a specific step, or set of steps, in the cell cycle is a cell-division-cycle gene (*cdc* gene). Circadian periodicity of cell-division rates has been demonstrated in cells derived from several tissues in the rat, and genes seemingly involved in the timing of cellular activity have been identified (see CIRCADIAN CLOCKS). The cell cycle is responsive to signal molecules, e.g., growth factors, which operate via SIGNAL TRANSDUCTION pathways (see ONCOGENES). It involves an ordered series of biochemical control switches overseeing the preparation for, and successful completion of, DNA replication and the segregation of replicated chromosomes (see Fig. 39).

DNA replication is confined to S-phase; G1 is the gap between M-phase (mitosis and cytokinesis) and S-phase, during which a cell may exit the cell cycle or continue dividing; G2 is the gap between S-phase and M-phase, during which a cell typically grows and prepares for division. The cell cycle machinery orchestrates the precise replication of the genome and cell division, the main driving force being the family of cyclin-dependent kinases (CDKs), enzymes that phosphorylate a number of cellular proteins that permit progression through the cycle's several sequential transitions. CDKs also coordinate this progression with timely responses to DNA-damaging events that threaten genomic integrity and that can result in diseases such as CANCER.

The most vulnerable stage of the cell cycle is S-phase, during which the nuclear genome must be faithfully replicated (see DNA REPAIR MECHANISMS). The cell grows continuously throughout interphase, during which two critical processes must be successfully completed before M-phase (MITOSIS) can begin: (i) replication of nuclear DNA, and (ii) duplication of the CENTROSOME. Monitoring systems termed *checkpoint pathways* ensure that unless this is achieved, progression is arrested. Checkpoints play a critical role in the DNA damage response system, providing an

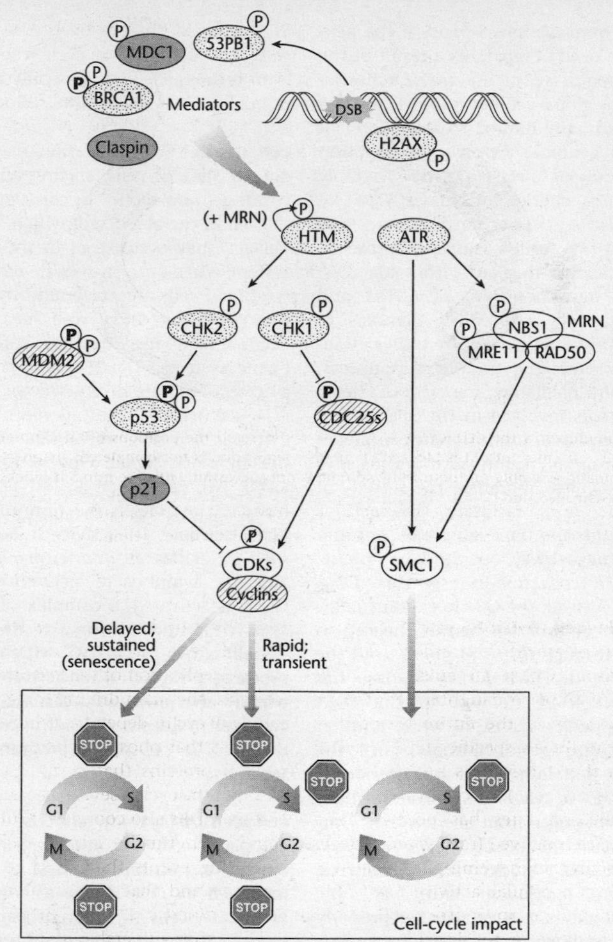

FIG 40. Simplified molecular details of CELL CYCLE checkpoints (3) → (1) shown in Fig. 40 in response to double-strand breaks (DSBs) in the nuclear DNA. Hatched ovals are proto-oncogene proteins; stippled ovals are tumour suppressor proteins. An encircled P indicates that a protein has been phosphorylated, either directly (bold P) via ATM and ATR, or indirectly via transducing kinases CHK2 and CHK1 (regular P). Dephosphorylation is indicated by a line struck through P. Arrows indicate positive influences; lines ending in T-junctions indicate inhibitory influences. Duration of the blockade on cell cycle progression mediated by p53 may be either transient, or permanent (senescence). The multiprotein complex formed from meiotic recombination 11 (MRE11) and RAD50/NBS1 (MRN) contribute to ATM activity after ionizing radiation. From M.B. Kastan and J. Bartek. *Nature* © 2004, with permission from Macmillan Publishers Ltd.

opportunity to monitor the appropriateness of cell suicide (apoptosis) over DNA REPAIR MECHANISMS and there are many points of contact between apoptotic and checkpoint programmes. There are also systems, *checkpoint controls*, ensuring that successful completion of a particular step occurs only once during a cell cycle. There is no progression from G1 into S-phase if the genome is in need of repair, as a result of the G1 checkpoint (see G1, P53 GENE). Orchestration of the global cellular response to DNA damage (usually limited to a few sites in the vast genome) involves cooperation between TUMOUR SUPPRESSOR PROTEINS, notably the proximal checkpoint kinases ATM (see ATAXIA TELANGIECTASIA MUTATED PROTEIN) and ATR, and two further classes of proteins, the checkpoint mediators (or, adaptors) and the transducer kinases, CHK1 and CHK2 (see Fig. 40). Most members of these two protein groups are either established or emerging tumour suppressors. Checkpoint mediators, the class of checkpoint factors involved in the spatio-temporal assembly of multiprotein complexes in the chromatin regions surrounding sites of DNA damage, modulate the activity of ATM/ATR and facilitate interactions between this and their substrates. Another checkpoint pathway, operating in S-phase, slows DNA replication in response to DNA damage. A third, the G2 checkpoint pathway (see G2), will not permit the cell to proceed through G2 to M-phase until the replication of DNA during S-phase has been completed, damage to the DNA blocking nuclear division, while another prevents entry into M-phase until the two DNA double helices forming the two chromatids of each chromosome have been disentangled from one another. During M-phase, the *spindle attachment checkpoint* control blocks sister-chromatid separation (anaphase) until all the chromosomes have been attached properly to the spindle apparatus (see MITOSIS). It is thought to derive from sensors on kinetochores themselves, which indicate either unattached kinetochores or those that are not under the proper tension resulting from bipolar attachment to kinetochore microtubules.

Very early embryonic cells are able to proliferate without receiving external mitogenic signals. This is probably because cyclin E levels are constantly high during the cell cycle of early embryos, effectively dispensing with the R point and allowing Cdk2 to initiate DNA replication in S-phase as soon as M-phase is over. Also in these cells, the centrosome cycle can operate even if the nucleus is physically removed, or nuclear DNA replication is blocked.

cell death Self-destruction may be the default state of cells, staying alive being contingent on receipt of constant SURVIVAL SIGNALS, in the absence of which 'death by neglect' may occur (e.g., in the immune system, when antigen-specific receptors of lymphoid cells are not bound by ligand). Alternatively, 'active cell death' may occur, involving death-receptor/death-ligand systems. 'Death receptors' include a subset of tumour necrosis factor receptors (TNF-Rs) whose 'death domain' on the cytosolic side of the membrane transduces the apoptotic signal. One such receptor, CD95 (whose natural ligand is CD95L), has a major role in the immune system and elsewhere. Its activity is boosted by such cytokines as interferon-γ and TNF and by lymphocyte activation. Once CD95 is activated, a complex of proteins associates with it to form a death-inducing signalling complex (DISC) which includes procaspase-8 which when activated in turn releases the proteolytic caspase-8, which activates procaspase-3 leading to completion of the cell death programme. See APOPTOSIS.

cell division (cytokinesis) The division of cytoplasm of a post-mitotic cell (see CELL CYCLE). After an initial period of cleavage, in which all cells divide at roughly the same rate, cells in different parts of the embryo start to divide at different rates. Rapid cell division in special parts of the lateral plate mesoderm give rise to limb buds, while adjacent regions divide more slowly and do not do so. The plane of cell division makes a big difference to the outcome: divisions perpendicular to the plane of an epithelial sheet cause the sheet to expand laterally by addition of new cells, but divisions in the same plane as the sheet will generate additional layers. If the cleavage plane is not medial, cells of different sizes are generated (see CELL GROWTH). Forty-four new cell generations (i.e., generations of cell division) are involved in the production of the human organism, although STEM CELLS and PROGENITOR CELLS continue to divide in the adult.

cell-division-cycle genes (*cdc* **genes**) Genes controlling a step, or set of steps, in the CELL CYCLE. They were originally discovered in yeasts.

cell growth The accumulation of macromolecular constituents within a cell which drives an increase in cell size. The phrase sometimes also implies the cell's subsequent mitosis and CELL DIVISION, the two processes together yielding cell proliferation. Animal cells, including human, can uncouple cell growth from cell division, and generally grow and divide only when stimulated to do so by signals from other cells, the cell size at which they do so depending to some extent on these signals. Cells such as neurons and muscle fibres continue to grow long after they have withdrawn from the CELL CYCLE. Production by a cell of proteins is a prerequisite for cell growth and proliferation, and of the many cellular signalling pathways that respond to MITOGENS, other GROWTH FACTORS and HORMONES have been elucidated, the most prominent being the *Ras* signalling PI3K–Akt–mTOR pathway (Akt, also known as PKB, and mTOR both being serine-threonine kinases; see MTOR, RAS SIGNALLING PATHWAYS). In its unphosphorylated state, the programmed cell death protein PDCD4 normally blocks protein synthesis at the level of TRANSLATION of mRNA on the ribosomes by inhibiting the RNA helicase activity of eIF4A and preventing its incorporation into the eIF4F complex required for translation of mRNA at the ribosome. PDCD4 is over-expressed in cell cycle-arrested cells but under-expressed in CANCER CELLS. In the absence of growth factors, PDCD4 remains unphosphorylated and inhibits eIF4A, protein synthesis and cell growth. However, in response to the activation of the mTOR pathway by growth factors, S6 kinase becomes phosphorylated and phosphorylates PDCD4, in which state PDCD4 is marked out for ubiquitination and so is degraded by the cell's proteasomes. This frees eIF4A to initiate cap-dependent initiation of mRNA translation and subsequent cell growth.

cell inclusion A largely organic intracellular cell product which is microscopically visible but not enclosed by a membrane and whose presence may be erratic; e.g. glycogen granules.

cell-mediated immunity A phrase indicating the prominent role of T CELLS in an adaptive immune response. These depend upon direct interactions between T lymphocytes and cells bearing the antigen they recognize, those of cytotoxic T cells being the most direct. Compare HUMORAL IMMUNE RESPONSE.

cell membranes Cell membranes are the major compartmentalizing systems in biology. Not only are cells limited and individuated by the plasma membrane (*the* cell membrane, interacting with the CYTOSKELETON and EXTRACELLULAR MATRIX), but the several membrane-bound organelles are permitted their unique identities because of cell membranes.

The fluid mosaic model of membrane structure developed by Singer and Nicholson in 1972 (see Fig. 41a) has become the standard conceptualization of biological membrane structure. Its important features include an unperturbed but fluid phospholipid bilayer, exposed on either side to the aqueous environment, and in which monomeric proteins are dispersed unencumbered and at low concentrations. Over subsequent years it has become apparent that molecules in the membrane have preferred associations and that membranes are typically crowded (see Fig. 41b), their bilayers varying considerably in thickness in a manner reflecting different degrees of orderliness of packing of the membrane's lipid components. Further, membrane proteins are usually oligomeric, most being hetero-oligomeric.

Sphingolipids are the second largest class of membrane lipids, and like phospholipids have a polar head group and two non-polar tails. But instead of glycerol, the long-chain amino alcohol sphingosine is the core, with a number of possible groups esterified to C1: phosphocholine or phosphoethanolamine (in sphingomyelins); glucose (in glycosphingolipids); a di-, tri- or tetrasaccharide (e.g., lactosylceramide); or a complex oligosaccharide (gangliosides). Both phospholipids and sphingolipids can be metabolized to LYSOPHOSPHOLIPIDS. A third class of membrane lipids is the glycerophospholipids (aka phosphoglycerides), in which two fatty acids are attached in ester linkage to C1 and C2 of glycerol while a highly charged polar (alcohol) or charged group is attached through a phosphodiester bond to C3. Each derivative is named for the head group alcohol, with the prefix 'phosphatidyl-'. Thus, phosphatidylcholine and phosphatidylethanolamine have choline and ethanolamine in their polar head groups. The common glycerophospholipids are DIACYLGLYCEROLS linked to head-group alcohols through a phosphodiester bond. Lipids also tend to group together, forming lipid–lipid and lipid–protein interactions (see CHOLESTEROL). Many experimental observations support patchiness of membranes as a principle, changes in lipid organization having significant effects on cell functions, such as SIGNAL TRANSDUCTION (e.g., PHOSPHATIDYLINOSITOL SIGNALLING SYSTEM) and membrane trafficking (see VESICLES).

Membrane lipids have both hydrophilic and hydrophobic poles that determine their positions within the bilayer. Their headgroups are the attachment sites for peripheral membrane proteins, crucial in cell–cell adhesion, signal transduction, and in generating membrane curvature. Different organelles within a cell have distinct lipid compositions and some regions of the plasma membrane have high cholesterol concentrations: the outer leaflet has especially high levels of sphingomyelin and glycosphingolipids, consistent with a highly ordered state. Such *ordered lipid phases* (l_o) are characterized by relatively elongated saturated fatty acyl side chains that pack tightly. Such dynamic LIPID RAFTS are relatively resistant to detergent, the outer leaflet forming 'microdomains'

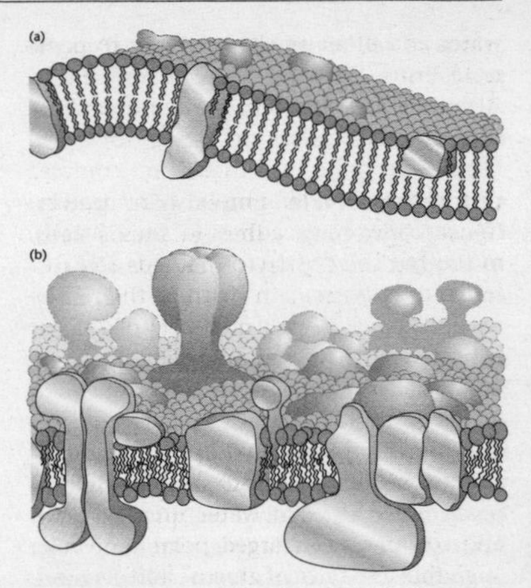

FIG 41. Diagrams indicating two 'fluid-mosaic' models of CELL MEMBRANE structure, in which a phospholipid bilayer is punctuated by protein molecules (shaded). (a) The earlier, Singer-Nicholson model; (b) an amended and updated version of this. From D.E. Engelman. *Nature* © 2005, with permission from Macmillan Publishers Ltd.

enriched with lipids additional to cholesterol (esp. sphingolipids), making it more rigid and impermeable. This also favours particular protein components, viz. those proteins attached post-translationally to lipids, such as to glycosylphosphatidylinositol (GPI). Receptor-associated Src-family kinases involved in immune-cell signalling are located in lipid rafts, as are many other signal transduction proteins – particularly those that cross-link. The antigen receptors of B and T cells migrate into lipid rafts but are stabilized there only after they cross-link, when they can associate with kinases.

At the opposite extreme, the endoplasmic reticulum (ER) is low in cholesterol, but has a large content of unsaturated lipids, consistent with a disordered membrane structure. Such *disordered lipid phases* (l_d) are characterized by rapid diffusion in the plane of the membrane, poorly ordered hydrophobic interiors and reduced membrane thickness. Lipids with unsaturated fatty acids, with kinks in their acyl chains, favour this disorganized tendency and the relatively loose packing of the atoms allows water and other small molecules to penetrate into the bilayer relatively easily. Although most membranes exhibit fluidity, rapid diffusion occurring for some lipids and proteins, there are complex restrictions on lateral mobility of proteins in particular (e.g., adhesion sites, attachments to the CYTOSKELETON). Lipids and proteins usually remain within their own monolayer, although members of the ABC TRANSPORTER FAMILY can 'flip' certain lipids from one side of the membrane to the other.

The membrane's hydrophobic interior acts as a barrier to ions, to small uncharged polar molecules (e.g., water, urea, glycerol) and to larger uncharged polar molecules (e.g., those the size of glucose and sucrose). However, this interior is permeable to small hydrophobic molecules such as O_2 and CO_2, and to lipid-soluble molecules such as alcohols and lipids (e.g., the lipid-soluble components of cholesterol, vitamin E and retinal pigments). The cytosol is a reducing environment, which decreases the probability that interchain disulphide (S–S) bonds will form between cysteine residues in proteins on the cytosolic side of the membrane; but these do form on the non-cytosolic side where they can influence stabilizing the folded structure of the membrane protein and its association with other polypeptides. Those portions of intrinsic and transmembrane proteins that lie within the hydrophobic membrane interior require hydrophobic amino acyl R groups to be positioned appropriately in their tertiary structures, often achieved post-insertionally. Hydrophobic and basic residues in these proteins often create a local positive electrostatic potential acting as a strong attraction for multivalent PIP_2, one of the most important membrane lipids. Ca^{2+}/calmodulin (see CALMODULINS) binds to such basic protein clusters and releases PIP_2 from the membrane (see PHOSPHATIDYLINOSITOL SYSTEM).

Membrane fluidity is essential for such cellular processes as membrane fusion (as in endocytosis, exocytosis and fertilization), and membrane folding is crucial in the formation of cilia and microvilli, and in membrane trafficking (see VESICLES, PHOSPHATIDYLINOSITOL SIGNALLING SYSTEM). At the very least, membrane lipids have permissive roles in these events; but more often, lipid composition is altered to accommodate protein and lipid movement from organelle to organelle in transport vesicles. Many organelles have specific GTPases that switch between inactive cytosolic and active membrane-associated forms, anchoring reversibly to the membrane by lipid groups. Rab GTPases and Arf GTPases (see ARF PROTEINS) have a central role in recruiting peripheral membrane proteins to organelles. Membrane proteins very often have ectodomains that bind signal molecules or other proteins/glycoproteins and without which the known functions of the neuroendocrine and immune systems, let alone development itself, would be utterly impossible.

Although the roles of cell membranes in DISEASE vary with the disorder (see Fig. 64), cell surface receptors involved in SIGNAL TRANSDUCTION are often implicated. In humans, one very significant factor affecting cell membranes is dietary uptake of cholesterol and fats, which are delivered to cells throughout the body by lipoproteins and subsequent receptor-mediated endocytosis.

cell shape See CYTOSKELETON.

cell signalling See RECEPTORS, SECOND MESSENGERS, SIGNAL TRANSDUCTION, SYNAPSES.

cell size See CELL GROWTH.

cell turnover The ratio of production to loss of cells varies with different tissues; it is low in mature muscle and nerve tissue, but high in proliferative tissues such as the mucosal epithelium of the alimentary canal, buccal cavity and epidermis. See CELL CYCLE.

cellular memory **1**. Heritable patterns of GENE EXPRESSION, passed on to daughter cells and cell lineages. Epigenetic mechanisms can ensure stable inheritance of a transcriptionally active state ('open' chromatin formation) for certain target genes or genome regions, or organize the chromatin of certain regions to adopt a highly condensed, transcriptionally inactive form. See EPIGENESIS. **2**. Cells may respond over the medium or long term to more transient signals, such as changes of SYNAPTIC STRENGTH, also sometimes regarded as a form of cell memory.

cellular oncogene (c-*onc*) See CANCER, CANCER CELLS, ONCOGENE.

cellular stress See STRESS (2).

cellulose A major constituent of plant cell walls, and therefore a major non-starch polysaccharide of the diet contributing to its fibre content. See GUT FLORA.

cement (cementum) Calcified tissue covering the root of a tooth.

centimorgan (cM) The distance unit employed in genetic (linkage) mapping. As a rule of thumb, 1 cM = ~1 Mb.

central canal (i) Microscopic canal running in the grey commissure component of the spinal cord and containing cerebrospinal fluid; (ii) a circular channel (Haversian canal) running through the centre of an osteon in compact bone, containing blood and lymphatic vessels and nerves.

central death rate See MORTALITY.

central nervous system (CNS) The BRAIN and SPINAL CORD. See NMDA (N-METHYL D ASPARTATE) RECEPTORS.

central pattern generators (CPGs, motor pattern generators, MPGs) Neural networks giving rise autonomously to rhythmic motor activity. The simplest pattern generators are single neurons whose membrane properties give them PACEMAKER properties. Widespread among invertebrates, CPGs have been shown to underlie many patterns of behaviour in vertebrates, e.g., alternate activation of flexors and extensors in walking, in swimming and the inspiratory–expiratory cycle of VENTILATION (see RESPIRATORY CENTRE). There is little evidence that any single neuron (i.e., a command neuron) in humans is able to exert control over a complex behaviour and its motor programmes. The circuit for the coordinated *control of walking*, or running, resides within the spinal cord and requires CPGs to generate the required rhythmic outputs to generate appropriate cycles of activity in which groups of muscles contract and relax in precisely controlled sequences. Each limb has an array of CPGs, each being an oscillator with two 'half-centres', one driving flexors, the other driving extensors. These motor reflexes do not operate in isolation but are continually modified by proprioceptor input from muscles and joints, and from the skin, enabling a smooth execution of motor commands (see INTENTIONAL MOVEMENTS, MOTOR CORTEX). According to one scheme, walking is initiated by activity in the MESENCEPHALIC LOCOMOTOR REGION (MLR), which projects to reticular nuclei in the medulla. These excitatory neurons release glutamate, producing a large depolarization of the CPG neurons, which then produce oscillating output for as long as the MLR input continues. This steady input along descending motor axons in the RETICULOSPINAL TRACTS (see SPINAL CORD, Fig. 168) excites pairs of interneurons synapsing with the α motor neurons controlling, respectively, FLEXOR MUSCLES and EXTENSOR MUSCLES. The interneurons respond to continuous input by alternately generating bursts of action potentials. The alternating rhythmicity arises through their reciprocal inhibition via other (inhibitory) glycinergic interneurons. GLUTAMATE is the main excitatory neurotransmitter in the spinal cord. Binding of one interneuron's NMDA RECEPTOR by glutamate causes entry of calcium ions, generating a quick burst of spikes. Rising intracellular $[Ca^{2+}]$ causes Ca^{2+}-activated potassium (K^+)-channels to open and K^+ ions to leave the cell. The resulting hyperpolarization stops further spikes and allows magnesium ions (Mg^{2+}) to enter and clog the NMDA channel, preventing further entry of Ca^{2+}, causing the K^+-channels to close and resetting the membrane for the next wave of oscillatory spikes in the presence of glutamate. Thus a burst of activity in one of the excitatory interneurons strongly inhibits the other, so that flexion on one side is accompanied by extension on the other. Addition of further interneuronal connections between the lumbar and cervical spinal segments can account for the swinging of the arms that accompanies walking. The basic locomotor rhythms of the CPGs are extensively modified by the supraspinal motor systems involving descending pathways in the CORTICOSPINAL TRACTS which generate INTENTIONAL MOVEMENT. The ventral SPINOCEREBELLAR TRACT carries information regarding the state of the spinal cord CPGs involved in locomotion to the CEREBELLUM. See GAIT, PARASOMNIAS, SLEEP (for sleepwalking).

central sulcus The sulcus in the cerebrum dividing the frontal lobe from the parietal lobe (see CEREBRAL HEMISPHERES, Fig. 47), separating the somatosensory from the motor cortex. The presence or absence of this structure distinguishes the brains of monkeys from those of prosimians, and its presence on the endocast of the Oligocene anthropoid *Aegyptopithecus* indicates that this fossil represents an early monkey.

central tolerance Tolerance of T cells to self-antigens. See T CELLS.

centre-surround receptive field A visual RECEPTIVE FIELD with a circular centre region and a surrounding region forming a ring around it whereby stimulation of the centre produces a response opposite to that generated by stimulation of the surround. See RETINA.

centric fusion See ROBERTSONIAN FUSION.

centrioles A hollow cylinder composed of nine sets of triplet MICROTUBULES held together by accessory proteins, each 300–500 nm long and 150 nm in diameter. Derived from the sperm cell at fertilization, the two members of a pair of centrioles each lie within a CENTROSOME, at right angles, near the nucleus and separate during early prophase. Centriole duplication begins in G1 and is completed by G2 of the CELL CYCLE.

centromere A constricted chromosomal region holding sister chromatids together until mitotic or 2nd meiotic anaphase. It is the site on the DNA where the kinetochore forms that captures microtubules from the mitotic/meiotic spindle. The surrounding chromatin consists of pericentric HETEROCHROMATIN.

centrosome The main microtubule-organizing centre (MTOC) in animal, and hence human, cells. It consists of a matrix of amorphous material surrounding a pair of CENTRIOLES (a mother and a daughter) lying in the cytoplasm close to the nuclear envelope and Golgi bodies. In interphase, microtubules point outwardly from the matrix, with their fast-growing plus (+) ends projecting to the cell's perimeter and their minus (–) ends associated with the centrosome. The matrix contains the important γ-tubulin ring complex which is the major factor responsible for nucleating microtubules. A procentriole is formed next to each existing centriole in S-phase, and during mitosis each centriole recruits pericentriolar material to form one of the poles of the spindle. The centrosome complex then splits, each centriole pair becoming part of a separate MTOC that nucleates a radial array of microtubules called an *aster* from opposite sides of the nucleus from which the spindle develops.

centrum The body of a vertebra (see VERTEBRAE).

cephalic flexure (mesencephalic flexure) A bend in front of the future midbrain region (between this and the forebrain) at a point opposite the front end of the notochord. It lies anatomically just anterior to the first SOMITE.

cephalic index The ratio of the maximum width of the head to its maximum length (i.e., in the horizontal plane, or front to back), sometimes multiplied by 100 for convenience. It was once widely used to categorize human populations, but is no longer used for that purpose except for describing individuals' appearances, and has no correlation with, or connotations concerning, intellect or behaviour. It is still sometimes used for estimating the age of foetuses for legal and obstetrical reasons, when differences in skull shapes between different populations are still of interest.

Cercopithecoidea Old World monkeys. See ANTHROPOIDEA.

cerebellum Developing from the metencephalon component of the hindbrain, the cerebellum comprises the cerebellar cortex and the deep cerebellar nuclei. It is involved in the learning and production of accurate and coordinated movements initiated by the MOTOR CORTEX. It integrates the originally separate motor components involved in learning new multijoint movement skills, such as juggling or throwing a ball accurately, into a seamless action that can be performed almost unconsciously – creating new motor programmes. It can operate in feedback or feedforward mode.

Although much smaller, the densely packed cerebellum contains as many neurons as both cerebral hemispheres combined. Its two hemispheres – not as obviously separated as the two cerebral hemispheres – comprise a thin superficial cortex of grey matter above a thicker zone of white matter (see Fig. 42) and are attached to the pons by three paired peduncles of white matter. The midline from which they emerge is recognizable merely as an enlarged 'bump', the *vermis*, forming the organ's longitudinal axis. The white matter consists mainly of afferent and efferent fibres running to and from the cortex, to which it sends irregular branched projections. Buried within the white matter are four pairs of deep cerebellar nuclei (e.g., the DENTATE NUCLEUS) which have important connections with the cerebellar cortex and nuclei of the brain stem and thalamus. Topographical representation of the body within the cerebellum is such that axial body parts are represented in the vermal region whereas the limbs and facial parts are represented in the intermediate zones, large parts of the lateral cerebellar regions being without topographical organization.

Neurologically, each hemisphere is divided into an intermediate zone and a lateral zone, the intermediate zone controlling muscle contractions in the distal portions of the limbs, the lateral zone participating with the cerebral cortex in the planning and timing of sequential motor movements. Unlike the cerebrum, each of the two cerebellar hemispheres controls its ipsilateral side of the body.

The three main functional and anatomical systems of the cerebellum are: the *spino-cerebellum* where output leaves via the vermis to the brain stem structures and contributes to the ventromedial pathways of the SPINAL CORD involved in control of axial musculature and posture (see RETICULOSPINAL TRACTS); the *ponto-cerebellum* in which the cerebellar hemispheres are involved in planning and coordination of limb movements by relaying visual input from the posterior parietal cortex to the motor cortical areas of the cerebrum and thence along the CORTICOSPINAL TRACTS; and the *vestibulo-cerebellum*, the only system of the three lacking any association with deep cerebellar nuclei, involved with posture and the control of eye movements (see VESTIBULOSPINAL TRACTS).

In superior aspect, two cerebellar lobes can be identified running across the entire cerebellum, the anterior and posterior (or medial) lobes making up the major part, while in inferior aspect the smaller, rostral, flocculonodular lobe is also visible. The anterior and posterior lobes are folded transversely into shallow ridges, the *folia*, greatly increasing the cortical surface area (~ 17 cm × 120 cm), and running uninterrupted from one side to the other. The most rostral portions of the flocculonodular lobes are small, lacking in folia, and receive information from the vestibular nuclei en route to the ipsilateral cortex.

Throughout, the cortex comprises three histological zones: an outer fibre-rich *molecular layer*; an intermediate *Purkinje cell layer*; and an inner *granular layer*, whose main cell type is the granule cell whose axons ascend into the molecular layer and branch out at right angles as *parallel fibres* forming excitatory glutamatergic synapses with dendrites of Purkinje cells (see Fig. 43). Each Purkinje cell receives only one synapse from each passing parallel fibre, but does so from as many as

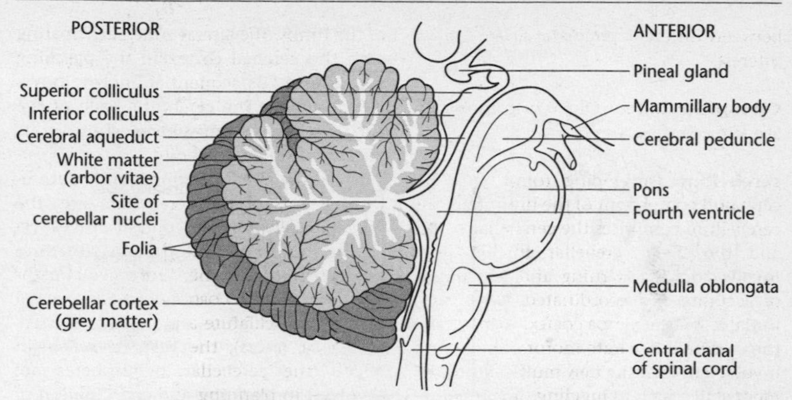

FIG 42. Diagram of midsagittal section of the CEREBELLUM and brain stem, from the right side. From *Principles of Anatomy and Physiology*, G.J. Tortora & S.R. Grabowski, © 2000. Reproduced with permission of John Wiley & Sons, Inc.

100,000 parallel fibres. Purkinje cells also receive numerous (>20,000) synaptic contacts with a single *climbing fibre* which arrives via the olivocerebellar tract from the inferior olivary nuclei and winds around the Purkinje cell's soma and dendrites. Climbing fibres fire with a frequency of 1–10 Hz, each time causing the Purkinje cell to discharge a complex spike. This local wiring, in which interactions take place between repeated modular units (parallel circuitry) is characteristic of the cerebellum, basal ganglia and the thalamus.

All afferent signals enter through one or other of the cerebellar peduncles. This includes afferent projections carrying proprioceptive information from the periphery of the body via the spinal cord, input from spinal cord MOTOR PATTERN GENERATORS, and acceleratory information from the vestibular and INFERIOR OLIVARY NUCLEI of the brain stem. Input from the higher centres (the contralateral motor, premotor and somatosensory cerebral cortices) arrives via the pontine nuclei (see PONS).

In feedback operation, the cerebellum compares motor intentions with motor performance, any discrepancy generating an error signal whose effect is to reduce the mismatch. Sensory signals caused by movement errors activate mossy fibres, which excite Purkinje cells and these inhibit deep cerebellar nuclei which drive the RED NUCLEUS and thalamus so that the movement error is prevented. For those movements that are too fast for feedback to be effective, the cerebellum operates pre-programmed (feedforward) sequences having predictable effects of motor function. Since most voluntary movements must be learnt, the cerebellum acquires by trial and error programmes specifying the motor commands required for a given movement. Sensory errors are translated into motor errors by the inferior olivary nucleus and sent to the cerebellum along olivocerebellar tract climbing fibres, which cause the Purkinje cells to become less responsive to the simultaneous mossy fibre input. Whenever the same input occurs, the Purkinje cells are less excited than they were before the learning (LONG-TERM DEPRESSION), which corrects the motor output. Hence, the cerebellum receives mossy fibre input representing the desired movement trajectory (from the motor cortex), sensory input (from visual cortex and proprioceptor pathways) and feedback from the inferior olive representing errors in motor performance.

Unlike the cerebral cortex, the cerebellar cortex lacks intrinsic long-range connections and relies exclusively on local short-range connections within its grey matter

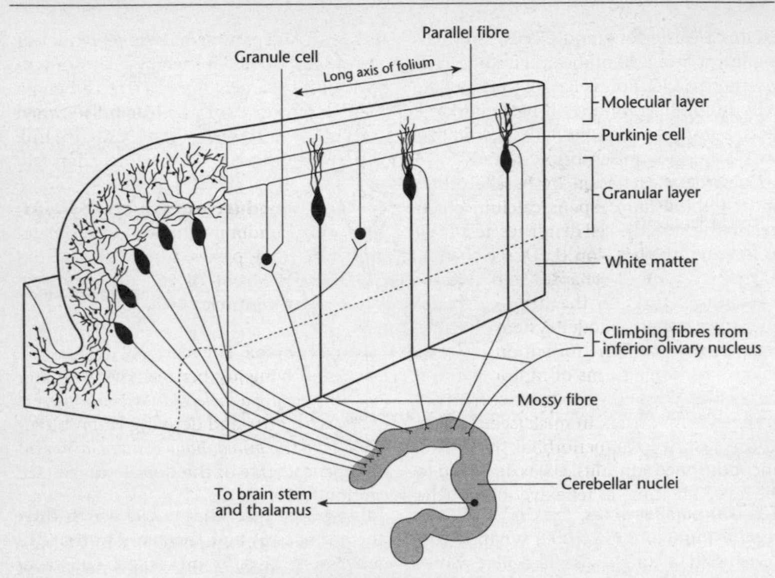

FIG 43. Diagram indicating main features of the cortex of the CEREBELLUM, showing connections between afferent (climbing fibre and mossy fibre) and efferent (axons of Purkinje cells) fibres. In such 'local wiring' a few neuronal types form individual 'modules' that are repeated, interaction between modules being restricted to neighbours (parallel circuitry). Most Purkinje cells are not shown in detail. From A.R. Crossman and D. Neary © 1995 *Neuroanatomy, An Illustrated Colour Text*; redrawn with permission from Elsevier Ltd.

forming an array of relatively simple NEURAL NETWORKS. Synaptic transmission of the main cerebellar cortical circuitry is mediated by excitatory GLUTAMATE and modulated by inhibitory GABA released by three interneuron classes: Golgi cells, basket cells and stellate cells. Its two excitatory inputs are provided by: (i) the *climbing fibres* which arise in the inferior olive of the medulla conveying proprioceptive information. Each climbing fibre (~15 million) makes synaptic contact with ~10 *Purkinje cells*. These show periodic activity of 1–10 Hz, debate centering on whether they provide a periodic clock for coordinating movements and/or provide signals for motor learning (see Fig. 44 and remaining entry); (ii) the *mossy fibres*, relaying proprioceptive information from the spinal cord and brain stem, and from the pontine-cerebellar cerebral neocortex conveying sensory and motor signals. Mossy fibres synapse on *granule cells* whose axons ascend into the outermost cerebellar cortex where they branch horizontally to extend as *parallel fibres*. Each Purkinje cell receives excitatory input from parallel fibres and from one climbing fibre that winds itself round the soma and dendrites, making ~300 powerful glutamatergic, excitatory synapses. Purkinje fibres form the single output system to the deep cerebellar nuclei (the cerebellum's main output neurons to the dentate and fastigial nuclei). Golgi cells receive excitatory synaptic connections from parallel fibres and in turn suppress granule cells using the transmitter GABA. Basket cells and stellate cells in the molecular layer are also GABAergic, receiving input from parallel fibres and sending axons at right angles to them to synapse with and inhibit Purkinje cells at specified (possibly modulatory) synapses, stellate cells mainly on Purkinje cell dendrites and basket cells on their somas and initial axon segments. Activation of a mossy fibre excites a cluster of granule cells, stimulating linear arrays of on-beam Purkinje cells via the parallel fibres. Basket and stellate cells inhibit surrounding off-beam Purkinje cells, a *surround antagonism* that resembles LATERAL INHIBITION in sensory systems.

Coincident excitation by parallel fibres and climbing fibres opens calcium channels in the Purkinje cell dendrites, resulting in long-term depression (LTD, see SYNAPTIC PLASTICITY) at these synapses and reduces inhibitory output to the deep cerebellar nuclei – a cellular model system that has been implicated in formation of the engram for some forms of motor learning such as the CLASSICAL CONDITIONING involved in the eyeblink reflex in mice (see Fig. 43). In this task, a weak periorbital shock (the unconditioned stimulus, US) is delivered to the eye, eliciting a reflexive blink (the unconditioned response, UR). When a neutral conditioned stimulus (CS) such as a tone is repeatedly paired with the periorbital shock so that they terminate simultaneously, the mouse learns to blink its eye in a carefully timed manner (the conditioned response, CR) such that the eyelid is lowered when the shock arrives. Neural activity evoked by the US and CS are conveyed by climbing fibres and mossy/parallel fibres, respectively, converging on the Purkinje cells. One favoured hypothesis is that the simultaneous firing of US-encoding climbing fibres and a CS-encoding mossy fibre–parallel fibre disynaptic relay causes LTD of excitatory parallel fibre–Purkinje cell synapses, especially of those active just before the US signal arrives. Repeated pairings result in LTD of the excitatory synapses between parallel fibres and Purkinje cells in which activity immediately precedes climbing fibre activation. This may reduce tone-evoked Purkinje cell activity, thereby attenuating inhibitory synaptic drive from Purkinje cells to deep cerebellar nuclei. Disinhibited populations of cells in the deep cerebellar nuclei begin to fire in the interval between CS onset and US onset, apparently driving the carefully timed, learnt, eyeblink. This suggests that the memory trace for associative eyelid conditioning is expressed in the firing rate and pattern of these deep nuclear neurons. This model thereby relates a synaptic event (parallel fibre LTD) to a behaviour (acquisition of the carefully timed eyeblink), although the causal link between them is not yet absolutely defined.

cerebral aqueduct (aqueduct of Sylvius) Duct in the midbrain through which cerebrospinal fluid passes from the choroid plexus in the roof of the 3rd ventricle into the 4th ventricle.

cerebral cortex The region of the CEREBRAL HEMISPHERES lying just beneath the pia mater and containing a *molecular layer* and layers of PYRAMIDAL CELLS and deriving embryologically from the *primordial plexiform layer* and the CORTICAL PLATE of the dorsal PALLIUM (see caption to Fig. 44).

Although the term 'cerebral cortex' includes the hippocampus, olfactory cortex (part of the palaeocortex) and neocortex (the cortices of the evolutionarily expanded cerebral hemispheres), it is commonly used to refer only to the latter. It develops from the pallium and its organization is laminar, comprising six cell layers, and the earliest step in its formation is specification of neuroepithelium to a cortical fate. Mouse genetic mosaics indicate that the LIM homeobox gene *Lhx2* (a vertebrate orthologue of the *Drosophila* selector gene *Apterous*) has been shown to act as a classic selector gene in determining cortical identity in that mammal. This gene is expressed in cortical precursor cells but not in the adjacent telencephalic dorsal midline (choroid plexus epithelium and the intervening cortical hem) which is a secondary source of bone morphogenetic proteins (see BMPS) and WNT SIGNALLING molecules.

The surface grey matter layer of the cerebral hemispheres, ~2–4 mm thick, overlies the cerebral white matter (see BRAIN). Because the grey matter enlarges during development faster than the white matter, the upper layers developing late, the cortical region is thrown into folds (gyri) between which are a few deep clefts, or fissures, and more numerous but shallower sulci. The human neocortex comprises almost 42% white matter, a peak value among primates. The longitudinal fissure separates the cerebrum into right and left halves, the two CEREBRAL HEMISPHERES, internally connected by the CORPUS CALLOSUM.

Each cortical layer possesses certain neuron types and its own characteristic input and projection patterns. Certain cell types, such as spindle cells (specialized deep-layer neurons), appear to be unique to primates. Layer I contains few cell bodies but many long white nerve fibres running horizontally interconnecting one area of cortex to another; Layers II and III also contain horizontal connections, commonly projecting from small pyramidal cells to neighbouring areas; Layer IV is where most input fibres terminate, containing large numbers of spiny stellate cells on which these projections terminate. It is particularly thick in the sensory cortex. Layers V and VI contain large numbers of pyramidal cells whose long descending axons project to subcortical areas, but there are also many neurons involved in intrinsic cortical circuits. Layer V is especially well developed in the motor cortex.

Most cortical neurons originate outside the cortex, in the 'proliferative zone' beneath what will become the VENTRICULAR ZONE (VZ) at the base of the cortex. Migrating neurons leave the VZ to form a single layer, the preplate. Between this and the VZ is an intermediate zone containing axons of the preplate neurons and axons growing into the cortex from developing subcortical structures such as the thalamus. During later phases of neurogenesis, a second region of dividing cells develops: the subventricular zone. This mainly generates small interneurons, continuing into the 2nd postnatal year in humans. To reach their final destination in the cortex, cells from the VZ migrate along glial cells, the laminar structure apparently resulting from the fact that each 'proliferative unit' in the subventricular zone produces about 100 offspring neurons which climb the same glial cell, each taking up residence on top of the last to arrive at the surface. All 100 offspring thereby contribute to the formation of one radial cortical unit and differentiate into six layers. This pattern of development has been described as 'inside-out', since cells migrate from deeper to more superficial locations. Mutations in the *LISI* gene may disrupt the timing and plane of division of cells in the proliferative zone, possibly leading to failure of neurons to migrate properly in the cortex. Layer acquisition by projection neurons and interneurons is highly coordinated for both cell types. Radial migration and laminar arrangement of projection neurons depends on Reelin, a secreted glycoprotein expressed near the pial surface by CAJAL-RETZIUS CELLS during embryogenesis. By contrast, layer acquisition by GABAergic interneurons does not seem to depend directly on Reelin signalling, these cells invading target layers after synchronically generated projection neurons reach their final destinations, possibly guided by cues released by them. There is speculation that Reelin may interact with the NON-CODING RNA molecule encoded by the *HAR1* region of the HAR1F GENE, although genomic data so far offer only provocative clues rather than direct answers to what caused the hominin lineage to undergo such rapid increase in brain size. Recent mRNA assays comparing gene expression levels in the cerebral cortices of humans, chimpanzees and macaques found 91 genes that have changed their activity level since the chimpanzee-hominin lineage split, ~80 of them being more active in the human brain.

There is growing support for the model that signals in the diencephalon and subcortical telencephalon regulate the location where axons from the different thalamic nuclei enter the developing cortex to create cortical circuits (see THALAMUS), their growth being influenced by axons growing in the opposite direction – from the cortex to the thalamus (the 'handshake' model). Thalamic axons wait in the deepest cortical layer (the *subplate*, SP; see Fig. 44) until Layer IV neurons have migrated and settled before they make contact with these target cells. The highly specific intralaminar and interlaminar connections (circuits) and networks

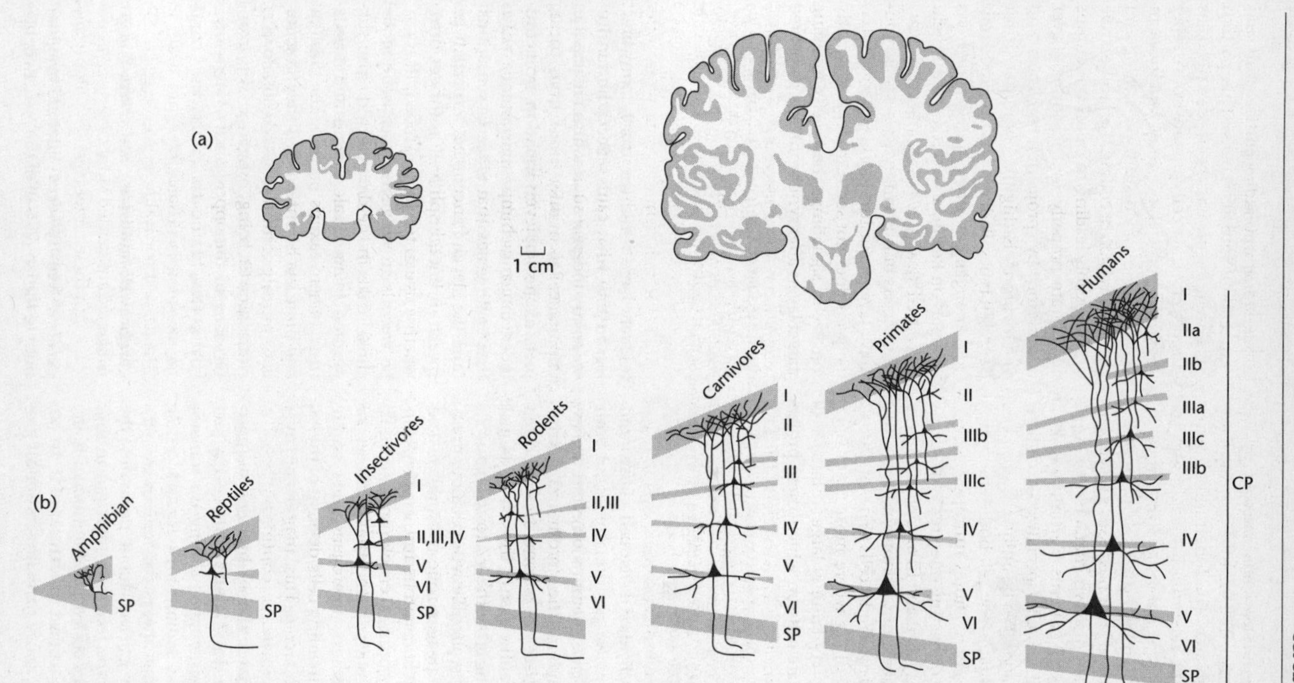

FIG 44. (a) Diagram of coronal sections through the CEREBRAL CORTEX of a chimpanzee (left) and of a human. (b) Diagram of the cellular compositions of vertebrate cerebral cortices. They derive from two cell populations during development: the primordial plexiform layer (PPL) and the cortical plate (CP). The PPL appears to be homologous to simple cortical structures in amphibians and reptiles, but during mammalian development is split by the cortical plate into two layers, layer I at the top and the subplate (SP) at the base. CP-derived cortical layers are then progressively elaborated in larger-brained mammals, humans having a larger proportion of these late-developing neurons which project locally, or elsewhere, within the cortex. From R.S. Hill and C A. Walsh, *Nature* © 2005, with permission from Macmillan Publishers Ltd.

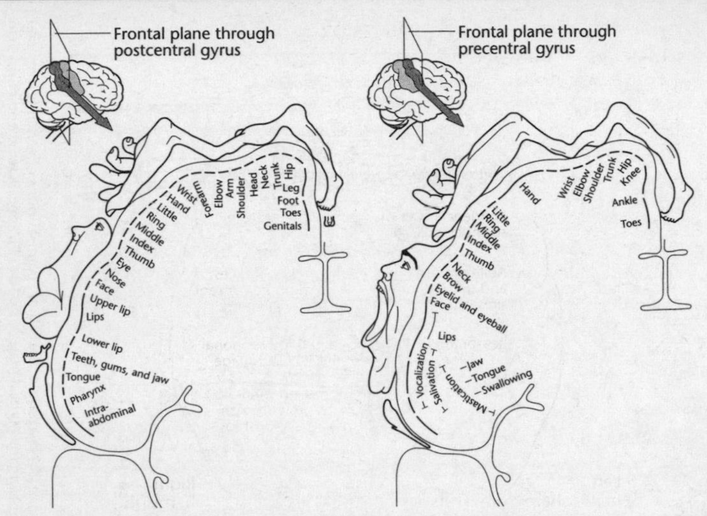

FIG 45. Diagrams of frontal sections of (a) the PRIMARY SOMATOSENSORY CORTEX (postcentral gyrus) and (b) the PRIMARY MOTOR CORTEX (precentral gyrus) of the right cerebral hemisphere, showing their respective somatotopic 'homunculi'. See CEREBRAL CORTEX. From *Principles of Anatomy and Physiology*, G.J. Tortora & S.R. Grabowski, © 2000. Reproduced with permission of John Wiley & Sons, Inc.

between individual cells, cell classes and columns involve positional cues, of both diffusible and membrane-bound types. Thus, semaphorin/neuropilin and ephrin/eph ligand/receptor proteins are implicated in orchestrating specific aspects of axonal and dendritic growth (see AXONAL GUIDANCE).

Once this scaffold of cell-specific connections and area-specific networks is laid down, activity-dependent mechanisms further shape connections between neurons. Activity rapidly alters the numbers of dendritic spines (see DENDRITES) and synaptic connections during development of the PRIMARY VISUAL CORTEX, particularly outside Layer IV.

The PRIMARY SOMATOSENSORY CORTEX (postcentral gyrus) and PRIMARY MOTOR CORTEX (precentral gyrus) are arranged somatotopically, as indicated in Fig. 45. These maps are not always continuous or symmetrical (e.g., the hand separates the head from the face in (a), but not in (b)). Nor are they scaled like the human body, resembling instead caricatures of it. In (a), the mouth, tongue and fingers appear grotesquely large while the trunk, arms and legs are tiny. In part this reflects the density of sensory receptors; but it also reflects the importance of sensory input from those parts of the body. Information from the index finger has more importance than that from the elbow; and tactile sensations from the lips and tongue are important in speech production, as well as for deciding whether a potential item of food is likely to be nutritious, or perhaps cause one to choke. Likewise, more cortical space is devoted to the face, tongue and hands than other regions because the variety and precision of movements they execute is so much greater. See ASSOCIATION CORTEX, ELECTROENCEPHALOGRAM, FRAGILE-X SYNDROME, MICROCEPHALY, NEURAL PLASTICITY, PREFRONTAL CORTEX, PREMOTOR CORTEX, PRIMARY VISUAL CORTEX, VISUAL ASSOCIATION CORTEX.

cerebral dominance See CEREBRAL HEMISPHERES.

cerebral hemispheres Bilateral evaginations of the lateral wall of the prosencepha-

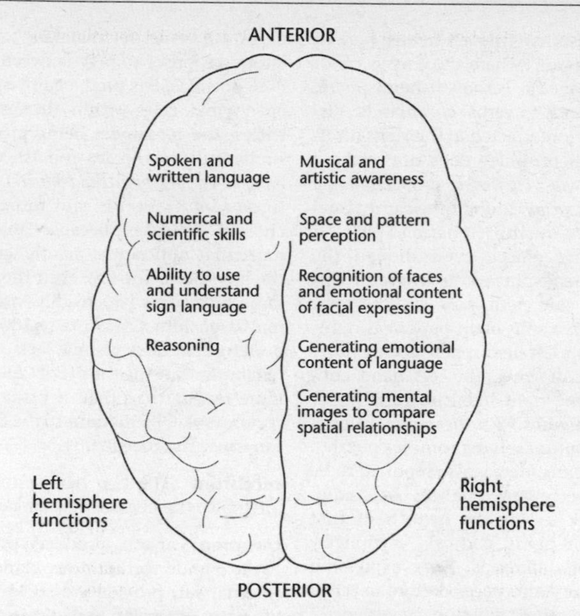

FIG 46. A simplified indication of the principal functional asymmetries between the left and right CEREBRAL HEMISPHERES (see entry). Precise localization is not to be implied. © 1990 John Wiley & Sons. *Principles of Anatomy and Physiology* by G.J. Tortora & N.P. Anagnostakos. Reprinted with permission of the publisher.

lon arising at the start of the 5th week of development, and comprising the bulk of the telencephalon. The basal parts grow rapidly by the 10th week and bulge into the lumen of the lateral ventricle, termed the corpus striatum on account of its striated appearance in cross-section. Where it attaches to the roof of the diencephalon, the wall of each hemisphere remains thin, comprising a layer of ependymal cells covered by vascular mesenchyme, together forming the choroid plexus and protruding into the lateral ventricle along the choroidal fissure. Above this, the wall of the hemisphere thickens to form the HIPPOCAMPUS, which bulges into the lateral ventricle. Expanding further, the hemispheres cover lateral aspects of the diencephalon, mesencephalon and cephalic portion of the mesencephalon (see BRAIN, Fig. 28). Part of the wall of each hemisphere, the corpus striatum, expands posteriorly as the dorsomedial caudate nucleus and the ventrolateral lentiform nucleus (see BASAL GANGLIA), these two separated by the internal capsule, made up of fibres of axons passing to and from the cortex. Continued growth of the hemispheres results in the formation of four lobes: frontal, parietal, temporal and occipital, named after the bones covering them (see BRAIN, Fig. 30).

Human left and right hemispheres are anatomically and functionally asymmetric (see Fig. 46), being specialized for different cognitive and behavioural functions. *Cerebral dominance* is the normal situation in which certain higher cognitive functions come to be dealt with largely by just one of the two cerebral hemispheres. This becomes established within a few years of birth and in most people the left hemisphere is dominant for language and mathematical reasoning while the right hemisphere performs better in SPATIAL AWARENESS and musical skills.

LANGUAGE functions are largely confined

to one hemisphere (the left hemisphere in most people) and where the corpus callosum is sectioned it is the left hemisphere, which responds to verbal commands. Visual information directed to the right (non-dominant) hemisphere does not evoke a verbal response. However, if such people are shown a word through the right visual field they can use the left hand to pick out the matching object, even though the dominant hemisphere is unaware of this transaction (see 'SPLIT-BRAIN' PHENOMENON). Even though a split-brain patient is right-handed, their left hemispheres being more practised at drawing, the left hand controlled by the right hemisphere is better at copying figures with three-dimensional perspective and at solving complex puzzles. The right hemisphere is also reported to be better at perceiving nuances of sound. Recent work testing the hypothesis that human left–right cortical asymmetry results from differential gene expression at early embryonic stages, before onset of organized cortical function, revealed 27 such genes consistently showing either left^{high} or left^{low} expression in 12-week human embryos. See CEREBRAL CORTEX, POSTERIOR TEMPORAL LOBE.

cerebrospinal fluid (CSF) Specialized clear, salty, EXTRACELLULAR FLUID (~150 cm^3) filling the central canal, the ventricles of the brain and subarachnoid space (see MENINGES), being secreted by the capillaries of the CHOROID PLEXUSES in the VENTRICLES and reabsorbed through the arachnoid villi. Its secretion depends mainly on active transport of Na^+, causing Cl^- to follow and leading to an osmotic passage of water from the blood. This Cl^- influx drives a Cl^--HCO_3^- antiport so that despite its low K^+ concentration, a small quantity of K^+ and HCO_3^- move from the CSF into the capillaries while small amounts of glucose move into the CSF. High levels of carbonic anhydrase in the choroid plexus also result in diffusion of HCO_3^- into the CSF via an apical ion transporter. The lack of protein buffer in CSF results in its pH changing more readily than that of tissue fluid (which may assist the NUCLEUS TRACTUS SOLITARIUS, adjacent to the 4th ventricle, in making rapid ventilatory responses to pH). Between 80% and 90% of the CSF is produced by specialized ependymal cells within the lateral ventricles, the remainder being produced by similar cells in the 3rd and 4th ventricles. A *blood–cerebrospinal fluid barrier* (part of the BLOOD–BRAIN BARRIER) is said to exist at the choroid plexuses because many large molecular substances hardly cross from the blood into the CSF even though these same substances pass readily into the normal tissue fluid. CSF contains 1000-fold less protein than does plasma, lack of this protein buffer causing the pH of CSF to change more readily than that of tissue fluid (see VENTILATION). Obstructions to the flow of CSF can cause hydrocephalus.

cerebrum The two hemispheres of the forebrain: the CEREBRAL HEMISPHERES.

cerumen Ear wax, produced by modified sweat glands (ceruminous glands) of the external ear, providing a sticky barrier to the entry of foreign bodies. See DEAFNESS.

ceruminous glands Accessory glands of the SKIN, being modified merocrine glands located in the ear canal (external auditory meatus) and contributing with sebaceous glands to the production of ear wax (cerumen). With hairs in the ear canal, cerumen protects the eardrum from dirt and small insects; but if it thickens and hardens it can block the ear canal and reduce hearing.

cervical cancer See PAPILLOMAVIRUSES, VIRUSES.

cervical flexure A sharp bend in the neural tube occurring during embryonic development between the brain and spinal cord.

cervical ganglia Three pairs of ganglia lying bilaterally on either side of the spinal cord in the cervical segments but receiving innervation from the first thoracic sympathetic ganglion. Some of their fibres innervate the radial muscles of the iris; others innervate the heart.

cervical plexus Nerve plexus formed from the anterior (ventral) rami of the first four cervical nerves (C1–C4), with contributions from C5 and supplying the skin and muscles of the head, neck and superior part of the shoulders and chest. See PHRENIC NERVES.

cervix The neck of any structure; but especially of the inferior cylindrical portion of the UTERUS.

CFCs See CHLOROFLUOROCARBONS.

C fibres Unmyelinated fibres (see AXON) with free nerve endings and of different diameters and conduction velocities, the smallest being selectively responsive to HISTAMINE. See A BETA FIBRES, A DELTA FIBRES, PAIN.

c-*fos* Gene encoding the transcription factor cFos. Regulated in the brain by CREB.

***CFTR* gene** See CYSTIC FIBROSIS.

CG islands (CpG islands) See CPG DINUCLEOTIDE.

cGMP See CYCLIC GMP.

cGMP-dependent protein kinase (protein kinase G) See PKG.

CGRP (calcitonin-gene-related peptide) See CALCITONIN.

Chagas disease Caused by the protozoan *Trypanosoma cruzei*, a neglected disease endemic in 18 countries in two ecological zones of the Americas: the Southern Cone, in which vector bugs (*Triatominae*) live inside human homes; and the Central American and Andean countries, where the vector lives both within and outside human homes. The parasite invades most organs, often causing heart, intestinal and oesophageal damage, killing up to one-third of those chronically infected. Vector control is the only current method of disease control. See TRYPANOSOMES.

channel proteins See ION CHANNELS.

chaperones Proteins that facilitate the folding of other proteins, those of vertebrates belonging to two classes of HEAT-SHOCK PROTEINS (hsp): hsp60 and hsp70. Mitochondria contain their own versions of these, distinct from those in the cytosol. Mitochondrial hsp70, e.g., is involved in the import of proteins across the inner mitochondrial membrane; and a special form of hsp70 assists protein folding in the ENDOPLASMIC RETICULUM. The hsp70 chaperones begin to act while proteins are still growing on the ribosomes, while hsp60-like proteins act later to fold completed proteins. Misfolded proteins get a chance to refold properly within a large barrel-shaped complex of hsp60-like proteins termed an 'isolation chamber', refolding being ATP-dependent. Proteins with a sizeable exposed patch of hydrophobic amino acids on their surfaces are usually abnormal and have probably misfolded, or suffered an accident that partially unfolds them. Many such proteins can form large aggregates, precipitating out of the cell solution and causing severe disease such as those involving NEURAL DEGENERATION, and the role of some chaperones is to recognize and remove or cover the hydrophobic patches. Failed attempts to refold a protein usually result in its delivery to a complex protease, the proteasome (see UBIQUITINS). Several chaperones bind to NUCLEAR RECEPTORS in the cytosol, usually being released when the receptor binds its ligand and enters the nucleus.

CHD See CORONARY HEART DISEASE.

checkpoints (cell cycle checkpoints), checkpoint controls See CELL CYCLE.

chemoaffinity hypothesis The notion that chemical markers on growing axons (especially their growth cones) are matched to complementary markers on their target destinations, thereby enabling precise connections to be made. See AXONAL GUIDANCE.

chemoattractants Substances (e.g., CHEMOKINES) that elicit cell motility in the direction of their increasing concentration gradients.

chemokines Inducible, low molecular mass chemoattractant and pro-inflammatory proteins (a subset of CYTOKINES) released by activated epithelial cells, macrophages, dendritic cells, eosinophils and mast cells, and stimulating the activation and migration from the bloodstream of phagocytes and lymphocytes with appropriate receptors. They fall into two groups: CC chemokines, with two adjacent cysteines near the amino terminus; and CXC chemokines, with the equivalent two cysteines separated by a single amino acid. CC chemokines bind to CC chemokine receptors (CCR1-9) while CXC chemokines bind to CXC chemokine receptors (CXCR1-6; see CXCRS). CC chemokines, e.g., macrophage chemoattractant protein-1 (CC chemokine ligand 2, or CCL2) promote migration of monocytes, lymphocytes and other cell types. Those CXC chemokines with a Glu-Leu-Arg motif just before the first cysteine (e.g., CXCL8) promote neutrophil migration, while those lacking this tripeptide (e.g., CXCL13) guide lymphocytes to their destinations in B-cell regions of the spleen, LYMPH NODES and gut. CXCL8 (formerly IL-8) induces neutrophils to leave the blood and enter the tissues (it is also released by the stomach mucosa in response to infection by *Helicobacter pylori*). CCL2 functions similarly with monocytes, promoting their differentiation into macrophages. CCL5 (RANTES) is released from $CD8^+$ T cells as a selective attractant for memory T cells, eosinophils, basophils and monocytes. Binding to CCR5 (a co-receptor for HIV) it is an HIV-suppressive factor.

Chemokines act with vasoactive factors to bring leukocytes close to the ENDOTHELIUM, and with cytokines such as TNF-α to induce the necessary adhesion molecules on the endothelial cells.

CCL3 (MIP-1α, macrophage inflammatory protein 1 alpha), in addition to its proinflammatory activities, inhibits the proliferation of haematopoietic stem cells *in vitro* and *in vivo*. CCL3 and CCL4 (MIP-1β) are produced by many activated cell types, including macrophages, dendritic cells and endothelial cells, and induce dendritic cells to release IL-12. CCL3L1 is a potent HIV-1-suppressive chemokine (see HIV).

chemoreceptors See CARDIOVASCULAR CENTRE, CAROTID BODIES.

chemotherapy Many chemotherapeutic agents act by damaging the chromosomes of cancer cells. These cells often lack G2/M checkpoint controls (see CELL CYCLE), so that drugs such as doxorubicin cause them to enter 'mitotic catastrophe' and eventual cell death, often by apoptosis. But oncoproteins are the most promising targets of anti-cancer therapies. The ectodomains of these receptors are potential targets, as with Herceptin (see BREAST CANCER), as are the signal transduction proteins immediately downstream of them (e.g., mTOR inhibitors, tyrosine kinase inhibitors, RAF inhibitors and MEK inhibitors; see RAS SIGNALLING PATHWAYS). Many chemotherapeutic agents act through their ability to alkylate guanine in DNA, a reaction which is reversed by the enzyme DNA methyltransferase. Gliomas and COLORECTAL CANCERS may in future be treatable by targeting this alkyltransferase. Chemotherapeutic drugs are believed to have selective toxicity for cancer cells because these cells lack fully competent DNA REPAIR MECHANISMS, preventing them from maintaining genomic stability. Those lacking BRCA1 or BRCA2 appear to get by with minimal genomic stability through the back-up repair mechanism afforded by PARP1 and its attendant repair proteins, which is important following the accidental breaks at replication forks during DNA REPLICATION. Cells lacking BRCA1 or BRCA2 prove to be sensitive to killing by inhibitors of PARP1 and, because these inhibitors have relatively little effect on normal cells they may prove therapeutically valuable. The multi-kinase inhibitor sorafenib targets BRAF, CRAF and the VEGF and platelet-derived growth factor receptor tyrosine kinases (see also LEUKAEMIAS, GASTROINTESTINAL STROMAL TUMOURS, MELANOMAS). Standard chemotherapy and radiotherapy may operate in part by inducing senescence within the tumour mass. There is rapid *in vivo* clearance of tumour cells that have undergone P53-triggered senescence.

chewing See MASTICATION.

Cheyne–Stokes breathing A periodic change in the frequency and tidal volume during VENTILATION, alternating between apnoea and mild hyperventilation. Breathing starts with slow shallow breaths, the frequency and tidal volume increasing to a maximum before returning to the earlier levels. It is sometimes observed in normal people, but especially in the terminally ill. During low ventilation, the arterial pCO_2 rises and stimulates the central chemoreceptors so that frequency and depth of breathing increase. The resulting fall in arterial pCO_2 then reverses the situation. There must be an abnormal delay between detection and response to CO_2 levels reflecting either reduced sensitivity to CO_2 or a sluggish blood flow to the brain.

chickenpox A systemic and highly infectious viral infection accompanied by a characteristic vesicular haemorrhagic rash, infection usually being symptomatic, but not generally severe, in immunocompetent children. The causative agent is varicella zoster virus (VZV), an enveloped VIRUS and member of the Herpesviridae, and five major glycoprotein antigen families (gp I–V) have been detected. Infection is commonest in children between 5 and 14 years. Recovery is followed by latency of the virus in the dorsal root ganglia (see SHINGLES) during which the virus switches off many of its genes, and it is not known what reactivates the virus. Specific treatment is rarely required in children, although adolescents and adults should be treated. Aciclovir is a synthetic acyclic purine nucleoside antiviral drug active against VZV, and is taken orally. Secondary staphylococcal pneumonia is common in adults. Live attenuated varicella vaccine is highly effective in prevention, particularly for health workers. Universal vaccination in early childhood has been introduced in the USA, where a combined MMRV vaccine is available.

chief cells See GASTRIC PITS.

chimaera, chimaerism (1) The presence in an individual of two or more genetically distinct cell lines derived from more than one zygote, i.e., of different genetic origin (compare MOSAICISM). *Dispersic chimaeras* result from double fertilization, two ova being fertilized by different sperm and the two resulting zygotes fusing to form one embryo, which can result in true HERMAPHRODITISM. *Blood chimaeras* result from exchange of cells, via the placenta, between non-identical twins while *in utero* (see MICROCHIMAERISM). (2) *Chimaeric genes* result from fusion of DNA sequences from different genes, often as a result of chromosomal translocations, so that a fusion protein is formed with altered biochemical properties from the original (see FUSION GENES).

chimpanzees There are two living species of chimpanzees (Hominidae): the common chimpanzee, *Pan troglodytes*, and the bonobo, *Pan paniscus*, whose lineages are believed to have split from a common ancestor 1–2 Mya. Chimpanzees adopt knuckle-walking when on the ground, but may brachiate while in trees. The last common ancestor of chimpanzees and humans is believed to have been quadrupedal, the oldest fossils resembling bipedal humans being dated to 6–7 Mya. The first discovered fossil chimpanzee, from the Middle Pleistocene Kapthurin Formation of Kenya, demonstrates that the East African Rift Valley did not present an impenetrable barrier to chimpanzee occupation of East Africa, as some had held, and that they were contemporaries of extinct species of hominins. Populations of common chimpanzees in Gabon and the Democratic Republic of the Congo were devastated by Ebola virus outbreaks between 2001 and 2003, and the world's largest protected area for great apes, its Sankuru reserve (30,500 km^2), is soon to be established there. It is being targeted to save the endangered bonobo, of which only an estimated 10,000 remain in the wild, all of them in the Congo.

Human and chimpanzee chromosomes differ by one chromosomal fusion, at least nine pericentric inversions, and in the distribution of constitutive heterochromatin. Human chromosome 2 results from fusion of two ancestral chromosomes that remained separate in the chimp lineage – chromosomes 2A and 2B in the revised nomenclature (formerly chimpanzee chromosomes 11 and 12). Chimpanzee and human genomes each contain ~40 Mb of species-specific euchromatic DNA sequence (so ~90 Mb in total), due mostly to insertions and deletions (indels), relative to the genome of the common ancestor. This difference amounts to ~3% of both genomes, dwarfing the 1.23% difference which is due to nucleotide substitutions, although the actual number of indel events that have occurred since lineage separation (~5 million) is far fewer than the number of substitution events (~35 million). It is believed that the major phenotypic differences between chimpanzees and modern humans are due more to differences in gene regulation than to differences in coding regions of genes, and the SVA retrotransposon is a possible source of regulatory differences between chimpanzees and humans (see TRANSPOSABLE ELEMENTS). Some of the issues involved in clarification of the relationships between humans and chimpanzees are discussed in the HOMINOIDEA entry. Other issues are raised in the entries on BRAIN SIZE, CHROMOSOMES, GENETIC DIVERSITY, GROWTH, INTELLIGENCE, MIRROR-SELF RECOGNITION, X CHROMOSOME and Y CHROMOSOME.

CHK1, CHK2 Checkpoint transducer kinases. See CELL CYCLE.

Chlamydia trachomatis Chlamydial bacteria have Gram-negative walls, yet lack peptidoglycan. They are primarily transmitted as airborne invaders of the respiratory system. *Chlamydia trachomatis* is the bacterium causing trachoma, the leading cause of blindness in humans, although other strains infect the urogenital tract, and chlamydial infections are one of the leading sexually transmitted diseases. Trachoma is a disease of crowding and poor hygiene, predominantly affecting poor, marginalized and displaced communities and is spread from eye to eye by unwashed hands and possibly by flies. Untreated, the infection (and repeated super-infections) may lead to a plaque of vascular inflammatory tissue (pannus) deforming the eyelid. Scarring and subsequent trichiasis damages the cornea, leading to scarring and blindness. Community-wide application of a single dose of oral azithromycin is as effective as 6 weeks of tetracycline eye ointment. There is urgent need of more reliable DNA-BASED DIAGNOSTIC TESTING instead of the former enzyme-immune assay. See NEISSERIA.

chloramphenicol First isolated from *Streptomyces venezualae*, an ANTIBIOTIC which inhibits protein synthesis by reversible binding to the 50s ribosomal subunit and preventing attachment of amino acyl tRNA. It acts against most Gram-positive and Gram-negative aerobic bacteria, and against spirochaetes, rickettsiae, chlamydiae and mycoplasmas. It is usually bacteriostatic, but bactericidal against *Haemophilus influenzae* and *Neisseria* spp.

chloride ion An ion with a reference daily intake value of 3.4 g and important in blood acid–base balance and hydrochloric acid production in the stomach.

chloride shift See RED BLOOD CELL.

chloroquine An antimalarial drug, probably acting by inhibiting the enzyme that polymerizes and detoxifies ferriprotoporphyrin IX. Retinal damage can follow prolonged high doses. It is the treatment of choice for non-falciparum MALARIA, although it retains a prophylactic role against *Plasmodium falciparum*. It exacerbates psoriasis, and is contradicted in people with this condition. Drugs such as chloroquine that raise the pH of endosomes, making them less acidic, inhibit the presentation of antigens (see ANTIGEN-PRESENTING CELLS).

choanae Openings of the nasal cavity into the pharynx.

cholecystokinin (CCK, CCK-PZ) A gastrin-like 33-amino acid peptide hormone and neurotransmitter released by the I cells in the duodenum and jejunum in response to products of fat digestion (see Fig. 47). Its multiple biologically active forms coordinate postprandial secretion of enzyme-rich fluid from the pancreatic

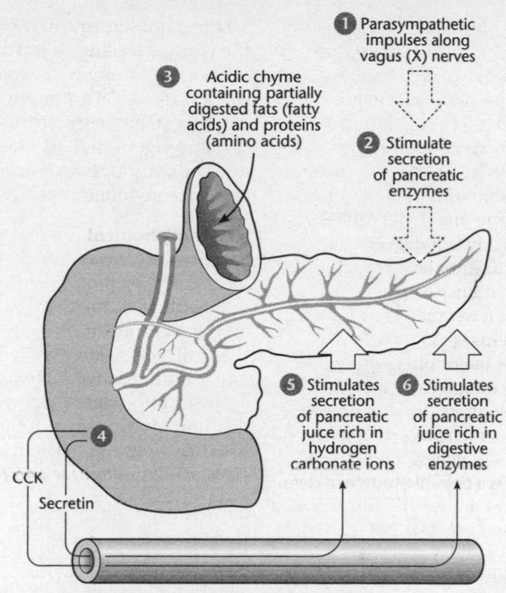

FIG 47. Diagram indicating the neural and hormonal enhancement of pancreatic juice output. Vagal stimulation and the presence of acidic chyme in the duodenum enhance SECRETIN production, while vagal stimulation and presence of fatty acids and amino acids in chyme stimulate release of CHOLECYSTOKININ. Large dashed arrows indicate neural stimulation; large complete arrows indicate endocrine stimulation. From *Principles of Anatomy and Physiology*, G.J. Tortora & S.R. Grabowski, © 2000. Reproduced with permission of John Wiley & Sons, Inc.

acinar cells (via the phosphoinositide/Ca^{2+} system), gall bladder contraction, relaxation of the sphincter of Oddi and inhibition of gastric emptying (delaying the passage of fat to the duodenum). It stimulates vagal afferents projecting to the nucleus tractus solitarius of the medulla, *en route* to the hypothalamus, signalling satiety (see APPETITE, Fig. 14; FEEDING CENTRE). See OXYNTOMODULIN.

cholera A disease caused by the O1 (and potentially the related O139) serotype of *Vibrio cholerae*. It is spread mainly through drinking faecally contaminated water although food (especially shellfish, which should be thoroughly cooked prior to eating) may also be a vehicle for the bacterium. Cholera is a pandemic infection and although rare in developed countries, is still an important disease world wide, occurring sporadically or as epidemics, large epidemics capable of occurring wherever safe drinking water and adequate sewage disposal facilities are lacking. Indeed, provision of these facilities is the most useful measure in preventing its spread. Incubation is usually 3–4 days and the symptoms of infection vary from a mild gastroenteritis-like disease to abrupt onset of severe highly infectious DIARRHOEA, fever not being prominent. Continuous fluid loss leads to shock. Cholera toxin is very closely related to the heat-labile toxin of *Escherichia coli*, and in time the same vaccine may be made effective against both. The target of the A_1 subunit of the enterotoxin activates adenylyl cyclase of the intestinal epithelial cells, causing active secretion of chloride and hydrogen carbonate ions, leading to loss of water and ions exceeding uptake by the colon, causing

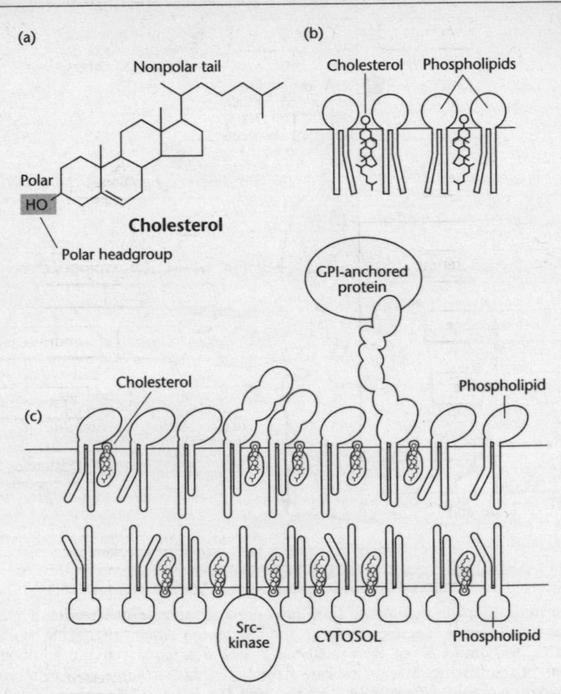

FIG 48. (a) The structure of CHOLESTEROL, indicating its polar hydroxyl group, rigid steroid ring structure, and non-polar hydrocarbon tail. (b) Diagram indicating cholesterol's preferential orientations with membrane phospholipids. (c) Diagram showing a small portion of a lipid raft, in which one protein (oval) is attached to the non-cytosolic surface of the membrane by a glycosylphosphatidylinositol (GPI) anchor, while another (see SRC-KINASES) is attached to the cytosolic surface. From F.R. Maxfield and I. Tabas, *Nature* © 2005, with permission from Macmillan Publishers Ltd.

DEHYDRATION (up to 10 L per day of water loss). The present cholera vaccines are of limited use, protection lasting only 6–24 months so that boosters are recommended after 2 years for those over 6 years. They have no effect on carriage of the bacterium and are not useful in preventing spread of cholera. See CAMPYLOBACTER, ORAL REHYDRATION THERAPY.

cholesterol A water-insoluble lipid, synthesized (much of it in the liver) by condensation of three molecules of acetyl-CoA by enzymes in the cytosol, followed by conversion of the resulting HMG-CoA to mevalonate by a monooxygenase reaction involving cytochrome P450 in the smooth ENDOPLASMIC RETICULUM (ER), and by subsequent reactions forming isoprene, squalene and ultimately cholesterol. It is delivered to organelles by transport within and without vesicles, bound to water-soluble protein transporters. It is also accessed by receptor-mediated ENDOCYTOSIS of LIPOPROTEINS (e.g., LOW DENSITY LIPOPROTEIN) followed by hydrolysis of their cholesteryl esters within late endosomes and lysosomes. It is a precursor of vitamin D production in the dermis and of the steroid SEX HORMONES, and has a role in synaptogenesis. Cholesterol synthesis requires just one oxidation step; but several such steps are needed for its degradation into bile acids and its conversion to steroid hormones. Introduction of an oxygen atom into cholesterol drastically reduces its half-life and directs the molecule to

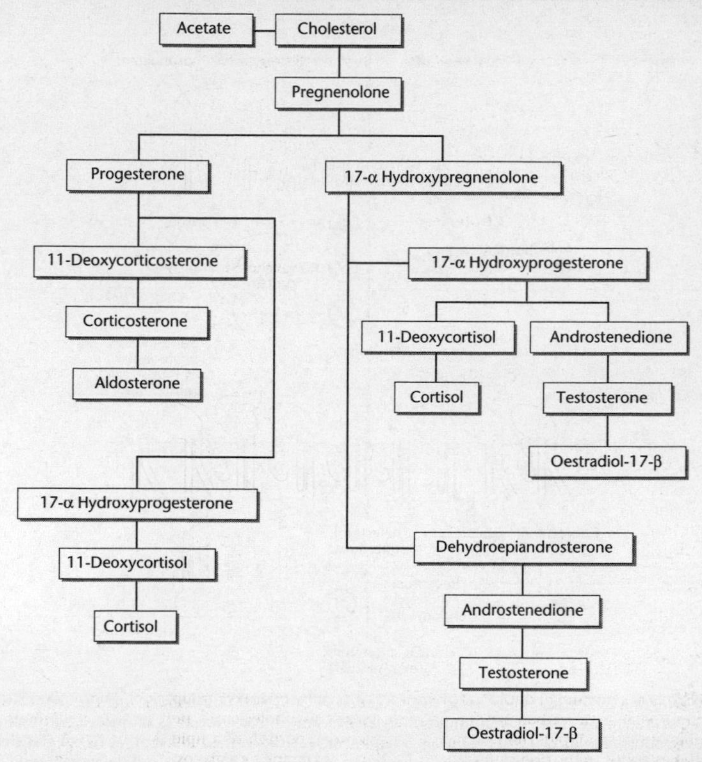

FIG 49. The pathways by which adrenal steroid hormones are synthesized from CHOLESTEROL. Most of these reactions involve the cytochrome P450 side-chain cleavage enzyme present in the endoplasmic reticulum and mitochondrial cristae of adrenal cortex cells. Most of the cholesterol involved in these reactions derives from LDL circulating in the blood. Redrawn from *Endocrine Physiology* © 2004, P.E. Molina (Lange Medical Books/McGraw-Hill) with permission of the McGraw-Hill Companies.

leave the cell. Such oxysterols are able to pass through lipophilic membranes much more rapidly than cholesterol. Liver and endocrine cells have the greatest ability to oxygenate cholesterol, and oxysterols are markers of oxidative stress (see ATHEROSCLEROSIS).

Cholesterol's polar head (far smaller than polar headgroups of other lipids, see Fig. 49) enables it to flip-flop readily between the two leaflets of the membrane and to lie in the aqueous medium surrounding the membrane, its remaining non-polar component lying vertically between the fatty-acyl tails of adjacent phospholipid molecules. When the polar head is situated next to the phospholipid ester carbonyl, cholesterol's relatively rigid fused ring structure can help to order these tails, which, along with cholesterol's ability to fill interstitial spaces, helps to increase membrane orderliness as in LIPID RAFTS (see CELL MEMBRANES, Fig. 41).

A major regulator of lipid organization, cholesterol's metabolism and abundance within cells are homeostatically controlled both by hormones and by several gene products, many of the latter being regulated by SREBP. Two of these are involved in transporting cholesterol out of the

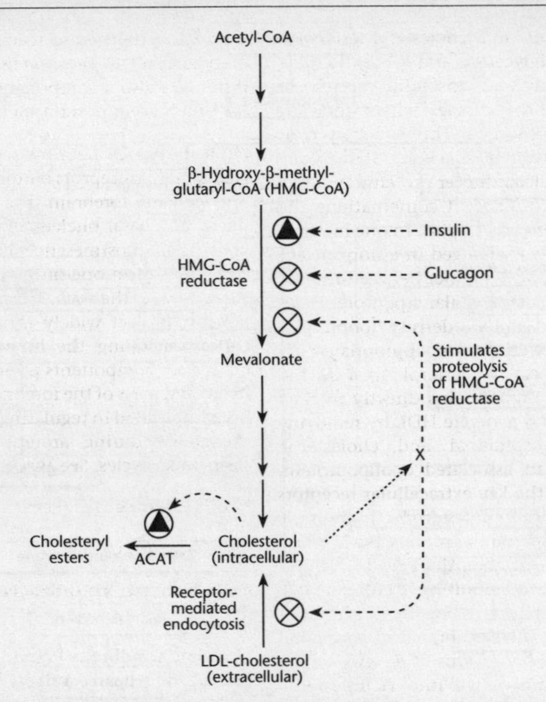

FIG 50. Diagram indicating the homeostatic regulation of CHOLESTEROL synthesis, and the central role of HMG-CoA reductase. The active (unphosphorylated) form of this enzyme is promoted by insulin, its inactive (phosphorylated) form by glucagon (and AMPK). Unidentified metabolites (X) of cholesterol, possibly cholesterol itself, promote rapid degradation of HMG-CoA reductase and downregulation of its gene's expression. Triangles represent positive regulation. © 2000 From *Lehninger Principles of Biochemistry* by D.L. Nelson and M.M. Cox, published by Worth Publishers. Redrawn with permission from the publishers.

endoplasmic reticulum in response to its low level elsewhere in the cell. At such times, the SREBP-escort protein and sterol sensor, SCAP, binds cholesterol, undergoes a conformational change, and escorts SREBP from the endoplasmic reticulum to the Golgi apparatus. Two proteases in the Golgi and a membrane-bound serine protease (SIP) then cleave the SREBP, the NH_2-terminal domain translocating to the nucleus and activating transcription by binding to certain sterol response elements (SREs) in the promoter/enhancer regions of multiple target genes including the LDL receptor and HMG-CoA reductase, the rate-limiting enzyme in cholesterol synthesis (see Fig. 50 and STATINS). Mammalian cells do not catabolize cholesterol and need to export it for its homeostasis at the cell and organismal levels. Increasing cholesterol levels in the cell, detected by SCAP, result in esterification of the excess by the cholesterol-sensitive ER enzyme ACAT and its storage in a more hydrophobic form in the cytosol as lipid droplets, whence it can be released on hydrolysis by neutral cholesterol ester hydrolases and used for cell membrane synthesis or – in steroidogenic cells – for steroid synthesis (see Fig. 49).

Depletion of liver cholesterol in mice activates SREBPS, increasing LDL receptor density; but it also increases production of the LDL receptor-destroying enzyme PCSK9.

Cholesterol and cholesteryl esters, as with triacylglycerols and phospholipids, are essentially water-insoluble yet must be carried to other tissues where they are stored or consumed. This is achieved by transport within plasma lipoproteins, complexes of specific carrier proteins, apolipoproteins, and different combinations and ratios of these other lipids. In one method, cholesterol is exchanged in a non-specific physicochemical manner between the cell surface and extracellular lipoproteins, its esterification on high density lipoproteins (HDLs) providing an appropriate net removal of cell cholesterol. In a second method, apolipoproteins directly interact with cells and generate HDL by removing cellular phospholipid and cholesterol. HDLs and an associated apolipoprotein, ApoA-I, are the key extracellular receptors for somatic cell cholesterol. The multiple membrane-spanning protein ABC TRANSPORTER FAMILY member ABCA1, which binds ApoA-I, is a rate-limiting factor in HDL assembly, promoting transfer of phospholipids across the membranes of peripheral cells to lipid-poor forms of ApoA-I, which are then transformed into HDLs in the blood by the action of lysolecithin-cholesterol acyltransferase (LCAT). Cholesterol enrichment of these particles is regulated independently. HDLs then bind to scavenger receptor-B1 (SR-B1) on liver cells and transfer their cholesterol and cholesteryl esters to them. The cholesterol is excreted into the bile either as free cholesterol or, after conversion, as BILE SALTS. Defects in the delivery of cholesterol and phospholipids to apolipoproteins lead to such diseases as ATHEROSCLEROSIS, the major human disease associated with cholesterol and lipid metabolism. See FAMILIAL HYPERCHOLESTEROLAEMIA.

choline An amino alcohol, member of the VITAMIN B COMPLEX, and of dietary origin, usually in the form of phosphatidylcholine but also on its own:

$$HO - CH_2 - CH_2 - N^+(CH_3)_3$$

It is converted via a CTP- (cytidine triphosphate-) requiring step to CDP-choline and then via a transferase reaction involving diacylglycerol to PHOSPHATIDYLCHOLINE (lecithin). It is also a component of ACETYLCHOLINE and is an important methyl donor.

cholinergic Of neurons secreting ACETYLCHOLINE, or of receptors binding it. The cholinergic basal forebrain (the medial septal nuclei and basal nucleus of Meynert) and brain stem (postmesencephalotegmental) complexes form one of the DIFFUSE MODULATORY SYSTEMS of the BRAIN. The forebrain components project widely into the cerebral cortex, including the hippocampus; the brain stem components project to the thalamus and parts of the forebrain. They have been implicated in regulating general brain excitability during arousal and general sleep–wake cycles. See ADRENERGIC.

cholinesterase See ACETYLCHOLINESTERASE.

chondrin See CARTILAGE.

chondroblast, chondrocyte See CARTILAGE.

chordae tendineae Tendon-like cords connecting heart valves to papillary muscles.

***chordin* gene** A gene which, activated by the transcription factor Goosecoid, antagonizes the activity of bone morphogenetic protein 4 (see BMPS) during dorsalization of cranial mesoderm into notochord, somites and somitomeres. See AXIS.

chorea Disorder producing involuntary jerky (choreatic) movements of the extremities and facial muscles; often resulting from disease of, or damage to, the BASAL GANGLIA. See HUNTINGTON'S DISEASE, PARKINSON'S DISEASE.

chorion Initially comprising the CYTOTROPHOBLAST and SYNCYTIOTROPHOBLAST, the chorion derives from the trophoblast and the extraembryonic mesoderm that lines it, eventually becoming the principal 'embryonic' component of the PLACENTA, while its inner layer eventually fuses with the amnion. Surrounding the embryo, and later the foetus, it produces HUMAN

CHORIONIC GONADOTROPHIN. See CHORIONIC VILLUS SAMPLING

chorionic villus sampling (CVS) Procedure in which a catheter is moved under ultrasonic guidance through the vagina and cervix to the chorionic villi, when ~30 mg of fluid are withdrawn and surveyed for chromosomal and other genetic disorders, as with AMNIOCENTESIS because the cells of the chorion and foetus have the same genome. The technique does not involve penetration of the abdomen, although it carries a slightly greater risk (1–2% spontaneous abortion) than does amniocentesis, however it can be performed as early as 8 weeks into gestation.

choroid The highly vascularized posterior portion of the middle layer of the eyeball (see EYES) lining most of the internal surface of the sclera and supplying the posterior surface of the retina with nutrients and oxygen, and removing wastes.

choroid plexuses Paired network of blood vessels, covered in specialized ependymal cells and support tissue, projecting into the lateral, 3rd and 4th ventricles of the brain, producing CEREBROSPINAL FLUID by filtration and forming a major component of the blood–cerebrospinal fluid (see BLOOD–BRAIN BARRIER). They form by invaginations of the vascular pia mater (see MENINGES), producing a vascular connective tissue core covered by ependymal cells. The endothelial cells of the CAPILLARIES involved have very dense tight junctions holding them together, preventing escape of large molecules. See CEREBRAL HEMISPHERES.

Chororapithecus abyssinicus A putative early gorilline primate from the Afar region of Ethiopia. See HOMINOIDEA.

chromaffin cells See ADRENAL GLANDS.

chromatic aberration See VISUAL ACUITY.

chromatid CHROMOSOMES consist of a pair of chromatids, joined at the centromere, during the period between S-phase of the cell cycle and separation of the chromatids at mitotic anaphase, or 2nd meiotic anaphase (see MITOSIS, MEIOSIS).

chromatin The complex of DNA and associated proteins, most of them HISTONES, of which eukaryotic CHROMOSOMES are composed. Assembly of the basic unit of chromatin, the NUCLEOSOME, is achieved during DNA REPLICATION, and involves chromatin-assembly factors (CAFs). Although interphase chromosomes are generally too extended and entangled for much clarity, they coil up and condense during mitosis. There is a steady process of condensation throughout prophase, both in mitosis and meiosis, by which interphase chromosome becomes compacted after replication into a metaphase chromosome that is ~50-fold shorter (although the DNA itself is ~10^4 times shorter than it would have been if unbound by histones). The final stage in this condensation only reduces the chromosome length 10-fold, but produces a dramatic change in their appearance (see Fig. 51). Integrated into the progression of the cell cycle, it requires a class of proteins, CONDENSINS, that couple ATP hydrolysis to chromosome coiling.

The subnuclear localization of chromatin in non-dividing cells is not random, specific genetic loci or entire chromosomes being localized within 'territories' (see NUCLEUS). DNA sequences within a chromosome are organized into either EUCHROMATIN or HETEROCHROMATIN, and in some cases large chromosomal loops containing active genes extend outside the defined territories. Changes in transcriptional activity are often coupled with changes in the subnuclear localization of chromosomes (see GENE EXPRESSION). Thus, in developing B cells and T cells, silent genes are repositioned in the nucleus at pericentromeric heterochromatin. Gene regulatory elements (e.g., locus control regions, enhancers and insulators) act by repositioning specific genetic loci to regions of active or silent transcription, while sequence-specific DNA-binding proteins may do similarly. See CHROMATIN REMODELLING, POLYCOMB GROUP, SYSTEMIC LUPUS ERYTHEMATOSUS.

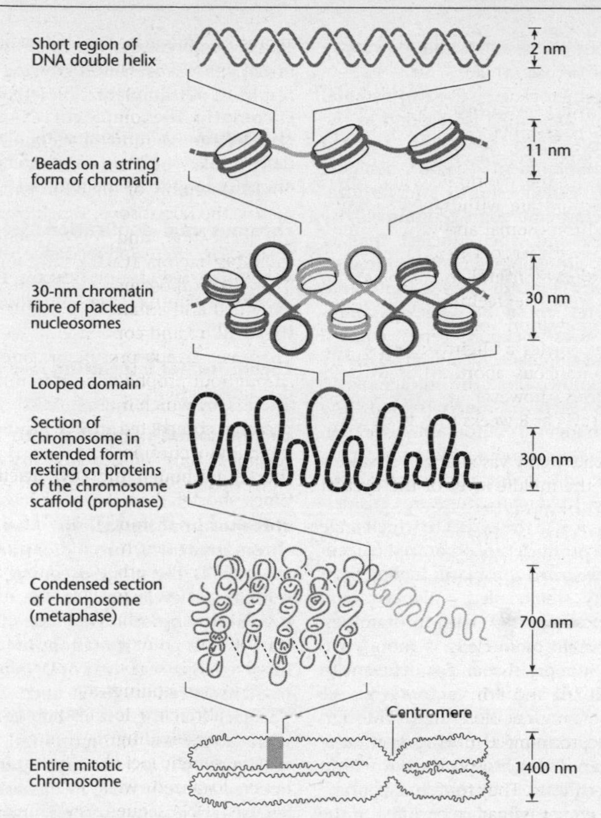

FIG 51. Diagram illustrating a widely accepted model for the hierarchy of CHROMATIN packaging, starting with the formation of nucleosomes ('beads on a string') as DNA is replicated. The final stage produces the fully condensed mitotic chromosome. It has been proposed that the formation of chromatin loops attached to a scaffold is the final stage of chromosome condensation. From *Molecular Biology of the Cell* (4th edition) by B. Alberts, A. Johnson, J. Lewis, M. Raff, K. Roberts and P. Walter. Reproduced by permission of Garland Science/Taylor & Francis LLC.

chromatin immunoprecipitation See DNA-BINDING PROTEINS.

chromatin remodelling Modification of CHROMATIN structure, as by changes to its structural proteins (especially HISTONES), by multi-subunit *chromatin remodelling complexes* (comprising >10 subunits which include ATP-dependent enzymes) and *cis*-regulatory elements (e.g., LOCUS CONTROL REGIONS, ENHANCERS) that are able to expose or redistribute NUCLEOSOMES, often in a tissue-dependent manner, and by RNA INTERFERENCE. They almost certainly enable DNA REPLICATION to proceed without displacing nucleosomes from the DNA. By remodelling chromatin, COREGULATORY PROTEINS (e.g., PGC-1), response element binding proteins (e.g., CREB-1) and TRANSCRIPTION FACTORS can access the DNA more easily (see GENE EXPRESSION). The result may be either the formation of a transcriptionally 'open' chromatin state, or the restoration of a transcriptionally inactive state when selected gene expression is shut down (see STEM CELLS, POLYCOMB GROUP).

During gametogenesis, the two parental genomes are formatted to respond to the oocyte's cytoplasm so as to proceed through development. Zygotes remodel the paternal genome shortly after fertilization, but before embryonic genome activity commences; and when nuclear transfer techniques are employed in mammalian cloning, any somatic nuclei transferred into oocytes must be quickly reprogrammed so as to express genes required for early development. Although relatively little is yet known about the initial molecular events involved, such remodelling involves changes in chromatin structure, DNA METHYLATION, GENOMIC IMPRINTING, TELOMERE length adjustment, and X CHROMOSOME INACTIVATION. In humans, there are considerable differences in the extent to which male and female pronuclei are organized epigenetically, the sperm chromatin having been sequentially remodelled, silenced and ultimately compacted with protamines; but the female pronucleus is more transcriptionally repressive and is deficient in generalized transcription factors (e.g., see TATA-BINDING FACTOR). DNA methylation is also more pronounced during spermatogenesis than during oogenesis, but within hours of fertilization, the paternal genome is actively demethylated in contrast to the maternal genome which appears to be passively demethylated during cleavage. By the blastocyst stage, the embryo's genome is hypomethylated but at the pre-gastrulation stage undergoes global *de novo* methylation, the extent varying with different cell lineages: somatic cell lineages are heavily methylated while TROPHOBLAST-derived cell lineages are hypomethylated. Early PRIMORDIAL GERM CELLS' genomic DNA remains largely unmethylated until gonadal differentiation has taken place. The expression of imprinted genes is set by parent-of-origin-specific methylation marks during late gametogenesis and, when lost, cannot be reset until passage through the germ line. Genome-wide disruption of imprinted gene expression, as occurs in uniparental embryos and nuclear transfer embryos derived from male primordial germ cells or non-growing oocytes, results in post-implantation lethality.

chromium A mineral with a reference daily intake of 120 μg and associated with enzymes in glucose metabolism.

chromosomal duplication See CHROMOSOMES (Fig. 52). For evidence of *genome duplication*, see GENE DUPLICATION. For aneuploids containing extra chromosomes, see NONDISJUNCTION.

chromosomal imprinting See EPIGENESIS, GENETIC IMPRINTING.

chromosomal instability (CIN) The most common form of genomic instability in CANCER CELLS. See GENOMIC INSTABILITY.

chromosomal mutations Alterations to chromosome structure and/or number that fall outside the range occurring during a normal cell cycle (see Fig. 81, GENETIC DISEASES AND DISORDERS). They are often contrasted with point mutations because they involve larger segments of DNA and often result in alterations visible microscopically. The fact that the risk of having children with abnormal chromosomes increases with maternal age indicates that primary oocytes are vulnerable to damage as they age (see AGE-RELATED DISORDERS, OOGENESIS).

Structural abnormalities result from misrepair of chromosome breaks (see DNA REPAIR MECHANISMS, GENOTOXIC STRESS) and from recombination between non-homologous chromosomes. Those involving a single break within one chromosome give rise to terminal deletions; those involving two breaks within a single chromosome include interstitial deletions, inversions, and duplication or deletion by unequal sister chromatid exchanges; while those involving three breaks within a single chromosome produce various rearrangements, such as inversion with deletion. Where two chromosomes are involved, those resulting from two breaks include reciprocal translocations, ROBERTSONIAN TRANSLOCATIONS, and duplication or deletion by unequal crossovers; while those resulting from three breaks include interchromosomal insertion, either direct or inverted.

Numerical abnormalities include polyploidy, autosomal aneuploidy and sex chromosome aneuploidy. Triploidy occurs in 1–3% of recognized human pregnancies and usually results from an egg being fertilized by two sperm. Few triploids survive to term, both this condition and the much rarer tetraploidy being incompatible with human life. Aneuploidy arises either through nondisjunction (failure of paired chromosomes to separate at 1st meiotic anaphase, or of sister chromatids to separate at 2nd meiotic anaphase) or through failure of a chromosome or chromatid to be included in one of the daughter nuclei as a result of its delayed movement during anaphase: 'anaphase lag'.

chromosome banding The genome is divisible into large replication time zones (see DNA REPLICATION), which gives rise to a distinctive banding pattern for each chromosome (see G BANDS). In Q-banding, CHROMOSOMES are stained with a fluorescent dye which binds preferentially to AT-rich DNA and viewed by UV fluorescence, fluorescing bands being known as Q bands and marking the same chromosomal segments as G bands. R-banding is essentially the reverse of the G-banding pattern, chromosomes being heat-denatured in saline prior to staining with Giemsa. This denatures AT-rich DNA, R bands being Q-negative. In T-banding, telomeres are identified as a subset of the R bands after severe heat treatment prior to Giemsa staining. In C-banding, thought to reveal constitutive heterochromatin, chromosomes are typically denatured with barium hydroxide solution prior to Giemsa staining. See CHROMOSOME PAINTING.

chromosome inactivation See DNA METHYLATION, HETEROCHROMATIN, X CHROMOSOME INACTIVATION.

chromosome map, c. mapping See GENOME MAPPING.

chromosome painting (chromosome *in situ* suppression hybridization) A method of finer-grained analysis of the organization of DNA than that provided by study of G BANDS, and a special application of FISH. It employs the same fluorescent dye on DNA probes to tag different DNA sequences from a single chromosome (or their amplified PCR products). DNA from different chromosomes can be tagged with different dyes. When hybridized back to a metaphase spread of chromosomes, tagged DNA derived from different loci spanning a whole chromosome gives a 'chromosome paint' in which the whole chromosome fluoresces. The procedure allows *de novo* recognition of chromosome rearrangements in clinical and cancer cytogenetics, but is also employed in comparative genome hybridization, where total DNA from two different, but related, sources are each tagged with a different dye. This can reveal differences involving gain or loss of subchromosomal regions, or of entire CHROMOSOMES.

chromosomes Thread-like structures, composed of chromatin, housed within the NUCLEUS of a cell, their DNA constituting the nuclear GENOME, two of its major functional features being the CENTROMERE and terminal TELOMERES (see also HETEROCHROMATIN, HUMAN ARTIFICIAL CHROMOSOMES). Chromosome morphology provides information about the evolutionary relationships (phylogeny) between primate species and each chromosome is a mosaic of different relationships (see GENETIC DIVERSITY, Fig. 82). Fig. 52 indicates some mechanisms involved in chromosomal duplication that occur during evolution, and synteny provides evidence for chromosomal rearrangements that have occurred during phylogeny. The number and physical appearance of mitotic chromosomes at late prophase, and at metaphase, determines the karyogram, while different CHROMOSOME BANDING methods reveal something of the underlying chromatin structure (see CHROMATIN, Fig. 51). The most obvious difference between human and great ape chromosomes is that humans have 23 pairs of chromosomes while the two chimpanzees, gorilla and orang-utan have 24 pairs.

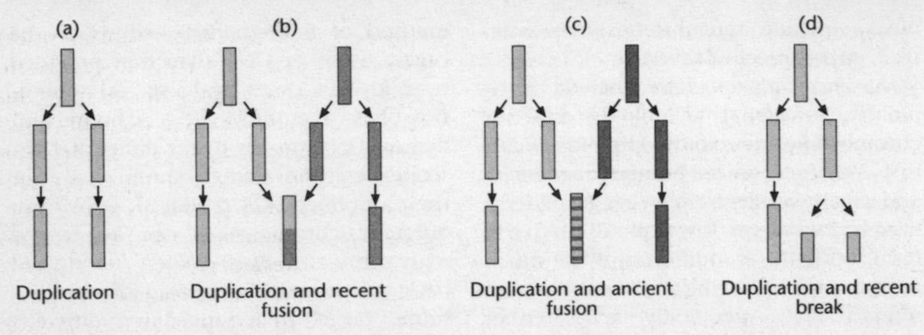

FIG 52. Diagram of a model of chromosome duplication followed by the four simplest chromosome rearrangements: (a) no rearrangement; (b) two different duplicate copies which fused recently; (c) two different duplicate copies which fused soon after the duplication and have had time for subsequent rearrangements; (d) a duplicate chromosome fragmented very recently. See CHROMOSOMES entry. From Hugues Roest Crollius, et al., © *Nature* 2004, with permission from Macmillan Publishers Ltd.

Chromosomes stained with Giemsa reveal light and dark bands (see G BANDS). Such G-banded karyograms (see Appendix III for the human karyogram) reveal a high degree of similarity between these four species, the difference in chromosome number apparently arising since the last common ancestor of humans with chimpanzees, by simple end-to-end fusion of two small chromosomes to form the large human metacentric chromosome 2. Fluorescent dyes can be used to distinguish chromosomes by using chromosome-specific DNA probes (see FISH, CHROMOSOME PAINTING). When comparisons are made between human chromosomes and those of other species, the first letter of the generic name and the first two of the trivial name prefix the chromosome, so that HSA18 indicates chromosome 18 of *Homo sapiens* (e.g., see SYNTENY). Nine chromosomes underwent major structural rearrangement during the evolution of the human lineage: chromosomes 1, 2, 5, 9, 12, 15, 16, 17 and 18.

Human chromosomes may be placed in the following groups: A (1–3), the largest chromosomes, 1&3 being metacentric, 2 being submetacentric; B (4&5), large and submetacentric, the two arms being of very different sizes; C (6–12, X), medium-sized and submetacentric; D (13–15), medium-sized and acrocentric, with satellites; E (16–18), small, 16 being metacentric; 17&18 submetacentric; F (19&20), small and metacentric; G (21,22, Y), small and acrocentric, with satellites on 21&22, but not on Y.

The International System for Cytogenetic Nomenclature (ISCN) arrived at a standardized nomenclature in 1971. Short arm locations are labelled p (*petit*) and long arm q (*queue*). Each arm is divided into regions, labelled p1, p2, etc., and q1, q2, etc. counting outwards from the centromere. These regions are delimited by specific landmarks, such as the centromere, the ends of the arms, and certain chromosome bands. Regions are then divided into bands labelled p11 (one-one), p12, etc., and sub-bands, e.g., p11.21, p11.22, in each case counting outwards from the centromere. Relative distance from the centromere is indicated by the words 'proximal' or 'distal'. Proximal Xq thus indicates the segment of the long arm of X closest to the centromere; distal 2p indicates the portion of the short arm of chromosome 2 most distant from the centromere, and so closest to the telomere. See CHROMOSOMAL MUTATION, HAPLOTYPE.

chromosome territories See NUCLEUS.

chronic In a chronic viral infection, $CD8^+$ memory T cells fail to respond to a new infection by the same pathogen (they are said to be in an 'exhausted' state), leading to a persistent infection. There is evidence that whereas functional T cells normally lose expression of the PD-1 antigen receptor protein they had during their acute response, memory T cells retain high levels of PD-1 expression during chronic infection. It may be that binding of PD-1 to the PD-L1 ligand normally expressed on the surfaces of chronically infected cells inhibits signalling through the T-cell antigen receptor and explains why exhausted T cells fail to divide in response to antigen or exhibit killer activities such as secretion of interferon γ and tumour necrosis factor α. Contrast ACUTE DISEASE.

chronic bronchitis An inflammatory disease characterized by excessive bronchial mucus secretion accompanied by a productive cough (i.e., in which sputum is raised) lasting at least three months of the year for two successive years. A chronic obstructive pulmonary disease, its leading cause is CIGARETTE SMOKING. Inhaled irritants lead to increased size and number of mucous glands and goblet cells, thickened and excessive mucus narrowing the airway and reducing ciliary effectiveness. See EMPHYSEMA.

chronic myeloid (myelogenous) leukaemia (CML) A LEUKAEMIA in which the nuclei of blood or bone marrow cells of 90% of patients contain a tiny ('Philadelphia', Ph^1) chromosome (see ONCOGENE) derived from a chromosome 22 by reciprocal translocation to and from the long arm of chromosome 9, and juxtaposes the *c-abl* oncogene from chromosome 9 with the *bcr* gene of chromosome 22, resulting in a chimaeric transcript derived from both genes that creates the oncoproteins BCR-ABL whose uncontrolled kinase activity is sufficient to cause clonal expansion of pluripotent haematopoietic STEM CELLS characteristic of leukaemia, making it an ideal therapeutic target (see TRANSFORMATION). By blocking this protein-tyrosine kinase, the drug imatinib (Gleevac) is a 'molecule-specific' therapeutic agent that induces nearly complete and sustained remissions in nearly all patients in the early stages of CML. See CHEMOTHERAPY.

chronic obstructive pulmonary disease (COPD) A respiratory disorder characterized by chronic and recurrent obstruction of airflow, increasing airway resistance. Its principal forms are chronic BRONCHITIS and EMPHYSEMA, and its most common cause is CIGARETTE SMOKING or 'passive smoking'. See α_1 ANTI-TRYPSIN.

chyle The milky-appearing fluid found in lacteals of the villi and SMALL INTESTINE after digestion.

chylomicrons LIPOPROTEIN aggregates, ~300 μm in diameter, comprising 1% protein and 99% lipid, transporting packaged triglycerides and cholesterol in the blood plasma. Formed by ENTEROCYTES of the small intestine, they enter the lacteals of the lymphatic system (since lymphatic capillaries lack a basement membrane), and thence the blood, in which they are carried to muscle and adipose tissues. In the blood capillaries of these tissues, their contained apolipoprotein C-II (ApoC-II) activates LIPOPROTEIN LIPASE on cell surfaces, this hydrolysing triglycerides into fatty acids and glycerol and enabling uptake of these molecules by muscle fibres and adipocytes. The remnant chylomicrons, lacking most of their triglyceride but containing cholesterol and apolipoproteins, remain in the plasma, bind to liver cell apolipoprotein receptors and are taken up by endocytosis where the cholesterol is released, the remainder being degraded in lysosomes. See BILE SALTS; compare MICELLES.

chyme The semi-fluid mixture of partially digested food and digestive secretions in the stomach and SMALL INTESTINE during food digestion. See CRYPTS OF LIEBERKÜHN.

chymotrypsin, chymotrypsinogen Chymotrypsin is a serine protease component of pancreatic juice, specific to the cleavage of peptide bonds adjacent to aromatic (hydrophobic) amino acid residues. It also catalyses the hydrolysis of small esters and amides. Its pro-enzyme (chymotrypsinogen) is acted on by trypsin to give the active enzyme.

	Lifelong non-smoker	*Smokers*			
Most recent number of cigarettes smoked per day before onset of disease	–	≥1 <5	≥ 5 <15	≥ 15 <25	≥ 25
Relative risk	1	8	12	14	27

TABLE 2 *Some of the early evidence linking lung cancer with* CIGARETTE SMOKING.

cigarette smoking The compelling association between cigarette smoking and lifetime risk of LUNG CANCER was presented by two groups of epidemiologists, in 1949–1950. The initial results from Richard Doll's group are given in Table 2 (the relative risk is compared with that of a non-smoker, which is set at 1).

It was data from the Framingham Heart Study, which now contains data from the past 60 years, that established smoking and cholesterol as risks for heart disease (see SINGLE NUCLEOTIDE POLYMORPHISMS). Cigarette smoke is rich in a variety of mutagenic carcinogens, including 3-methylchoanthrene (3-MC). Heavy smoking in fathers is reported to confer a fourfold increase in cancer risk to their children. Although ethanol has almost no mutagenic powers, the contribution of distilled alcoholic drinks to tumorigenesis seems to be due to their toxic effects on the epithelial cells of the mouth and throat so that stem cells underneath respond by regenerative division. If these cells already carry mutant alleles induced by tobacco tar, clonal expansion may enable their descendants to acquire further mutations leading ultimately to aggressive head-and-neck cancers (see LIFESTYLE).

Cigarette smoking influences cytokines, causing a dose-dependent increase in plasma ICAM-1 levels and a decrease in ADIPONECTIN levels, inducing endothelial dysfunction (see ENDOTHELIUM) – an early marker of ATHEROSCLEROSIS – observed in chronic smokers. It is not clear whether these effects are direct effects of NICOTINE, CARBON MONOXIDE or other components of tobacco smoke. The effects of free radicals in tobacco smoke may also enhance the oxidative stress already present (see MITOCHONDRION). Both hydrogen peroxide and nicotine suppress adipokine expression in ADIPOCYTES and smoking may directly regulate adiponectin concentration through lipolysis. Lower adiponectin levels found in chronic smokers confirm earlier findings that chronic smokers suffer from INSULIN RESISTANCE, which they share with those suffering from abdominal OBESITY. Meningococcal meningitis epidemiology reveals that one of its determinants is passive inhalation of cigarette smoke and chronic smokers have twice the risk of developing AGE-RELATED MACULA DEGENERATION. Smoking is the leading cause of CHRONIC BRONCHITIS, the cause of EMPHYSEMA, and is a factor in onset of OSTEOPOROSIS in women.

ciliary body The most anterior portion of the choroid (see EYES, Fig. 73), comprising: (i) the folded *ciliary process* containing blood vessels that secrete the aqueous humour and attaching to suspensory ligaments that connect to the lens; (ii) a circular band of smooth muscle, the ciliary muscle, forming its lateral aspect and altering the shape of the lens during ACCOMMODATION.

cilium (pl. cilia) Almost every vertebrate cell has a specialized non-motile cell surface projection, a *primary cilium*, which probes the extracellular environment for molecules recognized by its receptors. Primary cilia coordinate numerous intercellular signalling pathways regulating growth, survival and differentiation of cells during embryonic life, and maintenance of healthy tissues thereafter. Ciliary membranes are rich in cholesterol, the precursor of oxysterols.

All cilia (flagella are long cilia) are generated during interphase from a plasma membrane-associated foundation, the basal body, at the heart of which is a centriole. During interphase, the centriole moves to the plasma membrane and acts as a template for formation of the *axoneme*, the microtubular core of the cilium. These nine doublet microtubules, originating from the nine triplet microtubules of the basal body centriole, extend the length of the cilium at the ciliary tip. But because no protein synthesis occurs within cilia, its growth and maintenance require the intraflagellar transport (IFT) complex to move its structural components from the cell body to the ciliary tip, an anterograde movement driven by the motor protein Kinesis-2 (see HEDGEHOG GENE). The system is balanced between motors that build and those that dismantle a cilium, and if IFT fails, a cilium will break down. Most motile cilia have an additional central microtubular pair (the '9 + 2' microtubule arrangement); but non-motile cilia such as primary cilia lack this central pair (but the primary cilia of the PRIMITIVE NODE are an exception). The ciliary motor includes a left-right DYNEIN (see entry), a protein whose heavy chain in humans is encoded by the *DNAH5* gene. Mutations in this are associated with immotile cilia syndrome, an inherited disorder that includes mirror-image reversal of the internal organs in half of affected individuals (see SITUS INVERSUS). *Primary ciliary dyskinesia* (PCD) includes this and other inherited diseases in which mutations lead to defective ciliary proteins. Affected cilia are often shorter than normal and often immotile or move in a disorganized manner. Those with PCD tend to have an increased number of respiratory tract and sinus infections. In addition to the roles of the more familiar motile cilia in wafting mucus along the respiratory tract and in propelling spermatozoa (males with PCD have reduced fertility), non-motile primary cilia are crucial as sensory organelles in OLFACTION and vision (see ROD CELLS, CONE CELLS). In addition, they are involved in regulating the response of cells to Sonic hedgehog (see SONIC HEDGEHOG (SHH) GENE). See POLYCYSTIC KIDNEY DISEASE, STEREOCILIA. For kinocilium, see HAIR CELLS.

cingulate cortex Limbic region of the cortex, projecting to the HIPPOCAMPUS (see PAPEZ CIRCUIT). The right anterior cingulate cortex (ACC) is a relatively ancient part of the cortex and involved in many autonomic functions, motor and digestive, and in the regulation of heart rate and blood pressure. It is one region implicated in PAIN perception, and in the regulation of empathy and other emotional responses. It receives input from the hypothalamus via the mammillary bodies and relays signals from the amygdala to the frontal cortex. A 1999 report on SPINDLE CELLS in the ACC has aroused great interest. The ACC is active during demanding tasks involving judgement and discrimination, and signals from it have been detected in the frontal polar cortex (Brodman area 10), believed to be involved in regulation of cognitive dissonance (disambiguation of alternatives).

cingulate gyrus A deep cortical component of the LIMBIC SYSTEM (see CEREBRAL HEMISPHERES, Fig. 46) located along the inner surface of the longitudinal fissure, just above the corpus callosum. It is involved with the hypothalamus in causing a feeling of satisfaction associated with satiation (see APPETITE). The cingulate cortex projects to the HIPPOCAMPUS.

circadian clocks, circadian rhythms Circadian clocks are endogenous clocks residing in most body cells, whose major integrator are the circadian pacemakers located in the SUPRACHIASMATIC NUCLEI (SCN) of the anterior hypothalamus. The unexpected discovery that most peripheral tissues contain intrinsically independent circadian pacemakers pointed to the likely presence of a 'synchronization web' coordinating timing in all tissues. Many aspects of mammalian behaviour and physiology are subject to daily oscillations, or *circadian rhythms*, including SLEEP-WAKE CYCLES, activity, energy homeostasis, blood pressure, body temperature, renal activity, and liver metabolism. These are driven by a central circadian clock in the SCN, an intrinsic neural oscillatory pacemaker whose rhythm is sustained even when the neurons are isolated in cell culture. Generating cycles of approximately 25 hours, its phase must be adjusted each day to remain in resonance with geophysical time. *In vivo*, and in the absence of external cues, it fires with a sinusoidally varying frequency, peaking in the daytime. Circadian time (CT) is defined in hours since the onset (CT0) at hour 0 of a subjective light period. Circadian time in the SCN is entrained (synchronized) to the shifting solar day photoperiod (light:dark ratio) by glutamatergic innervation from the MELANOPSIN-containing retinal ganglion cells, mainly to the SCN core, via the RETINOHYPOTHALAMIC TRACT, and by MELATONIN secretion from the pineal gland during the dark cycle. Temperature, activity and food intake (e.g., glucose) modulate the rhythm. Most SCN neurons are GABAergic and co-release peptides that are assumed to inhibit their targets. The SCN synchronizes the peripheral clocks in other body cells through its neural and hormonal outputs to generate diurnal rhythms in sleep–wake cycles and in autonomic and endocrine functions. These appear to be tightly regulated in time as a result of daily waves of differential gene expression underpinning autonomous tissue-specific metabolic programmes. In circadian day, our physiology is given over to catabolic processes, whereas at night anabolic functions of growth and repair predominate (see CELL CYCLE). Most of the SCN output goes via axons in the subparaventricular zone (dorsal, dSPZ and ventral, vSPZ pathways) to the dorsomedial nucleus of the hypothalamus (DMH). Neurons of the vSPZ relay information required to coordinate daily sleep–wake cycles via the DMH to the ventrolateral preoptic nucleus (VLPO), while dSPZ neurons, projecting back to the medial preoptic area, control rhythms of body temperature. The DMH also projects to the paraventricular nucleus, controlling CRH release. The DMH thus integrates outputs from the SPZ with other inputs and drives circadian cycles of sleep, activity, feeding and corticosteroid secretion. Circadian rhythms are enabled to adapt to environmental stimuli, such as food availability (leptin and ghrelin acting on the ventromedial and arcuate nuclei), as well as sensory inputs, cognitive influences from the prefrontal cortex and emotional input from the limbic system.

Circadian clocks seem to operate through interlocking feedback loops of gene expression (see Fig. 53). Bmal1 and Clock proteins are the players of the positive loop, activating transcription of the cryptochrome (*Cry*) and period (*Per*) genes, which are major components of the negative loops. These loops are established by transcription activators operating as heterodimeric proteins (usually the basic helix-loop-helix Clock:Bmal1 complex, but the related Npas2:Bmal1 in brain regions), and by the transcriptional repressor of the Clock:Bmal1 complex, the Per/Cry heterodimer. E-boxes are DNA sequences within promoters that bind proteins belonging to the basic helix-loop-helix (bHLH) family of transcription factors. E-box-mediated activation of genes (e.g., *Per* and *Cry*) by Clock: Bmal in early circadian day is inhibited in late circadian day by nuclear accumulation of Per/Cry, closing an oscillatory negative feedback loop. The subsequent circadian expression cycle is initiated when Per/Cry levels decline as these protein products suppress their own gene expression. *Rev-erbα* is an orphan nuclear receptor and, by recruiting corepressor proteins, negatively regulates *Bmal1* transcription. *Rev-erbα* is expressed in phase with *Per* and *Cry*, driving *Bmal1*, in antiphase to the negative factors, thereby contributing to the initiation of a new cycle of gene expression. It is dispensable for basic oscillator function, but participates in determining period length and phase-shifting properties of the mammalian circadian timing system.

The striking finding that transcription of ≥10% of all cellular (metabolic) genes oscillates in a circadian manner illustrates how profoundly circadian transcriptional machinery influences a wide variety of cellular functions. At least 335 oscillating hepatic genes are known. Clock-controlled gene (CCG) expression is the ultimate regulator of metabolic rhythms throughout the

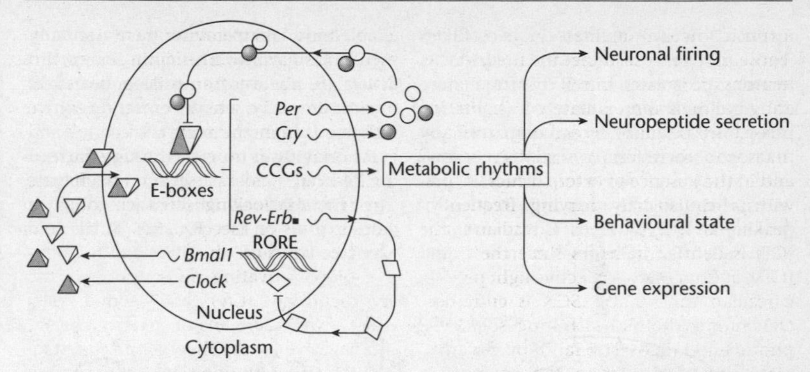

FIG 53. Diagram illustrating some important molecular feedback loops defining circadian time. *Per* = Period gene; *Cry* = Cryptochrome gene; ROREs are recognition sequences for members of the Rev-Erbα and ROR orphan nuclear receptor families. See CIRCADIAN CLOCKS entry for details. From M.H. Hastings, A.H. Reddy. © *Nature Reviews Neuroscience*, Macmillan Publishing Ltd. Redrawn with the permission of the publisher.

body, some of these genes being driven in phase with *Per*, while others are driven by RORE promoter recognition sequences and are sensitive to negative regulation by Rev-erbα in phase with *Bmal1*. Downstream signal cascade effects of some of these CCGs will orchestrate expression of further gene clusters to circadian phases. In 2007, work on mice indicated that the transcriptional coactivator PGC-1α (see PGC-1α) is strategically placed in the circadian control of energy metabolism. The ability of PGC-1α to activate *Bmal1* transcription is modulated by the relative abundance of the ROR and Rev-erb families of orphan receptors. Recent work shows that, in mice at least, PGC-1α is rhythmically expressed in liver and skeletal muscle, exerting only modest effects in these tissues on expression of *RORα* and *RORγ*, but promoting transcription of *Rev-erbα* and *Rev-erbβ*. In *PGC-1α* null mice, VO_2 is significantly higher in the dark than in wild-type mice, and higher in both light and dark phases, blunting the diurnal oscillation of VO_2.

A recent (2007) study on male mice given high-fat diets indicated a slight lengthening of their daily activity schedule. The link between circadian rhythm and metabolism may not be surprising since the two systems share many molecular signalling pathways. Expression patterns of some genes involved in lipid metabolism change in 24-hour cycles, while several nuclear receptors activated by sterols regulate clock-related genes. Conversely, a disrupted circadian rhythm leads people to crave high-fat foods, and there is evidence that children who lack sleep risk being overweight. Mutations in core clock genes correlate with altered stability, amplitude and/or length of activity cycles and, where tested, with altered circadian rhythms of electrical firing of the SCN *in vitro* and with sleep patterning in humans. The entrained oscillator loop imposes temporal order within and beyond the SCN through regulated expression CCGs that lie outside the loop but undergo periodic transcriptional activation and repression by Per/Cry complexes. There is evidence that the DNA-binding activity of the brain cell complex Npas2:Bmal1 is optimal when the ratios of reduced-to-oxidized NADP(H) and NAD(H) coenzymes are high; i.e. under anaerobic conditions. Changes in food intake affect cellular redox state, excess carbohydrate favouring a reducing environment. The Npas2:Bmal1 heterodimer activates the gene for lactate dehydrogenase (LDH), catalysing conversion of pyruvate to lactate in anaerobic conditions, influencing cellular redox potentials and thereby the effectiveness of its activator.

Such a negative feedback loop may contribute to circadian clock activity and help explain how the human forebrain oscillates between states of alertness and tiredness. As evening progresses, our body temperature falls and melatonin is secreted, facilitating sleep. Sleep onset is accompanied by increased secretion of GROWTH HORMONE (see SLEEP) and PROLACTIN, whereas predawn circadian activation of the corticotropin/cortisol axis prepares us for the demands of awakening. Alertness and core body temperature vary similarly, while plasma growth hormone and cortisol levels are highest during sleep, but at different times. Potassium excretion by the kidneys is highest during the day.

As individuals progress through the regular 24-hour cycle of sleep and wakefulness, their metabolism is adjusted accordingly in anticipation of the demands of the solar day (see Fig. 54).

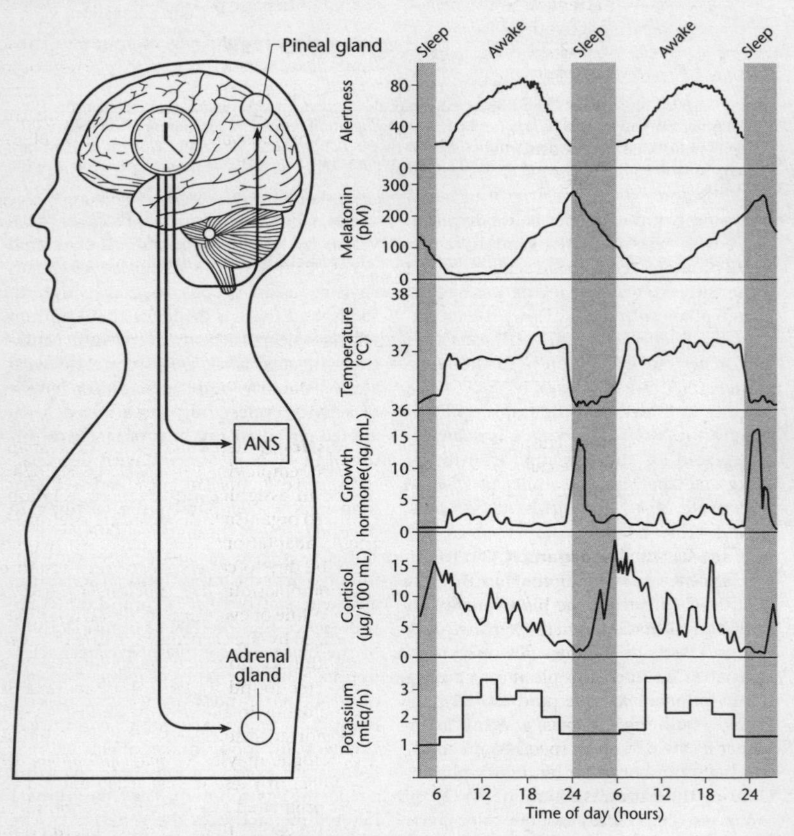

FIG 54. A current model of circadian organization in which a pacemaker in the suprachiasmatic nuclei (clock in diagram) projects through neural and endocrine pathways to drive and/or synchronize rhythms in peripheral physiology and behaviour (see text of CIRCADIAN CLOCKS). This ensures that as people progress through the 24-hour cycle of sleep (shading) and wakefulness, metabolism is adjusted to anticipate the demands and opportunities of the solar day. ANS = autonomic nervous system. Redrawn in part from © 2001 *Neuroscience* (2nd edition) by M.F. Bear, B.W. Connors and M.A. Paradiso, published by Lippincott, Williams & Wilkins. Reproduced with the kind permission of the publisher.

Circadian control of normal physiology, perhaps inevitably, results in its association with circadian variation in disease. Acute cardiovascular and cerebrovascular episodes (angina pectoris, intracerebral haemorrhagic stroke) show a pronounced morning peak associated with circadian changes in blood pressure, cardiac output, etc. The relative risk of acute myocardial infarction is 40% higher in the morning. Beta-blockers that attenuate the circadian surge in autonomic activation also stop the morning peak of CARDIOVASCULAR DISEASE (CVD). In the modern '24/7 society', social and commercial pressures opposing the internal temporal order are a growing cause of circadian stress, and are implicated in the aetiologies of chronic illnesses such as CVD and CANCER. See SEASONAL AFFECTIVE DISORDER.

circadian rhythms See CIRCADIAN CLOCKS.

circle of Willis (cerebral arterial circle) A ring of arteries forming an anastomosis at the base of the brain between the internal carotid and basilar arteries and arteries supplying the brain.

circulatory systems Humans have two circulatory systems: the CARDIOVASCULAR SYSTEM and the LYMPHATIC SYSTEM, between which lies the TISSUE FLUID.

circumventricular organs (CVOs) Brain midline organs lying along the third and fourth ventricles on the blood side of the BLOOD–BRAIN BARRIER, where capillaries are fenestrated and allow relatively free exchange between the plasma and tissue fluid. They include the posterior pituitary gland and choroid plexus, sealed off from the rest of the brain by specialized ependymal cells held together by tight junctions. Because this barrier is absent, oxytocin and vasopressin can be released directly into the systemic circulation. Other CVOs enable the brain to monitor the blood's water potential, ion concentrations and other selected molecules for homeostatic regulation. Thus the vascular organ of the lamina terminalis (OVLT), which contains OSMORECEPTORS, is a CVO. See VASOPRESSIN.

cirrhosis Destruction of LIVER parenchymal cells and their replacement with fibrous tissue that eventually contracts around the blood vessels, impeding flow of portal blood through the liver. It results most commonly from alcoholism (see ALCOHOL), but can result from poisoning (e.g., by carbon tetrachloride), viral diseases (e.g., infectious HEPATITIS), or obstruction and/or infection of the bile ducts. It is a common cause of hepatic jaundice.

***cis*-acting regulatory elements** Those REGULATORY ELEMENTS (PROMOTERS, promoter-proximal sequences, ENHANCERS, SILENCERS and LOCUS CONTROL REGIONS) that lie on the same chromosome as the regulated element.

cisternae See ENDOPLASMIC RETICULUM, GOLGI APPARATUS.

citric acid cycle See KREBS CYCLE.

CJD See CREUTZFELDT–JAKOB DISEASE.

clade All the descendants from a single node in a phylogenetic tree (see PHYLOGENETIC TREES). A monophyletic group. See CLADISTICS.

cladistics A method of biological classification employing inferred genealogies alone in assigning phylogenetic relationships to organisms, disregarding their phenetic adaptations (see PHENETICS), which may be due to convergence and therefore be homoplasious (i.e., not due to HOMOLOGY). A line of evolutionary descent is characterized by the occurrence of one or more evolutionary novelties (*apomorphies*). Any character found in two or more taxa is homologous in them if their most recent common ancestor also had it. Such a shared homologue may be *symplesiomorphous* (or *-morphic*) in these taxa if it is believed to have originated as a novelty in a common ancestor earlier than the most recent common ancestor, but *synapomorphous* (or *-morphic*) if not. See CLADE, HOMINIDAE, PHYLOGENETIC TREES.

classical conditioning (Pavlovian conditioning) Introduced into the study of learning by Pavlov at the start of the 20th century, classical conditioning is the learnt association between two stimuli: one of them ordinarily evokes a measurable reflex response, but the second either does not do so – or if it does, the response is either weak or unrelated to the first. After repeated pairings of the two stimuli, presentation of the second stimulus alone now elicits a new or different reflex response, the *conditioned response*, from any response it originally elicited.

The stimulus that originally brought forth a measurable response is termed the *unconditioned stimulus*, or US. The stimulus with which it is paired is termed the *conditioned stimulus*, or CS. After repeated pairing, the conditioned stimulus appears to become an anticipatory signal for the conditioned response. The relative timing of presentation of the two stimuli is important for successful conditioning to occur. The US and CS normally must either be presented simultaneously, or with the CS preceding the US by a brief interval. Indeed, classical conditioning develops best if, in addition to the contiguity of stimuli, there is an actual contingency between the conditioned and unconditioned stimuli (see CEREBELLUM and Fig. 43 for a simple form of conditioned motor learning). This prompts the suggestion that humans (and other animals) acquire classical conditioning because of the selective advantage gained by detection of causal relationships in their environments. If this is so, the neural mechanism underlying the behaviour would have evolved by NATURAL SELECTION.

The intensity, or probability, of occurrence of the conditioned response decreases if the conditioned stimulus is repeatedly presented without the unconditioned stimulus, a process termed EXTINCTION. Unlike classical conditioning, INSTRUMENTAL CONDITIONING can be regarded as the formation of a predictive relationship between a *response* and a stimulus rather than between two stimuli. For some neural correlates of classical conditioning, see LEARNING.

clathrin For clathrin and clathrin-coated vesicles, see VESICLES.

claustrum A thin sheet of grey matter parallel to and below part of the cortex; present in all mammals and of unknown function; two-way connections exist between the claustrum and most, if not all, parts of the cortex as well as subcortical structures involved in emotion.

clavicle (collarbone) S-shaped bone of the pectoral girdle, lying horizontally and bilaterally in the superior and anterior part of the thorax, superior to the first rib. Its broad, flat lateral end, the *acromial extremity*, articulates with the acromion of the SCAPULA. The medial end of the clavicle (the *sternal extremity*) is rounded and articulates with the sternum, forming the *sternoclavicular joint*. The *costal tuberosity* on the inferior surface of this end serves for attachment of the costoclavicular ligament.

CLC ion channels Proton-chloride exchangers that transport protons and chloride ions in opposite directions across cell membranes. See ION CHANNELS.

cleft lip, cleft palate A variable condition resulting from failure of midline of upper lip (25%) or of palate (25%), or both (50%) to close during gestation. Associated with loci on chromosomes 1, 6 and X, its heritability is 76%. Its percentage concordance in dizygotic twins is 5%, and 35% in monozygotic twins. Although often associated with syndromes (e.g., Van der Woude syndrome, in which it occurs with pits in the lower lip), its prevalence on its own is 0.1%. See BIRTH DEFECTS.

climacteric See MENOPAUSE.

climate forcing The view that climate change is a major promoter of macroevolutionary change. See HABITAT HYPOTHESIS.

cline A gradient (e.g., of gene frequencies) from one region to another.

clitoris One of the female EXTERNAL GENITALIA, usually <2 cm in length consisting of a shaft and a distal glans, well supplied with sensory receptors which initiate and intensify levels of sexual arousal. Its two erectile tissues are the corpora cavernosa, each expanding to form the crus of the clitoris. During sexual arousal, erectile tissue within the clitoris, and around the vaginal opening, become engorged with blood through parasympathetic stimulation. Erectile tissue corresponding to the corpus spongiosum of the PENIS lies deep, on either side of the vaginal orifice, each forming a bulb of the vestibule. On swelling, these cause narrowing of the vagina during coitus. The clitoris and penis become swollen during REM SLEEP.

clonal selection A central paradigm of ADAPTIVE IMMUNITY, clonal selection theory states that adaptive immune responses derive from individual antigen-specific lymphocytes (B CELLS, T CELLS) that are self-tolerant. These proliferate in response to antigen, and differentiate into: (i) antigen-specific effector cells that eliminate the antigen eliciting their response; and (ii) memory cells to sustain the immunity.

clone, cloning **1**. See DNA CLONING. **2**. Manipulated production of genetically identical cells or individuals. Legislatures in many countries have considered bills either seeking to ban every type of cloning, or that seek to make a distinction between reproductive cloning and therapeutic cloning. Monozygotic TWINS are clones of an unpremeditated kind, whereas *reproductive cloning*, which is banned, would aim to produce a cloned baby. The low success rates of all such mammalian cloning experiments, and the fact that many such animals have serious abnormalities, are probably due in large measure to our current inability to reprogramme epigenetic modifications of the donor cell (e.g., see GENOMIC IMPRINTING) and would make any attempt to clone a human grossly unethical. There is also the (Kantian) view that humans should be valued for themselves and not be used as a means to an end, and that therefore reproductive cloning could only ever be pursued for very questionable motives. Scenarios in which reproductive cloning is the only means a couple has of producing a child of their own could include the situation in which a woman with a serious mitochondrial disorder could not avoid passing the cause to her child through her oocytes, and thus she might benefit from reproductive cloning in which one of her partner's somatic cell nuclei was introduced into a donor enucleated oocyte.

In 1996, the mammalian cloning field was transformed when it was discovered that the cytoplasm of the unfertilized oocyte, unlike that of the zygote, could support reprogramming after nuclear transfer. Embryonic blastomeres produced by somatic cell nuclear transfer (see Fig. 55) ultimately led to the cloning of a sheep, Dolly, in 1997, since when cloning from adult cells has been achieved in numerous mammalian species.

The ability of mouse and bovine zygotes to support nuclear reprogramming has recently been reinvestigated, the conclusion being that in oocytes and zygotes activities crucial for nuclear transfer and/or reprogramming are lost to the oocyte cytoplasm after fertilization. It seems likely that factors critical to these processes are sequestered into the zygote pronuclei and therefore removed during the enucleation process, which would explain the failure of enucleated interphase zygotes to support development after nuclear transfer. It seems that breakdown of the pronuclear envelope on entry to the first cleavage division of the zygote releases critical factors into the cytoplasm, enabling chromosomes in zygotes arrested in mitosis to be removed without loss of any critical factors. When spindle-less chromosomes are microinjected from a donor cell into such an arrested zygote, a new spindle forms around them (see Fig. 56).

Therapeutic cloning involves the production of EMBRYONIC STEM CELLS in which the nucleus is derived from a somatic cell of the donor. The aim is to provide the donor with a source of transplantable cells, tissues or organs carrying the donor's tissue antigens to avoid immunological rejection. The ethical debate here centres on the creation of an embryo specifically for this purpose, or of using 'spare' eggs or embryos from IN VITRO FERTILIZATION clinics (see entry). The 'slippery slope' view holds that if therapeutic cloning is allowed, it will

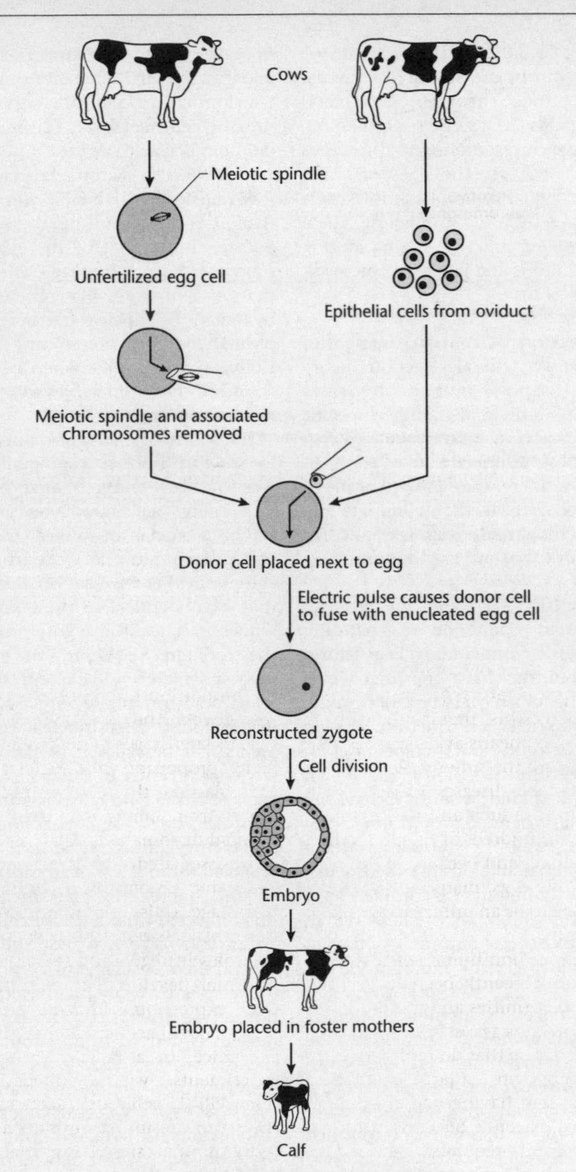

FIG 55. Outline of the method by which a differentiated cell from an adult cow, introduced into an enucleated egg from a different cow, can give rise to a calf. Different calves produced from the same differentiated cell donor animal are genetically identical and are therefore genetic clones of the donor, and of one another. See CLONE, CLONING. © 2002 From *Molecular Biology of the Cell* (4th edition) by B. Alberts, A. Johnson, J. Lewis, M. Raff, K. Roberts and P. Walter. Reproduced by permission of Garland Science/Taylor & Francis LLC.

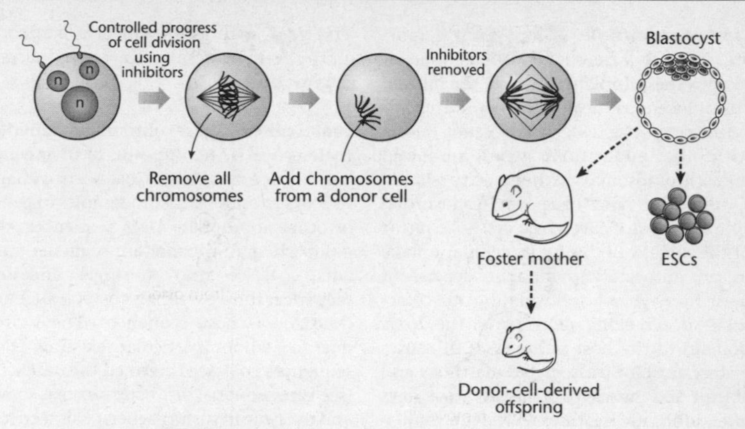

FIG 56. Following production of an abnormal mouse zygote by fertilization with two sperm cells (*n* indicates pronuclei), inhibitor use permitted progression through the cell cycle up to metaphase. The spindle was then mechanically removed and replaced by the condensed chromosomal content of a donor EMBRYONIC STEM CELL (ESC), similarly arrested at metaphase, a new spindle forming around the chromosomes. Removal of inhibitors permitted resumption of development to the blastocyst stage, which then gave rise either to live offspring or to new ESCs. The original mouse zygote was deliberately made aneuploid (3 pronuclei) to test whether it would support subsequent development, in case of possible use of discarded human aneuploid zygotes from IVF reactions for production of human ESC lines. See CLONE, CLONING.

pave the way for use of cloned embryos by unscrupulous operators for reproductive cloning. Against this, therapeutic cloning may be the only means available for eventually alleviating the suffering brought on by many incurable diseases (see SCREENING). Because cloning is such an inefficient process, requiring hundreds of eggs to produce one viable clone, and because of the scarcity of available eggs, many believe therapeutic cloning to be an unrealistic prospect (see TRANSPLANTS).

Efforts to clone non-human mammals by SCNT have until recently been dogged by a high rate of deformities and mortality (see SOMATIC CELL NUCLEAR TRANSFER entry for further details). Those that do result are commonly overgrown ('large offspring syndrome'), often having had a large and dysfunctional placenta. Successful cloning by this method requires considerable nuclear reprogramming, and it is important to determine the optimal cell cycle stage of the donor cell – which differs from mammal to mammal. Very few genes now seem to be required for reprogramming. In general, there is a decrease in the proportion of embryos becoming offspring when the donor nuclei come from cells in later stages of development (e.g., a higher proportion with nuclei from cells of early blastulas than late, and higher using nuclei from embryonic stem cells than from adult somatic cells). This is thought to be associated with the greater level of CHROMATIN REMODELLING required in the later-stage cells. Many imprinted genes (see GENETIC IMPRINTING, EPIGENESIS) have critical roles in regulating foetal growth and placental development. Perturbations in their expressions, linked to phenotypic defects, have been observed in sheep, cattle and mice. In animal (i.e., non-human) experiments, when cloned embryonic stem cells (ES cells) are injected into a blastocyst, the resulting animals appear perfectly normal suggesting that although reproductive cloning is clearly too unpredictable to consider for humans, ES cells derived by SCNT seem to be equivalent to normal ES cells so far as possible therapeutic uses are concerned.

Clostridium difficile A Gram-positive, sporing, obligate anaerobe of variable morphology, establishing itself where the normal anaerobic flora is disrupted. It produces two cytotoxins, A and B, whose genes (along with other minor toxin genes) are found on a chromosomal 'pathogenicity island'. Cytotoxin B and other large clostridial cytotoxins are cleaved by a cytosolic factor of the target cell during its cellular uptake in an autocatalytic manner dependent upon host cytosolic phosphoinositol cofactors, so achieving delivery of the toxic domain to the host cell cytosol. It causes watery antibiotic-associated diarrhoea and abrupt abdominal pain, usually 3–4 days after antibiotic treatment. Oral metronidazole (400 mg taken 8-hourly for a week) is often effective; but oral vancomycin (250 mg, 8-hourly for a week) is an alternative. Recently, *C. difficile* attracted attention because hypervirulent, ribotype 027 isolates, resistant to fluoroquinolones antibiotics, were reported in Canada and the USA. The bacterium's changing epidemiology is reflected in younger people encountering severe disease without prior antibiotic treatment. Formerly, only elderly immunocompromised individuals tended to suffer, mainly following hospitalization. There is a threat that strains such as 027 will spread worldwide and compromise antibiotic treatment. Antibiotic resistance in hospitals is an increasing problem (see ANTIBIOTICS, MRSA).

clot See BLOOD CLOT.

clusters of differentiation (CD) See CD ANTIGENS.

cM The centimorgan, a unit of physical linkage. See GENOME MAPPING.

c-*myc* See MYC.

CNS See CENTRAL NERVOUS SYSTEM.

CoA (coenzyme A) See COENZYMES.

coactivators COREGULATORY PROTEINS that can move among hundreds of different transcription factors and coordinately activate expression of appropriate subsets of genes involved in achieving a particular 'end goal' such as growth or a particular morphogenetic result. See CREB-1, NUCLEAR RECEPTOR SUPERFAMILY.

coalescence In evolutionary genetics, coalescence is the opposite of divergence (see PHYLOGENETIC TREES). Coalescence analysis seeks to establish, from samples of genes or other appropriate DNA sequences, the point when all the modern sequence variants coalesce into a single ancestral sequence: the MOST RECENT COMMON ANCESTOR (MRCA) of those sequences. The individual in whom particular existing DNA sequences coalesce is termed the *coalescent* (see MITOCHONDRIAL EVE, Y CHROMOSOME ADAM), and the time in history when coalescence is reached is the *coalescence time* (see TIME TO THE MOST RECENT COMMON ANCESTOR).

The coalescent approach is probabilistic, describing the hierarchical common ancestry of a sample of gene copies. The probability that two gene copies share a common ancestor (or coalesce) in the preceding generation is proportional to the reciprocal of the size of the entire population.

Not everyone who lived in the past had descendants and so different lineages can be traced to different MRCAs, but ultimately all the genetic variation now present at one human locus could be traced back to a coalescent. DNA sequences are much more easily analysed by coalescence methods if they pass only through one sex because they can then be treated as haplotypes. There is, in other words, no confounding recombination (see HOMINOIDEA, Fig. 95).

Because mitochondrial DNA and Y chromosomes are passed on matrilinearly and patrilinearly, respectively, a mother whose offspring are all sons will not contribute to the mitochondrial DNA of future generations. Similarly, a father whose offspring are all daughters will not contribute to the Y chromosomes of future generations. To illustrate the principle of coalescence, we can imagine (see Fig. 58a) a population that remains of constant size over the generations, and in which every individual has just two offspring, which may both be daughters (P = 0.25), sons (P

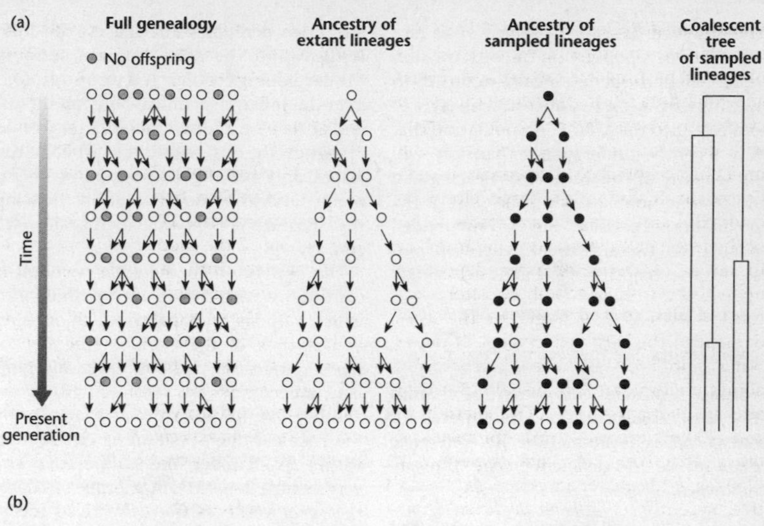

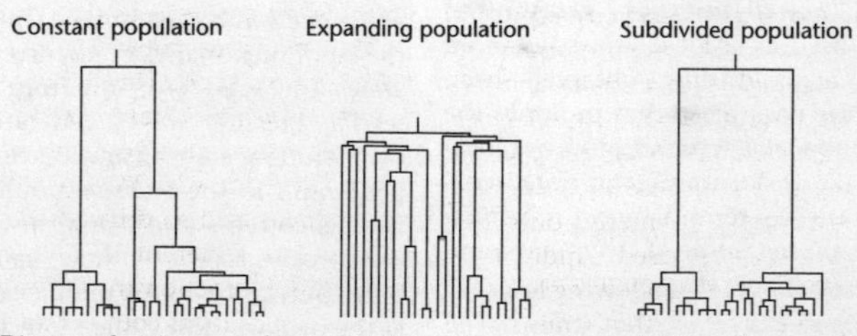

FIG 57. (a) Diagram showing genealogical relations among nine generations of a population with a constant size of 10 individuals (see COALESCENCE entry for details). Considering only those individuals who have contributed to the present generation reveals a most recent common ancestor of all lineages (left two lineages). If only a subset of lineages is sampled (third from left, sampled lineages in black) these share the same common ancestor as that shared by all extant lineages. (b) Model gene genealogies for populations experiencing different histories. © 2004 From *Human Evolutionary Genetics*, by M.A. Jobling, M. Hurles and C. Tyler-Smith. Reproduced by permission of Garland Science/Taylor & Francis LLC.

= 0.25), or one of each sex (P = 0.5). Starting with a population of 5000 parental pairs, it can be demonstrated that within 10,000 generations all the surviving Y chromosomes will have just one male MRCA, and all the surviving mitochondria will have just one female MRCA. Ideally, the entire MTDNA and Y CHROMOSOME sequences should be used in inferring the date of the human MRCA, and it is essential that data from a suitable outgroup (e.g., chimpanzee) are used to root the analysis (see PHYLOGENETIC TREES).

Because of recombination, different autosomal loci will coalesce in different MRCAs. And because recombination does not occur randomly throughout the genome, coalescence analysis of autosomal sequences provides greater challenges (see HAPLOTYPE BLOCKS, HOMO SAPIENS).

Coalescent simulations can be used repeatedly to test population models (e.g., neutral evolution, population demographies) concerning the origin of current human genetic diversity. Putative gene genealogies relating to a set of haplotypes are derived from such population models (null hypotheses) and then tested against the observed data for closeness of fit (see Fig. 57b). In a population of constant size, roughly half the gene genealogy consists in waiting for the last two ancestral lineages to coalesce into the MRCA. A population that has recently undergone expansion will show many of the lineages coalescing at a single time point marking the start of the population expansion. But in a subdivided population, two ancestral lineages will persist for the majority of the gene genealogy.

coated pits, coated vesicles See VESICLES.

cobalamins See VITAMIN B COMPLEX.

cobalt A mineral component of vitamin B_{12} (see VITAMIN B COMPLEX) and required for red blood cell production. Its reference daily intake and deficiency symptoms are unknown.

cocaine (benzoylmethyl ecogonine) Formerly used as a local anaesthetic, an addictive alkaloid drug obtained from leaves of the coca plant that prolongs the actions of DOPAMINE at its receptors by blocking reuptake of dopamine and noradrenaline into nerve terminals, with similar effects to amphetamine. It is sympathomimetic. See ADDICTION; PSYCHOACTIVE DRUGS, Fig. 144.

coccyx (adj. coccygeal) Triangular group of, usually, four vertebrae articulating with the fifth sacral vertebra. In males, it points anteriorly (i.e., is curved forwards); in females, it points inferiorly (i.e., is straighter). The coccygeal vertebrae usually fuse by 20–30 years of age.

cochlea The spiral part of the membranous labyrinth, ~32 mm in length and 2 mm in diameter but pea-sized *in situ*, embedded in the temporal bone anterior to the vestibule and containing the main organ of hearing, the spiral organ (organ of Corti). The cochlear duct (the *scala media*) arises as an outgrowth from the saccule during the 6th week, penetrating the surrounding mesenchyme in a spiral manner as the mesenchyme gradually develops into cartilage and vacuolates so that by the 10th week there are, additional to the cochlear duct, two perilymphatic spaces, the *scala vestibuli* and the *scala tympani*, forming the developing cochlea. It is wound around a central pillar, the conical bony modiolus, and at its base are two membrane-covered openings: the oval window attached to the stapes and bearing the tympanic membrane, and the round window housing the secondary tympanic membrane (see EARS, Fig. 68).

The perilymph in the scala vestibuli is continuous with that in the vestibule into which it opens, and apart from an opening at the apex of the cochlea (the helicotrema), the scala vestibuli and scala tympani are completely separated from one another for the length of the cochlea by the cochlear duct, containing endolymph within its chamber, the scala media. The very flexible Reissner's membrane separates the scala vestibuli from the scala media while the basilar membrane, upon which the basal membranes of HAIR CELLS rest, separates the scala tympani from the scala media. Pressure waves generated at the oval window are transmitted through the perilymph in the scala vestibuli, through the helicotrema and along the scala tympani to the round window, causing it to bulge outward into the middle ear. Motion at the oval window is therefore accompanied by motion at the round window. Since the endolymph within the scala media does not transmit pressure waves, pressure differences between it and the perilymph in the scala tympani cause the flexible basilar membrane to vibrate, moving the hair cells (resembling those of the vestibular apparatus) against the tectorial membrane which projects over them and is in contact with them. The resulting bending of the hair cells' stereocilia produces receptor potentials that can elicit nerve impulses in the afferent neurons of the cochlear branch of the VESTIBULOCOCHLEAR NERVE. A single row of 3,500 inner hair cells form ribbon synapses with myelinated axons of large bipolar cells (type I) in the spiral ganglion of the cochlear nerve. Each hair cell being innervated by ~10 axons, allowing signals to spread to many targets (divergence). About 12,000 outer hair cells, arranged in

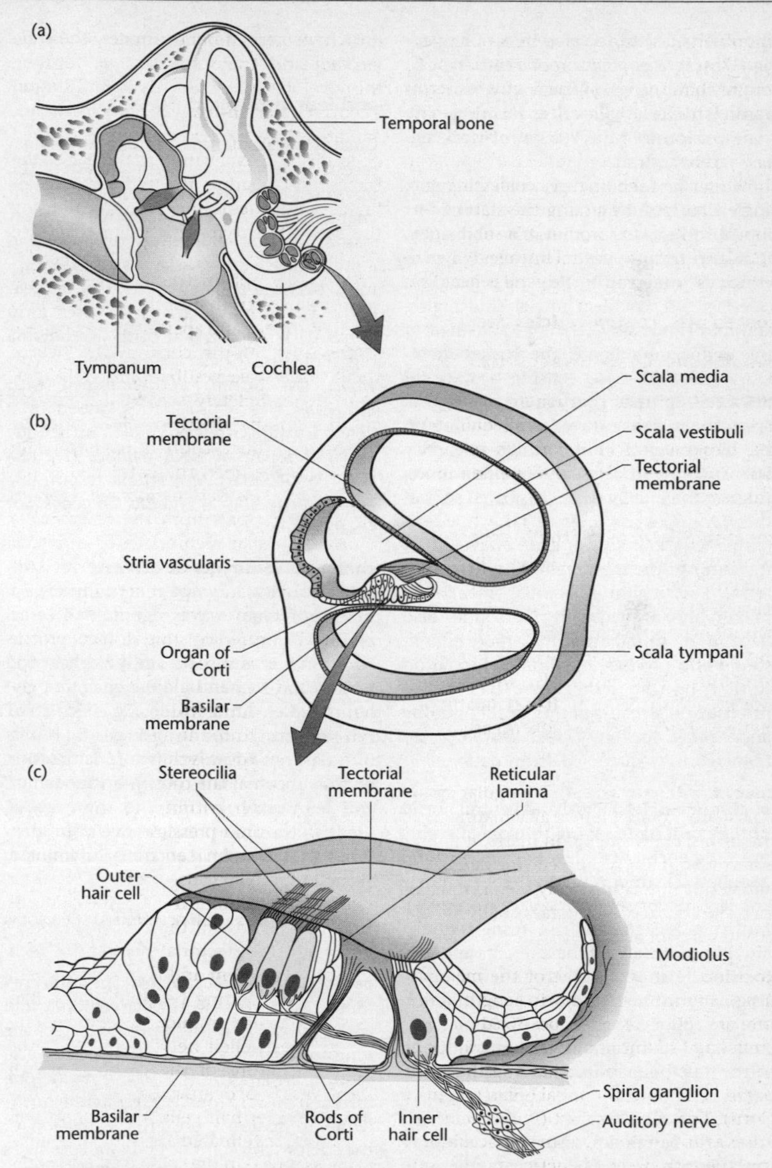

FIG 58. Diagrams of the COCHLEA, showing (a) its position within the inner ear; (b) enlargement of the three parallel chambers (the scalae) seen in section in (a); and (c) detail of the organ of Corti and its neural output. Redrawn in part from © 2001 *Neuroscience* (2nd edition) by M.F. Bear, B.W. Connors and M.A. Paradiso, published by Lippincott, Williams & Wilkins. Reproduced with the kind permission of the publisher.

three rows, are innervated by an unmyelinated axon from small bipolar cells (type II) in the spiral ganglion, each of which synapses with ten hair cells, allowing data compression and integration of weak signals (CONVERGENCE).

The basilar membrane is wider at its apex than at its base by a factor of ~5, while its flexibility increases from its base to its apex. It is also tonotopically organized, a map which is preserved in the spiral ganglion (see Fig. 58), cochlear nuclei, and higher auditory processing centres. Movement of the endolymph bends the basilar membrane near its base causing a wave of deformation that propagates towards its apex, the distance travelled depending on the frequency of sound: a high-frequency sound vibrates the basilar membrane more, causing the energy to be dissipated so that the wave does not progress far, while low-frequency sounds generate waves that travel up to the membrane's more flexible apex. The region of basilar membrane maximally displaced by a sound also depends on sound intensity: a more intense sound will produce maximal deformation further up the membrane than a less intense one. Because of this, both tonotopy and phase-locking (see HEARING) are involved in frequency discrimination.

cochlear nuclei Dorsal and ventral nuclei in the medulla receiving primary afferents from the COCHLEA (see Fig. 58) via the spiral ganglion. Distinct cell types process different features of sound (PARALLEL PROCESSING): bushy cells signal precise firing patterns and phase-locking of the afferents and project to the superior olivary nucleus enabling sound localization (see HEARING); stellate cells have a greater dynamic range and signal sound amplitude monotonically (sigmoid plots of sound level against firing rates). Axons from the ventral cochlear nucleus pass to the superior olivary nucleus (SON) on both sides, and to the contralateral INFERIOR COLLICULUS. Decussating auditory fibres crossing the pons form the trapezoid body. The dorsal cochlear nucleus projects directly to the contralateral nucleus of the lateral lemniscus, which projects to the inferior colliculus on that side.

cochlear transplant See COCHLEA.

Cockayne syndrome A rare autosomal recessive condition (named after the physician Edward Cockayne, 1880–1956) characterized by UV-sensitivity, short stature, mental retardation, deafness, and a premature ageing: it is a PROGERIA. Symptoms become apparent after the 1st year (Type I), although an early onset form (Type II) is apparent at birth. Xeroderma pigmentosum–Cockayne syndrome occurs when the disorder is combined with XERODERMA PIGMENTOSUM. It is caused by a defect in transcription-coupled nucleotide excision repair of DNA damage (see DNA REPAIR MECHANISMS), mutations responsible being in the *ERCC6* and *ERCC8* genes. See FANCONI ANAEMIA.

coeliac disease (celiac disease) A condition occurring in some genetically susceptible individuals after ingestion of cereal products containing the storage protein gluten, such as those from rye, wheat and barley, and treated by adherence to a gluten-free diet. Antigen-specific effector T_H1 cell response to the antigen gliadin results in a delayed hypersensitive inflammation of the upper small intestine (duodenum and jejunum), restricted to the mucosa. Predictive autoantibodies have been identified that target tissue transglutaminase. See ALLERGIES.

coeliac plexus (celiac plexus) See SOLAR PLEXUS.

coelom, coelomic cavities The adult coelomic cavities develop from a coelomic fold which grows past the posterior face of the lung and makes contact with the transverse septum, separating the *pleural cavity* (one for each lung) from the *peritoneal cavity*, while in males a posterior extension of the coelom through the body wall produces the scrotal pouch (scrotum). The pericardial cavity is also coelomic. See BODY CAVITIES, EXTRAEMBRYONIC COELOM.

coenzyme An organic molecule, often a

derivative of a mononucleotide or dinucleotide (and frequently derivatives of water-soluble VITAMINS), which is a soluble cofactor for an enzyme without being permanently joined as a prosthetic group. Coenzymes are often recycled vehicles for a chemical group that is either needed in a particular reaction or is a product of it. Many, such as ATP, NAD, NADP and FAD, are equipped with nucleotide components which are usually catalytically inactive although they act, and probably originally served, as handles by which enzymes attach to substrates. See ACETYL COENZYME A, COENZYME Q.

coenzyme Q (**ubiquinone**) A lipid-soluble benzoquinone which carries electrons in the membrane-associated electron-transfer chain on the inner membrane of the MITOCHONDRION. It can carry one electron, becoming the semiquinone radical ($QH^{\bullet}$), or two electrons, as ubiquinol (QH_2), acting at the junction between a 2-electron donor and a 1-electron acceptor. Freely diffusible in the lipid bilayer, it carries both electrons and protons, playing a central role in coupling electron flow to proton movements.

cofactor An ion, or coenzyme, required by an enzyme for its activity.

cognate Related, or connected. As of a ligand and its *cognate receptor*.

cognition Cognitive neuroscience is concerned with the biological basis of the ways in which individuals sense and interact with the world, and includes such fields as perception, learning and memory, decision making, problem solving and the 'executive' control of thought and action, speech and language and the recognition of self and others, in which the PREFRONTAL CORTEX and associated subcortical areas in the BASAL GANGLIA and midbrain play a disproportionately important role. However, the neurobiology of higher cognitive function (as with the control of emotion and of complex behaviour) is still a daunting frontier. Cognitive processes are in some respects analogous to computer programs in that both involve the processing, transformation, storage and retrieval of information (see NEURAL CODING). Cognitive functions are now being revealed by ELECTROENCEPHALOGRAM study and can be simulated by neural networks that employ parallel distributed processing.

Recent work has shown that it is possible to predict aspects of functional MAGNETIC RESONANCE IMAGING (fMRI) activation based on visual features of arbitrary scenes, using this to identify which of a set of candidate scenes an individual is viewing. Moreover, brain imaging studies demonstrate that different spatial patterns of neural activation are associated with thinking about different semantic categories of pictures and words (e.g., tools, buildings, animals). Recently, a computational model has been developed that predicts the fMRI activation of neurons associated with words for which fMRI data are not yet available. Once trained, the model predicts with highly significant accuracies fMRI activation for thousands of other concrete nouns already in the text corpus of a trillion words.

The capacity for ABSTRACTION is a major problem in cognitive neurobiology. One of the most frequent chromosomal rearrangements associated with cognitive deficits is the 22q13.3 microdeletion. The syndrome that results includes neonatal hypotonia; global developmental delay; normal-to-accelerated growth; absent, to severely delayed, speech; autistic behaviour; and minor dysmorphic features. Among the three genes located in the minimal telomeric region, *SHANK3* encodes a scaffolding protein found in excitatory synapses opposite the presynaptic active zone. Shank proteins are thought to function as master organizers of the postsynaptic density (see DENDRITES, SYNAPSES). Evidence suggests that abnormal gene dosage of *SHANK3* is associated with severe cognitive defects, including language and speech disorder and AUTISM spectrum disorder. See INTELLIGENCE, LANGUAGE, LANGUAGE AREAS, LEARNING, MEMORY, SELF-AWARENESS, THALAMUS.

cognitive ability See INTELLIGENCE.

cohesins Multisubunit ring-shaped complexes, deposited along the length of each sister chromatid as DNA REPLICATION occurs, binding together the two copies of a replicated chromosome until the start of mitotic anaphase (see MITOSIS), or 2nd meiotic anaphase. Meiosis-specific cohesins exist. Recent work indicates that cohesins can also occur in G2, independently of DNA synthesis, if DNA is damaged by double-strand breaks, forming new connections between sister chromatids. Cohesins must be degraded before chromatid separation can occur.

cohort A group of individuals of some specified age-range in a POPULATION. See CROSS-SECTIONAL STUDIES.

coincidence detectors These are referred to in entries on HEARING and NMDA RECEPTORS.

cold See COMMON COLD, STRESS RESPONSE, TEMPERATURE REGULATION.

cold receptors See THERMORECEPTORS.

collagenases Membrane METALLOPROTEINASES, being the only known mammalian enzymes capable of degrading triple-helical fibrillar collagens. Gelatinases are type IV collagenases.

collagens Major proline- and glycine-rich proteins of the extracellular matrix, secreted by connective tissue cells and others. As a major component of skin and bone they contribute up to 25% of BODY COMPOSITION in humans and other mammals.

Produced on rough endoplasmic reticulum (ER), collagen polypeptide chains are fed into the ER lumen as larger precursors, pro-α chains that have an amino-terminal signal peptide sequence as well as additional propeptide amino acid sequences at both N- and C-terminal ends. In the ER lumen, selected prolines and lysines are hydroxylated and some of the resulting hydroxylysines are glycosylated. Each pro-α chain combines with two others to form a hydrogen-bonded, triple-stranded, helical molecule termed *pro-collagen*, but in the absence of proline hydroxylation, e.g., in vitamin C deficiency, the defective pro-α chains fail to form a stable triple helix and are degraded within the cell. After secretion, the propeptides of the pro-collagen molecules are enzymatically removed to release α collagen molecules.

The *fibrillar collagens* are triple helices (superhelices) of three α collagen chains (termed *tropocollagen*), each ~1000 amino acids long, of which at least 25 forms are known, each encoded by a different gene. Different combinations of these genes are expressed in different tissues, although only ~20 different types of collagen have been identified. After secretion into the extracellular space, these ~300 nm long molecules assemble by covalent cross-links between lysine residues into thin, higher order, *collagen fibrils* many hundreds of micrometres in length. These may in turn aggregate into larger, cable-like bundles several micrometres in diameter and visible in the light microscope as *collagen fibres*. Certain other collagen types (fibril-associated collagens) are attached to the fibrils and probably link these to each other and to components of the EXTRACELLULAR MATRIX. Type IV collagens exist in several isoforms (comprising α1(IV) and α2(IV) forms) and form flexible sheet-like structures when polymerized. They are major components of basal laminae and are degraded by gelatinases (see COLLAGENASES).

Type I collagen fibres, the main type in skin and bone and forming 90% of human body collagen, bind to the proteoglycan decorin in connective tissues. Its fibrils comprise two α1(I) and one α2(I) molecules. The α1(I) collagen is encoded by the *COL1A1* gene which contains 41 exons encoding the part of the molecule that forms a triple helix: each exon encodes an integral number of copies (1-3) of an 18-amino acid motif itself composed of six tandem DNA REPEATS and originated by EXON duplication (see GENE DUPLICATION). See INSULIN-LIKE GROWTH FACTORS.

collaterals The branches of an axon, enabling a motor neuron to synapse with different target cells.

collecting ducts Tubules developing from

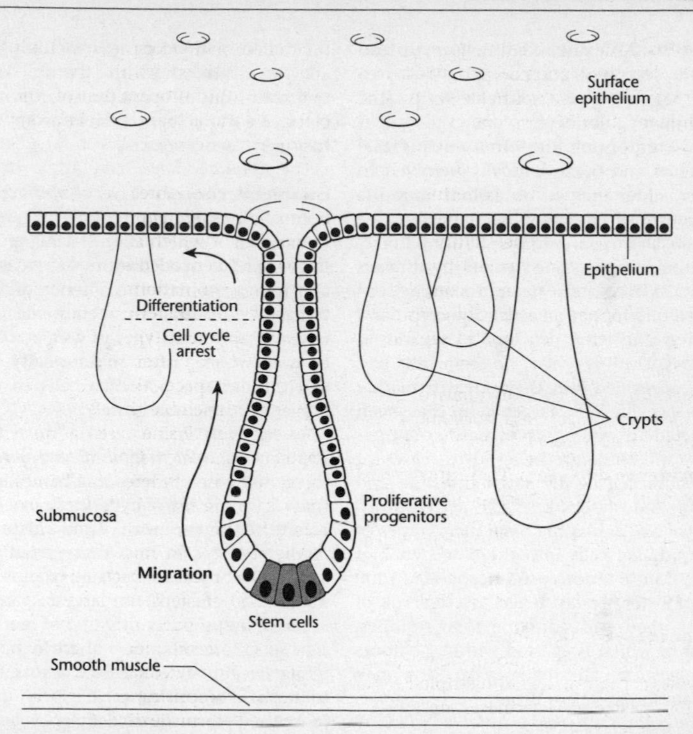

FIG 59. Diagram of the COLON epithelium and crypt. Putative stem cells (shaded) are found at the bottom of the crypt, proliferating progenitor cells occupy two-thirds of its length and differentiated cells populate the remainder of the crypt and the flat surface epithelium. From T. Reya and H. Clevers, © *Nature* 2005, with permission from Macmillan Publishers Ltd.

the ureteric buds and carrying urine from the cortex of the KIDNEY towards the renal papilla. By the time the tubular fluid reaches the end of the DISTAL CONVOLUTED TUBULE, 90–95% of the solutes in the glomerular filtrate have been reabsorbed and returned to the blood in the vasa recta. The collecting ducts are the main targets of VASOPRESSIN (ADH). Their water permeability is relatively low but can be dramatically increased within a few minutes by production of cAMP following binding of vasopressin to receptors in the basolateral membranes of the PRINCIPAL CELLS. This results in phosphorylation of aquaporin AQP2 and its insertion into the apical membrane, making the membrane more permeable to water. AQP3 and AQP4, located in the basolateral membranes, are not inducible by vasopressin and allow water through the ducts into the peritubular fluid of the renal medulla. Without AVP-mediated reabsorption of the remaining 10% of water from the glomerular filtrate, urine output would be ~18 L per day rather than the 1.5–2 L that it normally is.

colliculi See INFERIOR COLLICULUS, SUPERIOR COLLICULUS.

colon That part of the large intestine, ~1.2 m in length, into which 500 mL or more of chyme enters per day via the ileocaecal valve. Its mucosa (see ALIMENTARY CANAL,

Fig. 4) has a flat surface epithelium, instead of villi (see Fig. 59 and compare SMALL INTESTINE, Fig. 159), its epithelial cells (the absorptive colonocytes, and the goblet cells) confronting the often toxic faecal contents and being critically involved in the maintenance of intestinal immune homeostasis. The crypts of the colon (see Fig. 60) lack the nominal attribution to Lieberkühn; but the same signalling pathways (Wnt, BMPs, Notch) appear to be involved in cell differentiation along the crypt, and their malfunction can lead to COLORECTAL CANCERS. Until recently, no stem cells had been identified; but the activated marker gene *Lgr5*, a Wnt target gene, has been detected in cycling crypt base columnar cells, and these are believed to represent the stem cell of the small intestine and colon. Aldosterone-sensitive active transport of Na^+ across the basal membranes of its epithelial cells into the blood enables the colon to absorb osmotically 400–1000 mL of water per day. It acts as a reservoir of unabsorbed and unusable food residues, most of which is egested within 72 hours of ingestion, although up to 30% may remain for a week or more.

The colon's GUT FLORA is involved in many important aspects of nutrition (see also COLORECTAL CANCERS), but deficiency of the transcription factor NF-κB (see entry) has been shown in mice to lead to apoptosis of colonic epithelial cells, impaired expression of antimicrobial peptides and translocation of bacteria into the mucosa, triggering a chronic inflammatory response resembling inflammatory bowel disease (see ALLERGIES). This may be brought about by defective or ablated IKK subunits, essential for NF-κB activation.

colony stimulating factors (CSFs) Cytokine growth factors, e.g., GM-CSF (granulocyte-macrophage CSF), causing proliferation of white blood cells. GM-CSF is a product of T_H1, some T_H2 and some cytotoxic T cells, stimulating production of granulocytes and macrophages (myelopoiesis) and dendritic cells. CSFs are often used therapeutically in conjunction with anticancer drugs to counteract the latter's cytotoxic effect on dividing cells, including leukocytes. Patients with burns, other traumatic injuries, organ transplants or diabetes, are capable of benefit from such treatment. See CYTOKINES.

colorectal cancers Loss of APC activity occurs in most sporadic colorectal cancers, mutational inactivation leading to inappropriate stabilization of β-catenin and indicating that the absence of functional APC transforms epithelial cells through activation of the WNT SIGNALLING PATHWAY which, once mutationally activated, maintains adenoma cells in their progenitor state indefinitely.

Recent work has located a small DNA fragment (the 'Z' fragment) mapping to the gene-dense chromosome band 2q14.2 that is commonly hypermethylated in colorectal cancer, with three clusters of methylated CPG ISLANDS interspersed with and flanked by unmethylated islands. The methylated clusters, the largest spanning ~1 mbp, had significantly higher methylation in colorectal cancer than in normal tissue despite heterogeneity across these islands. Colorectal cancer has been linked to both dietary FOLATE deficiency and to variants in methylene-tetrahydrofolate reductase. In 1999, >25 years after the oncogene's discovery, mutant forms of the SRC GENE were found to be present in 12% of advanced human colon cancers. At all ages, women are less likely than men to develop colon cancer, and oestrogen hormone replacement therapy reduces its incidence in post-menopausal women and in several mouse colon cancer models. IL-6 secretion by KUPPFER CELLS requires NFκB activity. Ligand-binding of TOLL-LIKE RECEPTORS on these macrophages in mouse models induces MyD88 and NFκB activity, which is thought to promote tumour progression in the liver and colon. It appears that MyD88 activation in spontaneous colon cancer is driven by commensal GUT FLORA. In one model, MyD88 induced expression of the enzymes COX-2 and MMP-7, and of IL-1. MyD88 appears to regulate a tissue repair pathway that promotes, and is triggered by, growing malignancy. See ADENOMATOUS POLYPOSIS COLI PROTEIN (especially), CHEMOTHERAPY, FAMILIAL ADENOMATOUS POLYPOSIS, HEREDITARY NON-POLYPOSIS COLORECTAL CANCER, JUVENILE POLYPOSIS SYNDROME, P53.

colostrum See MILK.

colour blindness Anomalies in COLOUR PERCEPTION, sometimes due to retinal disease but most commonly resulting from lack of inheritance of a gene for one of the three CONE CELL pigments, or to inheritance of a 'hybrid gene' in which unequal crossovers within a gene result in its combining regions of the genes for the 'red'-absorbing and the 'green'-absorbing opsins. Any particular cone cell expresses just *one* of these pigment genes. The loci for the opsins of the red- and green-absorbing pigments lie close together on the X chromosome and are sex-linked, a tandem arrangement originating by tandem GENE DUPLICATION. The locus for the blue-absorbing opsin lies on chromosome 7.

In normal trichromatic vision, the cone cell population includes cells expressing genes for blue-absorbing, red-absorbing and green-absorbing opsins. In anomalous colour vision, the cone cell population either lacks cone cells producing one or more of these pigments (protanopia, deuteranopia, tritanopia), *or* contains cone cells expressing the following genes: (i) the normal gene for blue-absorbing pigment (~420 nm); (ii) *either* the normal gene for red-absorbing pigment (~563 nm) *or* that for the green-absorbing pigment (~532 nm); (iii) a hybrid gene encoding a visual pigment with maximal absorption intermediate between that of the normal red and green pigments. Red/green deficiency may be due either to *dichromatism* (complete red/green colour blindness) or to *anomalous trichromatism* (reduced sensitivity to either red or green). *Monochromatism* is rare, individuals lacking two or all three types of cone cell and having no perception of colour. Total absence of cones results in lack of photopic vision, individuals being effectively blind in daylight.

Dichromats have only two cone cell populations, some with a normal blue-absorbing opsin and some with a second visual pigment sensitive in the middle- to long-wave spectrum. This second pigment is usually either the typical green pigment (in protanopes, who lack L cones, causing severe protanomaly) or the typical red pigment (in deuteranopes, who lack M cones, causing severe deuteranomaly). *Anomalous trichromats* exhibit either deuteranomaly (4% of men having this reduced sensitivity to green), or protanomaly (1% of men having this reduced sensitivity to red). In mild anomaly (protanomaly or deuteranomaly) the anomalous hybrid pigment's absorption is slightly displaced, either from the normal green-absorption towards normal red-absorption (deuteranomaly), or from the normal red-absorption towards normal green-absorption (protanomaly). In rare cases, the blue pigment (S cones) may be lost, leading to *tritanopia*, an autosomal condition (prevalence 0.008% in both men and women), leading to inability to distinguish colours having a short-wavelength component (violet) from those without (yellow), both appearing grey. It may also be caused by genetic defects in the photoreceptors, by lesions in the neural pathways involved in colour perception, or by retinal disease. See ACHROMATOPSIA

colour constancy Our perception of the colour of an object as relatively constant despite great changes in the spectral composition of the incident light. This aspect of VISUAL PERCEPTION may be explained by the fact that there are many neurons in area V4 of the PRIMARY VISUAL CORTEX which fire only when particular colours are perceived, rather than when particular wavelengths are detected: changes in the wavelength composition of the incident light do not much affect the firing rates of these neurons. There is much we do not yet understand about how the cells in V4 make use of the information supplied by those in V1 (parvocellular-blob system) and V2 (thin stripes) to generate colour perceptions. There is evidence that the colour perceived depends not only on the two-dimensional image of an object projected onto the retina, but also on the object's three-dimensional shape.

colour opponency The acknowledgement that certain colours are never perceived in combination; thus we do not perceive reddish-green or bluish-yellow, although reddish-blue (mauve), greenish-yellow (chartreuse) and bluish-green (cyan) are perceived. Further, red and green lights can be mixed to give a white impression as can blue and yellow. Current theories to explain this colour opposition and cancellation hinge on the view that there are six primary colour qualities processed in opponent pairs of neural channels: red-green, blue-yellow and white-black. This is supported by study of COLOUR-OPPONENT CELLS. See SIMULTANEOUS COLOUR CONTRAST.

colour-opponent cells Retinal ganglion cells and cells of the LATERAL GENICULATE NUCLEI in whose sensory fields the inputs from different classes of CONE CELL are antagonistic. In P cells (see GANGLION CELLS) the antagonism is between 'red' (L) cones and 'green' (M) cones, while in non-M-non-P cells, the antagonism is between 'blue' (S) cones and 'yellow' cones: the RECEPTIVE FIELD centre receives inputs from one class of cone and the receptive field surround receives inputs from the other in such a manner that the response to one wavelength in the centre is cancelled by showing another wavelength in the surround (see RETINA, Figs 150 and 151). Such *concentric single-opponent cells* are to be contrasted with *concentric double-opponent cells* in the BLOB zones of the striate cortex (PRIMARY VISUAL CORTEX) which cells also have an antagonistic centre-surround receptive field although now the inputs from L and M cones are scattered throughout the field and vary depending upon whether they input to the centre or to the surround. Thus in some double-opponent cells the L inputs excite in the centre but inhibit in the surround, while the M cones have the opposite effect. In others the reverse is the case. Information from 'blue' (S) cones is processed by *coextensive single-opponent cells* in such a way that inputs from S cones oppose the combined inputs from L and M cones throughout the receptive field of the cell.

As an example, a cell with a red 'on' receptive field centre and a green 'off' surround has a centre fed mainly by R cones, so the cell responds to red light – even when this bathes the entire receptive field – by firing action potentials (the surround being relatively less sensitive to this wavelength). Only when green light is shone on the cell's receptive field surround is this response cancelled. Were white light to be shone onto the entire receptive field of this (L^+M^-) cell both centre and surround would be equally activated, so cancelling the cell's response. Blue-yellow colour-opponent cells operate in the same way. See COLOUR OPPONENCY, COLOUR PERCEPTION.

colour perception Four distinct cone cell opsin genes were present early in vertebrate evolution (~500 Mya), the visual pigments (iodopsins, 'visual purple') expressed by these having maximum sensitivities in the red, green, blue and violet/ultraviolet wavelengths. Probably as a result of a long nocturnal phase in their evolution, most mammals are dichromats, having lost the tetrachromacy of their reptilian ancestors. Trichromacy re-evolved in the primate lineage some 35 Mya after a duplicate of the gene encoding the long-wavelength-absorbing opsin arose by GENE DUPLICATION. Subsequent modification by mutation of the paralogous genes in the direction of shorter (green-wavelength) absorption by one of them seems to have been favoured by natural selection, possibly through enabling the animal to distinguish, visually, ripe (red) from unripe (green) fruits (but see VOMERONASAL ORGAN). The usual genetic arrangement in humans is a gene cluster on the X chromosome consisting of one 'red' gene and two 'green' genes. Mispairing of these regions on homologous chromosomes can lead to unequal cross-overs and result in the inheritance of COLOUR BLINDNESS.

Perception of different colours involves

light of different wavelengths exciting the three different types of CONE CELLS to different degrees. Perceived colour is based on comparison of the relative activities of those ganglion cells with receptive field centres containing red (L), green (M) and blue (S) cones (see COLOUR-OPPONENT CELLS; RETINA, Figs 150 and 151). S cones make up only 5–10% of the retina and are absent from the foveola. This is because the eye suffers from *chromatic aberration*, short-wave light being brought to focus at a different point from longer wavelengths. This would cause slight image blurring and compromise high VISUAL ACUITY. As a result, the central fovea is dichromatic. Further, the random distribution of M and L cones produces patches where only one type of cone occurs, making colour vision coarse-grained, failing to resolve fine detail.

Edwin Land has suggested that each cone type separately measures the brightness of each object in the field of view; the brightness of each object being given a value in relation to the brightest object in the field; and finally Land gives each object a brightness value for each cone type which is then used to predict the colours of object in the field using a formula of his devising. Although no single cone type's measurements are directly available to the cortex, it is possible that the theory can be successfully modified if the responses of the three classes of double-opponent cells are substituted for those of the cones themselves. Colour might be computed in some such fashion.

Ganglion cells convey a stream of information via the optic nerve, OPTIC CHIASMA (see Fig. 130) and optic tracts to the paired LATERAL GENICULATE NUCLEI (LGN) of the thalamus, where the different information deriving from the three classes of GANGLION CELL remain segregated (see PARALLEL PROCESSING). M-type ganglion cells (not to be confused with M-type cone cells) are 'colour blind'. The parvocellular pathway originating with the P-type ganglion cells projects to the parvocellular layers of the LGN (see PARVOCELLULAR-BLOB/INTERBLOB PATHWAY). These cells have small receptive fields (RFs) and colour single-opponency and two types are recognized: (i) red-green cells (in which the RF compares input from M and L cones); and (ii) blue-yellow cells, which combine inputs from M and L cones to give the sensation of yellow and lack a centre surround pattern. Blue-yellow cells are excited either by S cones and inhibited by a combined M-plus-L cone signal, or vice versa. Red-green cell response varies with the size of the light spot, and in general these cells are more wavelength-selective for large stimuli than for small ones. For example, a green-on/red-off cell is equally stimulated by a small green or white spot covering the RF centre because white light contains wavelengths exciting green cones. This same cell will be excited by a large green spot, but not by a large white one, because white light contains wavelengths that stimulate the red-off surround, while a large red spot will silence the cell. For small spots, red-green cells cannot distinguish red or green from white light, although they can signal brightness.

Area V4 of the EXTRASTRIATE CORTEX (see Fig. 72) receives input from the blob and interblob regions of the PRIMARY VISUAL CORTEX (or, striate) via relay in V2; and neurons in V4 have larger receptive fields than do cells in the striate. There is evidence that specialized colour processing takes place in the ventral pathway leading from the primary visual cortex to the temporal lobes (see ACHROMATOPSIA), so that the colours we perceive may be determined at a rather late stage of visual processing. See OLFACTORY RECEPTORS.

colour vision See COLOUR PERCEPTION.

commissure, commissural fibres Nerve fibres running from one cerebral hemisphere to the other, linking functionally related structures. See ANTERIOR COMMISSURE, CORPUS CALLOSUM. Compare ASSOCIATION FIBRES, PROJECTION FIBRES.

common cold (coryza) The common cold is an extremely infectious disease caused by numerous strains of rhinoviruses (Picornaviridae) and occasionally by coronaviruses (Coronaviridae; see VIRUSES), or by other viral respiratory infections. Spread by the droplet (aerosol) route, the viruses are also transmitted by hands and fingers. Treatment of the symptoms, which include mild fever, swelling of the mucosae of the nose, throat and conjunctiva, is by antipyretic analgesics, mild decongestants, and by bed rest in severe cases. Rhinoviruses encode a single polypeptide product which is cleaved to produce four main capsid proteins (VP1–4), the virus binding by its VP1 protein to host ICAM-1 (see CELL ADHESION MOLECULES). Icosahedrally symmetrical, the virus has a 'pocket' containing a strongly hydrophobic molecule which is a potential site for specific inhibition. New serotyped rhinoviruses are not emerging rapidly. The three main structural proteins encoded by coronaviruses are species-specific nucleocapsid proteins, the surface projection protein and the transmembrane (matrix) protein. These rhinoviruses differ from the SARS virus.

common-source outbreaks Outbreaks (epidemics) of disease traceable to a common source. They may be POINT-SOURCE OUTBREAKS or EXTENDED-SOURCE OUTBREAKS. Contrast PERSON-TO-PERSON OUTBREAKS.

communicable diseases An INFECTIOUS DISEASE capable of spreading from person to person.

compact bone Tissue forming the external layer of all bones and contributing the major part of diaphyses. It is characterized by the presence of HAVERSIAN SYSTEMS. Compare SPONGY BONE.

comparative genomics See HUMAN GENOME.

complement, complement system A heat-labile complex of plasma proteins (also widely distributed in body fluids) that interact with pathogens to opsonize them and mark them for destruction by PHAGOCYTOSIS – an activity said to 'complement' the antibacterial activity of antibodies (see Appendix XI). At sites of infection, complement activation also triggers INFLAMMATION. Several complement proteins are proteases circulating as inactive zymogens until proteolytically cleaved, usually by other complement factors. Three pathways are recognized, each triggered by a different molecular signal on the pathogen, but the 'early' steps of all three converging on the production of a protease, C3 convertase, bound covalently to the pathogen surface. Each pathway involves an enzyme cascade that hugely amplifies the response to the original signal, and many regulatory mechanisms prevent uncontrolled complement activation.

In the *classical pathway*, antigen:antibody complexes on the pathogen surfaces are the trigger. In the MANNOSE-BINDING LECTIN pathway (MB-L) the trigger is mannose on the pathogen surface, while the *alternative pathway* can be triggered by the binding of spontaneously activated C3 in the plasma to the pathogen surface in the absence of antibody. Plasma C3 is cleaved by C3 convertase on the pathogen surface to C3b (the major effector molecule in the system) and the minor inflammatory mediator C3a. C3b is deposited on pathogen surfaces, where it binds its receptor (CR1) on macrophages and neutrophils. C3b also binds C3 convertase to form C5 convertase which generates the peptide C5a, a major inflammatory mediator (see ANAPHYLACTIC SHOCK) which also activates macrophages to ingest bacteria already bound to CR1 receptors. C5 convertase also generates C5b, which initiates the 'late' complement activation steps which involve the interaction and polymerization of terminal complement factors to form a complex which creates lethal pores in some pathogen cell membranes. One molecule of C5b binds a C6 molecule, this complex then binding a C7 molecule and leading to exposure of a hydrophobic site on C7 which inserts into the pathogen's lipid bilayer. Similar hydrophobic regions are exposed on C8 and C9 (a protein related to perforin), enabling these, too, to enter the bacterial cell membrane. C8 (a protein complex) then induces polymerization of 10–16 molecules of C9 into a pore-forming structure termed the *membrane-attack complex* with a hydrophobic face in the lipid bilayer and a hydrophilic internal channel, ~10 nm in diameter, allowing free passage of water and solutes and leading to loss of cellular homeostasis, disrupting the bacterial proton gradient across the membrane. LYSOZYME can then penetrate and complete the destruction of the pathogen. Complement fragments (C5a, C3a and C4a) all use G PROTEIN-COUPLED RECEPTORS in their signalling. Factor D fulfils an essential role in the initiation and propagation of the alternative pathway of complement activation and in the amplification loop of C3 activation. Factor D is a serine protease of about 24 kDa that circulates in the blood as a constitutively active enzyme and is synthesized by fat cells and macrophages.

When C3b binds host cells, complement activation is prevented from proceeding by several plasma *complement-regulatory proteins* which interact with C3b and prevent C3 convertase from forming, or otherwise promote its rapid dissociation. Among these proteins are a membrane-attached protein, decay-accelerating factor (DAF); and complement receptor 1 (CR1), which competes with factor B for binding to C3b on the cell surface and can displace the active enzyme of C3 convertase if it has formed (see XENOGRAFTS). See ANAPHYLACTIC SHOCK; ANTIBODIES; B CELLS, Fig. 19; C-REACTIVE PROTEIN.

complementarity-determining regions (CDRs) Hypervariable regions (loops) of immunoglobulins and T-CELL RECEPTORS. There are three such loops (CDR1, CDR2, CDR3) in the variable (V) portion of each of the two chains in the heterodimer. See ANTIBODIES.

complementary DNA (cDNA) Double-stranded DNA formed by reverse transcription of (single-stranded) template mRNA followed by degradation of the original mRNA (using NaOH solution) and then using DNA polymerase 1 to synthesize the complementary DNA strand from the single-stranded DNA. It is often used in bacterial expression vectors for human gene products since it lacks the eukaryotic intron sequences that bacteria cannot splice out. cDNA banks and libraries, because they derive from mRNA, represent the DNA from protein-coding gene sequences (see EXPRESSED SEQUENCE TAGS).

complex cells of visual cortex See ORIENTATION-SELECTIVE CELLS.

complex diseases, complex disorders Diseases whose aetiologies include both environmental influences and an unknown number of genes that contribute to different degrees. Although there are some diseases that are clearly caused by defects in one gene (Mendelian disorders), susceptibility to common diseases is typically controlled by many genes (see ALLERGIES, DIABETES MELLITUS). Similar phenotypic outcomes can arise in genetically different families and ethnic groups (see GENETIC PREDISPOSITION), and detection of MODIFIERS of multigenic traits is difficult in segregating human populations because it is hard to distinguish the signals of disease genes against the genetic background of modifier genes (see MOUSE MODELS). Environmental influences contributing to complex disease PHENOTYPES include diet (e.g., in OBESITY and HYPERTENSION; deficiencies of FOLATE and METHIONINE), TOXINS and CARCINOGENS.

The reliability and validity of characterization of phenotypes of complex diseases (especially AFFECTIVE DISORDERS) are highly variable when compared with those of Mendelian disorders, or with those whose phenotypes include readily and accurately quantifiable tests (e.g., biopsies, insulin levels, glucose reactivity, immunoreactivity) and the limited ability to measure and characterize these phenotypes is often the rate-limiting step in aetiologic research.

As with any other character, it is necessary to prove claims of genetic determination. FAMILIAL RECURRENCE RISK, λ_R, indicates the degree of family clustering, and can be calculated for each type of relative.

There are MOUSE MODELS of several human complex diseases, and although genes involved in complex traits and disorders are identified either by biochemical and physiological identification of a causally involved candidate gene product or by locating that gene's position in a chromosome segment that is consistently co-inherited with the trait (genome-wide association studies), it is increasingly thought that variation in NON-CODING DNA, such as *cis*-regulatory sequences, rather than protein-coding sequences, may underlie many complex, non-Mendelian, diseases. One current view is that, instead of inheriting an array of multiple gene defects and then acquiring other complementary gene defects all of which add up to a disease phenotype, a person could inherit or acquire a defective allele encoding one or a few coactivators (or corepressors) for a relevant multigenic pathway and express that same disease phenotype through coordinated misregulation of a subset of target genes in multiple different pathways (see COREGULATORY PROTEINS). See COPY NUMBER.

Despite great effort during and since the final decade of the last millennium, the genetic basis of common human diseases remains largely unknown (2007). There is growing evidence that duplicated host immune response genes known to have dosage effects contribute to genetic susceptibility of some complex diseases (see SEGMENTAL DUPLICATIONS). Linkage and candidate gene association studies have, despite some successes, often failed to deliver clear-cut results; but genome-wide association studies have been assisted by three recent advances enabling detection of plausible effect sizes. These are: the International HapMap resource (see HAPLOTYPE, GENOMICS); the availability of dense genotyping chips (see DNA MICROARRAYS); and the assembly of appropriate large and well-characterized (properly controlled) clinical samples. The Wellcome Trust Case Control Consortium (WTCCC) brought together >50 UK groups researching the genetics of common complex human diseases of major public health importance: BIPOLAR DISORDER (BD), coronary artery disease (CAD; see CORONARY HEART DISEASE), Crohn's disease (CD; see ALLERGIES), HYPERTENSION (HT), RHEUMATOID ARTHRITIS (RA), type 1 DIABETES MELLITUS (T1D) and type 2 diabetes mellitus (T2D). See SINGLE NUCLEOTIDE POLYMORPHISMS for additional comment.

The British POPULATION (England, Wales, Scotland) on which the WTCCC work is being carried out is heterogeneous, having been shaped by waves of immigration from southern and northern Europe. The extent to which these events distort the findings is unclear, although individuals with non-Caucasian ancestry were excluded from the final analysis. Thirteen genomic regions showed strong geographical variation, the predominant pattern being along a NW/SE axis, thought to be due to marked geographical differences in NATURAL SELECTION, most plausibly in populations ancestral to those now present in the UK.

compliance The walls of structures are said to have high compliance if their walls stretch easily, or expand without tearing, in response to a small increase in pressure. This is normally true of arteries, the thoracic wall and lungs.

compulsive behaviour For obsessive-compulsive disorder, see FEAR AND ANXIETY.

computed tomography (CT) See CT SCAN.

computerized axial tomography (CAT) See CT SCAN.

COMT See CATECHOL-O-METHYLTRANSFERASE.

conchae (nasal conchae) The lateral projections of the ETHMOID BONE (superior and middle nasal conchae, or *turbinates*), and a third pair of separate facial bones (inferior nasal conchae), causing inhaled air to swirl and its particles to become trapped in mucus lining the nasal air passages before entering the rest of the respiratory tract. Air striking the mucous linings of the conchae is also warmed and moistened.

concordance Both members of a pair of twins are said to be *concordant* when either both are affected or neither is affected by a disease or disorder. Study of concordance (and its contrary discordance, in which one member of a twin-pair is affected and the other is not) offers partial resolution of the question of whether a trait is hereditary in the sense of being genetically determined (compare HERITABILITY). But since twins (especially monozygotic twins) tend to share similar environments more often than do unrelated members of their peer group,

TWIN STUDIES have been of limited use in human genetic research. See NATURE-NURTURE DEBATE.

condensins Large protein complexes that are a major structural component of mitotic chromosomes. They contain SMC proteins (structural maintenance of chromosomes) that use the hydrolysis of ATP to make large right-handed loops in DNA and drive coiling of interphase chromosomes to produce metaphase chromosomes (see CHROMATIN, Fig. 51).

conditioning See CLASSICAL CONDITIONING, INSTRUMENTAL CONDITIONING, LEARNING.

conduction velocity See AXON.

conduction zone The subdivision of the ventilatory system comprising the trachea and terminal bronchioles, whose functions include air transport, humidification, warming, particle filtration, vocalization and immunoglobulin secretion.

condyle Large, round protuberance on the end of a bone.

cone cells Colour-sensitive bipolar photoreceptor cells (~8 × 10^6 per eye), distributed throughout the RETINA (see Fig. 149) but the sole photoreceptor present in the fovea centralis and responsible for daylight vision (photopia; see LIGHT ADAPTATION). The cell's outer segment is the expanded tip of a primary CILIUM and contains double-layered membranous discs, slightly more numerous than those in ROD CELLS. As in rod cells, maintenance of their photoreceptor signalling machinery requires large amounts of IFT-mediated transport of proteins and lipids into the cilium (see RETINITIS PIGMENTOSA).

They are less sensitive to light than rod cells (lower signal amplification), and each cell contains less photopigment. But in the partially dark-adapted eye, cones receive signals from rod cells via gap junctions, augmenting acuity and colour perception. Cones provide higher temporal resolution than rods (resolving flickering at up to 55 Hz), being quicker to respond to light and having a shorter integration time. Cones are most sensitive to direct axial light rays. The cone system provides high VISUAL ACUITY, with less RETINAL CONVERGENCE than the rod system, allowing greater spatial resolution. The three types of cone cell have different photopigment molecules, *iodopsins*, sensitive to different components of the visible spectrum ('blue', 'green' and 'red'), mediating COLOUR PERCEPTION (see COLOUR BLINDNESS). More properly, these are short- (S), medium- (M) and long- (L) wavelength cones, respectively. Cones are concentrated in the FOVEA, although only L and M cones are present there (see COLOUR-OPPONENT CELLS). Cones all respond to photons of light, their differences being their probabilities of absorbing a photon of a particular wavelength, which produces different electrical responses in the three cone populations. Effectively, this registers different light intensities, which can be processed by the PARVOCELLULAR-BLOB PATHWAYS into colour perception.

In the dark, a cone cell is depolarized (–40 mV), so that voltage-gated calcium channels in its synaptic terminals are open and calcium enters the terminals triggering the release of the neurotransmitter glutamate, which excites (depolarizes) off-centre BIPOLAR CELLS and inhibits (hyperpolarizes) on-centre bipolar cells (each cone making contact with both types of bipolar cell). But on illumination the cone cell hyperpolarizes, causing closure of the calcium channels and reducing the quantity of glutamate released, which in turn depolarizes on-centre bipolar cells but hyperpolarizes off-centre bipolar cells (see RETINOBLASTOMA).

congenic strains Strains that are genetically identical, except for a specific locus. Produced by repeated backcrossing. Compare GENE KNOCKOUTS.

congenital Of a disorder present at birth. This does not imply a genetic cause (e.g., syphilis may be contracted at birth through contagious infection).

congenital adrenal hyperplasia (CAH) A group of disorders in which there are defects in testosterone synthesis in the testes, associated with excessive growth (hyperplasia) of the adrenal glands, where compensatory androgens are produced. Occurrence is ~1:12,000 in the USA, but varies globally with genetic history (e.g., 1:680 in Yupik Eskimos). When heterozygous, the gene appears to protect the Yupik against infections by *Haemophilus influenzae* B. Most individuals have a deficiency in the enzyme 21-hydroxylase needed to convert cholesterol to testosterone, one of the most common causes of ambiguous genitals in newborn females (see PSEUDOHERMAPHRODITISM). Exposure to atypically high levels of prenatal androgens results in masculinization of behaviour and ability patterns (see SEX DIFFERENCES) and females with CAH show 'tomboy' behaviour (but only a minority become homosexual). They also appear to perform better than normal females at INTELLIGENCE tests involving three-dimensional rotation. The late onset form of 21-hydroxylase deficiency may be the most common form of autosomal recessive disorder known in humans. NEWBORN SCREENING can pick up in excess of 90% of cases, the test involving immunoassay for the elevated steroid metabolite 17-hydroxyprogesterone. Contrast CONGENITAL ADRENAL HYPOPLASIA.

congenital adrenal hypoplasia Sex-linked condition caused by mutations in the *DAX1* gene, in which adrenal glands are underdeveloped. Dosage-dependent SEX REVERSAL results from a rare duplication of the locus.

congenital heart defects Considering the complexity of heart development and the devastating consequences of even slight perturbations in the process, it is not surprising that mutations in many genes can cause cardiac malfunction. Genes encoding signalling molecules (Jagged-1 in pulmonary stenosis), cell adhesion molecules, ion channels (e.g., HERG) and transcription factors (e.g., Nkx2.5 in cardiac looping, conduction defects and atrial septal defect; TBX5 in ventricular septal defect, atrial septal defect) are all implicated. One of the commonest defects is *interventricular septal defect* (or, 'hole-in-the-heart'), for which microRNAs have been implicated. Incomplete closure of the interventricular septum allows oxygenated blood to flow directly from left to right ventricles, resulting in its mixture with deoxygenated blood. Others include: *coarctation* (narrowing) *of the aorta*, leading to hypertension; *patent* DUCTUS ARTERIOSUS (its failure to close); *interarterial septal defect* (failure of the foetal FORAMEN OVALE to close); *tetralogy of Fallot*, combining four defects: interventricular septal defect, an aorta emerging from both ventricles rather than the left only, pulmonary semilunar stenosis, and an enlarged right ventricle. Analysis of families with supravalvar aortic stenosis (SVAS), characteristic of WILLIAMS–BEUREN SYNDROME and MARFAN SYNDROME, has helped understanding of aortic disease. SVAS is caused by mutation in the gene for elastin, while mutations in the fibrillin gene result in progressive ascending aortic aneurysms (dilations). Congenital heart block (see ARRHYTHMIAS) results from foetal damage due to placental transmission of autoantibodies when the mother has SYSTEMIC LUPUS ERYTHEMATOSUS. See BIRTH DEFECTS, DIGEORGE/VELOCARDIOFACIAL SYNDROME, DOWN SYNDROME, PHARYNGEAL ARCHES, SITUS INVERSUS.

congenital hypothyroidism Disease caused by inadequate production of the THYROID HORMONE thyroxine (T_4), failure to detect early enough leads to mental retardation, growth failure, constipation and slow metabolic rate. Its incidence is 1:3,600 to 1:5,000 in the USA. It is detectable by newborn SCREENING and successfully treated by hormone replacement therapy.

conjunctiva The thin, protective, mucous membrane external to the cornea, composed of stratified columnar epithelium with numerous goblet cells and supported by areolar connective tissue. Conjunctivitis caused by inflammation of the conjunctiva through bacterial colonization (e.g., by pneumococci, staphylococci or *Haemophilus influenzae*) is very contagious, but that caused by dust, smoke or atmospheric pollutants, is not.

connective tissues A variety of tissues derived from mesenchyme and the ground substance that these cells secrete. A characteristic cell type is the fibroblast, producing fibres of collagen and elastin that provide tensile strength and elasticity respectively. Another protein, reticulin, is associated with the polysaccharides in the basement membranes underlying epithelia and surrounding fat cells in adipose tissue. Loose connective tissue (areolar tissue) binds many other tissues (e.g., in capsules of glands, meninges, bone periosteum, muscle perimysium and nerve perineurium). Collagen fibres align along the direction of tension, as in tendons and ligaments. The viscosity of many connective tissues is due to the 'space-filling' hyaluronic acid. The peritoneum, pleural membranes and pericardium (serous membranes) are modified connective tissues, as are bone and cartilage. In addition to supportive roles, connective tissue is defensive, due largely to the presence of macrophages, which may be as numerous as fibroblasts. These and mast cells constitute part of the reticuloendothelial system. Connective tissues are often vascularized, and are permeated by tissue fluid.

connexins The major proteins and adhesion molecules of GAP JUNCTIONS. Distinct dominant mutations in the same connexins molecules (which have four transmembrane domains encoded by a large gene family) have been shown to underlie either skin disease or deafness, or both disorders. In response to specific chemical signals (e.g., increase in intracellular [Ca^{2+}]), connexins rotate so as to close the central pore. Connexin mutations also underlie peripheral neuropathy and cataract formation.

consanguinity A reproductive relationship between blood relatives (see RELATEDNESS) who have at least one common ancestor no more remote than a great-great grandparent. Their incidence may be 54% of marriages in Arab populations, where the commonest such marriages are between first cousins who are the children of two brothers, whereas in the Indian subcontinent uncle–niece marriages are the most commonly encountered. In these communities, the potential genetic disadvantages are held to be offset by the social advantages of greater family and marital stability. A recent study of genealogical data from Iceland, for 160,811 couples born in a 165-year period from 1800, revealed that women born between 1925 and 1949 who married third cousins had, on average, almost one more child and almost two more grandchildren than couples who were eighth cousins or more distantly related. First- or second-cousin marriages produced as many children as third-cousin unions, although the children died younger and had a lower probability of reproducing. See GENETIC COUNSELLING, INCEST, KIN DETECTION.

consciousness In reductionist approaches, consciousness is a property of (at least) the human central nervous system. A characteristic of human COGNITION, and especially of explicit MEMORY, it is generally regarded as being required for subjectivity, SELF-AWARENESS, intentionality, sentience, sapience, and the ability to perceive the relationship between oneself and one's environment. Several of these terms are becoming less intractable to reductionist biological research programmes. When ANAESTHETICS cause unconsciousness, they seem to do so by disconnecting functional communication within a brain complex in the POSTERIOR PARIETAL CORTEX, including the anterior and posterior CINGULATE CORTEX. The PAPEZ CIRCUIT, once thought to be the neural basis of conscious awareness of emotions and of cognitive effects on emotion, is now believed to be more restricted in its role and involved largely in explicit learning (that which can be consciously recalled) during emotional states. The pathway serving conscious visual perception includes the lateral geniculate nucleus of the thalamus and the primary visual cortex (the geniculocortical pathway), evidence for which comes from the phenomenon of BLINDSIGHT.

MIRROR CELLS, MIRROR SELF-RECOGNITION and SLEEP have undoubted biological relevance to consciousness, as have the INSULA, reticular activating system (see RETICULAR FORMATION), the AMYGDALA and HIPPOCAMPUS; and, in practical terms, so do ELECTROENCEPHALOGRAM recordings and brain imaging. The interaction and interference of multiple brain rhythms often seem like 'noise' in an electroencephalogram, and because local computation can be sensed by large parts of the cerebral cortex through its long-range connections (unlike the local circuits of the cerebellum which do not give rise to spontaneous or self-generated activity), 'local-global computation' captures the nature of cortical operations. We generate no subjective record of the local computations of the cerebellum, whereas it has been suggested that environmental inputs, perturbations, into the spontaneous activity of the cerebral cortex will, if of sufficient duration and involving a large enough population of neurons, be noticed: i.e., we become conscious of them.

A 23-year-old woman who sustained severe traumatic brain injury in 2005 and who remained unresponsive, but with preserved sleep–wake cycles, was assessed 5 months later by clinicians to satisfy all criteria for being in a vegetative state. A team used fMRI (see MAGNETIC RESONANCE IMAGING) to measure her neural responses during presentation of spoken sentences and compared her responses to acoustically matched noise sequences. Bilateral speech-specific activity was observed in the middle and superior temporal gyri, equivalent to that observed in healthy volunteers listening to the same stimuli. Sentences containing ambiguous words produced additional significant response from her left inferior frontal region, again similar to responses in volunteers and indicated the operation of semantic processes critical for speech comprehension. To address the question of whether such responses indicated her being in a state of conscious awareness, a second fMRI study involved her being given spoken instructions to perform two mental imagery tasks at specific times during the scan. One asked her to imagine herself playing a game of tennis, the other asked her to imagine visiting all the rooms in her home, starting at the front door. During the former episodes, significant activity was observed in the supplementary motor area, which contrasted with the significant activity observed in the parahippocampal gyrus, the posterior parietal cortex and the lateral premotor cortex during the latter – responses indistinguishable from those observed in healthy volunteers performing the same imagery tasks in the scanner. The conclusion drawn was that despite her vegetative condition, the patient retained the ability to understand spoken commands and to respond to them through her brain activity, though not through speech or movement. Her compliance with the research was also deemed to be indicative of clear intent, confirming her *conscious awareness* of herself and of her surroundings. See EMOTIONS.

consensus sequence Any DNA whose sequence has been multiply verified and collectively agreed upon by the sequencing community.

contraception Combined oral contraceptive pills contain synthetic oestrogen (ethinyl oestradiol, or mestranol) along with one of several synthetic progestogens, with a number of effects on the female reproductive tract. The major consequence is the inhibition of ovulation by preventing GnRH release and hence release of LUTEINIZING HORMONE. Combination oestrogen-progestin pills are taken over a 21-day period, followed by 7 days without; phasic oestrogen-progestin pills are taken for 21 days and vary in the relative proportions of the steroids, being followed by 7 days of placebo. Progestin-only pills are taken daily. Antiprogestins, such as *mifepristone*, block progesterone's effects by binding its receptors, disrupting the menstrual cycle, preventing normal uterine thickening in preparation for pregnancy and inhibiting gene action required for normal implantation. Oral contraceptives are believed to promote tumours of the liver. Intrauterine devices, which may cause additional bleeding and may not be tolerated, prevent blastocyst implantation by altering the endometrial lining; but some release progesterone. Barrier methods, such as condoms, foam and diaphragms, prevent fertilization by interfering with entry of sperm to the uterine cavity, or by destroying sperm in the vaginal cavity. The rhythm method relies on changes in mucus thickness and body temperature during the MENSTRUAL CYCLE, indicating a 'safe' period when intercourse should be possible without fertilization. Sterilization, which must be regarded as irreversible, disrupts the continuity of the fallopian tubes (tubal ligation, preferably by endoscopy) in women, and of the sperm ducts in men (vasectomy; diathermy of the vasa deferentia). Other male fertility control methods include weekly injections of testosterone ethanoate, which suppresses pituitary LH and FSH release: sperm disappear from the ejaculate after ~120 days, recovery of fertility occurring ~100 days after the last injection.

contractile ring A highly dynamic belt-like bundle of actin and myosin-II filaments, assembling just beneath the plasma membrane during anaphase and generating the force that pulls opposing membrane surfaces together prior to 'pinching-off' and completion of cytokinesis. Its components change by the minute, being regulated by the polo-like family of protein kinases. Contraction begins when Ca^{2+}-calmodulin activates myosin light chain kinase to phosphorylate myosin II. See CALCIUM/CALMODULIN-DEPENDENT PROTEIN KINASES.

contralateral Of structures on opposite sides of the MIDLINE. See ANATOMICAL PLANES OF SECTION.

contrast enhancement Amplification of the *differences* in the activities of neighbouring neurons. It is a general feature of information processing in sensory pathways, including the somatic sensory system, one general mechanism involved being feedforward LATERAL INHIBITION. See RETINA.

control elements See REGULATORY ELEMENTS.

convergence 1. Of NEURAL NETWORKS (e.g., in the HIPPOCAMPUS) in which many upstream neurons synapse with few downstream ones, allowing data compression and integration of weak signals (e.g., see RETINAL CONVERGENCE, COCHLEA, OLFACTION) and enabling more effective stimulation or inhibition of the postsynaptic neuron. The inputs may be from several different sources, as when a motor neuron that synapses with skeletal muscle fibres at neuromuscular junctions receives inputs from pathways originating in different brain regions. Contrast DIVERGENCE (2). **2**. See CONVERGENT EVOLUTION.

convergent evolution Evolution of the same character state by different routes. Such characters are analogous, or homoplasious.

convergent extension The coordinated intercalation of cells that narrows and lengthens a tissue. Regulated by the PLANAR CELL POLARITY pathway, a form of non-canonical WNT SIGNALLING PATHWAY, the processes of GASTRULATION and NEURULATION depend upon it.

co operative behaviour See ALTRUISM.

COP proteins Proteins (e.g., COPI, COPII) involved in aspects of trafficking between the Golgi apparatus and the rough endoplasmic reticulum. See VESICLES.

copy number, copy number variation The number of copies of a particular genome segment in a genome, whether nuclear or mitochondrial. Classically, variation in the copy number of the globin genes was shown to be responsible for haemoglobin disorders such as the α-THALASSAEMIAS. Copy number polymorphisms (CNPs) in populations can arise through deletions and duplications, giving copy number variants (CNVs). It may be that fixed segmental duplications, unstable regions and CNPs are manifestations of the same underlying process. The contribution of nuclear copy number polymorphisms to human GENETIC DIVERSITY is largely unknown, although recent estimates of large-scale CNPs (those ≥ ~100 kb) suggest that they contribute substantially to genomic variation between normal humans. A 2004 study involving 20 individuals from a variety of geographic backgrounds indicated that CNPs are widely distributed throughout the genome, some apparently located in 'hotspots' of copy number variation, some occurring within genomic regions where recurring *de novo* rearrangements are causes of developmental disorders, notably PRADER–WILLI SYNDROME, ANGELMAN SYNDROME and DIGEORGE/VELOCARDIOFACIAL SYNDROME, possibly reflecting instability of these genomic regions. Variable copy number of the *CCL3L1* gene, by SEGMENTAL DUPLICATION, has recently been linked to increased resistance to infection by HIV (see entry), differences in the copy number and dosage effects of immune response genes now being thought to provide a genetic basis for variability in responses to infectious diseases.

Work published in 2006 discovered 1,447 CNVs across 270 HapMap samples (see HAPLOTYPE), a 5- to 10-fold greater variation between any two randomly chosen genomes than previously suggested by studying SNPs alone (see SINGLE NUCLEOTIDE POLYMORPHISMS). Since more than half of these CNVs overlapped known annotated genes in the genome, it is likely that CNVs play a role in susceptibility to COMPLEX DISEASES. Observed differences in the copy number of genome segments between samples from two individuals should reflect germline differences or somatic variation. One interesting CNP in the 2004 study involved a triplication of the NEUROPEPTIDE Y receptor *PPYR1*, a gene directly involved in regulation of food intake and body weight. Preliminary work suggests that, as a class, CNPs are under negative selection. See DOSAGE COMPENSATION.

CoQ See COENZYME Q.

coracoid, coracoid process See SCAPULA.

co-receptors Cell-surface proteins and proteoglycans that increase sensitivity of an immune cell antigen receptor to antigen by binding associated ligands. CD8 and CD4 are MHC-binding T-cell co-receptors, while CD19 is part of a co-receptor complex on B cells. See ACCESSORY RECEPTORS, B-CELL CO-RECEPTOR, T CELLS.

coregulatory proteins Proteins, usually occurring as multiprotein complexes of 5–10 proteins (coactivators and corepressors) forming large regulatory machines within the nucleus that dock onto TRANSCRIPTION FACTORS and coordinately mediate their actions (see GENE EXPRESSION). While transcription factors bind DNA in a sequence-specific fashion, they usually lack the enzymatic activity required to modify chromatin, unwind DNA and recruit RNA polymerase II. Coregulators fulfil these roles, being recruited to transcription factors in response to cellular signals. Transcriptional coactivator proteins are very often the primary targets of hormonal control and signal transduction pathways (e.g., see CREB-1, NUCLEAR RECEPTOR SUPERFAMILY, PGC-1). A picture is emerging that, for any single physiological condition (e.g., diabetes mellitus, inflammatory diseases, cardiovascular disease) and other more general physiological processes (e.g., memory, learning and reproductive functions), a limited number of coregulatory molecules may control all the functionally relevant genes.

core sequence For core promoter, see PROMOTER.

core temperature The temperature of deep body structures, beneath the skin and subcutaneous layer. *Shell temperature* is the temperature near the body surface, and is normally 1–6°C below the core temperature (normally 37°C). See FEVER, TEMPERATURE REGULATION.

Cori cycle See GLUCONEOGENESIS, LACTIC ACID.

cornea The curved and transparent external surface of the EYE (see Fig. 73), continuous with the sclera and covering the pupil and iris. On its outer surface is the conjunctiva, a membrane folding back from the inside of the eyelids and attaching to the sclera. Lacking blood vessels, the cornea is supplied with nutrients by the aqueous humour behind it and by the tear film on its outer surface that is continuously replenished by blinking of the eyelids. The cornea performs most of the eye's refraction of light, so that it converges on the back of the eye (see LENS). The curvature of the cornea affects its focal length (or focal distance, the reciprocal of focal length in metres being the *dioptre*, converging lenses having positive values, diverging lenses having negative ones), its refractive power being ~42 dioptres, light being focused 0.024 m behind it – the distance between the cornea and the retina. Under water, this refractive power (which depends upon Snell's law) is abolished. For the *corneal reflex*, see EYE-BLINK REFLEX. See ACCOMMODATION.

cornification (keratinization) See EPIDERMIS.

corona radiata A covering of cumulus oophorus cells overlying the zona pellucida of the ovulated oocyte. It is only passed by capacitated spermatozoa. See FERTILIZATION, Fig. 76.

coronal Of a plane perpendicular to the ground and to the sagittal plane. See ANATOMICAL PLANES OF SECTION, Fig. 9.

coronary arteries The right and left coronary arteries exit the aorta close to the exit of aorta from the heart, the right being the smaller and carrying less blood. A major branch of the left coronary artery supplies most of the blood to the anterior part of the heart, another branch supplies the lateral wall of the left ventricle, while a third branch supplies much of the posterior wall of the heart. Branches of the right coronary artery supply the lateral wall of the right ventricle and the posterior and inferior parts of the heart. There are many anastomoses, however, and most parts of the myocardium receive blood from more than one arterial branch. This is a significant factor in helping some people to survive MYOCARDIAL INFARCTIONS. Aerobic EXERCISE tends to improve both the number of anastomoses and the density of capillaries, increasing the chances that a person will survive blockage of a small coronary artery. The CARDIAC VEINS drain the heart muscle. See BYPASS SURGERY, CORONARY HEART DISEASE.

coronary circulation See CARDIAC VEINS, CORONARY ARTERIES.

coronary heart disease (CHD, coronary artery disease, CAD) A COMPLEX DISEASE whose epidemic spread has ravaged Western populations in the 20th century, now ranking as the commonest worldwide cause of death in those >30 years of age. Aka coronary artery disease (CAD), or coronary ATHEROSCLEROSIS, it is a highly heritable chronic degenerative condition which may be clinically silent, or present with ANGINA PECTORIS or acute MYOCARDIAL INFARCTION resulting from the inflammatory and proliferative lesions of the atherosclerotic plaque. Pathogenesis is complex, with endothelial dysfunction, oxidative stress and inflammation contributing to instability of the plaque. HYPERTENSION, DIABETES and hypercholesterolaemia (see FAMILIAL HYPERCHOLESTEROLAEMIA) are known risk factors. There is evidence that a modest life-time reduction in LDL cholesterol, achievable by exercise, diet and/or drugs, can reduce the rate of CHD much more effectively than combating cholesterol at middle age by combined use of low-cost generic STATINS and other cholesterol-lowering drugs.

β_1-ADRENERGIC receptor antagonists are used in treatment of CHD and chronic heart failure. In addition to lifestyle and environmental factors, genome-wide linkage studies have identified two genes among several loci that may affect susceptibility: *ALOX5AP* (arachidonate 5-lipoxygenase-activating protein) and *LTA4H* (leukotriene A4 hydrolase). A recent genome-wide study found an association with CHD on chromosome 9p21.3, a region adjacent to the tumour suppressor genes *CDKN2A* and *CDKN2B*, which contains no annotated genes associated with established CHD risk factors such as plasma lipoproteins, hypertension or diabetes. Homozygotes for the risk allele make up 20–25% of Caucasians, and have a ~30–40% increased risk of CHD, while sibling recurrence risk (l_s) estimates for early myocardial infarction range from ~2 to ~7. Mutations in the gene encoding the protease PCSK9 are associated with a lowering of plasma LOW DENSITY LIPOPROTEIN (LDL). In one 2006 study of ~16,000 Caucasians and African-Americans from four US communities (average age 53 years at the initiation of the study in 1987), those African-Americans with a loss-of-function *PCSK9* mutation averaged plasma LDL-cholesterol (LDL-C) levels of 100 mg 100 cm^{-3}, lower by 38 mg 100 cm^{-3} (~40%) than the controls and associated with 88% reduction in CHD – despite the high incidence of hypertension (37%) and type 2 diabetes mellitus (13%). Caucasians with a missense *PCSK9* mutation lowering blood LDL-C levels by only 21 mg 100 cm^{-3} had a 47% reduction of CHD. Middle-aged individuals being treated with STATINS may reduce their plasma LDL-C levels by as much as 40%, although the relatively late treatment reduces CHD by only 23%. Surgical treatment includes ANGIOPLASTY and BYPASS SURGERY. See ApoB, BLOOD GLUCOSE REGULATION, CHOLESTEROL, CORONARY ARTERIES.

coronary thrombosis A blood clot blocking the coronary (arterial) circulation. See CORONARY HEART DISEASE, THROMBOSIS.

corpus albicans The scar left by a regressed CORPUS LUTEUM.

corpus callosum A large collection of commissural fibres (~2 × 10^8 axons) connecting corresponding regions of the NEOCORTEX other than the temporal fields (see ANTERIOR COMMISSURE). Appearing by the 10th week of development, its occasional absence does not necessarily lead to functional disorders, although it is sometimes associated with SAVANT SYNDROME. Severance is often beneficial in restoring a seizure-free life to epileptics. See PRIMARY SOMATOSENSORY CORTEX, SEX DIFFERENCES, SPLIT-BRAIN PHENOMENON.

corpus luteum A temporary endocrine organ forming upon rupture of the GRAAFIAN FOLLICLE from small (theca-lutein) but especially from large (granulosa-lutein) cells. It also contains fibroblasts, endothelial cells and immune cells. Maintained during the menstrual cycle by pituitary LH, it continues the secretion of oestradiol and is the main source of pre-placental progesterone, maintaining development during the first 2–3 months of pregnancy. Its survival is assured during this period by placental hCG, after which it regresses. Under hCG influence, it secretes progesterone, 17-hydroxyprogesterone, oestrogen, inhibin A and relaxin. In the absence of implantation, its regression marks the end of the menstrual cycle. This process of luteolysis involves an initial decline in progesterone secretion, followed by cellular changes leading to its gradual involution, a scar of connective tissue (the *corpus albicans*) remaining in the ovary.

corpus striatum Collective term for the NEOSTRIATUM (caudate nucleus and putamen) and palaeostriatum (globus pallidus). See BASAL GANGLIA, CEREBRAL HEMISPHERES.

cortex (pl. cortices) **1**. Any collection of neurons (e.g., cerebral and cerebellar cortices) forming a thin sheet; usually at the surface of the brain (see CEREBELLUM, CEREBRAL CORTEX). **2**. An outer layer of an organ (e.g., renal cortex, adrenal cortex).

cortical granules See CORTICAL REACTION.

cortical plasticity See CRITICAL PERIOD, NEURAL PLASTICITY.

cortical plate Cell layer of the immature cerebral cortex containing undifferentiated neurons but later contributing, with the primordial plexiform layer, to the mature cortex layers (see CEREBRAL CORTEX, Fig. 44, label CP).

cortical reaction On fusion of the sperm and oocyte plasma membranes at FERTILIZATION, there is a local uptake of calcium ions by the oocyte which spreads in waves through the cell. This initiates the cortical reaction in which Golgi-derived cortical granules, close to the oocyte membrane and containing proteolytic enzymes, release their contents by exocytosis. These alter the structure of the zona pellucida so that it is 'hardened' and will no longer bind sperm, which acts as a block to polyspermy.

corticobulbar tract Originating in the cerebral cortex, the direct descending motor tract controlling facial and head movements. Its axons terminate in the cranial nerve nuclei of the brainstem, decussation depending on the pathways involved (see CORTICOSPINAL TRACTS).

corticospinal tracts (pyramidal tracts) Bilateral glutamatergic descending motor tracts of the SPINAL CORD, the cell bodies of whose neurons originate in the cerebral cortex (see PREMOTOR CORTEX, PRIMARY MOTOR CORTEX). They are therefore *direct* motor tracts, forming a major component of the somatic motor system, and are especially involved in hand movements. Their axons pack tightly and pass through the internal capsule between the thalamus and lentiform nucleus, and descend into the pons. Here most medial fibres (*corticobulbar fibres*) cross the midline and pass to nuclei (trigeminal, facial, hypoglossal and accessory) of the relevant cranial nerves. These include those involved in intentional facial expressions. The remaining axons pass through the ventral pons and medulla oblongata, forming the two prominent *pyramidal tracts*. Roughly triangular in section, these collect at the base of the medulla, forming a bulge running down its ventral surface. Roughly 40% of fibres in these tracts originate in the primary motor cortex, ~30% in the premotor cortex and ~30% in the somatosensory cortex posterior to the central sulcus, or other regions of the parietal cortex. The corticospinal tract is therefore not simply the motor system by which the brain orders the muscles to contract: even this pyramidal motor system also influences the sensory input received by the cortex. About 75% of fibres decussate in the pyramids and form the contralateral *lateral corticospinal tract*, while 10–25% of pyramidal fibres remain ipsilateral and enter the *ventral (anterior) corticospinal tract*. About half of the neurons terminate at cervical levels, 20% at thoracic and 25% at lumbosacral levels. Most are fine myelinated or unmyelinated axons, with conduction velocities ~1–25 m s^{-1}, although faster-conducting myelinated axons of BETZ CELLS are also present.

The tracts are involved in the control of movement (see MOTOR CORTEX), especially novel movements, of muscle tone, spinal reflexes, spinal autonomic functions and transmission of sensory information to higher centres. In particular, in humans but not lower primates, they carry information for the control of voluntary, discrete, skilled movements (often termed 'fractionated movements'), such as those of individual fingers, that are usually the least likely to recover after STROKES involving the motor cortex (see also CEREBELLUM). The tracts are commonly excitatory to flexors, their axons synapsing directly with α motor neurons supplying distal limb muscles. They are commonly inhibitory to extensors, axons synapsing with Ia inhibitory interneurons innervating both these and axial muscles. Fusimotor (γ-efferent) neurons that require coactivation with α motor neurons in overriding the stretch reflex during voluntary movement are polysynaptically excited (see MUSCLE SPINDLES). During primate evolution, and human evolution especially, the indirect corticorubrospinal pathway has been largely replaced by the direct corticospinal pathway. For *extrapyramidal motor pathways*, see BASAL GANGLIA. Compare RUBROSPINAL TRACT, and see RED NUCLEUS for switching of pathways between the two descending motor pathways.

corticosteroids (corticoids) See GLUCOCORTICOIDS, MINERALOCORTICOIDS.

corticotrophs Cells constituting <20% of the anterior pituitary gland, secreting CORTICOTROPIN in response to CORTICOTROPIN RELEASING HORMONE.

corticotropin (ACTH) Synthesized as a large precursor, PROOPIOMELANOCORTIN, corticotropin is released in a pulsatile fashion from the anterior PITUITARY GLAND with ~7–15 episodes per day. Corticotropin binds G_s-coupled melanocortin 2 receptor and activates PKA, leading to phosphorylation of cholesterol ester hydrolase, release of cholesterol from cholesterol esters, and production of CORTISOL by the adrenal cortex ~15 minutes after a corticotropin surge. Its release is under the control of CORTICOTROPIN RELEASING HORMONE. In the absence of adequate corticotropin and glucocorticoid response, TUMOUR NECROSIS FACTOR is overexpressed in endotoxaemia.

corticotropin releasing hormone (CRH, CRF) A peptide transmitter synthesized by the parvocellular neurons of the paraventricular nucleus of the hypothalamus (PVN) and causing release of CORTICOTROPIN. Its own release is regulated by the effects of glucocorticoids on CRH gene expression: down-regulation in cells of the PVN and up-regulation in cells of the central nucleus of the amygdala and lateral bed nucleus of the stria terminalis. CRH acts locally, stimulating the oxytocinergic neurons that inhibit feeding. It is known to be anorexigenic (see ANOREXIA NERVOSA). There is considerable evidence that early traumatic life experiences, such as maternal neglect or child abuse, are linked to persistent, long-lasting hyperactivity of CRH neurons, one possible route by which such trauma increases vulnerability to adult depression (see AFFECTIVE DISORDERS). The hippocampus responds to cortisol release from the adrenal gland by inhibiting CRH release when activated. There is hope that CRH receptor antagonists may in the future be effective antidepressants and anxiolytics. Its gene is activated by CREB (see Fig. 60). CRH is also produced by the PLACENTA.

cortisol (hydrocortisone) The main human glucocorticoid, responsible for ~95% of their physiological activity (see GLUCOCORTICOIDS). A lipophilic hormone and neuromodulator, and a product of CHOLESTEROL metabolism (see Fig. 49), it is largely carried in plasma bound to α2-globulin (levels of which rise during PREGNANCY). It is released within 15 minutes of a surge in CORTICOTROPIN, which binds to the G_s-coupled plasma membrane melanocortin 2 receptor on plasma membranes of the zona fasciculata cells of the adrenal cortex. This signals activation of PKA, leading to release of cortisol from its esters. Activation of hippocampal glucocorticoid receptors by cortisol leads to feedback inhibition of the synthesis and release of hypothalamic CRH and corticotropin. Cortisol crosses cell membranes by passive diffusion and binds the type 2 glucocorticoid nuclear receptor before entering the nucleus and binding DNA as a transcription factor. Since most cells contain this nuclear receptor, most cells respond to corticotropin by transcribing selected genes. Its primary role in carbohydrate metabolism is a relatively minor one. It promotes GLUCONEOGENESIS, by inducing hepatic gene expression of the enzymes involved, in response to prolonged hypoglycaemia (see BLOOD GLUCOSE REGULATION). It also exerts permissive effects on hepatic gluconeogenesis by GLUCAGON and noradrenaline. Its anti-insulin effects in peripheral tissues (notably adipose and lymphoid tissues) are such as to promote the release of free fatty acids by lipolysis, indirectly reducing the depletion of liver glycogen and sparing glucose for use by the brain. See ANDROGENS, STRESS RESPONSE.

cortisone Inactive form of CORTISOL.

costal angle See RIBS.

costal breathing Shallow chest breathing. See VENTILATION.

costal tuberosity See CLAVICLE.

co-stimulatory molecules Those molecules expressed on the surfaces of ANTIGEN-PRESENTING CELLS, in response to binding of pathogen components to non-antigen-specific receptors, and which must be co-presented with antigen if lymphocyte proliferation is to result. B7.1 and B7.2 (CD80 and CD86) bind to CD28 on the T-cell surface. For B cells, CD40 ligand on the surface of an activated antigen-specific helper T cell, or common pathogen components such as lipopolysaccharide (from complement fragments), provide the co-stimulatory signal. For both macrophages and dendritic cells, co-stimulatory molecules are expressed in response to LIPOPOLYSACCHARIDE signalling through TLR-4 (see TOLL-LIKE RECEPTORS), but not during APOPTOSIS.

cotransporters (coupled transporters) Antiporters and symporters: membrane proteins in which movement of one ion species against its electrochemical gradient is powered by the downhill movement of another. They do not rely on ATP hydrolysis for their action, and energetically lie between ION CHANNELS and ION PUMPS and often mediate secondary active transport and facilitated diffusion (e.g., see ENTEROCYTES, PROXIMAL CONVOLUTED TUBULE).

coughing A long-drawn, deep inspiration followed by closure of the rima glottidis between the vocal cords, resulting in a strong expiration that opens the rima glottidis, forcing air out of the upper air passages; this reflex is usually in response to a foreign body lodged in the LARYNX, trachea or epiglottis. See WHOOPING COUGH.

countercurrent system See LOOP OF HENLE, VASA RECTA.

coupled transporters Antiporters and symporters. See COTRANSPORTERS.

Cowper's glands (bulbourethral glands) A pair of small compound mucous glands at the base of the penis which secrete mucus into the urethra under parasympathetic stimulation during sexual AROUSAL.

COX-1, COX-2 See CYCLOOXYGENASES.

coxal bones (hip bones) Each coxal bone is formed by the fusion during development of the ILIUM, ISCHIUM and PUBIS. See PELVIC GIRDLE.

Coxiella burnetti A mycoplasma (Rickettsiaceae), and cause of Q FEVER.

CpG dinucleotide This dinucleotide (read in 5′→3′ direction) occurs relatively rarely in human, mammalian, and indeed vertebrate, DNA. Such 'GC-rich' regions can be cut by 'rare-cutter restriction endonucleases', yielding quite long DNA restriction fragments from 20 Kb to several Mb in length that can be separated by pulsed field gel electrophoresis, a recent development of agarose gel electrophoresis.

During the course of vertebrate evolution, more than three out of four CpGs have been lost because of the manner in which enzymes of DNA REPAIR MECHANISMS operate: accidental deamination of unmethylated cytosine in DNA yields uracil, which is replaced by cytosine; but such deamination of 5-methyl cytosine yields thymine, which is not always distinguished from thymines that are normally present in DNA (see LI-FRAUMENI SYNDROME). CpG dinucleotides are ~10-fold more mutable than the genome average. The CpG sequences that remain are very unevenly distributed and present at several times their average density in so-called *'CG (or CpG) islands'* containing frequent unmethylated CpG dinucleotides that are 1000–2000 nucleotide pairs in length (an example of a DNA REPEAT). There are ~27,000 of these in the HUMAN GENOME and they often surround promoter regions of so-called 'housekeeping genes' – those essential for basic metabolic pathways and cell viability and expressed in most cells. They generally mark the 5′ ends of transcription units and provide a convenient way of identifying type II genes in the human genome (see DNA METHYLATION) since methylation of CpG islands is never erased during normal development (contrast GENOMIC IMPRINTING). Unmethylated CpG sequences are far commoner in bacterial than in mammalian DNA and are recognized by the TOLL-LIKE RECEPTOR TLR-9 (see SYSTEMIC LUPUS ERYTHEMATOSUS).

Mammalian cytosine methyltransferases show a strong preference for a hemimethylated DNA target (methylated on one strand only). Because the CpG sequence shows dyad symmetry, after DNA replication the newly synthesized DNA strands will receive the same CpG methylation pattern as the parental DNA so that the CpG methylation pattern can be stably transmitted to daughter cells. This *maintenance methylation* is carried out in mammalian cells by the Dnmt1 methyltransferase, and deficiencies in dietary METHIONINE and FOLATE lead to reductions in *S*-adenosylmethionine, the methyl donor for methylcytosine. Some tissue-specific genes are also associated with CpG islands, and an early observation in cancer epigenomics was that silencing of a gene can occur via aberrant 'focal' methylation of an adjacent CpG island (see CANCER CELLS). It is now believed that the aberrant CpG island methylation involved in some cancers may involve larger expanses of DNA (see COLORECTAL CANCER). The CpG island-containing SVA retrotransposon is a possible source of regulatory differences between chimpanzees and humans (see TRANSPOSABLE ELEMENTS). See TRANSCRIPTION, for HAIRPIN LOOPS.

CpG islands See CPG DINUCLEOTIDE.

cramps Painful, spasmodic, muscle spasms. Heat cramps occur as a result of profuse sweating that removes water and salt from the body, tending to occur in working muscles but during relaxation after the work has been done. Recovery usually results on drinking salty liquids.

cranial capacity The volume of the BRAIN. For fossils, it is often calculated from bone fragments, nowadays by computed tomography.

cranial cavity See BODY CAVITIES, Fig. 23.

cranial meninges See MENINGES.

cranial nerves It is usual to regard cranial nerves as nerves arising from nuclei in the brain stem, originating from RHOMBOMERES during development and innervating (mostly) the head. Some are components of the central nervous system; others belong to the somatic peripheral nervous system, while others are components of the AUTONOMIC NERVOUS SYSTEM. Sometimes, the OLFACTORY NERVE (CNI) and OPTIC NERVE (CNII) are included, these arising in the olfactory mucosa (see OLFACTORY TRACT) and RETINA respectively. Those originating in the midbrain are the oculomotor nerve (CNIII) and the trochlear nerve (CNIV); those originating in the pons are the trigeminal nerve (CNV), abducens nerve (CNVI), facial nerve (CNVII) and vestibulocochlear nerve (CNVIII, see NEUROFIBROMATOSIS); those originating in the medulla oblongata are the glossopharyngeal nerve (CNIX), the vagus nerve (CNX), the cranial portion of the accessory nerve (CNXI, whose spinal portion originates in the ventral grey horn of the first five cervical segments), and the hypoglossal nerve (CNXII). See entries on separate cranial nerves.

cranial vault The membrane (dermal) bones of the SKULL. See CALVARIUM.

cranium The skeleton of the SKULL (see Fig. 157) protecting the brain and organs of sight, hearing and balance; includes the frontal, parietal, temporal, occipital, sphenoid and ethmoid bones.

CRE Cyclic AMP response element. See CREB.

C-reactive protein (CRP) Opsonizing protein composed of five identical subunits that binds the phosphocholine component of certain bacterial and fungal cell wall LIPOPOLYSACCHARIDES and, by binding C1q, activating the COMPLEMENT cascade.

creatine An amino acid, not encoded by the genome, synthesized from glycine and arginine, with methionine as the methyl group donor. See CREATINE KINASE.

creatine kinase An enzyme reversibly phosphorylating creatine to PHOSPHOCREATINE in skeletal muscle fibres. ~4–6% is located on the outer mitochondrial membrane, ~3–5% in the sarcomeres of myofibrils and ~90% in the sarcoplasm. See MUSCULAR DYSTROPHIES.

creatine phosphate See ATP, PCR SYSTEM, PHOSPHOCREATINE.

creatinine A breakdown product of phosphocreatine in muscle, normally produced at a fairly constant rate. It is secreted by the PROXIMAL CONVOLUTED TUBULES of the KIDNEY.

CREB (cyclic AMP response element-binding protein) Localized within the nucleus, CREB is a member of the basic region leucine zipper proteins and is a transcription factor (TF) expressed in all brain cells and is best known for its involvement, as a synergist of NEUROTROPHINS, in neuronal survival, learning and MEMORY. Existing in three isoforms, it is one of several TFs that bind as dimers to the cAMP-response elem-

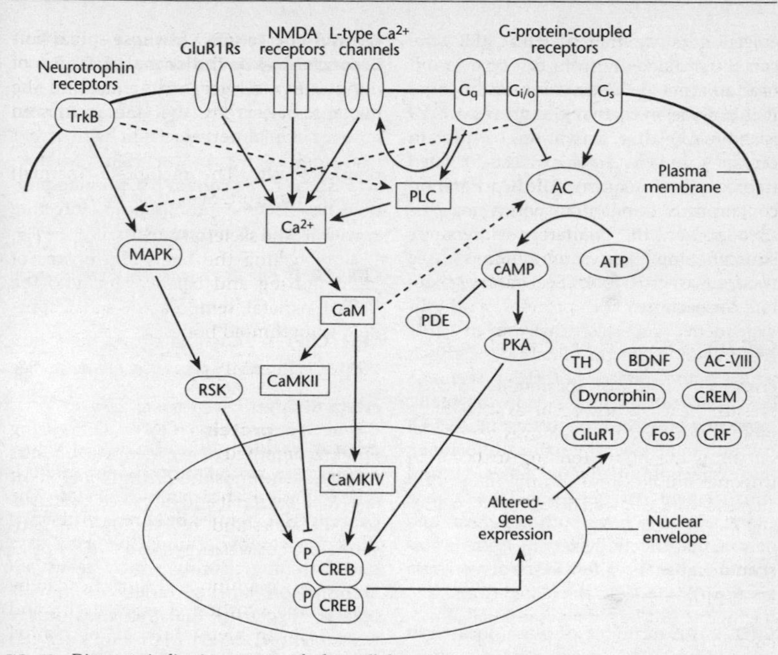

FIG 60. Diagram indicating some of the cellular events involved in the regulation of CREB. Neurotransmitters and neurotrophins bind cell surface receptors such as TrkB, AMPARs (GluR1Rs), NMDA receptors and G-coupled protein receptors, triggering intracellular signalling cascades culminating in phosphorylation of dimerized CREB at its two serine 133 sites, activating CREB-mediated gene expression (some of the protein products being indicated outside the nucleus to the right). CREB can be phosphorylated by calcium-calmodulin-dependent kinases (e.g., CaMKII, CaMKIV) on Ca^{2+} entry via NMDA and other receptors and channels. Many more interacting pathways than are shown (broken lines) occur in cells. Inhibitory pathways are indicated by T-junctions, stimulatory ones by arrows. CaM = calmodulin; PDE = phosphodiesterase; PLC = phospholipase C; TrkB = type 2 neurotrophin receptor tyrosine kinase receptor. From W.A. Carlezon Jr, R.S. Duman and E.J. Nestler, © 2005 *Trends in Neuroscience*, with permission from Elsevier Ltd.

ent (CRE) – a specialized DNA sequence occurring in several promoters and enhancers and containing the consensus nucleotide sequence TGACGTCA. A co-activator of numerous transcription factors (see RESPONSE ELEMENT, MECP2), CREB has numerous phosphorylation sites that regulate its activity in different ways and enable it to mediate in different pathways, including the experience-based learning neuroadaptations occurring after exposure to a maze (see HIPPOCAMPUS), to physical or emotional stressors (e.g., fear conditioning involving the AMYGDALA) and to addictive substances. It is activated by CALCIUM/CALMODULIN-DEPENDENT KINASES (e.g., CaMKIV) in response to Ca^{2+} entry to the cell. It is also a key player in FASTING responses, when it is coactivated by TORC2 as well as by its canonical coactivator, CBP. Once dimerized and phosphorylated at its two serine 133 sites (see Fig. 60), it initiates a cascade of events leading to formation of a transcription complex that promotes nearby CHROMATIN REMODELLING through histone acetylation (see Fig. 60). CREB-mediated gene expression is controlled by protein phosphatases (activated by CALCINEURIN) and by modulatory repressor proteins (CREMs) that bind to CREs and prevent CREBs from doing so.

The regulatory elements that CREB-1 targets differ markedly from cell type to cell type and include distinct subsets of genes within different brain regions. In LONG-TERM POTENTIATION, after activation by protein kinase A (PKA) and MAPK, CREB-1-related transcription is required for the formation of long-term memories, which may be encoded in the pattern of synaptic strengthening between neurons (see SYNAPTIC PLASTICITY), formation of new synaptic connections, or protein synthesis-dependent processes involved in their retrieval and reconsolidation. CREB-2, which is an inhibitor of CREB-1, seems to be a transcription repressor in the brain, repressing long-term memory. Since CREB is found in all neural circuits, influencing the expression of diverse genes, clinical enhancement of CREB-mediated gene expression may have both beneficial and detrimental effects, detracting from its possible therapeutic value (see DEPRESSION, RETT SYNDROME).

CREB-2 An inhibitor of CREB-1, located in the nucleus.

cremaster muscle A small band of skeletal muscle, a continuation of the internal oblique muscle, in the SPERMATIC CORD that elevates the testes when exposed to cold, and during sexual arousal.

CREMs (cyclic AMP response element modulator) See CREB.

crenated cells Cells whose membranes have become irregularly folded, usually as a result of being surrounded by a hypertonic fluid.

crest Prominent ridge or elongated projection on a bone surface, for attachment to connective tissue.

cretinism A condition, usually due to an underactive maternal THYROID GLAND in pregnancy, or in newborn children, characterized by MENTAL RETARDATION, and short stature through developmental abnormalities of the brain and skeleton. Early treatment with thyroid hormones overcomes the condition.

Creutzfeldt–Jakob disease (variant Creutzfeldt–Jakob disease, vCJD) A zoonotic disease, recently emerging in humans, caused by the human-adapted form of the PRION associated with BOVINE SPONGIFORM ENCEPHALOPATHY, (BSE). Uncharacteristically, the variant BSE prion responsible has become promiscuous, infecting species in addition to humans.

CRF (CRH) See CORTICOTROPIN-RELEASING HORMONE.

CRH (CRF) See CORTICOTROPIN-RELEASING HORMONE.

crista dividens See FORAMEN OVALE.

critical period In embryonic development, a time period during which intercellular communication alters a cell's fate. The concept was extended by Lorenz to describe the finite window of time (the first 2 days after hatching) during which geese can 'imprint' on a wide variety of moving objects, suggesting that this visual image was somehow 'etched' in the bird's nervous system, affecting subsequent behaviour. In neuroscience, a critical period became established as 'a limited period of time during which a particular aspect of brain development is sensitive to a change in the environment'. That synaptic activity can alter the fate of neural connectivity during the development of the CNS was enormously advanced by the work of Hubel and Wiesel on the establishment of OCULAR DOMINANCE in cats and primates. In macaque monkeys, occlusion of one eye for a few days during the first 2 weeks of postnatal life causes marked and permanent changes to the visual cortex such that few cells respond to stimulation of the previously occluded eye. Instead, the unoccluded eye establishes nearly complete ocular dominance, with expansion of ocular dominance columns at the expense of shrinkage of those of the occluded eye, so that the open eye 'takes control' of cortical territory normally occupied by the closed eye. These changes can be reversed simply by closing the previously open eye and opening that which was previously closed. In humans, binocular connections are formed and modified under the influence of visual stimuli during infancy and early childhood, the maintenance of binocular receptive fields depending on correlated patterns of activity arising from the two eyes (see NEURAL PLASTICITY). The plasticity of ocular dominance columns is not retained throughout life. As Hubel and Wiesel discovered, if deprivation starts later in life, the anatomical effects observed in layer IV of the primary visual cortex do not occur. The critical period lasts ~6 weeks in the macaque. There is also a critical period in LANGUAGE acquisition, and probably for INCEST-avoidance. See MENTAL RETARDATION.

critical thermal maximum (CT_{max}) A value often defined in practice by the onset of debilitating muscle spasms or loss of coordination rather than actual death of the animal. Very few animals can survive a value greater than 50°C.

Crohn's disease See ALLERGIES, COMPLEX DISEASES.

Cro-Magnon humans Anatomically modern human remains, discovered in 1868 at Cro-Magnon and Les Eyzies, France, and dated to 30 Kya. They exhibit a warm-adapted body stature, unlike the cold-adapted formula of Neanderthals, and are found with Upper Palaeolithic industries (the Cro-Magnon fossils being the first Upper Palaeolithic human remains to be published in detail). See HOMO SAPIENS.

crossing-over The chromosomal event of exchange of chromatid segments between homologous pairs of chromosomes during 1st meiotic prophase and, much more rarely, during mitosis, by breakage and reunion. Biologically, this is the sexual process. It is the result of the molecular events of a CROSSOVER, visible as a chiasma. It plays an important role in the generation of genetic diversity among the meiotic products, recombinant chromosomes containing DNA sequences that originally occurred on separate maternally and paternally derived chromosomes. *Unequal crossing-over* is a form of non-allelic homologous RECOMBINATION in which the crossover takes place between non-allelic sequences on non-sister chromatids of a pair of homologues. The sequences often show considerable sequence homology, presumably stabilizing the chromosome mispairing. *Unequal sister chromatid exchange* is analogous, involving unequal exchange between sister chromatids. See MEIOSIS, MUTATION.

crossover The breakage of two DNA molecules at the same relative position, and their rejoining in two reciprocal non-parental combinations (see RECOMBINATION). Multiple crossovers can involve more than two chromatids. It is a form of general recombination, resembling the process involved in DNA double-stranded break repair (see DNA REPAIR MECHANISMS). Recombination resulting in non-reciprocal transfer of DNA from one chromosome to another also occurs by general recombination, involving a limited amount of associated DNA synthesis (see GENE CONVERSION). The meiosis-specific heterodimer MSH4/MSH5 stimulates meiotic crossovers (see MSH2), during which two homologous non-sister chromatids are broken at the same relative position and rejoined in a new arrangement without loss or gain of genetic material by either. See CROSSING-OVER, INVERSION, TRANSLOCATION.

cross-sectional studies (cross-sectional analysis) A class of demographic, behavioural, etc., research methods involving simultaneous observation of some subset (e.g., cohort) of a population. Cross-sectional studies are used in most branches of science, in the social sciences and other fields as well. Compare LONGITUDINAL STUDIES.

crown of tooth The visible portion of a tooth, above the gums.

CRP See C-REACTIVE PROTEIN.

cruciate ligaments Ligaments within the articular capsule of the KNEE JOINT.

crying Inspiration followed by many brief expirations in which the rima glottidis remains open and vocal cords vibrate. Accompanied by facial expressions and tears.

cryobiology Transplant surgeons currently store organs perfused in special solutions at just above 0°C for a few days, although the temperature of liquid nitrogen (–196°C) would be preferred since molecular motion and decay are almost arrested, and since 1972 embryos have been so treated and successfully implanted. But organs fare far worse at such temperatures, water leaking from cells forming ice crystals in the intercellular spaces and destroying ducts and vessels. *Vitrification* is ice-free cryopreservation, organs being perfused with a viscous fluid which turns into a glassy (rather than crystalline) solid at low temperatures. Vitrification and conventional freezing methods confront the problem of avoiding ice crystal formation on thawing of material, and modification of the proteins that normally 'mop up' such crystals at marginally sub-zero temperatures so that they can operate well below zero is one line of research.

cryonics The (currently non-feasible) procedure of freezing the damaged body in the hope that when technology has advanced sufficiently it may be unfrozen and healed. See CRYOBIOLOGY.

cryptorchidism Disorder resulting from failure of the TESTES to descend from the abdominal cavity into the scrotum. If it persists into puberty it leads to arrest of spermatogenesis and infertility on account of the higher temperature.

Cryptosporidium parvum Opportunistic apicomplexan parasite associated with immunosuppressive conditions, including AIDS. Serotype A has its reservoir in humans; serotype B is excreted by cattle, but is found in other animals and is the commonest serotype causing human disease. Symptoms of cryptosporidiosis, after 3–6 days of incubation, are acute diarrhoea, often with abdominal pain and colic, lasting ~3 weeks. The organism attaches to the microvilli of the enterocytes and is transmitted by faecal–oral spread in nurseries. Surface waters containing infectious oocysts can contaminate water reservoirs and mains, leading to large outbreaks.

crypts of Lieberkühn Small glandular pits, 0.3–0.5 mm deep, between the villi of the SMALL INTESTINE (see Fig. 159) covered by an epithelium comprising a fair number of goblet cells among a large number of ENTEROCYTES. Crypts secrete a slightly alkaline fluid (pH 7.5–8.0, ~1.5 litres per day) which is rapidly absorbed by the villi, acting as a medium for the absorption of digestion products from the CHYME. Slow-cycling stem cells in the crypts (~1–6 per crypt) and PANETH CELLS (see SMALL INTESTINE, Fig. 159) escape the flow of epithelial cells up the crypts and villi. The crypts of the COLON (see Fig. 60) lack the nominal attribution to Lieberkühn.

crystallins A surprisingly diverse group of proteins, members generally being derived from, or identical to, metabolic enzymes or small heat shock proteins. α-crystallins are found in all vertebrate LENSES and to a lesser extent in other tissues, where they have non-refractive roles. β-crystallins and γ-crystallins form a structurally related crystallin superfamily. δ-crystallins are apparently restricted to birds and reptiles. αA- and αB-crystallins are distinct anti-apoptotic regulators. The small HEAT SHOCK PROTEIN αB-crystallin encoded by *CRYAB* also has neuroprotective functions (see MULTIPLE SCLEROSIS). It interacts with procaspase-3 to repress caspase-3 activation (see CASPASES). Human αA- and αB-crystallins prevent apoptosis by interacting with members of the Bcl-2 family (see APOPTOSIS). In certain conditions, α-crystallins prevent movement of Bax and $Bcl\text{-}X_S$ from the cytosol into mitochondria, preserving mitochondrial integrity, restricting cytochrome *c* release and blocking PARP degradation (see PARP1). Most, if not all of the vertebrate crystallin genes studied to date are regulated by a shared group of transcription factors, including CREB, Pax6, RETINOIC ACID receptors and Sox.

CSF See CEREBROSPINAL FLUID, COLONY STIMULATING FACTOR.

c-*src* See SRC GENE.

CT scan (computed tomography scan) Computed tomography (formerly computerized axial tomography, CAT), is a procedure in which a beam of X-rays traces an arc at multiple angles around a section of the body. This section, the CT scan, is reproduced on a monitor. It visualizes soft tissues and organs in greater detail than conventional radiography, different tissue densities being indicated by shades of grey. Multiple scans can be constructed to give a three-dimensional view of a structure. Compare MAGNETIC RESONANCE IMAGING, ULTRASOUND.

cumulus cells (cumulus oophorus) Follicle-derived cells surrounding the OOCYTE after its release from the follicle and posing a barrier to non-capacitated spermatozoa. Held together by hyaluronic acid (digested by sperm hyaluronidase). See ACROSOME, ZONA PELLUCIDA.

curare Arrow-tip poison and antagonist of ACETYLCHOLINE at nicotinic receptors, so causing paralysis. See α-BUNGAROTOXIN.

Cushing's syndrome Oversecretion of CORTISOL by the adrenal cortex, characterized by breakdown of skeletal muscle proteins and redistribution of body fat leading to thin legs and a 'moon' face, 'buffalo hump' on the back, and pendulous abdomen. Individuals bruise easily, and wound healing is poor. Causes include adrenal tumour and tumour of the anterior pituitary. High cortisol levels result in hyperglycaemia, type 2 diabetes mellitus, obesity, osteoporosis, weakness, hypertension and mood swings.

cusps The pointed or rounded projection(s) on the biting surface of a cheek tooth (premolar or molar). The premolars are bicuspid in humans. See DENTITION, SECTORIAL.

cuticle The outermost layer of cells of a HAIR, arranged like shingles on the side of a house.

C value paradox The originally puzzling observation that a single cell, such as an *Amoeba dubia*, can have a genome 200 times larger than that of humans (while that of a human is 200 times the size of a yeast cell's). The solution of the riddle turned out to be the large amount of REPETITIVE DNA present in some genomes.

CVDs See CARDIOVASCULAR DISEASES.

CVS See CHORIONIC VILLUS SAMPLING.

CXCRs CXC CHEMOKINE receptors, most others belonging to the CCR family. Ligands for both are known to be encoded by gene clusters in the human and chimpanzee genomes that are rapidly diverging. CXCR4 is a T-cell receptor for α-cytokines, in particular the chemokine CXCL12, and is a co-receptor bound by the lymphocyte-tropic variants of HIV. See GENE CLUSTERS.

cyclic AMP (adenosine 3′,5′-monophosphate, cAMP) An extremely important, and short-lived, intracellular SECOND MESSENGER synthesized by the enzyme adenylyl cyclase from ADENOSINE TRIPHOSPHATE (see entry for formula) by a cyclization reaction that removes pyrophosphate ($2P_i$), the reaction being favoured by hydrolysis of $2P_i$ to phosphate by a pyrophosphatase. cAMP is unstable in cells because it is degraded by phosphodiesterase type IV (PDE4) to 5′-AMP, modulating those intracellular pathways in which it is involved. Although it can activate some ion channels directly, cAMP generally functions by activating cAMP-dependent protein kinase (see PKA) which transfers a phosphate group to serines or threonines of target proteins, and by activating transcription of those target DNA sequences containing its response element, CRE (see also CREB). Signals that use cAMP as a second messenger include CORTICOTROPIN, CRH, DOPAMINE (D-1, D-2), the β-ADRENERGIC receptor system, FOLLICLE STIMULATING HORMONE, GLUCAGON, HISTAMINE, LUTEINIZING HORMONE, MELANOCYTE-STIMULATING HORMONE, PARATHYROID HORMONE, SEROTONIN (5-HT-1α. 5-HT-2) and THYROTROPIN. It is a strong inhibitor of fructose 1,6 bisphosphatase, a key regulatory enzyme in GLUCONEOGENESIS, and a strong positive modulator of phosphofructokinase-1, the key regulatory enzyme of glycolysis.

cyclic GMP (guanosine 3′,5′-monophosphate, cGMP) A second messenger produced from GTP by RECEPTOR GUANYLYL

CYCLASES, and of importance in olfaction (see OLFACTORY RECEPTOR NEURONS) and in phototransduction (see RHODOPSIN). Most of its effects are thought to be mediated by PKG, which phosphorylates Ser and Thr residues in target proteins when activated by cGMP. See GUANINE.

cyclin-dependent kinases (CDKs) Serine/threonine kinases that associate with regulatory protein subunits, cyclins, and in this combined form relay signals from the cell cycle clock (see CELL CYCLE) to numerous responding protein substrate molecules involved in the cell's growth-and-division cycle. Full activation of a cyclin-Cdk complex takes place when another kinase, Cdk-activating kinase, phosphorylates an amino acid close to the entrance to the Cdk active site. Cdk inhibitors (CdkIs, CKIs) block the actions of CDKs at various points in the cell cycle, particularly G1 and S-phase. See CYCLINS.

cyclins Proteins acting during the CELL CYCLE as regulatory subunits of CYCLIN-DEPENDENT KINASES (Cdks), to which they bind through a ~100-amino-acid-long domain, removing a blocking protein from the Cdk active site. There are four classes of cyclins, defined by the cell cycle stage in which they bind Cdks and function. Levels of the three D-type cyclins fluctuate in accordance with the levels of extracellular MITOGENS and 'inform' the cell of environmental conditions. But once synthesized in the cytoplasm they enter the nucleus and assemble in complexes with either Cdk4 or Cdk6, but after the G1/S transition D1 at least is exported to the cytoplasm. Levels of the G1/S cyclin, cyclin E (which binds Cdk2), increase after the R point at the end of G1 and peak at the start of S-phase, committing the cell to DNA REPLICATION. During S-phase it declines, reaching low levels at the start of G2. Levels of the S-cyclin, cyclin A (which binds Cdk2), increase during S-PHASE but fall from mid-S-phase to mid-G2. Levels of the M-cyclin, cyclin B (which binds Cdk1, aka Cdc2), increase gradually throughout G2 and M-phase (mitosis) as a result of increasing *M-cyclin* gene transcription, but fall dramatically as a result of ubiquitylation by the anaphase-promoting complex at the end of cell division. Cyclin E level increases after the R point at the end of G1 and peaks at the start of S-phase, during which it declines to reach low levels at the start of G2.

Apart from the D-cyclins, cyclin activation is independent of extracellular signals, in part because the cyclin-CDK complexes in one phase of the cell cycle are responsible for activation of those in the next phase and for shutting down those active in the previous phase. Cyclins D1 and E are ONCOGENES.

cyclooxygenases (COXs) Key enzymes in the conversion of arachidonic acid via PGH_2 to the two major groups of EICOSANOIDS (thromboxanes and prostaglandins). COX-1 is constitutively produced and protects the stomach lining from irritation. COX-2 only becomes active under specific conditions (notably, it is induced by NF-κB) to mediate wound-healing and inflammatory responses and is known to promote METASTASIS. Their activity is blocked by the NON-STEROIDAL ANTI-INFLAMMATORY DRUGS aspirin and ibuprofen, which often cause stomach irritation.

cyclopamine A teratogen that inhibits expression of *Sonic hedgehog* signalling pathway, blocking a receptor for the SHH protein, and resulting in cyclopia. See FOETAL ALCOHOL SYNDROME.

cyclosporin (ciclosporin) An immunosuppressive and non-cytotoxic drug, widely used in treatment of transplant patients. It is a cyclic decapeptide, is derived from a soil fungus (*Tolypocladium inflatum*), blocking T-cell proliferation by inhibiting the phosphatase activity of calcineurin, which transduces signals from the T-cell receptor to the nucleus. See IMMUNOSUPPRESSION.

cystic fibrosis (CF) An autosomal recessive disorder (chromosome site 7q31), affecting all exocrine epithelia, symptomatically pleiotropic, and the most common fatal GENETIC DISORDER in the western world. It occurs in ~1:2,000 births in Caucasians but is ten times rarer in the black population and almost non-existent in Asians (see NATURAL SELECTION, SALMONELLA). The gene responsible is the cystic fibrosis transmembrane conductance regulator (*CFTR*), whose protein product (an ABC TRANSPORTER protein of 1,480 amino acids) functions as a chloride-ION CHANNEL required for normal functioning in epithelia such as those lining the respiratory pathways, intestinal tract and exocrine ducts of the pancreas, and sweat glands as well as ducts of the testes. ~70% of the mutations in this gene result in omission of the amino acid phenylalanine at residue 508 in the protein leading it to be tagged for degradation. One diagnostic characteristic is a raised sodium and chloride ion level in the sweat, but failure to reach expected height and weight milestones in development are due largely to functional defects in enzyme output of the pancreas, which becomes fibrotic and fatty (see also PANCREATITIS). Supplementary digestive enzymes can be administered with food, but other exocrine glands become malfunctional later. Although a major contributory cause of death is the loss of airway function due to build-up of thick, viscous secretions by the lung, the bacterium *Pseudomonas aeruginosa* chronically infects lungs of >80% of CF patients and is the primary cause of morbidity and mortality in these patients.

Currently, treatments concentrate on CF's symptoms rather than its underlying cause. Advocates of GENE THERAPY aim to correct this by delivering functional copies of *CFTR* to a patient's cells. Studies indicate that 5–10% of normal activity levels of the gene would be sufficient to produce a good clinical response and initial trials employed adenovirus vectors to deliver the functional *CFTR* gene to airway epithelial cells. At high doses, this sometimes provoked inflammatory reactions and in one case resulted in death of the patient. Adenoviruses do tend to infect basal cells of the epithelium, where the highest levels of *CFTR* expression occur, but because these cells have only a 4-month lifespan (so requiring repeated administration of the treatment, with the problems of associated immune response) appropriate stem cells would be the ideal targets. Attempts are now being made to discover small molecules that might chaperone the mutant CFTR into the surface membrane and avoid its degradation. Recent approaches have used liposomes or adeno-associated virus vectors (non-pathogenic), but a significant problem is the physical barrier of mucus and glycocalyx covering the target epithelial cells. It has been suggested that heterozygotes may have reduced susceptibility to typhoid, and to cholera toxin; the variable association of meconium ileus with CF is controlled by a genetic variant unlinked to the *CFTR* gene (see MODIFIERS). One subset of mutations (nonsense mutations) in *CFTR*, resulting in a stop (termination) codon in the middle of the CFTR mRNA, initiates its NONSENSE-MEDIATED DECAY (NMD) and results in lack of CFTR; but there is hope that the drug PTC124 may enable ribosomes to 'read through' the premature stop codon and so produce some effective CFTR. Some 60% of Ashkenazi Jews with CF carry such a nonsense mutation (in most populations its frequency is 2–5%). CF patients are trying the drug in Phase 2 trials, both in the USA and Israel. See SCREENING.

cytochalasin A drug disrupting ACTIN filaments.

cytochrome *bc*$_1$ complex (complex III) A huge dimeric ubiquinone:cytochrome *c* oxidoreductase in the inner mitochondrial membrane, coupling transfer of electrons from UBIQUINONE to CYTOCHROME C to vertical transport of protons from the matrix into the intermembrane space. One subunit is encoded by the genome of the MITOCHONDRION (see MTDNA) and ten by the nuclear genome.

cytochrome *c* A soluble protein in the intermembrane space, whose single haem accepts an electron from complex III (the CYTOCHROME BC$_1$ COMPLEX) and then moves to complex IV (see CYTOCHROME OXIDASE) and donates it to a copper centre in that enzyme. Normally the cytochrome *c* channels in the outer mitochondrial membrane are kept closed by the anti-apoptotic Bcl-2 and Bcl-X_L (Bcl-2 also binds and inactivates Apaf-1) but the pro-apoptotic Bax, Bad, Bak and Bid act oppositely, prising it open. During APOPTOSIS, mitochondria swell, allowing cytochrome *c* to leak into the cytosol, where it binds Apaf-1 (apoptotic-protease-activating factor-1), a pro-apoptotic complex which binds an initiator CASPASE and activates it. The subsequent cleavage of I-CAD leads to DNA fragmentation. See MITOCHONDRION.

cytochrome oxidase (**cytochrome *c* oxidase**) Complex IV of the mitochondrial electron transport chain, carrying electrons from CYTOCHROME C to molecular oxygen. It is a dimer of two subunits, each containing 13 different polypeptide chains, 3 encoded by MTDNA and 10 by the nuclear genome. Oxygen has a high affinity for electrons, releasing a large amount of free energy when reduced to form water. Cytochrome oxidase holds onto oxygen at a special bimetallic (iron-copper) centre, where it remains between a haem-linked iron atom and a copper atom, until it has picked up a total of four electrons, the superoxide O_2^- being dangerously reactive (see MITOCHONDRION). Only then can the oxygen atoms in one oxygen molecule be safely released as two water molecules. A high concentration of the enzyme in cells is an indicator of intensive cellular activity, and as a result it is often stained in tissue sections (e.g., of brain). It is inhibited by CARBON MONOXIDE. See MITOCHONDRION.

cytochromes Proteins with characteristic strong absorption of visible light, owing to their iron-containing haem prosthetic groups. The MITOCHONDRION contains three classes, *a*, *b* and *c*, distinguished by absorption spectra and all components of the electron transport and oxidative phosphorylation. Unlike the haems of cytochromes *a* and *b*, that of cytochrome *c* is covalently attached. Cytochrome c_1 is a part of the CYTOCHROME BC_1 COMPLEX, and is embedded in the inner mitochondrial membrane, whereas CYTOCHROME C is soluble.

Cytochrome-P450 (or CYP) is a large family of monooxygenase enzymes, normally present in the smooth ENDOPLASMIC RETICULUM and microsomes rather than mitochondria, whose active sites contain iron atoms, each having different substrate specificities, but typically catalysing the conversion of C–H bonds to C–O bonds. Defined by the characteristic absorption (450 nanometres) of their ferrous carbon monoxide complex, they are involved in the production of cholesterol, steroids and other lipids, one member (P450c17) converting pregnenolone to dehydroepiandrosterone in the adrenal cortex. CYP is also important in the hydroxylation of many drugs, e.g., barbiturates and other xenobiotics, especially if hydrophobic and relatively insoluble. The organic substrate, RH, is hydroxylated to R–OH, incorporating one oxygen atom of O_2, the other oxygen atom being reduced to H_2O by NADH or NADPH. The carcinogen benzo(*a*)pyrene (BP) in cigarette smoke undergoes cytochrome-P450-mediated hydroxylation during detoxification. See CYTOCHROME OXIDASE, DRUG TARGETING.

cytoglobin The recently discovered cytoglobin gene *CYGB* is located on chromosome 17 (17q25.3) and expressed ubiquitously, but is of unknown function (see GLOBINS).

cytokine receptors A subfamily of enzyme-linked receptors, being the largest and most diverse set of receptors relying on cytoplasmic kinases (see SRC-KINASES) to relay signals into the cell. They are composed of two or more polypeptide chains, some very specific to a receptor and others found as components of several. They include receptors for cytokines, as well as receptors for certain hormones (e.g., GROWTH HORMONE and PROLACTIN). These receptors are stably associated with a class of cytoplasmic tyrosine kinases, Jaks, which activate latent gene regulatory proteins known as STATs (see JAK/STAT PATHWAY).

cytokines A generic term referring to any small, soluble, protein (~25 kDa) secreted by one cell and affecting the behaviour, or properties, of the secreting cell and/or of another cell. Those produced by leukocytes are termed *interleukins* (ILs) of which those produced by lymphocytes are termed *lymphokines* (see T CELLS). Cytokines fall into two major classes: the HAEMATOPOIETIN family (including growth hormones and many interleukins) and the TUMOUR NECROSIS FACTOR family, many of whose members are membrane-bound. They also include the INTERFERONS, those with chemoattractant effects on cells being known as CHEMOKINES (see entry for examples).

Cytokines may act in an autocrine-, paracrine- or endocrine-like manner and although stable cytokines can have hormone-like effects, the peptide hormones released by the endocrine system are usually excluded from cytokine membership. Those inducing proliferation in target cells are GROWTH FACTORS. Secreted by many cells, most famously by those of the immune system – including phagocytes of the innate immune system – they only modify the behaviour of those cells with appropriate cell-surface cytokine receptors. As expected, cytokine receptors are at least as diverse as are cytokines themselves and include many CD ANTIGENS and interleukin receptors (ILRs). Cytokines exert their signalling effects via the JAK/STAT PATHWAY, and sometimes induce proteins that inhibit cytokine signalling, via negative feedback loops, thereby ensuring temporal control of the signal. Pro-inflammatory cytokines produced by activated MACROPHAGES include IL-1β, IL-6 and TNF-α. Macrophage activation is linked to mechanisms controlling production of the cytokine IFN-γ, whose mRNA (like that of a variety of other cytokines) contains the $(AUUUA)_n$ sequence in its 3′ UTR, which greatly reduces the molecules' half-lives in the cell. T-cell activation apparently induces production of a protein promoting cytokine mRNA degradation.

IL-2 is produced by activated T_H1 cells, and some CD8 cells, inducing T-cell proliferation, B-cell growth and J-chain synthesis, and NK growth. IL-3 is produced by T_H1, T_H2 and some cytotoxic T cells and is a growth factor for progenitor haematopoietic cells (a multi-CSF; see COLONY STIMULATING FACTORS), operating via the JAK/STAT PATHWAY. IL-4 is produced by T_H2 cells, activating B cells and stimulating their growth and MHC class II expression and, with IL-13, provides the first signal switching B cells to IgE production. It also promotes growth and survival of T cells and growth of mast cells. IL-5 is also a T_H2 cell product, promoting eosinophil growth and differentiation. IL-6 belongs to the large family of haematopoietins and is an ENDOGENOUS PYROGEN released by macrophages as an acute-phase protein. It has been linked to tumour progression in some models of liver and colorectal cancer. IL-7 is essential for T-cell growth and development, and in maintaining the memory T-cell pool. IL-8, now known as the chemokine CXCL8, promotes neutrophil migration. IL-9 is a T_H2 cell product favouring development of T_H2 cells themselves, and has a causative role in allergic ASTHMA. IL-10 is produced by T_H2 and some T_H1 cells, upregulating MHC class II proteins in B cells and co-stimulating mast cell growth. IL-17 is produced by a subset of $CD4^+$ T cells. IL-23 is a heterodimeric cytokine receiving increasing attention because of its role in a wide range of chronic inflammatory diseases in MOUSE MODELS, including irritable bowel syndromes (see ALLERGIES), multiple sclerosis and arthritis (see AUTOIMMUNE DISEASES). IL-23 is closely related to IL-12, but differs from it functionally. Whereas IL-12 drives the classical T_H1 response characterized by IFN-γ production, IL-23 drives an autoreactive T-cell population characterized by the production of the IL-17-related cytokines IL-17A and IL-17F, promoting chronic inflammation dominated by IL-17, IL-6, IL-8 (CXCL8) and TUMOUR NECROSIS FACTORS as well as neutrophils and monocytes. See MIF.

cytokinesis The division of a cell's cytoplasm, which normally begins during anaphase of MITOSIS. A major feature is the formation of the CONTRACTILE RING.

cytomegalovirus (CMV) A herpes virus (see HERPES SIMPLEX VIRUS), and the commonest cause of congenital infection in the UK, 40% of adult populations in high income countries having evidence of immunity to the virus. Almost always asymptomatic in the immunocompetent, infection can occasionally cause mild fever and jaundice. Roughly 3% of the ~600,000 children born annually in England and Wales are congenitally infected, ~10% being symptomatic but ~10% having some permanent handicap (e.g., sensorineural deafness, neurological impairment, microcephaly). Microcephaly improves or disappears with growth in up to 50% of cases. Symptoms presenting at birth include prematurity, low birth weight, enlarged liver and spleen and prolonged jaundice. It can persist in a latent form. It is one of the most difficult infections to treat after a TRANSPLANT and more than half of patients who develop CMV pneumonitis post-transplant currently die from it. Ganciclovir is a favoured treatment and is often used in combination with other bone-marrow suppressive drugs (see IMMUNOSUPPRESSION), which are often toxic and oncogenic.

cytoplasm The entire cell, apart from its nucleus.

cytoplasmic inheritance Every human cell inherits cytoplasm as well as a nucleus, and to this extent all molecules inherited from the parental cell's cytoplasm are included. The term is usually intended to include the mitochondrial genome (see MTDNA).

cytosine A pyrimidine base, synthesized from aspartate and occurring in *cytidine* nucleotides, and in nucleic acids. Cytidine 5′-triphosphate is a metabolite of uridine 5′-triphosphate. Deamination of cytosine in DNA to uracil is probably the reason why DNA contains thymine and not uracil (see CPG DINUCLEOTIDE). Activation-induced cytosine deaminase (AID) and its conversion of cytosine to uracil are required for hypermutation in the production of ANTIBODY DIVERSITY. 5′-methylcytosine is deaminated to thymine (see DNA REPAIR MECHANISMS, DNA METHYLATION). Degradation of pyrimidines produces UREA.

cytosine deaminases See CYTOSINE.

cytosine methyltransferases Enzymes recognizing a CpG target sequence, showing a strong preference for a hemimethylated DNA target (one methylated on one strand only). The cytosine methyltransferase, Dnmt1, is the most important enzyme in maintaining the imprinted status of a gene. See CPG DINUCLEOTIDE, DNA METHYLATION, EPIGENESIS, GENOMIC IMPRINTING.

cytoskeleton A network of three major types of cytoskeletal filament (INTERMEDIATE FILAMENTS, MICROTUBULES and ACTIN filaments), whose subunits are held together by weak non-covalent forces of attraction so that their assembly and disassembly can occur rapidly and with little energy expenditure in response to the fluctuating needs of the cell. Much of it lies just beneath the CELL MEMBRANE and is a major factor in determining *cell shape*, in providing the cell's attachment capabilities, in enabling not only cell movements (e.g., ciliary movements, amoeboid locomotion and muscle contraction), but also phagocytosis and movements of material both within the cell and during endocytosis and exocytosis. Many additional proteins contribute to cytoskeletal properties (e.g., see RED BLOOD CELL, Fig. 146; MOTOR PROTEINS).

cytosol The gel-like component of the cytoplasm in which the organelles are suspended: the cytoplasmic matrix. It contains many enzymes, including those of glycolysis and, in liver and renal cortical cells, of gluconeogenesis.

cytotoxic (type II) reactions See HYPERSENSITIVITY.

cytotoxic T cells (T_C cells) See T CELLS.

cytotrophoblast Layer of properly individuated trophoblast cells (*contra* SYNCYTIOTROPHOBLAST), initially actively mitotic, lying between the inner cell mass and the SYNCYTIOTROPHOBLAST, with which it contributes to the CHORION. Within 8 days of fertilization, the inner cytotrophoblast cells proliferate to form the AMNION. See BLASTOCYST (Fig. 20), IMPLANTATION and SYNCYTINS.

D

δ- For these entries, see under headword prefixed by δ-.

D_1, D_2 receptors Subtypes of DOPAMINE receptor.

D-antigens See RHESUS BLOOD GROUP SYSTEM.

Daily Values (DVs) Dietary reference values which are appearing on food labels to help consumers plan a HEALTHY DIET. Based on two other sets of values (REFERENCE DAILY INTAKES, RDIs, and DIETARY REFERENCE VALUES, DRVs) it was thought that having two standards on food labels, one for vitamins and minerals and one for other nutrients, was confusing. Daily Values are based on a 2000 kcal reference diet approximating to the weight maintenance diets of postmenopausal women, women exercising regularly, teenage girls and sedentary men. Additional information is listed on large food labels for a daily intake of 2,500 kcal, adequate for young men. See ENERGY BALANCE.

dark adaptation (scotopia, scotopic vision) The increased visual sensitivity involved in adjusting to darkness after bright light, increasing by a value of ~1 million-fold; or, the adaptation of the eye for night vision (scotopia). In scotopic conditions, maximum sensitivity to light is at 507 nm, characteristic of rhodopsin, and therefore determined by ROD CELLS. Dark adaptation has two phases. Initially, we see 'nothing', but then visual acuity gradually improves, regaining maximum sensitivity after 30 minutes of darkness (e.g., starlight). In very dark conditions (e.g., a moonless night) CONE CELLS fail, despite signal boosting via gap junctions from adjacent rod cells. The increased influx of calcium ions into rod cells via the DARK CURRENT leads to closure of these gap junctions, so that relaying of the signal from rods is now via depolarizing (rod) bipolar cells which synapse with an amacrine cell population, leading to increased contrast sensitivity in the dark.

Cones adapt more rapidly than rods (usually completely within 8 minutes), accounting for the initial phase; but their final sensitivity is lower, i.e., their thresholds are higher. The second phase is due to rod cells. Regeneration of photopigments, depleted by bright light, accounts for much of the effect, since RHODOPSIN regenerates more slowly than iodopsins. Scotopic vision is achromatic, since all rod cells have the same spectral sensitivity. In the dark-adapted eye, GANGLION CELLS become much more sensitive to light through the shutting-off of dopamine SURROUND INHIBITION. Under photopic conditions (see LIGHT ADAPTATION), wavelength sensitivity is determined by cone cells and is maximal at 555 nm. The shift in spectral sensitivity between scotopic and photopic vision is the PURKINJE SHIFT, explaining why, as dusk falls, the last colour sensations to be lost are blues and greens. See MELANOPSIN.

dark current The current due to the inward flow of sodium ions into photoreceptors (rod cells and cone cells) in the dark, leading to hyperpolarization of the cell. The sodium ions enter through a cGMP-gated channel. See ROD CELLS.

'dark matter' Regions of the genome that

are transcriptionally active but which lie outside known or predicted exons and non-coding genes (see NON-CODING RNA). These regions lie within introns and in intergenic regions of the genome. Some of this may be due to novel polypeptide-coding genes, some to novel non-coding genes, and some to antisense transcription. See HUMAN TRANSCRIPTOME.

DARPP-32 (dopamine and cAMP-regulated phosphoprotein (molecular weight = 32 kD)) A protein which is highly concentrated in the neostriatum and nucleus accumbens and involved in signalling pathways mediating the opposing effects of the neurotransmitters DOPAMINE and GLUTAMATE (see Fig. 84). Phosphorylation of threonine-34 (T34) is required to mediate the actions of dopamine and inhibitor PROTEIN PHOSPHATASE 1 (PP1). It is capable of being phosphorylated on at least three additional sites, in response to activation of other signalling pathways. Glutamate results in phosphorylation of T75 and activation of PP1. In all cases, phosphorylation modulates the dopamine/D1 receptor/cAMP/PKA/phosphothreonine-34–DARPP-32/PP1 cascade. Drugs of abuse, and food reinforcement learning, activate this cascade and promote nuclear accumulation of DARPP-32 and inhibit its nuclear export. The resulting PP1 inhibition increases phosphorylation of histone H3, an important component of the nucleosome response. DARPP-32 is located in the only cells in the neostriatum which convey information away from this brain region (see BASAL GANGLIA). DARPP-32 gene knockout (deletion) studies in mice indicate that all behavioural, physiological, biochemical, pharmacological and toxicological responses to dopamine and anti-schizophrenic drugs seen in normal mice were either greatly diminished or abolished. Mutation of DARPP-32 serine-97 alters the behavioural effects of drugs of abuse and decreases motivation for food (see APPETITE).

Darwinism The theory formulated by Charles Robert Darwin (1809–1882) and Alfred Russel Wallace (1823–1913) which holds that species are not unchanging, or 'fixed', either in form or number; that new species continue to arise while others become extinct; that observed harmonies between an organism's structure and way of life are neither coincidental nor necessarily proof of the existence of a benevolent Creator, and that any apparent design is inevitable given: (i) the tendency of all organisms to produce more offspring than the environment can support (populations generally remaining stable); (ii) the variability between individuals in any measurable aspect of phenotype chosen; (iii) the tendency for some, at least, of this variance to be due to heritable differences between individuals. The inescapable conclusion drawn was that some, at least, of the phenotypic differences between individuals are not neutral with respect to the probability of an organism's surviving to the age of reproduction: that some of the variance has a positive effect such as survival while some has a negative effect. Since, on the whole, an organism's increased longevity favours its reproductive output, any component of phenotypic variance that favours an organism's reproductive output AND is heritable will tend to increase in frequency in subsequent generations in what could be termed a 'struggle for existence' between individuals, given the finite environmental resources available as population size increases.

The upshot was that any appearance of design in the living world could more simply be understood in naturalistic terms, as the result of the increasing adaptation over time of populations (and species) to their environmental circumstances, an adaptation that was the result of an increasing match between those circumstances and the population's genetic make-up. The mechanism that resulted in this match was termed NATURAL SELECTION.

DAX1 A locus on the short arm of the X chromosome, encoding a member of the nuclear hormone receptor family that downregulates *SF1* activity, so preventing development of Sertoli and Leydig cells.

Mutations can cause CONGENITAL ADRENAL HYPERPLASIA, and rare duplications cause dosage-sensitive SEX REVERSAL. See SRY.

***DCC* gene (deleted in colon carcinoma gene)** A tumour suppressor gene located on chromosome 18 and occurring in a mutated form in approximately 70% of colorectal cancers. See CANCER CELLS.

DDT A broad spectrum, persistent, organochlorine contact insecticide applicable both to soil and to leaves. Insoluble in water, but highly soluble in lipids, it tends to accumulate in fatty tissues. In the 1940s, DDT was discovered in the body fat of nearly all inhabitants of developed countries, reaching a plateau of ~10 ppm. There is no clear evidence that such levels are harmful to humans; but the related organochlorine DDD has been responsible for bird deaths. Pesticide resistance usually renders such pest control measures ineffectual, if used in isolation.

deacetylases See HISTONE ACETYLTRANSFERASES, SIRTUINS.

dead space (anatomic dead space) The non-exchanging components of the respiratory tract (the conducting zone) in which a portion of the inspiratory volume fails to reach the alveoli, remaining in the conducting airways of the nose, pharynx, larynx, trachea, bronchi, bronchioles and terminal bronchioles. It amounts to ~30% (~150 cm^3) of the tidal volume in an average adult. See RESPIRATORY ZONES.

deafness Significant or total loss of hearing. That caused by damage to the EARS may be of two kinds. *Sensorineural deafness* may be caused by impairment of the cochlea's HAIR CELLS or by damage to the cochlear branch of the vestibulocochlear nerve (often the result of reduced blood supply to the ears caused by atherosclerosis), through repeated exposure to loud noise, or through exposure to such drugs as aspirin or streptomycin. Mutations in the genes for CONNEXINS, components of GAP JUNCTIONS, have also been implicated. *Conduction deafness* arises through malfunction of the external or middle ear mechanisms for transmitting sound waves to the cochlea, sometimes through osteosclerosis (new bone deposition) around the oval window; by impacted ear wax (cerumen), or by damage to the eardrum, either through its thickening or through stiffening of the joints of the ear ossicles as an AGEING process. In marked contrast to the visual system, considerable auditory function is retained after unilateral lesions in the auditory cortex since both ears project to the cortex in both hemispheres. Damage to the PRIMARY AUDITORY CORTEX results in inability to detect the precise location of a sound (see HEARING), but frequency and amplitude discrimination are almost unaffected.

deamination Transamination of most AMINO ACIDS is the first step in their catabolism once they have reached the liver. The α-amino group is removed by transaminases (aminotransferases) to the α-carbon atom of α-ketoglutarate, generating the corresponding α-keto acid analogue of the original amino acid. No *net* deamination occurs, the effect being to collect the amino groups from many different amino acids in the form of GLUTAMATE. This can then serve as an amino group donor in biosynthetic pathways, or in excretion pathways leading to the production of nitrogenous wastes (see UREA). Glutamate is transported from the hepatocyte cytosol into the mitochondrion, undergoing oxidative *deamination* by glutamate dehydrogenase, in which either oxidized NAD or NADP can act as acceptor of the reducing equivalent. This produces the ammonium ion, NH_4^+ (see AMMONIA). The combined action of the aminotransferase and glutamate dehydrogenase is termed transdeamination. The prosthetic group of all aminotransferases is pyridoxal phosphate (see VITAMIN B COMPLEX). Different aminotransferases exist, each using a different amino acid as substrate. As a result of their activities, cells can mobilize different amino acids into the KREBS CYCLE for oxidation (see Appendix IId). Spontaneous deamination of the bases in nucleotides is a source of mutation, but rectifiable by nucleotide excision DNA REPAIR MECHANISMS (see ADENINE, GUANINE, CYTOSINE). See MALATE-ASPARTATE SHUTTLE.

death, death rate See CELL DEATH, MORTALITY.

death domains Protein domains originally discovered through their involvement in eliciting APOPTOSIS, but now known to be commonly involved in other protein–protein interactions. See TOLL-LIKE RECEPTORS.

death receptors Five families of death receptor proteins occur on various cell surfaces, binding of their ligands causing their trimerization, activating their cytoplasmic tails and thereby inactivating FADD (see APOPTOSIS) and promoting assembly of a death-inducing signalling complex (DISC) leading in turn to activation of a caspase cascade. See APOPTOSIS.

decalcification See CALCIUM and BONE REMODELLING, OSTEOCLASTS.

decidua The portion of the endometrium that becomes modified after IMPLANTATION, its different regions being named depending upon their positions relative to the site of the implanted blastocyst. The *decidua reaction* refers to the changes to the endometrium following implantation and, initially, restricted to its immediate vicinity but later affecting the whole inner uterine lining. Endometrial cells become polyhedral and loaded with glycogen and lipids while the extracellular spaces become filled with extravasate so that the tissue is oedematous.

decision-making Despite the importance of decision-making, and increased research interest, the specific underlying neural mechanisms remain unclear (see ACTION SELECTION, INTENTIONAL MOVEMENTS). One view is that at least two cortical pathways are involved in independently generating a decision concerning appropriate behaviour in given circumstances, these being extensions of the dorsal and ventral streams of the visual processing pathways (see EXTRASTRIATE CORTEX). In this scheme, (i) the parieto-premotor (extended dorsal) pathway makes decisions about motor actions in a mainly autonomous manner, whereas (ii) the temporo-ventrolateral prefrontal (extended ventral) pathway is mainly involved in deliberate decisions and inhibitory control through the inhibitory function of the ventrolateral prefrontal cortex.

Expectations guide our immediate behaviour; they can also be compared with actual outcomes to enable learning so that future behaviour can become more adaptive in terms of the pursuit of goals. Information about expected outcomes must be maintained in memory to allow it to be compared and integrated with information about internal state and current goals. The ORBITOFRONTAL CORTEX (see entry), a part of the FRONTAL CORTEX, is distinguished by its unique pattern of connections with subcortical structures involved in associative learning: the basolateral amygdala and nucleus accumbens, a position which enables it to use associative information to project into the future and to use the value of perceived or expected outcomes to guide decisions.

declarative memory (explicit memory) See MEMORY.

decussation The crossing of a fibre bundle from one side of the brain or spinal cord to the other (see AXONAL GUIDANCE); i.e., it crosses the MIDLINE. Some bundles only decussate partially, as do the optic nerve fibres at the optic chiasm: fibres from the nasal halves of the retinas cross while those from the temporal regions remain ipsilateral (same-sided). Visual stimuli from the left halves of the visual fields of both eyes excite the right visual cortex and vice versa. Several examples are to be found in the ascending and descending pathways of the SPINAL CORD (see Fig. 163). See CORPUS CALLOSUM, OPTIC CHIASMSA.

deep brain stimulation (DBS) Insertion of electrodes into the brain to hit a precise neuroanatomical region believed to be involved in psychiatric disorders. Connected to a battery-driven stimulator, the electrodes conduct pulses of current to the target neurons in efforts to normalize their activity. Controlled clinical trials have been promising in treatment of Parkinson's disease and are under way in patients with obsessive-compulsive disorder (see FEAR AND ANXIETY) and major depressive disorder (see AFFECTIVE DISORDERS).

defecation See EGESTION.

defensins Antimicrobial peptides believed to act by disrupting cell membranes of bacteria and fungi, and the envelopes of some viruses. A single intestinal cell can make as many as 21 different defensins with distinct activities, some active against Gram-positive and some against Gram-negative bacteria. The α-defensins (cryptidins) are made by PANETH CELLS while the related β-defensins are made by epithelia of the respiratory and urogenital tracts, the skin and tongue. Defensin genes are highly over-expressed in some cancers (e.g., lung and renal cancer), apparently creating a microenvironment unfavourable for the adaptive immune system.

deficiency diseases Diseases resulting from lack of sufficient dietary intake of micronutrients, typically VITAMINS (see vitamin entries by name), or occasionally where loss of a mineral exceeds intake (e.g., IRON, CALCIUM; see IODINE DEFICIENCY DISORDERS).

degenerative diseases See MUSCULAR DYSTROPHIES, NEURAL DEGENERATION.

dehydration Vomiting and DIARRHOEA cause a large water loss, and persistent diarrhoea can lead to severe dehydration even in mild gastroenteritis. Babies and young children are especially vulnerable because of their small total volume. The effect of CHOLERA toxin may be the loss of 10 L of water per day. Since this fluid loss also involves loss of solutes, intravenous infusion of sodium chloride would be a solution were it available in epidemic situations. ORAL REHYDRATION THERAPY is the best approach in these circumstances.

dehydroepiandrosterone (DHEA) A product of CHOLESTEROL (Fig. 50) metabolism by 17α-hydroxylation on pregnenolone by a microsomal cytochrome P450 enzyme in the adrenal cortex, and the most abundant adrenal ANDROGEN. It stimulates resting metabolic rate (see BASAL METABOLIC RATE) and lipid oxidation, and promotes glucose disposal by increasing the expression of adipocyte glucose transporters. See CYTOCHROMES.

deiodinases Enzymes catalysing progressive deiodination of thyroid hormones, requiring selenocysteine for optimal activity.

delayed early genes Genes expressed by a cell as a second wave, after IMMEDIATE EARLY GENES, in response to growth factors. Expression depends upon *de novo* production of proteins, notably transcription factors, that are products of the immediate early genes.

deletions The loss of one or more nucleotides from a DNA sequence, usually as a result of imperfect DNA REPLICATION. Deletions of one, two or more nucleotides (but not multiples of three) within a protein-coding sequence are frame-shift mutations and result in a shift to the left of the reading frame during mRNA translation on the ribosome, causing incorrect amino acids to be added to the growing chain from that amino acid onwards. It is also likely to result in generation of a stop codon, causing premature termination of polypeptide elongation. The effect of deletions in non-protein-coding sequences depends upon whether the sequence is informative. Deletions in regulatory elements may cause a change in GENE EXPRESSION, for example (see NON-CODING DNA).

Larger deletions affecting chromosome segments (see Fig. 61) result in partial or total gene deletions and may arise through unequal crossover between repeat sequences. They usually have serious consequences, often giving rise to a syndrome of phenotypic effects (e.g., see DIGEORGE/VELOCARDIOFACIAL SYNDROME, PRADER–WILLI SYNDROME, WILLIAMS–BEUREN SYNDROME). See GENOMIC IMPRINTING, MUTATION.

dementia Any disorder resulting in permanent or progressive loss of general intel-

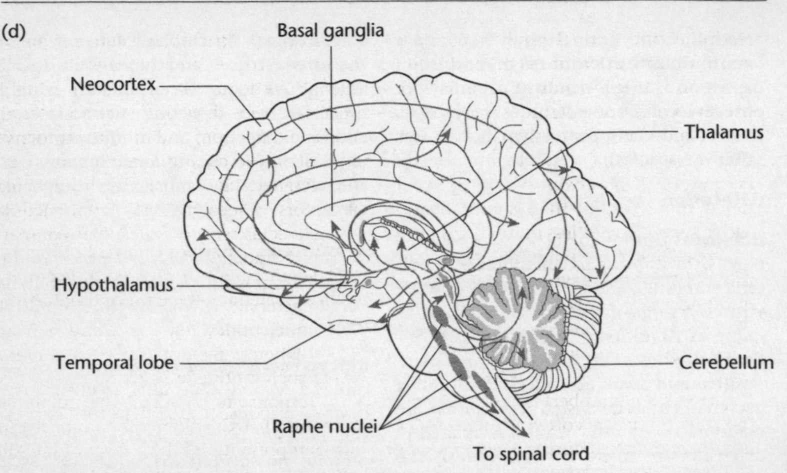

FIG 61. The chromosome 7 DELETION (grey band) typifying the WILLIAMS–BEUREN SYNDROME. In all, 28 loci have been identified within the region. The centromere region is hatched. From *Science* Vol. 310, No. 5749 (4 Nov. 2005), p. 803. Reprinted with permission from AAAS.

lectual faculties such as impairment of memory, judgement, abstract thinking and associated personality changes. ALZHEIMER'S DISEASE accounts for ~65% of cases of dementia. See NEURAL DEGENERATION.

deme A randomly mating subpopulation that exchanges migrants with other POPULATIONS, forming a meta-population. See EFFECTIVE POPULATION SIZE, ISLAND MODEL OF POPULATION STRUCTURE.

demographic stochasticity The variation evident in dynamics of small populations owing to the probabilistic nature of individual processes, such as birth, death, or pathogen transmission. See DISEASE.

demography See HUMAN DEMOGRAPHY.

dendrites Highly branched extensions of the cell bodies of NEURONS, the extent of such dendritic trees varying widely from one neuron to another, classification often referring to particular brain regions. In the cerebral cortex, two broad classes occur: stellate (star-shaped) cells and pyramidal (pyramid-shaped) cells. Cells with spiny dendrites are often called *spiny cells*, those lacking spines being *aspinous cells*. In the cerebral cortex, all pyramidal cells are spiny (see AMPARS), whereas stellate cells may be spiny or aspinous.

Since they contain the postsynaptic elements of excitatory and inhibitory SYNAPSES, dendritic spines are a critical locus of both physiological and anatomical change in the cerebral cortex. The extent to which each such synapse can be independently regulated is necessary if we are to explain the computational capacity of the brain. In cultured brain slices, LONG-TERM POTENTIATION (LTP) correlates with both enlargement of existing spines and addition of new spines while long-term depression (see SYNAPTIC PLASTICITY) correlates with spine shrinkage (dependent upon the degree of actin polymerization). After LTP induction, structural and functional plasticity is limited to the stimulated spine, although the threshold for LTP induction in synapses in spines in its immediate (10 μm) vicinity is reduced, an effect lasting for ~5 minutes. This effect has been found to occur not only in glutamate uncaging experiments by two-photon excitation, but also in response to electrical induction of LTP which more closely resembles the normal physiological situation. Co-activation of neighbouring synapses on a single dendritic segment does trigger local action potentials acting as spatially limited associative signals for LTP so that synapses that tend to fire together because of their proximity will be potentiated more readily than simultaneously active synapses with less mutual proximity. It is possible that, if potentiation increases the probability of a synapse's preservation during development, this may lead to spatial clustering of synapses with similar firing patterns ('cohorts', or 'cliques'; see HIPPOCAMPUS).

In vivo, spine numbers fall on neurons of the somatosensory cortex if their afferent inputs are removed during development or adulthood (see CEREBRAL CORTEX). Normal synaptic development, including maturation of dendritic spines, depends critically upon the environment during infancy and early childhood (see CRITICAL PERIOD, MENTAL RETARDATION). Hippocampal dendritic spine development is regulated by a brain-specific microRNA. Histone deacetylase inhibition has been shown to lead to increased sprouting of dendrites in mouse models (see NEURAL DEGENERATION). Shank proteins (see COGNITION) are thought to function as master organizers of the postsynaptic density (PSD), forming multimeric complexes with postsynaptic receptors, signalling molecules and cytoskeletal proteins present in dendritic spines and PSDs (see AUTISM). Mutations in the *FMR1* gene encoding the fragile-X protein FMRP also lead to abnormal dendritic spines (see FRAGILE-X SYNDROME). See ADDICTION and the APOE entry.

dendritic cells (DCs) One type of phagocytic ANTIGEN-PRESENTING CELL, with long branch-like processes, found in tissues and (when activated) in T-cell regions of lymphoid organs (e.g., see LYMPH NODES, Fig. 121) to which they migrate. They process and present antigens to T cells and B cells during ADAPTIVE IMMUNITY and activate other phagocytic cells (neutrophils and macrophages) during INNATE IMMUNITY (see Fig. 171, STEM CELLS). Many derive from LANGERHANS CELLS. Plasmacytoid DCs (pDCs) comprise a dendritic cell population highly specialized for detecting viral and certain bacterial infections and in contrast to myeloid dendritic cells uniquely express TOLL-LIKE RECEPTOR TLR-7 and TLR-9, intracellular receptors recognizing viral/bacterial nucleic acids within endosomal compartments. Normally, pDCs do not respond to self-DNA, possibly reflecting the fact that viral/bacterial DNA contains multiple CpG nucleotides that bind and activate TLR-9 whereas mammalian DNA contains fewer such motifs (see CPG DINUCLEOTIDE).

In response to microbial and inflammatory stimuli, DCs enhance their capacity for antigen presentation by a form of late differentiation termed *maturation* marked by cellular reorganization which includes an increase in, and redistribution of, class II MHC molecules from late endosomal and lysosomal compartments to the plasma membrane. Their growth and differentiation are stimulated by GM-CSF, immature DCs bearing self-antigens then migrating from the blood to reside in the tissues where they induce TOLERANCE. They patrol such sites as the skin and mucous membranes (e.g., vaginal epithelium) and upon encountering a pathogen, mature rapidly, express CO-STIMULATORY MOLECULES such as CD80 and CD86 and migrate to lymph nodes, where they lose their phagocytic ability but can now stimulate T cells. The binding of a combination of LPS and LPS-binding protein to DC cell-surface receptors CD14 and TLR-4 causes dendritic cells to mature into potent antigen-presenting cells. DCs eventually die in the lymph nodes, but not before they activate antigen-specific naïve T lymphocytes. Whether T CELLS attack or tolerate the antigens' DCs present depends to a degree upon whether or not other molecular 'danger signals' are present at the time, e.g., bacterial sugars or self-proteins released by host cell damage. In the presence of such co-stimulation, the DC rapidly matures and produces T-cell attractants; but in its absence, the DC may simply decommission any T cell with the receptor that recognizes its bound antigen, preventing an aggressive response.

Signals stimulating DCs to release IL-12 include the chemokines CCL3, CCL4 and CCL5 produced by many activated cell types, such as macrophages, other dendritic cells and endothelial cells. The C-type lectin DC-SIGN is specific to DC surfaces and its binding to ICAM-3 on naïve T cells in lymph nodes is unique to the interaction between these cells, findings which may have clinical application in developing procedures that improve tolerance to grafts.

dendritic spines See DENDRITES.

dengue fevers Once a disease restricted to poor people in tropical areas, this flavivirus infection (see YELLOW FEVER) now threatens middle-class urban populations in cities such as Singapore, infecting at least 50 million people per year in more than 100 countries. The viral vector is the culicine MOSQUITO *Aedes aegypti*. Four serotypes are widely distributed in tropical areas, regions with the highest disease burden being Southeast Asia and the Western Pacific, although outbreaks are occurring with increasing frequency in South America and the Caribbean. The feverish symptoms can be severe and debilitating, with ~2.5 billion people currently at risk, and ~500,000 cases of dengue HAEMORRHAGIC FEVER occurring annually. Mosquito management is currently the only prevention option; these breed in water that has collected in tyres and plastic containers. The *Aedes aegypti* genome has recently been published, hopefully leading to better understanding of the mosquito's biology and control.

dentate gyrus See HIPPOCAMPUS and Fig. 91.

dentate nuclei One of the two main output structures of the CEREBELLUM, which also receive sensory afferents. The PONS and dentate nuclei are closely linked to the neocortex in evolutionary growth, showing among the largest increases in primates – certainly in their surface areas.

dentine The main constituent of teeth, lying between the enamel and the pulp cavity. Secreted by odontoblasts (so mesodermal in origin), and of similar composition to bone, but containing up to 70% of inorganic material. See DENTITION.

dentition The teeth are positioned in the sockets of the alveolar processes of the maxillae (superior teeth) and the mandibles (inferior teeth) of the JAW. Adult Old World monkeys, apes and hominins (modern and fossil) have two INCISORS, one CANINE, two PREMOLARS and three MOLARS in each jaw quadrant, the dental formula being I2/2; C1/1; P2/2; M3/3. The teeth of later hominins lie in small rounded dental arcades, whereas those of the great apes are aligned in a U-shaped pattern with long, straight sides. Premolar teeth in *Homo* are not SECTORIAL, unlike those in monkeys and apes. In the latter two groups, but not in humans, the upper canine occludes with the first lower, sectorial, premolar. In all the later fossil hominins, and in modern humans, the first premolar is bicuspid, the two cusps being on the biting surface. Both upper and lower molars of Old World monkeys have four cusps linked in pairs by transverse ridges (*bilophodonty*) and contrast with the lower molars of great apes, which have five well-developed cusps, the rearmost of which has migrated outwards. The valleys between the five cusps (3 on the outside and 2 on the inside) of great apes form a Y-shape: the Y5 pattern.

Teeth of *Homo sapiens*, particularly the last of the cheek teeth, are relatively small compared with those of (other) apes; and tooth eruption is slow when compared with both these and early hominins. Deciduous teeth erupt relatively quickly after birth while the permanent teeth are growing within the jaws (see Fig. 62). Teeth that directly replace deciduous teeth are *successional* teeth and include all the permanent incisors, canines and premolars. Permanent molars grow in the jaws behind the deciduous dentition as available space arises. Apes and humans (hominoids) have two deciduous incisors, one deciduous canine and two deciduous premolars.

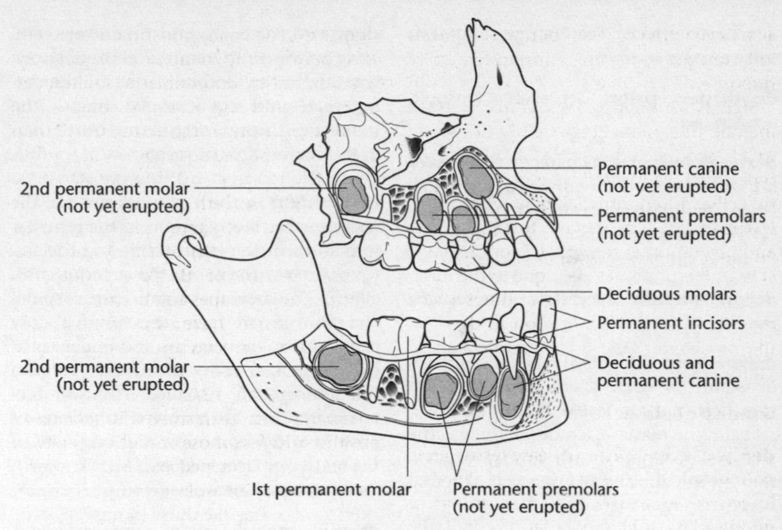

FIG 62. Diagram showing the replacement of deciduous teeth in a child of 8 or 9 years. Redrawn from *Langman's Medical Embryology*, T.W. Sadler. © 2004 Lippincott, Williams & Wilkins, with the permission of Wolters Kluwer.

While it is quite unusual to find 3rd molars missing in great apes, it is quite common to do so in modern humans. Two early *Homo erectus* specimens had congenitally missing third molars. For laser ablation studies of hominin teeth, see AUSTRALOPITHECINES.

deoxyribonucleic acid (DNA) The nucleic acid forming the genetic material of all cells (see GENOME), of mitochondria (see MTDNA) and of some VIRUSES. See DNA entries.

dependence See DRUG DEPENDENCE.

depression Currently a psychiatric, as opposed to a neurological, disorder believed to arise from abnormal activity in specific brain circuits, in the absence of a detectable lesion. There is presently no coherent neurochemical model for depression, although all treatments for the condition increase SEROTONIN neurotransmission. Muscarinic supersensitivity, an enhanced effect of acetylcholine on the muscarinic cholinergic receptors, in persons prone to depression and related conditions, has been suspected of having a causal role in signal-regulating brain imbalances for >30 years. Recently discovered links between the muscarinic receptor CHRM2, alcoholism and depression indicate a direct link between a specific gene and muscarinic hypersensitivity (see ALCOHOL). Neuroimaging studies employing fMRI and PET scanning may detect brain regions of interest; but their current spatial and temporal resolutions may not capture the real-time dynamics of brain function most relevant to mood and cognition. Voltage-sensitive dyes employed during optical imaging can identify changes in brain activity enabling correlations between real-time cellular activity and changing affective state. One recent (2007) mouse study of depression implicates the hippocampus in depression, and chronic or intense stressors (e.g., social defeat) result in behaviours akin to those characteristic of human depression. These down-regulate hippocampal neurogenesis and hippocampal expression of brain-derived neurotrophic factor (BDNF). Clinically effective ANTIDEPRESSANTS increase hippocampal neurogenesis in rodents, and blocking neurogenesis prevents their antidepressive effects. See AFFECTIVE DISORDERS, OBESITY, STRESS RESPONSE.

depth perception See BINOCULAR VISION, HYPERCOLUMNS.

derived character A character whose evolutionary origin distinguishes a new clade, whose members share it (synapomorphically), from a sister clade, whose members do not. Compare PRIMITIVE CHARACTER.

dermal papilla See DERMIS; HAIR FOLLICLE, Fig. 86.

dermatome The dorsal layer of dermatomyotome of each SOMITE, giving rise to mesenchymal connective tissue of the dorsal skin: the DERMIS. Each dermatome is the area of skin supplied with sensory innervation by the dorsal roots of a pair of spinal nerves. Compare MYOTOME.

dermis The tissue layer responsible for most of the structural strength of the SKIN. The dorsal dermis derives from DERMATOME while that of the remainder of the body derives from other mesenchyme. It comprises connective tissue containing fibroblasts, a few adipose cells, and macrophages. Collagen is its most abundant extracellular protein, although elastin and reticular fibres are also present. Compared with the underlying subcutis, with which it is continuous, the dermis contains relatively few adipose cells or blood vessels. But its free nerve endings include those for sensations of pain, itch, tickle, heat and cold, while MEISSNER'S CORPUSCLES, PACINIAN CORPUSCLES and RUFFINI ENDINGS are also present (see MECHANORECEPTORS, Fig. 120). Its more superficial, papillary, layer contains blood vessels supplying the overlying epidermis and is marked by projections extending towards the epidermis, the *dermal papillae* (see FINGERPRINTS). Its deeper reticular layer is composed of dense irregular connective tissue forming the bulk of the dermis. If overstretched, the dermis may rupture and form scar tissue (striae), or stretch marks.

descending motor tracts Axons from the brain descend through the SPINAL CORD along two main groups of pathways. The *direct pathways* (pyramidal pathways) originating in the cerebral cortex are the CORTICOSPINAL and CORTICOBULBAR tracts. The indirect pathways, originating other than in the cortex, are the RUBROSPINAL TRACT (origin in the red nucleus), the VESTIBULOSPINAL TRACT (origin in the vestibular nuclei), the RETICULOSPINAL TRACTS (origin in the reticular formation of the brain stem) and the TECTOSPINAL TRACT (origin in the superior colliculus). The lateral column (corticospinal and rubrospinal) tracts are involved mainly in voluntary movements and under cortical control; the ventromedial column (vestibulospinal, reticulospinal and tectospinal) tracts are involved in reflexes of posture and locomotion and originate in the brain stem (see POSTURAL REFLEX PATHWAYS). See ASCENDING SENSORY PATHWAYS.

desmosomes A form of INTERCELLULAR JUNCTION, typically occurring where cells require adhesion resisting shearing stresses that would otherwise shear them apart. They are rivet-like complexes in which the extracellular domains of cadherins form 'Velcro-like' attachments, their intracellular domains linking to a plaque of attachment proteins that in turn anchor intermediate filaments (especially keratin in most epithelial cells, and desmin in cardiac muscle fibres) to the CYTOSKELETON.

desquamation See EPIDERMIS.

detoxification See CYTOCHROMES, ENDOPLASMIC RETICULUM, LIVER.

development The epigenetic and genetic processes by which a fertilized ovum becomes an adult, the major stages of which are indicated in Appendices Va and Vb. Many anatomical entries include some information on the embryological derivation of the structure concerned. A remarkably small number of molecular signals – e.g., Wnt, Notch, Hedgehog, bone morphogenetic proteins and fibroblast growth factors (see entries) – is repeatedly employed in different developmental settings, often in association with appropriate cofactors, to orchestrate differentiation and patterning (e.g., see SOMITES, LIMBS). INSULIN-LIKE GROWTH FACTOR-I is the major growth-promoting factor during development, both before and after birth. See GENE EXPRESSION, GENE SILENCING, HOX GENES, PAX GENES.

DHA (docosahexaenoic acid) See OMEGA-3 FATTY ACIDS.

DHEA See DEHYDROEPIANDROSTERONE.

DHPLC (denaturing high performance liquid chromatography) A method for detecting mutations in which heteroduplexes, derived after denaturation by reannealment of DNA strands from different alleles, have an altered retention time on chromatography columns compared with normal allelic double-stranded DNA under near-denaturing conditions. PCR products from patients with heterozygous mutations will form heteroduplexes if denatured by heating and then cooled slowly. Homozygous mutations can be detected by mixing PCR products with normal sequence products to detect heteroduplexes. Putative SNPs discovered must be checked by direct DNA SEQUENCING. See DNA MICROARRAYS, SSCP.

diabetes insipidus Familial neurogenic diabetes insipidus, characterized by VASOPRESSIN deficiency, is due to mutations in the NEUROPHYSIN gene and improper targeting of the hormone to neurosecretory granules resulting in retention of the prohormone in the endoplasmic reticulum of the magnocellular neurons and leading to apoptosis. See KIDNEY.

diabetes mellitus The most common disease resulting from impaired pancreatic hormone release, both forms (type 1 and type 2) being characterized by impaired INSULIN release (see COMPLEX DISEASES). Because insulin favours glucose and lipid storage, and protein synthesis in muscle and other tissues, its deficiency leads to accumulation of blood glucose, decrease in lipid storage, and to protein loss (raising plasma amino acid levels) causing muscle wasting.

Type 1 diabetes (insulin-dependent diabetes mellitus, IDDM) is a chronic organ-specific AUTOIMMUNE DISEASE resulting from β-cell destruction (see PANCREAS), accounting for 2–5% of cases and usually presenting in younger people (hence 'juvenile-onset' diabetes), with a prevalence of <1% in childhood or adolescence. In the absence of insulin therapy, it is characterized by elevated levels of fasting plasma glucose, and of KETONE BODIES, sometimes enormously so, resulting in ketosis. It is polygenic, with >20 gene loci involved. Sibling recurrence risk (l_S) is ~15, twin data suggesting that >50% of phenotypic variance is due to genetic factors (5% concordance for dizygotic twins and 40% for monozygotic twins). Genes/regions for which strong associations are well known include the MHC, with the heterozygous genotype DR3/DR4 conferring ~25-fold increased relative risk compared with the general population. It is now known that this serotype is tightly linked to DQβ alleles that confer susceptibility. The MHC class II allele HLA-DR2 has a dominant protective effect, even in association with one of the susceptibility alleles. Genes encoding insulin, CTLA-4 (cytotoxic T-lymphocyte associated 4), PTPN22 (as with RHEUMATOID ARTHRITIS) and regions around the interleukin 2 receptor alpha and interferon-induced helicase 1 genes (*IFIH1/MDA5*) are also implicated. Recent genome-wide analysis has found strong associations with chromosomes 12 (12q13, 12q24) and 16 (16p13).

Improved prenatal survival seems to be the main benefit of genes encoding Type 1 diabetes, which was generally fatal until the advent of insulin replacement therapy. In affected families, normal foetuses are more likely to suffer MISCARRIAGE than those with diabetes-linked HLA antigens. Because only 20% of newborns with the Type 1 diabetogenic genotype actually become diabetic, 80% of the carriers reap the prenatal benefits without paying the postnatal price.

Type 2 diabetes (T2D), often a component of 'metabolic syndrome' (see INSULIN RESISTANCE), accounts for >90% of diabetes cases, with a prevalence of <10% in middle or old age. It is not a single disease, but rather a genetically heterogeneous group of metabolic disorders sharing glucose intolerance. Strongly associated with OBESITY, which in the long term predisposes to T2D, the waist-to-hip ratio is a good predictor of T2D (see ADIPOSE TISSUE). T2D is polygenic, resulting from loss of normal responsiveness of β-cells to glucose (see PANCREAS), followed later by net reduction in β-cell mass and decreased responsiveness of peripheral tissues to insulin action. It rarely leads to ketosis, usually presenting in obese adults (but increasingly in juveniles). Regardless of its aetiology (abnormalities in glucose transporters; abnormal insulin synthesis, processing, storage or secretion), the defect leads to excessive rise in plasma glucose which produces a compensatory increased insulin output (hyperinsulinaemia) which down-regulates insulin receptors, leading to insulin resistance. Percentage concordance is 25% for dizygotic twins and 95% for monozygotic twins. It is progressive, one of the main factors being a continued decline in pancreatic β-cell mass and function – e.g., a reduction in insulin output of ~75% during fasting hyperglycaemia. Even when BLOOD GLUCOSE LEVEL is within the normal range, β-cell function decreases progressively as fasting glucose level increases. Groups at increased risk of later developing diabetes exhibit β-cell dysfunction well before they would be considered to have reduced glucose tolerance, and include women with a history of gestational diabetes or polycystic ovarian syndrome, older hyperglycaemic subjects, and those with impaired glucose tolerance. First degree relatives of individuals with T2D also have impaired β-cell function although may not have impaired glucose tolerance. Longitudinal data from the Pima Indians, in whom the prevalence of T2D is higher than in almost any other group, support the link between insulin resistance and increasing loss of β-cell function, those not progressing to diabetes simply increasing their insulin output as sensitivity declined. Several large studies have implicated an allele of the gene for a β-cell zinc-transporting protein in non-obese diabetes. There are currently (2007) 10 non-Mendelian genetic risk factors for T2D, and the β-cell is the most likely source of this chronic disorder. The foetus of a diabetic mother with hyperglycaemia will also be hyperglycaemic, since glucose is in equilibrium between the foetal and maternal sides of the placenta. This leads to higher than normal birth weights for date and increased risk of foetal respiratory distress due to retarded lung maturation and surfactant production.

In OBESITY and T2D, expression of the GLUT4 glucose transporter is decreased selectively in adipocytes, while expression of the retinol-binding protein-4 (RBP4) is elevated in adipose tissue of mice with *Glut4* knockout, which induces expression of the gluconeogenic enzyme phosphoenolpyruvate carboxylase. Serum levels of RBP4 are elevated in both insulin-resistant mice and in humans with obesity and T2D, and mice transgenically overexpressing human RBP4, as well as those injected with recombinant RBP4, suffer insulin resistance. RBP4 is an ADIPOKINE that may contribute to the pathogenesis of T2D. PPAR-γ (see PPARS), PGC-1, the pancreatic β-cell KATP channel gene *KCNJ11*, and the transcription factor gene *TCF7L2*, have also been associated with diabetic phenotypes.

Maturity-onset diabetes of the young (MODY) is an autosomal dominant form of diabetes characterized by β-cell dysfunction, with a prevalence of $<0.01\%$, with age of onset in adolescence or early adulthood. It is monogenic, and mutations in any of at least seven genes may be responsible (see PANCREAS).

In diabetes and related metabolic syndromes, neural (polyneuropathy) and vascular dysfunction (microvascular disease) appear together clinically, reduced nerve conduction velocity being an early marker of neuropathy in diabetic MOUSE MODELS (see DIABETIC NEUROPATHIES). Like those forming in atherosclerosis, microvascular channels that develop in the diabetic retina as a result of neoangiogenesis may be particularly fragile and prone to micro-haemorrhage. Because NETRINS promote growth of both neurons and vasculature, they may be of clinical use in treatment of diabetes.

diabetic neuropathies A family of nerve disorders caused by DIABETES MELLITUS (DM) in time affecting nerves throughout the body. Symptoms include numbness and sometimes pain and weakness in the hands, arms, feet and legs. Problems may also occur in every organ system, including the digestive tract, heart and gonads. People with DM can develop nerve problems at any time, but the longer a person has diabetes, the greater the risk. Perhaps 50% of those with diabetes have some form of neuropathy, but not all with neuropathy have symptoms. The highest rates of neuropathy are among people who have had the disease for at least 25 years. It appears to be more common in people who have had problems controlling their blood glucose levels, in those with high levels of blood LDL and blood pressure, in overweight people, and in people over the age of 40. The most common type is *peripheral neuropathy*, also called distal symmetric neuropathy, which affects the arms and legs (see CONNEXINS). See NEURAL DEGENERATION.

diacylglycerol A glycerophospholipid second messenger of the PHOSPHATIDYLINOSITOL SYSTEM, remaining in the membrane after cleavage of PIP_2 and acting as a docking point and activator of protein kinase C (see PHOSPHOLIPASE C, Fig. 135). It can be cleaved to release ARACHIDONIC ACID, or used in the synthesis of other small messengers, the eicosanoids. See CELL MEMBRANES.

diagnostic screening See SCREENING.

diagnostic testing See DNA-BASED DIAGNOSTIC TESTING.

diamorphine Heroin. See OPIOIDS.

diapedesis The mechanism by which white blood cells, such as neutrophils and monocytes, enter the tissues from capillaries (extravasation), especially during INFLAMMATION. The initial step, commonly at a site of infection, involves the binding of carbohydrates on the leukocyte surface to SELECTINS (e.g., E-selectin) induced on the ENDOTHELIUM, an interaction that cannot anchor the cells against the flow of blood (hence the cells roll). However, this reversible binding allows stronger interactions to occur through induction by the chemokines CXCL8 and CCL2 of ICAM-1 on the endothelium and activation of its INTEGRIN receptors LFA-1 and CR3 on the leukocyte. This tight binding stops the rolling and enables the leukocyte to cross the capillary wall by squeezing between the endothelial cells, and involves traversing the basement membrane which it penetrates using a matrix metalloproteinase enzyme expressed on its surface. The leukocyte then migrates along a concentration gradient of chemokines (e.g., CXCL8, CCL2) secreted by cells at the site of infection.

diaphragm 1. The muscular and connective tissue partition, covered by a serous membrane, separating the pleural and peritoneal BODY CAVITIES and crucial in VENTILATION. The central tendon of the diaphragm develops from that portion of the transverse septum lying cranial to the developing liver. **2**. See CONTRACEPTION.

diaphysis (pl. diaphyses) The shaft, or body, of a BONE: its long, cylindrical and major portion.

diarrhoea The frequent passing of watery faeces, most often owing to inflammation of the small intestine or colon, often caused by viral gastroenteritis, bacteria (in food poisoning, bacterial dysentery), irritant drugs and poisons, and by allergic reactions. Estimates of the impact of diarrhoeal diseases worldwide range from 1 billion to 4 billion diarrhoeal episodes every year among children aged <5 years in developing countries, causing ~2.5 million deaths, of which 85% occur in the poorest parts of the world. In some countries, they account for >20% of all deaths in children aged <5 years. In more economically developed countries, diarrhoea is associated with <1% of deaths in young children. See DISEASE, Fig. 63; ORAL REHYDRATION THERAPY.

diarthrosis (pl. diarthroses) A freely movable JOINT; all of which are SYNOVIAL JOINTS.

diastema A gap between the lateral upper incisor and the upper CANINE teeth for the occlusion of the lower canine, present in all primates other than prosimians and modern humans.

diastole The relaxation phase of chambers of the heart during a CARDIAC CYCLE.

diastolic blood pressure The lowest pressure occurring in the arteries during ventricular relaxation (see CARDIAC CYCLE). Compare SYSTOLIC BLOOD PRESSURE.

diathesis Predisposition to a particular disease. See GENETIC PREDISPOSITION.

Dicer See RNA INTERFERENCE.

dichromatism See COLOUR BLINDNESS, COLOUR PERCEPTION.

dichrotic See CARDIAC CYCLE.

diencephalon The posterior part of the prosencephalon, rostral to the brainstem. Its two sides are separated by the lumen of the third ventricle, whose lateral walls they constitute, forming the epithalamus (including the pineal gland), thalamus, subthalamus and hypothalamus (see BRAIN, Fig. 30) and contributing to the formation of the pituitary gland.

diet See HEALTHY DIET.

Dietary Reference Values (DRVs) Values for recommended daily intake set for total fat, saturated fat, cholesterol, total carbohydrate, dietary fibre, sodium, potassium and protein. See DAILY VALUES, REFERENCE DAILY INTAKES.

diethylstilbestrol A synthetic oestrogen, formerly prescribed as an effective drug in treatment of metastatic prostate cancer until the more expensive GnRH agonist leuprolide was found to have fewer side-effects. It was also an approved drug in oestrogen-replacement therapy until it was found to be a teratogen. In the 1990s, the only approved indications for DES were treatment of advanced prostate cancer and treatment of advanced breast cancer in postmenopausal women.

diffuse modulatory systems Neural pathways of the central nervous system originating typically as a core of a few thousand neurons in the pons (LOCUS COERULEUS), the pons and medulla (RAPHE NUCLEI), the pons and midbrain tegmentum (the pontomesencephalotegmental complex), the midbrain (SUBSTANTIA NIGRA and ventral tegmental area) and the forebrain (basal forebrain complex). Their neurons are characterized by long, multibranched, axons, some of which make synaptic contact with 10^5 or more other neurons. These systems tend to release neurotransmitters into the extracellular fluid where they can diffuse and influence many neurons. They include cholinergic, dopaminergic, noradrenergic and serotonergic systems (see BRAIN, Fig. 30). Activation leads to increased vigilance (e.g., ATTENTION). Many abused drugs act directly on synapses of the diffuse modulatory systems (see PSYCHOACTIVE DRUGS, Fig. 144). Malfunctioning of the diffuse modulatory systems can be a cause of AFFECTIVE DISORDERS (see also SEPTUM **2**).

DiGeorge/velocardiofacial syndrome Deletion of a 3-Mb region of 22q11 on one copy of chromosome 22 is the most common genetic deletion known in humans, producing this syndrome. It involves neural crest abnormalities, including conotruncal and aortic arch defects, cleft palate and hypoplasia of the thymus and parathyroid glands. See CONGENITAL HEART DEFECTS, COPY NUMBER.

digestion See entries on regions of the ALIMENTARY CANAL.

digestive tract See ALIMENTARY CANAL.

digits Fingers of the hand (see FORE LIMB) and toes of the FOOT. See also PRIMARY SOMATOSENSORY CORTEX.

dihydrotestosterone (DHT) See TESTOSTERONE.

dilation Increased aperture of a structure. See PUPILLARY LIGHT REFLEX, VASODILATION.

dimeric, dimers Of molecules composed of two subunits. Many proteins (e.g., membrane receptors, transcription factors) only function in the dimeric state, be it as homodimers (identical polypeptide subunits) or heterodimers (non-identical subunits).

dinucleotide repeats See MICROSATELLITES.

dioptre The reciprocal of focal length. See CORNEA.

dioxin See TOXINS.

dipeptidase An enzyme in the brush borders (microvilli) of epithelial cells of the small intestine, hydrolysing dipeptides to single amino acids for transport across the membrane. See AMINOPEPTIDASE.

diphtheria Caused by the bacterium *Pseudomonas aeruginosa*, diphtheria was a serious childhood disease prior to widespread immunization and infections are common among immuno-compromised patients and developing countries still experience serious outbreaks. Diphtheria TOXIN belongs to a family of toxins that modify intracellular proteins, in this case the elongation factor eEF2 which catalyses the translocation of transfer and messenger RNAs on the ribosome. A toxoid is used in the vaccine.

direction-sensitive cells See VISUAL MOTION CENTRE.

directional selection A form of positive selection (see NATURAL SELECTION) in which a single mutation has a selective advantage over all other mutations, resulting in its rapidly reaching fixation (a frequency of 100%) in the population.

disaccharidases The enzymes sucrase, lactase and maltase are all components of the brush borders of epithelia of the mucosa of the small intestine converting, respectively, sucrose to glucose and fructose; lactose to glucose and galactose; and maltose and maltotriose to glucose. Glucose and galactose cross the brush borders through symports along with Na^+ (secondary active transport), fructose crossing by facilitated diffusion.

DISC (death-inducing signalling complex) See APOPTOSIS.

discordant twins Twin pairs in which only one member exhibits a given trait. See CONCORDANCE.

disease The classification of disease is an imprecise science. Categories of disease, disorders, or illnesses, might include the following: physical, mental, social, infectious, non-infectious, degenerative, inherited, self-inflicted and deficiency. These categories are not mutually exclusive, as illustrated by many COMPLEX DISEASES, AUTOIMMUNE DISEASES and AGE-RELATED DISEASES and diseases caused by NEURAL DEGENERATION. Molecules termed *predictive antibodies* often appear in the blood years before people show symptoms of some disorders, and tests to detect them could warn of the need to take preventive action.

An *infectious disease* is an illness caused by a pathogen which invades body tissues and causes damage, and not all are communicable (see INFECTION). It is now considered, on the evidence of DNA analyses, that many human bacterial diseases pre-date the rise of civilization (see SHIGELLA, TUBERCULOSIS, ANTHRAX). Emerging infections (EIs) can be classified into the following categories: 'newly emerging' (those not previously recognized in humans), re-emerging/resurging' (those that existed in the past but are now rapidly increasing either in incidence or in geographical or host range) and 'deliberately emerging' (those developed by mankind, usually for nefarious use such as bioterrorism). A key priority for infectious disease research is clarification of how pathogen (and vector) genetic variation, modulated by host immunity, transmission bottlenecks and epidemic dynamics, determine the wide variety of pathogen phylogenies observed at levels ranging from individual host to population (e.g., see HELICOBACTER PYLORI).

Endemic diseases are those occurring commonly at all times of the year (e.g., malaria in West Africa); an *epidemic*, or *outbreak*, of disease occurs when its incidence is clearly in excess of normal experience or expectancy; a *pandemic* is an epidemic that affects all or most world countries at the same time. A *point-source outbreak* occurs when a group of individuals is exposed to

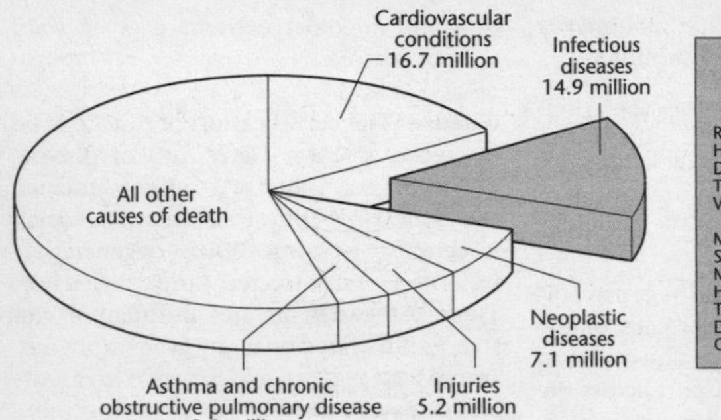

Infectious diseases	Annual deaths (million)
Respiratory infections	3.96
HIV/AIDS	2.77
Diarrhoeal diseases	1.80
Tuberculosis	1.56
Vaccine-preventable childhood diseases	1.12
Malaria	1.27
STDs (other than HIV)	0.18
Meningitis	0.17
Hepatitis B and C	0.16
Tropical parasitic diseases	0.13
Dengue	0.02
Other infectious diseases	1.76

FIG 63. Pie diagram indicating the leading causes of death worldwide in 2004. See DISEASE. From D.M. Morens, G.K. Folkers and A.S. Fauci, © *Nature* 2004, with permission from Macmillan Publishers Ltd. Figures from the World Health Organization.

a common source of infection at a defined point in time; an *extended-source outbreak* occurs when a group of individuals is exposed to a common source of infection over a period of time. Useful data can be found at http://earthtrends.wri.org

The *incidence* of a disorder is its rate of occurrence (e.g., 2 in 1,000 births are affected by neural tube defects). Its *prevalence* is the proportion of persons with the disorder in a given population at a given time. In 2004, the leading causes of death worldwide were as shown in Fig. 63, ~15 million (25%) being the direct result of infectious disease (see also MORTALITY, Fig. 126).

Although we know that genetics contributes to common diseases, we do not know much about which gene variants are responsible. If they are common, the HapMap project's toolkit should greatly accelerate their discovery (see HAPLOTYPE). The link between disease and signal transduction pathways is best validated for genetic disorders, four pathways in particular being associated with many complex diseases, such as DIABETES MELLITUS and CORONARY HEART DISEASE. These are the Wnt, Hedgehog (Hh), transforming growth factor-β and insulin/insulin-like growth factor (IGF) pathways (see Fig. 64). DNA sequence transposition can cause disease, as when genetic deficiency of gene product arises through insertional inactivation by retrotransposons (e.g., classic HAEMOPHILIA). Epigenetic causes may also be involved (see GENOMIC IMPRINTING), and both genetic and epigenetic factors contribute to CANCER.

See the Human Gene Mutation Database on www.hgmd.org/

DNA sequence transposition can cause disease. Genetic deficiency of gene product can arise through insertional inactivation by retrotransposons (e.g., classic HAEMOPHILIA).

dislocation (luxation) Detachment of a bone from a joint, with tearing of ligaments, tendons and articular capsules.

dissociation curves See HAEMOGLOBINS.

distal Farther from the attachment of a limb to the trunk; farther from the point of origin or attachment.

distal convoluted tubule (distal tubule, DCT) The cortical portion of a KIDNEY nephron lying between the thick ascending limb of the LOOP OF HENLE and the COLLECTING DUCT. Its epithelium contains both PRINCIPAL CELLS and INTERCALATED CELLS, both sensitive to ALDOSTERONE, and impermeable to both water and urea. The initial portion contains the MACULA DENSA. Principal cells reabsorb Na^+ and secrete K^+ (depending on dietary K^+ intake and body fluid level) while intercalated cells reabsorb K^+ (so that its intracellular level remains high) and HCO_3^-, and secrete H^+. K^+ leak channels present

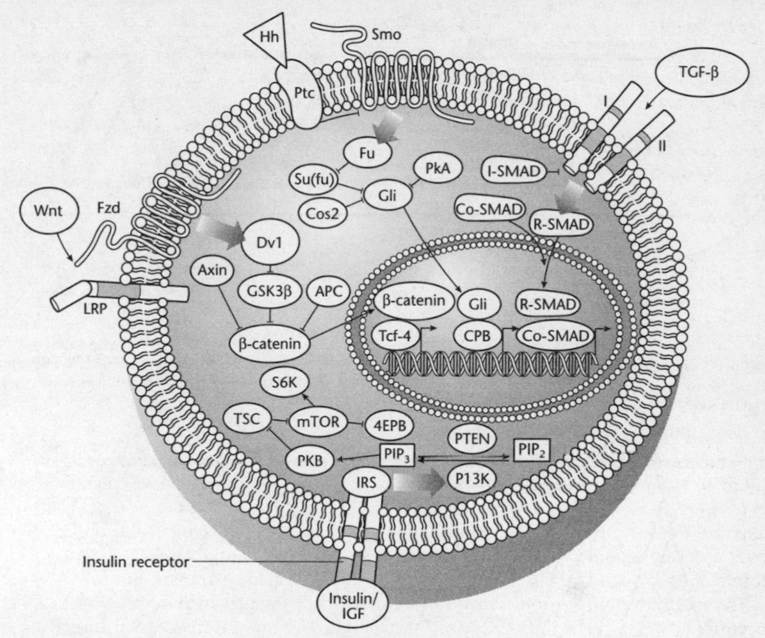

FIG 64. Diagram illustrating four signalling pathways linked to DISEASE. These are the Wnt, Hedgehog (Hh), transforming growth factor-β, and insulin/insulin-like growth factor pathways. Arrows indicate activation and T-junctions show inhibition of pathways. All four SIGNAL TRANSDUCTION pathways, and several of the proteins indicated within ovals, are referenced in the dictionary. From M.C. Fishman and J.A. Porter, © *Nature* 2005, with permission from Macmillan Publishers Ltd.

in both apical and basolateral membranes of the principal cells, so some K^+ diffuses down its concentration gradient into the tubular fluid where the K^+ concentration is very low, this being the main source of K^+ excreted in the urine. A variable amount of Ca^{2+} is also reabsorbed, depending upon the level of PARATHYROID HORMONE. Aldosterone increases Na^+ and water reabsorption and K^+ secretion by principal cells because it both increases the activity of existing Na^+ pumps and leakage channels and stimulates the synthesis of new ones. If aldosterone level is low, the blood K^+ concentration can rise, causing ARRHYTHMIAS and cardiac arrest at higher levels still. The DCT is not very permeable to water, so the solutes are reabsorbed with little accompanying water. See HYPERTENSION.

diuresis, diuretic Diuretics help turn excess body water into urine, relieving oedema and improving heart action because a smaller volume of blood has to be pumped. All diuretics act on the kidneys. See ANTIDIURETIC.

diurnal rhythms See CIRCADIAN RHYTHMS.

divergence **1**. In evolution theory, the genetic divergence between two species is the proportion of nucleotides that differ between representative individuals of the two species. This can be converted into a divergence time in terms of millions of years, but only if the differences have been accumulated at a constant rate as a result of new mutations (contrast COALESCENCE). Genealogical relationships (gene

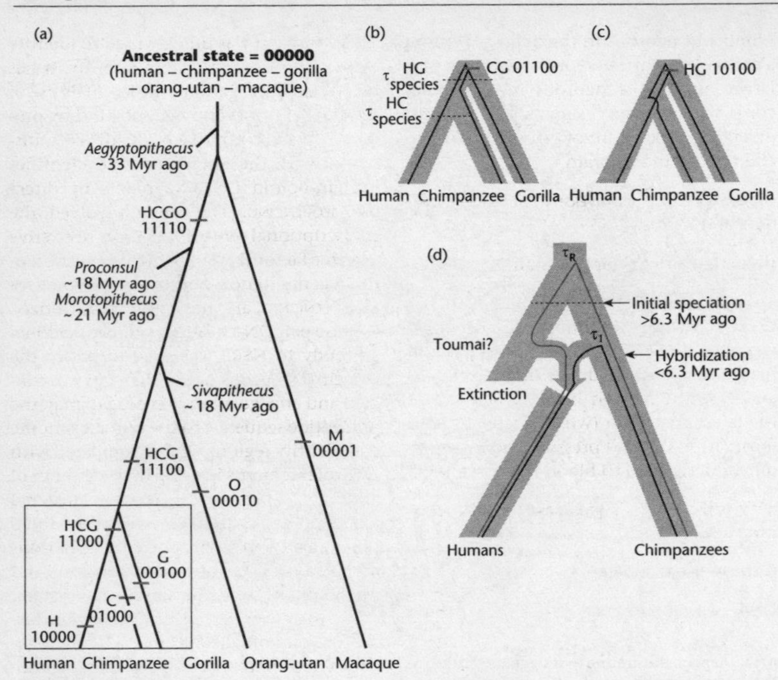

FIG 65. Reconstruction of genealogical DIVERGENCE patterns and times, in which divergent sites found by shotgun sequencing ~87 Mb of gorilla DNA are compared against human and chimpanzee genomes. Divergent sites are those in which two alternative alleles were observed across the aligned sequences of the species. Each divergent site class is designated by a string of 0s and 1s representing the bases seen in human–chimpanzee–gorilla–(orang-utan)–macaque, the macaque allele being defined as having state 0. (a) Historical relationships of these five taxa, with relationships to various fossils. Relative lengths of branches are estimated from the data by the number of divergent sites of each type. (b), (c) Species relationships (grey) can differ from genealogical relationships since, as here, humans and chimpanzees can share a common ancestor older than the gorilla speciation (see DIVERGENCE entry). (d) A revised model accounting for these results, which suggests that the first hominins became isolated from the chimpanzee ancestors >6.3 Mya, but hybridized back to the chimpanzee lineage, with chimpanzees and humans sharing a common ancestor around the time of hybridization in some regions (entire line) but before the initial speciation in others (pale line). The Toumaï reference is to SAHELANTHROPUS TCHADENSIS. From N. Patterson, D.J. Richter, S. Gnerre, E.S. Lander and D. Reich, © *Nature* 2006, with permission from Macmillan Publishers Ltd.

trees) do not always coincide with species relationships (species trees). This is because the genetic divergence time between species, $\tau(x)$, varies across the genome and is always greater than or equal to the speciation time, $\tau_{species}$. For example, humans and chimpanzees can sometimes share a common ancestor that pre-dates the gorilla speciation ($>\tau^{HG\ species}$) and although humans and chimpanzees are most closely related in most sections of their genomes ('HG' sites in Fig. 65; see HOMINOIDEA), regions exist in which chimpanzees and gorillas are most closely related ('CG' sites).

2. Of NEURAL NETWORKS (e.g., in the HIPPOCAMPUS) in which a single presynaptic neuron synapses with several postsynaptic neurons (or, elsewhere, with muscle fibres or gland cells), each of which it may influence (contrast CONVERGENCE). Thus a small number of neurons in the brain governing a particular body movement stimulate a far larger number of neurons in the spinal cord, while sensory signals feeding into diverging circuits are often relayed to several regions of the brain.

diversity See BIODIVERSITY, HUMAN GENOME DIVERSITY PROJECT.

diverticulum Sac or pouch in the wall of a canal or organ, especially in the colon.

diving reflex Stimulation of nasal receptors by water, causing cessation of breathing, bradycardia (slow heart rate) and vasoconstriction in all tissues except the heart and brain (which increases both mean arterial blood pressure and total peripheral resistance to blood flow).

dizygotic twins (fraternal twins, non-identical twins) See TWINS.

D loop See MTDNA.

DNA See DEOXYRIBONUCLEIC ACID.

DNA-based diagnostic testing See SCREENING.

DNA-binding proteins Proteins with DNA-binding motifs (e.g., BASIC HELIX-LOOP-HELIX MOTIF), involved in a wide variety of regulatory mechanisms including the control of TRANSCRIPTION, DNA REPLICATION, DNA REPAIR MECHANISMS and chromosome segregation. Both α-helices and β-sheets can be involved in DNA-recognizing motifs. Identification of their binding sites throughout the genome has been a high priority in recent years. This generally involves chromatin immunoprecipitation (ChIP) in which cells are treated with a reagent (e.g., formaldehyde) that cross-links protein and DNA. Chromatin is isolated, the DNA is fragmented, and an antibody added to precipitate the protein and associated DNA. The protein can then be analysed after removal of the DNA. Originally used to find out whether single genes were enriched by the precipitation, the classic genomic approach has been to analyse all precipitated DNA fragments using them as probes on a DNA MICROARRAY ('ChIP-chip'). This approach was quickly used to identify TRANSCRIPTION FACTORS. More recently, it has been used in genome-wide analysis of modified HISTONES and histone-binding proteins. A new approach ('ChIPSeq') dispenses with the microarray and identifies protein-bound DNA fragments by direct DNA SEQUENCING. The study involved the transcriptional repressor neuron-restrictive silencer factor (NRSF), which silences neuronal genes in non-neuronal cell types (see GENE SILENCING), and has a well-characterized 21-base pair DNA binding site. Monoclonal antibody to NSRF was used to purify the protein–DNA complexes, the DNA was isolated and an algorithm was used to map the nucleotide sequence to the human genome and identify regions where, compared with control fragments isolated in the absence of antibody, sequence reads clustered together. Genes that bound NRSF turned out to be highly enriched for functions involved in synaptic transmission and development of the nervous system, including those for 110 transcription factors, 22 microRNAs and five splicing regulators. See SINGLE-STRANDED BINDING PROTEINS.

DNA chip One form of DNA MICROARRAY.

DNA damage See DNA REPAIR MECHANISMS, GENOTOXIC STRESS.

DNA fingerprinting See DNA PROFILING.

DNA helicases Enzymes acting as motor proteins, breaking the hydrogen bonds between the base pairs and unwinding the double helix as they move along the DNA prior to DNA REPLICATION. They are also required in DNA REPAIR MECHANISMS, recombination and TRANSCRIPTION.

DNA hybridization See DNA PROBES.

DNA injection See DNA VACCINES for 'biolistics'.

DNA ligases Enzymes catalysing the formation of a phosphodiester bond in DNA when provided with an unattached but adjacent 3′ OH group and 5′ phosphate.

DNA markers Any variable (polymorphic) sites on a DNA molecule acting as

Mendelian characters and used as chromosomal reference points (or 'signposts') to help in mapping other DNA sequences or follow the passage of a chromosome segment through a pedigree. They include RESTRICTION FRAGMENT LENGTH POLYMORPHISMS, VARIABLE NUMBER OF TANDEM REPEATS, MICROSATELLITES and DNA MINISATELLITES. See HAPLOTYPE.

DNA methylation An important epigenetic mechanism, interacting with histone modification, in transmission of chromatin states that repress GENE EXPRESSION. It is the method of long-term silencing of transposons and imprinted genes (see GENOMIC IMPRINTING) and can only be erased in PRIMORDIAL GERM CELLS. Patterns of DNA methylation are somatically heritable largely through the action of DNMT1, the maintenance methyltransferase (see CPG ISLANDS); but they can be disrupted by environmental TOXINS. Dietary modification can also profoundly affect the process, in particular deficiencies in FOLATE and METHIONINE.

DNA sequences that are to be transcriptionally active need to be unmethylated (at least in the PROMOTER regions). In human (and other animal) cells, DNA methylation at CPG DINUCLEOTIDE residues by cytosine methyltransferases is one important mechanism for maintaining gene repression during development, the pattern of 5-methylcytosine distribution in the genome of differentiated somatic cells varying according to cell type (see GENE SILENCING, CHROMATIN REMODELLING, CYTOSINE) and the contribution of different methyltransferases to this *de novo* methylation is not yet resolved. Promoters of genes that undergo 'developmental' demethylation contain CpG islands, as do differentially methylated regions (DMRs) of imprinted genes. By contrast with the relative flexibility of histone methylation and demethylation, DNA demethylation seems to involve attachment through a carbon–carbon bond to the cytosine base and might not be so readily removed. Demethylation therefore has to pass through pathways involving base-excision or mismatch repair (see DNA REPAIR MECHANISMS).

An alternative view of the primary role of DNA methylation holds that it confers a form of genome protection by checking the spread of TRANSPOSABLE ELEMENTS, and although this may have been its original value, the dominant vertebrate role seems to be a mechanism for regulating expression of endogenous genes and reducing transcriptional noise by silencing genes whose activity is not required in a particular cell or cell line. See GERM LINE, ALKYLATING AGENTS.

DNA microarrays (DNA chips) Useful tools in the identification of gene sequence mutations that are having significant impact on GENOMICS, while such fields as drug discovery and toxicological research will benefit considerably from their use. Two variants of the process are commonly used: (i) In the technique originally known as *DNA microarray technology*, suitably amplified synthetic probe cDNA (500–5,000 bp in length) of known sequence is immobilized to a solid surface (the 'chip', commonly of glass or nylon) by high-speed robotics, in an orderly (often overlapping) arrangement termed a 'microarray'. A fluorophore-conjugated nucleotide is included in the reaction mixture so that the cDNA population is universally labelled. It is then exposed to a set of targets – free nucleic acid samples (e.g., DNA from a patient requiring TISSUE TYPING) whose identity/abundance is being detected, any hybridizing DNA then being identified. (ii) In the technique originally known as DNA chip technology, but now as a *high density oligonucleotide chip* (marketed as GeneChips) an array of oligonucleotides (20–80 bp in length), synthesized *in situ* and of known sequence, is exposed to labelled sample DNA, hybridized, and the identity/abundance of complementary sequences is determined. The unlabelled cDNA is used to prepare a labelled cRNA population by incorporation of biotin, later detected with fluorophore-conjugated avidin.

While DNA microarrays can be used for whole genome analysis, a sample of DNA being screened for a mutation is first amplified by PCR, fluorescently labelled, and hybridized with the oligonucleotides in the microarray. After hybridization, analysis of the colour pattern of the microarray generated by a computer enables automated mutation testing. Detection of known SNPs has proved to be very successful using the technique, although detection of unknown mutations has proved more difficult (see COMPLEX DISEASES). Custom-designed 20–25 bp (or longer) oligonucleotide sequences (probes) are synthesized for both the normal DNA sequence and/or possible single nucleotide substitutions of a gene and are attached to a 'chip', commonly of glass or nylon, by high-speed robotics, in an orderly arrangement termed a 'microarray' (see DHPLC, SSCP, PROTEOMICS).

Microarrays can be used to discover the transcriptome of a cell, as well as the tissue-specific expression pattern of a gene (see HUMAN TRANSCRIPTOME). They are termed *tiling microarrays* when the probes used are regularly spaced and (usually) overlapping. Probes can be synthesized oligonucleotides or PCR products, selected to be complementary to one strand or to both, and synthesized directly onto, or spotted onto, glass slides. Each spot on the array contains 10^6–10^9 copies of the same sequence, and is unlikely to be completely saturated when they are allowed to hybridize with fluorescently labelled cRNA or cDNA from cell samples. Regions of greater fluorescent intensity can reveal transcription within a large genomic region, and correlation of probe intensities in different tissues (co-expression analysis) can be used to identify probes that detect exons in the same transcript.

One technique of genome-wide analysis of those sequences to which silencers bind has dispensed with microarrays (see DNA-BINDING PROTEINS). See SEGMENTAL DUPLICATIONS.

DNA mismatch repair See MISMATCH REPAIR.

DNA polymerases (DNA pols) Protein complexes carrying out DNA REPLICATION of which more than 20 forms occur in mammalian cells. Most use an individual DNA strand as a template for synthesizing a complementary DNA strand (DNA-directed DNA polymerases), all synthesize in a 5′→3′ direction (although the template DNA is read in the opposite, 3′→5′ direction) and all require the 3′-hydroxyl end of a previously synthesized base-paired RNA primer strand as a substrate for initiating chain extension. This is provided by a PRIMASE enzyme. The incoming substrates are the deoxyribonucleotide triphosphates, dATP, dGTP, dCTP and dTTP. Three classes are usually recognized: (i) *classical high-fidelity DNA-directed DNA polymerases*, specific for the standard synthesis of chromosomal DNA from either the leading (DNA pol δ) or lagging (DNA pol α) strand, including DNA repair enzymes (β and ε), and a specialist mitochondrial DNA replication-and-repair enzyme, DNA pol γ; (ii) DNA polymerases with low fidelity (*error-prone DNA-directed DNA polymerases*) tend to be highly expressed in B cells and T cells and include DNA pol ι (involved in SOMATIC HYPERMUTATION); and (iii) *RNA-directed DNA polymerases* (aka reverse transcriptases) which synthesize DNA from an RNA template, being encoded by some highly REPETITIVE DNA and endogenous RETROVIRUS classes, and an activity also present in TELOMERASE. Retroviral reverse transcriptases can utilize either single-stranded RNA or DNA as template, enabling a duplex DNA molecule to be produced prior to integration into the host chromosome. They have no proofreading mechanisms.

Since DNA, unlike RNA, is the genetic material in humans and is passed as information to the next generation, DNA pols often include DNA proofreading and nuclease activities, the DNA pol 'looking back' to check for error, so that if a nucleotide has been misincorporated it will degrade the recently elongated strand in a 3′→5′ direction, using its EXONUCLEASE activity, then move forward synthesizing this stretch again (see DNA REPAIR MECHANISMS). Were DNA replication to proceed in a 3′→5′ direction, removal of the 5′-terminal deoxynucleoside triphosphate during proofreading would leave a new terminal nucleotide with no high energy bond to be cleaved, and so chain elongation would be impossible from the 5′-end. Compare RNA POLYMERASES.

DNA primases See PRIMASES.

DNA probes Nucleic acid hybridization is the ability of single-stranded nucleic acid molecules with sufficient base-complementarity to hybridize by base-pairing, forming double-stranded molecules. It is employed as a tool of molecular genetics (e.g., see DNA MICROARRAYS, FISH, POLYMERASE CHAIN REACTION, POSITIONAL CLONING), a standard technique being to use a single-stranded labelled nucleic acid probe to identify related DNA or RNA molecules within a complex mix of unlabelled nucleic acid molecules. This enables information to be gained about an imperfectly understood and typically complex mixture of nucleic acids (the 'target'). Conventional DNA probes are isolated either by cell-based DNA cloning (0.1–100s of kb long) or by PCR (usually <1 kb long), are usually double-stranded initially, and labelled by incorporation of labelled deoxyribonucleoside triphosphates (dNTPs) during an *in vitro* DNA synthesis process. The individual strands must be separated (denatured) by heating or alkaline treatment so that complementary sequences can anneal to the target, forming heteroduplexes. RNA probes are single-stranded RNA molecules, are typically a few 100s to several kb long, and synthesized using the four substrate ribonucleoside triphosphates (rNTPs), a phage RNA polymerase, and template DNA which has been cloned in a specialized plasmid vector with a phage promoter next to the multiple cloning site. Oligonucleotide probes are single-stranded and typically short (15–50 nucleotides long) and synthesized chemically. They are often labelled by incorporating a ^{32}P atom, or other isotopic group, at the 5′ end; but non-isotopic labelling could employ a fluorophore (a chemical group which fluoresces in a certain wavelength of light), or the coupling of a reporter molecule to which another molecule binds with high affinity (see DNA PROFILING, SCREENING).

Dot-blotting involves drying a spot of an aqueous solution of denatured target DNA on a nitrocellulose or nylon membrane and exposing it for an appropriate time to a solution containing single-stranded labelled probe sequences, decanting the probe solution, washing the membrane to remove excess probe, and drying it to reveal any heteroduplexes formed. Dot-blot assays are used with *allele-specific oligonucleotide* (ASO) probes. These are typically 15–20 nucleotides long and span a complementary DNA sequence containing a single nucleotide substitution (see SINGLE NUCLEOTIDE POLYMORPHISMS). They are used under hybridization conditions which permit heteroduplex formation only if there is complete complementarity between the probe and target sequences, so that a single mismatch renders the short heteroduplex unstable.

DNA profiling (DNA fingerprinting) An important method of genotyping individuals. After extraction of DNA fragments by appropriate restriction endonucleases, those highly polymorphic sequences containing MICROSATELLITES OF VNTRS are first amplified by PCR using primers that bracket the locus (if a *single locus probe* is to be used) or loci (if a *multi-locus probe* is used). SOUTHERN BLOTTING is then employed to reveal the bands occupied by the fragments. Because of their polymorphic nature, each locus produces a pair of bands, one each from the maternal and paternal chromosomes (see Fig. 66a), and although some individuals have several bands in common, the overall pattern is quite distinctive for each individual (see Fig. 66b). The band pattern therefore serves as a 'DNA fingerprint', identifying an individual almost uniquely. When examining the variability at 5–10 different VNTR loci, the probability that two random individuals would share the same genetic pattern can be ~$1:10^9$. The technique is used in forensic work and paternity testing. See SINGLE NUCLEOTIDE POLYMORPHISMS.

DNA proofreading See DNA POLYMERASES, DNA REPAIR MECHANISMS.

DNA repair mechanisms Life on Earth must cope with the constant exposure to DNA-damaging factors including the Sun's radiation, and all cells have multiple

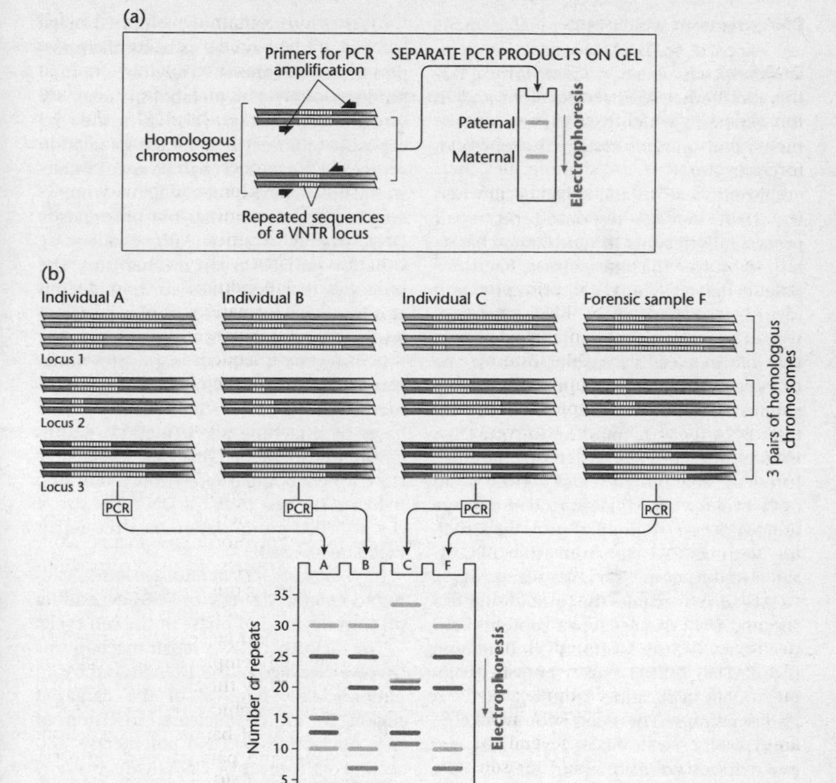

FIG 66. (a) Segments of a pair of homologous chromosomes containing different numbers of VNTR repeat DNA sequences (11 and 7 repeats, respectively) and the two bands they produce using agarose gel electrophoresis. (b) Three polymorphic VNTR loci and the banding produced from three 'suspects' and from a forensic DNA sample by DNA PROFILING (see entry for details). Individuals A and C can be eliminated from the enquiry, while individual B remains a clear suspect.

DNA repair mechanisms. Highly conserved DNA-repair and CELL CYCLE checkpoint pathways allow cells to rectify DNA damage from both endogenous and exogenous sources. If this is not possible, cells normally undergo APOPTOSIS. The genome of a typical mammalian cell accumulates many thousands of lesions during a 24-hour period, <0.1% of which become mutations. DNA POLYMERASES have proofreading functions that are an important part of DNA repair mechanisms (see also DNA REPLICATION).

Many human cells are terminally differentiated and will not divide again. Mechanisms repairing errors in DNA replication as well as errors in post-replicative DNA are therefore required since, if unrepaired, DNA damage can lead to chromosomal defects, inherited disorders and cancers.

The extents to which individual humans are exposed to DNA-damaging sources, and their cells can respond, are critical factors in whether a cancer develops and in the extent to which toxicities of current cancer and immune system therapies can be controlled.

Chromosome instability can occur when the DNA damage response and repair process fails, leading to syndromes characterized by growth abnormalities, haematopoietic defects, mutagen sensitivity and cancer predisposition. The tumour suppressor protein p53 (see P53 GENE) plays a key role in a cell's decision to arrest the cell cycle or undergo apoptosis following genotoxic stress. The enzyme PARP1 facilitates DNA repair by binding to such DNA breaks, attracting other repair proteins. Interplay between p53 and PARP1 is an early event when single-stranded damage to DNA occurs. A group of proteins associate with BRCA1 to form the BRCA1-associated genome surveillance complex (BASC). BASC includes the tumour suppressors and DNA damage repair proteins MSH2 (see entry), MSH6, MLH1, ATM, BLM, and the RAD50–MRE11–NBS1 protein complex. DNA replication factor C (RFC) is also a member. The association of BRCA1 with MSH2 and MSH6, required for transcription-coupled repair, may explain the role of BRCA1 in this pathway (see below). All members of this complex have roles in recognition of abnormal DNA structures, or damaged DNA, so that BASC may serve as a sensor for DNA damage. Mutations in ATM (see ATAXIA TELANGIECTASIA MUTATED PROTEIN), NBS1 (Nijmegen breakage syndrome), MRE11, BLM (see BLOOM SYNDROME) and WRN (see WERNER SYNDROME) are linked to a diverse group of disorders through interactions with BRCA1. BRCA2 interacts with RAD51 in repairing radiation-damaged DNA. Cancer cells lacking BRCA1 or BRCA2 function appear to survive by depending on PARP and its collaborators as a back-up system to maintain their genomes at some minimal level of integrity through homology-dependent repair, with implications for CHEMOTHERAPY.

DNA damage can result from:

(i) *Free oxygen radicals*, which can damage individual bases as well as break the phosphodiester backbone. They are normal products of cellular metabolism (e.g., see MITOCHONDRIA) but are also generated by exposure to an external ionizing radiation source (e.g., ULTRAVIOLET IRRADIATION). Double-strand breaks (DSBs) result when two breaks are close to one another but on opposite DNA strands, a particularly challenging situation for DNA repair mechanisms. The processes repairing DSBs are homologous end-joining and non-homologous end-joining. See GENOTOXIC STRESS.

(ii) ALKYLATING AGENTS, e.g., the mutagens nitrosoguanidine and ethylmethanesulphonate, which modify purine bases by attaching alkyl (methyl) groups covalently to them. This may destabilize the covalent bond to deoxyribose, resulting in loss of the base from the DNA. The size of the attached group determines the repair mechanism used.

(iii) *Inhibitors of DNA topoisomerases*, leading to enhanced single or DSBs depending on the enzyme and phase of the cell cycle.

The majority of DNA repair mechanisms involve cleavage of the DNA strand by an endonuclease, removal of the damaged region by an exonuclease, insertion of new nucleotides by DNA polymerase, and sealing of the break by DNA ligase. In error-free repair (*homologous end-joining*), the repair pathway either chemically repairs damage to a DNA base or deletes the damaged DNA and then uses an existing complementary sequence on the intact homologous chromosome as a template to restore the normal sequence on the damaged strand. In *non-homologous end-joining* (NHEJ, involving RAD51), the DSB is rejoined by DNA ligation, generally with loss of one or more nucleotides at the site of joining through trimming of the broken ends prior to ligation. Because so little of the genome consists of coding DNA, this error-prone emergency solution is tolerated for the sake of chromosome integrity (both deletions and short insertions can disrupt the informational content of coding sequences).

Pyrimidine dimers can be formed when

UV light unites adjacent pyrimidines in the same strand by C-6 and C-4 positions to form the 6-4 thymidine dimer, or when a cyclobutyl ring joins two adjacent pyrimidines to form a cyclobutane dimer. Both significantly perturb the local DNA shape and interfere with normal base-pairing. Chemical repair of such UV-induced photodimer lesions includes the action of chromophore-containing DNA photolyases which, in visible light (but not in the dark), bind the photodimer and split it to regenerate the original bases (*photoreactivation*).

Chemical repair to DNA can also be performed by DNA methyltransferases, which reverse alkylating lesions by removing certain alkyl groups added to the O-6 positions of guanine, transferring them to a cysteine residue on the enzyme (and inactivating it; see CHEMOTHERAPY). In the other error-free repair system, i.e., homology-dependent repair, two major pathways exist: excision repair and postreplication repair pathways, each depending on the antiparallel complementarity of the template strand to the strand being repaired. In excision repair, either an entire base or an entire nucleotide is removed and replaced. In *base excision repair*, DNA glycosylases cleave base-dribose bonds and produce a base-free sugar-phosphate backbone (see HUNTINGTON'S DISEASE). Another enzyme, AP endonuclease, then cuts the sugar-phosphate backbone and a deoxyribophosphodiesterase then removes a stretch of neighbouring sugar-phosphate residues so that a DNA polymerase can fill in the gap using the complementary strand as template. Spontaneous deamination of cytosine produces uracil, which is removed by uracil-DNA glycosylase. This prevents C→T transitions (see CYTOSINE), and the presence of thymine (which is 5-methyluracil) in DNA means that such spontaneous deamination can be recognized and repaired (see CPG DINUCLEOTIDE; but see also SOMATIC HYPERMUTATION). Deamination of adenine produces the base hypoxanthine, although much more rarely. In *nucleotide excision repair* mechanisms a multi-subunit repairosome recognizes distortions in DNA topology caused by the presence of an abnormal base, such as those caused by UV-induced pyrimidine dimers and addition of aflatoxin to guanine residues. It then makes a single-stranded 'bubble' by unwinding the DNA, then makes a 3′ and a 5′ nick in the damaged strand to excise ~30 nucleotides and then fills in the gap using the complementary strand as template. Because the repairosome preferentially repairs the DNA strand transcribed during gene expression, and because the damage leads to dissociation of the basal transcription apparatus and assembly of the repairosome, repair is carried out before transcription takes place. It is therefore called *transcription-coupled repair* (which requires BASC, as stated above) and is essential if repair is to be possible in the absence of DNA replication. Other proteins involved in this form of repair include ERCC8 and ERCC6 (see COCKAYNE SYNDROME).

Postreplication repair is needed when errors during DNA replication have escaped the 3′→5′ proofreading function of the replicating polymerase. The human *mismatch repair* system has not been fully analysed, but it recognizes mismatched base pairs, identifies the inappropriate base, excises it and repairs the strand. Sometimes the site contains an extrahelical loop generated by the slipped-mispairing mechanism (see MUTATIONS, Fig. 127). The inappropriate base is identified by hMutSα, which then recruits hMutLα to form a complex of proteins. An exonuclease then removes the mismatched segment and a DNA ligase joins the newly synthesized DNA to the original strand.

DNA repeats See REPETITIVE DNA.

DNA replication The entire genome is divided into large 'replication time zones', each programmed to replicate in a particular 'window' of time during S-phase of the cell cycle. It is this that gives rise to distinctive chromosome banding. Housekeeping genes seem to be arranged in early-replicating clusters, while many tissue-specific genes comprise miniature bands that are late-replicating in most tissues but become early-replicating in tissues where they are expressed. Since the basic unit of eukaryotic CHROMATIN (see Fig. 51) is the NUCLEOSOME, human DNA replication is associated with disassembly of the nucleosomes in the parental DNA strands, accurate copying of the parental DNA and reassembly of the original histones alongside assembly of new histones. The structure that accomplishes all this is the multienzyme *replisome complex* which contains at least 27 proteins, including a DNA polymerase. DNA replication uses a semiconservative mechanism employing leading- and lagging-strand synthesis, the region being replicated by the replisome complex moving progressively along the parental DNA, generating a Y-shaped structure termed a *replication fork*. DNA replication is initiated at specific points, *origins of replication*, of which there are many on each chromosome and possibly thousands in the human genome. These have AT-rich regions that melt (hydrogen bonds break so that the two DNA strands separate) when an initiator protein of the replisome complex binds to special binding sites adjacent to them. The preliminaries of DNA replication, in the form of pre-replicative protein complexes, have been assembled on each origin of replication during G1 of the cell cycle. During S-PHASE, the transcription factor E2F activates genes that enable assembly of the rest of the replisome complex (origin recognition complex, ORC). Replication proceeds *in both directions* from the multiple points of origin, the double helices produced eventually being ligated together so that two identical daughter DNA molecules are generated. The Mcm helicase component of the ORC opens up the DNA ahead of the replication forks and starts moving with each of them so that incoming deoxyribonucleoside triphosphates (DNTPs) can form base-pairs with the template strand, a process in which single-strand DNA-binding proteins contribute. DNA topoisomerase I is required to rotate the DNA ahead of the replication fork, which they do by adding themselves covalently to a phosphate backbone, reversibly breaking a phosphodiester bond so that the two sections of DNA helix on either side of the nick can rotate freely relative to one another. Any tension in the DNA drives the rotation in the direction relieving the tension. DNA topoisomerase II binds where the two double helices cross over each other. It uses ATP to nick both strands of the double helix (a double-strand break), creating a 'gate' through which the second double helix nearby passes, then reseals the break and dissociates from the DNA. These reactions prevent DNA tangling during replication.

Because of the antiparallel nature of the DNA double helix, and because all DNA POLYMERASES add nucleotides in a 5′→3′ direction, a replication fork has an asymmetric structure. One of the two daughter DNA strands (the *leading strand*) can be synthesized continuously, while its complementary strand (the *lagging strand*) must be synthesized in the opposite direction to the overall direction of DNA chain growth. Lagging strand synthesis is delayed, since it must wait for the leading strand to expose its template. It is synthesized in short sections (*Okazaki fragments*), ~100–200 nucleotides in length, incoming deoxynucleoside triphosphates carrying the energy required for chain elongation by the DNA polymerase. Once an Okazaki fragment is completed (taking a few seconds), DNA PRIMASE synthesizes RNA primers ~10 nucleotides in length, at intervals of ~100–200 nucleotides on the lagging strand. These are then elongated by the DNA polymerase to form a new Okazaki fragment, the RNA primer being erased by a DNA REPAIR MECHANISM and replaced by DNA, the 3′-end of the new complete Okazaki fragment being joined to the 5′-end of the previous one by DNA ligase. In one model, any stimulus or stress leading to an abnormal stretch of ssDNA, such as an arrested replication fork, would stimulate the binding of the site by replication protein A (an ssDNA-binding protein involved in DNA replication), which (in one model) then recruits the ATR-interacting protein (ATRIP) and its heterodimeric partner, ATR, to the ssDNA. Once there, ATR can phosphorylate key substrates, such as RAD17 and CHK1.

ATR has a critical role in most cellular stress responses that share inhibition of replication fork progression as a common mechanism.

As a replication fork advances, the replication apparatus somehow passes through parental nucleosomes without displacing them from the DNA, a process probably made possible by CHROMATIN REMODELLING proteins which destabilize the DNA–histone interface. Chromatin assembly factors (CAFs) are proteins that associate with replication forks and package the newly synthesized DNA with HISTONES as it emerges from the replisome complex, although not all histones are added in a DNA replication-dependent manner. On reaching the telomeres at the ends of the linear chromosomes there is nowhere to produce an RNA primer, and the TELOMERASE system completes DNA replication. See SPERMATOGENESIS.

DNA sequencing The 'gold standard' method for mutation screening employs the dideoxy chain termination method developed in the 1970s by Sanger. Originally employing radioactive labelling and manual data analysis, a new era of genome sequencing has dawned, making it affordable in small laboratories (see HIGH THROUGHPUT GENOTYPING). The inventions dispense with bacterial cloning and attach single-stranded DNA template (e.g., denatured PCR products) to aqueous beads encased in oil, where new complementary DNA is synthesized using a suitable DNA polymerase and an appropriate complementary primer. Four parallel reactions take place, each containing the four deoxynucleoside triphosphates (dNTPs), and a small proportion of one of the four analogous dideoxynucleotides (ddNTPs), each labelled with a different fluorescent dye, which serve as base-specific chain terminators. Lacking a hydroxyl group at the 3′ carbon position, these cannot form phosphodiester bonds, so that each reaction chamber will end up containing a mixture of DNA fragments of different lengths, each terminating in their respective dideoxynucleotide. Separating the reaction products on polyacrylamide

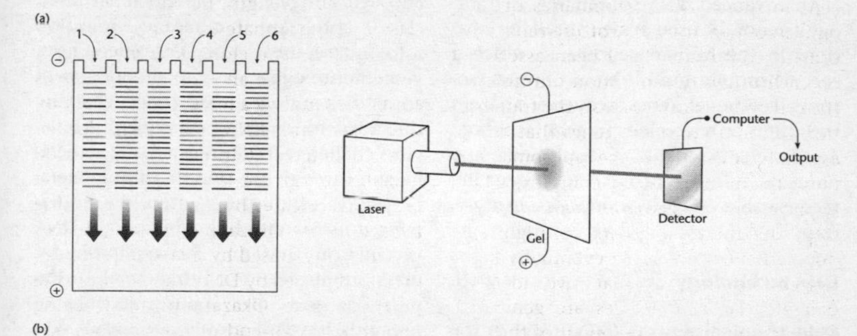

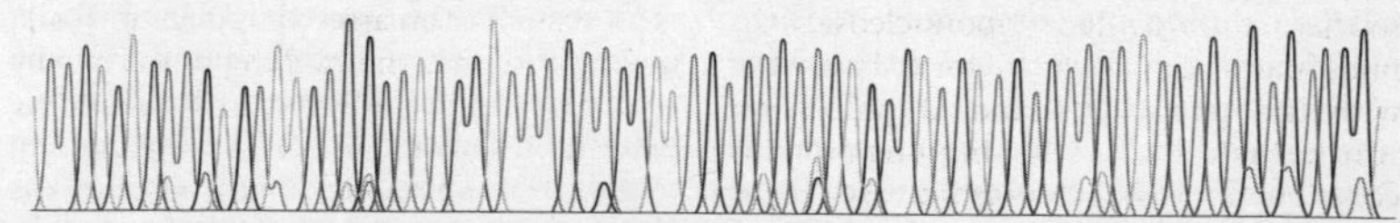

FIG 67. (a) The principles of automated DNA SEQUENCING (see entry for explanation). (b) An example of the DNA sequence output, in which the dye-specific (and therefore base-specific) intensity profiles are given for a sequence of cDNA for the human polyhomeotic gene, *PHC3*. From an original illustration in a paper by E. Tonkin, D.M. Hagan, W. Li and T. Strachan (2002), reproduced with the authors' kind permission.

gel (SOUTHERN BLOTTING) or capillary electrophoresis generates a ladder of DNA sequences of different lengths (see Fig. 67). A laser beam is focused at a specific constant region of the gel and as individual DNA fragments migrate past this the laser causes the dye to fluoresce. Maximum fluorescence occurs at different wavelengths for the four dyes. Either, fibre-optical techniques and flashes of white light are used to identify bases and 'read off' their sequence, or else bases are distinguished by bursts of fluorescent colours, a different colour for each base. High-speed charge-coupled device cameras are used to record the labelled bases. The DNA sequence complementary to the single-stranded DNA template used is generated by the computer software and the position of a mutation can be highlighted, again by using appropriate software. Such are the ease of use, increased throughput and accuracy of this technique that capillary sequencers, where DNA samples migrate through very thin long glass capillary tubes filled with gel, can sequence ~1 Mb per day.

An advanced DNA sequencing technology can now be used to identify all the locations in the human genome to which a specific protein binds (see DNA-BINDING PROTEINS). For novel genes, sequence analysis will often provide clues to gene function: motifs characteristic of transmembrane domains, tyrosine kinases, etc., are readily recognizable. See DNA PROFILING, GENOME MAPPING.

DNA technology See GENETIC MODIFICATION.

DNA topoisomerases (DTs) Any of the enzymes affecting the topology of DNA, e.g., coiling or relaxing it, and preventing tangling during DNA REPLICATION. All such modifications involve transient breakage of DNA strands, topoisomerase I producing single-strand nicks while topoisomerase II forms transient double-strand nicks. These breaks allow DNA to rotate into a relaxed molecule, and they finish by rejoining the strands. DNA gyrase is an example.

Inhibitors of DNA topoisomerases can lead to enhanced single- or double-strand breaks (DSBs) in DNA (see CHECKPOINTS, DNA REPAIR MECHANISMS, PROGERIAS).

DNA vaccines Although DNA can be used to raise immune responses against pathogenic proteins, those microbes with outer capsids that are composed of polysaccharides are more effectively countered by polysaccharide-bases subunit VACCINES. However, DNA vaccines offer a novel way of raising an adaptive immune response. For unknown reasons, injection of DNA into muscle leads to its expression and elicits antibody and T-cell responses to the protein encoded by the DNA. This was discovered when bacterial plasmids encoding proteins used in gene therapy elicited immune responses to the encoded proteins. DNA vaccines usually comprise a bacterial plasmid in which the expression unit is composed of promoter/enhancer sequences and downstream antigen-encoding and polyadenylation sequences required for plasmid amplification and selection. Host cells take up the foreign DNA, express the transgene, produce the encoded antigen and present it on MHC class I and class II molecules. Injection into mouse muscle of DNA encoding a viral immunogen leads to production of antibodies and cytotoxic T cells enabling the mice to reject a subsequent whole-virus challenge. DNA injected by a projectile gun through the skin on minute metal projectiles ('biolistics') enters the underlying muscle, affording a method of DNA vaccination that may be applicable for mass immunization. Intradermal, intramuscular and intravenous injections of plasmids have all proved successful, skin and mucous membranes being especially effective on account of their high densities of DENDRITIC CELLS. Bacterial DNA is also a potent inducer of innate immunity, the TOLL-LIKE RECEPTOR TLR9 recognizing a specific pattern in bacterial DNA. The first human trial of a DNA vaccine designed to prevent H5N1 avian INFLUENZA infection began in December 2006.

DNA-PK DNA-dependent protein kinase.

docosahexaenoic acid (DHA) See OMEGA-3 FATTY ACIDS.

dominant hemisphere See CEREBRAL DOMINANCE.

dominant-negative mutations Mutations in which a mutant allele in the heterozygous condition results in the loss of protein activity/function not only through its own malfunction but also by its interference with the function of the normal gene product of the normal allele. They are particularly common in genes for proteins that are dimers or multimers, such as structural proteins (e.g., collagens); but mutant *p53* alleles are also dominant-negative mutants.

dopamine (3,4-dihydroxyphenylethylamine, DA) A generally inhibitory monoamine NEUROTRANSMITTER with neuromodulatory roles, orchestrating motor behaviour and reward-driven learning. Derived from tyrosine, and a precursor in noradrenaline and adrenaline synthesis, it is inactivated by catechol-*O*-methyl transferase (COMT). Neurons that secrete or are stimulated by DA are termed *dopaminergic*. They are scattered throughout the central nervous system, but two groups of midbrain dopaminergic cells form DIFFUSE MODULATORY SYSTEMS of the BRAIN (see Fig. 30): the SUBSTANTIA NIGRA and the VENTRAL TEGMENTAL AREA (VTA). Degeneration of the substantia nigra causes depletion of dopamine, leading to PARKINSON'S DISEASE (treatment with L-dopa, which is converted to DA, activates DA receptors and alleviates disease symptoms). Dopamine and glutamate produce opposing physiological effects (see GLUTAMATE entry). DA potently augments the drive to obtain a rewarding stimulus (i.e., it increases the 'wanting' of a particular food, or drug; see REINFORCEMENT AND REWARD). Many pleasurable or addictive drugs (e.g., nicotine, morphine) produce their effects in part by initiating dopamine release in several brain regions. Destruction of ascending DA axons reduces food-seeking behaviour, and a variety of addictive drugs (heroin, nicotine and cocaine) enhance its efficiency, leading to its release in the nucleus accumbens. Pharmacological evidence links DA to reinforcement behaviour and self-stimulation by brain electrodes, and DA agonists such as amphetamines tend to increase the rate of self-stimulation (see PSYCHOACTIVE DRUGS, Fig. 144). DA release can produce a psychotic state in normal subjects. The mesocorticolimbic dopamine system is a common thread linking reinforcement, motivation for feeding (see APPETITE), and some kinds of drug addiction. Inactivation is by diffusion and reuptake into the presynaptic membrane by a high-affinity Na^+/Cl^--dependent dopamine transporter (DAT). Neurons releasing DA into the hypothalamic–pituitary portal system lack DAT. Competitive inhibition of DAT by amphetamines and cocaine results in potentiation of the effects of DA at the synapse, and may explain the reinforcing properties of these drugs. Inactivation of dopamine at the synapse is not dependent on enzyme degradation, although homovanillic acid is the main metabolite in primates. This cannot undergo reuptake, which requires action in turn by CATECHOL-O-METHYLTRANSFERASE and MONOAMINE OXIDASE present in synaptic membranes.

The dopaminergic neuronal system is active at waking levels throughout all states of SLEEP.

DA's usual effect is to open K^+ channels, the five receptors linked to these being of two subtypes D_1-like (D_1 and D_5) and D_2-like (D_2, D_3 and D_4). All are METABOTROPIC RECEPTORS, having the seven transmembrane-spanning domains characteristic of G-PROTEIN-COUPLED RECEPTORS. On binding DA they either activate adenylyl cyclase (D_1-like) or inhibit it (D_2-like). A dopamine-based gating mechanism has been suggested to account for the ability of the PREFRONTAL CORTEX to maintain neural representations of behavioural goals (mainly D_1 receptor-activated) while simultaneously destabilizing and updating them (mainly D_2 receptor-activated; see BISTABILITY).

D_1-like receptors (D_1 and D_5) respond to low DA levels, and are mainly involved in postsynaptic inhibition, most neuroleptic

drugs being antagonists of DA at these receptors and blocking its action (although this is not the reason for their antipsychotic action). These are coupled to G_s, and by activating D_1 receptors (which are extrasynaptic and postsynaptic), dopamine activates adenylyl cyclase and thereby the activity of PKA (see GLUTAMATE, Fig. 84), leading to phosphorylation of key substrate proteins, such as DARPP-32, leading in turn to inhibition in PPI and so decreasing dephosphorylation of its physiological substrates, one of which, indirectly, is PKA.

D_2-like receptors are largely synaptic, responding to only high levels of DA which occur only during phasic DA bursts. They are linked to G_i (inhibiting adenylyl cyclase) and involved in presynaptic and postsynaptic inhibition. The main subtype in the brain, they are involved in most of the DA effects currently understood. D_2 receptors occur in the LIMBIC SYSTEM (involved in mood and emotional stability) and in the BASAL GANGLIA (involved in the control of movement). Neuroleptic drugs blockade the D_2 receptors, which seems to underlie their antipsychotic action, although side-effects include some movement disorders.

D_3 and D_4 receptors are less abundant in the brain, but are found in the limbic system and may be involved in cognition and emotion. The D_3 receptors are presynaptic AUTORECEPTORS largely expressed in the NUCLEUS ACCUMBENS of the striatum. By closing presynaptic Ca^{2+} channels, they reduce dopamine release.

Dopamine can cause release of ENDOCANNABINOIDS. See SCHIZOPHRENIA.

dopaminergic Of nerve fibres that secrete, or are stimulated by, DOPAMINE.

dorsal The surface (posterior) in which the NEURAL TUBE forms, lying closest to the eventual spinal cord. Contrast VENTRAL.

dorsal aortae When the pharyngeal arches form, each arch receives its own artery. The AORTIC ARCHES (see Fig. 13) terminate in left and right dorsal aortae; but most of the right dorsal aorta, other than part of the right subclavian artery, disappears. See AORTA.

dorsal columns That part of the white matter of the SPINAL CORD (see Fig. 163) dorsal to the grey matter, comprising the FASCICULUS CUNEATUS and FASCICULUS GRACILIS. See DORSAL COLUMN–MEDIAL LEMNISCAL SYSTEM.

dorsal column–medial lemniscal system One of the main ascending pathways of the SOMATIC SENSORY SYSTEM in the SPINAL CORD (see Fig. 163), conveying proprioceptive, light touch and vibrational information from the skin and viscera via large diameter axons from muscle and joint receptors, and mechanoreceptors, to synapse in the *dorsal column nuclei* in the MEDULLA OBLONGATA. The cuneate nuclei receive input from cervical segments C1–C8 and the gracile nuclei receive inputs from thoracic segments T7–T12. Second-order fibres leave these nuclei as a discrete bundle (the MEDIAL LEMNISCUS), in the pons and midbrain, running anteriorly and decussating before reaching the ventroposterolateral nuclei (VPN) of the thalamus, whence they project to the PRIMARY SOMATOSENSORY CORTEX. Its general properties are: the considerable strength of its synaptic connections; the close similarity between the properties of its neurons and those of the sensory receptors supplying them, conserving dynamic features of the stimuli; and the somatotopic mapping which preserves localization at every stage in the pathway. See RELAY NUCLEI for enhancement of responses to tactile stimuli; SPINOTHALAMIC TRACT.

dorsal horn See SPINAL CORD.

dorsal ramus See SPINAL NERVES.

dorsal root ganglia See SPINAL NERVES.

dorsal visual pathway See EXTRASTRIATE CORTEX.

dorsiflexion Bending of the foot in the direction of its upper surface.

dosage compensation The amount of mRNA transcribed from a gene is generally directly proportional to the COPY NUMBER of the gene in the cell (the 'gene dosage effect'). Despite the presence of two X CHROMOSOMES in females, and only one in males, the X chromosome's housekeeping genes are expressed to approximately equal extents per cell in both sexes. This is because, in females, only one X chromosome is transcriptionally active, as a result of X CHROMOSOME INACTIVATION. See GENE DUPLICATION, SEX CHROMOSOMES.

dot-blot hybridization See DNA PROBES.

double-blind study A study in which neither the researchers nor the subjects know whether a given individual belongs to the control group (e.g., those receiving a PLACEBO) or to the experimental group until all the data are recorded. See DRUGS, RANDOMIZED CONTROLLED TRIALS.

double circulation A blood circulation which involves complete separation of deoxygenated and oxygenated blood by means of a pulmonary circuit and a systemic circuit (see HEART). This is found in the adult circulation; but the foetal circulation does not quite achieve this (see FOETUS).

double minutes (double minute chromosomes, homogenously staining regions) See ONCOGENE.

double-opponent cells Cells of the PRIMARY VISUAL CORTEX involved in processing colour. See COLOUR-OPPONENT CELLS.

double strand breaks (DSBs) See DNA REPAIR MECHANISMS.

down-regulation A cell's response to external signal molecules (or its absence) often involves repression (inhibition) of transcription of a particular gene, or gene set, or translation of their transcripts. This may, e.g., result in decrease in the amount of a particular receptor on the plasma membrane. These responses are termed 'down-regulation' of the product. See GENE EXPRESSION, TRANSCRIPTION FACTORS.

Down syndrome (Down's syndrome, DS) A disorder due to TRISOMY of chromosome 21 (or to translocation of parts of it). The overall incidence, after adjustment for the impact of antenatal screening (see PRENATAL DIAGNOSIS) is ~1:650–1:700. In the UK, ~50% of cases are detected prenatally. The rate of DS is at least 1:270 births for mothers over 35 years. The most common finding in the newborn is severe hypotonia (decrease or loss of muscle tone), coupled with small ears and a protruding tongue. Single palmar creases ('simian palm') are found in 50% of Down children, contrasting with 2–3% of the general population, and CONGENITAL HEART DEFECTS are present in 40–45% of Down babies, the three commonest being atrioventricular canal defects, ventricular septal defects and patent ductus arteriosus. Affected individuals show a wide range of intellectual ability (IQ range 25–75 in children; mean adult IQ 40–45). Most Down children are happy, very affectionate, and have relatively advanced social skills. Adult height is ~150 cm; and most adults develop ALZHEIMER'S DISEASE in later life: the *APP* gene is located on chromosome 21, those with DS producing higher levels of Aβ peptide from birth, amyloid deposits sometimes being found in their brains as early as 12 years. The additional chromosome 21 is of maternal origin in >90% of cases, usually as a result of NONDISJUNCTION, although ROBERTSONIAN TRANSLOCATIONS account for ~4% of all cases (in one-third of these one of the parents is a carrier). Children with MOSAICISM (1–2% of Down babies) are often less severely affected than are those with the full syndrome. The recurrence risk of the trisomy is related to maternal age, and usually of the order 1:200–1:100, a similar figure applying for the translocation if neither parent is a carrier. Female carriers of either a 13q21q or 14q21q translocation have a risk of ~10% of having a Down baby while for male carriers the risk is 1–3%.

dreaming, dreams Activity characterized by a set of related, cognitive features deficient compared with the waking state; notably: diminished self-awareness, diminished reality testing, poor memory, defective logic and the inability to maintain directed thought, decrease in capacity to orientate (times, changes and persons changing without notice) and associated increase in tendency to confabulate, giving cognitive defects an emotional salience regardless of illogicality. The deactivation of the PREFRONTAL CORTEX during dream states probably has much to do with the lack of temporal contiguity and inability to 'stay on task' that are characteristic of dreams. It is generally believed that the reason dreams are so perceptually intense, so emotional and hyperassociative, is that the brain regions supporting these functions are more active at these times; and similarly that the reason we lose track of time, place and person and cannot think critically is that the brain regions supporting these cognitive functions are less active. It is therefore of interest that the brain regions linked with these diminished psychological functions during REM sleep (see SLEEP) are known to be deprived of noradrenaline and serotonin – transmitters known to be necessary for attention, learning and memory (and, by implication, for orientation and active reasoning).

droplet infection Infection by agents carried in aerosols, as caused by coughing and sneezing.

drug addiction See ADDICTION.

drug dependence One component of drug ADDICTION, in which the biological changes brought about by a drug are such that normal functioning is only possible when the drug is present. It requires that an individual has experienced at least three of the following symptoms in the previous 12 months: tolerance for large doses, withdrawal reactions, loss of control over the substance's use, efforts to stop or cut this down, and continued use despite resulting physical or psychological problems. Dependence is two-fold: psychological dependence is a compulsion to use a drug repeatedly for its enjoyable effects; but physiological dependence (to a non-specific central nervous system depressant) is typified by a period of rebound hyperexcitability following a sedative's withdrawal, while withdrawal of an amphetamine (psychostimulant) is typified by fatigue, deep sleep, emotional depression and increased appetite.

drug discovery See DRUGS.

drug resistance RNA viral genomes are relatively unstable, on account of the lack of 'proof-reading' by reverse transcriptase. A single mutation may afford drug resistance (e.g., to the NRTI lamivudine and some NNRTIS; see HIV), while for resistance to zidovudine and some protease inhibitors up to three mutations are needed. Emerging resistance is being found in antiviral therapy to chronic HEPATITIS B. See MULTIDRUG RESISTANCE.

drugs Any substance taken medicinally to help recovery from illness, or to relieve symptoms (e.g., pain in ANALGESICS); but in a non-technical sense usually implying a PSYCHOACTIVE DRUG. In the former sense, drugs are rarely effective in 100% of patients to whom they are prescribed (see DRUG TARGETING). Among commonly prescribed drugs, 15–35% of patients have inadequate or no response to BETA-BLOCKERS; 7–28% of patients have an inadequate or no response to ANGIOTENSIN-converting enzyme inhibitors; 9–23% of patients respond inadequately to selective SEROTONIN re-uptake inhibitors; and 20–50% of patients respond inadequately to tricyclic ANTIDEPRESSANTS. Much of this individual variation in response depends on variations in the absorption, distribution, metabolism and elimination of drugs and variability in target drug receptors (see PROTEOMICS). These are probably accounted for by combinations of common polymorphisms in a limited number of genes, a more tractable goal of the HUMAN GENOME PROJECT than discovering genetic susceptibility factors to common diseases, and drug companies have invested heavily in GENOMICS to try to discover new drug targets. VITAMINS are on the borderline between drugs and components of a HEALTHY DIET. See ANTIBIOTICS, CHEMOTHERAPY, DOUBLE-BLIND STUDY, ORPHAN DRUGS, PLACEBO, RANDOMIZED CONTROLLED TRIALS (and see also http://www.nature.com/drugdisc).

drug targeting That an individual's particular genetic characteristics can influence drug effectiveness was revealed over 40 years ago when a gene variant leading to slowed metabolism of the anaesthetic succinyl-choline was found to explain why some patients awoke normally but remained paralysed after operation. Roughly 1:3,500 people are homozygous for this variant, placing them at risk of this side effect. Since then, a small but growing number of differences in drug metabolism have been explained by genetics, helping explain the variable effectiveness of certain drugs and the appeal of so-called 'race-related drugs' is growing. In 2004, 124 different genes were linked to resistance to four leukaemia drugs. There has been some hype and some commercial dishonesty in the drug-targeting story so far. Expanding knowledge of the human genome may eventually enable better targeting of drug treatments to those most likely to respond appropriately. Enhancement in effectiveness of drugs such as codeine, warfarin and mercaptopurine is caused by SINGLE NUCLEOTIDE POLYMORPHISMS in cytochrome-encoding genes (cytochrome P-450 2DS and cytochrome P-450 2C9) and the gene for thiopurine 5-methyltransferase. Genetic variants in the gene for vitamin K epoxide reductase control the dose response to warfarin. Polymorphism in the epidermal growth factor receptor gene confers susceptibility to the tyrosine kinase inhibitor gefitinib in non-small-cell lung cancer. But personalized medicine presents formidable challenges, especially with complex diseases, and remains a long way off. See TRYPANOSOMES, Fig. 149.

DSBs Double-stranded breaks in DNA; the most lethal type of DNA damage. See DNA REPAIR MECHANISMS.

DTI Diffusion tensor imaging. A modification of MAGNETIC RESONANCE IMAGING providing information on directionality and regularity of myelinated fibre tracts.

Duchenne muscular dystrophy (DMD) See MUSCULAR DYSTROPHIES.

ductus arteriosus Developing from the 7th left AORTIC ARCH (see Fig. 13b), a remnant of (i.e., homologous with) the ancient vascular connection between the pulmonary artery and systemic aorta, persisting in amniote embryos, including human, and serving as a shunt for most of the blood between the pulmonary artery and the descending aorta. Because of the high vascular resistance of the foetal lungs, only ~20% of blood leaving the right ventricle perfuses the lungs. Because of this, blood in the ascending aorta has a higher oxygen content than that in the descending aorta, resulting in comparatively good oxygenation of the foetal brain. Patent ductus arteriosus is a common feature of DOWN SYNDROME. See DUCTUS VENOSUS, FORAMEN OVALE, PLACENTA.

ductus deferens See VAS DEFERENS.

ductus venosus The opening of the umbilical vein into the foetal inferior vena cava, enabling ~80% of blood in the umbilical veins to bypass the foetal liver (the remaining ~20% travelling to the liver via the hepatic portal vein). As a result, oxygenated blood in the umbilical vein is mixed with the deoxygenated blood returning from the lower parts of the foetus via the inferior vena cava. However, the FORAMEN OVALE and crista dividens prevent further dilution of this blood with deoxygenated blood returning to the right atrium in the superior vena cava, causing most of the blood in the inferior vena cava to pass directly into the left side of the heart. See DUCTUS ARTERIOSUS, PLACENTA.

Duffy blood group The Duffy antigen was first detected serologically, and three main alleles of the Duffy gene (*FY*), A, B and O, encode a transmembrane chemokine RECEPTOR (Duffy antigen receptor for chemokines, DARC) expressed in many tissues but used by the malarial parasite *Plasmodium vivax* to enter red blood cells. A mutation in the promoter region (the *FY*O* allele) disrupts protein expression and confers protection against this parasite and is prevalent, even fixed, in many sub-Saharan African populations but virtually absent outside Africa (see MALARIA, NATURAL SELECTION, RACE). The near-fixation of the *FY*A* allele in some Asian populations raises the possibility that it, too, may have spread by selection, although more complex selective influences are suspected than in the *FY*O* case in Africa, possibly involving adjacent genomic regions.

duodenal gland See BRUNNER'S GLAND.

duodenum The most rostral part of the SMALL INTESTINE, ~25 cm in length and completing an almost 180° arc within the abdominal cavity. Its mucosa and submucosa (see ALIMENTARY CANAL, Fig. 4) form a series of *circular folds* bearing numerous villi, between which lie CRYPTS OF LIEBERKÜHN, and whose simple columnar epithelium contains ENTEROCYTES, goblet cells, PANETH CELLS and ENTEROENDOCRINE CELLS. The submucosa contains tubular coiled Brunner's glands (duodenal glands) opening into the crypts. Its mucosa is resistant to the effects of bile, but is unable to tolerate low pH. Consequently, gastric emptying which is too rapid may result in formation of duodenal (peptic) ulcers, which may also result from too much inhibitory stimulation of duodenal glands by the sympathetic system; while regurgitation of duodenal contents may result in gastric ulcers (the stomach mucosa not being resistant to bile). The presence of fatty acids or monoglycerides in the duodenum causes increased contractility of the pyloric sphincter, reducing entry of chyme into the duodenum (see STOMACH). The hormones CCK and GIP, produced by the duodenum, both delay gastric emptying.

duplicate genes, duplication See GENE DUPLICATION, SEGMENTAL DUPLICATIONS.

dura mater See MENINGES.

dwarfism See ACHONDROPLASIA, DYSPLASIAS, LARON DWARFISM.

dynactin See DYNEINS.

dynamic mutations Mutations involving trinucleotide (triplet) repeats that can change in size from one generation to the next, sometimes causing variability in phenotype and/or age of onset of an associated disorder. See TRIPLET REPEATS.

dynamins These GTPases (see GUANOSINE TRIPHOSPHATE) are involved in the final stage of budding of invaginating VESICLES from their membrane of origin. Dynamin forms a spiral around the tubular neck of the vesicle, extending lengthwise and constricting it as GTP is hydrolysed, pinching the vesicle off. Three different dynamin genes occur in mammals: Dynamin I is a neuron-specific GTPase required for endocytic recycling of synaptic vesicle membranes; dynamin II is expressed in most cell types, and dynamin III is expressed in the testis, heart, brain and lung tissues.

dyneins A group of MOTOR PROTEINS involved in a variety of cellular processes involving microtubules. The largest of the motor proteins, dyneins can be divided into two groups: the more ancient cytoplasmic dyneins (typically heavy-chain dimers with two large motor domains as 'heads'); and axonemal dyneins, also called ciliary or flagellar dyneins (see CILIUM), which include heterodimers and heterotrimers, with two or three motor heads. The latter are specialized for the rapid sliding of microtubules during ciliary movements. The base of the molecule binds tightly and ATP-independently to an A microtubule while the three large globular heads each have an ATP-dependent binding site for a B microtubule. On hydrolysing the ATP, the heads move towards the minus end of the B microtubule, producing the sliding force between adjacent ciliary or flagellar microtubules. Dynein molecules form bridges between neighbouring doublet microtubules of the axoneme. As the dyneins attempt to walk along adjacent microtubules, other linking proteins prevent this and the sliding is converted into a bending motion instead. Cytoplasmic dyneins are involved in behaviour of the mitotic and meiotic spindles (see MITOSIS) and in transport of a variety of cellular cargos rather rapidly (~14 μm.sec^{-1}), by 'walking' along cytoskeletal microtubules towards their minus ends (hence minus-end directed motors) which are usually oriented towards the centre of the cell. They must be activated by binding of *dynactin*, another multisubunit protein that is essential for mitosis. Dynactin includes components that weakly bind microtubules, others that bind dynein itself, and others that form small actin-like filaments composed of the actin-like protein Arp1. The latter may mediate attachment of the whole complex to membrane-bound organelles via a network of spectrin and ankyrin similar to that of the RED BLOOD CELL membrane. Dynactin may also regulate the activity of dynein. Compare KINESINS, MYOSIN.

dynorphin An endogenous ligand of (especially) κ OPIOID RECEPTORS (see CREB, Fig. 60) encoded by the dynorphin precursor gene (which also encodes leu-ENKEPHALIN). Increased dynorphin expression mediated by CREB seems to be linked with aversive or depressive-like effects, such as those accompanying drug withdrawal (see NUCLEUS ACCUMBENS).

dyskinesias Hyperkinetic disorders, resembling the motor dysfunction of HUNTINGTON'S DISEASE, which occur in 80% of PARKINSON DISEASE patients.

dyslexia Developmental dyslexia (DD), which is increasing in prevalence (~5–17% of the population), is characterized as a phonological disorder in which sensory and intellectual functions are intact, motivation and opportunities for acquiring reading ability are normal. It is diagnosed when children or adults fail to learn to read fluently despite normal intelligence, instruction and opportunities to do so. Evidence indicates the problem lies in the brain's ability to process and manipulate 'phonemes' – speech sounds enabling us to distinguish words, as in the initial consonants of 'pet' and 'bet'. Despite many recent functional imaging studies, controversy surrounds the nature of the cortical structures involved in DD. However, parieto-temporal language areas on the two sides are symmetrical in the dyslexic brain, lacking the normal left-side advantage. Development of the visual magnocellular system is impaired in dyslexics, development of the magnocellular layers of the lateral geniculate nucleus being abnormal (see MAGNOCELLULAR PATHWAY). Some work indicates that two pathways are involved in reading: one used by inexperienced readers and the other, faster, pathway which predominates in more skilled readers. Both involve three key areas in the left cerebral hemisphere: BROCA'S AREA, the parieto-temporal cortex and the occipito-temporal cortex; but novice readers appear to use the parieto-temporal region to dismantle words into phonemes in step-by-step phonological analysis, whereas experienced readers seem to rely more on the occipito-temporal region to recognize words in their entireties.

dyslipidaemia Elevated serum CHOLESTEROL or triglyceride levels. High levels of LDL and/or of VLDL, and low levels of HDL are features (see LIPOPROTEINS).

dysmenorrhea Painful menstruation.

dysplasias **1**. A cancerous tissue in which cells are aberrant cytologically. Most are caused by single-gene defects and associated with high recurrence risks for siblings and/or offspring. Some occur as BIRTH DEFECTS. Dysplastic growths exhibit alterations in the normal numerical ratios of cell types. It is considered to be a transitional state between completely benign growths and those that are premalignant. **2**. Abnormal skeletal growth, in which FGFs and FGFRs are involved; and *SOX9* in campomelic dysplasia (see SOX GENES). *FGFR1* and *FGFR2* are coexpressed in prebone and precartilage regions, including craniofacial structures, FGFR2 increasing proliferation and promoting osteogenic differentiation; *FGFR3* is expressed in cartilage growth plates of long bones and the occipital region, currently with unknown function. Mutations in these *FGFRs* have been linked to specific forms of craniosynostosis (FGFR1, FGFR2) and several forms of dwarfism (FGFR3; see ACHONDROPLASIA). In the skeletal dysplasia *thanatophoric dysplasia* (due to mutations in *FGFR3*) most parts of the body are affected. In *ectodermal* *dysplasia*, widely dispersed ectodermal tissues (hair, teeth, skin, nails) are involved.

dyspraxia *Developmental dyspraxia* covers a spectrum of developmental psychological disorders affecting the initiation, organization and performance of action even though there is no sensory or motor impairment. It may give rise to problems in learning basic movement patterns, development of a desired writing speed, acquisition of graphemes (letters of the alphabet) and numbers, establishment of the correct pencil grip, and an aching hand while writing. The motor coordination involved in walking, running and jumping are affected as well as judgement of distance. The condition often coexists with DYSLEXIA, dyscalcula (problems with mathematics), ATTENTION DEFICIT HYPERACTIVITY DISORDER, expressive language disorder (difficulties with verbal expression) and Asperger's syndrome.

dystonias These are HYPOKINESIAS seen in a variety of conditions, e.g., terminal HUNTINGTON'S DISEASE, strokes, and the effect of treatment with dopamine D2 receptor antagonists, such as metoclopramide.

dystrophin The human dystrophin gene comprises a total of >79 exons distributed over ~2.4 Mb of DNA located on chromosome Xp21. At least seven alternative PROMOTERS can be used, three of them located near the conventional transcription start site and used separately in cerebral cortex, striated muscle and Purkinje cells of the cerebellum. Their use results in large ISOFORMS (427 kDa) differing in their extreme N-terminal amino acid sequence as a result of using different parts of exon 1. At least four other alternative internal promoters can be used, generating smaller isoforms. See MUSCULAR DYSTROPHIES, NON-CODING DNA.

E

E2A (TCF3) A transcription factor involved in B-CELL differentiation and often mutated in acute lymphoblastic LEUKAEMIAS.

E2F A transcription factor (actually three, E2F1, E2F2 and E2F3) promoting expression of genes needed for G1/S progression in the cell cycle (see G1), including those encoding G1/S cyclins, S-cyclins and proteins required for DNA synthesis. It is transcribed in response to Myc protein (see MYC), and its activity is promoted when RETINOBLASTOMA PROTEIN (pRB) is inactivated by phosphorylation by G1-CDK in response to extracellular mitogens. Elevated levels of E2F indicate that pRB function has gone awry, E2F then driving activation of a number of genes encoding pro-apoptotic proteins (e.g., CASPASES 3, 7, 8 and 9 and Apaf-1).

eardrum (tympanic membrane) Thin semitransparent partition between the external auditory canal and the middle EAR (see Fig. 68). Its connective tissue interior is lined by cuboidal epithelium, the whole being covered in epidermis. It is aperiodic, its vibrations being conducted by ear ossicles to the oval window of the COCHLEA.

early-response genes See IMMEDIATE EARLY GENES.

ear ossicles (auditory ossicles) The incus, malleus and stapes (see EARS, Fig. 68; SKULL). The stapes attaches to the oval window of the inner ear; the incus and malleus are peculiar to mammals. The incus and malleus develop from Meckel's cartilage of the 1st PHARYNGEAL ARCH. The incus articulates with the stapes and with the malleus, which is attached to the tympanic membrane. Ligaments and skeletal muscles attach to the ossicles. The tensor tympani muscle, innervated by the mandibular branch of the trigeminal nerve (CNV), limits movement and increases tension on the eardrum, preventing damage to the inner ear from loud noises. The stapedius muscle, innervated by the facial nerve (CNVII), damps large vibrations of the stapes in these conditions.

ears The organs of HEARING and BALANCE. Although the ear forms one anatomical unit in the adult, it develops from three distinct components (see Fig. 68): (i) the sound-collecting external ear (the pinna and external auditory meatus); (ii) the middle ear, conducting vibrations from the external to the inner ear via the EAR OSSICLES; and (iii) the inner ear, transducing pressure vibrations into nerve impulses and registering changes in equilibrium. The 1st pharyngeal cleft and pouch give rise to the external auditory meatus and the middle ear cavity, respectively (see PHARYNGEAL POUCHES, Fig. 134).

By ~22 days of development, ectodermal thickenings on each side of the rhombencephalon, the otic placodes, invaginate to form *otic vesicles* (otocysts), each subsequently developing into a membranous labyrinth comprising ventral and dorsal components. The ventral part gives rise to the saccule and cochlear duct (see COCHLEA) while the dorsal component forms the utricle, semicircular canals and endolymphatic duct. See HEARING.

The semicircular canals also arise during

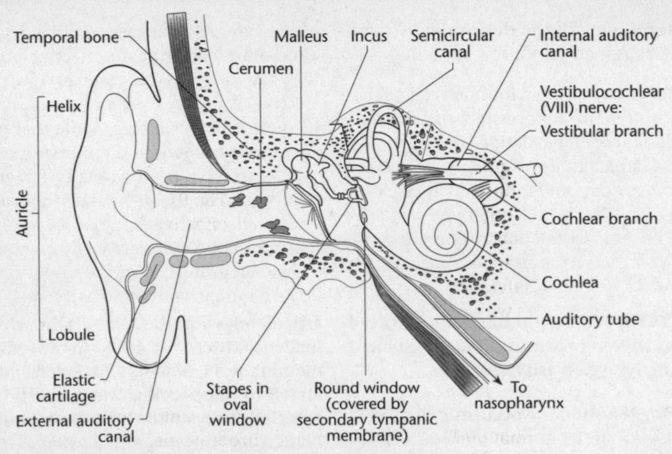

FIG 68. Diagram showing the three principal regions of the right EAR in frontal section. From *Principles of Anatomy and Physiology*, G.J. Tortora & S.R. Grabowski, © 2000. Reproduced with permission of John Wiley & Sons, Inc.

the 6th week, originally as flattened outpockets of the utricular portion of the otic vesicle. Functioning in BALANCE (esp. dynamic equilibrium), the canals lie at right angles to one another in three planes, two vertically and one horizontally. At one end of each canal is a swollen ampulla, each containing a crista (see VESTIBULAR APPARATUS).

eating Ingestion of food via the oral cavity. See ANOREXIA NERVOSA, BULIMIA NERVOSA, FEEDING CENTRE.

ebola viruses Single-stranded RNA viruses, with long filamentous morphology. Related to Marburg virus, it is the cause of Ebola haemorrhagic fever (in which death occurs within a few days), several outbreaks of which have occurred since the millennium in Gabon and the Republic of Congo. Human outbreaks consist of multiple simultaneous epidemics by different strains, each resulting from handling of a distinct gorilla, chimpanzee or duiker carcass. Recent outbreaks have devastated populations of the two primates, and fruit bats eaten by people are believed to act as reservoirs of the virus.

E-boxes DNA sequences within certain promoters, binding proteins belonging to the BASIC HELIX–LOOP–HELIX MOTIF (bHLH) family of transcription factors. The canonical E-box sequence for bHLH leucine zipper factors (bHLH-Zip) is CA[T/C]GTG. See CIRCADIAN CLOCKS.

eccrine sweat glands See SWEAT, SWEAT GLANDS.

ECG See ELECTROCARDIOGRAM.

eclampsia An exaggerated form of PRE-ECLAMSPIA, which has progressed to include convulsions, and sometimes coma.

E. coli See ESCHERICHIA COLI.

ecological goods and services Benefits derived from ecosystems, including water supply and regulation, erosion control, climate regulation (moderating weather extremes and mitigating such impacts as floods and droughts), production of food and raw materials, detoxification and decomposition of wastes, recycling and movement of nutrients, control of agricultural pests, pollination of crops and natural vegetation, and recreational opportunities. These are all threatened by human overpopulation and overconsumption.

ecstasy See MDMA.

ectoderm One of the three primary GERM LAYERS. See Appendix VI, ECTODERMIN.

ectodermin A protein found in cell nuclei and required in suppression of aberrant mesodermal differentiation in the ectoderm. If depleted of this protein, mesodermal and endodermal induction appears in the region that would otherwise become ectoderm. It is expressed abundantly in colorectal cancer cells and appears to have a causal role in the proliferation of cancer cells, where it interferes with TGF-β signalling.

ectomorph See BODY MASS, SIZE AND SHAPE.

ectopic Of a structure occurring in a location other than its normal one.

edema See OEDEMA.

EEG See ELECTROENCEPHALOGRAM.

effective population size, N_e The number of individuals of a given generation that contributes gametes to the subsequent generations. This abstract quantity depends on the breeding sex ratio, the number of offspring per individual, the type of mating system and the generation time per sex. It can be thought of as the size of an idealized (Wright–Fisher) population that experiences the same amount of genetic drift as the particular one under study. It measures the magnitude of genetic drift. The common simplifying assumptions in population genetics that population sizes are large, and that random mating is taking place, are often violated in real populations (see DEME, POPULATION). Not all individuals in a population might be of reproductive age; the sex ratio and population structure might be uneven. The 'effective population size' is a mathematical adjustment, developed by Sewall Wright, of the size of a census population to its equivalent size in a 'well-behaved' population. This can then be used as input into standard models, e.g., of the genetic drift experienced by different populations. In smaller populations, polymorphisms will be expected to reach fixation by drift more quickly than in larger populations. Early modern humans differed from chimpanzees, gorillas and the orang-utan in having an effective population size of ~10,000 compared with an effective size two to four times larger in the other three primates, while that of the ancestor of humans and chimpanzees also appears to have been similarly larger (see HOMINOIDEA, Fig. 95). It has been speculated that small effective population size, possibly associated with numerous expansions from small groups, was typical not only of modern humans but of many groups of the genus *Homo*.

Considering a single mating couple as a microcosm of a species with equal sex ratios, these two will have a total of four copies of each autosome, three copies of the X chromosome, two copies of mitochondrial DNA (mtDNA), only one of which will be inherited by the succeeding generation, and a single Y chromosome. So, given a 1:1 sex ratio, the N_e of the Y chromosome will be one-quarter that of the autosomes and one-third that of the X chromosome (assuming the reproductive variances of males and females are equal, as in the Wright–Fisher model). But because the Y chromosome is inherited paternally and mtDNA maternally, and because the X chromosome is inherited twice as often by females as by males, a higher reproductive variance of males than females reduces the N_e of the Y chromosome relative to that on mtDNA, the X chromosome and the autosomes, and increases the N_e of mtDNA and the X chromosome relative to that of the autosomes. But in cases of extreme male reproductive variance, it is possible for the N_e of the X chromosome to exceed that of the autosomes and for the N_e of the Y chromosome to reach as low as one-eighth that of the autosomes. This may explain why the Y chromosome displays such a high degree of population differentiation, although differences in generation times between the sexes also cause differences in their N_e values. It seems that, in humans, females have the shorter generation time, and this should lower the N_e of mtDNA relative to biparentally and paternally inherited loci, and the relative importance of these factors may differ between populations. See MAJOR HISTOCOMPATIBILITY COMPLEX, MOLECULAR CLOCK.

effective reproductive number (***R*~eff~**) The expected number of secondary cases caused by each infectious individual in a partially immune population. In well-mixed populations, $R_{eff} = sR_0$, where s is the fraction of the population that is susceptible and R_0 is the BASIC REPRODUCTIVE NUMBER. See EPIDEMIOLOGY.

effector cells Activated B CELLS and T CELLS of the immune system (see LYMPHOCYTES). Compare EFFECTORS.

effectors, effector organs Muscles and glands that are the targets of motor neurons.

efferent Leading, or carrying, away from a structure; used particularly of nerves, axons and blood vessels. Contrast AFFERENT.

efferent neurons A functional classification, referring to those neurons which project from one region of the nervous system to another, or to an effector organ. Those synapsing with skeletal muscles are MOTOR NEURONS. See MOTOR CORTEX.

egestion The removal of faeces from the ANUS (defecation). Distension of the rectal wall stimulates stretch receptors, initiating a defecation reflex. Sensory impulses travel along afferent fibres to the sacral spinal cord and motor impulses travel along parasympathetic nerves to the colon, rectum and anus. Contraction of longitudinal rectal muscles increases pressure within the rectum which, with voluntary contraction of the diaphragm and abdominal muscles and parasympathetic stimulation, open the internal sphincter.

EGF See EPIDERMAL GROWTH FACTOR.

egg cells Mature secondary OOCYTES, after ovulation.

eicosanoids A family of very potent but short-lived signalling molecules, derived from 20-carbon essential fatty acids, of which ARACHIDONIC ACID is the most common precursor. In response to hormonal or other stimuli, phospholipases A_2, C and D present in most mammalian cells, attack membrane phospholipids, releasing arachidonate from the middle carbon of glycerol. Enzymes in the smooth endoplasmic reticulum then convert arachidonate via PGH_2 to PROSTAGLANDINS and THROMBOXANES (see CYCLOOXYGENASE). Their synthesis is initiated in response to cell-specific stimuli, and over 16 types are known. See INFLAMMATION, PAIN.

ejaculation The culmination of the male sex act, associated with orgasm, in which semen is forced out of the urethra. The accumulation in the urethra of spermatozoa and secretions of the prostate gland and seminal vesicles (termed *emission*) is caused by sympathetic centres in the spinal cord (T12–L1) which are stimulated as sexual tension rises. The semen in the prostatic urethra produces sensory feedback along the pudendal nerves to the spinal cord. Integration of these impulses leads to both sympathetic and somatic motor output, the former causing constriction of the internal sphincter of the urinary bladder, the latter causing skeletal muscles of the urogenital diaphragm and the base of the PENIS to undergo rhythmic contractions forcing semen out as muscle tension increases throughout the body. SEMEN coagulates immediately after ejaculation through conversion of fibrinogen to fibrin; but fibrinolysis then occurs, releasing the sperm cells. See AROUSAL (2), EPIDIDYMIS.

elastase A protease component of the pancreatic juice, and released as proelastase which is converted to the active enzyme by TRYPSIN. It digests the elastin in connective tissues which holds meat together. See EMPHYSEMA.

elastic fibres See ELASTIN.

elastin A highly hydrophobic protein, the main component of elastic fibres and the dominant protein in the EXTRACELLULAR MATRIX of arteries (up to 50% of the artery's dry mass). It has a natural elastic recoil, enabling those tissues that are regularly deformed by pressure to regain their shapes (e.g., in skin and alveoli). Like COLLAGEN, it is rich in proline and glycine residues; but unlike collagen, it is not glycosylated and contains hydroxyproline but no hydroxylysine. Soluble *tropoelastin* is secreted into the extracellular space and assembled into elastic fibres close to the cell's plasma membrane, usually in cell-surface infoldings, after which it becomes highly cross-linked, forming an extensive network of fibres and sheets.

elbow, elbow joint The elbow joint is a compound hinge joint which acts as a third-class lever (see LEVERAGE, LEVERS, Fig. 113). It comprises the humeroulnar joint (between humerus and ulna) and the humeroradial joint (between humerus and radius), both of which are reinforced by ligaments. The shape of the trochlear notch and its association with the trochlea of the humerus limit movement of the joint to flexion and extension; but the rounded radial head rotates in the radial notch of the ulna and against the capitulum of the humerus, enabling pronation and supination of the humerus's head. The joint is surrounded by a joint capsule. When the flexors (biceps brachii) are extended, the short loop stretch reflex via 1a afferents and α motor neurons (see MUSCLE SPINDLES) causes them to produce tension by contracting. This requires relaxation of the antagonistic extensors (triceps brachii), the pathway for this *reciprocal inhibition* involving the same 1a afferents synapsing with inhibitory spinal cord interneurons which decrease the α motor neuron input to the extensors. If the flexors are sent motor commands to contract, the resulting stretch of the antagonistic extensors activates their stretch reflex, strongly resisting flexion. But the descending pathways activating the α motor neurons controlling the flexors also activate interneurons which inhibit these neurons supplying the extensors, overcoming the stretch reflex and permitting the action. This pathway is slower than the long cortical loop of the stretch reflex, and flexion and extension of the elbow joint also involves feedback from TENDON ORGANS.

electrical potential (voltage) Term used in the contexts of ACTION POTENTIAL and MEMBRANE POTENTIAL.

electrical synapses See SYNAPSES.

electrocardiogram (ECG) The summated record of the cardiac action potentials at a given time, detected by electrodes placed non-invasively on the body surface (compare ELECTROENCEPHALOGRAM). Each deflection in an ECG indicates an electrical event in the CARDIAC CYCLE that correlates with a subsequent mechanical event, being an extremely useful diagnostic tool in identifying a number of ARRHYTHMIAS (see for details) and other abnormalities.

A normal ECG (see Fig. 69) consists of a P wave, the result of action potentials causing depolarization of the atrial myocardium initiating atrial contraction; a QRS complex resulting from ventricular depolarization signalling onset of ventricular contraction; and a T wave, representing repolarization of the ventricles and preceding ventricular relaxation (repolarization of the atria cannot be detected because it takes place during the QRS complex). During the PQ (or PR) interval, which lasts ~0.16 seconds, the atria contract and start to relax and the ventricles begin to depolarize at its close. The QT interval lasts ~0.36 seconds, representing the approximate time for the ventricles to contract and start to relax.

electroencephalogram (EEG) Although a device for making electrical (voltage) recordings (electroencephalograms) from the surfaces of dog and rabbit brains had been constructed as early as 1875 by Richard Caton, the human EEG was first described in 1929 by the psychiatrist Hans Berger, and is of greatest use today in providing an ENDOPHENOTYPE that helps to diagnose some neurological conditions, notably EPILEPSY, and for researching into SLEEP (see Fig. 158). These electrophysiological patterns are highly heritable. Some two dozen electrodes are painlessly taped to the scalp in standard prescribed positions and connected to amplifiers, and small voltage fluctuations (μV) between

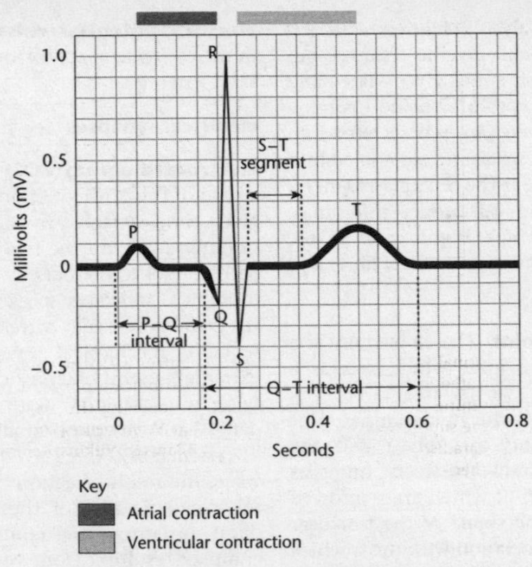

FIG 69. A diagram of a normal ELECTROCARDIOGRAM (see entry for details),, showing the P wave (atrial depolarization), QRS complex (onset of ventricular repolarization) and T wave (ventricular repolarization). See entry for details. From *Principles of Anatomy and Physiology*, G.J. Tortora & S.R. Grabowski, © 2000. Reproduced with permission of John Wiley & Sons, Inc.

chosen pairs of electrodes are registered using devices such as pen recorders (see Fig. 71). Because individual cortical neurons have too small an effect to be detected by this method, what is recorded is the *synchronous activity of groups of neurons* whose effects summate to produce a recordable voltage difference between the electrode pairs whose amplitude depends upon the number of cells generating simultaneous action potentials.

It is not certain what function neural oscillations serve; but they are known to facilitate SYNAPTIC PLASTICITY, influence reaction time, correlate with ATTENTION and perceptual binding (see BINDING PROBLEM), and possibly play a part in long-range coordination of distinct brain regions. By momentarily synchronizing the fast oscillations produced by different cortical areas, the brain may be binding together various neural components into a single perceptual construction and making them meaningful. Direct cortical recordings reveal ongoing rhythms encompassing a wide range of spatial and temporal scales: ultra-slow rhythms of <0.05 Hz coexist with fast transient oscillations of 500 Hz or more, the spatial coherence between such oscillations extending from several centimetres for the corticospinal tract to the micrometre scale for subthreshold membrane oscillations in a single neuron. Evidence for cross-frequency coupling, with one frequency band modulating the activity of another, is more abundant in non-human than in human data, but task-related changes in theta power have been detected in humans.

EEG recordings from normal awake and relaxed individuals exhibit a characteristic high-frequency, low amplitude (i.e., desynchronized), β activity (13–30 Hz). The β waveform is seen when a subject is alert or engaged in intense mental activity (see ATTENTION). It appears to be involved in mediating neural disinhibition and is influenced by variations in the $GABA_A$ receptor (see GABA). When the subject closes his/her eyes and becomes drowsy, a lower frequency, slightly higher voltage and synchronized α activity (8–13 Hz) occurs –

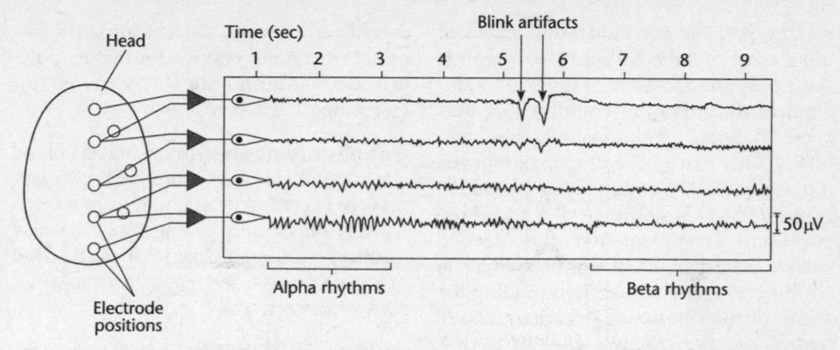

FIG 70. Diagram of a normal ELECTROENCEPHALOGRAM (see entry) with the subject awake and quiet. Recording sites are indicated at the left. The first few seconds indicate normal α activity (frequencies 8–13 Hz), the largest coming from the occipital regions. On opening the eyes halfway through the recording, α rhythms were suppressed. See SLEEP. © 2001 from *Neuroscience* (2nd edition) by M.F. Bear, B.W. Conners and M.A. Paradiso, published by Lippincott, Williams & Wilkins. Reproduced with the kind permission of the publisher.

resulting from simultaneous firing of many cortical neurons following thalamocortical activity (see THALAMUS). Delta rhythms (Δ rhythms) are quite slow (<4 Hz) but often of large amplitude, and are a hallmark of deep (NREM) sleep. Thalamocortical neurons become so hyperpolarized that they go silent and cortical neurons, now decoupled from the thalamus, fire with their own intrinsic rhythm. Desynchronized EEG activity is an EEG pattern lacking regular periodicity in its waveform as a result of blocking of α-wave activity through arousal or focused attention in the waking state, or blocking of Λ-wave activity through dreaming during sleep.

Theta rhythms (θ rhythms) have a frequency of 4–8 Hz. Theta rhythms occur during some sleep states, are the optimal protocol for generating LONG-TERM POTENTIATION, and are caused by regular firing of hippocampal neurons driven by the cholinergic pathway from the septum (see HIPPOCAMPUS). The phase of low-frequency theta rhythm modulates power in the high gamma (80–150 Hz) band, stronger modulation occurring at higher theta amplitudes. This transient coupling between low- and high-frequency brain rhythms coordinates activity in distributed cortical areas and may provide a mechanism for effective communication during cognitive processing. Different behavioural tasks evoke distinct patterns of theta/high gamma coupling across the cortex, indicating that transient coupling between low- and high-frequency brain rhythms coordinates activity in distributed cortical areas. This could provide a mechanism for effective communication during cognitive processing (see NEURAL CODING).

electron transport chain See MITOCHONDRION.

electrotonic conduction Direct flow of electric current, rather than action potentials, in the cytoplasm of a neuron, all the way from the point of excitation to the output synapse(s) (see ROD CELL). Electrotonic coupling is achieved by GAP JUNCTIONS.

Elephant Man disorder See NEUROFIBROMATOSIS.

elephantiasis A gross swelling of the legs and external genitals resulting from chronic blockage of lymph ducts by Bancroft's filarial worm (*Wuchereria bancrofti*).

elevation Movement of part of the body in a superior direction (towards the head, or upper part of the structure).

elongation, elongation factors Soluble cytosolic proteins (eEF1α, eEF1βγ and eEF2) making possible the elongation phase of TRANSLATION of mRNA during polypeptide synthesis on RIBOSOMES. Elongation also requires the initiation complex (see INITIATION FACTORS), amino-acyl-tRNAs and GTP. In the first phase of elongation, the initiating (5′)AUG codon being positioned in the P site, an amino-acyl-tRNA whose anticodon complements the second mRNA codon binds a complex of GTP-bound eEF1α, the whole then binding the A site of the ribosomal 70S subunit. GTP hydrolysis then releases the eEF1α-GDP complex, which is regenerated to eEF1α-GTP by eEF1βγ and GTP. In the second phase, a peptide bond is formed between the two amino acids still bound by their tRNAs to the A and P sites, the methionyl group being transferred from its tRNA to the amino-acyl-tRNA complex in the A site. Formation of the peptide bond is probably catalysed by 23S rRNA ribozyme. In the third and final step of the elongation cycle, the ribosome moves one codon toward the 3′ end of the mRNA, shifting the anticodon in the A site to the P site, and the uncharged tRNA is expelled from the P site. Movement of the ribosome along the mRNA requires eEF2, the energy needed coming from hydrolysis of another GTP molecule. A single eukaryotic ribosome adds ~2 amino acids per second to the growing polypeptide chain. For each amino acid correctly added to the growing polypeptide, two GTPs are hydrolysed as the ribosome moves along the mRNA from codon to codon in the 3′ direction. A proof-reading mechanism on the ribosome, involving eEF1α, ensures that the correct codon–anticodon pairing takes place (see TRANSLATION). Those tRNAs that have passed their amino acids to the growing polypeptide chain leave the ribosome via the E-site ('E' for exit). Elongation continues until the ribosome adds the last amino acid coded by the mRNA, termination being signalled by any of the three termination ('stop') codons: UAA, UAG, UGA. A single release factor, eRF, recognizes these and contributes to hydrolysis of the terminal peptidyl–tRNA bond so that the polypeptide is freed. The ribosome then dissociates into its large and small subunits, which are recycled for further polypeptide synthesis. Mg^{2+} is required for elongation to proceed.

embolism, embolus An *embolus* is a blood clot, air bubble or fat from broken bones, mass of bacteria, or other debris or foreign matter transported by the blood. It may lead to *embolism*, a closure or obstruction of a blood vessel. See BLOOD CLOTTING, TISSUE PLASMINOGEN ACTIVATOR.

embryo In its broadest sense, any developmental stage between fertilization and the time at which bone cells appear in the cartilage (~8 weeks in humans), after which the term 'foetus' applies. The embryo proper develops from some of the cells of the INNER CELL MASS, and a more restricted use would recognize the origin of these cells as the start of EMBRYONIC DEVELOPMENT.

embryoblast The non-polar cells lying interior to the trophoblast (see BLASTOCYST, Fig. 20) and which, as the blastocoele forms, congregate at one end of it to form the INNER CELL MASS.

embryo disc See EMBRYONIC DISC.

embryonic development In a general sense, development taking place in the ~8 weeks between fertilization and the origin of osteocytes (after which time the embryo is a foetus). The major superficial features of embryonic development are shown in Appendix V, and further details are provided in entries for many organs. A feature of the genetics of embryonic development is that polycomb proteins (see POLYCOMB GROUP) function as long-term repressors for embryologically expressed genes that are silenced in later development (see GENE SILENCING). Two major stages of embryonic development in terms of morphogenesis and organogenesis are GASTRULATION and NEURULATION. See GROWTH.

embryonic disc (germ disc) A flat disc of tissue separating from the inner cell mass during formation of the AMNIOTIC CAVITY (see Fig. 8b). It comprises an outer ectoderm adjacent to the amniotic cavity, and an endoderm on the side of the disc opposite the amnion. Initially almost round, it gradually elongates with a broad cephalic and a narrow caudal end. Growth and elongation of the cephalic end are caused by continuous migration of cells from the PRIMITIVE STREAK region in its direction, until by the end of the 4th week the primitive streak disappears. Cells at the anterior (cranial) margin form the *anterior visceral endoderm* (AVE) and express genes needed for formation of the HEAD, including transcription factors *OTX2*, *LIM1* and *HESX1*, and the secreted factor cerberus. These establish the cranial end of the embryo prior to GASTRULATION.

embryonic stem cells (ES cells, ESCs) Pluripotent STEM CELLS, with the promise of sourcing cells for therapeutic CLONING. They are currently derived, in humans, from cells isolated from the INNER CELL MASS of the preimplantation BLASTOCYST at ~5–6 days after conception, and more recently from post-implantation embryonic cells of the epiblast; but their specific embryonic origin is unclear. A blastocyst is currently required because earlier stages do not have an inner cell mass, and trophoblast cells are not stem cells. By the third week the GERM LAYERS have been laid down, most cells having lost their pluripotency and having begun to specialize.

Several genes required for early development (e.g., those encoding the pluripotency sustaining transcription factors OCT3/4, NANOG and SOX2; see below) are expressed by ESCs but silenced during cell differentiation. Their gene products are among those constituting an autoregulatory, pluripotent network the detailed components of which differ in different mammals, e.g., mouse and humans. Pluripotent cells are also characterized by epigenetic marks involved in the repression of developmental genes. These can be established by the PcG–protein repressive complex, PRC (see POLYCOMB GROUP), which marks genes with H3K27 and H3K4 methylation (see CHROMATIN REMODELLING). Later in development, the pluripotency-associated genes are repressed by histone and DNA methylations. The predisposition of ESC PRC2 targets to cancer-specific DNA hypermethylation indicates crosstalk between PRC2 and *de novo* DNA methyltransferases in an early precursor cell with a PRC2 distribution similar to that of ESCs. Such crosstalk may occur early in oncogenesis, when the PRC2 distribution resembles that of a stem cell.

Being pluripotent, ESCs can generate all the tissues of the body, most existing human ESC (hESC) lines being derived from unused embryos created for those seeking IN VITRO FERTILIZATION, a process which unfortunately destroys them. Prior to selection for use, an ESC must display characteristic epigenetic marks and undergo several rounds of cell division (passages) demonstrating its ability to provide a stable, or 'immortalized' cell line. Such ESCs usually recover from freezing and thawing and differentiate into various cell types in culture.

Different states impose different laws restricting the use of ESC lines, and many researchers are involved in developing methods for deriving ESCs which avoid destruction of human embryos. Some even attempt to source them from very young embryos that have already died through mitotic arrest during preparation for IVF treatment (see PREIMPLANTATION GENETIC DIAGNOSIS). But there are currently no clear criteria for deciding when an embryo is clinically dead. There is hope that work on mouse embryos, where ESCs can now be derived without loss of the embryo, may also be applicable to our source of human hESCs, and that it may be possible to create banks of hESC lines without additional risk to the embryos.

The most readily available human oocytes are aged ones that failed to be fertilized in IVF reactions. In contrast to recent success in mice, these have proved inadequate for SOMATIC CELL NUCLEAR TRANSFER (SCNT) and no human cell lines have currently been generated via SCNT. Although fresh unfertilized human oocytes are definitely preferred, there are logistical, medical and 'societal' reasons why these are not available in the numbers required. Frozen zygotes from couples who have completed their IVF treatment might be made available for stem cell research. But 3–5% of human zygotes prepared during IVF treatment contain abnormal numbers of pronuclei, and since human polyspermic zygotes can complete the first mitotic division, they could be arrested during mitosis and have their chromosomes removed prior to microinjection of donor chromosomes. Recent work in SCNT using mice (see CLONING, Fig. 56) raises the possibility of recycling such non-viable zygotes from IVF to produce ESC lines, thereby obviating the need for the oocyte donation, with all its ethical implications.

We now know that, in mice at least, chromosomes can be successfully transferred into zygotes arrested in mitosis and permit subsequent derivation of cloned mice. If human zygotes are to be used as recipient cytoplasts for producing genetically tailored hESC lines, they must be capable of reprogramming the genome of the donor adult somatic cell. Mouse meiotic oocytes and mitotic zygotes are able to do so; but artificial activation of unfertilized oocytes poorly mimics fertilization and evidence from cows suggests that it leads to poor clone development (neonatal respiratory failure and placental overgrowth being common phenotypes). This is very likely due to failures of unfertilized eggs to reprogramme somatic nuclei. Chromosome transfer into mitotic zygotes bypasses the need for artificial activation, without loss of reprogramming ability.

Until recently, inefficient reprogramming and poor embryonic development have characterized the results of SCNT in primates; but a modified SCNT approach producing Rhesus macaque blastocysts from adult skin fibroblasts has enabled ESC lines to be isolated from cloned embryos. DNA analysis confirmed that the nuclear DNA was identical to donor somatic cells, while mitochondrial DNA was confirmed as originating from oocytes, and key stem-cell markers were expressed. The cells differentiated into multiple cell types both *in vitro* and *in vivo*, a result which brings human therapeutic cloning a little closer to reality. Soon after that work was published (2007), two groups reported successful reprogramming of human skin cells into so-called *induced pluripotent cells*, iPCs, using slightly different combinations of four genes inserted by using a RETROVIRUS (akin to transfection). One group used *OCT3/4*, *SOX2*, *KLF4* and *c-MYC*, while the other used *OCT3*, *SOX2*, *NANOG* and *LIN28*. The next experimental step will be to reprogramme cells by switching on genes rather than inserting new copies, a technique that can induce mutation. It would, however, be premature to end research on human embryos, however desirable that goal. Producing an iPS cell line currently takes at least six months, and it is by no means certain that all such iPS cell lines are genetically equivalent. Since it is inadmissible to use human cells to create chimaeric animals, *contra* the situation with mouse cells, we will need to produce huge numbers of these cells from many different people and compare them in tests with hESCs to be sure. Because we currently have no alternative to retroviruses as vectors of the reprogramming genes there is always the risk of causing cancer. Adenoviruses would have the advantage of not inserting into the host chromosome; but another line of research is attempting to activate the small number of reprogramming genes by mimicking relevant signal transduction systems using small membrane-permeable signals.

Since their discovery in mice in the early 1980s, ESCs have raised hopes that they could be used therapeutically to replace lost or failing cells in diseases such as type I diabetes, Alzheimer's and other degenerative diseases. It is crucial to identify ESCs correctly, and to ensure their lack of contamination by other cells. We still need better knowledge of which identification cues, including possible cell surface markers, single out truly pluripotent ESCs. Their very pluripotency makes them too dangerous simply to inject into a patient in the hope that they will take cues from their surroundings and differentiate into just those cells requiring replacement, since there is currently too great a risk of their differentiating into inappropriate cell types (i.e. ectopic development). We do not adequately understand what such fate-determining cues are in particular cases, or whether they will be effective in a particular therapy; and unless they are genetically identical to the patient's own cells, ESCs have the same risk of immune rejection as an organ transplant (but see THYMUS). SCNT avoids this problem, and in 2004 this approach gave rise to the first hESC line.

Advances in developmental biology have informed the design of directed stepwise differentiation of stem cells that attempt to recapitulate the cell fate-determining decision processes in embryonic development that lead to lineage specification. Mouse motor neurons were some of the first differentiated cells to be derived from treating cells sequentially with embryonic signalling molecules: retinoic acid, and then an agonist of the hedgehog-mediated signalling pathway. Human ESCs have since been encouraged to differentiate in the following directions: neural precursor cells → subtype-specific neuronal progenitor cells → mature functional neurons; mesendoderm → mesoderm → cardiovascular precursor cells → immature cardiac cell types → mature cardiomyocytes; and mesendoderm → definitive endoderm → primitive gut tube → posterior foregut → pancreatic endoderm → endocrine precursor cells → hormone-producing endocrine cells. Ultimately, the goal is the discovery of efficient and well-defined medium conditions containing those small molecules that function synergistically with already-defined factors such as FGF2 (bFGF) and WNT proteins in increasing the homogeneity, functionality and yield of the intermediate progenitor cells or terminally differentiated cell types that will, it is hoped, enable successful therapeutic intervention in regenerative medicine.

emesis See VOMITING.

emission See EJACULATION.

emotional memory See AMYGDALA.

emotions These may be thought of as internally generated percepts that powerfully influence conscious experience, whether sleeping or waking, giving 'meaning' or 'significance' to those experiences. The experience and expression of emotion are a large part of 'being human', but for most non-human studies of emotion, their putative non-linguistic behavioural expressions, or manifestations, are all that we observe. Our understanding of the brain mechanisms underlying emotion derives from a combination of such non-human studies and clinical cases in our own species. Their selective value derives from the arousal and direction of ATTENTION to important features of a situation, as inducers to useful action, and in their social role in communication (emotions are expressed prior to acquisition of language).

Most stimuli evoking emotion derive from our senses, and result in expressed behaviour controlled by the somatic motor system, the autonomic nervous system and the secretory hypothalamus. The experiential components of emotions include perception of visceral changes due to autonomic and endocrine events, a motor component (especially involving facial muscles) and a conscious analytical (cognitive) component. The mechanisms underlying the experiential content of emotional experience (as of all other experience) are far harder to understand, although the cerebral cortex plays a key role (e.g., see INSULA, CINGULATE CORTEX). For strongly felt emotions at least, there is a close association with changes in physiological state; but it is not obvious which causes which. William James and Carl Lange in the 19th century each proposed that we experience emotion in response to physiological changes (the James–Lange theory) and that it is difficult to conceive of an emotion persisting in the absence of the physiological changes associated with it: crudely, the physiological changes *are* the emotion (we feel sad *because* we cry). In the 20th century, Walter Cannon and Philip Bard demonstrated that deprivation of sensory input (by spinal transactions) did not seem to abolish emotion, and that there was no one-to-one parity between an emotional experience and a particular physiological state. They proposed (the Cannon–Bard theory) that emotional experience can occur in the absence of emotional expression (you don't have to cry to feel sad).

We now know that emotions such as fear and rage are distinguishable physiologically, even though both involve the sympathetic division of the autonomic nervous system (see ANGER AND AGGRESSION, FEAR AND ANXIETY, REINFORCEMENT AND REWARD). Furthermore, emotion is sometimes affected by damage to the spinal cord affecting sensory input, while forcing oneself to express an emotion does sometimes result in one's experiencing it. For a time after its discovery in the 1930s, it was thought that the bidirectional nature of the PAPEZ CIRCUIT of the LIMBIC SYSTEM provided a suitable neurological basis for emotions, although the evidence was only suggestive. Its bidirectionality made it compatible with both the James–Lange and Cannon–Bard theories. The AMYGDALA is implicated in innate and learned fear responses (see CINGULATE CORTEX). It is known that D_2 DOPAMINE receptors in the LIMBIC SYSTEM are concerned with mood and emotional stability, and that emotions are modulated by the GLUCOCORTICOIDS. See also ANTERIOR COMMISSURE.

empathy The neurological basis of empathy is probably very complex, but some evidence points towards involvement of the CINGULATE CORTEX. See AUTISM, EMOTIONS.

emphysema Disorder characterized by destruction of the walls of the alveoli, caused by their chronic irritation and inflammation, and resultant coughing, producing abnormally large air spaces that remain filled with air during expiration. Causes include CIGARETTE SMOKING, air pollution and occupational exposure to industrial dust (see PNEUMOCONIOSIS). Since carbon dioxide diffuses across the respiratory membrane faster than does oxygen, the decreased surface area of respiratory membrane caused by the disease does not result in elevated arterial pCO_2 despite the low arterial pO_2. The elevated rate and depth of ventilation are largely due to the stimulatory effect of low pO_2 levels on the CAROTID BODIES and aortic bodies, although more extreme emphysema can result in elevated arterial pCO_2 levels. Over several years, the added ventilatory exertion increases the size of the chest ('barrel chest'). The inhalations from cigarette smoking cause decreased production by the liver of the protease inhibitor α_1 ANTITRYPSIN. Elastase in the alveoli is then free to damage the connective tissue, resulting in loss of alveolar elasticity. Treatment involves cessation of smoking and removal of other environmental respiratory irritants, carefully supervised exercise training, oxygen therapy and use of bronchodilators. See RESPIRATORY CENTRE.

enamel Extremely hard, acellular, covering of the exposed part of a tooth, comprising 97% inorganic material (two-thirds calcium phosphate crystals, one-third calcium carbonate), and 3% organic. The teeth of gorillas and chimpanzees have a thinner layer than those of fossil hominins, modern humans, orang-utans and even fossil apes (*Sivapithecus*). The molars of the robust australopithecines have the thickest enamel of any primate. In hominins, thick enamel may have evolved as an adaptation to a tough diet. See DENTITION.

ENCODE (Encyclopedia of DNA elements) See HUMAN GENOME.

end-diastolic pressure The pressure in the ventricles at the end of diastole; usually measured in the left ventricle as an approximation of the end-diastolic volume, or preload. See HYPERTENSION, STROKE VOLUME.

end-diastolic volume The volume of a heart ventricle just before it contracts. Compare END-DIASTOLIC PRESSURE, END-SYSTOLIC VOLUME; see CARDIAC CYCLE, STROKE VOLUME.

endemic disease Referring to a disease (e.g., MALARIA) that, where it occurs, occurs commonly at all times of the year. Compare PANDEMIC DISEASE.

endocannabinoids Lipids synthesized from components of neuron cell membranes when internal calcium ion levels rise, or when G-COUPLED PROTEIN RECEPTORS are activated as appropriate neurons fire in bursts. They include, at least, anandamide (from the Sanskrit 'ananda', meaning 'bliss') and 2-arachidonoyl glycerol (2-AG). Both bind endocannabinoid receptors CB_1, present mainly in the CNS but also in the PNS, and CB_2, present almost exclusively outside the nervous system on cells of the immune system. CB_1 and CB_2 are also activated by THC (see CANNABINOIDS). Anandamide, at least, has a role in guarding the nervous system against neuronal EXCITOTOXICITY (neuroprotection). Endocannabinoids differ profoundly in their neuroprotective effects, both chemically and in their activities, and the effects of some can worsen the situation by blocking release of GABA. Effects seem to include EXTINCTION of 'bad feelings', e.g., the fear and pain associated with memories of past experiences. Low numbers of cannabinoid receptors, or insufficient release of endocannabinoids, may be linked to post-traumatic stress syndrome, phobias and some forms of chronic pain. Some, at least, are associated with LONG-TERM DEPRESSION at appropriate synapses. Drugs targeting endocannabinoid degradation may be one therapeutic approach to treatment of PARKINSON'S DISEASE. See FEAR AND ANXIETY, RETROGRADE INHIBITION, SYNAPTIC PLASTICITY.

endocardium The innermost layer of the HEART. Comprising a thin layer of endothelium overlying a thin layer of connective tissue, it provides a smooth lining to the heart chambers and also covers the heart valves.

endocasts (endocranial casts) Plaster or latex casts of the insides of skulls, from which it is possible to determine the shape of the brain of an extinct species due to changes in shape of the dura mater that reflect those of the brain. Thus, cortical surface details such as gyri and sulci sometimes leave their mark on the inside of the skull, although since these foldings become tighter as the brain enlarges, they are more difficult to interpret because the dura mater then becomes thicker and less pliable. See BIPEDALISM, CRANIAL CAPACITY, SYLVIAN FISSURE.

endochondral ossification Bones formed by OSSIFICATION of pre-existing cartilaginous tissue. Normally there is no ossification in the human foetus until after the 4th month of pregnancy. Neural crest forms endochondral bones in the head, while mesoderm forms endochondral bones in the skeleton and trunk.

endocochlear potential See ENDOLYMPH.

endocrine glands Glands releasing signal molecules into the blood and affecting distant target organs. See ENTEROENDOCRINE CELLS; compare AUTOCRINE SIGNALLING, PARACRINE SIGNALLING.

endocytosis The uptake of macromolecules, particulate substances, and even cells, by progressive enclosure by a small portion of the cell surface membrane, which first invaginates and then pinches off to form an *endocytic vesicle*, within which compartment the molecules or particles are contained. One form, PHAGOCYTOSIS, involves ingestion of large particles (e.g., microorganisms, dead cells) via large vesicles: *phagosomes*. The other form is pinocytosis, by which fluids and solutes are enclosed by much smaller *pinocytic vesicles*. Whereas endocytic vesicles produced by pinocytosis are small and uniform (~100 nm in diameter), those produced by phagocytosis vary in diameter with the size of the ingested particle (usually >250 nm in diameter). Unlike pinocytosis, which is a constitutive process of cell membrane, phagocytosis is a triggered process in which transmembrane cell surface receptor proteins must be activated by binding specific ligands (*receptor-mediated endocytosis*). These ligands include ANTIBODIES that have bound infectious microorganisms, some COMPLEMENT components, and Gram-negative antigens such as LIPOPOLYSACCHARIDE. Many cells take up CHOLESTEROL by this process, and if the process is blocked cholesterol can accumulate in the blood as LDL. The ligand/receptor complexes then accumulate in coated pits and enter the cell in clathrin-coated VESICLES. Endocytosed molecules are initially delivered by endocytic vesicles to early endosomes, some being selectively recycled to the cell membrane prior to lysosome fusion while others pass on into late endosomes. Here they meet lysosomal hydrolases delivered to the late endosome by the GOLGI APPARARUS (see LYSOSOMES).

endoderm One of the primary GERM LAYERS. It originates during gastrulation, from cells which displace the hypoblast. It covers the ventral surface of the early embryo and forms the roof of the yolk sac. The main organ system derived from it is the ALIMENTARY CANAL. See Appendix VI.

endogenous Originating within the system concerned; within these pages, within the human body, its cells, etc.

endogenous clock, e. rhythm Rhythms of electrical activity arising in cells and tissues in the absence of external influences (although the latter may modify them). Examples occur in the SA NODE of the heart, in CENTRAL PATTERN GENERATORS, and in the SUPRACHIASMATIC NUCLEUS and preoptic nuclei of the HYPOTHALAMUS.

endogenous pyrogens Cytokines released by macrophages, notably the acute phase proteins TNF-α, IL-1β and IL-6 that raise body temperature during FEVER in response to bacterial surface components such as LIPOPOLYSACCHARIDE, which is therefore an exogenous pyrogen. See TEMPERATURE REGULATION.

endogenous retroviral sequences See LONG TERMINAL REPEATS.

endolymph The fluid filling the scala media of the COCHLEA (see Fig. 58) with an ionic composition similar to that of tissue fluid. Secreted by a specialized epithelium, the stria vascularis, lining the outer wall of the cochlear duct, its K^+ concentration is ~160 mM and its Na^+ concentration is ~2 mM. Specialized marginal cells in this epithelium have Na^+, K^+-ATPases on their basolateral membranes enabling them to concentrate K^+ for secretion into the endolymph, generating an electrical potential 80 mV more positive than that of the PERILYMPH (the endocochlear potential). See HAIR CELLS.

endometriosis A disease defined by the presence and/or growth of endometrial tissue outside the uterus, a common disorder mainly occurring in women of reproductive age (6–10%). Because ectopic endometrial implants respond to natural or induced decreases in oestrogen level, the condition is considered 'oestrogen-dependent'. The untransformed epithelial cells have been found in the pelvic cavity, pericardium, pleura, aorta and brain – arguably as metastases (see METASTASIS). Symptoms include reduced fertility and several types of pain (e.g., severe dysmenorrhea (excessive menstrual pain), deep dyspareunia (pelvic pain during intercourse), dyschezia (pelvic pain during defecation) and chronic pelvic pain). GnRH agonists are often prescribed in treatment.

endometrium The mucous membrane lining the UTERUS. The endometrial stroma adjacent to the site where implantation has occurred is oedematous and richly vascularized (see RELAXIN). This region, the decidua (stratum functionalis), is modified during PREGNANCY and shed after the baby's BIRTH. The basal layer (stratum basalis) is permanent and gives rise to a new stratum functionalis after each menstruation. From puberty until menopause, the endometrium undergoes cyclical changes of ~28 days' duration during which it passes through the follicular (proliferative) phase, the secretory luteal phase and the menstrual phase (see MENSTRUAL CYCLE). Endometrial (uterine) glands develop as invaginations of the lumenal epithelium and extend almost to the myometrium. See ENDOMETRIOSIS.

endomorph See BODY SIZE AND SHAPE.

endomysium Invagination of the perimysium, separating each individual muscle fibre.

endoneurium Connective tissue, continuous with the perineurium, that wraps around individual nerve fibres in a nerve.

endonucleases Enzymes capable of beginning degradation of a nucleic acid strand or molecule at any site compatible with the specificity of their active site. Some only degrade single-stranded nucleic acid and some, e.g., the restriction endonucleases used in gene technology, target particular nucleotide sequences. See DNA POLYMERASES, EXONUCLEASES.

endophenotype Phenotypic traits or markers that can represent intermediate forms of expression between the output of genes involved in a disease and the broader disease phenotype. Their identification may lead to more rapid success in identifying susceptibility genes for the disorders. The brain's electrical activity patterns (see ELECTROENCEPHALOGRAM) are a form of endophenotype. See EXPRESSIVITY, PENETRANCE, SCREENING.

endoplasmic reticulum (ER) A cytoplasmic organelle comprising a vast network of continuous membranous stacks, cisternae, continuous with the outer nuclear membrane. It contains no genome, but nevertheless is self-replicating. A ribosome-free compartment of the ER (smooth ER), continuous with the cisternae, projects into the cytosol and pinches off COPII-coated transport vesicles. These transport material from the ER, to their target sites, as dictated by their COPII coat proteins, complementary SNAREs, and the Rab proteins in the target membrane (see VESICLES). Smooth ER of liver cells contains enzymes of the cytochrome P450 family (see CYTOCHROMES) which detoxify lipid-soluble drugs and their metabolites (which would otherwise remain in membranes) by direct reduction of carbonyl groups (>C=O) to hydroxyl groups (>HC–OH) and by conjugation of these with sulphate or glucuronic acid, rendering them water-soluble and excretable. Extra smooth ER made during a time of such drug administration seems to be removed afterwards by autophagosomes.

The ER membrane also bears HMG-COA REDUCTASE, the rate-limiting enzyme in CHOLESTEROL synthesis, and cytochrome P450 (see CYTOCHROMES). Within the ER lumen are enzymes and chaperones enabling secretory and membrane proteins to be assembled into their secondary and tertiary structures after translation on the ribosomes which are attached by docking proteins to the cytosolic ER surfaces (forming 'rough ER'; see SIGNAL RECOGNITION). Post-translational modification (PTM) of proteins includes hydroxylation and glycosylation (e.g., see COLLAGENS, PROTEOGLYCANS), enzymatic cleavage and disulphide bond formation (e.g., see INSULIN). One of the commonest PTMs is addition of O-LINKED B-N-ACETYLGLUCOSAMINE to serine/threonine residues. Cyclooxygenase in the ER lumen is instrumental in the conversion of arachidonic acid to prostaglandins and thromboxanes and is inhibited by NSAIDs such as aspirin and ibuprofen. Proteins failing either to fold or to be assembled properly after they enter the ER are detected by an ER-based surveillance system and retro-translocated back into the cytosol for degradation by the 26S PROTEASOME system.

Cells have evolved elaborate mechanisms to ensure the accuracy with which proteins are folded and assembled prior to export or transport to the cell surface, and stringent quality control is imposed by the ER, which has its own form of hsp70 CHAPERONE, BIP, which helps protein folding. Only properly folded proteins are permitted to leave the ER, misfolded proteins being degraded. Accumulation of unfolded proteins by the ER, fluctuations in energy and nutrients, hypoxia, toxins, viral infections and increased demand on synthetic machinery give rise to *ER stress* when the ER activates a complex response system (the *unfolded protein response*) to restore ER integrity. This is mediated via inositol-requiring enzyme 1(IRE-1), PKR-like ER kinase (PERK) and activating transcription factor 6 (ATF6). This can lead to activation of genes serving to protect cells from such ER stress by increasing the ER's protein-folding capacity through increasing the quantity of ER, enhancing degradation of misfolded proteins and/or decreasing the synthesis of new proteins. However, if homeostasis cannot be achieved, UPR signalling eventually leads to a cell's APOPTOSIS – an effective way of protecting the individual from rogue cells and dysfunctional expression of signalling molecules.

The two principal inflammatory pathways that disrupt INSULIN action, involving JNK and IKK signalling, are linked to activity of the ER stress sensors IRE1 and ATF6 (cell-protective, but both attenuated during persistent ER stress) and PERK (a more persistent pathway leading to translational inhibition, and induction of the pro-apoptotic transcription regulator *Chop*). The varied time courses mediated by these sensor responses probably influence the cell's ultimate fate in response to ER stress. When IRE1 activity was artificially sustained, cell survival was enhanced, indicating a causal link between the duration of the type of UPR signalling and life or death of the cell after ER stress. Experimental cell systems indicate that ER stress has an important role in mediating insulin resistance in obesity in animal models, probably related to the fact that its membranes bind inactive sterol regulatory element-binding proteins (SREBPS) prior to their proteolytic release. Its CALCIUM pumps sequester Ca^{2+} ions (see PHOSPHOLIPASE C, Fig. 135), which can be released as a second messenger within cells. It is also a major source of reactive oxygen species (see MITOCHONDRION), and therefore of OXIDATIVE STRESS in cells. Enzymes involved in oxidation of fatty acids from the omega (ω) carbon atom (that most distant from the carboxyl group) are uniquely located in the ER of liver and kidney cells and include alcohol dehydrogenase and aldehyde dehydrogenase.

β-endorphin Derived from PROOPIOMELANOCORTIN, OPIOIDS with larger molecules than ENKEPHALINS, and containing a met-enkephalin or leu-enkephalin sequence. Expressed mainly in those neurons of the hypothalamus projecting to the thalamus or brain stem. It is also secreted by the adrenal medulla as part of the overall response to STRESS.

endosomes Membrane-bound organelles carrying newly internalized materials from endocytic vesicles (see ENDOCYTOSIS). Some of the endocytosed molecules are selectively recycled to the cell surface membrane, while others pass on into late endosomes. Endosomal compartments form a varied set of membranous tubes extending from the periphery of the cell to the perinuclear region, often close to the GOLGI APPARATUS and nucleus. Early and late endosomes differ in their membrane protein compositions, but their internal environment is kept acidic (pH ~6) by membrane vacuolar H^+ATPases, and later endosomes are generally more acidic than earlier ones. Early endosomes form a compartment acting as the main sorting station in the endocytic pathway, endocytosed ligands that dissociate from their receptor in the low pH are usually degraded along with other soluble contents. Any ligands remaining attached share the fate of the receptor, most of which are recycled, returning to the plasma membrane (at the same or a different site from their origin) in transport vesicles that bud off from the endosome. Others progress to lysosomes and are degraded, a route followed by EGF receptors (receptor down-regulation). Late endosomes develop into mature endosomes as they receive lysosomal hydrolases delivered in vesicles from the Golgi apparatus. The mature endosome is now a LYSOSOME. See PHAGOSOME, VESICLES.

endosteum Membranous lining of the medullary cavity and the cavities of spongy bone.

endothelins A family of 21-amino acid peptides, all with two disulphide bonds holding them in a conical spiral shape. Endothelin-1 (ET-1) is produced by endothelial cells, and by neurons and astrocytes of the CNS. It produces a relatively long-lasting vasoconstriction, although its physiological significance remains unclear. Its increased production in preeclampsia may result from endothelial dysfunction. Human endothelin-1 gene has five exons encoding pre-proendothelin-1, a 212-

amino acid peptide which undergoes further post-translational modification by an endopeptidase to proendothelin-1, which is further processed to the active peptide by endothelin-converting enzyme (ECE). ET-1 is involved in several pathological conditions, including sepsis, asthma and anaphylactic shock. Mouse studies indicate that ET-1 is degraded by MAST CELLS, protecting them from some of its adverse effects. ET-2 and ET-3 are produced mainly by the kidney and intestine. They appear to act locally in a paracrine or autocrine manner.

endothelium The innermost monolayer of thin, flattened cells, held together by tight junctions, lining the entire cardiovascular system (heart and blood vessels) and surrounded by a BASEMENT MEMBRANE. It is the only cell layer component of blood capillaries. Arteries and veins develop from small vessels constructed solely of an endothelium and the basement membrane (see VASCULOGENESIS). In liver sinusoids and kidney glomeruli, endothelial cells are specialized to form a sieve-like structure. Fenestrae in the plasma membranes, created in the manner of nuclear pores, allow water and most solutes to pass more easily (see CAPILLARIES).

Endothelial cells are crucially important in organ development. The primitive embryonic endoderm is induced by several factors, including those from adjacent endothelial cells, to become endodermal buds that eventually form the liver and pancreas, an association sustained in the mature organ. There is evidence that endothelial cells can be recruited from small numbers of bone marrow progenitor cells that circulate in the blood. A growing endothelium has terminal tip cells bearing filopodial extensions, behind which are stalk cells (see ANGIOGENESIS).

Endothelia influence circulating blood supply by releasing substances (e.g., NITRIC OXIDE and prostacyclins) into the blood which keep platelets inactive and unaggregated; but damage to endothelia reduces the concentrations of these inhibitory substances and promotes PLATELET aggregation (see CARDIOVASCULAR DISEASES (CVDS)). Acetylcholine released by vagal efferents in arteriole walls binds muscarinic receptors on endothelial cells and activates nitric oxide synthase (to which Hsp90 is normally bound), causing them to release NITRIC OXIDE (NO). This diffuses to any underlying smooth muscle and binds and activates guanylyl cyclase to produce cyclic GMP, triggering a response that causes the smooth muscle to relax, enhancing blood flow. Nutrient excess (see OBESITY) has deleterious effects on the endothelium, causing inflammation through impairment of the endothelial IRS-PI3K signalling pathway (see INSULIN) and reducing NO production. This might offer a plausible link between nutrient excess and CVDs. The endothelial glycocalyx, prostacyclin and thrombomodulin are ANTICOAGULANTS, and after an injury damaged endothelia slowly release TISSUE PLASMINOGEN ACTIVATOR, involved in removing remaining blood clots.

In lymphoid organs, the endothelial cells appear larger than those found elsewhere and are termed *high endothelial venules*. Normally, the endothelium resists firm adhesion of leukocytes, but the cytokines TNF-α and TNF β, both secreted by activated T_H1 cells, alter the endothelial cell membrane (*endothelial cell activation*) at sites of infection so that macrophages adhere to them (a process helped by the cytokine-immobilizing effects of heparan sulphate proteoglycans). TNF-α and chemokines released by activated macrophages by mechanisms of INNATE IMMUNITY, such as production of COMPLEMENT factors C3a and C5a, or HISTAMINE, also activate nearby endothelial cells. Activated endothelial cells trigger both the complement cascade and the coagulation cascade of BLOOD CLOTTING and release Weibel–Palade bodies, causing the adhesion molecule P-selectin to be delivered to the cell surface (see SELECTINS). Transcription and translation of RNA-encoding E-selectin are also stimulated, so that this also appears on the cell surface along with the adhesion molecule VCAM-1 which binds monocytes and neutrophils to the endothelium, leading eventually to recruitment of activated T CELLS to the site of infection by DIAPEDESIS. However, monocytes also the leukocyte classes found in nascent atheroma (see ATHEROSCLEROSIS). See BLOOD–BRAIN BARRIER.

endotherm (adj. endothermic) An animal whose body heat is generated mainly by its own metabolism rather than received as incident heat from its environment (as in ectotherms). Such animals (e.g., birds and mammals) must consume food more quickly, and with a greater energy content, than ectotherms. See TEMPERATURE REGULATION.

endotoxins See TOXINS.

end-plate potential (EPP) See NEUROMUSCULAR JUNCTION.

end-stopped cells of visual cortex See ORIENTATION-SELECTIVE CELLS.

end-systolic volume (residual systolic volume) The volume of blood remaining in a ventricle after systole. Compare END-DIASTOLIC VOLUME; see CARDIAC CYCLE, STROKE VOLUME.

endurance running (ER) See GAIT.

endurance training See EXERCISE.

energy balance (energy homeostasis) Humans use a variable mix of glucose and fat as energy-releasing substrates, depending upon food intake. Total daily energy expenditure is attributable to: resting metabolic rate (60–70%; see BASAL METABOLIC RATE), thermic effects of eating (~10%), and physical activity and recovery (15–30%). The DAILY VALUES for energy-releasing nutrients are given as a percentage of daily kilocaloric intake: 60% for carbohydrates, 30% for total fats (of which 10% for the total for unsaturated fat), and 10% for proteins. Excess dietary carbohydrate can be converted into triglycerides in the liver and exported as VLDLs (see LIPOGENESIS, LIPOPROTEINS); but although humans and other animals can synthesize fats from carbohydrates via acetyl CoA, they cannot synthesize carbohydrates from fatty acids. Energy homeostasis is controlled by long-term adiposity hormones (see ADIPOKINES, INSULIN) and short-term gastrointestinal signals, reflecting the overall state of energy stores and individual meal intake, respectively. When energy intake exceeds expenditure, the resulting nutrient excess can trigger responses in several cell types that could generate metabolic dysfunction (oxidative stress, unfolded proteins in the endoplasmic reticulum) that promote INFLAMMATION which can limit further exposure to nutrients by blocking insulin action. These are cells of the ENDOTHELIUM, hepatocytes, muscle cells, ADIPOCYTES, monocytes and MACROPHAGES.

During waking and feeding hours, we use glucose as an efficient energy source (see BLOOD GLUCOSE REGULATION) while under FASTING conditions, as during sleep, we primarily use fats. After an overnight fast, almost all the liver's glycogen, and most skeletal muscle glycogen, has been depleted. As blood glucose begins to fall, INSULIN secretion slows as GLUCAGON and GROWTH HORMONE secretion are stimulated. The fat stores (triglycerides) are mobilized as VLDLs for muscle and liver (triglyceride being their primary fuel) and are broken down in these tissues by fatty-acid β-oxidation (see FATTY ACID METABOLISM), a process inhibited by insulin. It is thought that distinct switches control glucose and fat metabolism in the liver and that regulation of fatty acid oxidation is much more sensitive to insulin than is gluconeogenesis. Glucose for the brain is provided by liver GLUCONEOGENESIS.

Although the two opposing sets of pathways (anabolic and catabolic) share many reversible reactions, each has at least one enzymatic reaction unique to it that is essentially irreversible, allowing the organism to respond to different needs for energy (see RESPIRATORY QUOTIENT). Maintaining blood glucose levels becomes a challenge during prolonged STARVATION CONDITIONS or high-energy endurance EXERCISE, when muscle and liver glycogen reserves decline rapidly. A shift in fuel usage from glucose to fats and ketone bodies by peripheral tissues is critical for maintenance of systemic glucose homeostasis. Opposing pathways (e.g., those of gluconeogenesis and glycolysis) are integrated by key regulatory enzymes, usually early in the pathways, so that stimulation of a biosynthetic pathway (anabolic) is accompanied by inhibition of the corresponding degradative (catabolic) pathway.

For people in highly industrialized states, 40% or more of the daily energy requirement is supplied by dietary triglycerides (neutral fat; see BALANCED DIET) while ~10% of energy here may come in the form of short-chain fatty acids produced by the GUT FLORA.

Studies in cultured liver cells and *in vivo* indicate that nuclear PGC-1α is sufficient to coactivate nearly all aspects of the *hepatic fasting response*, including gluconeogenesis, fatty-acid β-oxidation, ketogenesis and bile-acid homeostasis (see PGC-1 GENES). This it does by coactivating key hepatic transcription factors (e.g., HNF4α, PPARα, GR, FOXO1, FXR and LXR) and directly activating transcription from their target promoters. Also in liver, PGC-1β binds to the lipogenic SREBPS and LXR and has been shown to activate cholesterol and triglyceride synthesis and to promote their export to the bloodstream. PGC-1β also coactivates FOXA2 and promotes fatty-acid β-oxidation (see FATTY ACID METABOLISM).

Several hormones involved in the regulation of energy intake and metabolism include such satiety signals as glucagon, cholecystokinin and adrenaline (acting via beta receptors) – all decreasing appetite and increasing energy expenditure; and appetite-promoting opioids, growth hormone-releasing factor (GHRF), glucocorticoids and adrenaline (acting via alpha receptors), insulin, progesterone and somatostatin – all reducing energy expenditure. Many neural satiety signals are transmitted to the medulla via afferent fibres of the vagus nerve that synapse in the nucleus tractus solitarius in the hindbrain, participating in gustatory, satiety and visceral sensation (see APPETITE, Fig. 14; see TASTE for the role of primary and secondary taste neurons in feeding behaviour), while several gut peptides target the medulla and arcuate nucleus of the HYPOTHALAMUS.

The *adiposity negative feedback model* of energy homeostasis, also involving the hypothalamus, was invoked in the mid-20th century, its central premise being that circulating signals inform the brain of changes in body fat mass (see ADIPOSE TISSUE), which responds by initiating adaptive homeostatic adjustments of energy balance to stabilize fat reserves. The negative feedback loops that were proposed were that a chemical signal would circulate in proportion to body fat content; that it would promote weight loss through neuronal mechanisms, and that blockade of these would lead to increase in food intake and body mass. Although insulin appears to have been a key negative-feedback signal and regulator of body fuel stores in invertebrates, LEPTIN may have been added to the signal repertoire more recently in animal evolution, although cross-talk between these two mammalian hormones in the neuronal and signal transduction pathways they invoke indicates a shared evolutionary past. It is not known whether the hyperphagia and weight gain induced in animal models by reduced neuronal signalling by insulin and leptin contribute to the pathogenesis of common obesity in humans. Plasma levels of the food intake-reducing PYY_{3-36} (see entry), a physiological gut-derived satiety signal, decline in advance of meals while those of the food intake-promoting GHRELIN rise shortly before meals and fall rapidly on feeding. Cells of the mediobasal hypothalamus respond to ghrelin by activation of AMP-ACTIVATED PROTEIN KINASE (AMPK) which leads to increased food intake and body weight, inhibition of this enzyme (as by leptin and insulin) having the opposite effect.

In addition to hormones, nutrient-related signals are also involved in energy balance. Free fatty acids exert insulin-like effects in key brain areas including the ARCUATE NUCLEUS (ARC) of the hypothalamus, an important appetite-regulating centre, probably through raising intracellular long chain fatty acyl-CoA accumulation, thought to signal nutrient abundance.

During any 24-hour period, two distinct phases relating to the ingestion of a meal alternate: the fed state (reflecting overall anabolic metabolism) and the fasted, or catabolic, phase, characterized by mobilization of energy sources. These alternating phases maintain adequate glucose supply to the brain and maintain TEMPERATURE REGULATION, food digestion and physical activity.

engram The means by which a MEMORY is stored.

enhancers *Cis*-acting REGULATORY ELEMENTS (or sequences) that markedly increase the basal level of transcription of a neighbouring gene, and are capable typically of operating over considerable distances upstream or downstream of their gene (~50 kb), and in either orientation (unlike the core promoter sequence). They regulate not only the rate but also the tissue specificity of expression. Enhancers are only capable of gene activation of specific gene loci and cannot overcome the repressive effect of heterochromatin when linked to recombinant transgenes at such sites. Enhancers and promoters can be identified by the modification state of the nucleosomal histones in which they reside. See LOCUS CONTROL REGION.

enkephalins Pentapeptide OPIOIDS produced mainly in short interneurons in such brain regions as the periaqueductal grey matter, limbic system, thalamus and within the substantia gelatinosa of the spinal cord. The enkephalin precursor gene encodes either tyrosine-glycine-glycine-phenylalanine-methionine (met-enkephalin) or tyrosine-glycine-glycine-phenylalanine-leucine (leu-enkephalin). See ENDORPHINS.

enteric Relating to the ALIMENTARY CANAL.

enteric nervous system An interconnected network of ~10^8 neurons, often regarded as the 'third' division of the AUTONOMIC NERVOUS SYSTEM (ANS) forming part of the peripheral nervous system and serving the alimentary canal (see Fig. 4). It is organized into two cylindrical sheets along the length of the gut: the myenteric plexus (Auerbach's plexus) between the longitudinal and circular smooth muscle layers of the muscularis, and the submucosal (Meissner's) plexus lying in the submucosa from the pylorus of the stomach to the anus. It can act autonomously to coordinate gut motility (peristalsis and SEGMENTATION) and secretion, although its activity is modulated by inputs from both divisions of the ANS. Its sensory neurons are multipolar and phasic, responding either to stretch or else having neurites in the mucosa responding to chemical signals, or to distortion of the mucosa associated with presence of chyme. Typically, in response to stretch by a bolus of food or chyme, a sensory neuron stimulates contraction of circular-layer smooth muscle on the oral side but inhibits it on the anal side, whereas longitudinal smooth muscle layers are relaxed on the oral side but contracted on the anal side. The effect of this coordinated muscle contraction is peristalsis.

enterocytes Cells of the epithelium of the SMALL INTESTINE (see Fig. 159 and ALIMENTARY CANAL, Fig. 4), other than enteroendocrine cells and goblet cells. They cover the CRYPTS OF LIEBERKÜHN and villi, developing from stem cells in the crypts (see SMALL INTESTINE, Fig. 163), secreting large quantities of water and electrolytes that are reabsorbed by the villi with the products of digestion. While part of the epithelium, enterocytes engage in both transcellular and PARACELLULAR TRANSPORT, their connecting TIGHT JUNCTIONS being variably effective barriers to the passage of molecules in-between the cells. They have a brush border of microvilli with which micelles fuse, enabling lipid transfer into their cytoplasm. Although small amounts of short-chain fatty acids are absorbed by simple diffusion, the majority of triglycerides are reformed in the smooth endoplasmic reticulum (SER) by re-esterification of monoglycerides; phospholipids are resynthesized, and much of the cholesterol undergoes re-esterification. The lipids accumulate in the SER to form chylomicrons which are released at the basolateral surface by exocytosis. These enter lacteals and leave the intestine via the lymph, bypassing the liver in the short term (see CHYLOMICRONS). The apical microvilli of enterocytes lining the villi contain disaccharidases which complete digestion of carbohydrates, while carrier proteins take up dipeptides in conjunction with hydrogen ions (H^+). Intracellular dipeptidases then convert the dipeptides to amino acids. After a meal, in response to the increased activity of their sodium-dependent glucose transporter SGLT1, the microvilli incorporate GLUT2 GLUCOSE TRANSPORTERS. Active transport of Na^+ across the basal surfaces lowers the enterocyte's internal Na^+ concentration, favouring Na^+ entry from the intestinal lumen across the microvilli. Symporters in the brush borders only allow Na^+ entry in combination with either glucose or galactose (via the SGLT1 cotransporter), or an amino acid – secondary active transport which also leads to osmotic uptake of water. At least ten separate transporters have been identified for amino acids. Fructose is absorbed by facilitated diffusion via the GLUT5 carrier, and by GLUT2 when this has become incorporated in the membrane. Potassium ions (K^+) are absorbed passively along a concentration gradient set up by this water absorption, anions in general following the electrical potential generated by the active sodium transport. These operations are very similar to those occurring in the cells of the PROXIMAL CONVOLUTED TUBULE. Glucose and fructose are transported by facilitated diffusion across the basal surface of the enterocyte by GLUT2, diffusing into the blood and taken to the liver via the hepatic portal vein. As the enterocytes move up the villus and are released into the intestinal lumen, they break down and release their brush border enzymes to engage in extracellular digestion. See COLON.

enteroendocrine cells (APUD cells) Peptide hormone-producing cells of the gut's mucosal epithelium, having a relatively short lifespan and being continually replaced by progenitor cells. Also known as APUD cells (amine precursor uptake and decarboxylation), which relates to their role in hormone synthesis. In the small intestine, their differentiation is affected by the position along the crypt-to-villus axis and by interactions with neighbouring cells. The gastrointestinal tract is the body's largest endocrine organ and releases >20 different regulatory peptide hormones, many of them sensitive to gut nutrient content, and believed to mediate short-term feelings of hunger and satiety. Cells in the stomach produce GASTRIN and GHRELIN; those in the duodenum produce CHOLECYSTOKININ, SECRETIN, GASTRIC INHIBITORY PEPTIDE and motilin, while L cells of the small intestine produce a precursor protein, preproglucagon, processed further to produce GLUCAGON, GLUCAGON-LIKE PEPTIDE-1, glucagon-like peptide-2, peptide YY (see PYY_{3-36}) and oxyntomodulin, and N cells produce NEUROTENSIN. Gut hormones can activate circuits in the hypothalamus and brain stem responsible for the regulation of ENERGY BALANCE (e.g., by modulating circadian rhythms of feeding) and a number also act as brain neurotransmitters.

enterogastric reflex Collective term indicating all hormonal and neural mechanisms involved in the control of gastric emptying. See STOMACH.

enterokinase Misnomer for ENTEROPEPTIDASE.

enteropeptidase An enzyme attached to the microvilli of the epithelial cells of the small intestine and converting trypsinogen to trypsin.

entorhinal cortex Neocortical region forming part of the hippocampal complex, adjacent to the HIPPOCAMPUS (see Fig. 91), to which it sends information via a bundle of axons termed the *perforant path*. These axons synapse with granule cells in the dentate gyrus, or with pyramidal cells in the CA3 region of the cornu ammonis (CA). The dorsocaudal medial entorhinal cortex contains a directionally oriented, topographically organized neural map of the spatial environment, whose key unit is the 'grid cell', activated whenever the individual's position coincides with any vertex of a regular grid of equilateral triangles spanning the surface of the environment. The entorhinal cortex may support some spatial computations previously attributed to the hippocampus. See also ALZHEIMER'S DISEASE, CINGULATE GYRUS, PARAHIPPOCAMPAL GYRUS.

environmental variance See VARIATION.

eosinophils Granulocytic leukocytes originating in bone marrow, their granules containing arginine-rich basic proteins stained bright orange by eosin. Believed to be particularly important in defence against parasites such as worms (e.g., *Schistosoma*), their production and release are increased by activation of T_H2 cells and release of IL-5. In the absence of infection or other immune stimuli very small numbers usually occur in the circulation, most being found in tissues, notably in the connective tissues underneath the epithelia of the alimentary canal, bronchial and urogenital systems (see ALLERGIES). CC chemokines induce their chemotaxis and activation. On activation, they release highly toxic proteins and free radicals from their granules, causing degranulation of BASOPHILS and MAST CELLS. Activation also causes their synthesis of prostaglandins, leukotrienes and cytokines, amplifying inflammation by activating epithelia and recruiting further eosinophils and leukocytes.

ependymal cells Epithelial cells lining the VENTRICLES of the brain and central canal of the spinal cord, and held together by TIGHT JUNCTIONS. This effectively seals off the CIRCUMVENTRICULAR ORGANS from the rest of the brain (see BLOOD–BRAIN BARRIER). Those surrounding the lateral ventricle are polarized with oriented bundles of motile cilia whose action creates a direction to the flow of cerebrospinal fluid (see SUBVENTRICULAR ZONE). See NEURAL TUBE, NEUROGENESIS.

ephrins Membrane-bound signalling ligands that bind and cross-link their cell-surface receptors and stimulate angiogenesis and axon migration. Type A ephrins are linked to the membrane by a glycosylphosphatidylinositol anchor; type B ephrins are transmembrane proteins. Ephrins and Eph receptors (the most numerous RECEPTOR TYROSINE KINASES) can simultaneously act as both ligand and receptor. On binding to an Eph receptor, some ephrins not only activate the receptor but also become activated themselves to transmit signals into the interior of the ephrin-expressing cell. An ephrin protein on one cell can thus lead to bidirectional reciprocal signalling that changes both cells' behaviour and is required in keeping cells in particular parts of the brain from mixing with cells in neighbouring parts. For instance, one 'repulsive' chemical signal marker on cell membranes of the posterior optic tectum of amphibians, helps to guide growing temporal retinal axons away from the posterior optic tectum and towards the anterior tectum. Ephrins are expressed in a gradient across the tectal surface, their highest levels being found on the posterior tectal cells. When the ephrin receptors on the membranes of the growing axons encounter ephrin, axon growth stops. See AXONAL GUIDANCE, LUNG CANCER, SEMAPHORINS, VASCULOGENESIS.

epiblast A layer of high columnar cells of the embryoblast, being pluripotent derivatives of the INNER CELL MASS (see BLASTOCYST, Fig. 21) within which a cavity forms by the 7-day blastocyst stage and becomes the AMNIOTIC CAVITY (see Fig. 8). With the hypoblast, the epiblast initially forms a flat disc (see GERM DISC) next to the cytotrophoblast within the blastocyst. Epiblast cells lining the cavity and lying next to the hypoblast develop into ectoderm and all cell types of the adult while those forming the roof of the cavity develop into the amnion. See EMBRYONIC STEM CELLS, PRIMITIVE STREAK.

epicardium (visceral pericardium) The thin outer layer of the heart wall, comprising serous tissue and mesothelium. See PERICARDIUM.

epicondyle Projection on a bone, above a condyle; for connective tissue attachment.

epidemics See DISEASE.

epidemiology The study of the distribution and determinants of DISEASES of populations. This commonly involves descriptions of disease in terms of time (day, month, year of onset of symptoms), person (age, sex or socioeconomic circumstances) and place (region or country). The effects of disease determinants (the disease aetiology) are usually identified by analytical case-control or cohort studies. Seasonal cycles of infectious disease are common (e.g., measles has a 2-year cycle while mycoplasma pneumonia typically peaks every 3–4 years). Gradual changes in incidence of an infectious disease over many years (long-term secular trends) may be due to demographic, social, behavioural or nutritional changes in a host population, to climatic or environmental change, or to public health intervention. See BASIC REPRODUCTIVE NUMBER (R_0), CANCER, CIGARETTE SMOKING, COMPLEX DISEASE, DEMOGRAPHIC STOCHASTICITY, EFFECTIVE REPRODUCTIVE NUMBER (R_{eff}), SINGLE NUCLEOTIDE POLYMORPHISMS.

epidermal growth factor (EGF) The first GROWTH FACTOR to be discovered, EGF binds at its specific EGF receptor (EGF-R) to surfaces of cells that have already been stimulated to grow. This receptor has a Src-like tyrosine kinase in its cytoplasmic domain, activating cell growth and division when EGF binds the receptor (see RECEPTOR TYROSINE KINASES). In addition to its role as a mitogen, EGF is a survival factor, blocking activation of the apoptosis programme by activating a cascade that leads directly or indirectly to activation of the BCL 2 family of proteins.

As with FGF receptors, EGF-Rs can be affected by point mutations or small deletions affecting the cytoplasmic domain and generating truncated and constitutively active receptors, sending growth-stimulatory signals into the cell and therefore acting as ONCOPROTEINS. About one-third of GLIOBLASTOMAS examined involve such faulty EGF-Rs. It can cause dramatic changes in cell shape. See MARFAN SYNDROME, TRANSFORMING GROWTH FACTOR-α.

epidermis The thin outermost portion of the SKIN, superficial to the DERMIS, and consisting of stratified keratinized squamous epithelium containing four major cell types: keratinocytes (~90%), MELANOCYTES (~8%), Langerhans cells (immune cells), and Merkel cells specialized for detecting light touch in association with nerve endings (see Fig. 71). Mitotic division in the *stratum basale* (the *Malpighian layer*; see KERATINOCYTES) pushes older cells to the surface, where they are lost (desquamation). These cells become increasingly keratinized (see KERATIN) with their age, providing both resistance to abrasion and a permeability barrier. Wrinkling of the skin is due to inability of epidermal cell replacement to keep up with loss, and to maintain integrity of connective tissue (see WERNER SYNDROME). While in the *stratum spinosum*, keratinocytes accumulate lamellar bodies containing lipid; in the *stratum granulosum* they release the lipid sealant from the lamellar bodies, and die; in the *stratum lucidum*, if present, the dead cells contain dispersed keratohyalin, while in the *stratum corneum* cells are cornified (contain keratin), have a tough protein casing and are surrounded by lipids.

Thick skin has all five epithelial strata, the stratum corneum alone having several layers. It is found in skin areas subject to pressure or friction (e.g., palms of hands, soles of feet, fingertips). *Thin skin*, covering the rest of the body, is more flexible and contains fewer cell layers, the stratum lucidum being absent and the stratum granulosum comprising one or two cell layers (see SKIN for possible confusion in terminology). See HAIR, HAIR FOLLICLE, SEBACEOUS GLANDS.

epididymis (pl. epididymides) A coiled tube between the rete testis and vas deferens, comprising a head region, body and tail region. Efferent ductules in the head region empty into a long convoluted tube (the duct of the epididymis, several metres in length) located in the body, which ends in the tail region. Final sperm maturation (spermateliosis) takes place in the epididymis: further loss of cytoplasm and maturation of the ACROSOME occur, along with ability to bind the zona pellucida. Sperm in the head region are incapable of becoming motile and cannot fertilize the second-

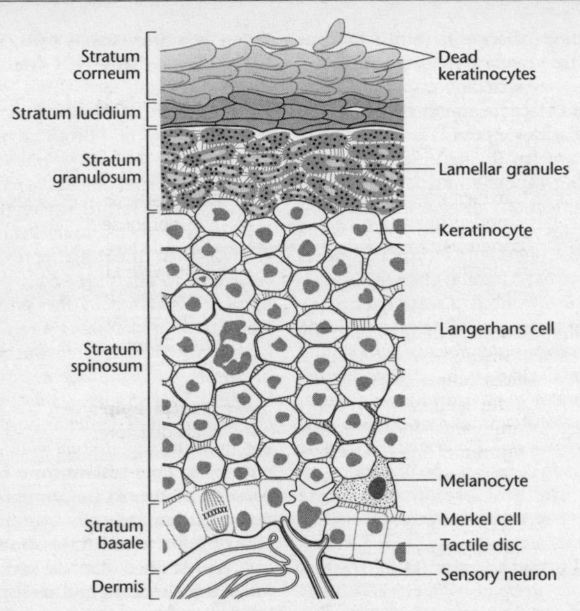

FIG 71. Diagram indicating the principal cell types and layers of the EPIDERMIS. From *Principles of Anatomy and Physiology*, G.J. Tortora & S.R. Grabowski, © 2000. Reproduced with permission of John Wiley & Sons, Inc.

ary oocyte, whereas those in the tail region are capable of becoming motile and can engage in fertilization. It takes 12–16 days for sperm to pass along the epididymis into the VAS DEFERENS.

epidural anaesthesia Injection of anaesthetics into the epidural space, between the walls of the vertebral canal and the dura mater of the spinal cord, anaesthetizing the spinal nerves. It is often given to women during childbirth (see BIRTH).

epigenesis (adj. epigenetic) In its broadest sense, any developmental process involving interactions not only between genes and their cellular environments, but ultimately between genes and the environments of the organisms of which they are components (see EPIGENETICS, NATURE–NURTURE DEBATE). Such interactions include heritable changes (from cell to daughter cell, or from parent to offspring) that do not depend upon changes in DNA sequence, and may involve CHROMATIN REMODELLING and are particularly clear in CANCER CELLS. Epigenetic regulation of gene expression includes the enzymatic modification of HISTONES, which are in turn recognized by other structural proteins and enzymes and may together stabilize the pattern of GENE EXPRESSION (see DNA METHYLATION, GENOMIC IMPRINTING, PRADER–WILLI SYNDROME, RETT SYNDROME). The next great frontier in the epigenetics of human disease is to determine its possible role in common non-neoplastic human diseases.

epigenetics Coined by Waddington in 1947, the causal study of development; or, study of the causal interactions between genes and their products which bring the phenotype into being. Today, it might be regarded as the study of heritable changes in GENE EXPRESSION that are not due to changes in DNA sequence (sometimes referred to as the 'epigenome'). Such mechanisms are crucial for the development and differentiation of an organism's different cell types (see CHROMATIN REMODELLING, DNA METHYLATION), as well as for X-CHROMOSOME INACTIVATION in female mammals. See EPIGENESIS, NATURE–NURTURE DEBATE.

epiglottis An unpaired cartilage (composed of elastic cartilage), attached to the thyroid cartilage, and projecting as a free flap towards the tongue. It covers an opening of the larynx during SWALLOWING, preventing food from entering it.

epilepsy A condition, caused by abnormal electrical activity in the brain, in which a person experiences repeated seizures, affects 7–10% of the general population during their lifetime, and is not so much a disease as a symptom of one. Its causes, when discoverable, include tumours, trauma, metabolic dysfunction, infection and vascular disease. Many forms of epilepsy show a genetic predisposition, the genes involved encoding many protein families including ion channels, transporters and receptors. Abnormal neuronal activity in the limbic system is implicated in epileptogenesis, involving signalling through various receptors including not only the ionotropic NMDA RECEPTOR and AMPAR, but also $GABA_A$, neuronal nicotinic ACh receptors and metabotropic glutamate receptors. The central histaminergic neuron system inhibits epileptic seizures, probably via H_1 and H_3 receptors. Proteins that have been implicated lie in the postsynaptic density in brain SYNAPSES and include three that bind PSD-95, a major scaffolding protein. These are stargazin, LG11 and ADAM22. Removal of part of the medial temporal cortex, including the amygdala, the anterior two-thirds of the hippocampus, and the overlying cortex from both hemispheres, has been known to alleviate seizures, although causing anterograde amnesia. See CORPUS CALLOSUM.

epimysium The fibrous envelope of connective tissue surrounding a skeletal muscle.

epinephrine See ADRENALINE.

epineurium Connective tissue encapsulating one or more fasciculi (see FASCICLE) in the nervous system.

epiphyseal growth plate Layer of hyaline cartilage in the metaphysis of a growing bone (i.e., just medial to an epiphysis) by whose activity the bone elongates (see BONE, Fig. 27). Between the ages of 18 and 25 years the epiphyseal plates 'close' as the epiphysial cartilage degenerates. The cartilage cells stop dividing and are replaced by bone, leaving a thin epiphyseal line (see ADOLESCENT GROWTH SPURT). No further increase in length of the diaphysis is then possible.

epiphysis (pl. epiphyses) The proximal or distal ends of a BONE.

epistasis The ability of one mutation to override a mutation at another locus in a double mutant. The overriding mutation is *epistatic*; that which is overridden is *hypostatic*. It results in situations where the two loci are involved in the same biochemical pathway, the epistatic mutation being 'upstream' of the hypostatic one. Compare SUPPRESSION.

epithalamus The small region superior and posterior to the thalamus, comprising the PINEAL GLAND and HABENULAR NUCLEI.

epithelium (adj. epithelial) Any sheet or tube of firmly coherent cells (linked, e.g., by desmosomes; see INTERCELLULAR JUNCTIONS) with minimal material between them and resting on a BASAL LAMINA which itself rests on connective tissue. Of ectodermal or endodermal origin, lining cavities and tubes and covering surfaces, often those exposed to toxic substances or otherwise unfavourable conditions, so that the apical surfaces of the cells are free.

epitopes (antigenic determinants) Either a site on an antigen which is recognized by an antigen receptor (and hence immunogenic) or by antibody. T-cell epitopes are peptide fragments, derived from a protein antigen, which bind appropriate MHC molecules and are recognized by a particular T cell. B-cell epitopes are antigenic determinants recognized by B cells.

epizootics A DISEASE outbreak occurring in animal populations.

EPP See END-PLATE POTENTIAL.

EPSP (excitatory post-synaptic potential) See SYNAPSE.

Epstein–Barr virus (EBV) A DNA virus of the Herpesvirus family and cause of BURKITT'S LYMPHOMA (cancer of B lymphocytes) and nasopharyngeal carcinoma. ~15% of clinically recognized cases of infection have clinical hepatitis. It infects B cells by binding to CR2 (CD21), a component of the B cell co-receptor complex for complement, and enters latency as episomes in B cells after a primary infection that often passes without diagnosis. In some people, the initial infection causes infectious mononucleosis (GLANDULAR FEVER). EBV is shed in pharyngeal secretions, transmission also occurring via shared eating utensils. In some human populations, the virus evades host immune responses by evolving proteins whose peptide fragments cannot bind MHC molecules. A virally encoded cytokine homologue of IL-10 inhibits T_H1 cells, reducing IFN-γ production.

ER See ENDOPLASMIC RETICULUM.

***erb* gene, Erb protein** See CANCER.

erectile tissue See CLITORIS, PENIS.

erection See PENIS.

erector (arrector) pili muscles See HAIR FOLLICLE.

ERK (extracellular signal-regulated kinase) An enzyme (classical MAP-KINASE) involved in one of the kinase pathways induced by *Ras* activation (see RAS SIGNALLING PATHWAY).

erythroblastosis foetalis (haemolytic disease of the newborn, HDN) Disease occurring when a Rh-negative mother makes IgG antibodies specific to the rhesus blood group antigen expressed on the red blood cells of her foetus. Foetal and maternal blood are normally kept apart during pregnancy, but an Rh-negative mother can be sensitized to the Rh antigen inherited from the paternal parent, usually at the time of delivery. Anti-Rh antibodies will cross the placenta and cause haemolysis in any subsequent Rh-positive foetus. HDN is prevented by giving all Rh-negative women an injection of anti-Rh antibodies (anti-Rh gamma globulin, RhoGAM) soon after every delivery so that they bind to and inactivate the foetal Rh antigens before they can sensitize the mother.

erythrocytes See RED BLOOD CELLS.

erythropoietin (EPO) A cytokine growth factor and haematopoietin family member, synthesized by kidney cells under low blood oxygen levels (see ALTITUDE), acting through the JAK/STAT PATHWAY and having its principal effect on bone marrow by stimulating red blood cell precursors to survive, proliferate and differentiate. EPO is synthesized and released in response to local arterial hypoxia and initiates red blood cell formation within 15 hours of altitude ascent. Produced by recombinant DNA technology, it is used therapeutically by millions of anaemia patients, often those with kidney disease, reducing the need for repeated blood transfusion. It seems also to have general tissue-protecting effects, notably in the nervous system, eye, heart and kidney. It does so by targeting components common to many different neurodegenerative pathways (e.g., anti-apoptotic, antioxidant, glutamate-inhibitory, anti-inflammatory, neurotrophic, stem cell modulatory and angiogenic). A modified form of EPO, carbamyl-EPO (CEPO), seems to have neuroprotective properties but without the erythropoietic effects of EPO and seems a suitable candidate for clinical trials of chronic neurological and psychiatric conditions requiring long-term treatment.

Escherichia coli A facultatively anaerobic Gram-negative bacterium occupying the colon (see GUT FLORA), fermenting a wide range of sugars, including lactose, and producing acid and gas. It is now possible to distinguish commensal from pathogenic serotypes. Pathogenicity factors for *E. coli* include capsular K antigens (K1), fimbriae, lipopolysaccharide, enterotoxins (heat-stable and heat-labile), verocytotoxin (shiga-like toxin; Shiga toxigenic *E.coli* (STEC); see SHIGELLA).

essential amino acids Amino acids that cannot be synthesized by an animal and must therefore be present in its diet or be absorbed from the GUT FLORA. Those for humans are underlined in the list of AMINO ACIDS. See HEALTHY DIET.

essential fatty acids Those fatty acids that humans cannot synthesize and which must be present in the DIET. An intake of 1–2% of the total dietary energy requirement should be in this form in order to avoid deficiency symptoms (e.g., fragile red blood cells, skin lesions, kidney damage, hair and weight loss). The polyunsaturated linoleic and linolenic acids, present in fish oils, are precursors of the EICOSANOIDS prostaglandins and leukotrienes, which are important in the regulation of several physiological processes including lowering of blood cholesterol and reducing thrombus formation. This is because linoleate can be converted into some other polyunsaturated acids, such as ARACHIDONIC ACID, which can only be synthesized from this source. See ASPIRIN, OMEGA-3 FATTY ACIDS, OMEGA-6 FATTY ACID.

essential hypertension (idiopathic hypertension) See HYPERTENSION.

estrogens See OESTROGENS.

ESTs See EXPRESSED SEQUENCE TAGS.

ethanol (ethyl alcohol) See ALCOHOL.

ethmoid bone An unpaired bone of the neurocranium (see SKULL) forming part of the anterior cranial floor, the medial wall of the orbits, and the superior portion of the nasal septum, dividing the nasal cavity into right and left sides. The perpendicular plate and cribriform plate (containing olfactory foramina) lie superiorly and anteriorly forming respectively the superior portion of the nasal septum and roof of the nasal cavity. Branches of the olfactory nerve (CNI) enter the cranial cavity from the nasal cavity through the olfactory foramina. Its *ethmoid sinus* (ethmoid labyrinth) consists of a maze of interconnected ethmoid 'air cells', giving the bone a sieve-like appearance. Its lateral masses contain two thin, scroll-shaped projections, the superior and middle nasal turbinates. See CONCHAE.

ethology The approach, associated particularly with the work of Konrad Lorenz and Niko Tinbergen, which seeks to interpret animal behaviour in terms of its individual fitness and adaptiveness to the organism. Lorenz championed the comparative method in tracing the evolution of animal behaviour. Reaching its zenith in the 1960s and early 1970s it has been replaced as a methodology by interpretations relying on the notions of INCLUSIVE FITNESS and SELFISH GENE THEORY and the rise of behavioural ecology. Opponents of ethology took issue on two major points. They argued (i) for interactionism, taking issue with the notion that instinctive behaviour is innate; (ii) that genetic determinism of behaviour and the innate-learning dichotomy of ethology omitted individual development (ontogeny; e.g., social deprivation) as a causal factor in the maturation of behaviour. This acceptance of epigenetic interaction in the development of behaviour, which Tinbergen adopted, meant that no element in an animal's behaviour could be assumed to be purely instinctive or learned. Lorenz argued that learning is almost always adaptive in outcome because learners arrive in the world with innate structures that determine the adaptive outcome of LEARNING. See NATURE–NURTURE DEBATE, SOCIOBIOLOGY.

euchromatin Those regions of CHROMATIN that are decondensed and believed to represent gene loci that are transcriptionally active. Genes in euchromatic regions replicate early, the chromatin containing hyperacetylated histones which stain poorly in the nucleus. Active genes in this region are enriched for methylation at lysine 4 (Lys4) of histone H3 (H3-K4), H3-K36 and H3-K79. See HETEROCHROMATIN. Silenced euchromatic genes contain methylated H3-K9, H3 tri-methyl K27 and H4 mono-methyl K20.

eukaryote, eukaryotic Cells which either contain a NUCLEUS, or are derived from one that did.

Eustachian tubes (pharyngo-tympanic tubes, auditory tubes) Canals linking the middle ear cavities to the PHARYNX and closed externally by the eardrum. They develop from the second pharyngeal pouches (but the first to be seen in human development), lying on either side between the maxillo-mandibular and hyoid arches, and permit equilibration of pressure on the two sides of the eardrums (e.g., during swallowing).

evoked potential (evoked response) Electrical signals in the cerebral cortex representing the summed activity of thousands of neurons, recorded by metal electrodes or electrolyte-filled glass capillaries. The latency between stimulus and evoked potential gives an indication of the number of relay stations in the pathway.

evoked response See EVOKED POTENTIAL.

evolution **1**. *Microevolution*: the change in appearance, or gene frequencies, of populations and species over generations. Many of the most rapidly evolving proteins function in host defence and immunity (e.g., genes of the MAJOR HISTOCOMPATIBILITY COMPLEX) and some of those involved in reproduction. Increasing evidence indicates that it results from NATURAL SELECTION, sexual selection and GENETIC DRIFT. **2**. *Macroevolution* (phyletic evolution): the origins and extinctions of species and grades.

excitatory post-synaptic potential (EPSP) See SYNAPSE.

excitement For sexual, see AROUSAL.

excitotoxicity Over-excitation of neurons by GLUTAMATE, which is stored in large quantities in brain synapses. If the glutamate concentration outside neurons rises to that within them (>3 mM), they die in minutes. Disease states leading to neuronal death (e.g., cardiac arrest, stroke, brain trauma, seizures and oxygen deficiency) can all result in glutamate leakage. Insufficient ATP is generated for calcium pumps, membranes depolarize and Ca^{2+} leaks into neurons, triggering glutamate release. This further depolarizes the neurons, releasing further Ca^{2+} and causing further glutamate release. Glutamate receptors allow excessive entries of Na^+, K^+ and Ca^{2+}, the latter entering through NMDA RECEPTORS. Neurons swell osmotically, as the Ca^{2+} activates intracellular enzymes that degrade proteins, lipids and nucleic acids. This has been implicated as one possible factor in ALZHEIMER'S DISEASE and AMYOTROPHIC LATERAL SCLEROSIS, and one role of ENDOCANNABINOIDS appears to be to reduce excitotoxic damage. See TOXINS.

excretion The removal from the blood (usually) of waste products of metabolism. Excretory organs include the KIDNEYS, LIVER, LUNGS and SKIN. Contrast EGESTION.

exercise During any form of exercise, cardiac contractility and output are increased, respiration is enhanced and blood flow redirected to provide adequate supply to the brain and skeletal muscle. MINUTE VENTILATION commonly increases to 100 L or more (~17- to 20-fold larger than resting value). The sympathetic nervous system is activated, growth hormone is released and the hypothalamic-pituitary-adrenal axis is activated, with resultant rise in cortisol, suppression of insulin release and stimulation of glucagon release. These responses stimulate lipolysis and hepatic and muscle glycogenolysis, while muscle lactate output provides a substrate for hepatic gluconeogenesis.

We are appreciating the role of such coactivator proteins as PGC-1α in coordinating the multiple pathways involved in complex biological programmes such as the onset of exercise (see PGC-1 GENES). PGC-1α functions as a sensitive 'rheostat' in SKELETAL MUSCLE, responding to neuromuscular input and the ensuing contractile activity. Its expression is readily inducible by exercise training in rodents and humans and has been shown to be intimately linked with the CIRCADIAN CLOCK.

Initial increased oxygen consumption during exercise always lags behind energy expenditure, but after several minutes of submaximal exercise, liberation of 'hydrogen atoms' from glycolysis, fatty acid oxidation and the Krebs cycle becomes proportional to exercise intensity, and VO_2 attains a steady rate (see RESPIRATORY QUOTIENT). In light exercise, the steady rate is reached rapidly; moderate-to-heavy aerobic exercise requires longer, creating a larger OXYGEN DEFICIT and taking longer for oxygen consumption to return to the resting level in recovery. Also in light exercise, lipolysis and fatty acid release by adipose tissue increase ~threefold, mainly through increased β-ADRENERGIC stimulation, the energy for muscular work being derived from ATP produced by the aerobic pathway involving oxidation of fatty acids and ketones. The power generated solely by fatty acid oxidation is about one-half that obtained if carbohydrate is the sole energy source (see FATTY ACIDS). A rate limit exists for fatty acid use by active muscle at about 65% VO_2 max, aerobic training (see later) enhancing this limit, which is probably caused by reduced fatty acid delivery from adipose tissue to muscle. As exercise intensity increases from mild to moderate and intense, energy substrate selection switches from lipid to carbohydrate dependence, growth hormone and cortisol contributing only minimally to exercise-induced rise in liver glucose output. In more vigorous exercise, glycogen is oxidized aerobically. At higher work rates still, ATP demand is supplied by the anaerobic pathway, producing ATP more rapidly through conversion of glycogen to lactic acid. Depletion of muscle glycogen, as occurs in prolonged exercise (e.g., marathons), consecutive days of heavy training or low-carbohydrate diets, decreases a muscle's maximum aerobic power output and is probably the cause of 'peripheral' or local muscle fatigue during exercise. The contribution of amino acid oxidation to total energy expenditure is small during short-term intense exercise, accounting for ~3–6% of total ATP supplied during prolonged exercise. However, the intermediary metabolism of several amino acids, such as glutamate, alanine and branch-chain amino acids, influences the availability of KREBS CYCLE intermediates. Sustained dynamic exercise especially stimulates branch-chain amino acid oxidation, with resulting ammonia production. If intense enough, such exercise causes a net loss of muscle protein (through decreased protein synthesis, increased breakdown, or both). Although some of these amino acids are oxidized, others supply substrates for gluconeogenesis and possibly for acid–base regulation; but gluconeogenesis cannot replenish or even maintain glycogen reserves without regular carbohydrate consumption.

Typical *aerobic training* programmes (see OVERLOADING OF MUSCLES) take place 3 days per week, usually alternating with a rest day. Whereas 3- to 5-minute daily aerobic exercise periods produce training effects in some poorly conditioned people, 20- to 30-minute sessions achieve more-optimal results if intensity reaches at least 70% of maximal heart rate. To achieve meaningful weight loss through exercise, each session should last at least 60 minutes at a sufficient intensity to expend 300 kcal or more. *Endurance training* increases total skeletal blood flow by enlargement of the diameters of the main arteries and veins and 10% increase in capillary supply. It lowers blood lactic acid levels and extends the level of exercise required before the start of lactate accumulation in the blood (the LACTATE THRESHOLD, see Fig. 108). It also causes skeletal muscle fibres to have larger and more numerous mitochondria, a ~twofold increase in aerobic system enzymes (within 5–10 days of training) coinciding with increased mitochondrial ATP production. Such typical increases in mitochondrial protein greatly exceed the 10–20% increases in VO_2 max achieved by endurance training and probably reflect a person's ability to sustain a high percentage of aerobic capacity without significant blood lactate accumulation. The enhanced fatty acid mobilization and oxidation increases during endurance training conserve glycogen stores and result from greater flow of blood through trained muscle, increased synthesis of fat-mobilizing and fat-metabolizing enzymes, enhanced muscle fibre mitochondrial respiratory capacity and reduced catecholamine release for the same absolute power output compared with untrained persons. Also, carbohydrate oxidation is greater during maximal exercise, although the reduced carbohydrate use during submaximal exercise reduces muscle glycogen turnover. Trained and untrained individuals attain similar steady-state oxygen consumption values, but the endurance-trained person reaches steady state rapidly with a smaller OXYGEN DEFICIT than the untrained or sprint-power (anaerobic) athlete. Although the basic type of SKELETAL MUSCLE fibre does not change to any great extent, highly trained endurance athletes have larger slow twitch fibres than fast twitch fibres in the same muscle. It is not clear whether any training schedule affects muscle myoglobin content, although myoglobin content is large in slow twitch fibres with high capacity to generate ATP aerobically. Endurance athletes generate ~20 × their resting energy release during a 2.5-hour marathon, recycling as much as 80 kg ATP in the process (see ADENOSINE TRIPOPHOSPHATE, PHOSPHOCREATINE). The heart's mass and volume increase greatly with long-term aerobic training, increasing the number of anastomoses in the CORONARY ARTERIES, enlarging CARDIAC MUSCLE fibre diameter and enlarging the left ventricular cavity, with modest thickening of its walls. Training also decreases the intrinsic firing rate of SA NODE tissue, explaining the small resting and large submaximal exercise bradycardia (by 12–15 beats min^{-1}) in highly conditioned endurance athletes and in sedentary individuals who train aerobically. This is normally associated with a compensatory increase in maximum STROKE VOLUME and cardiac output, representing the most significant adaptation in cardiovascular function through aerobic training.

Regular aerobic training blunts the typical age-related decline in static and dynamic lung functions (e.g., RESIDUAL VOLUME). Pulmonary adaptations of aerobic exercise reflect a breathing strategy that minimizes respiratory work at a given exercise intensity. Maximal exercise ventilation increases (TIDAL VOLUME and VENTILATION rate increase) and several weeks of aerobic training reduce the VENTILATORY EQUIVALENT for oxygen during submaximal exercise, reducing the fatiguing effects of exercise on ventilatory muscles and allowing more oxygen to become free for locomotor muscle activity. Exercise training also increases inspiratory muscle capacity to generate force and sustain a given level of inspiratory pressure, reducing the muscle's lactate production, prolonging high intensity exercise. The ventilatory muscles also metabolize lactate better. Regular physical exercise enhances hippocampal synaptic plasticity and neurogenesis, and is neuroprotective.

Anaerobic exercise is exercise of short duration but high intensity (e.g., a 100-metre sprint, a fast 25-m swim or heavy weightlifting) and requires an immediate supply of energy provided largely by the intramuscular HIGH-ENERGY PHOSPHATES ATP and phosphocreatine, PCr (see ADENOSINE TRIPHOSPHATE, Fig. 2). The maximum rate of energy transfer from these sources exceeds by 4–8 times the maximum energy transfer from aerobic respiration. Even during a world-class 100-m sprint, there is insufficient existing phosphagen to maintain speed consistently and a runner slows down during the last few seconds, so that the quantity of phosphagens available significantly affects performance. Athletes highly trained in anaerobic power have fast twitch fibres occupying a much greater proportion of a muscle's cross-sectional area than do those undergoing endurance training.

The ability to sustain exercise beyond a brief period and recover from previous exertion requires additional ATP production, derived largely by substrate-level phosphorylation from GLYCOLYSIS, with resulting lactate formation. This 'buys time' while the oxygen supply is still inadequate for the muscle to resynthesize ATP aerobically (see OXYGEN DEBT).

The highest exercise level that can be performed without a significant change in blood pH is the *anaerobic threshold*. Work rates in excess of this level result in greater production of lactic acid by skeletal muscles, a rise in its plasma concentration and a fall in pH (due largely to lactate buffering and pCO_2 increase; see LACTIC ACID). This stimulates the carotid bodies, resulting in increased rate and depth of ventilation (see RESPIRATORY CENTRE) so that blood pCO_2 can fall below resting levels and pO_2 rise above resting levels.

At the pO_2 in the lungs, haemoglobin is 98% saturated with oxygen, while at the pO_2 of active tissues haemoglobin is 25% saturated with oxygen, so that 73% of the oxygen absorbed in the lungs is unloaded in the tissues. The Bohr effect and raised temperature of blood in exercising tissues contributes to this unloading (see HAEMOGLOBIN).

Prolonged exercise increases the activity of the GLUCOSE TRANSPORTER in muscle cell membranes, which is the main reason why glycogen synthesis increases during the hours following prolonged exercise. This *glycogen supercompensation* only occurs in muscles where glycogen levels have been reduced to low values and is responsible for the good correlation between pre-exercise muscle glycogen concentrations and cycling time to exhaustion. Eating a high-carbohydrate diet or drinking high-carbohydrate drinks after heavy exercise (carbo-loading) increases muscle glycogen content by up to 90% and improves endurance capacity by up to 50%. Glycogen-loading diets would require consumption of roughly 1 g carbohydrate kg^{-1} body mass 1 hour prior to exercise, or 5 g carbohydrate kg^{-1} body mass if taken 4 hours prior to exercise. Alternatively, it could involve a 15–25% carbohydrate drink or solid high-carbohydrate supplement taken with each meal, stabilizing one's body weight during all phases of training by matching energy consumption to training energy demand, which also helps to maintain carbohydrate reserves.

Fatigue during exercise is not solely the result of reduced muscle glycogen levels and glucose availability. Dehydration caused by SWEAT production brings on fatigue earlier than muscle glycogen depletion. Well-formulated sports drinks containing sodium and potassium salts and 5–8% carbohydrate (glucose or dextrins) if drunk throughout prolonged heavy exercise delay onset of severe dehydration and improve performance by decreasing the rate of glycogen breakdown. During prolonged exercise, an increase occurs in plasma protein-bound levels of the amino acid tryptophan. Plasma fatty acids displace this, the free tryptophan entering the brain where some of it will be converted to SEROTONIN, causing fatigue. But drinking a carbohydrate solution raises insulin levels and so decreases the mobility of fatty acids from adipose tissue so that more tryptophan binds albumin and onset of fatigue is delayed. See OVERTRAINING SYNDROME.

exocoelomic cavity The primitive YOLK SAC.

exocrine gland Gland secreting material onto an epithelial surface via a specialized duct (e.g., salivary glands, sweat glands and the exocrine pancreas).

exocytosis The mode by which large molecules are exported from a cell, and by which membrane budding occurs. See VESICLES.

exogenous Originating outside the system concerned. Contrast ENDOGENOUS.

exogenous pyrogens Pathogen-associated substances, such as lipopolysaccharide, which cause FEVER. Compare ENDOGENOUS PYROGENS.

exon shuffling The copying and insertion into other genes of exons or exon groups encoding whole protein domains. See EXONS.

exons Those segments of eukaryotic genes, usually adjacent to an INTRON, comprising nucleotide sequences that will be represented by complementary sequences in mRNA (and amino acid sequences in the translated polypeptide), or in the final tRNA or rRNA, after RNA PROCESSING in the nucleus. Sex-specific exons occur in the cytosine methyltransferase gene *Dnmt1* (see GENOMIC IMPRINTING).

Although many of these exons are 'expressed sequences', not all exons are coding exons (e.g., see UTRS). *Symmetrical exons* have a nucleotide number that is an exact multiple of the triplet codon sequence, *non-symmetrical exons* having a number of nucleotides not exactly divisible by three. The retrotransposition machinery has made important contributions to gene evolution by making copies of exons or exon groups and then shuffling them from one location to another within the genome (*exon shuffling*), a process usually limited to symmetrical exons since it avoids frameshift mutations. The mechanism is most likely to involve LINE element-assisted RETROTRANSPOSITION. The presence of the numerous introns on either side of exons in eukaryotic DNA facilitates this, allowing genetic recombination to combine the exons of different genes without interrupting their DNA sequences, enabling genes for new proteins to evolve speedily. Indirect evidence for this is provided by the numerous different proteins that contain the same protein domains. Exon shuffling at the DNA level has been exploited in ANTIBODY engineering.

Exon duplication (see GENE DUPLICATION) has enabled the evolution of proteins with similar repeating domains covalently linked in series (e.g., FIBRONECTIN). *Exon-splicing enhancer* (ESE) sequences have recently been identified. These purine-rich sequences are required for the correct splicing of exons with weak splice site consensus sequences (see RNA PROCESSING). See MUSCULAR DYSTROPHIES for a possible therapeutic approach involving ESE blocking.

exonucleases Enzymes degrading a nucleic acid from one end of the molecule, many operating only in the 5´→3´ or the 3´→5´ direction. A few degrade only single-stranded molecules. See DNA POLYMERASES, ENDONUCLEASES.

exotoxins See TOXINS.

experience-based neuroadaptations See DENDRITES, LEARNING, MEMORY, NEURAL PLASTICITY.

expiration See VENTILATION.

expiratory reserve volume The volume of air, additional to the expiratory TIDAL VOLUME, that can be expired. See FORCED VITAL CAPACITY, SPIROMETER, VENTILATION.

explicit memory (declarative memory) Memories requiring conscious recall, concerned with people, places, objects and events. They require specialized anatomical organization such as that found in the MEDIAL TEMPORAL LOBE and HIPPOCAMPUS. See MEMORY.

expressed sequence tags (ESTs) ES-tagging is a variant of SEQUENCE-TAGGED SITE MAPPING, and of more limited application. Sequence-specific primers at the untranslated 3´ ends of cDNA clones (see COMPLEMENTARY DNA) are designed to identify sequences of expressed genes, which can be assigned a specific chromosomal location by methods including SOMATIC CELL HYBRIDIZATION, FISH and POLYMERASE CHAIN REACTION analysis of gridded arrays of yeast artificial chromosomes. ESTs have applications in the discovery of new human genes, GENOME MAPPING, and in identification of coding regions in genomic sequences. Their mapping location and cell or tissue of origin enables them to provide a rich source of partial sequences of a large number of positional candidate genes for inherited human diseases; however it is relatively slow and cannot usually identify mRNAs expressed at low levels. See DNA MICROARRAYS, DNA SEQUENCING.

expression See EMOTIONS, EXPRESSIVITY, GENE EXPRESSION.

expressivity A measure of the degree to which a given allele is expressed in the phenotype; a measure of the intensity of the phenotypic trait. As with penetrance, expressivity is central to the concept of the reaction norm. It may be influenced by MODIFIERS and by the individual's environment, and is a complicating factor in genetic counselling. See screening.

extended-source outbreak See DISEASE.

extension Increase in the angle between two bones. Compare FLEXION; see EXTENSOR MUSCLES.

extensor muscles Muscles whose contraction results in EXTENSION. They include the 'antigravity' muscles, particularly those of the limbs, involved in maintaining posture. The overall effect of the VESTIBULOSPINAL TRACTS is extensor activation. RETICULOSPINAL TRACT inhibition and excitation of extensor motor neurons tend to cancel net extensor activity, while reticulospinal inhibition of interneurons mediating reflex contraction of FLEXOR MUSCLES tends to enhance net extensor activity. For involvement in walking, see CENTRAL PATTERN GENERATORS, GAIT.

external auditory meatus (auditory canal) Canal of the external EAR, located in each temporal bone and leading from the pinna to the tympanic membrane. About 25 mm long and 7 mm in diameter, it is derived from the 1st pharyngeal cleft (see PHARYNGEAL POUCHES, Fig. 134). Just posterior to it is a large inferior projection, the MASTOID PROCESS. See HEARING.

external ear The pinna and external auditory meatus of the EAR.

external genitalia (primary genitalia) The externally visible sex organs, in males the PENIS and scrotum, in females the CLITORIS and VULVA, including the genital cleft, LABIA MAJORA and external LABIA MINORA. In the 3rd week of development, mesenchyme cells from the primitive streak migrate around the cloacal membrane and form a pair of slightly elevated cloacal folds, which unite cranially to form the medial *genital tubercle*. During this indifferent stage (see SEX DETERMINATION), the folds are subdivided caudally into urethral folds anteriorly and anal folds posteriorly, and are surrounded by another pair of swellings, the *genital swellings*. In males, these form the scrotal swellings, while in females they become the labia majora, although there is no distinction until the 7th week.

Before they differentiate at about this time, the penis of males and the clitoris of females are represented by the genital tubercle. But under the influence of androgens, this elongates rapidly in males and pulls the urethral folds forward to form the urethral groove extending along the caudal aspect of the penis and folding over to form the canal of the penile URETHRA. The scrotal swellings arise in the inguinal region and move caudally to form the scrotum, whose two halves are divided by the scrotal septum. In females, OESTROGENS cause the genital tubercle to elongate slightly to form the clitoris (it is actually larger than the penis until the end of the 4th month), but the urethral folds do not fuse. Instead, they develop into the labia minora, while the genital swellings enlarge as the labia majora. The urogenital groove between the urethral folds remains open as the vestibule and will house the opening of the urethra and the entrance to the vagina. SEBACEOUS GLANDS are present, unassociated with hair.

external intercostal muscles See INTERCOSTALS.

external nares The external nostrils; the openings into the nasal cavity from the exterior.

external respiration See GASEOUS EXCHANGE.

exteroceptor Any receptor responding to stimuli outside the body.

extinction 1. In CLASSICAL CONDITIONING, the decrease in intensity, or probability of occurrence, of a conditioned response after the conditioned stimulus has been repeatedly presented without the unconditioned stimulus. Rather than being a form of 'forgetting', it seems to be a kind of 'learning to unlearn' – e.g. learning that the conditioned stimulus now signals that the unconditioned stimulus will not occur. This principle underlies therapies designed to overcome fear conditioning. See ENDOCANNABINOIDS. **2.** See HABITAT HYPOTHESIS.

extracellular Strictly, of those molecules and spaces outside a cell's plasma membrane. In reality however, many molecules of the EXTRACELLULAR MATRIX are rooted in that membrane.

extracellular fluid (EF) All the fluid outside cells, comprising ~20% of total body mass. It includes BLOOD plasma (confined to extracellular spaces within blood vessels), TISSUE FLUID (occupying extracellular spaces outside blood and lymph vessels; see Fig. 174), LYMPH (confined to lymph vessels), CEREBROSPINAL FLUID (confined to the central canal, ventricles of the brain and subarachnoid space) and SYNOVIAL FLUID (confined to articular capsules). Regulation of EF volume is due largely to homeostatic mechanisms responding to changes in BLOOD PRESSURE (see entry).

extracellular matrix (ECM) A complex network of macromolecules lying between the cells of tissues and colonies, largely comprising (i) the polysaccharide chains of glycosaminoglycans (GAGs), normally covalently linked to proteins as PROTEOGLYCANS, and (ii) largely locally secreted proteins including the structural GLYCOPROTEINS collagen and elastin, adhesive fibronectin and such signal molecules as LAMININS. It forms the BASAL LAMINA between an epithelium and its underlying connective tissue stroma, to which cells adhere via hemidesmosomes, focal adhesions and non-junctional mechanisms (see INTERCELLULAR JUNCTIONS, Fig. 106). The anchorage dependence of cells cultured *in vitro*, by secretion of ECM components that adhere to the glass or plastic, reflects the need for normal cells to be tethered to ECM components in order to survive and proliferate (see INTEGRINS). The ECM may contain bound growth factors and matrix METALLOPROTEINASES (MMPs), secreted and membrane-anchored proteinases which, by loosening up the dense meshwork of matrix molecules, may promote both normal physiological and tumorigenic invasion of the connective tissue stroma (see METASTASIS). The ECM is crucial in NEURAL REGENERATION. See STEM CELLS.

extraembryonic coelom The chorionic cavity. Developing in the EXTRAEMBRYONIC MESODERM, it surrounds the primitive yolk sac and amniotic cavity except where the germ disc attaches to the trophoblast. See BLASTOCYST (Fig. 20), COELOM.

extraembryonic mesoderm Developing from cells of the primitive yolk sac and from cells that pass through the most caudal part of the PRIMITIVE STREAK, a fine loose connective tissue that eventually fills all the space between the trophoblast externally and the amnion and exocoelomic membrane internally (see BLASTOCYST, Fig. 20). Large cavities develop in it, coalescing to form the EXTRAEMBRYONIC COELOM whose inner mesoderm is the extraembryonic splanchnopleuric mesoderm (chorionic plate) and whose outer mesoderm is the somatopleuric extraembryonic mesoderm (connecting stalk) which, when supplied with blood vessels, becomes the UMBILICAL CORD.

extrafusal muscle fibres See MUSCLE SPINDLES.

extrapyramidal motor pathways See BASAL GANGLIA.

extrastriate cortex (V2, V3 and V4) This comprises ~30 distinct areas of the cerebral cortex receiving projections from the primary visual cortex (area V1), and also from the RETINAS, whose contributions to visual processing are still vigorously debated. The macaque monkey has been the source of much of our knowledge, although it is thought that most regions of the extrastriate cortex have counterparts in humans. Internally classified on the basis of its cytoarchitecture (e.g., cytochrome oxidase distribution), connections and physiology, it includes Brodmann areas 18 and 19 in the occipital cortex, and areas of the temporal and parietal cortex. Two large-scale cortical streams (see PARALLEL PROCESSING) project from layers 2, 3 and IVC of the PRIMARY VISUAL CORTEX (see Fig. 141) to the extrastriate cortex (see Fig. 72):

one projecting dorsally towards the PARIETAL CORTEX and involved in analysis of visual motion – the 'where' stream (see VISUAL MOTION CENTRE); the other projecting ventrally towards the INFEROTEMPORAL CORTEX (ITC) and FUSIFORM GYRUS, serving object recognition – the 'what' stream (see PROSOPAGNOSIA).

Most output from V1 goes to V2, the *secondary visual cortex*, occupying part of area 18, which has a characteristic pattern of thin and thick stripes when stained with cytochrome oxidase. Thus the magnocellular and parvocellular pathways from the LATERAL GENICULATE NUCLEUS extend through and beyond areas V1 and V2 (see BLINDSIGHT). Cells in the thick stripes derive from IVCα via IVB, are part of the magnocellular pathway (dorsal stream), and are motion-sensitive and binocular. They project in turn via V3 to the MEDIAL TEMPORAL CORTEX (MT), V5, and from there to the medial superior temporal cortex (MST), and POSTERIOR PARIETAL CORTEX. The interstripe region of V2 receives interblob inputs from layers 2 and 3 of area V1, and also projects to V3 and then to V4. However, many cells in V3, and some of those in V4, are orientation-selective and are a continuation of the parvocellular–interblob pathway for form perception. The parvocellular–blob path from layers 2 and 3 of area V1 projects to the thin stripe of V2, and thence to area V4 and the ITC. This ventral stream is crucial for object recognition, thin stripe V2 cells and some V4 cells being wavelength-sensitive and showing colour constancy. This is probably why loss of colour vision, ACHROMATOPSIA, occurs in patients with damage to V4. Extensions of the dorsal and ventral streams have been implicated in DECISION-MAKING. See ATTENTION.

extravasation Escape of fluid (e.g., blood, lymph) from a vessel into the tissues; or exit of leukocytes from the circulation to the tissues by DIAPEDESIS. See METASTASIS.

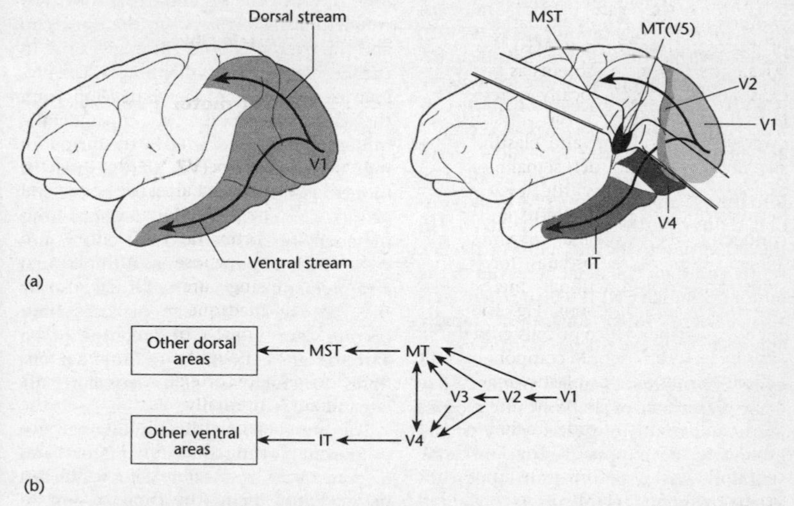

FIG 72. (a) Diagram showing the macaque monkey's dorsal and ventral visual processing streams projecting from the primary visual cortex (V1) to the EXTRASTRIATE CORTEX (see entry). The extrastriate visual areas (MST, MT) under the overlying temporal cortex are shown exposed. (b) Representation of the flow of information in the dorsal and ventral pathways. © 2001 from *Neuroscience* (2nd edition) by M.F. Bear, B.W. Connors and M.A. Paradiso, published by Lippincott, Williams & Wilkins. Reproduced with the kind permission of the publisher.

extremities The limbs. See APPENDICULAR SKELETON.

extrinsic Not contained in or belonging to a body; external; outward. In anatomy, attached partly to an organ or limb and partly to some other part, e.g., certain groups of muscles (e.g., extrinsic eye muscles). Opposed to *intrinsic*.

extrinsic pathway See BLOOD CLOTTING.

exudate Escaping fluid or fluid-like material when oozing from a space; may contain serum, pus and cellular debris.

***ex vivo* gene therapy** Transfer of genes to a patient's recipient cells cultured in the laboratory, after which they can be replaced into the patient. It is used for cells that can be accessed for initial removal, and can be induced to engraft and survive for a long time after replacement. Such cells include haematopoietic cells, skin cells, etc. See GENE THERAPY.

eye-blink reflex The eyelids are closed by relaxation of the levator palpabrae muscles, supplied by the oculomotor nerve (CNIII), coupled with contraction of the orbicularis oculi muscles, supplied by the facial nerve (CNVII). If the reflex occurs when the cornea is irritated (the *corneal reflex*), the afferent fibres run in the trigeminal nerve (CNV). Very bright light shone directly into the eye also leads to >99% eyelid closure. In this case, the afferent signals arise in the retina, pass to the midbrain visual reflex centre and then collateral fibres run to the motor nucleus of CNVII. This *dazzle reflex* is also elicited when an object rapidly approaches the eye (the *threat reflex*).

eyes Eye rudiments appear in the 22-day embryo as a pair of shallow optic grooves on the side of the forebrain, which rapidly expand to form optic vesicles on optic stalks and invaginate to form optic cups as the overlying ectoderm is organized into lens placodes. The transcription factor gene *PAX6* is the master gene for eye development, usually being expressed in a central band of the anterior neural ridge. *Sonic hedgehog* (*SHH*), secreted by the prechordal plate, inhibits *PAX6* expression in the midline and up-regulates *PAX2* expression instead. *PAX2* then regulates differentiation of the optic stalk as *PAX6* is expressed in the optic cup and in the overlying ectoderm forming the lens (see PAX GENES). Cells of the surface ectoderm, initially in contact with the optic vesicle, elongate and form the *lens placode*. This subsequently invaginates to form the lens vesicle and during the 5th week this loses contact with the ectoderm, lying in the mouth of the optic cup. FGFs from the ectoderm promote differentiation of the neural (inner) layer of the optic cup and induce expression of *CHX10*, forming the retina. By the 7th week, the outer layer of the optic cup has become the pigmented layer of the retina while the more complex inner, neural, layer develops rods and cones (in its posterior four-fifths) while the adjacent mantle layer gives rise to the neurons and supporting cells of the RETINA (see Fig. 149). The anterior fifth of the inner layer remains one cell thick and divides to form the inner layer of the IRIS and CILIARY BODY. The ectodermal region between the optic cup and the overlying epithelium gives rise to the muscles of the iris diaphragm. The pigmented outer layer of the iris develops from the external layer of the optic cup under the influence of TGFβ secreted by the surrounding mesenchyme and *MITF* which it induces (see MELANOCYTES, WAARDENBURG SYNDROME). The ciliary muscle develops from mesenchyme adjacent to the ciliary process, which originates from a thin neural and pigment layer of the retina. Also by the 7th week, the lens placodes have become lens vesicles and consist of an external lens epithelium and internal lens fibres, to which secondary lens fibres are later added (see LENS).

The mesenchyme surrounding the eye primordium at the 5th week differentiates into an inner layer equivalent to the pia mater of the brain (the CHOROID) and an outer layer, the SCLERA, comparable to the dura mater. However, the mesenchyme overlying the anterior chamber of the eye vacuolates and splits into an inner layer in front of the lens and iris, and an outer layer

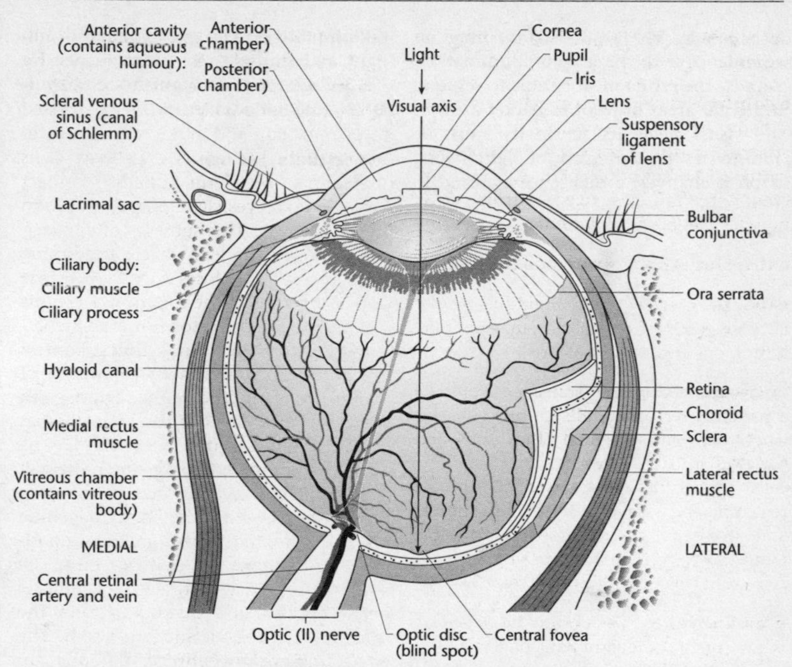

FIG 73. Superior view of transverse section of the right eyeball, showing its gross anatomy (see EYES). From *Principles of Anatomy and Physiology*, G.J. Tortora & S.R. Grabowski, © 2000. Reproduced with permission of John Wiley & Sons, Inc.

continuous with the sclera: the central substantia propria of the cornea. This becomes covered with internal and external epithelia to form the eventual CORNEA. The vacuolated anterior chamber becomes filled with AQUEOUS HUMOUR. The vitreous body (posterior chamber) develops from mesenchyme that invades the inside of the optic cup, where it forms hyaloid vessels supplying the inner surface of the retina and a fibrous network between this and the lens. The interstitial spaces between these vessels later fill with the transparent gelatinous VITREOUS HUMOUR, the vessels leaving behind the hyaloid canal.

Six extrinsic eye muscles are involved in eye movements and are innervated by axons originating in the MIDBRAIN and located in the trochlear nerve (CNIV), abducens nerve (CNVI) and the oculomotor nerve (CNIII). Four of them (the superior, inferior, medial and lateral rectus muscles) run roughly anteroposteriorly, two (the superior and inferior oblique muscles) inserting on the sclera at an angle to the globe of the eye. The superior oblique muscle runs anteroposteriorly but loops through a pulley-like loop, the trochlea, before insertion. In addition to their role in voluntary control of eye movements, these nerves and muscles respond reflexly to rotation of the head so that the retinal image is kept stationary (gaze stabilization). They are also involved in gaze shifting: rapid SACCADES, smooth pursuit of a moving object, and disjunctive movements enabling both eyes to remain directed towards an approaching or receding object during BINOCULAR VISION. They also generate the MICROSACCADES which prevent sensory adaptation of rods and cones.

The eye can alter its sensitivity to light by

as much as 500,000–1,000,000 times in accordance with the level of illumination. Because the retina must be able to respond to lighter areas but not to darker areas in the visual field, its sensitivity must be adjusted as the intensity of light falling upon it changes, a difference of some 10 billion-fold between the extremes of sunlight and starlight, for example. See Fig. 73. See BLINDNESS, CIRCADIAN CLOCKS, EYE-BLINK REFLEX, PUPILLARY LIGHT REFLEX.

eye sockets See ORBITS.

F

Fab fragment A disulphide-linked heterodimer (the fragment antigen-binding regions) of an immunoglobulin molecule, joined by the hinge region to the F_C region in intact ANTIBODY and separated from it on the amino-terminal side of the disulphide bonds of the hinge by papain digestion. Each of the two chains contains one V domain and one C domain, their juxtaposition forming the antigen-binding region (see ANTIBODIES, Fig. 11; T-CELL RECEPTORS, Fig. 117). Another protease, pepsin, cuts the antibody on the carboxy-terminal side of the disulphide bonds to release an $F(ab')_2$ fragment with a few more amino acids. See RECEPTOR CLUSTERING.

face By the end of the 4th week of development, facial prominences (maxillary and mandibular prominences) appear, formed mainly by the first pair of pharyngeal arches which surround the stomodaeum, and consisting mainly of neural crest-derived mesenchyme. Five embryonic mesenchymal structures, termed prominences or processes, fuse during its formation. The frontonasal prominence forms the forehead, nose, and midportion of the upper jaw and lip. The lateral nasal prominence and medial nasal prominence are components of this (see Fig. 74). Two maxillary processes form the lateral parts of the upper jaw and lip, and two mandibular prominences form the lower jaw and lip (see PHARYNGEAL POUCHES, Fig. 134).

facet A smooth, flat articular surface on a bone; involved in a JOINT (e.g., intervertebral joints).

facial bones See SKULL.

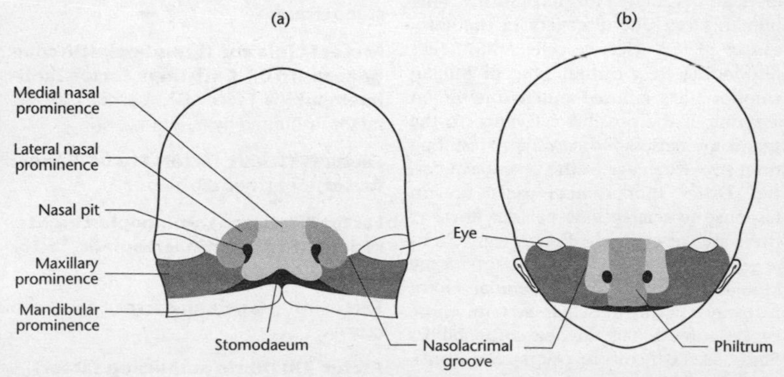

FIG 74. Diagrams of (a) a 7-week embryo and (b) a 10-week embryo, showing the formation of the FACE (see entry). The nasal placodes have already invaginated to form nasal pits, creating the ridges of the nasal prominences. The NOSE forms from five facial prominences. Redrawn from *Langman's Medical Embryology*, T.W. Sadler. © 2004 Lippincott, Williams & Wilkins, with permission of Wolters Kluwer.

facial nerve (cranial nerve VII) Nuclei in the pons receive sensory information from taste receptors and provide motor impulses leaving via the facial nerves and regulating secretion of SALIVA and tears (see LACRIMAL GLANDS), and contraction of muscles involved in facial expression.

facial recognition The use of functional MAGNETIC RESONANCE IMAGING with humans indicates that there is a small area in the brain that is more responsive to faces than to other stimuli, a finding consistent with electrophysiological studies on AREA IT in monkeys. At approximately 2–3 months of age, human infants begin to vocalize and smile at others spontaneously in a face-to-face social-communicative context. It has been proposed that a subcortical visual pathway involving the superior colliculus controls tracking moving face stimuli in the first month of life, and that at around 2 months there is a shift in processing from the subcortical visual pathway to a second pathway, involving the AMYGDALA, mediating plastic cortical visual pathways and enabling recognition of characteristic *facial expressions* (see IMITATION). Whereas a 6-month-old human infant is equally good at recognizing facial identity in both human and non-human primates, 9-month-old infants and adults show a marked advantage for recognizing only human faces. The discovery in the HIPPOCAMPUS of face-selective cells, and of cells responding to a certain type of human emotion, has aroused much interest on account of the possible relevance to the syndrome PROSOPAGNOSIA, usually resulting from stroke damage to the EXTRASTRIATE CORTEX. Other hippocampal cells fire in response to one specific person, however they are recognized. Facial expressions (e.g., smiling) engendered by EMOTIONS are brought about by extrapyramidal motor pathways in the reticular system; those evoked intentionally involve corticobulbar fibres (see CORTICOSPINAL TRACTS). See AUTISM, FUSIFORM GYRUS, INFERIOR TEMPORAL CORTEX.

facilitated diffusion Carriage of an ion or molecule across a membrane by a carrier protein, but not directly involving ATP hydrolysis (although this may ultimately be the energy source enabling transport, as in secondary active transport). See SYMPORTERS.

facilitation A relatively simple form of SYNAPTIC PLASTICITY in which increase in synaptic efficiency occurs either when a presynaptic action potential occurs soon after a previous one, or when the amount of transmitter released by an axon terminal is increased by its synapsing with another neuron. It occurs at many excitatory synapses.

Factor I (fibrinogen) See BLOOD CLOTTING.

Factor II (prothrombin) See BLOOD CLOTTING.

Factor III (tissue factor, tissue thromboplastin) See BLOOD CLOTTING.

Factor IV (calcium) See BLOOD CLOTTING.

Factor V (proaccelerin, labile factor, Ac-globulin (Ac-G)) See BLOOD CLOTTING.

Factor VII (serum prothrombin conversion accelerator (SPCA), proconvertin, stable factor) See BLOOD CLOTTING.

Factor VIII (antihaemophilic factor (AHF), antihaemophilic globulin (AHG), antihaemophilic factor A) See BLOOD CLOTTING.

Factor IX (plasma thromboplastin component (PTC), Christmas factor, antihaemophilic factor B) A BLOOD CLOTTING factor, inhibited by α_1 ANTI-TRYPSIN.

Factor X (Stuart factor, Stuart–Prower factor) See BLOOD CLOTTING.

Factor XI (plasma thromboplastin antecedent (PTA), antihaemophilic factor C) See BLOOD CLOTTING.

Factor XII (Hageman factor) See BLOOD CLOTTING.

Factor XIII (fibrin-stabilizing factor) A protein product of platelets during BLOOD CLOTTING.

FAD See FLAVINE ADENINE DINUCLEOTIDE.

FADD (Fas-associated death domain protein) See APOPTOSIS.

faeces Material egested from the anus (see EGESTION). Roughly 100–150 g is normally eliminated per day, 30–50 g of which is solid (the rest being water). The solid portion consists largely of cellulose, epithelial cells shed from the mucosa, bacteria (see GUT FLORA), some salts, and the brown pigment stercobilin. It contains ~5% fat, mostly derived from bacteria. Bacteria in faeces and the mucus overlying the intestinal epithelium come mainly from the Cytophaga-Flavobacterium-Bacteroides divisions (e.g., the genus *Bacteroides*), and the Firmicutes (e.g., genera *Clostridium* and *Eubacterium*) contributing ~30% each.

fainting See SYNCOPE.

falciform ligament Ligament attaching liver to anterior abdominal wall and diaphragm.

Fallopian tubes See OVIDUCTS.

familial adenomatous polyposis (FAP, APC) A dominantly inherited form of COLON cancer, affecting ~1:5,000 people who accumulate benign polyps (adenomas, because glandular) in the colon lining that may progress to malignancy (see METASTASIS, Fig. 124) and COLORECTAL CANCER. It is caused by mutation in the tumour suppressor gene *APC* (ADENOMATOUS POLYPOSIS COLI). In the absence of a Wnt signal (see WNT SIGNALLING PATHWAY), the *APC* protein product, APC, fosters degradation of the oncogene product β-catenin and prevents its entry into the nucleus where it would otherwise act as a transcriptional co-activator with transcription factors. Without an active degradation pathway involving ubiquitinylation, β-catenin accumulates leading to tumour formation. See HEREDITARY NON-POLYPOSIS COLON CANCER.

familial hypercholesterolaemia (FH) Inherited in an autosomal dominant manner, FH is the best-known disorder of lipid metabolism. Persons with FH have high blood cholesterol levels due to deficient or defective function of their LDL receptors (or lack of endocytosis of the LDL:receptor complex), leading to increased levels of endogenous cholesterol synthesis and increased risk of developing premature ATHEROSCLEROSIS and coronary artery disease (see CORONARY HEART DISEASE) and an early death by heart attack. It is estimated that one person in 20 presenting with early coronary artery disease is heterozygous for a mutation in the LDL receptor gene, and that such mutations are heterozygous in 1:500 of the general population (see LOSS-OF-FUNCTION MUTATIONS).

familiality Phenotypic character states, or traits, are familial if members of the same family have them in common, *for whatever reason*. By contrast, character states are heritable only if the similarity arises from having genotypes in common. See FAMILIAL RECURRENCE RISK, HERITABILITY.

familial recurrence risk, λ (λ_R) The ratio of risk of a character's appearance in relatives of affected individuals (probands) to: (i) the population PREVALENCE; or (ii) relatives of controls, as calculated from controlled family studies of first degree relatives (i.e., an individual's closest relatives: parents, offspring and siblings, each sharing on average 50% of their genes with the affected individual). Sibling recurrence risk, l_s, is the ratio of disease manifestation, given that a non-twin sibling is affected, compared with disease prevalence in the general population. See COMPLEX DISEASES, FAMILIALITY, GENETIC RISK, PHENOTYPE, TWIN STUDIES.

Fanconi anaemia, F.a. groups A–G Fanconi anaemia is an autosomal recessive disorder associated with UV-sensitivity, decreased blood cells, short stature, limb, heart and kidney malformations, patchy skin pigmentation, mental retardation and cancers (esp. LEUKAEMIAS). Multiple chromosome breaks are observed in cultured cells (see GENOMIC INSTABILITY) on account of impaired interstrand DNA cross-link repair (see DNA REPAIR MECHANISMS). Of the five groups, or subtypes, the commonest (A) maps to chromosome 16q24. See BRCA2.

farsightedness (**hypermetropia**) An abnormality of refraction of light entering the eye, in which the image focuses behind the retina. See ACCOMMODATION.

Fas (**FAS, CD95**) A cell-surface 'death receptor', and member of the TNF receptor (TNFR) family, expressed on many cells, but especially on activated lymphocytes. On binding its ligand (FasL, CD178, a member of the TNF family of membrane-associated cytokines), Fas is activated, causing adaptor proteins to bind to its cytoplasmic 'tail' (containing a 'death domain') and activate a CASPASE cascade leading to APOPTOSIS. Both Fas and FasL are normally induced during the course of an adaptive immune response, the apoptotic FasL being released by some METALLOPROTEINASES. See AUTOIMMUNE DISEASE, IMMUNOLOGICALLY PRIVILEGED SITES.

fascia The fibrous membrane covering, supporting and separating muscles.

fascicle, fasciculus A small bundle, especially of nerve or muscle fibres, surrounded by the connective tissue of the perineurium, or perimysium. A nerve may be one of several fasciculi, all encapsulated by epineurium.

fasciculations Disorganized muscle contractions, caused by motor neuron diseases (see NEURAL DEGENERATION) and by such depolarizing drugs as succinylcholine (a nAChR agonist).

fasciculus cuneatus The more lateral of the two dorsal columns of the SPINAL CORD carrying proprioceptive and discriminative touch information along primary afferent axons entering the cord at the upper thoracic and cervical dorsal roots. See DORSAL COLUMN–MEDIAL LEMNISCAL SYSTEM, MULTIPLE SCLEROSIS.

fasciculus gracilis The more medial of the two dorsal columns of the SPINAL CORD, carrying proprioceptive and discriminative touch information along primary afferent axons entering the cord at the sacral, lumbar and lower thoracic levels. See DORSAL COLUMN–MEDIAL LEMNISCAL SYSTEM.

fasting The fasting, or catabolic, phase of energy metabolism is the period during which endogenous energy sources are mobilized. Overnight fasting is a feature of normal life; but lack of ingested food over a 24-hour period will result in falling BLOOD GLUCOSE concentration. The transition from fed to fasted states involves drastic metabolic changes in the liver during its adaptation to nutrient deprivation (see AMP-ACTIVATED PROTEIN KINASE), ensuring an adequate fuel supply to the brain, which can normally only metabolize glucose as a fuel. Fasting has been shown to promote the nuclear localization of hepatic FOXA2, where it promotes glycolysis, fatty-acid β-oxidation, and the synthesis and secretion of KETONE BODIES, along with a reduction in GLUCONEOGENESIS and hepatic fat. The CREB-regulated transcription coactivator 2 (aka TORC2) is induced by fasting (in particular by the presence of pancreatic glucagon) to become dephosphorylated at Ser 171 and enter the nucleus in liver cells and coactivate CREB, an important factor in triggering the activation of gluconeogenesis. As a preliminary for this, the body's less important proteins will be degraded and the carbon skeletons of deaminated glucogenic amino acids will be converted into pyruvate or intermediates of the KREBS CYCLE. Insulin disrupts TORC2 activity by inducing Ser/Thr kinase SIK2, which undergoes activation by PKB-mediated phosphorylation at Ser 358, leading to Ser 171 phosphorylation and cytoplasmic translocation of TORC2. See ENERGY BALANCE, SMALL INTESTINE, STARVATION CONDITIONS.

fast inhibitory post-synaptic potentials See SYNAPSE.

fast oxidative-glycolytic fibres Intermediate in diameter between SLOW TWITCH FIBRES and FAST TWITCH FIBRES, these skeletal muscle fibres contain a rich capillary supply and much myoglobin and so, like slow twitch fibres, also appear dark red. With high intracellular glycogen levels, they can also generate ATP by glycolysis. The ATPase in the myosin heads hydrolyses

ATP rapidly (hence 'fast' fibres), making their contraction velocity (less than 100 msec) faster than in slow twitch fibres, causing tension to be reached more quickly.

fast pain See PAIN.

fast synapses See SYNAPSE.

fast synaptic transmission See SYNAPSE.

fast twitch fibres (fast phasic fibres, fast glycolytic fibres) With the largest diameter of all SKELETAL MUSCLE fibres, containing the greatest number of myofibrils, fast twitch fibres generate the most powerful contractions. Their high glycogen and low myoglobin content, relatively scanty blood supply and small numbers of mitochondria indicate their adaptation to generate ATP by glycolysis. The ATPase in their myosin heads hydrolyse ATP rapidly, adapting them for anaerobic activities of short duration (contrast SLOW TWITCH FIBRES). Anaerobic exercise leads to increase in mass of these fibres by hypertrophy.

fat See ADIPOSE TISSUE, BODY COMPOSITION.

fat-free body mass (FFM) See BODY COMPOSITION.

fatigue Muscle fatigue is the failure to maintain the required force or power output (see MOTOR UNITS). In the 'central fatigue concept' prolonged EXERCISE raises blood tryptophan concentration, leading via SEROTONIN to fatigue (by an unknown mechanism). Fatty acids stimulate this effect, but high blood glucose concentrations oppose it, delaying fatigue. Autonomic failure is a fairly rare condition occurring in older people, and more commonly in men than women. Clinical signs are dizziness, fatigue and blackouts after meals or during exercise. The common feature is persistent low blood pressure.

fats See LIPIDS.

fatty acid metabolism The oxidation of long-chain fatty acids to acetyl-CoA is a tightly controlled central energy-yielding pathway (see ENERGY BALANCE), occurring in skeletal muscle especially. There are three stages in the complete oxidation of fatty acids, starting from the carboxyl end of the molecule: (i) their oxidation to 2-carbon fragments in the form of acetyl-CoA, generating the reduced coenzymes $FADH_2$ and NADH + H^+; (ii) the oxidation of acetyl-CoA to CO_2 in the KREBS CYCLE; and (iii) the transfer of electrons from the resulting reduced coenzymes to the electron transport chain in the inner mitochondrial membrane, which drives ATP synthesis. If acetyl-CoA is not completely oxidized via the KREBS CYCLE, it may be converted (as in the liver) to KETONE BODIES.

Free fatty acids entering the cytosol from the blood plasma are first converted to fatty acyl-CoA by acyl-CoA synthetases in the outer mitochondrial membrane. This is then converted to fatty acyl-carnitine by carnitine acyl-transferase I on the outer surface of the inner mitochondrial membrane (see FATTY ACIDS). This enters the mitochondrial matrix by facilitated diffusion via the acyl-carnitine/carnitine transporter of the inner mitochondrial membrane, the acyl group finally being transferred from carnitine to coenzyme A in the mitochondrial matrix by carnitine acyltransferase II on the inner face of the inner mitochondrial membrane. The single bond between the methylene (–CH_2–) groups in fatty acids is relatively stable, and activation of the carboxyl group on carbon-1 by its attachment to CoA during β-oxidation overcomes this, allowing step-wise oxidation of the fatty acyl group at the C-3 (or β) position. One complete pass through the β-oxidation sequence generates one molecule of acetyl-CoA, two pairs of electrons and four protons (H^+), reducing the length of the fatty acid chain by two carbon atoms. The shortened fatty acid can then go through further 2-carbon removals until it is degraded. The $FADH_2$ molecules generated are oxidized via the electron-transferring flavoprotein of the inner mitochondrial membrane, providing ~1.5 molecules of ATP per electron pair. Each molecule of NADH delivers a pair of electrons to the inner membrane's NADH dehydrogenase, generating ~2.5 molecules of ATP per pair. The complete oxidation of one molecule of the 16-carbon palmitic acid from palmitoyl-CoA can release sufficient energy to generate a total of 108 molecules of ATP:

$$\text{Palmitoyl} - \text{CoA} + 23O_2 + 108P_i + 108\text{ADP} \rightarrow \text{CoA} + 108\text{ATP} + 16CO_2 + 23H_2O$$

In metabolically active tissues degradation of fatty acids in the KREBS CYCLE continues only if sufficient oxaloacetate and other cycle intermediates combine with acetyl-CoA formed during β-oxidation, and pyruvate generated by glycolysis plays an important role in maintaining oxaloacetate levels. Fatty acid breakdown in such tissues therefore depends somewhat on a continual background level of carbohydrate catabolism.

Oxidation of saturated fatty acids and those with even numbers of carbon atoms is slightly less complex than is oxidation of unsaturated fatty acids and of those with odd numbers of carbon atoms. Two additional enzymes are required to break the *cis*-configuration double bonds of most common unsaturated fatty acids during oxidation. One enzyme, an isomerase, is needed for monounsaturated fatty acids, such as oleic acid, while an additional reductase is needed for polyunsaturated fatty acids (PUFAs), such as linoleic acid. Odd-numbered fatty acids require three extra enzymatic steps and coenzyme B_{12} (see VITAMIN B COMPLEX, Fig. 180). PEROXISOMES can also carry out β-oxidation of fatty acids. Oxidation of fatty acids from the omega end of the molecule (the carbon atom furthest from the carboxyl group) occurs in liver and kidney endoplasmic reticula.

Although fatty acid oxidation within mitochondria produces acetyl-CoA, there is no net synthesis of glucose possible in humans because the pyruvate dehydrogenase and pyruvate kinase reactions proceed irreversibly. Acetyl-CoA cannot therefore be used to form pyruvate and then glucose simply by reversing glycolysis. Rather, the acetyl group released from acetyl-CoA enters the Krebs cycle for further oxidation. Because fatty acids released by LIPOLYSIS from triglycerides within muscle and adipose tissue fail to provide gluconeogenic substrates, these tissues rely on protein catabolism for energy when the body's glycogen reserves are depleted. After a meal, insulin stimulates two separate relay systems in liver cells, via ISR1 and ISR2 (see INSULIN). The enzyme Akt is a key player in both signalling pathways, both of which shut down gluconeogenesis and fatty acid oxidation – the latter far more effectively.

Inborn errors in fatty acid metabolism may involve the long-chain-, medium-chain- or short-chain-specific acyl-CoA dehydrogenase enzymes, the principal defect being the accumulation of mitochondrial acyl-CoA intermediates during fasting (e.g., during sleep) with consequent reduced CoA availability for other mitochondrial reactions. Among northern Europeans, the frequency of carriers of a recessive mutation in the gene encoding medium-chain acyl-CoA dehydrogenase (MCAD) is ~0.025, homozygous individuals being unable to oxidize fatty acids of 6–12 carbon atoms in length. This could lead to hypoglycaemia, sleepiness, vomiting and cerebral oedema, and is rectified by intake of frequent low-fat, high carbohydrate meals.

fatty acids Aliphatic and usually unbranched carboxylic acids, often of considerable length, which: (i) are constituents of biological membranes; (ii) are energy storage molecules and (iii) second messengers; (iv) may be added post-translationally to modify the structures of proteins; and (v) may modulate GENE EXPRESSION. Tissues that actively synthesize fatty acids (see LIPOGENESIS), such as mammary gland, adrenal cortex, adipose tissue and liver, use NADPH (much of it generated by the PENTOSE PHOSPHATE PATHWAY) to reduce double bonds and carbonyl groups of intermediates. The *de novo* synthesis of fatty acids by liver cells, as in conditions of excess carbohydrate, involves the stepwise addition of activated 2-carbon precursor units in a conserved set of biochemical reactions different from

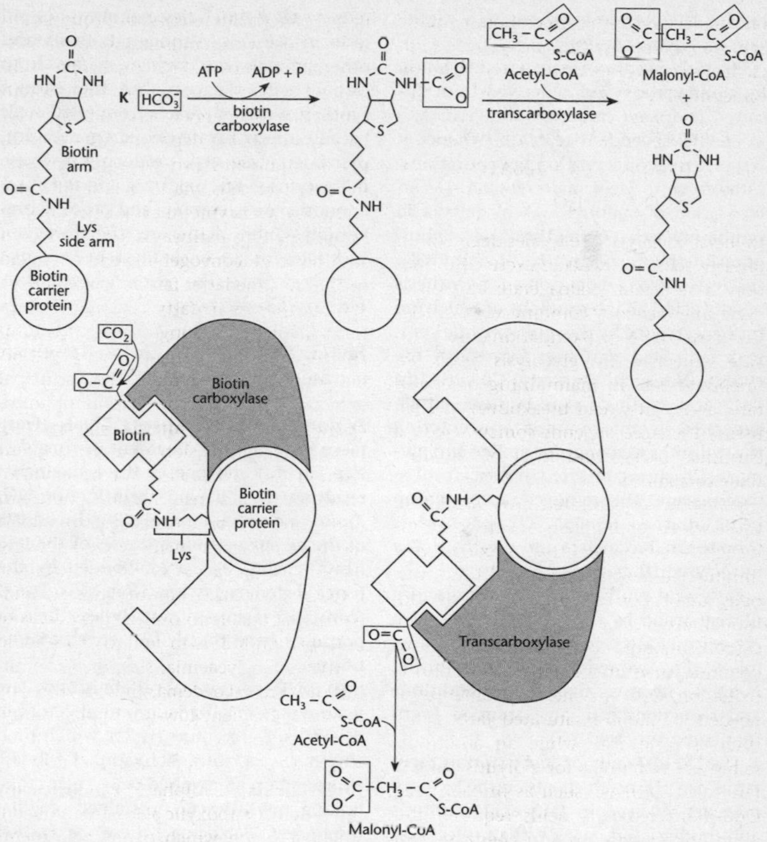

FIG 75. Diagram indicating the role of acetyl-CoA carboxylase in the synthesis of malonyl-CoA during the production of long-chain FATTY ACIDS. Its three functional regions are: a biotin-carrier protein; biotin carboxylase (activating CO_2 by attaching it in an ATP-dependent reaction to a nitrogen atom in the biotin ring); and transcarboxylase, transferring activated CO_2 from biotin to acetyl-CoA, producing malonyl-CoA. The active enzyme in each step is shown stippled. Biotin's flexible arm transfers the activated CO_2 between the two enzyme components. © 2000 From *Lehninger Principles of Biochemistry*, by D.L. Nelson and M.M. Cox, published by Worth Publishers. Redrawn with permission from the publishers.

those by which fatty acids are oxidized. The long and highly reduced alkyl hydrocarbon chains of the constituent fatty acids of neutral fats make them valuable storage fuels, the yield of ~38 kJ g^{-1} on complete oxidation being more than twice that for the same mass of carbohydrate or protein (see FATTY ACID METABOLISM).

Malonyl-CoA is the first intermediate in the biotin-dependent cytosolic biosynthesis of long-chain fatty acids from acetyl-CoA (see Fig. 75), and its concentration increases whenever there is plentiful dietary carbohydrate (see GLUCOSE). In this generally anabolic environment, the inhibition of carnitine acyltransferase I (see FATTY ACID METABOLISM) by malonyl-CoA prevents oxidation of fatty acids and favours triglyceride synthesis (see CARDIOVASCULAR DISEASE, OBESITY).

The carboxyl group transferred to biotin by biotin carboxylase is derived from dissolved hydrogen carbonate ions (HCO_3^-) in an ATP-dependent reaction, whence it is transferred to acetyl-CoA by biotin transcarboxylase to yield malonyl-CoA. Decarboxylation of malonyl-CoA by fatty acid synthase (which requires the cofactor phosphopantetheine, a derivative of vitamin B_5) still leaves two carbon atoms for synthesis of fatty acids, all the reactions of which are further catalysed by the multienzyme complex fatty acid synthase, located in the cytosol. Nearly all the acetyl-CoA used in fatty acid synthesis derives from citrate generated in mitochondria and entering the cytosol in exchange for malate and pyruvate. Citrate is converted in the cytosol to oxaloacetate, the removed acetyl group being attached to CoA. Oxaloacetate is then reduced to malate and returns to the mitochondrial matrix.

Fatty acid synthase activity results in a net extension by two carbon atoms of a pre-existing acetyl group on the synthase, followed by two reduction steps requiring NADPH. Iterative two-carbon additions and reductions generate saturated even-numbered fatty acids.

The SREBP family of transcription regulators has significant effects on the levels of proteins involved in fatty-acid synthesis and modification, and on their effects on cholesterol synthesis.

Cells can derive fatty acids from three main sources: from ingested dietary fats, from fats stored in cells as lipid droplets (see LIPIDS), and from FREE FATTY ACIDS (NEFA) synthesized by the liver from excess dietary carbohydrate and mobilized for use elsewhere, bound to plasma serum albumin. Most fatty acids in triglycerides and phospholipids are unsaturated, the double bonds being in the *cis*-configuration. SREBPS, LXRs, PPARs and hepatocyte nuclear factor are all regulated by polyunsaturated fatty acids. This affects the amount of fatty acids in the cell and their level of unsaturation, which in turn can alter the saturation of fatty acids incorporated into phospholipids as well as the cholesterol:phospholipid ratio in the cell. Although humans and other animals can synthesize fats from carbohydrates via acetyl CoA, they cannot synthesize carbohydrates from fatty acids because the pyruvate dehydrogenase and pyruvate kinase steps producing pyruvate in the cytosol, and oxidizing it in the mitochondria, are exergonic, and proceed irreversibly. Fatty acids are therefore not substrates for GLUCONEOGENESIS. See ESSENTIAL FATTY ACIDS, FREE FATTY ACIDS, OMEGA-3 FATTY ACIDS, OMEGA-6 FATTY ACIDS.

favism (fabism) See GLUCOSE-6-PHOSPHATE DEHYDROGENASE.

F_C domains and receptors The F_C (Fragment crystallizable) fragment of ANTIBODIES (see Fig. 10) and one of three fragments resulting from papain digestion of IgG. An IMMUNOGLOBULIN (see Fig. 99), it consists of the carboxy-terminal halves of the two heavy chains, disulphide-bonded by the hinge region (see ISOTYPE SWITCHING). The C (constant) regions of antibodies (their F_C portions) have three main effector functions:

(i) the F_C portions of certain isotypes are ligands for specialized immune effector cell receptors, F_C receptors (F_CRs), which then trigger PHAGOCYTOSIS. Activation $F_C\gamma$Rs are expressed on all myeloid cells, their cross-linking resulting in sustained cellular responses. Balancing these activation receptors (A) is the inhibitory $F_C\gamma$RIIB receptor (I) which, when coligated to activation receptors, dampens the cellular response. The coexpression of the two opposing receptor classes establishes a threshold for cellular triggering by IgG antibodies. The A/I ratios differ by several orders of magnitude between different IgG subclasses, for which different F_CRs display significantly different affinities. The F_C portions of IgG1 and IgG3 antibodies bind the F_C receptors of phagocytic cells such as neutrophils and macrophages, which consequently bind and engulf pathogens coated with these isotypes. The F_C portion of IgE binds to a high affinity $F_C\varepsilon$ receptor on mast cells, basophils and activated eosinophils, which then release inflammatory mediators on binding specific antigen. The $F_C\gamma RIII$ (CD16) receptor of NK cells recognizes the IgG1 and IgG3 subclasses;

(ii) the F_C portions of antigen:antibody complexes can bind to COMPLEMENT, initiating the complement cascade;

(iii) by engaging specific cell receptors, the F_C portion enables active transport of antibody to body compartments it would not normally reach without this help. These places include mucous secretions, tears and milk (IgA) and the foetal circulation (IgG).

Many kinds of microorganism appear to have responded to the destructive potential of the F_C portion by evolving proteins that bind to it or cleave it proteolytically (e.g., Protein A and Protein B of *Staphylococcus* and Protein D of *Haemophilus* spp.). F_C receptors bind the F_C portions (antibody constant regions) of Ig isotypes and are found on mast cells, macrophages, monocytes and NK cells. The bacterial genus STAPHYLOCOCCUS is able to bind F_C portions of immunoglobulins, associated with their antiphagocytic ability.

fear and anxiety Fear is normally an acute EMOTION with behavioural and physiological response components crucial to adaptation and survival during times of perceived threat and danger. Chronic fear, in which symptoms continue well past their usefulness, can develop into pathological states associated with anxiety and depressive disorder (see AFFECTIVE DISORDERS). Fear-related behavioural responses are orchestrated by several neurochemical signals, in particular the regulation in the brain of corticotropin-releasing hormone by glucocorticoids in the HPA axis (see STRESS RESPONSE). Chemical destruction of serotonergic neurons in animals reduces behaviours associated with anxiety, and several anxiolytic drugs are SEROTONIN receptor ligands. It is probable that serotonergic cells (see BRAIN, Fig. 30d) comprise an anxiogenic system that is important in learning about aversive situations, akin to behavioural SENSITIZATION. The exaggerated expression of fear during rabies infection caused the neurologist James Papez (see PAPEZ CIRCUIT) to link this disease symptom to the pathological cytoplasmic bodies found within neurons of the HIPPOCAMPUS. Work on mice indicates that lack of the endocannabinoid receptor CB_1 leads to a failure to lose the fear of a sound which they had learnt to associate with a mild shock to the feet after the sound had been coupled to the shock. This suggests that endocannabinoids may be involved in extinguishing the bad feelings and pain triggered by reminders of past experiences and that reduced numbers of their receptors may be associated with post-traumatic stress syndrome, phobias and some forms of chronic PAIN.

Recent (2007) functional MAGNETIC RESONANCE IMAGING (fMRI) evidence involving humans and simulated low and high intensity threat from a distant or near 'virtual predator' indicates that knowledge that a chase is about to commence increases blood flow in the frontal cortex (anterior cingulate, orbitofrontal and ventromedial prefrontal cortices), possibly indicating planning of subsequent navigation during the chase. At the start of the chase, the cerebellum and PERIAQUEDUCTAL GREY MATTER were activated. When the predator was remote, the ventromedial prefrontal cortex and lateral AMYGDALA were activated; but close proximity of the predator brought a shift to signals from the central amygdala and periaqueductal duct. This shift from forebrain to midbrain may have been associated with increased fear as the predator approached, and activities of the periaqueductal grey and nearby dorsal raphe nucleus were highly correlated with subjective fear of the predator and the level of confidence that the subject could escape. Activation of the prefrontal cortex by distal, unpredictable threats may foster anxiety, while activation of the periaqueductal duct by proximal threats predictive of PAIN may elicit panic.

Anxiety disorders have been linked to hyperactivity of the amygdala and diminished activity of the hippocampus, while both these and prefrontal cortex dysfunction are associated with depressive disorders and CREB-related NEURAL PLASTICITY has been implicated. In rodents, anxiety-like effects are induced not by CREB enhancement in the amygdala, but by its reduction within the NUCLEUS ACCUMBENS. See WILLIAMS–BEUREN SYNDROME.

Obsessive-compulsive disorder (OCD) is a spectrum of neuropsychiatric conditions affecting ~2% of humans. Frequently familial, it is believed to involve dysregulation of frontostriatal circuitry, the orbitofrontal cortex being central to neurobiological models. A lack of animal models has hampered research into the condition, which is characterized by obsessions (recurrent, or persistent, thoughts, impulses, or images experienced during the disturbance in which the person attempts to ignore or suppress them, there being a tendency to haggle over fine details); and compulsions (ritualized, repetitive behaviours, or mental acts, which a person feels driven to perform in response to an obsession, these behaviours being aimed at reducing or preventing the distress caused). OCD is often regarded as an anxiety disorder and often begins in early life, running a chronic and relapsing course which causes significant distress and disability. Diagnosis requires there to be either obsessions, or compulsions, alone; or obsessions and compulsions. OCD sufferers must, by definition of the disorder, recognize that these obsessions/compulsions are unreasonable or excessive (see DEEP BRAIN STIMULATION, TOURETTE SYNDROME). Accumulating, if circumstantial, evidence implicates abnormalities in neural circuits spanning the frontal, striatal and thalamic brain regions (see NEOSTRIATUM). fMRI is now commonly used in association with psychotherapy. Based on cognitive and emotional tasks, it also points to abnormalities in frontal-striatal-thalamic circuits. Abnormalities in synaptic function here could result in unintended behaviours, or perhaps unintended thoughts. A mouse model in which the gene encoding SAPAP3 (a scaffolding protein which locates to excitatory glutamate-responsive synapses and is expressed solely in the mouse neostriatum) is absent results in excessive grooming behaviour and hair loss. However, like other major psychiatric disorders, OCD seems to be heterogeneous, with complex underpinnings. It probably arises through an interaction between developmental and environmental factors, probably involving several genes. These may then impinge on the circuitry currently being studied in mice. See DEEP BRAIN STIMULATION.

Anxiety disorders and depression often coexist, and as with the former, exaggerated activity of the HPA axis is associated with severe depression. Continuous exposure to cortisol, as during chronic stress (e.g., during post-traumatic stress disorder), can cause a decrease in volume of the hippocampus. Hyperactivity of neurons that release CRH has been linked to anxiety disorders, and in the future anxiolytic drugs may be found to block CRH receptors. Currently, most *anxiolytics* are BENZODIAZEPINES (BDZs) which act on $GABA_A$ receptors, and SEROTONIN-selective reuptake inhibitors (see SSRIS). BDZs induce sleep when given in high doses at night, and sedation when given in low doses during daytime. A reduction in anxiety is likely to explain, in part, the widespread social use of ALCOHOL, and anxiety disorders and alcohol abuse often go hand in hand.

The genes contributing to depression and anxiety disorders are still unknown (but see SEROTONIN), although the recently discovered single nucleotide polymorphism (Val66Met) in the brain-derived neurotrophic factor (BDNF) gene may be related to mood and anxiety disorders common in human populations.

feedback In neural networks, signals whose direction flows from higher-order cells to lower-order cells. A neuron may be the source of signals to itself, through *recurrent connections*. Feedback connections may be excitatory, but are more commonly made by inhibitory interneurons (the neural equivalent of NEGATIVE FEEDBACK), as when RENSHAW CELLS enable motor systems to correct errors during a movement. See POSITIVE FEEDBACK.

feedforward In NEURAL NETWORKS, pathways

in which signals flow downstream from lower-order to higher-order neurons (directionally, i.e., from input to output). The signals may be excitatory or inhibitory. Feedforward inhibition, in which lower order cells excite inhibitory interneurons projecting forward to dampen activities of neighbouring higher-order cells, operates in surround inhibition in sensory pathways, and in the reciprocal inhibition operating in motor reflexes. See FEEDBACK.

feeding centre Because lesions in the ventrolateral hypothalamus (VLH) inhibit feeding (aphagia), while lesions in the ventromedial hypothalamus (VMH) cause overeating, it was long thought that the hypothalamus contained two centres involved in feeding: a 'hunger' or 'feeding centre' in the VLH and a 'satiety centre' in the VMH. It is now believed that VLH lesions exerted their effect through damage to the nigrostriatal pathway from the substantia nigra to the dorsal striatum. In addition to aphagia, VLH lesions produce akinesia and sensory neglect (also features of PARKINSON'S DISEASE), and it is now thought that it is the destruction of dopaminergic neurons that reduces motivation to eat rather than its having specific effects on a feeding mechanism. Progressive inanition (marked weight loss, muscle weakness and decreased metabolism) are the consequence. Also in contrast to earlier views, it is noticeable that although VMH lesions cause hyperphagia and obesity, they do not affect the size of individual meals. This seems to be because these lesions destabilize the autonomic and endocrine regulation of metabolism, leading to an increase in plasma insulin output (see PANCREAS) and more rapid clearance of plasma glucose and amino acids after meals. In the *glucostatic model*, a transient fall in plasma glucose triggers feeding in rats and, because of the elevated insulin level in animals with VMH lesions, feeding reoccurs earlier after a meal. In this model, VMH lesions reduce the intervals between meals and are responsible for meal timing rather than meal size (see APPETITE). See ENERGY BALANCE, LEPTINS.

female pseudohermaphroditism See CONGENITAL ADRENAL HYPERPLASIA, PSEUDOHERMAPHRODITISM.

females See SEX DETERMINATION.

femoral Pertaining to the FEMUR; e.g. femoral arteries.

femur (thighbone) The longest, heaviest and strongest bone in the body, articulating proximally with the acetabulum of the hip bone and distally with the tibia and patella. Because the shafts of the femurs are angled inwardly (medially), the knee joints are brought nearer to the midline, which is more pronounced in females since their pelves are broader. The ligament of the head of the femur connects it to the acetabulum of the hip bone (see PELVIC GIRDLE, Fig. 132), while its greater and lesser trochanters are projections serving as points of attachment for the tendons of some of the thigh and buttock muscles (e.g., see GLUTEUS MAXIMUS). See SKELETAL SYSTEM, Fig. 156.

fenestrated capillaries See CAPILLARIES.

ferritin An acute phase protein, metalloprotein, and the major store of IRON in the liver and ENTEROCYTES. The iron–transferrin complex carries iron from the liver to the haematopoietic tissue in the red BONE MARROW.

fertility The strategy of limiting fertility by techniques such as delaying marriage and sexual abstinence within marriage have been practised by rural and urban peoples, known to have been first used by African pastoralists relying on slow-breeding livestock, and by pre-industrial German farmers attempting to maintain their farm integrities by reducing the number of inheritance claims. In the late 19th and early 20th centuries, urbanizing Britain reduced its fertility in 70 years, from 6–8 children per family being common, to families with >4 children being rare. Despite local departures, this fertility decline has become global, first in Europe and by the end of the 20th century in the Americas and Asia also. In the West, this decline has continued long after MORTALITY rates were greatly reduced. In Africa, this fertility decline is happening fastest in urban areas (see HUMAN DEMOGRAPHY), and began (e.g., in Ethiopia) before the HIV epidemic.

Male infertility is a long-standing enigma of significant medical concern and the integrity of sperm chromatin is a clinical indicator of male fertility and IN VITRO FERTILIZATION potential as chromosome aneuploidy and DNA decondensation or damage are correlated with reproductive failure. Recently, several evolutionarily conserved genes for DNA compaction, chromosome segregation and fertility have been revealed by RNA interference in nematode worms and mice, as well as unexpected roles in spermatogenesis for genes involved in other processes. These discoveries provide an opportunity to identify causes of male INFERTILITY as well as possible targets for male contraceptives.

fertilization Adhesion and subsequent fusion of the plasma membranes of an oocyte and a capacitated sperm cell leading to close contact between the female and male pronuclei followed by loss of their nuclear envelopes so that the diploid chromosome number of *Homo sapiens* (2n = 46) is restored (see Fig. 76). The region of the sperm membrane which fuses to the oocyte (see OOGENESIS) is that covering the posterior portion of the sperm head since the membrane covering the acrosomal head cap disappears during the ACROSOME reaction. The head, midpiece (containing mitochondria, which are eliminated) and tail of the spermatozoon enter the oocyte cytoplasm although the plasma membrane is left on the oocyte surface. The oocyte then releases cortical granules containing lysosomal enzymes which make its membrane impenetrable to other sperm (see CORTICAL REACTION) and the ZP-1, ZP-2 and ZP-3 glycoproteins of the ZONA PELLUCIDA polymerize to form a gel preventing further sperm from binding and penetrating (which would result in polyspermy; see CHIMAERA). The sperm delivers a molecule (PLCζ) that triggers calcium (Ca^{2+}) oscillations in the egg membrane (the *activation reaction*), involving repetitive Ca^{2+} pulses, lasting several hours and triggering development by activating enzymes controlling

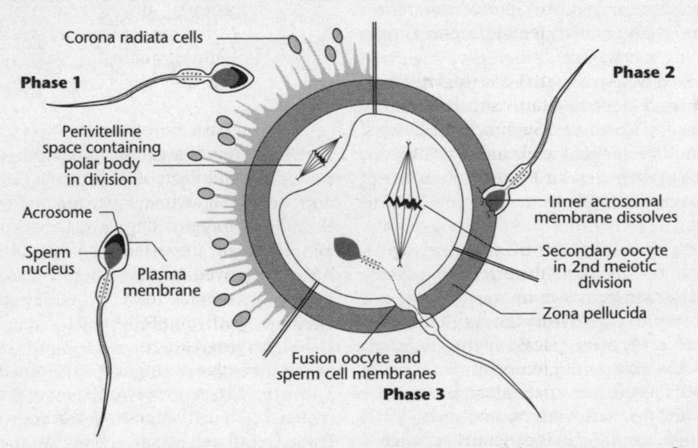

FIG 76. Diagram showing three phases of oocyte penetration by a sperm cell during FERTILIZATION. In phase 1, spermatozoa pass through the corona radiata; in phase 2, one or more spermatozoa penetrate the zona pellucida; in phase 3, a single spermatozoon penetrates the oocyte membrane, losing its own plasma membrane in the process. The lines separating the three phases are artistic licence. Redrawn from *Langman's Medical Embryology*, T.W. Sadler. © 2004 Lippincott, Williams & Wilkins, with permission of Wolters Kluwer.

the cell division cycle. The sperm also contains several kinds of mRNA, some coding for proteins needed for early embryo development, and microRNAs. Immediately after fertilization, the oocyte completes its second meiotic division and the second polar body is extruded. The haploid chromosomes of the mature oocyte (ovum) become surrounded by a nuclear envelope to form the female pronucleus. This and the sperm nucleus (male pronucleus) approach each other but do not fuse. Instead, their membranes break up and the two haploid chromosome sets are released so that the first (diploid) mitotic division of the zygote can now take place (these cleavage divisions produce the BLASTOCYST). Distinct genome-wide epigenetic reprogramming occurs immediately after fertilization and during early pre-implantation development. Many sequences in the paternal genome become suddenly demethylated shortly after fertilization (see SPERM). The sperm provides a centriole (lacking in the ovum) and this enters along with the sperm nucleus and axoneme of the tail, a centrosome forming around it. It replicates and helps form the spindle of the first cleavage division. Spermatozoa may remain alive in the female reproductive tract for several days.

It is estimated that 3–5% of fertilized human zygotes contain supernumerary sets of chromosomes. Such zygotes always fail to develop and are excluded from clinical use in centres for IN VITRO FERTILIZATION. See SEMENOGELINS.

fever Development of a higher-than-normal body temperature resulting from infection or invasion by certain foreign substances. Lymphocytes, neutrophils and macrophages release pyrogens that raise the set-point temperature of the hypothalamus (see TEMPERATURE REGULATION) so allowing temperature and metabolic rate to increase. Fever is generally beneficial to host defence since most pathogens grow better at lower temperatures, and adaptive immune responses are more intense at elevated temperatures. Furthermore, host cells are better protected from the effects of the tumour necrosis factor (TNF-α) at raised temperatures. See ENDOGENOUS PYROGENS.

FGFs See FIBROBLAST GROWTH FACTORS.

fibre **1**. See NON-STARCH POLYSACCHARIDE. **2**. See MUSCLES. **3**. See nerve AXONS.

fibrillar adhesions See FIBRONECTINS.

fibrin See BLOOD CLOTTING.

fibrinogen (Factor I) See BLOOD CLOTTING.

fibrinolysin See PLASMIN.

fibrinolysis Fibrinolytic agents, dissolving fibrin in unwanted blood clots, include STREPTOKINASE and recombinant TISSUE PLASMINOGEN ACTIVATOR. Dispersion of venous THROMBOSES relies on such agents, and on anticoagulants such as HEPARIN.

fibroblast growth factors (FGFs) and FGF receptors (FGFRs) In humans, nine members of the FGR family, and four receptors (FGFR1–FGFR4 with multiple isoforms of each), occur. They regulate such cellular events as division, differentiation and migration. Signalling to cells is mediated by the receptors, all RECEPTOR TYROSINE KINASES, each having three extracellular immunoglobulin domains, a transmembrane segment and the cytosolic kinase domain.

FGF2 (aka basic fibroblast growth factor, bFGF) is secreted by cardiac mesoderm and blocks the inhibition of liver-specific genes by notochordal factors in the region of the embryonic foregut destined to become LIVER. It is an important component of media employed in maintaining cultures of human EMBRYONIC STEM CELLS. FGF10 initiates LIMB growth, and FGF4 and FGF8 maintain the progress zone of the apical ectodermal ridge during LIMB development. FGF8 from the PRIMITIVE STREAK is involved in induction of cell migration through the streak and in cell specification into mesoderm. It is also expressed by the PRIMITIVE NODE (see entry), and involved in segmentation of SOMITES, and in SEX DETERMINATION, and is antagonized by RETINOIC ACID.

FGFR-1 (promoting osteogenic

differentiation) and FGFR-2 (increasing proliferation) are co-expressed in prebone and cartilage regions, including craniofacial structures. FGFR-3 is expressed in cartilage growth plates of long bones, but is of uncertain function. As with EGF receptors, FGFRs can be affected by point mutations or small deletions affecting the cytoplasmic domain and generating constitutively active receptors. Very specific missense mutations in these receptor genes have been linked to a series of skeletal dysplasias (achondroplasia A, hypochondroplasia H and craniosynostosis syndromes (e.g., Apert's syndrome)). FGFR-8 promotes outgrowth of the TELENCEPHALON and regulates its rostral regionalization. Unlike most other FGFs, the osteoblast-derived FGF23 acts as a hormone and helps maintain homeostasis in phosphate and vitamin D metabolism by the kidneys. FGF21 has recently been accepted as a key metabolic hormone involved in adaptation to STARVATION CONDITIONS (see FASTING). The quadrupled regions for FGFRs in the human genome, on chromosomes 4, 5, 8 and 10, are evidence favouring the hypothesis of tetraploidy in the evolution of vertebrates.

fibroblasts The characteristic cell type of vertebrate connective tissue, responsible for the synthesis and secretion of extracellular matrix materials such as tropocollagen (see COLLAGENS). They migrate during development and give rise to mesenchymal derivatives.

fibrocartilage See CARTILAGE.

fibroids (leiomyomas) Benign tumours of the wall of the uterus.

fibronectins Encoded by an IMMEDIATE EARLY GENE, large dimeric glycoproteins of the extracellular matrix, with multiple modular domains for binding to other matrix macromolecules such as heparin and collagen. Unlike collagens, fibronectin molecules only assemble into fibrils on certain cell surfaces, the two long polypeptide chains are linked by disulphide bonds near their C-terminal ends. As with COLLAGENS, IMMUNOGLOBULINS and ALBUMINS, each domain is encoded by a separate exon within the gene, suggesting that the gene evolved by multiple exon duplications (see GENE DUPLICATION). Each isoform is variably produced by exon shuffling, the main type of module being the type III fibronectin repeat, which binds to integrins. They include the soluble *plasma fibronectin*, which probably enhances blood clotting, wound healing and phagocytosis. Other isoforms are cell-surface molecules, requiring cell-surface fibronectin-binding integrins in order to form fibrils. Some are involved in cell guidance during developmental migrations, and in rounding up of endothelia. On fibroblasts, fibrils form at sites termed *fibrillar adhesions* (distinct from focal adhesions), where transmembrane integrins link intracellular actin to extracellular fibronectins, the contractile actomyosin cytoskeletal complex pulling on the fibronectin matrix, generating tension and exposing fibronectin self-binding sites for further fibril formation and further integrin-binding sites, so promoting formation of the EXTRACELLULAR MATRIX.

fibrous joints See JOINTS.

fibrous plaque See ATHEROSCLEROSIS.

fibula Bone of lower leg, smaller than the tibia but parallel and lateral to it and attached to it by an interosseous membrane. The distal projection of the fibula, the *lateral malleolus*, articulates with the talus bone of the ankle and forms the prominence commonly called 'the ankle bone'.

Fick's Law A law governing the rate of diffusion across a membrane or other rate-limiting barrier. Assuming the membrane is at all permeable to the diffusing substance, the law states that at a given temperature the rate of diffusion by the substance across it will be directly proportional to the surface area across which diffusion occurs and to the diffusion gradient (concentration difference) of the substance across the membrane but inversely proportional to the thickness of the membrane.

fight or flight response The acute setting of the STRESS RESPONSE.

filopodia Dynamic ACTIN-supported receptor-bearing extensions (small ones are termed microspikes) of a cell membrane, up to ~50 μm long and 0.1 μm wide, protruding from the surfaces of migrating and growing cells (e.g., axon growth cones, capillary tip cells). They have sensory roles and grow and retract rapidly as a result of actin polymerization and depolymerization.

filtering Most of the THALAMUS is concerned with controlling which sensory information is sent to the cortex, a major factor in ATTENTION. Sensory filtering of a kind could also be said to occur in HABITUATION (see also SENSORY ADAPTATION).

fimbriae See OVIDUCTS.

fingerprinting See DNA FINGERPRINTING.

fingerprints Parallel curving ridges of the epidermis of thick skin of fingers and toes reflecting the same structure of the underlying papillae of the DERMIS. They increase friction and improve the grip of hands and feet. First used in criminal investigation by Henry Faulds in 1880, those of identical twins are readily distinguishable on close examination.

fingers See LIMBS, FORE LIMB.

first-order neuron The neuron synapsing with a primary sensory afferent in the dorsal horn of the SPINAL CORD (see Fig. 163).

FISH (fluorescence *in situ* hybridization) A method by which complex abnormalities in chromosome structure can be resolved that would be impossible or very difficult to detect by standard histological preparations using brightfield microscopy (see CHROMOSOME PAINTING). A DNA PROBE is labelled by addition of a reporter molecule (e.g., a small protein or fluorescent dye) and allowed to hybridize *in situ* with target DNA, such as nuclei or chromosomes that are already fixed on a slide. Excess dye is washed off and, if the reporter is a protein, a fluorescent antibody to it is then added and the preparation incubated. Hybridized slides are then studied with a fluorescence microscope, when a fluorescent signal indicates the presence and position of the target material. Different fluorophores emit in different wave bands, so that different target sections of chromosomes may be 'painted' different colours. See G BANDS, POSITIONAL CLONING, SEGMENTAL DUPLICATIONS.

fissures 1. Narrow slits between adjacent parts of bones, through which blood vessels or nerves pass. **2**. The deepest grooves between folds of the CEREBRAL CORTEX (i.e., deeper than sulci).

fitness 1. Darwinian, or biological, fitness (selective value): the factor describing the difference in reproductive success of an individual (or genotype) relative to another. **2**. See EXERCISE, HEALTH.

fits See SEIZURES.

flagellum See CILIUM.

flat bones Bones composed of two almost parallel plates of compact bone with a layer of spongy bone between. Include the cranial bones, the sternum (breastbone), ribs and shoulder blades.

flavine adenine dinucleotide (FAD) A flavine nucleotide coenzyme acting as an electron carrier, yielding a semiquinone form if one electron is accepted and $FADH_2$ if two are carried (as in dehydrogenations). It is a covalently bound component of Complex II in the inner mitochondrial membrane, electrons passing from succinate to FAD, and then via the Fe–S centres to COENZYME Q. See MITOCHONDRION.

flavonoids The most numerous and widely distributed group of plant phenolics, concentrated mainly in the epidermis of leaves and skins of fruits. Also present in such derived products as tea, chocolate, wine and olive oil. They account for two-thirds of the current 0.3 g per day dietary phenol intake (phenolic acids accounting for the rest). Interest in their health effects have centred upon their antioxidant potential – notably of LDL (see ATHEROSCLEROSIS) and DNA. Scope has widened to include anti-inflammatory properties and their ability to influence smooth muscle relaxation. Platelet activation and aggregation also seem to be affected, depending upon the flavonoids, so it is pertinent that red wine and chocolate contain different flavonols and that the alcohol content of red wine (in moderation) may be more important in health promotion.

flexion Muscle contraction reducing the angle between two articulating bones. Contrast EXTENSION; see FLEXOR MUSCLES.

flexor muscles Muscles whose contraction results in FLEXION. For involvement in walking, see CENTRAL PATTERN GENERATORS, GAIT; for involvement in forearm movement, see ELBOW, HAND. Compare EXTENSOR MUSCLES.

floor plates Midline portions of the developing NEURAL TUBE, lacking neuroblasts and serving primarily as a pathway for nerve fibres crossing from one side to the other. Compare ROOF PLATES.

'flu See INFLUENZA.

fluid mosaic theory See CELL MEMBRANES.

fluorescence *in situ* hybridization See FISH.

fluoxetine An ANTIDEPRESSANT drug, also known as Prozac.

fMLP Any peptide fragment with the amino-terminal amino acid *N*-formylated methionine, being a potent chemotactic factor for inflammatory cells such as NEUTROPHILS. *N*-formylated methionine is the amino-terminal amino acid present on proteins produced by all bacteria, being encoded by the bacterial 'start' codon AUG (in eukaryotes AUG encodes methionine).

fMRI See MAGNETIC RESONANCE IMAGING.

foam-cells See ATHEROMA, ATHEROSCLEROSIS.

focal adhesion Anchoring INTERCELLULAR JUNCTION which, like a HEMIDESMOSOME, binds a cell to its extracellular matrix, but in which the adhesion proteins are INTEGRINS. Formation of focal adhesions may cause the cytoplasmic domains of integrins to activate signalling pathways evoking such cellular responses as cell migration, division and survival. Compare fibrillar adhesions (see FIBRONECTIN).

focusing See ACCOMMODATION.

foetal alcohol syndrome (FAS) A constellation of developmental abnormalities occurring in children born to alcoholic mothers (see ALCOHOL). A 1996 survey of American women suggested that 20% of women who drink alcohol continue to do so while pregnant and that as a result 0.1% of infants have FAS. Symptoms include facial abnormalities, stunted growth (see INTRA-UTERINE GROWTH RETARDATION), and learning and memory problems. These effects of alcohol appear to be mediated via receptors for two brain neurotransmitters, the glutamate NMDA RECEPTOR and the GABA receptor, killing rat brain neurons and causing abnormal neuron growth. See HOLOPROSENCEPHALY, MENTAL RETARDATION.

foetal stem cells Originally derived from umbilical cord cells, early reports of pluripotency have not been confirmed. See EMBRYONIC STEM CELLS.

foetus An EMBRYO becomes a foetus when bone cells appear in its cartilage (at ~8 weeks). Intrauterine life imposes environmental conditions very different from those experienced by the adult. Foetal adaptations to this environment include foetal HAEMOGLOBIN, the DUCTUS ARTERIOSUS, DUCTUS VENOSUS and FORAMEN OVALE, much of which relates to its organ of gaseous exchange being the PLACENTA and not the lungs. See also AMNIOTIC FLUID, BIRTH.

folate A B vitamin derivative of folic acid (pteroylglutamic acid). Its reduced form, tetrahydrofolate (H_4 folate), is formed by a two-step reaction catalysed by dihydrofolate reductase (DHFR) and serves as a donor of one-carbon units in a variety of metabolic reactions. The donated carbon often comes from the conversion of the amino acid serine to glycine. In transcriptionally quiescent cells, one of the *DHFR* gene's two promoter regions, the major promoter, is bound by an interfering RNA encoded by the minor promoter. This causes dissociation of the preinitiation complex (see TRANSCRIPTION) and prevents the recruitment of transcription factors to the major promoter. As a source of methylene tetrahydrofolate, it is required in the synthesis of thymidine nucleotides (see THYMINE). It is photodegraded by ultraviolet radiation, the possible adaptive response being to favour dark skins in regions where there is high exposure to UVR (see SKIN COLOUR). Its deficiency in preparation for, and during, pregnancy is responsible for neural tube defects such as SPINA BIFIDA and ANENCEPHALY;

Deficiency in folate and METHIONINE leads to aberrant genomic imprinting of IGF2; and colorectal cancer risk is linked to both dietary folate deficiency and to variants in methylene-dihydrofolate reductase, which has a critical role in directing the folate pool towards remethylation of homocysteine to methionine. See CPG DINUCLEOTIDES, FRAGILE SITES.

folic acid (pteroylglutamic acid) See FOLATE, VITAMIN B COMPLEX.

follicle See GRAAFIAN FOLLICLE.

follicle stimulating hormone (FSH) A very large glycoprotein and GONADOTROPIN synthesized and secreted in both sexes by gonadotrophs of the anterior pituitary gland in response to GONADOTROPIN RELEASING HORMONE (GnRH; see also MENSTRUAL CYCLE for pulsatile effects) and by circulating hormones or their own metabolites (see TESTES, Fig. 172). FSH and LH share a common α-subunit, but have different β-subunits, and the ratio of FSH:LH production depends upon the frequency of GnRH pulses. The FSH β-subunit is synthesized in response to low-frequency GnRH pulses and suppressed at higher pulse rates.

FSH circulates unbound in the plasma and has a half-life of 1–3 hours. As with LH, it produces its effects by binding to plasma membrane $G\alpha_s$ protein-coupled receptors located in the Leydig and Sertoli cells of the testis, activating adenylyl cyclase and raising cAMP levels in the cells, in turn activating a protein kinase A cascade (see PKA). Playing an important role in testis development by controlling SERTOLI CELL division and seminiferous tubule growth, it accounts for 80% of the size of the testis. FSH is involved in initiation of spermatogenesis in puberty by regulating production of androgen-binding protein (see SEX HORMONE-BINDING GLOBULIN) by Sertoli cells and is responsible for production and release of inhibin B (see INHIBINS). Inhibin B suppresses FSH secretion by repressing production of its β-subunit. At the male pituitary, locally produced ACTIVIN interacts with inhibin B regulation, stimulating FSH synthesis. See TESTOSTERONE.

follicular cells **1**. Thyroxine-secreting cells of the THYROID GLAND. **2**. A single layer of cells surrounding and supplying nutrient to the oocyte in an immature ovarian follicle, developing into the GRANULOSA CELLS. Follicular cells remaining at the time of ovulation develop into the CORPUS LUTEUM under the influence of LH. See OVARIES.

follistatin A product of the PRIMITIVE NODE. Also a later product of GRANULOSA CELLS; its mode of action here is not fully understood, although it appears to neutralize the effect of activin on steroid production through autocrine or paracrine signalling. See INHIBINS.

fomites Inanimate environmental objects (e.g., hand towels, face flannels) that passively transfer pathogens from source to host.

fontanel, fontanelle Space between SKULL bones at birth; bridged by dense fibrous connective tissue.

food See DIET, ENERGY BALANCE, FASTING.

food aversion See REINFORCEMENT AND REWARD.

food poisoning Food poisoning is caused mainly by the rod-shaped bacterium SALMONELLA typhimurium, although SHIGELLA and CAMPYLOBACTER are also sometimes involved. *Staphylococcus aureus* can also cause food poisoning, notably in the USA, through its several enterotoxins (an exotoxin), preformed TOXINS produced during replication of the bacteria in prepared food

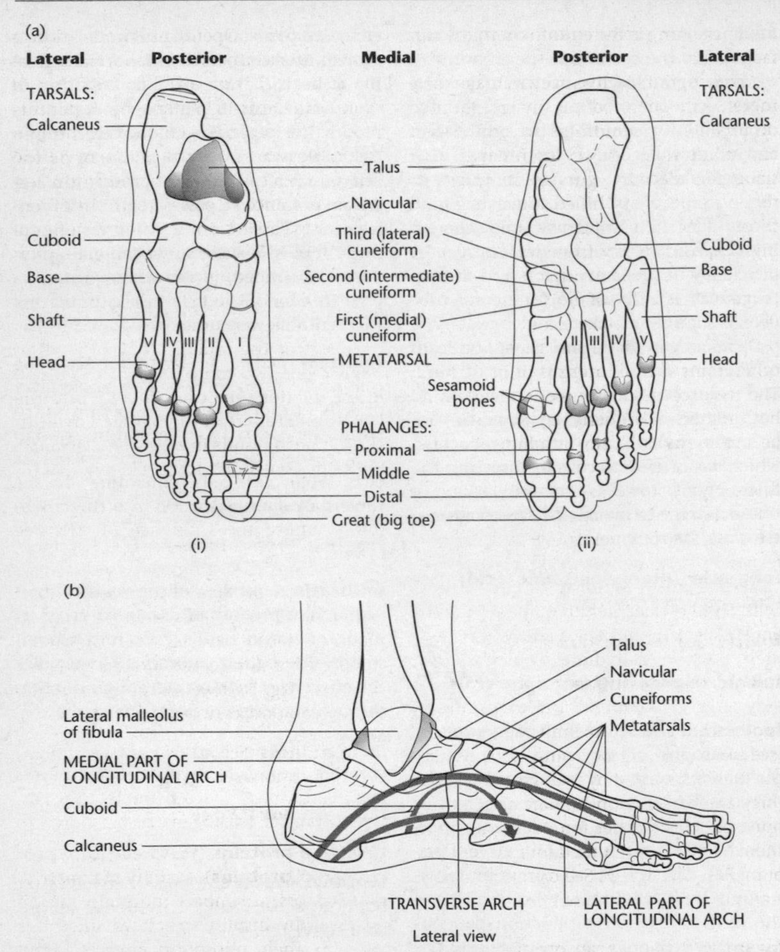

FIG 77. Diagrams showing (a) the skeletal anatomy of the right FOOT (see entry) in (i) superior view and (ii) inferior view, and (b) the right foot in lateral view, showing its three arches (thick arrows). From *Principles of Anatomy and Physiology*, G.J. Tortora & S.R. Grabowski, © 2000. Reproduced with permission of John Wiley & Sons, Inc.

and acting on intestinal neurons to induce vomiting. *Clostridium perfringens*, present in a variety of cooked and uncooked meat, poultry and fish, also produces an endotoxin eliciting diarrhoea and intestinal cramps, with nausea and vomiting in one-third of cases. Enterotoxins produced by *Escherichia coli* of the O157:H7 strain causes at least 20,000 cases and 250 deaths per year. *Salmonella* is found in the GUT FLORAS of many farm animals, especially those reared intensively on highly concentrated feeds. Most chickens in deep litter houses have salmonella in their gut, and increasingly these are finding their way into unpasteurized milk and eggs and hence into foods made with raw eggs. The cause of the gut irritation in salmonella food poisoning is the endotoxin in the surface wall of the bacterium.

About 10 million live bacteria have to be ingested in our food for an infection to occur and for symptoms to arise. Meat and poultry become contaminated after slaughter, when the guts are removed. As meat is frozen or chilled after slaughter, there is little bacterial growth until defrosting. Bacteria can pass from defrosting meat in a fridge or display to other foods, particularly if there is a film of moisture to swim in. People handling chicken, for example, can pass the organisms from their hands to other items such as utensils or other food. The most common source is meat which has not been properly cooked and has been left in warm surroundings, or food which has come into contact with bacteria since it was cooked.

Diarrhoea and vomiting are common symptoms of food poisoning, and antibiotics are usually ineffective because of difficulty in getting the drug to the infected cells. Fluid replacement is of major importance, especially in babies and young children, who dehydrate quickly. See REINFORCEMENT AND REWARD.

foot Feet develop from limb bud footplates (see LIMBS) and follow a similar pattern to the hands, retarded by 1–2 days. The great (big) toe of all non-human primates is opposed to the other four, set apart from them by a broad cleft used for grasping tree branches. See Fig. 77. See GAIT, HAND, SACRAL PLEXUS.

foramen In bones, an opening through which blood vessels, nerves or ligaments can pass.

foramen magnum A major foramen in the inferior part of the occipital bone of the SKULL (see Fig. 157), through which the spinal cord, accessory nerves and vertebral arteries pass. Within it, the medulla oblongata connects to the spinal cord. Its positioning relative to the orbital plane has been used as a measure of the degree of BIPEDALISM in hominin evolution (see SAHELANTHROPUS TCHADENSIS, Fig. 154).

foramen ovale Opening between left and right atria of the foetal heart, normally closing at birth (failure so to do resulting in 'hole in the heart'). While open, it permits most of the oxygenated blood returning to the foetal heart from the placenta via the inferior vena cava to pass across to the left atrium, whence it passes to the left ventricle, bypassing the non-functional lungs. On its superior surface, the foramen ovale is bounded by the *crista dividens*, a modification of the incomplete heart septum. On its lower surface, it is bounded by the valve of the foramen ovale. Between them, these two route most of the blood in the inferior vena cava directly into the left atrium. Failure of the foramen ovale to close at birth results in *blue-baby syndrome*. See DUCTUS ARTERIOSUS, PLACENTA.

forced vital capacity See VITAL CAPACITY.

fore limb (upper limb, arm) Developing from the embryonic limb buds (see LIMBS for some details), the fore limbs rotate 90° laterally during the 7th week of gestation so that the extensor muscles lie on the lateral and posterior surface, the thumbs lying laterally (see Fig. 78). See ELBOW; LEVERAGE, Fig. 113; PECTORAL GIRDLE; WRIST, Fig. 182.

forgetfulness See AMNESIA and, for memory suppression, see MEMORY.

forkhead box P2 See FOX P2.

Forkhead proteins (Forkhead box proteins, FOX proteins) A family of transcription factors, members normally being repressed by insulin signalling since this leads to their phosphorylation by PKB and their retention in the cytoplasm. They are involved in cell growth, division, differentiation and longevity, and some interact with HOX proteins (see HOX GENES). The protein PHA-4, a forkhead protein of the nematode worm *Caenorhabditis elegans*, which is very similar to the mammalian FOXA proteins, and has been shown to confer extended survival under dietary restriction (see LIFE EXPECTANCY). The closest mammalian homologue of the nematode daf-16 transcription factor, whose gene targets are involved in metabolism and stress resistance, is Foxo3 (see FOXO). See ENERGY BALANCE, FOXA2, FOXP2, NATURAL SELECTION.

form perception Perception of the outlines and orientations of images. See EXTRASTRIATE CORTEX, INFEROTEMPORAL CORTEX, PARVOCELLULAR–BLOB/INTERBLOB PATHWAY, PRIMARY VISUAL CORTEX.

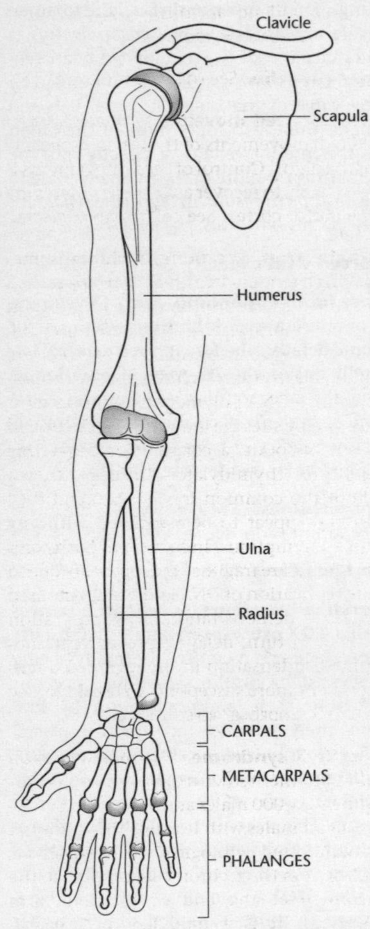

FIG 78. Diagram showing the right FORE LIMB, in anterior view. From *Principles of Anatomy and Physiology*, G.J. Tortora & S.R. Grabowski, © 2000. Reproduced with permission of John Wiley & Sons, Inc.

fornix The large bundle of axons leaving the HIPPOCAMPUS (see Fig. 91), looping around the thalamus and projecting to the HYPOTHALAMUS.

fossa Shallow depression on a bone's surface.

founder effect A form of GENETIC DRIFT, occurring when a small group of individuals becomes isolated reproductively from the main population, remains a small population through several subsequent generations, and founds a new colony. If the population increases, drift will continue to occur but at a slower rate. The founder effect is thought to be responsible for the discrepancy in GENETIC DIVERSITY between native African human populations and that of the remainder of present humanity, and for the loss of genetic variation resulting from POPULATION BOTTLENECKS, such as are thought to have occurred in the colonization of the New World (e.g., the lack of blood group B). See MIGRATION, OUT-OF-AFRICA MODELS.

fovea (fovea centralis, pl. foveae) A small depression in the centre of the MACULA LUTEA of the retina, ~1.5 mm in diameter and representing 5° of the VISUAL FIELD and the area of greatest VISUAL ACUITY (resolution). The cell layers present are the pigmented epithelium, the photoreceptors (exclusively tall and slender CONE CELLS), the outer nuclear layer and the outer plexiform layer (see RETINA, Fig. 149). The ganglion cell axons leaving the fovea are directed to the temporal region of the optic disc and so do not obstruct light on its way to the photoreceptors. See FOVEOLA.

foveola The central portion of the FOVEA, 0.4 mm in diameter, in which cones do not converge on bipolar cells. It lacks short-wavelength (S) CONE CELLS, improving VISUAL ACUITY.

FOXA2 (forkhead box A2, hepatic nuclear factor-3) A member of the family of Forkhead transcription factors and, in addition to developmental roles, a key switch regulating GLUCAGON gene expression by the pancreas and consequent FATTY

ACID METABOLISM in the LIVER during FASTING (see FATTY ACID METABOLISM). The nematode PHA-4 is very similar to mammalian FOXA proteins and triggers hormones coordinating physiological responses to dietary restriction (see LIFE EXPECTANCY). See FORKHEAD PROTEINS.

FOXO A family of Forkhead transcription factors. The SIRTUIN SIT1 and FOXO3 form a complex in cells in response to oxidative stress. SIRT1 increases FOXO3's ability to induce cell cycle arrest, and resistance to oxidative stress, but inhibits FOXO3's ability to induce cell death. As a result, the Sir2 proteins may increase organismal LIFE EXPECTANCY by directing FOXO-dependent responses away from apoptosis and towards stress resistance. See FORKHEAD PROTEINS.

***FOXP2* (forkhead box P2)** A gene on human chromosome 7 encoding a Forkhead transcription factor whose involvement in a SPEECH and LANGUAGE disorder (SPECIFIC LANGUAGE IMPAIRMENT) was discovered in 2001 and aroused great interest because individuals with the relevant mutations have no overt impairment but rather a lesion in the neural circuitry affecting a language process. Human FOXP2 protein differs from the gorilla and chimpanzee amino acid sequence at just two residues, and from orang-utan and mouse sequences at just three or four residues – a fairly typical number. However, the unusual excess of rare alleles and of high-frequency alleles at the human *FOXP2* locus is indicative of a SELECTIVE SWEEP consistent with positive selection acting on the locus, apparently within the last 200 kyr. As yet, no evidence indicates that the amino acid replacements in the human lineage have functional significance in themselves, although such changes in a transcription factor could have such import. But the genetic architecture typically underlying complex traits such as speech and language abilities makes it most improbable that this was the only locus under positive selection in their evolution.

Subtle differences in *FOXP2* sequence between humans and non-humans show evidence of positive evolutionary selection by K_A/K_S ratio (see MUTATION) and the locus exhibits unusually low DNA sequence diversity in diverse human populations; i.e., many such populations share a common ancestral *FOXP2* sequence, which is evidence for a 'selective sweep' after the appearance of HOMO SAPIENS. However, genome-wide empirical studies indicate weaker evidence for this (see NATURAL SELECTION). See FORKHEAD PROTEINS.

FOX proteins See FORKHEAD PROTEINS.

fractionated movements Precise, often skilled, movements of the limbs, especially of the digits. Control of these movements is least likely to recover after STROKES involving the motor cortex. See CORTICOSPINAL TRACTS.

fragile sites Locations on chromosomes having a tendency to break or to appear as a gap or constriction when cells are grown in culture appropriately. Most common fragile sites (occurring in most people) are inducible in culture in the presence of aphidicolin, a DNA polymerase inhibitor; most rare fragile sites are induced by reduction in levels of FOLATE, a cofactor for converting UMP to thymidylate. Little is known about the common fragile sites, and they do not appear to be associated with any disease symptoms. Induction of both common and rare fragile sites is probably due to late replication of DNA and the agents used in their demonstration. Late replication would, in turn, delay or prevent chromosome condensation in the region and render them more susceptible to breakage. See FRAGILE-X SYNDROME, REPETITIVE DNA.

fragile-X syndrome The most common form of inherited MENTAL RETARDATION, occurring in 1:5,000 males and accounting for 4–8% of all males with learning difficulties. It is manifested cytologically by a FRAGILE SITE (*FRAXA*) in the X chromosome, close to the telomere at the end of the long arm (Xq27.3). It is a modified, or atypical, example of X-linked inheritance. A fragile site is a non-staining gap usually involving both chromatids at which a chromosome is liable to break, and is more difficult to detect cytologically in female carriers than in males. The FRAXA mutation results from expansion of the trinucleotide CGG in the 5′ untranslated region (5′ UTR) of the *FMR-1* gene (see REPETITIVE DNA), whose protein product is required for proper synaptogenesis in the 1st meiotic prophase, whose absence or mutation leads to formation of abnormal dendritic spines (see DENDRITES). In normal individuals, there are 10–50 copies of this triplet, but a small increase to 50–200 copies renders this repeat sequence unstable (*premutation*). Such a male (a 'normal transmitting male') is at increased risk of a late-onset neurological condition 'fragile X tremor/ataxia syndrome'. All inheriting daughters are of normal intelligence; but their sons will have a high risk of further mutation of the premutation during meiosis, leading to a full mutation (see ANTICIPATION). The premutation is also vulnerable to expansion during mitosis. A normal allele can be identified by PCR, although Southern blotting is required to detect full mutations (see DNA PROFILING). Full mutations suppress transcription of *FMR-1* by hypermethylation, abnormal methylation of the *FMR1* promoter preventing binding of nuclear respiratory factor (NRF1) and inhibiting the gene's transcription. This is believed to be responsible for the clinical features. This gene contains 17 exons encoding a cytoplasmic protein with important roles in development of neurons of the CEREBRAL CORTEX, a protein that can be detected in plasma using monoclonal antibodies. 50% of carrier females have mild learning difficulties. A second fragile site *FRAXE* (Xq28), close to *FRAXA*, also involves CGG triplets; but these are expanded much less frequently than in *FRAXA*. Some males with these expansions have mild learning difficulties. A third site, *FRAXF*, also lies close to *FRAXA*, but does not cause clinical abnormalities. See GENETIC COUNSELLING.

frameshift mutation See MUTATIONS.

Frank–Starling mechanism See STROKE VOLUME.

fraternal twins Non-identical TWINS.

freckles See MELANINS, SKIN COLOUR.

free energy See ADENOSINE TRIPHOSPHATE for a consideration on free energy of hydrolysis.

free fatty acids (non-esterified fatty acids, NEFA) Fatty acid molecules released from triglycerides by the action of lipases. They are transported in blood plasma bound to albumins. Insulin-resistant individuals are unable to suppress release of NEFA from ADIPOSE TISSUE. See FATTY ACIDS.

free radicals (radicals) Highly reactive chemical species produced when a covalent bond breaks evenly so that the two atoms in the bond each receive one of the two shared electrons. Each is symbolized by a dot before the first chemical symbol or before the ionic charge value. See MITOCHONDRION, VITAMIN E.

frequency-dependent selection A form of BALANCING SELECTION in which the frequency of a genotype determines its fitness. If a genotype has a higher fitness when rare than when common, an intermediate equilibrium value will be reached over time. See NATURAL SELECTION, MAJOR HISTOCOMPATIBILITY COMPLEX, POLYMORPHISM.

frontal One of the ANATOMICAL PLANES OF SECTION (see Fig. 9). See 'frontal' entries.

frontal bone One of the membrane bones of the SKULL (see Fig. 157).

frontal cortex See FRONTAL LOBE, PREFRONTAL CORTEX.

frontal lobe One of four major subdivisions of each cerebral hemisphere (see BRAIN, Fig. 29), lying anterior to the central sulcus, under the frontal bone (see SKULL, Fig. 157). It has a role in controlling movement and in associating the functions of other cortical areas, and lesions in the left frontal lobe may cause BROCA'S APHASIA. Little theory supported frontal lobotomy (or prefrontal lobotomy; see PREFRONTAL CORTEX) in treating severe human behavioural disorders, a technique developed By Egas Monitz, for which he won the 1949 Nobel Prize in medicine. However, frontal lesions had been observed to have a calming effect on chimpanzees. The effect of surgery was described as 'relief from anxiety' and 'escape from unendurable thoughts'. Despite having little effect on intelligence test scores, or memory, it appeared to blunt emotional thought components and responses associated with the LIMBIC SYSTEM, giving rise to socially inappropriate behaviour and leading to difficulties in planning goal-related behaviour. Nowadays, drug therapy has replaced frontal lobotomy in treating serious emotional disorders. See AFFECTIVE DISORDERS.

fructose A hexose carbohydrate and (with GLUCOSE) a breakdown product of sucrose. Its absorption into the apical microvilli of cells of the small intestine (see ENTEROCYTES) is by facilitated diffusion using the GLUCOSE TRANSPORTER GLUT-5, although its exit from the basal surfaces employs GLUT-2. In the polyol pathway, which occurs chiefly in the retina, lens, Schwann cells, kidneys and aorta, glucose is converted to sorbitol which is then slowly oxidized to fructose which may then be a substrate in GLYCOLYSIS.

fructose 1,6-bisphosphatase A Mg^{2+}-dependent enzyme generating fructose-6-phosphate from fructose 1,6-bisphosphate, the allosteric activator of pyruvate kinase. See GLUCONEOGENESIS.

FSH See FOLLICLE STIMULATING HORMONE.

F_{ST} calculations See GENETIC DISTANCE, MOLECULAR CLOCK.

functional brain imaging Methods for detecting regional changes in blood flow in the brain. See PET and MAGNETIC RESONANCE IMAGING.

functional genomics See GENOMICS.

functional magnetic resonance imaging (fMRI) See MAGNETIC RESONANCE IMAGING.

functional residual capacity (FRC) The volume of gas remaining in the lungs at the end of a tidal expiration, comprising the expiratory reserve volume and residual volume. See SPIROMETER.

funiculi Dorsal, lateral and ventral columns of white matter into which the SPINAL CORD is sometimes considered to be divided.

fusiform gyrus An extremely long convolution running over the inferior aspect of the temporal and occipital lobes of the BRAIN (see Fig. 30), separated medially by the collateral sulcus from the lingual gyrus and the anterior parahippocampal gyrus, and laterally by the inferior temporal sulcus from the inferior temporal gyrus (see INFERIOR TEMPORAL CORTEX). It is part of the EXTRASTRIATE CORTEX and has face-selective areas and its activity is statistically predictive of a subject's responses, suggesting that the conscious perception of faces could be made explicit in this area. Lesions in the fusiform gyrus tend to produce PROSOPAGNOSIA.

fusion genes (chimaeric genes) Genes whose sequences have become joined, usually as a result of chromosomal events such as TRANSLOCATIONS, or non-homologous CROSSOVERS. Their gene products may thereby include amino acid sequences from both genes, as in some *fusion proteins* (see ABL GENE). Almost all fusion genes have been found in haematopoietic malignancies such as LEUKAEMIAS and lymphomas, although 'hybrid genes' are a common cause of COLOUR BLINDNESS. Mutations in some tyrosine kinase receptors convert them into oncoproteins (see ONCOGENES) as a result of gene fusion, resulting in truncation of sequences encoding the receptor ectodomains and alteration of the remaining cytoplasmic domain so that it is fused with the protein product of the other member of the fusion gene. Since these are proteins normally prone to dimerize or oligomerize this results in a ligand-independent dimerization of these receptors and explains the oncogenic powers of these fusion proteins. Several oncogenes that cause misexpression of a protein are associated with chromosomal translocations diagnostic of various B-cell tumours. In follicular lymphoma, 85% of patients have a translocation between chromosomes 14 and 18 causing a transcriptional enhancer near the chromosome 14 breakpoint to be fused to the BCL-2 gene on chromosome 18. The resulting enhancer–*bcl-2* fusion causes large amounts of Bcl-2 protein to be expressed in B cells, effectively blocking apoptosis and giving them a long life in which to accumulate mutations promoting cell proliferation. See CHIMAERA (2).

fusion proteins Proteins resulting from gene fusion (see FUSION GENES), either naturally occurring or by employing recombinant DNA techniques. Also, those proteins deliberately tagged to another for research purposes.

***FY* gene** See DUFFY BLOOD GROUP.

G

γ- For these entries, see the headword prefixed by γ-.

***g* (general cognitive ability)** See INTELLIGENCE.

G0, G_0 (G zero) Phase of the CELL CYCLE in which growth of the cell and progression through the cycle are arrested (see G1). This quiescent state is brought about by absence of sufficient mitogenic growth factors in a cell's environment or the presence of growth-inhibitory substances, blocking expression of IMMEDIATE EARLY GENES. See MYOD.

G1 (or G_1) Gap I phase, the earliest period of interphase of the complex CELL CYCLE, following G0, and the period between the end of cytokinesis and the start of DNA REPLICATION. It may last from 12 to 15 hours, during which time a cell accumulates RNA and proteins and makes critical decisions about its growth versus its quiescence and whether, as a quiescent cell, it will differentiate. Centriole duplication starts in G1 and is completed by G2.

Early embryonic cells have constantly high cyclin E levels and lack G1 and G2, and many signals influence whether such a cell decides whether to self-renew, differentiate (become specialized and function) or die. Many mechanisms operate in G1 to suppress CDK activity and hinder entry into S-phase. The *Myc* gene (see MYC) is upregulated by MITOGEN binding, increasing transcription of IMMEDIATE EARLY GENES encoding G1 CYCLINS (D cyclins) and inducing transcription of E2F genes. By associating with a second transcription factor, Miz-1, Myc can also function as a transcription repressor, repressing expression of genes encoding $p15^{INK4B}$ (inhibitor of CDK4) and $p21^{Cip1}$ ($p21^{Waf1}$), CDK inhibitors which shut down the activities of CDK4/6 and CDK2. By preventing their expression, Myc resists the growth-inhibitory actions of TRANSFORMING GROWTH FACTOR-β. Without active cyclinD-CDK4/6 complexes, the cell is unable to progress through early and mid-G1 and reach the R point (see below). During much of G1, the similarly acting CDK4 and CDK6 associate with the D1, D2 and D3 cyclins, and diverse signal transduction pathways initiated by metabolic, stress and environmental cues are integrated and interpreted during this period. The decision to proceed into late G1 is made at a transition termed the *restriction point*, or R point (aka the G1 checkpoint) and is taken several hours before the G1/S-phase transition (blocked by the nuclear protein p53). Once formed, cyclinD-CDK4/6 complexes usher a cell up to, and possibly beyond, the R point, and the remaining G1-cyclins behave in a pre-programmed manner through to the end of M-phase independently of extracellular signals. The E-type cyclins E1 and E2 associate with CDK2 and allow the DNA replication in S-PHASE to progress. The prereplicative complex contains an origin of replication complex (ORC) which formed on the chromosome at the end of the previous M-phase (mitosis), and in addition contains a hexameric DNA helicase (Mcm) and a helicase loading factor (Cdc6). It is this that licenses the ORC to initiate DNA REPLICATION during S-phase, in response to E2F. A large body of evidence suggests that deregulation of the R-point decision-making machinery accompanies the formation of most, if not all, types of CANCER CELLS.

G2 Gap 2 phase of the CELL CYCLE, lasting from 3 to 5 hours, during which a cell prepares itself for entry into M-phase and cell division. At the end of G2, a cell has accumulated a stock-pile of phosphorylated M-Cdk (the M-cyclin–Cdk1 complex). The major event of G2, on completion of successful DNA replication, is the removal of these inhibitory phosphates by the protein phosphatase Cdc25 (see CYCLINS). At the same time, the inhibitory activity of the kinase Wee1 on M-Cdk is also suppressed (by positive feedback) so that M-Cdk activity increases abruptly. Remarkably, M-Cdk triggers all the events of early M-phase including assembly of the mitotic spindle, chromosome condensation, attachment of chromosomes to the spindle and breakdown of the nuclear envelope.

G6PD See GLUCOSE-6 PHOSPHATE DEHYDROGENASE.

Gα The subunit of heterotrimeric G PROTEINS (see entry) which dissociates on ligand-binding of the G protein-coupled receptor and mediates the ligand's effects.

$G_i(G_I)$ See G PROTEINS.

G_{olf} See G PROTEINS.

G_q See G PROTEINS.

G_s See G PROTEINS.

GABA (γ-aminobutyric acid) A largely inhibitory amino acid neurotransmitter of the central nervous system, especially the higher centres (contrast GLYCINE), derived from GLUTAMATE and released at perhaps one-third of synapses (said to be GABAergic). It helps maintain overall balance in the CNS by damping neurons' ability to respond to excitatory signals from other cells. On release into a synaptic cleft, GABA can activate two types of receptor: $GABA_A$ (ionotropic, linked to and opening a chloride channel) and $GABA_B$ (metabotropic, linked to G proteins activating K^+ channels). Both hyperpolarize the cell and inhibit action potential initiation. GABA transmission by cortical interneurons is vital for cortical plasticity (see NEURAL PLASTICITY), and seems to be involved in timing of the critical period of plasticity in the visual system. It is potentiated by very low concentrations of BENZODIAZAPINE tranquillizers (e.g., Valium) and barbiturates such as phenobarbital. The *GABR1* gene, encoding the ligand-gated ion channel ($GABA_A$ receptor, beta 1), and involved in GABA neurotransmission, is implicated in BIPOLAR DISORDER. See ANAESTHETICS, CANNABINOIDS, GLYCINE, NMDA RECEPTOR, PAIN.

GABAergic (Adj.) Of neurons that secrete GABA, and of membranes and membrane receptors sensitive to it.

gain-of-function mutations Those MUTATIONS resulting in either increased levels of gene expression, or development of a new function(s) in the gene product. They are dominantly inherited, those rare occurrences in the homozygous state resulting in the more severe phenotypes, as in homozygous achondroplasia. The expanded triple repeat mutations causing HUNTINGTON'S DISEASE and mutations in cellular ONCOGENES responsible for many forms of CANCER are examples. Compare LOSS-OF-FUNCTION MUTATIONS.

gait Walking (including KNUCKLE-WALKING), running, climbing and swimming. In comparison with apes, human legs have many long spring-like tendons connected to short muscle fascicles that generate force economically. In all mammals, humans included, walking (see CENTRAL PATTERN GENERATORS) employs an 'inverted pendulum' where the centre of mass vaults over a comparatively extended leg during the stance phase and efficiently transfers potential and kinetic energy, out-of-phase with every step. Optimal speed of walking, considered energetically, is ~1.3 $m.s^{-1}$ and is largely a function of leg length. Most humans switch to running at ~2.4 $m.s^{-1}$, employing a mass-spring mechanism exchanging kinetic and potential energies very differently. Tendons and ligaments in the leg are collagen-rich and store elastic strain energy through recoil in the propulsive phase, the legs flexing more during running than in walking, flexing and extending at the knee and ankle during the mid-stance support phase.

In what follows, italicized numerals refer to Appendix VIII.

During running, the elastic structure of the plantar arch of the FOOT (*1*; see also Fig. 79) helps maintain mid-tarsal rigidity for powered plantar flexion during toe-off (toes short, *2*) and absorbs some impact force after heel strike. The elastic structures of this arch function as a spring during running and return ~17% of the energy generated during each stance phase (see Fig. 79). For this arch to be effective as a spring, the transverse tarsal joint must also restrict rotation between the hind foot and the anterior tarsals, and in modern humans this is achieved by a projecting medial flange on the proximal cuboid.

During human evolution, such a foot feature is first apparent in the OH 8 specimen, usually attributed to *Homo habilis* (see HOMO). Skeletal strength is also a major factor in running, peak vertical ground reaction forces approaching 3–4 times body weight at higher speeds. The enlarged iliac pillar (*15*) and shorter femoral neck (*4*) reduce stress, while enlarged areas of articular surfaces (*5*) relative to body mass in most lower body joints of *Homo* compared with *Australopithecus* and *Pan* are adaptations spreading such forces (see BIPEDALISM, VALGUS ANGLE). *Homo* also has several derived features enhancing trunk stabilization during running,

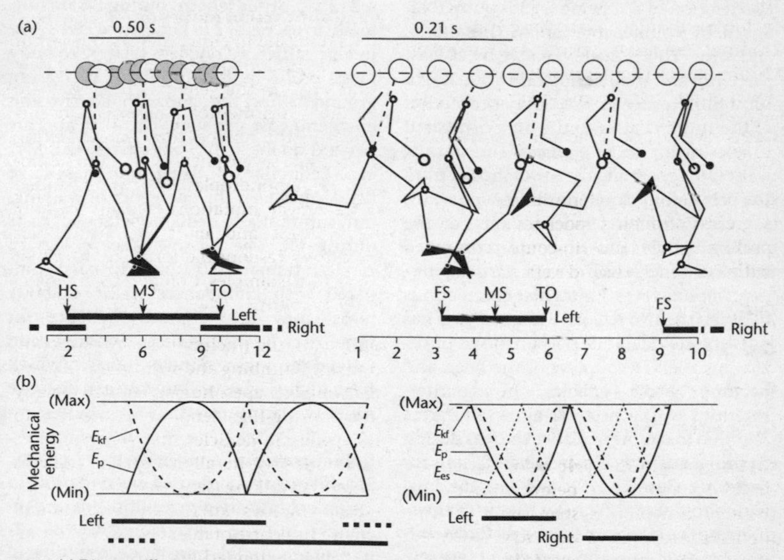

FIG 79. (a) Diagram illustrating some postural comparisons during walking and running GAIT, with relative centres of gravity indicated by circle positions at the top. FS = foot-strike; HS = heel-strike; MS = mid-stance; TO = toe-off. (b) Biomechanical contrasts between the two gaits shown in (a). When walking, an inverted pendulum mechanism exchanges forward kinetic energy (E_{kf}) for gravitational potential energy (E_p) between heel-strike and mid-stance, this exchange being reversed between mid-stance and toe-off. While running, a mass-spring mechanism causes E_p and E_{kf} to be in phase, both energies declining to minima between FS and MS. Tendons and ligaments in the leg partially convert decreases in E_p and E_{kf} to elastic strain energy (E_{es}) during the first half of the stance, this subsequently being released through recoil between MS and TO (see GAIT). From D.M. Bramble on D.E. Lieberman, © *Nature* 2004, with permission from Macmillan Publishers Ltd.

when there is a tendency to pitch forward – especially during heel strike. These include: expanded attachment areas on the sacrum and posterior iliac spine (*6*) for large erector spinae muscles, and a greatly enlarged gluteus maximus (one of the most distinctively human features), and surface area for its attachment (*7*); and the transverse processes of the sacrum are also relatively larger than in *Australopithecus*, indicating a more mechanically stable sacroiliac joint.

Greater leg length (*8*) enables greater stride length and is favoured during endurance walking and endurance running, as is the long ACHILLES' TENDON (*9*) which not only stores energy but also acts as a shock absorber. During the 'aerial' phase of running in *Homo*, inertially induced trunk rotation about its vertical axis generated by forward acceleration of the swing leg is counterbalanced by counter-rotation of the thorax and arms. This is made possible by at least three hip and shoulder modifications: (i) a substantially greater autonomy of rotation of the trunk relative to the hips compared to apes owing to an elongate and narrow waist (*10*) separating the lower thorax from the pelvis, fully developed in *H. erectus*; (ii) a greater structural independence of the pectoral girdle and head (*11*) compared with *Pan*, and possibly with *Australopithecus* (*Pan* has several muscular connections between the two regions whereas *Homo* has only the cleidocranial portion of the trapezius muscle). Decoupling of the head and pectoral girdle enables the counter-rotations of the pectoral girdle and arms required to counter-balance the legs during running, and may also be advantageous for throwing (see Fig. 79); and (iii) the low, wide shoulders (*12*) characteristic of *Homo* increase these counterbalancing forces during arm-swinging and may permit energy-saving reductions in forearm mass (*13*), being 50% less massive relative to total body mass in *Homo* than in *Pan*. The head stabilization needed during running in *Homo*, with its vertically positioned head, is achieved by a combination of decreased facial length and occipital projection behind the foramen magnum. From all this, and the relatively much larger radius of the posterior semicircular canal in *Homo* than in *Pan* or *Australopithecus*, which probably increases the sensitivity to head pitching while running compared with walking, and the enlargement of the nuchal ligament (*14*) in *Homo* (first evident in KNM-ER 1813, generally attributed to *H. habilis*), a convergent feature of cursorial mammals, it appears that *Homo* has indeed evolved derived features for bipedal running.

Endurance running (ER) may be regarded as running many kilometres over extended time periods using aerobic metabolism. No primates other than humans are capable of ER, although some such as patas monkeys can sprint for short distances. Human ER speeds range from ~2.3 $m.s^{-1}$ to as much as 6.5 $m.s^{-1}$ in elite athletes, and recreational joggers average between 3.2 and 4.2 $m.s^{-1}$. Stride lengths during ER are normally in excess of 2 m but can exceed 3.5 m in elite athletes. Long legs relative to body mass probably favour ER by increasing ground contact time and with it decreasing energetic cost per step. This first appears beyond doubt with *Homo erectus* ~1.8 Myr ago. Reductions of distal limb mass have little impact on the energetics of walking, but substantially reduce metabolic costs during ER. The relatively heavy legs of modern humans (30% of body mass) compared with chimpanzees (18% of body mass) may favour greater reliance on more slowly contracting, oxidative and fatigue-resistant muscle fibres, which have higher relative abundance in distance runners than sprinters.

galactosaemia A recessive trait (see MENDELIAN INHERITANCE) caused by mutation in chromosome site 9p resulting in a deficiency in the enzyme galactose 1-phosphate uridyl transferase resulting in inability to convert galactose to glucose. Begins in newborn period and results in vomiting when fed milk, jaundice and predisposes to Gram-negative meningitis. Resolved by removal of milk from diet (galactose is a digestion product of lactose). Tests for galactosaemia are included in many NEWBORN SCREENING schedules.

galactose A breakdown product of lactose, milk sugar (see GALACTOSYLTRANSFERASE), which can be phosphorylated by galactokinase in a Mg^{2+}-dependent reaction involving UDP as a coenzyme-like carrier of hexose groups. The net reaction is the conversion of galactose 1-phosphate to glucose 1-phosphate. The surface antigen α-Gal on XENOTRANSPLANTS (xenografts) is one reason for their rejection by humans. See GALACTOSAEMIA.

galactosyltransferase An enzyme, present in most tissues, that contributes to glycoprotein biosynthesis by promoting transfer of the activated galactose residue of UDP-galactose to the monosaccharide *N*-acetylglucosamine, already attached to protein. After birth, the specificity of the enzyme in the mammary gland epithelium changes so that it transfers the galactosyl group of UDP-galactose to glucose at a high rate, forming lactose. This change in specificity is due to the enzyme's association with α-lactalbumin, produced postpartum. Its synthesis is regulated by high prolactin and low progesterone, leading to the formation of the α-lactalbumin-galactosyltransferase complex ('lactose synthase').

galanin An inhibitory transmitter of the ventrolateral preoptic nucleus of the hypothalamus. See RETICULAR FORMATION, SLEEP–WAKE CYCLES.

gall bladder Developing as an outgrowth of the embryonic bile duct, the gall bladder is a thin-walled muscular sac, ~10 cm in length, on the inferior margin of the LIVER. Its wall is folded and can accommodate the BILE not required immediately for digestion, this being diverted into the gall bladder because of the relatively high tone and resistance of the sphincter of Oddi. It concentrates bile by up to 20-fold by absorbing Na^+, Cl^-, HCO_3^- and water. The sphincter of Oddi relaxes under the stimulus of CHOLECYSTOKININ (CCK) and initial parasympathetic vagal stimulation in response to fats in the duodenum followed by that of CCK cause contraction of its wall, forcing bile towards the duodenum. Sympathetic stimulation suppresses emptying of the gall bladder. For *gallstones*, see BILE.

GALP See GLYCERALDEHYDE 3-PHOSPHATE.

gamete intra-Fallopian transfer (GIFT) An early and simple form of assisted reproductive technology (ART), devised in the 1980s, to help couples with INFERTILITY problems to achieve pregnancy (and the only ART fully acceptable to the Roman Catholic Church). Gametes (i.e., ova and sperm) are washed and inserted via a catheter into the woman's OVIDUCT, so that fertilization occurs within her body (contrast IN VITRO FERTILIZATION, which is more popular). Fluid containing the ova (unfertilized eggs) is placed in a laboratory dish and inspected microscopically. A nutrient culture is added to one or two selected ova, and then sperm and eggs are loaded sequentially into a catheter and introduced into the patient's oviduct via laparoscopy or a small lower abdominal incision. Fertilization takes place in the oviduct and achieves a 50% pregnancy rate per retrieval cycle in younger women, and lower in older women (but if an older woman can provide many eggs the pregnancy rate can be as high as in younger women). If zygotes produced by IVF are introduced into an oviduct, the technique is termed ZIFT.

ganglion (pl. ganglia) A collection of peripheral nervous system neurons, or of their cell bodies (e.g., dorsal root ganglia). The only central nervous system ganglia are the BASAL GANGLIA.

ganglion cells The only cells of the retina which generate action potentials, this as a result of information received from bipolar cells. They are involved in establishment of OCULAR DOMINANCE before birth. Their axons leave by way of the optic disc, becoming myelinated and collectively forming the OPTIC NERVE.

Two classes of retinal ganglion cells exist, each with on-centre and off-centre cells (see RECEPTIVE FIELD; RETINA, Figs 151 and 152). These classes are: M (for *magni*, or large; ~5% of the ganglion cells) and P (for *parvo*, or small; ~90% of the ganglion cells). The remaining 5% of ganglion cells (nonM-nonP cells) are less well characterized. M (magnocellular) cells have large receptive fields, conduct impulses more rapidly along the optic nerve and show a transient burst of action potentials in response to sustained illumination. Responding to large objects, they appear to be involved in analysis of gross features of stimuli, their orientation and movement. P (parvicellular) cells have small receptive fields, respond with a sustained discharge as long as the stimulus is present, are wavelength sensitive and involved more in the analysis of detailed stimulus form (contrast and contours) and colour than are M cells (which are colour-insensitive). P cells receive their input from single cones, or from several cones with the same wavelength sensitivity. The commoner 'red-green cells' have a receptive field which compares input from M and L cones. 'Blue-yellow cells' lack a centre surround pattern but are either excited by S cones and inhibited by a combined M- and-L signal, or vice versa. M cells are 'broad band' cells, getting combined input from M and L cones (but not S cones), and from rods.

Information from photoreceptor cells is segregated into distinct on-centre and off-centre pathways, a transformation which helps the higher centres (LATERAL GENICULATE NUCLEUS, PRIMARY VISUAL CORTEX) detect weak contrasts and rapid changes in light intensity. Different ganglion cells are specialized for processing different aspects of the visual image, such as the motion of a stimulus, or fine spatial details and colour (see COLOUR PERCEPTION). Most of the wavelength-sensitive P cells and nonM-nonP cells are COLOUR-OPPONENT CELLS, having receptive fields excited by one type of cone cell but inhibited by another.

gap junctions Most human cells, other than mature skeletal muscle fibres and blood cells, communicate with their neighbours via gap junctions, in which adjacent cell membranes are separated by a gap of ~2–4 nm (among the largest diameter pores) spanned by channel-forming proteins, CONNEXINS, which may close the pore. Six connexins combine to form a channel (a connexon) which usually allows ions to pass in both directions between the cells' cytosols, discriminating only on the basis of size. A gap junction consists of two closely applied pairs of connexons, one pair in each membrane. The gate of the gap-junction channel acts like the iris diaphragm of a camera. Because electrical current can pass through gap junctions (in both directions), cells so connected are said to be *electrotonically coupled*, the gap junctions producing electrical SYNAPSES (see entry). During brain development, gap junctions are thought to enable neighbouring cells to share both electrical and chemical signals that may assist their growth and maturation. See CARDIAC MUSCLE, ION CHANNELS, LIGHT ADAPTATION.

GAPs (GTPase-activating proteins) See GUANOSINE TRIPHOSPHATE.

gaseous exchange For the human system, this phrase usually refers to the net movement in opposite direction of molecules of oxygen and carbon dioxide at the exchange surfaces afforded by the capillaries in contact with alveoli in the lungs. However, gaseous exchange also occurs across capillary surfaces in general, surrounded as they are by tissue fluid. But the organs normally considered to have specific gaseous exchange functions are the lungs and the placenta.

Fick's law governs the rate of gas diffusion across the alveolar membrane, and the diffusion constant (D) relates directly to gas solubility (S) and inversely to the square root of the relative molecular mass of the gas (RMM): ($D \alpha S \div \sqrt{RMM}$). Despite their fairly similar RMMs, on a per-molecule basis, CO_2 diffuses ~20 × faster than O_2 through thin membranous tissues such as the respiratory membrane (see ALVEOLUS, Fig. 5), because of its higher solubility.

gastric glands See GASTRIC PITS.

gastric inhibitory peptide (glucose-dependent insulinotropic polypeptide, GIP) A 43-amino-acid polypeptide, somewhat resembling secretin, produced by K cells of the mucosa lining the duodenum and upper jejunum in response to the presence of fat, amino acids and glucose in the upper intestine. It promotes INSULIN release from the pancreas and inhibits gastric secretion and motility. It also enhances insulin-stimulated glucose uptake, stimulation of fatty acid synthesis and their incorporation into triglycerides. Inhibition of this activity may provide a useful anti-obesity strategy.

gastric lipase A lipase secreted in very small amounts by the gastric glands, being stable at the low pH levels of gastric juice. It is most effective in hydrolysing short-chain triglycerides found in milk, so probably more important in children than in adults.

gastric pits Deep invaginations of the mucosa of the STOMACH (see this entry for details of secretion) into which *gastric glands* open. The histology of the gastric glands differs with the region of the stomach, but they contain four types of cell: mucous neck cells situated near the gland's opening into the gastric pit and secreting a specialized mucus; chief cells, located in the basal regions of the glands and secreting PEPSINOGENS; PARIETAL CELLS, scattered among the chief cells and secreting hydrochloric acid and intrinsic factor; and enteroendocrine cells (G cells), secreting GASTRIN.

gastrin A peptide hormone secreted by G cells (ENTEROENDOCRINE CELLS) of the antrum of the stomach under vagal stimulation. In the cephalic phase of gastric secretion (see STOMACH, Fig. 166), anticipation and sight of food increase vagal output which is prolonged in the gastric phase by stretching of the stomach wall. Gastrin reaches the gastric glands via the blood, stimulating them to secrete mucus, acid and pepsinogens. Peptides and free amino acids also increase gastrin output. Vagal activity and gastrin stimulate histamine release from mast cells, this acting on parietal cells via H_2 receptors to stimulate H^+ secretion.

gastrointestinal hormones Hormones produced by the ENTEROENDOCRINE CELLS of the gastrointestinal mucosal epithelium.

gastrointestinal stromal tumours (GIST) Relatively uncommon tumours displaying aberrant tyrosine kinase activation for which few therapeutic options have been available.

Imatinib (Gleevec) is a promiscuous tyrosine kinase inhibitor, blocking activity of additional tyrosine kinases additional to BCR-ABL for which it was originally used (see CHRONIC MYELOID LEUKAEMIAS). The c-Kit receptor and the platelet-derived growth factor receptor (PDGFR) are two such, and a subset of GISTs displays mutations in the c-*Kit* gene so that patients with such tumours, who are generally unresponsive to conventional chemotherapy, show excellent response to imatinib. Almost one-third of patients with GISTs lacking mutations in c-*Kit* have mutations in the *PDGFRA* gene, for which imatinib is, at least initially, a robust agent (see CANCER, Fig. 34; STROMA).

gastrointestinal tract The portion of the ALIMENTARY CANAL from the stomach to the colon, inclusive, responsible for the bulk of food digestion and absorption.

gastrula, gastrulation The most important event occurring during the 3rd week of gestation, and a particularly sensitive time for teratogenic insult and for disruption by genetic abnormalities (e.g., CAUDAL DYSGENESIS). It is the process that establishes the three GERM LAYERS (ectoderm, endoderm and mesoderm; see Appendix VI; STEM CELLS) in their appropriate spatial distributions and begins with the formation of the PRIMITIVE STREAK. Gastrulation, like subsequent NEURULATION, is dependent upon the coordinated cell movements of convergent extension, itself dependent upon the non-canonical WNT SIGNALLING PATHWAY. It continues in caudal segments while cranial structures are differentiating, so that the embryo develops cephalocaudally. See ECTODERMIN.

gatekeeper genes See TUMOUR SUPPRESSOR GENES.

'gay gene' See SEXUAL ORIENTATION.

G bands Bands visible on mitotic CHROMOSOMES (see Appendix III) after mild trypsin treatment and staining with Giemsa stain (see CHROMOSOME BANDING). Light (G-light) bands alternate with dark (G-dark) bands, the dark bands once thought to have a relatively low GC content, and so relatively gene-poor (see GC-RICH SEQUENCES); but the banding is now thought to reflect the CHROMATIN packing density, densely packed G-dark chromatin taking up more of the stain. G-light bands are early-replicating and are probably active (i.e., their DNA is being transcribed; however, mitotic metaphase is not a time when much transcription takes place). G-dark bands replicate late in S-phase. See KARYOTYPE.

GC-rich sequences See REPETITIVE DNA.

GCTs Germ cell tumours. See CANCER.

gene Usually regarded as the smallest physical unit of heredity encoding a molecular cell product. Type I genes encode large ribosomal RNAs and 5.8S RNA; Type II genes include all polypeptide-encoding genes, those encoding snoRNAs and some snRNAs; Type III genes encode transfer RNAs, 5S RNA, some snRNAs and some other small RNAs. This classification is reflected in the RNA POLYMERASE involved in transcription. The genes of eukaryotes also include introns and regulatory sequences, notably a PROMOTER, immediately upstream of the coding sequence. See GENE EXPRESSION, HUMAN GENOME, TRANSCRIPTION.

gene chips See DNA CHIPS.

gene clusters The organization on the same chromosome of genes within a GENE FAMILY, often closely linked and thought to have arisen by tandem GENE DUPLICATION (e.g., see HOX GENES). When not quite tandemly repeated, the genes are said to be closely clustered, as opposed to tandemly clustered. The α- and β-GLOBIN gene clusters are closely clustered and regulated by a single control region. Compound gene clusters occur where the physical proximity of members of a gene family is less close on the chromosome, members being separated by genes of unrelated sequence and function (e.g., the HLA COMPLEX on chromosome 6p21.3; see Table 3 and MUTATIONS). One proposal is that the expression of

Location (human)	*Cluster*	*Median* K_A/K_I
1q21	Epidermal differentiation complex	1.46
6p22	Olfactory receptors and HLA-A	0.96
20p11	Cystatins	0.94
19q13	Pregnancy-specific glycoproteins	0.94
17q21	Hair keratins and keratin-associated proteins	0.93
19q13	CD33-related Siglecs	0.90
20q13	WAP domain protease inhibitors	0.90
22q11	Immunoglobulin-λ/breakpoint critical regions	0.85
12p13	Taste receptors, type 2	0.81
17q12	Chemokine (C-C motif) ligands	0.81
19q13	Leukocyte-associated immunoglobulin-like receptors	0.80
5q31	Protocadherin-β	0.77
1q32	Complement component 4-binding proteins	0.76
21q22	Keratin-associated proteins and uncharacterized ORFs	0.76
1q23	CD1 antigens	0.72
4q13	Chemokine (C-X-C motif) ligands	0.70

TABLE 3 *Some of the known* GENE CLUSTERS *in the human and chimpanzee genomes that are rapidly diverging. The right column gives the maximum median* K_A/K_I *ratios, if the cluster is stretched over more than one window of ten genes (see* MUTATIONS*). From* Nature *(2005). © 2005 Nature Publishing Group. Reproduced with permission from Macmillan Publishers Ltd.*

genes in some gene clusters is coordinated by a LOCUS CONTROL REGION located some distance upstream of the cluster (e.g., see HAEMOGLOBINS). Imprinted genes (see GENOMIC IMPRINTING) are usually found to be organized as gene clusters. See MUTATION, Fig. 127.

gene conversion Non-reciprocal transfer of genetic information from one chromosome to another. It involves alteration of one strand of heteroduplex DNA to make it complementary with the other strand at any position(s) where mispaired bases occur. It occurs in association with homologous RECOMBINATION events in meiosis (and more rarely in S-phase prior to the mitotic cell cycle). A donor DNA sequence, a few hundred bases or ~1 kb in length, is transferred from one gene to another with substantial sequence homology, the donor sequence being repaired back to its original form. The ousted sequence on the recipient chromosome is usually degraded enzymatically. See CANCER, Y CHROMOSOME.

gene dosage See COPY NUMBER, DOSAGE COMPENSATION.

gene duplication Duplicated genes are common in complex organisms and arise relatively rapidly in evolutionary time: on average once per 100 Myr per gene. Some duplications are of limited scope, involving single genes; others involve segmental duplications, or even whole genome duplications. Such events alter the COPY NUMBER of a gene. Approximately 10% of human genes contain duplicated sequences produced by EXON duplication, duplicating a particular protein domain in the encoded product. Proteins such as IMMUNOGLOBULINS and ALBUMINS, and those with major structural roles, e.g., COLLAGENS and FIBRONECTIN, often contain repeating motifs. Such intragenic tandem duplication, commonly the result of unequal sister chromatid exchange, allows coding sequences to expand and, often, to diversify. Unequal crossing-over can result in *tandem duplication* of entire genes, and lead eventually to GENE CLUSTERS (e.g., the α- and β-globin gene clusters; see Fig. 80 and PARALOGOUS GENES). Quite large-scale duplications originating in this way may occur, involving chromosome segments containing several genes. But these are probably rare on account of gene dosage effects (see DOSAGE COMPENSATION). Exon shuffling by RETROTRANSPOSITION has also provided an important mechanism

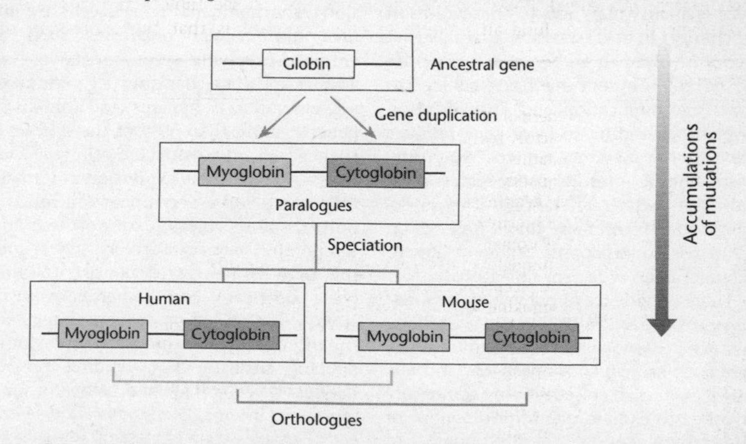

FIG 80. Diagram indicating how a tandem gene duplication event can be the source of paralogous genes within a species, each of which is orthologous to its homologue in a different species after a speciation event. Such putative events can explain the existence of gene clusters and gene families, including the origins of the myoglobin and cytoglobin genes, ~500 Mya. © 2004 From *Human Molecular Genetics* (3rd edition), by T Strachan & AP Read. Reproduced by permission of Garland Science/Taylor & Francis LLC.

by which some genes have been duplicated, although it has often led to production of PSEUDOGENES. Gene duplication may sometimes be the result of retrotransposition, whereby a cDNA copy (intronless) of an RNA transcript is inserted into a new chromosomal location. Such events have been very important in organizing the human genome, since ~40% of its DNA sequence consists of retrotransposon-derived repeats, a small percentage of which are actively transposing (see HUMAN GENOME).

Gene duplication is recognized as a major source of new genes and functions, and until recently it was usually assumed that duplicate genes were free to evolve new functions. But several recent case studies and comparisons of genome content suggest that most new duplicate genes do not have new functions but that paralogous gene pairs are often 'sub-functionalized' with two or more functions being partially or completely subdivided between the two genes.

Tandem gene duplication normally results in duplication of both the coding and regulatory sequences, and subsequent sequence divergence in *cis*-acting regulatory elements may lead to the evolution of changes in GENE EXPRESSION. It seems that genes involved in an organism's response to its environment are often subject to extensive duplication and diversification over time, leading to large gene families such as the OLFACTORY RECEPTOR repertoire, sometimes as a result of SEGMENTAL DUPLICATION (see entry's references). The subsequent 'death' of these duplicates, which gives rise to PSEUDOGENES, often relates to changes in an organism's life history.

There is evidence of polyploidy, or *whole genome duplication* (WGD), in vertebrate history (see MAJOR HISTOCOMPATIBILITY COMPLEX, FIBROBLAST GROWTH FACTOR RECEPTORS and HOX GENES), two such episodes being commonly invoked to explain quadrupled regions of the human genome. The high number of single-copy genes in the human genome is not necessarily evidence against this view since there is also evidence that extensive gene loss of duplicates has occurred. Two distinctive features of WGD are (i) that genes on one chromosome segment have a strong tendency to have copies (paralogues) on a single other chromosome, and (ii) when compared with a genome that has not undergone WGD, the duplicated genome shows 2:1 mapping with interleaving – i.e., with conserved synteny and local gene order as the orthologous region of the unduplicated genome but with the duplicated genome genes interleaving in alternating stretches (doubly conserved synteny, DCS) due to subsequent differential gene loss between sections of the paralogous (sister) chromosomes of the duplicated genome. It seems that ~80% of orthologous loci shared between the pufferfish *Tetraodon nigroviridis* and humans have this arrangement (see HUMAN GENOME, Fig. 97).

gene expression Genes are expressed when they are transcribed into RNA, during which usually one strand of the DNA double helix is used as a template for production of an RNA transcript by one of the RNA POLYMERASES (see TRANSCRIPTION). While most genes in unicellular organisms are in a state of perpetual expression, gene repression is the dominant theme in cells of complex multicellular organisms such as animals, involving CHROMATIN REMODELLING and EPIGENETIC mechanisms (e.g., see GENOMIC IMPRINTING). In humans, and animals in general, more than 50% of the GENOME is silenced in any particular cell type (see GENE SILENCING). Gene expression in somatic cells needs to be very finely controlled – both tissue-specifically (see ISOFORMS) and temporally (see CIRCADIAN CLOCKS). Non-coding DNA sequences (REGULATORY ELEMENTS) lying upstream and sometimes downstream of the coding sequences regulate the timing of gene expression by providing docking sites for DNA-binding general TRANSCRIPTION FACTORS and activator and repressor proteins. Their *combinatorial interaction* enables gene expression levels to be adaptable enough to change during development, across different tissues, and in different environments (see HUMAN TRANSCRIPTOME). Gene expression is exquisitely regulated by SIGNAL TRANSDUCTION pathways, sometimes involving changes in membrane voltage (see CALCIUM, GONADOTROPIN RELEASING HORMONE). Tandem GENE DUPLICATION of coding sequences and regulatory elements followed by sequence divergence of the regulatory elements may enable the evolution of altered gene expression, both temporal and spatial.

Through studies in epigenetics, it is now clear that gene expression control is not as simple as an on/off switch (see BISTABILITY), but is more like a 'dimmer switch' where an almost infinite array of interrelating gene expression states is possible. Some of these multistable states may be rarely occupied. We know that complex processes such as metabolism, morphogenesis, cell growth and inflammation are coordinately regulated, and yet the genes that encode the components of the numerous signalling pathways involved in each sub-programme of these processes are widely scattered across the genome and activated by different transcription factors. In the late 1960s, Britten and Davidson proposed a scheme in which the co-expression of such blocks of genes might lie under the control of a 'master gene' (see HOX GENES). Since the early 1990s, evidence has accumulated that genes encoding COREGULATORY PROTEINS (co-activators and co-repressors) might fulfil the 'master gene' role. Particularly relevant here are those genes and REGULATORY ELEMENTS (e.g., LOCUS CONTROL ELEMENTS) involved in chromatin remodelling.

Activator proteins can bind to an ENHANCER thousands of nucleotides from a gene's promoter, either upstream or downstream of it. In addition to their DNA-binding motif, these proteins have an activation domain which binds the mediator of the RNA polymerase II holoenzyme. The simplest model for this 'action-at-a-distance' is that the DNA between the enhancer and the promoter loops out, through binding of architectural binding (looping) proteins. This enables the activators bound to the enhancer to contact proteins, such as Pol II, bound to the promoter and initiate transcription (see REGULATORY ELEMENTS, TRANSCRIPTION FACTORS).

In the 'looping-scanning' model of gene regulation, a locus control region regulates several genes. Several proteins that bind here scan through large portions of DNA, looping the intervening region out until they locate the relevant gene. Genes destined for expression are then moved to a region of the nucleus full of activating proteins (euchromatic regions) while those destined for silencing are moved into a nuclear region full of repressive factors (heterochromatic regions). Genes from different chromosomes might come into contact when the CHROMATIN containing them loops out from their chromosome 'territory' (see NUCLEUS). Activator proteins can also remodel chromatin, by binding to and recruiting histone acyl transferases and producing patterns of histone acetylation recognized, e.g., by the general transcription factor TFIID. They can also serve to recruit additional activator proteins.

King and Wilson proposed that altered gene regulation might solve the paradox of how few genetic changes drove the considerable anatomical and behavioural gulf between humans and chimpanzees. Mutational changes in regulatory sequences are a putative major cause of such differences (see MUTATION), but there are formidable obstacles to achieving the goals of such research, which is only in its infancy. Not least is the difficulty in obtaining experimental material from great apes, whose study should follow intellectual and ethical guidelines generally similar to those for research on human subjects. Nor will adult tissues alone give us the information needed to interpret gene expression and its consequences during development, probably the time when many of the crucial differences between humans and great apes are expressed, as King and Wilson appreciated.

Some research on humans and chimpanzees indicates that the evolution of gene expression patterns both within and between species conforms to the predictions of the NEUTRAL THEORY OF MOLECULAR EVOLUTION – that most expression differences observed are selectively neutral or nearly so, and largely a function of the time since the two species shared a common ancestor. That said, there is evidence that, since the divergence of human and chimpanzee lineages, more gene expression changes have occurred in the human lineage than in the chimpanzee's, brain tissue having the greatest acceleration in expression change of the tissues studied.

Each tissue appears to be under different levels of evolutionary constraint with respect to the expression of genes within it. Brain seems to be under more constraint than liver, despite its relatively greater change in gene expression, and genes that are expressed in many tissues are subject to more constraints than those expressed in few. The ratio of differences in gene expression between tissues in chimpanzees and humans to diversity in expression within these species is higher in the testis than in any other tissues studied up to 2005. This is particularly true of genes on the X CHROMOSOME that are expressed in the testis, suggesting positive selection on these expression changes (see NATURAL SELECTION, SPERMATOGENESIS).

It is becoming increasingly clear that changes in gene expression, largely resulting from changes in sequences of NON-CODING DNA (particularly the *cis*-regulatory regions to which transcription factors bind, but also interfering RNAs: see RNA INTERFERENCE, FOLATE), can impact hugely upon an organism's phenotype (see NUCLEAR RECEPTOR SUPERFAMILY). The organization of the genome within the nucleus, and the effects this has on gene expression, are attracting increased attention (see NUCLEUS). Not only do they probably contribute to the evolution of species novelties, they also contribute to disease susceptibility (especially COMPLEX DISEASES) and to differences in response to drug treatment. Cancer drugs are now designed to target specific cell SIGNAL TRANSDUCTION pathways, and it is hoped that choice of drug type and timing of use may be informed by the gene *expression signature* of the tumour (see RAS SIGNALLING PATHWAY).

Actively transcribed genes within the interphase nucleus localize at distinct regions ('transcription factories') where Pol II is concentrated, multiple genes – both *cis* and *trans* – apparently sharing the same factory. Temporary inactivation of 'active' genes, involved in their 'pulsing' action, seems to require location away from the factory, so that gene mobility appears to be an important factor in their expression. Transient interactions between regulatory elements and genes in the nucleus have been associated with coordinately regulated gene expression, as between the interferon-γ (*Ifng*) gene and the locus control region of the T_H2 cytokine locus in naïve $CD4^+$ T CELLS.

gene families and superfamilies Genes with considerable DNA sequence similarity (homology), at least in their exons, often originating through tandem GENE DUPLICATION. Three broad classes of gene families are often distinguished: (i) *classical gene families*, whose members have a high degree of DNA sequence homology over most of their gene lengths, or at least over their coding regions and include those for HISTONES, α-globins and β-globins; (ii) gene families whose gene products contain large, highly conserved sequences (domains) while other regions have low sequence homology. These genes often encode developmentally important proteins, such as transcription factors (where the product must recognize a specific DNA sequence in the target gene); (iii) gene families whose products share very short conserved sequences (motifs) indicating some shared general function.

The members of gene superfamilies are more distantly related in evolutionary terms, the gene products sharing a common general function but having weak sequence homology, without shared motifs, yet having certain structural features in common. These include the IMMUNOGLOBULIN superfamily, globin superfamily and G PROTEIN-COUPLED RECEPTOR superfamily. Gene families that have undergone local expansion since the ancestor of humans and chimpanzees include those for olfactory receptors and immunoglobulins. See GENE CLUSTERS.

gene fusion See FUSION GENES.

gene knockouts Techniques of REVERSE GENETICS providing selective gene deletion, usually in embryonic stem cells, in order to identify gene function. Such techniques initially resulted in all cell types inheriting the deletion and sometimes caused lethal developmental abnormality; but in *conditional knockout* techniques, the deletion can be restricted to a specific cell or tissue type by employing a tissue-specific promoter attached to the transgene. This enables the function of a gene within a specific cell type to be analysed in the absence of its developmental or pleiotropic effects. Compare CONGENIC STRAINS, TRANSGENICS.

gene locus The region of a chromosome occupied, in a population, by alternative forms (alleles) of the same GENE.

gene markers See GENETIC MARKERS.

gene modification See GENETIC MODIFICATION.

gene mutation See MUTATION.

gene over-expression Transcription of a gene caused either by a gain-of-function mutation, insertional mutation (e.g., by a mobile genetic element, such as a retrovirus), gene amplification or by chromosomal translocation. See ONCOGENES.

gene pool The total genetic profile (the loci present, the alleles and their frequencies) present in a discrete interbreeding population.

gene probe See DNA PROBE.

general cognitive ability (*g*) See INTELLIGENCE.

generalization The ability to recognize objects from a wide variety of perspectives (vantage points). See VISUAL PERCEPTION and references there.

general recombination See RECOMBINATION.

generator potential A membrane voltage change resulting from stimulation of a receptor cell (see RECEPTOR POTENTIAL) that reaches the threshold sufficient for an action potential to be conducted along the axon. A prolonged generator potential can trigger further action potentials until the stimulus, and the generator potential, come to an end. See ADAPTATION (2).

genes As the term is generally employed in molecular terms, those DNA sequences that are transcribed into RNA, along with any control region lying immediately upstream of the transcribed sequence. Alternative RNA splicing (see RNA EDITING) and EXON SHUFFLING have somewhat complicated this definition. Such sequences are not equivalent in every respect to Mendelian 'factors', which were distinguished purely on the basis of their effects on phenotype. See GENE LOCUS.

gene sequencing See DNA SEQUENCING.

gene silencing GENE EXPRESSION can be repressed prior to TRANSCRIPTION by epigenetic methylation of DNA or histones, while TRANSLATION of transcripts can be prevented by RNA INTERFERENCE. There also appear to be regions of the NUCLEUS into which chromatin containing genes to be repressed is moved. Techniques employing gene silencing (e.g., RNA INTERFERENCE) are instances of REVERSE GENETICS.

Over 50% of human nuclear genes are silenced in any differentiated cell type. Long-term silencing of transposons and imprinted genes by DNA METHYLATION is stably maintained from the gametes, throughout development, and into the adult (see GENOMIC IMPRINTING, TRANSPOSABLE ELEMENTS), and X CHROMOSOME INACTIVATION is another example of relatively stable epigenetic silencing (see also HETEROCHROMATIN). Most genes with tissue-specific expression patterns are set up to be repressed at an early stage of development and are only reactivated in their tissues of expression; a silencing programme seems to be established in a global manner, being stably maintained through the many cell cycles that occur during development. Several genes required for early development (or purely during germ cell development) are known to undergo a defined kinetics of acquisition of repressive histone modifications and DNA methylation, being expressed by EMBRYONIC STEM CELLS but silenced during the somatic gene-expression programme. These genes include those encoding the pluripotency sustaining transcription factors OCT4 and NANOG, permanent silencing of which probably safeguards against accidental expression in somatic cells, which would lead to dedifferentiation and predisposition to CANCER.

Although very little is known about gene repression in the early embryo, genes that are required later are transiently held in a flexibly reversible repressed state by epigenetic histone modifications. Each embryo also inherits the DNA-methylation pattern from the gametes, although this is erased in the zygote and early morula (see CHROMATIN REMODELLING, DNA REPLICATION). As a cell differentiates, genes required for the pluripotent state are switched off by histone modifications and DNA methylation, markings which also silence many of these genes in mature germ cells. These marks probably require rapid reversal after fertilization so as to permit their re-expression in the next generation.

In differentiated cells, however, genes involved in development are repressed by the POLYCOMB GROUP (PcG) protein system, which methylates histone H3K27, an inactivating mark (see HISTONES). Special epigenetic regulation is required in PRIMORDIAL GERM CELLS.

Transcription factors, in addition to their role in positive gene expression, can sometimes act as repressors of expression – e.g. Polycomb proteins in subsequent long-term repression of genes that are only expressed during embryonic development. The transcriptional repressor NSRF silences expression of neuronal genes in non-neuronal cell types and in neuronal progenitor cells (see DNA-BINDING PROTEINS). During development of pluripotent cells (e.g., EMBRYONIC STEM CELLS) pluripotency-associated genes are repressed, potentially permanently, by DNA methylation. Subsequently, gene silencing may be at the level of TRANSCRIPTION, or it may prevent translation of mRNA.

An early observation in CANCER epigenomics was that silencing of a gene can occur via aberrant tumour-specific transcriptionally repressive chromatin or methylation of an adjacent CpG island. See EPIGENETICS.

gene testing/screening See SCREENING.

gene therapy Any treatment that involves the direct genetic modification of a patient's cells in order to achieve a therapeutic goal. In *germ-line therapy* (banned in many countries) modification of a gamete, zygote or early embryo produces a permanent and transmissible alteration. *Somatic cell gene therapies* are attempts to modify specific cells or tissues that cannot be transmitted by the patient to their offspring and constitute the basis of all current gene therapy trials and protocols. The approach used in somatic cell gene therapies might involve *gene supplementation* (gene augmentation) whereby a functioning copy of a gene is introduced to provide a normal gene product to compete with the defective gene product. *Gene replacement* would aim to replace a defective gene (particularly GAIN-OF-FUNCTION MUTATIONS) by a functioning copy, and even to correct a defective gene *in situ*. Clinical trials using recombinant adeno-associated viruses (rAAV) for treating a retinal degenerative disorder, Leber's congenital amaurosis (LCA), were begun in 2007 after the technique restored photoreceptor function in blind Briard dogs in 2001. This is an attractive approach, since the eye is easily accessible and allows local application of vectors with reduced risk of systemic effects, and the retina is relatively immune-privileged. It is hoped the approach might be used in treatment of macula degeneration and retinosa pigmentosa (see RHODOPSIN), and encouraging reports of two small studies in 2008 (six human LCA patients) indicate that viral introduction of a working copy of the gene *RPE65* (retinal pigment epithelium 65), required for converting vitamin A into a form used in the production of rhodopsin, into retinal cells did significantly improve light perception in four patients, and enabled two to read several lines of an eye chart whereas before they could only detect hand motions. *Targeted inhibition of gene expression*, especially relevant to infectious diseases, aims to silence selected pathogen functions or activated oncogenes, down-regulate autoimmune responses, or even silence gain-of-function mutant alleles in genetic disorders. The *targeted killing of specific cells* would be a line of treatment applicable in cancers. The possible therapeutic use of HUMAN ARTIFICIAL CHROMOSOMES (see entry) as vectors would avoid some of the disadvantages of viral vectors.

Disorders that are possible candidates for gene therapy include both genetic and non-genetic diseases, treatment of genetic disease not being equivalent to genetic treatment of disease. Among the candidate single-gene (Mendelian) disorders are adenosine deaminase deficiency, haemoglobinopathies, cystic fibrosis (gene supplementation), haemophilia A and B and Duchenne muscular dystrophy. Candidate multifactorial diseases (see COMPLEX DISEASES) include CANCERS (gene supplementation), CORONARY HEART DISEASE, peripheral vascular disease and AUTOIMMUNE DISEASES. See RETROVIRUSES.

genetic code The table of correspondence between (i) all possible triplet sequences (codons) of messenger RNA and (ii) the amino acid which each triplet causes to be incorporated into a polypeptide during its synthesis on a ribosome. The code is almost uniform from prokaryotes to eukaryotes and was discovered in the mid-1960s by Niremberg, Khorana, and others. See MTDNA for mitochondrial genetic code differences.

genetic counselling Services introduced in the 1960s, the common theme of which is that there is a process of communication and education addressing concerns relating to the development and/or transmission of a hereditary disorder. The counsellor tries to ensure that the consultand (who seeks counselling) is given information enabling understanding of (i) the medical diagnosis and its implications in terms of prognosis and possible treatment; (ii) the mode of inheritance of the disorder and the risk of developing and/or transmitting it, and (iii) the choices or options available for dealing with the risks. Armed with this information, consultands should be able to reach their own fully informed decisions without undue pressure or stress. Alternative reproductive approaches to conception may be mentioned, such as artificial insemination using donor sperm (AID), use of donor ova and preimplantation genetic diagnosis (see IN VITRO FERTILIZATION, SCREENING). Such approaches may be useful if one partner is infertile (see INFERTILITY), as in the case of KLINEFLETER SYNDROME and TURNER SYNDROME, or in order to bypass transmission of a parent's disadvantageous gene(s) to the baby. Bayes' Theorem is widely used in genetic counselling to provide an overall probability of an event, or outcome, including carrier status.

It cannot be assumed that the concept of risk is as well understood by prospective parents as it is by the counsellor, and it is important to explain that a risk applies to each pregnancy, and that chance does not have a memory. Likewise, a couple faced with a probability of 1:25 that their next baby will have a neural tube defect should be aware that there is a 24:25 probability that it will not. It can be helpful, but also alarming, that ~1:40 babies has a congenital malformation or handicap of some sort (e.g., see GENETIC DISEASES AND DISORDERS, Fig. 81). An additional risk of 1:50 might therefore be regarded, upon reflection, as relatively low.

A common cause of learning difficulties, FRAGILE-X SYNDROME, poses difficulties for counsellors. For a woman carrying a full mutation, there is a 50% probability of each of her sons being affected, and of each of her daughters being carriers (any daughter having a probability of 0.25 of having mild learning difficulties). Prenatal diagnosis can be offered based on DNA analysis from chorionic villi; but in the case of a female foetus with a full mutation, accurate phenotype prediction cannot be made. Disputed paternity is also a difficult problem for which the help of a clinical geneticist is sometimes sought, limitations of earlier approaches having been overcome by the development of DNA PROFILING. However, EXPRESSIVITY and PENETRANCE of genes, and delayed age of onset, are always complicating factors to be taken into account, as is ANTICIPATION. Special problems that can arise in genetic counselling include CONSANGUINITY and INCEST, consanguineous marriages being widespread in many parts of the world.

genetic determinism Our grasp of GENE EXPRESSION continues to reveal increasingly complex genetic and epigenetic pathways involved. Most phenotypic traits of individuals with the same genotype (e.g., identical twins) can only be predicted to be concordant within specific confidence limits (see PHENOTYPE). Factors responsible for this include genetic PENETRANCE, EPIGENETIC factors, and lifetime experiences. See CRITICAL PERIOD, GENETIC PREDISPOSITION, NATURE–NURTURE DEBATE.

genetic diseases and disorders Many inherited diseases and disorders are caused by mutations in a single gene, and are often termed Mendelian disorders (see *Online Mendelian Inheritance In Man* website: www.nchi.nlm.hih.gov/omim/). In others,

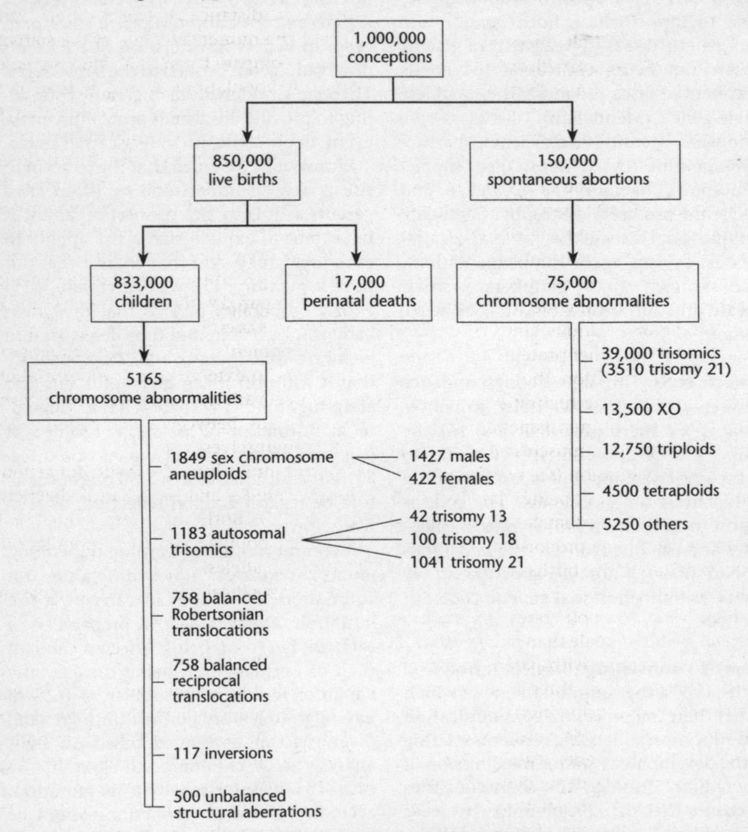

FIG 81. Diagram illustrating the likely fates of a million human zygotes.© 2005, W.H. Freeman & Co. Redrawn from *Introduction to Genetic Analysis* (8th edition), A.J.F. Griffiths, S.R. Wesler, R.C. Lewantin, W.M. Gelbart, D.T. Suzuki and J.H. Millar, with permission from the publisher.

an individual needs to carry mutations in two or three genes before any symptoms appear; while in COMPLEX DISEASES, several or many genes are involved (e.g., see ALZHEIMER'S DISEASE). The various forms of DIABETES also exemplify the complexity of disease genetics and aetiology. Elderly at-risk individuals who are unaffected by their disease genes may point in the direction of a solution to the problem of inherited disease (see MODIFIERS). Mutant alleles of certain TRKs can be transmitted in the human germ line, explaining the origins of a number of familial CANCER syndromes in which family members show greatly increased risks of contracting one.

It is estimated that one-third of people who inherit a genetic disorder do so as a result of either frameshift or nonsense mutations – base substitutions in DNA that alter the reading frame of the complementary mRNA or cause it to contain a 'stop' codon in place of a normal amino acid-coding triplet sequence. Nonsense mutations may cause the ribosome translating the mRNA into polypeptide to terminate its synthesis prematurely. Release of the incomplete fragment into the cytosol could cause various kinds of damage, e.g., binding to other proteins and causing them to malfunction (but see NONSENSE-MEDIATED DECAY). The highest known mutation rate for a human genetic disorder is that for Duchenne muscular dystrophy at ~1:10^4 per generation (see MUSCULAR DYSTROPHIES), most rates being much lower (~1:10^7). Chromosomal abnormalities account for a large proportion of the genetic problems in live births (see Fig. 81, and GENOMIC DISORDERS).

Observed MUTATION rates in humans appear higher in male than female gametes and often increase with paternal age. It is likely that there are three main classes of gene mutation causing human genetic disorders: (i) nucleotide substitutions scattered along the gene, usually associated with age- and sex-dependent effects; (ii) small insertions and deletions (especially deletions) lacking age effects but with a slight maternal excess; and (iii) 'hot-spots' occurring almost exclusively in males and increasing steeply with age (see, e.g., APERT'S SYNDROME). It seems that high levels may arise from infrequent mutational events, suggesting that such mutations may confer a selective advantage on the spermatogonia in which they arise. See AGE-RELATED MACULA DEGENERATION, FAMILIAL RECURRENCE RISKS, GENETIC COUNSELLING, GENETIC PREDISPOSITION and entries for particular disorders.

genetic distance 1. The distance between two loci as determined by recombination (see GENOME MAPPING). **2**. Measures allowing comparison of the relatedness of populations, or of molecules: the greater the evolutionary distance between them, the greater the numerical value of the statistic used to compare them. If a value is greater between populations A and B than it is between C and D, we conclude that C and D are more closely related than are A and B. With such pair-wise comparisons it becomes possible, given certain assumptions, to convert distance measures to an evolutionary time-scale enabling judgements to be made concerning the relative recency of a common ancestor between A and B compared with that between C and D (see MOLECULAR CLOCK).

Developments in DNA sequencing enable us to extend earlier work on allele frequency variation between populations to calculate between-allele genetic distances themselves.

Two commonly used statistics for genetic distance are F_{ST} and Nei's standard genetic distance, D. Both vary from values of 0 (identical populations) to 1 (populations sharing no alleles).

When specifically employed as a measure of genetic distance,

$$F_{ST} = V_p \div p(1 - p)$$

where p and V_p are the mean and variance of gene frequencies between the two populations, respectively, which gives added weight to alleles that are almost fixed (p ~100%) or barely polymorphic (p ~0%).

Nei's standard genetic distance (D) relates the probability of drawing two identical alleles from the two different populations ($\sum x_i y_i$) to the probability of drawing identical alleles from the same population, $\sum(x_i^2$ and $y_i^2)$, by the equation

$$D = -\ln\left(\sum x_i y_i \div \left(\sum x_i^2 (y_i^2)\right)^{\frac{1}{2}}\right)$$

With certain assumptions taken about the processes driving population divergence (e.g., genetic drift, mutation rate, population bottlenecks, immigration/emigration, etc.) this relationship can be used to generate a version of the statistic that is linear with respect to evolutionary time. This is a useful property of a genetic distance measure, but so is its variance, and if the sample size required from a population is very great in order for it to be representative, this will seriously affect its usefulness.

As for genetic distances between molecules, if they are DNA sequences then two homologous alleles, or mini- or micro-satellites, will differ at a discrete number of sites, or repeat units, respectively. Assuming these accumulate in stepwise fashion, these values in themselves constitute a simple genetic distance, although back-mutations and parallel mutations will lead to an underestimate of the number of mutational steps distancing the molecules. But corrections can be made for this.

Many different approaches are used in PHYLOGENETICS to construct phylogenetic trees and networks from genetic data, each applicable to a particular type of data.

genetic diversity In 1972, Lewontin published research in which variation at 17 genetic loci (for blood groups, serum proteins and red blood cell enzymes) was assessed by the immunological and electrophoretic methods then available, from POPULATIONS classified into seven so-called 'races': Caucasians, Black Africans, Mongoloids, South Asian Aborigines, Amerinds, Oceanians and Australian Aborigines, based on morphological, linguistic, historical and cultural criteria. The mean proportion of genetic variation within populations was 85.4%, that between populations *within* the 'races' was 8.3% while only 6.3% was found *between* the 'races'. The unmistakable conclusion was that most genetic variation lies *within* populations and that 'races' have no genetic reality (see Table 4). Claims about fixed genetic differences between races have proved to be the result of insufficient sampling (e.g., of populations separated by huge geographical distances), a conclusion corroborated by subsequent analyses using DNA markers for genetic diversity in 'continental groups' (rather than 'races'). Analyses of MTDNA and the Y CHROMOSOME do indicate proportionately less of the variation within groups and more of it between groups, although this is to be expected from their respective matrilineal and patrilineal inheritances (i.e., smaller EFFECTIVE POPULATION SIZES). Scientists work hard to ensure that the biological knowledge emerging from the HUMAN GENOME PROJECT is not used inappropriately to make socially constructed racial categories appear as though they were grounded in biology (e.g., through a misunderstanding of the biological basis of DRUG TARGETING).

The pattern of human genetic variation holds considerable information about selection events that have shaped our species (see NATURAL SELECTION).

The main pattern of human genetic variation across the globe is one of gene frequency gradients. The genome of any particular individual is more usefully thought of as a mosaic of haplotype blocks. As far as a given HAPLOTYPE BLOCK in the

	Within population	*Between populations within continental groups*	*Between continental groups*
Autosomal	~83–88%	~3–8%	~8–13%
mtDNA	~72–81%	~4–6%	~13–22%
Y-chromosomal	~35–80%	~12–25%	~16–50%

TABLE 4 *Selected studies of apportionment of human genetic diversity* GENETIC DIVERSITY.

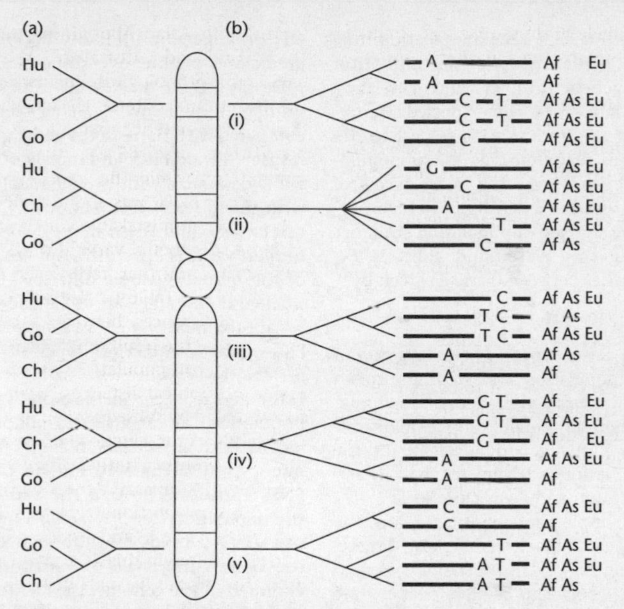

FIG 82. Diagram using a hypothetical chromosome to indicate: (a) that DNA sequences of different regions reveal different between-species relationships for human, chimpanzee and gorilla, most regions indicating that humans are most closely related to chimpanzees; and (b) that the among-human relationships for these same regions (but now on five different chromosomes, i–v) reveal that although most DNA variants are found in people from all the three continents mentioned, a few variants are found on only one continent, most often Africa. For region (ii), the level of sequence variation within humans is low relative to that observed at this region between species, giving a 'star-shaped' branching pattern of relatedness on account of the high frequency of nucleotide variations (A, T, C, G) unique to single chromosomes. Such regions may contain genes contributing to traits distinguishing humans from their closest relatives. Af = Africa; As = Asia; Eu = Europe. From S. Pääbo, © *Nature* 2003, with permission from Macmillan Publishers Ltd.

genome is concerned, a person from, e.g., Europe is often more closely related to a person from Africa or Asia than to another person from Europe sharing his or her complexion (see Fig. 82b). Each of us contains ~30% of the entire haplotype variation of the human gene pool. Recent work on COPY NUMBER variants indicates greater variation than has been indicated by study of SINGLE NUCLEOTIDE POLYMORPHISMS.

Many of the genetic differences between humans and other primates are a result of large duplications and deletions. Comparisons of within-species variation among humans and among the great apes (common chimpanzee, *Pan troglodytes*, bonobo, *Pan paniscus* and gorilla, *Gorilla gorilla*) have shown that humans have less genetic variation than they (see Fig. 82a). This strongly supports the view that our ancestors survived a population 'bottleneck' in which a larger, genetically more diverse, population was reduced to a smaller more homogeneous one. Comparative studies of X chromosomes in these primates, and in orang-utans, also indicate that humans have much less variation in non-coding DNA regions where mutations accumulate at a steady rate, providing evidence that the bottleneck occurred 190-160 Kyr BP. See HOMO, MTDNA, MIGRATION (2).

A population may not be randomly mating because of its internal subdivision into smaller sub-populations, or demes. The demes within such a metapopulation may undergo partial genetic differentiation through genetic drift, in which case they may deviate from the expected Hardy-Weinberg equilibrium with respect to the frequencies of heterozygotes, subpopulation divergence resulting in an excess of homozygotes and a deficiency of heterozygotes. The statistic of population structure measuring this Wahlmund effect is F_{ST} (which varies between 0 and 1) given by

$$F_{ST} = (H_T - H_S) \div H_T$$

where H_T is the expected heterozygosity of the metapopulation and H_S is the mean expected heterozygosity across subpopulations. When subpopulations are highly differentiated, the genetic diversity of the metapopulation is far greater than in any subpopulation and F_{ST} is close to 1.

In the simplest model of gene flow, the *island model* due to Sewall Wright, a metapopulation is split into 'islands' of equal size, N, between which genes are exchanged at the same rate per generation, *m*. Given the model's assumptions of no geographical substructure, each population persisting indefinitely, no mutation or selection, equilibrium between mutation and GENETIC DRIFT, and migrants being a random sample of the source 'island', then $F_{ST} = 1/(1 + 4Nm)$.

Geographical substructure is included in the *stepping-stone model*, adding the proviso that gene exchange is assumed to occur only between adjacent discrete subpopulations, between which equal rates of migration are assumed to occur. Both models have been used to demonstrate that even very low migration rates can retard population genetic differentiation.

Sewall Wright also devised the *isolation by distance* model, in which migration is modelled as occurring within the population's overall range but where mating choice is limited by distance (typically less than the whole population range, e.g., distance between birthplaces of parent and offspring). In this model, once an equilibrium has been established between gene flow and the differentiating effect of genetic drift, genetic similarity declines in a predictable manner over distance.

Despite their worldwide distribution, humans show a low F_{ST} value even when compared with those in the geographically restricted waterbuck and impala of Kenya, while compared with two mammal species with much wider distributions (N. American coyotes and Eurasian grey wolves) the human value is lower still. Any two copies of the human genome differ by ~0.1% of nucleotide sites (i.e., are ~99.9% identical), or approximately one variant per 1,000 bases. The commonest types of variant are SINGLE NUCLEOTIDE POLYMORPHISMS (SNPs). Estimated to occur in the world's human population at about 10 million sites where both alleles are present at a frequency ≥ 1%, these 10 million common SNPs constitute 90% of the variation in the population (see DNA FINGERPRINTING, GENOME MAPPING, GENOMICS). Although SNPs are the dominating type of variation so far explored in the genome, the identification of genome-wide large-scale COPY NUMBER polymorphisms is virtually untouched.

genetic drift Statistically significant alteration in population gene frequencies resulting not from NATURAL SELECTION, emigration or immigration, but from causes operating randomly with respect to the fitnesses of the alleles concerned. Its effects are expected to be greatest in small populations, where alleles may easily go to fixation or extinction by chance alone. See FOUNDER EFFECT, MTDNA, NEUTRAL THEORY OF MOLECULAR EVOLUTION.

genetic fingerprinting See DNA PROFILING, SINGLE NUCLEOTIDE POLYMORPHISMS, VNTRS.

genetic imprinting (chromosomal imprinting, parental imprinting) See GENOMIC IMPRINTING.

genetic lesions Mutations leading to loss or malfunction of a protein in a metabolic or signal transduction pathway.

genetic mapping See GENOME MAPPING.

genetic markers Variant alleles (polymorphisms) which are used to 'label', and so keep track of, some other biological structure, phenotype or process, throughout the course of a genealogy, or of an experiment. A marker allele may be either a sequence of DNA which encodes a cell product (e.g., a globin chain in haemoglobin), or else a non-coding sequence (e.g., an *Alu* sequence). Increasingly, markers are particular molecular variants in either DNA or protein (e.g., allozymes), rather than morphological.

Marker gene segregation is one method used to distinguish heritability from familiality of a character trait. This depends upon showing that genotypes carrying different alleles of certain marker genes also differ in their average phenotype for a quantitative character under investigation. The marker genes themselves make no contribution to the variance of this character; but if they vary in relation to it then the presumption is that they are linked to genes that do influence the character and its variation (see HAPLOTYPE). See BIOMARKERS, DNA MARKERS, SCREENING.

genetic predisposition To say of any aspect of phenotype that it is 'inherited' implies that its determinants are passed from parent to offspring, and the cardinal feature by which inherited predispositions are recognized clinically is family history. Normally, genetic predisposition to a disease is understood as a tendency to incur that disease on account of inheritance, by each cell, of one altered allele at a locus when two such altered copies of it are required for the cell to develop abnormally. Predisposition for the eye cancer RETINOBLASTOMA is genetic and involves receiving one deletion or mutated copy of the *RB1* gene at conception. Such a predisposition would require mutation of the second chromosome in a retinal cell before development of the cancerous condition occurred. *Genetic susceptibility* to disease covers any inheritance of a genetic disease determinant other than one for a single-gene disorder (i.e., typically multigene, or COMPLEX DISEASES). Not all inherited determinants are properly described as 'genetic' in the classical sense of that term, because: (i) it is not just nuclear genes that are inherited but also the genetic and non-genetic components of the (egg) cytoplasm; (ii) certain epistatic modifications of DNA are also inherited (see EPISTASIS); and (iii) factors affecting embryonic development in the uterus, properly described as *congenital*, are also inherited in the broad sense given above. See FAMILIAL RECURRENCE RISKS, FRAGILE-X SYNDROME, GENETIC DISEASES AND DISORDERS.

genetic probe See GENE PROBE.

genetic risk See COMPLEX DISEASES, GENETIC DISEASES.

genetic screening See SCREENING.

genetic susceptibility to disease See COMPLEX DISEASES, GENETIC DISEASES.

genetic variance See VARIATION.

genetic variation See GENETIC DIVERSITY, VARIATION.

gene trees See PHYLOGENETIC TREES.

geniculocortical pathway The pathway serving conscious visual perception, projecting from the LATERAL GENICULATE NUCLEI to the primary visual cortex and extrastriate cortex. See RETINOFUGAL PROJECTION.

genital ducts Both sexes originally have two pairs of genital ducts: mesonephric (Wolffian) ducts and paramesonephric (Müllerian) ducts, although only the former persist in males and the latter in females (see SEX DETERMINATION).

genital ridges (gonadal ridges) A pair of longitudinal ridges, appearing on the lateral surface of the mesonephros and formed by proliferation of the epithelium and condensation of the underlying mesenchyme. They are invaded by primordial germ cells during the 7th week of development, after which they develop into either OVARIES or TESTES. See SEX DETERMINATION.

genital swellings Swellings in either side of the urethral folds which form the scrotal swellings in the male and the labia majora in the female. See EXTERNAL GENITALIA.

genital tubercle See EXTERNAL GENITALIA.

genome Very approximately, the total genetic material within a cell of an organism. In an individual eukaryote, such as an individual human, it approximates to the complement of nuclear (chromosomal) and mitochondrial DNA (see MTDNA) present at FERTILIZATION. With the genomes of dozens of species, including human, chimpanzee and rhesus macaque, already in hand, and with powerful bioinformatics techniques for analysing and comparing them, research into the molecular basis of human evolution has increased explosively.

Like all eukaryotic cells, human cells house their genome in a dedicated intracellular compartment, the NUCLEUS. The inner membrane of this compartment plays a pivotal role in genome organization, such spatial organization being a key contributor to its function. As with all eukaryotic genomes, the human nuclear genome is packaged into the nucleosome particles (see CHROMATIN, Fig. 51) of CHROMOSOMES, preventing access by most DNA-binding proteins.

A genome is less like a static library of information than an active and adapting computer operating system for the organism, being remodelled over the course of evolutionary time by numerous processes. These include MUTATIONS of widely differing size, genome duplication (see GENE DUPLICATION), SEGMENTAL DUPLICATIONS and RETROTRANSPOSITION. It is, in essence, a sequence of DNA base pairs, distributed unevenly among chromosomes whose number and content continue to evolve. All organisms with a DNA-based genome share the unifying characteristic of the genetic code: rules for translation of nucleotide triplets into the amino acid sequences of proteins. But genomes also contain diffusely encoded information, including the sequence motifs bound by regulatory proteins which, unlike the genetic code, are capable of non-lethal variability enabling many variations of a motif to be bound by a given protein. Genomes encode, by a diffuse set of sequence-based rules, a 'nucleosome positioning code', an intrinsic nucleosome organization that accounts for much nucleosome positioning and, by implication, facilitates such specific chromosome functions as transcription factor binding, transcription initiation, and even some CHROMATIN REMODELLING.

Comparatively little of the DNA in a eukaryotic genome (in humans less than 2%) consists of protein-coding genes, and even a large proportion of these DNA sequences consist of non-coding INTRONS. Regulatory DNA sequences (see REGULATORY ELEMENTS), also non-coding, lie upstream of the coding sequences and regulate the timing of GENE EXPRESSION, expression levels which must be adaptable enough to change during development, across different tissues, and in different environments. There are also genes that encode RNA molecules as end-products, such as genes for transfer and ribosomal RNAs, directly involved in protein synthesis. In addition, there are DNA sequences that encode species of RNA, including microRNAs, not *directly* involved in protein synthesis but of whose biological importance we are increasingly aware. Then there are the vast stretches of DNA between these genes, so-called 'junk DNA' or perhaps more aptly 'genomic dark matter', much of which comprises either autonomous or non-autonomous TRANSPOSABLE ELEMENTS. Recent experiments have found that a significant fraction (e.g., 74% in brewer's yeast, and 90% in fission yeast) of this intergenic DNA is actively transcribed. Some of these intergenic stretches include PSEUDOGENES, derived either by SEGMENTAL DUPLICATION or by reverse transcription of mRNA and subsequent insertion of the DNA 'copy' (retrotransposition), a process in which LINE elements are involved. Not all of these are quite 'dead', in that a number of transcriptionally active intergenic areas overlap with pseudogenes. Unexpectedly long regions of highly conserved DNA (ultraconserved elements) were first identified in 2004 by comparing the human, mouse and rat genomes. Most are noncoding regions that have undergone little or no change since mammal and bird lineages diverged ~300 Mya. Many may function as distant enhancers of neighbouring developmental genes, and recent (2007) work suggests that selection coefficients involved may actually be much stronger than those for protein-coding regions, pointing to a functional role. There is increasing evidence that genome-wide association (GWA) studies represent a powerful tool in identification of genes involved in common human diseases (see GENOMICS, COMPLEX DISEASES). See GENE DUPLICATION, HUMAN GENOME PROJECT.

genome mapping Although large-scale DNA sequencing can be performed by piecing together random sequenced fragments that overlap, most sequencing of large genomes involves genetic (large-scale) mapping and physical (small-scale) mapping prior to sequence analysis. This requires access to large clone banks (e.g., provided by BACTERIAL ARTIFICIAL CHROMOSOMES) and a proficient system for recording the data recovered. A *genetic map* (linkage map) gives positions of genes relative to one another in the genome – locating them on their correct chromosome and indicating the relative (but not physically exact) order of the genes along their chromosome (genetic distances). It is normally arrived at by studying the proportion of recombinant offspring produced from parents whose genotypes are such that one is heterozygous at two loci whose relative positions on the chromosomes are being determined while the other is doubly recessive at those loci. LINKAGE DISEQUILIBRIUM indicates that the two loci are on the same chromosome, a larger proportion of recombinant offspring from these matings indicating larger relative separation between the loci on the chromosome. The principles of linkage mapping were worked out using the fruit fly *Drosophila*, but since people cannot be bred to order, a human geneticist prior to the 1970s had to ferret out linkage information from the medical literature, hospital records or population data, and more often than not the early studies on linkage in humans involved a blood group locus or other polymorphic gene in which heterozygotes are not rare.

Fortunately, in the 1980s, the discovery and use of restriction enzymes, which digest DNA at specific nucleotide sequences, revealed the highly polymorphic nature of human DNA at the molecular level (see HAPLOTYPE). Use of such DNA MARKERS enabled human linkage studies to be carried out in the absence of experimental breeding and in 1983 the gene for HUNTINGTON'S DISEASE was mapped by looking for recombinants among children from parents heterozygous for the disease gene and for a nearby RFLP site. The gene mapped near the tip of the short arm of chromosome 4, and subsequent discovery of nearby DNA markers on the other side of the gene (flanking markers) enabled the disease gene to be isolated and cloned. The most recent high-level resolution maps of the human genome have polymorphic markers that are, on average, spaced at intervals of <1 cM.

A *physical map* differs from a genetic map in that the relative positions of the genes are not merely inferred from recombination data but are directly located on real objects – DNA segments – that can be isol ated, sequenced, and then arranged in the actual order and orientation found in the organism's genome. It builds on genetic mapping, with a further level of detail. Two approaches are employed: the whole genome shotgun approach and the ordered clone approach.

In the *whole genome shotgun* (WGS) sequencing, sequencing precedes mapping. Randomly selected DNA clones are taken from a whole genome library (a shotgun library), read and assembled into a consensus sequence covering the entire genome by matching homologous sequences shared by clones which overlap (*clone contigs*, or *sequence contigs*). Where gaps occur, the two ends of a cloned sequence can be used as primers to sequence further into uncloned fragments (*primer walking*). WGS is good at producing draft-quality sequences of complex genomes, where repetitive DNA is abundant. Paired-end reads are used, in which pairs of 'sequence-reads' from opposite ends of a genomic insert in the same clone can span the gap between two sequence contigs and provide their relative orientation automatically.

In *ordered clone* sequencing, mapping precedes sequencing and proceeds by assembling contiguous clones of the organism's DNA, identified by similarities in restriction enzyme recognition sites, into overlapping clone contigs. These will eventually extend the whole length of a chromosome and are then fully sequenced using the logic of the whole genome shotgun approach. Once the physical map is complete, the ordered clones can be sequenced in detail to provide an agreed consensus sequence for the entire genome.

Low resolution physical mapping techniques include SOMATIC CELL HYBRIDIZATION and radiation hybrids, and fluorescence *in situ* hybridization (see FISH). High-resolution physical mapping techniques employ long-range restriction mapping (using enzymes that cut infrequently in the DNA) by pulsed field gel electrophoresis, high resolution FISH and DNA SEQUENCING (see also EXPRESSION SEQUENCE TAGS, HIGH THROUGHPUT GENOTYPING, SEQUENCE TAGGED-SITE MAPPING).

Difficulties in achieving consensus arise when DNA sequences are present in more than a single copy (often on more than one chromosome), especially if they are longer than the DNA sequences that can be conveniently cloned in standard vector systems (see SEGMENTAL DUPLICATIONS), and when they occur in inverted alignment (e.g., see Y CHROMOSOME).

As an example, in a 2005 study of the Zebrafish *golden* phenotype, linkage analysis of 1126 homozygous *gol*^{b1} embryos revealed a single crossover between *golden* and microsatellite marker z13836 on chromosome 18 giving a genetic map distance of 0.044 centimorgans (cM) corresponding to ~33 kilobases of a physical map. Comparison of *golden* cDNA to genomic sequences showed that the wild-type gene contained nine exons encoding 513 amino acids and computer matching revealed close similarities of the protein to sodium/calcium exchangers with highest similarities to mouse *Slc24a5* and human *SLC24A5*. Shared intron/exon structure and gene order between fish and mammals increased support for the view that the Zebrafish *golden* gene and human *SLC24A5* are ORTHOLOGUES and injection of the human *SLC24A5* mRNA into *golden* Zebrafish embryos rescued normal melanin pigmentation in the fish and demonstrated functional conservation of the gene. This work is potentially of considerable human import, demonstrating the remarkable theoretical and practical significance of research (see SKIN COLOUR) whose original direction lay in a completely different direction. See HUMAN GENOME PROJECT, PHYLOGENETIC FOOTPRINTING.

genome sequencing See CANCER, DNA SEQUENCING, GENOME MAPPING.

genomic disorders Conditions which arise from DNA rearrangements that bring about changes in regional genomic architecture and are therefore recombination-based. They contrast with conventional monogenic disorders (due to specific mutations within a gene, reflecting errors in DNA replication and/or repair), with digenic disorders (due to heterozygous mutations at two different gene loci), and with conditions resulting from faulty GENOMIC IMPRINTING. The DNA rearrangements involved lead to loss or gain of a dosage-sensitive gene (or genes) or to gene disruption, and are usually caused by non-allelic homologous recombination between region-specific low-copy repeats (LCRs), genomic DNA repeats spanning ~10–400 kb and sharing ≥ 97% sequence identity. The Genomic Medicine Database (GMED; webpage: gmed.bu.edu) currently allows one to survey chromosomes for possible associations between a limited number of traits, such as hypertension and cholesterol levels and up to 100,000 genetic markers (SNPs). See GENETIC DISEASES AND DISORDERS.

genomic duplications See GENE DUPLICATION.

genomic imprinting (gametic imprinting, parental imprinting) Genes which, although inherited from both parents, behave as though only one allele were present, i.e., hemizygously. The silenced allele is methylated, usually on the cytosine base of a CPG DINUCLEOTIDE in differentially methylated regions (DMRs) that overlap a promoter; but more often imprinted genes occur in clusters where a single DMR is methylated in the germ line (see PRIMORDIAL GERM CELLS) and regulates GENE SILENCING in the whole cluster. It is a particularly interesting example of epigenetic modification (see EPIGENESIS), involving genes whose expression is controlled, often tissue-specifically, in a manner dependent on the parent of origin. In tissues where the imprinted gene is expressed, the expression of either the paternally inherited allele or the maternally inherited allele is consistently repressed, resulting in *monoallelic expression* (equivalent to hemizygous inheritance). The pattern of such expression can be faithfully transmitted to daughter cells following cell division. Such imprinted genes tend to occur in those GENE CLUSTERS harbouring imprint control regions, ICRs, and containing both maternally and paternally imprinted genes. Often, both the gene's DNA strands can be transcribed, although the antisense strand (often >100 kb in length) prevents transcription of the other (sense) strand. When malfunctions in the imprinting mechanism occur, or when the parental origin is not the expected one, human imprinting diseases can arise (e.g., PRADER–WILLI SYNDROME, ANGELMAN SYNDROME, BECKWITH–WIEDEMANN SYNDROME, HUNTINGTON'S DISEASE, early onset myotonic dystrophy, FRAGILE X SYNDROME). Girls who inherit only one X chromosome (see TURNER SYNDROME) more frequently exhibit behavioural socialization problems if they lack paternal X chromosome rather than the maternal one, and a candidate imprinted gene has been identified that may be responsible (see AUTISM, BIPOLAR DISORDER). *Loss of imprinting* indicates that the imprinting pattern has been reversed. Considering the situation in Beckwith-Wiedemann syndrome (BWS), two closely-linked loci are involved. A normal maternal chromosome 11 has an inactive *IGF2* locus and an active *H19* locus with an unmethylated *H19* ICR, while a normal paternal chromosome 11 has an active *IGF2* locus, a methylated *H19* ICR and an inactive *H19* locus. In BWS, the maternal chromosome has reversed its imprinting and methylation pattern so that no growth-suppressing RNA is produced.

As chromosomes pass through the male and female GERM LINES they must acquire imprints, signalling the difference between maternal and paternal chromosomes. One key component, at least in maintaining an imprinted status, is allele-specific DNA METHYLATION. All imprinted genes are characterized by CpG-rich regions of differential methylation (see CPG DINUCLEOTIDE) for which cytosine methyltransferase is required. The cytosine methyltransferase DNMT1 is the most important enzyme in maintaining the imprinted status of a gene and the *Dnmt1* gene in mice is known to have sex-specific exons. Although DNMT1s (DNA methyltransferase 1s) can maintain methylation of imprinted-gene DMRs (differentially methylated regions), how particular genes are selected for *de novo* methylation during OOGENESIS or SPERMATOGENESIS is not known, although it possibly involves pre-existing histone marks (see also FERTILIZATION). See PRIMORDIAL GERM CELLS, CLONE.

genomic instability An almost universal feature of CANCER CELLS, resulting from loss of spindle CHECKPOINT, ability to proceed through the cell cycle despite DNA damage, and ability to keep dividing beyond cell SENESCENCE. It is also a hallmark of AGEING, and may vary significantly between tissues. Chromosomal instability is the usual form in cancer cells, although MICROSATELLITE instability is a feature of some tumours. A major contributor to genomic instability is loss or mutation of *TP53* (see P53 GENE). The BRCA1 gene product is an important component of the BASC complex maintaining genomic stability. See AUTOPHAGOSOMES, CARETAKER GENES, CHEMOTHERAPY, GENOTOXIC STRESS.

genomics The study of the DNA sequences and properties of entire GENOMES. This includes discovery and identification of all the genes present, followed by the discovery of the functions of all their products at the biochemical, cellular and organismal levels. In the past decade, there has been a shift in the direction of research from the reductionist approach of studying single genes and their products to a more holistic one in which many, or even all, such products in a cell or tissue are studied simultaneously. Such global analysis of gene function is the rationale of *functional genomics*, a central theme of which is a description of a cell's transcriptome (see HUMAN TRANSCRIPTOME). From this, the cell's proteome can be established (see PROTEOMICS). Mutating a gene can sometimes affect expression profiles of others, suggesting links that may include functional pathways and networks. *Genome-wide association studies* compare the distribution of ~500,000 SINGLE NUCLEOTIDE POLYMORPHISMS (SNPs) simultaneously in hundreds or thousands of people with and without a particular disease, enabling correlations to be found between SNPs and particular symptoms. This provides an indication of the increased risk of a COMPLEX DISEASE associated with a particular SNP. See DNA-BINDING PROTEINS, DNA MICROARRAYS, HUMAN GENOME, TRYPANOSOMES.

genotoxic stress *Genotoxic influences* are those whose effect is to cause DNA damage (mutation) by either single or double-strand breaks. It is *genotoxic stress* of this kind to which DNA REPAIR MECHANISMS respond. Double-strand (ds) breaks (DSBs) are often created by ionizing radiation, such as ultraviolet light and X-rays. A single DSB occurring anywhere in the genome seems sufficient to induce a measurable increase in p53 levels (see P53 GENE). The identities of the proteins detecting such breaks remain unclear, but they transfer the signal to the ATM kinase. DNA damage often triggers cell death by APOPTOSIS. DNA is highly reactive and is easily altered by cell processes such as oxidation: one estimate is that the human genome undergoes about 10^5 modifications each day, each with a finite probability of residual damage. Damage to DNA during its replication in S-phase of the cell cycle prevents fresh origins of replication from being initiated.

SIGNAL TRANSDUCTION pathways activated following DNA damage are those involving phosphatidylinositol-3-OH-kinase-like kinases (PIKKs), and the proximal checkpoint kinases ATM and ATR (see CELL CYCLE, Fig. 40). The ATR-dependent checkpoint pathway recognizes and signals the presence of multiple DNA damage events and stalled DNA REPLICATION forks during S-phase. The ATM kinase seems to be activated primarily following DNA damage, whereas the ATR kinase and checkpoint pathway recognizes and signals the presence of multiple DNA damage events and stalled DNA REPLICATION forks during S-phase, helping cells deal with cellular stresses affecting DNA or chromatin structure. The higher order chromatin structural changes associated with double-strand DNA breaks induce a conformational change in ATM so that its homodimeric form present in unstressed cells dissociates to its monomeric form which can then phosphorylate nucleoplasmic proteins (e.g., p53) and migrate to the DSB and phosphorylate other substrates (e.g., NBS1, BRCA1 and SMC1).

A break occurring during G1 of the cell cycle will result in a chromosome break if unrepaired during S-phase; but a break occurring during G2 will, if unrepaired, result in a chromatid break. Repairs either involve joining two broken ends together, or adding a telomere to a broken end. An estimated ~50 DNA double-strand breaks occur during a typical mammalian cell division cycle; non-homologous end-joining is the principal mechanism for DSB repair, especially during G0 and G1 phase, and there is evidence that impairment of NHEJ is associated with increased chromosomal aberrations, including TRANSLOCATIONS. If damage is incorrectly repaired in such a way that there are no free broken ends, it is possible for abnormal chromosomes (e.g., those in which the wrong broken ends are joined together) to pass through the cell cycle CHECKPOINTS.

Cells differ greatly in their response to DNA damage. Those in the post-mitotic epithelium of adult intestinal crypts of Lieberkühn are resistant to apoptosis in response to ionizing radiation, but the dividing cells just a few hours younger, of the same lineage and just one cell deeper in the crypt, are acutely sensitive to both radiation- and drug-induced apoptosis. On the other hand, small $CD4^+/CD8^+$ lymphocytes of the thymus cortex, although past their last mitosis, still remain sensitive to apoptosis after DNA damage and other stimuli.

The Gram-negative bacterial genotoxin CDT (cytolethal distending toxin) produced by, e.g., *Escherichia coli, Salmonella typhi, Shigella dysinteriae, Helicobacter spp.* and *Campylobacter spp.* share a common nuclease-encoding operon whose products (CdtA, CdtB and CdtC) create DNA lesions and induce cellular distensions in epithelial cells and death by apoptosis in many lymphocytes. CdtA and CdtC are both lectin-type structures homologous to the plant toxin ricin.

genotype The complete genetic complement of a nucleus, cell or individual. Genotypes of the latter two include both nuclear and mitochondrial DNA. See GENOTYPING, PHENOTYPE.

genotyping Methods used in detecting and recording DNA variation, including HIGH THROUGHPUT GENOTYPING methods. See also DNA PROFILING.

germ cells Diploid cells giving rise to the gametes, either by OOGENESIS or SPERMATOGENESIS. They are derived by mitosis from PRIMORDIAL GERM CELLS.

germ disc See EMBRYONIC DISC.

germ layers The main layers, or groups of cells, distinguishable in the embryo during and immediately after GASTRULATION. Very approximately, *ectoderm* gives rise to epidermis (in the presence of BMP4) and nerve tissue (in the presence of the BMP4-inactivating proteins Noggin, Chordin and Follistatin); *endoderm* gives rise to the alimentary canal and associated glands, and *mesoderm* (see PRIMITIVE STREAK) gives rise to blood cells, connective tissue, kidney and muscle. Cartilage derives from more than one lineage. See Appendix VI; PRIMITIVE NODE; STEM CELLS.

germ line The GERM CELLS and the cells that give rise to them (PRIMORDIAL GERM CELLS), contributing cells to the next generation of individuals. Other cells of an individual constitute the *soma*. In vertebrates, new methyl-to-cytosine mutations can be transmitted to the next generation only if they occur in the germ line, in which cells the DNA is inactive and highly methylated. See CpG DINUCLEOTIDE, DNA METHYLATION, GENE THERAPY.

germinal epithelium See OVARY.

gestation The duration of PREGNANCY.

ghrelin A 28-amino acid hormone synthesized by numerous cells of the gastrointestinal tract, but especially in the fundus of the stomach, with a probable involvement in meal initiation. Acylated by addition of an octanoyl group at the third serine residue, its powerful appetite-stimulating and growth hormone-releasing activities are mediated by the growth hormone secretagogue receptor (GHS-R). In its acylated form, it can cross the BLOOD–BRAIN BARRIER and activate neuropeptide Y/AgRP-expressing neurons of the hypothalamus (see APPETITE, Fig. 14; CIRCADIAN CLOCKS). Plasma levels of ghrelin rise during fasting and immediately before meals, falling within an hour of food intake. It opposes the hypothalamic actions of leptin, insulin and PYY_{3-36}, suggesting that the effect of weight loss in increasing ghrelin levels may contribute to weight gain, so that ghrelin agonists might be useful in the treatment of specific patient groups with ANOREXIA. See ENERGY BALANCE; compare LEPTIN.

giantism Abnormal increase in long bones during childhood caused by hypersecretion of human GROWTH HORMONE, hGH. Compare ACROMEGALY.

Giardia lamblia An important protist (Diplomonadida) parasite of the human intestine, related to free-living and parasitic treponemas, and the most prevalent protist parasite in the USA (incidence up to 0.7%). Worldwide giardiasis is common among people with poor faecal-oral hygiene, while modes of transmission include contaminated water supplies and sexual activity. It is microaerophilic.

Giemsa A stain that produces dark bands in AT-rich chromosome regions. See G BANDS.

GIFT See GAMETE INTRA-FALLOPIAN TRANSFER.

glandular fever (infectious mononucleosis) A systemic disease cause by EPSTEIN–BARR VIRUS (EBV), often detected by presenting as a sore throat. After 6–8 weeks' incubation fever, sore throat and a white creamy tonsil exudate occurs; gross pharyngeal swelling may make swallowing difficult. Most convalesce steadily, but persisting *chronic fatigue syndrome* is a rare complication.

glans penis See PENIS.

glaucoma Abnormally high intraocular pressure due to buildup of AQUEOUS HUMOUR within the anterior chamber of the EYE. The fluid compresses the lens into the vitreous humour, putting pressure on the retina. Persistent pressure leads progressively from mild visual impairment to irreversible retinal damage. It occurs more commonly with advancing age, affecting over 67 million people worldwide (primary open-angle glaucoma, POAG, affecting 33 million), and is a leading cause of BLINDNESS. Glaucoma is genetically heterogeneous, at least 8 loci having been associated with the disorder.

glenoid cavity The shallow depression, inferior to the acromion of the SCAPULA, receiving the head of the HUMERUS.

glenoid fossa See MANDIBULAR FOSSA.

glial cells (neuroglia) Non-conducting nerve cells. There are three types in the central nervous system: ASTROCYTES, OLIGODENDROCYTES and MICROGLIAL CELLS and one type in the peripheral nervous system: the SCHWANN CELL. They have many of the same voltage-gated ION CHANNELS of axons, but communicate by means of chemical signalling. They have a large repertoire of membrane receptors, some of which respond to neurotransmitters. See NEURONS, EPENDYMAL CELLS.

glioblastomas A type of human brain tumour caused commonly by the cells' faulty receptor for epidermal growth factor (a chromosomal deletion results in absence of part of its extracellular domain) so that even in EGF's absence they produce a stimulatory signal.

globins A family of porphyrin-containing proteins that bind oxygen reversibly and are therefore important features of the human respiratory system. They are classified into four functional classes (haemoglobins (Hb), myoglobin, neuroglobin and cytoglobins). Sequence homology between the globin polypeptides suggests that the family has evolved through a series of GENE DUPLICATION events, some ancient and some more recent. Polypeptides encoded by the different gene clusters are generally distantly related to one another, although the α-globin and β-globin families are clearly more closely related than is either to the other classes. The α-globin gene cluster (16p13.3) contains two human α-globin genes (*HBA1* and *HBA2*) which actually encode identical products, indicating that the duplication responsible is very recent. The two foetal γ-globin genes (*HBG1* and *HBG2*) are located in the β-globin cluster (11p15.5) and encode polypeptides differing by a single amino acid, indicating recent duplication. The *HBD* gene encoding the δ-globin of adult A_2 is also in the β-globin cluster (11p15.5), as is *HBE1* encoding ε-globin. The phylogeny of the β-globin cluster has been useful in tracing the history of HOMO SAPIENS; see GENE FAMILIES AND SUPERFAMILIES. Gene expression in the α- and β-globin gene clusters can be controlled by common LOCUS CONTROL REGIONS (LCRs) located upstream of the clusters. Without their respective LCRs, globin gene expression is negligible, although we do not fully understand how an LCR operates. The 'open' active chromatin domain resulting from transgenic LCR action makes the DNA more susceptible to cleavage by DNase I. The β-globin LCR seems to comprise short sequences at five major DNase I-hypersensitive (HS) sites located ~50–60 kb upstream of the β-globin gene of erythroid DNA (but not of DNA from cells that do not express globin genes), while the α-globin LCR is located 60 kb upstream of the α-globin gene. Although a transgenic β-globin LCR can induce chromatin opening at ectopic locations, it does not appear to do so when at its original location in the β-globin gene cluster. Other DNase I hypersensitive sites are located in promoters of the globin genes but show developmental stage specificity, this 'haemoglobin switching' in globin gene expression reflecting the competition between the globin genes for interaction with their respective LCR and stage-specific activation of gene-specific silencer elements. See HAEMOGLOBINOPATHIES.

α- and β-globulins See GLOBULINS.

globulins Produced by hepatocytes and plasma cells, globulins account for ~38% of the total plasma protein mass. The α- and β-globulins are liver products and transport lipids (including fat-soluble vitamins), carbohydrates, hormones (see BINDING PROTEINS), iron and copper, while γ-globulins are the ANTIBODIES produced by plasma cells. Fibrinogen is often grouped with the globulins.

globus pallidus See BASAL GANGLIA.

glomerular filtrate The product of net filtration from a KIDNEY glomerulus into the surrounding BOWMAN'S CAPSULE. See also MACULA DENSA, PROXIMAL CONVOLUTED TUBULE.

glomerulus **1**. Fenestrated capillaries of the filtration unit of the renal cortex (see KIDNEYS). **2**. Spherical zones, nodes, of OLFACTORY BULBS. **3**. Synaptic complexes of the cortex of the cerebellum, containing the swollen terminal of a single mossy fibre and its synaptic connections with the dendrites of 4–5 granule cells.

glossopharyngeal nerve (cranial nerve IX) A mixed CRANIAL NERVE, whose cell bodies lie in nuclei in the medulla (rhombomeres 6 and 7), containing afferents from the tongue (see taste) and efferents to the pharyngeal muscles and parasympathetic secretomotor fibres to the parotid salivary glands.

glottis A pair of mucous membranous folds, the vocal cords, in the larynx, separated by a space, the rima glottidis. See LARYNX, SWALLOWING.

GLP-1 See GLUCAGON-LIKE PEPTIDE-1.

glucagon A 29-amino acid single-chain polypeptide hormone synthesized by α-cells of the PANCREAS and secreted in response, *inter alia*, to low plasma glucose levels (see BLOOD GLUCOSE REGULATION), although this is more pronounced when INSULIN is absent. Its release is powerfully stimulated by certain amino acids, notably arginine and alanine, by CCK, and by division of the autonomic nervous system. In common with insulin, it has a half-life of ~6 minutes in the plasma. Insulin appears to inhibit glucagon release directly.

The glucagon receptor, as with the GLUCAGON-LIKE PEPTIDE 1 receptor, is a G PROTEIN-COUPLED RECEPTOR, expressed on liver cells (its principal target organ), pancreatic β-cells, kidney, adipose tissue, heart and vascular tissues, some brain regions, stomach and adrenal glands. Receptor binding activates adenylyl cyclase, leading to raised intracellular [Ca^+] and to activation of PKA (see entry). This favours glycogenolysis (see GLYCOGEN), and the release of plasma glucose by liver cells. The glucagon-receptor complex undergoes endocytosis and glucagon is then degraded. During FASTING conditions, FOXA regulates glucagon release.

Glucagon also promotes GLUCONEOGENESIS (see entry), but decreases GLYCOLYSIS (see entry). In addition to its effects on phosphofructoskinase-2 (PFK-2), favouring gluconeogenesis, it mediates (i) increased expression of glucose-6-phosphatase, enabling free glucose to enter the bloodstream; (ii) suppression of glucokinase, decreasing entry of glucose into glycolysis; (iii) phosphorylation of glycogen phosphatase, stimulating glycogenolysis; (iv) inhibition of glycogen synthase, decreasing glycogen production; and (v) stimulation of the expression of phosphoenolpyruvate carboxylase, stimulating gluconeogenesis. It also promotes LIPOLYSIS.

glucagon-like peptide-1 (GLP-1) See GLUCAGON-LIKE PEPTIDES.

glucagon-like peptides (GLP-1, GLP-2) Polypeptides (formerly known as enteroglucagons), secreted by the ileum and colon and acting as circulating insulin- and satiety-promoting hormones (i.e., incretins) in response to high glucose level in the gut lumen. Produced by the same enteroendocrine L cells that produce PYY, GLP-1 binds to membrane receptors (e.g., on pancreatic β-cells), activating PHOSPHOLIPASE C and adenylyl cyclase, raising cAMP levels and activating cAMP-dependent protein kinase A (see PKA). It exists in several forms, and amplifies INSULIN release, promoting its exocytosis in response to a glucose load. See APPETITE, Fig. 14; LEPTIN.

glucocorticoids Steroid hormones (and transcription factors) produced by metabolism of CHOLESTEROL (see Fig. 50) and released by the adrenal cortex (see ADRENAL GLAND) in response to CORTICOTROPIN. The expression of as many as 1% of the genes in the genome may be regulated by glucocorticoids. CORTISOL is responsible for the bulk of their effects in humans, e.g., promotion of gluconeogenesis and hepatic glycogen synthesis, fat mobilization from adipocytes and nitrogen excretion and degradation of muscle protein (at high levels) and decrease of glucose and amino acid utilization. They down-regulate pro-inflammatory cytokine synthesis, reduce the number of mast cells (reducing histamine release), and up-regulate anti-inflammatory cytokine synthesis. At high levels, they depress immune responses (see IMMUNOSUPPRESSION). They stabilize LYSOSOME membranes, slowing release of their enzymes, and decrease capillary permeability and depress phagocytosis. They modulate perception and emotion (e.g., FEAR AND ANXIETY) and increase ATTENTION. Chronic high levels of glucocorticoids enhance retention of fear-conditioned memory (see CORTICOTROPIN RELEASING HORMONE). CIRCADIAN RHYTHMS of secretion are generated by the dorsomedial nucleus of the hypothalamus under input from the suprachiasmatic nucleus, among others.

The classic glucocorticoid receptor (GR), a NUCLEAR RECEPTOR expressed within virtually all cells and complexed with heat-shock proteins in the cytosol, is glucocorticoid receptor-α, one of the superfamily of steroid, thyroid, retinoid and orphan receptors operating as ligand-activated transcription factors in the regulation of GENE EXPRESSION. Binding its ligand causes the GR to translocate to the nucleus, where it binds specific hormone response elements, leading to increased or decreased transcription of target genes. However, glucocorticoids also bind mineralocorticoid receptors with high affinity, although this does not lead to permanent maximal occupancy of the ALDOSTERONE receptor. Plasma glucocorticoids bind glucocorticoid-binding protein and albumin in the plasma, allowing only ~10% of unbound hormone to cross cell membranes, and aldosterone-sensitive cells convert glucocorticoids to inactive forms with less affinity for the mineralocorticoid receptor (see MINERALOCORTICOIDS).

glucogenic Referring to non-carbohydrates that can be substrates for glucose synthesis. Used especially in the cases of certain AMINO ACIDS (e.g., alanine). See KETOGENIC.

glucokinases (hexokinases) Regulatory enzymes, in several isoforms (isozymes), catalysing the conversion of glucose to glucose 6-phosphate and initiating its entry into the glycolytic pathway (see GLYCOLYSIS). The chromosomal location of the *GLK* gene is 7p15. They require ATP as a substrate, in its $MgATP^{2-}$ form. Muscle hexokinase is allosterically inhibited by its product. Hexokinase D (glucokinase) is the predominant isozyme of liver, being directly regulated by blood glucose concentration and inhibited by fructose 6-phosphate. Glucokinase is present in β-cells of the PANCREAS, where it is often described as the pancreatic glucose sensor because it catalyses the rate-limiting step of glucose metabolism in β-cells, and is the primary regulator of glucose-controlled insulin secretion. Mutations in the glucokinase gene cause mild hyperglycaemia, often stable throughout life and treatable by diet alone.

gluconeogenesis The production of glucose from such non-hexose sources as pyruvate, lactate, glycerol and the glucogenic amino acids, all favoured when a cell's energetic requirements are being adequately met, and when plasma GLUCAGON levels are therefore high (see FASTING for some details). In humans, it occurs mainly in LIVER hepatocytes, and to a lesser extent in the renal cortex. Because the pathways of GLYCOLYSIS (see Appendix IIc) and gluconeogenesis (see Appendix IIb) both occur in the cytosol they must be reciprocally and coordinatedly regulated. Although much of the LACTATE generated by very active skeletal muscle is used by the liver and other organs as a respiratory substrate, some of the OXYGEN DEBT is believed to result in blood glucose production by the liver. It is thought that muscle fibres then use it to replenish their glycogen, completing what is known as the *Cori cycle*. FATTY ACIDS do not provide gluconeogenic substrates. During feeding, increases in circulating pancreatic INSULIN inhibit hepatic gluconeogenesis through activation of the serine/threonine kinase PKB (aka AKT) and subsequent phosphorylation of the forkhead transcription factor FOXO1.

Seven of the ten enzymatic steps in gluconeogenesis are identical to those in glycolysis and have their directions determined by mass action. But three glycolytic steps are irreversible, having a large negative free energy change, and cannot be used in gluconeogenesis. Instead, they are by-passed by three separate sets of enzymatic steps ('by-pass reactions'; see Appendix IIb), favoured by CORTISOL. The first of these, converting pyruvate to phosphoenolpyruvate, is achieved by reactions involving the cytosol and mitochondria, the alternative pathways being determined by the availability of lactate and the cytosolic requirements for NADH. When the cell has a high ATP:ADP ratio (see ATP), NADH accumulates and inhibits the KREBS CYCLE so that acetyl-CoA accumulates, inhibiting the NADH dehydrogenase complex and promoting gluconeogenesis by activating mitochondrial pyruvate carboxylase. This converts pyruvate to oxaloacetate, which is then reduced to malate by mitochondrial malate dehydrogenase, the malate entering the cytosol and being converted by phosphoenolpyruvate carboxykinase to phosphoenolpyruvate in a GTP-dependent process.

The second glycolytic step that cannot be a part of gluconeogenesis is the highly exergonic phosphorylation of fructose-6-phosphate by the regulatory enzyme phosphofructokinase-1, which is allosterically modulated by fructose 2,6-bisphosphate. Generation of fructose-6-phosphate from fructose 1,6-bisphosphate is catalysed by the Mg^{2+}-dependent enzyme fructose 1,6-bisphosphatase, which hydrolyses the C-1 phosphate. This enzyme is inhibited by AMP, which would be indicative of low ATP levels. A high ATP:ADP ratio (and NADH level) also favours the conversion of fructose 6-phosphate, a key substrate of both glycolysis and gluconeogenesis, to fructose 2,6-bisphosphate by phosphofructokinase-2 (PFK-2). Fructose 2,6-bisphosphate is not an intermediate in either glycolysis or gluconeogenesis and its concentration is determined by its rates of synthesis by PFK-2 and breakdown by its phosphatase FBPase-2, both parts of the same enzyme complex and regulated in a reciprocal way by GLUCAGON, whose level in the blood rises when blood glucose falls. Rising glucagon levels activate, via cAMP, the FBPase-2 enzyme component, while inactivating the PFK-2 enzyme component. This reduces fructose 2,6-bisphosphate levels in the cell, inhibiting glycolysis but stimulating gluconeogenesis and enabling the liver to replenish blood glucose.

The third bypass step is the reversal of the

GLUCOKINASE (aka hexokinase) reaction of glycolysis, catalysed by the Mg^{2+}-dependent GLUCOSE-6-PHOSPHATASE on the lumenal surface of the endoplasmic reticulum of liver and renal cortex cells but absent from muscle and brain.

Gluconeogenesis is energetically expensive, six high-energy phosphate groups being required for conversion of each molecule of glucose formed from pyruvate:

$$2\ \text{Pyruvate} + 4ATP + 2GTP + 2NADH + 4H_2O \rightarrow \text{glucose} + 4ADP + 2GDP + 6P_i + 2NAD^+ + 2H^+$$

Nevertheless, under intracellular conditions the overall free energy change is still negative (though less so than glycolysis) so that both are essentially irreversible processes in cells.

glucose Biosynthesis of the monosaccharide glucose is an absolute requirement in mammals because embryonic cells, the central nervous system, erythrocytes, testes and renal medulla require blood glucose as their sole, or main, energy source (see ENERGY BALANCE). The human brain alone requires >120 g glucose daily and low levels of plasma glucose trigger mobilization of triglycerides through the action of adrenaline and glucagons on adipocyte adenylyl cyclase (see BLOOD GLUCOSE REGULATION). Excess glucose that cannot be stored as glycogen, or oxidized, is converted into FATTY ACIDS in the cytosol for storage as triglycerides (see LIPOGENESIS), and generally indicates an anaerobic cellular environment (high NADPH:NADP and NADH:NAD ratios). Such changes in glucose levels can modulate and entrain circadian oscillations in cells grown in culture (see CIRCADIAN CLOCKS). See PANCREAS.

glucose intolerance The existence of higher than normal, but sub-diabetic, levels of blood GLUCOSE – in either the fasting state (impaired fasting glycaemia (IFG)), or the non-fasting state – with its consequent appearance in the urine (glycosuria). While an increase in mortality is associated with IFG, a much greater and more significant risk has been identified with IGT, together with likelihood of progression to DIABETES MELLITUS, the microvascular and macrovascular consequences of which are a major cause of premature mortality. In view of the risks associated with glucose intolerance and diabetes, and the benefits of their early detection, the WORLD HEALTH ORGANIZATION has defined an approach for opportunistic screening for these conditions in symptomatic individuals, and asymptomatic individuals in high risk categories. This involves an initial random or fasting plasma glucose measurement and confirmatory diagnosis, where necessary, with a standardized oral glucose tolerance test (OGTT) in which a blood sample for plasma glucose determination is taken after an overnight fast. The patient is then given 75 g glucose dissolved in water orally and a further blood sample taken at 2 hours post glucose load for further plasma glucose determination. Diabetes is confirmed if the fasting glucose is ≥7.0 mmol/L and/or a 2-hour glucose ≥7.8 and <11.1 mmol/L. IFG is identified by a fasting glucose ≥6.1 and <7.0 together with a 2 hour glucose <7.8 mmol/L. See INSULIN RESISTANCE.

glucose-6-phosphatase (Glc-6-Pase) An enzyme hydrolysing glucose-6-phosphate, releasing a phosphate group and free glucose, after which glucose may be exported from the cell via GLUCOSE TRANSPORTERS. The reaction completes the final step in GLUCONEOGENESIS and GLYCOGENOLYSIS and therefore plays a key role in BLOOD GLUCOSE REGULATION (see PANCREAS). See GLUCOKINASES.

glucose-6-phosphate dehydrogenase (G6PD) An enzyme found in all cells (a 'housekeeping enzyme') catalysing the first step in the PENTOSE PHOSPHATE PATHWAY, deficiency in which may lead to the unpleasant condition *favism*, which can be fatal in children. Complete deficiency is unknown, and probably lethal.

The G6PD locus lies on the X chromosome in a genomic region (Xq28) with low single nucleotide polymorphism density in HapMap data (see HAPLOTYPE). G6PD is the only enzyme present in red blood cells

GLUT-1	Ubiquitous, but prominent in erythrocytes and endothelial linings of brain blood vessels	Uptake of glucose by skeletal muscle and adipocytes under basal conditions
GLUT-2	Present in pancreatic β-cells, hepatocytes, enterocytes of intestine, and kidney tubules; low affinity for glucose	Part of the glucose-monitoring system of β-cells and hepatocytes, ensuring glucose absorption only when plasma glucose concentration is high. Also transports fructose
GLUT-3	Principally in neurons	With GLUT-1, allows glucose across the blood–brain barrier to enter neurons
GLUT-4	Principally in skeletal muscle and adipocytes. Not primarily located in plasma membranes, but recruited to them on binding of INSULIN to its membrane receptor	The main insulin-sensitive transporter
GLUT-5	In spermatozoa and enterocytes of small intestine	Mainly a fructose transporter
SGLT1	Present in microvilli of enterocytes	Cotransporter with Na^+

TABLE 5 *The distributions and roles of some* GLUCOSE TRANSPORTERS.

that can produce NADPH. In G6PD-deficient individuals, NADPH production is reduced and detoxification of HYDROGEN PEROXIDE is inhibited so that cellular damage results from lipid peroxidation, leading to red blood cell membrane rupture. Geographically, G6PD deficiency tends to occur in regions where MALARIA is (or has been) most prevalent, *Plasmodium falciparum* being sensitive to oxidative damage, its growth being inhibited in G6PD-deficient erythrocytes. Epidemiological studies indicate that in heterozygous females and hemizygous males, the risk of severe malaria is reduced by ~50%.

Although most individuals with the deficiency are asymptomatic, the ~400 million sufferers undergo lysis of their red blood cells and release of free haemoglobin soon after breathing in pollen of (or more commonly 24–48 hours after ingesting) fava (or faba) beans (the broad bean *Phaseolus faba*), which is a traditional ingredient of falafel, an important food source since antiquity in Mediterranean and Middle Eastern countries.

glucose transporters The membrane glucose transporter of red blood cells (GLUT-1) is specific for D-glucose and enables it to enter the cell by facilitated diffusion at a rate ~50,000 times greater than the unaided diffusion rate. A different transporter, GLUT-2, transports glucose out of liver cells when glycogen is hydrolysed to replenish blood glucose. Muscle and adipose tissue have the GLUT-4 transporter which is distinguished by its stimulation by insulin. See Table 5.

glucosinolates See ALKALOIDS.

glucuronic acid See BILE PIGMENTS, LIVER.

glutamate The AMINO ACID which is the major product of transamination in the liver (see DEAMINATION) and the major excitatory neurotransmitter in the brain and spinal cord, exciting almost all central neurons by activating several types of receptor, the major classes being the ionotropic NMDA, AMPA, kainate receptors, and the metabotropic receptors. The first three subtypes were identified through pharmacological approaches, AMPA, NMDA and kainate being synthetic glutamate agonists and naturally occurring toxins not found in the brain. Glutamate is a transmitter of photoreceptors in the retina and inhibits some BIPOLAR CELLS, but is excitatory for others. GABA is a derivative of glutamate.

Ionotropic glutamate receptors, iGluRs (see Fig. 83), are linked directly with a membrane cation channel. They include AMPA and NMDA RECEPTORS, the former (see AMPARS) mediating moment-to-moment signalling, the latter mediating SYNAPTIC

PLASTICITY (see Fig. 169). The iGluRs are relatively non-selective, although NMDA receptors and some AMPARs favour calcium permeation. NMDA receptors show slow depolarization kinetics and are important in temporal summation and CENTRAL PATTERN GENERATORS. Unlike the recycling AMPARs, NMDA receptors do not recycle and are much more stable in the membrane. *Metabotropic glutamate receptors* are linked via G PROTEINS to ion channels and second messenger pathways (see PHOSPHOLIPASE C, Fig. 135; CREB, Fig. 60), and at least eight subtypes have been identified. Some cause depolarization by inhibition of postsynaptic K^+ channels; others are involved in presynaptic autoinhibition. Mutant glutamate receptors are implicated in EPILEPSY.

Advances in technology now permit delivery of stimuli to postsynaptic termini in various spatio-temporal patterns by methods that bypass the presynaptic membrane. Addition of a large light-absorbing side group to glutamate inactivates, or 'cages', it. Light pulses (e.g., 2-photon excitation) can then be used to break down the cage and release the active neurotransmitter. This delivery of glutamate to individual dendritic spines (see DENDRITES) deep within brain tissue can mimic the time course of normally released glutamate.

One major mechanism by which glutamate and DOPAMINE produce opposing physiological effects appears to involve a positive feedback loop amplifying their mutually antagonistic actions. The loop has three components: PKA, protein phosphatase 2 (PP2A), and the threonine-75 (T75) residue of DARPP-32. Dopamine causes both increased activity and decreased inhibition of PKA; but glutamate activates CDK5 (cyclin-dependent kinase 5), which phosphorylates threonine-75 of DARPP-32, inhibits PKA, which reduces PP2A activity and so maintains DARPP-32 phosphorylated. This pathway amplifies either the effects of glutamate or dopamine, whichever is the more effective transmitter at any given time (see Fig. 84).

Decarboxylation of glutamate gives rise to γ-aminobutyrate, the inhibitory transmitter GABA.

The numbers of one form of glutamate receptor, the AMPA receptor (AMPAR), in one domain of the dendritic postsynaptic membrane vary as it moves rapidly in to and out from this membrane, under the influence of the chaperone Stargazin. Stargazin also controls the overall flux of ions through the water-filled AMPAR pore and enables the receptor to function even in the presence of glutamate. AMPAR concentration in the membrane is a major factor controlling the strength of excitatory transmission between neurons and perhaps also the storage of memories in the brain (see SYNAPTIC STRENGTH). Oversecretion of glutamate can lead to EXCITOTOXICITY. Glutamate is an important carrier of AMMONIA to the liver. See ASTROCYTES, GLUTAMATE DEHYDROGENASE, SCHIZOPHRENIA.

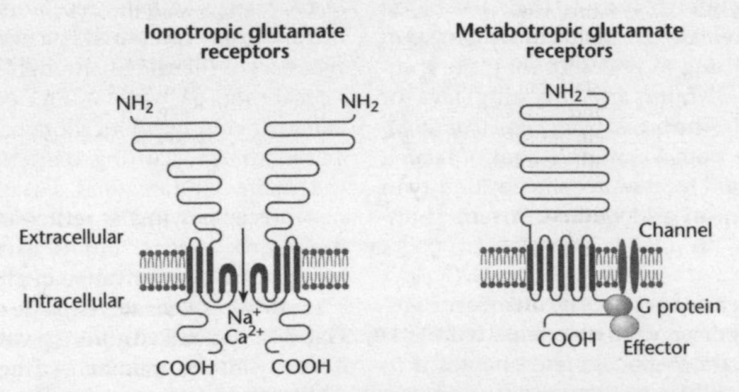

FIG 83. Diagram showing the structures of two classes of GLUTAMATE receptor: ionotropic receptors (e.g., NMDARs, AMPARs) and metabotropic receptors. From S Nakamishi, © 2005 *Trends in Neurosciences*, with permission from Elsevier Ltd.

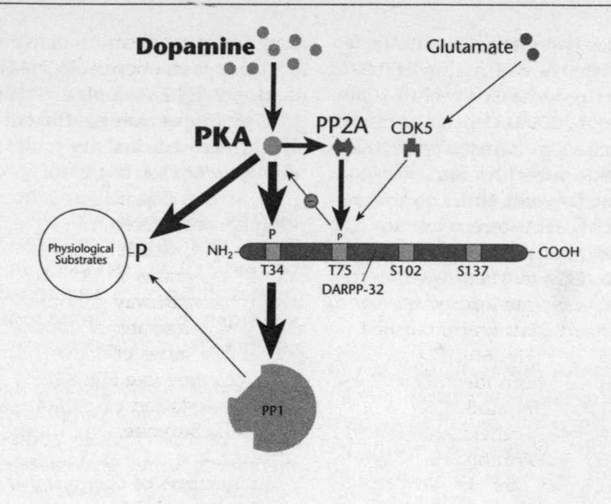

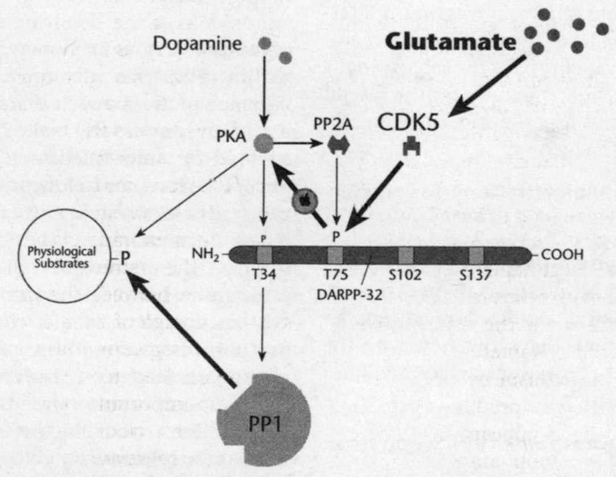

FIG 84. A diagram representing some of the opposing signalling pathways involved in mediating the biological effects of the neurotransmitters glutamate and dopamine (see GLUTAMATE). The larger the arrows and lettering, the higher the levels, and activities, of the component in the pathway. Thus, P indicates a phosphorylated state; but the relative degree of phosphorylation is indicated by the size of the letter. The horizontal dark and cross-hatched structure is the protein DARPP-32 (see entry). PP1 = PROTEIN PHOSPHATASE 1; PP2A = protein phosphatase 2A; PKA = protein kinase A (see PKA). From *Science* Vol. 294, No. 5544, p. 1029. Reprinted with permission from AAAS.

glutamate dehydrogenase An allosteric enzyme operating at an important intersection of carbon and nitrogen metabolism. Its six identical subunits are positively modulated by ADP and negatively by GTP. See AMMONIA.

glutamate receptors See GLUTAMATE.

glutamatergic Of fibres, receptors and synapses employing or responding to GLUTAMATE.

glutathione (GSH when reduced; GSSG when oxidized) An amino acid derived from glycine, glutamate and cysteine. Its antioxidant properties probably help maintain sulphydryl groups of proteins in their reduced state (making cystine formation less likely) and the iron of haem in the ferrous (Fe^{2+}) state. It is a major 'sink' for newly synthesized NITRIC OXIDE and is also involved in deoxyribonucleotide synthesis. Its redox function is also used protectively to remove toxic peroxides (e.g., HYDROGEN PEROXIDE) formed by cells under normal aerobic conditions (see MITOCHONDRION). The enzyme glutathione *S*-transferase (GST) is often used to tag a target protein so that the protein can be adsorbed onto glutathione-bearing beads and purified.

gluteus maximus (buttock) The largest and heaviest of the three gluteal muscles moving the femur and one of the largest muscles of the body – a distinctive human feature (see GAIT). It is the chief extensor of the femur, extending the thigh at the hip joint and rotating it laterally. Its origins lie in the iliac crest, sacrum, coccyx and aponeurosis of the sarcospinalis while its insertion is mainly on the lateral part of the linea aspersa under the greater TROCHANTER of the femur.

glycans An alternative term for polysaccharides. See GUT FLORA.

glyceraldehyde 3-phosphate (GALP) A triose phosphate intermediate in GLYCOLYSIS (see Appendix IIc), oxidized by glyceraldehyde 3-phosphate dehydrogenase with the production of 3-phosphoglycerate and NADH (see NAD).

glycerate 3-phosphate (GP) See GLYCEROL.

glycerol A component of triglycerides, released by lipase action and usually phosphorylated by glycerol kinase. The resulting glycerol-3-phosphate is oxidized to dihydroxyacetone phosphate during GLYCOLYSIS. In skeletal muscle and brain tissue, dihydroxyacetone phosphate accepts two reducing equivalents from NADH in a reaction catalysed by cytosolic glycerol 3-phosphate dehydrogenase. An isozyme of this enzyme, bound to the outer face of the inner mitochondrial membrane, channels two reducing equivalents (electrons) into the respiratory chain by reducing ubiquinone (compare the MALATE–ASPARTATE SHUTTLE). Without the need for membrane transport systems, this delivers reducing equivalents from cytosolic NADH, via ubiquinone, into Complex III rather than Complex I, providing sufficient energy for synthesis of 1.5 ATP molecules per pair of electrons. See DIACYLGLYCEROL.

glycine After GABA, the most widespread inhibitory neurotransmitter in the central nervous system, especially in the lower centres (brain stem and spinal cord), its ionotropic receptors linked to and opening chloride channels. See NMDA RECEPTOR.

glycocalyx A carbohydrate-rich region at the surface of most cells, deriving mainly from oligosaccharide components of membrane-bound glycoproteins and glycolipids. See CELL MEMBRANES, EXTRACELLULAR MATRIX, LEPTINS.

glycogen The major storage carbohydrate within cells, especially those of LIVER and SKELETAL MUSCLE. A much-branched glycan, its many terminals make it especially adapted to rapid breakdown by glycogen phosphorylase, and elongation by glycogen synthase, while as a reserve of glucose it does not greatly affect the water potential of the cell. Branching also makes the molecule more soluble. The initial step in its synthesis (*glycogenesis*) is the conversion by phosphoglucomutase of glucose 6-phosphate to glucose 1-phosphate, UDP-glucose pyrophosphorylase then uses UTP to synthesize UDP-glucose, which is the immediate substrate for glycogen synthesis by the active (unphosphorylated) form of glycogen synthase, glycogen synthase *a*. This enzyme favours the transfer of the glucose residue of UDP-glucose to a non-reducing end of a branched glycogen molecule. The enzyme requires a primer ($\alpha 1 \rightarrow 4$) polyglucose chain, or branch, having at least 8 glucose units. The ($\alpha 1 \rightarrow 6$) bonds are formed by a glycogen-branching enzyme, glycosyl-($4 \rightarrow 6$)-transferase which transfers a terminal sequence of 6–7 glucose residues from the non-reducing end of a branch having ≥11 residues to the C6 hydroxyl group of a more interior glucose residue of the same or of a different chain. Since glycogen synthase requires a primer, this is provided by the enzyme *glycogenin*, which catalyses the covalent attachment to itself of a glucose residue at its Tyr^{194} residue. Gylcogenin then forms a tight association with glycogen synthase, during which time up to 7 more glucose units are added autocatalytically (mediated by glycogenin's glucosyltransferase activity). Once this is achieved, glycogen synthase *a* can continue glycogen synthesis, eventually being released, although the glycogenin remains within the glycogen particle.

The balance between glycogen synthesis and breakdown (*glycogenolysis*) is controlled by the hormones GLUCAGON and INSULIN (see entries). Glycogenesis is one of the early cytosolic effects of binding of insulin to its receptor, whereas the early effects of glucagon are mediated by PKA (see entry). ADRENALINE also stimulates hepatic glycogenolysis, but mainly when glucagon is deficient. its main target being SKELETAL MUSCLE. Enhanced sympathetic activity promotes hepatic glycogenolysis via noradrenaline. See EXERCISE.

glycogenesis See GLYCOGEN.

glycogenin See GLYCOGEN.

glycogen-loading diets See EXERCISE.

glycogenolysis The hydrolysis of GLYCOGEN, promoted by ADRENALINE and GLUCAGON.

glycogen phosphorylase A PHOSPHORYLASE of muscle and liver cells catalysing the conversion of glycogen and inorganic phosphate to glucose 1-phosphate:

$$\underset{\text{Glycogen}}{(\text{Glucose})_n + \text{P}_\text{i} \rightarrow (\text{glucose})_{n-1}} + \text{glucose 1-phosphate}$$

It exists in two forms: the more active and phosphorylated phosphorylase *a*, and the less active and dephosphorylated phosphorylase *b*. There are in fact nine separate sites in five designated regions of the molecule susceptible to phosphorylation by one or other cellular protein kinases, so that the activity of the enzyme is capable of modulation in response to a variety of signals, many of them arising through signal transduction pathways and resulting in fine-tuning of its activity.

glycolipids Lipids to which mono-, di-, or oligosaccharides have been covalently attached during post-translational modification in the ENDOPLASMIC RETICULUM and GOLGI APPARATUS. They are located especially, but not exclusively, in the outer leaflet of the plasma membrane (see CELL MEMBRANES), where they include glycosphingolipids, galactocerebrosides and gangliosides.

glycolysis The major anaerobic pathway by which all human cells oxidize glucose to yield ATP and pyruvate (see Appendix IIc), occurring in the cytosol and therefore *outside* mitochondria. For red blood cells, lacking mitochondria, it is the default ATP synthesizing pathway, while for skeletal muscle fibres during vigorous exercise it may be thought of as 'buying time', allowing substrate-level phosphorylation during times when the muscle's oxygen supply is inadequate and/or energy demands outstrip the muscle's capacity to resynthesize ATP aerobically. Glycolysis and GLUCONEOGENESIS are reciprocally and coordinately regulated (see Appendix IIb): if pyruvate is converted to acetyl-CoA, it fuels the mitochondrial KREBS CYCLE; and if it is converted to oxaloacetate, it can initiate gluconeogenesis. Phosphofructokinase-1, a key regulatory enzyme of glycolysis, is positively modulated by AMP and ADP but negatively modulated by ATP and citrate (see ATP). Most glycolytic enzymes require Mg^{2+} for their activity, the substrate binding sites being specific for Mg^{2+} complexes with ADP, ATP and the glycolytic intermediates. The reducing equivalents of NADH produced by glycolysis are available either for production of LACTIC ACID (which regenerates NAD for glycolysis in anaerobic conditions), or for mitochondrial ATP synthesis via two alternative shuttles (see MALATE–ASPARTATE SHUTTLE, GLYCEROL) in aerobic conditions. See ENERGY BALANCE, PYRUVATE CARBOXYLASE.

glycoproteins Proteins to which simple or complex sugar units have been added to the N-terminal end during post-translational modification in the ENDOPLASMIC RETICULUM and GOLGI APPARATUS. They contain 1–60% carbohydrate (compare PROTEOGLYCANS) and include, *inter alia*, ANTIBODIES, other IMMUNOGLOBULINS (many of them CELL ADHESION MOLECULES), many receptor molecules, and some hormones (e.g., LH, FSH). See GLYCOLIPIDS.

glycosaminoglycans (GAGs) Strongly anionic unbranched polysaccharide chains composed of repeating disaccharide units, at least one of the two sugar units always being an amino sugar and usually sulphated, giving the molecule a strong negative charge. Components of MUCINS, the simplest is HYALURONIC ACID, which is also the only GAG not to be covalently attached to protein in the form of PROTEOGLYCANS.

glycoside hydrolases Enzymes digesting glycosidic bonds in carbohydrate molecules. In addition to amylases and disaccharidases of human origin, they include important enzymes produced by the mutualistic bacteria of the colon (e.g., see *Bacteroides thetaiotaomicron*; see GUT FLORA).

glycosylphosphatidylinositol (GPI) A glycolipid attached to some proteins post-translationally, anchoring them in the outer leaflet of the plasma membrane. It is a candidate toxin produced by the malarial parasite *Plasmodium falciparum* (see MALARIA). See CELL MEMBRANE.

GM-CSF Granulocyte-macrophage colony-stimulating factor, acting through the JAK/STAT PATHWAY. A haematopoietin cytokine produced by T cells and macrophages and stimulating growth and differentiation of myelomonocytic lineage bone marrow cells, especially DENDRITIC CELLS.

GNEFs (guanine nucleotide exchange factors) See GUANOSINE TRIPHOSPHATE.

GnRFs, GnRHs See GONADOTROPIN RELEASING HORMONES.

goblet cells Pear-shaped specialized columnar epithelial cells, unicellular glands, present from the nasal passages to the small bronchi; in the necks of gastric pits; among the enterocytes of the intestinal mucosal epithelium; and in other mucous membranes (e.g., urogenital tracts), cells secreting mucoprotein-containing fluid. As with other epithelial cells of the gut mucosa, the transcription factor Math I, a component of the NOTCH SIGNALLING PATHWAY, is required for their differentiation. The mucus produced is an important protectant (see STOMACH) and lubricant, but overproduction occurs during CHOLERA and CYSTIC FIBROSIS. See BRUNNER'S GLANDS.

Golgi apparatus (G. body, G. complex) A dynamic organelle, usually close to the nucleus and centrosome, consisting of a collection of flattened membrane-enclosed cisternae (the *Golgi stack*, usually 4–6 cisternae per complex) sandwiched between two complex networks of tubules, the *cis* Golgi network (on the nuclear side of the complex) and the *trans* Golgi network (on the plasma membrane side), many stacks usually being interconnected by tubular connections forming a single complex. The *cis* network is formed by fusion of COPII-coated vesicular clusters arriving from the ENDOPLASMIC RETICULUM (ER), proteins and lipids arriving at the *cis* and exiting from the *trans* Golgi network, destined for targeted membranes. These networks are the main sorting station in the biosynthetic-secretory pathway. Oligosaccharide and protein processing is achieved in sequence by different enzymes in different cisternae. Removal of mannose residues and addition of *N*-acetylglucosamine occur in the middle compartment while the addition of galactose and sialic acid occur in the *trans* compartment (see PROTEOGLYCANS). According to the *vesicular transport model*, transport through the Golgi apparatus, molecules in transit move through the cisternae in sequence, carried by transport vesicles. These budding vesicles remain tethered by filamentous proteins restricting their movement, making easier fusion with the correct target membrane. In the *cisternal maturation model*, the cisternae themselves move through the Golgi stack, the distribution of enzymes being explained by retrograde flow.

In general, peptides that are destined to become neurotransmitters are synthesized by the rough ER and cleaved in the Golgi apparatus to form the active transmitter.

COPI-coated packages bud from the pre-Golgi compartments and Golgi cisternae and deliver their cargo to target membranes in accordance with complementary SNAREs, and Rab proteins in the vesicle membrane (see VESICLES). Secretory vesicles bud from the *trans* network and release their contents by exocytosis at the plasma membrane. The biosynthetic-secretory pathway involves movement of coated vesicles from the ENDOPLASMIC RETICULUM towards the Golgi apparatus and cell surface membrane. Vesicles leaving the Golgi for the plasma membrane must not include those proteins required to stay in the Golgi.

Golgi tendon organs See TENDON ORGANS.

gomphosis (pl. gomphoses) A fibrous joint with a cone-shaped peg fitting into a socket. The articulations of the roots of the teeth with the sockets of the maxillae and mandible are the only examples in humans.

gonadotrophs Cells constituting ~5–10% of anterior pituitary gland cells, secreting gonadotropins. Most (~60%) produce both LH and FSH; 18% secrete LH only and ~22% produce FSH only, synthesis and release being under the influence of GnRF from the hypothalamus. See GONADOTROPIN RELEASING FACTOR.

gonadotropins (gonadotropic hormones) Among the largest hormones known, these glycoproteins are secreted by gonadotrophs of the anterior PITUITARY GLAND in response to GONADOTROPIN RELEASING FACTORS. Heterodimeric, each consists of a common α-subunit conferring receptor-binding ability, and a unique β-subunit conferring biological specificity. See FOLLICLE STIMULATING HORMONE, LUTEINIZING HORMONE. Although not a gonadotropin, THYROTROPIN is a member of the same group of glycoproteins.

gonadotropin releasing hormone (GnRH, GnRF) Synthesized and secreted in both sexes by neurons of the preoptic region of the hypothalamus by a pulse generator whose structure and activity are not fully understood, but whose rhythms of GnRH secretion are perhaps more evident in the control of the MENSTRUAL CYCLE (see entry for details) than in the male. Until PUBERTY, GnRH-releasing neurons are inactive through tonic GABAergic inhibition. Male activity is regulated by testosterone and inhibin B, although output of GnRH in both sexes is dependent upon several other factors: positive influences from noradrenaline, neuropeptide Y and leptin; negative influences from β-endorphin, interleukin-1, dopamine and GABA; and both stimulatory and inhibitory influences from 17β-oestradiol (see TESTES, Fig. 178). The negative feedback exerted by testosterone is achieved via its local conversion to 17β-oestradiol.

GnRH binds a specific G PROTEIN-COUPLED RECEPTOR on gonadotrophs of the anterior pituitary gland, activating PHOSPHOLIPASE C (see Fig. 135), leading to phosphoinositide turnover and Ca^{2+} mobilization and influx. This leads to increased transcription of the FSH and LH α- and β-subunit genes, and increases LH and FSH release. GnRH is also called LHRH since it causes a much greater release of LH than of FSH. Pulsatile GnRH release stimulates pulsatile release of LH and FSH, but the ratio of LH and FSH production is set by the frequency of GnRH pulses. Activin interacts with inhibin B to increase FSH β-subunit synthesis (highest in response to low-frequency GnRH pulses) and a synthesis which is suppressed at the higher frequency pulses and amplitude that stimulate LH β-subunit synthesis. See ACTIVINS, INHIBINS.

gonads Collective term for OVARIES and TESTES, developing from the paired GENITAL RIDGES.

gonochorism Having a chromosomally determined sex, which is either male or female. Contrast hermaphroditism (see SEX DETERMINATION).

gonorrhoea See NEISSERIA.

goosecoid A transcription factor whose name derives from the two Drosophila homeodomain proteins to which its homeodomain is similar: gooseberry and bicoid. See PRIMITIVE NODE.

Gorilla The largest living PRIMATE (Subfamily Gorillinae), being a ground-dwelling omnivore inhabiting forests of central Africa. Populations in Gabon and the Republic of the Congo were devastated by Ebola virus outbreaks between 2001 and 2003, and human encroachment and agriculture make the remaining populations very endangered. Until recently, three species of gorilla were recognized: Western Lowland, Eastern Lowland and Mountain Gorillas. Two species tend to be recognized today: the Western Gorilla (*Gorilla gorilla*), now restricted to the Congo basin, with two subspecies (Western Lowland Gorilla, *G. g. gorilla*, and Cross River Gorilla, *G. g. diehli*); and the Eastern Gorilla (*Gorilla beringei*), now found only in Rwanda, Uganda and the eastern Democratic Republic of the Congo, with two subspecies (Mountain gorilla, *G. b. beringei*, and Eastern Lowland Gorilla, *G. b. graueri*). In 2007, the IUCN Red List moved *Gorilla gorilla* from 'endangered' to 'critically endangered', and it is believed that *Gorilla beringei* will follow suit when its population survey is completed. Some of the issues involved in clarification of the relationships between gorillas, humans and CHIMPANZEES are discussed in the HOMINOIDEA and GENETIC DIVERSITY entries. See GROWTH.

Gorillinae The subfamily of hominids (Family Hominidae) containing *Gorilla* and its closest relatives since the separation of its lineage from the Homininae. One such putative species is *Chororapithecus abyssinicus* (see HOMINOIDEA). See GORILLA.

gout Painful disorder resulting either from excessive production of uric acid or inability to remove it adequately from the blood. Uric acid then reacts with plasma sodium to form sodium urate, crystals of which accumulate in soft tissues such as the kidneys and in cartilage of the ears and joints (gouty ARTHRITIS). In the latter, crystals irritate and erode the cartilage, causing inflammation and pain. If untreated, the ends of the articulating bones fuse, making the joint immovable. It is rarer in women, possibly related to the menstrual cycle.

GPCRs See G PROTEIN-COUPLED RECEPTORS.

G protein-coupled receptors (GPCRs) 'Serpentine' SEVEN-MEMBRANE-SPANNING RECEPTORS that are METABOTROPIC RECEPTORS. On binding their extracellular ligands, which include CATECHOLAMINES, they activate one or more G PROTEINS. The senses of sight, smell and much of taste are initiated by ligands that bind GPCRs, which can differ profoundly in their membrane trafficking (see VESICLES). GPCRs are the largest family of neural signalling receptors, representing ~1% of transcripts encoded by the human genome and targets of the majority of therapeutic drugs currently in clinical use. CHEMOKINE receptors are also examples, as are the fMLP receptor and receptors for COMPLEMENT fragments C5a, C3a and C4a. Activation by agonist binding of GPCRs controls most SIGNAL TRANSDUCTION pathways, some (e.g., OPIOID RECEPTORS) allowing direct coupling of G proteins to ion channels.

Although GPCRs span the cell membrane they are not considered to be voltage-sensitive, although recent evidence suggests the muscarinic ACh receptors m2R and m1R display charge-movement-associated currents analogous to 'gating currents' of voltage-gated channels. The cannabinoid receptor CB_1 is one of the most abundant GPCRs in the brain. GPCRs share several features, in addition to receptor and G protein resemblances. The ligand binding site is buried deep in the membrane and involves residues from several transmembrane segments, the conformational change which results on binding (or light absorption) induces a conformational change in the receptor that enables it to interact with a specific G protein. When activated, a GPCR undergoes reversible phosphorylation of several Ser or Thr residues. This is reversed by ARRESTINS, which therefore inhibit their activity. Heterotrimeric G proteins activate or inhibit effector enzymes (e.g., adenylyl cyclase, phosphodiesterases or phospholipase C), changing the concentration of a SECOND MESSENGER. In hormone-detecting systems, an activated protein kinase regulates a particular cellular process; in sensory systems, the output is a change in membrane potential which signals to another neuron in a pathway to the CNS. All such systems self-inactivate (see G PROTEINS). See GENE FAMILIES AND SUPERFAMILIES, GUANOSINE TRIPHOSPHATE, OLFACTORY RECEPTORS, ROD CELLS, TASTE.

G proteins (GTP-binding proteins, guanosine nucleotide-binding proteins) Cytoplasmic heterotrimeric proteins, sometimes membrane-bound, with GTPase activity, activated on binding GUANOSINE TRIPHOSPHATE. The intrinsic GTPase activity eventually converts the GTP to GDP, which inactivates the protein so that it acts as a molecular 'switch': self-inactivation. Two families occur: heterotrimeric G proteins (components Gα, Gβ and Gγ), and monomeric G proteins (e.g., Ras; see RAS SIGNALLING PATHWAYS). When stimulated by a G PROTEIN COUPLED RECEPTOR, heterotrimeric G proteins release the Gα subunit to activate a number of distinct cytoplasmic enzymes, such as ADENYLYL CYCLASE and phospholipase C-β (see PHOSPHOLIPASE C, and Fig. 139) generating the SECOND MESSENGERS cAMP, and diacylglycerol and PIP_2, respectively. As with the monomeric Ras proteins, activation of the Gα subunit occurs when it binds GTP. The GβGγ dimers can activate their own effectors, e.g., PI3Kγ, PLC-β and Src. Important heterotrimeric examples include: G_i G_t), involved in transduction pathways inhibiting adenylyl cyclase, lowering cAMP levels in the cell; G_q, involved in coupling cell membrane receptors to phospholipase C and thence to the inositol 1,4,5-triphosphate second messenger pathway. It is linked to α_2-ADRENERGIC receptors, D2 and D4 DOPAMINE receptors, $GABA_B$, SEROTONIN ($5\text{-}HT_1$), mGlu, types II and III GLUTAMATE, M2 and M4 MUSCARINIC, and μ, δ, and κ OPIOID RECEPTORS; G_{olf}, an olfactory-specific G protein which activates adenylyl cyclase when the G_{olf}-linked olfactory receptor binds its odorant ligand; G_s, activating adenylyl cyclase and linked to β-ADRENERGIC receptors, D1 and D2 DOPAMINE receptors, H2 HISTAMINE receptors and melanocortin-2 receptors (e.g., in the adrenal cortex). See ALZHEIMER'S DISEASE, Fig. 6; GUANOSINE TRIPHOSPHATE, RHODOPSIN (for transducin).

Graafian follicle A mature, antral, ovarian follicle in which a secondary oocyte is produced and which ruptures at ovulation. Folliculogenesis involves signalling between germ and somatic cells and involves two phases: a gonadotropin-independent (preantral) phase, and a gonadotropin-dependent (antral, or Graafian) phase. In the former, primordial follicular growth up to 0.2 mm takes place during foetal life and infancy. Primordial follicles are composed of an outer layer of flattened epithelial FOLLICULAR CELLS and a small oocyte, both enveloped by a basal lamina. The pool of primordial follicles in the ovary reaches its maximum ~20 weeks into gestation, decreasing logarithmically throughout life until completely depleted at MENOPAUSE. At puberty, primary follicles form when the follicular cells become cuboidal and start to divide mitotically, becoming granulosa cells. When 2–5 mm in diameter, roughly half the follicles that are growing are rescued from apoptosis and enter the growth phase, which begins ~85 days before ovulation and in the luteal phase of the preceding MENSTRUAL CYCLE. In the preantral follicular growth phase, granulosa cells proliferate under their own ACTIVIN stimulation, which also up-regulates their FSH and steroid hormone receptor expression. Activin also antagonizes INHIBIN, stimulating more pituitary FSH synthesis and release. Theca cells from the ovarian stroma associate with the follicle and granulosa cells to produce a secondary follicle. At this stage, any androstenedione received by granulosa cells from thecal cells is converted to testosterone. The antral stage that follows is the longest, tertiary follicles forming as thecal cells hypertrophy, the theca interstitial cells expressing FSH and LH receptors. These Graafian follicles have an external fibrovascular coat connected to the surrounding stroma of the ovary by a network of blood vessels. Blood-borne FSH stimulates thecal cells to convert cholesterol to progesterone and then to androstenedione, this passing to the granulosa cells which convert it to 17β-oestradiol by aromatase activity and release it into the albuminous antral follicular fluid and blood circulation. As FSH stimulates LH receptor expression in what becomes the dominant follicle, ≥25 mm in diameter, granulosa cells surrounding the oocyte remain intact and form the cumulus oophorus (see CUMULUS CELLS). When the follicle is mature, a surge in LH induces the pre-ovulatory growth phase. LH induces genes involved in OVULATION (e.g., progesterone receptor, cyclooxygenase-2) and luteinization (e.g., cell cycle inhibitors, steroidogenic acute regulatory protein, protein kinases, etc.). The 1st meiotic division is completed, producing a secondary oocyte and polar body (see OOGENESIS). The 2nd meiotic division is arrested in metaphase approximately 3 hours prior to ovulation, when the follicle ruptures and releases the secondary oocyte. See FERTILIZATION.

grafts, graft rejection Tissue grafts can be from one site to another on the same individual (*autografts*), or between unrelated, or allogeneic, individuals (*allografts*), in which case the graft is accepted initially but rejected ~10–13 days after grafting. This *first-set rejection* depends on a T-CELL response in the recipient (nude mice, lacking T cells, do not reject such grafts; but they do when given normal T cells). A recipient later given a second graft from the same donor rejects it more quickly: a *second-set rejection*. This rapid response can be transferred to normal or irradiated recipients by T cells from the initial recipient, explicable by its being caused by a memory-type immune response from clonally expanded and primed T cells specific for the donor tissue. Such responses are a major barrier to effective tissue transplantation, mediated by responses of CD8 T cells, CD4 T cells, or both, to allogeneic MHC molecules expressed by the grafted tissues. In most tissues, these will be MHC class I antigens. Even grafts between HLA-identical siblings invariably incite a slow rejection, because of differences in their minor histocompatibility antigens, unless the siblings are identical twins. The corticosteroids used in transplant recipients are often given in combination with other drugs in an attempt to keep their toxic effects to a minimum, and often in combination with cytotoxic immunosuppressive drugs (see IMMUNOSUPPRESSION).

Bone marrow transplantation was invented to enable physicians to increase chemotherapy and radiotherapy to myeloblastic doses in the hope of eliminating endogenous cancer cells. Unfortunately, they contain T cells, which encounter and respond to host antigens in almost all tissues, leading to multi-system *graft-versus-host syndrome*; yet haematopoietic cells often fail to engraft if T cells are eliminated.

Grafts also contain other 'passenger leukocytes': antigen-presenting cells (APCs). In *direct allorecognition*, donor APCs leave the graft and migrate to LYMPH NODES where their MHC molecules then activate host T cells bearing corresponding T-cell receptors. These alloreactive effector T cells are carried back to the graft and attack it. But the recipient's APCs also take up allogeneic proteins, presenting them to T cells by self-MHC molecules. Graft-derived peptides include those from minor histocompatibility proteins and from the foreign MHC molecules. In addition, cytotoxic T-cell attack will be made by cells recognizing the graft MHC molecules. Antibody responses by pre-existing alloantibodies to blood group and polymorphic MHC antigens can result in a COMPLEMENT-dependent reaction within minutes of transplantation, avoidable by ABO-matching and cross-matching with the latter to discover whether recipient antibodies react with donor white blood cells. See THYMUS, TRANSPLANTS.

graft-versus-host disease (GVHD) A complication that sometimes arises in

BONE MARROW transplant recipients, when white cells derived from the donated bone marrow attack the recipient's tissues. See GRAFTS, THYMUS.

Gram-negative bacteria (Gram –ve b.) Bacteria which, because of the thin peptidoglycan layer in their cell walls, do not retain the crystal violet-iodine complex when decolourized by acetone or alcohol in the Gram stain technique. The lipid membrane of their cell walls contains LIPOPOLYSACCHARIDE as a major component. Gram –ve cocci include *Neisseria gonorrhoeae* and *N. meningitidis*; Gram –ve bacilli include *Salmonella typhi*, *Shigella flexneri*, *Campylobacter jejuni*, *Vibrio cholerae*, *Yersinia pestis*, *Haemophilus influenzae*, *Legionella pneumophilus* and *Bordetella pertussis*. See ANTIBIOTICS.

Gram-positive bacteria (Gram +ve b.) Those bacteria that, because of the thick peptidoglycan layer in their cell wall, retain the crystal violet-iodine complex when decolourized by acetone or alcohol in the Gram stain technique. Gram +ve cocci include *Staphylococcus aureus*, *Streptococcus pneumoniae* and *S. pyogenes*; Gram +ve bacilli include *Listeria monocytogenes*, *Bacillus anthracis* and *Corynebacterium diphtheriae*. There are some bacteria that, despite having cell walls of the Gram +ve kind, do not retain the crystal violet-iodine complex and appear as Gram-negative cells, while some archaebacteria appear Gram +ve although lacking peptidoglycan. Bacteria, parts of whose cell walls appear Gram +ve and parts Gram –ve, are described as Gram-variable. See ANTIBIOTICS.

grand mal See EPILEPSY, SEIZURES.

granule cells Cell type in the cortex of the CEREBELLUM.

granulocytes Those WHITE BLOOD CELLS with granular cytoplasm: NEUTROPHILS, BASOPHILS and EOSINOPHILS.

granulosa cells Cuboidal and low-columnar cells developing from FOLLICULAR CELLS and organized in six or seven layers surrounding the primary oocyte; secrete follicular fluid and OESTROGENS under FSH stimulation and INHIBIN under FSH and LH regulation, and convert androstenedione produced by the theca cells into 17β-oestradiol. They also secrete FOLLISTATIN. See OOGENESIS.

Graves disease An AUTOIMMUNE DISEASE resulting in hyperplasia of the THYROID GLAND and autonomous thyroid hormone secretion through thyrotropin receptor stimulation by the thyroglobulin-like antibody, immunoglobulin G. THYROGLOBULIN levels are elevated. Due to thyroid-reactive T cells, it is eight times as common in women as in men, and its incidence peaks in the third to fourth decades of life. Hashimoto disease is another autoimmune disease of the thyroid gland.

Grb2 (growth factor receptor-bound protein 2) Pronounced 'grab two', a growth factor receptor-binding protein in the RAS SIGNALLING PATHWAY cascade, forming either a direct intermediary between the receptor and Sos, or a downstream of Shc, linking Shc to Sos. Free in the cytosol, it contains two SH3 groups and one SH2 group, the SH3 domains having affinity for two distinct proline-rich sequences present in Sos, while its SH2 domain associates with the phosphotyrosines present on the C-terminus of many ligand-activated receptors. In consequence, Sos moves from the cytosol to become tethered via the Grb2 linker to the receptor. See SH2 AND SH3 DOMAINS.

grey matter Generic term for a collection of neuron cell bodies, containing much water, in the central nervous system, and appearing grey when freshly cut. Contrast WHITE MATTER; see ADOLESCENCE.

growth Usually regarded as irreversible increase in dry mass, some aspects of growth are included in the entries, and references included there, on BONES, BRAIN, BRAIN SIZE, GROWTH FACTORS, GROWTH HORMONE and INSULIN-LIKE GROWTH FACTORS.

The pattern of prenatal growth is of great importance for an individual's future well-being, the rate of increase in body length being at a maximum around 16–20 weeks of GESTATION. A mother's preparation for

PREGNANCY (see entry) should include appropriate additional components of her diet, and she should be aware of the possible damage that ALCOHOL, NICOTINE and other drugs can do to her unborn child. During the previous EMBRYONIC DEVELOPMENT, growth rate is usually less, with greater emphasis on differentiation of cells. Until weeks 26–28 of gestation, increase in foetal mass is due largely to accumulated protein, as cells enlarge and proliferate; but during the final ~10 weeks the foetus starts to accumulate FAT (up to 400 g), distributed subcutaneously and in deep depots, and peak increase in foetal mass is around 34 weeks of gestation, declining somewhat towards the time of BIRTH. Growth rate continues to be high in neonatal life, declining significantly through early childhood, with a spurt at ADOLESCENCE.

It is well established that the period of maturation of *Homo sapiens* is nearly twice as long as in *Gorilla gorilla* and *Pan troglodytes*, an extension thought to be important for transmission of the numerous additional learnt behaviours of modern humans over the African apes, and recent analysis of dental and femoral development in *Australopithecus* indicate its similarity to the African apes in these respects. However, the corresponding developmental timescales of *H. ergaster* and *H. neanderthalensis* are closer to those of *H. sapiens* than to those of the two African species. See also EMBRYONIC DEVELOPMENT.

growth cone A dynamic and sensory ACTIN-supported extension of a developing AXON, bearing filopodia and involved during AXONAL GUIDANCE in reaching the axon's synaptic target.

growth factors (GFs) A collective term for relatively small extracellular signal proteins promoting CELL GROWTH, proliferation (see MITOGENS), differentiation or survival. They stimulate cells to take up excess nutrients and use them for anabolic processes. They also provide many of the signals that unite the cells within a tissue into a single community, all members of which are in continual communication with their neighbours. In some cases, one GF stimulates cell growth and cell division; in other cases, these are brought about by separate GFs. All GFs bind cell-surface receptors that are enzyme-linked (mitogens, especially, binding RECEPTOR TYROSINE KINASES), and act as local mediators (paracrine signals) at very low concentrations (~10^{-9}–10^{-11} M), initiating slow responses (in the order of hours), and usually involving many steps in an intracellular kinase cascade that eventually lead to a nuclear signal resulting in changes in GENE EXPRESSION (see IMMEDIATE EARLY GENES, SIGNAL TRANSDUCTION), in e.g., various ways of helping cells to emerge from the quiescent G0 state into an active growth-and-division cycle. They include the *Hedgehog* and *Wnt* products (see HEDGEHOG GENE, WNT SIGNALLING PATHWAYS). Many growth factors activate RAS SIGNALLING PATHWAYS while the interleukins IL-2, IL-3, IL-4, IL-5, IL-6 and GM-CSF, trigger JAK/STAT PATHWAYS, their CD receptors being located on specific leukocyte classes. ERYTHROPOIETIN and THROMBOPOIETIN also use the JAK/STAT pathway. VEGFs (vascular endothelial growth factors) are secreted glycoproteins that stimulate growth of cells lining blood vessels (see ANGIOGENESIS, VASCULOGENESIS) and lymphatic vessels (VEGF-C, VEGF-D). The latter two have been implicated in responses of the lymphatic system to tumours, notably breast cancer and melanomas, inducing lymph vessels to branch and invade the tumour, assisting in their METASTASIS, some interacting importantly with PROTEOGLYCANS in the extracellular matrix.

Antibodies to VEGFs have proved useful in control of colorectal, lung and breast cancers, anti-Neuropilin potentiating this effect. See EPIDERMAL GROWTH FACTOR, FIBROBLAST GROWTH FACTOR (FGF), INSULIN-LIKE GROWTH FACTOR, PLATELET-DERIVED GROWTH FACTOR, TRANSFORMING GROWTH FACTOR-B SUPERFAMILY.

growth hormone (GH, hGH, somatotropin) The most important growth-promoting factor during post-natal life (but see PLACENTA), being released in pulsatile bursts (most frequent during ADOLESCENCE) by somatotrophs of the anterior pituitary in response to hypothalamic GHRH. GHRH binds to G_S-coupled somatotroph receptors and activates the catalytic subunit of adenylyl cyclase. This ultimately leads to phosphorylation of CREB and transcription of the gene encoding pituitary-specific transcription factor (Pit-1) which in turn activates the GH gene. The pulsatile GH bursts are inhibited by SOMATOSTATIN and by raised blood glucose and free fatty acids. GH secretion shows a definite circadian rhythm, with marked elevations in output during deep NREM SLEEP (bursts of secretion every 1–2 hours). Decreased blood glucose (as in fasting), anxiety, pain, fever and strenuous exercise all promote GH release, as do amino acids (especially arginine) indirectly, by decreasing somatostatin release. Age-related sex steroid mechanisms, adrenal glucocorticoids, thyroid hormones and renal and hepatic functions all influence pulsatility of GH release, although the pulse-generator is not yet identified. High plasma GH levels inhibit further GH release.

A neuromodulator, with structural similarity to PROLACTIN, GH binds its receptor (GHR, a CYTOKINE RECEPTOR) and activates the JAK/STAT PATHWAY, inducing production of IGF-I (see IGFS). Many of its anabolic effects (e.g., longitudinal growth) are mediated mainly through IGF-I, and start to be important during the 1st and 2nd years of postnatal life, peaking during the ADOLESCENT GROWTH SPURT at puberty. Before the epiphyses of the long bones have fused, GH stimulates chondrogenesis and widening of the epiphyseal plates, probably the result of stimulation of chondrocyte precursor differentiation, and subsequent bone matrix deposition. In adults, it promotes BONE REMODELLING, stimulating proliferation and differentiation of osteoblasts, bone protein synthesis and growth. To a lesser degree, it promotes bone resorption.

GH has short-term insulin-like effects, such as lipogenesis and anabolic glucose and amino acid utilization, favouring muscle protein synthesis. After ~2 hours, these are followed by anti-insulin glucose-sparing effects (lipolysis, gluconeogenesis, hyperglycaemia and hyperinsulinaemia). GH inhibits adipocyte differentiation, reduces triglyceride accumulation and stimulates lipolysis and fatty acid oxidation. Overall, GH counters the effects of insulin on lipid and glucose metabolism (as with CORTISOL). It also influences the immune system by promoting B CELL responses and antibody production, macrophage activity, and T CELL function. It also modulates mood and behaviour.

The human growth hormone gene cluster includes five genes clustered within 67 kb on chromosome 17q22-24, while growth hormone receptor mRNA exhibits at least eight different 5′ UTRS as a result of alternative splicing (see RNA PROCESSING).

Genetic disruption of the growth hormone axis in MOUSE MODELS leads to metabolic changes, including low serum glucose and insulin, increased insulin sensitivity, increased expression of *Ppar-γ* (see PPARS), decreased β-oxidation of fatty acids, increased gluconeogenesis, enhanced antioxidant defences and stress resistance.

growth hormone-inhibiting hormone (GHIH) See SOMATOSTATIN.

growth hormone-releasing hormone (GHRH) See GROWTH HORMONE.

growth plate See EPIPHYSEAL GROWTH PLATE.

growth rate See GROWTH, and references there.

GTP See GUANOSINE TRIPHOSPHATE.

GTPases (GTP-binding proteins) See GUANOSINE TRIPHOSPHATE.

guanine A purine base, synthesized *de novo* from 5-phosphoribosyl 1-pyrophosphate (PRPP) by several steps including incorporation of amino acid moieties (glutaminyl, glycyl, aspartyl) via inosinate (IMP) to yield guanylate (GMP). Guanosine nucleotides include CYCLIC GMP and GUANOSINE TRIPHOSPHATE. Deamination of guanine to the base xanthine occurs relatively rarely compared with that of cytosine to uracil (see DNA REPAIR MECHANISMS). Degradation of purines produces URIC ACID.

guanine nucleotide exchange factors (GNEFs) See GUANOSINE TRIPHOSPHATE.

guanosine diphosphate (GDP) See GUANOSINE TRIPHOSPHATE.

guanosine monophosphate (GMP) See CYCLIC GMP, PURINES.

guanosine triphosphate (GTP) A nucleoside triphosphate generated by carriage of phosphoryl groups from ATP to GDP by nucleoside bisphosphate kinase. GTP is produced during the KREBS CYCLE within mitochondria and is energetically equivalent to ATP. It is present in large excess over GDP in cells, being required during GLUCONEOGENESIS, the elongation stage of translation (see ELONGATION FACTORS), and by all G PROTEINS. GTPases (GTP-binding proteins) are a large family of proteins with the same globular GTP-binding domain. When tightly bound, GTP is hydrolysed to GDP, when the domain undergoes a conformation change that inactivates it (see G PROTEINS). GTPases are regulated by proteins that determine whether GTP or GDP is bound. GTPase-activating proteins (GAPs) increase the rate of hydrolysis of bound GTP to GDP by GTPases, thereby inactivating them (e.g., down-regulating *Ras* gene activity; see RAS SIGNALLING PATHWAYS). *Guanine nucleotide exchange factors* (GEFs), such as Sos, promote exchange of bound nucleotide by stimulating dissociation of GDP and subsequent uptake of GTP from the cytosol. This is the method by which GTPases (e.g., Ras; see RAS PROTEINS), ARF PROTEINS and RAB PROTEINS are activated. The roles of GAPs and GEFs are equivalent to those of a protein phosphatase and a protein kinase, respectively. Monomeric GTPases include Ras while trimeric GTPases include those associated with receptors (see G PROTEIN-COUPLED RECEPTORS), and both forms are involved in membrane trafficking (see VESICLES). One molecule of GTP binds each α-tubulin monomer and links it to β-tubulin in a MICROTUBULE, but is not hydrolysed in the process and may be considered an integral part of the TUBULIN heterodimer structure. See SIGNAL RECOGNITION.

guanylyl cyclase (guanylate cyclase) Enzyme converting GTP to CYCLIC GMP. It may be a component of a RECEPTOR GUANYLYL CYCLASE; but a distinctly different type is cytosolic, bound by the heat-shock protein Hsp90, and bearing a tightly-associated HAEM group to which NITRIC OXIDE can bind, activating cGMP production. In the heart, cGMP brings about less forceful contractions by stimulating the ion pump(s) that expel calcium (Ca^{2+}) from the cytosol.

gubernaculum A mesenchymal band extending from the caudal pole of the testis to the inguinal region. As the testis begins to descend, an extra-abdominal portion of the gubernaculum forms and grows from the inguinal region toward the scrotal swellings and contacts the scrotal floor during descent. The gubernaculum also develops in females, but normally remains rudimentary. See EXTERNAL GENITALIA.

gustation See FEEDING CENTRE, TASTE.

gut-associated lymphoid tissues (GALT) Peripheral lymphoid organs (see LYMPH NODES) associated with the gastrointestinal tract, including the tonsils, adenoids, appendix and PEYER'S PATCHES. The latter are the most organized and important of these, antigen being collected and transported across the intestinal epithelium by specialized endothelial cells, the multi-fenestrated cells (M cells, often targeted by pathogens), the follicles comprising a large central dome of B cells with deeper T-cell zones. The B cells give rise to immunoglobulin-A (IgA)-secreting cells. Dendritic cells in GALT induce T-cell-independent expression of IgA and gut-homing receptors on B cells. See INFECTION.

gut flora (g. microbiome, g. microbiota) The adult human gastrointestinal tract contains microorganisms from all three domains of life: bacteria, archaea and eukarya. Gut bacteria attain the highest cell densities recorded for any ecosystem, although diversity is low (e.g., only 8 of the 55 known bacterial divisions, of which 5 are rare). The two dominant divisions, whose members comprise >90% of phylogenetic types, are the Bacteroidetes (e.g., *Bacteroides*) and Firmicutes (e.g., *Clostridium* and *Eubacterium*). Some, such as *Clostridium perfringens* and *Bacteroides fragilis*, are anaerobic (anaerobes dominate the normal gut);

others, such as *Enterobacter aerogenes*, are aerobic or facultatively anaerobic. The Human Microbiome Project is comparing people's microbiota on a species level, while an initiative termed Metagenomics of the Human Intestinal Tract (MetaHIT) aims to discover differences in microbial genes and expressed proteins, regardless of their source, and to examine associations between bacterial genes and human phenotypes (e.g., OBESITY, INFLAMMATORY BOWEL DISEASES). Inflammation may boost oxygen levels and favour enterobacteria.

Although densities of cells are huge (10^{13}–10^{14} cells), biodiversity is low – apparently the result of strong host selection and coevolution – although it contains at least 100 times as many genes as our own genome. Metagenomic analysis reveals that although two people may each have 1,000 kinds of bacteria living in their guts, they may have only 10 species in common. Most reside in the distal gut (where strains of *Escherichia coli* comprise the major facultative anaerobes), out-competing potentially pathogenic bacteria and synthesizing essential amino acids and vitamins and processing dietary components such as the many plant polysaccharides that are common components of dietary fibre and rich in xylan-, pectin- and arabinose-containing components. These glycans are indigestible on account of the human genome's lack of appropriate enzyme-encoding genes, but one bacterium with an extraordinary capacity for digesting these is *Bacteroides thetaiotaomicron*, a highly adaptable glycophile and prominent mutualist of the distal intestinal habitat of adult humans. Indeed, at least 81 different glycoside hydrolase families are represented in the human gut microbiome, many of them absent from the human 'glycobiome'.

The fucose-containing mucus lining the distal gut appears to encourage bacterial growth. The gut supports a trophic chain of microorganisms in which primary fermenters process glycans to short-chain fatty acids (SCFAs; mainly ethanoate, propionate and butyrate) absorbed by the host and contributing ~10% of energy in a Western diet, plus the gases H_2 (converted to methane by mesophilic archaeons) and CO_2. The bacterial product butyryl-coenzyme A is the principal energy source for epithelial cells of the colon. Our gut microbiome synthesizes isopentenyl pyrophosphate (IPP) from pyruvate and glyceraldehyde 3-phosphate via deoxyxylulose 5-phosphate (DXP) and 2-methyl-D-erythritol 4-phosphate (MEP). This MEP pathway has been proposed as a target for new antibiotics; but this may be detrimental to the microbiota and, as a result, to humans. Conversion of the BILE PIGMENT bilirubin into urobilinogen by colon bacteria leads to some absorption into the blood. On average, individuals on a 'British diet' egest 15–20 g dry bacteria per day.

The vitamins B_1 (thiamine) and B_6 (pyridoxal form) are also bacterial products of DXP. Vitamin B_{12} is also a product of colonic bacteria.

In mice, colonization by microbiota produces substantial increase in serum glucose and insulin, both of which stimulate hepatic LIPOGENESIS and fat storage. An individual's microbiotal composition is influenced by host genotype, and the relative abundance of Bacteroidetes:Firmicutes increases as obese individuals lose weight on either a fat- or carbohydrate-restricted low-calorie diet, work on mice indicating that the 'obese microbiome' has an increased capacity to harvest energy from the diet. See FAECES, FOOD POISONING, HELICOBACTER PYLORI.

Guthrie test See PHENYLKETONURIA.

gynecomastia A rare male condition, sometimes occurring during puberty, in which the breasts become enlarged. It is due to the influence of oestrogen. It is also a frequent occurrence in KLINEFELTER SYNDROME.

gyrus (pl. gyri) Convolution (fold) of the cerebral cortex between two of which occurs a sulcus. Some are consistently located, marking functional areas. See BRAIN, Fig. 31.

HAART (highly active antiretroviral therapy) See HIV.

habenular nuclei Small-celled (medial) and larger-celled (lateral) nuclei in the EPITHALAMUS influenced by the sense of smell and involved in emotional and visceral responses to odour. Both nuclei receive fibres from basal forebrain regions (SEPTUM, BASAL NUCLEI, and the hypothalamic lateral preoptic nucleus). The lateral habenular nucleus receives an additional projection from the medial segment of the globus pallidus. Both nuclei project by way of the retroflex fasciculus to the interpeduncular nuclei which project to the brain stem and hypothalamus, and to a medial zone of the midbrain TEGMENTUM.

habitat hypothesis The theory that extinction and speciation are driven by major climatic fluctuations ('climate forcing'), and that evolutionary change is confined to short bursts in small populations at these times. If true, it is a pattern that should be apparent in all species' groups (see TURNOVER-PULSE HYPOTHESIS). The issue is debated, some research suggesting that the correlation between climate change and extinction rates is greater than that between climate change and speciation rates and that speciation depends more on changes in local conditions than global ones.

habituation A decrease in response to a repeated benign stimulus, and an example of non-associative learning. As an example whose neural correlates have been well studied, the mollusc *Aplysia* withdraws its gill and siphon into the mantle cavity if a jet of water is squirted onto the siphon, but this gill withdrawal reflex habituates after repeated presentation of the jet (see MEMORY). Compare SENSORY ADAPTATION.

haem An iron-containing prosthetic group of haemoglobins, myoglobin and tightly but not covalently bound in cytochromes (except in *c*-type cytochromes) and cytosolic GUANYLYL CYCLASE. Synthesized within the MITOCHONDRION, its iron atom has six coordination bonds: four to nitrogen atoms forming part of the porphyrin ring system and two perpendicular to the porphyrin. The coordinated N-atoms help prevent conversion of the haem iron in haemoglobin and myoglobin to the ferric state. Iron in the ferrous (Fe^{2+}) state binds reversibly with oxygen. The transcription coactivators PGC-1α and PGC-1β (see PGC1) are known to be involved in regulation of genes involved in haem biosynthesis. See NITRIC OXIDE.

haemagglutination The clumping of red blood cells caused by any substance (e.g., agglutinin antibodies anti-A and anti-B) binding to and cross-linking their cell-surface antigens. It is the principle behind blood typing. Some enveloped viruses, such as those causing INFLUENZA and MEASLES, have haemagglutinin molecules that bind to glycoproteins on host cells and initiate the infection process.

haemagglutinins See HAEMAGGLUTINATION.

haemangioblasts The common precursor cell for blood vessels and blood cell formation in embryos. Those lying within blood islands form haematopoietic stem cells, precursors of all blood cells; those at the periphery of blood islands develop into angioblasts, precursors of blood vessel tissue.

haematocrit (packed red cell volume) The proportion of the blood volume composed of RED BLOOD CELLS, determined by centrifuging whole blood in a haematocrit tube. In men it averages 0.47 (47%) and in adult females 0.42 (4.2%). Its value decreases with longer-term acclimatization to ALTITUDE.

haematoma Swelling resulting from bleeding into the tissues.

haematopoiesis (haemopoiesis) Blood cell formation. The first HAEMATOPOIETIC STEM CELLS appear in distinct extraembryonic and embryonic sites: the aorta-gonad-mesonephros (AGM) region, major vitelline and umbilical vessels, and the placenta. The site of formation then moves from the AGM to the foetal liver. Finally, during the last two months of intrauterine life, haematopoiesis relocates to the BONE MARROW – again through stem cell migration. After birth, haematopoiesis takes place in red BONE MARROW only. See COLONY STIMULATING FACTORS, GM-CSF, HAEMATOPOIETINS.

haematopoietic stem cells (HSCs) The pluripotent cells that give rise to blood cells, both lymphoid and myeloid, throughout the human lifespan (see HAEMATOPOIESIS) although, in adults, it is quite a rare cell population. Possibly the best understood of our STEM CELLS, they circulate freely but seem to have little function outside specific anatomical locations (their niches) from which they derive those cues enabling them to persist and to change in number, fate and function. HSCs develop in the context of a stromal cell niche (see STROMA), consisting, at least in part, of the osteoblast lineage. Thus, *stem cell factor* (Kit ligand, steel factor), a transmembrane protein on bone marrow stromal cells, binds to c-Kit, a signalling receptor carried on developing B cells and other developing leukocytes. Developing mainly in trabecular (spongy) BONE, at the endosteal marrow adjacent to the bone's surface, HSCs are believed to migrate inward in the central marrow, differentiating into PROGENITOR CELLS away from a possible gradient of self-renewal cues which probably include Notch, Hedgehog and Wnt signalling pathways (see entries), ligands to which they are known to have appropriate cell-surface receptors. In the absence of the protein PTEN (see AKT SIGNALLING PATHWAY), not only is there a gradual loss of haematopoietic stem cells, but their differentiation characteristics allow LEUKAEMIAS to develop in which large numbers of cells behave as CANCER stem cells.

haematopoietins A major family of CYTOKINES made by CD4 (T_H2) effector T CELLS. They include IL-3, IL-4, IL-5, IL-13 and GM-CSF. They bind closely related receptors (class I cytokine receptors). They signal through the JAK/STAT PATHWAY. ERYTHROPOIETIN (Epo) is also a haematopoietin.

haemoglobinopathies The disorders of human HAEMOGLOBIN (Hb) are divisible into two main groups: structural globin chain variants (e.g., sickle-cell disease) producing electrophoretically detectable abnormalities, and disorders of synthesis of the globin chains (thalassaemias). See GLOBINS.

Sickle-cell disease is an autosomal recessive disorder in which a missense mutation causes an abnormality in the β chains of haemoglobin. Sickle-cell haemoglobin (HbS) contains valine instead of glutamate at position 6 of the β-globin, so that there is no electric charge at this position. HbS therefore has two fewer negative charges than HbA, resulting in a 'sticky' hydrophobic point of contact at position 6 of the β chain, which is on the outer surface of the molecule. These sticky spots cause deoxy-HbS molecules to associate abnormally, causing a decrease in their solubility and the formation (particularly in homozygotes, who have sickle cell *disease*) of long, fibrous aggregates within the erythrocytes that cause erythrocytes to sickle in low pO_2. Without treatment, such individuals are likely to die in early childhood. Heterozygous individuals, with the sickle cell *trait*, usually lack symptoms unless oxygen tensions are very low and have a selective advantage in regions where MALARIA is endemic, giving rise to a balanced polymorphism (see NATURAL SELECTION). Foetal DNA can be screened for the presence of the sickle-cell gene, which involves one base substitution, from T to A (see SCREENING). Missense mutations can also give rise to HbC and HbE; deletion mutations can lead to Hb variants lacking one or more amino acids (e.g., Hb Freiburg); insertion and frameshift mutations and 'fusion genes' resulting from unequal cross-overs also generate variant Hbs. HbC (glutamate-to-lysine at position 6 in the β-globin) protects West African children against malaria caused by *Plasmodium falciparum* in both the homozygous (CC) and heterozygous (AC) states.

Thalassaemias are genetic disorders arising mostly from reduced production of the α- or β-globin chains of haemoglobin so that unequal ratios of the two globins are produced. Unpaired free globin chains accumulate and precipitate in red blood cell precursors, causing haemolysis (*haemolytic anaemia*) and consequent hyperplasia of the bone marrow. The β-thalassaemias account for hundreds of thousands of childhood deaths per year and result from any of about 100 different mutations. Some of these affect the promoter region of the gene, while stop mutations (see NONSENSE-MEDIATED DECAY), frameshift mutations and others affecting RNA splicing are known. Heterozygous individuals have thalassaemia minor and suffer mild anaemia; homozygotes have thalassaemia major and often die before 10 years of age from ANAEMIA, a reduction in the level of haemoglobin needed for oxygen transport, abnormalities of bones and bone marrow and spleen enlargement. The mutations are in the same β-globin gene as that causing sickle cell disease, although in thalassaemias the red cells do not sickle. Some 20% of individuals in parts of Italy are carriers of a thalassaemia gene and the geographical distribution of some thalassaemias is similar to that of malaria. The α-thalassaemias result from underproduction of the α-globin chains and are most frequent in persons from Southeast Asia. Severe α-thalassaemia, in which no α-globin chains are formed, is associated with foetal death in utero. In the milder form, compatible with life, the underproduction of the α-globin chains leads to excess of β-globin chains and formation of the β-globin tetramer HbH, whose oxygen affinity is close to that of myoglobin so that it does not release O_2 in the peripheral tissues. The α-thalassaemias are mostly due to deletions of one or more α-globin gene, unequal crossovers giving rise to variations in the COPY NUMBER of the gene. Individuals homozygous for δβ thalassaemia produce no δ- or β-globin chains, but are usually only mildly anaemic because of increased γ-globin production.

haemoglobins (Hbs) Allosteric proteins with quaternary structure, produced within cells derived from proerythroblasts in the bone marrow, and functioning in mature RED BLOOD CELLS to load oxygen in the adult lungs and foetal placenta and unload it in the respective adult and foetal body tissues.

Each of the four non-covalently bound globin molecules in any haemoglobin molecule (2α, 2β in adult A_1; 2α, 2δ in adult A_2; 2α, 2γ in foetal HbF; and 2ζ, 2ε;2α, 2ε and 2ζ, 2γ in consecutive embryonic Hbs) has an attached Fe^{2+}-iron-porphyrin prosthetic group, HAEM, to which one molecule of oxygen can bind reversibly, making four oxygen molecules per haemoglobin molecule. Adult haemoglobin consists mainly of type A_1, with 2–3% of A_2, although HbF makes up ~0.5% of Hb in the blood of normal adults. The two pairs of identical globin chains are encoded by the *HHA* and *HHB* genes, members respectively of the α-globin gene cluster on chromosome 16 and the β-globin gene cluster on chromosome 11 (compare MYOGLOBIN). In early embryonic development, haemoglobins are made in the yolk sac; but the liver becomes the major site of synthesis in the foetus, before giving way to the bone marrow in the adult. This developmental sequence is accompanied by switching genes on and off in each of the two major globin GENE CLUSTERS (see also GLOBINS) generating slightly different forms of haemoglobin (*haemoglobin switching*). At ~8 weeks of gestation, the foetal liver synthesizes predominantly HbF and a little HbA. In the newborn, ~70% of haemoglobin is HbF. But by the end of the first year HbF has declined to ~1% as HbA predominates.

The differences in the molecular compositions of Hb account for their different oxygen-haemoglobin dissociation curves (see Fig. 85). These plot the sigmoidal relationship between the percentage O_2-saturation of Hb for a given partial pressure of oxygen (pO_2) in the blood. Increase in blood temperature, pCO_2, H^+ and quantity of 2,3-bisphosphoglycerate (BPG) bound to the β-globin all decrease the affinity of Hb for oxygen and shift this curve to the right, favouring unloading of bound O_2 (the *Bohr effect*), while the opposite changes shift the curve to the left. As erythrocytes break down glucose they produce 2,3-bisphosphoglycerate (BPG), which greatly reduces the affinity of Hb for oxygen, shifting the oxygen-haemoglobin dissociation curve to the right. Its production rate increases at high altitude, resulting in greater O_2 unloading in the tissues but without affecting O_2-binding in the lungs.

The steepness of the central portion of

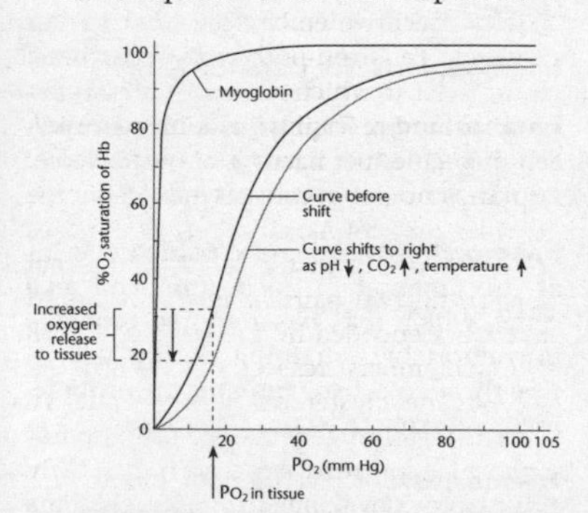

FIG 85. Oxyhaemoglobin dissociation curves, illustrating the Bohr effect (see HAEMOGLOBINS). © 2006, Oxford University Press. Redrawn from *Human Physiology, The Basis of Medicine*, G. Pollock & C.D. Richards, with permission of the publisher.

O_2-Hb dissociation curves results in a greater volume of oxygen loading and unloading over a given change in blood pO_2 than would have been the case had the relationship between % Hb-O_2-saturation and pO_2 been linear. This is the result of *cooperative binding* of oxygen. The first oxygen molecule to bind a haem causes a slight allosteric shift in the molecule's shape and exposes the second haem, making binding of the second O_2 molecule easier. Similarly, binding of the second and third O_2 molecules to their haem groups facilitates binding of the third and fourth.

The greater O_2-affinity of foetal Hb than adult Hb is essential if foetal Hb (HbF) is to be oxygenated by maternal HbO_2 in the placenta. It also has to unload its bound O_2 at lower pO_2 than does adult Hb. In the foetus, γ-subunits of Hb are synthesized rather than β-subunits, the $\alpha_2\gamma_2$ product having a much lower affinity for BPG and a much greater affinity for O_2 than the adult $\alpha_2\beta_2$ Hb. See HAEMOGLOBINOPATHIES, NITRIC OXIDE.

haemolysis Red blood cell rupture followed by release of haemoglobin. It occurs in haemolytic anaemia, transfusion reactions, ERYTHROBLASTOSIS FOETALIS and MALARIA.

haemolytic anaemia A disorder in which red blood cells either rupture, or are destroyed, at an excessive rate. It can result from inherited defects within red blood cells leading to weakening of the membrane, but also from damage to the red cells by drugs, snake venom, artificial heart valves, autoimmune disease and ERYTHROBLASTOSIS FOETALIS. It can also result from thalassaemias (see HAEMOGLOBINOPATHIES) and from Rhesus-incompatibility between mother and foetus. See ANAEMIA.

haemolytic disease of the newborn See ERYTHROBLASTOSIS FOETALIS.

haemophilia Classic haemophilia (haemophilia A) is due to a recessive mutation in the X-linked gene encoding plasma FACTOR VIII, one of many proteins involved in BLOOD CLOTTING. Affecting about 1:10,000 males, its effects vary from mild to severe, symptoms including bleeding into joints on first walking, with consequent inflammation and pain. A substantial proportion of female carriers are predisposed to a bleeding tendency since their mean Factor VIII concentration is about half normal due to *lyonization* of the X chromosomes (see X-CHROMOSOME INACTIVATION). Linkage analysis using polymorphic intragenic markers has superseded blood testing for carrier detection. Prenatal diagnosis is sometimes requested and can usually be achieved by mutation or linkage analysis (see SCREENING).

The Factor VIII gene locus is near the tip of the long arm of the X chromosome, and is large (186 kb, with 26 exons), the product being post-translationally modified into two polypeptides. Deletions account for 5% of cases, tend to occur in the female germ line, and usually cause complete absence of Factor VIII expression. Frameshift, nonsense and missense mutations tend to occur in the male germ cells, but a 'flip' inversion accounts for 50% of all severe cases, is about ten times more frequent in the male germ cells, and results from recombination between a small gene (A) located within intron 22 of the Factor VIII gene and other copies of the A gene located upstream near the telomere. Very occasionally, malfunction of the gene is the result of a LINE insertion.

Haemophilia B (Christmas disease) is also an X-linked recessive, but non-allelic, disorder, caused by mutation in the gene (Xq27.1-q27.2) encoding Factor IX protein. Over 800 different point mutations, deletions and insertions have been reported in the gene, and by analysing 2.2 kb of the 34 kb gene (which contains 8 exons) it is possible to detect the mutation in 96% of all patients. The rare variant haemophilia B Leyden, caused by promoter mutations, exhibits the unusual characteristic of age-dependent expression. It is very severe in childhood, with Factor VIII levels <1% of normal, but after puberty its level rises to 50% of normal and the condition then becomes asymptomatic. As with Factor VIII in haemophilia A, Factor IX levels are rarely used in carrier detection, linkage or direct mutation analysis having replaced the method.

Both forms of haemophilia are treated using plasma-derived Factor VIII or Factor IX, although these preparations did not originally exclude blood-borne viruses such as HIV and hepatitis B virus, with disastrous consequences for haemophiliac males. Recombinant Factor VIII became available in 1994 although expense has limited its worldwide use. Adeno-associated viral vectors expressing Factor IX, injected into skeletal muscles, have not been successful in the long term although haemophilia remains a prime target of GENE THERAPY.

Haemophilus influenzae A fastidious bacterial genus whose growth requires NAD and haematin normally found in the blood. Six capsular endotoxin serotypes (a–f) occur, type b readily colonizing unimmunized infants below the age of 5 (Hib immunization is readily available in developed countries). Manifestations of Hib disease include pneumonia, meningitis, septic arthritis and facial cellulites, all commonly accompanied by bacteraemia. Hib isolates are increasingly resistant to amoxicillin and erythromycin; clarithromycin and quinolone are generally effective against most isolates. It can cause MENINGITIS.

haemopoiesis See HAEMATOPOIESIS.

haemorrhage Rupture of a blood vessel. See BIRTH, BLOOD CLOTTING, HAEMOPHILIA, STROKES, VIRAL HAEMORRHAGIC FEVERS.

haemorrhoids Distended (varicose) veins at the junction of the rectum and anal canal, which may bleed and be generally uncomfortable. Straining during defecation through constipation may contribute, but is unlikely to be the sole cause.

haemostasis Prevention of blood loss, including vascular spasm, formation of PLATELET plugs, and BLOOD CLOTTING.

hair A product of the EPIDERMIS (see HAIR FOLLICLE), comprising a shaft (above the skin surface) and root (beneath the skin surface). Most of the root and shaft comprise columns of dead keratinized cells arranged in three concentric layers: medulla (central hair axis with soft keratin), cortex (the bulk of the hair and containing hard keratin) and cuticle (a single layer of cells containing hard keratin, whose edges overlap resembling tiles on a roof). Reduced hair increases convection rates and heat dissipation – an advantage for long-distance walking in hot environments, and endurance running (see GAIT), which may have been important in pursuit of prey by early modern humans. See HAIR COLOUR.

hair cells Mechanosensory cells held in an elaborate epithelial structure of supporting cells within the organ of Corti of the COCHLEA (auditory hair cells, see Fig. 58) and the maculae of the utricles and saccules of the VESTIBULAR APPARATUS (see EARS, Fig. 68). Each has at its apical surface a characteristic bundle of 70 or more *stereocilia* and a single motile *kinocilium* (a conventional CILIUM anchored to its basal body) longer than the longest stereocilium. Stereocilia decrease progressively in length away from the kinocilium, defining the *axis of polarity* (or, orientation) for the hair cell. A fine filament (tip link) runs from the tip of each shorter stereocilium and attaches at a higher point on its taller adjacent neighbour. The orientation of hair cells differs in different parts of the vestibular apparatus. In the ampulla, they are orientated so that they all have their kinocilium pointing in the same direction; but in the utricle and saccule their orientation is more complex. Coupled with the orientation of the semicircular canals, utricle and saccule, this enables the vestibular apparatus to encode any movement of the head.

Hair cells have a resting potential of about –60 mV, but as the bundle of stereocilia tilts in the direction of the kinocilium in response to sound or changes in acceleration, mechanosensory TRP RECEPTORS in the ciliary membranes open and allow the influx of positive charge (K^+) from the endolymph that results in depolarization of the hair cell, producing a generator potential which may then lead to formation of an action potential being transmitted along the vestibular afferent fibres. Tilting in the opposite direction closes the channels and leads to repolarization. As a result, hair cells release greater or lesser amounts of neurotransmitter at their synapses with first-order sensory neurons in the vestibular branch of the VESTIBULOCOCHLEAR NERVE. As the potential of the endolymph is ~80 mV positive, the effective voltage across the apical surface of a hair cell is ~140 mV which, together with the sizeable chemical gradient of K^+, makes them extremely sensitive. Efferent neurons also synapse with hair cells and regulate their sensitivity.

hair colour The MELANOCYTES of the HAIR FOLLICLE are responsible for hair colour, and the melanocyte stem cell population seems to be required for continued production of pigment. Age-related greying in both mice and humans is accompanied by bulge region melanocyte attrition. Human melanocortin receptor 1 gene (*MC1R*) is highly polymorphic (see MELANOCORTINS) and three of its alleles are associated when homozygous with red hair, defined also by fair skin, inability to tan and propensity to freckle. Mutations in the *POMC* gene (see PROOPIOMELANOCORTIN) also produce the red-haired phenotype. The inheritance of red hair seems to be autosomal and recessive, although heterozygotes show different tanning ability compared to wild-type homozygotes suggesting a gene dosage effect. See MELANINS, SKIN COLOUR.

hair follicle Complex appendix of the squamous interfollicular epidermis of the SKIN, originating embryologically as skin placodes and deriving from multipotent epidermal stem cells in the bulge region (see Fig. 86). They originate embryologically as skin placodes, appearing as solid proliferations of the EPIDERMIS which penetrate the underlying dermis. Their termini, the *hair buds*, invaginate to form a *dermal papilla* which rapidly fills with mesoderm in which blood vessels and nerve cells

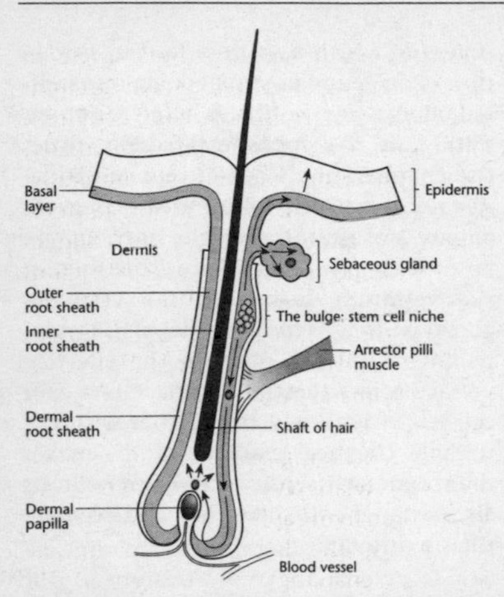

FIG 86. Diagram showing a HAIR FOLLICLE (see entry) with its hair, as present in the newborn. Arrows indicate the directions of cell migration (and loss from the epidermis). From T. Reya and H. Clevers, © *Nature* 2005, with permission from Macmillan Publishers Ltd.

develop. The WNT SIGNALLING PATHWAY controls the fate of cells derived from the bulge region in established hair follicles. In particular, different members of the Tcf/Lef family are expressed in different lineages. Bulge region cells generate several cell lineages, giving rise to the hair itself, the sebaceous gland, and the basal layer of the interfollicular epidermis (where they differentiate into slow-cycling keratinocyte stem cells). The more peripheral cells of the hair bud become cuboidal and give rise to the epithelial (outer) root sheath. To form a hair, cells flow downwards from the bulge through this outer root sheath and enter a transit amplifying compartment, the germinative matrix, at the base of the follicle. On leaving this, cells undergo terminal differentiation in the precortex compartment, becoming spindle-shaped and keratinized, forming the hair shaft. Continuous proliferation of cells at the shaft base pushes hair upwards. The dermal root sheath forms from surrounding mesenchyme and a small muscle, the arrector pili muscle, is usually attached to this. Cells from the bulge can also flow upwards through the outer root sheath to produce sebocytes (see SEBACEOUS GLANDS), or seed the basal layer of the epidermis between the follicles with slowly proliferating stem cells which eventually proceed to terminal KERATINOCYTE differentiation. The first hair to form, *lanugo hair*, is shed at roughly the time of birth, being replaced by coarser hair from new hair follicles. Work on mice indicates that if a healed skin wound is larger than ~0.5 cm in diameter, new hairs form at the centre of the wound, skin sections revealing changes resembling those during embryonic hair-follicle development. New hair follicles grow and pass through the hair cycle, becoming indistinguishable from neighbouring hair. The size of the healed wound appears to be critical.

Hair follicle receptors (hair end organs) branch out from the dendritic tree at the distal end of a sensory axon and wrap themselves around follicles (see MECHANORECEPTORS, Fig. 120). They are sensitive to light touch, but since the field of follicles innervated overlaps those of adjacent axons the sense of light touch is not highly localized despite being highly sensitive on account of CONVERGENCE.

hairpin loops Nucleic acid molecules, or parts of them, that fold back on themselves by base-pairing. Hairpins are involved in TRANSCRIPTION termination. See INVERTED REPEATS, MICRORNAS, RNA INTERFERENCE, TRANSFER RNA.

hair receptors For *hair end organs*, see HAIR FOLLICLE.

hallucinations See HALLUCINOGENIC DRUGS, SCHIZOPHRENIA.

hallucinogenic (psychedelic) drugs Lysergic acid diethylamide (LSD) and related drugs (psilocin, mescaline, dimethyltryptamine) induce dramatic states of altered perception, vivid and unusual sensations, and feelings of ecstasy. Sometimes, unwanted effects include panic, frightening delusions and hallucinations, which usually fade but may recur ('flashbacks'). Serotonergic systems may be involved in LSD action, as it inhibits firing of 5HT-containing neurons in the raphe nuclei (probably by inhibiting firing of $5HT_2$ receptors (see SEROTONIN)). But there is no withdrawal syndrome. See MDMA, PSYCHOACTIVE DRUGS.

Hamilton's rule The prediction that genetically determined behaviour which benefits another individual, although at some cost to the agent with the allele(s) responsible for the behaviour, will spread by NATURAL SELECTION when the relation rB>C is satisfied, where r = the degree of RELATEDNESS between agent and recipient, B = the improvement of individual fitness of recipient caused by the behaviour, and C = the cost to the agent's individual fitness as a result of the behaviour. This is the concept of *inclusive fitness*, and is an explanation of kin altruism.

hand The termini of the FORE LIMBS (see Fig. 78; see also LIMBS). Tendons of the forearm muscles attach to the WRIST (see Fig. 182) or continue into the hand, in association with blood vessels and nerves, being held close to bones by strong fascial structures. Such *extrinsic muscles* moving the hand and digits are innervated by nerves derived from the BRACHIAL PLEXUS. The *anterior compartment muscles* originate on the humerus, typically inserting on the carpals, metacarpals and phalanges, functioning as FLEXOR MUSCLES. The *posterior compartment muscles* also originate on the humerus, inserting on the metacarpals and phalanges and functioning as EXTENSOR MUSCLES. In both compartments, the muscles are grouped as superficial and deep. The intrinsic muscles in the palm produce weak, but intricate and precise, movements of the digits characterizing the human hand, their origins lying within the hand. The general activities of the hand are free motion, power grip (forcible movement of fingers and thumb against the palm), precision grip (change in the position of a handled object requiring exact control of positions of fingers and thumb, as in threading a needle) and pinch (compression between thumb and index finger, or thumb and first two fingers). Because the thumb is positioned at a right angle to the other digits, its movements in different planes are crucial to precise hand activities. Its movements include flexion (medially, across the palm), extension (laterally, away from the palm), obduction (anteroposteriorly, away from the palm), adduction (anteroposteriorly, towards the palm) and opposition (across the palm so that the tip of the thumb meets a finger). The mobility of the hand is far greater in humans than in other primates. Although opposability of the thumb is a characteristic of Old World primates, including the Early Miocene *Proconsul*, it is relatively restricted in apes (and absent from living apes and monkeys) compared with the later hominin lineage where, by enabling the precision grip, it has had major significance in the development of tool production. See HANDEDNESS.

handedness Approximately 90% of the human population is naturally more skilled with the right hand than with the left, a motor asymmetry strongly correlated with language dominance (see LANGUAGE AREAS). It is due to functional asymmetry in the cerebral cortex, and chimpanzees do not show such marked handedness. It is interesting that humans are different from non-human primates in regard to handedness as well as language, but while animals of many species show consistent preferences for using one hand over the other, there are typically equal numbers of left-handed and right-handed individuals.

handplates See LIMBS.

haplogroups Well-resolved subclusters of HAPLOTYPE lineages within a (human) population.

haploid A cell, or nucleus, containing an unpaired chromosome set. In humans, these cells arise during meiosis, do not divide further, and give rise to the gametes. See OOGENESIS, SPERMATOGENESIS.

haploinsufficiency See LOSS-OF-FUNCTION MUTATIONS.

haplorhine Having nostrils surrounded by

hairy skin (i.e., the rhinarium is absent), as in living tarsiers and lorises (see PRIMATES). Contrast strepsirhine (see PRIMATES).

haplotype A particular combination of alleles at adjacent SINGLE NUCLEOTIDE POLYMORPHISMS (SNPS) that is usually inherited as a unit, or block (see HAPLOTYPE BLOCKS), because of LINKAGE DISEQUILIBRIUM (LD) attributable to the fact that RECOMBINATION during meiosis tends to occur in certain regions of the chromosomes more often than in others. Many empirical studies indicate highly significant LD in the human genome, frequently with strong associations between nearby SNPs. Under positive selection (see NATURAL SELECTION), a favoured allele may rise in prevalence rapidly enough for crossing-over not to break down the association with alleles at nearby loci on the ancestral chromosome, an indication that it arose from a single historical mutation event.

Most single nucleotide variation in the human genome results from such single historical mutation events, the rate of these mutations being very low (~10^{-8} per site per generation) relative to the number of generations since the MOST RECENT COMMON ANCESTOR of any two humans (~10^4 generations). Each such new mutation (allele) is therefore initially associated (physically linked) with the other alleles that happen to occupy the same chromosome. This specific set of alleles on a particular chromosome, or part of one – the particular combination of alleles at nearby SINGLE NUCLEOTIDE POLYMORPHISMS (SNPs) – is known as a haplotype (see Fig. 87b). New haplotypes are formed by additional mutations, and by recombination between maternal and paternal chromosome segments during meiosis, resulting in a chromosome that is a mosaic of the two parental chromosomes. In many chromosome regions there are only a few haplotypes and these account for most (~90%) of the variation among people in those regions (see Fig. 87 and GENETIC DIVERSITY, Fig. 82). This is helpful, since genotyping only a few carefully chosen SNPs in the region will supply enough information to

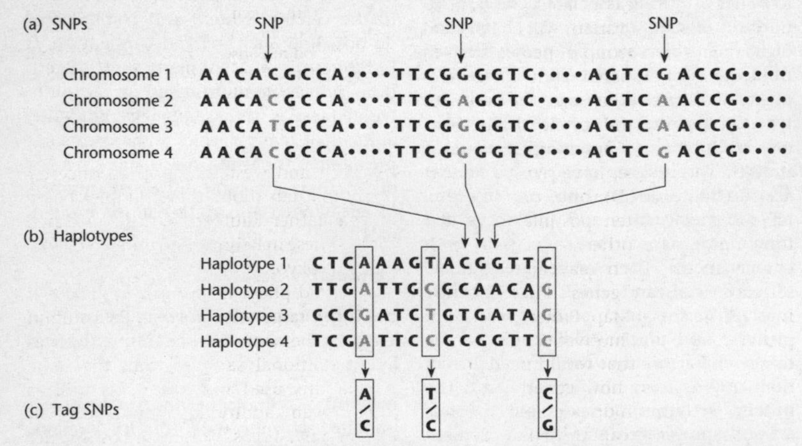

FIG 87. Diagram illustrating single nucleotide polymorphisms (SNPs) and tag SNPs. (a) Three SNPs from four versions of the same chromosome region. In the first SNP, two alleles occur: C- and T-containing. (b) Observed HAPLOTYPES (see entry), extending across 6 kb of DNA, only the variable bases (SNPs) being illustrated. Three of these (arrowed) are those indicated in (a). Most of the population had haplotypes 1–4; but genotyping only the three tag SNPs (rectangular boxes and (c)) is sufficient to identify these four haplotypes uniquely. If, for instance, a chromosome had the pattern A–T–C at these three tag SNPs, this pattern matches that determined for haplotype 1. From *the International HapMap Consortium*, © *Nature* 2003, with permission from Macmillan Publishers Ltd.

enable prediction of many of its other SNPs. Only a few of these so-called 'tag' SNPs are therefore needed to identify each of the common haplotypes in a chromosome region (see Fig. 87c).

If, e.g., people suffering from some disease (e.g., type 1 DIABETES) also turned out to carry an allele of some GENETIC MARKER having nothing to do with diabetes more often than would be expected on the assumption that their association in the population were random, that would be evidence that diabetes is influenced by a gene on the same chromosome as the marker gene – strong LINKAGE DISEQUILIBRIUM between the two loci indicating that the diabetes-related gene lay fairly close to the marker.

On average, a single SNP has 3–10 other SNPs associated with it in population samples, depending on the population. A haplotype with a particular set of sequence variants in the genome can therefore serve as a GENETIC MARKER to detect association between a particular genomic region and susceptibility to a particular disease, even if the marker has no functional effects and the disease-associated variant is uncommon (a rare variant will be enriched in frequency in a group of people affected by a disease compared with a group of unaffected controls, as was the case with the *CFTR* gene and cystic fibrosis). In other words, since polymorphisms often 'travel together' (i.e. have significant LINKAGE DISEQUILIBRIUM, LD), once one has been identified it is often possible to predict the form of many others.

Such an approach may accelerate the search for 'disease genes'. Phase I of the International HapMap Project was completed in 2005 and motivated by the desire to develop a tool that would make it economical to assess how genetic variation influences common diseases (see COPY NUMBER). Its aim was to capture SNPs with a frequency ≥5% in 269 individuals of four reference populations (i.e., four ethnic origins) and to identify a minimum set of SNPs ('tagging SNPs'; see Fig. 88) sufficient to express all common variants in our genome. Populations sampled were from the Yoruba in Ibadan, Nigeria; populations of

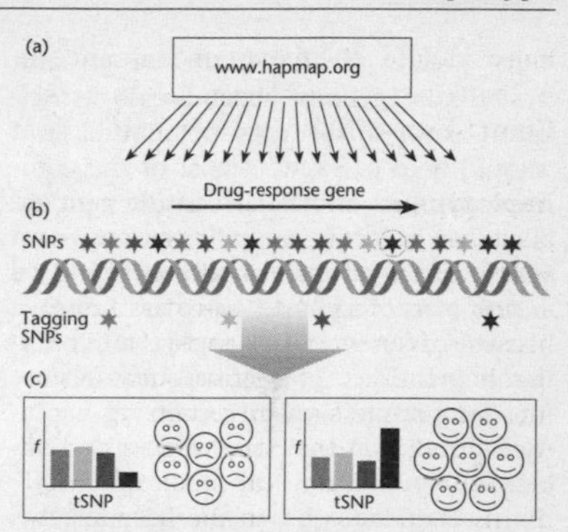

FIG 88. Diagram indicating the use made of the HapMap website (see HAPLOTYPE). (a) Sequence data for an interesting gene (or region) are downloaded from the HapMap website. (b) Those SINGLE NUCLEOTIDE POLYMORPHISMS associating closely with each other have the same level of shading and are represented in (c) by correspondingly shaded rectangles. They can be compressed to a subset of tagging SNPs, shown as one from each box type. The tagging SNPs are genotyped in a population sample in which individuals vary for some character of interest, such as a good or bad reaction to some therapeutic approach. A tagging SNP correlating with a particular reaction indicates that one of the SNPs with which it is associated influences that reaction. In (c), one such SNP (circled) causes a good response. From D.B. Goldstein and G.L. Cavalleri, *Nature* © 2005, with permission from Macmillan Publishers Ltd.

northern and western European ancestry living in Utah (both these samples being mother–father–adult child trios); unrelated Han Chinese in Beijing; and unrelated Japanese in Tokyo.

A second phase of the project (Phase II HapMap) targeted a further 2.1 million SNPs in the same individuals, as well as from additional samples from the same populations used in Phase I, as well as from seven additional populations in Kenya, Italy, Texas, Colorado, Los Angeles and people of African ancestry in the southwestern United States. Published in 2007, it provides an SNP density of approximately 1 per kb and is estimated to contain ~25–35% of all the 9–10 million common SNPs (minor allele frequency ≥0.05) in the assembled human genome. These data have greatly facilitated the design and analysis of genome-wide disease association studies (see COMPLEX DISEASES) and should help to capture most of the common patterns of variation in the genome (e.g., see SKIN COLOUR) and associate them with particular disease susceptibility, to better protect against illness and improve the effectiveness of drug application to target populations. Two striking features are the long-range similarities among haplotypes, and SNPs that show almost no linkage disequilibrium with any other SNP. Both HapMap phases provide insights into the selective forces acting on SNPs in coding regions of genes (see NATURAL SELECTION).

Because of the ethical implications of this work, appropriate protocols must be observed in the authorization of DNA sampling and the release of data obtained. The approach promises to have profound implications for our understanding of human biology and human health. The geographical distribution of haplotypes reflects human history, with a loss of haplotype diversity as distance increases from Africa (see MIGRATION). See HLA COMPLEX, NATURAL SELECTION, NESTED CLADISTIC ANALYSIS.

haplotype blocks The genomes of living humans are ~99.9% identical to each other, DNA sequence variation along a chromosome being structured in 'blocks' where nucleotide substitutions are associated in HAPLOTYPES. These blocks are stretches of DNA sequence in which 3–7 SNPs that occur at frequencies >5% account for most of the GENETIC DIVERSITY (see Fig. 82) found among humans, and the catalogue of haplotypes for every block constitutes the haplotype map (HapMap) of the HUMAN GENOME. The human autosomal genome appears to be a mosaic of 'haplotype blocks', each block having its own pattern of variation. Delineation of blocks depends on the methods used to define them, but they are typically 5–200 bp in length. The two parts of the human genome where no recombination occurs at all, the mitochondrial genome (see MTDNA) and the Y CHROMOSOME, can be regarded as haplotype blocks. See RECOMBINATION.

HapMap See HAPLOTYPE.

haptens Small organic molecules of simple structure (e.g., phenyl arsonates and nitrophenyls) that bind antigen-binding sites of ANTIBODIES or T-CELL RECEPTORS but cannot by themselves elicit an adaptive immune response. However, antibodies can be raised against them if they are attached covalently to an immunogenic protein carrier and the manner of binding of haptens by these 'anti-hapten antibodies' has been influential in defining the precision of antigen binding by antibodies. See PENICILLIN.

haptic touch See KINESTHESIA.

HARs (human accelerated regions) DNA sequences that have experienced a significantly accelerated rate of base substitution in the human lineage since the time of the common ancestor of humans and chimpanzees. In their 2006 paper, Haussler and colleagues discovered 49 HARs, 96% of which are in non-coding DNA segments. The 118-bp region HAR1, in the last band of chromosome 20q, showed the most dramatically accelerated change. It is part of a pair of overlapping divergently transcribed genes, containing a two-exon *HAR1F* gene and an alternatively spliced *HAR1R* gene with isoforms of two and three exons. See NATURAL SELECTION.

***HAR1F* (human accelerated region 1 forward) gene** A gene expressed specifically in CAJAL–RETZIUS CELLS in the developing human neocortex (see CEREBRAL HEMISPHERES), a critical period for cortical neuron specification and migration. One of about 10 genes to have emerged between 2002 and 2006 as having played a significant role in the expansion of the cerebral cortex during hominin evolution, it lies in the last band of chromosome 20q. It is very active in the developing brains of 2-month to 5-month-old human embryos and part of its sequence (the 118-bp HAR1 region) shows the most dramatically accelerated base-pair change since the human-chimpanzee ancestor, with an estimated 18 substitutions compared with the expected 0.27 substitutions on the basis of the slow rate of change in this region in other amniotes (see NATURAL SELECTION). It encodes an apparently stable RNA molecule and is co-expressed with REELIN, a known regulator of cortical development, and may well regulate the expression of other genes.

***hAT* elements** See TRANSPOSABLE ELEMENTS.

Haversian system (osteon) Anatomical unit of compact BONE (see Fig. 26). Approximately cylindrical, each consists of concentric layers (lamellae) of hard calcified bone secreted by osteocytes which occupy spaces (lacunae) within this matrix. A central canal (Haversian canal) containing blood vessels and nerves runs longitudinally through the centre of each osteon and connects via lateral branches to the periosteal blood vessels and nerves which enter the diaphysis (shaft) through pores leading into canals (Volkmann's canals). In compact bone, osteons run parallel, enabling the bone to resist bending stresses and as the directions of these stresses change during life the directions in which new osteons are constructed alter accordingly. Haversian systems appear to be a feature of endothermic vertebrates.

hay fever See ALLERGIES.

HCG (hcg, hCG) See HUMAN CHORIONIC GONADOTROPIN.

HDL (high density lipoprotein) See LIPOPROTEINS.

head 1. The superior part of the human body, cephalic to neck, originating from ectoderm and head mesenchyme (see Appendix VI, EMBRYONIC DISC). It is the location for the BRAIN and major organs of sense (e.g., EARS, EYES). Tilting or rotation of the head and body as a unit activates motor neurons to muscles maintaining the head vertical with respect to gravity (see POSTURAL REFLEX PATHWAYS). See SKULL, FACE and HOX GENES. **2**. Rounded articular projection supported on the neck of a bone and involved in a JOINT.

health Formerly regarded simply as the absence of disease, or lack of disease symptoms, it is now usual to extend the definition of 'good health' to be a state of physical and mental well-being. A person who feels healthy and lacks obvious signs of disease may still be storing up trouble for the future by leading a lifestyle that incurs avoidable RISK FACTORS, increasing numbers of which decrease a person's health accordingly. Maintaining good health requires an awareness of 'good life practice'. Many of the cellular processes that are vital for human health are regulated by the INSULIN/INSULIN-LIKE GROWTH FACTOR signalling pathway, including glucose homeostasis, fat metabolism, cell growth and differentiation, and ageing.

Physical health refers to the state of the body, and may be measured by checking temperature, blood pressure, blood glucose level, and many other such variables. Abnormalities in their values may be symptomatic of compromised physical health, even in the absence of more obvious symptoms of illness. Physical health, measured by the level of physical FITNESS, depends upon properly functioning homeostatic mechanisms: when these control circuits become deregulated, illness usually results. Emotional well-being, not least an ability to cope with STRESS, is also a major component of health. See EXERCISE, HEALTHY DIET, LIFESTYLE.

healthy diet A properly balanced diet, containing all the nutritional requirements with none in deficiency or excess (see DAILY VALUES, MALNUTRITION). The phrase 'optimal diet' could be used in preference. However, there is no one optimal diet but an indefinite number of ways of obtaining it. A reasonable excess of most dietary constituents never does anyone harm, but a gross excess of some may be harmful. High-fat diets predispose to OBESITY. Deficiencies of any dietary components are, in the long run, extremely harmful and express themselves, perhaps after some years, in general lassitude and liability to develop anaemia and infectious diseases such as tuberculosis with, for example, consequent changes in maternal and infantile mortality rates.

Peoples in different geographical regions and of different ages and physiological conditions will vary in both dietary requirements and intakes. In western Europe, for instance, only about 10% of energy is obtained from dietary protein while the Inuit may obtain 44% of their energy from this source (see ENERGY BALANCE). Compared with North Americans, people in many tropical countries obtain less of their protein from meat and much more of it from pulses (e.g., beans). Furthermore, all people require different nutrients at different times of their lives, be they in the womb, in early post-natal existence, childhood, adolescence or adult life, because different parts of the body, with their different chemical constituents, grow at different rates at different times (see OMEGA-3 FATTY ACIDS, OMEGA-6 FATTY ACIDS).

Optimal diets contain adequate levels of protein, carbohydrate, fats, vitamins and mineral salts, water and roughage (or 'fibre', non-starch polysaccharide). Lipids required in addition to fats include cholesterol and the fat-soluble vitamins A, D, E and K. In addition to the above-mentioned fatty acids, the B vitamin derivative FOLATE is required in larger amounts by a woman in preparation for, and during, her PREGNANCY (see entry). Certain VITAMINS and minerals are essential to the control of cell biochemistry, either as aids to enzyme activity or as components of hormones, so that few of them are present in large quantities at any time. Calcium and phosphorus are the major exceptions in that both are crucial structurally (in BONES, hydroxyapatite being a form of calcium phosphate), while the former is involved in regulation of a huge variety of biochemical processes as a SECOND MESSENGER, and the latter is a major constituent of nucleic acids (DNA and RNA), ATP, NADP and cell membranes as well as having a major role in enzyme activation and inactivation (see SIGNAL TRANSDUCTION, PHOSPHOINOSITIDES). Because cows' milk has more phosphate than human milk it can lead to loss of blood calcium and magnesium in the baby's blood unless it is diluted first. Excess sodium in the diet is a risk factor in HYPERTENSION, which tends to be offset by magnesium, calcium and potassium when sodium intake is reduced (see DISTAL CONVOLUTED TUBULE). Iodine is a requirement for THYROID HORMONE production. See also CALCIUM, IRON, MAGNESIUM, PHOSPHATE, ZINC. For potassium, see DISTAL CONVOLUTED TUBULE.

In temperate climates, people can live for a year on just meat, fish and eggs; vegetarians can survive on cheese, eggs, milk, nuts, fruits and vegetables (a diet just as good but no better nutritionally), while vegans eliminate all animal products (including milk products and eggs) but may run short of one of the B vitamins and develop pernicious anaemia as a result.

Animal protein is very concentrated. Most of the animal we eat is muscle ('meat'), containing high concentrations of muscle proteins, the major connective tissue protein collagen (vitamin C is required for synthesis of collagen) and blood proteins. Protein is much less concentrated in the plant material we eat, and larger quantities of plants need to be eaten to obtain the animal equivalent in terms of protein mass. *First-class proteins*, including meat, fish, soya beans, milk and eggs, contain all the AMINO ACIDS in the proportions required to maintain health. *Second-class proteins* do not contain all the essential amino acids in the correct proportions and are mainly of vegetable origin (e.g., cereals and pulses).

Protein synthesis (on ribosomes) is halted if any of the 20 amino acids is unavailable. Some of these, the essential amino acids, must enter in our food, as we cannot synthesize them. Those whose diet contains a staple (i.e., any food item contributing 10% or more of the average daily protein intake) which is low in a particular amino acid run a risk of essential amino acid deprivation. Beans (and other legumes) are poor in tryptophan and other sulphur-containing amino acids, whereas they are rich in lysine (an essential amino acid); while rice is rich in the sulphur-containing amino acids. So beans and rice eaten together are mutually complementary resources. Maize and wheat are poor in isoleucine and lysine but rich in other amino acids scarce in legumes such as beans.

Another potentially serious problem for vegetarians is the low organic iron content of plant food. Meat contains iron in organic compounds which is more easily assimilated (taken across the ileum and into cells) than is the iron found in plants. So vegetarians need to supplement the iron content of their diets. This is especially true for women in general (menstrual blood flow is a major iron loss) but especially for pregnant women (the baby needs a lot of iron for its own haemoglobin; see PREGNANCY).

If maize forms the major staple, the vitamin nicotinic acid (essential for respiration) may be bound to cell wall material and difficult to absorb across the ileum unless treated first with alkali. In Mexico, lime water is used in preparation of foods containing maize, so reducing the prevalence of the deficiency disease pellagra – common in other parts of the world where maize is a staple and lime water is not used. See DAILY VALUES.

hearing Central auditory pathways process three features of sound input in parallel (see PARALLEL PROCESSING): tone, loudness and timing. From loudness and timing, the brain calculates the location of a sound in space. Parallel processing begins in the COCHLEAR NUCLEI. Primary auditory afferents within the cochlear nerve bifurcate to terminate in the paired ventral and dorsal cochlear nuclei of the medulla. Axons run from the ventral cochlear nuclei to the superior olivary nuclei (SON), both ipsilaterally and contralaterally, and to the contralateral INFERIOR COLLICULUS (IC) of the midbrain. All ascending auditory pathways converge on the IC. Fibres that cross the medulla contralaterally constitute the *trapezoid body*. The SON compares input from the two ears to compute the whereabouts of a sound, whose coordinates are analysed in terms of elevation in the vertical plane and azimuth in the horizontal plane (see below). It projects to the nuclei of the lateral lemniscus in the pons which in turn projects to the IC, which projects to the MEDIAL GENICULATE NUCLEUS of the thalamus. This sends its output via the internal capsule in an array termed the *auditory (or acoustic) radiation*, which projects to the PRIMARY AUDITORY CORTEX. This is the pathway for conscious auditory perception, including SPEECH, all components of which retain the TONOTOPY of the basilar membrane of the COCHLEA. Neurons in this pathway also have different temporal response patterns: some have a transient response to a brief sound; others have a sustained response. Some are also intensity-tuned, giving a peak response to a particular sound intensity (see NEURAL CODING).

Information about *sound elevation* is due to the asymmetrical shape of the pinna. Sound waves arriving from different directions in the vertical plane are reflected by the pinna differently, leading to time delays between arrival at the tympanum along the direct (unreflected) and reflected paths. Information about *sound azimuth* involves comparison between the inputs of the two ears (binaural sound localization) and enables horizontal position to be computed to ~1° of arc, and two methods are used: (i) Analysis of interaural sound *amplitude* differences as low as ~1 dB involves neurons of the lateral superior olivary nuclei (LSO), which receive inputs from both ipsilateral and contralateral cochlear nuclei. Two kinds of binaural neuron are present. One type (EE) is moderately excited by sound presented to either ear but giving maximal response when both ears are stimulated. The other type (EI) is excited by sound in one ear but inhibited by sound in the other (therefore inhibited by sound in both ears). The contralateral route is via a glycinergic inhibitory neuron, equal sound amplitude in both ears resulting in overall inhibition of the LSO neuron. Increasing the amplitude in the contralateral ear augments this inhibition; but increasing amplitude in the ipsilateral ear results in firing of the LSO neuron, maximum firing rate occurring when the amplitude difference between the two ears is 2 dB or more. The same sound produces reverse responses in the opposite LSO, each of which projects via the lateral lemniscus to the central nucleus of the inferior colliculus. Each LSO has a *tonotopic map*, restricted largely to high-frequency input, and projects to the deep layers of the SUPERIOR COLLICULUS to form an auditory space map in register with the retinotopic map. (ii) Analysis of interaural sound *time* differences depends on the phase differences in the arrival of low-frequency waves (<3 kHz) at the two ears. There are analysed by neurons in the cochlear nuclei (bushy cells) capable of *phase-locking* (consistent firing of a cell at the same phase of a sound wave) in response to low-frequency stimuli, differences as short as 20 μs being detectable (see COCHLEA). This makes for easy determination of sound frequency: it is the same as the frequency of the neuron's action potentials. Cells in the medial superior olivary nucleus (MSO) act as *coincidence detectors* and, if a phase difference between the two ears occurs, the bushy cells corresponding to the further ear fire slightly later. The details of how azimuth information is derived from these phase-lock responses are known for birds, but not as yet for mammals.

See DEAFNESS, EARS.

heart The four-chambered organ lying in the pericardial cavity and receiving venous deoxygenated (right side) and oxygenated (left side) blood, pumping it via arteries to the lungs (from the right ventricle) and to the remainder of the organs of the body (from the left ventricle) via the aorta. The adult heart is thus a major component of a double adult circulation, with pulmonary and systemic circuits, four non-return heart valves preventing back-flow of blood. Foetal heart tissue relies upon glycolysis for its energy metabolism, but there is a switch to mitochondrial biogenesis and oxidative metabolism early in neonatal life (see CARDIAC MUSCLE). The foetal heart has adaptations to intrauterine life, such as the FORAMEN OVALE.

The cardiovascular system appears during the third week of development from cardiac progenitor cells lying in the epiblast, lateral to the primitive streak. These migrate to the splanchnic layer of the LATERAL PLATE MESODERM (see Appendix VI) and are induced by underlying pharyngeal endoderm to form cardiac myoblasts. Blood islands appear in the mesoderm on either side of the midline, forming vessels and blood cells. Each of these paired cardiac primordia unites to form an endothelial-lined tube (endocardial tube) surrounded by myoblasts, while the pericardial cavity

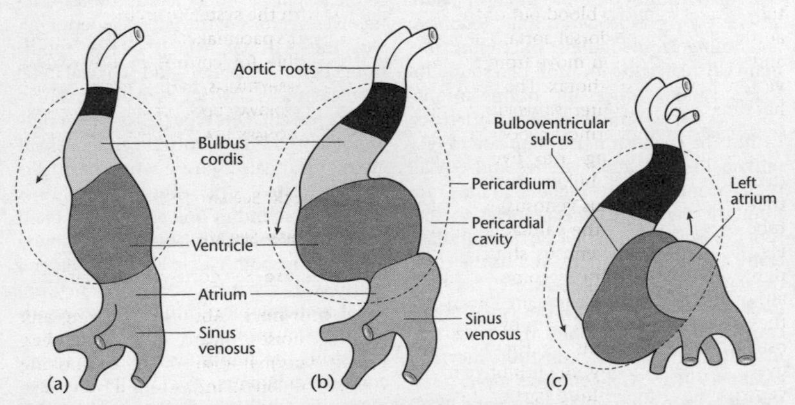

FIG 89. Diagrams in which arrows show the movements involved in the formation of the cardiac loop (see HEART). (a) 22 days; (b) 23 days; (c) 24 days. Redrawn from *Langman's Medical Embryology*, T.W. Sadler. © 2004 Lippincott, Williams & Wilkins, with permission of Wolters Kluwer.

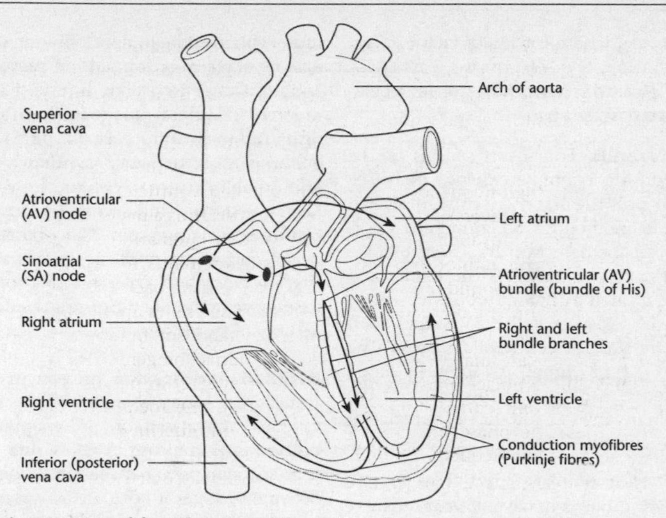

FIG 90. Anterior view of frontal section of the HEART, indicating major anatomical features and the direction of action potentials through the conduction system controlling the CARDIAC CYCLE. From *Principles of Anatomy and Physiology*, G.J. Tortora & S.R. Grabowski, © 2000. Reproduced with permission of John Wiley & Sons, Inc.

develops coelomically within the mesoderm. A pair of longitudinal dorsal aortae (see AORTIC ARCHES, Fig. 13) develops from other blood islands and the two endocardial tubes unite to form a continuous expanded tube (the primitive heart tube), except that their caudal ends remain separate. These receive venous drainage as the tube begins to pump blood out of the first aortic arch into the dorsal aorta. The heart and pericardial region move from the cervical region into the thorax. The heart wall has three layers: an outer EPICARDIUM, middle MYOCARDIUM and innermost ENDOCARDIUM.

After differentiating into five regions (truncus arteriosus, bulbus cordis, ventricle, atrium and sinus venosus) the heart tube curves to form the cardiac loop (see Fig. 89). The sinus venosus shifts to the right and the atrium becomes divided although the two atria remain joined by the ostium primum and later by a second ostium which remains as the foramen ovale. By the 30th day, the primitive right ventricle has formed from part of the bulbus cordis. The direction of cardiac looping is determined by an asymmetric axial signalling system that also affects the position of the lungs, liver, spleen and gut (see SITUS INVERSUS, CONGENITAL HEART DEFECTS).

By the end of the 2nd month, the ventricles have become divided by a septum and the truncus arteriosus divided by a spiral fold so that the right ventricle communicates with the pulmonary arch and the left with the systemic arch.

The heart's pacemaker circuit (see Fig. 90) is responsible for control of the CARDIAC CYCLE. See ARRHYTHMIAS, CARDIAC CYCLE, CARDIAC OUTPUT, CARDIOVASCULAR CENTRE, CORONARY ARTERIES, CORONARY HEART DISEASE, ELECTROCARDIOGRAM.

heart attack See MYOCARDIAL INFARCTION.

heart bypass See BYPASS SURGERY.

heart disease See CORONARY HEART DISEASE.

heart murmurs Abnormal rushing and gurgling noises, heard before, between and after normal HEART SOUNDS, or masking them, most often indicating a valve disorder.

heart rate The number of CARDIAC CYCLES

per minute, which is a major factor in CARDIAC OUTPUT. See ARRHYTHMIAS, BAINBRIDGE REFLEX, CARDIOVASCULAR CENTRE (for its regulation), EXERCISE, SA NODE.

heart sounds Four heart sounds can be heard during a complete CARDIAC CYCLE (see Fig. 36), only the first and second, indicated in the figure, being loud enough in a normal heart for detection by a stethoscope. The first (S1, or 'lubb') is the louder and longer, created by blood turbulence associated with closure of the AV valves soon after the start of ventricular systole. The second (S2, or 'dupp') is caused by blood turbulence associated with closure of the semilunar valves at the start of ventricular systole.

heat Heat increasing the temperature above 45°C burns tissue and heat-sensitive ion channels in NOCICEPTOR membranes open at this temperature. Non-nociceptive THERMORECEPTORS are involved in sensations of warmth when skin is heated between 37 and 45°C and recruit different pathways from nociceptive stimuli. See HEAT-SHOCK PROTEINS.

heat-shock proteins (HSPs) A group of proteins whose expression increases when a cell is exposed to raised temperatures or other STRESSORS such as infection, inflammation, exposure to toxins (alcohol, arsenic, ultraviolet light), starvation, hypoxia and emotional stress. The numerical suffix (e.g., '70' in Hsp70) refers to the relative molecular mass of the protein in kDa. Cells express HSPs when their temperature rises above a set point and shuts down synthesis of the other proteins they were synthesizing. HSPs include the numerous CHAPERONES that are present within cells in the absence of stress conditions, monitoring normal protein folding and refolding. Bacterial HSPs, and their degraded fragments on ANTIGEN-PRESENTING CELLS, are potently immunogenic, and HSPs are of interest to oncologists as potential 'cancer vaccines'. The small HSP αB-crystallin (see CRYSTALLINS) has neuroprotective functions. Hsp20 is involved in preventing platelet aggregation, in development of the smooth muscle phenotype in development, in cardiac myocyte function, and in preventing apoptosis after ischaemic injury. It is also involved in skeletal muscle function and its response to insulin. Hsp90 binds both endothelial NITRIC OXIDE synthase (NOS) and soluble GUANYLYL CYCLASE, as well as NUCLEAR RECEPTORS and several transcription factors, stabilizing them. Many more HSPs occur, and many multiple splice variants (see RNA PROCESSING, EXON SHUFFLING) provide tissue specificity. They are implicated in the activity of THERMORECEPTORS.

Hebbian modification Increase in effectiveness of a synapse caused by the simultaneous activation of presynaptic and postsynaptic neurons. Hebb's rule states that all synapses between two neurons become stronger if both are activated simultaneously. Synapses with this type of plasticity are said to be Hebbian, and can mediate associative LEARNING because they act as coincidence detectors that associate firing of the presynaptic and postsynaptic cell. See NMDA RECEPTORS, SYNAPTIC PLASTICITY.

***Hedgehog* gene (*Hh*)** The Hh family members, encoded by *Hh*, are morphogens, secreted lipoprotein GROWTH FACTORS that regulate tissue patterning, cell proliferation, and many other biological processes (see WNT SIGNALLING PATHWAYS for similarities with Wnt proteins). In a manner universally employed by metazoans (animals), graded distribution of Hh protein induces differential cell fate in a dose-dependent way. Hh-mediated signalling leads to the expression of genes encoding GLI-family factors, which are transcription regulators, and its defectiveness can cause human birth defects and cancer. Hh signalling (see CANCER CELLS, Fig. 35) culminates in the conversion of Gli transcription factors from repressors to activators, a conversion in which the transmembrane proteins Smoothened (Smo) and the Hh receptor Patched (Ptch) are crucial. In the absence of Hh signals, Ptch maintains Smo in an inactive state, and Gli transcription factors are converted to their repressor forms; but upon binding Hh, Ptch no longer represses

Smo and Gli transcription activators are produced, executing the Hh transcriptional programme. Several components of the Hh pathway, notably Smo and Gli, are present in the primary cilia (see CILIUM). There is evidence that both the 'on' and 'off' states of the mammalian cellular Hh pathway depend on the presence of such a primary cilium, in the absence of which the required intraflagellar transport proteins are lacking. Hedgehog-signalling components are apparently tethered or released from the internal scaffolding of cilia in response to signalling, moving in or out of cilia as part of the signalling process. See DISEASE, Fig. 64; SONIC HEDGEHOG.

hedonic experience See OPIATE RECEPTORS.

helicases See DNA HELICASES.

Helicobacter pylori An orally acquired Gram-negative pathogenic bacterium of the gastric mucosa (see STOMACH), to which it adheres and causes a local inflammatory response (see INFECTION). The chemokine CXCL8 (formerly IL-8) is released as a result, leading to an influx of leukocytes. Children acquire it as infants, predominantly within families and usually from their mothers. Although there is no overt disease in most people, up to 5% of those infected respond very differently. In some, excess gastrin is released which is associated with gastric and duodenal ulcers. Others experience a chronic inflammatory response leading to atrophy of the stomach, reduced acid production and increased risk of carcinoma of the STOMACH (an estimated 9% of all cancer deaths worldwide arise because of chronic infections by the bacterium) and unusual tumours of the MUCOSA-ASSOCIATED LYMPHOID TISSUE. Some of these tumours, despite being monoclonal proliferations with a transformed phenotype, depend for survival and growth on antigenic or other stimulation by *H. pylori*, but may regress if the infection is effectively eliminated by antibiotics. Its populations are genetically diverse, due to high rates of recombination. Seven populations and subpopulations exist, with distinct geographical distributions. These derive their gene pools from ancestral populations that arose in Africa, Central Asia, and East Asia, subsequent spread being attributable to human migratory fluxes (e.g., the prehistoric colonization of Polynesia and the Americas, and Bantu expansion within Africa). The bacterium can be used to trace patterns of human MIGRATION, its nucleotide polymorphisms being evidence of the ancestry of the strain and of the stomachs through which it has passed. See GUT FLORA.

helicotrema An opening connecting the scala tympani and scala vestibuli of the COCHLEA.

helium dilution method See RESIDUAL VOLUME.

helix-loop-helix See BASIC HELIX-LOOP-HELIX MOTIF.

helper T cells (T_H cells) See T CELLS.

hemidesmosomes Resembling 'half-desmosomes', these anchoring junctions connect the intermediate filaments of the basal surface of an epithelial cell to its underlying basal lamina, distributing shearing forces through the epithelium (see INTERCELLULAR JUNCTIONS, Fig. 103). Extracellular domains of INTEGRINS mediating the adhesion bind a laminin protein in the basal lamina, while an intracellular domain binds a plectin, which in turn binds to keratin intermediate filaments. Compare FOCAL ADHESIONS.

hemo- For entries with this prefix, relating in general to blood, see entries prefixed by HAEMO-.

Hensen's node See PRIMITIVE NODE.

heparin An anticoagulant, normally present in low concentrations in the blood. A highly negatively charged conjugated polysaccharide, its anticoagulant effects are mobilized on binding antithrombin III (see ANTICOAGULANTS), whose effectiveness is increased by two or three orders of magnitude. Other clotting factors removed by the heparin:antithrombin III complex include Factors XII, XI, IX and X. Heparin is also an inflammatory mediator, promotes smooth muscle contraction, and is toxic to parasites. Produced by several cell types, heparin is released in especially large amounts by basophilic MAST CELLS and by BASOPHILS. Dispersion of venous thromboses relies on such anticoagulants, and on fibrinolytic agents (see FIBRINOLYSIS).

hepatic portal vein The portal vein conducting blood, and non-lipid digestion products, from the small intestine to the LIVER (see Fig. 116). Its wall contains OSMORECEPTORS.

hepatitis Disease caused by at least six unrelated viruses that infect and inflame the liver. Hippocrates (2400 yr BP) mentioned it. *Hepatitis A* is caused by a picornavirus (HAV), a small RNA virus. Although the virus is not cytopathic, hepatic damage occurs after the feverish stage of illness and is probably immunologically mediated. A highly infectious disease, so spreading especially easily among children, most cases are subclinical, although moderate morbidity can result. Overall, its mortality is <0.1%, although ~1% of hospital admissions suffer more severe disease. In conditions of poor hygiene, children are most affected although immunity is life-long. In high-income countries, adults are more commonly infected. Epidemics occur explosively in common-source outbreaks and are commonly associated with food contaminated by an infected handler.

Hepatitis B usually involves acute hepatitis and is caused by a small enveloped virus (HBV), the genome consisting of partially double-stranded DNA whose four major genes each encode more than one polypeptide (the surface protein gene encoding three proteins). The virus seems not to be cytotoxic, and the mechanism of its binding to hepatocytes is uncertain. Viral DNA is integrated into the host genome, although the viral genes are mostly truncated or inactivated. As a result of this integration, liver carcinoma is an important complication (100-fold excess risk compared with the general population). In up to 10% of patients, persistent infection (probably due to failure of the initial immune response) causes chronic hepatitis progressing to cirrhosis after ~30 years. There are ~350 million chronic carriers of HBV worldwide. Where carrier numbers (carriage rates) are high, acute infection occurs mainly in infants and young children, mostly via intrapartum and horizontal transmission (some owing to skin disease and biting arthropods). In low-prevalence countries, most infections are sporadic and involve adult needlestick injuries, shared syringes or sexual contact. Treatment is with antiviral drugs and aims at eliminating the viral e-antigen, derived by cleavage of the major nucleocapsid antigen HBcAg and cause of e-antigenaemia, but emerging resistance to these drugs is increasingly encountered. Prevention includes hepatitis B vaccination, involving a course of three injections containing HBsAg, the major capsid protein, which probably provides life-long immunity. Use of condoms, and needle exchange schemes for intravenous drug users are also important measures. Passive protection can be provided by the HBIG immunoglobulin, which must be given within one week of exposure and should be supplemented by vaccine. Chimpanzee research was crucial in the development and trials of vaccines.

Hepatitis C virus (HCV) is an enveloped positive-sense RNA virus (Hepacivirus) with different amounts of the same three surface envelope proteins as HBV. Discovered in 1986, hepatitis C infected 170 million people worldwide in 2001. Many cases are subclinical and liver failure is extremely rare in immunocompetent individuals, although may occur in those who are immune-compromised or have agammaglobulinaemia. Transmission is by exposure to blood and body fluids and needlesharing by intravenous drug users has made transmission easier (37% of users are HCV-positive). HIV/HCV co-infection is common, affecting 7–15% of HIV cases in different communities. Sexual transmission occurs much less readily than with HBV. The genome has a single large open reading frame encoding the core and envelope proteins and exists in six major genotypes with varying potential for causing persistent viraemia and for response to antiviral treatment. INNATE IMMUNITY is important in limiting the initial infection. Prevention methods are similar to those for hepatitis B. No vaccine or specific immunoglobulin is currently available, but protease inhibitors are being investigated as therapeutic agents.

hepatocyte growth factor (HGF, scatter factor) A GROWTH FACTOR whose production increases during liver damage and stimulating hepatocytes to divide in culture. It is produced by mesenchyme during development, stimulating growth and branching of the ureteric buds, where its tyrosine kinase receptor is MET. Hereditary papillary renal cancer is due to inherited point mutant alleles in the *met* gene encoding the receptor for HGF, resulting in constitutive, ligand-independent firing of the receptor.

hepatocytes Liver cells, roughly hexagonal in section, forming the hepatic cords between the hepatic sinusoids and the bile canaliculi (see LIVER, and Fig. 116). They are metabolically extremely active, with a relatively large smooth, and extremely abundant rough, endoplasmic reticulum, and numerous Golgi apparatus and mitochondria.

hepcidin A liver-derived peptide and central regulator of body iron metabolism. It acts on small intestinal enterocytes, macrophages and other body cells to limit IRON entry into the plasma.

HER2 receptor An EGF receptor-related protein. See BREAST CANCER.

Herceptin Chimaeric anti-HER2/Neu monoclonal antibody with murine antigen-combining (variable) domains with a human constant domain. See BREAST CANCER.

herd immunity The phenomenon whereby disease can be excluded from a population in spite of the presence of some susceptibles because the proportion of individuals that is immune is sufficient to ensure that the effective reproductive number is less than one. See EPIDEMIOLOGY.

hereditary non-polyposis colorectal cancer (HNPCC) Occurring in about 15% of COLORECTAL CANCERS, and some other cancers, this autosomal dominant and highly penetrant disorder appears to be the result of a mutation which itself increases mutability by reducing the fidelity of replication of certain REPETITIVE DNA sequences (e.g., dinucleotide repeats such as CACACA ...) scattered through the genome, leading to early onset of the disease. Genes involved have been mapped to 2p15–p22 and 3p21.3, the first of which is a homologue of the bacterial gene *mutS* (see MISMATCH REPAIR) and has an ancient evolutionary lineage. Unlike FAMILIAL ADENOMATOUS POLYPOSIS, there is no preceding phase of polyposis.

hereditary predisposition See GENETIC PREDISPOSITION.

hERG (human ether-a-go-go-related gene) potassium channels Potassium (K^+) ION CHANNELS found in many tissues and cell types (e.g., neural, smooth muscle and tumour cells) but especially in cardiac muscle fibres where they are responsible for the slow delayed rectifier K^+ current in the cardiac action potential. Mutations in the *HERG* gene can cause cardiac ARRHYTHMIAS, with long QT (Brugada syndrome). See ELECTROCARDIOGRAM.

Hering–Brewer reflex (inflation reflex) A reflex limiting the degree to which inspiration proceeds. Stretch receptors in the walls of the bronchi and bronchioles initiate impulses when the lungs are inflated, the information being sent via the vagus to the medulla oblongata where they inhibit the RESPIRATORY CENTRE and result in expiration. Expiration results in reduced stretch receptor firing and less inhibition of the respiratory centre, allowing inspiration to begin again. In infants the reflex prevents overinflation of the lungs, while in adults it is important only when the tidal volume is large, as in EXERCISE.

heritability Members of the same human family tend not only to have genes in common, but also share similar environments. Familial similarity in a phenotypic character does not, therefore, entail its heritability: shared linguistic characteristics between family members are familial, but of themselves genetically uninterpretable (see FAMILIALITY).

In order to determine whether a human trait is heritable, we need to eliminate the possible contributions of covariance between genotype and environment (see VARIATION). The quantitative assessment of heritability of a character is given by the broad heritability, H^2, that part of the total phenotypic variance, ${s_p}^2$, that is due to genetic variance, ${s_g}^2$:

$$H^2 = \frac{s_g^2}{s_p^2} = \frac{s_g^2}{s_g^2 + s_e^2}$$

H^2 tells us what component of the population's variation is due to variation in genotype. It tells us nothing about those components of an *individual's* phenotype that can be ascribed to their genotype and environmental background – which would be an inappropriate way of asking questions about heritability. A major issue will be to tease apart heritability that results from the interaction of genes, maternal effects and culture (see NATURE–NURTURE DEBATE). Public information (PI) may provide an effective tool for such studies.

hermaphroditism See SEX DETERMINATION.

heroin (diamorphine, diacetylmorphine) More lipid-soluble than MORPHINE, and therefore with a more rapid onset of action when injected, the higher peak levels resulting in a greater sedative effect. It is used increasingly in small amounts epidurally to relieve pain. It has a high misuse potential because of the intense euphoric experiences when taken intravenously; but TOLERANCE develops quickly. Substitution of oral long-acting methadone reduces the harm of heroin ADDICTION.

herpes simplex virus (HSV), herpes zoster A cytolytic DNA virus with a linear double-stranded genome encoding 84 proteins. These colonists and invaders of mucosal surfaces can also infect skin if this is accessible to entry, causing degeneration and 'ballooning' of epidermal prickle cells, leading to vesicle formation. Infections of the mouth and upper body have been caused typically by type 1 viruses, while genital infections were usually caused by the genetically distinct type 2 viruses. But this distinction is nowadays less clear, with both sites being susceptible to both forms of virus. The vesicle roof in mouth and other mucous membrane infections is unstable, and is sloughed off, but in the skin keratinized cells form the roof. HSV-1 and HSV-2 have ~50% homology of their genomes, these serotypes differing in the repeats of sequences of both terminal ends of the genome. The virus spreads between cells and up sensory axons, where it establishes latent infection of the sensory ganglia. It does not integrate its genome into that of the host, latency being a dynamic equilibrium, with intermittent viral replication, movement down the neuron and production of a new lesion. Some HSV latency genes interfere with the host's immune recognition, and others prevent apoptosis of the infected cell. Aciclovir is a synthetic acyclic purine nucleoside antiviral drug active against both HSV serotypes. All herpes VIRUSES have the property of latency. Cells they infect are subject to programmed cell death when targeted by NK CELLS and cytotoxic T CELLS. For varicella zoster virus (VZV), see CHICKENPOX; for herpes zoster, see SHINGLES. See also CYTOMEGALOVIRUS, EPSTEIN–BARR VIRUS.

herpes viruses See HERPES SIMPLEX VIRUS.

HERVs Human endogenous retroviruses. The human genome has been colonized by thousands of copies of HERVs, related to RETROTRANSPOSONS, which are now permanently integrated in the genome. Many have been associated with teratocarcinoma and leukaemia, while others are involved in multiple sclerosis, schizophrenia and diabetes. Some copies have been co-opted by the genome and now perform cellular functions. The syncytin gene corresponds to part of an endogenous retrovirus and is essential to the normal development of the placenta. The HERV-K group (see TRANSPOSABLE ELEMENTS) is of particular interest.

heterochromatin Originally defined as that portion of the genome retaining deep staining with DNA-specific dyes during the return from interphase to metaphase in the cell cycle, heterochromatin actually has a constellation of properties although exceptions in every instance make definition difficult. Highly compacted CHROMATIN with regions of silenced DNA (see CHROMOSOME BANDING), it is gene-poor, contains repetitious sequences, replicates late in S-phase, does not engage in meiotic recombination, and contains hypoacetylated histones and high levels of DNA METHYLATION. It is classified into pericentric (constitutive) heterochromatin and non-centromeric (facultative) heterochromatin (see Appendix III for its distribution). Recent work implicates RNA INTERFERENCE mechanisms in targeting and maintaining heterochromatin. In the fission yeast *Saccharomyces pombe*, the densely packed structure of heterochromatin is dissolved at mitosis, followed by binding of RNA polymerase II at S-phase, coinciding with production of increasing amounts of small interfering RNAs (siRNAs) derived from these transcripts. These silence transcription of centromeric RNA by RNA interference.

The 33 constitutive heterochromatic regions of the human genome, devoid of genes and not targeted by the Human Genome Project, are highly repetitive and largely refractory to current cloning and sequencing methods (but see GENOME MAPPING). They all tend to be highly polymorphic in length in the human population. There are four types: the 24 centromeres comprising ALPHA SATELLITE REPEATS; three secondary constrictions adjacent to the centromeres on arms 1q, 9q and 16q; the five acrocentric chromosome arms 13p, 14p, 15p, 21p and 22p encoding the 5S, 18S and 28S ribosomal RNA genes; and the single large region on distal Yq. As a result of histone methyltransferase activity, these regions contain H3 tri-methyl K9 (tri-methylated on lysine 9) and mono-methyl K27, and H4 tri-methyl K20, and are irreversibly silenced throughout cell cycles (compare EUCHROMATIN). C-banding is a method of staining chromosomes specific for constitutive heterochromatin. Chromosomes are treated successively with dilute acid, alkali (barium hydroxide), warm saline and stained with Giemsa. Alkali hydrolyses the DNA, already depurinated by fixation in alcohol-acetic acid and by the dilute acid treatment. Saline incubation extracts the facultative heterochromatin more easily than the more compact constitutive DNA, which stains more strongly. Facultative heterochromatin can become transcriptionally active again, as in X CHROMOSOME INACTIVATION, where the chromatin is characterized by the presence of H3 tri-methyl K27 and di-methyl K9, and mono-methyl K20. See HISTONES.

heteroduplex DNA Double-stranded DNA in which there is some mismatch between the two strands in terms of base-pairing. Such DNA has been important in the detection of mutation (see DNA SEQUENCING) since it has an abnormal mobility on polyacrylamide gels, abnormal denaturing profile, and the mismatched bases are sensitive to cleavage by enzymes. The approach is simple and cheap, but of limited sensitivity (it does not reveal the position of change, and only sequences <200 bp can be used) and other approaches to GENOME MAPPING for mutations are preferred. See GENE CONVERSION.

heterogametic sex The sex (male in humans) which produces two chromosomally distinct forms of gamete.

heteroplasmy The occurrence of two or more distinct organellar (e.g., mitochondrial) haplotypes within an individual, there being multiple copies of these genomes within each individual.

heterozygous Of a gene locus within an individual's genome that is occupied by different alleles. See HETEROZYGOUS ADVANTAGE.

heterozygous advantage The situation

in which an individual who is heterozygous at a particular gene locus has a greater fitness than either of the two homozygotes. It is a form of BALANCING SELECTION. See MAJOR HISTOCOMPATIBILITY COMPLEX, MALARIA (for genetic resistance), SINGLE NUCLEOTIDE POLYMORPHISMS, TAY–SACHS DISORDER, TUBERCULOSIS.

***HEXA* gene** See TAY–SACHS DISORDER.

hexokinases See GLUCOKINASES.

HFE A MHC class IB gene product with no known immunological role but which is expressed on the surfaces of epithelial cells of the intestinal mucosa and functions in regulating uptake of dietary iron, possibly through interactions with the TRANSFERRIN receptor.

HGPPS (familial horizontal gaze palsy with progressive scoliosis) A congenital eye movement disorder associated with hypoplasia of abducens nuclei, uncrossed ascending dorsal column–medial lemniscal sensory pathways, and uncrossed descending corticospinal motor pathways. Affected individuals lack horizontal smooth pursuit, SACCADES, optokinetic nystagmus or vestibuloocular responses (see VESTIBULAR APPARATUS). These phenotypes are consistent with linkage to the 11q23-q25 locus, affected individuals being homozygous in this candidate region, homozygous mutations in the ROBO3 protein being identified. ROBO3 blocks repulsion by the Slit protein of axons crossing the midline of the spinal cord, by blocking the functions of ROBO1 and ROBO2. ROBO3 is normally downregulated after an axon crosses the midline, enabling ROBO1 and ROBO2 to have their repulsive effects during AXONAL GUIDANCE.

HHA, HHB See HAEMOGLOBIN.

HIF1 (hypoxia-inducible factor 1) A heterodimeric transcription factor combining HIF1α, a basic helix-loop-helix protein induced and stabilized by HYPOXIA, and HIF1β, the constitutively expressed aryl hydrocarbon receptor nuclear translator protein (ARNT). Over-expression of a natural antisense transcript (aHIF, of which there are two isoforms) has been shown to be associated with non-papillary renal carcinomas.

high altitude See ALTITUDE.

high density lipoproteins (HDLs) See LIPOPROTEINS.

high-energy phosphates (phosphagens) Groups of phosphorylated compounds transferring chemical energy required for cell work, depending upon their tendency to donate their phosphate group to water (to be hydrolysed) as indicated by their standard free energies of hydrolysis (the more negative the value, the greater the tendency to be hydrolysed). Phosphate-bond energy (not bond energy, the energy required to break a bond) indicates the difference between free energies of reactants and products, respectively, before and after hydrolysis of a phosphorylated compound. Some of the important high-energy phosphates are given in the entry for ADENOSINE TRIPHOSPHATE, Fig. 2.

high endothelial venules (HEVs) Venules entering peripheral lymphoid organs, having specialized endothelial cells that appear to be larger than any found elsewhere in the body. See LYMPH NODES.

high throughput genotyping Techniques enabling rapid testing for presence or absence of a known DNA sequence change (i.e., GENOME MAPPING), as during (i) diagnoses of diseases with limited allelic heterogeneity, (ii) diagnosis within a family (other family members being tested for the mutation), (iii) screening a panel of normal control samples for the presence of a change suspected of causing a disease phenotype, and especially (iv) testing a DNA sample for a pre-defined SINGLE NUCLEOTIDE POLYMORPHISM (SNP). The underlying method, which can be readily adapted to a DNA MICROARRAY format, is usually PCR-amplification of the relevant sequence by primer extension (minisequencing) followed by the Sanger DNA SEQUENCING method, in which the 3′ end of the sequencing primer is immediately upstream of the nucleotide to be genotyped. The reaction mix contains DNA polymerase and four differently labelled dideoxy nucleoside triphosphates (ddNTPs), the test DNA acting as a template for addition of a single ddNTP to the primer. The added nucleotide is then identified using any of a variety of methods, allowing the base at the position of interest to be identified. See PROTEOMICS, TRIPLET (TRINUCLEOTIDE) REPEATS.

highly repetitive DNA See REPETITIVE DNA.

hind limb (lower limb, leg) Developing from the embryonic limb buds (see LIMBS), the lower limb rotates ~90° medially during the 7th week of gestation, placing the extensor muscles on the anterior surface and the big toe in a medial position. See GAIT, LEVERAGE, SKELETAL SYSTEM. For superficial musculature, see Appendix X.

hindbrain The region of the BRAIN (see Fig. 29) developing from the RHOMBENCEPHALON, including the CEREBELLUM, PONS and MEDULLA OBLONGATA. See RETINOIC ACID.

hip bone (coxal bone) See PELVIC GIRDLE.

hippocampus, hippocampal complex, h. formation The hippocampus, a region of the medial temporal lobe of the brain, is part of the palaeocortex and has a three-layered structure rather than the six layers of the neocortex. They are bilaterally paired structures adjacent to the olfactory cortex, linked by hippocampal commissures and forming prominent bulges, the parahippocampal gyri, in the floors of the lateral ventricles. The hippocampus itself, as part of the hippocampal complex (a major component of the LIMBIC SYSTEM; see Fig. 91), plays important roles in the consolidation of new episodic memories into long-term memory, in learning, and in creating a cognitive spatial map of the physical environment (although recent work with rats indicates that their storage and recall of spatial memory can occur independently of the hippocampus; see ENTORHINAL CORTEX). It is often considered a single giant cortical module with rich recurrent excitatory connections (NEURAL NETWORKS). However, with the help of inhibitory neurons, these excitatory signals can also be rerouted within the entire hippocampal complex. Damage to the hippocampus (e.g., in anoxia/ischaemia, epilepsy and Alzheimer's disease) can cause both retrograde and anterograde amnesia (see MEMORY), but leaves procedural learning unaffected. The hippocampus contains numerous glucocorticoid receptors responding to cortisol, and its activation down-regulates the HPA system by suppressing CRH release (see STRESS RESPONSE). The hippocampal formation is often damaged in anoxia/ischaemia, epilepsy and Alzheimer's disease.

The four subregions of the hippocampal complex, the first two forming the *hippocampus* itself, are: (i) the *dentate gyrus* (strictly, the dentate nucleus within the dentate gyrus) populated mainly by *granule cells* and receiving its major input from the entorhinal cortex via the perforant pathway (see below); (ii) the *cornu ammonis*, with four distinct subfields, CA1–CA4. Each subfield comprises one main cell layer, the *pyramidal cell layer*, the apical and basal dendrites of whose cells leave the hippocampus separately, some of their axons coalescing to form the *fimbriae* within the *fornix* and projecting to the mammillary bodies and diencephalon. The CA1 region receives inputs from many brain regions and sensory systems. CA1 axons also project to the subiculum and entorhinal cortex, the recurrent collaterals of the CA1-to-CA3 subfields (see Fig. 94) being particularly important in acquisition of context-dependent episodic MEMORY (see below).

A major input to the hippocampus from the entorhinal cortex comes via the lateral and medial *perforant pathways*, axons of which synapse with granule cells of the dentate gyrus or with pyramidal cells in the CA3 subfield. The perforant pathway is glutamatergic and excitatory. Axons of the granule cells, termed mossy fibres, also synapse with CA3 pyramidal cells. CA3 pyramidal cell axons branch, giving rise to commissural fibres passing to the opposite hippocampus, efferents leaving the hippocampus via the fornix and terminating mainly in the hypothalamus (see PAPEZ CIRCUIT) or thalamus, and collaterals which turn back to form synapses with themselves and neighbouring CA3 cells (recurrent collaterals) or with pyramidal cells in CA1. A looped circuit is thus completed: entorhinal cortex–dentate gyrus–CA3–CA1–entorhinal cortex (the 'trisynaptic' pathway is underlined; see p.405). The recurrent excitatory circuitry of the CA3 subfield exemplifies a network using 'random' connections, with a more or less equal probability of connecting local, intermediate or distant neurons. Its recursive connection matrix is thought to function as a large 'autoassociator', allowing reconstruction of entire episodes from remembered fragments. (iii). The *subiculum* (subicular complex) is transitional cortex between the hippocampus proper and the six-layered neocortex. With the hippocampus, it forms the *hippocampal formation*. (iv). The *entorhinal cortex* is the hippocampal region providing the main cortical

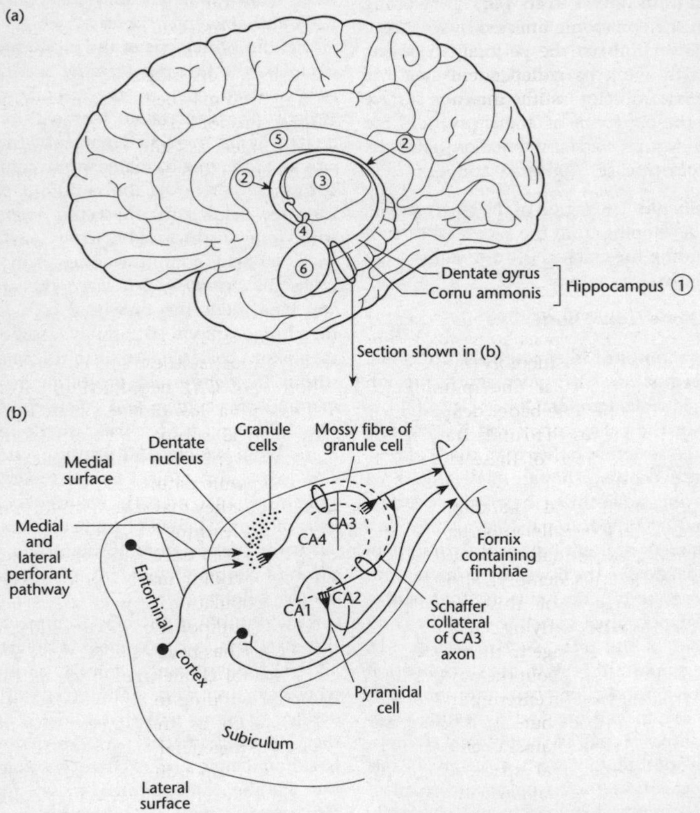

FIG 91. (a) The position of the hippocampal formation and related structures of the human brain. (1) Dentate gyrus and cornu ammonis; (2) fornix-fimbriae; (3) thalamus; (4) mammillary nucleus; (5) corpus callosum; (6) amygdaloid body. The main body of the fornix becomes progressively thinner as it bends posteriorly upwards towards the corpus callosum. (b) Frontal section through the hippocampus shown in (a), showing the cornu ammonis (CA) of the hippocampus, its subfields (CA1–CA4), and associated regions. © 2001 Redrawn from *Neuroscience* (2nd edition) by M.F. Bear, B.W. Connors and M.A. Paradiso, published by Lippincott, Williams & Wilkins. Reproduced with the kind permission of the publisher.

input to the hippocampus. It has undergone the greatest regional and laminar differentiation during primate evolution, is six-layered, and organized into columns rather similar to those of the cerebral cortex.

As E.R. Kandel and W.A. Spencer had realized by 1968, the unique functions of the hippocampus arise not so much from the intrinsic properties of pyramidal cells but from the pattern of functional interconnections of these cells and how those interconnections are affected by learning. In 1972, work on the perforant pathway revealed the activity-dependent synaptic plasticity, a long-lasting increase in the synaptic strength of transmission, now known as LONG-TERM POTENTIATION (LTP; see Fig. 120). In 1973, Lømo and Bliss discovered that brief high-frequency stimulation of excitatory synapses in the hippocampus could produce LTP. Subsequently, a use-dependent synaptic weakening termed LONG-TERM DEPRESSION (LTD) was also found to occur in the hippocampus. Damage to the hippocampus was recognized as linked to memory impairment and as the molecular underpinnings of LTP and LTD became better defined, neuroscientists attempted to understand memory storage in terms of these changes in synaptic strength.

Hippocampal and parahippocampal structures are critically important in the acquisition of information about temporal events (declarative memory), although the long-term storage of this information occurs within the overlying cerebral cortex, as indicated by damage to these areas. We have much to learn about the way information is transformed on entering and leaving the hippocampus, but it is generally accepted that it contains a cellular representation of personal space. Originally thought to have general navigational properties, it is now clear that the hippocampus can differentiate information about space into a multitude of context-specific representations that can be retrieved independently for subsequent upstream processing by algorithms integrating self-motion information into metric and directionally oriented representations valid in all contexts. When rats explore a maze, their EEG has a theta (θ) rhythm with a frequency of 4–10 Hz reflecting periodic firing of hippocampal neurons driven by the cholinergic septohippocampal pathway. During θ discharge, PLACE CELLS are firing in phase although all other hippocampal cells are silenced by increased activity of GABAergic inhibitory cells. This ensures that only cells involved in learning a particular environment are active. Coupled with the fact that one of the most effective protocols for *in vitro* LTP production closely resembles a θ discharge, this suggests that the θ rhythm may be the brain's endogenous stimulus for learning.

With mice as subjects, a large-scale neuron-ensemble recording technique, in conjunction with powerful decoding algorithms, has revealed (2006) that various robust experiences (e.g., free-fall within a plunging elevator; a sudden flow of air to the back; or an earthquake-like shake of the cage) can evoke different changes in the firing of certain CA1 cells (see NEURAL CODING). Various startle-triggered ensemble responses of CA1 neurons form distinct patterns in a low-dimensional-encoding mathematical subspace, whose spontaneous post-event reactivations indicate the highly dynamic nature of the process and suggest that they may be crucial for immediate post-learning fixation of newly formed memory traces (episodic memory). Functional coding units, or ensembles, in the CA1 population, termed *neural cliques*, have similar response properties and selectivity. What has been dubbed the 'general startle neural clique' comprises cells capable of responding to all types of startling stimuli, such as the three above-mentioned. 'Sub-general startle cliques' are cell groups responding to a combination of two such events, but not all. Additionally, neuron groups exist which exhibit high specificity towards one type of startling event, e.g., elevator-drop (the drop-specific neural clique), sudden air-blows (the air-specific neural clique) or earthquakes (earthquake-specific neural clique). Individual neurons of a given clique

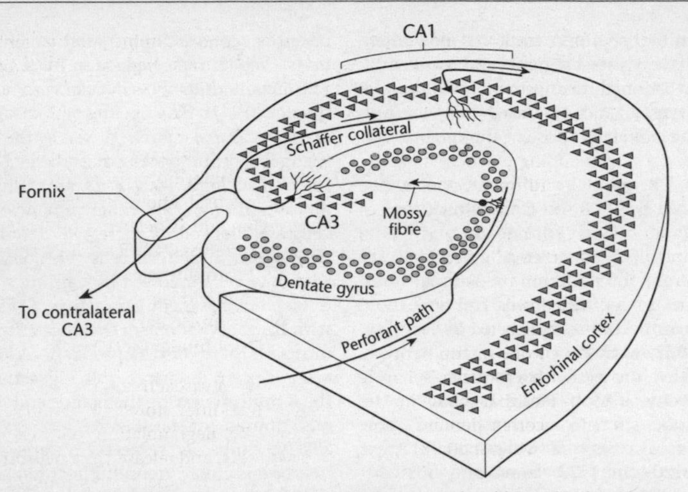

FIG 92. Diagrammatic representation of the conjectured categorical and hierarchical organization of a memory-encoding neural clique assembly into a feature-encoding pyramid. See HIPPOCAMPUS. From L. Lin, R. Osnan and J.Z. Tsien © 2006 *Trends in Neurosciences*, with permission from Elsevier Ltd.

exhibit 'collective co-spiking' temporal dynamics (see DENDRITES). Responses of the individual member neurons vary from trial to trial, but the consistency and specificity of the collective co-spiking of the clique responses is evident. This robust co-spiking of member neurons is preserved over time-scales of 20–30 ms, a time window matching the membrane time constant of CA1 pyramidal cells. Clique neurons can be further grouped into four major subtypes on the basis of their spike characteristics: (i) transient increase; (ii) prolonged increase; (iii) transient decrease; (iv) prolonged decrease. Each clique assembly is organized in a categorical hierarchy which can be modelled as a 'feature-encoding pyramid', whose base is the most general clique and whose apex is the most specific clique. Each pyramid can then be thought of as one face of a 'polyhedron' representing all events of a given category, such as 'all startling events' (see Fig. 92).

The structure of the feature-encoding pyramid affords, through a combinatorial and self-organizing process, a network mechanism for creating possibly unlimited numbers of unique internal patterns that encode the behavioural episodes encountered during a mouse or human life. Such parallel-binding networks afford possibilities not only for vast memory storage but also for cognitive functions of the brain, such as ABSTRACTION and generalization. This is congruent with anatomical evidence that virtually all sensory input to the hippocampus derives from higher-order, multimodal, cortical areas, and also with its known high level of sub-regional DIVERGENCE and CONVERGENCE. See FACIAL RECOGNITION.

There is a prominent ELECTROENCEPHALOGRAM wave pattern, 'theta activity' that seems to coordinate information as it enters each of the links of the hippocampal circuit. Synaptic strengths at each synapse in the trisynaptic pathway (underlined above) can be modulated in a manner partly dependent on the NMDA GLUTAMATE receptor. It is believed that altering the synaptic strength of individual synapses may enable neurons to integrate into ensembles that, when activated, might represent salient features of the environment. Place cells, only active when the animal is at a particular position in space, could identify its current spatial location and, with other ensemble neurons, track its movement. Studies on hippocampal NEURAL CODING indicate that small changes in environmental or contextual features bring about changes in the correlated clique (ensemble) activities ('rate remapping') whereas larger changes lead to recruitment of different cliques, especially in CA3 ('global remapping'), the former but not the latter being disrupted in mice genetically engineered to lack NMDA receptors in the dentate gyrus. Such mice were also slower than controls in learning to discriminate between two similar contexts in which one of the environments was paired with a footshock. This is consistent with the idea that synaptic plasticity at entorhinal cortex–dentate gyrus synapses is important when spatial context is used to identify the appropriate memory and making the appropriate behavioural response (whether or not to become immobile). If this is relevant to the ability to recall 'what' happened 'where' (spatial memory), then it could be of significance to episodic memory in humans. Ability to learn to navigate a water maze, however, was unaffected by the NMDA receptor-mutant mice. They could learn to discriminate between which arms were rewarded and which were not (i.e., they acquired 'spatial reference memory'). However, the mutant mice were unable to recall which arms of the water maze they had already visited, making more errors than controls in their spatial 'working' memory. A role for the hippocampus in processing the temporal context associated with an event has been described in rodents, and it appears that NMDA receptor-dependent plasticity of dentate gyrus synapses, and possibly the associated rate remapping, contribute to temporal information processing. Further work will unravel the roles of different forms of synaptic plasticity and of different subregions of the hippocampus (dorsal-posterior and ventral-anterior) in behavioural regulation. In this context, it is of interest that the dorsocaudal region of the medial entorhinal cortex (dMEC) represents location accurately prior to the hippocampus and, in rats at least, has an extraordinary microstructure; and that hippocampal dendritic spine development has been shown to be regulated by a brain-specific microRNA. See ASTROCYTES, ALZHEIMER'S DISEASE.

Hirschsprung disease A disease involving failure of migration of ganglionic cells to the submucosal and myenteric plexuses of the colon and associated in 50% of familial cases with loss-of-function mutations in the RET proto-oncogene on hsa10q11.2. Affected individuals suffer from colon obstruction and enlargement on account of the lack of hindgut motility.

histamine A vasoactive amine and derivative of histidine, stored in granules and released by degranulation of MAST CELLS and BASOPHILS, and released as a neurotransmitter by histaminergic neurons of the anterior HYPOTHALAMUS (loss of histaminergic neurons leads to a loss of vigilance). Histamine exerts its effects via H_1–H_4 METABOTROPIC RECEPTORS (G-PROTEIN-COUPLED RECEPTORS). H_1 is located on smooth muscle, endothelium and in the central nervous system, causing vasodilation, bronchoconstriction and PAIN and itching at insect stings, producing some of the symptoms of immediate HYPERSENSITIVITY reactions. H_1 antagonists (antihistamines) produce sedation and have antiemetic functions. H_2 is located on PARIETAL CELLS, regulating gastric secretion. H_3 is involved in presynaptic inhibition in the CNS, decreasing release of acetylcholine, noradrenaline and serotonin (see SLEEP–WAKE CYCLE). Several isoforms of H_3 receptor are known. H_4 is located in the thymus, small intestine, spleen, on basophils and in bone marrow. The central histaminergic neuron system inhibits epileptic seizures, probably via H_1 and H_3 receptors (see EPILEPSY). H_3 antagonists increase wakefulness. Histamine is released from mast cells in the genitals as part of orgasm. Low levels in the blood are associated with symptoms of SCHIZOPHRENIA. Supplements of FOLATE and niacin (see VITAMIN B COMPLEX) taken together can increase blood levels of histamine. See ALLERGIES.

histocompatibility Term denoting whether tissue grafts and organ transplants are tolerated or rejected. Rejection results from differences between donor and recipient in their major and minor histocompatibility antigens. See H-Y ANTIGENS, MAJOR HISTOCOMPATIBILITY COMPLEX.

histology The study of tissues; in particular their microscopic structure and morphology.

histone See HISTONES.

histone acetylases See HISTONE ACETYLTRANSFERASES.

histone acetyltransferases (HATs, histone acetylases) Enzyme catalysing addition of an acetyl group to specific lysine residues in HISTONES, leading to CHROMATIN REMODELLING and stimulating GENE EXPRESSION. Several HAT-containing proteins, such as CREB-binding protein (CBP), p300 and SRC-1, are bound by PGC-1 coactivators (see PGC-1 GENE).

histone chaperone Protein escorting HISTONES to DNA for deposition. See CHAPERONES.

histone code A term indicating that particular stretches of chromatin, through different covalent modifications of the histone tails, gives a particular 'meaning' to the proteins in its immediate environment. Different types of marking might signal that the stretch of chromatin is newly replicated, or that gene expression should not take place along it. Because histone tails are extended, and probably accessible even when chromatin is condensed, they provide a convenient format for conveying messages to effector proteins nearby. See HISTONES.

histone deacetylase inhibitors (HDACIs) Small molecular inhibitors of specific histone deacetylases. They have been targeted in possible therapies for a number of cancers and several HDACI cancer therapies are undergoing clinical trials. See HISTONES.

histone deacetylases (HDACs) DEACETYLASE enzymes, some of them SIRTUINS, removing acetyl groups from HISTONES. Some cancers have been linked to malfunctions in the HDACs, these enzymes proliferating and disproportionately down-regulating the genes controlling the normal process of apoptosis. If apoptosis is inhibited in cancerous cells, they will multiply and form a tumour. Histone deacetylation and DNA METHYLATION have been shown to act synergistically to promote onset of some tumours. Increased histone acetylation by inhibition of histone deacetylases has been shown to induce recovery of LEARNING and MEMORY in mouse models (see NEURAL DEGENERATION).

histone demethylases Enzymes mediating reversal of histone methylation by HISTONE METHYLASES, recent discovery of numerous such enzymes suggesting a central role for them in regulating histone methylation dynamics (see HISTONES). The human histone lysine demethylase LSD1 (aka AOF2) has been shown to use some non-histone substrates (see *p53* GENE). The human JmjC-domain-containing proteins UTX and JMJD3 demethylate trimethylated Lys 27 on histone H3, and ectopic expression of JMJD3 results in strong decrease in H3K27me3 levels and delocalization of POLYCOMB GROUP proteins *in vivo*. Such decrease is associated with HOX GENES during differentiation.

histone lysine methyltransferase Enzyme catalysing transfer of methyl groups to the ε-amino residue of lysine residues in HISTONES. See HISTONE METHYLASES.

histone methylases Enzymes (e.g., histone methyltransferases) adding methyl groups to HISTONES, usually to their amino- or carboxy-terminal tails. Lysine is a key substrate residue, often occurring several times (mono-, di- or tri-methylation) on one residue. As with histone ubiquitylation, histone methylation has variable effects, depending upon the precise residue and context. Trimethylation of lysine 4 in H3 (H3K4me3) occurs at the 5´ ends of open reading frames as genes become activated; but H3K9me3 occurs in compact pericentromeric heterochromatin, transcriptionally inert. Histone H3K27 trimethylation has been linked to polycomb-group-protein-mediated suppression of *Hox* genes and animal body patterning, X-chromosome inactivation, and possibly to maintenance of embryonic stem cell identity, while imbalance of its methylation through over-expression of the methylase EZH2 is implicated in metastatic prostate cancer and aggressive breast cancers. Histone lysine methylation has recently been shown to be reversible (see HISTONE DEMETHYLATION).

histones Very stable basic DNA-BINDING PROTEINS (see this entry for ChIP-chip technology) of major importance in the packaging of eukaryotic chromatin into the NUCLEOSOMES that form the bulk of the chromosome, and in the establishment of chromatin domains (euchromatin and heterochromatin) and other effects crucial to the control of GENE EXPRESSION (see also GENE SILENCING). Most histone synthesis occurs during S-PHASE of the cell cycle, histone mRNA being very unstable and degenerated within minutes of synthesis.

Each nucleosome contains an octomeric core of the four major (canonical) histones: H2A, H2B, H3 and H4. A fifth type of histone, H1, occurs in eight isoforms and links neighbouring nucleosomes. H1, H2A and H2B are lysine-rich while H3 and H4 are arginine-rich. All four core histones share a structural motif, the *histone fold*, and during nucleosome assembly these bind to each other to form very stable H3–H4 and less stable H2A–H2B dimers, the former combining to form tetramers. Each tetramer then combines with two H2A–H2B dimers to form the compact octamer core around which the DNA is wound. In addition to the histone fold, each of the core histones has a long N-terminal amino acid 'tail' extending out from the DNA-histone core. These tails are available for several post-translational covalent (and sometimes non-covalent) modifications (PTMs), which control a great deal of chromatin structure (see HISTONE CODE). Histone H1 consists of a globular core with two extended tails. It is larger than the core histones and much less well conserved in sequence, occurring in several distinct forms. A single H1 molecule binds to each nucleosome, making contact with both the DNA and the protein, and changes the direction of the linker DNA as it leaves the nucleosome. It is required to pull the nucleosomes together into a 30-nm fibre (see CHROMATIN, Fig. 51), although the mechanism is uncertain, and the histone tails may also be involved in this.

Members of the histone GENE FAMILY (61 human genes) are found in modest-sized clusters at a few locations, including two large clusters on chromosome 6, but also occur as dispersed orphan genes. Apart from these orphan genes whose expression continues throughout the CELL CYCLE, histone gene expressions are tightly regulated, each histone being deposited onto DNA in a process that is coupled to DNA REPLICATION in S-phase. *Chromatin assembly factors* are proteins that associate with replication forks and package the newly synthesized DNA on emergence from the replication machinery. The acetylated H3–H4 tetramer is deposited first (deacetylation following incorporation into chromatin), followed by the H2A–H2B dimers, processes requiring histone chaperones. These function alongside CHROMATIN REMODELLING factors to mediate the accurate positioning of nucleosomes on the DNA template. See SYSTEMIC LUPUS ERYTHEMATOSUS.

ATP-dependent protein machines termed *chromatin remodelling complexes* can change nucleosome structure temporarily so that the DNA becomes less tightly bound to the core particle, possibly resulting from movement of the H2A–H2B dimers in the core. The freeing-up of DNA in this way allows access by other proteins involved in gene expression, DNA replication and repair, and different remodelling complexes are probably involved in each of these processes.

Covalent modifications of the histone tails include acetylation and methylation of lysine (K) residues, methylation of arginine (R) residues and phosphorylation of serine (S) residues. The ATR kinase is involved in at least some of the latter. Histone acetylation tends to destabilize chromatin structure, possibly by making it more difficult to neutralize the negative charges on DNA as chromatin is compacted, while deacetylated histones are associated with transcriptionally inactive chromatin (see HISTONE ACETYLTRANSFERASES and other histone-modifying enzyme classes listed above). The two bromodomains of the general transcription factor TFIID are spaced apart at the correct distance to bind acetylated lysines that are six or seven residues apart on the H4 N-terminal tail, associated with transcriptionally active chromatin. POLYCOMB GROUP (PcG) repressive complexes 2 and 3 (PRC2/3), involved in transcriptional silencing, contain a HISTONE LYSINE METHYLTRANSFERASE (Ezh2) targeting different lysine residues on H1 or H3. Ezh2 levels are increasingly elevated in prostate cancer progression, as are other PRC2/3 components in CANCER CELLS. A surprising discovery is that a single mark recruits numerous proteins whose regulatory functions are not only activating, but also repressing. Possibly an intricate 'dance' of associations takes place over time, positive-acting complexes involved during TRANSCRIPTION initiation and elongation, negative-acting complexes binding during attenuation. In pluripotent cell types (e.g., EMBRYONIC STEM CELLS), the PRC system marks the histones associated with genes required during development and differentiation by inducing methylation of the lysine at position 27 of histone H3 (H3K27), a mark which usually occurs outside the context of DNA methylation (see RNA). Epigenetic silencing by PRCs might be mitotically heritable.

All histone post-translational modifications (PTMs) are removable. Acetylation and phosphorylation are 'dynamic marks' that are reversible by nuclear deacetylases and phosphatases. Two classes of lysine demethylase have now been identified: the LSD1/BHC110 class (removing H3K4me1 and me2 in euchromatin), and the jumonji class (removing H3K4me2 and me3, H3K9me2 and me3, and H3K36me2 and me3), ending the debate about the reversibility of histone lysine methylation, and whether it is the only 'true' epigenetic histone PTM. LSD1 (aka AOF2) catalyses removal of its target methyl groups using a flavin-adenine-dinucleotide-dependent oxidation, whereas JmjC-domain-containing proteins remove them from histones through a hydroxylation reaction requiring α-ketoglutarate and $Fe_{(II)}$ as cofactors. Changes in substrate specificities of factors that irreversibly modify the histone components of chromatin are likely to have profound effects on gene expression.

In addition to the canonical histones of the core, four different isoforms of H3, and four of H2A, have been discovered. The canonical H3.1 and H3.2 histones are expressed during S-phase and deposited during replication by a DNA replication-dependent nucleosome assembly pathway. Another H3 variant (CENP-A) is only present at the centromere. The DNA replication-independent remodelling complex FACT ('facilitates chromatin transcription') can separate the H2A and H2B dimer and replace H2A with the variant H2AZ. It can also replace H3 with H3.3. The H2AZ and H3.3 nucleosomes seem to contribute to the establishment of euchromatin, and are more transcriptionally competent than the original (but see SPERMATOGENESIS). On the other hand, the chromatin-assembly factor-1 (CAF1) has been shown to interact with the heterochromatin-binding protein HP1, which itself interacts with the H3-K9 methyltransferase, resulting in H3 methylation and attachment of HP1 to the methylated residue. The stimulation of tri-methylation of H4-K20 by further methylating enzymes, which depends on the presence of HP1, would then follow, leading to the formation and propagation of pericentric heterochromatin.

HIV (human immunodeficiency virus) The human immunodeficiency viruses HIV-1 and HIV-2 are enveloped retroviruses belonging to the lentivirus subfamily Retroviridae. They cause lifetime infection of many tissues, particularly of $CD4^+$ T CELLS (CD4 helper lymphocytes), dendritic cells and macrophages (whose mannose receptor binds HIV). Around 14,000 people are infected per day and, if untreated, develop profound helper T-cell deficiency and consequent immunodeficiency (*acquired immune deficiency syndrome*, AIDS) over a period of several years, dying of uncontrolled opportunistic or HIV disease. Possession of ≤200 $CD4^+$ T cells per mL of blood (normal persons have 600-1200) is considered to constitute AIDS. Infection is capable of disrupting whole communities. Carried in infected CD4 T cells, dendritic cells and macrophages, and as the free virus in blood, semen, vaginal fluid or mother's milk, it is most commonly spread by sexual intercourse (across mucosal surfaces in which dendritic cells reside), and by contaminated needles or blood products (until most countries started screening blood donations). Perinatal infection also occurs and, at present, NEWBORN SCREENING is anonymous and used solely for epidemiological purposes. HIV viraemia can be controlled, and the onset of AIDS prevented, by multidrug antiviral treatment; but this requires a considerable investment in drugs and healthcare which poorer states cannot afford.

HIV-1, of which the simian immunodeficiency virus (SIV) of chimpanzees is the ancestor, has a world-wide distribution, producing a more severe and rapidly progressive disease than the less aggressive and mostly West African HIV-2. HIV-1, the much more common form of the virus, appears to have crossed the species barrier from chimpanzees to humans ~1930, infecting ~1% of the world's adult population by 2007. It is *not* now thought that the virus was introduced into humans within polio vaccines. HIV-2, the other major form of the virus, is closely related to a virus infecting sooty mangabey monkeys, from which it apparently jumped host to humans when these were kept as pets or hunted for food.

The spread of HIV-1, and the absence of an effective vaccine, are in large part due to its ability to evade antibody-mediated neutralization, largely because of the remarkable diversity, glycosylation and conformational flexibility of its envelope. The initial infection with HIV usually occurs after the transfer of body fluids from an infected person to an uninfected one.

The different variants of HIV infect different cell types as a result of the expression of specific receptors for the virus by those cells. The virus enters by means of its envelope spikes, complexes of three copies of each of two non-covalently associated viral glycoproteins, gp120 and gp41 (see Fig. 93 and VIRUS, Fig. 177b). The gp120 portion binds to the cell-surface CD4 molecule, altering its conformation in the process to reveal a site for attachment to a seven-span chemokine receptor. Variants of HIV associated with primary infections use the CC chemokine receptor CCR5 as co-receptor (hence 'R5 viruses'), requiring only low levels of CD4 on the cells they infect. R5 viruses are those preferentially transmitted by sexual contact, and infect dendritic cells, macrophages and T cells *in vivo*. 'Lymphocyte-tropic' variants of HIV infect only $CD4^+$ T cells *in vivo*, use CXCR4 as co-receptor ('X4 viruses') and require high levels of CD4 on their host cells. Binding to the co-receptor mediates fusion of the viral, and host cell, membranes and entry of the viral core particle into the host cell cytosol. The core particle must disassemble to release the RNA genome and allow reverse transcription into DNA. Unidentified host factors are believed to mediate this uncoating of the core particle. The discovery that the protein TRIM5α (tripartite motif-containing protein 5-α) causes rapid uncoating, premature disassembly and loss of infectivity, encourages hope that this may in time offer a prophylactic approach that bypasses the need for a vaccine.

In common with other RNA viruses, HIV has an inaccurately replicated genome (~1 mutation per replication), host RNA POLYMERASES lacking a proof-reading capability. This probably increases its antigenic variability, helping it to evade effective immune responses. Two copies of its RNA genome enter the host cell along with viral reverse transcriptase and integrase. The HIV-1 genome is flanked by long terminal repeats (LTRs) involved in viral integration. Using the RNA as a template, reverse transcriptase transcribes the viral RNA into a double-stranded complementary DNA (cDNA) copy which then migrates to the nucleus along with the integrase and Vpr protein and is enzymatically integrated (as a provirus) into the host cell's genome. In activated CD4 T cells, viral replication is subsequently initiated by transcription of the provirus; but HIV can establish a *latent infection* lasting from 18 months to 15 years or more, during which the provirus remains quiescent. Latency seems to occur in memory CD4 T cells and dormant macrophages, which appear to act as an important reservoir of infection for host cells. HIV also becomes trapped in the form of immune complexes on the surfaces of follicular dendritic cells and although not infected these cells may also act as a store. Production of infective virus from the integrated proviral DNA is initiated by the cellular transcription factor NFκB (see entry) present in all activated T cells. $NF_{\kappa}B$ binds promoters in cellular and viral LTRs, initiating transcription of viral RNA by host RNA polymerase.

HIV RNA transcripts are multiply and alternatively spliced (see ALTERNATIVE SPLICING), producing mRNAs for viral proteins. The genome can be read in three frames, enabling several proteins to be encoded by a small genome. Several of the genes overlap in different frames. Not all the gene product functions are understood. Thus, the matrix protein monomers are encoded by *gag*; the reverse transcriptase, protease and integrase enzymes by *pol*; the transmembrane glycoproteins by *env*; the reverse transcriptase, protease and integrase enzymes by *pol*; the viral infectivity factor by *vif*; the viral protein R, enabling transport of viral DNA to the nucleus and initiating host cell cycle arrest, by *vpr*; the viral protein U, promoting intracellular degradation of CD4 and release of virus from the cell membrane, by *vpu*; the negative-regulation factor, which augments viral replication, decreases CD4, MHC class I and II expression, by *nef*; the transactivator positive transcription regulator required for viral replication by *tat*; the regulator of viral expression, allowing

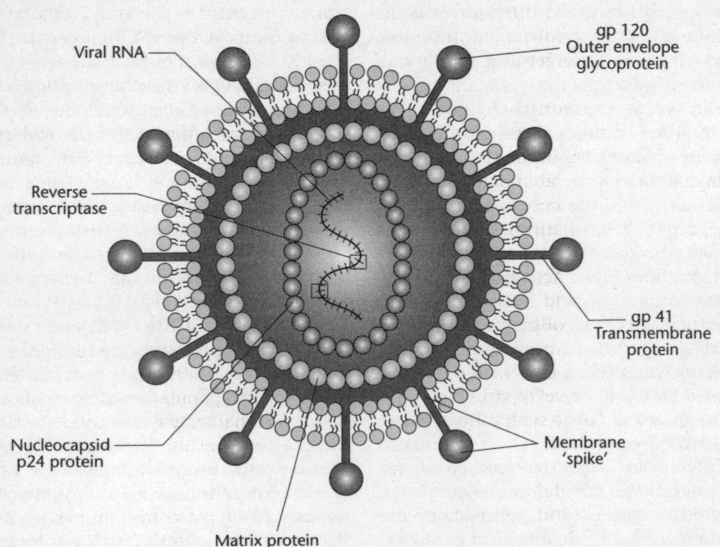

FIG 93. Diagram of the structure of the human immunodeficiency virus, HIV. Each membrane spike comprises three molecules of both gp120 and gp41. © Blackwell Publishing. Redrawn from *Infection (Microbiology and Management)*, (3rd ed), by B. Bannister, S. Gillespie and J. Jones, with permission of the publisher.

export of unspliced and partly spliced transcripts from the nucleus, by *rev*. CD4 adheres to freshly made gp160 proteins, preventing them from assembling with other viral proteins into new viruses. HIV has evolved the protein product of *vpu*, Vpu. This binds both CD4 and a complex containing the E3 ubiquitin enzyme, causing CD4 to be ubiquitinated and destroyed by the proteasome.

The exposed proteins of HIV are heavily glycosylated by enzymes of the infected host cell so that they are recognized by the immune system as 'self-antigens'. Genetic, immunological and structural studies of the HIV-1 envelope glycoproteins have demonstrated extraordinary diversity, manifest in a variety of immunodominant loops and multiple overlapping mechanisms of humoral evasion. The latter include self-masquerading glycan and conformational masking, evolutionarily improved barriers of diversity and evasion that have confounded traditional vaccine development. But recent studies indicate that these glycans, and the human glycosylation pathway involved in their biosynthesis, can be targets for drugs and vaccines.

Primary infection, of macrophages and T-lymphocytes, causes an influenza-like illness in up to 80% of individuals, and although there is an abundance of virus in peripheral blood (acute viraemia) and a large increase in circulating CD4 T cells, the diagnosis is usually missed although suspicion may be aroused. Activated CD8 T cells kill infected cells and this is probably important in controlling virus levels, which decline after a peak, and antibody is produced. The best current indicator of future disease is the level of virus persisting in the plasma after the symptoms of acute viraemia have subsided.

Although most HIV-infected people go on to develop acquired immune deficiency syndrome (AIDS) and ultimately die through opportunistic infection or cancer, some do not. Genetic variation in either the virus (e.g., natural attenuation) or the host (notably in HLA type) can alter the rate of disease progression, and a period of apparent quiescence of the disease (clinical latency) of between 2 and 15 years is usual, during which there is persistent replication of the virus and a slow decline in the number of $CD4^+$ T cells until these are too few to prevent illness. It is increasingly clear that virus levels are but one part of the reason for T-cell loss. Many suspect that immune activation is important: HIV stimulates the immune system to such an extent that it eventually triggers T-cell death. Those HIV-infected individuals who do not progress to AIDS (non-progressors) have low concentrations of circulating virus and are under intensive study in an effort to understand how the virus is being controlled. Individuals who are seronegative despite high exposure to HIV have specific cytotoxic lymphocytes and T_H1 lymphocytes indicating this exposure and are the focus of attention of those involved in vaccine development.

Strong evidence that CCR5 is the major macrophage and T-lymphocyte co-receptor used by HIV in establishing primary infection *in vivo* offers the possibility that antagonists of this receptor might block this primary infection. Low molecular weight inhibitors of CCR5 have been shown to block infection *in vitro*, and may be the model for development of prophylactic drugs that could be taken by mouth. There are significant interindividual and interpopulation differences in the copy number of a SEGMENTAL DUPLICATION on human chromosome 17q (Hsa17q) encompassing the gene encoding CC chemokine ligand 3-like 1 (CCL3L1, aka MIP-1αP), a ligand for CCR5. CCL3L1 is a potent dominant HIV-1-suppressive chemokine, and possession of a COPY NUMBER of *CCL3L1* lower than the population average is associated with markedly enhanced HIV/AIDS susceptibility. This susceptibility is enhanced in those who also possess disease-accelerating *CCR5* genotypes. Increasing *CCL3L1* copy number is positively associated with CCL3/CCL3L1 secretion and negatively associated with the proportion of $CD4^+$ T cells expressing CCR5. African populations possess a significantly greater copy number of *CCL3L1* genes (median value = 6) than do non-Africans (median value = 3). The duplicated region has an ancestral correlate in chimpanzee (median value = 9). Work suggests that it is not the absolute copy number per se, but the gene dose relative to the average copy number in each population that confers HIV/AIDS susceptibility.

A few individuals carry mutated chemokine receptors and are resistant to HIV infection. Experimental work indicates that homozygosity for a 32-base-pair deletion from the coding region of the gene for the CCR5 cell-surface chemokine receptor, leading to frameshift truncation of the translated protein, is responsible for this resistance. Some advantage also seems to be conferred in the heterozygous condition. The frequency of this allele in Caucasoid populations is 0.09 (i.e., ~10% are heterozygous carriers of the allele and ~1% are homozygous). It has not so far been detected in Japanese or in black Africans from Western or Central Africa; but variation in the CCR5 promoter region has been found in Caucasian and African Americans, resulting in different rates of disease progression.

The aims of highly active antiretroviral therapy (HAART) are to minimize viral replication, minimize emergence of resistance to antiviral drugs, reduce infectiousness, halt progress of immunological damage and to enable reconstitution of immune function. Antiretroviral drugs include nucleoside-related reverse transcriptase inhibitors (NRTIs) such as AZT and ZDV (zidovudine), abacavir (ABC) and lamivudine (3TC), while tenofovir disoproxil is a nucleotide reverse transcriptase inhibitor. Some new antiretrovirals in production (non-nucleoside reverse transcriptase inhibitors, NNRTIs) target reverse transcriptase by alternative mechanisms. Viral protease inhibitors (PIs) include atazanavir, amprenavir, saquanivir (best tolerated), ritonavir, idinavir (most potent) and nelfinavir, all of which interact with cytochrome CYP450, which causes adverse reactions. Their combined use is often problematic, and rifampicin (used in treatment of tuberculosis) reduces their blood levels. Much work has been done in attempts to develop genetically engineered vaccines to HIV and trial protocols for genetic manipulation of host cells to make them resistant to HIV usually employ the principle used to treat X-linked SCID: haematopoietic stem cells are transfected with a gene that will hopefully inhibit HIV replication, usually introduced in a retroviral vector, before adoptive transfer back into the patient. For HIV-1 it has proved difficult to generate vaccines capable of inducing antibodies neutralizing the broad range of virus strains found in patients.

The possibility of effective HIV gene therapy is some distance in the future; but recently effective gene-therapy techniques are extending a range of novel strategies in clinical trials. These include disabling the HIV gene *tat* (see above) using a mouse retroviral ribozyme vector introduced into $CD34^+$ stem cells that produce the $CD4^+$ target cells of HIV, and using HIV itself as a vector to insert an 'antisense' gene for an envelope protein, *env*.

HLA complex, HLA system (human lymphocyte (leukocyte) antigen system) The human analogue of the MAJOR HISTOCOMPATIBILITY COMPLEX (HMC) of mice and other mammals.

HMG-CoA reductase The complex regulatory enzyme catalysing the rate-limiting reaction in which β-hydroxy-β-methylglutaryl-CoA (HMG-CoA) is converted to mevalonate during synthesis of CHOLESTEROL (see Fig. 50).

HNPCC See HEREDITARY NON-POLYPOSIS COLON CANCER.

HnRNPs (heterogeneous RNPs) Proteins involved in packaging newly formed RNA prior to its export from the nucleus via nuclear pore complexes. They also promote various steps in SPLICEOSOME assembly and bind splicing enhancer sequences that enhance splice site recognition.

Hodgkin's disease A lymphoma in which neoplastic cells account for 99% of the cells in a tumour mass, and associated with the EPSTEIN–BARR VIRUS. There are five histologic types. Vinblastine, from *Catharanthus*, is one drug employed in therapy. Disease starts in a single group of lymph nodes and spreads progressively and predictably.

holoprosencephaly (HPE) A spectrum of disorders resulting from damage to the anterior midline of the germ disc, giving rise to a child with a small forebrain (the two lateral VENTRICLES often merging into a single ventricle) and eyes close together (hypotelorism). Mutations in the SONIC HEDGEHOG GENE, which regulates establishment of the ventral midline of the CNS, are sometimes responsible, as is defective CHOLESTEROL biosynthesis in the foetus. Other loci are also implicated. HPE can be caused by excessive alcohol intake during the first two weeks after fertilization, alcohol acting as a teratogen. It is responsible for 1:250 pregnancies ending in MISCARRIAGE.

homeobox genes Members of the HOX GENE family, all containing a short DNA sequence, 180 base-pairs in length, specifying a 60-amino acid DNA-binding sequence (the *homeodomain*).

homeodomain See HOMEOBOX GENES, HOX GENES.

homeostasis Neural and endocrine systems and circuits tending to maintain the constancy of the internal environment through NEGATIVE FEEDBACK. The HYPOTHALAMUS is central to many, but not all, homeostatic mechanisms, notably those involving fluid balance (see OSMORECEPTORS), TEMPERATURE REGULATION and ENERGY BALANCE.

Hominidae (hominids) In evolutionary taxonomy, a hominoid family. Two principal taxonomies exist today: *evolutionary classification* and *cladistic classification*. Both make use of phylogenies (whose constructions constitute the field of systematics) in naming and classification; but they differ in the extent to which overall phe-

A: Evolutionary taxonomy, or 'systematics'

Superfamily	*Family*	*Subfamily*	*Representative genera*
Hominoidea			
	Hylobatidae		
	Pongidae		
		Ponginae	*Pongo*
		Gorillinae	*Gorilla*
		Paninae	*Pan*
	Hominidae		
			Ardipithecus, Sahelanthropus, Australopithecus, Paranthropus, Kenyapithecus, Homo

B: Cladistic taxonomy, or 'cladistic systematics'

Superfamily	*Family*	*Subfamily*	*Tribe*	*Representative genera*
Hominoidea				
	Hylobatidae			
	Pongidae			
		Ponginae		*Pongo*
		Gorillinae		*Gorilla*
		Homininae		
			Panini	*Pan*
			Hominini	*Ardipithecus, Sahelanthropus, Australopithecus, Paranthropus, Kenyapithecus, Homo*

TABLE 6 *Two approaches to classification of the hominids.* See HOMINIDAE.

netic similarity between organisms contributes to the process. It is therefore not a debate about what the 'real' world is like: it is about how we should name and classify species; but it is reflected in the different contents of the taxon Hominoidea that these two procedural approaches arrive at. Cladistic taxonomy ignores apparent phenetic similarities and relies entirely on the genealogical relationships between organisms. It therefore does not support construction of a distinct family for humans (see CLADISTICS).

In evolutionary classification (A in Table 6), the Pongidae contains the subfamilies Ponginae, Gorillinae, and Paninae, while a distinct family, the Hominidae, is created for humans (see HOMINOIDEA for discussion). In cladistic classification (B in Table 6), the Hominoidea contains the families Hylobatidae and Pongidae and their closest relatives since their separation from the clade containing *Proconsul*, and dispenses with the Hominidae. Humans are placed within the Pongidae ('apes'), which contains the subfamilies Ponginae, Gorillinae and Homininae. See HOMININI.

Homininae (hominines) In cladistic taxonomy, the pongid subfamily including the tribes Panini and Hominini and their closest relatives since departure of their lineage from the Gorillinae. See HOMINIDAE.

Hominini (hominins) In cladistic taxonomy, the tribe containing all taxa on the human lineage subsequent to its split from its common ancestor with the Panini (or, chimpanzee lineage). One of the main criteria for inclusion in the tribe is the presence of features indicative of BIPEDALISM. It includes *Sahelanthropus*, the AUSTRALOPITHECINES and members of the genus HOMO. The approximate time depths between members of the taxon, based on sequence analysis of mitochondrial DNA and one nuclear DNA locus are indicated in Fig. 94.

Estimates of hominin body mass, using postcranium-based regression equations, are as follows, using averaged male and female values for *H. sapiens* and values given incrementally (kg): *H. habilis*, 34; *H. africanus*, 36; *Paranthropus robustus*, 36; *P. africanus*, 37; *P. boisei*, 44; *H. sapiens*, 53; *H. erectus*, 57; *H. ergaster*, 58; *H. heidelbergensis*, 62; *H. neanderthalensis*, 76. See HOMINIDAE.

Hominoidea The primate superfamily comprising apes and humans (if humans are not regarded as apes): the hominoids. The absence of a tail is the evolutionary novelty that distinguishes the taxon from the Anthropoidea; but also significant is the structure of the elbow region, which enables its greater mobility. In apes, the humerus is rounded where it articulates with the similarly rounded head of the radius, enabling the radius to rotate around it. But being deeply keeled and ridged, the humerus restricts the ridged ulna to movement in only one direction. The shape of the hominoid humerus, and the heads of the radius and ulna, are therefore highly diagnostic (see FORE LIMB, Fig. 78). Hominoid taxonomy has undergone substantial revision in the past 40 years and there is no consensus; but three families used to be recognized: Hylobatidae (gibbons and siamang – the 'lesser apes'), Pongidae (orang-utans, chimpanzees and gorillas – the 'great apes') and Hominidae (humans). Current thinking holds that humans and only a subset of the traditional Pongidae share a most recent common ancestor – making the traditional 'Pongidae' paraphyletic, along with the term 'ape' if humans are excluded from it. Today, depending on the classification employed (see HOMINIDAE), either three or two families of hominoids are recognized: Hylobatidae (4 genera and 12 species of gibbons), Pongidae and Hominidae; or Hylobatidae and Pongidae. Some anatomical features distinguishing living apes from Old World monkeys relate to their locomotory behaviour. While monkeys climb with their gripping hands and feet on branch tops, using their tails for support and balance, and usually feed in a seated position on fleshy ischial callosities, apes typically brachiate from branch to branch (see GAIT) and have comparatively shorter trunks, broader chests, longer arms, more flexible shoulder joints; they also lack tails, and the great apes lack ischial callosities.

The hominoid fossil record in Africa between 10 and 5 Mya is very sparse (see PRIMATES), hindering resolution of the exact relationships between gorilla, chimpanzees and humans by morphological data alone. Phenetic considerations alone suggest that chimpanzees are more closely related to gorillas than to humans; but most measures of GENETIC DISTANCE suggest that chimpanzees are more closely related to humans than to gorillas (although some such data suggest the opposite). A *trichotomy* would involve simultaneous DIVERGENCE of all three lineages.

First, a definitive phylogeny has to be obtained by examining, from as many loci as possible, gene trees that are single copy (not duplicated), orthologous and selection-neutral (so that polymorphisms do not cross species boundaries but go to fixation within the lifetime of a species). These data can then be analysed in several ways (see PHYLOGENETIC TREES). In terms of gorilla–chimpanzee–human phylogeny, eight out of ten independent data sets established by 2002 supported a most recent common ancestor between chimpanzees and human, two between chimpanzees and gorilla, and none between gorilla and human. The chimpanzee–human (*Pan–Homo*) clade is also supported by chromosome breakpoint analyses showing that a 0.1-Mb autosomal fragment must have been transposed onto the Y chromosome in the common ancestor of humans and the two chimpanzees, excluding the gorilla. Once a definitive phylogeny has been obtained, and branch-points dated (not without controversy), the proportion

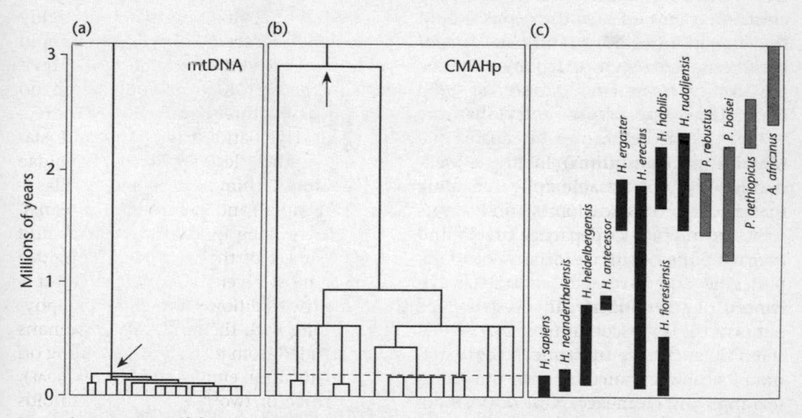

FIG 94. A comparison of the time depths of genealogies from different components of the genome. (a) A time-scaled mitochondrial DNA (mtDNA) gene tree. (b) A time-scaled gene tree for the cytidine monophosphate-*N*-acetylneuraminic acid hydroxylase pseudogene (CMAHp) on human chromosome 6. These trees can be compared with the reconstruction of hominin evolution as inferred from the fossil record, and shown in (c). Taxa in the genus *Homo* are shown in black; *Paranthropus* and *Australopithecus* are shown in grey. Although deep lineages of mtDNA are found in some hunter-gatherer populations, such as the Khoisan or the Mbuti, these lineages trace back only to the time of the global most recent mtDNA common ancestor (that is, <170,000 years ago; arrow in (a)). By contrast, polymorphisms on the X chromosome and autosomes, used in (b), contain information about human population structure well before the emergence of anatomically modern humans (dotted line in (b), corresponding to the top of the *H. sapiens* bar in (c)), with a subset of loci characterized by the segregation of two highly divergent sequence haplotypes that coalesce before the Pliocene–Pleistocene boundary (~1.8 million years ago), when early forms of *Homo* and *Paranthropus* were extant (arrow in (b)). Redrawn from 'Reconstructing human origins in the genomic era' (D. Garrigan and M.F. Hammer), *Nature Reviews Genetics*, with kind permission of the authors and publishers.

of gene trees not supporting this phylogeny can be used to estimate the EFFECTIVE POPULATION SIZE of the ancestral population between the splitting of the first lineage from the remaining lineage, and the subsequent splitting of the second and third species from that (see DIVERGENCE and Fig. 65). If this ancestral population does correspond to the chimpanzee–human common ancestor, its effective population size appears to have been at least five times greater than that of the descendant human lineage (see Fig. 95). It may prove impossible to explain this population BOTTLENECK satisfactorily.

Homo Although established by Linnaeus in 1758, great uncertainty confounds the origin of genus *Homo*. In part this is due to the relative paucity of hominin fossils from the relevant period, 2–3 Myr BP; but it is also due to the fact that there is currently no clear agreement on how the genus should be defined. Until quite recently, the general consensus had been that any putative member of *Homo* should have, at ≥600 cm^3, a larger BRAIN SIZE (see entry) than the 'robust' australopithecines; be capable of a modern precision grip involving a well-developed and opposable pollex enabling manufacture of stone tools; and a less robust mandible, smaller cheek teeth and narrower premolars than any australopithecine; and have had LANGUAGE (as evidenced by endocranial casts). See INTELLIGENCE, 'LESS IS MORE' HYPOTHESIS.

But these criteria are not sufficiently rigorous. Absolute brain size does not translate into any clear biological disposition; language functions do not so simply equate either with gross brain appearance or with structures on the cerebral surface; there is too much ignorance about early hominin functional hand structure; and there is no certainty that *Homo* did produce the earliest stone tool culture (2.6–2.3 Myr ago) because at least one AUSTRALOPITHECINE genus, *Paranthropus*, was also present at the scene.

Our understanding of the biogeography of the genus *Homo* has changed considerably in the past twenty years. It is now clear,

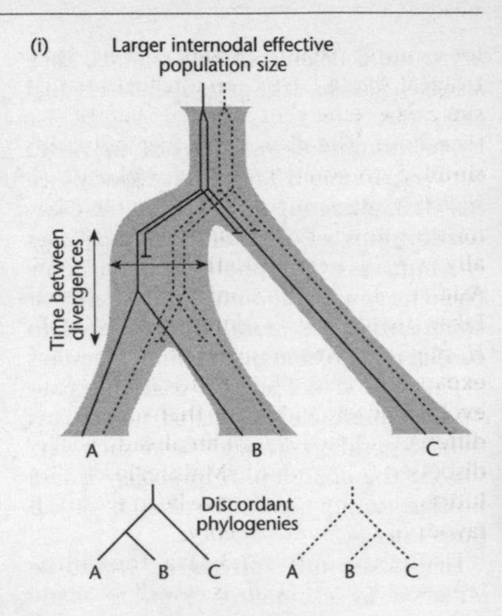

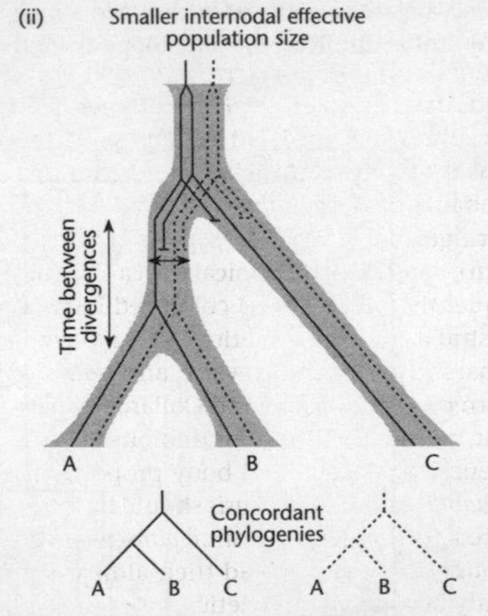

FIG 95. Diagram indicating the consequence of effective population size on the frequency of discordant gene trees (see HOMINOIDEA entry). The time between speciation events is the same in scenarios (i) and (ii). Two *gene trees* are shown within each *species tree* (grey outline) by the solid and dashed lines. In (i), the greater width of the species ancestral to A and B indicates a larger effective population size, allowing a polymorphism in the dotted gene tree to persist between the two speciation events. Interpretation of the lineage assortment in (i) therefore leads to discordant (species) phylogenies. © 2004 From *Human Evolutionary Genetics*, by M.A. Jobling, M. Hurles and C. Tyler Smith. Reproduced by permission of Garland Science/Taylor & Francis LLC.

for example, that dispersals of *Homo* from tropical Africa have been a regular feature since the emergence of the genus (see MIGRATION). The view that other hominins simply 'evaporated' in the wake of *H. sapiens* as it advanced is no longer tenable: modern humans overlapped geographically with other populations of *Homo* in Asia and Europe for considerable periods. Likewise, the debate centering on whether *H. sapiens* evolved once prior to range expansion (the Out-of-Africa view), or evolved separately more than once from different populations of an already widely dispersed *H. erectus* (the Multiregional Evolution view), seems to have been settled in favour of the former theory.

Dating to possibly 2.3 Mya, the earliest representative of *Homo* is usually regarded as being *H. habilis*. *H. erectus* dates from 1.9 Mya, evolving by ~600 ka into a larger-brained form, *H. heidelbergensis*, that occupied Africa and Europe. *H. erectus* persisted in Central Java until ~30 ka, but by ~195 ka (fossils at Omo, Ethiopia) the African descendants of *H. heidelbergensis* had become distinguishable as *H. sapiens* (modern man), spread across tropical Africa and subsequently (~80 ka later) colonized Eurasia, Australia and the Americas ~45 ka ago – apparently via the Levant and via the Horn of Africa. Wood and Collard propose that, on the basis of their dubious distinctiveness of skeletal and body proportions, *H. habilis* and *H. rudolfensis* should be transferred to the genus *Australopithecus*, even though the genus would then almost certainly become paraphyletic.

Not all adaptations of a primate to homeostasis, acquisition of food and reproduction can be reliably reconstructed from the fossil record. Arguably of those that can are BODY MASS and shape, the skeletal features of locomotor behaviour (see BIPEDALISM, GAIT, VALGUS ANGLE), relative brain size, the rate and pattern of development, and the relative size of the apparatus for chewing food prior to swallowing. Human and chimpanzee genomes differ in 1.23% of the nucleotides that can be aligned between species (i.e., 35 million point mutations), and in addition differ in ~5 million insertions, deletions, duplications and inversions – genome rearrangements varying in scale from a few nucleotides to whole chromosomes (see GENETIC DIVERSITY). But these are differences of degree and kind to be expected for a pair of primate species sharing a common ancestor 5–7 Mya (see NEUTRAL THEORY OF MOLECULAR EVOLUTION), and only a small fraction of the DNA differences between humans and chimpanzees are non-neutral and relevant to functional differences. See all entries for *Homo* species, below.

Homo antecessor ('Pioneer Man') A species named in 1997, on the basis of material dating to ~800 kya from the Gran Dolina cave, Atapuerca, northern Spain, establishing the human settlement of Western Europe during the early Pleistocene. Fossils include the bones of the forehead and face of a child, jaw and tooth fragments, a patella and fragments of arm and foot bones, mostly cut by stone tools. The authors decided that the material could not be accommodated within any existing *Homo* species, eventually rejecting the possibility of its being a primitive form of *H. heidelbergensis*, but that it probably evolved from *H. ergaster*. The species, if accepted, is a possible ancestor to Neanderthals and *Homo sapiens*. Such a classification would represent a greater degree of subdivision of later *Homo* than some scholars believe is warranted, some assigning the material to early *H. neanderthalensis*, and it is usually considered not to have left descendant species. Some of the original authors now recognize in the face, jaw and teeth, close resemblances to *H. erectus* of China, and suggest that *H. antecessor* was replaced by populations of *H. heidelbergensis* originating in Africa and bearing handaxe tools, and that *heidelbergensis*, rather than *antecessor*, was the ancestor of Neanderthals. The teeth of *antecessor* resemble those from similarly dated jawbones from the Algerian site of Tighenif (formerly Ternifine). Since these latter fossils have been referred to a species named *Homo mauritanicus*, the Spanish material (if it has specific status) may have to be assigned to that species.

A hominin mandible (ATE9-1) and an isolated lower P4 (see PREMOLARS) provisionally assigned to *H. antecessor*, has more recently been discovered among Mode 1 lithic tools (but lacking hand axes and cleavers) at the cave site of the Sima del Elephante, Atapuerca, close to the sites of Gran Dolina, Galéria and Sima de los Huesos. It has been dated, mainly by faunal associations, to the Early Pleistocene (1.2–1.1 Mya), making this the oldest and most accurately dated record of human occupation in Europe. The authors suggest that the most parsimonious interpretation of these fossils is that western Europe was settled during the Early Pleistocene by a hominin population originating in the east, and that it may be related to an 'early' demographic expansion of hominins out of Africa (see the entries HOMO ERECTUS and HOMO ERGASTER).

Homo erectus Introduced in 1892, the first fossils attributed to this species came from Trinil, Indonesia. Its distinctive large brow ridges, low cranial vault and sharply angled occipital region have been found elsewhere in Indonesia, in mainland Eurasia and in Africa. The earliest specimens may be ~1.5 Myr old and the youngest possibly <100 Kyr old – both of Javanese origin. Chinese specimens span a similar age range, from ~1.5 Mya to possibly ~250 Kya. During this time span, there was a modest increase in cranial capacity of the specimens. Despite uncertainty concerning putative fossils of this species dating to ~2 Mya from southwestern China, the conservative view is that *H. erectus* descended anagenetically from *Homo habilis/rudolfensis* in Africa nearly 2 Mya. Against this view, recent finds from the Koobi Fora formation at Ileret, east of Lake Turkana, northeastern Kenya, indicate that *H. habilis/rudolfensis* and *H. erectus* were sympatric for ~500 kyr. More recently, many experts have sided in favour of the view that *H. erectus* descended from *H. ergaster* in Asia, after the latter species had migrated out of Africa.

In line with this last opinion, although the species name *H. erectus* was formerly applied to all Middle Pleistocene hominins it is now restricted by many to a small group of Asian fossils. This is because the oldest specimen formerly attributed to this, *H. erectus*, KNM-ER 3733 from the Koobi Fora on the east side of Lake Turkana (dated radiometrically to >1.5 Mya), is now placed by many in a separate taxon, HOMO ERGASTER (see this entry). In the early 1990s a spectacular skeleton (KNM-WT 15000) of a juvenile male, from Nariokotome on the western side of Lake Turkana and dating to 1.5 Mya, came to be representative of the view that the species had a very modern body: tall (160–185 cm), large (50–70 kg), with long legs, and otherwise only subtly different postcranially from anatomically modern humans. Until the mid-1990s, no specimen thought to be earlier than ~1 Mya (Modjokerto, Java) had been discovered outside Africa and it was generally believed that the species arose in Africa ~2 Mya and spread into Asia after a delay of ~1 Mya (the 'Out-of-Africa 1' model). In 1994, redating of the Javan material to ~1.8 Mya (Mojokerto) and ~1.6 Mya (Sangiran) implied either that *H. erectus* spread across Asia soon after its African origin, or that the species originated in Asia, close to ~2 Mya. In 1992, the discovery of a *H. erectus* mandible at Dmanisi, Georgia, dated by faunal associations to ~1.6–1.8 Mya, and subsequent fossils of similar age at that site, have strengthened opinion that *H. erectus* left Africa much earlier than previously held, expanding its range into Asia soon after its origin there ~2 Mya without entering Europe (see Fig. 96). This despite the belief by some that a fossil from Ceprano, Italy, recently dated to ~800 kya has affinities with *H. erectus* (but is sometimes attributed to *H. antecessor*).

The new Ileret fossils include KNM-ER 42700, a small late subadult calvaria of ~1.55 Mya and endocranial capacity estimated at 691 mL, indicates that *H. erectus* showed substantial size variation throughout the early Pleistocene. It has been argued from this that the Dmanisi crania cannot be seen as primitive, or transitional between *H. habilis* and *H. erectus* on size alone. KNM-ER 42703 is a right maxilla estimated to be 1.44 Mya. Multivariate analysis of maxillary molar dimensions clearly sep-

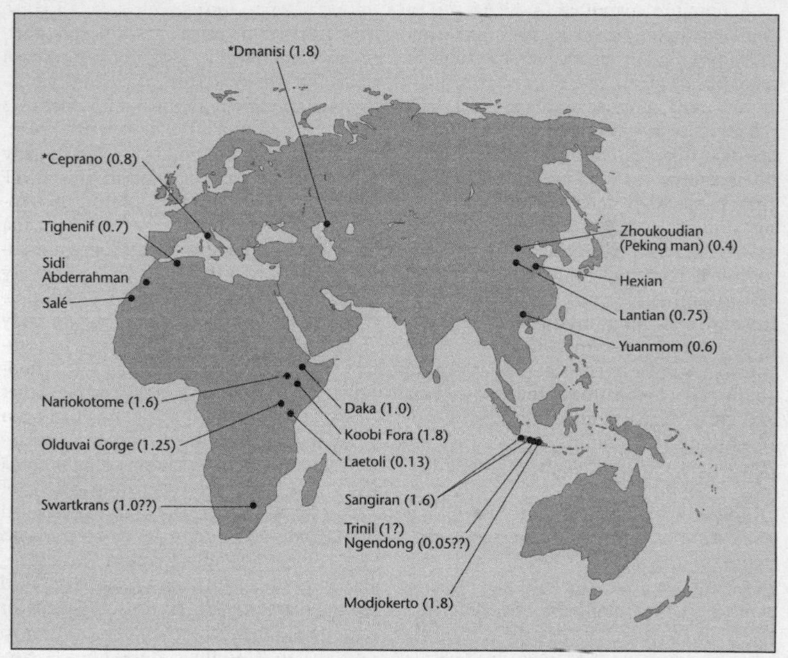

FIG 96. Distribution map of sites where *Homo erectus* and similar forms have been located, bracketed numbers being their estimated ages (Mya). Recent redating of fossils from Java, if correct, provides evidence that this species expanded its range beyond Africa soon after its origin. *For the Ceprano and Dmanisi material, see the HOMO ERECTUS entry. © 2004 Blackwell Publishing. Redrawn from *Principles of Human Evolution*, R. Lewin and R.A. Foley, with permission of the publishers.

arates *H. erectus* from *H. habilis*, the former having smaller molars overall, and a mesiodistally shorter 3rd molar, and KNM-ER 42703 lies in the *H. habilis* cluster. This dating now indicates that *H. habilis* and *H. erectus* occurred together in eastern Africa for nearly 0.5 Mya (see HOMO HABILIS).

Full description of the Georgia material should reveal whether the bones all belong to one species, *H. erectus*, or to two or more. Despite burned stone tools and clay associated with the Koobi Fora material, there is no unambiguous evidence that *erectus* cooked its food. See ACHEULIAN CULTURE.

Homo ergaster A taxon introduced in 1975, in response to the widely held belief that early African specimens of HOMO ERECTUS (see entry) deserved specific status. The two species are commonly thought to have an ancestor-descendant relationship, *ergaster* originating in Africa ~2 Mya, quickly expanding its range into Asia and giving rise to *erectus*, either there or in Africa. It is therefore believed by many to have been the first hominin to migrate out of Africa. The lower jaw and associated ACHEULIAN CULTURE discovered in 1991 at Dmanisi in the Caucasus Mountains of the Republic of Georgia, and two complete crania discovered in 1999, are very similar to African specimens of *H. ergaster* and date to ~1.7 Mya. Many anatomical features of *H. ergaster* are shared with *H. erectus*, the main differences from *erectus* being *ergaster*'s higher cranial vault, less prognathic face, smaller jaws and teeth (and relatively smaller molars), thinner cranial bones and absence of a sagittal keel. These derived characters may reflect a greater dependence on biting and tearing, the large brow ridges and occipital torus buttressing against associated stresses. The absolute BRAIN SIZE of *ergaster* has been estimated as 854 cm^3. While this is substantially larger than in earlier 'ape-like' hominins, it may be a result of *ergaster*'s larger body size. The skeleton of a young male discovered in 1984 at Koobi Fora on the west side of Lake Turkana (see Fig. 96, HOMO ERECTUS) provides a very good picture of *ergaster* anatomy, including as it does the ribs and vertebrae. Judging by modern human developmental standards, this boy died at age ~11 years; but enamel growth lines of the teeth indicate that this specimen (KNM-WT 15000) developed comparatively quickly and may only have been ~8 years at death. It stood 1.625 m tall, and would probably have reached 1.9 m as an adult. The skeleton is robust and would have been heavily muscled. By contrast with earlier hominins, *ergaster* had relatively shorter arms, but long legs, narrow hips and narrow shoulders – much like the body proportions of people inhabiting tropical savannas today. These features would have enabled *ergaster* to run well over several kilometres (see GAIT), a probable adaptation to life in savannas where the large body surface area would tend to reduce overheating. The ratio of the female *ergaster* birth canal to that of the newborn's head is comparable to that of modern humans, suggesting that *ergaster* infants had extended dependence on parental support. This has been associated with a reduction in sexual dimorphism, increased paternal investment and monogamous mating systems. However, estimates of the rate of *ergaster*'s tooth enamel growth suggest that postnatal development, although slower than in australopithecines, was faster than it is in modern humans.

Homo floresiensis The discovery in 2003 of this small hominin (Tribe Hominini) from the Liang Bua cave on the island of Flores, Indonesia, was unexpected. The original description of the holotype (LB1) as a new species has been disputed, some claiming that the remains might be those of a dwarfed *H. erectus* (a unique example for *Homo* of island dwarfism), a pygmy, a pathological modern human, or remnants of a much more ancient hominid lineage. If it represents a non-sapiens hominin, then its 18 kya date would indicate just how recently *Homo sapiens* became the sole representative of its genus.

The description is of a small-bodied bipedal hominin, with endocranial volume and stature (body height) similar to, or smaller than, *Australopithecus afarensis*, lacking AUSTRALOPITHECINE masticatory adaptations and with reduced facial height and prognathism, and smaller postcanine teeth, with superior cranial vault bone thicker than in australopithecines but similar to those of *H. erectus* and *H. sapiens*. Its supraorbital torus arches over each orbit, and does not form a flat bar as in Javan *H. erectus*. Its first and second molar teeth are of similar size and the mandibular symphysis lacks a chin. The ilium has a marked lateral flare; the neck of the femur is long relative to femur head diameter; the long axis of the tibia is curved, with the midshaft having an oval cross-section. According to eight standard measurements, the LB1 femur resembles those of early hominins such as *H. habilis* or even *A. afarensis*. But there is no other evidence anywhere indicating such an early exodus of a primitive hominin out of Africa. The nearly complete left and partial right foot bones of LB1 suggest that the feet were 70% as long as the very short femur (compared with 55% for living humans), that the great toe was very short, and the foot probably not arched. The three complete carpals (trapezoid, scaphoid and capitate) from the left wrist are well preserved, show no pathology, and lack the shared derived features of the modern human and Neanderthal carpals thought to have evolved by at least 800 kya. Thus, the trapezoid is wedge-shaped with a narrow palmar tip and wide dorsal base which is reflected in the shapes and articular configurations of the carpals articulating with it. This wrist morphology may help falsify or support hypotheses regarding the phylogenetic position of *H. floresiensis*. Apart from a partial lunate from Zhoukoudian, China, carpals are absent from fossils of *H. erectus* sensu lato. The discovery of hominin carpals dating between 1.8 and 0.8 Mya would help clarify the relationship between *H. floresiensis* and *H. sapiens*.

Homo habilis Introduced in 1964 in classifying material from Olduvai Gorge in Tanzania, additional specimens have since been found in East and, more controversially, southern Africa. It is a putative ancestor of *H. erectus* (see HOMO), dating to at least 1.9 Myr and possibly 2.3 Myr ago, the most recent material (OH 13) dating to 1.65 Myr ago. However, recent discovery in 2000 from Ileret, east of Lake Turkana, Kenya, of a right maxilla dating to 1.55 Mya includes part of the palatine and zygomatic processes, most of the alveolar process and teeth C to M^3. If correct, this date precludes *H. habilis* from being directly ancestral to *H. erectus* (see HOMO ERECTUS).

The taxon includes dozens of specimens, with a wide range of anatomical variation, and several scholars have favoured splitting the material into two species: *Homo habilis* and HOMO RUDOLFENSIS. If so, the *rudolfensis* material has a flatter, broader face, and broader postcanine teeth with thicker enamel and more complex crowns and roots. The hand bones associated with OH 7, the type specimen, have suggested an ape-like climbing ability, and the relatively long arms of OH 62 and KNM-ER 3735 also support such ability; so *H. habilis* may have been capable of both terrestrial bipedalism and effective arboreal activity.

Homo heidelbergensis Introduced in 1908 in describing the Mauer jaw (discovered 1907), which with related material was subsequently incorporated in a grade-based (rather than clade-based) taxon 'archaic *H. sapiens*', the taxon has recently been revived to include material from a number of European and African Middle Pleistocene sites. It represents the first occupation by *Homo* of central and northern Europe, an occupation (judging from associated fauna and vegetation) likely to have been limited to mild climatic episodes. By 600 kya, a large-brained population of African *H. heidelbergensis* (cranial volumes 1,200–1,300 cm^3 compared with ~1,000 cm^3 for *H. erectus*) had already begun the 'evolutionary trajectory' in the direction of anatomically modern humans, now dated to ~190 kyr from Omo in Ethiopia (see HOMO SAPIENS). A different lineage is believed to have evolved in Europe and western Asia while *Homo erectus* persisted in eastern Asia. The early stages of this transition to Neanderthals are found in specimens from Sima de Los Huesos, Atapuerca, northern Spain, which are dated to ~400 kya (see HOMO ANTECESSOR). A tibia found in 1993 at Boxgrove, West Sussex, UK, and two front lower jaw teeth, have also been allocated to this species and dated to ~500 kya, representing the earliest *physical* evidence of humans known from the British Isles.

Homo neanderthalensis A species introduced in 1864, but only widely used quite recently for fossils having a projecting face, a large rounded cranial vault, and robust limb bones. Taurodont molars are common. Fossils with Neanderthal traits appeared in Europe and western Asia ~400 kya and vanish ~30 kya. Over this period they evolved morphological traits making them more distinct from the ancestors of modern humans that were evolving in Africa. Neanderthals are believed to have evolved from European *Homo heidelbergensis*, early signs of this transition evident in specimens from the pit, Sima de Los Huesos, Atapuerca, Spain, ~400 kya.

Few museum specimens have yielded DNA, and the risk of contamination with modern human DNA is great. But in 1997 the first Neanderthal mtDNA was sequenced and suggested that the Neanderthal lineage split from the common lineage with modern humans ~500 kya, so that the lineage may not have contributed any mtDNA to modern humans. Mitochondrial data provide no access to the gene and gene regulatory sequences of the nuclear DNA that might reveal biological features unique to each. However, DNA recovered from a ~45 kyr male specimen from Vindija

Cave, Croatia, has provided ~10^6 bp of exceptionally uncontaminated nuclear DNA and indicates that the Y chromosome was substantially more different from human and chimpanzee Y chromosomes than were other chromosomes. Although several assumptions are involved in the calculations, this material has enabled a provisional MOST RECENT COMMON ANCESTOR for Neanderthals and modern humans at 706 kya (95% confidence interval of 468 kya–1,015 kya) and suggests that these lineages split ~370 kya. Some research is focusing on obtaining DNA from Neanderthal remains from caves in southern France, where geographical overlap with modern humans probably occurred for ~12 kyr. This might help resolve the issue of whether the populations interbred. No divergent (i.e., Neanderthal) mtDNA types have been found in any survey of modern humans (see MIGRATION (2)). Some recent work suggests that this absence of Neanderthal DNA sequences is still compatible with a maximum of 120 admixture events between the two populations during their coexistence and that maximum interbreeding rates of the two populations could have been <0.1%, consistent with ecological evidence for differences in habitat exploitation between Neanderthals (heterogeneous vegetation mosaics) and modern humans (open plains habitats). The question of cultural transfer between Neanderthals and modern humans remains open. The implications of stone tool technology that is apparently transitional between the Middle Palaeolithic industries of Neanderthals and those of modern humans during the Upper Palaeolithic of western Europe, at the time of overlap, is hotly debated. But it is generally accepted that Neanderthals were capable of behaviour that is attributed to modern humans.

***Homo rudolfensis* (*Australopithecus rudolfensis*)** Introduced in 1986, similarities between the KENYANTHROPUS PLATYOPS fossils and the larger-brained and later fossils exemplified by KNM-ER-1470 (~1.8 Myr) from Koobi Fora in Kenya previously attributed to *Homo* or *Australopithecus rudolfensis* have led to *rudolfensis* being assigned to *Kenyanthropus*; but although the specific name 'rudolfensis' is widely used, their generic attributions are controversial. The KNM-ER 1470 lectotype has anteriorly placed and forwardly sloping zygomatic processes. See HOMO ERECTUS, HOMO HABILIS.

***Homo sapiens* (anatomically modern humans)** In 1967, two hominin skulls were unearthed in the Omo Valley, southwest Ethiopia. Originally dated to 135,000 kya, but now dated to 195,000 kya, one of these skulls (Omo 1) is accepted as evidence of the oldest discovered anatomically modern human (AMH). Characteristic features distinguishing early modern *Homo sapiens* from 'Archaic' *H. sapiens* (see HOMO entry; HOMO HEIDELBERGENSIS) are: a less robust skeleton; small or absent brow ridge; shorter, higher, skull (larger forehead); and shorter jaws, a developed chin, slightly smaller teeth, and slightly larger BRAIN SIZE (see entry for correlates).

Using palaeontological and archaeological records and the patterns of neutral DNA polymorphism in the human genome, the evolution of *H. sapiens* can be divided into two epochs: (i) processes occurring in the lineage up to the emergence of AMH and (ii) processes, particularly demographic ones, accompanying the global diaspora of AMH subsequent to its origin in Africa. Of particular interest is the resolution of whether or not AMH completely replaced archaic forms of *Homo* already present in Asia and Europe without interbreeding, because of its relevance to interpretation of current human genetic diversity, and with it attempts to target clinical approaches to health and disease.

Early attempts to make use of genetic data in inferring human history relied on allele frequencies of classical polymorphisms, such as the ABO blood group system, to reconstruct population phylogeny. But with DNA SEQUENCING data to hand, inferences about population history could come from reconstruction of gene trees (see PHYLOGENETIC TREES). However, the coalescent approach (see COALESCENCE) has proved especially useful in inferring the history of changes in human population size and structure: its demography. Because gene trees constructed from both mtDNA and the non-recombining Y chromosome are PARAPHYLETIC with respect to modern African populations, Africa was inferred as the place of the MOST RECENT COMMON ANCESTOR of all modern humans – i.e., their trees are rooted in Africa – and autosomal and X-linked loci usually support this (see PSEUDOGENES).

Research into the two haploid compartments of the human genome (see MTDNA and Y CHROMOSOME) played a decisive role in the palaeontological debate over human origins (see MITOCHONDRIAL EVE). Restriction enzyme digestion of human mtDNA in a 2006 work suggested a root in Africa and a time depth of ~200 kyr (see Appendix VIIa). It promotes a general consensus that AMH originated in a small, isolated African population during the Late Pleistocene, one which completely replaced archaic forms of *Homo* as it increased in size and range (see MIGRATION (2)). The deep branches within the African lineages and 'starlike' structure within non-African lineages indicate that population size in the African lineages has remained roughly constant, contrasting with a rapid population expansion in non-African lineages (see COALESCENCE, Fig. 55; MIGRATION (2)). Although less detailed than the mtDNA studies, current phylogeny of the Y chromosome has also greatly contributed to the interpretation of the origins of modern humans (see Appendix VIIb). It, too, reveals a complete separation of African and non-African lineages, its first two branches leading exclusively to African lineages while its third contains both African and non-African lineages. The time to the most recent common ancestor (TMRCA; see COALESCENCE, MOST RECENT COMMON ANCESTOR) for the complete phylogeny is estimated as 59 (40–140) kyr (assuming 25 years per generation), the TMRCA for the branch containing African and non-African lineages (marked with an asterisk in Appendix VIIb) being 40 (31–79) kyr. In 2008, Excoffier and colleagues provided the first formal statistical evaluation of the likelihood for the various schemes explaining modern human origins, concluding that a recent expansion from a single African origin is best supported by current geographical spread of genes (see MULTIREGIONAL MODEL). Probably the best supporting evidence for hybridization between modern and archaic humans is the length of time it takes to reach the MRCA of some genes (predating the origin of *H. sapiens*). But Excoffier et al. demonstrate that such cases may also arise if humans had a recent, and single, African origin.

Although autosomal and X-chromosomal phylogenies are complicated by recombination (e.g., see DIVERGENCE, PHYLOGENETIC TREES), analysis of very closely linked polymorphisms (~10 kb apart or less) reveals useful haplotypes capable of giving gene trees. An early example allowed Harding and colleagues in 1997 to construct a haplotype tree from 326 different sequences, from widely different geographical samples, found in the 2.67 kb β-globin region of chromosome 11 (see GLOBINS). Once again, the tree rooted in Africa, some lineages being exclusively African but many others containing both African and non-African lineages. The TMRCA for the entire phylogeny, as expected for the larger EFFECTIVE POPULATION SIZE of an autosomal locus, was 750 (400–1300) kyr (assuming 20 years per generation), with some lineages dating to >200 kyr and exclusively to Asia! While this might be interpreted as indicating the existence of lineages arising in Asia prior to 200 kyr and surviving today through interbreeding between ancient and modern populations, it is also possible that these lineages exist in Africa but have not so far been discovered there. Then again, β-globin has been a target of balancing selection; so the assumption of neutrality required to calculate MRCAs is questionable. The few other studies of haplotypes of autosomal loci point almost unanimously to their African origins; but there is some evidence of ancient non-African contributions to the modern gene pool.

Since the publication of results of the HUMAN GENOME PROJECT, on HAPLOTYPES and on SINGLE NUCLEOTIDE POLYMORPHISMS, there are now additional data on DNA polymorphisms from two other compartments of the HUMAN GENOME: the X CHROMOSOME, and autosomes. These have given us new insights into both of the above-mentioned epochs, some of them in conflict with earlier thinking. The difference in ploidy of these new compartments means that they provide different levels of resolution of the evolutionary and demographic history of our species. Traits distinguishing humans from other primates originated in human-specific DNA sequence changes (see MUTATIONS). Given the similarities, long established, between the protein-coding components of chimpanzee and human genomes, gene regulatory and other functional non-coding elements (see NON-CODING DNA) are now thought to have played a major part in bringing about the observable differences between the species, a view for which evidence is accumulating. See HOMO entry.

homocysteine A metabolite in the synthesis of cysteine by mammals, during which it is condensed with serine by cystathione β-synthase to form cystathione and then deaminated, in a reaction requiring pyridoxal phosphate (see VITAMIN B COMPLEX), to α-ketoglutarate and cysteine. See ALZHEIMER'S DISEASE. See HOMOCYSTINURIA.

homocystinuria A Mendelian autosomal recessive (chromosome site 21q) caused classically by deficiency in the enzyme cystathione β-synthase and associated inability to break down the amino acid methionine. Incidence 1:50,000–1:150,000 in the USA. If untreated, it leads to mental retardation in 65–80% of cases, with seizures and psychiatric disturbances common. Vitamin and/or dietary treatment resolves the disorder, for which there are NEWBORN SCREENING schedules similar to PHENYLKETONURIA.

homogenously staining regions (HSRs, double minutes) See ONCOGENES.

homologous genes, chromosomes and chromosome segments See HOMOLOGY, where the same principles that apply to DNA sequences also apply to genes, chromosomes and chromosome segments. See SYNTENY.

homologous recombination (general recombination) Recombination in which exchange occurs between a pair of homologous DNA sequences, usually located on two adjacent copies of the same chromosome, although other chromosomes sharing the same nucleotide sequence can also participate. It is essential for every proliferating cell since accidents occur in nearly every round of DNA REPLICATION, normally only occurring during the S and G2/M phases of the CELL CYCLE. The details of the process are not fully understood, but are variations on the homologous end-joining involved in the initiation of replication forks. In meiotic nuclei it is initiated by double-strand DNA breaks, and its successful completion is essential for accurate chromosome segregation during 1st meiotic anaphase. See DNA REPAIR MECHANISMS.

homology Similarity due to descent from a common ancestor. Two DNA sequences are homologous if they are derived from the same DNA sequence in a common ancestor. Homology is an all-or-nothing concept. Thus all sequences in a multigene family, such as the histone sequences, are homologous, although divergence between any two sequences gives paralogous sequences (if gene duplication was involved) or orthologous sequences (if speciation was involved). Contrast ANALOGOUS; see HOMOLOGOUS GENES, HOMOLOGOUS RECOMBINATION.

homoplasies Characters with the same state, but generated by different means (convergent evolution).

horizontal Of a plane parallel to the ground. See ANATOMICAL PLANES OF SECTION, Fig. 9.

horizontal cells Small interneurons of the outer plexiform layer of the RETINA (see Figs 149 and 150). Like photoreceptors and some amacrine cells, they respond by graded changes in membrane potential and do not produce action potentials. Horizontal cells have large dendritic trees, but short processes, and relay information from distant rods or cones to on- or off-centre bipolar cells, which they influence through LATERAL INHIBITION in the RECEPTIVE FIELDS of both BIPOLAR CELLS and GANGLION CELLS. They are extensively interconnected by gap junctions and form a network of reciprocal connections at triad synapses between neighbouring photoreceptor cells, spanning an area of retina termed the S space. Horizontal cells in the S space provide the signal for *surround inhibition*, hyperpolarizing the receptive field surround of BIPOLAR CELLS in the indirect pathway, this signal being proportional to a measure of mean luminescence over a fairly large retinal area. In the dark, they are excited by glutamate release from photoreceptors; but since they release GABA, this tends to inhibit photoreceptors (e.g., a central cone cell). Light hyperpolarizes surrounding photoreceptors, causing them to release less glutamate. This reduces horizontal cell excitation, causing less GABA output and so, by feedback on the central cone cell, more glutamate release. Depending on the type of bipolar cell the central cone cell synapses with, this will cause it either to hyperpolarize (if it is a depolarizing (ON) bipolar cell), or cause it to depolarize (if it is a hyperpolarizing (OFF) bipolar cell). The ganglion cell response will be commensurate.

hormone replacement therapy (HRT) See CONGENITAL HYPOTHYROIDISM, MENOPAUSE, OESTROGENS.

hormone response elements (HREs) See RESPONSE ELEMENTS.

hormones Signal molecules released by ductless glands (endocrine glands; sometimes single cells) into the bloodstream, in which they travel and, on entering the tissue fluid, either bind their cognate receptors on plasma membranes (amide and peptide hormones) and activate SIGNAL TRANSDUCTION pathways, or pass through the plasma membrane and bind members of the NUCLEAR RECEPTOR SUPERFAMILY in the cytosol or nucleus before binding DNA and modifying gene expression (steroid and thyroid hormones, retinoids, and calcitriol, the hormonally active form of vitamin D). Several are released rhythmically, some in response to CIRCADIAN CLOCKS. Many are transported bound to PLASMA PROTEINS. It is common nowadays to include, where they travel via the blood, GROWTH FACTORS and CYTOKINES within the broader 'hormone' category, even though these usually act locally (in paracrine or autocrine fashion) rather than at a distance. Peptides form the largest hormone group, varying considerably in size: thyrotropin has only three amino acid residues, whereas growth hormone and follicle-stimulating hormone have almost 200. Amides, derivatives of amino acids, include the catecholamines and thyroid hormones. Steroid hormones are derivatives of cholesterol (see Fig. 49) and include sex hormones, hormones of the adrenal cortex and calcitriol. Non-classical endocrine tissues include brain tissues (especially the HYPOTHALAMUS), the heart (atrial natriuretic peptide), kidneys (erythropoietin), liver (insulin-like growth factors), gastrointestinal tract (gastrin, secretin, cholecystokinin, gastric inhibitory peptide), platelets (platelet-derived growth factor, transforming growth factor-β), lymphocytes (interleukins), adipose tissue (leptins) and non-specific sites (epidermal growth factor, transforming growth factor-α. Most are removed from the blood by the liver. For further details, see under entries for particular hormone.

housekeeping genes Genes expressed in virtually all somatic cells (e.g., genes for enzymes involved in respiratory pathways, fatty acid metabolism, protein synthesis and many mitochondrial proteins). They are especially concentrated in G–C-rich regions of the genome (see CPG DINUCLEOTIDE). They replicate very early in S-PHASE (see DNA REPLICATION), tending to occur in less condensed chromatin.

***Hox* genes (homeobox genes)** A family (or complex) of 38 genes, and 214 orphan homeobox genes, containing a 180-base pair sequence (the *homeobox*) which encodes a 60-amino acid DNA-binding motif, the *homeodomain*. *Hox* genes encode transcription factors or co-activators, proteins that activate cascades of genes regulating such key developmental events as segmentation (see SOMITES) and body AXIS formation. Their involvement in expression of body plan was revealed by classical genetics experiments in the fruit fly *Drosophila melanogaster* and the laboratory mouse. Some of the non-protein-coding transcripts generated in Hox clusters also encode microRNAs (see entry). *Hox* gene expression is modulated by other gene products during embryogenesis (see FORKHEAD PROTEINS), and by BMP, Notch-ligand- and hedgehog-mediated signalling pathways. Extreme homeotic transformations are lethal at early stages of development. The 38 human *Hox* genes are collected on chromosomes into GENE CLUSTERS in arrangements such that those affecting the more cranial structures lie at the 3′ end of the DNA and are expressed first, while those controlling more posterior developmental regions are expressed progressively towards the 5′ end – a feature shared with the fruit fly *Drosophila* (see LIMBS). *Hox* genes are not expressed in the fore- and midbrain, but are important for anteroposterior patterning from the hindbrain posteriorly. The most anterior *Hox* genes have their anterior boundaries in the hindbrain, and combinations of activity are such that each RHOMBOMERE carries a unique 'code' of expression (see PATTERN FORMATION, RETINOIC ACID). As development proceeds, 'head' HOX proteins specify the cell arrangements and structures and result, for instance, in chewing organs; 'thoracic' HOX proteins specify, *inter alia*, LIMBS, while 'abdominal' HOX proteins specify genital and excretory organs. Recent evidence (2006) links the functions of non-coding Hox cluster transcripts and CHROMATIN REMODELLING proteins in Hox regulation (see RNA). The unfinished HUMAN GENOME PROJECT identified three large regions that had been quadrupled (i.e., undergone a four-fold duplication), one of these being the Hox cluster regions found on chromosomes 2p, 7, 12 and 17 (see GENE DUPLICATION) – strong, but not definitive, evidence for two rounds of whole genome duplication during vertebrate evolution. See LEUKAEMIAS.

HPA axis The hypothalamic-pituitary-adrenal axis, regulating the secretion of cortisol from the adrenal cortex. See AFFECTIVE DISORDER, STRESS RESPONSE.

hPL See HUMAN PLACENTAL LACTOGEN.

hsa A prefix indicating the position of a gene on a chromosome of *Homo sapiens*. See CHROMOSOMES.

human accelerated regions (HARs) See HARS, NATURAL SELECTION.

human artificial chromosomes (HACs) First created in 1997, HACs are minichromosomes that are useful in gene expression studies and as gene transfer vectors. Crucial in their construction was the creation of functional centromeres and their incorporation into HACs along with telomeres and genomic DNA. Inside cells, they remain independent and function as accessory chromosomes during repeated rounds of the cell cycle. They have potential for use in GENE THERAPY, whereas most such systems under development at present utilize viral vectors which often require integration of the therapeutic gene into an existing chromosome, which can interfere with normal gene expression (rather like a TRANSPOSON). Viral vectors can also induce immune responses. HACs have the potential to avoid these disadvantages.

human chorionic gonadotropin (HCG, hCG, hcg) A large heterodimeric glycoprotein secreted by the syncytiotrophoblast, it reaches the mother's ovaries and causes the corpus luteum to remain functional and producing progesterone until placental production takes over. It is detectable in serum at day 6–8 after implantation, its levels peak at 60–90 days of gestation, decreasing after this to a low level. With structural and functional similarities to LH, it has a much longer half-life but exerts its effects through the LH receptor. It is believed to play an important role in initial development of Leydig cells, and hence in the regulation of embryonic testosterone production and male sex organ development.

human cytomegalovirus (HCMV) A very large DNA herpes virus with more than 200 genes in its genome. Virions also contain at least four types of mRNA molecule transcribed from its genes after infection, translated in the host cell cytoplasm and incorporated into the virion during viral assembly.

human demography Demographic processes include changes in population size, distribution and structure. Demographic variables include age- and sex-related disease and mortality, migration, socioeconomic grouping and fecundity, a few of which are referred to in entries on AGE-RELATED DISEASES, GENETIC DIVERSITY, EFFECTIVE POPULATION SIZE, EPIDEMIOLOGY, FERTILITY, MIGRATION, MORTALITY and POPULATION.

Cities are centres of concentrated production, consumption and waste disposal that generate changes in land use and global environmental problems. They are also centres of improvisation and imagination, which may provide timely solutions to the problems they generate. The urban implications of climate change are that large coastal city populations are likely to be put at risk of flooding, storm surges and other extreme weather conditions. The United Nations Population Division estimates that by 2030 each of the major regions of the developing world will hold more urban than rural dwellers, and that by 2050 two-thirds of their people (95% of the net increase) will probably live in urban areas. Although the rural-to-urban migration of populations that started in the late 20th century is widespread, and perhaps the greatest human-environment experiment of all time, the importance of the natural increase of urban populations tends to be under-emphasized. This unprecedented rate of urban growth has taken place on <3% of the global terrestrial surface, yet its impact has been global: 78% of carbon emissions, 60% of residential water use and 76% of wood used by industry is attributable to cities. The excessive demands of a highly stratified urban elite have led to previous collapses of otherwise successful societies (e.g., the salinization resulting from irrigation practices in 3rd millennium BC Mesopotamia). Deforestation of New World tropical rainforests is largely the result of demand for beef by Western Hemisphere societies. The negative correlation between wealth and FERTILITY that regularly emerges in modern nation-states is a challenge for evolutionary demographers, since a basic tenet of evolutionary biology is that greater individual resources and fertility are usually positively correlated. It has been suggested that the emergence of post-industrial life, largely free as it is from fear of early child mortality, may have generated a situation in which a runaway process of ever-increasing levels of investment in children drives fertility ever lower. See CROSS-SECTIONAL STUDIES, LONGITUDINAL STUDIES.

human diversity See GENETIC DIVERSITY, HUMAN GENOME DIVERSITY PROJECT.

human embryonic stem cells (hESCs) See EMBRYONIC STEM CELLS.

human gene therapy See GENE THERAPY.

human genome (HG) A human zygote obtains the vast majority of its DNA from the two pronuclei that pool their chromosomes at FERTILIZATION, a small remainder being present in the mitochondria derived from the egg cytoplasm. The phrase 'human genome' is usually understood as referring to the human nuclear GENOME (see entry), distributed over 23 pairs of CHROMOSOMES. But although 'genome' suggests that 'genes' are involved, 45% of human nuclear DNA comprises dispersed repeat elements (LINES and SINES) with copy numbers ranging from hundreds to thousands, amounting to ~20 times as much DNA as encodes human proteins (the usual sense of 'genes'). Moreover, as much as 8% of this genome derives from endogenous retroviral genomes, although only several of the ~40,000 retrovirus-derived segments have ever been shown to be genetically intact and capable of specifying infectious retrovirus particles (see TRANSPOSABLE ELEMENTS). It has been customary to search for protein-coding genes by equating each with an OPEN READING FRAME. It is currently (2007) thought that the human genome contains 19,209 protein-coding genes (with perhaps ~100 more to be discovered), although there has been an explosion in the discovery of RNA 'genes', DNA sequences that do not encode proteins but do have a biological function (see NON-CODING DNA, NON-CODING RNA). There is growing, sometimes indirect, support for the view that more than half of the *functional* complement of the HG is non-protein-coding and that microRNAs play a large part in its regulation through RNA INTERFERENCE (see HUMAN TRANSCRIPTOME). It is the aim of researchers working in the ENCODE (Encyclopedia of DNA Elements) Project Consortium to identify every DNA sequence with functional properties (see GENE EXPRESSION). The results up to 2007, involving analysis of just 1% (30 Mb) of the DNA, reveal: (i) that the HG is, in fact, pervasively transcribed (including into non-protein-coding transcripts), and that these extensively overlap one another; (ii) new understanding about transcription start sites; (iii) a more sophisticated view of chromatin structure, and its involvement in DNA replication and transcription regulation; and (iv) an integration of these new sources of information relevant to mammalian evolution and to the functional landscape of the HG.

In addition to the above, the HG also contains an abundance of large DNA segments that have duplicated over the past 40 Myr. Such SEGMENTAL DUPLICATIONS represent ≥5% of the human genome, and are frequently located near centromeres and telomeres (see SUBTELOMERES).

Unexpectedly low numbers of genes in humans (the fruit fly *Drosophila* has almost half as many as humans, and humans have just 1.5 times as many as the nematode worm *Caenorhabditis*) has led to the belief that alternative PROMOTERS, alternative splicing (see RNA PROCESSING) and post-translational modification of proteins generate very many more distinct protein species than are apparent from counting the number of genes in the HG.

What may be termed 'the HUMAN GENOME', however, incorporates global genetic variation (see HUMAN GENOME PROJECT, VARIATION). Discovery of new human genes is becoming an increasingly rare event. Study of the pufferfish (*Tetraodon*) genome in 2004 suggested an additional ~900 predicted genes in the human genome, and although considerable uncertainty remains about the precise number, it is thought to be no more than ~22,000. It is recognized that genome complexity does not always parallel biological complexity (e.g., see C-VALUE PARADOX), and alternative splicing may help to explain small human gene number.

The euchromatic portion of the nuclear genome comprises ~3,000 Mb, while heterochromatic regions comprise ~200 kb. Although there is a genome-wide average of 41% GC base combinations in euchromatin, the proportion of the CPG DINUCLEOTIDE is conspicuously under-represented in the nuclear genome, and is unevenly distributed both between chromosomes and along their lengths (see G BANDS). But certain small regions of transcriptionally active DNA (CpG islands) have the expected CpG density and, significantly, are unmethylated (see DNA METHYLATION).

Protein-coding genes are not evenly distributed on the chromosomes, as was first heralded by the uneven distribution of CpG islands. The draft sequence of the HG confirmed that gene density is comparatively high in subtelomeric regions, and that some chromosomes (e.g., 19 and 22) are gene-rich, while others (e.g., X and 18) are gene-poor.

Comparative genomics can provide powerful evidence for reconstructing the common ancestral karyotype from which two or more (often distantly) related species' karyotypes have evolved. Analysis of the pufferfish (*Tetraodon nigroviridis*) genome in 2004 made it possible to infer the basic structure of the ancestral vertebrate genome, apparently comprising 12 chromosomes, and to reconstruct much of the evolutionary history of ancient and recent

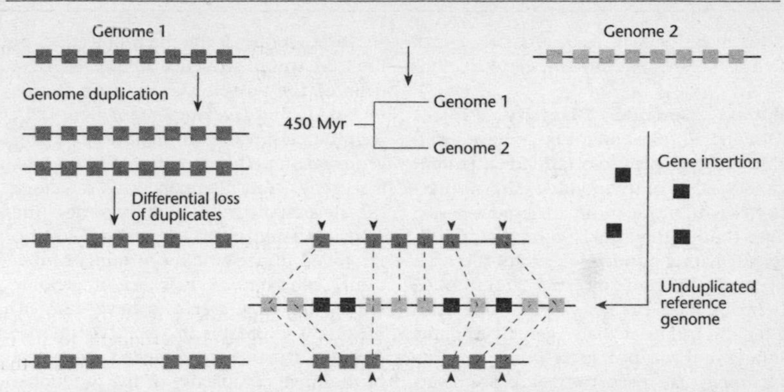

FIG 97. An illustration of the comparative genomic approach, used to reveal whole genome duplications in the pufferfish, *Tetraodon*, and to reveal paralogons in the HUMAN GENOME. Genome 1 (*Tetraodon*) undergoes a duplication event, creating two identical sets of chromosomes and genes, followed by differential gene loss in two duplicated chromosomes (left side). Genome 2 (e.g., human) experiences some gene insertions, but no wholescale duplication. In most such unduplicated reference genomes, large regions of 'double conserved synteny' are detectable, such that every chromosome in Genome 2 maps to two chromosomes in Genome 1 in an interleaving pattern (middle lower). Genes that have remained in two copies (arrowheads) would be anchor points to identify a paralogon. This approach has detected 'double conserved segments' in a genome that underwent a whole genome duplication ~200–300 Mya, separating from its reference genome ~450 Mya. See also GENE DUPLICATION. From G. Panopoulou and A.J. Poulstka, © 2005 *Trends in Genetics*, with permission from Elsevier Ltd.

chromosome rearrangements leading to the modern human karyotype. As described in the GENE DUPLICATION entry, one of the distinctive features of *whole genome duplication* (WHD) is double conserved synteny in the genome that has not undergone WHD. It seems that ~80% of orthologous loci shared between the pufferfish *Tetraodon nigroviridis* and humans have this arrangement (see Fig. 97).

However, comparisons of the human and chimpanzee genome have not yet offered major insights into the genetic elements underlying bipedal locomotion, a large brain, linguistic abilities, elaborated abstract thought, or any other seemingly unique aspect of human phenotype. This is indicative of the now widely held view that interpretation of DNA sequences requires functional information from the organism that cannot be deduced from sequences alone (see GENE EXPRESSION, GENOMICS, HUMAN TRANSCRIPTOME).

Searching the human chromosomes for duplicated genes that are located next to each other on more than a single location revealed that the *Hox*-gene-bearing chromosomes (located on chromosomes 2, 7, 17 and 12) do indeed indicate conserved SYNTENY with members of several gene families. Similarly, the human MAJOR HISTOCOMPATIBILITY COMPLEX cluster and its paralogous regions provide evidence for chromosomal duplication early in vertebrate evolution.

Comparison of the human HUNTINGTON'S DISEASE gene, *HTT*, with the equivalent gene in pufferfish revealed that both have exactly 67 exons with almost perfect conservation of INTRON positions. Recent (2008) fine-scale mapping of eight human genomes from individuals of diverse geographical ancestry revealed that half of the 1,695 structural variants located (insertions, deletions and inversions > 6 kbp in length) were observed in more than one individual, and that half also lay outside regions of the genome previously described as structurally variant. They included 525 new insertion sequences not present in the human reference genome, many of them being variable in COPY NUMBER (i.e., polymorphic) between individuals. See MUTATIONS; PHYLOGENETIC FOOTPRINTING; SYNTENY, Fig. 170.

Human Genome Diversity Project Although an ideal might be to have available the evidence from individual patient genotypes in order to guide clinical practice, such information is decades away. Even though there are far more genetic differences among individuals than there are between different ancestral groups (i.e., between modern 'races'), this does not alter the fact that many genetic variants influence how people respond to drug treatment (see DRUG TARGETING). See SINGLE NUCLEOTIDE POLYMORPHISMS.

Human Genome Project (HGP) HGP was launched in 1990, its aim being the accurate sequencing of the vast majority of the euchromatic portion of the human genome. For greater understanding of inherited disease, the mouse and human genomes were both mapped, while sequencing of the smaller genomes of several other organisms was begun for better interpretation of the human genome.

The two groups involved, the International Human Genome Sequencing Consortium (IHGSC) involving 26 centres in six countries, and Celera Genomics, each published draft human sequences in 2001, enabling gene identification, regional differences in genome composition, the distribution and history of transposable elements, distribution of polymorphism, and the relationship between genetic recombination and physical map distances between gene loci.

By 2004, the ~150,000 gaps remaining in the human sequence (at least 10% of the euchromatic genome) had been reduced to a mere 341, most involving segmental duplications that require improved methods of resolution, with an error rate of ~1 per 10^6 bases. Containing 2.85 billion (10^9) base pairs, this amounted to ~99% of the euchromatic genome, and a series of papers describing individual chromosomes with annotation of their genes and other features was either in preparation or in press.

Although one of the main objectives of the HGP was to sequence all 3×10^9 base pairs of the HUMAN GENOME, other factors included: the development of new DNA technologies; the construction of human gene maps and the mapping of human 'disease genes' and pharmacogenetic factors; the development of bioinformatics; the study of comparative genomics; and an improved understanding of human functional genomics. A new era of genome sequencing now seems to have dawned, making it affordable in small laboratories (see DNA SEQUENCING). Growing awareness of the huge complexity of the functional organization of the human genome in terms of GENE EXPRESSION within the NUCLEUS offers the prospect that, e.g., inappropriate levels of a patient's particular (often inducible) liver enzyme could be remedied by combining detailed knowledge of the individual's genome with a schedule of drug delivery peculiar to that individual's needs. Work on human GENETIC DIVERSITY indicates that there are far more genetic differences among individuals than there are between different ancestral groups (i.e., between modern 'races'), many genetic variants influencing how people respond to drug treatment (see DRUG TARGETING). Although one ideal might be to have the evidence from individual patient genotypes in order to guide clinical practice, such information is still decades away. As the transition to genome-wide genotype–phenotype associations occurs, the challenge will be to replicate positive findings in order to separate true associations from all the false positives attained.

Before sequence completion of the human genome, the gene number estimate was in the region of ~70,000, but this has been reduced to ~22,000 (in part through comparison with the genome of the pufferfish *Tetraodon nigroviridis*). However, the largest estimates of human gene numbers based on expressed sequence tags (ESTs), identified by sequencing cDNA libraries, have not been similarly reduced from their original prediction of 100,000–200,000 human genes. This discrepancy appears to have been resolved since the discovery in 2005 that an amazing 62% of the mouse genome is transcribed and that a large proportion of mRNA transcripts (the transcriptome) of a cell, tissue or organism, consists of NON-CODING RNA, a high proportion of which overlap to some extent with a transcript from the opposite DNA strand.

One of the major findings of the HGP has been the extent to which the genes encoding AXONAL GUIDANCE proteins have expanded during vertebrate evolution, perhaps especially so in humans.

See CANCER, GENETIC DIVERSITY, HAPLOTYPE, HUMAN GENOME DIVERSITY PROJECT, LEUKAEMIAS, NATURAL SELECTION, SINGLE NUCLEOTIDE POLYMORPHISMS.

human immunodeficiency virus See HIV.

human kinome The HUMAN GENOME sequence reveals 518 distinct genes encoding protein kinases which, as a group, have been called 'the human kinome'. Of these, 90 phosphorylate tyrosine residues, the remainder phosphorylating serine or threonine residues of substrate proteins. Because they show clear structural relatedness to each other, indicating their common evolutionary descent with modification, it complicates the design of drugs that target them selectively.

human lymphocyte antigens See HLA COMPLEX.

human papillomaviruses (HPVs) Causative agents of genital warts, cervical cancer and anal cancer. See VIRUSES.

human placental lactogen (hPL) A powerfully somatotropic hormone produced by the syncytiotrophoblast and secreted into both maternal and foetal circulations after the 6th week of pregnancy. It exerts a negative feedback effect on growth hormone and may increase the risk of gestational diabetes in susceptible women. In the foetus, it activates lactogenic receptors, modulating development, regulates intermediary metabolism and stimulates production of insulin-like growth factors, insulin, adrenocortical hormones and pulmonary surfactant. With oestrogens, it stimulates proliferation of maternal MAMMARY GLAND tissue.

human somatotropin See GROWTH HORMONE (hGH).

human transcriptome The human TRANSCRIPTOME is the total set, and relative amounts, of different RNA transcripts present in a human cell or tissue. The sensitivity of tiling microarray techniques (see DNA MICROARRAYS) enables abundant and rare transcripts to be identified, many lines of evidence now indicating that more of the HUMAN GENOME is transcribed than was thought at the start of the millennium. In 2004, microarray tiling through human chromosomes 20 and 22 (hsa20 and hsa22) indicated 47% of positive probes lay outside their currently annotated exon positions, 22% within introns and 25% within intergenic regions. Using tiling arrays of 5-nucleotide (nt) resolution, a recent study of polyadenylated RNAs >200 nt (long RNAs, lRNAs) and whole-cell RNAs <200 nt (short RNAs, sRNAs) over the entire non-repetitive portion of the human genome reported the potential biological function of a significant proportion of long unannotated transcripts to serve as precursors for sRNAs, and that sRNAs specifically localized to gene boundaries (5′ and 3′), termed 'promoter-associated sRNAs' and 'termini-associated sRNAs'. The third class of RNAs, lRNAs, overlapped the 5′ boundaries of polypeptide-coding genes, but although including the promoter and the first exon and intron regions did not include most of the other exons. In this study, ~64% of detected poly(A)$^+$ transcription did not align with previous annotations, and ~41.8% of transcribed sequences seemed to remain exclusively in the nucleus while the remainder were transported into the cytosol. See HUMAN GENOME PROJECT.

human variation See GENETIC VARIATION, PHENOTYPE.

humans See HOMININS.

humerofemoral index The ratio of humerus to femur length, expressed as a percentage. For modern humans, it is 71.4% for males and 69.8% for females. For male common chimpanzees, it is 101.1% for males and 102% for females, while for the pygmy chimpanzee it is 102.2% for both sexes.

humerus The longest and largest bone of the upper limb, articulating proximally with the glenoid cavity of the scapula and distally with the radius and ulna, forming the elbow joint. See SKELETAL SYSTEM, Fig. 156.

humoral immune response Because body fluids were formerly known as 'humours', these are immune responses mediated by ANTIBODIES and therefore perhaps more obviously reliant on the services of B CELLS (see Fig. 19) than of T cells.

hunger centre See FEEDING CENTRE.

hunger dependency A reduction in the response of neurons as more food is eaten. See TASTE.

Huntington's disease (HD, Huntington's chorea) Clinically, HD is characterized by chorea, psychiatric disturbances and dementia, and pathologically by loss of long projection neurons in the cortex and striatum, leading to progressive mental disturbances, physical degeneration and eventually death – usually within 15–20 years of initial symptoms. A genetically dominant, late onset disorder, HD is one of several neurodegenerative diseases caused by gain-of-function tandem repetition of a codon for glutamine in the coding region of the *HTT* (*HD*) gene near the tip of the short arm of chromosome 4, this being the first (1983) disease-causing gene to be mapped against DNA MARKERS (see GENOME MAPPING). Normal individuals have 6–39 of these CAG repeats, but individuals with HD have 36–180 repetitions, the greater the expansion being associated with earlier disease onset (see TRIPLET REPEATS). The protein encoded by the normal *HTT* allele is a 3144-amino acid protein, *huntingtin*, apparently implicated in synaptic plasticity through association with clathrin-coated membranes along dendrites and involved in both presynaptic and postsynaptic functions. Evidence also suggests that mutant huntingtin affects cortical neurons producing brain-derived neurotrophic factor (BDNF), which is required for survival of striatal neurons. In cultured cells, additions of glutamine to huntingtin weakens its interaction with the protein Hip1 with which it normally binds, so that free Hip1 then interacts with Hippi and sets off an apoptotic cascade involving caspase-8. Although caspase-3 can fragment huntingtin and cause a piece to self-aggregate and form inclusions in the cytoplasms and nuclei of cultured cells, it is uncertain whether these are required for neuronal death in HD, although such amyloid-like fibrils occur *in vitro* in patients with HD. Rather, it now appears that inclusion-body formation in HD does not harm neurons but instead protects them against the mutant huntingtin protein. Even so, evidence from ALZHEIMER'S DISEASE and HD models has implicated these abnormal proteins in the disruption of glutamatergic neurotransmission and calcium signalling and drugs aimed at their removal or neutralization may improve neuronal survival. Recent evidence from mice indicates that the neural toxicity derives from expansion of inherited CAG repeat tracts at mid-life which serve as templates for synthesis of increasingly toxic HD protein. The disease is associated with degeneration of the neostriatal regions of the basal ganglia, and dysfunctional NMDA RECEPTORS have been implicated.

Various lines of evidence implicate mitochondrial dysfunction in HD and there are several mechanisms by which mutation in *HTT* could achieve this, in one of which huntingtin interacts with the outer mitochondrial membrane of brain cells, lowering its membrane potentials and causing depolarization at lower calcium loads than normal mitochondria. NMR spectroscopy reveals increased lactate in the cortex and basal ganglia and biochemical studies in humans reveal decreased activities of complexes II and III of the electron transport chain of the inner membrane of the MITOCHONDRION. Recent work with mice indicates that the level and accumulation of oxidative damage to DNA correlate well with the degree of expansion of the CAG repeat, and that the age-dependent somatic repeat expansion actually arises in the process of removing oxidized base lesions. It occurs in germ cells during differentiation and in the striatum of disease patients – although expansion cannot be monitored in humans. Both *in vivo* and *in vitro* work favours the view that a single base excision repair enzyme, 7,8-dihydro-8-oxoguanine-DNA glycosylase (OGG1), is involved (see DNA REPAIR MECHANISMS). It initiates an escalating oxidation-excision cycle leading to progressive age-dependent CAG expansion, providing a direct molecular link between oxidative damage and toxicity in post-mitotic neurons through an error-prone response to single-strand DNA breaks.

Although symptoms may appear earlier or later, they often present in the thirties and forties. Ubiquinated derivatives of the mutant proteins are found in intranuclear inclusions and may induce APOPTOSIS of striatal cells, although it is not yet clear whether this is a cause of the motor symptoms (see DYSKINESIAS). Neuronal dysfunction rather than cell death may be responsible for early neurological deficits, but the probability of neuronal death is constant with time and nucleation of the affected huntingtin molecules into aggregates (a random mechanism) appears to be the cause of neuronal death. A non human primate (rhesus macaque) model of HD involving transgenesis of a polyglutamine-expanded *HTT* gene has recently been developed.

H-Y antigens Minor histocompatibility antigens encoded by Y-chromosome-specific genes and therefore not expressed in females. Female anti-male minor histocompatibility responses occur as a result, but male anti-female responses do not because both sexes express X-chromosome genes. The Y-chromosome gene *Smcy* encodes a protein from which peptides occur on MHC class I proteins. The minor histocompatibility antigens encoded by autosomes have not been characterized, although at least 10 have been identified.

hyaluronic acid (hyaluronan, hyaluronate) The simplest GLYCOSAMINOGLYCAN, and the only one that is neither sulphated nor covalently attached to a protein to form a glycoprotein. Located in many tissues, it is involved in cell migration, in resisting compressive forces in joints and tissues, and in tissue repair. When secreted from the basal surface of an epithelium, it can create a space into which other cells can migrate. It also serves as a lubricant in synovial fluid. Excess hyaluronic acid is usually degraded by *hyaluronidase*, a component of some snake venoms.

hybrid genes See FUSION GENES.

hybridoma A clone resulting from proliferation of a hybrid cell resulting from artificial fusion of a normal antibody-producing B cell with a B-cell tumour cell. It is a technique involved in production of monoclonal antibodies.

hydrocortisone See CORTISOL.

hydrogen peroxide A highly reactive oxidant (H_2O_2) generated by the electron transport chain during aerobic respiration in the MITOCHONDRION and through the action of drugs such as the antimalarial primaquine and natural products such as divicine, the toxic ingredient of fava beans (see GLUCOSE-6-PHOSPHATE DEHYDROGENASE). It is normally detoxified by reduced GLUTATHIONE and glutathione peroxidase, but is also broken down by catalase in peroxisomes, both these detoxification processes requiring NADPH. Activated MACROPHAGES and neutrophils produce H_2O_2, further chemical and enzymatic reactions generating from it other toxic ions, including the hydroxyl radical, hypochlorite and hypobromite ions. See PKC.

17α-hydroxypregnenolone A metabolite of CHOLESTEROL (Fig. 49) in the synthesis of cortisol, 17β-oestradiol and testosterone.

17α-hydroxyprogesterone An intermediate in cortisol synthesis from progesterone. See CHOLESTEROL, Fig. 49.

hyoid bone A horseshoe-shaped bone supporting the tongue and held between the mandible and the larynx by ligaments and muscles (see SKULL, Fig. 157). It does not articulate solidly with any other bone, and is regarded as a sesamoid bone. It comprises a horizontal *body* attached to two pairs of processes, the *cornua*, resembling the horns of a bull. The muscles inserting on it attaching it to the tongue and larynx are used in SPEECH, SWALLOWING and other tongue movements; other muscles inserting on it are involved in MASTICATION. See AUSTRALOPITHECUS AFARENSIS.

hypercapnia An abnormal increase in the carbon dioxide concentration (pCO_2) in the blood, bringing about vasodilation and increased blood flow (but not in the lung). It leads to increased pCO_2 in the cerebrospinal fluid. It can result from reduced VENTILATION. Its occurrence often coincides with HYPOXIA, as following ischaemia and within necrotic tumours.

hypercholesterolaemia Elevated blood CHOLESTEROL, contributing to atherosclerosis and cardiovascular disease. Two fungal products, compactin and lovastatin, inhibitors of HMG-CoA reductase, were the first statin drugs developed to treat hypercholesterolaemia.

hypercolumns Cells in the PRIMARY VISUAL CORTEX (see Fig. 142) responsive to stimuli with the same retinal position and axis of orientation are organized into columns running from the surface next to the pia mater through to the white matter – a 2 mm thickness. As with the PRIMARY SOMATOSENSORY CORTEX, these functional columns include cells from all six cell layers (see CEREBRAL CORTEX, Fig. 45). Each hypercolumn contains a complete set of orientation-specific columns, representing 360°, a set of OCULAR DOMINANCE COLUMNS (see later) and several BLOBS, each occupying about 1 mm^2 of cortical surface. Each orientation column (30–100 μm wide) contains in layer IVC stellate cells with concentric RECEPTIVE FIELDS, just above and below which lie simple cells. All the cells in a given column have fields centred on almost the same retinal positions, although fields differ in size (involved in BINOCULAR VISION and depth perception). This generates receptive fields with particular axes of orientation, an example of ABSTRACTION over the circular receptive fields of retinal and lateral geniculate cells. Columns with different responses to stimulus axis of orientation lie side by side, the axis shifting by about 10° from one column to the next. Termed *orientation columns* by Hubel and Wiesel, these are interspersed by 'BLOB' regions, columns of cells concerned with colour discrimination. Another columnar system in a hypercolumn is the *ocular dominance column*, which receives input from one or other eye (ipsilateral, I, or contralateral, C), those for the two eyes alternating in a regular manner (see PRIMARY VISUAL CORTEX, Fig. 143). Columns of cells with similar functions are linked by horizontal connections (collaterals) between cells in a single layer. Thus, such groups of cells can fire simultaneously in response to a stimulus with a specific orientation and direction of movement. Within the ocular cortex, CREB plays a key role in establishing ocular dominance. See ORIENTATION-SELECTIVE CELLS.

hyperglycaemia Raised average blood glucose concentration, associated with DIABETES MELLITUS and impaired mitochondrial function (see MITOCHONDRION). See BLOOD GLUCOSE REGULATION.

hyperkinesias Motor disorders arising from dysfunction of the basal ganglia in which overactivity occurs giving rise to frequent, random, twitch-like or writhing movements.

hypermutation See ANTIBODY DIVERSITY, SOMATIC HYPERMUTATION.

hyperphagia Voracious eating. See FEEDING CENTRE.

hyperplasia Increase in the number of cells in a tissue, rather than increase in the volume of cells already present (HYPERTROPHY). Hyperplasia is thought to be far less significant than hypertrophy in muscle enlargement. *Hyperplastic growths* are cancers that differ minimally from normal in terms of tissue structure apart from the excessive numbers of cells involved.

hypersensitivity A state of heightened sensitivity to an innocuous allergen leading to symptomatic reactions on reexposure to it. Type I (anaphylactic) hypersensitivity reactions (immediate hypersensitivity reactions, or ALLERGIES) involve IgE antibody triggering of mast cells and histamine release; type II hypersensitivity (cytotoxic) reactions involve IgG antibodies against cell-surface or matrix antigens, usually leading to activation of COMPLEMENT; type III (immune-complex) reactions involve antigen–antibody (IgA or IgM) complexes, and complement; type IV (cell-mediated) reactions (delayed-type hypersensitivity), occurring hours to days after antigen is injected, are mediated by CD4 T_H1 cells (see T CELLS).

hypertension Acute or chronic elevation of BLOOD PRESSURE (see entry) to values higher than 140 mm Hg systolic pressure and 90 mm Hg diastolic pressure. Vascular pathology tends to produce, or exacerbate, high blood pressure, and is an important risk factor in CARDIOVASCULAR DISEASE (see http://www.who.int/whr/2002/en/).

Arteriosclerosis can lead to narrowing of the vessels, increasing total peripheral resistance. Coarction of the aorta (see CONGENITAL HEART DEFECTS) can also increase systolic blood pressure, notably to the upper body. In polycythaemia (where red blood cell count is raised), blood pressure may rise through increased blood viscosity. As arterial pressure increases, often through STRESS, the operation of the Starling mechanism (see CARDIAC MUSCLE, STROKE VOLUME) leads to raised left ventricular END-DIASTOLIC PRESSURE, although this is minimized by hypertrophy of the cardiac muscle. But this only partially compensates for the increased work load put on the heart, and left ventricular failure may eventually occur. It is twice as common in African Americans as in Caucasians.

Chronic hypertension is a dysregulated state, resulting in large measure from chronic activation of the HPA axis (see STRESS RESPONSE) and sympathetic nervous system. It has adverse effects on the heart and circulation, leading to hypertrophy of cardiac muscle (esp. the left ventricle), increasing the risk of heart failure (see CARDIAC MUSCLE). It also increases the rate at which arteriosclerosis develops, which in turn increases the risk of thrombosis. Associated problems include stroke, coronary infarction, renal haemorrhage, and retinal haemorrhage leading to poor vision.

Contributory factors include decrease in functional kidney mass, excess aldosterone or angiotensin release, and increased resistance to blood flow in the renal arteries – all increasing total blood volume, leading to increased CARDIAC OUTPUT, increased blood flow through tissue capillaries and consequent contraction of precapillary sphincters – again increasing peripheral resistance and blood pressure. *Idiopathic*, or *essential*, hypertension, one of the most common disorders of the modern world, is a COMPLEX DISEASE. Over 30 years ago, Guyton and colleagues suggested that defective handling of sodium by the kidney (see DISTAL CONVOLUTED TUBULE) is a requisite, final common pathway in hypertension pathogenesis. Powerful support for this view has since emerged from the genetic studies of Lifton and colleagues showing that virtually all Mendelian disorders with major effects on blood pressure derive from genetic variants affecting salt reabsorption within a circumscribed portion of the distal convoluted tubule. Linkage studies are consistent with susceptibility genes of modest effect size.

β_1-ADRENERGIC receptor antagonists (beta-blockers such as propranolol) and α_2-adrenergic agonists are used in treatment of hypertension; and ANGIOTENSIN and ALDOSTERONE levels can be reduced by drugs blocking angiotensin-converting enzyme (ACE). See ACETYLCHOLINE, PREECLAMPSIA.

hyperthyroidism Excessive functional activity of the THYROID GLAND. Occurring in ~2% of women and 0.02% of men. There are several possible causes, including Graves disease, toxic nodular goitre, excess iodine intake and TSH-producing tumour of the hypophysis.

hypertrophy Increase in mass of a tissue

by cell enlargement (contrast HYPERPLASIA). In striated muscle, it usually occurs in response to contraction at maximal or almost maximal force, especially if the muscle is loaded during contraction.

hypervariable microsatellite sequences See VNTRS.

hypervariable (HV) regions See ANTIBODIES, ANTIBODY DIVERSITY, SOMATIC HYPERMUTATION.

hyperventilation Rapid deep breathing, allowing exhalation of more carbon dioxide (CO_2) until its pressure in the blood, and blood pH, are restored to normal. If plasma CO_2 level falls significantly (HYPOCAPNIA), e.g. to 40 mm Hg, central and peripheral CO_2 receptors are not stimulated and so the inspiratory reflex reverts to its endogenous (lower) level until CO_2 levels rise again. Competitive swimmers are no longer encouraged to hyperventilate before diving in since their blood oxygen level can fall to dangerously low levels before the CO_2 receptors kick in, causing fainting. The dizziness often experienced after hyperventilation is due to the reduced cerebral blood flow accompanying reduced dissolved CO_2 (pCO_2) levels. See OXYGEN TOXICITY, RESPIRATORY CENTRE.

hypoblast A layer of small cuboidal cells adjacent to the blastocyst cavity which, with the epiblast, forms the embryoblast (termed the inner cell mass until the blastocoele develops).

hypocapnia Low blood CO_2 concentration (pCO_2) between 40 and 20 mm Hg, inducing considerable vasoconstriction. Compare HYPERCAPNIA.

hypocretinergic, hypocretins See OREXINS.

hypoglossal nerve (cranial nerve XII) Originating in nuclei in the medulla, the nerve containing fibres controlling tongue movements during SPEECH and SWALLOWING.

hypoglycaemia Low blood glucose concentration. See BLOOD GLUCOSE REGULATION.

hypokinesias Disorders in which motor activity is reduced, the prototypical example being PARKINSON'S DISEASE. In animal studies, lesions of the globus pallidus result in abnormal co-contraction of agonist and antagonist muscles in the contralateral limbs, increasing limb stiffness and causing rigidity and slow movement (bradykinesia). Bilateral lesions in animals typically result in adoption of abnormal flexed positions that are apparently irreversible (see DYSTONIAS).

hypophysis The PITUITARY GLAND.

hypotension Low BLOOD PRESSURE.

hypothalamus A small but enormously important component of the LIMBIC SYSTEM, being that part of the diencephalon (see BRAIN, Fig. 28) consisting mainly of nuclei positioned in and around the floor of the third VENTRICLE, just ventral to the THALAMUS. Developing under the influence of SHH, product of the SONIC HEDGEHOG GENE, and subsequent expression of *NKX2.1* (a homeobox gene whose expression is restricted in mice to two domains within the forebrain: the basal telencephalon and the hypothalamus), its nuclei are implicated in sleep, appetitive behaviours, and the control of homeostatic autonomic and endocrine functions (largely the responsibility of the *paraventricular nuclei*, PVN), playing a key role in the regulation of body temperature, fluid balance and energy balance. It receives input from the HIPPOCAMPUS via the subiculum and postcommissural fornix, whose fibres project largely to the MAMMILLARY BODIES. Input from the AMYGDALA arrives via the stria terminalis (following a similar looping course to that of the fornix) and the amygdalofugal pathway. Many noradrenergic and serotonergic axons synapse with hypothalamic neurons (see BRAIN, Fig. 30), but dopaminergic neurons simply traverse it. Output from the mammillary bodies goes via the mammillothalamic tracts to the anterior thalamic nuclei, thence to the cingulate cortex and along the mammillotegmental tracts to the midbrain.

Many hypothalamic neurons, but not those of the lateral hypothalamus, are organized into *nuclei*, clusters of cells whose projections (axons) reach into other brain regions and are classified on the basis of their location and neuropeptide secretion (although nuclei may contain more than one neuronal cell type). Three longitudinal zones (see the coronal section, Fig. 98b) are often recognized: the *periventricular zone* (including the suprachiasmatic, paraventricular, arcuate and posterior hypothalamic nuclei); the *medial zone* (including the medial preoptic, anterior, ventromedial, dorsomedial and medial mammillary nuclei); and the *lateral zone* (including the lateral preoptic, supraoptic and lateral mammillary nuclei, and the lateral hypothalamic area). It also has four divisions in the rostro-caudal axis (see midsagittal section, Fig. 98a): the *preoptic*, *anterior*, *tuberal* and *mammillary* subdivisions.

At the rostral end of the hypothalamus lies the optic chiasma, while its most caudal region is the mammillary bodies. It is ideally located for coordinated visceral and somatic motor responses, being anatomically and physiologically linked through the infundibulum to the PITUITARY GLAND (see Fig. 137) while linking the hippocampus and the anterior thalamic nuclei. The hypothalamic-pituitary-adrenal axis (HPA axis) is a major part of the neuroendocrine system controlling the STRESS RESPONSE. Small (parvocellular) neurosecretory cells in the ventromedial nuclei secrete hormones (including CORTICOTROPIN RELEASING HORMONE, CRH) into the thickened floor of the 3rd ventricle, the median eminence, whence they are carried via the hypothalamic-pituitary portal circulation to the anterior pituitary. Large (magnocellular) neurosecretory neurons in the PVN and SUPRAOPTIC NUCLEI (SON) synthesize neurohormones in their rough endoplasmic reticula, modify them in Golgi complexes, carrying them in secretory vesicles along microtubules to be released from the axon terminals into capillaries in the posterior pituitary in response to membrane depolarizations, as at ordinary SYNAPSES. These peptides include VASOPRESSIN (ADH) from the SON, and OXYTOCIN from the PVN. The PVN also secretes corticotropin releasing hormone (CRH) and thyrotropin releasing hormone (TRH), initiating the humoral response to excessive adiposity and high leptin levels. Leptin receptors are present on neurons of the ARCUATE NUCLEUS, ventromedial and dorsomedial nuclei, regions regulating feeding behaviour. These neurons project axons to the lower brain stem and preganglionic neurons of the sympathetic division of the spinal cord, while others project to the lateral hypothalamus and inhibit feeding in response to leptin. As a result, the hypothalamus is critical in the regulation of food intake (see APPETITE) and ENERGY BALANCE. The SON also initiates neural circuits responsible for the sensation of THIRST.

The hypothalamus integrates the actions of the AUTONOMIC NERVOUS SYSTEM, with which it is linked via the anterior thalamic nuclei. The INAH-3 (interstitial nuclei of anterior hypothalamus) is about twice the size in men as in women, and half the size in gay men as in straight men. Although interesting, this tells us little by itself about the origins of SEXUAL ORIENTATION. But the preoptic nuclei, usually considered part of the anterior hypothalamus, secrete GONADOTROPIN RELEASING HORMONE, critical in reproduction, and receive input from thermoreceptors in the skin, mucous membranes and the hypothalamus itself, all involved in TEMPERATURE REGULATION. The ventrolateral preoptic nuclei (VLPO) are primarily active during SLEEP and involved in suppression of the ascending reticular system (see RETICULAR FORMATION) which is needed for initiation of non-REM sleep. VLPO release neurotransmitters galanin and GABA, inhibiting the monoaminergic cell groups of the locus coeruleus, the raphe nucleus and the tuberomammillary nucleus (see RETICULAR FORMATION). The dorsomedial nucleus is one of the largest sources of input to the ventrolateral supraoptic nucleus and orexin-containing neurons, and is crucial for conveying information from the SUPRACHIASMATIC NUCLEUS (SCN, see entry) to the sleep–wake regulatory system (see SLEEP–WAKE CYCLES). The SCN is responsible for MELATONIN secretion by the

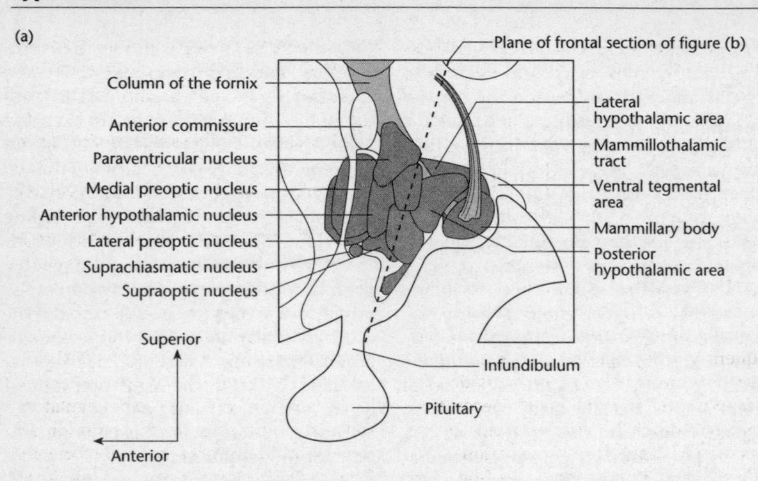

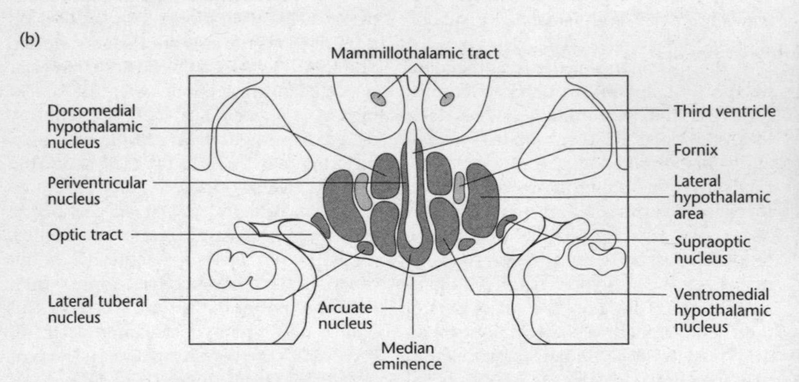

FIG 98. Diagrams to show the nuclei of the HYPOTHALAMUS, some of which are only visible in coronal section. (a) Midsagittal (medial) section; (b) coronal (frontal) section along the plane shown by the hatched line in (a). From *Essentials of Neural Science and Behaviour*. E.R. Kandel, J.H. Schwartz and T.M. Jessell © 1995 by Appleton & Lange (Simon & Schuster). Reproduced with permission of The McGraw-Hill Companies.

pineal gland. The tuberomammillary nucleus (TMN) has histaminergic neurons which make excitatory synapses with thalamic relay neurons and are a part of the ascending RETICULAR FORMATION involved in waking. Animal experiments involving removal of the posterior hypothalamus indicated that this region may have a role in the expression of anger and aggression. Cutting the connections between the hypothalamus and ventral tegmentum alleviated, but not entirely, the sham anger elicited by non-aggressive stimulation. The medial hypothalamus projects to the periaqueductal grey matter, electrical stimulation of which can produce affective aggression and lesions which can disrupt it. The amygdalofugal input to the hypothalamus from the basolateral nuclei of the amygdala is also involved in affective aggression, while lesions in the corticomedial nuclei of the amygdala, and stria terminalis, increase a cat's predatory aggression (the corticomedial nuclei inhibiting this). See AMYGDALA.

hypothermia The condition in which the core body temperature falls below 35°C resulting, progressively, in mental confusion, impaired TEMPERATURE REGULATION (32° C), cessation of temperature regulation (30°C) and cardiac ARRHYTHMIAS (~28°C).

hypothyroidism Condition resulting from insufficient THYROID HORMONE action, occurring in ~2% of adult women but less frequently in adult men. Primary hypothyroidism (95% of cases) is due to disease of the thyroid gland. Secondary hypothyroidism is characterized by a decrease in thyrotropin secretion and reduced thyroid hormone secretion. See CONGENITAL HYPOTHYROIDISM.

hypovolaemia Low blood volume. See BLOOD PRESSURE.

hypoxanthine A purine produced by the deamination of ADENINE.

hypoxia Reduced oxygen tension (pO_2) in a tissue, which may be acute or chronic. If chronic, it may weaken cells without killing them, and is often caused by ischaemia (see HYPERCAPNIA). The mechanisms by which hypoxia triggers increases in levels of the nuclear protein p53 are poorly understood, although up-regulation of HEAT-SHOCK PROTEINS does occur. In response to hypoxia, HIF1α protein accumulates (see HIF1) owing to its decreased interaction with VHL, and activates the hypoxic response elements (HREs) of target gene regulatory sequences, resulting in transcription of genes implicated in the control of metabolism and ANGIOGENESIS, and of apoptosis and cellular stress. Some of its targets include ERYTHROPOIETIN, VEGF, GLUCOSE TRANSPORTERS and multiple glycolytic enzymes.

I

iatrogenic Of diseases that are hospital-borne and spread.

ibuprofen An analgesic non-steroidal anti-inflammatory drug.

ICAMs See INTERCELLULAR ADHESION MOLECULES.

I cells See INTERCALATED CELLS.

ICSH (interstitial cell stimulating hormone) See LUTEINIZING HORMONE.

IDDM Type 1, insulin-dependent, DIABETES MELLITUS.

identical twins TWINS developing from the same fertilized egg.

idiopathic In medicine, indicating a condition which arises spontaneously or from an obscure or unknown cause.

idiopathic haemochromatosis (IHC) A recessively inherited disorder in which the absorption of dietary iron is increased, resulting in serious organ damage in later life (e.g., cirrhosis of the liver). The ratio of affected males:females is 5:1; but the condition is of benefit to women, who routinely experience high iron loss during menstruation. The *HFE* gene responsible is linked to the HLA region on chromosome 6p21. Two variants occur: *C282Y* (frequency in Northern Europe ~0.1, for which 85–100% of affected individuals are homozygous); and *H63D*, which is more common in the general population and only moderately increases the risk of haemochromatosis when homozygous. Genetic testing is common in affected families. Mutation in the transferring receptor 2 gene and the ferroportin gene *SLC11A3* is also implicated in the phenotype.

IFT system The intraflagellar transport system of eukaryotic cilia and flagella. See CILIUM.

Ig See IMMUNOGLOBULINS, and antibody isoforms listed below (IGA, IGD, etc.).

IgA The principal antibody class in epithelial secretions, such as tears, SALIVA, MILK, respiratory and intestinal secretions, where complement and phagocytes are not normally present. Its molecules are characterized by α heavy chains and IgA-secreting plasma cells are found predominantly in the LAMINA PROPRIA where IgA functions mainly as a neutralizing antibody. See F_C DOMAIN.

IgD The immunoglobulin class characterized by δ heavy chains. Of unknown function, it appears as a surface immunoglobulin on mature naïve B cells.

IgE The immunoglobulin class characterized by ε heavy chains and produced by plasma cells in lymph nodes. The cytokines IL-4 and IL-13 provide the first signal in ISOTYPE SWITCHING of B cells to IgE production. Unlike other immunoglobulins, it is predominantly located in tissues and in the absence of bound antigen binds the specific Fc receptor FcεRI on MAST CELLS (just beneath the skin and mucosa, and along blood vessels in connective tissue), BASOPHILS and activated EOSINOPHILS. It is always monomeric, its hinge region being replaced by a third C region. IgEs are involved in ALLERGIC REACTIONS. See F_C DOMAIN.

IGFs, IGFBPs See INSULIN-LIKE GROWTH FACTORS.

IgG The immunoglobulin class characterized by γ heavy chains, and the most abundant immunoglobulin isotype in blood plasma and extracellular fluid. It is always monomeric, diffuses easily into the tissues from the blood, is an efficient opsonizer of pathogens and activator of COMPLEMENT. See F_C DOMAIN.

IgM The immunoglobulin class characterized by μ heavy chains, found mainly in blood plasma, and to a lesser extent in lymph. IgM monomers contain a third C domain in place of the hinge region and form pentamers, whose ten antigen-binding sites can bind multivalent antigens such as bacterial capsular polysaccharides. They are the first immunoglobulins to appear on surfaces of B CELLS, because they can be expressed without isotype switching, and the first to be secreted – before B cells have undergone SOMATIC HYPERMUTATION. They are good activators of COMPLEMENT.

IHC See IDIOPATHIC HAEMOCHROMATOSIS.

IKK signalling See NF-κB.

IL(s) Interleukin(s). See CYTOKINES.

ileum The final, and longest, region of the SMALL INTESTINE, joining the colon at the ileocaecal sphincter.

iliac crest The superior border of the ileum, ending anteriorly in a blunt anterior superior iliac spine and posteriorly in a sharp posterior superior iliac spine.

ilium The largest of three bones which, after birth, fuse to form the hip (or coxal) bone. See PELVIC GIRDLE.

imipramine An ANTIDEPRESSANT drug.

imitation Learning of a non-genetic behaviour similar to that performed by another individual. Since it is difficult to identify whether an observed action is totally novel, imitation has also been defined as performance of a similar behaviour after observing the behaviour of another individual, regardless of whether the behaviour is already in the animal's repertoire. Debate surrounds the issue of whether *neonatal imitation* in humans should be interpreted as imitation (see FACIAL RECOGNITION). For a limited period between 5 and 10–11 weeks of age, an infant female chimpanzee raised by humans in a nursery produced gestures of tongue protrusion, mouth opening and lip protrusion more frequently than any other gesture when the experimenter specifically demonstrated these gestures. Recent work suggests that wild chimpanzees only rarely use body movements as population-specific gestural signals to communicate with another individual, and chimpanzees appear to have difficulty processing information about body movements. This limited capacity for whole-body imitation in chimpanzees may limit characteristics of social behaviours, for a behaviour must be recognized and performed by both partners for it to have a signal function. By contrast, a human can anticipate and imitate another human's actions by processing information about body movements, a process in which MIRROR NEURONS may be involved. It is nowadays thought that imitation of novel tool behaviours is a rarer event than had been assumed in the past. Whereas in *Homo sapiens*, and to an unknown degree in ancestral hominins, any tool-using behaviour can rapidly disseminate by imitation, this may only occur in circumscribed circumstances in other hominoids, and may be almost absent in monkeys. LANGUAGE development in childhood involves transformation of auditory signals in the auditory centres in the temporal lobes into verbal output from the motor cortex. See INTELLIGENCE.

immediate early genes (IEGs, immediate-early response genes) Genes whose transcription is induced, within less than ~1 hour, by binding of GROWTH FACTORS to the cell's plasma membrane, even if protein synthesis is blocked by drugs such as cycloheximide. Multiple non-convergent signal transduction pathways appear to be involved in control of IEG expression, including some activated by depolarization of the plasma membrane. Some have been implicated in LONG-TERM POTENTIATION (see SYNAPTIC PLASTICITY). The total IEG number is variously estimated to be 50–100, and they encode a number of interesting proteins, including β-actin, γ-actin, tropomyosin, fibronectin, and cell surface GLUCOSE TRANSPORTERS. IEGs encoding transcription factors (TFs) include MYC, *fos*^{b}, *junB* and *egr-1* (encoding a zinc finger TF), while others encode G_1-cyclins (promoting passage through the restriction point in late G1) and cytokines. Within an hour of IEG mRNA appearing, a second wave of gene activity occurs (blocked by cycloheximide). Expression of these delayed early genes (late response genes) depends on *de novo* protein synthesis, many of these proteins being transcription factors synthesized in the wake of IEG transcription.

immediate-response genes See IMMEDIATE EARLY GENES.

immune suppression See IMMUNOSUPPRESSION.

immune system The MYELOID TISSUES and LYMPHOID TISSUES which enable the responses of INNATE IMMUNITY and ADAPTIVE IMMUNITY.

immune tolerance Failure of the immune system to respond to an antigen. An important feature of a properly functioning immune system is tolerance to self-antigens, which develops during lymphocyte development (see B CELLS, T CELLS, THYMUS). Such self-tolerance has two facets: central (or thymic) tolerance and peripheral (non-thymic) tolerance. The former involves negative selection of T cells and the generation of T_{reg} cells. Peripheral tolerance results from inappropriate co-stimulation once T cells have left the thymus and leads to T-cell anergy instead of T-cell activation. Tolerance to allografts involves MAST CELLS.

immunity See ADAPTIVE IMMUNITY, INNATE IMMUNITY, PASSIVE IMMUNITY, and Appendix XI.

immunization One of the most significant public health achievements of the 20th century, the introduction of immunization programmes led to the control of smallpox, polio and diphtheria in those parts of the world where appropriate programmes were implemented. It may be achieved actively by use of a VACCINE, or passively by introduction of an IMMUNOGLOBULIN preparation. See ADAPTIVE IMMUNITY.

immunoglobulin-like domains (Ig-like domains) Protein domains resembling the Ig domains characteristic of IMMUNOGLOBULINS, but less closely related to them.

immunoglobulins (Ig) A very large gene superfamily (the Ig superfamily; see GENE FAMILIES AND SUPERFAMILIES) encoding plasma and cell-surface proteins (including transplantation antigens) involved in antigen recognition, and cell–cell interaction in the immune and nervous systems (e.g., ICAMs, and NCAM; see CELL ADHESION MOLECULES). The positioning of the two major INTRONS has been well conserved, suggesting that they integrated into an ancestral globin gene >800 Mya. All have at least one immunoglobulin domain (Ig domains), or immunoglobulin-like domains, characterized by an antiparallel sandwich of two β-sheets (β-barrels) held together by a disulphide bond, the so-called *immunoglobulin fold*. These domains are of two types: C domains and V domains, the amino acids in each domain usually being encoded by a separate exon.

Molecules of the Ig superfamily (see Fig. 99) include antibodies, B- and T-cell antigen receptors, MAJOR HISTOCOMPATIBILITY COMPLEX (MHC) proteins, CD4, CD8 and CD28 co-receptors, most of the invariant polypeptide chains which associate with B- and T-cell receptors, and the various F_C lymphocyte receptors (and those on other leukocytes). Many of these proteins are dimers or higher oligomers in which Ig or Ig-like domains of one chain interact with those of another. It is likely that the entire superfamily evolved by exon and gene duplications from a gene encoding a single Ig-like domain such as the β_2-microglobulin of MHC class I proteins, or the mouse Thy-1

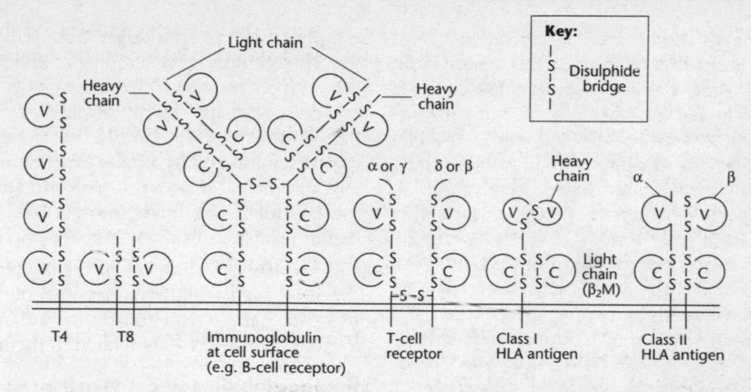

FIG 99. An illustration showing a few examples of the very large Ig superfamily (see IMMUNOGLOBULINS). Many are dimers formed from extracellular variable domains (V) located at the N-terminal ends, and constant domains (C) at the C-terminal ends closer to the membrane. The light chain of Class I HLA antigens, β_2-microglobulin, has a single C domain which does not span the membrane but associates with the heavy chain, which does. This has two V domains and one C domain, the overall structure resembling that of a Class II HLA antigen. © 2004 From *Human Molecular Genetics* 3, T. Strachan & A.P. Read, © 2004 Garland Science. Reproduced by permission of Taylor & Francis Books UK.

protein, which may have mediated cell–cell interactions prior to the origins of vertebrates.

immunologically privileged sites Those sites, including the brain (see BLOOD–BRAIN BARRIER), in which allogeneic tissue does not elicit rejection. It results both from presence of a physical barrier to cell (e.g., naïve lymphocyte) and antigen migration and to the existence of soluble immunosuppressive mediators (e.g., certain anti-inflammatory cytokines, such as TGF-β). Tissue fluid in these sites does not pass through conventional lymphatics although antigens leaving such sites can have immunological effects but induce tolerance or some other non-destructive response. The expression of FAS ligand by involved tissues may also provide further protection by inducing apoptosis of Fas-bearing lymphocytes that enter. The antigens sequestered in such sites are often the targets of autoimmune attack.

immunological memory See ADAPTIVE IMMUNITY.

immunological tolerance See IMMUNE TOLERANCE.

immunoprecipitation For chromatin immunoprecipitation, see DNA-BINDING PROTEINS.

immunoproteasomes A form of PROTEASOME found in cells exposed to interferons and containing a third subunit which displaces a constitutive unit of the usual proteasome.

immunosuppressant See IMMUNOSUPPRESSION.

immunosuppression Compounds that inhibit adaptive responses are termed *immunosuppressive drugs* and are used mainly in the pre-emption, and treatment, of graft rejection, and in severe autoimmune disease. Some graft recipients do not reject their graft when they stop taking immunosuppressants, for which there is no clear explanation (see TRANSPLANTS). Drugs employed to suppress the immune system inhibit both its protective and harmful effects and are of three kinds: synthetic anti-inflammatory GLUCOCORTICOID drugs of the corticosteroid family (e.g., prednisone); cytotoxic drugs (e.g., azathioprine (Aza) and cyclophosphamide) that interfere with DNA synthesis and have their major effects on tissues with dividing cells; and fungal and bacterial derivatives (e.g., CYCLOSPORIN A, tacrolimus and rapamycin).

Normal DNA bases do not absorb significantly at UVA wavelengths, but the thiopurines (Aza, 6-MP and 6-TG) used as cancer therapeutic and immunosuppressive agents do so. These are all prodrugs requiring metabolic activation into thioguanine nucleotides that are precursors for 6-TG incorporation into DNA. This has been associated with selective UVA photosensitivity and reactive oxygen species generation in cultured cells with 6-TG substituted DNA and may partly explain the prevalence of skin cancer in long-term survivors of organ transplantation.

Cyclosporin A is a cyclic decapeptide that blocks T-cell proliferation by inhibiting the phosphatase activity of a Ca^{2+}-activated enzyme CALCINEURIN and reducing expression of several cytokine genes normally induced on T-cell activation. Along with another immunosuppressive drug *tacrolimus*, used in ~80% of kidney and ~90% of liver and pancreas transplants, it thereby reduces the expression of several cytokine genes normally induced on T-cell activation, including IL-2 (an important growth signal for T cells).

Its effectiveness in preventing rejection of allogeneic organ grafts depends on careful control of its concentration: high doses are needed at the time of grafting, but once the graft is established the dose must be reduced to permit useful protective immune responses. *Rapamycin* blocks the late stages of T-cell activation after T-cell receptor engagement, arresting cells in G1 of the cell cycle so that they die by apoptosis. It allows activation of lymphocytes by antigen but blocks their proliferation driven by growth factors such as IL-2, IL-4 and IL-6.

Antibodies against cell-surface molecules can be used to remove specific lymphocyte subsets, or to inhibit their function. This non-toxic approach either triggers lymphocyte destruction *in vivo* (*depleting antibodies*) or blocks the function of a target protein without killing the cell (*non-depleting antibodies*). IgG MONOCLONAL ANTIBODIES causing lymphocyte depletion target these cells to the F_C receptors of macrophages and NK cells which then engulf the lymphocytes by phagocytosis. Many antibodies are being tested for their ability to inhibit allograft rejection and to modify expression of autoimmune diseases. Unfortunately, monoclonals are made most easily by using mouse cells and humans rapidly develop resistance to mouse antibodies which not only blocks their action but can also lead to allergic reactions and, if treatment is continued, to anaphylaxis.

T_R (T_{reg}) cells are a dedicated lineage of regulatory T CELLS (see entry) making up 2–10% of the total $CD4^+$ T-CELL population of the thymus and secondary lymphoid organs in mice, rats and humans. *In vitro* work suggests that these cells can suppress $CD4^+$ and $CD8^+$ T-cell responses by an undefined contact-dependent (but IL-10/TGF-β-independent) mechanism. It may involve 'reverse' signalling by B7 co-stimulatory molecules upon cross-linking on the dendritic cell or activated T-cell surface by a high-affinity receptor (CTLA4) expressed at a high level by T_R cells. High levels of IL-9 (a mast cell growth and activation factor) are produced by activated T_R cells, seemingly the link through which they recruit and activate MAST CELLS to mediate regional immune suppression, since IL-9 greatly accelerates allograft rejection in tolerant mice (see IMMUNE TOLERANCE). A subset of T cells also express receptors found on natural killer cells and are known as NKT cells, and natural killer cells themselves can exert a regulatory function. See LYMPH NODES, PLACENTA, THYMUS.

implantation An inflammation-like process of attachment of the BLASTOCYST (see Fig. 21) to the uterine epithelium. The blastocyst enters the uterus 3 days prior to implantation, at which time it 'hatches' by squeezing out from the zona pellucida, which is enzymatically punctured. Implantation only occurs during a short period when the endometrium is receptive, between days 20 and 24 of the MENSTRUAL CYCLE (see entry), during which the uterine epithelial cells are columnar and bear microvilli, having lost their progesterone receptors, and when stromal cells are proliferating. The blastocyst must also be 'implantation competent'. In humans, implantation beyond the time of endometrial receptivity leads to MISCARRIAGE. The blastocyst normally implants along the anterior or posterior wall of the body of the uterus, becoming embedded between the openings of the uterine glands. Trophoblast cells divide rapidly to form an inner CYTOTROPHOBLAST, while the outer multinucleated cell layer forms the SYNCYTIOTROPHOBLAST which begins to invade the connective tissue of the uterus. Low implantation rates are common in women undergoing assisted reproduction. Experiments on mice have shown that lipid molecules, PROSTAGLANDINS, generated by the enzyme COX-2 are essential for implantation; and circulating RELAXIN also seems to be required. Absence of LPA_3 receptors (see LYSOPHOSPHOLIPIDS) and cytoplasmic phospholipase $A_{2\alpha}$ in mice produce delayed implantation, embryo crowding, retarded foetal development and placenta-sharing by embryos and both appear to be required for prostaglandin production and implantation. It remains to be seen whether this has relevance to the human condition *placenta praevia*, in which the placenta is attached close to, or covers, the cervix; or to multiple pregnancies that have tended to arise from assisted reproductive technologies.

implicit memory Memory for perceptual and motor skills, expressed through performance, without conscious recall of past episodes. See MEMORY.

imprint control elements (imprinting control regions, ICRs) *Cis*-acting regulatory elements acting over long distances, occurring in gene clusters subject to GENOMIC IMPRINTING. See NUCLEUS.

imprinting 1. See GENETIC IMPRINTING. **2**. See CRITICAL PERIOD.

impulses See NERVE IMPULSES.

***in situ* gene therapy** See GENE THERAPY.

inbreeding A term indicating that two individuals who are breeding partners are more closely related than two randomly chosen individuals. Its result amounts to the probability that two alleles at any locus within a given individual are identical by descent from a common ancestor, which will be a function of population size (see EFFECTIVE POPULATION SIZE). When recessive deleterious alleles segregate homozygously in the offspring, a fitness cost (inbreeding depression) results. The most extreme form of human inbreeding is INCEST.

incest Incestuous relationships are those between first-degree relatives (e.g., brother/sister; parent/child) and associated with a very high risk of abnormality in offspring (less than half being entirely healthy). Severe (25%) or mild (35%) mental retardation, autosomal recessive disorders (10–15%) and congenital malformations (10%) are the most serious consequences. Marriage between such relatives is forbidden, on religious and/or legal grounds, in almost every culture. Incest-aversion between those reared in the same family appears to depend on a CRITICAL PERIOD in childhood. See CONSANGUINITY, GENETIC COUNSELLING, INBREEDING, KIN DETECTION.

incidence (of diseases, genetic disorders) The proportion of people (e.g., 2 per 10,000) who develop a characteristic, e.g., a DISEASE or inherited disorder, over a particular period – commonly a year. Contrast PREVALENCE; see FAMILIAL RECURRENCE RISK.

incisors Incisor teeth of prosimians evolved from simple conical teeth and in the aye-aye (*Daubentonia madagascarensis*) grow throughout life. In simians, incisors are blade-like or shovel-shaped (spatulate), used for cutting, paring, peeling or stripping food. Throughout the Old World monkeys, apes and hominins there is a trend for reduction in relative incisor size, and in modern humans they are small compared to those of great apes and monkeys. See DENTITION.

inclusive fitness The evolutionary principle in population genetics that an individual may promote future representation of its own genes, even if it leaves no offspring itself, by contributing to the biological fitness of close relatives. See ALTRUISM, HAMILTON'S RULE, SELFISH GENE THEORY.

incomplete penetrance See PENETRANCE.

incubation period The time between INFECTION and the appearance of disease symptoms. See LATENCY.

incus See EAR OSSICLES.

indels Collective term for genomic INSERTIONS and DELETIONS. See MUTATIONS.

indifferent gonad The developing gonad in both sexes up to the time of arrival of the primordial germ cells and before acquisition of distinctive male or female features in the 7th week of development. See SEX DETERMINATION.

indifferent stage The developmental period, up to about the 7th week, during which it is impossible to distinguish the sexes anatomically.

inducible enzyme An enzyme whose concentration increases in response to a molecular signal, usually through increased expression of the gene(s) encoding it (see GENE EXPRESSION). Many liver enzymes are inducible, and the phenomenon of inducibility extends to many other kinds of molecular cell product. Repression and repressibility are usually involved in control of enzyme and of other cell productions (e.g., see TRANSCRIPTION FACTORS).

induction In development, the production by one tissue of a new cellular property in a dependently differentiating second tissue, where the inducing tissue neither exhibits the resulting property nor changes its own development as a result of the interaction.

infarction Death of tissue due to interruption of its blood supply. Myocardial infarction is 'heart attack'. Compare ISCHAEMIA.

infection Invasion of tissues by a pathogen. About 15 million (>25%) of 57 million annual deaths worldwide are thought to be related directly to infectious diseases, the heaviest burden falling on developing countries (e.g., see HIV, MALARIA). This figure does not include the millions of deaths from complications associated with chronic infections (e.g., liver failure and hepatocellular carcinoma in those affected with hepatitis B or C viruses). Infant and child deaths from malaria and diarrhoeal diseases alone stand at >3 million annually (see DISEASE, Fig. 63). There is growing support for the view that differences in the dose of immune response genes may constitute a genetic basis for variability in responses to infectious diseases.

Local defences (see Appendix XI) are always active in preventing infection. For a pathogen to invade the human body, it must first bind to or cross an epithelium. On encountering a potential human host, a pathogen usually first makes contact with the host's mucosa or skin, to which it may then adhere prior to crossing the epithelium. The dry keratinized epithelium of the skin is one local defence which must normally be broken for infection to occur by this route. Pathogen entry via the internal epithelia is far more common, where colonization may continue for some time without adverse consequence to the host. During this time, immunity to the pathogen may develop (as occurs with *Haemophilus influenzae* and *Neisseria meningitides*). Antibodies developed in this way to some mild pathogenic strains may provide immunity to more powerful pathogens. In the early stages of infection, responses provided by INNATE IMMUNITY (e.g., the ACUTE PHASE RESPONSE) predominate, setting in train T-CELL activation and initiation of the responses of ADAPTIVE IMMUNITY.

When an infection eludes or overwhelms local defences and innate immune mechanisms it generates a threshold level of circulating non-self antigen. Some pathogens are permanently extracellular (e.g., the bacteria *Escherichia coli*, *Streptococcus pneumoniae*, *Helicobacter pylori*, *Neisseria gonorrhoeae* and *Mycoplasma* spp.), but most have an extracellular phase in which

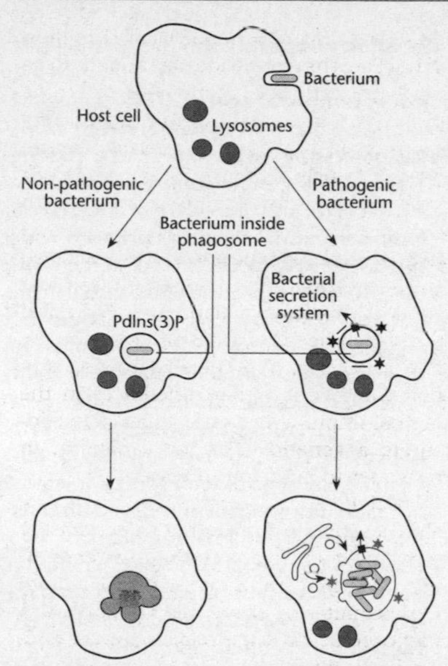

FIG 100. Diagram showing how many pathogenic bacteria can evade the usual digestion within a phagosome (left) by using their secretion systems (right) to transfer virulence factors into the host cytosol. These factors arrest pathways that generate the proteins enabling organelle identity (stars, centre right) and the normal endocytic pathway. Once this pathway has been evaded, further proteins are secreted (stars, bottom right) setting up interactions with the secretory pathway that attract delivery of host-cell proteins and lipids while the bacteria replicate. PdIns(3)P = phosphatidylinositol-3,4,5-triphosphate. See INFECTION. From R. Behnia and S. Munro, *Nature* © 2005, with permission from Macmillan Publishers Ltd.

their antigens are accessible to circulating antibody in the interstitial spaces, blood and lymph. However, intracellular phases of pathogen life cycles, often within MACROPHAGES (see Fig. 119), are not accessible to antibody and can only be targeted by T cells. Infected cells usually up-regulate their production of HEAT-SHOCK PROTEINS. Bacterial pathogens enter cells either by receptor-mediated endocytosis and replicate within ENDOSOMES (as with the bacteria *Mycobacterium* spp., *Salmonella typhimurium*, *Yersinia pestis*, some *Listeria* spp. and the protoctists *Trypanosoma* spp. and *Leishmania* spp.), or force their way inside non-phagocytic cells by using 'secretion systems' to inject virulence factors across the host cell's plasma membrane. These virulence proteins trigger localized alteration of the cytoskeleton and engulfment of the pathogen. But irrespective of the route of entry, most bacteria enter the endocytic pathway and the endosome becomes a mature lysosome in which the pathogen is degraded unless they have evolved mechanisms preventing fusion with lysosomes by altering endosomal identity markers (see VESICLES). During internalization of pathogenic bacteria by PHAGOCYTOSIS, the pathogens often capitalize in several ways on phosphoinositide-mediated signalling to invade, and thrive in, host cells (see Fig. 100). Some bacteria (e.g., SALMONELLA, *Mycobacterium tuberculosis*) deliver phosphoinositide phosphatases into the host cell cytoplasm that then help reprogramme cell function to allow invasion and/or to determine the appropriate fate of pathogen-containing vacuoles (e.g., see TUBERCULOSIS).

Some pathogens, e.g., viruses, the bacteria *Chlamydia* and *Rickettsia* spp., SHIGELLA, LISTERIA MONOCYTOGENES and some protozoa, replicate within the cytosol. Such infected cells display on their surfaces the epitopes of antigenic peptide fragments derived from their parasite's proteins, which are normally recognized by two sets of highly variable cell surface receptors: antigen-specific T-CELL RECEPTORS (TCRs) and B-cell IMMUNOGLOBULINS. *L. monocytogenes* and some mycobacteria can escape from cell vesicles and enter the cytoplasm, avoiding macrophage activation. But their presence can be detected by CD8 T cells, which can release them by killing the cell. They can then be taken up by newly recruited macrophages capable of activation to antimicrobial activity.

T cells can only recognize and act on antigens if they are delivered to the cell surfaces of antigen-presenting cells (APCs), or of infected cells, and these only arrive there by virtue of the conjoined glycoproteins of the MAJOR HISTOCOMPATIBILITY COMPLEX, the MHC molecules. However, many pathogenic bacteria can maintain themselves within cell compartments by perturbing the cytoplasmic GTPases and lipids in their membranes, subverting normal membrane trafficking (see VESICLES).

When the infection is due to pathogens of the intestinal mucosa, such as *Salmonella typhi* (cause of typhoid fever) or *Vibrio cholerae* (cause of cholera), the adaptive immune response occurs in GUT-ASSOCIATED LYMPHOID TISSUES. Gut epithelia are not passive victims of infection, but express surface molecules (such as the Toll-like receptor TLR-5) and intracellular receptors (e.g., see NOD PROTEINS) that detect bacterial invasion and activate the NF-κB pathway, releasing inflammatory cytokines and neutrophil-attracting chemokines (e.g., IL-8) and CC chemokines (attracting monocytes, eosinophils and T cells) as a result (see HELICOBACTER PYLORI).

infectious diseases See DISEASE, INFECTION.

inferior Ventral in position. Contrast SUPERIOR.

inferior colliculus A part of the MIDBRAIN tectum (see CEREBELLUM, Fig. 42), receiving sensory input from the ear via the lateral lemniscus and acting as an important relay station for this information *en route* to the thalamus. See HEARING.

inferior olivary nucleus (inferior olive) A nucleus of the rostral MEDULLA, dorsolateral to the pyramid, from which excitatory climbing fibres project via the olivocerebellar tract to the cerebellar cortex. It integrates proprioceptive information from muscles, one of its roles being to detect and correct errors in motor performance. Work with non-human primates indicates that there, at least, it is involved in switching control of learnt and novel motor patterns between the corticospinal and rubrospinal tracts (see RED NUCLEUS). See CEREBELLUM.

inferior parietal cortex See POSTERIOR PARIETAL CORTEX.

inferior temporal cortex See INFEROTEMPORAL CORTEX.

inferior temporal lobe See TEMPORAL LOBE.

inferotemporal cortex (ITC) An association area, part of the ventral stream of visual processing (see VISUAL CORTEX and Fig. 178), receiving input from V4 of the EXTRASTRIATE CORTEX and involved in a later stage of form perception than the primary visual cortex (see ATTENTION). It is concerned with the 'what' identification of the visual image (see OBJECT RECOGNITION). Its cells are not organized retinotopically, but almost all their large and usually bilateral RECEPTIVE FIELDS (RFs) include the foveal region, occasionally including the whole VISUAL FIELD. Cells of the ITC have very large RFs, usually bilateral, are sensitive for form and colour but fairly unresponsive to object size, retinal position or orientation. Many of them respond to specific objects, including hands or faces. Lesions in the ITC of macaques cause visual AGNOSIA and inability to learn tasks dependent upon object recognition, but leave visuospatial tasks unaffected. See FUSIFORM GYRUS.

infertility, infertility treatments The impact of environmental (e.g., anti-androgenic) TOXINS and the innate inadequacy of human SPERM are compounded by effective contraception and assisted reproductive technologies. It has been suggested that this lifting of the selection pressure on fertility will lead to a loss of advantage to those endowed with genes for high fecundity over those without, and that future generations may experience a further decline in SEMEN quality and human fertility. All human dominant mutations appear to arise in the male germ line, and 5–15% of severely infertile men have wholesale Y CHROMOSOME deletions. Even microdeletions destabilize sperm transmission, often causing it to be lost during gamete production. The mitochondrial and genomic DNA of spermatozoa are especially vulnerable to oxidative stress since sperm generate reactive oxygen species, are transcriptionally inactive, and are rich in targets for oxidative attack; they are also deficient in antioxidants and DNA repair mechanisms. Interest is growing around the sperm proteome, and a recent (2004) study found at least 20 proteins present in significantly different quantities in sperm of fertile and infertile men. Oversecretion of PROLACTIN is a common cause of female infertility (see also IMPLANTATION). See FOLIC ACID and impaired spermatogenesis. See GENETIC COUNSELLING, IN VITRO FERTILIZATION, KLINEFELTER SYNDROME, TURNER SYNDROME.

inflammasome See INFLAMMATION.

inflammation A local protective response to microbial invasion or to injury, characterized by PAIN, heat and swelling, and normally occurring adaptively within hours of INFECTION or wounding. It is also a key maladaptive feature of OBESITY, type 2 DIABETES MELLITUS, ATHEROSCLEROSIS and inflammatory bowel diseases (see ALLERGIES). Inflammatory-like processes are also involved in ovulation and implantation. In the adaptive form, it is initiated by MACROPHAGE release of pro-inflammatory cytokines (e.g., TNF-α, IL-1 and IL-18), and lipid mediators of inflammation (prostaglandins, leukotrienes and platelet-activating factor) that are produced by enzymatic degradation of membrane phospholipids. Inflammation may ultimately result from damage to membrane lipids by reactive oxygen species (ROS) such as peroxide, superoxide anion, hydroxyl radical and singlet oxygen (see MITOCHONDRION), and affected cells up-regulate their production of HEAT-SHOCK PROTEINS. Local increases in TNF-α cause the cardinal clinical signs of inflammation: heat, swelling, pain and redness. COMPLEMENT cleavage product C5a is also a potent inflammatory mediator. Cytokines produced during an inflammatory response increase local small blood vessel diameter and modify the cells of their ENDOTHELIUM which, with C5a, lead to extravasation by DIAPEDESIS of neutrophils and monocytes.

The endothelial cells of capillaries become more permeable, and more loosely attached (also a response to C5a), allowing plasma proteins and fluid to leak into the tissue and accounting for oedema and pain. Local mast cells are stimulated by C5a to release the inflammatory molecule histamine and TNF-α. The increased entry of plasma into infected tissues leads to activation of tissue DENDRITIC CELLS (which then migrate to LYMPH NODES) and drainage of the increased tissue fluids into the lymph. The adaptiveness of inflammatory responses lies in the resulting delivery of additional effector immune cells to the site, the provision of a physical barrier to further pathogen spread through blood coagulation which blocks minor blood vessels, and promotion of tissue repair. In a typical 'successful' inflammatory response, the duration and magnitude of TNF-α release is limited, and it is not released systemically (for its adverse effects, see TUMOUR NECROSIS FACTORS). High sympathetic activity and resulting catecholamine release stimulate β-adrenergic-receptor-dependent release of the anti-inflammatory IL-10 which, with TGF-β, specifically inhibits the release of TNF and other proinflammatory mediators. On the other hand, adrenal glucocorticoids, adrenaline, α-melanocyte-stimulating hormone and other classical 'stress hormones' inhibit inflammatory cytokine synthesis and intracellular signal transduction. In addition, a cholinergic anti-inflammatory pathway involving efferent activity of the vagus nerve exists, its ACh release in organs of the reticuloendothelial system in the heart, liver, spleen and gastrointestinal tract, interacting with α-bungarotoxin-sensitive nicotinic receptors on tissue macrophages and inhibiting the release of TNF, IL-1, HMGB1 and other cytokines. Prostaglandins and leukotrienes play a complex role in regulating the inflammatory response to injury and infection (see ENDOTHELIUM).

inflammatory bowel diseases (IBDs) Two main types of chronic IBD occur: Crohn's disease and ulcerative colitis (UC), in which patients suffer from chronic diarrhoea and weight loss, abdominal pain, fever and fatigue. Mucosal ulceration is patchy in Crohn's disease but continuous in UC. Coeliac disease, although an IBD, is not chronic, but a response to gluten. See ALLERGIES for some details.

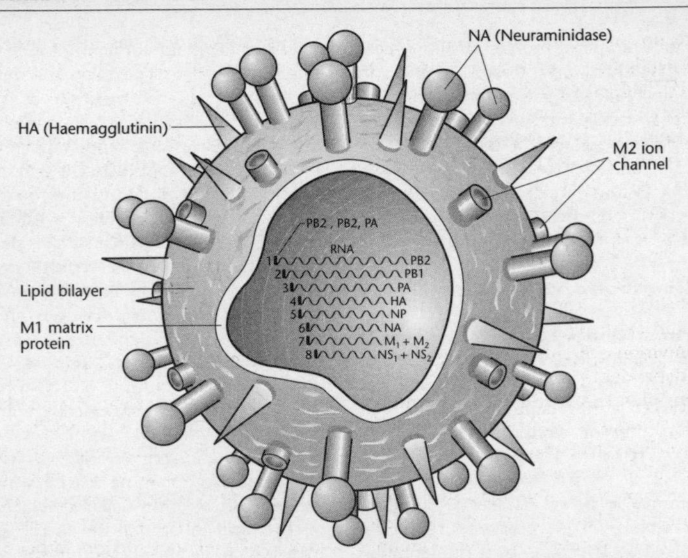

FIG 101. Diagram showing the envelope, matrix proteins and segmented genome of an INFLUENZA virus. A universal 'flu vaccine would target such conserved proteins as the envelope's M2, or the inner protein NP. From *Science* Vol. 312, No. 5772 (21 Apr. 2006), p. 380. Reprinted with permission from AAAS.

influenza ('flu, flu) Epidemic and occasionally pandemic respiratory disease caused by orthomyxoviruses known as human influenza viruses. Three types of 'flu virus exist: type A (responsible for regular outbreaks, including pandemics, and subdivided into subtypes); type B (causing sporadic outbreaks and not divided into subtypes); and type C (common, but seldom causing disease symptoms).

Orthomyxoviruses are enveloped viruses, the lipid envelope surrounding the nucleocapsid in which the viral RNA exists as separate minus-strand pieces in influenza virus (see Fig. 101). Once the virion has bound to a cell membrane of the host's respiratory tract (see VIRUS, Fig. 177), the nucleocapsid enters the cytoplasm and then the nucleus, where viral genome replication takes place after the nucleocapsid coat has been removed. The viral mRNA molecules are then transcribed in the nucleus using oligonucleotide primers cut from the 5′-ends of newly synthesized capped cellular mRNAs and the poly-A tails are then added so that the viral mRNAs can leave the nucleus and enter the host cell cytoplasm.

The influenza A virus genome has eight RNA segments encoding a total of ten viral proteins, some of which form 'spikes' on the outside of the envelope. The two proteins whose antigenic properties are of the greatest importance both form such spikes: *haemagglutinin* (H, or HA) by which the virus binds human cells, and the enzyme *neuraminidase* (N, or NA). Both of their genes mutate easily, a property responsible for the *antigenic drift* that makes each season's 'flu strain slightly different antigenically from the previous year's so that annual vaccines must be tailor-made. Distinct antigenic properties of different H and N molecules are used to classify influenza type A viruses into subtypes: 16 for H (H1–H16) and 9 for N (N1–N9), numerous combinations of which are found in bird species. 'Flu pandemics occur when a new virus that is transmitted easily between people arises in a human population where no one is immune. The 'Spanish 'flu' pandemic of 1918–19 caused by H1N1 killed between 20 and 100 million people, and its genome has been reassembled from preserved tissues and from the body of an Alaskan victim. The pandemic of 1957 was caused by H2N2 and that of 1968 by H3N2.

Most influenza vaccines promote host antibody production to the viral haemagglutinin, so that the viral HA protein is coated before it can adhere to host cells. In addition to neuraminidase, the influenza A virion contains two other enzymes: RNA-dependent RNA polymerase, transcribing plus RNA strands using the virion's minus strands as templates, and an RNA endonuclease which cuts a primer from capped mRNA precursors.

The N1 and N2 neuraminidases of viruses currently circulating in humans belong to two phylogenetically distinct groups. Group-1 contains the subtypes N1, N4, N5 and N8 while Group-2 contains N2, N3, N6, N7 and N9. The bird 'flu strain that is currently causing outbreaks in human populations of south-east Asia is of strain H5N1; but cases of human infection have also been caused by H9N2 and H7N7. As yet, these avian viruses do not spread easily among humans; but the fear is that humans (or pigs, or cats) simultaneously infected by human and bird 'flu strains might engender a novel recombinant strain of the virus (*antigenic shift*) that carries antigens to which no one is immune and which is easily transmissible between humans. Segmentation of the RNA genome makes antigenic shift more likely.

Unfortunately, bird 'flu viruses kill the chicken eggs in which vaccines to 'flu are normally grown. A solution to this may be 'reverse genetics', whereby the haemagglutinin gene is deliberately mutated to reduce its pathogenicity so that growth in eggs becomes possible (an alternative to using eggs is to produce vaccines by using cell culture in fermenters). Such an attenuated strain of H5N1 was produced by two teams in 2004 – a 'phenomenal advance'. But making such a candidate vaccine is just the first step, for it has to be tested in humans. Further, an appropriate and cheap adjuvant (e.g., another protein, such as core hepatitis B virus protein) then needs to be included to make the vaccine more immunogenic so that less of the antigen is needed – important, since people require two doses of vaccine for immunogens to which they have not been previously exposed.

Ideally, a 'flu vaccine would protect against any strain of the virus for years or even a lifetime. The M2 ion channel protein that barely protrudes from the virion's envelope is required for uncoating of the virus within the host cell and is remarkably constant in amino acid sequence from one human 'flu strain to another and would be an ideal target for a universal 'flu vaccine. However, M2 antibodies appear to operate by binding to already-infected cells and promoting their destruction rather than preventing viral attachment in the first place – so people would probably still suffer 'flu symptoms. Conserved internal viral proteins (i.e. those exhibiting little variation) such as nucleoprotein (NP), or DNA encoding such proteins (see VACCINES), might enable a vaccine to elicit a cytotoxic T cell response rather than one based on antibodies. Work on a combined NP/M2 DNA vaccine, and on others where the inner proteins act as adjuvants, is in progress.

In addition to vaccines, two classes of anti-'flu drugs are now available for chemoprophylaxis and treatment: (i) the neuraminidase inhibitors oseltamivir (brand name Tamiflu) and zanamivir (brand name Relenza), effective against all N subtypes; and (ii) the M2 inhibitors amantadine (several brand names) and rimantadine (brand name Flumadine), which are only effective against influenza A virus. Unfortunately, high-level resistance of the virus to these inhibitors is conferred by single amino acid substitutions in both M2 and N proteins and since 2000 appears to be growing.

Work on crystal structures has revealed significant conformational differences between group-1 and group-2 neuraminidases, notably a large cavity adjacent to the active site in group-1 and absent from group-2 molecules. Armed with such information, the rational design of modified inhibitors of viral enzymes leading to an effective group-specific inhibitor should be possible, and be of value against the currently circulating H1N1 viruses, the H3N8 viruses that repeatedly cause 'flu in equines and are now causing widespread disease in canines, and the avian H5N1 viruses currently threatening human populations (see DNA VACCINES, http://www3.niaid.nih.gov/healthscience/healthtopics/Flu/default.htm).

information processing See NEURAL CODING. For processing of genetic information, see ENDOPLASMIC RETICULUM, GENE EXPRESSION, TRANSCRIPTION, TRANSLATION.

infundibulum A downward extension of the diencephalon giving rise to the pituitary stalk and posterior pituitary (pars nervosa). See PITUITARY GLAND, Fig. 137.

ingestion Entry of food via the mouth (see ORAL CAVITY). See FEEDING CENTRE.

inguinal Pertaining to the groin.

inguinal canal An oblique passageway, ~4–5 cm long, in the anterior abdominal wall through which pass the spermatic cord and ilioinguinal nerve in males and the round ligament of the uterus and ilioinguinal nerve in females. The TESTES pass through it during their descent, normally completed by 28 weeks of development.

inheritance In addition to transmission of genetic information (see HUMAN GENOME), the cytoplasm of the ovum and a centriole from the sperm (see FERTILIZATION), each human also inherits other non-genetic information. Such non-genetic inheritance can affect human traits as much as does genetic inheritance, and the wide range of transmissible phenotypic characteristics include lifestyle, habits, wealth, employment, education, land quality and tenure, and social status. See PHENOTYPE.

inherited predisposition See GENETIC PREDISPOSITION.

inhibins Heterodimer glycopeptide hormones (α-β_A and α-β_B), and members of the TRANSFORMING GROWTH FACTOR-β SUPERFAMILY, exerting both endocrine and paracrine responses. Both A and B forms contribute to regulation of LH and FSH release by negative feedback at the anterior pituitary. The B-form (the physiologically important form in males) is produced and released from the Sertoli cells of the TESTES (see Fig. 178) in response to FSH stimulation, and suppresses FSH release from the pituitary in a classic negative-feedback loop in which it binds to a membrane-spanning serine/threonine kinase receptor – but only if Sertoli cells are proliferating, and spermatogenesis is taking place (plasma levels correlating with total sperm count and testicular volume). In males, inhibin B action is antagonized by ACTIVIN, resulting in FSH release. In females, inhibin production by granulosa cells of the GRAAFIAN FOLLICLE is regulated by FSH and LH, and locally by autocrine and paracrine signalling by growth factors (EGF, TGF) and hormones (androstenedione, ACTIVINS and FOLLISTATIN). Under FSH stimulation, preantral follicles secrete inhibin B exclusively (suppressing FSH secretion) whereas inhibin A is a marker of corpus luteum function and is released under LH stimulation. See FOLLICLE-STIMULATING HORMONE, LUTEINIZING HORMONE.

inhibition See GENE SILENCING, LATERAL INHIBITION.

inhibitory post-synaptic potential (IPSP) See SYNAPSE.

initiation factors (eIFs) The nine or more proteins specifically required for initiation of eukaryotic TRANSLATION of mRNA on the ribosomes. In cap-dependent translation, after the addition of a 5′ cap to the mRNA (see RNA PROCESSING) and in the absence of the protein 14-3-3σ, a complex of proteins termed eIF4F binds to the cap through the eIF4E component (other proteins of eIF4F including eIF4G, which binds poly(A) binding protein; and eIF4A, which has RNA helicase activity, removing the RNA's secondary structure and permitting binding of mRNA to the 40S ribosomal subunit). eIF4F associates with eIF3, and through this binds to the 40S subunit. The initiation complex, facilitated by eIF4B, binds the mRNA and locates the initiating (5´)AUG codon (the 'start' codon) by scanning from the 5´ end until the first AUG is encountered, signalling the start of the reading frame (see TRANSLATION). The initiating (5´)AUG codon is positioned in the P site of the ribosome. For some genes, alternative choices are made for the initiating codon, yielding ISOFORMS differing in their N-terminal sequence. Mg^{2+} is required for assembly of the 40S and 60S ribosomal subunits. The serine-threonine kinase mTOR (see entry MTOR) is required for phosphorylation of proteins in signalling pathways which respond to growth signals by translation of mRNA. See ELONGATION FACTORS.

initiation sites For initiation sites for TRANSCRIPTION, see PROMOTER; for those involved in TRANSLATION, see that entry.

innate-like lymphocytes (ILLs) Cells that are really a part of the adaptive immune system, though behaving as though a part of the INNATE IMMUNITY system. They contribute to rapid innate responses to pathogens but use a limited set of antigen-receptor gene segments to make IMMUNOGLOBULINS and T-CELL RECEPTORS, but only if RAG-1 and RAG-2 proteins are present (see SOMATIC RECOMBINATION).

innate immunity (non-specific immunity) An arsenal of permanent barriers and responses initiated in the first few hours of INFECTION by a pathogen (see Appendix XI). These prevent or deter infection, clear the infection, or contain it while responses of ADAPTIVE IMMUNITY develop. They include the physical barrier provided by the surface epithelia themselves, whose cells are held together by tight junctions. Tears and saliva contain the antibacterial enzymes lysozyme and phospholipase A. The acidic pH of the stomach and the enzymes, bile salts, fatty acids and lysolipids of the upper gastrointestinal tract are all microbicidal or microbistatic and unfavourable to infection. Saliva contains the ANTIMICROBIAL PEPTIDES histatins, while Paneth cells of the epithelium of the small intestine secrete DEFENSINS, and other antimicrobial peptides are produced by other epithelia, notably of the respiratory tract (see ALVEOLUS). Epithelia also support a flora of non-pathogenic bacteria that not only produce antimicrobial substances (e.g., lactic acid and bacteriocins in the case of vaginal lactobacilli) but also out-compete potential pathogens (see GUT FLORA). Ciliary movement of mucus upwards along the respiratory tract traps dust-laden pathogens, which are then swallowed and subjected to stomach acid.

If a microorganism does cross an epithelial barrier and begins to replicate in the tissues it will be recognized and bound by a variety of receptors on MACROPHAGES (see Fig. 119) which normally results in its PHAGOCYTOSIS and death. NEUTROPHILS will similarly engulf and kill pathogens in the blood. If it binds, the MANNOSE-BINDING LECTIN will initiate the COMPLEMENT cascade. Resulting activation of neutrophils and monocytes, and their adherence to the ENDOTHELIUM, elicits inflammatory responses. Also an innate immune response is activation of DENDRITIC CELLS and MACROPHAGES, resident in most tissues, by lipopolysaccharide and peptidoglycans of Gram-negative bacteria (see PATTERN RECOGNITION RECEPTORS, TOLL-LIKE RECEPTORS, ACUTE-PHASE RESPONSE). This causes INFLAMMATION through macrophage release of such CYTOKINES as TNF-α, IL-β1 and IL-18, and chemokines. Inflammation generated by these innate mechanisms causes release of E-selectin and P-selectin onto the cell surfaces of the vascular ENDOTHELIUM, causing neutrophils and monocytes to adhere to, and roll on, the endothelial surface. This interaction acts as a focus for further NEUTROPHIL and macrophage recruitment. Other cytokines released by activated macrophages include IL-1 and IL-6, raising body temperature and initiating the ACUTE PHASE RESPONSE. In the case of viral infection, the response involves expression of numerous cytokine genes (e.g., IL-12 by dendritic cells, which in turn activates NK cells to produce IFN-γ). When bound to pathogen molecules, some of the receptors on macrophages and dendritic cells induce their expression of

CO-STIMULATORY MOLECULES so that these phagocytic cells can then initiate the responses of ADAPTIVE IMMUNITY. During the innate immune response to a pathogen, antigen-presenting cells are stimulated to express the B7 proteins which act as costimulatory proteins in activating T cells. Mouse models indicate that innate immunity seems to be essential for effective host defence, those engineered to lack macrophages and polymorphonuclear leukocytes suffering uncontrolled infection.

inner cell mass (ICM) Cells forming off-centre within the BLASTOCYST (see Fig. 20) by aggregation from the inner non-polar cells (embryoblast) and giving rise to all the cells of the future organism as well as those of the yolk sac, amnion and allantois (see EMBRYO; EMBRYONIC STEM CELLS).

inner ear The part of the ear lying within the temporal bone and including the VESTIBULE, COCHLEA and SEMICIRCULAR CANALS.

innervation The nerve supply, e.g., to a skeletal muscle or blood vessel.

inositol 1,4,5-triphosphate (IP_3) Second messenger released on activation of phosphatidylinositol 4,5-bisphosphate. Most IP_3 receptors in most cells are in the endoplasmic reticulum and, when bound, these open IP_3-gated Ca^{2+}-release channels causing a rapid rise in the concentration of Ca^{2+} in the cytosol. IP_3 is rapidly dephosphorylated by specific phosphatases to form IP_2. See PHOSPHATIDYLINOSITOL SIGNALLING SYSTEM.

inositol phospholipids See PHOSPHATIDYLINOSITOL SIGNALLING SYSTEM.

insertion **1**. A form of mutation. An alteration in the nucleotide sequence of the genetic material by inclusion of one or more nucleotides that were previously absent. Insertion of one, two or more nucleotides (but not multiples of three) within a protein-coding sequence will shift the reading frame of the translation process to the right and result in incorrect amino acid inclusion for the rest of the translation process, and will probably produce a stop codon that prematurely ends polypeptide elongation. The consequence of insertions into a non-protein-coding DNA sequence depends upon whether or not the sequence is informative. Insertions into a REGULATORY ELEMENT could well change GENE EXPRESSION. Large insertions can also result from unequal crossover (see TRIPLET REPEATS) or the insertion of mobile TRANSPOSABLE ELEMENTS. See NEUROFIBROMATOSIS. **2**. For a muscle's insertion, see MUSCLES.

insight learning See INTELLIGENCE, LEARNING.

insomnia The most common sleep disorder, in which there is failure to obtain the required amount or quality of sleep for normal functions during the day. See SLEEP–WAKE CYCLES.

inspiration See VENTILATION.

inspiratory capacity The maximum volume of gas inspired following tidal expiration. It averages 3.6 L in men and 2.4 L in women. See SPIROMETER, Fig. 165.

inspiratory centre See VENTILATION.

inspiratory reserve volume The volume of air additional to the inspiratory TIDAL VOLUME that can be inhaled. Its average value is 3 L for men, and 1.9 L for women. See INSPIRATORY CAPACITY, SPIROMETER, Fig. 165; VENTILATION.

instrumental conditioning (operant conditioning, trial-and-error learning) Associative learning in which an animal learns to associate a response, a particular motor act, with some meaningful stimulus – typically a reward, such as food – presented shortly after performance of the act. The animal learns to predict the consequences of its own behaviour (see LAW OF EFFECT). Originally introduced by Thorndike, and systematized by Skinner and others, its paradigm example is that of a hungry rat that randomly presses a lever protruding from a wall in a test chamber and promptly receives food as a result. Its subsequent rate of lever-pressing increases above the spontaneous rate and, if sufficiently regularly rewarded by food, the rat is more likely to press the lever in future whenever it is hungry. A response need not evoke a rewarding stimulus: instrumental conditioning will sometimes occur if the response prevents the occurrence of an aversive stimulus (see REINFORCEMENT AND REWARD, INTELLIGENCE).

Unlike CLASSICAL CONDITIONING, in which learning involves altered reflex responses to selected stimuli, operant conditioning may be thought of as the formation of a predictive relationship between a *response* and a stimulus. The behaviours whose frequencies are altered as a result of operant conditioning are termed *operants* and originally occur spontaneously, or without apparent eliciting stimuli and so they are sometimes regarded as emitted rather than elicited. As with classical conditioning, however, timing is crucial. For successful instrumental conditioning, the reinforcer (reward) must be given soon after the operant response. If it is delayed, or if it occurs prior to the operant response, learning is generally poor. As with classical conditioning, high predictability, or contingency, of the learnt relationship improves its durability, and there is also an optimal time interval between response and reinforcement which varies with the specific task and with the species involved.

The Thorndike rat mentioned above was motivated by hunger to feed, and operant conditioning is used to investigate MOTIVATION. For some neural correlates of instrumental conditioning, see LEARNING. See BIOFEEDBACK.

insula (adj. insular) A 'fifth cerebral lobe' deep within the lateral fissure, forming the floor of the lateral sulcus and separating the temporal lobe from the rest of the cerebrum (see BASAL GANGLIA, Fig. 18). It is often considered a limbic-related cortex (see LIMBIC SYSTEM), projecting to the thalamus and amygdala, and there is increasing interest in its role in body representation and subjective emotional experience. Brodmann's area 43 (TASTE area I) is located at the junction of the insula and the dorsal wall of the lateral sulcus, adjacent to the somatotopic mapping of the tongue, and believed to be involved in the conscious perception of taste, while taste area II is in the anterior insula and appears to be involved in the affective aspects of TASTE, OLFACTION and viscero-autonomic inputs, as well as having important roles in the experience of PAIN, anger, fear, disgust, happiness and sadness. The frontoinsular cortex is closely connected to the insula, and together they form part of the ORBITOFRONTAL CORTEX in which circuitry associated with spatial awareness and sense of touch are located. This region in the right hemisphere is implicated in navigation and the perception of three-dimensional rotations.

The insular cortex contains MIRROR CELLS and is responsible for planning the articulatory movements required for SPEECH on the basis of verbal working memory input from the prefrontal cortex, and it projects to BROCA'S AREA. The anterior insular cortex also contains SPINDLE CELLS, and at a higher density than is found in the anterior CINGULATE CORTEX (ACC). Significant olfactory and gustatory roles of the ACC and frontoinsular cortex appear to have been supplanted, during recent evolution, apparently in the interests of higher cognitive functions.

Insulators Two kinds of *cis*-acting REGULATORY ELEMENTS. *Enhancer-blocking insulators* block the action of an ENHANCER on a promoter when placed between the two, but not otherwise, while *barrier insulators* block the extension of a heterochromatic region into a euchromatic region when placed at the junction between the two.

insulin A protein hormone produced and secreted by β-cells of the pancreas, regulating both metabolism and gene expression. In its active form, insulin consists of two polypeptides, A and B, connected by disulphide bonds. It is encoded by separate parts of a single insulin gene that encodes preproinsulin, a longer polypeptide which includes a signal sequence directing its passage into secretory vesicles. In the endoplasmic reticulum, preproinsulin is proteolytically cleaved by removal of 23 amino acids (the signal sequence) at the amino terminal end and three disulphide bonds are formed, producing proinsulin which is stored in secretory granules in β cells of the pancreas. When elevated blood glucose (see BLOOD GLUCOSE REGULATION, PANCREAS) triggers insulin secretion, proinsulin is converted to insulin by prohormone convertases that cleave two peptide bonds to remove the C peptide (required during earlier folding of the protein) and form the mature 51-amino acid insulin molecule (see Fig. 102), composed of A and B chains.

In response to rising plasma levels of glucose and amino acids, release of insulin by β-cells is pulsatile and rhythmic, two identifiable rhythms occurring with periods of 5–10 and 60–120 minutes. This appears to be critical in suppression of liver glucose production and in insulin-mediated glucose disposal by adipocytes. The specific effects of insulin on skeletal muscle dominate its action. Although insulin mediates ~40% of glucose uptake by the body, ~80–90% of this occurs in skeletal muscle. Glucose uptake by body cells is mediated through GLUCOSE TRANSPORTERS in the cell membrane, the GLUT-4 transporter being unique in being recruited to the cell surface of fat cells (adipocytes) and skeletal muscle fibres from Golgi-derived cytosolic vesicular compartments when insulin binds its membrane receptor (see INSULIN RESISTANCE).

On binding its receptor, insulin triggers receptor autophosphorylation on tyrosine residues in the cytoplasmic domain (the β-chain). This opens up the active site, allowing the activated receptor to use ATP to phosphorylate several *insulin receptor substrates* (IRS-1, −2, −3, −4) on multiple tyrosine residues. The ARNO/cytohesin family of Arf guanine nucleotide exchange factors (see GUANOSINE TRIPHOSPHATE) seem to be involved in this pathway. Phosphorylated IRS-1 becomes a nucleation centre for a complex of proteins serving as a scaffold for the recruitment of downstream signalling proteins that relay the signal from the insulin receptor to targets in the cytosol and the nucleus.

The *early cytosolic effects* of insulin are largely metabolic and are mediated by the PHOSPHATIDYLINOSITOL-3-KINASE (PI_3K) pathway (see this entry, and DISEASE, Fig. 64), activation of PI_3K being through the SH2 domain of IRS-1. The IRS-PI_3K signalling pathway promotes glucose transport, GLYCOLYSIS, GLYCOGEN synthesis, and the regulation of protein synthesis. Activated PI_3K converts the membrane lipid PIP_2 to PIP_3 which indirectly activates PKB (Akt), causing it to be attracted to the plasma membrane.

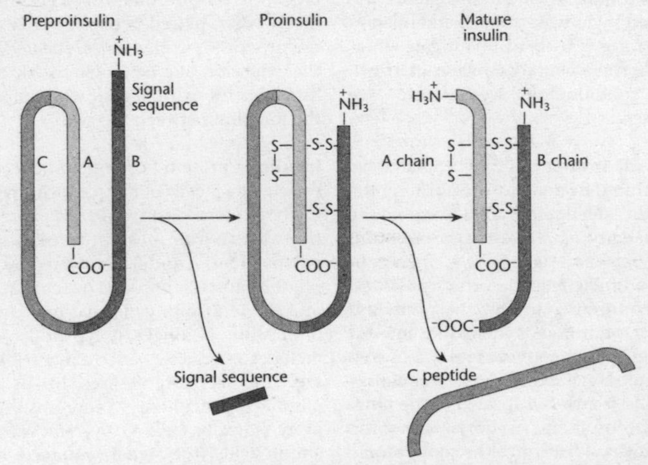

FIG 102. Diagram indicating the processing of preproinsulin to the functional INSULIN molecule (see entry). Disulphide bonds are represented by –S–S–. © 2000 From *Lehninger Principles of Biochemistry* by D.L. Nelson and M.M. Cox, published by Worth Publishers. Redrawn with permission from the publishers.

After further phosphorylation by another protein kinase (PDK1), PKB then phosphorylates serine or threonine residues on its target proteins, which include glycogen synthase kinase 3 (GSK3). Since active GSK3 phosphorylates and inactivates glycogen synthase, the cascade of protein interactions initiated by the binding of insulin to its receptor has the effect of stimulating glycogen synthesis. PKB is thought to be involved in the movement of GLUT-4 from internal vesicles to the plasma membrane, promoting glucose uptake. Some of the other pleiotropic effects of insulin receptor signalling via PIP_3 and the PI_3K pathway include promotion of protein synthesis and cell growth, sending a strong antiapoptotic signal, thereby promoting cell survival. In conditions of nutrient excess (see OBESITY) reduced IRS-PI_3K signalling and inflammation occur widely (see ENDOPLASMIC RETICULUM, MITOCHONDRION). If this were to occur in pancreatic β cells, as has been suggested, impaired insulin secretion could result which, combined with insulin resistance, leads to type 2 DIABETES since a functional IRS-PI_3K pathway is essential for β cell survival.

Insulin's *intermediate cytosolic effects* are mediated by similar phosphorylation or dephosphorylation of enzymes, dephosphorylation of HMG-CoA reductase promoting CHOLESTEROL synthesis. In adipocytes, insulin inhibits LIPOLYSIS and ketogenesis by promoting dephosphorylation of hormone-sensitive lipase, inhibiting the rate-limiting step in release of fatty acids (FA); i.e., the breakdown of triglycerides to FA and glycerol. Fewer FA are thereby available for ketogenesis. In liver cells, the forkhead transcription factors FOXA2 and FOXO1 lie downstream of activated PKA but are affected differentially by insulin. FOXA2 is inhibited by PKA more than is FOXO1, resulting in a greater repression of fatty acid oxidation than of GLUCONEOGENESIS, which are both effects of insulin (see also FASTING). By mediating dephosphorylation of inhibitory sites on hepatic acetyl-CoA carboxylase, insulin promotes LIPOGENESIS in adipocytes through its activation. This increases the production of malonyl-CoA, thereby reducing the rate at which FA can enter hepatic mitochondria for oxidation and ketone body production (see FATTY ACID METABOLISM). By antagonizing catecholamine-induced lipolysis through the phosphorylation and activation of phosphodiesterase, insulin leads to a decrease in cAMP levels and consequently of PKA activity. Prolonged insulin stimulation enhances the synthesis of lipogenic enzymes, and the repression of gluconeogenic enzymes. The MAPK pathway is less involved in metabolic effects of insulin; but its signalling cascades do mediate some of its mitogenic and differentiation roles.

Once insulin has bound its surface receptor, the ligand–receptor complex is internalized into endosomes, whence insulin receptor tyrosine kinase activity probably reaches less accessible targets than those near the cell surface. After acidification of the endosomal lumen, insulin breaks away from its receptor and is degraded by acidic insulinase, the receptor being recycled to the plasma membrane.

insulin-like growth factors (IGFs, somatomedins) Small peptide hormones synthesized primarily by the liver and circulating at higher concentrations than INSULIN (which they resemble structurally, although retaining the C peptide). Acting as mitogens, IGFs stimulate DNA, RNA and protein synthesis (see DISEASE, Fig. 65) for its signalling pathway. Although both IGFs (IGF-I and IGF-II) are important in embryonic development, IGF-I is then the major growth-promoting factor (see PLACENTA). Postnatally, IGF-I is found in nanomolar concentrations, exerting its effects mainly after induction by GROWTH HORMONE (GH), when it is the main mediator of GH's anabolic and linear growth-promoting effects (e.g., on skeletal muscle, bone, cartilage, liver, kidney, skin, nerves and lungs; see AGEING). IGF-II, by contrast, is minimally GH-dependent. Some extrahepatic secretion of IGF-I occurs (e.g., by bone and cartilage cells, and the reproductive system), where it acts paracrinally. In cartilage cells,

IGF-I acts synergistically with GH. IGF-I stimulates bone formation, paracrinally and possibly autocrinally, by increasing proliferation of osteoblast precursors and by promoting production, and inhibiting degradation, of type I COLLAGEN.

Both IGFs bind specifically to two high-affinity membrane receptors, both ligand-activated tyrosine kinases. These then autophosphorylate and phosphorylate substrates which include insulin receptor substrate 1 (see INSULIN RECEPTOR, IR). The signal cascade that ensues is one of the most potent natural activators of the AKT (PKB) signalling pathway (see PKB), which includes activation of PHOSPHATIDYLINOSITOL 3-KINASE, Grb2, Raf, MAPK and others involved in mediating growth and metabolic responses. Although IR and the IGF-I receptor are similar, the former is largely involved in metabolic processes in healthy individuals, while IGF-I receptors mediate mainly growth and differentiation.

Both IGFs can bind any of six different high-affinity binding proteins (IGFBPs), produced by a variety of tissues through complex regulatory processes. IGFBPs usually inhibit IGF-I action through competitive binding, almost 80% of IGF circulating bound to IGFBP-3 (increasing its half-life). Serine proteases cleave IGFBPs; but by binding the extracellular matrix, IGFBPs potentiate IGF action. IGF-II's physiological role is unclear, despite its occurrence in quantities threefold higher than IGF-I. In bone, IGF-I production is under GH, parathyroid hormone and sex steroid control, while in sex organs the sex steroids are the main promoters. Its concentrations parallel those of GH, being low at birth, increasing during childhood and puberty and beginning to decline in the third decade (as does IGF-II). Although IGF-I exerts its effect on binding the IGF-I receptor, some effects are expressed on binding IGF-II and insulin receptors. The *IGF2* gene is associated with BECKWITH–WIEDEMANN SYNDROME.

insulin receptor (IR) A tyrosine-specific protein kinase, IR is a heterotetrameric glycoprotein membrane receptor (2α- and 2β-subunits linked by disulphide bonds), the α-chain lying extracellularly and binding insulin, the β-chain having a short extracellular component, a transmembrane segment and a tyrosine kinase-bearing cytosolic segment activated by binding of insulin to the α-chain. On binding insulin, an IR autophosphorylates on tyrosine residues in the β-chain, leading to phosphorylation of several insulin-receptor substrates on multiple tyrosine residues (see INSULIN). IR is a member of the insulin-receptor family, along with INSULIN-LIKE GROWTH FACTOR receptor and the insulin-related receptor, all of which are involved in cell division, metabolism and development.

insulin resistance (insulin insensitivity) A reduced sensitivity of peripheral target tissues, including β-cells of the PANCREAS (see entry), to insulin, manifesting as reduced insulin-stimulated BLOOD GLUCOSE clearance and hyperglycaemia, intimately associated with weight gain and often associated with CARDIOVASCULAR DISEASE risk factors such as lipid abnormalities, glucose intolerance and high blood pressure. 'Metabolic syndrome' (syndrome X, insulin resistance syndrome) is a condition characterized by hypertension, hyperinsulinaemia, insulin resistance, glucose intolerance, hypertriglyceridaemia, lowered plasma HDL, and central obesity, of which type 2 DIABETES MELLITUS (T2D) is often a part. It is a disorder associated with impaired mitochondrial function and increased oxidative stress (see MITOCHONDRION). Chronic CIGARETTE SMOKING is associated with insulin resistance, and LIFESTYLE variation such as giving up smoking, increased EXERCISE and increased carbohydrate intake, are associated with increased insulin sensitivity. Fluctuations in insulin sensitivity occur during puberty and pregnancy, and with ageing. Its role in the aetiology of the atherogenic process is controversial.

In insulin-resistant states, the orderly phosphorylation of insulin receptor substrates is damaged (see INSULIN), preventing the correct regulation of glucose and fat metabolism. As a result, insulin-resistant individuals exhibit hyperglycaemia, in part due to elevated GLUCONEOGENESIS by the liver, the insulin signal being transmitted preferentially to the fatty-acid-oxidation switch rather than to the glucose switch so that insulin-resistant individuals are not only hyperglycaemic but also accumulate triglycerides in the liver instead of breaking them down. Insulin resistance syndromes, including OBESITY, T2D and syndrome X, are associated with what has been defined as *atherogenic lipoprotein phenotype*: the co-existence of moderately increased plasma triglyceride levels, an increased preponderance of small, dense LDL particles, and reduced HDL cholesterol concentrations. This phenotype results from insulin resistance having opposite effects on the activities of two different lipases. In adipose tissue, lipolysis by hormone-sensitive lipase is uncontrolled, the resulting increased mobilization and delivery of free fatty acids to the liver increasing hepatic triglyceride synthesis and VLDL production. Together with reduced activity of endothelial lipoprotein lipase as a consequence of insulin resistance, this results in impaired clearance of these lipoproteins and hypertriglyceridaemia. Hypertriglyceridaemia promotes cholesterol ester transfer protein- (CETP-)mediated transfer of cholesterol ester from LDL to HDL particles to triglyceride-rich lipoproteins, in exchange for triglyceride. The activity of hepatic lipase on these cholesterol-depleted and triglyceride-enriched LDL and HDL particles produces (i) atherogenic small dense LDL particles with poor receptor affinity that readily penetrate the vascular intima and are susceptible to oxidation and macrophage uptake; and (ii) small dense HDL particles that are rapidly removed from the circulation and are consequently less cardioprotective. The cholesterol ester-enriched chylomicron and VLDL remnants resulting from increased CETP activity are more susceptible to macrophage uptake and therefore to developing atheromatous plaque.

Rare human mutations in the gene encoding PPAR-γ (see PPARS) are associated with severe insulin resistance and type 2 diabetes. Located in nuclei of many cells, especially adipocytes, when it binds its ligand (which may include non-esterified fatty acids, NEFAs), PPAR-γ complexes with another transcription factor retinoid X receptor (RXR) and activates target gene transcription. If OBESITY alters the availability of NEFAs, it could reduce PPAR-γ signalling and produce insulin resistance.

In obese and insulin-resistant individuals, the expression of GLUT-4, an insulin-regulated GLUCOSE TRANSPORTER, is greatly reduced in adipocytes (fat cells), but not in muscle cells. Levels of circulating RETINOL BINDING PROTEIN-4 (RBP4) have been found to be coordinately regulated by changes in adipocyte GLUT-4 levels, and raised in several mouse models of insulin resistance, and in obese humans. This induces insulin resistance through reduced PHOSPHATIDYLINOSITOL 3-KINASE (see entry) signalling in muscle and enhanced expression of the enzyme phosphoenolpyruvate carboxykinase (favouring GLUCONEOGENESIS) in the liver through a retinol-dependent mechanism. By contrast, ADIPONECTIN acts as an insulin sensitizer, stimulating fatty acid oxidation in an AMPK and PPAR-α-dependent manner. Mice lacking adiponectin suffer severe insulin resistance in the liver. Chronic exposure to high insulin levels, obesity and excess growth hormone all lead to down-regulation of insulin receptors.

integrases Enzymes that cut out retroviral DNA and insert it elsewhere into chromosomes. They are very similar to transposases.

integrins Heterodimeric transmembrane cell surface ADHESION receptors (αβ dimers), by which cells bind either to each other (see INTERCELLULAR ADHESION MOLECULES) or the EXTRACELLULAR MATRIX (ECM). In the latter case stable interaction zones are formed within the plasma membrane by means of which the ECM can be linked, mainly through the β subunit, to the cytoskeleton (e.g., via talin, filamin and α-actinin). These zones then act as *focal adhesions* for protein complexes which recruit adaptor proteins and kinases and affect both cytoplasmic enzymes and signalling pathways. Among many proteins recruited into these junctions is the cytoplasmic TYROSINE KINASE, *focal adhesion* kinase, which binds the cytoplasmic tail of one of the integrin subunits. Cross-phosphorylation of these molecules creates phosphotyrosine docking sites for SRC-KINASE to bind. Responses include cell migration, proliferation and survival. Integrins cause release of anti-apoptotic signals, reducing the likelihood of anoikis, and are therefore critical during development of tumours. Many GROWTH FACTOR signalling pathways are modulated by integrins. Integrins on neurite filopodia bind LAMININS in the ECM and are important in AXONAL GUIDANCE.

intelligence Cognitive ability, often held to be of a kind enabling choice of a successful action without the need to evaluate all states of the intervening movements. Many of the conceptual difficulties associated with intelligence in the NATURE–NURTURE DEBATE arise because it is a faculty that drives behavioural change within the lifetime of an individual. Indeed, a major distinctive feature of human intelligence is flexibility. An animal's motor flexibility, its play, the technologies it develops, all are reflections of its motor repertoire. Baboons, despite observation of chimpanzees, never develop any technologies for obtaining termites. Chimpanzees can not only reproduce playful acts, they can also simulate or image actions using mental representations to guide their problem-solving. Shown a fruit overhead, a chimpanzee can picture placing two sticks together (play), obtain them, and use them to knock down the fruit (despite putting them at 90° to one another). Such simulation of motor acts (involved in 'insight' LEARNING) is held by many to represent a pre-existing capacity, the final development of which appears only in human primates: recursive LANGUAGE which enables recombination of mental elements; mixing features of one object with those of another; producing 'ghosts', flowers with faces, etc. Humans can represent what they imagine, whereas chimpanzees represent what they perceive. This makes counterfactuals a natural step, leading to art and science (e.g., laws about ideal gases). Several have noticed that biological evolution and intelligence are the only flexible processes on Earth capable of producing endless solutions to problems confronted by living beings.

Unlike pure achievement tests, conventional intelligence tests attempt to exclude items that rely on material explicitly taught at school, but include items that call for problem-solving, reasoning and ABSTRACTION. Non-verbal intelligence tests require no specific knowledge of language. The results of one such widely used test (Raven Progressive Matrices) suggest, when appropriately controlled, that SEX DIFFERENCES in intelligence tests that test such *general cognitive ability* (*g*) are negligible or non-existent. Tests designed to draw on *specific cognitive abilities* reveal moderately interpretable differences between the sexes: females do better on tests of verbal fluency (e.g., saying in one minute as many words as possible beginning with 'R') and perceptual speed (e.g., crossing out all the words beginning with 'e' in a long string of words). Achievement tests reflecting written language, grammar and arithmetic computation also show a female advantage. On the other hand, there is a male advantage at the highest levels in mathematical reasoning tests, and also at the highest level in SAT verbal tests, although males are more variable in performance than females. Males also perform better than females on tests of spatial visualization, especially when three-dimensional mental rotation is required (similar differences are found in rodents in maze-learning tests). Multivariate genetic analysis indicates that genetic correlates of specific cognitive abilities are very high: close to 1. Performance on such tests is positively affected by low female sex hormone levels, while performances on which females normally score better (e.g., speed in articulation; manual coordination) are improved by higher levels of these hormones (see CONGENITAL ADRENAL HYPERPLASIA, CAH). The effect may not be a direct hormonal one, however. Females with CAH may, however, have interest and activity patterns more like those of the other sex, providing test-relevant experiences influencing spatial visualization. The situation requires evaluation. But the lower spatial visualization scores of males with idiopathic hypogonadotropic hypogonadism seem to reinforce the role of pubertial testosterone in the male spatial visualization advantage. Even as children, females with TURNER SYNDROME, who have low oestrogen levels, show a significant deficit in spatial and mathematical abilities in spite of normal verbal intelligence; but if there is a hormonal mechanism involved, it has not been clarified. See ARTIFICIAL INTELLIGENCE, PREFRONTAL CORTEX.

intentional (voluntary) movements Those movements whose command patterns originate in the MOTOR CORTEX, enabled by the innate and learned microcircuitry of the striatum. These project to the thalamus and to subthalamic nucleus, providing basal ganglia output for limb and facial movements. The thalamus projects back to the same region of the cortex giving rise to the striatal inputs, closing a loop. Two pathways in the BASAL GANGLIA (see entry) have opposing effects on firing of thalamic and cortical neurons, and both are activated when the motor cortex initiates a specific movement. The direct pathway activates thalamic neurons, allowing movement sequences to occur, while the indirect pathway inhibits thalamic neurons and suppresses unwanted movements (see ACTION SELECTION). Modulatory input from the dopaminergic nigrostriatal tract (see SUBSTANTIA NIGRA) enhances the direct pathway but suppresses the indirect pathway, so that dopamine neurotransmission enables movements to occur. Knowledge of limb position is achieved by KINESTHESIA, requiring corollary discharge of motor efferents to the sensory cortex (providing information about the intended movement) and sensory feedback from muscle spindles, tendon organs, and other proprioceptors informing the sensory cortex directly of the progress of movement. Nicotinic ACh receptors (see ION CHANNELS) in skeletal muscle membranes at neuromuscular junctions mediate all voluntary movement. See CEREBELLUM, CORTICOSPINAL TRACTS, LANGUAGE, MOTOR CORTEX, REINFORCEMENT AND REWARD.

interblobs Regions of the primary visual cortex (V1) staining weakly for cytochrome oxidase and interspersed among BLOBS. A component of the pathways involved in perception of shape/form. See PARVOCELLULAR–BLOB/INTERBLOB PATHWAYS.

intercalated cells (I cells) Kidney cells, of two types, in the ends of the DISTAL CONVOLUTED TUBULES and in the collecting ducts that help regulate the pH of body fluids (see KIDNEYS). They are sensitive to ALDOSTERONE. Compare PRINCIPAL CELLS.

intercalated discs See CARDIAC MUSCLE.

intercellular adhesion molecules (ICAMs) Cell-surface ligands of the immunoglobulin superfamily (see IMMUNOGLOBULINS) for the leukocyte INTEGRINS and crucial for the tight binding of lymphocytes and other leukocytes to the ENDOTHELIUM and to ANTIGEN-PRESENTING CELLS (APCs). ICAM-I and ICAM-2 are expressed on endothelial as well as APC surfaces and enable lymphocytes to pass through capillary walls (extravasation by DIAPEDESIS), plasma ICAM-1 levels increasing in a dose-dependent manner with CIGARETTE SMOKING; ICAM-3 is only expressed on leukocytes and is probably involved in T-cell adhesion to APCs, notably dendritic cells. All three bind to the T-cell integrin LFA-1, and ICAM-3 also binds the dendritic cell lectin DC-SIGN.

intercellular junctions Cell junctions (see Fig. 103) occurring at points of cell–cell and cell–matrix contact in every tissue. Functionally, they include *occluding junctions* (e.g., TIGHT JUNCTIONS) that prevent even small molecules from crossing from one side of the cell sheet to the other; *anchoring junctions* attaching cells mechanically to neighbouring cells or to the extracellular matrix and involving either ACTIN filaments (e.g., ADHERENS JUNCTIONS, FOCAL ADHESIONS) or INTERMEDIATE FILAMENTS (e.g.,

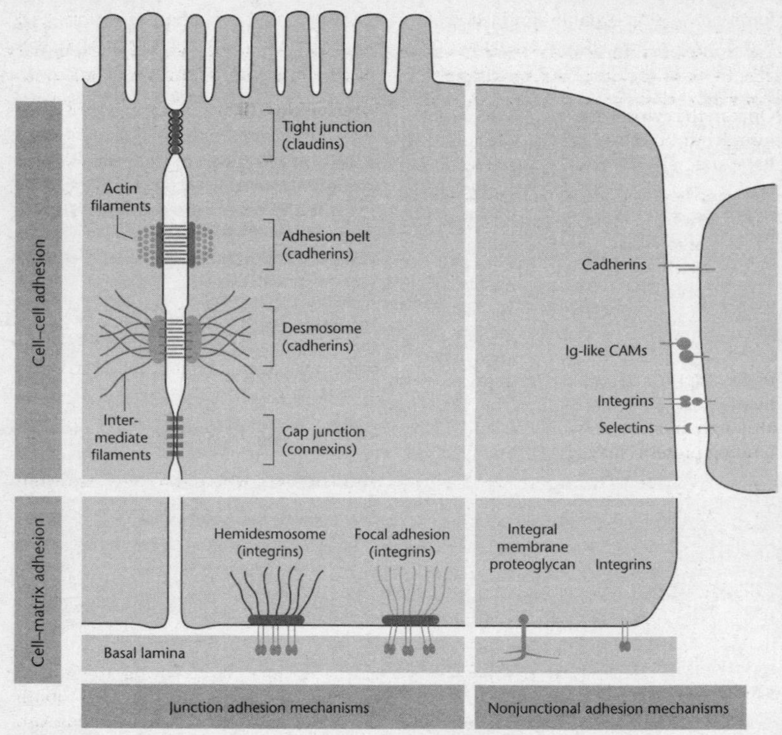

FIG 103. Some of the INTERCELLULAR JUNCTIONS, and non-junctional adhesive mechanisms, linking cells to each other, and linking cells to the extracellular matrix. © 2002 From *Molecular Biology of the Cell* (4th edition), by B. Alberts, A. Johnson, J. Lewis, M. Raff, K. Roberts and P. Walter. Reproduced by permission of Garland Science/Taylor & Francis LLC.

DESMOSOMES and HEMIDESMOSOMES); and *communicating junctions* (e.g., GAP JUNCTIONS and SYNAPSES) that allow passage of chemical or electrical signals from one involved cell to a partner.

intercostals Muscles attached to the RIBS, by whose contractions and relaxations (along with those of the diaphragm) the shape and volume of the thoracic cavity are altered during breathing. External intercostals originate on the inferior border of the rib above and insert on the superior border of the rib below. On contraction, they elevate ribs during inspiration and increase lateral and anteroposterior dimensions of the THORAX. Internal intercostals originate on the superior border of the rib below and insert on the inferior border of the rib above. On contraction they draw adjacent ribs together during forced expiration, decreasing the lateral and anteroposterior dimensions of the thorax.

interferons (IFNs) Pleiotropic signalling CYTOKINES, which are induced by viral infection. They not only interfere with viral replication and probably block the spread of viruses to uninfected cells, but are also growth-inhibitory, gating proliferative responses to MITOGENS. In some cells, free viral nucleic acids can be internalized by endocytosis and detected by TOLL-LIKE RECEPTORS within endosomes, leading to interferon production. Double-stranded RNA is not a normal product of mammalian cells but forms the genome of some viruses. It is recognized by the TOLL-LIKE RECEPTOR TLR-3. Once in the cytosol of a host cell (especially non-dendritic cells), viral double-stranded RNA, and indeed viral single-stranded RNA, is specifically recognized by 'RIG-like' RNA helicase receptors encoded by retinoic acid inducible gene 1 (RIG-1) and melanoma differentiation-associated gene 5 (MDA5). This ability to recognize 'non-self' RNA lies in part in the recognition of the 5′-triphosphate (ppp) moiety of dsRNA and ssRNA (although 7SL RNA self-RNA is also present in the cell and has this ppp attachment and is not recognized by RIG-1), although nucleotide sequence and RNA-binding proteins may also be involved.

The antiviral effector molecules interferon-α (IFN-α, a family of several closely related proteins) and interferon-β (IFN-β, product of a single gene) are collectively known as 'type 1 interferon', or IFN-α/β, and are distinct from interferon-γ (IFN-γ). IFN-α/β is produced by expression of gene in response to binding by transcription factors (IRF-3, NF-κB) which are themselves activated to bind by RIG-1-activation. IFN-α/β secreted by an infected cell binds a common cell-surface receptor, the *interferon receptor*, on both the infected cell and its uninfected neighbours, which is coupled to a Janus-family tyrosine kinase through which it signals (see JAK/STAT PATHWAY). This leads to expression of genes encoding interferons.

IFN-γ, secreted by T_H1 and cytotoxic T CELLS, is not directly induced by viral infection but is produced later and mediates pathways in both innate and adaptive immune responses to intracellular pathogens, increasing expression of MHC class I and MHC class II molecules – even by cells that do not normally express them (see MAJOR HISTOCOMPATIBILITY COMPLEX). It is also involved in the control of activated macrophages (see CYTOKINES).

IFN anti-proliferative effects result in part from their suppression of phosphorylation of retinoblastoma protein and from their inhibitory effects on cyclin-dependent kinases (CDKs), induction of various CDK inhibitors and suppression of c-Myc. IFNs also activate NATURAL KILLER CELLS.

interleukins (ILs) Those CYTOKINES produced by leukocytes, the generic term 'cytokine' often being used instead. Interleukins are individually numbered (although this nomenclature will be replaced in time by one based upon cytokine structure). See CYTOKINES for examples.

intermediary proteins Proteins linking proteins of the CYTOSKELETON either to one another or to integrins in the plasma membrane. They include α-actinin, talin, vinculin and tensin.

intermediate filaments One of three major types of cytoskeletal filament (see CYTOSKELETON), whose proteins are composed of smaller subunits formed from α-helical coiled coils that are themselves elongated and fibrous, forming rope-like fibres ~10 nm in diameter. They are not polarized, and have no associated motor proteins. A pair of parallel dimers associates in an antiparallel manner to form a staggered tetramer, representing the soluble subunit analogous to the αβ-tubulin dimer, or ACTIN monomer. Different families are expressed in different cell types, and there is far greater sequence variation in their subunit isoforms than occurs in isoforms of actin or tubulin (see MICROTUBULES). One group (lamins A, B and C) forms the nuclear lamina lining the inner face of the nuclear envelope; others extend across the cytoplasm providing mechanical strength (e.g., those spanning the cell from one INTERCELLULAR JUNCTION to another). The KERATINS occur in epithelial cells; VIMENTIN-LIKE PROTEINS occur in cells of mesenchymal origin, and neurofilament proteins (NF-L, NF-M and NF-H) occur in axons.

intermediate mesoderm Cells originating by migration through the midstreak region of the PRIMITIVE STREAK and briefly connecting the lateral plate mesoderm and paraxial mesoderm, giving rise to the urogenital system. See MESODERM and Appendix VI.

internal capsule A thick band of PROJECTION FIBRES (both sensory and motor) connecting the cerebral cortex with the brain stem and spinal cord.

International HapMap Project See HAPLOTYPE.

interneurons Any neurons of a neural network synapsing between a sensory and a motor neuron, as in a typical polysynaptic reflex arc. Excitatory or inhibitory, they are confined to the grey matter of the central nervous system, enabling cross-connectivity between neural pathways and resultant integration of reflexes, and learning. Most human neurons (perhaps 90%) are interneurons. See PURKINJE CELLS, PYRAMIDAL CELLS, RENSHAW CELLS.

interorbital pillar The bones forming the bridge of the nose, including the paired and medial nasal bones and those parts of the maxillae forming part of the lateral wall of the nasal cavity. During chewing (mastication) in strepsirhines, it has been suggested that the interorbital region can be modelled as a simple beam under bending or shear. During unilateral biting, or mastication, the face is twisting on the brain case and the interorbital pillar transmits axial compressive forces from the toothrow to the braincase. Strain data reveal that the interorbital region cannot be modelled as an anteroposteriorly oriented beam bent superiorly in the sagittal plane during incision or mastication, providing some support for the view that the interorbital region is a member in a rigid frame subjected to axial compression during mastication.

interphase The stage in the CELL CYCLE preparatory to nuclear division, subdivided into G1, S-PHASE and G2. Chromosomes are located in distinct territories in the interphase nucleus. All cilia are generated during interphase (see CILIUM).

intersex See CONGENITAL ADRENAL HYPOPLASIA, SEX REVERSAL.

interspersed repeat sequences At least 45% of the human genome comprises interspersed repetitive non-coding DNA in the form of LINES, SINES and LTRS (see TRANSPOSABLE ELEMENTS). A further ~2% is occupied by simple sequence repeats (see MICROSATELLITES). The Y CHROMOSOME contains additional IRSs.

interstitial cells See LEYDIG CELLS.

interstitial fluid See TISSUE FLUID.

interstitium That part of the body, about one-sixth in all, consisting of spaces between cells. A large part of this is occupied by collagen bundles and peptidoglycan filaments, entrapping the interstitial fluid so as to produce a 'tissue gel' and restricting movement of fluid components to diffusion. But some rivulets of free fluid movement are also present, amounting to perhaps 1% of the interstitial volume in most tissues.

intervertebral discs Fibrocartilage rings containing a pulpy and highly elastic material (the *nucleus pulposus*) forming strong joints between the bodies of adjacent VERTEBRAE. Under compression, they flatten and broaden, absorbing vertical shock and bulging from their intervertebral space. Superior to the sacrum, discs account for a quarter of the length of the vertebral column. A herniated ('slipped') disc occurs most often in the lumbar region, and follows injury to or weakening of the anterior or posterior ligaments, when the nucleus pulposus may herniate (protrude) posteriorly, often towards the spinal cord and spinal nerves, causing acute pain. Parts of the laminae of the vertebrae and intervertebral disc may be removed to relieve the pressure.

intima (tunica interna) The endothelium, basement membrane and (in arteries and arterioles) underlying elastic lamina of blood vessels larger than capillaries.

intraflagellar transport complex (IFT) See CILIUM.

intrafusal muscle fibres See MUSCLE SPINDLES.

intra-uterine device (IUD) See CONTRACEPTION.

intra-uterine growth retardation (IUGR) A condition defined as having less than 10% of predicted foetal weight for gestational age. It is most commonly caused by inadequate maternal–foetal circulation, leading to reduced foetal growth, and may result in significant foetal morbidity and mortality unless diagnosed. Its actual incidence is only ~4–7% since roughly one-quarter of infants below the 10th percentile have a normalized birth weight when corrected for low maternal weight, paternal phenotype, or residence at higher altitudes. Conditions associated with IUGR include *medical*: chronic hypertension, preeclampsia early in gestation, diabetes mellitus, systemic lupus erythematosus, chronic renal disease, inflammatory bowel disease; *maternal*: cigarette smoking, alcohol use, cocaine use, warfarin, malnutrition, prior history of pregnancy with IUGR, and residence at altitude > 1800 metres; *infectious*: syphilis, cytomegalovirus, toxoplasmosis, rubella, hepatitis B, HSV-1 or HSV-2, and HIV-1; and *congenital*: trisomies of chromosomes 21, 18 and 13, and TURNER'S SYNDROME. It may also result from mutations in the *IGF-I* gene and persist after birth.

intra-uterine insemination (IUI) Formerly termed *artificial insemination*, those procedures by which sperm is introduced without sexual intercourse into the vagina, cervix or oviducts. Sperm may be from the male partner or from a donor. This may be an approach suitable for couples whose INFERTILITY arises from a variety of causes in the male partner (or from his HIV infection); for couples wishing to avoid genetic disorders in their child which may be passed on via the male partner; for women with blocked oviducts, and for lesbian couples who desire a child.

intrinsic factor A glycoprotein secreted by parietal cells of gastric glands, required for the absorption of vitamin B_{12} across the mucosal epithelium of the lower ileum. See VITAMIN B COMPLEX.

introgression Infiltration of genes of one species population (conspecific or otherwise) into the gene pool of another, often by hybridization and subsequent backcrossing. In crop production, often involves a small, polymorphic genomic region being introduced, transgenically or otherwise, into a given variety through repeated backcrossing, so that all progeny (or kernels) from an individual in which a (trans)gene has been introgressed will possess that gene.

introns Heterogeneous DNA sequences, with different functional capacities, lying between the coding sequences of most genes, creating 'split genes' (a small minority of genes in complex organisms lack introns). Exceptionally high conservation of intron positions can be found, as in the IMMUNOGLOBULINS and the Huntington's disease gene (see HUMAN GENOME), although other intron locations (e.g., those for actins, myosins and tubulins) are more variable and may have evolved more recently. Introns are classifiable into the different intron groups below.

(i) *Spliceosomal introns* probably arose from group II introns, and are transcribed into RNA in the primary mRNA transcript and are excised from this by spliceosomes (see RNA PROCESSING). Only a few short sequences, those at or near the splice junctions and at the branch site, and splice enhancers and silencers, appear to be important in gene function. Because of this, they are often very long, mainly because they can tolerate insertion of mobile genetic elements, and facilitate exon shuffling.

(ii) *Group I and II introns* are self-splicing introns, found mainly in rRNA and tRNA genes (see SNORNAS entry), and may act as mobile elements. Mobile group II introns encode a reverse transcriptase-like activity strikingly similar to that of LINE1 elements.

(iii) *Archaeal introns* have only been found in archaeal tRNA and rRNA genes and so do not occur in human genes.

Other than pseudogenes, introns are among the most rapidly evolving sequences in eukaryotes and are therefore of great value in studies of population structure. They permit diversification by exon shuffling and exon duplication (see GENE DUPLICATION). See NORTHERN BLOTTING.

invariance Visual ability to recognize a given object from different perspectives, e.g., from front, or side; in light or shadow, and so on. In view-invariant theories, this is due to the visual cortex performing some sort of neural calculation transforming the visual structure of different images into a common format. In view-invariant theories, temporal associations between different views are learned and stored in the memory as such.

inversin A primary cilium-associated protein (see CILIUM), mutations in whose gene disrupt left–right axis formation in the mouse and cause one form of human POLYCYSTIC KIDNEY DISEASE. It targets the cytosol pool of the protein Dishevelled for degradation, inhibiting the canonical WNT SIGNALLING PATHWAY.

inversions Double-break rearrangements within a single chromosome in which a segment is reversed in position (alignment). If an inversion involves a centromere, it is a pericentric inversion; if it involves only one arm, it is a paracentric inversion. A pericentric inversion of chromosome 7 occurs as a common structural variant (polymorphism). While it is of no functional significance, other inversions may give rise to gene imbalance in offspring, with clinical consequences. There is a tendency for those with mental retardation to have balanced TRANSLOCATIONS or inversions. See SPECIATION for chimp/human evolution.

inverted repeats Palindromic DNA strands, which fold back on each other to give HAIRPIN LOOPS when denatured. DNA-only transposons, which do not occur in the human genome, have short inverted repeats at each end. MicroRNAs (see entry) are derived from larger ~70-nucleotide-long precursors containing an inverted repeat, permitting double-stranded hairpin RNA formation.

***in vitro* fertilization (IVF)** A technique of assisted reproductive technology (ART) developed primarily as a treatment for human INFERTILITY. Typically, a woman is superovulated using GnRH analogues, exogenous gonadotropins (containing FSH and/or LH) and synthetic hCG. Oocytes are collected 34–36 hours afterwards by puncture and aspiration of the follicles. Initially done using laparoscopy, endovaginal ultrasound and aspiration avoids general anaesthesia, and is often preferred. Eggs are microscopically examined for maturation, a metaphase II oocyte either being available or obtained by further incubation. Oocytes are fertilized either with sperm from a recent semen sample or by cryopreserved samples from a sperm donor or from the male partner of the infertile couple. Intracytoplasmic sperm injection (ICSI) was developed to circumvent the inability of the sperm from some men to fertilize an egg (e.g., lack of sperm motility). It is important that sperm are separated from the rest of the semen as quickly as possible to reduce the potentially destructive effects of oxidative chemicals in the semen. Normally achieved by centrifugation, this method takes up to an hour, may damage the sperm, and runs the risk of contamination by sperm from other donors. Separation of sperm by an electric current permits sperm to be taken directly from the testis, involves fewer steps than centrifugation, and avoids the risk of oxidation and contamination of sperm. It is not known why negatively charged sperm have the most intact DNA, but sialic acid added during the final stages of sperm maturation may be responsible, possibly indicating that everything has gone absolutely normally for sperm to have got to that point.

Oocytes are assessed for fertilization 12–18 hours postinsemination, the presence of two pronuclei and/or extrusion of the 2nd polar body confirming identification. Embryos are transferred to growth media, quality being determined by the cleavage rate and by the shape of the blastomeres. Embryo culture conditions can affect expression of imprinted genes (see GENOMIC

IMPRINTING), associated with higher than expected frequencies of both ANGELMAN SYNDROME and BECKWITH-WIEDEMANN SYNDROME through this method of conception. Blastocysts, which are increasingly employed in transfer, are much more active metabolically than are early cleavage-stages and use glucose rather than pyruvate/lactate. The culture conditions result in a stress response manifested by changes in gene expression (e.g., of heat-shock proteins), apoptosis and metabolism.

Embryo transfer is performed transcervically to the uterine cavity by catheter. Pressures to avoid multiple gestation have led to transfer of only one or two embryos in an IVF cycle in some countries. Multiple pregnancies are associated with higher incidence of complications and often with intensive care of premature babies, low birth weight and learning disabilities. Single embryo transfer (SET) raises the issue of identification of the 'best' embryo.

Progesterone supplementation is required to generate a suitable intrauterine environment for IMPLANTATION. During IVF, well over half of human embryos arrest through failure of mitosis and are therefore unsuitable for transfer to women. Such embryos, which are discarded, may be an ethically sound source of EMBRYONIC STEM CELLS (see also PREIMPLANTATION GENETIC DIAGNOSIS). An alternative to IVF includes INTRA-UTERINE INSEMINATION using donor sperm (AID). See GAMETE INTRA-FALLOPIAN TRANSFER.

***in vivo* gene therapy** See GENE THERAPY.

iodine deficiency disorders (IDDs) The most familiar IDD is goitre, a swelling in the neck due to enlargement of the THYROID GLAND. A healthy adult contains 15–20 mg of iodine, of which ~75% is present in the thyroid gland, which has a very active iodide-trapping mechanism. Iodine deficiency interferes with thyroid hormone synthesis, blood T_4 levels falling but T_3 levels rising. Resulting increased thyrotropin output increases thyroidal iodide uptake and hormonal turnover and cells of the thyroid follicles hypertrophy and multiply, forming the swelling. This is easily reversible with iodide supplements. Iodine deficiency in the foetus leads to greater incidence of stillbirths, miscarriages and congenital abnormalities. See CRETINISM.

iodopsins The visual pigment of CONE CELLS, consisting of retinal combined with a photopigment protein, opsin. Three major forms of opsin exist, each differing slightly in their amino acid composition from that of rod cells (see RHODOPSIN) and responding to a narrower range of wavelengths, but behaving very similarly to it in vision. See COLOUR BLINDNESS, COLOUR PERCEPTION, LIGHT ADAPTATION.

ion channels Large transmembrane proteins found in virtually every cell, whose evolution, along with transporters and pumps, enables ions to be transported across the ion-impermeable lipid barrier of CELL MEMBRANES. Opening and closing (*gating*) of most channels is regulated by biological signals, e.g., binding of intracellular or extracellular ligands (*ligand-gated channels*), changes in membrane potential (*voltage-gated channels*), changes in temperature, or mechanical stress. In many cases, gating is also influenced by biochemical reactions (termed *modulation*) such as phosphorylation. Also included here are *ion pumps*, channels whose conformational changes require associated biochemical reactions involving release of free energy in order to make reversibility difficult. The free energy may be provided by hydrolysis of ATP or GTP, or collapse of a concentration gradient (e.g., a proton gradient); but as with all ion pumps that create ion gradients they can function in reverse if there is a steep enough gradient of the ion they transport across the membrane. Many of these channels are co-transport channels, either symporters or antiporters. Accessory proteins are tightly associated with most channels, some forming part of much larger macromolecular complexes.

Over 340 human genes are thought to encode ion channels, and mutations in over 60 such genes are known to cause human neurological disease ('channelopathies'; e.g., see MYASTHENIA GRAVIS, EPILEPSY), although inherited forms of these are very rare. The channels are composed of one or more pore-forming subunits. Accessory subunits specify the location and abundance of ion channels in the membrane, modulating their properties and fine-tuning their sensitivities to ligands and pharmacological agents. Many such proteins act as chaperones involved in trafficking channels from one membrane to another (see VESICLES).

Whereas GAP JUNCTIONS discriminate only on the basis of ion size, some channels, such as the acetylcholine receptor AChR, permit a variety of cations through (rings of negatively charged amino acid residues in the pore channels excluding anions). Others, notably members of the pore loop family, possess rings of positively charged residues and are anion-selective (e.g., GABA and glycine receptors). Some, such as the Cys-loop receptors (see below), seem to have evolved with multicellularity in the animal lineage when cell–cell interactions, such as those of the nervous system, became important. Ion channels are gated pores permitting passive flow of ions down their electrochemical gradients. Behaving as molecular nanoswitches, their often restricted tissue expression, membrane location, and complex structural heterogeneity make them attractive targets for drug design and therapy in anaesthesia and sedation, anxiety and depression, diabetes, and infectious disease. Channels can be subdivided on the basis of either structure or function.

Structure: The central pore through which the ions pass is formed by four or five transmembrane (TM) α-helices fitting together like the staves of a barrel. These pore-forming helices are often formed from separate modular subunits, so that the channel is tetrameric or pentameric. But voltage-gated Ca^{2+} (Ca_V) and Na^+ (Na_V) channels are composed of a single subunit containing four similar repeated domains, and some K^+ channels are dimers in which each subunit is composed of two repeated domains (see CALCIUM for calcium channels involved in signalling). *Pore-loop channels* are the largest ion channel group, members possessing a pore loop, a protein region that loops back into the membrane to discriminate very selectively the ion species that can penetrate. The tetrameric nature of most of these channels, which include GLUTAMATE receptors, hERG and inwardly rectifying K^+ channel (Kir), contributes to ion-channel diversity with closely related subunits associating to form heteromeric channels with correspondingly diverse properties.

Non-pore-loop channels are found in several families with no obvious evolutionary relationships. The pentameric ligand-gated Cys-loop receptors (e.g., the anionic $GABA_A$, $GABA_C$ and GLYCINE receptors, and the cationic 5-HT_3 serotonin and nicotinic ACh receptors) contain four TMs, one helix of each of the five subunits contributing to the pore. They can assemble from five copies of a single subunit type, but more commonly they contain several different subunits. The ClC chloride (Cl^-) channels have many helices that go only partway across the membrane, and include some proteins which are channels and others which are proton-coupled ion pumps. It is a large family, with isoforms expressed in nearly every cell (the human genome has nine). They blur the classical distinction between channels and transporters, since the two forms of ion transport share many common features. The CFTR (see CYSTIC FIBROSIS) channel is the only member of the ABC TRANSPORTER FAMILY member that is an ion channel. The gate that regulates chloride-ion flow through the membrane-spanning pore in CFTR is thought to be opened when ATP binds to its two cytoplasmic nucleotide-binding domains (NBD1 and NBD2) and, acting as a molecular glue, holds them together. Hydrolysis of one of the ATPs disrupts this interaction, closes the channel, and terminates anion flow. CFTR's closest relatives in the human genome are the multidrug-resistance-related proteins (MRPs) which are ATP-driven efflux pumps, and the sulphonylurea receptors (SURs) which associate with inward rectifier potassium-ion channels to form nucleotide-sensitive K^+

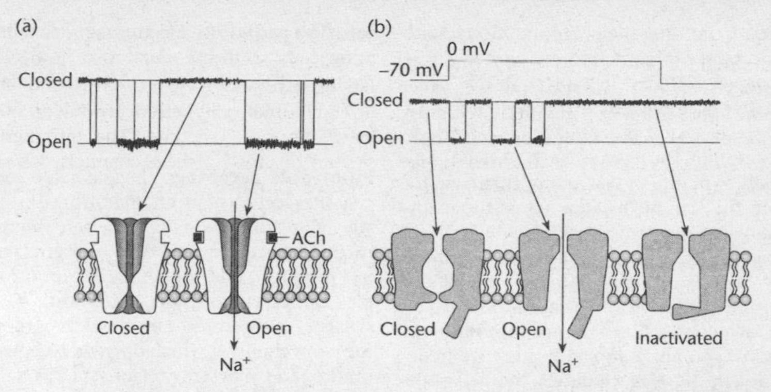

FIG 104. Diagrams illustrating the opening and closing of ION CHANNELS, with associated single-channel voltage recordings. Despite the randomness of these events, their frequencies can be influenced by (a) ligand-binding, or (b) appropriate transmembrane voltage. Following opening, some voltage-gated channels remain in an inactivated (non-conducting) state in which they are refractory to subsequent depolarization. From F.M. Ashcroft, *Nature* © 2006, with permission from Macmillan Publishers Ltd.

conductance. Non-pore-loop Ca^{2+}-release channels (ryanodine channels; see CALCIUM) occur in endoplasmic and sarcoplasmic reticula.

Function: *Ligand-gated ion channels* are IONOTROPIC RECEPTORS, typically located in the postsynaptic membranes of SYNAPSES, binding one or more specific activator molecules, or ligands (e.g., NEUROTRANSMITTERS). Flexible loops (gating loops) link the ligand-binding domains to the pore. In as little as tens of microseconds, this triggers the opening of an integral membrane pore and allows ions to flood across into the postsynaptic cell, changing the membrane voltage and, as a consequence, the cell's activity. Ligand-gated ion channels can therefore be considered as transducers of chemical signals into electrical output. Their opening and closing regulate information flow throughout the brain, examples including the ionotropic AMPARS and NMDA RECEPTORS (see GLUTAMATE, Fig. 83), and they are often modulated by allosteric ligands, channel blockers, ions or the membrane potential. They are likely to be the major sites at which ANAESTHETIC agents have their effects. One family of K^+ channels, the Kir channels, stabilize the resting membrane potential and act as a brake on excitability. One member, the K_{ATP} channel, is inhibited by the non-hydrolytic binding of ATP, but activated by interactions with Mg^{2+}-bound nucleotides at separate sites. The inhibitory effect dominates (channel closed) when cellular phosphorylation potential is high; but as metabolism declines the channel opens, providing a unique electrical transducer of the metabolic state of the cell.

Voltage-gated ion channels (see Fig. 104) are opened or closed by changes in transmembrane voltage, and are often modulated through phosphorylation by protein kinases. Following opening, some voltage-gated channels enter an inactivated (non-conducting) state in which they are refractory to subsequent depolarization (see REFRACTORY PERIOD). Action potentials depend on the activities of voltage-gated Na^+, Ca^{2+} and K^+ (Kv) channels (see CALCIUM for ryanodine channels). See COTRANSPORTERS.

Stretch-activated ion channels are mechanosensitive (MS; see MECHANORECEPTORS) and occur in many membranes.

Ion pumps are membrane proteins (see above) that are substrates for protein kinases and use free energy release to transport ions and molecules against their electrochemical gradient (active transport), as is required in restoration of ionic equilibrium after a burst of neuronal activity. Ion pumps that alter their conformational state by autophosphorylation are termed *P-type transport ATPases*. In typical animal cells, passive ion movements contribute most to the electrical potential across the membrane; but the P-type Na^+K^+-ATPase (the sodium pump, or Na^+ pump) helps maintain the osmotic balance across the membrane by keeping the intracellular $[Na^+]$ low. The excess of anionic charge generated by organic (mainly protein) molecules within the cytosol is balanced mainly by K^+, which is actively pumped into the cell by the sodium pump and can move freely through K^+ leak channels (see MEMBRANE POTENTIAL). On its cytosolic surface, the pump hydrolyses ATP and becomes phosphorylated and undergoes a conformational change enabling it to pump Na^+ out. Binding of K^+ on the extracellular surface then leads to dephosphorylation returning the protein to its original conformation during which phase it transfers K^+ across the membrane and releases it in the cytosol. There are thought to be three Na^+- and two K^+-binding sites reflecting the ratio of ions transported across the membrane, and more conformational states occur than are indicated here. The P-type transport ATPase family also includes CALCIUM pumps (Ca^{2+}-ATPase) that remove Ca^{2+} from the cytosol after its entry resulting from signalling events, and the proton (H^+-K^+) pumps of stomach PARIETAL CELLS.

The CFTR and sulphonylurea (SUR) receptor belong to the ATP-binding cassette (ABC) family of pumps that transport nutrients, drugs and solutes across the membrane. The CFTR is a chloride channel, its ATPase activity being used to drive the protein between open and closed confirmations. The ABC protein SUR is neither a pump nor an ion channel, but a channel regulator.

ionic regulation The maintenance of the relative proportions of various ions in the body fluids, important roles being played by the adrenal cortex (especially by secretion of ALDOSTERONE) and components of the kidney nephrons (see KIDNEY and cross-references). Compare OSMOREGULATION.

ionizing radiation Electromagnetic radiation causing the production of ionized, chemically reactive molecules. It includes ULTRAVIOLET IRRADIATION and X-IRRADIATION. See MUTAGENS.

ionotropic receptors Ligand-gated ION CHANNEL receptors for certain NEUROTRANSMITTERS, falling into two classes: the pentameric nicotinic receptors (subtype of ACETYLCHOLINE receptor), and the tetrameric ionotropic GLUTAMATE receptors. Mediating *fast synaptic transmission* (see SYNAPSES), they form ion channels which open on receptor binding and allow entry of ions such as Na^+, K^+, or Cl^- into the cell. The process from transmitter release to the postsynaptic response takes <1 ms, making receptor-coupled channels ideal for rapid communication within the central and peripheral nervous systems (compare METABOTROPIC RECEPTORS). The *nicotinic receptor family* includes nAChR, $GABA_A$, $GABA_C$, glycine and serotonin 5-HT_3 receptors. The glutamate receptor family includes GluR1–GluR4 (AMPARS), GluR6 (kainite receptors) and NMDA RECEPTORS.

ion pumps See ION CHANNELS.

IP_3 (inositol 1,4,5-triphosphate) A second messenger in the PHOSPHATIDYLINOSITOL SIGNALLING SYSTEM. See PHOSPHOLIPASE C, Fig. 135.

IPSP (inhibitory postsynaptic potential) See SYNAPSE.

IQ repeats Calmodulin-binding motifs of proteins.

iris (pl. irides) A contractile, coloured, structure, part of the vascular tunic of the EYE (see Fig. 73), lying in front of the lens, between the posterior and anterior chamber. It consists mainly of smooth muscle, surrounding an aperture, the pupil. It is attached at its outer margin to the ciliary processes. Its major role is to alter the aperture of the pupil during the PUPILLARY LIGHT REFLEX. Blue irides are due not to pigment, but from light scattering by the tissues of the iris which overlay a deeper layer of black melanin. Brown irides contain brown melanin.

iris reflex See PUPILLARY LIGHT REFLEX.

iron Present in haemoglobins, myoglobin, cytochromes and non-haem proteins (iron-sulphur proteins), a 75-kg human male contains ~4 g iron, a 55-kg female ~2.1 g, males storing more in ferritin and haemosiderin than women. It is stored in the liver by FERRITIN and carried in the plasma (to red bone marrow) by TRANSFERRIN. Because of this efficient recycling of iron, the need for dietary iron in adults is largely due to loss by bleeding and the death of intestinal cells. Optimum daily allowance is 20 mg, recommended daily allowance is 14 mg, increasing in preparation for PREGNANCY (although debate surrounds the need for additional iron intake during pregnancy if the mother has adequate vitamin C and meat intake). Haem iron absorption from meat and meat products (~5–10% of iron intake in most industrialized countries) is ~25% efficient, and is influenced very little by the iron status of the subject. The low pH of the stomach lumen solubilizes both ferrous and ferric iron salts. Non-haem iron is the main form of dietary iron, ingested in cereals, vegetables, pulses, beans, fruits, etc. Its absorption is greatly influenced by ligands (e.g., phytates, phenolics) from which it is difficult to remove, although the presence of vitamin C, and prolonged fermentation, are effective. High fibre content in the diet reduces iron absorption on account of its phytate content. Calcium interferes with iron absorption, dairy products having an adverse effect. Meat, fish and seafood all improve non-haem iron absorption. Non-haem iron, being in the reduced (Fe^{3+}) state, requires prior oxidation to Fe^{2+} (e.g., by vitamin C, or ferric oxidoreductase in the microvilli) before it can be absorbed. Haem iron (e.g., in cytochromes) not ingested with meat is poorly absorbed. Fe^{2+} is taken up with H^+ into ENTEROCYTES by the DMT1 transporter and stored in its reduced form to iron-binding proteins including ferritin (normally ~two-thirds is stored this way). Iron complexed with haem is taken up by a separate pathway. Absorbed iron is released into the blood in its reduced form across the basolateral enterocyte membrane through the ferroportin transporter, where it combines with plasma transferrin for transport to the liver (see HEPCIDIN).

Iron requirements during pregnancy increase; but iron absorption then improves – particularly in the 2nd trimester (50% increase) and last trimester (fourfold increase). Even so, with prevailing industrialized diets there will usually be a total iron deficit of ~0.5 g during pregnancy (although the loss due to menstruation will not occur), and during delivery there is an average additional blood loss of ~250 mg iron. Inherited disorders of iron homeostasis are an important class of human diseases, the most prominent being the primary iron loading conditions known collectively as haemochromatosis (see IDIOPATHIC HAEMOCHROMATOSIS). See ANAEMIA, GLUTATHIONE.

irregular bones Bones with complex shapes and variable amounts of spongy and compact bone. They are classifiable as long bones, short bones or flat bones. They include vertebrae and certain facial bones.

irritable bowel syndrome (IBS, spastic colitis) A disease of the entire alimentary canal in which reaction to stress leads to development of cramps and abdominal pains associated with alternating diarrhoea and constipation, often with loss of appetite.

irruptive population curves ('j' curves) Graphs of population density against time in which density is normally fairly stable but occasionally explodes (irrupts) exponentially to a peak on account of some factor (e.g., temperature) that briefly increases the carrying capacity, but then crashes to a stable lower level.

ischaemia Obstruction of the blood supply to the part of the body, as during myocardial infarction or THROMBOSIS. Ischaemia of walls of the heart causes reduced blood flow and results in an elongation of the PR interval (see ELECTROCARDIOGRAM). The HEAT-SHOCK PROTEIN Hsp20 is involved in preventing apoptosis after ischaemic injury. Compare INFARCTION; see CARDIOVASCULAR DISEASE, HYPOXIA, MITOCHONDRION; PAIN.

ischium Bone fusing with ilium and pubis after birth and forming the more posterior portion of the hip bone. Contributes to the acetabulum and bears the ischial tuberosity (see PELVIC GIRDLE).

island model of population structure A commonly used model to describe gene flow in a subdivided population in which each subpopulation of constant size, *N*, receives and gives migrants to each of the other subpopulations at the same rate, *m*. Under the island model, $F_{ST} = 1/(4Nm + 1)$. See GENETIC DISTANCE, MOLECULAR CLOCK.

islets of Langerhans Patches of endocrine tissue lying scattered within the more abundant exocrine tissue of the PANCREAS (see entry). They develop from parenchymatous tissue in the 3rd month of foetal life.

isoforms Different forms of a protein, being alternative expression products of the same gene but generated by use of alternative PROMOTERS, splice sites and polyadenylation (see RNA PROCESSING). Isoforms can provide tissue-specificity (e.g., see PGC-1 GENES, ACTIN), developmental stage specificity (e.g., the INSULIN-LIKE GROWTH FACTOR II gene), differential subcellular localization (e.g., soluble or lipid-bound isoforms), differential functional capacity (e.g., the progesterone receptor) and sex-specific gene regulation. See ISOZYMES, NORTHERN BLOTTING.

isometric contraction Skeletal muscle contraction in which the load is too great to allow a change of muscle length. The muscle does no work and no power is generated since the load is not moved through a distance (power = force × velocity). Compare ISOTONIC CONTRACTION.

isotonic (isosmotic) Term used of any medium containing the same concentration of osmotically active molecules as, e.g., a cell or tissue immersed in it. Isotonic solutions provide an osmotic buffer or support for cells. See SWEAT.

isotonic contraction Skeletal muscle contraction in which a muscle shortens against a constant load. Compare ISOMETRIC CONTRACTION.

isotopes The entry on AUSTRALOPITHECINES gives an indication of the value of isotopes in hominin research.

isotype switching The switching of antibody by activated B CELLS from IgM to another class, altering the antibody effector function by switching the F_C DOMAIN of the molecule, in response to cytokines – especially those released by armed effector helper T cells – enabling the adaptive immune response to adjust appropriately to different forms of pathogen infection. Although no further V(D)J recombination is involved, isotype switching does involve irreversible SOMATIC RECOMBINATION events mediated by a special form of non-homologous DNA recombination guided by stretches of repetitive DNA recombination sites termed *switch regions* (S_μ, S_γ, S_ε and S_α) located just 5′-upstream of the heavy-chain C-region (C_H) genes C_μ, C_γ, C_ε and C_α, respectively (see ANTIBODY DIVERSITY), there being four different C_γ sequences ($C_{\gamma 3}$, $C_{\gamma 1}$, $C_{\gamma 2b}$, $C_{\gamma 2a}$) each with its own S_γ switch region, needed for production of the four different IgG subtypes. One interaction essential for all isotype switching, in addition to stimulation by a particular cytokine (interleukin), is a co-stimulatory interaction between CD40 ligand on the T-cell surface and the CD40 on the B-cell surface. Those with the X-linked hyper IgM syndrome have CD40 deficiency and produce no IgG, IgA or IgE.

When a B cell switches from co-expression of IgM and IgD to expression of an IgG subtype, DNA recombination takes place between the S_μ and S_γ switch regions to generate that IgG subtype, a 'looping-out' event deleting the intervening C_μ and C_δ coding regions. This links the VDJ region to the $C_{\gamma 3}$, $C_{\gamma 1}$, $C_{\gamma 2b}$, $C_{\gamma 2a}$, C_ε and C_α sequences which would give rise to the IgG3 subtype of IgG. The formation and splicing of RNA transcribed from the switch recombination sites are initiated by cytokine release (e.g., IL-4) from T_H2 cells, alternative recombination between the remaining splice sites and looping out of intervening C regions giving rise to the generation of further Ig subtypes. TGF-β induces switching to IgG2b and IgGA (the latter by removal of all four C_γ and the C_ε sequences; see MUCOSAL-ASSOCIATED LYMPHOID TISSUES), whereas the T_H1 cytokines, such as IFN-γ, result in IgG2a, 2b and 3 switching. Switching is also strongly influenced by the nature of the stimulating antigen (see ANTIBODIES).

isozymes (isoenzymes) Enzyme ISOFORMS, often differing in the kinetic or regulatory properties, in the cofactor used, or in their subcellular distributions (e.g., soluble or membrane-bound). One of the first discovered was lactate dehydrogenase.

IT (area IT) See INFERIOR TEMPORAL CORTEX.

itch receptors A class of C fibre (see AXONS, PAIN) responding to histamine released by mast cells.

IUGR See INTRA-UTERINE GROWTH RETARDATION.

IUI See INTRA-UTERINE INSEMINATION.

IVF See IN VITRO FERTILIZATION.

J

JAK/STAT pathway The signalling pathway employed by more than 30 cytokines and hormones. JAKs, or Janus kinases, are soluble tyrosine kinases which bind activated CYTOKINE RECEPTORS (see Fig. 105), such as those for interferons (α- and γ-), ERYTHROPOIETIN and THROMBOPOIETIN, and transduce signals via the STAT proteins which have SH2 DOMAINS. The cytokine or hormone activates its receptor by dimerizing it, leading to reciprocal autophosphorylation of the associated JAKs at appropriate tyrosine residues leading to phosphorylation of the receptor itself. The phosphotyrosines attract inactive STAT transcription factors, which bind the SH2 domains and become phosphorylated by the JAKs. Mutated JAK2 enzyme is the cause of a blood disorder often linked to chronic myeloid LEUKAEMIA.

The STATs then dimerize as their SH2 domains bind the phosphorylated tyrosine of their partner, and translocate to the nucleus where they operate as active transcription factors to activate key gene expression, such as genes encoding class I MHC proteins and interferons. Other signalling proteins acting via this pathway include GROWTH HORMONE and PROLACTIN. In addition to phosphorylating STATs, JAKs can also phosphorylate substrates that activate other mitogenic pathways, including the RAS SIGNALLING PATHWAYS. Several STATs are constitutively activated in a number of human cancers (e.g., Stat3 in myelomas). STATs can be activated in the cytoplasm by tyrosine kinases other than receptor-associated JAKs (e.g., by Src; see SRC GENE).

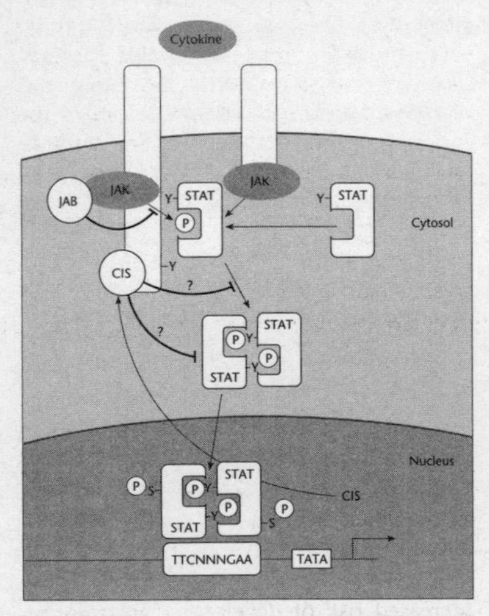

FIG 105. Involvement of the JAK/STAT PATHWAY in cytokine signalling, with CIS (cytokine-induced Src-homology 2 protein) associating with the receptor (vertical bars through the membrane) and regulated by Janus kinases (JAKs) and signal transducers and activators of transcription (STATs). The protein JAB interacts with JAKs, inhibiting their action (indicated by a bar at the end of an arm).

James–Lange theory See EMOTIONS.

jaws The jawbones are the MANDIBULAR BONE (upper jaw) and MAXILLARY BONE (lower jaw). See MASTICATION.

jejunum That region of the ALIMENTARY CANAL extending from the duodenum to the ileum, and therefore a part of the SMALL INTESTINE. Approximately 1.5 m in length, it is usually empty at death (*jejunum* being Latin for 'empty'). A sharp bend, the duodeno-jejunal flexure, forms its junction with the duodenum, but although the mucosal wall thickness decreases progressively along the small intestine, there is no clear anatomical distinction between the jejunum and the ileum.

JmjC-domain-containing proteins (jumonji C-domain-containing proteins) See HISTONE DEMETHYLASES, HISTONES.

JNK signalling pathway The Jun amino-terminal kinases (JNKs) are members of the MAPK family (see MAPK PATHWAY) and activated by a variety of environmental stresses, inflammatory cytokines, growth factors and G PROTEIN-COUPLED RECEPTOR agonists. Stress signals are delivered to this cascade by members of the Rho family (see RAS SIGNALLING PATHWAYS), and as with other MAPKs, the membrane-proximal kinase is a MAPKKK that phosphorylates and activates MKK4 or MKK7, the SAPK/JNK kinases (stress-activated protein kinases/ JNK kinases) which translocate to the nucleus and regulate transcription of multiple transcription factors. See STRESSORS.

joint kinesthetic receptors Proprioceptive receptors located within a joint and stimulated by the latter's movement (see PROPRIOCEPTION). Several types occur around the articular capsules of synovial joints, including free nerve endings and type II cutaneous mechanoreceptors (see RUFFINI ENDINGS); small PACINIAN CORPUSCLES in connective tissue around articular capsules, and receptors, similar to TENDON ORGANS, within articular ligaments. See MECHANORECEPTORS.

joints (articulations, arthroses) Any regions of contact between two bones, between bone and cartilage, or between bone and teeth. At such regions skeletal elements are said to articulate, enabling movement (see Fig. 106). Classifications of joints tend to be either structural or functional. The main considerations in structural classification are whether or not a space (synovial cavity) occurs between the articulating elements and the type of connective tissue binding the elements together. Functional classifications relate to the degree of movement a joint permits.

Structural types include: (i) *fibrous joints,* lacking a synovial cavity and held together by fibrous connective tissue rich in collagen fibres. They include sutures (thin dense connective tissue uniting bones of the SKULL), syndesmoses (in which the bones united are further apart than in a suture and connected by a greater quantity of fibrous connective tissue, either a bundle (ligament) or sheet (interosseous membrane)), gomphoses (in which a fibrous peg fits into a socket, as with teeth in the maxillae and mandibles); (ii) *cartilaginous joints,* also lacking a synovial cavity, bones being held together by cartilage. They include synchondroses (in which the connecting material is hyaline cartilage) and symphyses (in which the ends of the articulating bones are connected by a broad, flat disc of fibrocartilage, as in intervertebral joints (see VERTEBRAE)). SYNOVIAL JOINTS do have a synovial cavity and are held together by the dense irregular connective tissue of an articular capsule, and often by accessory ligaments. On functional grounds, *synarthroses* (sing. synarthrosis) are immovable joints; *amphiarthroses* are slightly moveable; and *diarthroses* are freely moveable (all of the latter being synovial joints). See ARTHRITIS, LEVERS, PACINIAN CORPUSCLES, RHEUMATOID ARTHRITIS, VERTEBRAE.

jugulars Veins draining the head, comprising paired left and right internal and external jugular veins. The two external jugular veins join the appropriate SUBCLAVIAN VEIN before joining their respective internal jugular veins to form the left and right brachiocephalic veins.

juvenile polyposis syndrome (JPS) An autosomal condition in which patients bear 50–200 polyps, usually in the rectosigmoid region of the gut, and have an increased risk of developing gastrointestinal polyps and COLORECTAL CANCER. Inactivating mutations in *Smad4* and BMP receptor type 1A account for up to 50% of cases, implying that loss of intestinal BMP signalling is the primary cause and suggesting a role for the BMP pathway in epithelial homeostasis. See COLON, SMALL INTESTINE.

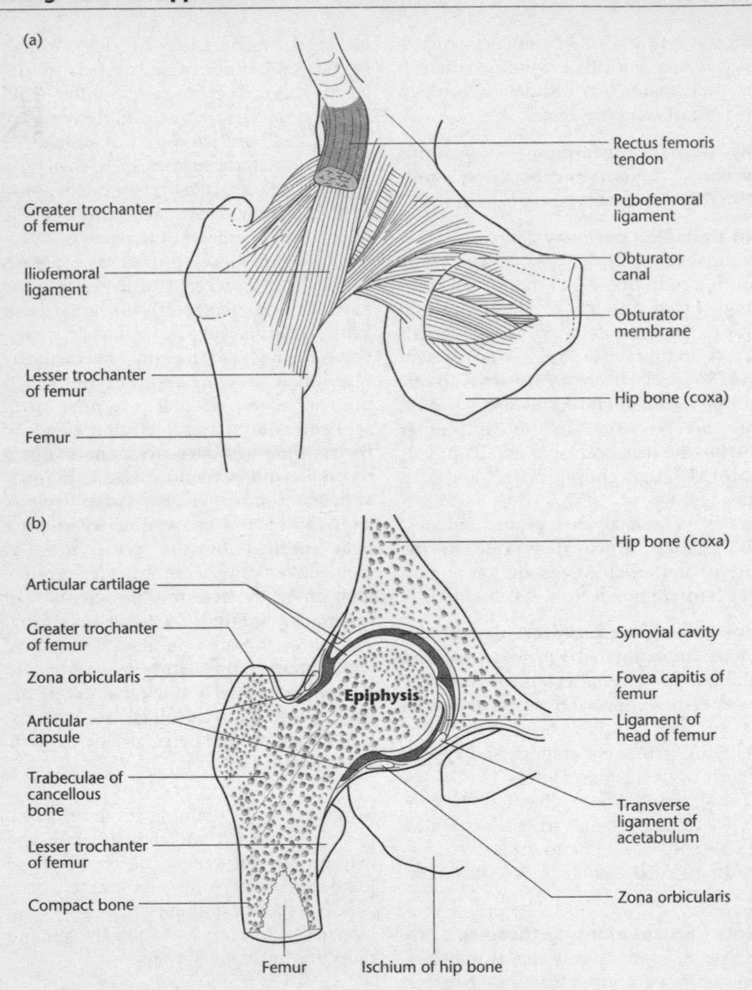

FIG 106. (a) Anterior view of right hip JOINT, showing the head of the femur articulating with the hip bone, ligaments (shaded) and tendon (stippled). (b) Frontal section through (a). From *Principles of Anatomy and Physiology*, G.J. Tortora & S.R. Grabowski, © 2000. Reproduced with permission of John Wiley & Sons, Inc.

juxtaglomerular apparatus Juxtaglomerular cells (granular cells) are specialized smooth muscle cells lining the afferent arterioles entering Bowman's capsules of the KIDNEY cortex, and secreting the enzyme renin when BLOOD PRESSURE falls. Renin has an important role in aldosterone secretion by the adrenal cortex and consequently in maintenance of sodium balance. In association with specialized cells of adjacent ascending limbs of loops of Henle (the MACULA DENSA), they constitute the juxtaglomerular apparatus. See ANGIOTENSINOGENS.

K

K_A/K_S and K_A/K_I ratios See MUTATION.

kainate receptors A subtype of GLUTAMATE-receptor ionotropic ion channel with major modulatory roles in pre- and post-synaptic sites.

karyotype The chromosome constitution of a species, arrived at by *karyotyping*. This states the total number of chromosomes, including the sex chromosome constitution. Normal karyotypes for human females and males are 46,XX and 46,XY, respectively. Chromosomes may be displayed in the form of a karyogram (see Appendix III for the human karyogram). See CHROMOSOME, FISH, G BANDS, HUMAN GENOME.

Kell blood group system (Kell–Cellano blood group system) This antigen system is determined in part by the XK locus which encodes the Kell blood group precursor substance (Kx, or XK) and is X-linked, mapping to Xp21.2-p21.1. The Kell peptide antigens, however, are determined by a locus at 7q33, the gene products being located on the human red blood cell surface and important determinants of blood type as targets for autoimmune or alloimmune disorders that destroy blood cells. Those without the Kell antigen (K_0) must be transfused with blood of type K_0 to prevent haemolysis. The two commonest alleles are K_1 (Kell) and K_2 (Cellano), the Kell protein containing the Kell antigens bound to XK being a 93-kd transmembrane protein and zinc-dependent endopeptidase responsible for cleaving endothelin-3. Individuals lacking a specific Kell antigen may develop antibodies against it when transfused with blood with the antigen. Absence of the XK protein leads to marked reduction in Kell antigens, but the reverse is not the case.

Kenyanthropus platyops The name given to fossils discovered in 1999 from the western shores of Lake Turkana in Kenya and dating to the period 3.2–3.5 Myr ago, contemporary with *Australopithecus afarensis*. Demonstrating a combination of primitive and derived features, the most distinctive feature being its flat face, it has been placed in a genus of its own. Similarities with the later, larger-brained, KNM-ER-1470 fossils have led to these latter 'rudolfensis' specimens also being attributed to *Kenyanthropus*. See HOMININI.

keratin The keratins form the major and most diverse family of vertebrate INTERMEDIATE FILAMENT proteins, with ~20 isoforms located in different human epithelial tissues and ~10 more restricted to hair and nails. Type I keratins are acidic, while type II keratins are basic, each keratin filament being a composite of an equal ratio of the two chains, forming heterodimers that join in pairs to form the tetrameric keratin subunit. These tetramers are then cross-linked by disulphide bonds to form tough coverings as in the skin EPIDERMIS, and in hair and nails. Several genetic disorders are due to mutations in the keratin genes.

keratinocytes Cell type forming ~90% of cells of the skin EPIDERMIS (see Fig. 71), deriving from stem cells in the bulge region of HAIR FOLLICLES, and producing the internal protein KERATIN that inserts into desmosomes and strengthens the cells, protecting the skin and underlying tissues from heat, microbes and chemicals. While in the stratum basale (the *Malpighian layer*), they undergo mitotic division roughly every 19 days, one daughter cell becoming a new stratum basale cell while the other is pushed towards the surface. While part of the stratum granulosum, keratinocytes contain lamellar granules from which they release a water-proofing lipid sealant and die. Cells take 40–56 days to reach the surface and desquamate. They may derive pigment from MELANOCYTES (see also MELANINS). Some keratinocytes have THERMORECEPTOR membrane channels.

ketogenic Referring especially to those AMINO ACIDS (e.g., glycine) that yield ketone bodies when deaminated and are either used in triglyceride synthesis or are catabolized in the KREBS CYCLE. They are not GLUCOGENIC.

ketone bodies Despite their name, these are water-soluble fuels, formed (ketogenesis) during FATTY ACID METABOLISM in the liver and exported to the brain and other organs for use in the KREBS CYCLE when glucose is in short supply. They include ACETOACETIC ACID, β-hydroxybutyric acid and acetone. Ketone bodies in the blood and urine of untreated diabetics can reach very high levels (90 mg/100 mL), a condition termed *ketosis*, or *ketoacidosis* if the levels of the mentioned carboxylic acids overcome the capacity of the blood's hydrogen carbonate buffering system and result in a lowering of blood pH – a life-threatening condition (see AMMONIA, DIABETES MELLITUS). See KETOGENIC, STARVATION CONDITIONS.

ketosis (ketoacidosis) See KETONE BODIES.

kidney The permanent kidneys are each roughly the size of a clenched fist. Developing from somatic mesoderm, they lie behind the peritoneum on the posterior abdominal wall, either side of the vertebral column, extending over the last thoracic (T12) to the third lumbar (L3) vertebrae and partially protected by the ribcage. Each develops from the foetal METANEPHROS and ureteric bud, an outgrowth of the mesonephric duct close to its entrance to the cloaca. The ureteric bud gives rise to the ureters, the renal pelvises, major and minor calyces and ~1–3 million collecting ducts (collecting tubules). The metanephric mesoderm gives rise to the excretory units, or nephrons. The ureteric bud penetrates metanephric tissue, which becomes moulded over its distal end to form a cap. The bud later dilates, forming the primitive renal pelvis, splitting into cranial and caudal portions, the future renal calyces. These each develop two new buds, which in turn subdivide as they penetrate further into the metanephric tissue forming 12 or more generations of tubules. By the start of the 6th month, coalescence of third and fourth generation tubules takes place to form the 8–20 minor calyces of the renal pelvis, as elongation of collecting tubules of the fifth and later generations is accompanied by their convergence on the minor calyces to form the renal pyramids into which urine flows to the URETER.

Each newly formed collecting tubule has a cap of metanephric tissue, each of which develops under signals from the tubule to form a small renal vesicle which develops into an S-shaped tubule with an invaginated pocket at one end. Capillaries grow into this pocket to form a knot, the *glomerulus*, as the pocket deepens to form a BOWMAN'S CAPSULE (the two combined being one renal corpuscle, or *filtration unit*). The glomerulus and tubule together comprise an excretory unit, or *nephron*, of which there are ~1.3 million per kidney. At the nephron's other end, a connection is made with a collecting tubule. Continuous lengthening of the nephron results in formation of the proximal convoluted tubule (PCT), loop of Henle and distal convoluted tubule (DCT). A frontal section (see Fig. 107) reveals an outer cortex and inner medulla. A renal artery gives rise first to segmental arteries and then to interlobar arteries, which ascend within renal columns towards the cortex. Near the base of each pyramid these diverge to form the arcuate arteries, from which interlobular arteries project into the cortex, giving rise to

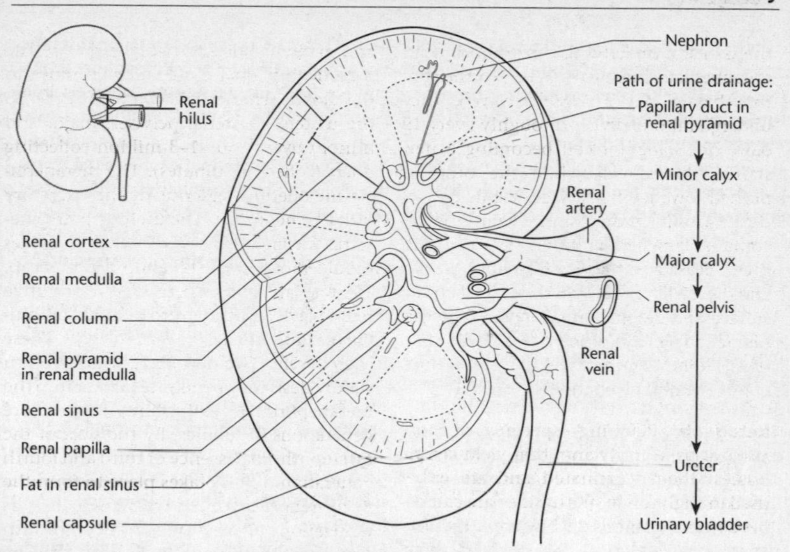

FIG 107. Frontal section of right KIDNEY. From *Principles of Anatomy and Physiology*, G.J. Tortora & S.R. Grabowski, © 2000. Reproduced with permission of John Wiley & Sons, Inc.

afferent arterioles from which arise the glomeruli of the filtration units. These offer huge resistance to blood flow, accounting in large part for the need for a relatively high left ventricular pressure compared with that of the right (see BLOOD PRESSURE for the kidneys' role in its maintenance). The efferent arterioles leaving the glomeruli of the superficial nephrons give rise to the peritubular capillaries, while those of the juxtamedullary nephrons give rise to the VASA RECTA passing deep into the medulla and providing its blood supply.

The volume of blood flowing through the kidneys per minute (renal blood flow) is the product of the CARDIAC OUTPUT and RENAL FRACTION, amounting to ~1,200 mL min^{-1} in healthy individuals, or ~650 mL plasma min^{-1}. Sympathetic neurons, releasing noradrenaline, innervate the blood vessels of the kidneys and their stimulation, as during shock or intense EXERCISE, constricts the small arteries and afferent arterioles, decreasing renal blood flow and filtrate formation. Because the afferent arteriole giving rise to the glomerulus is wider than the efferent arteriole leaving, hydrostatic pressure builds up in the glomerulus. The glomerular capillaries are fenestrated (the endothelial cells have pores in them), easing the loss of fluid and solutes into a Bowman's capsule, although the surrounding basement membrane prevents loss of all but the smaller proteins, while the podocytes of the visceral layer of Bowman's capsule that lie on the basement membrane permit fluid to leave only through filtration slits between their pedicel 'feet'. On average, 19% of plasma is filtered into Bowman's capsule, the glomerular filtration rate (GFR) being the volume of filtrate entering all the Bowman's capsules of both kidneys min^{-1} (650 mL $min^{-1} \times 0.19 \approx 125$ mL filtrate min^{-1}). The smooth muscles in the walls of the afferent and efferent arterioles can alter vessel diameter and glomerular filtration pressure, or NET FILTRATION PRESSURE, in the renal corpuscle: dilation of afferent arterioles increases filtration pressure and glomerular filtration.

About 65% of the filtrate volume is reabsorbed back into the peritubular capillaries during its passage through the PROXIMAL CONVOLUTED TUBULE (PCT, see entry for details). UREA enters the glomerular filtrate at the same concentration it has in the plasma, and the thick ascending limb of the LOOP OF HENLE, the DISTAL CONVOLUTED TUBULES and cortical portion of the COLLECTING DUCT are impermeable to it. However, the descending limb is permeable to it, and some diffuses out as its concentration increases on account of water loss in the tubular fluid. But as the volume of urine in the *medullary* collecting ducts decreases on account of VASOPRESSIN (ADH), the concentration of urea in the ducts increases so that it diffuses out into the interstitial fluid of the deep medulla, where the ducts are permeable to it. As urea concentration rises in the medulla, further water is removed osmotically from the collecting ducts, and some urea diffuses back into the tubular fluid in the descending and thin ascending limbs of loops of Henle, also permeable to urea. Repeated transfer of urea between the permeable portions of the nephron and the interstitial fluid (*urea cycling*) leads to ~50% of the original plasma urea being lost passively per transit through the kidneys. Uric acid, the end-product of purine metabolism, circulates in the plasma as the anion urate, very little of which is bound to plasma proteins. It is freely filtered through the glomerulus and largely reabsorbed in the PCT, only ~5–10% being excreted per circuit. Its clearance is ~6–9 mL min^{-1}, amounting to 0.5–1 g per day. Roughly 5% of all kidney stones are deposits of uric acid or its salts (see GOUT).

The renal cortex is metabolically extremely active. Two of its enzymes are involved in the conversion of VITAMIN D to calcitriol, under the influence of FGF23, and its cells can synthesize glucose by GLUCONEOGENESIS. Most filtered Ca^{2+} ions are reabsorbed passively in the PCT, but a combination of active and passive absorption occurs in the thick ascending limb while absorption in the DCT is stimulated by binding of parathyroid hormone to PTHR1 receptors and is a three-step process involving passive entry across the apical cell surface, cytosolic diffusion assisted by vitamin D-dependent calcium-binding proteins (calbidins), and active transport of Ca^{2+} across the basolateral membrane via a high-affinity Ca^{2+}-ATPase ion pump and Na^+/Ca^{2+} ion exchanger. Na^+-Cl^- symporters in the DCT apical membranes reabsorb these ions in a concentration-dependent manner, while passive reabsorption of Na^+ into capillaries and passive secretion of K^+ into the tubular fluid (ALDOSTERONE-related) occurs by principal cells of the DCT. In the collecting duct, most of the intercalated cells secrete H^+ into the tubular fluid through H^+-ATPases so that urine can be up to 3 pH units more acidic than blood. Intercalated cells also contain carbonic anhydrase, so that HCO_3^- produced by dissociation of carbonic acid crosses their basolateral membranes through Cl^-/HCO_3^- antiporters and diffuses into peritubular capillaries. Other intercalated cells have H^+ pumps in their basolateral membranes and Cl^-/HCO_3^- antiporters in their apical membranes, and between them these cells help maintain the pH of body fluids by excreting excess H^+ when pH is too low and by excreting excess HCO_3^- when pH is too high. The extent of Na^+, Cl^- and water reabsorption and K^+ secretion by renal tubules is regulated by four hormones: angiotensin II (AT_2), aldosterone, vasopressin (ADH) and atrial natriuretic peptide. See HEPATOCYTE GROWTH FACTOR (for hereditary papillary renal cancer), OSMOREGULATION (where aquaporins are dealt with), POLYCYSTIC KIDNEY DISEASE, RENAL FAILURE.

kidney machine The artificial KIDNEY (renal dialysis machine) allows arterial blood to flow through tubes made of a selectively permeable membrane surrounded by dialysis fluid containing the same concentration of solutes as blood plasma, except for those of metabolic wastes. Pores in the membrane are too small to allow proteins to pass into the surrounding fluid and only smaller solutes (e.g., urea) with a higher concentration in the plasma than in the dialysis fluid diffuse across. Compressed CO_2 is used to force the fresh dialysis fluid through the machine, in a countercurrent direction to the blood.

killer T cells (T_C cells) See T CELLS.

kinases Enzymes that phosphorylate substrates. Many phosphorylate proteins (see PROTEIN KINASES), but some phosphorylate lipids (e.g., see PHOSPHATIDYLINOSITOL 3-KINASE). See SIGNAL TRANSDUCTION.

kin detection In 1891, Westermarck proposed that mutual exposure during childhood weakens sexual attraction among adults. Relatively little work has been done regarding the existence or design of human kin detection mechanisms; but one such (2007) tested whether sibling ALTRUISM and aversion to personal engagement in sibling INCEST and opposition to third party sibling incest behaved in accordance with current socioecological and population biological theory that ancestral human foragers would have been subject to inbreeding depression and KIN SELECTION. The evidence indicated that there is a kin detection system in humans, and that it uses two distinct, ancestrally valid cues to compute relatedness: the familiar other's perinatal association with the individual's biological mother, and the duration of sibling co-residence. If a potential sibling is younger, we note how much time he/she spends with our mother; if older, we assess how much time we ourselves have spent with that person.

kin recognition See KIN DETECTION.

kinesins Humans have at least 40 forms of this superfamily of MOTOR PROTEINS (see entry), which includes at least 10 families of kinesin-related proteins (KRPs). Kinesins move along a microtubule toward its plus end (hence '*plus-end directed motors*'), carrying, e.g., membrane-bound organelles attached to their tail region. They are involved in the behaviour of the mitotic and meiotic spindles (see MITOSIS). Their motor domains are smaller than those of myosins, and they track along different filaments (myosin binds actin) and have different kinetic properties; but they share very similar core structures and appear to have shared a common evolutionary origin. The fastest movements kinesins achieve are ~2–3 μm sec^{-1}, considerably slower than is achievable by the DYNEINS. Small movements of switch loops at the NTP-binding site regulate the docking and undocking of the motor head domain. This is attached by a linker region to the long coiled-coil middle region whose dimerization is driven by the association of regularly spaced hydrophobic amino acids (see MYOSIN). The C-terminal tails attach to a cargo. When ATP binds to a head region its linker domain amplifies the motor domain conformational change and docks, throwing the 2nd head forward to a fresh attachment site 8 nm closer to the microtubule plus end than the docked 1st head. Kinesin forms a relatively lengthy rigor-like tight association with a microtubule when ATP is bound, releasing this attachment on ATP hydrolysis. Nucleotide hydrolysis cycles are so coordinated that the two-headed motor can walk 'hand-over-hand'. Most kinesins have their motor domain at the N-terminus of the heavy chain and move towards the microtubule plus end; but one family of KRPs has the motor domain at the C-terminus and walks towards the minus end of a microtubule. Kinesins are responsible for fast axonal transport of mitochondria, secretory vesicle precursors and components of synapses towards nerve terminals.

kinesthesia The sense enabling knowledge of the positions of one's limbs even when one is blindfolded, and in which MUSCLE SPINDLES play an important part. Normally, a person can accurately replicate the position of one arm by moving the other to the same angle; but if both arms are initially raised at 90° to the horizontal and a muscle, such as the biceps, of one arm is stimulated by vibration, the blindfolded person will move the other to a new and incorrect perceived position. This *vibration illusion* occurs because the vibration stimulus to the stretch receptors, causing increased afferent discharge, is interpreted by the central nervous system as indicating that the vibrated muscle is longer than it is. Active exploration of an object to determine its shape and texture is termed *haptic touch*, and is employed to great effect by the blind.

kinins Any of various structurally related autacoid polypeptides, such as BRADYKININ and kallikrein, produced from the extracellular peptide kininogen. They act locally to induce vasodilation and contraction of smooth muscle. The kinin system is an enzymatic cascade of plasma pro-enzymes, triggered by tissue damage, producing several inflammatory mediators, such as bradykinin. Kinins are responsible for many effects in leukocytes including the release of other inflammatory mediators, such as cytokines, prostaglandins, leukotrienes and reactive oxygen species. The resulting increase in vascular permeability promotes influx of plasma proteins and neutrophils to the site of tissue injury, and causes PAIN. In leukocytes, their effects are mediated by kinin receptor subtypes B_1, B_2, or both, depending on species and cell type. In contrast to the other leukocytes, neutrophils contain the complete system for the synthesis and release of bioactive kinins.

kinocilium A conventional solitary CILIUM, anchored to its basal body, found on the apical surface of HAIR CELLS.

kinome See HUMAN KINOME.

kin selection Selection favouring genetic components of any behaviour by one individual which has beneficial consequences for another, and whose strength is proportional to the relatedness of the two individuals. See ALTRUISM.

Klinefelter('s) syndrome A relatively common condition (incidence ~1:1000 male live births) due to the presence of an additional X CHROMOSOME as a result of NON-DISJUNCTION (compare TURNER SYNDROME). Childhood symptoms include clumsiness, self-obsessiveness, some mental retardation and mild learning difficulties, especially in relation to verbal skills. Adults tend to be slightly taller than average, with long lower limbs, ~30% of individuals having moderately severe gynaecomastia. Individuals are usually infertile, with small, soft testes, and treatment with testosterone from puberty onwards is beneficial for development of SECONDARY SEXUAL CHARACTERISTICS and long-term prevention of OSTEOPOROSIS. A very few individuals do produce some sperm, and testicular sperm aspiration and intra-cytoplasmic sperm injection has resulted in fertility. Cases of maternal inheritance of the additional X chromosome increase with advancing maternal age, and a small proportion of individuals show mosaicism (46,XY/47, XXY). Very rarely, two additional X chromosomes are found, such individuals showing more marked Klinefelter characteristics, and being severely mentally retarded. See GENETIC COUNSELLING

knee-jerk reflex See MUSCLE SPINDLES.

knee joint The most complex, and largest, JOINT in the human body. It comprises three joints within a single synovial cavity: an intermediate patellofemoral joint; a lateral tibiofemoral joint and a medial tibiofemoral joint. No complete independent articular capsule unites the bones, the ligamentous sheath surrounding the joint comprising mostly muscle tendons or their expansions. But capsular fibres connect the articulating bones. The patellar ligament strengthens the anterior surface of the joint and houses the PATELLA; the oblique poplitial ligament and tendon of the semimembranous muscle strengthen the posterior surface; the arcuate poplitial ligament strengthens the lower lateral part of the posterior surface; the tibial collateral ligament is crossed by the sartorius, gracilis and semitendinosus muscles, all strengthening the medial aspect; the fibular collateral ligament strengthens the lateral aspect and is covered by the tendon of the biceps femoris muscle; and the intracapsular ligaments connecting the tibia and femur are the anterior and posterior cruciate ligaments. Two fibrocartilage discs between the tibial and femoral condyles help compensate for the bones' irregular shapes. The joint can engage in flexion, slight medial rotation, and lateral rotation of the leg in flexed position.

knockouts See GENE KNOCKOUTS.

knuckle-walking A semiquadrupedal

form of GAIT adopted by gorillas, especially the heavier male, when on the ground. The more arboreal chimpanzees, having similar limb proportions and upper body shape, also adopt this posture.

Koch, Robert Robert Koch (1843–1910) developed criteria for the definition of a pathogen during his work on the isolation and identification of bacteria such as *Mycobacterium* tuberculosis and *Bacillus anthracis*. Koch's 'postulates', indicating when such an identified bacterium is a pathogen, are: (i) the bacterium can always be identified in cases of the disease; (ii) the bacterium is only found in the presence of the disease; and (iii) when the bacterium is cultured in pure growth outside the body and then re-introduced into a healthy host, it will produce the same disease. Although these criteria operate reasonably well for many bacteria, they do not adequately define the host–pathogen interaction for agents such as viruses and prions, discovered later.

koniocellular layers (k. zones) Interlaminar layers, containing very small cells, lying ventral to each of the six principal layers of the LATERAL GENICULATE NUCLEI. They receive input from the small, slowly conducting, nonM-nonP types of retinal ganglion cells with large dendritic trees and large receptive fields, which signal average illumination. See also PRIMARY VISUAL CORTEX.

Krebs cycle (citric acid cycle, tricarboxylic acid cycle, TCA cycle) The cyclical metabolic pathway (see Appendix IIc) occurring in the mitochondrial matrix and generating reduced coenzymes (NADH, FADH) which are subsequently oxidized by the inner mitochondrial membrane during ATP synthesis (see MITOCHONDRION). The main substrate for the cycle is acetate, derived by decarboxylation and dehydrogenation from the pyruvate entering the mitochondrion from the cytosol as a product of GLYCOLYSIS. The cycle itself generates a little GTP, which may transfer its phosphoryl group to ADP to synthesize ATP; but it is the main source of the blood's CARBON DIOXIDE. The Krebs cycle is linked to the UREA cycle.

Kupffer cells MACROPHAGES resident along sinusoids of the LIVER. With macrophages of the red pulp of the SPLEEN, they remove large numbers of dying cells from the blood daily. They express little MHC class II protein and no TLR-4 and are therefore unlikely to elicit an autoimmune response, although they produce large quantities of self peptides in their endosomes and lysosomes. The production of IL-6 has been implicated in progression of liver cancer (see COLORECTAL CANCER).

kwashiorkor A severe protein deficiency disorder of children in which arms and legs become emaciated owing to muscle protein degradation as an energy source. The abdomen swells (oedema) as a result of TISSUE FLUID build-up, blood protein levels being too low to draw water back through capillaries. It typically develops when a child is weaned and its diet shifts from protein-rich breast milk to low-protein starchy foods. See MALNUTRITION.

L

λ The symbol notation for FAMILIAL RECURRENCE RISK of disease. See COMPLEX DISEASES.

labia majora (sing. labium majus) Two longitudinal folds of skin, inferior and posterior to the mons pubis covered by pubic hair and containing an abundance of adipose tissue, sebaceous and sweat glands and homologous to the scrotum. See EXTERNAL GENITALIA.

labia minora (sing. labium minus) Medial to the labia majora, two smaller skin folds devoid of pubic hair and adipose tissue and containing numerous sebaceous glands but fewer sweat glands than them. They are homologous to the spongy (penile) urethra. See EXTERNAL GENITALIA.

labour See BIRTH.

lacrimal glands (tear glands) Almond-sized glands, being accessory structures of the eye and situated between the upper eyelids and the eyebrow region. Their secretion, tears, drain through 6–12 ducts onto the surface of the conjunctiva of the upper eyelids, passing medially over the anterior surface of the eyeball and draining via two lacrimal canals into a lacrimal sac and ultimately into the nasal cavity. Tears are a watery solution containing salts, some mucus and the antibacterial enzymes LYSOZYME and PHOSPHOLIPASE A. Irritation of the conjunctiva causes the lacrimal glands to oversecrete, as does inflammation of the nasal mucosa. Their secretion is controlled from the PONS via parasympathetic outflow along the facial nerves, sensory return to the pons being via the trigeminal nerve.

lactacid oxygen debt See OXYGEN DEBT.

lactase See DISACCHARIDASES, LACTOSE INTOLERANCE.

lactate See LACTIC ACID.

lactate dehydrogenase (LDH) An enzyme of the cytosol catalysing the reduction of pyruvate to lactate, using NADH as reductant, during build-up of an OXYGEN DEBT. It may also be an important component in CIRCADIAN CLOCK mechanisms by influencing cellular redox potentials.

lactate threshold The highest oxygen consumption or exercise intensity generating less than a 1.0 $mM.dm^{-3}$ increase in blood lactate concentration (see Fig. 108). Its value is higher in trained than untrained athletes (see EXERCISE), and its value is affected by such factors as low tissue oxygen, the reliance on glycolysis, activation of fast-twitch SKELETAL MUSCLE fibres and reduced lactate removal. When its value rises to a level of 4.0 $mM.dm^{-3}$ (or *onset of blood lactate accumulation*) this may indicate the maximum prolonged exercise intensity that a person can sustain.

lactation The synthesis and secretion of milk by a mother's MAMMARY GLANDS, usually occurring after parturition (BIRTH) and continuing for 2–3 years, or longer, providing that suckling occurs regularly. During PREGNANCY, oestrogens and progesterone cause branching of the ducts and expansion of the secretory units of the mammary glands, leading to their enlargement. Oestrogen is mainly responsible for breast enlargement during pregnancy, while progesterone causes development of the secretory alveoli; but growth hormone, prolactin,

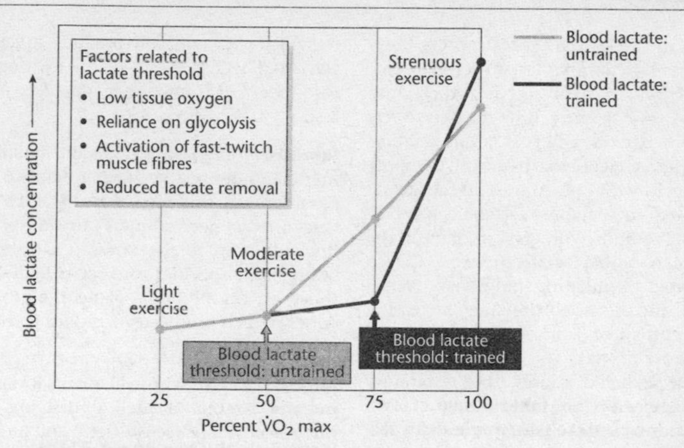

FIG 108. Line graphs indicating the blood lactate concentrations and lactate thresholds for untrained (solid) and trained (hatched) subjects at different levels of exercise and plotted against percentage of maximal oxygen consumption. © 2001 Redrawn from *Exercise Physiology* (5th edition) by W.D. McArdle, F.I. Katch and V.L. Katch, published by Lippincott, Williams & Wilkins. Reproduced with the kind permission of the publisher.

thyroid hormones, glucocorticoids and insulin all play a part, as do human placental lactogen and human somatotropin. During most of a pregnancy, oestrogen and progesterone inhibit the effect of prolactin on the mammary glands; but before parturition, high oestrogen levels stimulate greater production of PROLACTIN (see entry). After parturition, the levels of all three hormones decrease; but freed from its inhibition, prolactin can then stimulate milk production. During suckling, sensory impulses are sent from the nipples to the secretory oxytocinergic neurons of the hypothalamus, which release OXYTOCIN. Oxytocin causes release of milk into the mammary ducts via the milk ejection reflex through myoepithelial contraction in the lactiferous ducts, sinuses and breast tissue alveoli. Milk flow begins within 30 seconds to 1 minute of suckling (milk let-down), which persists throughout lactation. Mammary gland secretion increases during the first 36 hours after birth, reaching ~500 mL d^{-1} four days after birth, and during this time they produce colostrum, which contains little fat and less lactose than milk. True milk appears around the fourth day, both colostrum and milk containing maternal antibodies.

lacteal The modified lymph vessel within VILLI of the small intestine, opening into the local lymphatic circulation, carrying CHYLOMICRONS produced by ENTEROCYTES of the mucosal epithelium.

lactic acid, lactate The product of pyruvate reduction by LACTATE DEHYDROGENASE in those cells deprived of sufficient oxygen to support aerobic oxidation of the pyruvate generated by GLYCOLYSIS or – as in the case of red blood cells – by cells lacking mitochondria.

Lactate does not accumulate in the plasma at all levels of exercise, as indicated in Fig. 108 (see LACTATE THRESHOLD). Its concentration is expressed in mM dm^{-3} of whole blood (vol %). In strenuous exercise, the glycogen within *inactive* tissues can become available to supply the needs of active muscle. Such active glycogen turnover through the *exchangeable lactate pool* provides a gluconeogenic precursor for carbohydrate synthesis via the Cori cycle in liver and the kidney cortex. Lactate produced in fast-twitch SKELETAL MUSCLE fibres can circulate to other fast-twitch fibres for conversion to pyruvate and then into acetyl-CoA for entry into the KREBS CYCLE where it is oxidized. This *lactate shuttling* enables glycogenolysis in one cell to supply other cells with fuel. Almost all the lactate generated in anaerobic metabolism is buffered by sodium hydrogen carbonate in the plasma to sodium lactate, the excess CO_2 so generated stimulating pulmonary ventilation and disproportionately increasing the ventilatory equivalent (see EXERCISE, RESPIRATORY CENTRE). The additional carbon dioxide exhaled causes the respiratory exchange ratio to exceed 1.00. Heart, liver, kidney and skeletal muscle all use lactate as an energy source during exercise and recovery, the liver being the main organ removing it from the blood for GLUCONEOGENESIS. See OXYGEN DEBT.

lactogenesis See LACTATION, MAMMARY GLANDS, MILK.

lactose intolerance A condition resulting from insufficient lactase production by the intestinal mucosa. Lactase activity normally disappears after childhood in most human populations, excepting those originating in northern Europe, but a mutation in a regulatory region near the gene for lactase (*LCT*) allows lactose tolerance to persist into adulthood and apparently underwent positive selection in parts of Europe after the domestication of cattle. It exhibits large population differentiation between Europeans and non-Europeans (see NATURAL SELECTION). Undigested lactose in the chyme causes fluid retention in the faeces, bacteria in the gut flora fermenting it and producing gases. Symptoms include diarrhoea, gas, bloating and abdominal cramps after consumption of milk and other dairy products. Treatment is by dietary supplements to aid lactose digestion. See DISACCHARIDASES.

lamina propria Loose vascular connective tissue and smooth muscle of the tunica intima of larger blood vessels. It also lies between the mucosal epithelium and muscularis mucosa of the digestive tract, where it is filled with antibody-producing plasma cells that secrete 3–5 g of the IGA isotype into the gut lumen per day (see GUT-ASSOCIATED LYMPHOID TISSUES).

laminins Large, multidomain heterotrimeric and permissive glycoproteins of the basal lamina, to which as many as 12 distinct INTEGRIN heterodimers can bind, enabling adhesion to the EXTRACELLULAR MATRIX. Many tissue-specific forms occur, including those on the filopodia of neurite growth cones, which bind integrins during AXONAL GUIDANCE.

lamins Proteins whose filaments form the nuclear lamina, a shell underlying the nuclear envelope (see NUCLEUS) and providing its structural rigidity. Direct phosphorylation by M-Cdk (see G2) results in their depolymerization.

Landsteiner, Karl Born 1868, died 1943, an Austrian biologist and physician noted for his development in 1901 of the modern system of classification of blood groups from his identification of the presence of agglutinins in the blood. With Alexander S. Wiener, he identified the Rhesus factor in 1937 (see RHESUS BLOOD GROUP SYSTEM).

Langerhans cells Immature dendritic cells, a relatively scarce cell type in the epidermis of the SKIN, arising from red bone marrow and participating in the innate immune response to a bacterial infection. In the immature state they are highly phagocytic and macropinocytic but cannot activate T lymphocytes. But on activation through the Toll signalling pathway by lipopolysaccharide, they migrate into afferent lymphatics and mature into DENDRITIC CELLS in regional lymph nodes. See MUCOSAL-ASSOCIATED LYMPHOID TISSUES.

language The capacity of the human brain for language is one of its most important specializations, and linguists regard speaking, signing and language comprehension as primary faculties of language (i.e., innate and biologically determined), but regard reading and writing as secondary abilities (see NATURE–NURTURE DEBATE). The major linguistic factors, which certainly interact, are: (i) at the word level, phonology and lexico-semantics, and (ii) at the sentence and discourse level, sentence comprehension and syntax (see LANGUAGE AREAS). There is some evidence that sign language, which also involves facial expression (chimpanzees do not sign to each other) may pre-date speech in the human lineage, and BROCA'S AREA is involved in both speech and gesture. It is probably no coincidence that humans express language through accompanying hand movements rather than the voice, and imitation probably played a causal role in the evolution of brain structures involved. Language development in childhood involves transformation of auditory signals in the auditory centres in the temporal lobes into verbal output from the motor cortex. Human communication is intentional, and this would involve imitation of a role model's motor action (see MIRROR CELLS). Language dominance shows a strong correlation with HANDEDNESS, and the PREMOTOR CORTEX is involved in syntactic analysis and coordination of the two hands. Sign language is a true language, using arbitrary signs and possessing an internal grammar as sophisticated as speech. The native, or first, language (L1) is acquired during the first years of life (see BRAIN, SPEECH); in contrast, reading and writing are learned with much conscious effort and repetition. It may be that the subsystems that support language are probably not, for the most part, unique to language, and we may ultimately understand language as a powerful new combination of old elements. However, the existence of developmental dyslexics indicates that reading ability requires specific neural mechanisms. Syntax (rules governing the internal structure of sentences), grammar (rules governing the use of language), recursion (the embedding of sentences within sentences, and of acts within acts) and inflexion (the modification or marking of a word to reflect grammatical, i.e., relational, information, such as gender, tense, number or person) may all pre-date the evolution of anatomically modern HOMO SAPIENS; but they may all have been achieved first with hands and not the voice.

Recursion is a prominent feature of human social cognition: child A watches child B watch child C watch the teacher. With chimpanzees, we may observe animal A watch its mother, animal B watch its mother, etc. – an iteraction of acts all on the same level. Although recursion is not a necessary condition for theory of mind (the attribution of mental states), and there is no evidence for recursion in chimpanzees, chimpanzees will choose photographs showing a human carrying out an act that fulfils a goal (solves a problem). Although chimpanzees provide evidence of goal attribution to other individuals, the recursive thought 'A thinks that B thinks that D thinks that E's goal is to beat A', is beyond them (see AUTISM, INTELLIGENCE).

Ability with a second language (L2) can be mastered at any time of life, although L2 ability (including accent) rarely becomes comparable to that of L1 if it is acquired beyond the hypothesized 'sensitive', or CRITICAL PERIOD, from early infancy until PUBERTY (~12 years): there is a loss of flexibility for reorganization of the cerebral cortex due to acquired APHASIA after puberty, while proficiency in L2 acquisition (English) declines after 7 years when Chinese or Korean speakers move to the USA. Just as spoken pidgins can be turned into fully grammatical creoles only when learned by a generation of children, so the same has proved true of sign languages. It is possible that different linguistic abilities are acquired during their own developmental courses whose critical periods differ in duration and timing. Children brought up in multilingual households can discriminate individual languages early and speak them with a perfect accent. Prior to ~7 years, brain loci for learning other languages overlap extensively with those for the native language; but later the brain regions involved in learning L2 do not overlap those for L1. The effects of focal lesions reflect this. See COGNITION.

language areas A considerable volume of brain is involved in language. Language function is predominantly localized to a distributed neuronal network in the left perisylvian cortex in 97% of right handed people and ~60% of left handed people, the left cerebral hemisphere being dominant in 90% of people and the right hemisphere in 7.5% (see CEREBRAL HEMISPHERES, Fig. 46). However, the anatomical asymmetries are less marked than the functional ones, and brain imaging suggests that language areas are less lateralized in women than in men. Electrical stimulation indicates that language areas in the brain are more extensive than BROCA'S AREA and WERNICKE'S AREA, and include other cortical areas as well as parts of the thalamus and primary visual cortex. Within Broca's and Wernicke's areas, there may be specialized regions, possibly on a scale of functional columns, as in the somatosensory cortex or ocular dominance columns.

Adults who learn sign language use the left hemisphere; but both hemispheres are employed by those for whom sign language is L1 (see LANGUAGE). Excision of the left hemisphere in infants with neurological disease does not prevent speech fluency, although it does decrease speech comprehension, but adults so treated suffer permanent, near total loss of language, reflecting the huge NEURAL PLASTICITY of the infant brain. Much of our understanding of the cortical areas involved in processing written and auditory information into speech comes from studies of aphasiacs (see APHASIAS) and through brain scanning (functional MAGNETIC RESONANCE IMAGING and PET SCANNING).

In the Wernicke–Geschwind model of language processing (see Fig. 109), the key functional elements are BROCA'S AREA, WERNICKE'S AREA (larger on the left side in 60% of people), the arcuate fasciculus linking them, and the ANGULAR GYRUS bordering

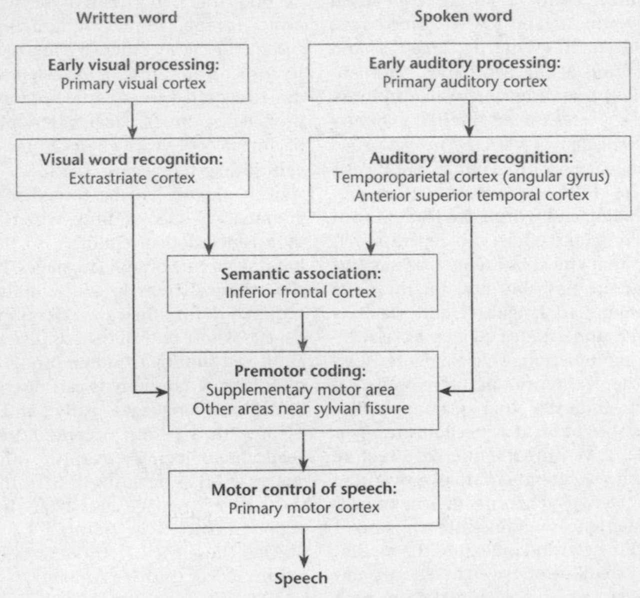

FIG 109. One model of LANGUAGE processing during verbal repetition of a written or spoken word. Below each stage is the location in the (typically) left cerebral hemisphere where task-specific activity is observed with PET scans. © 2001 Redrawn from *Neuroscience* (2nd edition) by M.F. Bear, B.W. Connors and M.A. Paradiso, published by Lippincott, Williams & Wilkins. Reproduced with the kind permission of the publisher.

Wernicke's area. The model also involves sensory and motor areas involved in HEARING language and in SPEECH. According to this model, the auditory system (see COCHLEA) processes sounds and signals reach the PRIMARY AUDITORY CORTEX. These signals are not understood as meaningful until they are processed by Wernicke's area, and in order to repeat words in speech, word-based signals are passed to Broca's area via the arcuate fasciculus. The signals are then processed into neural coding for the muscle movements involved in speech and sent to nearby MOTOR CORTEX required for movement of the lips, tongue, larynx, etc.

In reading written text aloud, the model holds that visual information from the retina and lateral geniculate nucleus is further processed by the PRIMARY VISUAL CORTEX and EXTRASTRIATE CORTEX. It is then passed to the angular gyrus, where it is assumed a transformation occurs such that output evokes the same pattern of activation in Wernicke's area as if the words were spoken rather than written, after which the pathway is the same as for hearing words.

In fact, read words do not have to be processed in the angular gyrus. Instead, their signals can pass straight to Broca's area from the visual cortex. In addition, subcortical structures (e.g., the thalamus and caudate nucleus) are involved in language processing (as work on APHASIAS has revealed), although these regions are not included in the model. Recovery of language function after a stroke suggests alternative pathways can compensate to some extent for the damage. And since most aphasias involve both comprehension and speech difficulties, the clear-cut functional differences outlined by the model cannot be entirely correct. However, the SECONDARY AUDITORY CORTEX and extrastriate cortex do not respond to visual stimuli that are not words and may be required for encoding of words that are either seen or heard.

Sounds must acquire semantic meaning for them to be used linguistically. A phoneme is the smallest sound unit uttered in speech, whereas a morpheme is the smallest linguistic unit to have semantic meaning. PET scans and fMRI indicate that auditory phonological processing is associated with activity in the superior temporal gyrus (STG) whereas lexico-semantic processing is more typically associated with activation of the left extra-sylvan temporoparietal regions, including the angular gyrus and supramarginal gyrus (AG/SMG). Processing of syntactic structures plays a critical role in selective integration of lexico-semantic information into sentence meaning (see LANGUAGE), and evidence suggests that the left inferior frontal gyrus (IFG) region is involved in the selection and integration of semantic information, which is separable from simple lexico-semantic processing. This and the left lateral PREMOTOR CORTEX (possibly the putative 'grammar centre') are activated during analysis of syntactic structures and have no counterparts in non-human primates. Severe language impairment has long been known to have a familial basis. The discovery of the FOXP2 locus in 2001 caused much debate. Twin studies have revealed genetic factors for brain structure, notably the language areas in the left hemisphere and longitudinal studies of brain structure and function may reveal further developmental tendencies, as well as individual differences. See COGNITION, INTELLIGENCE, SPLIT-BRAIN PHENOMENON.

large intestine The portion of the ALIMENTARY CANAL extending from the ileocaecal valve to the anus, comprising the appendix, caecum, COLON (see entry), rectum and anal canal. See EGESTION.

Laron dwarfism Condition resulting from mutations in the gene encoding the GROWTH HORMONE receptor, GRH, resulting in growth retardation (but not *in utero*), midfacial hypoplasia, blue sclera and limited elbow extension.

larynx Developing from the embryonic lung bud (see LUNGS), the larynx is located in the anterior part of the throat; the larynx consists of an outer casing of nine cartilages (the largest being the thyroid cartilage, or Adam's apple) interconnected by muscles and ligaments, and connected by membranes and/or muscles to the HYOID BONE superiorly, to the TRACHEA inferiorly. The thyroid and unpaired cricoid cartilage keep the air passage open for air movement during ventilation. One of the unpaired cartilages is the EPIGLOTTIS, which projects as a free flap towards the tongue. Two pairs of ligaments attach the paired arytenoid cartilages to the thyroid cartilage, the superior ones being covered by a mucous membrane called the vestibular folds (false vocal cords). When these come together during swallowing, they prevent air from entering the larynx and leaving the lungs (as during holding of breath). The inferior ligaments, also covered by a mucous membrane, are the vocal folds (true vocal cords) which, with the opening between them (the *rima glottidis*), are called the *glottis* (see COUGHING). The vocal cords are the primary source of sound production and the greater their vibration the greater the sound amplitude, while the greater the frequency of vibration the higher the sound pitch, such sounds being produced when only the anterior parts of the folds vibrate. Lower pitches are produced when progressively longer sections vibrate. Modulation of the sound is produced by the tongue, lips, teeth and other structures in vocalization during SPEECH. Skeletal muscles move the arytenoid and other cartilages, changing the position and lengths of the vocal folds. The larynx is innervated by the superior laryngeal branch of the vagus nerve (CNX), routed via the inferior vagal ganglion. See SWALLOWING.

latency An extended interval between infection and development of disease symptoms. See CHICKENPOX, HIV.

lateral corticospinal tract A descending tract of motor axons in the lateral part of the white matter of the SPINAL CORD, ~75% of its fibres having decussated in the pyramids of the medulla, 10-25% remaining ipsilateral, although eventually most of these also decussate near their termination. See CORTICOSPINAL TRACTS.

lateral fissure (lateral sulcus) See SYLVIAN FISSURE.

lateral geniculate nucleus (LGN) Subcortical visual centre of the THALAMUS (Fig. 173a); 'geniculate' meaning 'bent like a knee'), and the principal such site for VISUAL PROCESSING, receiving visual projections from the retina and relaying them to the visual cortex. Before birth, spontaneous bursts of retinal activity (see RETINA) are sufficient to direct axons to their correct 'addresses' in both the LGN and the PRIMARY VISUAL CORTEX; but input from the visual environment is critically important after this (see NEURAL PLASTICITY). Only about 10–20% of the presynaptic connections in each LGN come from the retina. Some come from the reticular formation, concerned with activity related to alertness and attentiveness; but ~80% of its excitatory synapses project back from the primary visual cortex in the corticofugal feedback pathway. The LGN also receives projections from the brain stem, modulating the magnitude of its responses to visual stimuli. It is the first location in the ascending visual pathway where what we see can be influenced by how we feel.

As with cells in the retina, cells in the LGN have small, circular, RECEPTIVE FIELDS. As a consequence of the OPTIC CHIASMA (see Fig. 130), the right LGN receives projections from the right hemiretina of each eye bearing information from the left VISUAL FIELD (see Fig. 179), while the left LGN receives projections from the left hemiretina of each eye, with information from the right visual field. Axons from the retinal ganglion cells terminate in a topographically organized manner in each LGN, each having a retinotopic representation of the contralateral visual field (see Fig. 110).

About half the mass of each LGN represents the FOVEA. Each LGN has six layers of cell bodies, each separated by layers of axons and dendrites. This layering suggests that various types of retinal information are being treated in parallel pathways, and indeed axons arising from M-type, P-type and nonM-nonP ganglion cells in the two retinas synapse on cells in different

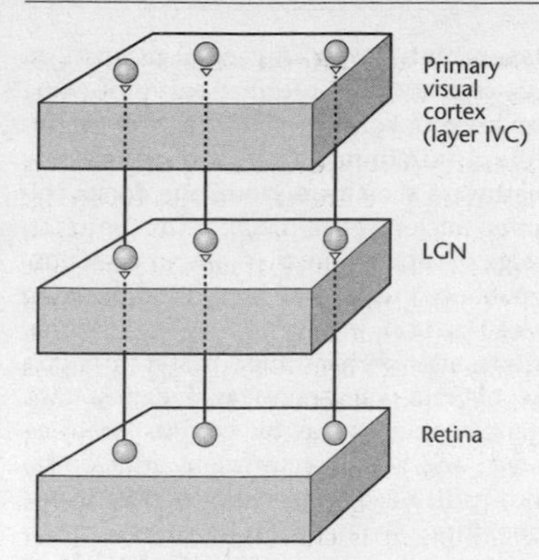

FIG 110. Diagram indicating that retinotopy is preserved in the projection of the LATERAL GENICULATE NUCLEUS (LGN) to the primary visual cortex. © 2001 Redrawn from *Neuroscience* (2nd edition) by M.F. Bear, B.W. Connors and M.A. Paradiso, published by Lippincott, Williams & Wilkins. Reproduced with the kind permission of the publisher.

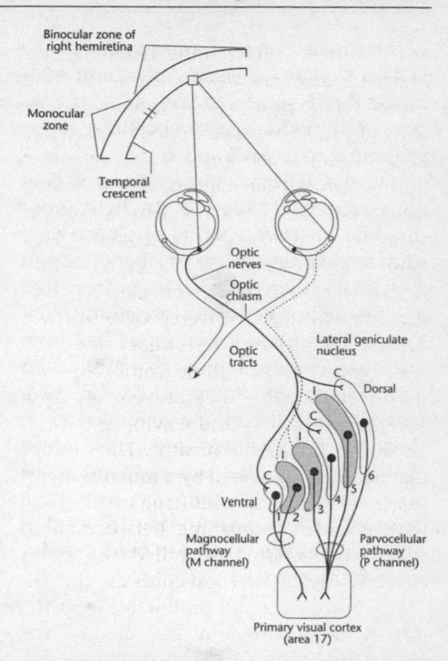

FIG 111. Diagram illustrating the six cell-body layers of the LATERAL GENICULATE NUCLEUS, in each of which (left and right nuclei) fibres from the contralateral (C) nasal hemiretina terminate in layers 1 and 4; fibres from the ipsilateral (I) temporal hemiretina terminate in layers 2, 3 and 5. Layers 1 and 2 are the magnocellular layers, layers 3 through to 6 being the parvocellular layers. All these neurons project to the primary visual cortex, each layer containing a neural representation of the contralateral hemifield, mapping in such a manner as to preserve the spatial arrangement of receptors in the retina (retinotopy). The ipsilateral pathway is shown by a pale grey line and its cell-body layers are also shown pale grey.

layers of the LGN. The two most ventral layers, containing relatively large cells, are the magnocellular layers, with a major input from the M-type cells. The four dorsal layers (parvocellular layers) receive input from the P-type cells. The six LGN layers receive inputs from both the ipsilateral eye (layers 2, 3 and 5) and the contralateral eye (layers 1, 4 and 6). Despite segregation of the left-eye and right-eye fibres the retinotopic maps occur in perfect vertical register so that a needle inserted perpendicularly would pass through cells responding to the same points in visual space but viewed alternately by the two eyes (see Fig. 111).

In addition, numerous tiny neurons are sandwiched just beneath each of the six LGN layers, and form the six koniocellular layers (interlaminar layers). This is where the nonM-nonP ganglion cells of the retinas synapse. The magnocellular system processes information relating to movement, texture and stereoscopic vision; the parvocellular system deals with information relating to colour, shape (form) and contour.

Microelectrode studies demonstrate that the visual receptive fields of LGN cells are all but identical with those of the retinal ganglion cells projecting to them: magnocellular cells respond just like M-type ganglion cells and are insensitive to differences in wavelength, while parvocellular cells behave just like P-type cells, and many of them exhibit red-green colour opponency. Cells in the koniocellular layers beneath layers 3 and 4 exhibit blue-yellow colour opponency, but cells in other koniocellular layers lack colour sensitivity. On-centre and off-centre cells are interspersed in all LGN layers.

The single major synaptic target of the LGN is the PRIMARY VISUAL CORTEX and both M and P cells project to layer 4C of the PRIMARY VISUAL CORTEX, the koniocellular layers projecting to layers 2 and 3.

Wiesel and Hubel showed that in neonatal mammals, reducing the light input into one eye over a period of several days leads to reduction in the number of cells in the visual system receiving input from that eye, a reduction persisting into adulthood – long after normal visual input has been restored. They also showed that monocular light deprivation during a 'critical' neonatal period also caused a reduction in thickness of LGN innervated by the closed eye, a thinning that others found to be accelerated if the animals were also deprived of REM SLEEP during the critical period of susceptibility. See OPTIC CHIASMA, SUPERIOR COLLICULUS.

lateral inhibition A general mechanism of FEEDFORWARD information processing underlying contrast enhancement by surround inhibition, common in sensory pathways such as found in the dorsal column nuclei of the medulla, in the RETINA (e.g., see HORIZONTAL CELLS) and in OLFACTION, enabling greater sensory discrimination (see Fig. 112). In general, sensory information is altered every time it passes through a set of brain synapses (see RELAY NUCLEI). It is a general mechanism for CONTRAST ENHANCEMENT, and a similar surround antagonism occurs in relation to Purkinje cells in the CEREBELLUM. See RECIPROCAL INHIBITION.

lateral intraparietal region (LIP) A subregion of the parietal lobes known to be centrally involved in visuo-spatial ATTENTION, motor planning and decision-making and, with the connected MIDDLE TEMPORAL REGION, involved in visual motion

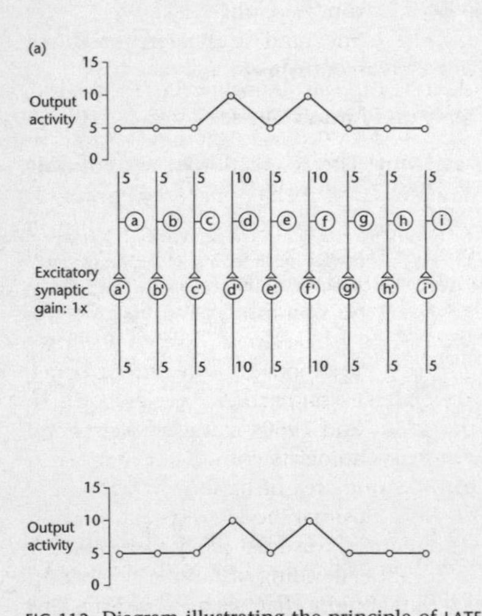

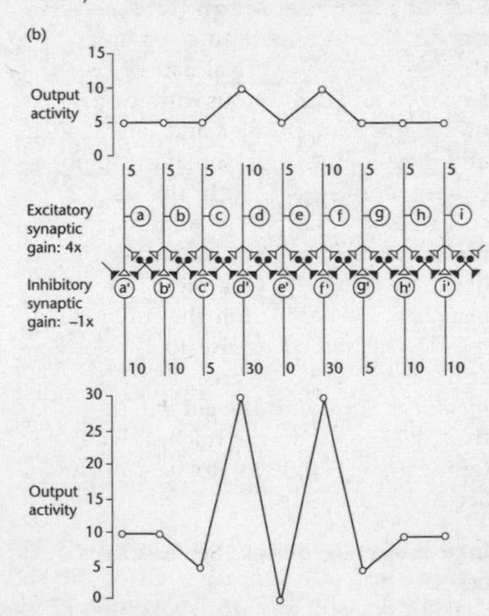

FIG 112. Diagram illustrating the principle of LATERAL INHIBITION underlying contrast enhancement in sensory pathways. In (a), dorsal root ganglion neurons a to i relay information to dorsal column nucleus neurons a′ to i′. Assuming the output of the dorsal column nucleus cells is simply the presynaptic input times a synaptic gain factor of 1, then if the input activity of cell a′ is 5, its output activity is also 5. This does nothing to enhance the difference between the more active neurons d and f, and the others. By contrast, in situation (b), added inhibitory neurons project laterally to inhibit each cell's neighbour. If the synaptic gain of the excitatory synapses (open triangles) is 4, while the synaptic gain of the inhibitory synapses (black triangles) is –1, multiplying the input to each synapse by its synaptic gain then summing the effect of all synapses on each cell a′ to i′ results in significant contrast enhancement, the difference in the activities of cells d and f and their neighbours having been greatly amplified in the outputs of cells d′ and f′. © 2001 Redrawn from *Neuroscience* (2nd edition) by M.F. Bear, B.W. Connors and M.A. Paradiso, published by Lippincott, Williams & Wilkins. Reproduced with the kind permission of the publisher.

processing (see VISUAL MOTION CENTRE). Recent work with rhesus monkeys (*Macaca mulatta*) suggests that LIP neurons encode directions of motion according to their category membership (see CATEGORIZATION) and that the region may be an important nexus in transforming visual direction selectivity to more abstract representations encoding the behavioural relevance, or meaning, of stimuli.

lateral lemniscus A tract in the brainstem between the superior olivary nucleus and the inferior colliculus, conveying auditory information. See HEARING.

lateral motor pathways Neural pathways involved in voluntary motor control. They are the CORTICOSPINAL TRACT and RUBROSPINAL TRACT. Compare MEDIAL MOTOR PATHWAYS.

lateral plate mesoderm (lateral mesoderm) Cells originating by migration through the more caudal part of the PRIMITIVE STREAK and continuous with the paraxial mesoderm, but thinner and lateral to it, and developing into somatic mesoderm (where continuous with the mesoderm covering the amnion) and splanchnic mesoderm (where continuous with mesoderm covering the yolk sac). The *somatic mesoderm* develops from the intermediate mesoderm and gives rise to the kidneys and reproductive organs. The *splanchnic mesoderm* surrounds the gut and its derivatives, giving rise to the trachea, lungs and (along with ectoderm) to SMOOTH MUSCLE. See Appendix VI.

late response genes See IMMEDIATE EARLY GENES.

laterodorsal tegmental nuclei See TEGMENTUM.

lathyrism See TOXINS.

law of effect The empirically simple idea, governing much of voluntary behaviour, that those behaviours that are rewarded tend to be repeated at the expense of behaviours that are not, whereas behaviours followed by aversive consequences tend not to be repeated. See INSTRUMENTAL CONDITIONING, LEARNING, REINFORCEMENT AND REWARD.

LBP Lipopolysaccharide-binding protein. An ACUTE PHASE PROTEIN that binds lipopolysaccharide, transfers it to CD14 on phagocytes, and then interacts with the TOLL-LIKE RECEPTOR TLR-4 resulting in release of the pro-inflammatory transcription factor NF-κB.

LCAT See LYSOLECITHIN-CHOLESTEROL ACYL TRANSFERASE.

L cells Enteroendocrine cells throughout the length of the gut, secreting PYY.

LCT See LACTOSE INTOLERANCE.

LDL See LOW DENSITY LIPOPROTEIN.

L-DOPA (laevo-dopa, *l*-dihydroxyphenylalanine) An intermediate in the pathway from phenylalanine to noradrenaline, and immediate precursor of the neurotransmitter DOPAMINE.

lean body mass See BODY COMPOSITION.

learning The acquisition of reproducible alterations of behaviour as a result of experience, the storage of which over time is MEMORY. Both are ultimately dependent upon NEURAL PLASTICITY.

It may be non-associative (HABITUATION, BEHAVIOURAL SENSITIZATION) or associative (CLASSICAL CONDITIONING, INSTRUMENTAL CONDITIONING; see REINFORCEMENT AND REWARD). In the 1950s and 1960s many biologists and most psychologists considered that learning was one area of biology in which use of simple animal models, especially invertebrate ones, was least likely to contribute to an understanding of human behaviour: it was thought that such organisms had neural organizations qualitatively too simple to be of relevance. But work by E.R. Kandel and others on simple learning in the giant marine snail *Aplysia* showed this to be incorrect.

If a jet of water is squirted onto *Aplysia*'s siphon, the siphon and gill retract (see MEMORY and Fig. 122). Repeated application of the jet results in progressively weaker responses (habituation), and pioneering work by Kandel and colleagues in the 1960s revealed that one of the motor neurons (L7) in the mollusc's abdominal ganglion, that receives monosynaptic input from the siphon, innervates the muscles producing gill withdrawal. By a process of elimination, Kandel showed that the cellular events underlying gill withdrawal reflex habituation occurred at the synapse between the sensory and motor neuron. Repeated stimulation of the sensory neuron proved sufficient to cause progressive decrease in the size of the excitatory postsynaptic potential (EPSP). He and his colleagues then showed that, given the same stimulus intensity, repeated stimulation led to fewer and fewer synaptic vesicles releasing their transmitter into the synaptic cleft: habituation in this instance was therefore caused by a presynaptic loss of SYNAPTIC STRENGTH. Short-term habituation, just like short-term sensitization, proved to result from covalent modification of pre-existing proteins; but long-term habituation involved a loss of synaptic connections, whereas long-term facilitation resulted in an increase.

It has been said that there is 'no learning without memory', and SEROTONIN is as critical in learning as it is in memory. Some influential models of learning and memory involve our current understanding of events leading to SYNAPTIC PLASTICITY, such as LONG-TERM POTENTIATION, long-term depression (e.g., see CEREBELLUM for motor learning) and the modification of dendritic processes (see DENDRITES). Different forms of learning, such as classical conditioning and spatial learning, recruit different modulatory neurotransmitters acting via second messenger pathways that may promote neuronal growth and long-term memory. The CREB transcription factor is a crucial mediator of experience-based neural plasticity (neuroadaptation). Marking of specific synapses for involvement in such long-term processes favours the ATTENTION required for memory retention and recall (e.g., 'insight learning': see INTELLIGENCE). *Implicit learning* is learning in which the responses cannot be consciously generated (as in fear learning by the AMYGDALA). See COREGULATORY PROTEINS, ETHOLOGY.

lecithin See PHOSPHATIDYLCHOLINE.

lecithin-cholesterol acyl transferase See LYSOLECITHIN-CHOLESTEROL ACYL TRANSFERASE.

lectins Highly specific cell-surface carbohydrate-binding proteins, including many CELL ADHESION MOLECULES which bind glycoproteins or glycolipids. Many are involved in interactions between immune cells. See HAEMAGGLUTINATION (for haemagglutinin), MANNOSE-BINDING LECTIN, RICIN, SELECTINS.

legs See GAIT, HIND LIMB, LEVERAGE, SKELETAL SYSTEM.

legumes Members of the plant family Fabaceae, or their fruits. Important nitrogen-fixing plants on account of mutualistic root bacteria (*Rhizobium*), they are usually rich protein sources in the diet. See GLUCOSE-6-PHOSPHATE DEHYDROGENASE for favism.

Leishmania *Leishmania* spp. cause a spectrum of human diseases in tropical and subtropical regions. Leishmaniasis is currently a low priority endemic disease in 88 countries (2005) on 4 continents caused by over 20 species of the TRYPANOSOME protozoan *Leishmania*, transmitted to humans by sandflies. >90% of self-healing cutaneous cases (e.g., caused by *L. major*, or *L. mexicana*) occur in the Middle East, Brazil and Peru. >90% of severe, life-threatening visceral cases (e.g., caused by *L. donovani s.l.*) occur in Bangladesh, Brazil, India and Sudan. After injection, the parasite loses its flagellum and undergoes phagocytosis by MACROPHAGES, many surviving there by inhibiting the cell's production of NITRIC OXIDE. It proliferates in the phagolysosomes, leading to lysis and serial infection of other macrophages. NATURAL KILLER CELLS are major immune cells in defence. The genome of *L. major* was sequenced in 2005. See TRYPANOSOMIASIS.

leishmaniasis See LEISHMANIA.

lemniscus A TRACT that meanders through

the brain, in the manner of a ribbon. See LATERAL LEMNISCUS, MEDIAL LEMNISCUS.

lens Transparent, biconvex structure (more convex on its posterior surface) of the EYE (see Fig. 73) consisting of cuboidal epithelial cells anteriorly and long columnar cells (lens fibres) posteriorly. The latter lose their nuclei, accumulate the refracting proteins CRYSTALLINS, and become covered by an elastic transparent capsule. With the CORNEA, the lens is involved in ACCOMMODATION. Similar REGULATORY ELEMENTS and transcription factors (including Pax6) seem to unify lens expression of crystallin genes in vertebrates and molluscs, although the combinations of regulatory sequences within these elements differ. Crystallins comprise ~80–90% of the water-soluble proteins of the transparent lens, packing closely together and reducing concentration fluctuations in the cytoplasm so that incident light-scattering is minimized. With increasing age, the near point of vision (the distance at which blurring occurs) increases. This presbyopia is due to the lens becoming sclerotic and less flexible. See VISUAL FIELD, Fig. 179.

lentiform nucleus Collective term for the globus pallidus and putamen. See BASAL GANGLIA, CEREBRAL HEMISPHERES.

leprosy An indolent disease, caused by the slow-growing and poorly infective and unculturable bacterium *Mycobacterium leprae*. It mainly affects the skin, nerves and mucosae, occasionally the eyes; but its choice environment is SCHWANN CELLS, where it causes demyelination. Nerve damage in leprosy is widely thought to be secondary to the immune response to *M. leprae* itself. This tuberculoid disease (TT) is used in those mounting a strong immune attack and is characterized by hypopigmented skin lesions and inflammation of peripheral nerves. Spread is by close contact, especially within families. Peripheral nerve lesions lead to paralysis, anaesthesia of limbs and neuropathic joint disease. Untreated, there is no way of preventing these effects. Recommended treatment is multidrug therapy with rifampicin, clofazimine and dapsone. It has an exceptionally stable genome, and evidence from comparative genomics indicates the disease may have originated in the Indian subcontinent and that all extant cases are attributable to a single clone whose dissemination worldwide seems to have originated in Eastern Africa or the Near East, and spread with successive human MIGRATIONS.

leptin A small protein hormone encoded by the *ob* gene and produced by adipocytes. Released in proportion to BODY FAT, it binds receptors on cells of the HYPOTHALAMUS known to regulate feeding behaviour and decreasing appetite by inhibiting production of hypothalamic NEUROPEPTIDE Y (see APPETITE, Fig. 14; CIRCADIAN CLOCKS; compare GHRELIN). Leptin receptors are also found at much lower levels on cells of the adrenal cortex and β-cells of the pancreas. Defective receptors curtail its signalling function and induce leptin resistance, which is common among obese individuals and reduces its potential as a therapeutic agent. Leptin may be a factor in the onset of PUBERTY.

The leptin receptor is a class 1 cytokine receptor that regulates gene transcription via activation of the JAK/STAT PATHWAY of signal transduction. Genes under its control include those encoding NEUROPEPTIDE Y, CORTICOTROPIN-RELEASING HORMONE and PROOPIOMELANOCORTIN. In some cells, notably hypothalamic neurons, leptin also activates the insulin receptor substrate–PHOSPHATIDYLINOSITOL-3-OH-KINASE pathway. AMP-ACTIVATED PROTEIN KINASE decreases activity of the nutrient-sensing enzyme mTOR (mammalian target of rapamycin), implicated in hypothalamic leptin action. Leptin also stimulates the sympathetic nervous system, increasing blood pressure, heart rate and thermogenesis by uncoupling electron transfer from ATP synthesis in mitochondria of adipocytes. See ENERGY BALANCE, GLUCAGON-LIKE PEPTIDE-1, GLYCOCALYX, INSULIN.

Lesch–Nyhan syndrome A genetic disorder, almost always occurring in male children, involving lack of hypoxanthine-guanine phosphoribosyltransferase, an enzyme involved in recycling PURINES (see entry). As a result, free PRPP levels rise and purines are overproduced by the *de novo* pathway, resulting in high levels of uric acid and gout-like damage to tissues. Affected children usually manifest the condition by 2 years, being poorly coordinated, mentally retarded, and often with self-mutilating and aggressive tendencies. See ADA DEFICIENCY.

lesion Disruption of a pathway, malfunction of a normally occurring process, or a symptom of such malfunctioning. (i) Anatomical, e.g.: those occurring in tooth decay (caries) and classified on the basis of tissues affected; plaques of abnormal protein characteristic of ALZHEIMER'S DISEASE; or the fatty streaks characteristic of ATHEROSCLEROSIS; ablations and resections (surgical removals) may generate lesions deliberately or unavoidably. (ii) Genetic, in which neither allele of a gene is functional so that normal gene product is absent. (iii) Biochemical, in which an enzyme-controlled step in a metabolic pathway is non-functional – often as a result of (ii).

'less-is-more' hypothesis One of three prevalent hypotheses to account for the evolution of 'humanness traits', favouring the view that loss-of-function genome changes have given rise to major derived features of the human lineage compared with prototypical apes (e.g., lack of body hair, preservation of certain juvenile traits into adulthood, and expansion of the cranium). See BRAIN SIZE.

let-down reflex See LACTATION, OXYTOCIN.

leukaemias (leukemias) Malignancies of any of a variety of haematopoietic cell types, including lineages leading to lymphocytes and granulocytes, in which tumour cells are non-pigmented and dispersed throughout the circulation (i.e., they do not form solid tumour masses, lymphomas). As with parallels between adenomas and progenitor cells in the intestinal crypt, the influence of the WNT SIGNALLING PATHWAY components on haematopoietic stem and progenitor cells suggests that the same pathway may be dysregulated in leukaemias. Lymphocytic (lymphoid) leukaemias arise in both B-cell and T-cell lineages. Multiple myelomas are malignancies of plasma cells. Acute and chronic myelogenous (myeloid) leukaemias derive from precursor cells of the lineage that forms the various granulocytes, and monocytes (which give rise to macrophages). Acute leukaemia risk is increased by exposure to the CARCINOGEN benzene.

Human leukaemias often harbour recurrent chromosome translocations. The 'mixed lineage leukaemia' gene (*MLL*), a vertebrate homologue of a fruit fly gene that maintains expression of the *Hox* genes, is repeatedly found fused to various gene partners, such as *AF9*, in acute myeloid leukaemias. Myeloid PROGENITOR CELLS are produced by bone marrow STEM CELLS but do not self-renew. However, in mice, those with the *MLL-AF9* fusion gene express a set of 'stem cell genes' making them self-renew and develop into cancer cells capable of initiating, maintaining and propagating the cancer and subverting stem-cell properties in non-stem-cell (i.e., progenitor) populations. Expression of this gene set is hierarchical, with a specific set of genes expressed as an early response to *MLL-AF9* expression. Part of this mouse stem-cell self-renewal signature is also expressed in human cells from patients with leukaemia associated with *MLL* FUSION GENES. The enzyme mTOR (see MTOR) is inhibited by the drug rapamycin, which not only prevents leukaemia from developing in mice but also inhibits progression of established mouse leukaemias by reducing the number of cancer STEM CELLS.

Chronic myeloid leukaemias (or chronic myelogenous leukaemias, CMLs) are almost always associated with a reciprocal chromosome translocation that fuses the *bcr* and *abl* genes (Philadelphia, Ph^1, chromosome), a genetic lesion occurring in the haematopoietic STEM CELLS, but which expresses itself in more committed myeloid progenitor cells (although Ph^1 chromosomes may also be present in lymphoid cells). CML precursors may be dependent on Wnt signalling for growth

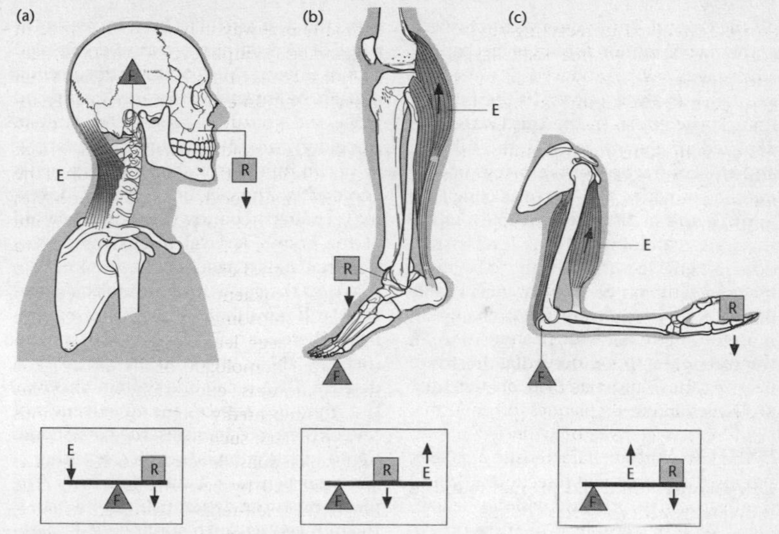

FIG 113. Diagrams illustrating examples of different human LEVER systems. In first-class levers (a), the fulcrum is between the effort and the resistance; in second-class levers (b) the resistance is between the fulcrum and the effort; in third-class levers (c) the effort is between the fulcrum and the resistance. E = effort; F = fulcrum; R = resistance. From *Principles of Anatomy and Physiology*, G.J. Tortora & S.R. Grabowski, © 2000. Reproduced with permission of John Wiley & Sons, Inc.

and renewal. Mutated JAK2 enzyme (see JAK/STAT PATHWAYS) in bone marrow is responsible for polycythaemia vera, a blood disorder often linked to chronic myeloid leukaemia.

Acute lymphoblastic leukaemias (ALLs) are highly aggressive, undifferentiated, lymphoid malignancies derived from a progenitor cell in which 80% of cases arise from the B-cell and 20% of cases from the T-cell lineage. Chromosomal aberrations are a hallmark of ALLs, but fail to induce leukaemia alone. Genome-wide analysis of leukaemic cells from 242 paediatric ALL patients revealed deletion, amplification, point mutation and structural rearrangement in genes encoding principal regulators of B-cell development and differentiation in 40% of B-progenitor cases. The *PAX5* gene was the most frequent target of somatic mutation (31.7% of cases), resulting in reduced levels of PAX5 protein, although deletions were also detected in *E2A* (aka *TCF3*).

leukocytes, leucocytes See WHITE BLOOD CELLS.

leukocytosis An increase in the numbers of circulating NEUTROPHILS, as is brought about by release of the cytokines TNF-α, IL-1 and IL-6.

leverage, levers A lever is a rigid shaft capable of turning about a hinge, or fulcrum (F), and transferring a force, or effort (E), applied at one point along the lever to a weight, or resistance (R). Three classes of levers occur, based on the relative positions of these components (see Fig. 113). In the first-class lever depicted, the fulcrum extends through several cervical vertebrae as the trapezius muscle contracts to elevate the head, a lever system which is capable of being a force multiplier if the fulcrum is closer to the resistance than to the effort. In the second-class lever depicted, the ball of the foot is the fulcrum as the gastrocnemius muscle contracts to pull the heel up off the ground. This system sacrifices speed and range of motion for force, mechanical advantage being produced because the resistance is always closer to the fulcrum than is the effort. In the third-class level depicted, the ELBOW JOINT is the fulcrum and the contraction of the biceps brachii muscle provides the force flexing the joint. Being a distance multiplier rather than a force multiplier, this lever system does not enable a great weight to be lifted; but a weight can be lifted a considerable distance by a relatively small change in muscle length because the insertion of the biceps brachii on the radial tuberosity is only a short distance from the fulcrum. See GAIT, MUSCLES.

levers See LEVERAGE, LEVERS.

Lewy bodies Amyloid plaques found in individuals with PARKINSON'S DISEASE.

Leydig cells (interstitial cells) Testosterone-secreting cells of the TESTES, lying between the seminiferous tubules and comprising about 20% of adult testis mass. Their differentiation requires SRY expression, the 'master gene' for testis development, and is possibly under hCG influence, since onset of testosterone production in the embryo precedes LH secretion. Numerous in the newborn male for a few months, Leydig cells are absent during childhood until puberty. When present, they secrete large amounts of TESTOSTERONE. See ANDROGENS, SEX DETERMINATION.

LH See LUTEINIZING HORMONE.

life expectancy Life expectancy at age *x* is the average number of years of life remaining for a population of individuals, all of age *x*, and all subject for the remainder of their lives to the observed age-specific death rates corresponding to the current life table. A *life table* is a statistical method of translating central death rates for the population into the probability of a person of a given age dying during the following year, and can be used to calculate life expectancy.

Since 1840, in industrialized countries throughout the world, maximum life expectancy at birth has been increasing at a rate of ~3 months per year, and shows no sign of slowing. Most of this is due to elimination of infectious diseases tending to strike the young; but mortality rates in the elderly are also declining, due largely to declining rates of CARDIOVASCULAR DISEASE and CANCER. The genetic control of longevity is indirect, coming through regulation of the body's survival mechanisms – the essential maintenance and repair processes that slow the build-up of the molecular and cellular lesions that will probably end our lives. These control components of systems such as those involved in antioxidant defence, DNA repair and protein turnover. If a mutation reduces the effectiveness of one of these mechanisms, the corresponding lesions accumulate earlier, resulting in a shorter period of assured longevity. The most consistent determinant of lifespan is the mitogenic growth hormone/IGF1 pathway, prolonged damping of which by genetic manipulation or by *caloric restriction* promotes longevity in animal models, whereas persistent up-regulation shortens life (see STARVATION CONDITIONS, TEMPERATURE REGULATION). In the nematode worm *Caenorhabditis elegans*, the forkhead protein PHA-4 has been shown to confer extended survival under dietary restriction, and is very similar to the mammalian FOXA proteins (see SIRTUINS), while in many organisms decreased activity of mTOR (see entry) is associated with increased lifespan.

It is well known from animal models that genes influence lifespan; but human studies are confounded by necessary reliance on observational rather than experimental methods, and by the requirement to tease out non-genetic contributions. Lifestyle, habits, wealth, employment, education, land tenure and social status are all phenotypic variables that can be transmitted to offspring. The HERITABILITY of human longevity is low, estimated to be from 0.22 to <0.33 from twin studies (it only weakly 'runs in families'), and there are good *a priori* grounds for supposing that many genes may be involved. Some, but not all, work suggests greater heritability of longevity for daughters than for sons. One recent study in pre-industrial Scandinavian society (1640–1870) indicates that the greater the number of sons delivered, the lower the expected maternal longevity, whereas the greater the number of daughters raised, the higher the expected maternal longevity.

The AGEING process is a malleable contributory factor in longevity, including as it does antioxidant defences and DNA REPAIR MECHANISMS. Various data (from mice, fruit flies and nematode worms) indicate that reduced growth hormone/IGF1 signalling, involving expression of the sirtuins, is a contributing factor in increased longevity. Several cell signalling pathways associated with these factors converge on the Forkhead/FOXO family of transcription factors regulating expression of a range of stress response proteins affecting antioxidant capacity, cell cycle arrest, DNA repair and apoptosis, among others (see MITOCHONDRION). See PROGERIAS.

lifespan See LIFE EXPECTANCY.

lifestyle Lifestyle choices have considerable consequences for the probability of developing some of the commonest COMPLEX DISEASES. One's choices of diet (see HEALTHY DIET), level of EXERCISE, and use of 'recreational' drugs (e.g., see ALCOHOL, CIGARETTE SMOKING) can profoundly affect one's health, often impacting on OBESITY. Lifestyle factors (1990 data for the USA) known or suspected to be causal for human cancers by percentage of total cancer cases were as follows: 1–2% occupational exposures; 34% tobacco-related (lung, bladder, kidney); 5% diets low in vegetables, high in nitrates or salt (e.g., stomach); 37% diets higher in fat, lower in fibre (bowel, pancreas, prostate, breast); 2% tobacco- and alcohol-related (mouth, throat). In 1839 an Italian physician reported that BREAST CANCER was present in nunneries at rates six times higher than those among women in the general population who had given birth multiple times. The greatly increased incidence of lung cancer in miners of the St. Joachimsthaler mines was eventually traced to the high levels of radioactivity in the ores they dealt with. See http://www.who.int/whr/2002/en/; CORONARY HEART DISEASE, INSULIN RESISTANCE, LIFE EXPECTANCY.

Li–Fraumeni syndrome An unusual and rare familial susceptibility to a wide variety of cancers due to mutant *p53* germ-line alleles (see P53 GENE). In ~70% of such multicancer families, the mutant *p53* allele is inherited in a Mendelian fashion, and family members have a high probability of developing some form of malignancy early in life, although the age of onset varies: adrenocortical carcinomas (~5 years); sarcomas (~16 years); brain tumours (~25 years), breast cancer (~37 years) and lung cancer (~50 years). A variety of point mutations in the *p53* reading frame may be transmitted in the germ line, predominantly involving G:C→A:T transitions at CpG sites, as would occur if a 5-methylcytosine underwent spontaneous deamination and was replaced by a thymidine (see CPG DINUCLEOTIDE). See ATAXIA TELANGIECTASIA MUTATED PROTEIN, WILMS' TUMOUR.

ligaments Connective tissue connecting bone to bone. *Yellow elastic ligaments* mainly comprise elastic fibres and form relatively extensible ligaments, as in those joining vertebrae, and the vocal cords of the LARYNX. *Collagenous ligaments* largely comprise parallel bundles of collagen, resisting extension. See GAIT.

ligamentum nuchae (nuchal ligament) See GAIT, OCCIPITAL BONE, OCCIPITAL CREST.

ligand-gated ion channels See ION CHANNELS.

ligases Enzymes catalysing condensation of any two molecules where hydrolysis of ATP or another such triphosphate is involved. They are involved in many polymerizations, including DNA REPAIR MECHANISMS.

light adaptation (photopia, photopic vision) Daytime vision, involving reduced sensitivity (adaptation) of the EYES to light that occurs over a matter of minutes or hours of exposure to bright light and during which photoreceptors become less sensitive, allowing them to respond to levels of illumination that can vary by as much as four orders of magnitude. It is due to the low concentrations of photochemicals that remain in the rods and cones, and to the fact that much of the retinol will have been converted to vitamin A, leading to a maximum sensitivity to light of 555 nm, characteristic of the combined absorption spectra of the three CONE CELL pigments. Prolonged illumination of the ROD CELLS causes their cGMP levels to fall to the level where their response to light is saturated, additional light causing no further hyperpolarization. Daytime vision therefore depends entirely on the CONE CELLS, whose photopigments need more energy to bleach. The light-evoked closure of the rod cell cation channels reduces Ca^{2+} influx, causing the $[Ca^{2+}]$ in the outer segment to fall. Since Ca^{2+} normally inhibits the activity of guanylyl cyclase, this fall in $[Ca^{2+}]$ increases the production of cGMP, offsetting its destruction by light. Far more rapid is the neural adaptation brought about by the effect of the neurotransmitter DOPAMINE in reducing the gap junction communication between one class of retinal neurons in response to increase in light intensity, helping the retina switch from using rod photoreceptors to cone photoreceptors. Most of the signals transmitted by bipolar cells, horizontal cells, amacrine cells and ganglion cells of the RETINA decrease rapidly in bright light and account for a fewfold decrease in sensitivity to light. Pupillary diameter can be reduced by up to 30-fold in a fraction of a second in bright light, reducing the intensity of light striking the retina. Light adaptation is very much faster than DARK ADAPTATION. See MELANOPSIN, PURKINJE SHIFT.

limbic system A group of mammalian cortical structures (see Fig. 114) first recognized by Broca in 1878, on the medial surface of the cerebrum and distinct from the surrounding cortex. They form a ring, or border (*limbus* = border), around the CORPUS CALLOSUM. It may be thought of as the entire neuronal circuitry controlling emotional behaviour and motivational drives and attaching a behavioural significance and response to a stimulus – notably its emotional content (see AMYGDALA, EMOTIONS). Limbic-related areas of the cortex include the ORBITOFRONTAL CORTEX and INSULA.

The limbic system comprises a collection of areas lying mainly along the medial aspect of the temporal lobe, and includes the mammillary bodies and some other regions of the HYPOTHALAMUS (see also CIRCADIAN RHYTHMS), the cingulate gyrus, parahippocampal structures, entorhinal cortex, hippocampal complex, septal nuclei (see SEPTUM) and amygdala. The core of the limbic system is the *affective striato-thalamo-cortical circuit*, and the connections this makes with the amygdala. The affective

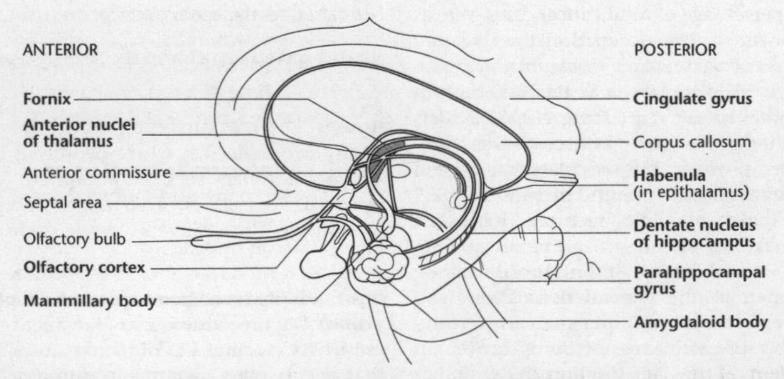

FIG 114. Medial view of midsagittal section of the LIMBIC SYSTEM (bold labels) and associated structures in the right cerebral hemisphere. Redrawn from *Anatomy and Physiology*, R.R. Seeley, T.D. Stephens and P. Tate. © 2006 McGraw-Hill, with the permission of The McGraw-Hill Companies.

basal ganglia loop has much in common with the motor loop (see BASAL GANGLIA) except that its striatal component is the NUCLEUS ACCUMBENS, which projects to the ventral pallidum (globus pallidus). This relays via the mediodorsal thalamus to the anterior CINGULATE CORTEX and medial orbitofrontal cortex. Connections from the cortex back to the nucleus accumbens close the loop, which also has reciprocal connections with the amygdala, responsible for fear learning; but it is modulated by the dopaminergic mesolimbic system involved in reward learning (see REINFORCEMENT AND REWARD). See SLEEP.

limbs The FORE LIMBS and HIND LIMBS. Limb development PATTERNING is initiated by FGF10 secreted by lateral plate mesoderm in the limb-forming regions. Further development is regulated by overlapping craniocaudal expression of HOX GENES, some of which differ in their cranial limits of expression. That of *HOXB8*, for instance, is at the cranial border of the fore limb, misexpression altering limb position. The *HOX* genes regulate the types and shapes of bones of the limb. *HOX* A and *HOX* D clusters are the main determinants, variations in their combined patterns of expression probably accounting for differences in FORE LIMB and hind limb structures (see GAIT, LEVERAGE, SKELETAL SYSTEM). Factors determining fore limb versus hind limb are the transcription factors *TBX5* (fore limbs) and *TBX4* with *PITX1* (hind limbs). Patterning of the anteroposterior axis of the limb is regulated by the *zone of polarizing activity* (ZPA), a cluster of cells at the limb bud's posterior border. Producing RETINOIC ACID, these initiate expression of SONIC HEDGEHOG, whose protein product regulates the anteroposterior axis. By the end of the 4th week of development, limb buds are visible as outpocketings from the ventrolateral body wall (see Appendix Vb). Initially consisting of mesenchyme derived from the LATERAL PLATE MESODERM and eventually forming the appendicular skeleton of the limb, ectoderm at the distal borders of the limb buds thickens to form the *apical ectodermal ridge* (AER). This produces FGFs and has an inductive influence on the adjacent mesenchyme, causing it to proliferate rapidly but to remain undifferentiated. But cells distal to the AER do begin to differentiate into cartilage and muscle in a proximodistal progression. By the early 6th week, the terminal portions of the limb buds have flattened to become *handplates* and *footplates*, separated by a circular constriction from the proximal segment, which itself becomes separated into two segments within which the hyaline cartilages of the tibia and fibula (hind limb) and radius and ulna (fore limb) develop. Similar hyaline cartilage 'models' of the other limb bones are also present by the 7th–8th weeks, selective cell death by apoptosis in the AER separating the AER into five distal segments in each limb bud, which give rise to the fingers and toes. Further development of these is dependent on signals from the AER ectoderm. BMPs from the ventral ectoderm regulate the dorsoventral axis. Joints form in the cartilage as chondrogenesis is arrested and a joint interzone is induced, cells in the joint cavity dying by apoptosis. The first indication of limb musculature occurs in the 7th week as mesenchyme derived from dorsolateral cells of the somites migrates into the limb buds. As the limb buds elongate, this muscle tissue splits into flexor and extensor components, the segmental arrangement being replaced by fusion so that limb muscles are derived from several segments. By the end of the embryonic period (the 12th week) endochondral OSSIFICATION has begun, primary ossification centres being present in all long bones of the limbs. For superficial musculature, see Appendix X.

LINEs (long interspersed nuclear elements) Autonomous non-retroviral RETROTRANSPOSONS with a long evolutionary history that can encode reverse transcriptase and move via an RNA intermediate that is often produced from a neighbouring promoter. Of the three human LINE families (LINE1, LINE2, LINE3), only LINE1 (L1) contains any actively transposing members. L1 is also the most abundant family, comprising ~17% of the human genome. L1 sequences are ~6.1 kb in length and encode two proteins: an RNA-binding protein, and a protein with endonuclease and reverse transcriptase functions, the former having a TTTT↓A target site. As a result the integration site is AT-rich, which therefore tends to avoid gene-rich regions. There is an internal promoter in the 5´ untranslated region, and after translation the LINE1 RNA assembles with its own proteins and moves into the nucleus, where the endonuclease cuts one strand of a DNA duplex, leaving the 3´ OH group to act as a primer for reverse transcription. Indeed, LINE1 machinery is responsible for most of the reverse transcriptase activity in human cells. See PSEUDOGENES, SINES.

linkage The simultaneous occurrence of two or more gene loci on the same chromosome. See HAPLOTYPE, LINKAGE DISEQUILIBRIUM. Compare SEX LINKAGE INHERITANCE.

linkage disequilibrium When a new mutation of a gene arises in a population, it must occur as a single event on a chromosome, that chromosome also having a particular combination of alleles at all its other loci. If the mutant allele *a* arose at the *A* locus on that chromosome, a chromosome that also carried the allele *b* at the *B* locus, then unless recombination (crossing-over) occurs at some time between the *A* and *B* loci, all gametes carrying the new mutant allele *a* will also carry the allele *b* as the new *ab* HAPLOTYPE. Normally, repeated recombination between the loci will, in time, randomize the combinations of alleles of different loci such that if the frequency of *a* at the *A* locus were 0.2 and the frequency of *b* at the *B* locus were 0.4, the frequency of the *ab* haplotype would be $(0.2 \times 0.4) = 0.08$, a situation termed linkage equilibrium.

However, the initial situation described above for the *ab* haplotype is one in which the distribution of *a* and *b* alleles in the population is non-random – a situation referred to as linkage disequilibrium. This association decays from generation to generation at a rate governed by the amount of recombination between the *A* and *B* loci, which is dependent upon the distance between the loci: the further apart the loci are on the chromosome, the more rapidly will linkage equilibrium be achieved. The intermediate levels of partial linkage disequilibrium can be informative in dating population divergence times (e.g., see MIGRATIONS). See NATURAL SELECTION, SELECTIVE SWEEP.

linkage map See GENOME MAPPING.

linoleic acid See OMEGA-6 FATTY ACIDS.

linolenic acid See OMEGA-3 FATTY ACIDS.

lipases Enzymes digesting triglycerides to free fatty acids, monoglycerides, diglycerides and glycerol, the bulk being secreted by the pancreas where its effectiveness in the small intestine is greatly enhanced by BILE SALTS (see CHYLOMICRONS). Minor amounts of lingual lipase are swallowed with food in the oral cavity, digesting ~10% of fats in the stomach, while the stomach also secretes very small amounts of gastric lipase. Hormone-sensitive triacylglycerol lipase is a glucagon- and adrenaline-sensitive enzyme of ADIPOCYTES activated by the intracellular messenger cyclic AMP. See PHOSPHOLIPASE C.

lipid bilayer See CELL MEMBRANES.

lipid rafts Small cholesterol-rich areas in the CELL MEMBRANE in which the outer leaflet of the bilayer is enriched with particular lipids, notably sphingolipids and cholesterol, making it more rigid than the surrounding membrane. Glycophosphatidylinositol-linked proteins (GPI-linked proteins) tend to be found in lipid rafts, as do intracellular proteins attached to the membrane by fatty-acid or prenyl links. Receptors found outside lipid rafts can migrate into them once they have been oligomerized by ligand-binding. Thus, antigen receptors of B CELLS and T CELLS associate with lipid rafts when cross-linked by antigen. See CHOLESTEROL, Fig. 48).

lipids Any organic compound that is soluble in organic solvents. They include triglycerides (triacylglycerols, neutral fats), oils, waxes, phospholipids, steroids and sterols, intracellular messengers (e.g., phosphoinositides), the fat-soluble vitamins (A, D,E and K), pigments (retinal) and anchors for membrane proteins (e.g., covalently attached fatty acids). The synthesis of these water-insoluble molecules usually starts with water-soluble precursors such as acetate (ethanoate) and is endergonic and reductive, requiring ATP and a reduced electron carrier such as NADPH as a reductant (see LIPOGENESIS).

Their insolubility in water causes them to aggregate in lipid droplets within the cytosol so that they are osmotically inactive; and their unsolvated state makes them far better energy reserves than polysaccharides in terms of minimizing overall body mass (water of solvation can account for up to two-thirds of the overall mass of glycogen reserves). However, dietary triglycerides must be emulsified before they can be digested by the water-soluble enzymes of the alimentary canal; and their mobilization from storage depots requires their being bound to plasma proteins. Triglycerides provide more than half the energy requirement of organs such as the liver, heart and resting skeletal muscle, as well as of the renal cortex (see ENERGY BALANCE).

lipogenesis The synthesis of triglycerides from glucose and amino acids, taking place mainly in liver cells and ADIPOCYTES when glycogen stores are filled (see FATTY ACIDS). Promoted by INSULIN, mainly through modification of transcription factors such as PPAR-γ (see PPARS) and activation of lipogenic and glycolytic enzymes, it is opposed by GROWTH HORMONE and LEPTIN. GLUCOSE molecules can be used as a source of acetyl-CoA and glyceraldehyde-3-phosphate (then converted into glycerol), amino acids being converted into acetyl-CoA. The latter are then joined together to form fatty acid chains, three of which are then esterified to glycerol to form triglycerides. Recent studies have implicated PGC-1β as a key regulator of hepatic lipogenesis and lipoprotein secretion in response to dietary intake of saturated fats, when its expression level increases (see PGC-1). PGC-1β docks and coactivates the SREBP and LXR families of transcription factors, coupling triglyceride and cholesterol synthesis, lipoprotein transport, and VLDL secretion. Short-chain fatty acids generated by microbial fermentation also induce lipogenesis (see GUT FLORA).

lipolysis Breakdown of triglycerides (triacylglycerols) to fatty acids and glycerol, often intended to refer to the process within ADIPOCYTES, where it is stimulated by β_2-adrenergic GLUCOCORTICOIDS, sympathetic nervous stimulation, and by GLUCAGON, and inhibited by INSULIN. It is a major feature of SKELETAL MUSCLES, although these do not export the products. See ENERGY BALANCE.

lipopolysaccharide (LPS) The dominant cell wall component of GRAM-NEGATIVE BACTERIA, these molecules having in common a lipid region (Lipid A), a core polysaccharide, and a highly variable O-specific chain which usually distinguishes the bacterial serotype. The Lipid A complex is responsible for the toxicity of these endotoxins (see TOXINS) while the polysaccharide complex makes them molecule water-soluble. Receptors recognizing LPS are located on dendritic cells and macrophages and associate with the signalling receptor TLR-4, activating the transcription factor $NF_{\kappa}B$ (see TOLL-LIKE RECEPTORS, Fig. 175). The receptors mediate ENDOCYTOSIS on binding their ligand (see PHAGOCYTOSIS, INFECTION). See ACUTE PHASE PROTEINS, INNATE IMMUNITY, LECTINS.

lipoprotein lipase An enzyme present on appropriate cell surfaces which, when activated by apolipoprotein ApoC-II, hydrolyses triglycerides in lipoproteins and blood plasma to fatty acids and glycerol. This occurs, *inter alia*, within the capillaries of adipose tissue, heart, skeletal muscle and mammary glands. It is the enzyme involved in clearing triglyceride-rich CHYLOMICRONS and VLDL particles from the bloodstream, providing free fatty acids (NEFAs) to adipose tissue for storage, and to cardiac and skeletal muscle for oxidation and energy release. Its activity in adipocytes is reduced by GROWTH HORMONE.

lipoproteins (lipoprotein particles) Spherical complexes of apolipoprotein and lipids, mediating lipid transport between different tissues via blood plasma and lymph. Their water-facing surfaces consist of a monolayer of phospholipids and free CHOLESTEROL molecules (the polar heads of all these facing the outside) and associated APOLIPOPROTEINS which, in addition to their structural role, act as receptor ligands and enzyme modulators. Their interiors consist of varying mixtures of cholesterol, cholesteryl esters and triglycerides (triacylglycerols). Different combinations of lipid and protein produce particles of different densities and having different functions, from CHYLOMICRONS (99% lipid) and very low density lipoproteins (VLDLs; 92% lipid), via LOW DENSITY LIPOPROTEINS (LDLs, 75% lipid), to high density lipoproteins (HDLs, 55% lipid).

Exogenous cholesterol and fatty acids, released by hydrolysis of dietary fat in bile salt micelles in the small intestine, are absorbed into intestinal mucosal cells where they are re-esterified to form cholesterol esters and triglyceride and then packaged with free cholesterol, phospholipid, Apo B-48 and apoA apoproteins to form nascent CHYLOMICRONS. These large triglyceride rich lipoproteins are secreted into the lacteals and transported via the thoracic duct into the blood where they acquire C and E apoproteins from HDL. On reaching adipose and muscle cells in the peripheral tissues, lipoprotein lipase on the luminal surface of capillary endothelial cells, activated by apoC-11 on the chylomicron surface, rapidly hydrolyses triglyceride in the core of the particle enabling fatty acids to be taken up by these tissues for energy production or storage. The chylomicron reduces in size and surplus surface components: cholesterol, phospholipid, A and C

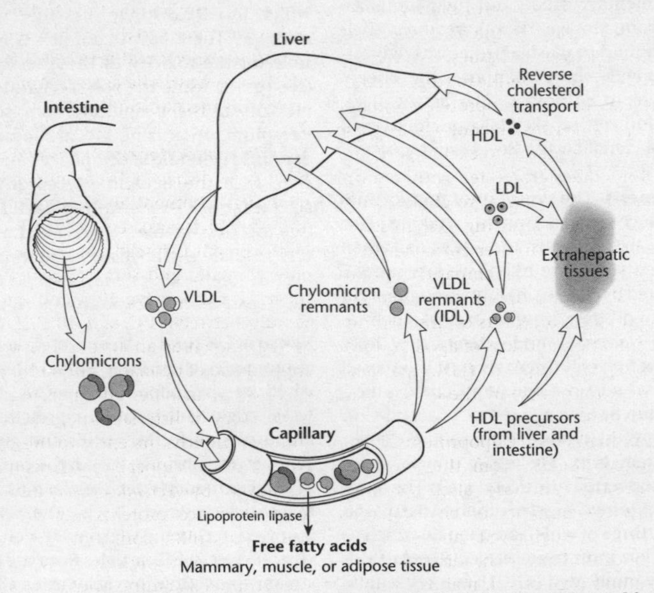

FIG 115. Diagram indicating some of the various lipoproteins transported in blood plasma. Most dietary lipids are packaged into chylomicrons, which unload non-esterified fatty acids to high-demanding tissues. Higher density remnants are removed by the liver and returned to the plasma as VLDL. Excess cholesterol from extrahepatic sources is transported to the liver as HDL. See LIPOPROTEINS and its references. © 2000 From *Lehninger Principles of Biochemistry* by D.L. Nelson and M.M. Cox, published by Worth Publishers. Redrawn with permission from the publishers.

apoproteins leave the chylomicron and join the pool of nascent HDL particles in circulation. Further triglyceride is then transferred from the chylomicron core remnant to mature HDL particles in exchange for cholesterol ester by cholesterol ester transfer protein (CETP). The resulting triglyceride depleted, cholesterol ester enriched chylomicron remnant is of intermediate density and contains apo B-48 and apo E. The latter on the surface of the particle is recognized by hepatic chylomicron remnant receptors enabling this potentially atherogenic particle to be removed from the circulation.

In an analagous manner, endogenous fatty acids surplus to oxidative requirements in the liver are esterified to form triglyceride and packaged with free cholesterol, phospholid and apoproteins B-100, C, E and A and then secreted into the blood as *very low density lipoproteins* (VLDL). After acquisition of more C and E apoproteins from HDL the VLDLs interact with LPL in peripheral tissues, to lose fatty acids from core triglyceride to adjacent cells and surface lipid and apoprotein components to the nascent HDL pool. CETP reacts with the core remnant, to exchange triglyceride for cholesterol ester with mature HDL particles. In contrast to chylomicron remnants the resulting triglyceride depleted, cholesterol ester enriched intermediate density VLDL remnant, containing apo B-100 and apo E can either be taken up by LDL receptors or undergo further delipidation and loss of apo E to form low density lipoprotein (LDL) particles before receptor uptake by the liver or peripheral tissues.

Nascent high density lipoproteins (HDL) are secreted directly from the liver and small intestine and are also generated from surface remnants of chylomicrons and VLDL as described above. These small, phospholipid rich, discoidal particles contain cholesterol, mainly in unesterified form, A, C and E apoproteins and the enzyme LCAT (see LECITHIN-CHOLESTEROL ACYL TRANSFERASE) They play a critical role in the mobilization of cholesterol from peripheral tissues, its esterification and transport to the liver for disposal, either directly or after transfer to intermediate density lipoproteins. Unesterified transferred to these particles via an ATP binding casette transporter (ABCA1) from peripheral cells is esterified by LCAT and transferred to the core of the particle allowing it to mature to a larger, less dense, spherical particle which can receive more cholesterol via an ABCG1 transporter. This mature HDL particle can interact directly with a hepatic scavenger receptor class B type 1 (SRB1) which removes free and esterified cholesterol from the particle without endocytosis. The cholesterol depleted particle can then return to the circulation. Alternatively, as described above, cholesterol esters can be transferred from mature HDL particles to chylomicron and VLDL remnants by CETP with the potential of their subsequent uptake by remnant or LDL receptors (see OESTROGENS). HDL thus plays a critical role in the transport of cholesterol from the peripheral tissues to the liver (reverse cholesterol transport) and is believed to play a major protective role against atherosclerosis by this mechanism.

Listeria monocytogenes A small, Gram-positive rod-shaped bacterium, common in the environment and found in human and animal faeces, sewage slurry and on land where it is spread, on vegetables, soft cheeses, pâtés and some prepared chilled meats. It can survive 60°C and multiplies rapidly between 4°C and 40°C. *Listeriosis* is best known as an INFECTION of pregnant women and of neonates, although ~40% of cases occur in older children and adults. Most cases of listerial MENINGITIS occur in summer months. Its entry to the host cell is by the 'zipper mechanism', also employed by *Yersinia pseudotuberculosis*. Several surface proteins, e.g. internalins (InlA and InlB), contribute to its entry into non-phagocytic cells, which requires the terminal 35 amino acids of E-cadherin. After contact and adherence, involving receptor clustering, actin polymerization and membrane extension, phagocytosis ensues. Listeriolysin is a cysteine-based toxin homologous to streptolysin O, and appears to act by allowing the bacterium to escape from phagolysosomes into the host cytosol, where it replicates freely. Cell-to-cell transfer involves 'filopods', protecting the organism from exposure to macrophages and complement. *L. monocytogenes* is sensitive to penicillin, ampicillin and tetracycline, but chloramphenicol is not to be used. NATURAL KILLER CELLS and CD8 T cells are the major defensive immune cells. See SHIGELLA.

liver Appearing in the 3rd week as a diverticulum of the endodermal epithelium at the distal end of the foregut (duodenum), the liver is influenced by fibroblast growth factors (FGF2) secreted by the cardiac mesoderm and BMPs secreted by the transverse septum. The diverticulum invades the mesentery of the transverse septum between the pericardial cavity and the stalk of the yolk sac, as the connection between it and the foregut narrows to form the bile duct. BMPS secreted by this septum enhance the liver endoderm's response to FGF2. Endothelial liver cords mingle with vitelline and umbilical veins, which form hepatic sinusoids, and differentiate into parenchymatous hepatocytes forming the linings of biliary ducts. Hepatocyte and biliary cell lineages are at least partly under the influence of the hepatocyte nuclear transcription factors HNF3 (FOXA2) and HNF4. The mesoderm of the transverse septum gives rise to haematopoietic cells, KUPFFER CELLS and connective tissue cells. The transverse septum becomes membranous, forming the lesser OMENTUM and falciform ligament (collectively termed the ventral mesentery) and contributing to the central tendon of the diaphragm. Its connective tissue encapsulates the liver, except for the *bare area* on its surface adjacent to the diaphragm. By the 10th week, the liver has ~10% of the embryo's total mass, a result to some extent of its role in HAEMATOPOIESIS, a role that subsides during the last two months *in utero*. Bile begins to be formed by hepatocytes by the 12th week, as the hepatic and cystic ducts converge to form the bile duct.

The mature liver is an accessory organ of the gastrointestinal tract, comprising two major lobes (left and right) and two minor lobes (quadrate and caudate). A *porta* is on the liver's inferior surface, where various vessels, ducts and nerves enter and exit the organ. Those entering include the hepatic portal vein, hepatic artery and a small nerve plexus. Those exiting include lymphatic vessels and two hepatic ducts, one each from the left and right lobes. The connective tissue capsule branches at the porta into the body of the liver to provide its main support, dividing it into hexagonal lobules, with a portal triad at each corner (see Fig. 116). Central veins at the centre of each lobule unite to form hepatic veins which unite on exiting from the liver to enter the inferior (posterior) vena cava. Radiating out from the central vein of each lobule are hepatic cords composed of *hepatocytes*, the metabolically active cells of the liver. Capillary-like hepatic *sinusoids* lie between the hepatic cords and consist of very thin, sparse, endothelial cells and KUPFFER CELLS, but lack mural cells and a basement membrane. A bile canaliculus lies between the cells of each cord.

The liver as a homeostatic organ

(i) Carbohydrate homeostasis. By storing GLYCOGEN, hepatocytes (liver cells) are able to release glucose into the blood plasma in response to glucagon and adrenaline presence. They also absorb excess plasma glucose in response to insulin – a major factor in BLOOD GLUCOSE REGULATION. Hepatocytes are also the major site of GLUCONEOGENESIS from lactate and amino acids.

(ii) In fat homeostasis, hepatocytes are the major site of *de novo* synthesis of fatty acids and production of VLDLs (see LIPOPROTEINS). They are important regulators of plasma lipoprotein levels and of appropriate interchange of free fatty acids between the plasma and ADIPOCYTES. Triglycerides exported by the liver into the circulation are taken up by adipocytes through a lipoprotein lipase-mediated process. By producing BILE SALTS, hepatocytes promote emulsification of dietary fat – required for effective lipase action and uptake of fatty acids and glycerol, and fat-soluble vitamins, by the ileum epithelium. Hepatocytes convert excess sex hormones to CHOLESTEROL and produce cholesterol as a normal product of fat metabolism. They are therefore important in plasma cholesterol maintenance. See LXRS.

(iii) By amino acid transamination (see DEAMINATION) and production of UREA, hepatocytes remove excess amino acids from the plasma.

(iv) By secreting PLASMA PROTEINS, liver cells maintain the blood colloidal osmotic potential – essential for capillary function and water balance (see EXTRACELLULAR FLUID). These proteins include prothrombin and fibrinogen – essential in the homeostatic process of blood clotting. The liver is also the source of the proteins involved in the ACUTE-PHASE RESPONSE. Other protein products include hormone BINDING PROTEINS, thereby increasing hormonal half-lives.

(v) By destroying damaged red blood cells, the phagocytic Kupffer cells in the liver sinusoids help recycle IRON, which hepatocytes store as ferritin.

(vi) The liver contributes to heat generation, and is responsive to THYROID HORMONES when core body temperature falls (see TEMPERATURE REGULATION).

(vii) Liver cells detoxify the plasma, releasing its wastes into the bile (the smooth endoplasmic reticulum is large in liver cells). Ammonia (see UREA), ethanol, antibiotics and psychoactive drugs are dealt with here.

(viii) Hepatocytes are short-term stores of fat-soluble vitamins (A, D, E, K) and carry out the first step in the conversion of VITAMIN D to the hormonally active form calcitriol. Vitamin B_{12} and copper are also stored in the short-term.

(ix) By conjugating certain hormones to

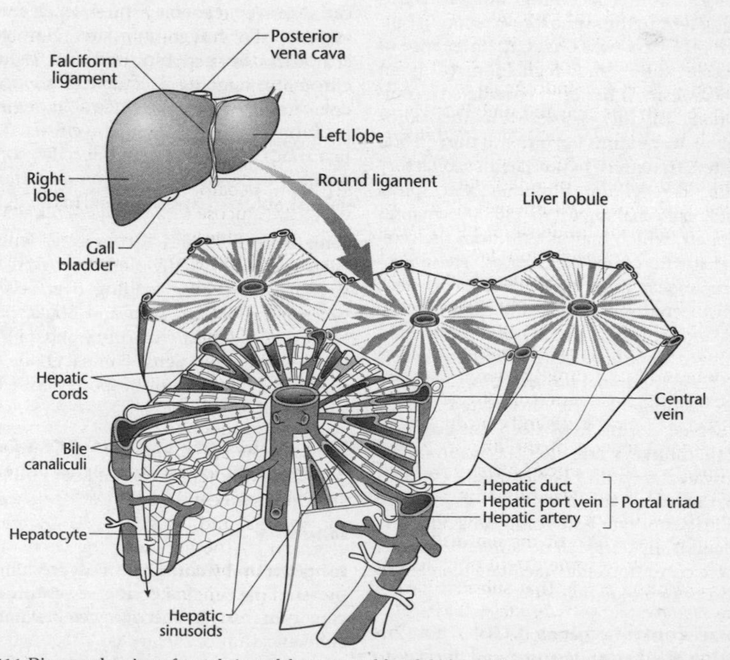

FIG 116. Diagram showing a frontal view of the LIVER and four lobules, with some histology of one of them. Redrawn from *Anatomy & Physiology*, R.R. Seeley, T.D. Stephens and P. Tate. © 2006 McGraw-Hill, with the permission of the McGraw-Hill Companies.

sulphate or glucuronic acid, the liver decreases their potency and increases their rate of excretion in the urine and bile.

(x) B cells, NK cells and macrophages all develop in the foetal liver.

liver lobules See LIVER.

liver X receptors See LXRS.

locomotion See BIPEDALISM, CENTRAL PATTERN GENERATORS, ELBOW, GAIT, INTENTIONAL MOVEMENTS, MUSCLE SPINDLES.

locus 1. See GENE LOCUS. **2**. A small and well-defined group of cells (e.g., the LOCUS COERULEUS).

locus coeruleus (LC) Nucleus of noradrenergic neurons, located in the PONS, whose efferents form the noradrenergic DIFFUSE MODULATORY SYSTEM of the BRAIN (see Fig. 30). It is implicated in promoting AROUSAL (information concerning arousal status being sent to the AMYGDALA), ATTENTION, regulation of SLEEP–WAKE CYCLES, maintenance of muscle tone, and regulation of brain metabolism. It has been considered a component of the MESENCEPHALIC LOCOMOTOR REGION. Its neurons fan out into the cerebral cortex, thalamus, hypothalamus, olfactory bulb, cerebellum, midbrain and spinal cord, each making up to 250,000 synapses *en route*, which may extend from the cerebral to the cerebellar cortices. These neurons appear to be best activated by new, unexpected and non-painful external stimuli, and are least active in non-vigilant individuals. Some of its neurons are responsible for reducing spinal motorneuron excitability, producing the associated atonia observed during REM sleep (see SLEEP). In both monkeys and rats, LC neurons are activated within behavioural contexts that require a cognitive shift – as involved in interruption of on-going behaviour and adaptation. Release of noradrenaline under these conditions may facilitate reorganization of neural networks.

locus control regions (LCRs) Those *cis*-acting REGULATORY ELEMENTS, additional to PROMOTER regions, that engage in both intra- (adjacent) and interchromosomal (long-distance) activation of GENE EXPRESSION, as in developmentally regulated interchromosomal interaction of related genes. They are defined by their ability in transgenic assays, the LCR being coupled to a test gene or gene cluster, then transfected and allowed to integrate into another genome. LCRs can then direct high-level expression of the linked genes at all integration sites. They are often composed of transcriptional activators and INSULATOR elements.

Contra ENHANCER regions, LCRs overcome repressive effects of condensed heterochromatin on the expression of genes, have a dominant effect, and appear to act by CHROMATIN REMODELLING at a higher order of complexity, establishing an 'open' active chromatin domain and direct gene activation (see GLOBINS). When enhancer regions are transgenically introduced in the same genetic systems the linked gene can show very variable expression, depending on the integration site. However, although the β-globin LCR can induce chromatin opening at ectopic locations, it does not appear to do so when at its original location in the β-globin gene cluster. The fact that LCRs confer tissue-specific, copy-number dependent expression on a transgene should perhaps be viewed in this light.

long bones Bones with greater length than breadth, usually slightly curved for strength (spreading loading over several points); e.g. femur, tibia and fibula, phalanges (toes), humerus, ulna and radius. Consist mainly of dense compact bone tissue with a fair amount of spongy bone tissue. Contrast SHORT BONES.

long chain fatty acyl-CoA (LCFA-CoA) An important intracellular signal of nutrient abundance and ENERGY BALANCE.

longevity See LIFE EXPECTANCY.

longitudinal fissure The most prominent FISSURE of the cerebral cortex, separating the cerebrum into left and right cerebral hemispheres.

longitudinal studies Demographic, behavioural, etc., research which provides data about the same individual at different points in time. Change can also be studied using repeated CROSS-SECTIONAL STUDIES collected from different samples. These reveal how the population as a whole has changed, but they do not reveal the complex pattern of changes at the individual level which lead to these changes.

long sight (farsightedness) See ACCOMMODATION.

long terminal repeats (LTRs) Nucleotide sequences repeated at both ends of some integrated retrovirus (LTR virus) particles. See DNA REPEATS, TRANSPOSONS.

long-term depression Weakening of synaptic connections between neurons, and an important form of neural plasticity underlying learning and memory. It is known to occur in the cerebral cortex, hippocampus, cerebellum and spinal cord. See SYNAPTIC PLASTICITY.

long-term memory See MEMORY.

long-term potentiation (LTP) A form of SYNAPTIC PLASTICITY by which neuronal activity leads to a persistent increase in the efficacy of excitatory SYNAPSES in the brain. It is one of the best-studied models of MEMORY at the cellular level, and a particularly good candidate for declarative memory. Theta rhythms (see ELECTROENCEPHALOGRAM) occur during some sleep states and are the optimal protocol for generating LTP, which may be either associative (Hebbian) or non-associative. Brain-derived neurotrophic factor (see NEUROTROPHINS) operates in synergy with electrical activity of neurons in producing LTP.

LTP was first demonstrated during work on the rabbit hippocampus in the early 1970s. Similar work was done on the perforant pathway synapses of neurons of the dentate gyrus, although LTP is also now known to occur in the neocortex, amygdala and elsewhere in the nervous system. Most work is performed on the CA3 Schaffer collateral (Sc) synapses on CA1 pyramidal neurons in brain slice preparations of HIPPOCAMPUS (see Fig. 117). In response to brief, low-frequency stimulation of the Scs, CA1 cells show a brief excitatory postsynaptic potential (EPSP) as a result of release of GLUTAMATE (see also DENDRITES); but if a brief tetanic burst of high frequency is given (e.g., ~100 Hz for 0.5 s), subsequent low-frequency stimulation elicits a larger EPSP in the CA1 cells (see Fig. 117).

Increase in synaptic strength, indicated by evocation of a much greater EPSP on subsequent test stimulation, is the key feature of long-term potentiation (see SYNAPTIC PLASTICITY), and in the case of CA1 neurons this can last many weeks, even a lifetime. But it is now apparent that such high-frequency stimulation is not a prerequisite for LTP. What is required is that synapses are active at the same time that the postsynaptic neuron is strongly depolarized, as it would be in cases of temporal or spatial summation of EPSPs (see HEBBIAN MODIFICATION). If an EPSP is followed by a postsynaptic action potential within ~50 msec, the synapse potentiates. The action potential or EPSP alone do not produce this; LTP only results from the precise temporal relation of the action potential and the EPSP. Action potentials generated in the soma can actually back-propagate to the dendrites of some cells. This is possible because NMDA RECEPTORS have a high affinity for glutamate, which remains bound to the receptor for many tens of milliseconds. Timely arrival of an action potential causes these dormant receptors to eject the magnesium ions clogging the channel, so that if glutamate is still attached, Ca^{2+} will enter the cell and trigger LTP. The rise in internal Ca^{2+} concentration activates PKC and CaMKII (see CALCIUM/CALMODULIN-DEPENDENT PROTEIN KINASES). It also stimulates an isoform of adenylyl cyclase, so that cAMP concentrations rise in LTP. This leads to persistent activation of PKA, promoting CREB-mediated gene expression (e.g., see IMMEDIATE EARLY GENES). Evidence implicates both presynaptic and postsynaptic changes during LTP which, in the hippocampus, include: increased glutamate release as a result of the RETROGRADE MESSENGER effect of nitric oxide; sudden addition of AMPA channels to the postsynaptic membrane in large groups; their

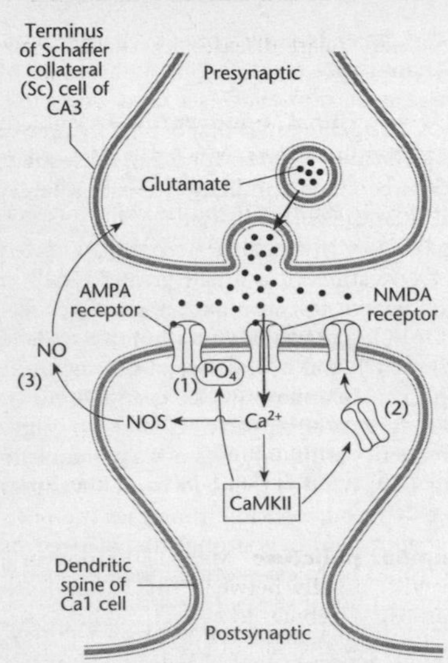

FIG 117. Diagram showing some of the pre- and postsynaptic changes accompanying LONG-TERM POTENTIATION in the CA1 subfield of the hippocampus. CA1 cells have AMPA and NMDA glutamate receptors. Calcium ions entering through the NMDA receptor activate protein kinases, leading to phosphorylation (1) and increased conductance of existing postsynaptic AMPA receptors, to insertion of already-prepared AMPA receptors (2), and to release of the retrograde messenger nitric oxide (NO), enhancing glutamate release (3). CaMKII = Ca^{2+}-calmodulin-dependent kinase II. © 2001 Redrawn from *Neuroscience* (2nd edition) by M.F. Bear, B.W. Connors and M.A. Paradiso, publisher by Lippincott, Williams & Wilkins. Reproduced with the kind permission of the publishers.

phosphorylation, leading to increased conductance; a delayed increase in the number of synaptic vesicles released (see Fig. 117); and the appearance of dendritic spines from new synaptic contacts within axons (see DENDRITES).

It was shown quite early that drugs that block LTP, given to an animal before it learns a new task, prevent new memories from being formed. If LTP is required for hippocampal MEMORY formation, it should be possible to observe LTP in the hippocampus when an animal learns something. Early in 2006, such an observation was reported in mice conditioned to blink upon hearing a tone, and later that year a team showed that abolishing LTP *after* rats had learnt to avoid a 'shock zone' (by blocking an enzyme required to sustain hippocampal LTP) eradicated both LTP and memory of the shock zone's location. If it is true that LTP is a molecular mechanism of memory, it is unknown how the many forms of LTP identified in brain tissue relate to different kinds of memory. See ADDICTION, LONG-TERM DEPRESSION.

long-term secular trends See EPIDEMIOLOGY.

loop of Henle That portion of a KIDNEY nephron between the PROXIMAL CONVOLUTED TUBULE and DISTAL CONVOLUTED TUBULE. Fluid enters the loops at a rate of ~40–45 mL min^{-1}, its composition being different from that of blood plasma and the original glomerular filtrate, although its osmolarity is still close to that of blood. Its descending limb is very permeable to water but quite impermeable to solutes other than urea. As fluid passes along the descending limb of a medullary nephron, water moves out by osmosis into the peritubular fluid so that the tubular fluid increases in osmolarity, reaching 1,200 mOsm dm^{-3} at the hairpin bend. The thin ascending limb is impermeable to water but permeable to solutes, which diffuse out into the more dilute peritubular fluid and then into the descending VASA RECTA. The thick ascending limb is also impermeable to water, but its epithelial cells actively transport Na^+ out across their basal membranes into the peritubular fluid and vasa recta, causing Na^+, K^+ and Cl^- to be reabsorbed from the tubular fluid through Na^+-K^+-$2Cl^-$ symport channels in the apical membranes. Cl^- diffuses through leak channels in the basolateral membranes, but much of the K^+ brought in by symports diffuses out again into the tubular fluid through numerous K^+ leak channels in the apical membranes so the main effect of the symports is to reabsorb Na^+ and Cl^-. The K^+ returning to the tubular fluid gives the peritubular fluid and blood a net negative charge relative to the fluid in the ascending limb, drawing Na^+, K^+, Ca^{2+} and Mg^{2+} from the tubular fluid into the vasa recta.

loss-of-function mutations Those MUTATIONS resulting in either reduced activity, or complete loss, of the gene product. In the heterozygous state, such mutations would result, at worst, in half-normal levels of protein product. Because the catalytic activity of the normal allele's product is normally more than adequate, loss-of-function mutations will tend to be inherited in an autosomal or sex-linked recessive manner. However, loss-of-function mutations in the heterozygous state resulting in half-normal product levels are termed *haploinsufficiency mutations*. The phenotypic effect of such mutations is often the result of gene dosage effects in genes encoding receptors, or more rarely in enzymes, which are rate-limiting (e.g., see FAMILIAL HYPERCHOLESTEROLAEMIA). Homozygous mutations in such genes often have more severe phenotypic effects. Compare GAIN-OF-FUNCTION MUTATIONS. See CANCER, 'LESS-IS-MORE' HYPOTHESIS, TUMOUR SUPPRESSOR GENE.

Lou Gehrig's disease See AMYOTROPHIC LATERAL SCLEROSIS.

low copy repeats (LCRs) See GENOMIC DISORDERS, SEGMENTAL DUPLICATIONS.

low density lipoprotein (LDL) See LIPOPROTEINS. LDL receptors (under thyroid hormone regulation) are found on plasma membranes of cells that need to take up cholesterol, binding initiating LDL endocytosis and entry of the LDL/receptor complex into the cell within an endosome. Fusion with a lysosome containing enzymes hydrolysing cholesterol esters releases cholesterol and fatty acids into the cytosol. The causal relation between low-density lipoprotein cholesterol (LDL-C) and CORONARY HEART DISEASE is now well established in studies using multivariate analysis. Loss-of-function mutations in the gene encoding PCSK9, a secreted enzyme of the serine protease family, are associated with lifelong low LDL levels and dramatically reduced prevalence of coronary heart disease. See FAMILIAL HYPERCHOLESTEROLAEMIA, STATINS.

lower critical temperature Ambient temperature below which the BASAL METABOLIC RATE becomes insufficient to balance heat loss, resulting in falling body temperature. See TEMPERATURE REGULATION.

LTRs See LONG TERMINAL REPEATS.

lumbar plexus Nerve plexus formed from the anterior (ventral) rami of SPINAL NERVES L1–L4 and (unlike the brachial plexus) lacking intricate intermingling of fibres. It supplies the anterolateral abdominal wall, external genitalia and parts of the lower limbs.

lumbar puncture Medial insertion of a needle, usually between the 3rd and 4th lumbar vertebrae, to collect cerebrospinal fluid from the sac of dura (spinal canal) for diagnosis of meningitis and other disorders affecting the brain and spinal cord, or to inject drugs such as antibiotics, or anaesthetics (in operations on the lower half of the body).

lumbar vertebrae (L1–L5) The largest and strongest elements in the vertebral column, supporting as they do the greatest amount of body weight. See SOMITES.

lunate sulcus See BIPEDALISM.

lung cancer The leading cause of CANCER deaths in the USA, several subtypes of lung cancer exist, each with its own characteristic GENE EXPRESSION signature (or profile). Comparative genomic hybridization and expression profiling offer hope of progress towards understanding of the critical genomic alterations that characterize each subtype. Mouse models have indicated the possible stem cell origin of lung cancer as the bronchioalveolar stem cell (BASC), whose anatomical location and ability to self-renew and differentiate into multiple lung cell types are features consistent with those predicted for a lung stem/progenitor cell. Adenocarcinomas and squamous cell carcinomas differ histopathologically but, surprisingly, have nearly identical genomic signatures and may arise from such a common stem/progenitor cell. Mutations in the *Ras* oncogene (see RAS SIGNALLING PATHWAY) are found in 40% of human lung cancers. Human small cell carcinomas of the lung tend to show amplification of the oncogenes c-*MYC*, of the related N-*MYC* and L-*MYC*, and of the *Hdm2* gene (see P53 GENE). It is thought that human lung cancer may be due in part to a structural retroviral tumour-causing protein (the coat protein, Env) similar to that of JSRV (jaagsiekte sheep retrovirus) that killed the first cloned mammal, Dolly. See CIGARETTE SMOKING, METASTASIS.

lungs A pair of elastic dome-shaped structures lying in the thoracic cavity, separated by the heart and other structures in the mediastinum. At ~4 weeks, the respiratory diverticulum (lung bud) grows out from the ventral wall of the foregut, its location determined by expression of the transcription factor gene *TBX4*. TBX4 not only induces lung bud formation but also the continued growth and differentiation of the lungs. The epithelium of the inner lining of the larynx, trachea, bronchi and lungs is entirely of endodermal origin. The cartilaginous, muscular and connective tissue components derive from splanchnic mesoderm.

The lungs extend from the diaphragm to slightly superior to the clavicles, lying against the ribs anteriorly and posteriorly. The broad inferior portion of each lung is concave, fitting over the convex area of the diaphragm. The right lung is larger than the left, with an average mass of 620 g, that of the left being ~560 g. In the human lung, air is distributed into >130,000 terminal bronchioles by 16 generations of a dichotomous bronchial tree, each terminal bronchiole further subdividing six more times into thin-walled alveolar ducts and alveoli, where gas exchange between capillary blood and alveolar gas occurs (see ALVEOLUS). All branches of the tree up to and including the terminal bronchioles constrict (see ASTHMA). The high compliance of the lungs is due to their large surface area and low surface tension. An increase in lung compliance occurs in EMPHYSEMA. See LUNG CANCER; PLEURAL MEMBRANE; SPIROMETER, Fig. 165; VENTILATION.

lupus An AUTOIMMUNE DISEASE in which B cells produce antibodies against the individual's own nucleic acids and associated proteins.

luteinizing hormone (LH, ICSH) A very large glycoprotein and gonadotropic hormone produced and released by gonadotrophs of the anterior pituitary gland under the influence of GONADOTROPIN RELEASING HORMONE (see also MENSTRUAL CYCLE details of pulsatile release). It circulates unbound in the plasma and has a half-life of 30 minutes and uses CYCLIC AMP as a second messenger in its signalling. In adult males, LH causes the LEYDIG CELLS of the testis to secrete testosterone and its rate of release increases when testosterone level in the blood is low and decreases when it is high. LH acts on mature Graafian follicles to terminate the programme of gene expression associated with the follicular phase of the menstrual cycle, increased cAMP levels turning off transcription of genes for IGF-I, FSH, oestrogen receptors and cyclin D2 as well as those for enzymes for steroid synthesis. LH induces genes involved in ovulation (e.g., progesterone receptor, cyclooxygenase-2) and luteinization (e.g., cell cycle inhibitors, steroidogenic acute regulatory protein, protein kinases, etc.). The GnRH agonist LHRH-A has been used for many years in treatment of prostate cancers, precocious PUBERTY and ENDOMETRIOSIS.

LXRs (liver X receptors) Nuclear receptors that coordinate hepatic lipid metabolism, their ligands being oxysterols. They form functional heterodimers with retinoid X receptors (RXR) and appear to be transcriptional switches integrating hepatic glucose metabolism, and cholesterol and fatty acid synthesis. Glucose and glucose-6-phosphate are direct agonists, and probably ligands, of both LXR-α and LXR-β, inducing expression of LXR target genes. LXR/RXR signalling inhibits cell proliferation, and induces apoptosis, in β-cells of the PANCREAS.

lyases Enzymes catalysing the breaking of various chemical bonds by means other than hydrolysis and oxidation, often forming a new double bond or a new ring structure. They differ from other enzymes in that they only require one substrate for the reaction in one direction, but two substrates for the reverse reaction.

lymph An EXTRACELLULAR FLUID defined by its location within lymph vessels. See LYMPHATIC SYSTEM.

lymphatic system First discovered by Hippocrates ~400 BC; but although early 20th century anatomists mapped much of the lymphatic system, understanding of its vital role in immunity only began in 1937 when Florey (ultimately responsible for the mass production of PENICILLIN) noted lymphatic proliferation in inflammatory conditions (see INFLAMMATION). The lymphatic vessels originate in *lymph capillaries*, which drain TISSUE FLUID, and have valve-like structures in their walls by which immune cells enter and leave the lymph and also acting as a series of one-way valves, preventing lymph back-flow. Resembling blood capillaries, their simple squamous epithelial cells slightly overlap, are attached more loosely, and have no basement membrane. They join to form larger vessels, which resemble VEINS in structure but have thinner walls and more valves. These join to form lymphatic trunks, each draining a major portion of the body: jugular trunks draining the head and neck; subclavian trunks draining the FORE LIMBS, superficial thoracic wall and mammary glands; bronchiomediastinal trunks draining the thoracic organs and deep thoracic wall; intestinal trunks draining most of the abdominal organs; and lumbar trunks draining the hind limbs, pelvic organs and the remaining abdominal organs (gonads, kidneys and adrenal glands). On the right side, the jugular, subclavian and bronchiomediastinal trunks typically separately join the right thoracic vein, but in 20% of people they join to form a short right lymphatic duct prior to this. But inferior to this on the right side, and on the left side entirely, the trunks drain into the thoracic duct, the largest lymph vessel (38–45 cm long). The left jugular and subclavian trunks join the thoracic duct just before it enters the left SUBCLAVIAN VEIN.

Only in 1999 was a protein marker (LYVE-1) of lymph tissue discovered, since when lymphatic research has taken on a new lease of life. There is convincing evidence that it is the interplay between tumours and the lymphatic system which creates the main route by which solid tumours spread. Besides the vessels, the lymphatic system includes LYMPHOID ORGANS. Lymphocytes can move from lymph nodes via the thoracic duct and enter the blood. See GROWTH FACTORS.

lymph node Occurring throughout the body, a form of peripheral (or secondary) LYMPHOID ORGAN located where lymph vessels converge. As sites of intersection of the blood and lymphatic systems, they are admirably positioned to initiate adaptive immune responses (see ADAPTIVE IMMUNITY). They are larger and more numerous adjacent to likely places of infection such as the throat, lungs, gastrointestinal tract and exposed parts of the limbs. Most are embedded in adipose tissue, where very fine connecting lymph ducts pass adjacent adipocytes on entering and leaving the nodes so that glycerol and fatty acids can be released to immune cells as required. They include BRONCHIAL-ASSOCIATED LYMPHOID TISSUE, GUT-ASSOCIATED LYMPHOID TISSUES and MUCOSAL-ASSOCIATED LYMPHOID TISSUE.

Lymph nodes (see Fig. 118) consist of an outermost cortex and an inner medulla. Lymph draining the tissues enters the cortex through afferent lymphatic vessels, and lymph node stroma and high endothelial venules release the chemokines CCL19, CCL20 and CCL21, which attract those antigen-presenting cells (APCs), the DENDRITIC CELLS and MACROPHAGES, bearing foreign antigens. These antigens activate naïve T cells that have entered the peripheral lymphoid organs from the blood by crossing the walls of the high endothelial venules. Oligosaccharides on the endothelial cells also bind L-selectin on lymphocytes, causing them to loiter. The architecture of these organs almost guarantees that a foreign antigen will encounter the antigen-specific T-cell receptor on those rare T cells that bear them. Some of the APCs will then respond and trigger an adaptive immune response. The outer cortex is composed of B cells organized into lymphoid follicles and deep, paracortical, areas composed largely of T cells and dendritic cells. During an immune response, the central areas of some of the follicles become occupied by proliferating B cells. These centres are termed *germinal centres* (see T CELLS) and eventually become senescent, the follicles being termed secondary lymphoid follicles. The lymph node enlarges as a result, a person being said to have 'swollen glands'.

lymphocytes Collective term for B CELLS and T CELLS. The consensus view is that lymphocytes develop from 'common lymphoid progenitors' in the BONE MARROW, themselves descendants of pluripotent haematopoietic STEM CELLS. Small lymphocytes that have matured in the bone marrow and thymus but have not yet encountered their specific antigen, referred to as naïve lymphocytes, circulate continually from the blood into peripheral lymphoid tissues, entering the latter by crossing high endothelial venules. During an infection, lymphocytes recognizing the antigens of the pathogen are arrested in the lymphoid tissue and proliferate, enlarge, and differentiate into effector cells.

lymphoid organs Organs of the LYMPHATIC SYSTEM containing large numbers of lymphocytes within a framework of non-lymphoid cells with which they interact during their development and as a prelude to initiating the responses of ADAPTIVE IMMUNITY. Central (or primary) lymphoid organs (BONE MARROW and THYMUS) are the sites of lymphocyte generation. Secondary (or peripheral) lymphoid organs are the sites of initiation of lymphocyte maintenance and initiation of adaptive immune responses and include the SPLEEN and LYMPH NODES, TONSILS and PEYER'S PATCHES. Lymphatics drain tissue fluid from the peripheral tissues, through the lymph nodes and on to the thoracic duct and left subclavian vein. In lymphoid organs, endothelial cells express oligosaccharides recognized by

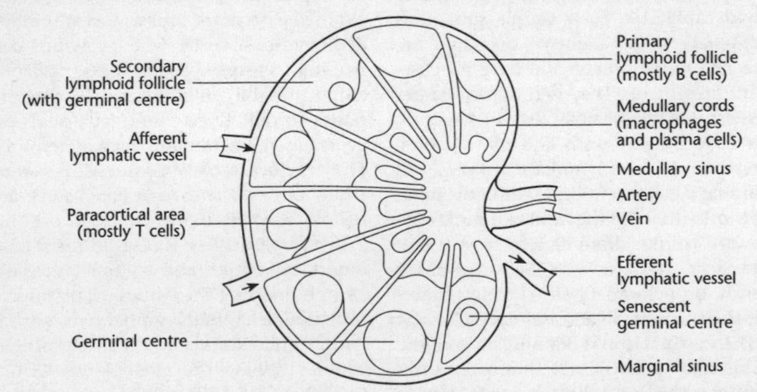

FIG 118. The structure of a LYMPH NODE. Antigen-presenting dendritic cells migrate from the tissues to the paracortical areas, where passing naïve T cells are trapped if their receptors bind peptides that are presented. B cells enter the paracortical areas by passing through the high endothelial venules located there. In the presence of the appropriate antigen, B cells with antigen-specific receptors are trapped by dendritic cells at the border of the paracortical areas and primary lymphoid follicles, where they are most likely to encounter armed helper T cells that can activate them. On receipt of these key signals, B cells migrate to a medullary cord and undergo clonal expansion, their daughter cells maturing into plasma cells or memory B cells. In the absence of T-cell signals, the antigen-stimulated B cells die within 24 hours (see B CELLS).

L-selectin on lymphocytes, causing these to become trapped. See LYMPHOID TISSUES.

lymphoid tissues Tissues in which the common lymphoid progenitor cells occur (i.e., bone marrow) and in which their cellular offspring (B CELLS, T CELLS and NATURAL KILLER CELLS) mature and reside. These latter tissues are typically in LYMPHOID ORGANS. Dendritic cells very similar to those derived from the common myeloid progenitor (see MYELOID TISSUE) are also produced by the common lymphoid progenitor cells. See GUT-ASSOCIATED LYMPHOID TISSUES; STEM CELLS.

lymphokines Those CYTOKINES produced by lymphocytes.

lymphomas Tumours of lymphocytes which grow in lymphoid tissues (especially secondary LYMPHOID ORGANS) and other tissues, but do not enter the blood in large numbers.

lymphotoxin See TUMOUR NECROSIS FACTORS.

lysolecithin-cholesterol acyl transferase (LCAT) The enzyme catalysing the formation of cholesterol esters from lecithin (see PHOSPHATIDYLCHOLINE) and cholesterol. It occurs on the surfaces of newly formed HDLs (see LIPOPROTEINS, CHOLESTEROL).

lysophospholipids (LPs) Polar metabolites of such activated membrane phospholipids as sphingomyelin and phosphatidylcholine. Glycerol-based and sphingosine-based phospholipids, abundant structural components of cell membranes, are metabolized into eicosanoids and LPs. Originally recognized as intracellular second messengers, G PROTEIN-COUPLED RECEPTORS for some of them have since been found for them on the cell surface. LPs include lysophosphatidic acid (LPA) and phosphatidic acid (PA, a precursor of DIACYLGLYCEROL), which are interconverted by lysophosphatidic acid acyl transferase and phospholipase A_2 activity, respectively. These are involved in curving the membrane in opposite directions. LPA uses four different receptors (LPA_1–LPA_4) to produce different effects. Other LPs include lysophosphatidylcholine (LPC), lysophosphatidylinositol (which activates the polymodal K^+-channel of some MECHANORECEPTORS even better than does LPC), sphingosylphosphoryl choline (SPC) and sphingosine 1-phosphate (S1P). They are implicated in tumorigenesis, angiogenesis, immunity, atherosclerosis, neuronal survival and IMPLANTATION. Lysolipids are found in the upper gastrointestinal tract, creating a barrier to infection.

lysosomes Membrane-bound organelles, 0.2–0.5 μm in diameter, forming from late ENDOSOMES (see ENDOCYTOSIS) and containing ~40 types of hydrolytic enzymes, all ACID HYDROLASES, delivered to them by a route leading outward from the ENDOPLASMIC RETICULUM via the *cis* and *trans* networks of the GOLGI APPARATUS. In the *cis* network, the lysosomal hydrolases pick up a unique marker group, mannose-6-phosphate (M6P), added to their N-linked oligosaccharides. This is recognized by transmembrane M6P receptor proteins present in the *trans* network, which bind to the hydrolases on the lumenal side of the membrane and to adaptins involved in assembling the clathrin coats on the cytosolic side. When budded from the *trans* network, these clathrin-coated VESICLES (see entry) target the late endosome. Initially bound to the M6P receptor protein on the lumenal surface, the hydrolases dissociate in late endosomes as the pH drops to 6. Their diverse morphology reflects the variety of digestive functions carried out by their proteases, nucleases, glycosidases, lipases, phospholipases, phosphatases and sulphatases, which would not function if they escaped into the cytosol, with its pH~7.2. The lysosomal lumen is maintained at an internal pH of ~5.0 through activity of an ATP-dependent membrane proton pump. Lysosomes are involved in breakdown of intracellular debris (autophagy) and phagocytosed microorganisms (see INFECTION), transport proteins in the membrane allowing the products of digestion to be transported to the cytosol as nutrients. Most of the membrane proteins are highly glycosylated, helping to protect the membrane from digestion by the lysosomal enzymes. Normally, membrane components delivered to lysosomes and late endosomes are hydrolysed inside these organelles; but several *lysosomal storage disorders*, including TAY–SACHS, NIEMANN–PICK type A or B (involving sphingomyelinase) and Sandhoff diseases, result from impaired hydrolysis or trafficking of lipids, lysosomes and endosomes acquiring MULTIVESICULAR BODIES. Proteins destined for the lysosome have a mannose-6-phosphate SIGNAL SEQUENCE added to them while in the *cis*-compartment of the GOLGI APPARATUS and recognized by a receptor protein in the *trans*-compartment. See MANNOSE-BINDING LECTIN.

lysozyme An antibacterial enzyme which dissolves the wall of some Gram-positive bacteria. It is secreted in tears (see LACRIMAL GLANDS) and SALIVA, and released from phagocytes on ingestion of microorganisms.

lytic enzymes Enzymes involved in rupturing a cell membrane during escape of an intracellular (e.g., viral) pathogen.

macroevolution A term most commonly applying to evolutionary processes and events that occur at the level of taxa higher (i.e., more inclusive) than species. Whereas some hold that factors extrinsic to biological systems (e.g., environmental catastrophes) are required to explain such evolution, others (Neodarwinians) seek explanations for macroevolutionary change solely in microevolutionary processes, in which organisms interact with each other and with their physical environments.

macronutrients Substances, such as carbohydrates, proteins and fats, required in large amounts in the DIET.

macrophages Long-lived mononuclear phagocytes that mature continuously from monocytes that have left the blood and entered the tissues (see PHAGOCYTOSIS). IL-3 and GM-CSF produced by activated T_H1 cells induce their differentiation in the bone marrow, being found in large numbers in connective tissues, in the submucosa of the gastrointestinal tract, in the interstitium and alveoli of the lung, along liver sinusoids (see KUPFFER CELLS), and throughout the spleen. The T_H1 CHEMOKINE CCL2 prompts MONOCYTES to leave the bloodstream and enter tissues to become tissue macrophages, where they amplify immune responses. They are located in many regions of a LYMPH NODE (see Fig. 118) but especially at the afferent marginal sinus and in the medullary cords, where lymph collects before leaving via efferent vessels. They cannot take up soluble antigen efficiently, but ingest particulate antigens, preventing their entry into the blood. Although they are not such specialized ANTIGEN-PRESENTING CELLS as dendritic cells, those in the T-cell areas of LYMPH NODES stimulate immune responses to many sources of infection. T_H1 cells (see T CELLS) are specialized to activate macrophages that have ingested, or been infected by, pathogens. Such activation involves secretion of IFN-γ and other effector molecules by T_H1 cells, while their CD40 ligand binds its receptor on the macrophage (see INFECTION). Other cell-surface receptors (see Fig. 119) involved in macrophage activation include those for bacterial glucans, mannose and lipopolysaccharide (see TOLL-LIKE RECEPTORS), scavenger receptors for anionic polymers (their expression augmented by macrophage colony stimulating factor, M-CSF), acetylated low-density lipoproteins and other pathogen-derived molecules. Ligands binding TLRs cause secretion of pro-inflammatory cytokines (e.g., TNF-α) and lipid mediators of INFLAMMATION (e.g., CXCL8, activating cells of the ENDOTHELIUM) and cause other macrophages (and dendritic cells) to produce cell-surface CO-STIMULATORY MOLECULES that lead, importantly, to the induction of ADAPTIVE IMMUNITY.

Macrophage activation promotes the cells' microbicidal behaviour and amplifies the immune response. The number of molecules of MHC class II, B7, CD40 and the TNF receptor all increase on the cell surface, making the cell more effective at presenting antigen to naïve T cells and more responsive to CD40 ligand and to TNF-α. The macrophage products IL-1, IL-6 and TNF-α all induce the ACUTE PHASE RESPONSE

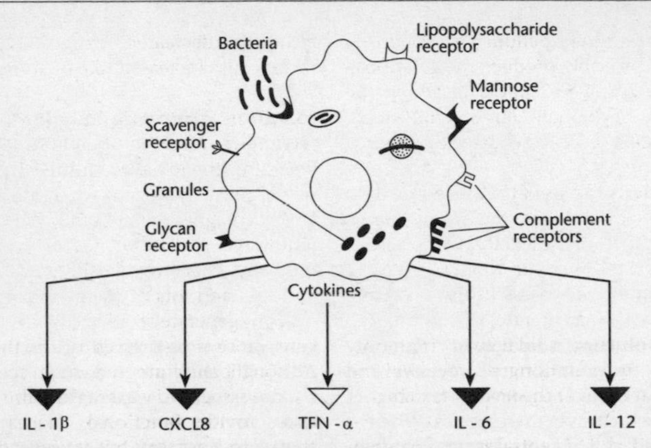

FIG 119. Diagram showing important cytokines secreted by MACROPHAGES in response to bacterial products. IL-1β activates the vascular endothelium and lymphocytes; TNF-α induces local inflammation by activating the vascular endothelium, increasing vascular permeability; IL-6 activates lymphocytes, increasing antibody production (all three are endogenous pyrogens, generating fever); CXCL8 is a chemokine recruiting neutrophils, basophils and T cells to sites of infection; IL-12 activates NK cells and causes CD4 T cells to differentiate into T_H1 cells.

in the liver, while IL-12 directs differentiation of naïve CD4 T cells into T_H1 effector cells (see T CELLS). Their lysosomes fuse more efficiently to phagosomes, exposing intracellular or recently ingested microbes to a variety of lysosomal enzymes. During phagocytosis, activated macrophages, like NEUTROPHILS, produce antimicrobial peptides, proteases (effective against extracellular parasites), NITRIC OXIDE, the superoxide ion and HYDROGEN PEROXIDE, all directly toxic to bacteria. During this process, termed '*respiratory burst*', the large membrane-associated NADPH oxidase of macrophages produces superoxide, requiring a brief increase in oxygen consumption. The superoxide is then converted into hydrogen peroxide by superoxide dismutase, from which other enzymes and reactions involving the Fe^{2+} produce microbicidal hypochlorite and hydroxyl radicals. In severe inflammation, this response can damage the endothelium. Tight regulation of macrophages by T_H1 cells minimizes both this damage and energy consumption. This involves not only down-regulation of cytokine production (see CYTOKINES) but also inhibition of macrophage activation by cytokines such as TGF-β and IL-10. Several of these inhibitory cytokines are produced by T_H2 cells, and induction of this class of T cell is an important route for controlling macrophage effector functions. Macrophages play an important part in atheroma formation during ATHEROSCLEROSIS.

Pathogens which proliferate within macrophage VESICLES pose special problems to host defence (see INFECTION), when their activation by armed T_H1 cells expressing CD40 and secreting IFN-γ is essential (see T CELLS). Activated T_H1 cells can express the Fas ligand and thereby kill macrophages that express Fas, preventing their acting as a reservoir of infection if infection results in their failure to become activated. See TUBERCULOSIS.

macula (statolith organ, pl. maculae) **1**. A thickening of the wall of both the utricle and saccule of the VESTIBULAR APPARATUS, lying perpendicular to one another and housing sensory mechanoreceptor HAIR CELLS for static, and some components of dynamic, equilibrium. They provide sensory information on the position of the head in space, essential in maintenance of static posture and balance. Between the hair cells of the sensory epithelium lie supporting cells that probably produce the gelatinous glycoprotein of the cupula in which the stereocilia of the hair cells are embedded. **2**. See MACULA LUTEA. **3**. See MACULA DENSA.

macula densa Cells of the ascending limb of the DISTAL CONVOLUTED TUBULES of kidney nephrons (part of the JUXTAGLOMERULAR APPARATUS) which release renin when blood NaCl concentration increases. It also increases glomerular filtration rate in response to reduced volume of fluid delivery to the distal tubule by decreasing resistance of the afferent arterioles. See ANGIOTENSINOGENS.

macula lutea The central RETINA, ~6 mm in diameter and representing 15° of the visual field and yellowish in hue, whose own central region is the FOVEA, ~1.5 mm in diameter and representing 5° of the visual field. See AGE-RELATED MACULA DEGENERATION.

magnesium A major mineral component of the DIET, routinely ingested in insufficient amounts. In adults, ~50% is present as Mg^{2+} salts in bone matrix, ~45% as Mg^{2+} within cells (the 2nd most abundant intracellular cation), and ~1% in the extracellular fluid (~50% free and the rest bound to plasma proteins). It is a cofactor for enzymes involved in glycolysis, gluconeogenesis and protein synthesis (ribosome assembly; initiation and elongation during translation), and for the Na^+/K^+ ATPase sodium pump. It is required in neuromuscular activity, nerve impulse transmission, cardiac muscle contraction and for PTH secretion. Deficiency is thought to be a risk factor in onset of diabetes, high blood pressure, difficulties during pregnancy (e.g., migraines, high blood pressure, miscarriages, stillbirths and low birth-weight babies) and cardiovascular disease. Addition to the drinking water of rats can eliminate hypertension, while addition to the diet of rabbits reduces plasma lipid levels and plaque formation in blood vessels. Free Mg^{2+} enters the glomerular filtrate, most of these being absorbed in the loop of Henle and only ~10–15% entering the urine. Control of reabsorption is unclear, although decreased extracellular Mg^{2+} levels cause increased rate of absorption.

magnetic resonance imaging (MRI) A technique revealing brain structure by detecting microwaves emitted by atomic nuclei (e.g., hydrogen) when aligned in a strong magnetic field and pulsed with radiowaves. The nuclear magnetic moments that provide the signal in MRI are tiny, and lots of them are required in order to generate an image, but it can provide more soft-tissue contrast in cross-sectional brain images than a CT SCAN, without the associated ionizing radiation. It can also provide functional images (fMRI), similar to a PET SCAN but safer because it is non-invasive. It is also faster and, at 3 mm^3, has greater resolution. Blood-oxygen-level-dependent (BOLD) MRI, developed in the 1990s, is the usual approach, measuring haemodynamic changes including alterations in blood flow, blood volume or intravascular magnetic susceptibility, and capable of being linked to EEG data to reveal functional meaning of activation patterns associated with MRI. Since higher rates of flow are associated with increased metabolic activity, fMRI makes use of the fact that oxyhaemoglobin has a different magnetic resonance from that of deoxyhaemoglobin. More active regions of the brain receive more blood, which then donates more of its oxygen, and fMRI detects the location of increased neural activity by measuring the ratio of oxyhaemoglobin to deoxyhaemoglobin. In a recent cause célèbre, fMRI scans of a young woman with severe head injuries revealed evidence of cognition that could not have been anticipated from standard MRI scans. Acquisition of molecular and physiological information by nuclear magnetic resonance imaging, using detectable nanoparticles, is starting to help in diagnosis of cancers at a premalignant stage. Recent techniques (Xenon MRI) that increase the sensitivity of MRI using hyperpolarized ^{129}Xe gas may extend the use of MRI so as to provide diagnostic information at the molecular as well as at the functional and anatomical levels. Real-time fMRI is finding uses in psychotherapy, although in studies of ALZHEIMER'S DISEASE fMRI is still a step or two behind PET. False negative findings in fMRI are common, even in healthy volunteers. See CONSCIOUSNESS, LANGUAGE, ULTRASOUND, WILLIAMS–BEUREN SYNDROME.

magnocellular cells (M cells) **1**. One class of retinal GANGLION CELL, projecting to the magnocellular layers of the LATERAL GENICULATE NUCLEUS. They exhibit fast and phasic responses, particularly to movement, and are involved in location and movement of objects in the visual field. See MAGNOCELLULAR PATHWAY; compare PARVOCELLULAR-BLOB/INTERBLOB PATHWAYS. **2**. One of two classes of neurosecretory cell in the hypothalamus, predominating in the paraventricular and supraoptic nuclei, and producing large amounts of oxytocin and vasopressin. Compare PARVOCELLULAR CELLS (2).

magnocellular pathway One of the three parallel processing pathways involved in visual perception, specialized for detecting motion and spatial relationships. It is apparently largely 'colour blind', relying on brightness cues alone. Originating in the magnocellular (M-type) GANGLION CELLS of the retina, it passes through the magnocellular layers of the LATERAL GENICULATE NUCLEI (see Fig. 111), each of whose cells are activated by only one eye, hence *monocular*, then via the IVCα layer of the PRIMARY VISUAL CORTEX to the thick stripes of V2 and finally on to the middle temporal region (V5) concerned with depth and motion (see EXTRASTRIATE CORTEX, Fig. 72). This 'dorsal' pathway is concerned with *where* objects are, and for timing visual events during reading (see DYSLEXIA). It signals any visual motion that occurs if the observer's unintended movements lead to images moving of the fovea (retinal slip), and these signals are then used to bring the eyes back on target. Compare PARVOCELLULAR-BLOB/INTERBLOB PATHWAYS. For ocular dominance columns, see HYPERCOLUMNS, VISUAL CORTEX.

major histocompatibility complex (MHC; human HLA complex) A vertebrate gene complex, located in humans on chromosome 6 and spanning 3.6 Mb it contains ~1620 allele sequences at the 19 loci, among the most polymorphic in the human genome. Ancient gene duplication events have led to many pseudogenes within the MHC region. Many of the human alleles predate the human–chimpanzee lineage bifurcation, such trans-species polymorphisms being explicable only in terms of the operation of NATURAL SELECTION in preventing fixation of alleles over time (see DIVERGENCE, NEUTRAL THEORY OF MOLECULAR EVOLUTION). Much of the variation is concentrated in the exons encoding the antigen-binding groove of the protein, supporting the operation of selection. Two alternative selection pressures have been invoked to explain MHC diversity: heterozygous advantage (whereby an individual benefits from a broader range of antigen-binding specificities), and frequency-dependent selection (whereby low-frequency alleles are favoured if pathogens have evolved mechanisms enabling evasion of immune detection in individuals carrying higher-frequency alleles).

MHC genes control a person's TISSUE TYPE (class I and class II transplantation antigens); i.e., they determine how strongly or weakly a graft of one person's skin or kidney will provoke an immune response in a recipient. One or more MHC proteins is expressed on every human nucleated cell, most densely so on the surfaces of haematopoietic cells, their distributions reflecting the different functions of the effector T cells that recognize them. Known as the *human leukocyte antigen (HLA) complex* in humans, and H-2 in mice, many of their protein products are involved in the development of adaptive immunity through presentation of bacterial and viral antigen peptide fragments to T CELLS. No two humans (apart from identical twins) have the same tissue antigen constitution, this MHC diversity probably having evolved in response to selection pressure from pathogens (e.g., see EPSTEIN–BARR VIRUS, MALARIA). Since outbred human populations present a greater variety of peptides from each pathogen, not all individuals in the population will be equally susceptible to a given pathogen, one effect of this is to limit the pathogen's spread. In some mammals, MHC antigens play a part in controlling inbreeding (see later).

MHC proteins are peptide-binding IMMUNOGLOBULINS (see Fig. 99), whose expression by cells is regulated by cytokines, especially interferons, produced during immune responses. Falling into two classes, MHC class I and MHC class II, they differ from one another in their distinct subunit structures and, as a consequence, in the functional classes of T cell they interact with (see T CELLS). MHC class I molecules present peptides from viral pathogens to CD8 cytotoxic T cells and, since viruses can infect any nucleated cell, most nucleated cells express MHC class I. MHC class II molecules bind peptides derived from pathogens taken up into endocytic vesicles of macrophages, immature dendritic cells, B cells and other antigen-presenting cells and present them to CD4 T cells. Vesicles containing newly synthesized MHC class II molecules fuse with the acidified endosomal vesicles and bind peptides of the fragmented antigen on their way to the cell surface. Here they are recognized by T_H1 and T_H2 helper T cells, whose major function is to activate other cells of the immune system.

MHC class I molecules are heterodimers of a membrane-spanning α chain and a non-covalently linked β_2-microglobulin that is not encoded within the MHC locus. The α chain folds into three domains (α_1–α_3), of which α_1 and α_2 form a peptide-binding cleft that is open at only one end. This cleft binds short peptides 8–10 amino acids in length by both ends.

MHC class II molecules are non-covalent complexes of two transmembrane glycoprotein chains, α and β each with two domains (α_1, α_2 and β_1, β_2), the α_1 and β_1 domains contributing to the peptide-binding cleft, which is open at both ends. The length of the peptide that can be bound is not limited. There are three human class I α-chain genes (HLA-A, -B and -C) and three pairs of MHC class II α- and β-chain genes (HLA-DR, -DQ and -DP). In many people, the HLA-DR cluster also contains an extra β-chain gene, with its own product. Together, these genes enable any one individual to present a much broader range of peptide fragments than were only one MHC molecule of each class expressed on the cell surface. The human HLA-A gene cluster is believed to be rapidly diverging from that of the chimpanzee (see GENE CLUSTERS, Table 3). There are many loci in the MHC complex, termed MHC class IB genes, whose products resemble MHC class I-type molecules but show little polymorphism. Like MHC class I products, their products associate with β_2-microglobulin on the cell surface, although not all have obvious immunological functions (e.g., see HFE).

Genetic diversity of the MHC class I and II loci is essential for accurate recognition of a range of foreign antigens and immunological signal molecules. They are rapidly evolving loci, influenced by recombination, gene conversion, duplications and deletions, transposable element and retroviral insertions and Darwinian selection. Chimpanzees have greater overall diversity than humans in the class I loci; but both chimpanzee and human lineages have lost certain MHC II allelic lineages, sequence variability depending on the locus examined. This will reflect different population structures (e.g., EFFECTIVE POPULATION SIZE) and histories as well as responses to their environments, pathogen-mediated selection being one possible cause (possibly accounting for different human and chimpanzee responses to HIV-1/SIVcpz, hepatitis B and C, the malarial parasite *Plasmodium falciparum* and other pathogens).

The debate as to whether or not the human MHC cluster and its paralogous regions arose through chromosomal duplication at the origin of the vertebrates has been resolved in favour of *en bloc* duplication of a proto-MHC region. Application of chromosome painting (see FISH) in the cephalochordate *Branchiostoma* (amphioxus) indicates that a single chromosome in this animal bears orthologues to 31 genes in the human genome (on chromosomes 1, 6, 9 and 19), and that this proto-MHC region has been duplicated paralogously. The MHC paralogous region on human chromosome 9 even retains the ancestral organization. In mice and some other vertebrates, MHC peptide ligands can function as chemosignals and have been shown to influence maternal recognition of young, and mate choice. Whether such MHC-linked olfactory-mediated cues influence mate preferences in humans is debated; but they have probably been a factor favouring maintenance of MHC polymorphism and production of disease-resistant progeny in other vertebrates.

Alleles at adjacent MHC loci are often involved in strong linkage disequilibrium (LD), although little is known about the mechanisms responsible for this. It has recently (2006) been shown that the human MHC HLA-DR2 haplotype, which predisposes to multiple sclerosis, displays more extensive LD than other common Caucasian HLA haplotypes in the DR region and seems likely to have been maintained through positive selection (see MULTIPLE SCLEROSIS).

malaria A debilitating and often fatal disease caused by any of four species of the apicomplexan protist genus *Plasmodium*: *Plasmodium falciparum* (malignant tertian malaria); *P. vivax* and *P. ovale* (benign tertian malaria); *P. malariae* (benign quartan malaria), the only consistent clinical features being fever and rigors. *Plasmodium falciparum*, whose genome was sequenced in 2002, causes the most virulent form of human malaria (~200–300 million infections and 1–3 million deaths annually). Fevers tend to occur every three days (tertian fever), or every four days (quartan fever). Abdominal pain, headache, diarrhoea, dysuria and, commonly, sore throat and cough, are often present. Hypoxia, hypoglycaemia and convulsions are complicating factors. Severe and complicated forms of falciparum malaria include *cerebral malaria*, which presents with encephalopathy and largely affects children under four years, and *blackwater fever* in which severe intravascular haemolysis produces haemoglobinuria, with acute renal failure quickly developing in untreated cases.

Benign malarias should be treated quickly with chloroquine, to which they are rarely resistant. Uncomplicated falciparum malaria may be treated with quinine, followed by Fansidar (if quinine-resistant), clindamycin or Malarone. In many countries, severe and complicated falciparum malaria is treated intravenously with quinine and dextrose; but artesunate is now often preferred. Chemoprophylaxis for benign malarias tends to involve use of chloroquine or proguanil; for falciparum malaria it usually involves mefloquine, Malarone or doxycycline, depending on the region. It is very important to seek medical advice well before visiting regions where malaria is endemic or even sporadic, as drug effectiveness varies; and advice for pregnant mothers and neonates may also vary. Most of the ~2,000 cases in Britain annually affect travellers who have not taken prophylaxis.

An infected female *Anopheles* mosquito will inject 10–30 sporozoites of *Plasmodium* from her salivary glands into the human bloodstream during a blood meal. Within 30–60 minutes, these enter liver cells and are clinically silent, dividing asexually to form thousands of merozoites, which rupture the hepatocytes within a few days and attach to, and enter, RED BLOOD CELLS by adhering to surface antigens such as blood group determinants. It may be that because red blood cells, lacking nuclei, do not express MHC class I molecules, that *Plasmodium* is able to gain access to this privileged environment. *P. falciparum* can enter red blood cells at all stages of red cell development whereas *P. vivax* infects only reticulocytes and *P. malariae* prefers senescent cells. Once inside a red blood cell, the merozoites proliferate by binary fission, using haemoglobin as a protein and energy source and generating a waste haem pigment (haemozoin) that is deposited within special vacuoles within the parasite, in the endothelium and in surrounding tissues. This causes intense inflammation and platelet and complement activation. The haem group is photoactive and easily detectable by direct laser-desorption mass spectrometry and is a promising diagnostic marker as it builds up inside the parasite. This erythrocytic stage of the life cycle persists as daughter merozoites burst their host cells and infect new ones, haemolysis being increased by an immune response to antigens on the surface of infected cells. Glycosylphosphatidylinositol (GPI; see CELL MEMBRANE), originating in the parasite, has the properties predicted of a toxin, and is a candidate for anti-toxic vaccines. Fever, chills and progressive anaemia ensue, death often occurring from occlusion of blood vessels in the brain, lungs and other organs. In pregnancy, infected placentas restrict supplies of oxygen and glucose to the foetus. Eventually, some merozoites develop into male and female gametocytes that can be ingested by a previously uninfected mosquito (the primary host), in whose intestine they develop into gametes. After fertilization, the oocyst embeds in the gut wall, further asexual proliferation occurs, and sporozoites travel to the salivary glands for injection into a human (the secondary host).

DNA analysis suggests that *Plasmodium vivax* jumped from a Southeast Asia primate to humans about 1 Myr BP, and that although *P. falciparum* was probably infecting the earliest African hominins, the all but complete absence of SNPs polymorphisms combined with analysis of its microsatellite DNA suggests that all existing human strains emerged from a bottleneck that may have occurred as recently as 5 kya, coincident with the establishment of slash-and-burn agriculture in the African rainforest when mosquitoes and humans would have undergone population expansions (see NATURAL SELECTION). Most of its amino acid replacements are associated with drug therapy or immunological status of the human host. That *Plasmodium* infection can select for the expression of particular alleles of the MAJOR HISTOCOMPATIBILITY COMPLEX is supported by the strong association of the HLA-B53 allele with recovery from *falciparum* malaria in West Africa where the disease is endemic. The allele is rare elsewhere, where lethal malaria is uncommon. For both the *G6PD* (GLUCOSE-6-PHOSPHATE DEHYDROGENASE) and CD40 ligand (*TNFSF5*; see ANTIGEN PRESENTING CELL) gene loci, haplotypes carrying putative protective mutations show a combination of high frequency and long-range linkage disequilibrium, confirming positive selection ~10 kya, in the Neolithic era. Other evidence for the selective impact of *Plasmodium* spp. comes from studies on the HAEMOGLOBINOPATHIES, notably sickle-cell anaemia and thalassaemias.

Genes that evolve quickly are probably least suitable as targets for antimalarial drugs because they will probably continue to evolve quickly. Genes that have been relatively constant for millions of years and produce a good immune response are probably more rewarding. The presence of an unpigmented chloroplast remnant, the apicoplast, in many parasitic apicomplexans such as *P. falciparum* is a promising target for antimalarial drug design since animal cells lack plastids. See DUFFY BLOOD GROUP, GLUCOSE-6-PHOSPHATE DEHYDROGENASE, THALASSAEMIAS.

New insecticides are being developed, current mosquito bednets using one of the pyrethroid group. But the main mosquito vectors, *Anopheles gambiae* and *A. funestus*, are already developing resistance to these (there has been intercontinental spread of pyrimethamine-resistant malarial parasites). Nevertheless, evidence indicates that insecticide-treated mosquito bednets, spraying and more effective drugs have had a direct and major effect on reducing the malaria burden in Kenya and in Tanzania's Zanzibar archipelago, although there are very few studies on the impact of such control programmes. Insecticides exploiting mosquito and parasite genome sequences to target either the vector or thwart resistance to existing pesticides are being developed. Introducing transgenic mosquitoes into the wild is proving difficult, since *A. funestus* is still impossible to engineer. RNA interference (RNAi) could be developed to silence expression of specific mosquito genes, the C-type lectin gene family of *A. gambiae* being possible targets, having undergone recent evolutionary expansion.

The drug *artemisinin*, originating from

the Chinese herb *Artemisia annua* (sweet wormwood), has been used there for >1,500 years as a herbal fever remedy, has been very successful in Southeast Asia and cures 90% of patients in Africa, within three days; but it is at least ten times as expensive as established malaria treatments to which *Plasmodium* has become resistant. Hence, it is not readily available where most needed in Africa. It kills all blood-borne stages of the parasite, but is only effective for a short time so the dosage regimen can lead to patient non-compliance and subsequent failure. Its effective endoperoxide bridge, protected by bulky chemical rings, is a challenge to reproduce in candidate drugs, and most of those produced so far have had poor anti-malarial activity or are not amenable to scaling up without loss of stereoselectivity. Development of a water-soluble drug, OZ277, that can cross the gastrointestinal tract is a very promising prospect, and seems to operate by producing free radicals within *Plasmodium*-infected erythrocytes.

Chloroquine resistance in *P. falciparum* is due to amplification of the gene encoding an ABC TRANSPORTER protein that pumps the drug out of the cell. In the borders of Thailand, drug-resistant forms of *P. falciparum* are rife. Treatment here usually consists of mefloquine and artesunate given together; but resistance is a growing problem. Parasites bearing extra copies of the *P. falciparum* multidrug-resistance gene-1 (*pfmdr1*) were most likely to be resistant to mefloquine when this was given alone, or combined with artesunate.

Numerous attempts have been made to develop an anti-malarial vaccine, work which is a slow and lengthy sequence of steps rather than involving a single flash of brilliance. The problem is not intractable. Injection of birds with irradiated sporozoites (which cannot divide) can prime their immune system to target normal sporozoites, and can do so for at least 10 months in 90% of experimental human subjects. However, simple syringe injection of sporozoites raises practical problems of reduced infectivity, and the number of mosquito bites required to achieve the dose needed is in the thousands. The risk of infection by insufficiently irradiated parasites is also an issue. In rodent experimental systems using *P. berghei* and *P. yoelii*, both the T cells (mostly $CD8^+$) that target intra-hepatocytic stages, and antibodies recognizing proteins on the sporozoite surface and prevent their entering hepatocytes, are involved in protection. IFN-γ, IL-12 and NITRIC OXIDE (the latter being an effective killer of intra-hepatocytic *P. falciparum*) are crucial. Much is still unclear about how sporozoite irradiation generates protection; but although the cells do not die after entering the endosomal compartment, they do not proliferate in the hepatocytes and their continued presence there is required for protection. This suggests that continued synthesis of parasite antigens is important, and development of vaccines specifically targeting infected hepatocytes is one promising line of research.

The erythrocytic stage is the only stage that can be genetically manipulated. Knocking out the *UIS3* gene in *P. berghei* within rodent erythrocytes does not prevent the sporozoites from entering hepatocytes but does prevent them from proliferating once there. Mice immunized by ~30,000 knockout *P. berghei* sporozoites were protected against wild-type sporozoites injected one month after immunization. This raises the question of whether genetically modified *P. falciparum* might be used as a live vaccine for humans (they would need to be prevented by a 'double-crossover' strategy from genetic reversion). In the quest for a live attenuated malaria vaccine, it may become possible to produce infectious sporozoites from erythrocytic stages without requiring mosquitoes; but human erythrocytes would still be required for parasite multiplication.

Malaria diagnosis is most needed in rural areas, where there is often no electricity to power mass spectrometers. The 'gold standard' for diagnosis today is the thick and thin blood smear, which involves counting parasites in stained blood smears by microscopy; but this low-tech approach is often relatively insensitive. Small mass spectrometers can operate in inhospitable field conditions and could provide a diagnosis in seconds, although it would not discriminate the species of parasite. More finance is needed for development of sensitive and discriminative portable field-based mass spectrometers.

malate–aspartate shuttle A shuttle for transporting reducing equivalents from cytosolic NADH into the mitochondrial matrix, employed in the liver, kidneys and heart. NADH in the cytosol (and intermembrane space of the mitochondrion) passes two reducing equivalents to malate, which is transported across the inner mitochondrial membrane by the malate–α-ketoglutarate transporter. In the matrix, malate passes these reducing equivalents to NAD^+ for oxidation by the electron transport chain. Oxaloacetate formed from malate cannot pass into the cytosol directly, but is first transaminated to aspartate (see DEAMINATION), which can leave via the glutamate–aspartate transporter. Oxaloacetate is regenerated in the cytosol, which completes the cycle. See GLYCEROL for the glycerol 3-phosphate shuttle operating in skeletal muscle and brain.

males See SEX DETERMINATION.

malignant tumours Once a cell has become immortalized (see CANCER), it may undergo malignant transformation. Only partially understood at present, this process involves host and tumour cells acting in concert to produce an invasive, spreading, entity that ultimately kills its host. It must instruct its host body to produce new blood vessels (angiogenesis), evade host immune recognition and modify its microenvironment so as to seed and survive in less-than-optimal conditions. Invasion by epithelial cancer cells of the connective tissue stroma involves disruption of the processes which normally prevent intermixing of cells from different tissues. This seems to require remodelling of the cancer cell extracellular matrix.

It is becoming clear that a cancer cell must experience sequential genetic modification, overcoming the considerable growth-inhibitory mechanisms that normally prevent tumour progression (see METASTASIS, Fig. 124). Chromosome aberrations are particularly common in malignant cells, which often show marked variation in chromosome structure and number – often due to translocations – leading to novel chimaeric genes with novel biochemistry, or level of proto-oncogene activity (see ONCOGENE). In several malignant tumours, receptor dimerization has occurred through chimaeric fusion of genes encoding GROWTH FACTOR receptors to unrelated genes that happen to specify proteins that normally dimerize (see RECEPTOR TYROSINE KINASES). Dimerization of receptors is often the prelude to growth factor activation of a cell, so these tumour cells have been made constitutively active as though they were being stimulated by growth factor when in fact there is no growth factor present.

malleus See EAR OSSICLES.

malnutrition Many undernourished people have too little intake of energy and protein in their diet (see HEALTHY DIET). Protein-energy malnutrition causes weakness, weight loss, impaired immunity and other symptoms in adults; but its effects are especially devastating in children. KWASHIORKOR (swollen belly) is one result; but MARASMUS is another. In both cases, the damage to children is often too severe for complete recovery even when adequate dietary intake is restored, and those who do survive are unlikely to function at normal physiological or mental levels. Vitamin deficiency diseases include: BERIBERI (due to vitamin B_1 deficiency), PELLAGRA (niacin deficiency), XEROPHTHALMIA (night blindness) due to vitamin A deficiency; scurvy (due to VITAMIN C deficiency), RICKETS and OSTEOMALACIA (both due to VITAMIN D deficiency).

Secondary immunodeficiency and its consequences are common effects of malnutrition. See CALCIUM, IRON, MAGNESIUM, ZINC, and other mineral entries.

Malpighian layer The stratum basale of the EPIDERMIS, where mitotic division of KERATINOCYTES occurs.

maltase See DISACCHARIDASES.

mammary glands Modified sudoriferous (sweat) glands that produce milk, lying over the pectoralis major and serratus anterior muscles and attached to them by a layer of deep fascia of irregular connective tissue. Most mammary gland development takes place postnatally, under the influence of oestrogen, GH, IGF-I and epidermal growth factor. It involves branching and extension of ducts and 15–20 lobules into a fatty stroma. At the onset of puberty, oestrogen stimulates development of the nipple, some duct elements and an underlying fat pad (whose size determines the size of the breasts). Progesterone and prolactin (synergistically) and thyroid hormone promote ductal branching and alveolar budding. After puberty, during the luteal phase, progesterone promotes sprouting of alveolar structures from the ducts. During PREGNANC,, PROLACTIN, progesterone and HUMAN PLACENTAL LACTOGEN cause expansion of the terminal duct alveolar lobules and their secretory function, causing synthesis of milk proteins, enzymes involved in lactogenesis, and an increase in fat droplet size. Lactose presence is dependent upon the enzyme system comprising α-lactalbumin and GALACTOSYLTRANSFERASE (see entry for details), an enzyme present in the Golgi membranes of the mammary gland alveolar cells. Lactose is packaged together with proteins in granules that are released by exocytosis into the alveolar lumen. Elevated progesterone levels during pregnancy prevent milk production; but prolactin levels rise during the 3rd trimester, and at BIRTH progesterone levels fall and prolactin maintains continuous milk production. See BREASTS, LACTATION.

mammillary bodies Two distinct, rounded, eminences in the posterior HYPOTHALAMUS (see Fig. 98 and CEREBELLUM, Fig. 42). They are components of the LIMBIC SYSTEM and contain the mammillary nuclei, receiving afferents from the HIPPOCAMPUS, mainly via the fornix, and projecting to the anterior nuclei of the thalamus and brain stem. See PAPEZ CIRCUIT.

mandibular bone (mandible, lower jaw) A facial bone of the SKULL (see Fig. 157), developing from the 1st PHARYNGEAL POUCH (see entry) and articulating posteriorly with the temporal bone. Its two major portions are: the body, extending anteroposteriorly, and the ramus, extending superiorly from the body towards the temporal bone. Each ramus has a posterior condylar process that articulates with the mandibular fossa of the zygomatic process of the temporal bone (part of the zygomatic arch), while the coronoid process, separated from the condylar process by a depression, the mandibular notch, is the insertion of the TEMPORALIS MUSCLE. The alveolar process of the mandible contains the sockets for the lower teeth (see DENTITION, Fig. 62). The mandible is inferior to the MAXILLARY BONE. See MASTICATION.

mandibular fossa (glenoid fossa) In the temporal bone, the depression in the zygomatic process into which fits the condylar process of the MANDIBULAR BONE, forming the jaw joint (see SKULL, Fig. 157). It is bounded, in front, by the articular tubercle and behind by the tympanic part of the bone, which separates it from the external auditory (acoustic) meatus. It is divided into two parts by a narrow slit, the petrotympanic fissure.

mania See AFFECTIVE DISORDERS.

mannose-binding lectin (MBL) A pathogen-binding protein and a trigger for the MB-lectin COMPLEMENT cascade. Found in normal blood plasma at low levels, it is produced by the liver in increased amounts during an ACUTE PHASE RESPONSE, acting as an opsonin for monocytes. Mannose-6-phosphate receptors are cell-surface receptors that normally target a molecule to the lysosomal compartment of cells (see LYSOSOME), and binding by granzyme B seems to be crucial to rendering cells susceptible to apoptosis by cytotoxic T CELLS. See LECTINS.

mantle layer Layer of the embryonic spinal cord (see NEURAL TUBE), surrounding the neuroepithelial layer, and containing neuroblasts which give rise to its grey matter. Compare MARGINAL LAYER.

MAOs See MONOAMINE OXIDASES.

MAPK (mitogen-activated protein kinase) pathway Many MITOGENS activate the small GTPase Ras, which activates a MAP kinase cascade leading to increased levels of the gene regulatory protein Myc (see entry). A MAP-kinase must be phosphorylated at both a threonine and a tyrosine residue, which are separated by just one amino acid. The enzyme catalysing both of these phosphorylations is MAP-kinase-kinase (MAPKK, aka MEK). This is itself activated by MAP-kinase-kinase-kinase which, in the mammalian Ras signalling pathway, is known as Raf. Raf is activated by activated Ras. Neurotrophin receptor activation uses this pathway, activating CREB (see Fig. 60). The adapter protein Shc, an upstream regulator of the MAPK pathway, is one of the targets of c-Src (see SRC GENE).

MAP-kinases See MAPK PATHWAY.

maple syrup (maple sugar) urine disease A rare autosomal recessive disorder (chromosome site 19q) caused by defective enzyme branched-chain keto-acid dehydrogenase. Results in lethargy, irritability and vomiting, progressing to coma and death in the first 1–2 months if untreated. Urine smells sugary. Occurrence <1:250,000 in the USA. Treatment involves dietary restriction in branch-chain amino acids (leucine, isoleucine, valine). The NEWBORN SCREENING procedure is as for phenylketonuria, although leucine can be measured by the bacterial inhibition assay.

marasmus A body-wasting disease of very young children resulting from protein-energy MALNUTRITION, common in bottle-fed infants of poor families in which parents dilute prepared formula milk. Symptoms include low weight, muscle wasting, dry hair and skin, and retarded growth and mental development.

MARCKS (myristoylated alanine-rich C-kinase substrate) A natively unstructured (unfolded) protein, with a high fraction of negatively charged residues and few hydrophobic residues, involved in control of free PHOSPHATIDYLINOSITOL-4,5-BISPHOSPHATE (PIP_2) levels in the plasma membrane and associated peripheral regions of the cell. Despite its negative charge, it binds to the negatively charged phospholipid membrane by two membrane anchors. In the process, it draws three membrane PIP_2 molecules into close proximity beneath it. But at higher cytoplasmic Ca^{2+} concentrations, when calcium/calmodulin binds to the protein, MARCKS is pulled off the membrane, freeing and exposing the PIP_2 molecules. It is a major substrate for protein kinase C (PKC) in most mammalian cells, its continued detachment from the membrane being favoured by PKC phosphorylation of its terminal residues. See MYRISTOYL GROUP.

marfanoid physique Describing a person who is tall and thin, with the subjective features of long limbs, toes and fingers. Associated with HOMOCYSTINURIA. See MARFAN SYNDROME.

Marfan syndrome (MFS) Clinicians consider diagnosis of Marfan syndrome for any patient with MARFANOID PHYSIQUE. A disorder of fibrous connective tissue, in particular a defect in the glycoprotein type 1 fibrillin encoded by the *FBN1* gene. Compared to family members, affected individuals have laxity of the joints, a span:height ratio >1.05, a reduced upper to lower segment body ratio, pectus deformity, lens subluxation (ectopia lentis) and scoliosis. Dilatation of the ascending aorta (see CONGENITAL HEART DEFECTS), a life-threatening complication, can lead to its dissection but can be alleviated by β-adrenergic blockade (if tolerated) although surgical replacement may be needed. Pregnancy is a risk factor for a woman with MFS who already has some aortic dilatation. Diagnosis of MFS requires careful clinical assessment and may involve lumbar MRI in doubtful cases.

The disorder has an autosomal dominant inheritance pattern, most associated with lesion in the *FBN1* locus on chromosome

15q21, a large gene with 65 exons spanning 200 kb and five domains. The largest domain (~75% of the gene) comprises ~46 epidermal growth factor repeats. See SCREENING.

marginal layer The outermost layer of the developing SPINAL CORD, surrounding the mantle layer and basal plate, and giving rise to its white matter.

marijuana See CANNABINOIDS.

marker genes See GENETIC MARKERS.

masseter muscle Muscle running from inside the zygomatic arch and inserting on the outside and rear of the lower jaw, pulling jaws forward and up (protraction), and assisting in side-to-side motions. Compare TEMPORALIS MUSCLE.

mast cells Large myeloid cells found in high concentrations in vascularized connective tissues lying just beneath epithelia, notably those of the gastrointestinal and respiratory tracts. When activated by binding circulating monomeric IgE at their FcεRI receptors, and subsequent cross-linking the IgE by multivalent antigen, they release the inflammatory mediators HISTAMINE, cytokine TNF-α and HEPARIN from their cytoplasmic granules (degranulation) and secrete the lipid mediators of INFLAMMATION (prostaglandin D_2 and leukotriene C4). The resulting increase in blood flow and permeability of the vascular ENDOTHELIUM leads to accumulation of fluid and blood proteins (e.g., antibodies) in the local tissue. Their armoury also includes protein-degrading enzymes (e.g., chymase and tryptase). TNF-α leads to recruitment of lymphocytes and other leukocytes, including eosinophils and T_H2 lymphocytes, to the site of inflammation, linking the innate and adaptive immune systems. TNF-α also leads to migration of dendritic cells to LYMPH NODES, enlargement of these organs, and recruitment of lymphocytes to them so that encounters between lymphocytes and APCs are encouraged. IL-9 is a mast cell growth and differentiation factor, and is of importance in the role of mast cells in the development of immune tolerance, where they serve as enforcers for T_R cells (see IMMUNOSUPPRESSION). In mice, at least, they also neutralize toxins from snake bites and bee stings. Among the proteins mast cells degrade is endothelin-1, which can produce sepsis under some circumstances. Mouse models indicate that this can protect them from the toxic effects of endothelin-1, enabling them to avoid septic shock. Mouse mast cells also protect them from the toxins in rattlesnake and bee venoms, carboxypeptidase A contributing to the former. See ALLERGIES, ASTHMA.

mastication (chewing) The muscles of mastication include the TEMPORALIS, MASSETER and medial pterygoid muscles which close the mouth (elevating the mandible) and grind the food between the teeth (medial and lateral excursion of the mandible) in which all muscles of mastication participate. Opening of the mouth, other than by relaxation of the elevators of the mandible and the force of gravity, requires contraction of the mandibular depressors: the lateral pterygoids, digastric, geniohyoid and mylohyoid muscles. The hyoid muscles attach to the HYOID BONE and are also involved in movement of the mandible. The trigeminal nerve innervates the main elevators and the trigeminal and hypoglossal nerves innervate the hyoid muscles.

mastoid process A knob-like process of the temporal bone (see SKULL, Fig. 157) forming an attachment surface for neck muscles rotating the head and for a hyoid muscle. It is not solid bone, but is filled with cavities (mastoid air cells) connected to the middle ear.

maturation See GROWTH.

maxillary bone (maxilla) Developing from the 1st PHARYNGEAL POUCH (see entry), one of the two bones forming the upper jaw (the other being the premaxilla). The premaxilla bears the incisors, and in humans the maxilla covers the premaxilla. It articulates directly with every bone of the face other than the mandibular bone, forming the floor of the ORBIT, part of the lateral walls and floor of the nasal cavity, and most of the hard PALATE. Each contains a large maxillary sinus emptying into the nasal cavity, its alveolar process containing the sockets for the upper teeth (see DENTITION).

maximal oxygen consumption ($\dot{V}O_{2max}$, aerobic capacity, maximal oxygen uptake, maximal aerobic power) The plateau of oxygen consumption (mL kg^{-1} min^{-1}) when this rate is plotted against gradually and successively increasing treadmill gradient (work rate). Performing more-intense EXERCISE results solely from energy transfer from glycolysis, with resulting lactate accumulation. It provides a measure of a person's capacity for aerobic ATP resynthesis, an important determinant in a person's ability to sustain high-intensity exercise for >4–5 minutes. See CARDIOVASCULAR DISEASE.

***MC1R* gene** See SKIN COLOUR.

M-Cdk see cyclins.

Mdm2, MDM2 A protein initially identified as one encoded by double-minute chromosomes in mouse sarcoma cells (hence mouse double minutes). The human homologue of the *mdm2* gene (*MDM2*) was subsequently found to be amplified in sarcomas, notably in human lung tumours. It is a major agent of both p53 and pRB destruction (see P53 GENE, RETINOBLASTOMA PROTEIN) and can act as an oncoprotein (see CELL CYCLE, Fig. 40).

MDMA (3,4-methylenedioxymethamphetamine, 'ecstasy') An amphetamine-like drug with mixed stimulant and HALLUCINOGENIC effects, the latter possibly being due to SEROTONIN release. There is increasing evidence that long-term use destroys serotonin receptors or serotonergic terminals and increases the risk of psychiatric disorders. See PSYCHOACTIVE DRUGS.

MDMX A protein, structurally related to MDM2, both of which act as antagonists of the p53 signalling pathway.

ME (myalgic encephalomyelitis) Often combined with chronic fatigue syndrome (CFS) as CFS/ME, ME is not a conventional illness, lacks diagnostic tests, has no treatment, and tends to be over-diagnosed; some practitioners at least recognize it as a genuine distressing, debilitating condition. An extensive literature exists on non-specific but possible causes, including up-regulation of anti-viral pathways, immune abnormalities, disruption to the HPA AXIS, neuropsychological impairments, oxidative stress and dysfunction of the autonomic nervous system. A significant body of evidence points towards vascular dysfunction in the peripheral circulation of CFS/ME patients, with orthostatic intolerance (a cluster of symptoms, e.g., dizziness, altered vision, nausea, fatigue, headache and sweating, associated with standing upright) being a characteristic of many patients.

measles A systemic infection caused by an enveloped paramyxovirus (see VIRUSES), 120–200 nm in diameter, of a single serological type which only infects primates. Its helical nucleocapsid contains one single-stranded RNA molecule, coated with a protein and associated with an RNA-dependent RNA polymerase. Encoding six structural genes, the genome encodes three major envelope antigens: M, H and F proteins. The H glycoprotein antigen is genetically the most variable and responsible for HAEMAGGLUTINATION and adsorption of the virus with host cell surface receptors. The F glycoprotein mediates fusion with the host cell membrane and haemolysis. No neuraminidase is encoded. Viral replication in the respiratory mucosa is followed by a primary viraemia when the virus enters the reticuloendothelial cells, causing secondary viraemia and suppressing the immune system, making those already weakened by hunger, poverty or disease susceptible to pneumonia, diarrhoea and acute encephalitis. It can invade most cell types although infection of the respiratory mucosa makes this liable to secondary bacterial infections. Measles encephalitis is thought to result from hypersensitivity to the virus leading to demyelination, gliosis and infiltration of fat-laden macrophages around cerebral blood vessels.

A highly infectious systemic viral infection, and notifiable disease, the main features of measles are respiratory disease and rash. In unprotected populations it tends to spread in large epidemics, affecting mainly children, and causes devastating outbreaks in refugee camps and other populations lacking organized healthcare and vaccination: it will emerge and spread when immunization rates are <95%. Where available, live attenuated measles vaccines are offered to children at 12–15 months as part of the measles/mumps/rubella (MMR) preparation, prior to which age there is poor immune response. Although MMR is contraindicated in children with immunodeficiency disorders, children who are HIV-positive should be vaccinated since risks from measles are greater than those from the vaccine. >95% of recipients of MMR require only a single dose for protective antibody production, minor side-effects (fever, loss of appetite and sometimes a rash) occurring in 5–10% of recipients; but febrile convulsions may occur. Very rarely (~1:10^7 doses), subacute sclerosing panencephalitis (SSPE) occurs after vaccination; but the risk of neurological complications is 10- to 100-fold greater after the disease than after the vaccine. Globally, deaths from measles fell by 39% from 1999 to 2003 (from 873,000 to 530,000), and by 46% in Africa, largely through improved immunization. The WHO and Unicef have set a new target to cut deaths by 90% by 2010 from the 2000 level.

Secondary bacterial infection is common and frequently severe, while post-measles encephalitis is life-threatening. After an incubation of 8–13 days, clinical features include fever and respiratory symptoms, extreme irritability and febrile convulsions being common in the early stages. Conjunctival inflammation, haemorrhagic chest rash, running eyes and nose and persistent cough and, commonly, diarrhoea follow. There is no specific treatment for acute measles, but secondary bacterial infection is responsible for most of the 5–20% fatality rate in developing-world outbreaks. See DISEASE.

meatus A tube-like opening in a bone.

mechanoreceptors Sensory receptors (see Fig. 120) responding to mechanical forces and involved in the senses of HEARING (see HAIR CELLS of the COCHLEA) and TOUCH, in the control of balance (see VESTIBULAR APPARATUS), as well as in PROPRIOCEPTION (see entry), the detection of osmotic pressure gradients (see OSMOREGULATION) and visceral stretch (see STRETCH RECEPTORS). Transient receptor potential (TRP) membrane channel proteins sense vibration, touch and osmotic membrane stretch. At least one associated ion channel, the polymodal K^+ channel, can be activated by both force and osmolarity, and by the lysophospholipid LPC. Cutaneous low-threshold mechanoreceptor primary afferent AXONS (see entry) relay skin mechanoreceptor signals via the dorsal roots of SPINAL NERVES to synapse with dorsal horn cells in laminae III–VI (see SPINAL CORD, Fig. 157, for route taken by primary afferents to the spinal cord). Mechanoreceptors present in the skin include MEISSNER'S CORPUSCLE, the HAIR FOLLICLE receptor, MERKEL CELLS, the PACINIAN CORPUSCLE and the RUFFINI END ORGAN.

Meckel syndrome A human disorder associated with neural tube defects, caused by mutations in genes implicated in the functions of cilia (see CILIUM).

Meckel's cartilage See PHARYNGEAL ARCHES.

MeCP2 (methyl-CpG binding protein 2) Initially identified in 1992 through its ability to bind methylated DNA, MeCP2 (product of the *MeCP2* gene) was believed to be a transcription repressor – disrupting mutations which were expected to increase expression of target genes. One *MeCP2* mutation gives rise to a spectrum of related, but distinct, postnatal neurodevelopmental disorders including RETT SYNDROME and AUTISM spectrum disorders.

Its action is now thought to depend on the methylation state of target promoter sequences, and that it may also have a transcription activating role. Work on mice indicates that thousands of genes in the

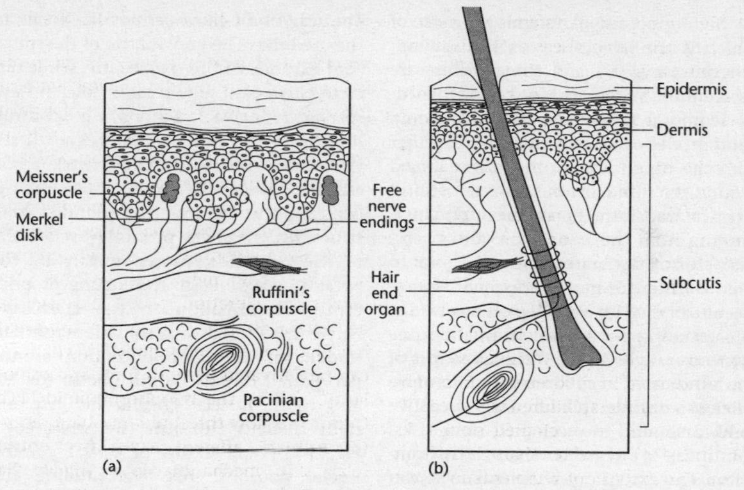

FIG 120. The distribution of various MECHANORECEPTORS in, and underlying, (a) smooth, and (b) hairy skin. © 2006, Oxford University Press. Redrawn from *Human Physiology: The Basis of Medicine*, G. Pollock & C.D. Richards, with permission from the publisher.

hypothalamus of *MeCP2*-deficient, wild-type and *MeCP2*-overexpressing animals undergo small but important changes in expression, often in opposite directions with deletion or over-expression of the gene. Promoters of genes bound by MeCP2 are enriched for CpG islands (see CPG DINUCLEOTIDE) compared with those repressed by MeCP2, the activated promoters being lightly methylated compared with those repressed. This suggests that MeCP2 repressor/activator function may depend on the DNA METHYLATION status of the relevant target sequence. The transcriptional activator CREB is associated with MeCP2 on the chromatin of promoters it activates, and both proteins are phosphorylated in response to neuronal activation – at least in the mouse hypothalamus. Activity-dependent transcription coordinates the processes of postnatal synaptic development and NEURAL PLASTICITY, which may be disrupted in disorders of cognitive function.

medial Of structures closer to the MIDLINE than others; contrast LATERAL.

medial geniculate nucleus (MGN) A nucleus of the thalamus receiving input from the INFERIOR COLLICULUS and projecting to the PRIMARY AUDITORY CORTEX.

medial lemniscus A midbrain tract of fibres of the ascending sensory pathway, in which axons of second-order neurons from the medulla pass through the pons and midbrain before terminating in the ventroposterolateral nucleus of the thalamus. It conveys contralateral somatosensory information. See DORSAL COLUMN-MEDIAL LEMNISCAL SYSTEM; SPINAL CORD, Fig. 163. Compare LATERAL LEMNISCUS.

medial motor pathways Comprising the VESTIBULOSPINAL TRACT and RETICULOSPINAL TRACT, pathways originating in the medulla and terminating in the intermediate and ventral grey matter of the SPINAL CORD. They are components of POSTURAL REFLEX PATHWAYS (contrast LATERAL MOTOR PATHWAYS, which mediate voluntary movements).

medial prefrontal cortex (MPFC) See PREFRONTAL CORTEX.

medial temporal cortex (MT, V5) The medial and underside of the temporal lobe of the BRAIN (see Fig. 29) is occupied by the hippocampal formation, most of which is the HIPPOCAMPUS and subiculum, but the amygdala and rhinal fissure are adjacent to this region. Monkeys with medial temporal lesions appear to provide a good model of human amnesia. Together with the hippocampus, the cortex in and around the rhinal fissure seems to perform a critical transformation of the information arriving from the association cortex, possibly storing declarative memories prior to their transfer for more permanent storage (see MEMORY). MT is also a region involved in late-processing of visual information (see ATTENTION). Cells in the V2 thick stripes of the EXTRASTRIATE CORTEX (see Fig. 72) have motion-sensitive and binocular receptive fields and send most of their output, via V3, to the MT visual cortex, V5. This constitutes an extension of the magnocellular pathway, and lesions in V5 cause akinetopsia (see VISUAL MOTION CENTRE). MT visual cortex projects to the medial superior temporal (MST) cortex and the posterior parietal cortex via the 'dorsal stream'. See BASAL GANGLIA, Fig. 18.

mediastinum The broad, median partition between the pleurae of the lungs, separating the thoracic cavity into two anatomically distinct chambers. It extends from the sternum to the vertebral column of the thoracic cavity and from the neck to the diaphragm and contains the heart, thymus, trachea, oesophagus as well as blood vessels and nerves. The lungs lie either side of the mediastinum. See BODY CAVITIES.

medulla The central part of an organ, typically where the outer part is termed the *cortex*, as in the adrenal gland and kidney (see LYMPH NODE). Often used as an abbreviation for the MEDULLA OBLONGATA.

medulla oblongata The region of the hindbrain caudal to the PONS, ~3 cm in length. Bilateral swellings, the pyramids, on its ventral surface are caused by the fibres of the paired CORTICOSPINAL TRACTS (often termed the pyramidal tracts, being broader towards the pons and narrower towards the spinal cord), 85% of whose fibres decussate in the caudal medulla. The origins of CRANIAL NERVES IX–XII lie in the medulla. The grey matter of the spinal cord extends through its centre, while discrete clusters of grey matter (nuclei) have specific reflexive functions, e.g., controlling heart rate (see CARDIOVASCULAR CENTRE), VENTILATION (see RESPIRATORY CENTRE), BLOOD PRESSURE, SWALLOWING, VOMITING, hiccupping, COUGHING and sneezing. Two rounded, oval, structures, the olives, protrude just laterally to the pyramids, caused mostly by the INFERIOR OLIVARY NUCLEI, conducting proprioceptive information to the CEREBELLUM. Nuclei associated with somatic sensations (touch, vibration, proprioception) are the two paired dorsal column nuclei: the left and right nucleus gracilis and left and right nucleus cuneatus (see SPINAL CORD, Fig. 163), and axons of dorsal column nuclei neurons cross the midline and ascend in the medial lemniscus to terminate in the ventroposterolateral thalamus en route to the somatosensory cortex (see DORSAL COLUMN–MEDIAL LEMNISCAL SYSTEM). Two other nuclei, the COCHLEAR NUCLEI and SUPERIOR OLIVARY NUCLEI, are both involved in HEARING. The NUCLEUS TRACTUS SOLITARIUS receives axons of primary afferent cell neurons involved in TASTE and project either to the hypothalamus or thalamus, while reticular nuclei receive input from the mesencephalic locomotor region (see CENTRAL PATTERN GENERATORS), whence axons run in the RETICULOSPINAL TRACTS.

megakaryocytes Large, highly polyploid, cells of red bone marrow, produced from erythrocyte/megakaryocyte progenitor cells, and giving rise to PLATELETS by a pinching-off process (cellular autotomy). Lying close to endothelial cells of a sinus, each sends long processes between adjacent cells from which each buds off ~10^4 platelets. Thrombopoietin is thought to be the growth factor inducing their proliferation. See STEM CELLS.

meiosis The nuclear division, occurring during OOGENESIS and SPERMATOGENESIS, which halves the chromosome number and produces genetic variation by crossing-over between maternal and paternal chromosomes (see Fig. 121) and by the independent assortment of non-homologous chromosomes at first anaphase. The meiosis-specific heterodimer MSH4/MSH5 stimulates meiotic crossovers (see MSH2), a form of general RECOMBINATION, which begins with the breakage of both strands of the double helix in one of the recombining chromosomes. The two new ends are subject to degradation, exposing two single strands with overlapping 3′ ends, which seek out a region of unbroken duplex DNA having the same nucleotide sequence, thereby producing a four-stranded structure termed a Holliday junction. New DNA synthesis using the two unpaired strands of the homologous chromosomal DNA as templates occurs, and two more DNA breaks and rejoinings are required so that the original maternal and paternal chromosomes end up having different DNA sequences from those with which they started. The genetically varied haploid nuclei are then prepared for possible FERTILIZATION.

Meissner's corpuscles (tactile corpuscles) Rapidly adapting MECHANORECEPTORS (see Fig. 120) of the skin, ~1/10th the size of PACINIAN CORPUSCLES and lying just beneath the Malpighian layer in the dermal papillae of the DERMIS of glabrous skin (e.g., the raised parts of fingertips). Two-point discrimination (fine TOUCH), the ability to discriminate simultaneous stimulation at two points on the skin, differs between regions of the body surface, Meissner's corpuscles being numerous and close together in the tongue and fingertips but less numerous and less densely distributed in other regions, such as the back. They respond best to pressure vibrations of ~50 Hz and contribute to the sense of a surface's texture.

MEK (mitogen-activated protein kinase kinase) A MAPK enzyme in the *Ras* signalling pathway, directly activated by the *Braf* gene product (see RAS SIGNALLING PATHWAY, Fig. 145).

melanin-concentrating hormone (MCH) A cyclic peptide and orexigenic neurotransmitter of a group of neurons of the lateral HYPOTHALAMUS (see Fig. 98) whose inputs include projections from leptin-sensitive cells of the arcuate nucleus. These neurons are well placed to inform the cortex of plasma leptin levels and may contribute to motivation of food-seeking behaviour. MCH and orexin levels in the brain rise when plasma leptin level falls. MCH and α-melanocyte stimulating hormone (α-MSH) have antagonistic effects on several physiological functions, and whereas the former increases feeding in mammals, the latter reduces it (see APPETITE). In teleost fish, MCH causes lightening of skin pigment cells by causing pigment aggregation.

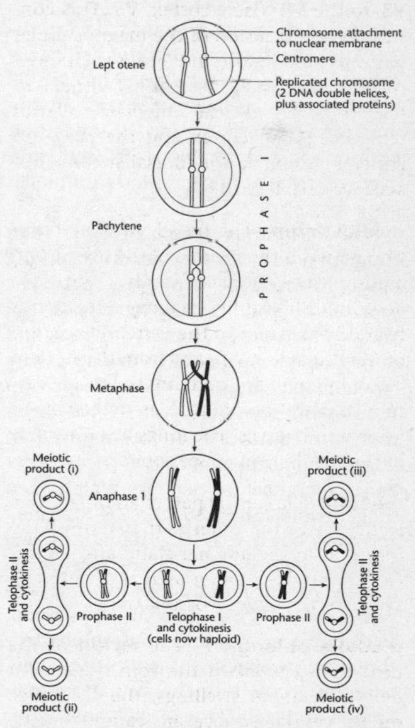

FIG 121. Diagram showing the behaviour of one pair of homologous chromosomes during MEIOSIS. One cross-over results in an observable chiasma, visible at metaphase I. From that stage onwards, the two chromosomes are shown black or white to indicate the origins of segments.

melanins Derivatives of the tyrosine metabolite dopaquinone, melanins are the most noticeable pigments in hair and skin, being insoluble and polymeric: *eumelanin* (dopaquinone polymerized, then conjugated to protein) is brown-black, *phaeomelanin* (dopaquinone combined with thiol-containing cysteine or GLUTATHIONE, then polymerized and conjugated to protein) yellow/red, responsible not only for skin colour, but also for pigmentation of hair and, to a large extent, the iris of the eye. Multiple signalling pathways affect production of melanins, which are important in protecting DNA in the epidermis and dermis from ultraviolet radiation, and in maximizing VISUAL ACUITY by controlling light scatter in the RETINA. Melanin is packaged and delivered to KERATINOCYTES by melanosomes, the protease-activated receptor-2 (PAR2), a serpentine receptor on keratinocytes, having a central role. Once inside keratinocytes, melanosomes are distributed, and in response to UV-irradiation positioned strategically over the 'sun-exposed' sides of nuclei to form umbrella-like caps.

Phaeomelanin is more photolabile than eumelanin and can then produce, among other by-products, hydrogen peroxide, superoxide and hydroxyl radicals, all known to trigger oxidative stress and cause possible further DNA damage. Only one human gene, *MC1R* (for melanocortin 1 receptor; see MELANOCYTES), has been shown unequivocally to be involved in normal human pigmentation. The receptor binds a number of peptides which then trigger eumelanin production within melanocytes; but inactivating mutations lead to phaeomelanin production. Most melanocytes produce both melanins, the ratio being affected by enzyme expression, availability of tyrosine, and sulphydryl-containing reducing agents (e.g., glutathione) in the cell. The formation, maturation and trafficking of melanosomes is crucial to pigmentation. Different variants of the *MC1R* gene are more likely to be found in people with red hair and fair skin, who tan poorly and should stay out of the sun; and variants are also associated with melanoma and non-melanoma skin cancer – possibly independently of their role in determining fair skin. *MC1R* is also the major 'freckle gene' (see SKIN COLOUR), 60% of cases of freckles (ephelides) being associated with variations in this gene, whatever the skin type or hair colour. Neuromelanin is a dark iron-binding pigment, possibly a byproduct of monoamine neurotransmitter synthesis, which accumulates with age in neurons of four deep nuclei of the brain where these monoamines are produced: the substantia nigra, locus coeruleus, the dorsal motor nucleus of the vagus nerve and the median raphe nucleus of the pons. See ALBINISM, Α-MELANOCYTE STIMULATING HORMONE, MELANOMAS, NEURAL DEGENERATION, PARKINSON DISEASE, TANNING.

melanocortins Derived from PROOPIOMELANOCORTIN, these include Α-MELANOCYTE STIMULATING HORMONE, β-MSH and γ-MSH. The highly polymorphic melanocortin-1 receptor (MC1R) is a G PROTEIN-COUPLED RECEPTOR with at least 30 allelic forms, whose agonists activate adenylate cyclase (see MELANINS, SKIN COLOUR). Resulting cAMP increase leads to phosphorylation of CREB, genes then activated including the microphthalmia transcription factor gene, *MITF*, which is pivotal in the expression of numerous pigment enzymes and differentiation factors. Agonists of MC1R include corticotropin and α-MSH, causing increased MELANIN production through raised cAMP levels. Human *MC1R* is highly polymorphic, with at least 30 allelic forms. The melanocortin-4 receptor (MC4R) is located in the paraventricular nucleus of the hypothalamus, where it modulates energy metabolism and is mutated in 4% of cases of severe OBESITY (see APPETITE, ARCUATE NUCLEUS).

melanocytes The pigment cells of amniote skin, originating in the embryonic NEURAL CREST and migrating to the skin at around 12 weeks as melanoblasts, where they either differentiate into melanocytes or remain as melanocyte stem cells (MSCs). Various signalling pathways and transcription factors regulate melanocyte migration very precisely, mutations in the genes involved leading to hypopigmentation arising from lack of melanocytes rather than lack of pigment in viable melanocytes. Key genes in these pathways include *PAX3*, *SOX10*, *MITF*, endothelin-3 and endothelin receptor B (*EDNRB*). Melanocytes produce MELANINS (see entry) contained within granules called *melanosomes* which can pass into the other cells of the skin (e.g., hair follicle cells and KERATINOCYTES), although they remain within melanocytes in cells of the iris. Located around hair bulbs and in the basal layer of the EPIDERMIS (see Fig. 71); the iris; retina; inner ear and membranes lining the mouth and protecting the brain and spine. MSC self-renewal appears to require the microphthalmia-associated transcription factor (MITF); but MSCs in HAIR FOLLICLES which carry the amplified *MITF* gene (or other key mutations) may evade conventional chemotherapy by surviving in a dormant state and recurring as lethal metastatic MELANOMA. Hair follicle MSCs have important roles in both normal hair pigmentation and senile hair greying, and specific genetic defects have shed light on the survival properties of this cell population. Melanocytes are needed in the normal development of the cochlea, and there is an association between deafness and one form of pigmentation abnormality caused by mutation in the *PAX3* gene (see WAARDENBURG SYNDROME). Dysregulation of melanocyte growth or survival is a very common occurrence. Most people harbour multiple MOLES that only rarely (if ever) transform into invasive melanomas. See SKIN COLOUR.

α-melanocyte stimulating hormone (α-MSH) An anorectic (and melanocortin) peptide neurotransmitter produced from PROOPIOMELANOCORTIN and released by neurons of the ARCUATE NUCLEUS in response to presence of leptin from adipose tissue. It decreases food intake by binding the melanocortin receptor MC4 receptors (MC4R) in the lateral hypothalamus (see APPETITE, OBESITY). It also binds to the MC1 receptor (MC1R) on the human MELANOCYTE membrane, activating the tyrosinase required for production of MELANINS (see SKIN COLOUR, TANNING).

melanomas Extremely aggressive malignant tumours of MELANOCYTES in which these cells pass through several pathological stages, associated with sequential mutations. Mutated *BRAF* (see RAS SIGNALLING PATHWAY) is found in 60% of melanomas and is associated with formation of naevi (MOLES), while loss of p^{INK4a} product is associated with formation of melanoma stem cells and invasiveness. These seem to be prior stages to a melanoma becoming metastatic and invading other organs (e.g., lung) via the blood. Melanocytes have a master regulating gene product, MITF (microphthalmia-associated transcription factor), whose gene (probably an oncogene) is amplified in about 10% of primary melanomas and 20% of metastatic melanomas. Its targets include Bcl-2 (an inhibitor of apoptosis), cyclin-dependent kinase 2 (Cdk2), a cell-cycle regulator, and the cytokine receptor gene *c*-Met. Together, they probably contribute to MITF's dual role in melanocyte differentiation and survival/proliferation, while some probably serve as surrogate drug targets for the MITF oncogenic pathway. Another pathway emerging as important in melanoma is the PI(3)K pathway (see PHOSPHATIDYLINOSITOL-3-OH-KINASE), and function of the tumour-suppressor gene PTEN is lost in 5–10% of melanomas. Treatment is difficult on account of resistance to cytotoxic drugs (arising from highly motile cells that have enhanced survival properties). Most drugs operate by inducing apoptosis in malignant cells. Postoperative adjuvant therapies involve IFN-α, high-dose IL-2 (toxicity a problem). Dacarbazine (DTIC) is the approved chemotherapeutic agent for advanced melanoma, while several single-agent drugs and immunotherapies have been approved, although contributing little to overall patient survival. NRAS and BRAF are validated chemotherapeutic targets in melanoma, and the multi-kinase inhibitor sorafenib targets BRAF and when combined with carboplatin and paclitaxel shows encouraging results. See CANCER, CHEMOTHERAPY, SKIN CANCER, ULTRAVIOLET IRRADIATION.

melanopsin Discovered in 1998 in light-sensitive cells of frog skin, a photopigment demonstrated two years later to be present in a subtype of ganglion cell in the mammalian (and human) RETINA where it is involved in the regulation of circadian rhythms (see CIRCADIAN CLOCKS). These cells are intrinsically photosensitive and adjust their sensitivity in accordance with the recent history of both light and DARK ADAPTATION. Targets of the melanopsin signalling system include the suprachiasmatic nucleus and the olivary pretectal nucleus involved in the PUPILLARY LIGHT REFLEX. Light entrainment is less effective in mice with gene knockout for melanopsin.

melanosomes See MELANOCYTES.

melatonin An amphiphilic hormone derivative of SEROTONIN. It is produced by the PINEAL GLAND under stimulation from the SUPRACHIASMATIC NUCLEUS and secreted during darkness (production being inhibited by daylight), a pattern controlled by sympathetic stimulation regulated by light signals from the retina carried via the retinohypothalamic tract. Transported across the blood–brain barrier, feedback effects of melatonin modify activity of the suprachiasmatic nucleus, inducing drowsiness and loss of alertness (see SEASONAL AFFECTIVE DISORDER) and affecting SLEEP–WAKE CYCLES. As an antioxidant, it is five times more effective than vitamin C and twice as effective as tocopherols (vitamin E), and its metabolites are also strong antioxidants. Photoperiodic control of melatonin secretion is apparently unimportant in human reproduction. The amount of melatonin synthesized decreases with age, and women with breast cancer tend to have reduced melatonin levels. This is paralleled by statistics on other cancer subtypes; but it is not clear whether the low melatonin raises free radical levels and increases cancer risk, or whether it is a consequence of the cancer. Animal studies indicate that the onset of PUBERTY can be influenced by melatonin suppression – e.g., by constant light regimes. A melatonin-based contraceptive pill for women, B-Oval, seems to reduce LDL cholesterol levels by 10–20%. See CIRCADIAN CLOCKS.

membrane attack proteins See COMPLEMENT.

membrane bones (dermal bones) Bones developing directly from mesenchyme by intramembranous ossification rather than from pre-existing cartilage (cartilage bones). In humans and other vertebrates, they are largely confined to the flat bones of the cranial vault (see SKULL) and MANDIBLE.

membrane metalloproteinases (MMPs) Cell surface enzymes which degrade EXTRACELLULAR MATRIX molecules. See TUMORIGENESIS.

membrane potential See RESTING POTENTIAL.

membranes Human cells have a variety of membranes: the cell membrane, membranous organelles, and VESICLES.

membrane trafficking The movement of proteins and lipids between different cell compartments. See MEMBRANES, VESICLES.

membranous labyrinth A paired series of sacs and tubes, all lined by epithelium and containing endolymph, and comprising most of the organs of the inner EAR (e.g., COCHLEA, VESTIBULAR APPARATUS, see Fig. 58). Each lies within the bony labyrinth of the temporal bone, surrounded by perilymph.

memory The molecular mechanism by which information gained through LEARNING is stored, acquisition of which can only be assessed through some external measure of modified behavioural performance. For this reason, memory research is bound up with the study of learning. Human memory is a polygenic cognitive trait and a fundamental brain function which appears to be capable of study independently of other higher cognitive abilities. It is ultimately made possible by neural plasticity, as described below. Heritability estimates of ~50% indicate that naturally occurring genetic variability has an important impact on its effectiveness, most contributory genes as yet identified being involved in the same signal transduction pathways employed for non-memory purposes in other cells. It is likely that protein ISOFORMS exist, specific to memory-storing structures in the brain such as the HIPPOCAMPUS (see also COREGULATORY PROTEINS). Memory loss is a feature of many neurodegenerative diseases, but inhibition of HISTONE DEACETYLASES in a mouse model has been shown to induce recovery of learning and memory (see NEURAL DEGENERATION). This work also showed that another known memory-boosting manipulation – environmental enrichment through exposing the animals to a variety of experiences over their lifetime – improves the memory of the genetically engineered mice by increasing the levels of histone acetylation in their hippocampi.

In 1894, the great neuroanatomist Ramón y Cajal (and subsequently Hebb) expressed the view that memory is stored in experience-dependent changes in synaptic strength (neural plasticity), notably in the growth of new connections between neurons. More recent research, especially from 1973, has vindicated his view. It is now believed that learning results from changes in the strengths of synaptic connections between very precisely interconnected cells (see AXONAL GUIDANCE, SYNAPSES), and that the NMDA RECEPTOR, which possesses the synaptic coincidence-detection function, is the key molecular switch for the acquisition, consolidation and storage of memories in the mammalian brain. Experience alters the strength and effectiveness of these chemical connections via signal transduction pathways that involve somewhat complex phosphorylations and dephosphorylations of intracellular and membrane proteins. The ultimate goal of memory research is to produce a comprehensive model of memory storage that flows from molecules to behaviour, with all intermediate steps (e.g., neural plasticity) defined. Such reductionist understanding does not yet exist for any form of memory in any model organism. We are also unable to explain the correspondence (if any) between neural events and the contents of our experiences. Understanding of the molecular mechanisms underlying memory consolidation is still elementary, although non-REM SLEEP seems to be involved in at least some forms.

The early model for the biochemical study of learning and memory involved modifications to the very precisely wired, and apparently invariant, neural circuitry of the gill withdrawal reflex in the giant nudibranch mollusc *Aplysia* studies by Eric Kandel and colleagues; more specifically, the synaptic connections between sensory neurons from the gill and siphon, and motor neurons in the abdominal ganglion (see Fig. 122). One key experimental discovery with *Aplysia* was that the animal remembers an aversive stimulus (a shock to its tail) and learns to enhance its defensive reflex responses ('gill withdrawal', in which both the gill and siphon retract) to a *variety* of subsequent stimuli such as a gentle stimulus to the siphon below the strength needed to elicit gill withdrawal on its own (termed BEHAVIOURAL SENSITIZATION; see also CONDITIONING, HABITUATION). The memory's duration turned out to be a function of the number of repetitions of the aversive stimulus. Whereas one shock leads to a *short-term memory* (sensitization) lasting only minutes and does not require cells to synthesize new proteins, four or five spaced shocks to the tail generate a *long-term memory* lasting several days, which does require new protein synthesis. Further training of four brief bouts per day for four days generated even longer-lasting memory that survived for weeks. This long-term memory endures by virtue of growth of new synaptic connections. In *Aplysia*, a typical sensory neuron has ~1,200 synaptic connections; but following long-term sensitization this increases to ~2,600. With time, the number returns to ~1,500 as the memory fades, the strength of the synaptic connections between the precisely interconnected cells weakening.

From these simple cases of memory

storage involving reflex behaviour (procedural memory), several pointers to learning mechanisms in mammals arose.

(i) Even the short-term memory of sensitization involved synaptic changes, although no new proteins were synthesized. These involved increase in cyclic AMP (cAMP) levels in the abdominal ganglion, and more specifically in the presynaptic termini of serotonergic modulatory neurons synapsing with the terminals of the sensory neurons. Serotonin release by the modulatory neurons was found to act on specific receptors in the presynaptic terminals of sensory neurons, leading to greater release of their own neurotransmitter onto the motor neurons, termed *presynaptic facilitation*. The transmitter raised cAMP levels in the presynaptic cytoplasm, activating the enzyme PKA whose effects included phosphorylation of the presynaptic K^+ channels. The resulting altered membrane voltage caused more Ca^{2+} entry through voltage-gated channels, causing a faster exocytosis of transmitter from synaptic vesicles and increased firing of motor neurons leading to gill withdrawal. Such increased release of transmitter at the synapse as a result of modulatory neuron transmitter release onto the presynaptic neuron is an example of *synaptic strengthening* (see also LEARNING).

(ii) In a cell culture containing a single sensory neuron making synaptic contact with a single motor neuron, research showed that memory storage has two phases: the transient memory just described, which lasts minutes (short-term sensitization), and an enduring memory that lasts days or longer. Repeated stimulation of the synapse leading to long-term facilitation caused synthesis of new proteins and growth of new synaptic connections (see SYNAPTIC PLASTICITY). In the 1970s, these slow synaptic events were found to depend upon activation by neurotransmitters of metabotropic receptors that in turn activated second-messenger pathways (see SYNAPSE, Fig. 168). These operated via a transcription cascade involving the selective activation of *memory-enhancer genes* and selective inactivation of *memory-suppressor genes*. Notwithstanding involvement of these nuclear events,

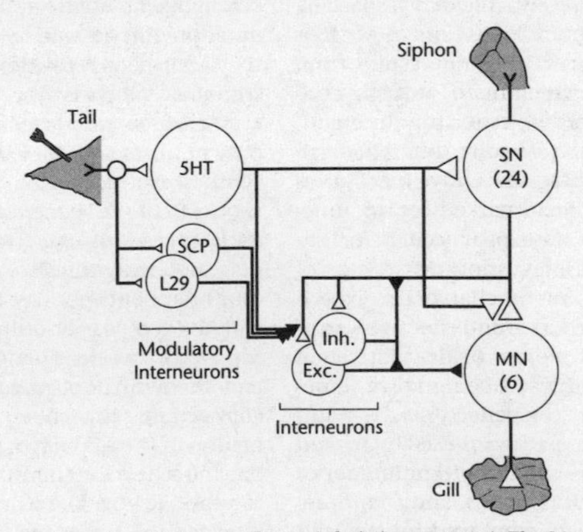

FIG 122. Circuit diagram of the gill-withdrawal reflex in *Aplysia* (see MEMORY). The 24 sensory neurons (SN) innervating the siphon connect directly with six motor neurons (MN) and with populations of excitatory and inhibitory interneurons that connect in turn with the motor neurons. Tail stimulation activates three classes of modulatory neurons: serotonergic (5HT) neurons, neurons releasing the small cardioactive peptide (SCP) and L29 cells. These act on terminals of the sensory neurons and on those of excitatory neurons; but the serotonergic action is the most important, and blocking their action blocks the sensitizing stimuli. From *Science* Vol 294, No. 5544, p. 1032. Reprinted with permission from AAAS.

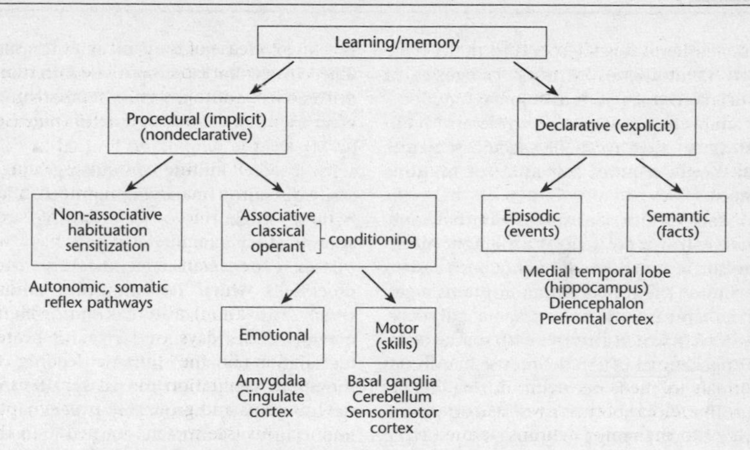

FIG 123. One classification of the forms of MEMORY. Structures with major responsibility for a particular category are shown below the respective rectangles. Modified from *Neuroscience* (2nd edition) by M.F. Bear, B.W. Connors and M.A. Paradiso, publisher by Lippincott, Williams & Wilkins. Reproduced with the kind permission of the publishers.

long-term changes in synaptic structure and function were confined to those synapses involved in presynaptic facilitation and not to any other synapses made by the same sensory neuron. In long-term facilitation, therefore, the function of a synapse was determined not only by its history of usage but also by the state of the transcriptional machinery, notably the transcription factor CREB, in the cell's nucleus – a situation quite different from short-term facilitation. Only those synapses that had been marked by SEROTONIN made use of proteins produced by CREB-mediated gene expression for synaptic growth and a persistent change in synaptic strength involved in long-term memory.

Many classifications of learning and memory exist, Fig. 123 being fairly representative. There is evidence that the cellular and molecular pathways used by *Aplysia* for storing short- and long-term memories are conserved in mammals, and that these same pathways are employed by them in both implicit and explicit memory storage (the latter memories requiring conscious recall and typically concerned with places, objects and events). *Implicit (motor) memory*, or *procedural memory* ('knowing how'), is memory for such skills as learning to walk, ride a bicycle or play a musical instrument (the latter also requiring declarative memory for the musical notation), in which improvement occurs gradually through regular rehearsal, and its performance is neither forgotten nor involves conscious recall. It involves LEARNING to produce a motor response to a particular stimulus, and may be non-associative (HABITUATION and SENSITIZATION) or associative (CLASSICAL CONDITIONING, INSTRUMENTAL CONDITIONING). It is likely that emotional memories (see EMOTIONS) are not stored in the AMYGDALA, but in the associated cerebral cortex. *Explicit*, or *declarative, memory* is memory for facts, and declarative learning is fast, requires few trials, involves conscious recall, and may be readily forgotten. Its two components are episodic memory and semantic memory. *Episodic memory* is memory for specific events, establishing associations at a specific time and place, and in humans involves the HIPPOCAMPUS (see for details) and PREFRONTAL CORTEX. Spatial navigation learning (most studied in rats) is a special form of this. *Semantic memory* ('knowing that') is memory of facts unrelated to events, and in humans involves the anterior temporal lobes (especially that of the left hemisphere), different categories of such memories, such as colours, functions, names, being stored in different brain areas, so that recall of specific semantic memories requires activation of multiple sites.

Many neuroscientists suspect that LONG-TERM POTENTIATION (LTP) is a memory mechanism, although proof has not been easy to acquire. The hippocampus contains a cellular representation of extrapersonal space, lesions in which interfere with spatial tasks. Hippocampal LTP in the mouse has phases similar to those occurring during *Aplysia* facilitation: a short burst of stimuli producing effects lasting 1–3 hours and involving no new protein synthesis, whereas repeated trains of electrical stimuli in the CA1 region of the mouse hippocampus produce a *late phase of LTP* which persists for at least a day and involves PKA, MAPK and CREB-mediated transcription and translation and appears to lead to growth of new synaptic connections. Mutant mice with faulty hippocampal PKA exhibit a decrease in late LTP and related deficits in hippocampus-generated long-term memory involving extrapersonal space, even though learning and short-term memory remain unimpaired. In the mouse, as in humans, a key feature in the stabilization of PKA- and protein synthesis-dependent explicit memory is ATTENTION.

Over-expression of endogenous CALCINEURIN can act as a negative regulator of synaptic plasticity, learning and memory through its protein dephosphorylating activity. Whereas the short-term synaptic changes required for both implicit and explicit memory, e.g., WORKING MEMORY, involve covalent modification of existing proteins and lead to modification of pre-existing synaptic connections, long-term synaptic changes require activation of gene expression, new protein synthesis and formation of new synaptic connections, and are subject to constraints such as protein dephosphorylation. Just as it does in the storage of implicit memory (short-term presynaptic facilitation) in *Aplysia*, PKA plays a major role in the storage of explicit memory of extrapersonal space in the hippocampus, transforming declarative and procedural short-term memory (STM) into long-term memory (LTM). STM is temporary, limited in capacity, needs continued rehearsing and is easily disrupted by conflicting input. LTM (remote memory) is long lasting and appears to have no obvious limit.

Memory consolidation refers to the gradual process of neural reorganization during which new memories become remote memories. Learning of facts and events (declarative memory) initially depends on the hippocampus. As time passes, after initial learning, the hippocampus declines in importance as a more permanent memory is established over a few years in regions of the neocortex, possibly via intermediate processing in the MEDIAL TEMPORAL CORTEX. Serial models of the process suggest that before STM decays, elements may be selected by ATTENTION and arousal mechanisms prior to consolidation. The parallel processing alternative holds that input goes in parallel to both STM and LTM but that consolidation acts to save selected material in LTM. The 'schema' concept of memory consolidation, positing acquired knowledge structures into which newly acquired information can fit more easily than in their absence, is an influential one in psychology. Recently, rats which had learned associations between flavours and their spatial locations have been shown to learn new associations within their familiar testing arena far more rapidly than rats without such regular training. Their neocortices were shown to be able to incorporate information from the hippocampus unexpectedly rapidly, and it is tempting to take this as evidence that memory consolidation proceeded this rapidly because the new information was fully compatible with what had been learned already.

Amnesia, loss of memory, may be anterograde or retrograde, depending on whether memories are lost for events and facts acquired before (retrograde) or after (anterograde) the brain damage.

Memory suppression has been a controversial issue for decades; but recent work

(2007) provides evidence for an active suppression mechanism for emotional memories, involving at least two pathways with staggered phases of modulatory influence. The first involves the right inferior frontal gyrus over sensory components of memory representation. This is consistent with models in which the thalamus is a critical means of gating WORKING MEMORY information. A second involves cognitive control by the right medial frontal gyrus over memory processes and emotional components of memory representation via modulation of the hippocampus and amygdala, both of which are needed for retrieval of emotional memory.

Memory appears to be a distinctive cognitive function that can be studied independently of other higher cognitive abilities. Most genes as yet identified as affecting memory are involved in the same signal transduction pathways employed for non-memory purposes in other cells, although it is possible that isoforms specific to memory-storing structures in the brain, such as the HIPPOCAMPUS, exist. Work on the fruit fly *Drosophila* suggests that signalling via the cyclic AMP pathway is involved.

memory cells Immune cells involved in a secondary immune response to antigen. See B CELLS, T CELLS.

menarche The onset of MENSTRUATION. It is thought that it requires either a critical body mass of ~47 kg, or possibly a critical ratio of fat to lean mass. Years since menarche are used as one measure of biological maturation. See BREAST CANCER, PUBERTY.

Mendelian disease See MENDELIAN INHERITANCE.

Mendelian inheritance Any phenotypic trait, including disease (e.g., metabolic disorders), whose inheritance behaves as though it were controlled by a single gene locus. The *Online Mendelian Inheritance In Man* website is: http://www.ncbi.nlm.nih.gov/entrez/query.fcgi?db=OMIM. See NATURAL SELECTION; contrast COMPLEX DISEASES.

meninges (sing. meninx) Collective term for the three connective tissue membranes overlying the brain and SPINAL CORD (see Fig. 161) and lying between them and the overlying bones of the skull and vertebral column, respectively. The toughest and outermost is the leather-like *dura mater*, with venous sinuses running through it. Just under this lies the *arachnoid membrane* (or arachnoid mater, superficially resembling the network of a spider's web), separated from the dura by the subdural space, which is traversed by cerebral veins entering the venous sinuses in the dura. Small herniations in the arachnoid mater, arachnoid villi, protrude through the dura into the venous sinuses. It is here that bulk flow of CEREBROSPINAL FLUID (CSF) into blood takes place via mesothelial tubes, valves in the tubes preventing reflux of blood into the CSF. The thinnest membrane, and closest in proximity to the nerve tissue, is the *pia mater*, separated from the arachnoid by a fluid-filled space, the subarachnoid space, containing CSF and housing blood vessels. See MENINGITIS.

meningitis Inflammation of the MENINGES, meningism being the group of symptoms accompanying, but not necessarily indicating, this: a global headache, neck stiffness, nausea, vomiting and photophobia. *Viral meningitis* is usually self-limiting and rarely lasts >1 week, viruses causing it including echovirus, coxsackievirus, poliovirus, mumps virus, HERPES SIMPLEX type 2, influenza type A or B, and arboviruses (usually meningoencephalitis) including tick-borne, St. Louis, and West Nile encephalitis. Drugs that inhibit the binding of enteroviruses to their host-cell receptor are being developed and are intended for treatment of viral meningitis (notably herpes simplex type 2). *Bacterial meningitis* is a medical emergency on account of the high mortality of untreated or late-treated cases. Empirical antibiotic treatment should be given prior to bacterial diagnosis and prior to lumbar puncture if diagnosis is strongly suspected, not possible or delayed. Causative bacteria include *Neisseria meningitidis* (see NEISSERIA), the cause of meningococcal meningitis (the commonest form in the UK), *Streptococcus pneumoniae*, LISTERIA MONOCYTOGENES, HAEMOPHILUS INFLUENZAE, MYCOBACTERIUM TUBERCULOSIS, other GRAM-POSITIVE BACTERIA, ESCHERICHIA COLI and MYCOPLASMA PNEUMONIAE. There are at least 13 serotypes of *Neisseria meningitidis*, based on the outer membrane protein (OMP), the most commonly implicated being A, B, C and W135. The outer membrane contains a lipid A-containing lipopolysaccharide causing activation of macrophages and release of TNF-α and a mediator of SEPTIC SHOCK in severe cases. A purpuric rash appears in most patients. Among ANTIBIOTICS, chloramphenicol is effective against meningococcal meningitis, most now being resistant to benzylpenicillin. In hospital, the treatment of choice is cephalosporin (cefotaxime or ceftriaoxone)

menopause (climacteric) Progressive failure of the female reproductive organs, usually between 45 and 55 years. The number of OOCYTES declines through complete depletion of primary follicles (see GRAAFIAN FOLLICLES) through atresia and declining responsiveness of the ovaries to GONADOTROPINS. Menstrual cycles often become irregular before finally terminating. Levels of pituitary FSH and LH are high in postmenopausal women since negative feedback control by oestrogen is lost; but LH surges no longer occur. Many somatic and emotional changes accompany loss of ovarian steroids. The uterine muscle becomes fibrous, the vagina may become dry, and there is a loss of breast tissue. Bone strength often declines through increased bone resorption (see OSTEOPOROSIS). Most of these changes can be successfully treated by hormone replacement therapy if required.

menses See MENSTRUATION.

menstrual cycle (ovarian cycle) A ~28-day hormonally controlled cycle in women (although varying between 18 and 40 days), occurring from PUBERTY to MENOPAUSE, during which a mature oocyte is produced by the OVARIES and the uterus is prepared for implantation, in the absence of which a new cycle begins. Its rhythm can be upset, or its occurrence prevented, by neural influences during stress, anxiety and emotional upset, and by anorexia nervosa. The first cycles are usually anovulatory, the first ovulatory cycles occurring 6–9 months later, becoming fully ovulatory after 2–3 years. The cycle is divisible into follicular and luteal phases.

By convention, the *follicular phase* (corresponding to the proliferative phase of the endometrial cycle) begins on day 1 of the cycle, the first day of menses (menstruation), during which a dominant GRAAFIAN FOLLICLE grows and develops. It is characterized by oestradiol-induced epithelial cell proliferation of the endometrium in preparation for implantation, and up-regulation of oestradiol and progesterone receptor expression, which reaches a peak by the time of ovulation. The ratio of production of LH and FSH by the pituitary is determined by the frequency of GnRH pulses from the hypothalamus (see GONADOTROPIN RELEASING HORMONE). During the follicular phase, FSH output stimulates the dominant follicle to release high levels of 17β oestradiol (see OESTROGENS), and inhibin B. At mid-cycle, the rising oestradiol level acts on the hypothalamus to generate low-amplitude, high-frequency (90-minute) pulses of GnRH, resulting in an LH surge at ~day 14. This is the result of positive feedback resulting from increased sensitivity of the pituitary gonadotroph cells to GnRH, triggered by rising oestradiol, and from increased gonadotroph GnRH receptor density. Inhibin B levels fall during the LH surge, while inhibin A levels rise during the late follicular phase, peaking as LH and FSH surge. Elevated LH levels stimulate follicular development, resumption of the 1st meiotic division by the primary oocyte, cause decreased FSH secretion, and increasing PROGESTERONE synthesis by the granulosa cells so that oestrogen output falls slightly. At the time of OVULATION, under LH influence, the cells of the stalk attached to the oocyte dissociate and the follicle ruptures, 36 hours after the LH surge. It is suspected that the switch from oestrogen to progesterone production is somehow responsible for this.

During the follicular phase, under oestrogen stimulation, the uterine mucosa has undergone relative hypertrophy, its glandular structures have increased, and its epithelial cells have acquired a high density of progesterone receptors. During the *luteal phase* (corresponding to the secretory phase of the endometrial cycle) which begins after ovulation, the corpus luteum produces high levels of progesterone, reducing the surges of GnRH to 1 pulse per 3–4 hours and suppressing the frequency and amplitude of LH release. Progesterone also downregulates the density of GnRH receptors on gonadotrophs and decreases transcription of the α- and β-subunits of LH and FSH. During this time, FSH can build up in the gonadotrophs ready for release if the corpus luteum undergoes atrophy (which it does from ~day 21 in the absence of hCG). The luteal phase is characterized by progesterone-induced endometrial epithelial cell differentiation into secretory cells as uterine glands and arteries become coiled and the tissue becomes succulent. The glands start to secrete a thick fluid rich in sugars, amino acids and glycoprotein, favouring IMPLANTATION and placenta formation. Granulosa cells secrete inhibin A, its concentration falling in line with those of oestradiol and progesterone during luteal regression. Toward the end of the luteal phase, glandular expression of oestrogen receptors declines as a result of suppression by rising progesterone levels. In the absence of fertilization, the corpus luteum regresses after 10–12 days, steroid output quickly declines, and MENSTRUATION ensues. During the transition from the luteal to the follicular phase, pulses of GnRH release occur every 90–120 minutes and FSH release predominates again, initiating a new cycle.

menstruation (menses) Menstrual period. Shedding of dead tissue and blood of the uterine endometrium as blood flow is restricted and ischaemia occurs resulting from decline in oestradiol and progesterone secretions at the end of the non-fertile MENSTRUAL CYCLE. Severe obesity, malnutrition and strenuous physical exercise may prevent menses; but a woman who has not had a menstrual period by the completion of PUBERTY or by age eighteen is said to have primary AMENORRHOEA.

mental retardation Condition said to have occurred if a disruption of brain development results in below average cognitive functioning that impairs adaptive behaviour. Of its many causes, CONGENITAL HYPOTHYROIDISM is one of the most common preventable forms, affecting ~1:4,000 infants. The most severe effects arise from genetic disorders, such as phenylketonuria and Down syndrome. It may arise through distress during pregnancy, such as from infections with German measles (rubella), or foetal asphyxia; or from poor nutrition (see FOETAL ALCOHOL SYNDROME). But perhaps the majority of cases arise though inadequacies in general nutrition, socialization and sensory stimulation during infancy (see CRITICAL PERIOD). DENDRITES of mentally retarded children have far fewer spines, and those present are unusually long and thin, resembling those of the normal foetus. Large chromosomal disorders causing mental retardation include those responsible for DOWN SYNDROME. A syndrome identified in 2006 and expected to explain ~1% of undiagnosed cases of mental retardation is caused by a very small deletion in chromosome 17 and flanked by DNA sequences typical of other areas prone to deletion. It carries about six genes, one of which is the so-called tau gene, *MAPT*, that is mutated in some neurodegenerative disorders, including ALZHEIMER'S DISEASE. See also CRETINISM, FRAGILE-X SYNDROME, HOMOCYSTINURIA, KLINEFELTER SYNDROME, PHENYLKETONURIA.

Merkel cells, Merkel discs A TOUCH receptor, being the least abundant neuronal cell type in the *epidermis* of the skin (see MECHANORECEPTORS, Fig. 120), where they are located in the deepest epidermal layers and synapse with the flattened process of a non-neural epidermal cell to form the Merkel disc, the type 1 cutaneous mechanoreceptor. The epithelial cell may be the mechanically sensitive part, since it makes a synapse-like junction with the nerve terminal. Like RUFFINI ENDINGS, they are slowly adapting, generating a more sustained response during a long stimulus than do PACINIAN CORPUSCLES or MEISSNER'S CORPUSCLES.

mesencephalic locomotor region (MLR) An alleged 'locomotor region' of the mammalian midbrain and projecting to the medullary reticular nuclei. It includes the periaqueductal grey matter, the cuneiform nucleus, the pedunculopontine nucleus and the LOCUS COERULEUS, different subsets of which appear to be activated during locomotion in different behavioural contexts. See CENTRAL PATTERN GENERATORS.

mesencephalon The primary brain vesicle (see BRAIN, Fig. 28) giving rise to the MIDBRAIN. See ALAR PLATES, BASAL PLATES.

mesenchyme Embryonic MESODERM, deriving from the sclerotome of each SOMITE (see entry) and comprising widely scattered tissue giving rise to CONNECTIVE TISSUES and blood cells.

mesenteries Double layers of PERITONEUM enclosing an organ (e.g., much of the abdominal alimentary canal) and connecting it to the body wall. Kidneys are covered on their anterior surfaces only. See OMENTUM.

mesoderm Originally cells of the mesodermal GERM LAYER, form a thin layer either side of the notochord in the midline of the embryonic disc. This layer arises after formation of definitive endoderm by migration of epiblast cells through the PRIMITIVE STREAK, and is under the control of FGF8 (see FIBROBLAST GROWTH FACTORS) and BMP4 (see BONE MORPHOGENETIC FACTORS). By roughly the 17th day, at the onset of NEURULATION, cells close to the midline on either side proliferate to form a thickened plate, the PARAXIAL MESODERM, which becomes segmented to form the SOMITOMERES. Cells at the dorsolateral portion of each SOMITE (see entry) migrate as precursors to limb and body wall skeletal musculature. More laterally, the mesoderm remains thinner, forming the LATERAL PLATE MESODERM, which eventually gives rise to smooth muscle (see Appendix VI). Intermediate mesoderm, temporarily connecting the paraxial and lateral plate mesoderm, eventually gives rise to the kidneys and gonads. Other mesodermal cells (see SCLEROTOME) give rise to BONE, CARTILAGE (although some cartilage originates from the neural crest) and other connective tissue.

mesolimbic system Dopamine-containing cell clusters in the ventral tegmentum of the midbrain (see BRAIN, Fig. 28), and the NUCLEUS ACCUMBENS to which they project.

mesomorph See BODY SIZE AND SHAPE.

mesonephros The 'second kidney', derived from intermediate mesoderm between the upper thoracic and upper lumbar segments and developing in the 4th week as the PRONEPHROS degenerates, the excretory tubules of the mesonephros lengthen and acquire a tuft of capillaries (glomerulus). The tubules grow around the glomeruli to form Bowman's capsules as the tubules join the common collecting duct, the mesonephric (Wolffian) duct, which is continuous with the remains of the pronephric duct. By ~6th week, the mesonephric kidneys form large bilateral ovoid organs and, with the developing OVARIES or TESTES lying on their medial sides, form the *urogenital ridges*. By the end of the 8th week, the majority of the mesonephric tubules have disappeared, a few caudal ones persisting in the male to form the vasa efferentia and linking the rete testis to the mesonephric duct which persists as the vas deferens.

messenger RNA Produced by TRANSCRIPTION and subsequent RNA PROCESSING, a molecular intermediate in the deciphering of the protein-coding genes into protein. However, most messenger RNAs do not encode proteins (see HUMAN GENOME PROJECT, NON-CODING RNAS).

metabolic disorder A disorder resulting from a biochemical lesion, very often exhibiting MENDELIAN INHERITANCE.

metabolic pathway See METABOLISM.

metabolic rate See BASAL METABOLIC RATE.

metabolic reactions See METABOLISM.

metabolic syndrome See INSULIN RESISTANCE.

metabolism The totality of enzyme-mediated biochemical pathways occurring within the body, both anabolic (build-up) and catabolic (breakdown). These pathways may be linear (e.g., GLYCOLYSIS, GLUCONEOGENESIS), and may branch frequently. They may also be cyclical (e.g., KREBS CYCLE). The enzymes involved tend to be encoded by HOUSEKEEPING GFNES, and the pathways are usually highly regulated by regulatory proteins ensuring cellular homeostasis. See BASAL METABOLIC RATE.

metabotropic receptors A subtype of receptor in the CELL MEMBRANF and in the membranes of VESICLES. All are monomeric proteins with seven transmembrane domains (serpentine receptors) with their N-terminus on the external membrane surface and their C-terminus on the internal surface. They can be classified into RECEPTOR TYROSINE KINASES and G PROTEIN-COUPLED RECEPTORS, and have neurotransmitters as ligands; but in contrast to IONOTROPIC RECEPTORS, they do not form an ION CHANNEL pore but are indirectly linked with ion channels via G PROTEINS or by SIGNAL TRANSDUCTION pathways involving SECOND MESSENGERS. They mediate *slow synaptic transmission* (see SYNAPSES), channel opening or closing involving activation of a number of other molecules downstream of the receptor, taking longer than is the case with ionotropic receptors; but they stay open longer (from seconds to minutes) and have longer-lasting effects, which can also be more widespread within the cell. They include metabotropic GLUTAMATE receptors, $GABA_B$ receptors, most SEROTONIN receptors, and receptors for NORADRENALINE, ADRENALINE, HISTAMINE, DOPAMINE, neuropeptides and ENDOCANNABINOIDS.

metalloproteinases Zinc-dependent endopeptidases which are mediators of tissue (uterus, mammary gland, foetal membranes, birth canal) growth and remodelling, and often located within the EXTRACELLULAR MATRIX, ECM (*matrix metalloproteinases*, MMPs). Some are released in response to RELAXIN in preparation for birth and lactation; others are involved in tumour cell migration and in angiogenesis to supply tumours (see METASTASIS). Collectively, MMPs degrade all kinds of ECM proteins, and are involved in cleavage of cell surface receptors, release of apoptotic ligands (e.g., FAS ligand), and chemokine activation and inactivation.

metalloproteins Metal-containing proteins, including alcohol dehydrogenase (zinc), calmodulin (calcium), dinitrogenase (molybdenum), ferritin (iron) and plastocyanin (copper) and membrane METALLOPROTEINASES.

metanephros The third urinary organ to develop, appearing in the 5th week and forming the permanent KIDNEY. Its excretory units develop in the same manner as those of the MESONEPHROS, but the development of its duct system differs from that of the two earlier kidney systems.

metaphase The phase of MITOSIS and of both divisions of MEIOSIS when chromosomes are aligned at the equator of the spindle.

metaphase spread A single cell's condensed chromosomes at metaphase, displayed on a slide. Can then be stained, e.g., by Giemsa, FISH, etc.

metaplasia A cancerous tissue in which one type of normal cell layer is displaced by cells of another type not normally encountered at that location within a tissue. Metaplasias are most frequent in epithelial transition zones at the borders between epithelia (e.g., at the junction of the cervix and the uterus, or between the oesophagus and stomach).

metapopulation A population of populations.

metastasis (verb, metastasize) A complex, multi-step process by which primary tumour cells invade adjacent tissue, enter the systemic circulation (intravasate), translocate via the vasculature, halt in distant capillaries, extravasate into the surrounding tissue parenchyma, and then proliferate from microscopic growths (micrometastases) into macroscopic secondary tumours (see CANCER). Few cellular processes have been identified that are of greater complexity than tumour metastasis, and most deaths from cancer are due to its occurrence (see Fig. 124). The leading edge of the tumour cell protrudes following cytoskeletal alteration and adhesion to the extracellular matrix while the trailing edge detaches from it. We do not yet know how many separate gene-expression programmes are initiated in simple cell motility, and there are probably many additional requirements for metastasis. Expression of non-canonical Wnt5a (see WNT SIGNALLING PATHWAYS) correlates with metastatic melanoma invasiveness in humans, and exposure to this protein (but not to other Wnts) promotes invasiveness in cultured melanoma cells. Recent work on mouse metastatic breast cancer cells implicates microRNAs (see entry) in the coordination of these complex programmes. The transcription factor Twist binds to the E-box upstream of the DNA sequence encoding this miR-10b, inducing it. Of the potential binding sites of miR-10b, one inhibits mRNA translation of the transcription factor HOXD10, affecting its downstream targets. One such target is the RHOC protein, which is a component of signalling cascades mediating metastasis. RHOC expression is increased in response to miR-10b and consequent reduced HOXD10 levels, and this increase promotes tumour cell motility.

The initial metastatic step is localized invasion of a CARCINOMA across the epithelial basement membrane, followed by intravasation into blood or lymph microvessels. At some possibly distant anatomical site, commonly the first capillary bed after entering the heart (i.e., the lungs), the cells, ~20 μm in diameter, become trapped. A microthrombus forms as platelets attach to the cancer cell, which then pushes aside an endothelial cell to reach and attach to the basement membrane. Its proliferation leads to extravasation and formation of dormant micrometastases, some of which may then acquire the ability to colonize the surrounding connective tissue. While travelling in these vessels, cancer cells are susceptible to death by shearing forces of the blood and by ANOIKIS, and any subsequent colonization involves cross-talk between cancer cells and their stroma of local fibroblasts and endothelial cells (see EXTRACELLULAR MATRIX). This interaction leads to growth of new blood vessels (see Fig. 124; ANGIOGENESIS), evasion of host immune recognition and alteration of the microenvironment so that metastases form in spite of unfavourable growth conditions. The gene *p53* must be inactivated or bypassed in order for tumour progression to occur (see LI-FRAUMENI SYNDROME). Cancer invasion may be a deregulated form of the normal physiological invasion process required for neuronal growth and wiring in the embryo, tissue remodelling, angiogenesis and wound healing. The angiogenesis associated with a tumour often provokes pro-inflammatory signals which can convert immune tolerance to activation by induction of dendritic cells, which can ingest tumour antigens and carry them to lymph nodes. Alternatively, tumour cells may migrate directly to lymph nodes and activate T cells by this route. Immune tolerance of a tumour may arise through 'ignorance' of its foreign nature because it does not generate a new antigen during its pro-inflammatory phase, or by down-regulating the MHC or antigen-processing machinery, T-cell anergy or deletion. The alpha-defensins are antimicrobial peptides whose genes are highly over-expressed in some cancers (e.g., lung and renal cancer), apparently creating a microenvironment unfavourable for the adaptive immune system.

Discovery of gene sets ('signatures') whose expression profiles are associated with general and organ-specific metastasis has led to clinical screening programmes and the search for molecular targets amenable to therapy. Discovery since 2003 of four genes promoting not only growth of primary tumours but also intravasation, colonization of the lung, extravasation

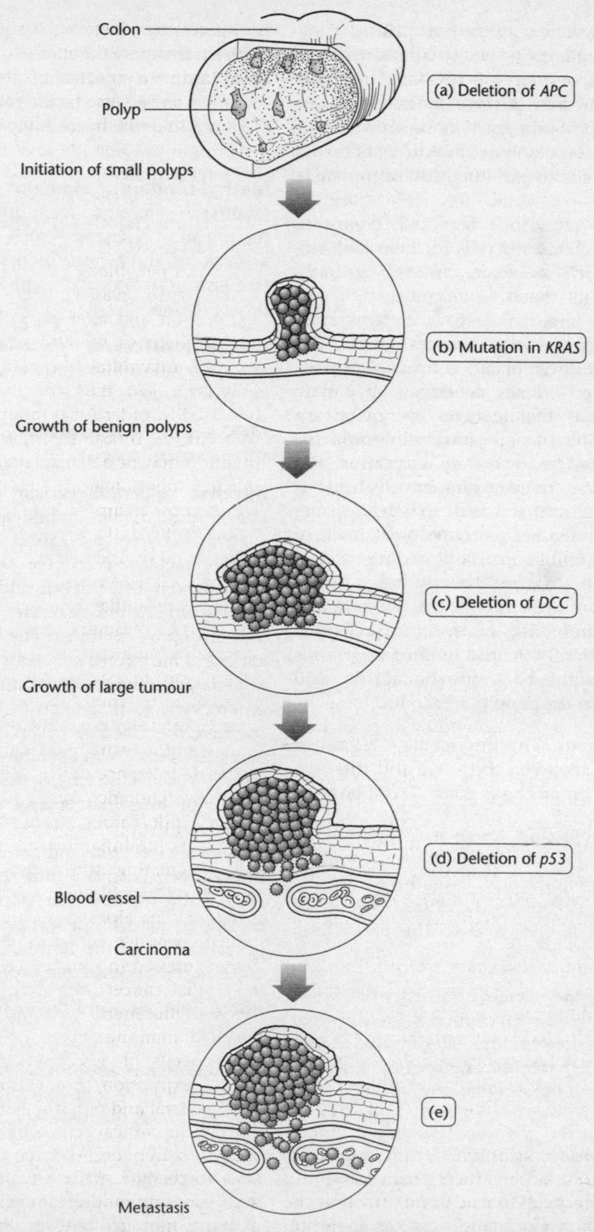

FIG 124. Diagram indicating the role of a stepwise sequence of mutations in proto-oncogenes and tumour suppressor genes during the progression of a non-inherited colon tumour (see METASTASIS). This case involved (a) a deletion of the *APC* tumour suppressor gene on chromosome 5; (b), a mutation in the *KRAS* proto-oncogene on chromosome 12; (c) deletion of *DCC* tumour suppressor gene on chromosome 18; (d), deletion of *p53* tumour suppressor gene on chromosome 17; and (e) subsequent genetic changes. From *Basic Human Genetics* (2nd edition), A.P. Mange and E.J. Mange. © 1999 Sinaur Associates Inc. Redrawn with permission of the publisher.

and metastatic outgrowth is exciting. They encode epiregulin, two METALLOPROTEINASES, and cyclooxygenase-2 (COX-2). Inactivation of all four of these genes was found to abrogate primary tumour growth and metastasis (notably extravasation) of breast cancer cells to the lung, increasing their apoptosis.

A new clone of cancer cells competes with neighbouring cells for food and services, such as waste removal, initially within its natal environment. When tumours have grown to a certain small size, their growth becomes limited by their ability to obtain cellular resources, and they normally develop a 'glycolytic phenotype' that involves energy release by glycolysis in a hypoxic environment, interpreted by some as an adaptation: the metabolic activity of tumour cells leads to local acidosis that is toxic to neighbouring normal cells, and promotes breakdown of the extracellular matrix providing tumour cells with a competitive advantage fostering growth and invasion. This anoxic adaptation often involves a specific nucleotide substitution in the *p53* tumour suppressor gene leading to a specific amino acid change in the protein it encodes.

metatarsus The intermediate region of the foot, between the ankle and the phalanges, comprising five metatarsal bones.

metencephalon A BRAIN VESICLE, part of the rhombencephalon (hindbrain) rostral to the myelencephalon, extending from the pontine flexure to the rhombencephalic isthmus. It gives rise to the cerebellum and pons.

methionine One of the essential AMINO ACIDS, necessary for normal biosynthesis of *S*-adenosylmethionine, the methyl donor for methylcytosine. Deficiency leads to aberrant GENOMIC IMPRINTING of IGF2 (see FOLATE).

methylation Choline is an important methyl donor in alkylating reactions, such as those involved in DNA METHYLATION in which methyltransferases are involved (not properly enzymes if the methylation is irreversible). HISTONE METHYLASES and methyltransferase are also of critical importance in organizing the genome. Methylation can also be the result of ALKYLATING AGENTS, which are important mutagens.

methyl-binding domain proteins (MBPs) Proteins recruiting histone-modification enzymes, such as HISTONE DEACETYLASES and histone methyltransferase (see HISTONE METHYLASES), leading to histone deacetylation.

methyltransferases See DNA METHYLATION, DNA REPAIR MECHANISMS, HISTONE METHYL LYSINE METHYLTRANSFERASES.

MHC genes, MHC proteins See MAJOR HISTOCOMPATIBILITY COMPLEX.

micelles Spherical particles of colloidal size forming when amphipathic lipids, such as BILE SALTS, are mixed with water. Dietary fats are converted by bile salts into mixed micelles of bile salts and triglycerides which also sequester cholesterol and fat-soluble vitamins. The hydrophilic surface of the micelle enables it to enter the aqueous layer surrounding the microvilli of enterocytes, so that micelles enormously increase the fraction of lipid molecules accessible to water-soluble intestinal LIPASES. Micelles attach to the cell membranes of intestinal enterocytes, whereupon monoglycerides, free fatty acids, cholesterol, lecithin and fat-soluble vitamins diffuse passively into the duodenal cells while the bile salt portion of the micelle remains in the gut lumen until it is reabsorbed in the terminal ileum. See CHYLOMICRONS, ENTEROCYTES, LIPOPROTEINS.

microarray analysis See DNA MICROARRAYS, GENOMICS.

microbiome, microbiota The totality of microbial life living in, or on, the body. For the gut microbiota, see GUT FLORA. See also INFECTION.

microcephaly Usually, an abnormality in which the cranial vault is smaller than normal, the brain failing to grow as the skull fails to expand. Primary microcephaly

(MCPH) is characterized by a 70% reduction in BRAIN SIZE, comparable to that of early hominins, but normal stature. Its causation varies, some forms being inherited in a recessive autosomal manner (see ASPM), others being due to prenatal infection or exposure to certain drugs and other teratogens. The recessive condition has an incidence of 4–40 per million live births in western states and is the commonest known cause of MCPH. Many children with the condition are severely retarded. The cranial vault is smaller than normal, and as the size of the cranium depends on growth of the brain, the primary defect lies in brain development. Foetal infection by toxoplasmosis can result in microcephaly, as it can in cerebral calcification, mental retardation and hydrocephalus. Radiation exposure early in development can also be a cause. MCPH is associated with certain genes encoding centrosome proteins including, *ASPM/Asp, SAS4/CENPJ, Centrosomin/CDK5RAP2* and *MCPH1*. One *ASPM* variant which appeared ~37 kya, roughly when Europeans began to show symbolic behaviour, was originally thought to be a target of positive selection; but later work indicated that its frequency did not differ significantly from those of alleles for other genes elsewhere in the genome, arguing against positive NATURAL SELECTION for the allele. See ANENCEPHALY.

microchimaerism Condition in which cells from a foetus live on, possibly for decades, in the mother's body, or vice versa, after pregnancy. There is a growing body of evidence suggesting that these cells might be a cause of some forms of AUTOIMMUNE DISEASES and other clinical conditions.

microevolution Evolutionary processes occurring at the level of species populations rather than at the level of whole species or higher taxa.

microglial cells Small macrophages of the central nervous system, comprising ~10% of its cells, acting immunologically as scavengers, proliferating and migrating to a site of brain damage, removing cell debris, and initiating inflammatory responses. They continually extend and retract cell processes whose filopodia-like protrusions form transient bulbous tips. They derive from bone marrow precursor cells – unlike the macroglia, which derive from ectodermal precursors in the nervous system. Compare ASTROCYTES.

β_2-microglobulin See IMMUNOGLOBULIN-LIKE DOMAINS, MAJOR HISTOCOMPATIBILITY COMPLEX.

micronutrients Substances, organic (e.g., vitamins) and inorganic (e.g., cobalt), required in minute or small amounts in the diet compared with the main components, the macronutrients.

microprojectiles For 'biolistics', see DNA VACCINES.

microRNAs (miRNAs, miRs) Abundant, small, single-stranded, naturally occurring non-coding RNAs (encoded by the genome) that enter the RNA INTERFERENCE pathway to regulate protein-encoding genes by controlling TRANSLATION of target mRNAs, and thereby GENE EXPRESSION, during critical aspects of development. To repress translation or facilitate degradation of mRNAs, miRNAs bind to the latter's 3′ untranslated regions (see UTRS) and have been implicated in regulating diverse cellular pathways. Some HOX GENES encode them, and they derive from 'hairpin' RNAs, RNAs that are effectively double-stranded by virtue of base-pairing between self-complementary base sequences (inverted repeats). These are excised by the enzyme Drosha to form precursor miRNAs. In flies, further cleavage of these sequences by Dicer and Dcr1 and its cofactor Loqs gives rise to mature miRNAs of ~22 nucleotides. These are then incorporated into the RISC protein complex, containing the effector protein Ago1 (a member of the Argonaut family). It is possible that miRNAs regulate translation of upward of a quarter of the HUMAN GENOME.

There is emerging evidence that some miRNAs can function as oncogenes or tumour suppressors. A 2006 genetic screen of all known human miRNAs identified two that are expressed in all tested samples of testicular germ-cell tumours that had non-mutated *p53* genes. By creating mice deficient in a muscle-specific miRNA (miR-1-2) one study has shown its importance in ventricular myocyte hyperplasia, and revealed a role for it in one of the commonest forms of CONGENITAL HEART DEFECT: ventricular septal defects ('hole-in-the-heart'); a second has associated over-expression of miR-1 in adult mice with development of cardiac ARRHYTHMIA; a third focuses on the muscle-specific miR-133, with the discovery that it is a negative regulator of CARDIAC MUSCLE hypertrophy. Another miR with a modulatory role in controlling this hypertrophic response is miR-208, transcribed from a sequence embedded in the non-coding region of the gene for the heavy chain of the α-myosin protein. It has recently been shown (2007) that miR-10b is highly expressed in metastatic breast cancer and positively regulates cell migration and invasion (see METASTASIS). Hippocampal dendritic spine development is regulated by a brain-specific miRNA.

microsaccades Eye movements that occur even when fixating the gaze, being the largest form of such fixational eye movements. Each movement causes a given group of photoreceptor cells to intercept light from a different part of the field of view, so changing its responses. They carry an image linearly across dozens to hundreds of retinal photoreceptors and increase the rate of impulses generated by both the lateral geniculate nucleus and the primary visual cortex by ushering stimuli in and out of a neuron's receptive field. Work published in 2006 demonstrated for the first time that they engender visibility when subjects try to fix their gaze on an image, and that the larger and faster microsaccades are most effective at preventing image fading through neural adaptation. Smaller fixational eye movements include drift (wavy paths) and tremor (zig-zag paths superimposed on drift). Abnormal fixational eye movements occur in AMBLYOPIA, in severe forms of which excessive drift and too few microsaccades can cause objects and even large portions of the visual scene to fade away during fixation. Evidence indicates that the direction of a subject's microsaccades is biased towards their true point of focus even though the gaze is directed elsewhere, indicating and even being directed by a person's covert thoughts. Further, the frequency of microsaccades is indicative of something attracting a person's ATTENTION in the peripheral field of view. Compare NEGLECT SYNDROME.

microsatellites Tandemly repeated short mono-, di- or tri-nucleotide DNA sequences (see REPETITIVE DNA), often polymorphic in length, and originally discovered on the basis of high-speed centrifugation. Often defined as regions containing eight or more tandem repeats of a sequence 1–8 bp in length they do not, unlike VNTRS, tend to be restricted to the ends of chromosomes but are more evenly distributed (interspersed) throughout the genome and, after PCR-amplification, are often used in DNA PROFILING. Some human microsatellites have significantly longer repeat lengths than those of orthologous chimpanzee sites, although the latter often contain interrupted repeat sequences and thus differ fundamentally from those of human sites. Microsatellite SNPs also occur, often within introns, probably resulting most often from replication slippage (see MUTATIONS, Fig. 127). Microsatellite instability (MIN) is a DNA-level instability observed in a few tumours, especially some colon carcinomas, and probably enables a cell to amass enough mutations. It may even be the result of cell selection (see CANCERS). The microsatellite D9S1120 has recently been of value in theories of human MIGRATION into the Americas. See DNA MARKERS, INTERSPERSED REPEAT SEQUENCES.

microtubule organizing centres (MTOCs) See CENTROSOME.

microtubules Hollow cylindrical cytoskeletal structures formed from 13 parallel protofilaments, each comprising alternating α-TUBULIN and β-tubulin molecules (the αβ-tubulin dimer is the soluble subunit, analogous to the actin monomer or staggered tetramer of intermediate filaments).

Because the microtubule has a distinct polarity, α-tubulins are exposed at one end and β-tubulins at the other; elongating microtubules can help push out membranes, while shrinking microtubules of the spindle apparatus can help pull sister mitotic/meiotic chromatids apart during anaphase; each tubulin subunit has an exposed GTP-binding site, at the interface between it and the adjacent subunit, which can hydrolyse 1 molecule of GTP, a reaction that is speeded up when they are incorporated into filaments. Free subunits are usually in the GTP-form (T-subunits), but shortly after addition to a filament the GTP is hydrolysed to GDP (forming D-subunits), much of the free energy release being stored in the polymer lattice. The protein TAU is involved in stabilizing microtubules. The taxol group of drugs, which kill dividing cells and are used in treating breast cancer, also have this effect. Microtubules form tracks along which proteins and membranous organelles can be moved by the motor proteins DYNEINS and KINESINS. See CENTROSOME, CILIUM.

microvilli Minute finger-like projections of the apical cell membranes of many absorptive epithelial cells (e.g., of nephrons and the small intestine), increasing the surface available for absorption 25-fold. A dense covering of microvilli is termed a *brush border*. Roughly ~1 μm long × 0.1 μm diameter, they often contain enzymes, specialized transport proteins and membrane pumps. When several thousand strong, they form the *brush borders* observed in electron micrographs. About 40 actin microfilaments (see ACTIN) run along its length, supported by accessory proteins (e.g., α-actinin, fimbrin, integrins). See STEREOCILIA.

micturition Reflex emptying of the urinary bladder. Its neural integration occurs in the sacral region of the spinal cord, modulated by centres in the pons and cerebrum. Stretch receptors in the wall of the bladder send impulses via sensory neurons to the sacral spinal cord via the pelvic nerves, whereupon parasympathetic motor fibres cause smooth muscle in the bladder wall to contract in pulses while decreased somatic motor impulses relax the skeletal muscle of the external urinary sphincter. Urine flows from the bladder when pressure is sufficient. After the age of 2–3 years, this reflex can be modified by descending fibres from the pons and cerebrum allowing voluntary control of urination, which may also involve voluntary control of abdominal muscles causing increased abdominal pressure.

midbrain (mesencephalon) A relatively undifferentiated primary vesicle of the BRAIN (see Fig. 28) lying between the forebrain and hindbrain. Its dorsal surface (the *tectum*) differentiates into paired superior and INFERIOR COLLICULI, while the floor (the *tegmentum*) contains the paired SUBSTANTIA NIGRA and RED NUCLEI, involved in the control of voluntary movements. The SUPERIOR COLLICULI (aka the optic tecta in non-mammalian vertebrates) receive direct inputs from the retinas prior to relaying these to the LATERAL GENICULATE NUCLEI of the thalamus. The inferior colliculi receive input from the ears prior to relaying these to the thalamic medial geniculate nuclei. Axons innervating the extrinsic muscles of the EYES and belonging to cranial nerves III and IV originate here, while other axons involved in regulation of consciousness, mood, pleasure and pain project widely to the central nervous system.

middle ear See EARS.

middle temporal region (MT, V5) This region and the interconnected lateral intraparietal region are known to be involved in visual motion processing (see EXTRASTRIATE CORTEX, Fig. 72; VISUAL MOTION CENTRE).

midline The midline of the embryo, giving rise to the left–right body AXIS, originates during the generation of BILATERAL SYMMETRY. A plane passing through the midline of the body or of an organ is the *midsagittal plane*. Structures closer to the midline are termed *medial*, while those further from it are termed *lateral*; structures and pathways which do not cross the midline are termed *ipsilateral*, while those (e.g., neural pathways; see DECUSSATION) which do cross the midline are termed *contralateral*. See ASYMMETRY, AXONAL GUIDANCE.

midsagittal A plane passing through the MIDLINE of the body, or of an organ. See ANATOMICAL PLANES OF SECTION (see Fig. 9).

MIF (macrophage migration inhibitory factor) A CYTOKINE acting as a key regulator of innate and acquired immunity, overriding the suppressive effects of glucocorticoids. Although it induces inflammation in response to bacterial invasion, viruses, etc., it also regulates the activation of macrophages and T cells and insulin release from the pancreas. It may also allow cells to bypass p53-induced apoptosis. MIF binds and inhibits the cytoplasmic protein Jab1, and may be taken up during endocytosis. Its effects seem to be dependent upon its concentration, inhibiting inflammation and cell growth at high levels and promoting them at low levels.

mifepristone (RU-486) Antiprogesterone sex hormone (PROGESTERONE antagonist), initiating menstruation, causing abortion if administered within eight weeks of the previous menses. See BRCA1.

migration Here restricted to population migration, the major feature of current human migration is the trend from rural to urbanized life (see HUMAN DEMOGRAPHY). In addition to the use of archaeological remains, human migrations have been tracked either by studying the genetic similarities and differences among modern populations, or by using such 'proxy' measures of genetic similarity as 'linguistic phylogeny'. More recently, study of human pathogens and parasites has also contributed.

The two main hypotheses for the evolution of modern humans agree that *Homo erectus* migrated from Africa ~2 Myr ago (but see the HOMO ERECTUS and HOMO ERGASTER entries for modifications of this view). The '*Recent African Origin*' hypothesis (or *Uniregional Hypothesis*) holds that anatomically modern humans originated in Africa ~100–200 Kyr ago and spread to the rest of the world, replacing *H. erectus* with little or no genetic mixing. The *Multiregional Hypothesis* proposes that transformation to modern humans occurred from *H. erectus* at different times and in different parts of the world, supporting evidence coming from the alleged continuity between fossils, and cultural and morphological data between archaic and modern humans outside Africa. The debate between these two alternatives seems to have been settled in favour of the former.

Non-Africans generally carry a small subset of African GENETIC DIVERSITY. Their MTDNA and Y CHROMOSOME phylogenies stem from a common ancestor some distance from the root in the molecular phylogenies (see PHYLOGENETIC TREES), most simply explained by a common origin of all non-African peoples from a small subset of African ones. The geographical distribution of HAPLOTYPES reflects human history, with a loss of haplotype diversity as distance increases from Africa. Based initially on estimates of haplotype variation in human mitochondrial DNA, coalescence to the MOST RECENT COMMON ANCESTOR (the so-called MITOCHONDRIAL EVE) currently indicates an East African individual living >130 Kya. Once a phylogenetic tree is reconstructed, and, e.g., an ancestral mtDNA type of interest is identified (possibly an expansion type, a founder type or a disease type), the job of dating ancestors can start. Starlike clusters (see HOMO SAPIENS) in a phylogenetic tree are not necessarily in themselves informative for counting and dating mtDNA types arriving in new regions. The cluster's age is obtained by equating average branch length with elapsed time, measured in numbers of mutations (for which standard errors can be calculated, reflecting the number of branches available for computation). 'Mutational time' can then be converted to absolute time by multiplying it with the mutation rate. Mutation rate is the 'Achilles heel' for any DNA chronology, and efforts have been made to determine these for mtDNA.

Current interpretations of the human fossil record indicate that fully modern humans emerged in sub-Saharan Africa by ~195 Kya. By ~35 Kya, modern humans were living all across Eurasia, even into Australia. The route(s) taken in achieving this distribution is still unclear, and although the salient view for the past 20 years has been that such migration could have occurred as early as ~100 Kya, recent genetic evidence suggests that the spread out of Africa took place closer to 60–50 Kya. The earliest mtDNA lineages, known as L1 types, are restricted to Africa and (with frequencies <1%) adjacent Arabian and Mediterranean regions (see Appendix IV(i)a). Those who made this migration across the Sahara and then via the Levant (~120 Kya), with remains at Qafzeh and Skhul in the Middle East (120–80 Kya), appear neither to have left any subsequent trace there nor to have led to an expansion into Europe. Their rugged morphology, including pronounced brow ridges, is suggestive of Neanderthals (which inhabited nearby Tabun at ~50 Kya); but their genetic dating indicates that they were an evolutionary dead-end.

At ~60–80 Kya, a renewed expansion repopulated Africa with L2 and L3 mtDNA types, the original L1 types becoming a minority other than in the ancestors of Khoisan (Bushmen) and Biaka (West Pygmies). This re-expansion resulted in the first, and apparently only, successful modern human migration out of Africa, 54 ± 8 Kya. All present-day non-Africans are descended from an L3 type that gave rise to two founder populations: M and N (see Appendix IV(i)b). The founding population was probably very small, since only this L3 type survived. Two exodus routes have been favoured: (i) down the Nile and then by land across the Sinai Peninsula, or (ii) from the Horn of Africa, across the Bab el Mandab Straight to Yemen. The migration route soon split (see Appendix IV(i)c), the tropical 'southern' coastal route to Australasia and New Guinea being the quickest. An arrival in Australia by ~40 Kya (modern human burials at Niah Cave, Lake Mungo, in southeastern Australia) is supported by MTDNA and Y chromosome evidence and would have involved a minimal eastward migration of ~300 m per year. Colonization of western Eurasia, probably from an offshoot of the 'southern' route, followed a lengthy pause (possibly in the Persian/Arabian Gulf) until climate improved (desert extending from North Africa to Central Asia), migration into Europe then taking place via the Levant. A clade of M known as M1 (see Appendix VIIa) is present in high frequencies in the Horn of Africa and is largely African-specific. The possibility that it might represent a back-migration into East Africa is supported by one of the numerous clades of U, U6, found mainly in northern Africans but also present in eastern Africans. This is corroborated by current phylogeography of the Y chromosome (albeit with lower molecular resolution than mtDNA). The dispersal of Levantine people to Europe and North Africa ~40 Kya was apparently marked by, respectively, the mtDNA haplogroups U5 and U6/M1. Similarities in the blade technologies of the Levantine 'Aurignacian' and northern African 'Dabban' support this genetic evidence.

In the period 60–30 Kya, migrants into northern Eurasia would have encountered harsher climatic conditions, and the presence of Neanderthals. Their M and N mtDNA mutated into descendant haplogroups that remain continent- or region-specific today. When Ice Age climatic conditions stabilized ~30 Kya, human populations outside Africa grew strongly (as evidenced by the star-like mtDNA clusters), expanding into Europe and northern Asia and driving *Homo neanderthalensis* to extinction. Archaeological evidence of modern human presence increasing at this time includes Gravettian culture in Europe and Zhoukoudian remains in China. Settlement of the Americas (Alaska) ~25 Kya was probably associated with the Beringia land bridge connecting America and Asia (sea level being 120 m lower than today). Founder analysis of mtDNA (see Appendix IV(ii)d) indicates that a small population from northern Siberia migrated across Beringia, bearing the mtDNA subtypes A, B, C and D found today in tribes across the Americas.

Apparently, all of these tribes speak an ancestral Amerind language.

As the last glacial maximum ~20 Kya forced humans south, proposed glacial refuges in Beringia and Iberia led to concomitant narrowing of mtDNA diversity among survivors to A/D and H/V types, respectively. Subsequent loss of B, C and X types within the Beringia group would have produced the ancestors of the Eskimo-Aleut and Dene-speaking peoples, the larger Amerind group subsequently colonizing the Americas (Meadowcroft, Pennsylvania, by ~18 Kya and Monte Verde, Chile, by ~14 Kya). Such distances would have involved a spread of ~1,000 m per year (see Appendix IV(ii)e). The post-glacial warm phases (15–13 Kya) enabled recolonization of northern latitudes, the almost 100% presence today of the A-type in Eskimo-Aleut and Na Dene speakers, the very high frequencies of H and V in northwestern Europeans (who probably came from Iberia or southern France), and the complete loss of X and B in northern Asia, all testifying to the reduced mtDNA diversity of the migrants (see Appendix IV(ii)f). Magdalenian culture in Europe and Clovis culture in North America are associated with these expansions north (a revised time range for the latter, 11–10.8 Kya, suggesting that it may not have been the first culture to arrive in the Americas). The climatic stability from 11.4 Kya onwards enabled agricultural practices to develop and the spread of the major proto-languages. The Holocene has seen dispersals into Madagascar (~2 Kya; see Appendix IV(iii)g) and the Pacific (by 1 Kya) by ocean-going Austronesians, and the recolonization of Greenland from Alaska by neo-Eskimos, whether Yupik- or Inuit-speaking (~2 Kya). An earlier (4.5 Kya) recolonization of Greenland by palaeo-Eskimo peoples apparently fell foul of the 2.6 ky Arctic cooling cycle. Recently, a study on the extent of variation in the microsatellite D9S1120 in Native Americans has supported the theory that the Americas were colonized by a small founding population. It appears in roughly one-third of more than a dozen native North and South American groups, but very rarely elsewhere except in two groups in northeastern Siberia.

The bacterium HELICOBACTER PYLORI has probably accompanied modern humans since their origins, and has thrown light on their migrations. Sequences of seven of its housekeeping genes and of one virulence-associated gene, differ according to the continent of origin. As a result of recombination between different strains, the partial LINKAGE DISEQUILIBRIUM existing between polymorphic nucleotides within genes has proved informative in genetic analysis of *H. pylori* populations. Five ancestral populations emerged from analysis: Africa1, Africa2, EastAsia, Europe1 (AE1) and Europe2 (AE2). An ancestral population tree suggests that Africa2 evolved before the other populations split, EastAsia and AE1 diverging from each other most recently. Four modern populations (hp), and five subpopulations (hsp), were identified. *H. pylori* strains within hpEurope are recombinants between AE1 bacteria from central Asia and AE2 bacteria from the Near East and Africa, indicating human population fusion. The relative contributions of AE1 and AE2 nucleotides to modern European populations are consistent with the entry of Neolithic farmers into Europe from the Near East. Speakers of Austronesian languages (Maoris and other Polynesians) are believed to have arrived in New Zealand after sequential island-hopping, likely to have led to repeated population bottlenecks (consistent with low variation and strong implied genetic drift in hspMaori bacteria). Supporting this, the hspMaori subpopulation of hpEastAsia was isolated exclusively from Maoris and other Polynesians in New Zealand. Almost all *H. pylori* strains isolated from East Asia (hpEastAsia) were assigned to the hspEastAsia subpopulation. Discovery of hpEastAsia nucleotides (hspAmerind strains) in Inuits, and Amerinds in North and South America, can be explained by their being carried during the colonization of the Americas ≥12 Kya. The absence of any sign of strong drift in the sequences indicates either that *H. pylori* accompanied these migrants in larger numbers and/or that it was introduced on multiple occasions. Known human migrations can also explain the spread of hpEastAsia and hpAfrica1 populations. The hspSAfrica and hpAfrica2 populations were found only in South Africa, where they formed the majority of strains. Strong similarity between hspSAfrica and hspWAfrica strains reflects the low genetic distance between speakers of the Niger-Congo languages and is consistent with the hspWAfrica strain being carried to southern Africa in the rapid expansion of Bantu farmers from central West Africa. The hspWAfrica strains were at low frequency in South Africa but high frequency in West Africa and in the Americas, especially among African-Americans in Louisiana and Tennessee, consistent with the modern migration of slaves from West Africa to the Americas. See LEPROSY.

milk The composition of milk produced by the mammary glands varies with the time since parturition. 'Mature milk' is not secreted for 2–3 weeks, prior to which the fluids vary in composition. For the 1st week or so after delivery, *colostrum* is secreted at ~40 mL per day. This sticky yellowish fluid is low in fats, lactose and some B vitamins, but rich in protein (casein, α-lactalbumin and lactoblobulin), minerals, and the fat-soluble vitamins A, D, E and K (see MAMMARY GLANDS for details). Casein has many phosphoserine groups that bind calcium, and is an important source of amino acids, phosphate and calcium. Milk also contains significant quantities of IGA antibodies, providing the baby with passive immunity to some infections. During the 2nd and 3rd weeks post-partum, the proportion of proteins, including immunoglobulins, decreases, but the milk becomes increasingly rich in fats and sugars so that its calorific content increases. By the 3rd week, the mature milk composition is attained, high in fats (especially short-chain triglycerides digestible by gastric lipase in children), sugars (e.g., lactose) and essential amino acids. The medium-chain fatty acids caprylic (octanoic) and capric (decanoic) acids are produced specifically by mammary gland and no other tissue. Iso-osmotic with plasma, its calorific content is ~75 kcal per 100 mL. See LACTATION, PROLACTIN (for hormonal influence on milk composition).

milliosmole See OSMOLARITY.

mineralocorticoids Steroid hormones, above all ALDOSTERONE, that are products of CHOLESTEROL (see Fig. 49) metabolism and released from the adrenal cortex (see ADRENAL GLANDS) under the stimuli of angiotensin II and extracellular K^+ (and to a lesser extent by corticotropin). Mineralocorticoid receptors (MRs), type I GLUCOCORTICOID receptors (GRs), are not ligand-selective, having a high affinity for both glucocorticoids and mineralocorticoids, so that both aldosterone and the glucocorticoid hormones bind mineralocorticoid receptors with similar high affinity. MRs and GRs are NUCLEAR RECEPTORS, occurring in the cytosol complexed with heat-shock proteins which act as stabilizing molecular CHAPERONES. They descend from a GENE DUPLICATION event deep in the vertebrate lineage but now have distinct signalling roles. In most vertebrates, GRs are specifically activated by the stress hormone cortisol and regulate metabolism, inflammation and immunity whereas MRs are activated by aldosterone to control electrolyte homeostasis and other processes. The presence of a cortisol-clearing enzyme in many MR-expressing tissues makes the receptor largely aldosterone-specific.

minerals Inorganic ionic substances, the importance of which are indicated in specific entries (e.g., IRON, PHOSPHATE), in the HEALTHY DIET entry, and references included there.

minisatellite DNA A site on DNA with a variable number of one fairly long repeat sequence dozens to hundreds of nucleotides in length. See DNA MARKERS, MICROSATELLITE DNA, VNTRS.

minute ventilation ($\dot{V}_E$, minute volume) The volume of air breathed per minute, increased by either the rate or depth of VEN-

TILATION, and derived by the equation:

$$\dot{V} = \text{breathing rate} \times \text{tidal volume}$$

At rest, if 0.5 dm^3 of air is breathed in per breath, and the ventilation rate is 12 breaths per min, the minute ventilation is 6 dm^3 per min. See EXERCISE; RESPIRATORY CENTRE.

minute volume See MINUTE VENTILATION.

miRNAs, miRs See MICRORNAs.

mirror cells (mirror neurons) Neurons discovered in 1991 by Rizzolatti in parts of the brain (including the SUPERIOR TEMPORAL SULCUS) that represent both an action and a vision of that action. That is, e.g., a neuron involved in controlling the 'reach-for-peanut' in a macaque will fire when the monkey performs the action and when it observes a fellow monkey, or a researcher, perform the same action – i.e. the neurons are components of both sensory and motor pathways. Later work revealed further mirror neurons that were active during the observation and the imitation of a particular, very specific, action. Brain imaging techniques later showed such neurons to exist in corresponding parts of the human brain, supporting the theory that mirror neurons, or the networks they are part of, not only send motor commands but enable monkeys and humans to determine the intentions of others by mentally simulating their actions (including, in humans, complex intentions), a prerequisite for behavioural imitation. Human brain regions with mirror neurons include the premotor cortex, inferior frontal gyrus, anterior cingulate and insular cortices. More recently, neurons have been found to fire not only when a certain action is enacted and observed, but also when the same action (but not others) is heard. In birds, a class of brain neurons has been identified that is active both when a bird hears a song and when it replies by singing a similar song. Such mirror neuronal activity could contribute to IMITATION. See PLACE CELLS.

mirror self-recognition On the basis of early work in the mid-20th century, a three-stage developmental sequence of reactions of infants to their own image in a mirror has been described. At ~6–12 months, they show surprise at their image and react by reaching out and laughing as if the image were real. At ~12–24 months, they start to show interest in the relationship between their own movements and the reflection. After ~24 months, they begin to play with their image, apparently aware that it is not real. Such work, in so far as it revealed whether or not the infant showed self-recognition, depended on the infant's use of LANGUAGE. In 1970, Gallup reported that young chimpanzees recognized themselves in a mirror, which included grooming parts of their bodies visually inaccessible without the mirror image, picking food from between the teeth while watching the mirror image, making faces at the mirror, and so on. After 10 days of such exposure, the chimps were anaesthetized and marked with a dye above one eyebrow ridge and on the top half of the opposite ear. On regaining consciousness, the chimps touched the marks more frequently in the presence of the mirror than in its absence and made mark-directed visual inspections and hand movements, touching the marked area and examining the fingers even though the dye had dried. This approach provided an objective test of whether nonverbal human infants and non-human animals can recognize themselves in a mirror. However, not even all primate species are suited to the approach; and it is not clear that self-directed movements indicate possession of a sense of self, or a sense of its own body. That said, there seems to be a qualitative gap in self-recognition between hominoids and other primates. See SELF-AWARENESS.

mirror visual feedback A technique devised by Vilayanur Ramachandran, enabling patients to perceive real movements in a phantom limb. It consists of a virtual reality box made by placing a vertical mirror inside a cardboard box with its lid removed, the front having two holes in it through which the patient inserts the 'good limb' (say the right one) and moves into a position that would 'insert' the phantom limb (the left). With the mirror in the middle of the box, the right limb is now on the right side of the mirror with the phantom limb on the left. The patient is asked to view the reflection of the normal limb in the mirror and to move it about slightly until the reflection appears to be superimposed on the felt position of the phantom limb, thus creating the illusion of observing two limbs when only the mirror reflection of the intact limb is actually being seen. If the patient then sends motor commands to both limbs to make mirror symmetric movements, as in clapping, they will 'see' their phantom limb moving too. Receiving this visual feedback that the phantom limb is moving correctly in response to commands, the brain apparently 'unblocks' inhibiting signals freezing the phantom limb and enabling it to be 'moved' to a position that may relieve REFERRED PAIN. See BIOFEEDBACK, PHANTOM LIMB SYNDROME.

miscarriage (spontaneous abortion) Unassisted premature termination of PREGNANCY. It has been estimated that ~50% of all human conceptions are either lost before implantation or shortly afterwards, before the mother realizes that she is pregnant (see GENETIC DISEASES AND DISORDERS, Fig. 81). For recognized pregnancies, ≥15% end in spontaneous miscarriages prior to 12 weeks' gestation, severe structural abnormalities apparently being present in 80–85% of these. Chromosomal abnormalities such as trisomy, monosomy or triploidy (see ANEUPLOIDY) are detected in ~50% of all miscarriages, this figure rising to 60% when a major structural abnormality is present. These affect 15–20% of clinical pregnancies. Submicroscopic or *de novo* single-gene disorders probably account for a proportion of the remainder. Induction or injection of cytokines such as IFN-γ and IL-12, which promote T_H1 responses in experimental animals, promotes foetal resorption (the equivalent of spontaneous abortion in humans). Excess VITAMIN A during the first trimester can be teratogenic, leading to abnormalities of the foetal nervous system and heart and some of these miscarry. HOLOPROSENCEPHALY is responsible for 1:250 pregnancies resulting in miscarriage. See DIABETES MELLITUS, IMPLANTATION, RELAXIN, TURNER SYNDROME.

mismatch repair (MMR, mismatch proofreading) A DNA REPAIR MECHANISM, in eukaryotes independent of DNA METHYLATION, and involving different protein heterodimers (see MSH2, MSH3, MSH6), by which incorrect nucleotide (or base) insertions and small insertion or deletion loops (IDLs) introduced by DNA POLYMERASES are recognized and corrected on the newly synthesized strand. Such newly synthesized strands are known to be preferentially nicked. Once the appropriate MSH heterodimer has bound DNA at a mismatched base pair (detected by the kink it induces in the DNA), another heterodimer formed from MLH1 and PMS2 (the principal homologues of bacterial MutL) binds the MSH dimer and scans the nearby DNA for a single-stranded nick. The sequence between the nick and the mismatch is then cut out and replaced. MMR also counteracts recombination between homologous but diverged DNA sequences and is important in preventing accumulations of mutations in cancer-related genes during the several stages involved in tumorigenesis (see CANCER CELLS). Inherited defects in one allele of the genes encoding MSH2, MLH1 or occasionally PMS2, result in the common cancer hereditary predisposition non-polyposis COLORECTAL CANCER.

***MITF*gene** A gene encoding microphthalmia transcription factor (MITF), a member of the Myc-related family of BASIC HELIX–LOOP–HELIX MOTIF transcription factors and conserved in all vertebrates. See MELANOCYTES, WAARDENBURG SYNDROME.

mitochondrial DNA See MTDNA.

mitochondrial Eve The name often used to denote the female who is the MOST RECENT COMMON ANCESTOR of all surviving human mitochondrial DNA (see MTDNA); i.e., its coalescent. This ancestor was not the only living human female at the time, and so the term 'Eve' has misleading connotations. Rather, it is the manner in which mitochondrial DNA is lost from the human gene pool and the peculiarities of its genetics and inheritance that enable us to pinpoint this individual both geographically and temporally (see COALESCENCE). The original 1987 'Eve' paper of Cann, Stoneking and Wilson was based on results of mtDNA RESTRICTION MAPPING and used only part of the genetic information in this molecule. Even so, it was immediately evident that modern human mtDNA is much less varied than is that of chimpanzees (ten times less). Although the 147 human donors were drawn from five continents, native Africans were under-represented. However, the conclusion reached was that 'all these mitochondrial DNAs stem from one woman who is postulated to have lived about 200,000 years ago, probably in Africa.' Although it proved impossible to root the mitochondrial tree unequivocally in Africa, work this century which answered earlier criticisms and, *inter alia*, sequenced the entire mtDNA genome, confirmed the original work's main conclusion that modern humans did originate in Africa, maybe even within the last 100,000 years, and at least one study supports an origin of half this time period (see HOMO SAPIENS). Comparison with mtDNA from the African apes confirmed the restricted nature of human mtDNA variation, lending support to the view that we have had very different evolutionary histories. See Y CHROMOSOME ADAM.

mitochondrion (pl. mitochondria) The cytoplasmic organelle that is the main source of cellular ATP, playing essential roles of fatty acid β-oxidation, phospholipid biosynthesis, calcium signalling, production of reactive oxygen species and apoptosis. It is also the site of porphyrin synthesis in mammals, required when chelated to iron in the formation of the HAEM prosthetic group of cytochrome *c*, myoglobin, haemoglobin, catalase and peroxidase. They also synthesize steroid hormones from CHOLESTEROL in steroidogenic tissues. Mitochondria are key regulators of cell survival and death, have a central role in AGEING, and interact with many of the specific proteins implicated in genetic forms of neurodegenerative diseases. There are strong links between aerobic capacity (MAXIMAL OXYGEN CONSUMPTION), mitochondrial function and the full range of cardiovascular symptoms (see CARDIOVASCULAR DISEASE).

Human mitochondria generate ~65 kg ATP per day. In a liver cell, a mitochondrion has, on average, a lifespan of ~10 days, being digested by AUTOPHAGOSOMES (see entry) after their fusion with LYSOSOMES (autophagy). The outer membrane is readily permeable to small molecules and ions, which pass through transmembrane channels termed *porins*. The inner membrane is impermeable to most small molecules and ions, including protons, other than through specific transporters. The most notable inner membrane transporters include those carrying pyruvate, fatty acids, amino acids (or their α-keto-acids) into the mitochondrial matrix. ADP and inorganic phosphate (P_i) are taken in, in exchange for ATP produced by the KREBS CYCLE. The inner membrane also bears the components of the electron transport chain and enzymes involved in oxidative phosphorylation.

The mitochondrial matrix contains the pyruvate dehydrogenase complex, and is the site of the KREBS CYCLE and β-oxidation of fatty acids (see FATTY ACID METABOLISM).

There are four unique electron-carrying complexes in the inner mitochondrial membrane (IMM). Complexes I (NADH DEHYDROGENASE COMPLEX) and II catalyse electron transfer to COENZYME Q (ubiquinone) from the electron donors NADH and succinate respectively. Complex III (cytochrome bc_1 complex) carries electrons from UBIQUINONE to cytochrome *c*, and Complex IV (CYTOCHROME OXIDASE) completes the transfer by passing electrons from cytochrome *c* to molecular oxygen, O_2, prior to water formation. Once an oxygen molecule has picked up one electron to form a superoxide ion ($O_2^{\bullet-}$) it becomes dangerously reactive and rapidly takes up an additional three electrons from wherever it can find them. Complex IV holds onto oxygen at a special bimetallic centre where it remains trapped between a haem-linked iron atom and a copper atom until it has picked up a total of four electrons, after which the two oxygen atoms of an oxygen molecule can be safely released as two molecules of water. But this four-electron reduction of O_2 releases incompletely reduced intermediates such as HYDROGEN PEROXIDE and hydroxyl FREE RADICALS that can damage cellular components, the inhibition of cytochrome oxidase activity increasing free-radical generation:

$$O_2 + e^- \rightarrow O_2^{\bullet -} \leftarrow + e^- \rightarrow H_2O_2$$
$$+ e^- \rightarrow {}^{\bullet}OH + e^- \rightarrow H_2O$$

The signalling protein $p66^{Shc}$ plays a critical role in the mitochondrial role in energy metabolism and ROS production, being required for early mitochondrial responses to an oxidative H_2O_2 challenge. Phosphorylation of $p66^{Shc}$ on the Ser^{36} residue, and inhibiting PKC β protects mouse cells against the challenge while its over-expression reproduces the mitochondrial fragmentation and Ca^{2+}-signalling defect in cells expressing $p66^{Shc}$ but not in cells lacking it. Phosphorylation enables $p66^{Shc}$ to enter the intermembrane space, where it interacts with cytochrome *c* to produce H_2O_2. This can promote opening of the mitochondrial permeability transition pore leading to cytochrome *c* entry to the cytosol and APOPTOSIS of the cell. It is possible that AGEING may be associated with impaired autophagy of impaired mitochondria (see AUTOPHAGOSOMES), and $p66^{Shc}$-dependent fragmentation of such mitochondria may localize and attenuate mitochondrial damage where *reactive oxygen species* (ROS) are produced. See SIRTUINS.

Mismatches between NADH production and the use of ATP can stress the electron transport chain (Complexes I–IV) and modulate production of these reactive intermediates, of which mitochondria are the primary cellular source (the ENDOPLASMIC RETICULUM is another), especially when proton re-entry to the matrix is uncoupled from ATP production and heat is released (see TEMPERATURE REGULATION, ADIPOCYTES). ROS, which may ultimately be the trigger for INFLAMMATION, may leak out into the cytosol and enter other organelles, including the nucleus, producing *oxidative stress*, since they are generally regarded as a cellular hazard that can damage proteins (e.g., protein kinases and phosphatases), lipids and nucleic acids (see AMYOTROPHIC LATERAL SCLEROSIS).

Oxidative damage to DNA and RNA generates 8-hydroxy-2′-deoxyguanosine, while lipid peroxidation generates F2-isoprostanes. Certain oxidized phospholipids and short-chain aldehydes arising from lipoprotein oxidation can induce transcriptional activation of the VCAM-1 gene via NF-κB (see ATHEROSCLEROSIS). ROS generation is greatly elevated following metabolic upsets after ISCHAEMIA, reperfusion of the heart, stroke in the brain, in a wide range of disorders involving NEURAL DEGENERATION, and is now being implicated in METASTASIS.

However, ROS are required in a regulatory role to transduce growth signals via some receptor tyrosine kinases and in regulation of ATP production via activation of uncoupling proteins. Protection from ROS is normally provided by NADPH and GLUTATHIONE, the balance between ROS production and detoxification being controlled by a set of cellular enzymes which includes superoxide dismutase (SOD), catalase, and those involved in glutathione synthesis. The superoxide anion ($O_2^{\bullet -}$) is converted by superoxide dismutase to hydrogen peroxide, subsequently being reduced to water by glutathione peroxidase (see PARKINSON DISEASE). As PGC-1 (see PGC-1 GENES), which is regulated by the sirtuin SIRT1, stimulates mitochondrial biogenesis and mitochondrial-based respiration, it also increases expression of *SOD*, glutathione peroxidase and the enzymes of glutathione synthesis. ROS production by mitochondria is believed to be greatly reduced by PGC-1-stimulated expression of the uncoupling protein-encoding genes *UCP2* and *UCP3* (see UCPS) whose products dissipate the proton gradient across the inner mitochondrial membrane, reducing ATP production. Non-esterified fatty acids induce expression of UCP-2, which in addition decreases insulin secretion.

All mammalian genomes have retained just 37 genes from the genome of the original free-living bacterium from which mitochondria evolved (see MTDNA). Yeasts and plants have retained more. Hundreds, perhaps thousands, of these ancestral genes have been transferred to the nuclear genome during the evolution of the stable mutualism that has evolved between the bacteria and the present organelle so that ~99% of human mitochondrial proteins are now encoded by nuclear genes so that more than 80% of human mitochondrial diseases fail to follow maternal inheritance patterns. Currently, we are very ignorant of what these nuclear genes are (termed *nuclear insertions of mitochondrial genes*, or numts). Simultaneous stimulation of 'mitochondrial genes' encoded by the two genomes (nuclear and mitochondrial) by PGC-1α and PGC-1β coactivators results in increased capacity for fatty acid β-oxidation, Krebs cycle and oxidative phosphorylation. They also induce expression of genes involved in mitochondrial haem biosynthesis, ion transport, translation of proteins and protein import, and can stimulate mitochondrial biogenesis – even of mitochondria with different metabolic characteristics. There is evidence that respiration stimulated by PGC-1β is more highly coupled to ATP production than is that induced by PGC-1α, the relative activities of the two coactivators in a cell possibly enabling fine-tuning of mitochondrial function to metabolic needs (e.g., see ADIPOCYTES). A protein destined for the mitochondrion from the cytosol requires an N-terminal SIGNAL SEQUENCE, typically containing many hydrophobic amino acids and several positively charged ones, all forming an amphipathic helix with charged groups on one surface and nonpolar groups on the other.

Mitochondria have been thought to contribute to AGEING through accumulation of mtDNA mutations and net production of reactive oxygen species (see NEURAL DEGENERATION). It is of interest that protective mtDNA lineages with the UJKT mtDNA haplogroup may decrease the risk of developing Parkinson's disease compared with haplogroup H. The protective haplogroup seems to have arisen in areas requiring cold adaptation where relative uncoupling of electron transport from ATP synthesis to increase heat generation may increase longevity and decrease ROS production.

mitogens Those GROWTH FACTORS stimulating cell proliferation (see CELL GROWTH), operating through pathways activated by their binding RECEPTOR TYROSINE KINASES (RTKs), aka *mitogen-activated protein kinases*. Other signalling pathways rely almost exclusively on serine and threonine kinases. Almost all animal mitogens, including human, control the rate of cell division by acting in G1 of the cell cycle. During G1, many mechanisms act to suppress CDK activity, hindering entry to S-phase of the CELL CYCLE, and mitogens act to release the brakes on CGK activity. Several operate via the PHOSPHATIDYLINOSITOL 3-KINASE pathway (see PKB) muting the effect of the CDK inhibitor, $p21^{Cip1}$, and/or $p27^{Kip1}$. Within minutes of mitogen binding, transcription of IMMEDIATE EARLY GENES is detected (e.g., levels of *myc* mRNA (see MYC) rise on binding, falling after mitogen removal). Mitogens include the steroid hormones oestrogens, progesterone and testosterone. Some signals, such as transforming growth factor-β, may overrule mitogenic signals and interrupt proliferation. See INTERFERONS, RAS SIGNALLING PATHWAY.

mitosis (M-phase) The phase of the CELL CYCLE, triggered by M-Cdk (see CYCLINS) during which the nucleus is replicated, the final stages of which overlap with cytokinesis (division of the cytoplasm, or cell division), both being required for increase in cell number (see SELF-RENEWAL). A dynamic sequence of cyclical events, mitosis is usually analysed into five stages: prophase, prometaphase, metaphase, anaphase and telophase (see Fig. 125).

The mitotic spindle starts to self-assemble in the cytoplasm during prophase and each replicated chromosome nucleates its array of MICROTUBULES. The relatively few, long, microtubules of interphase rapidly give way to a larger number of shorter microtubules surrounding each CENTROSOME which start to form the mitotic spindle. These changes are initiated by the cyclin-dependent kinase M-Cdk which phosphorylates the proteins controlling microtubule dynamics. Microtubules growing from the centrosome at one pole meet (overlap) and engage with those growing from the other. Microtubules that attach to chromosomes become stabilized and do not depolymerize. They attach end-on at the chromosome's kinetochore, a protein complex that assembles on the centromere during late prophase. Polymerization of tubulin, elongating microtubules, occurs at the kinetochore. If a chromosome attaches successfully by its kinetochore to a microtubule from each of the two centrosomes it is tugged back and forth until it assumes a position, the metaphase plate, equidistant from the two spindle poles. Here it awaits the checkpoint signal to split its centromere, delivered only when all chromosomes have a bipolar microtubule attachment (see CELL CYCLE). At least seven families of KINESIN-related motor proteins have been localized to the vertebrate mitotic spindle, the balance between plus-end-directed and minus-end-directed motor proteins apparently determining spindle length. When two microtubules from opposite centrosomes interact in the microtubule overlap zone, plus-end-directed motor kinesin-related motor proteins cross-link them and tend to drive them in directions pushing the centrosomes apart, a process in which minus-end-directed dynein motors are also thought to be involved.

About half the time spent in a mitotic cycle is spent in metaphase, after which kinetochores are believed to pull the chromosomes toward the poles, overcoming a 'push' in the opposing direction from the aster microtubules. Anaphase then begins with the abrupt release of the cohesin linkage holding sister chromatids together. The anaphase-promoting complex, a large proteasome complex, degrades anaphase inhibitors such as securin, and the mitotic cyclin M-cyclin (cyclin B), inactivating M-Cdk as a result. The transition from metaphase to anaphase initiates dephosphorylation of many proteins that were phosphorylated at prophase. It is not until mitotic kinase (M-Cdk) is switched off that phosphatases can act unopposed. With securin gone, the protease separase is activated and cleaves the COHESIN complex, ungluing the chromatids.

The initial poleward movement of chromosomes (anaphase A) is mainly the result of shortening of kinetochore microtubules at their centromere attachment and depends on motor proteins at the kinetochore that somehow cause depolymerization of tubulin. In human cells, anaphase B begins shortly after anaphase A and stops when the spindle is about twice its metaphase length. In one model, motor proteins cross-link the overlapping antiparallel microtubules and slide them past each other, pushing the spindle poles apart, while minus-end-directed motor proteins bind both the cell cortex and astral microtubules pointing away from the spindle and pull the poles apart. By the end of anaphase, daughter chromosomes (formerly sister chromatids) have arrived at the opposite poles of the cell and begin to decondense. This stage, telophase, is completed when a nuclear envelope reassembles around each genome, associating with the surfaces of individual chromosomes and eventually coalescing, to form two daughter interphase nuclei. Nuclear pore complexes are incorporated into the envelope and dephosphorylated lamins reassociate to form the nuclear lamina. Two genetically identical nuclear genomes have then been formed (see NUCLEUS). Compare MEIOSIS; see CELL GROWTH.

mitotic recombination See RECOMBINATION.

mitral valve (**bicuspid valve**) The atrioventricular valve on the left side of the HEART.

mixed nerves Nerves containing axons of both sensory and motor neurons.

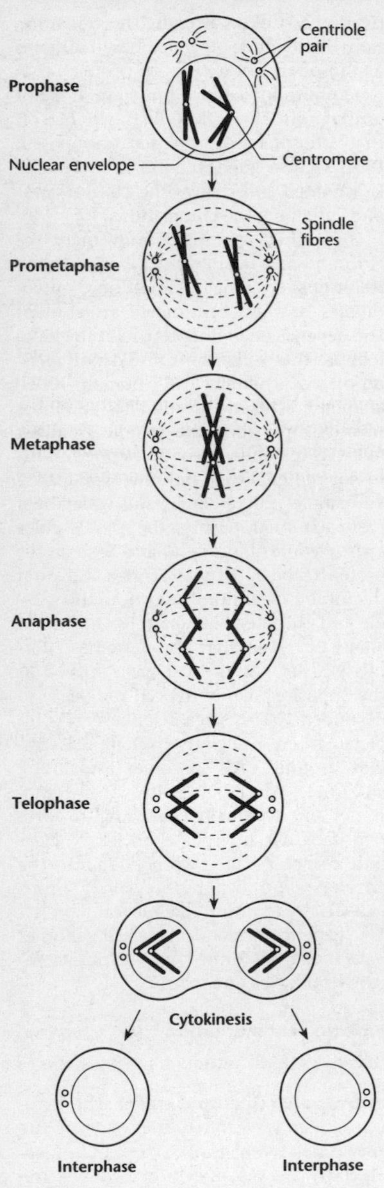

FIG 125. Diagram showing the behaviour of one pair of homologous chromosomes during MITOSIS.

MMR vaccine See MEASLES, VACCINES.

Mode 1, Mode 2 ... Mode 4/5 technologies A scheme for comparing stone tool industries devised by J. Desmond Clark, according to which the crude flaked pebble tools of the Oldowan industry (see OLDOWAN CULTURE) are regarded as Mode 1 technology; bifaces (the most common examples being hand axes), present in the Acheulian industry (see ACHEULIAN CULTURE), are categorized as Mode 2 technology; the gradual replacement of hand axe culture by the production of sizeable flakes, which were further shaped or retouched to produce sizeable regular and symmetrical tools, represents Mode 3 technology (some flakes being hafted to produce spears, first associated with remains of HOMO HEIDELBERGENSIS); tool kits emphasizing blades (long, thin and flat retouched flakes with a sharp edge) are indicative of Mode 4/5 technologies (see AURIGNACIAN CULTURE).

modifiers Variants of one gene that alter the phenotypic expression of another ('target') gene at a different locus in the genome. Selection for such modifiers can establish a genetic background within which a new mutation at a locus will be expressed to a greater or lesser extent (see PENETRANCE). The tendency for health to persist in spite of the presence of susceptibility genes (see GENETIC RISK) could be explained by the presence of modifier genes and protective alleles conferring genetic resistance to disease (see RESISTANCE GENES). A classic example is a genetic variant on human chromosome 19q (and chromosome 7 in mice) that controls the association of meconium ileus with CYSTIC FIBROSIS. Phenotypic variation among individuals sharing identical mutations is both a complication in conventional genetic studies and a signal that modifier genes are involved. In multigenic models of human disease (see COMPLEX DISEASES), variants in genes such as dead-end and TRP53 modulate susceptibility to testicular cancer; apolipoprotein E influences susceptibility to heart disease, and carboxypeptidase E, low density lipoprotein receptor and leptin affect susceptibility to obesity.

molars In primates, other than tarsiers, there has been a trend to add a fourth cusp on the inside of the rear end of the upper molar to the basic mammalian tricuspid molar, and to lose the leading cusp on the main 'triangle' of the lower molar, leaving only four to five cusps. Among Old World simians, both upper and lower molars of Old World monkeys have four cusps linked in pairs by transverse ridges (bilophodonty); but the lower molars of the great apes have five well-developed cusps, the rearmost having moved outwards. The valleys between these five cusps (three outside and two inside) create a Y-shaped pattern, the lower molars of great apes therefore having a Y-5 pattern. Fossil hominins either have massive molars, or reduced molars as in early *Homo* (see AUSTRALOPITHECINES). See DENTITION, TAURODONT.

molecular clock The hypothesis that molecular (e.g., DNA sequence) evolution occurs at a sufficiently constant rate for divergence between two sequences to be accurately related to the time since they split from a common ancestor. The dating of molecular divergencies has built-in uncertainties of unknown magnitude since calibration is based on palaeontological data.

Inferences of such times from calculations of GENETIC DISTANCE (see also PHYLOGENETIC TREES, Fig. 140) between two populations rely on the principle of constancy of a molecular clock, although the processes underlying such genetic distances do not always show linear relationships with time. Consequently, modifications of these statistics have been proposed.

The processes just alluded to include genetic drift, mutation, recombination and natural selection. When drift is believed to be the sole influence, the rate of population divergence will depend only on their EFFECTIVE POPULATION SIZES, N_e, and assume these have remained constant (an approach required when calculating F_{ST} values). But if it is believed that mutation rate is driving the molecular clock then Nei's standard genetic distance, D, is employed. See MOST RECENT COMMON ANCESTOR.

moles (melanocytic naevi) Benign, senescent neoplasias produced when MELANOCYTES in the skin escape their tight regulation by keratinocytes, proliferating and spreading. This can be restricted to the epidermis (junctional naevus), the dermis (dermal naevus) or overlapping components of both (compound naevus). Some naevi are dysplastic, with morphologically atypical melanocytes. Although generally benign, they can progress to the radial-growth-phase MELANOMA, an intraepidermal lesion that can involve some local microinvasion of the dermis. These cells can also progress to the more dangerous vertical-growth-phase, in which they have metastatic potential, invading the vascular and lymphatic systems.

monoamine oxidases (MAOs) Enzymes degrading CATECHOLAMINES and SEROTONIN, especially in the liver and in axon terminals, and located on the outer membranes of mitochondria. There are five allelic variants of the MAOA gene encoding monoamine oxidase A, one of which results in low MAOA levels. A combination of severe childhood maltreatment and low-activity MAOA has been found to correlate strongly with male adult antisocial behaviour. High-sensation seeking behaviour appears to be correlated with low MAO levels. MAO inhibitors (MAOIs), which cause a marked MOOD elevation, include phenelzine (see ANTIDEPRESSANTS, Fig. 12).

monoamines The neurotransmitters NORADRENALINE, DOPAMINE and SEROTONIN, all produced by the diffuse modulatory systems of the BRAIN (see Fig. 30). See AFFECTIVE DISORDERS, DEPRESSION, MEMORY, MOOD.

monoclonal antibodies (mAbs, monoclonals) Because B cells have only a few days' life in culture they are unsuitable for commercial antibody production. Although antisera contain antibodies, these are of polyclonal origin and therefore heterogeneous. But in 1975 the development of a special type of hybrid cell line enabled pure antibodies to be produced on a commercial scale. An antibody-producing B cell from an appropriately immunized mouse was fused to an appropriate mutant human tumour cell, which produces no antibody itself, to produce a heterokaryon (containing two or more nuclei) that eventually formed a hybrid cell (hybridoma) that could be cultured. Cultures screened for the desired antibody could then be subcloned from single cells to give antibody molecules that have identical structures, including their antigen-binding site and isotype. These are now used routinely in serological assays, as diagnostic probes and as therapeutic agents.

The problem of using monoclonals as therapeutic agents is that humans readily develop antibodies to mouse antibodies. Exon shuffling at the DNA level has been exploited in antibody engineering, where antibodies with novel combinations of protein domains can be artificially constructed. Such work produced *humanized antibodies* in which a lesser or greater part of the rodent mAb is replaced by the human equivalent. Mice lacking their own immunoglobulin genes can then be made transgenic for human heavy- and light-chain genes by use of yeast artificial chromosomes so that their T- and B-cell receptors are encoded by human genes. But because the B cells develop in mouse bone marrow, these receptors are not tolerant to most human proteins. These mouse cells can then be used to induce human monoclonals against epitopes on human cells or proteins.

Antibody-like molecules that are not recognized as foreign can also be developed by cloning sequences for *human* antibody heavy- and light-chain V regions from a DNA library into a filamentous phage so that each phage expresses one heavy-chain and one light-chain V region as a surface fusion protein with antibody-like properties (see PHAGE DISPLAY LIBRARY). Then, using a surface coated with antigen, those that bind the desired protein can be selected. The unbound phages are washed away, the bound phage recovered, multiplied in bacteria, and again bound to antigen. Each phage isolated in this way will produce a monoclonal antigen-binding particle analogous to a monoclonal antibody, and the technology may replace the hybridoma construction.

monocular deprivation See OCULAR DOMINANCE.

monocular zone (temporal crescent) That portion of the VISUAL FIELD (Fig. 179) occupied by light from the lateral or temporal portion of the hemifield and projecting only onto the medial or nasal hemiretina of the eye on the same side (the nose blocking this light from reaching the eye on the opposite side). See OPTIC CHIASMA, Fig. 130.

monocytes Myeloid leukocytes, 10–20 μm in diameter, with a kidney-shaped nucleus (see STEM CELLS), produced by haematopoietic progenitors (monoblasts) in bone marrow. While in the blood (3–8% of leukocytes), they carry out phagocytosis of foreign particles using intermediary (opsonizing) proteins such as antibodies or complement that coat the pathogen, as well as by binding to the microbe directly via pattern-recognition receptors that recognize pathogens. They carry F_C receptors and can kill infected host cells via antibody (antibody-dependent cell-mediated cytotoxicity). They give rise to tissue macrophages after extravasation from capillaries by DIAPEDESIS. The chemokine CCL2 initiates this migration, and this and CCL7 and CCL13 act on the monocyte receptor CCR2 to suppress their production of IL-12. Their migration into atherosclerotic plaques is largely responsible for atheroma production (see ATHEROSCLEROSIS).

monogenic disorders Those GENETIC DISORDERS involving a single locus. See MENDELIAN INHERITANCE.

monooxygenases Abundant enzymes of complex action, catalysing reactions in which only one of the two oxygen atoms of O_2 is incorporated into an organic substrate, the other being reduced to H_2O. Two substrates are involved, the main substrate accepting one of the two oxygen atoms while a cosubstrate furnishes hydrogen atoms (some using $FMNH_2$ or $FADH_2$, others using NADH or NADPH) to reduce the other to H_2O. Since the main substrate is hydroxylated, they are hydroxylases. The enzyme which hydroxylates the phenyl ring of phenylalanine to give tyrosine, and which is defective in phenylketonuria, is a monooxygenase, as is the haem protein CYTOCHROME P-450.

monophyletic A grouping of taxa, the common ancestor of which is unique and universal, so that they fall into a single CLADE and do not exclude any members of that clade. Only such groupings are regarded as coherent evolutionary lineages. See PHYLOGENETICS. Compare PARAPHYLETIC, POLYPHYLETIC.

monosomy The condition in which a nucleus contains one fewer chromosomes than the normal diploid number, often resulting from nondisjunction, from translocation or from loss of a chromosome ('anaphase lag') during its movement to the pole of the cell during meiosis. Autosomal monosomies are almost invariably incompatible with survival to term. See ANEUPLOIDY, TURNER SYNDROME.

monosynaptic reflex A REFLEX ARC involving a synapse between the sensory and motor neurons, without the involvement of a relay neuron.

mono-unsaturated fatty acids Those FATTY ACIDS containing one double bond.

monozygotic twins 'Identical' twins. See TWINS.

mood The term 'affect' refers to emotional state or mood, and AFFECTIVE DISORDERS are those conditions in which an individual's health is affected by their disturbance (e.g., see FEAR AND ANXIETY). The monoamine hypothesis of mood disorders attempts to explain depression by depletion of either noradrenaline (NA) or SEROTONIN neurotransmitters at synapses of the appropriate DIFFUSE MODULATORY SYSTEMS in the BRAIN (see Fig. 31). However, no simple equation of mood with transmitter level is credible, since elevation of NA level by drugs (e.g., cocaine) is not as effective as prevention of its reuptake by antidepressants, and the time delay of a few weeks in the relief of depressive symptoms by antidepressants hinted at the importance of other factors. D_2 DOPAMINE receptors occur in the limbic system, concerned with mood and emotional stability.

morbidity Ill-health.

morpheme The smallest linguistic unit that has semantic meaning (usually, a word). In spoken LANGUAGE, morphemes are composed of PHONEMES, the smallest linguistically distinctive units of sound.

morphine An alkaloid, with very similar structure to codeine and HEROIN (diamorphine). Like other OPIOID analgesics, it produces a range of central effects including analgesia, euphoria, sedation, depression of ventilation, depression of the vasomotor centre, stimulation of the nucleus of the oculomotor nerve, nausea and vomiting. Peripheral effects include constipation and biliary spasm. It may cause release of histamine. Both physical and psychological DRUG DEPENDENCE on opioid analgesics gradually develops, sudden termination of administration precipitating withdrawal.

morphogens Secreted signalling molecules which organize a field of surrounding cells into patterns (see PATTERN FORMATION). Emanating from a localized source, each morphogen's distribution forms a concentration gradient determining the arrangement and fate of responding cells in accordance with the perceived concentration. Examples include fibroblast growth factor-8 (FGF-8) produced by the PRIMITIVE NODE, several members of the TRANSFORMING GROWTH FACTOR-β SUPERFAMILY, and RETINOIC ACID.

mortality The *death rate* (officially, the *central death rate*) is the number of deaths in a population of a given age, ethnicity or sex, divided by size of the same population

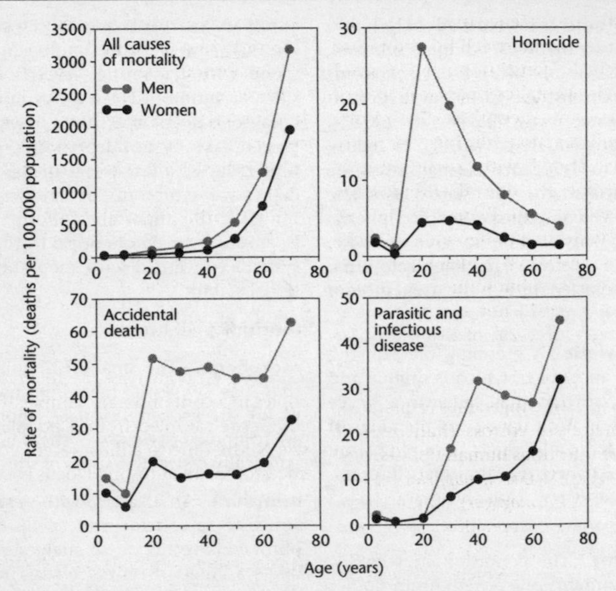

FIG 126. Graphs indicating sex differences in MORTALITY in the 1997 USA population. Although the three specific mortality rates shown indicate higher rates in men than in women, the timing of onset of male-biased mortality varies across causes, those for death through homicide and accidental causes increasing immediately after puberty. For parasitic and infectious diseases, the disparity occurs much later. From *Science* Vol. 297, No. 5589 (20 Sept. 2002), p. 2008 (since corrected). Reprinted with permission from AAAS.

at risk and then multiplied by a constant (as per 100,000). This gives the number of deaths expected for a population of that size. These figures may then be converted into a probability of dying (see LIFE EXPECTANCY). Underlying causes of death vary widely with each such population (see Fig. 126); but the risk of death is highest at birth, declines to its lowest levels by 13 years, and increases exponentially thereafter. Since the 1960s, extremely rapid declines in mortality have occurred in the developed world among the major fatal degenerative diseases (e.g., cardiovascular disease, stroke and some cancers). Most of the gains in life expectancy at birth that occurred in the 20th century are attributable to declines in infant, child and maternal mortality. See HUMAN DEMOGRAPHY.

morula Solid mass of 8–16 cells (blastomeres) resulting from cleavage divisions of the zygote. These initially rounded cells, with uniformly distributed microvilli, adhere to one another by E-cadherin. They then flatten out to maximize contact and undergo compaction, during which cell polarity is introduced, microvilli becoming restricted to the apical surfaces and E-cadherin to the basolateral surfaces. The cells form tight junctions with their neighbours, cytoskeletal elements becoming reorganized. By the 16-cell stage, the morula consists recognizably of external polarized cells (TROPHOBLAST) and internal non-polar cells (embryoblast), and as the latter increase in numbers the two cell types begin to communicate via gap junctions. The blastocoele develops at about the 32-cell stage, when the structure is termed a BLASTOCYST. Human morulae implant in the uterine endometrium five days after fertilization (see PREIMPLANTATION GENETIC DIAGNOSIS).

mosaicism The presence in an individual or tissue of two or more cell lines differing in their genetic constitution but derived from a single zygote (compare CHIMAERISM). Chromosome mosaicism usually results from nondisjunction during an early embryonic mitosis with persistence of more than one cell line. If chromosome 21 were involved, the cell line with the trisomy 21 would survive (the embryo being expected to have ~33% mosaicism for trisomy 21) while the monosomic line with 45 chromosomes would not survive. Mosaicism accounts for 1–2% of cases of DOWN SYNDROME.

mosquitoes Two-winged flies (dipterans), whose females are VECTORS (italicized) of some major infectious human DISEASES: DENGUE FEVERS (*Aedes*), MALARIA (*Anopheles*), YELLOW FEVER (*Aedes* and *Haemogogus*), chikungunya (*Aedes*), Japanese encephalitis (*Culex*) and WEST NILE FEVER (*Culex*). Major difficulties in their control, some of which are discussed in the above entries, include their rapid aquatic life cycle, varied ecologies and evolving resistance to insecticides. The Asian tiger mosquito (*Aedes albopictus*) is especially troublesome because, unlike those of many other tropical mosquitoes, its eggs can survive cold winters by diapausing, facilitating its spread into the USA, Central and South America, and Europe, since the early 1980s. The used tyre trade seems to have been largely responsible, the pools of rainwater collecting in the tyres providing breeding sites while the containers in which the tyres are shipped provide a sheltered journey. According to the World Health Organization, spread of dengue since the 1970s has placed 2.5 billion people at risk, with an estimated 50 million cases per year. Malaysian authorities are considering field trials involving release of millions of male *Aedes aegypti* that have been genetically modified to produce offspring that die in the larval stage in the absence of tetracycline. But there has been resistance to the proposal from local environmentalists.

mossy fibres 1. Axons of fibres of the dentate gyrus synapsing on cells in division CA3 of the HIPPOCAMPUS (Fig. 91). **2**. Axons of a neuron arising in a number of brainstem cell groups, including the pons (pontine nuclei), relaying information from the cerebral neocortex and synapsing on cerebellar granule cells (see CEREBELLUM).

most recent common ancestor (MRCA) In evolutionary genetics, the species (or DNA sequence) into which any two (usually extant) species or sequences coalesce (see COALESCENCE). The time to the most recent common ancestor (TMRCA) is the number of generations back in time when a single gene copy gave rise to all the gene copies in a contemporary sample. If n gene copies are sampled from a population of size N, the time to a most recent common ancestor for an autosomal locus is expected to be $4N(1 - 1/n)$ generations. See HOMO SAPIENS.

motion-selective cells One kind of neuron (*component direction-sensitive*) in the primary visual cortex (V1) responds to motion of a stimulus only in a direction perpendicular to the plane of orientation for which it is specialized; but others (*pattern direction-selective*) respond to motion of complex patterns for which there is no single plane of orientation, an ability dependent upon receipt of inputs from a group of direction-sensitive neurons. Pattern direction-sensitive neurons respond to approximately what we normally understand by motion, i.e., the 'global' motion of a particular object. See VISUAL MOTION CENTRE.

motivational system See REINFORCEMENT AND REWARD.

motor cortex Reciprocally interconnected and somatotopically organized areas of the frontal lobe of the BRAIN (see Fig. 29) involved in the planning and execution of voluntary movements. The PREMOTOR CORTEX (Area 6) is divisible into the premotor area (PA) involved especially in planning movements requiring sensory cues, and the supplementary motor area (SMA) which forms a motor loop with the

BASAL GANGLIA. Area 6 sends output to the striatum which projects back to the SMA via connections with the globus pallidus and the thalamic ventral posterior nucleus (VPN). The PRIMARY MOTOR CORTEX (Area 4, or M1) forms a motor loop, via the pons, with the CEREBELLUM. Axons project from the cerebellum to the THALAMUS, which projects back to Area 4 and PA, forming other motor loops. The thalamus is therefore central to information flow between the motor cortex, the cerebellum and the basal ganglia.

Projections from the PREFRONTAL and POSTERIOR PARIETAL CORTICES converge on the premotor cortex, which therefore lies at an interface where signals encoding *what* actions are desired are converted into signals specifying *how* those actions will be executed in the form of *motor commands* which result in appropriately coordinated sequences of muscle contractions (for the control of walking, see CENTRAL PATTERN GENERATORS). Between them, Areas 4 and 6 contribute over half of the axons of the CORTICOSPINAL TRACTS (see SPINAL CORD, Figs 162 and 163), the major pathway for the execution of INTENTIONAL MOVEMENTS, especially hand movements. Roughly 40% of the remaining axons in these tracts originate in the SOMATOSENSORY CORTEX or other parietal regions.

Area 6 is involved, with the posterior parietal cortex, in planning spatial and temporal features of voluntary movements, especially those involved in response to external stimuli. Neurons here fire up to 800 ms before a voluntary movement begins. Simple movements increase cerebral blood flow (cbf) in Area 4 alone; more complex tasks also involve increased cbf in Area 6; but mentally rehearsing a complex task in the absence of its execution increases cbf in Area 6 alone. The same motor tasks can be performed by different synaptic combinations (motor equivalence), executed through the two descending lateral motor pathways. Several somatotopic maps exist in the motor cortex (see CEREBRAL CORTEX, Fig. 44) and their topography are preserved in the orderly distribution of fibres in the corticospinal tracts: e.g., axons synapsing with lower motor neurons in the legs are lateral while those synapsing with lower motor neurons in the arms are medial. Such mapping is not a one-to-one mapping to individual muscles or movements.

The set of neurons in Area 4 that is activated during execution of an intentional movement involving a particular muscle depends, *inter alia*, on the movement's direction and force. Recordings from single cells in Area 4 of conscious macaque monkeys during intentional limb movements show that they correlate with force, rate of change of force, velocity, acceleration or joint position. Direction of movement is not encoded well by single Area 4 cells, but the average firing of a few hundred cells (population coding) is (see NEURAL CODING). None of these variables maps in an orderly way to the cortex, however. The premotor area (PA) and supplementary motor area (SMA) contain neurons which fire in a manner correlated with the direction and force of a movement, and both control distal limb muscles via Area 4. In addition, the SMA directly controls proximal limb muscles via the corticospinal tract, while the PA neurons synapse with reticular neurons in the brain stem which project to both axial and proximal limb muscles.

It may be surprising that many Area 4 neurons have *sensory* receptive fields (RFs). However, this region receives substantial input from the somatosensory cortex, and Area 4 neuron RFs are located in those positions where the neuron's activity is likely to cause movement and where there are likely to be sensory consequences (neurons in caudal Area 4 respond mostly to cutaneous mechanoreceptor input, those more rostral respond to proprioceptor input, especially from MUSCLE SPINDLES). See GAIT, SEIZURES.

motor end-plate The region of a muscle fibre's surface membrane (often invaginated), and the adjacent axon terminal which innervates it, at a NEUROMUSCULAR JUNCTION.

motor neuron diseases (MNDs) See NEURAL DEGENERATION.

motor neurons (motoneurons) Normally referring to those efferent neurons whose axons synapse with skeletal muscle fibres (see NEUROMUSCULAR JUNCTIONS), all of which release acetylcholine as their neurotransmitter; but the phrase is sometimes applied to projection neurons in motor pathways generally, even if they do not synapse with muscles. For example, *upper motor neurons* originating in the motor cortex, red nucleus and cerebellum pass down descending tracts (e.g., the CORTICOSPINAL TRACTS), eventually synapsing with interneurons in the grey matter of the SPINAL CORD (see Fig. 162). These synapse with *lower motor neurons* which then synapse with effectors. Although myelination of nerve fibres in the spinal cord begins at approximately four months of intrauterine life, some motor fibres descending from the higher brain centres do not become myelinated until the first year of postnatal life. For motor neuron diseases, see NEURAL DEGENERATION.

motor pathways See CORTICOSPINAL TRACTS, MOTOR CORTEX, RUBROSPINAL TRACT.

motor pattern generators See CENTRAL PATTERN GENERATORS.

motor proteins (molecular motors, motor molecules) Proteins whose changes in shape (conformational changes) are used to drive chemical systems away from thermodynamic equilibrium and generate directed movements of molecules and organelles within cells. They must utilize free energy changes from some other source, usually the large negative free energy release of nucleoside triphosphate (NTP) hydrolysis (usually ADENOSINE TRIPHOSPHATE, or GTP), which makes for irreversibility within cells. Binding such a NTP shifts a motor protein conformation from state 1 → state 2, hydrolysis of the NTP → NDP + P_i shifts the conformation from state 2 → state 3, and release of the bound NDP + P_i shifts the protein back to state 1. Motor proteins that generate directional movement by these processes include MYOSIN (which 'walks' along actin filaments, including those of skeletal and other muscle fibres and produces contractile force), KINESIN and cytoplasmic DYNEIN proteins (which walk along microtubules). Interactions between these cytoskeletal motor proteins and their sites of attachment on the filament track must repeatedly form and break and they have a 'head' region (motor domain) which not only binds and hydrolyses the NTP but also identifies the track and the direction of movement along it. Cargo identity is determined by the 'tail' of the motor protein. Axonemal dynein causes sliding of microtubules in the axonemes of cilia and flagella. See MITOSIS.

motor unit A motor neuron plus all the skeletal muscle fibres it stimulates (see NEUROMUSCULAR JUNCTION). One motor neuron makes contact with an average of 150 muscle fibres (sometimes 1,000 or more) causing their simultaneous contraction or relaxation. The greater the precision and control of movement required, the smaller are the motor units involved and the greater their number. All muscle fibres innervated by a single motor neuron are of the same basic type, and there are correspondingly three basic types of motor unit – fast twitch type I units, fast twitch type II units and slow twitch units – corresponding to the characteristics of the muscle fibres innervated (see SKELETAL MUSCLE). Muscle fatigue is the failure to maintain the required force or power output. Fatigue-resistant (FR) motor units can sustain moderate force for ~5 minutes before a steady decline sets in over several minutes. Fast-fatigue motor units achieve great force, but repetitive stimulation leads to precipitous drop in force production after ~30 seconds. See EXERCISE, MOTOR CORTEX.

mouse, mouse models It is our shared evolutionary heritage with other living species that underwrites our belief that studying them will deliver insights into our own biology and pathology. Our closest relatives, the great apes, have not proved very useful as human disease models but other mammals, notably mice, have.

Unlike primates, mice can be bred easily and have a short generation time, so that breeding programmes can produce recombinant inbred strains and CONGENIC STRAINS. Sequencing the mouse genome has enabled identification of more genes in the human genome than has sequencing of the human genome alone, since their sequences tend to be conserved in both (see HUMAN GENOME PROJECT, PHYLOGENETIC FOOTPRINTING). In order to distinguish them, the names of mouse genes that are homologous to human genes are often not fully capitalized, and their human homologues often begin with '*h*'. There are mouse models for all the major classes of human disease – inherited disorders, cancers, infectious diseases and autoimmune disorders – and even of such COMPLEX DISEASES as ALZHEIMER'S DISEASE, ATHEROSCLEROSIS, DEPRESSION and other AFFECTIVE DISORDERS, and ESSENTIAL HYPERTENSION. So also for diseases associated with mutations in GLUTAMATE receptors. Recently (2005) transchromosomic mice have been employed in models of DOWN SYNDROME (DS). In addition to their normal diploid chromosome complement, such chimaeric mice also carry an extra human chromosome 21 (Hsa 21) because they develop from a blastocyst into which was microinjected a mouse EMBRYONIC STEM CELL containing the extra chromosome. This human chromosome can now be transmitted almost complete in the mouse germline and results in mice that recapitulate certain features of humans with DS, including changes in behaviour, synaptic plasticity, cerebellar neuron number, heart development and mandible size. Indeed, mice have a medial temporal lobe system, including a HIPPOCAMPUS, resembling those of humans, and use their hippocampus to store MEMORY of places and objects, just as we do. Neuroanatomical differences between rodent and human brains caution against jumping from studies of rodent behaviour to human psychotherapy. Animal models are woefully inadequate representations of human-specific psychiatric disorders. Classical animal tests for psychiatric disorders are based on responses to clinically proven drugs, and these do not necessarily reflect the cause or biological basis of the disorder they are supposed to mimic. Human psychiatric disorders are complex amalgams of affective, cognitive and behavioural abnormalities and even though we may be able to model aspects of one such dimension in rodents, this can only provide clues to an understanding of the complete human disorder (see AFFECTIVE DISORDERS, ADDICTION, REINFORCEMENT AND REWARD). However, being very social animals, mice are a good choice for modelling symptoms of AUTISM. It is widely held that the time is ripe for applications of recent discoveries about the human brain to creating more useful mouse models, often by deletion or addition of mutant candidate susceptibility genes.

Orthologous gene mutations (see ORTHOLOGUES) are more likely to generate similar phenotypes in humans and mice than in humans and invertebrates such as worms or flies; and mouse and human exon sequences, and exon–intron boundary structures, are usually well conserved between mice and humans. The great benefit of using transgenic/gene targeted mice is that specific disease models can be constructed to order and, provided that the relevant gene clones are available, mice can be generated with a relevant alteration in a specific target gene. Once a candidate disease gene has been identified in humans (see POSITIONAL CLONING), mouse mutants can then be constructed to allow functional analysis (e.g., see ALDOSTERONE) – although it should never be assumed that gene expression patterns will be the same in mice and humans. GENE KNOCKOUT studies in 1995 suggested that lack of a gene for NITRIC OXIDE synthase (NOS) makes mice more aggressive. But many researchers warn against carrying out, on such bases, studies of NOS levels among custodial populations serving sentences for aggressive offences unless there is a very strong intellectual case for doing so.

Discoveries in mice often guide studies in humans, as in the case of a family of five siblings affected with sensorineural hearing loss (see DEAFNESS). Based on comparisons of genetic and sequence maps for humans and mice, linkage in this family led to identification of a candidate gene (*CDH23*), encoding cadherin 23, and subsequent validation in this family. Variability in the severity of hearing loss in the five siblings raised the possibility of involvement of MODIFIERS, confirmed by mouse studies. Variants in a plasma-membrane calcium pump gene modulate the severity of hearing loss in mice with *DCH23* mutations, and research showed that variants in this same calcium pump protein (ATP2B2) accounted for the variability in hearing loss in the five siblings.

A project completed in 2006 that identified 8.3 million SNPs in 15 mouse strains should allow detailed examination of the way in which combinations of environmental and genetic factors influence the risk of diseases as well as the ill effects of toxins and other chemical species.

Mousterian culture Under a scheme of classification developed by de Mortillet in 1867, and named after Le Moustier Cave containing remains from the early Cave Bear-Mammoth Age, a Middle Palaeolithic Neanderthal flake tool human culture containing small hand-axes with rather flat butts. Most of these industries belong to the last glaciation, from ~100→40 kya, often described as the Mousterian or Acheulian culture (see ACHEULIAN CULTURE). Characteristic of the Neanderthals of western Europe, the Middle East, and as far east as Lake Baikal, Siberia, its most salient features are simple discoid or Levallois flaking in which flakes produced are suitable for conversion into side-scrapers (with retouch along one or more edges). See OLDOWAN CULTURE.

mouth (oral cavity, buccal cavity) Developing from the embryonic stomodaeum, the mouth is formed by the cheeks, hard and soft palates, and TONGUE.

M-phase The phase of the cell cycle (see MITOSIS) during which the nucleus is replicated.

MRCA See MOST RECENT COMMON ANCESTOR.

MRI See MAGNETIC RESONANCE IMAGING.

mRNA Messenger RNA. See RNA, TRANSCRIPTION, TRANSLATION.

MRSA (methicillin-resistant *Staphylococcus aureus*) Strains of *Staphylococcus aureus* (see STAPHYLOCOCCUS) were formerly mostly susceptible to penicillin; but as penicillin usage became widespread, penicillinase-producing strains began to predominate. Introduction of methicillin and, later, flucloxacillin solved the problem initially; but latterly strains of methicillin-resistant *S. aureus* have become a major problem in hospitals globally. It developed by acquisition of the *mecA* gene encoding the low-affinity penicillin-binding protein PBP2 enabling the bacteria to synthesize its cell wall in penicillin's presence. MRSA is commonly resistant to several other antimicrobials and some strains ('E', for epidemic) spread readily in hospitals. Strains of MRSA can be identified by real time PCR tests within a few hours. *Lactobacillus fermentum*, smeared on surgical wounds, has been used to prevent resistant *Staphylococcus aureus* infection. *L. fermentum* secretes a protein that seems to prevent adhesion of *S. aureus* to collagen. The staphylococci which cause hospital-acquired and childhood pneumonia do not carry plasmids, although they can incorporate plasmid DNA into their genomes, and this is thought to explain the worldwide spread of MRSA. Patient screening for MRSA, and staff screening if continued transmission occurs, are important control measures; as is visitors' use of alcoholic hand gels for hand-washing before and after entering a ward; and similar hand-washing by staff before and after each patient contact. See CLOSTRIDIUM DIFFICILE, MULTIDRUG RESISTANCE.

α-MSH See α-MELANOCYTE-STIMULATING HORMONE.

MSH2, MSH3, MSH6 Three eukaryotic MutS-related proteins (related to the bacterial mismatch-repair MutS proteins) which interact to form heterodimeric complexes (e.g., MHS2/MHS6, MHS2/MSH3) which are very sensitive to DNA damage of several types (see DNA REPAIR MECHANISMS). The meiosis-specific heterodimer MSH4/MSH5 stimulates meiotic crossovers.

MT For area MT of the cerebral cortex. See EXTRASTRIATE CORTEX, Fig. 72; MEDIAL TEMPORAL CORTEX.

mtDNA (mitochondrial DNA) Although sperm mitochondria do enter the egg at fertilization, they are eliminated. Mitochondria, and mitochondrial DNA, are therefore inherited, by both sexes, matrilinearly via the egg cytoplasm, in contrast to the patrilineal inheritance, in males only, of the Y CHROMOSOME. The quantity of mtDNA per mitochondrion is small compared with the quantity of nuclear DNA per nucleus, and is one of the two haploid compartments of the human genome (see HAPLOTYPE BLOCKS); but it is more mutable than nuclear DNA, probably because mtDNA replication is more error-prone, and because the number of replications required to produce the thousands of mitochondria per cell is far greater. MtDNA is replicated by polymerase-γ (POLG), which has 3´-to-5´ exonuclease (proofreading) activity in addition to its 5´-to-3´ polymerase activity. If the proof-reading activity of POLG is eliminated and the polymerase activity preserved, mtDNA mutations accumulate through uncorrected errors in replication.

Mitochondrial data provide no access to the gene and gene regulatory sequences of the nuclear DNA.

Sanger's group at Cambridge, UK, first sequenced human mtDNA in 1981, since when pathogenic mutations have been discovered in 30 of the 37 mtDNA genes, all of which play roles in either oxidative phosphorylation or mitochondrial protein synthesis. Although a mitochondrially inherited condition can affect both sexes, it is only passed on by affected mothers, producing a recognizable pedigree pattern. Mitochondria contain 2–10 copies of this 16.569 kb circular, double-stranded, molecule, one strand of which (the 'heavy' strand) is relatively rich in G bases while the other (the 'light' strand) is relatively rich in C bases. Although principally double-stranded, a small section of mtDNA (the displacement (D) loop region) has a triple-DNA strand structure due to repetitive synthesis of a short heavy strand segment, 7S DNA. In humans, mtDNA genes (all of which are intronless) encode a small number of proteins, 13 in all, all of which are crucial in electron transport (e.g., CYTOCHROME C OXIDASE) and ATP synthesis by the inner mitochondrial membrane. In addition, it contains genes for tRNA (22) and rRNA (2) involved in the translation stage of mitochondrial protein synthesis. This small functional load has allowed the mitochondrial genetic code to drift from the 'universal' genetic code. There are 60 sense codons, one fewer than in the nuclear genetic code, and four stop codons, AGA and AGG being additional to the UAA and UAG of the nuclear code, while UGA encodes tryptophan instead of being a stop codon, and AUA specifies methionine not isoleucine.

Mitochondrial diseases can be caused by mutations in both mtDNA and nuclear DNA (see MITOCHONDRION), although most mtDNA mutations are selectively neutral. These mutations may be maternally inherited or acquired, the former being widespread in most tissues while the latter arise throughout life and remain confined to the cells in which they originated, although the high mtDNA copy number provides protective genetic redundancy. mtDNA deletions have been detected in various regions of the human AGEING brain, particularly in the substantia nigra, the proportion of deleted mtDNAs increasing with age, perhaps contributing to the age-dependence of PARKINSON'S DISEASE. Nigral neurons containing >60% of deleted mtDNA molecules (each neuron containing just one of several kinds of deletion) showed a striking loss of cytochrome *c* oxidase, whereas hippocampal neurons from aged individuals did not contain similar high levels of mtDNA deletions. Increased point-mutation levels in a cytochrome oxidase gene (*CO1*) correlates negatively with cytochrome oxidase activity. Many maternally inherited and incurable neuromyopathies are caused by mutations in mitochondrial tRNA genes (see SOMATIC CELL NUCLEAR TRANSFER).

Point mutations and deletions in mtDNA accumulate in tissues with age, correlating with decline of mitochondrial function. They have been implicated in contributing to AGEING – notably age-related hearing loss, hair loss and greying, thymic involution, testicular atrophy (with depletion of sperm count), loss of bone mass, loss of intestinal crypts, progressive decrease in circulating red blood cells, and weight loss, especially loss of muscle mass.

Mitochondrial DNA studies strongly supported the 1960s work of Sarich on blood proteins, that the time of separation between chimpanzee and human lineages, based on genetic distance, is 5–7 Myr rather than the previous estimate of ~15 Myr, based on fossils alone. Because of their rate of production and general neutrality, mtDNA mutations are useful in the study of human evolution (see MIGRATION; COALESCENCE; HOMININI, Fig. 94; HOMO NEANDERTHALENSIS; HOMO SAPIENS; MITOCHONDRIAL EVE).

Although the overall size and organization of human, chimpanzee and gorilla mitochondrial genomes are identical, the mitochondrial D-loop (involved in initiation of mtDNA replication) is the most rapidly evolving region, with considerable allelic heterogeneity – commonly used to measure intraspecific genetic diversity. The extensive divergence of D-loop sequences between western and lowland gorilla subspecies is of interest. Chimpanzees of both species, depending on the population studied, have 3–10 times more mitochondrial sequence variation than do humans. Differences in human and African great ape mitochondrial sequences could reflect differences in male and female migration patterns, population subdivisions, population bottlenecks (see GENETIC DRIFT), and/or selective pressures between the species; but the data tend to parallel human and African great ape population diversity estimates based on X chromosome sequence analysis. See HOMININI, Fig. 94.

MTOCs (microtubule organizing centres) See CENTROSOME.

mTOR (mammalian target of rapamycin) A serine-threonine kinase component of the most prominent signalling pathway in CELL GROWTH (the growth factor/hormone–PI3K–Akt–mTOR pathway; see PKB). This pathway allows cells and organisms to adapt to changes in nutrient availability. It is activated in the presence of abundant nutrients and inactivated under STARVATION CONDITIONS. When activated, it transduces extracellular growth signals to the cell's translation machinery by phosphorylation of targets that initiate the translation process. It is a critical inhibitor of autophagy (see AUTOPHAGOSOMES) and enhances mitochondrial metabolism and the production of reactive oxygen species (see AGEING, MITOCHONDRION). In mammalian cells, mTOR appears to be an important regulator of mitochondrial metabolism. Upstream regulators of TOR include PTEN, AKT-1 and serine/threonine kinase 11 (STK11), all frequently altered in human tumours.

M-type ganglion cells See GANGLION CELLS.

mucins (mucoproteins) Jelly-like, sticky or slippery glycoproteins formed by complexing a glycosaminoglycan to a protein, the former contributing most of the mass. Some provide an intercellular bonding material, others lubrication.

mucosa (pl. mucosae; mucous membrane) General term for a moist epithelium and its underlying connective tissue (LAMINA PROPRIA). Applied especially in the context of the lining of the gut and the respiratory and urogenital tracts. The epithelium is usually simple or stratified, often ciliated, and usually contains GOBLET CELLS that secrete mucus. They are often sites of INFECTION, but are usually moistened by tissue fluids containing lysozymes capable of destroying bacterial peptidoglycan.

mucosal-associated lymphoid tissue (MALT) Lymphoid tissues associated with mucosal epithelia lining the mouth and intestine, respiratory tract and reproductive tract. The epithelia of these surfaces are interspersed with a network of dendritic cells (those in the skin are termed LANGERHANS CELLS). The stromal cells of MALT, and particularly those of GUT-ASSOCIATED LYMPHOID TISSUES (a subset of MALT), secrete TGF-β, which induces IgA ISOTYPE SWITCHING. During foetal life, waves of γ:δ T CELLS leave the thymus and migrate to these epithelial barriers, where they are abundant and may have a role in oral tolerance. Unexpectedly, they present foreign antigens.

mucous membrane See MUCOSA.

mucus A slimy secretion containing mucins (mucoproteins) and secreted by GOBLET CELLS of mucous membranes.

Müllerian ducts (paramesonephric ducts) One of the two pairs of embryonic GENITAL DUCTS, persisting in females as uterine tubes (oviducts), opening into the abdominal cavity in a funnel-like structure and fusing in the midline to form the uterine canal. See SEX DETERMINATION.

Müllerian inhibiting hormone (M.i.substance, MIH, MIS) Hormone produced by SERTOLI CELLS of the male human foetus causing Müllerian ducts to regress by apoptosis. Only produced if foetal cells contain a Y chromosome. See ANDROGEN INSENSITIVITY SYNDROME, PSEUDOHERMAPHRODITISM, SEX DETERMINATION.

multidrug resistance (MDR) DRUG RESISTANCE to hydrophobic cytotoxic compounds made possible by the presence in the cell membrane of a multidrug resistance protein, which is a serious impediment to improved healthcare. The BLOOD–BRAIN BARRIER is able to exclude a wide variety of lipophilic compounds that are potentially neurotoxic. Structures of multidrug transporters are now known from four distinct families: ABC TRANSPORTER FAMILY; the MFS (major facilitator superfamily) family, energized by an electrochemical proton gradient; the RND (resistance-nodulation-division) family, found in prokaryotes and eukaryotes, energized by proton movement down the transmembrane electrochemical gradient; and the SMR (small multidrug resistance) family, found only in prokaryotes. Their superabundant presence is often due to amplification of the coding gene. Proteins conferring resistance to vinblastine, actinomycin D, mitomycin, taxol, colchicine and puromycin are known, although such cytotoxic drugs actually up-regulate multidrug transporters in bacterial and mammalian cells, often as part of a general stress response. See CANCER, MRSAS, TUBERCULOSIS.

multifactorial inheritance, m. disorders Inheritance of a phenotype in which two or more gene loci are involved (see POLYGENIC INHERITANCE) and in which environmental factors also play a part. Multifactorial disorders are discussed in the entry on COMPLEX DISEASES. See also BIRTH DEFECTS, FAMILIAL RECURRENCE RISKS.

multigene diseases See COMPLEX DISEASES.

multilocus probe See GENE PROBE.

multiple allelism The simultaneous occurrence in a population of more than two alleles of a single gene locus.

multiple endocrine neoplasia (MEN) Two forms of this familial cancer occur: MEN1, involving the parathyroid, thyroid, anterior pituitary, pancreatic islets and adrenal, is caused by mutation at 11q13; and MEN2, involving the medullary thyroid (phaeochromocytoma), caused by mutation of the *RET* locus at 10q11.2. Both forms are inherited in an autosomally dominant manner, seemingly through mutation arising in the male germ line. Screening for MEN1 should begin from 8 years, repeated annually up to 50 years.

multiple sclerosis (MS) A progressive COMPLEX DISEASE, believed by many to be an AUTOIMMUNE DISEASE, in which multiple plaques of demyelination occur in the CNS, resulting in defects in impulse propagation (see OLIGODENDROCYTES). The optic nerves, brainstem and cervical spinal cord are especially affected. Despite their presence in an immunologically privileged site, brain autoantigens such as myelin basic protein are targeted in MS. Clearly, this antigen does not induce the tolerance attributable to clonal deletion of the self-reactive T CELLS. Specific damage to the FASCICULUS CUNEATUS of the cervical spinal cord leads to loss of proprioceptive information from the hands and fingers. This causes profound loss of dexterity and inability to identify the shape and nature of objects by touch alone. Remissions probably reflect repair and re-myelination. Prevalence increases with distance from the Equator, and environmental factors during childhood appear to play an important role in development of MS in later life. People immigrating from low- to-high-prevalence zones (e.g., from the tropics to northern Europe) come to have the same risk of developing the disease as the high-prevalence zone natives if they arrive before 15 years of age; but immigrants older than 15 years of age retain the susceptibility of the home country. Incidence is commoner in women than in men, and the sibling recurrence risk is ~4%. For identical twins, the likelihood that the second twin may develop MS if the first twin does is ~30%; for fraternal twins (who do not inherit identical genomes), the likelihood is closer to that for non-twin siblings, or ~4%.

In mice with experimental autoimmune encephalomyelitis, a model of MS, local catabolism of the amino acid tryptophan (Trp) by indoleamine 2,3-dioxygenase (IDO) seems to reduce pro-inflammatory cytokine (e.g., IFN-γ) production by T_H1 cells and suppresses their proliferation, and such Trp catabolites and their derivatives may offer a strategy for dealing with T_H1-mediated autoimmune diseases such as this. The small heat shock protein αB-crystallin (see CRYSTALLINS) encoded by *CRYAB* has anti-apoptotic and neuroprotective (anti-inflammatory) functions and is the most abundant gene transcript present in early active multiple sclerosis lesions, whereas these transcripts are absent from normal brain tissue. CRYAB is a major target of T_H1-mediated immunity to the myelin sheath from MS brain. Antibody to CRYAB has been detected in cerebrospinal fluid from MS patients, and in sera of mice with autoimmune encephalomyelitis (EAE). Administration of recombinant CRYAB ameliorated EAE in mice. It seems that an immune response against the anti-inflammatory CRYAB would exacerbate inflammation and demyelination, and may have a role as a therapeutic agent.

The human MHC HLA-DR2 haplotype predisposes to MS. Work on two MS-associated HLA-DR alleles at separate loci, using humanized mice, indicates that the linkage disequilibrium (LD) between the two alleles may be due to functional epistatic interaction in which one allele modifies the T-cell response activated by the second allele through activation-induced cell death, associated with a milder form of MS-like disease.

multipolar neurons Neurons with more than three neurites (i.e., axons and dendrites); commonly those within a restricted tissue family.

multipotent stem cells Nowadays, these are termed PROGENITOR CELLS.

multiregional model The multiregionalist view of the mode of human evolution was the first comprehensive model of the origin of modern humans to be elaborated, dating back to Weidenreich's 1940s hypothesis that modern human anatomical diversity owes its existence to the evolution of distinctive traits, by natural selection and genetic drift, in different geographical regions originally established in early populations of *Homo erectus* and persisting by regional continuity since then. Gene flow through mating between the geographical populations was thought to be limited; but many modern multiregionalists take the view that gene flow has played a significant role in establishing the geographical pattern of present-day human phenotypes (the multiregional evolution hypothesis). Proponents of the model regard East Asian hominin fossils as showing a continuous record of remains from ~1 Mya (at Lantian in northeastern China), through unquestionable *H. erectus* fossils of Zhoukoudian (500–200 kya) to fossils displaying a mixture of *H. erectus* and more derived traits (e.g., a 200 kya partial skeleton from Jinniu Shan and a similarly aged skull at Dali generally treated as archaic *sapiens*) to allegedly transitional material between *erectus* and *sapiens* (dated to ~20 kya) in the Upper Cave of Zhoukoudian, through to material from the historical period and the present day. If the attribution and dating of two crushed crania from Yunxian, east-central China, to archaic *sapiens* at ~350 kya are correct, this would be difficult to reconcile with the transition envisaged by multiregional evolution hypothesis since these fossils would be almost twice as old as some of the youngest *H. erectus* specimens. For the Recent African Origin model, see HOMO SAPIENS, MIGRATION.

multivesicular bodies Late endosomes and lysosomes containing large amounts of invaginated cell surface membrane, often as internal vesicular compartments within the organelle, and in the process of migrating in the cytosol along microtubules, shedding tubules for recycling to the plasma membrane. Impairment of sphingolipid hydrolysis in these membranes through enzyme or activator protein deficiency is the cause of inherited lysosomal storage disorders (see LYSOSOME, TAY–SACHS DISORDER).

mumps A moderately infectious, systemic, paramyxovirus (see VIRUSES) and notifiable infection, often regarded as epidemic parotiditis, causing significant morbidity but few fatalities. Only one serotype occurs, and transmission is by droplet spread. The virus (120–200 nm) contains a single linear RNA strand associated with an RNA-dependent RNA POLYMERASE forming a helical nucleocapsid. The genome encodes six major proteins. The outermost envelope contains glycoproteins and haemagglutinins, with neuraminidase and cell fusion activity; the middle envelope consists of the host membrane, and the innermost envelope contains glycosylated viral structural proteins. Clinical features can appear in any order and include parotiditis in >70% of cases, generally bilateral, subsiding in 7–10 days; orchitis is common in adult men and if severe can lead to testicular atrophy (although rarely bilaterally); meningitis and meningoencephalitis occur in ~15% of cases, often being mild and overlooked. Rarer features include mastitis, cochlear infection and hearing impairment, oophoritis in women, and arthritis (often in the 2nd week). Myocarditis can occur in adult cases, especially in the rare fatalities. Mumps VACCINE is live and attenuated (usually as part of the combined MMR vaccine) and is usually prescribed at 12–15 months. It is highly effective and >95% of recipients develop immunity. A sustained coverage level of ≥85% is required to prevent transmission and to eliminate the disease. A mild, self-limiting meningoencephalitis occurs after ~1:11,000 vaccinations with the Urabe Am9 strain, which is therefore not used in the UK; the incidence of this complication with the Jeryl Lynn strain is very low, while the incidence following natural infection is far higher, at 1:200 cases. See DISEASE, INFECTION.

mural cells The cells surrounding the endothelium in blood vessels larger than a capillary, and including pericytes and smooth muscle fibres.

muscarinic receptor A subtype of ACETYLCHOLINE receptor that is metabotropic and coupled to a G PROTEIN (see G PROTEIN-COUPLED RECEPTORS).

muscle fibres See CARDIAC MUSCLE, SKELETAL MUSCLE, SMOOTH MUSCLE.

muscle pump See VEINS.

muscles Three types of muscle cells (or fibre) occur in humans: CARDIAC MUSCLE, SKELETAL MUSCLE and SMOOTH MUSCLE. Skeletal muscles are organs which normally produce movements by exerting force on tendons which, in turn, pull on bones or other structures (see LEVERS). A skeletal muscle ordinarily has its *origin* on a stationary bone via a tendon, while its *insertion* is on a moveable bone, also via a tendon (the extrinsic eye muscles insert on the sclera). Most movements require the coordinated actions of several skeletal muscles, and most skeletal muscles are arranged at joints in opposing (or antagonistic) pairs, e.g., flexors/extensors and abductors/adductors. The desired action is caused by contraction of the prime mover (or *agonist*), while the opposing muscle (the *antagonist*) is relaxing. Most movements also involve *synergists*, which often serve to steady a movement; and some of these stabilize the origin of the muscle and are termed *fixators*. See Appendix VI and Appendix X.

muscle spindles Specialized organs (3–10 mm in length and 100 μm in diameter) within skeletal muscles, involved in maintaining a muscle's length as it is stretched. They often work in opposition to TENDON ORGANS, which maintain muscle tension. Each consists of a spindle-shaped fluid-filled capsule of connective tissue containing ~7 modified muscle fibres which are interspersed among, and oriented parallel to, the normal skeletal muscle fibres, with their ends anchored in endomysium and perimysium. Responding to changes in muscle *length* they are important PROPRIOCEPTORS, being stretched when the adjacent skeletal muscle fibres are stretched, as in load bearing or as a result of contraction of an antagonistic muscle.

Muscles involved in complex movements contain more spindles per gram than those performing gross movement patterns. Each spindle contains a group of two kinds of *intrafusal muscle fibres*: the fairly large *nuclear bag fibre* (~2 per spindle), containing numerous nuclei packed centrally throughout its diameter; and the *nuclear chain fibre* (~4–5 per spindle), containing many nuclei along its length, each attaching to the surface of a nuclear bag fibre. The central region of each intrafusal fibre lacks actin and myosin, but its contractile ends are innervated by the fusimotor fibres of *γ motor neurons*. Around the muscle spindle are more normal skeletal muscle fibres, the *extrafusal muscle fibres*, innervated by large-diameter efferent *α motor neurons*. Cell bodies for both motor neuron types lie in the dorsal horn of the SPINAL CORD, directly activating motor neuron cell bodies without brain involvement (the *short loop*). Stimulation of one α motor neuron can result in up to 3,000 skeletal muscle fibres contracting, interneurons coordinating the motor response through appropriate inhibitory and excitatory motor neurons. This stretch reflex (see later) functions to oppose sudden changes in muscle length, and in this way stretch reflexes in general act as self-regulating compensatory mechanisms, enabling a muscle to adjust, without the brain's involvement, to different loads.

Two sensory afferent fibres service the spindle. A primary afferent (Group Ia) fibre entwines around the mid-region of the bag fibre, responding to stretch of the spindle by increasing its rate of firing (spindle shortening decreases the rate). A second group of smaller sensory nerve endings (flower-spray endings) makes contact mainly with the chain fibres, but also with the bag fibres, its secondary afferent (Group II fibre) also passing to the dorsal horn. Stimulation of both by sudden stretching of the muscle leads to reflex activation of the motor neurons to the spindle, causing the muscle to contract more forcefully (the stretch reflex), reducing the stretch stimulus from the spindle. It also contributes to a *long (cortical) loop* of the stretch reflex circuitry by means of which Area 4 of the MOTOR CORTEX can modify, via the corticospinal tracts, the γ motor neurons of the spindles in response to unexpected loads. This long loop modifies muscle contraction on a timescale slower than that of the stretch reflex, but faster than voluntary movements (see ELBOW).

A *stretch reflex* (myotatic reflex) has two components: a phasic component brought about by the dynamic activity of the Ia afferents (as occurs when the tendon of a muscle is tapped, or during a sudden jolt); and a tonic component, which is much the more important in maintaining posture, being a sustained contraction brought about by the static activity of the Ia afferents and of the secondary, group II, afferents. Even if the length of the entire muscle does not change (isometric contraction), contraction of the end portions of the intrafusal fibres by higher brain centres will stretch the middle portions so that they can respond to stretch. This activation regulates the spindle length independently of the overall muscle length, preparing the spindle in advance of other lengthening actions even though the muscle itself remains shortened. There are two categories of γ motor neurons that can be activated independently by the CNS, their firing rates being raised when particularly complex movements are being performed. Stretch reflexes must be overridden to permit movement, since the muscle must contract isotonically and shorten. Overriding the stretch reflex during voluntary movement is achieved by descending motor pathways in which γ efferent neurons are coactivated with α motor neurons innervating the extrafusal fibres. This is achieved polysynaptically via the CORTICOSPINAL TRACT and enables the intrafusal fibres to be sufficiently taut to respond to stretch. Adjustment of γ efferent activation enables the spindles to monitor continuously the length of the muscle within which they lie.

In the misleadingly named *patella tendon stretch reflex* (the 'knee-jerk reflex'), the simplest reflex arc involving only one synapse (monosynaptic), spindles in the extensor *muscles* of the thigh become stretched as the hammer strikes the patella tendon. Neurological examinations often include eliciting such reflexes as absent or abnormal reflexes often relate to the level of damage to the nervous system.

muscular dystrophies GENETIC DISORDERS resulting in loss of muscle tissue and weakness through myopathy (disorder of the muscle tissue itself) rather than neuropathy (disorder of nerves supporting the muscle). There are currently no cures.

Duchenne muscular dystrophy (DMD) is an X chromosome-linked recessive myopathy (see SEX LINKAGE) affecting primarily males (~1:3,500 newborn males) and leading to progressive muscle deterioration, beginning in infancy, with death in the second or third decade. Discovery of the gene was an early success, in 1986, for the procedure of POSITIONAL CLONING. The *DMD* gene is the largest in the human genome (see NON-CODING DNA for its possible significance in human evolution), providing many targets for mutation, and its mutation rate (~10^{4}) is the highest for any known human genetic disorder – about one-third of which are *de novo* mutations. Early detection by newborn SCREENING involves measuring the level of creatine kinase in a dried blood specimen taken after 4 weeks, and is currently limited to pilot schemes. It is also possible, using molecular techniques, to detect carrier mothers and affected offspring prenatally with >95% accuracy. Mothers have serum creatine kinase levels above those of the general public, and in conjunction with pedigree risk information and results of linked marker DNA this can help calculations of her being a carrier. Most cases involve mutation in a gigantic X chromosome locus and even a mother who is not a carrier may produce the mutation in her germ cells (perhaps >12% of DMD cases are attributable to this). This high mutation rate, in males as well as females, may be due to the huge size of the locus, two so-called 'hot spots' in the locus being especially prone to breakages and deletions. Some cases are the result of insertion of SINES OR LINES near or within the locus. All DMD patients lack significant levels of the protein DYSTROPHIN in their muscle, one of the major cytoskeletal muscle proteins. A milder form of muscular dystrophy, *Becker muscular dystrophy*, is caused by mutations in the same hotspots but does not result in severe dystrophin deficiency. Abnormal dystrophin is produced, but there is later and more progressive onset of symptoms. In both forms of the disorder, one or more exons are deleted in ~60% of cases. In one *in vitro* experiment involving muscle cells from two patients with DMD, an antisense oligonucleotide was used to block an exon-splicing enhancer (see EXONS). The reading frame was restored and significant levels of dystrophin in the muscle cells confirmed the therapeutic potential of this approach. It seems that restoring ~20% of normal dystrophin gene expression in muscle would benefit DMD patients; but the huge size of the gene and targeting its correct delivery into both skeletal (including diaphragm) and cardiac muscle cells each poses problems for gene therapy approaches. Even dystrophin cDNA is too large for many vectors. One approach attempts to up-regulate the expression of utrophin, a molecule that binds actomyosin to the muscle cell membrane at the neuromuscular junction. The utrophin locus is intact in DMD boys, being encoded at a separate autosomal locus. Cell therapy involving transplantation of myoblasts into muscles of affected boys has not yet led to clinical improvement, but the possibility that transplanted bone marrow stem cells might migrate to muscle and act as muscle stem cells would offer hope of treating DMD in a manner similar to the methods that have cured some SCID patients.

Myotonic dystrophy is a dominantly inherited muscular dystrophy with variable phenotypic presentation even in the same family. Patients may show severe weakness when newborn (congenital myotonic dystrophy); others show mental retardation in infancy; some develop 'hatchet-faced' dysmorphologies when adult; others only have cataracts in middle age. The disorder tends to get more severe in successive generations (ANTICIPATION). Its cause is a trinucleotide repeat expansion (see TRIPLET REPEATS) in the gene for cyclic AMP kinase, located on chromosome 19 and may be due to precipitation or other malfunction of its RNA transcript in some nuclei or to abnormal behaviour of the repeat DNA regions. The longer a repeat is, the more probable it is that it expands between generations.

Spinal and bulbar muscular atrophy (Kennedy disease) is an X-linked disorder of motor neurons, characterized by progressive muscle weakness and atrophy resulting from amplification of the trinucleotide CAG in exon 1 of the gene (*AR*) encoding the androgen receptor.

muscularis The muscular tunic of the ALIMENTARY CANAL (see Fig. 4), controlled by the ENTERIC NERVOUS SYSTEM.

mutagens Environmental agents that cause mutations. Ionizing radiation includes electromagnetic waves of very short wavelength (X-rays and γ rays), high-energy particles (α particles, β particles and neutrons) and ULTRAVIOLET IRRADIATION. X-rays, γ rays and neutrons have great penetrating power, but α particles penetrate soft tissues to a depth of a fraction of a millimeter, while β particles penetrate up to a few millimeters. The rad is a measure of ionizing radiation absorbed by tissues, 1 rad being equivalent to 100 ergs of energy absorbed per gram of tissue. Humans can be exposed to a mixture of radiation, and the rem (roentgen equivalent for man) is a unit allowing all to be compared in terms of X-rays. 1 rem of radiation is that absorbed dose producing in a given tissue the same biological effect as 1 rad of X-rays. In SI units, 100 rem is equivalent to 1 sievert (Sv), and 100 rad is equivalent to 1 gray (Gy), sieverts and grays being roughly equivalent. The average natural radiation experienced by a human per month is 0.2 millisieverts (mSv). At its height, atmospheric nuclear testing raised this by 0.01 mSv. In the UK, the peak due to Chernobyl was six times smaller than this. On the other hand, the background radiation per month in Cornwall (UK) is far higher, at 0.6 mSv, on account of the increased radioactivity in the granite. It has been recommended that occupational exposure should not exceed 50 mSv per year. UK regulations limit increased doses to the public to less than 0.08 mSv per month (1 mSv yr^{-1}) – all small compared to the lethal dose for humans of ~6,000 mSv. At Chernobyl, 20 out of 21 who received this dose died within a few weeks; but of the 55 workers who received 2,000 mSv–4,000mSv, only one died.

In humans, chemical mutagens may be more important than radiation in producing genetic damage. Mustard gas and other ALKYLATING AGENTS, benzene, some basic dyes and food additives, and aflatoxin are all mutagenic. Free radicals produced by normal metabolic processes are also mutagens (see MITOCHONDRION). For more, see CARCINOGENS, DNA REPAIR MECHANISMS.

mutations Alterations in the arrangement, or amount, of genetic material in a genome, tending to occur during DNA REPLICATION in S-phase of the cell cycle, and often involving TRANSPOSABLE ELEMENTS (see MUTAGENS). A *point mutation* is a minor sequence change involving single base-pair INSERTIONS, DELETIONS and SUBSTITUTIONS. Mutations occur in both coding and NON-CODING DNA, and many of them are repaired by DNA REPAIR MECHANISMS, but the significance of those escaping detection depends upon where and when they occur in the genome, and only those occurring in the germ line are capable of being transmitted to the next generation. Of all mutation processes, point substitutions are the most prevalent, with indels (insertions and deletions) 10-fold less frequent. Some mutations are responsible for cell transformation of CANCER CELLS; but most are either selectively neutral (e.g., those that do not alter coding sequences; see NEUTRAL THEORY OF MOLECULAR EVOLUTION), deleterious (see NATURAL SELECTION) or occasionally beneficial. Mutations can be detected by GENOME MAPPING methods.

Larger genomic alterations, *macromutations*, tend to have microscopically observable effects on chromosomes (see CHROMOSOMAL MUTATIONS) and include chromosomal inversions, deletions, duplications and nondisjunctions. These are often either deleterious or incompatible with life, indels being highly disruptive of protein-coding sequences, for instance. But those with little adverse effect have often been of enormous significance in genome evolution (e.g., see SEGMENTAL DUPLICATIONS).

~175 mutations are estimated to appear in every human individual each generation, of which ~1.6 are estimated to be deleterious. Some gene mutation rates are much higher than others (e.g., see NEUROFIBROMATOSIS). Comparisons of human and chimpanzee PSEUDOGENE sequences give an estimated deleterious mutation rate of at least three per generation – very high for species with low reproduction rates. Mutations at many loci show increasing rates with parental age; and when the parent of origin of a *de novo* mutant chromosome is determined, a strong bias towards paternal origin is often observed (see SPERMATOGENESIS). Microsatellite mutations occur three to five times more frequently in fathers than in mothers, and all dominant mutations in our species (e.g., those causing ACHONDROPLASIA, multiple endocrine neoplasia and APERT'S SYNDROME) seem to arise in the male germ line (see INFERTILITY). CpG dinucleotides are ~10-fold more mutable than the genome average.

Mutations have been classified (non-mutually exclusively) into four categories, (i)–(iv).

(i) *Changes in DNA sequence*. SUBSTITUTIONS are replacements of a single nucleotide by another and are the most common form of mutation; DELETIONS involve the loss of one or more nucleotides, while INSERTIONS (= additions) involve the addition of one or more nucleotides.

(ii) *The structural effect (if any) a mutation has on an encoded protein*. If a mutation does not alter the polypeptide product of the gene, this is termed a *synonymous* (or silent) mutation. Because of the degeneracy of the genetic code, a single nucleotide pair (or, 'base pair') substitution, especially one occurring in the 3rd position of a codon, will often result in a different triplet encoding the same amino acid, with no consequent alteration in the protein's properties. Mutations leading to change in the encoded polypeptide are *non-synonymous* mutations and can occur in three main ways: the single nucleotide pair substitution encodes a different amino acid from the original (*missense* mutations); the substitution leads to generation of a stop codon (*nonsense* mutations; see NONSENSE-MEDIATED DECAY, THALASSAEMIAS); or the mutation involves insertions or deletions of nucleotides that are not multiples of three, so disrupting the reading frame during translation (*frameshift* mutations) and often resulting in a stop codon downstream of the mutation (see THALASSAEMIAS). These often occur at repeated bases, where the prevailing model (see Fig. 127) proposes that indels arise when loops in single-stranded regions are stabilized by the 'slipped mispairing' (*replication slippage*) of short tandem repeats, especially during replication. This occurs when normal pairing between two homologous complementary strands of a double helix is altered through either one or more repeats of the parental strand looping out, leading to deletion in the newly synthesized strand, or to one or more repeats of the newly synthesized strand looping out, leading to an insertion within it. The loop-outs result in incorrect pairing of repeats. Repeated slippage has been proposed as the cause of DNA repeat expansion and the formation of MICROSATELLITES (see REPETITIVE DNA; compare unequal CROSSING-OVER).

The K_A/K_S ratio is the ratio of non-synonymous to synonymous substitutions in protein coding regions of two compared genomes – expected to be $\ll 1$ if the region has been subjected to strong selective constraints after genome splitting on account of the substantial proportion of amino acid changes that could only have been eliminated by purifying selection. The K_A/K_S ratio will be expected to be >1 if the region is subject to weak selective constraints, or continued positive selection and for the human–chimpanzee lineage as a whole is 0.23. This implies that 77% of amino acid alterations in genes in the Homininae are sufficiently deleterious to be eliminated by NATURAL SELECTION. Recent work seems to indicate that a substantial fraction of the genes with the highest K_A/K_S ratios had roles in brain development or function and that there is a small subset of neural genes for which there has been strong positive selection since the split in the chimpanzee/human lineages. The K_A/K_S ratio in the 5,802 proteins identified as orthologues in all pairwise comparisons between human, mouse, and the two pufferfish species *Tetraodon nigroviridis* and *Takifugu rubripes*, indicates that the ratio is much higher between the two pufferfish than between human and mouse, suggesting that evolution is taking place more rapidly along the pufferfish lineage (possibly positively correlated with the higher rate of neutral mutation).

The K_A/K_I ratio (the median protein sequence divergence) is the ratio of the number of non-synonymous nucleotide substitutions per non-synonymous site to the number of substitutions per site in interspersed DNA repeats in 250-bp regions around each gene being compared. Genes expressed solely in the brain show significantly lower K_A/K_I ratios than those of other tissue-specific genes, and ubiquitously expressed genes show lower K_A/K_I ratios than genes expressed in single tissues. The chromosome with the most extreme mean K_A/K_I ratio among extant

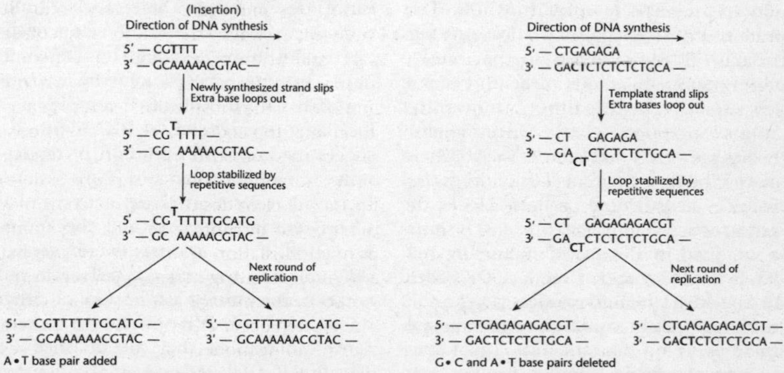

FIG 127. The prevailing model for 'slipped mispairing' replication errors leading to indel (insertion and deletion) MUTATIONS in *E. coli*, in which frameshifts can occur. Inserted and deleted bases are shown in bold. © 2005, W.H. Freeman & Co. Redrawn from *Introduction to Genetic Analysis* (8th edition), A.J.F. Griffiths, S.R. Wesler, R.C. Lewontin, W.M. Gelbart, D.T. Suzuki and J.H. Miller, with permission from the publisher.

hominines is chromosome X (0.32). See GENE CLUSTERS, Table 3.

(iii) *The functional effects of mutations on their encoded protein.* These are brought about either by LOSS-OF-FUNCTION MUTATIONS, GAIN-OF-FUNCTION MUTATIONS, or by DOMINANT-NEGATIVE MUTATIONS.

(iv) *Mutations occurring in* NON-CODING DNA. These include changes in gene regulatory regions (e.g. promoters) and regulatory proteins (e.g. transcription factors) affecting the level of GENE EXPRESSION. Mutations of the highly conserved splice donor (G:T) and splice acceptor (A:G) sites result in aberrant splicing (see RNA PROCESSING) and may lead to frameshift mutations (see (ii) above) through retention of intronic sequence or loss of coding sequence (exon skipping). Non-coding DNA differences are considered as possible major factors responsible for phenotypic differences between chimpanzees and humans. This is currently difficult to test since we need more information from microarray expression studies. See 'LESS-IS-MORE' HYPOTHESIS and the Human Gene Mutation Database on http://www.hgmd.cf.ac.uk/ac/index.php.

myasthenia gravis There appear to be two forms of this disease: (i) An acquired AUTOIMMUNE DISORDER in which autoantibodies are produced against the α-chain of the nicotinic ACETYLCHOLINE receptor (nAChR). This leads to increased internalization and degradation of the receptors so that muscle fibres (especially cranial muscles) become less responsive to ACh, the only transmitter at NEUROMUSCULAR JUNCTIONS. Since the antibodies are polyclonal there is variability in susceptibility to them. A class of drugs that includes neostigmine partially blocks the action of acetylcholinesterase and is sometimes used in treatment, increasing Ach levels in the synaptic cleft. The DR3 allele of the MHC (=HLA) confers a ~2.5-fold increased risk of contracting the disease, there being no sex ratio bias. (ii) A non-immune, heritable, disorder in which antibodies to ACh receptors are not involved but in which there is a reduction in functional nicotinic ACh receptors in motor end-plates due to mutation (a 'channelopathy'; see ION CHANNELS). Another myasthenic disease, Lambton–Eaton myasthenic syndrome, results from autoantibodies against presynaptic voltage-gated Ca^{2+} channels in motor neuron terminals, leading to reduced ACh release.

Myc (c-Myc) Oncoprotein encoded by genes of the *Myc* (or c-*Myc*) ONCOGENE family on chromosome 8, an IMMEDIATE EARLY GENE whose product (Myc) acts in the nucleus as a signal for cell proliferation, has a half-life of only 25 minutes, and whose expression is tightly controlled in normal cells by MITOGEN availability (see RAS SIGNALLING PATHWAYS). It is usually expressed in a deregulated or elevated manner in tumour cells, commonly as a result of gene amplification (see CANCER), such overexpression inhibiting cell differentiation, which itself contributes to tumour development (compare RETINOBLASTOMA PROTEIN). *Myc* is one of the genes activated by β-catenin (see WNT SIGNALLING PATHWAY), and mutations in the *APC* gene inhibit the APC protein's ability to bind β-catenin so that it is not degraded and accumulates in the nucleus. This stimulates transcription of *c-Myc* and other Wnt target genes, even in the absence of Wnt signalling, resulting in uncontrolled cell proliferation which promotes cancer development. The biological functions required by Myc to induce oncogenesis remain elusive, although it is known to have a role in DNA replication by interacting with the pre-replicative complex and localizing to early sites of DNA synthesis, suggesting a role in initiation of replication. Over-expression of *Myc* causes increased replication origin activity with subsequent DNA damage and checkpoint activation. If *Myc* alone is overexpressed, cells are driven round the cell cycle but without deregulated expression of other oncogenes. Such cells tend not to become cancerous because they are destined for APOPTOSIS. But if the *Myc* gene is translocated so that it comes under the control of a strong promoter, it may result in tumour (see BURKITT'S LYMPHOMA).

Mycobacterium tuberculosis The bacterial cause of TUBERCULOSIS.

Mycoplasma pneumoniae Mycoplasmas are very small microorganisms lacking the rigid cell wall of other bacteria, and are therefore completely resistant to β-lactam ANTIBIOTICS. *M. pneumoniae* is a pathogenic bacterium which is not demonstrable by conventional culture or Gram-staining whose importance lies in it being a cause of respiratory morbidity in children, and less often of adults of working age, and because of its tendency to occur in outbreaks. Its reservoir of infection is infected members of the population. Mycoplasma pneumonia is commonest in children of school age, epidemics occurring in the UK at 3- to 4-year intervals. It competes with STREPTOCOCCUS PNEUMONIAE as the commonest cause of community-acquired pneumonia. It is also a cause of bacterial MENINGITIS. Treatment involves orally taken erythromycin, oxytetracycline or azithromycin. As with mycoplasmas generally (e.g., see Q FEVER), PCR-based diagnosis is now the standard technique.

MyD88 An adaptor protein in the TOLL-LIKE RECEPTOR signalling pathway.

myelencephalon The secondary brain vesicle (see BRAIN, Fig. 28) giving rise to the medulla oblongata, differing from the spinal cord in having inverted lateral walls but like it in having alar and basal plates. Latero-medially, the alar plates give rise to three groups of sensory nuclei: somatic afferent, special visceral afferent and general visceral afferent. The basal plates similarly give rise in a latero-medial direction to three groups of motor nuclei: general visceral efferent, special visceral efferent and somatic efferent.

myelinated neurons See AXONS, MYELIN SHEATH, NEURONS.

myelination See MYELIN SHEATH.

myelin sheath, myelination Formed by glial cells (SCHWANN CELLS in the peripheral nervous system, and OLIGODENDROCYTES in the central nervous system), a myelin sheath consist of many layers (up to 100) of their plasma membrane wrapped in a tight helix around a nerve AXON, preventing leakage of ions (current) across the axon membrane (except at NODES OF RANVIER). These membranes contain large amounts of *sphingomyelins*, lipids containing sphingosine instead of glycerol, with phosphocholine or phosphoethanolamine attached to C1. In addition to protein (30%), phospholipid (30%) and cholesterol (19%), the membranes contain galactolipids and plasmalogens, initial stages of the latter's synthesis taking place in peroxisomes. Schwann cells myelinate peripheral nerves at about the time they begin to function, whereas from about the 4th month of intrauterine life, oligodendroglia (oligodendrocytes) myelinate axons of the spinal cord although some of the motor fibres descending from higher brain centres do not become myelinated until the 1st year of postnatal life. Myelination increases the transmission of NERVE IMPULSES by 5- to 50-fold, and conserves energy for the axon since only the nodes depolarize. The excellent insulation afforded by myelination of large axons, and the 50-fold decrease in membrane capacitance, also allow repolarization to occur with little ion transfer. After an action potential, when sodium channels begin to close, repolarization occurs so quickly that many of the potassium channels have not started to open. Conduction of an impulse in myelinated axons is therefore accomplished almost entirely by ion conduction through voltage-gated sodium channels at NODES OF RANVIER. See LYSOPHOSPHOLIPIDS, MULTIPLE SCLEROSIS, NEUROLEMMA.

myeloid leukaemia See LEUKAEMIAS.

myeloid tissue The common myeloid PROGENITOR CELLS, found in red BONE MARROW and produced from a HAEMATOPOIETIC STEM CELLS, is the precursor of granulocytes (neutrophils, basophils, eosinophils), monocytes (and their mature form, macrophages), dendritic cells, and mast cells of the innate immune system. Compare LYMPHOID TISSUE; see STEM CELLS.

myelomas A malignancy of the antibody-producing cells of BONE MARROW, often termed multiple myeloma in advanced disease. Often detected by BENCE JONES PROTEINS.

myenteric plexus (Auerbach's plexus) See ENTERIC NERVOUS SYSTEM.

myoblasts Precursor muscle cells, whose differentiation into SMOOTH MUSCLE, CARDIAC MUSCLE or SKELETAL MUSCLE is dependent upon a pattern of gene expression which includes MYOD. Smooth and cardiac muscle myoblasts develop from splanchnic mesoderm (a derivative of LATERAL PLATE MESODERM), whereas all skeletal muscle myoblasts develop from PARAXIAL MESODERM.

myocardial infarction A 'heart attack', resulting from death of heart tissue because of interrupted blood supply, resulting in ischaemia (see CORONARY HEART DISEASE). Dead tissue is replaced by scar tissue, the heart muscle losing strength. It may also disrupt the heart's conducting system involved in control of the CARDIAC CYCLE, causing sudden death by ventricular fibrillation (see ARRHYTHMIAS). Most acute heart attacks are caused by a blood clot forming on an atherosclerotic plaque in a coronary artery and then growing by positive feedback until the artery is blocked. Treatment may involve injection of a thrombolytic agent to disrupt blood clots (e.g., see TISSUE PLASMINOGEN ACTIVATOR). The same Y402H variant of the complement factor H (*CFH*) gene that is a strong susceptibility factor for age-related macular degeneration is also a susceptibility factor for myocardial infarction. See CORONARY ARTERIES.

myocardium The middle layer of the heart wall, lying between the epicardium and endocardium and composed of cardiac muscle tissue and forming the bulk of the organ. Its arteries have characteristic anastomoses.

myocyte A general term for any type of MUSCLE cell.

MyoD A BASIC HELIX–LOOP–HELIX protein with a key role in regulating SKELETAL MUSCLE differentiation (see SOMITES), being one of the earliest markers of myogenic commitment. It commits mesoderm cells to a skeletal lineage, and then to regulate that process. In addition to MyoD, the other regulatory proteins Myf5, myogenin and Mrf4 are also synthesized, directly activating transcription of muscle structural genes as well as the *MEF2* gene encoding an additional regulatory protein which activates further muscle structural genes. MyoD also removes cells from the CELL CYCLE (i.e., halts proliferation) by enhancing the transcription of p21. MyoD is inhibited by cyclin dependent kinases (CDKs); but since CDKs are in turn inhibited by p21, MyoD enhances its own activity in the cell.

myofibrils Threadlike structures consisting largely of groups of thick (MYOSIN) and thin (ACTIN) *myofilaments* running longitudinally within striated muscle fibres and forming the functional units of contraction (composed of sarcomeres) within the fibre.

myofilaments See MYOFIBRILS.

myogenic Of muscle tissue capable of rhythmic contractions independently of external nervous stimulation, contractions arising from the properties of the muscle itself. See PACEMAKER; contrast NEUROGENIC.

myoglobin A monomeric protein encoded by a gene located on chromosome 22 (22q13.1, compare HAEMOGLOBINS) and found in muscle cells, assisting diffusion of oxygen into the mitochondria.

myometrium The smooth muscle layer of the UTERUS.

myopia See NEARSIGHTEDNESS.

myosins A protein superfamily which includes *myosin II*, the first MOTOR PROTEIN to be discovered, members of which interact with ACTIN filaments. The human genome contains ~40 myosin genes. Myosin molecules are either monomers or dimers when functional. Each has a heavy chain which generally begins with a recognizable force-generating myosin motor domain at the N-terminus, which is usually associated with one or more regulatory light chains. The motor domains, although much larger than those of KINESINS, share structural similarity with them. They are also related to the ras proteins. The heavy chain, which usually moves towards the plus end of an actin filament, is attached by a neck, or hinge, region to a C-terminal domain by a variable-length middle region (sometimes absent altogether). Myosin I is monomeric, while myosin II is dimeric, comprising two heavy chains and four light chains (two per myosin head). Dimerization occurs when the two long α-helices of the heavy chains wrap around one another to form a coiled-coil structure, driven by the association of regularly spaced hydrophobic amino acids. The coiled-coil tail of myosin II bundles itself with those of others to form large bipolar 'thick' myosin filaments with several hundred myosin heads oriented in opposite directions at the two ends of the filament. Myosin VI is apparently unique in moving towards the actin minus end. In non-muscle cells, myosin activity can be regulated by phosphorylation and dephosphorylation of multiple sites on both the heavy and light chains and, when activated by Rho (see RAS SIGNALLING PATHWAYS), calcium-dependent myosin light-chain kinase (MLCK) causes myosin II to assume an extended state promoting assembly into a bipolar filament prior to cell contraction. In muscle cells, the myosin II filaments are much longer than in non-muscle cells and force generation takes place when Ca^2+, released from the sarcoplasmic reticulum, shifts tropomyosin to expose myosin-binding sites on adjacent actin filaments. ATP hydrolysis occurs on the myosin head while it is detached from the actin filament and the deformation of the hinge causes the head to assume a 'cocked' or 'primed' position, ADP and P_i remaining attached to it. When the head attaches to a myosin-binding site on actin, P_i and then ADP are released, triggering the power stroke that moves the actin filament relative to the motor protein. This shortens sarcomeres in skeletal muscle fibres and results in forceful contraction. Myosins can achieve movements of 0.2–60 μm sec^{-1}, faster than KINESINS. This is largely due to the fact that many myosin motor heads interact with the same actin filament, a set of linked myosins moving the actin a total distance of 20 steps during the time it takes a myosin head to go through one ATP cycle. In this time, kinesin has moved the equivalent of only two steps. See MICRORNAS.

myotatic reflex (stretch reflex) See ELBOW, MUSCLE SPINDLES.

myotome Deriving during development from the dorsomedial portions of SOMITES, segmentally arranged groups of cells giving rise to the vertebral muscles spanning the vertebrae and allowing the back to bend. Compare DERMATOME; see Appendix VI.

myotonic dystrophy See MUSCULAR DYSTROPHIES.

myristoyl group A C_{14} lipid attached to a glycine residue added to the extreme N-terminus of a protein during post-translational modification and enabling the protein to interact with a membrane receptor, or with the lipid bilayer of a membrane. See MARCKS.

N

N-acetyltransferases Liver enzymes catalysing the transfer of acetyl groups from acetyl-CoA to a wide specificity of arylamines, including SEROTONIN. These enzymes are involved in the breakdown of various drugs and carcinogens. Two human acetyltransferases genes, *NAT1* and *NAT2*, have had different histories. The *NAT1* sequence has remained fairly constant in 13 different geographical populations studied, while that of *NAT2* is polymorphic with allele frequencies differing significantly from group to group. One variant is far more prevalent in people from western and central Eurasia, possibly associated with survival value in the agricultural lifestyles developed there. People who inactivate isoniazid (see TUBERCULOSIS) slowly (50% in the USA and Western Europe) are homozygous for an autosomal recessive of *NAT2*, in contrast to the Japanese, who are mainly rapid inactivators. The *NAT2* polymorphism has been associated with risk modifications in bladder, colorectal, breast and lung cancers. See HISTONE ACETYLTRANSFERASES, O-LINKED B-N-ACETYLGLUCOSAMINE.

NAD, NADH (nicotinamide adenine dinucleotide) COENZYME 1, of which vitamin B_3 (niacin) is a component. It exists as NAD^+ in its oxidized form when it is capable of accepting a hydride ion ($:H^-$, the equivalent of a proton and two electrons) to become the reduced form, NADH. The half-reaction and general reaction are thus:

$$NAD^+ + 2e^- + 2H^+ \rightarrow NADH + H^+$$

or

$$AH_2 + NAD^+ \rightarrow NADH + H^+$$

in which NAD^+ does not indicate a net charge on the molecule, merely that the nicotinamide ring is in its oxidized form with a positive charge on its nitrogen atom. As with NADP, any given enzyme catalyses a reaction transferring the hydride ion to either side of the nicotinamide ring: the front (A side) or the back (B side), and are termed type A dehydrogenases or type B dehydrogenases, respectively. The coenzyme acts as a readily soluble electron carrier between metabolites, readily diffusing from one enzyme to another, and there is no net gain or loss since it is recycled between oxidized and reduced forms. The combined concentrations of NAD^+ and NADH in most tissues is $\sim 10^{-5}$ M, while that of $NADP^+$ + NADPH is ~10-fold lower, and in many cells the NAD^+:NADH ratio is high. NAD is used mostly in oxidation reactions such as GLYCOLYSIS and the KREBS CYCLE, high NADH levels occurring when there is a high ATP: ADP ratio in the cytoplasm, a situation which favours GLUCONEOGENESIS over glycolysis. NADH is the reductant used by lactate dehydrogenase in reducing pyruvate to LACTIC ACID. See SIRTUINS, whose activity is favoured by low NADH:NAD^+ ratios. See MALATE–ASPARTATE SHUTTLE, NADH DEHYDROGENASE COMPLEX.

NADH dehydrogenase complex (complex I) The largest of the respiratory enzyme complexes of the inner mitochondrial membrane, containing >40 polypeptide chains. It accepts electrons from NADH generated by the KREBS CYCLE and passes them through a flavin and at least seven iron-sulphur centres to UBIQUINONE (coenzyme Q). See MITOCHONDRION.

NADP, NADPH (nicotinamide adenine dinucleotide phosphate) COENZYME 2, of which vitamin B_3 (niacin) is a component. Unlike NADH (see NAD entry for this and other relevant information), NADPH is generally present in cells in greater amounts than $NADP^+$, and NADPH is generally used as the reductant in anabolic reactions, e.g., those involved in the synthesis of FATTY ACIDS (see SREBPS). The half-reaction and general overall reactions are:

$$NADP^+ + 2e^- + 2H^+ \rightarrow NADPH + H^+$$

or

$$A + NADPH + H^+ + \rightarrow AH_2 + NADP^+$$

The PENTOSE PHOSPHATE PATHWAY is an important mode of NADPH production.

Na^+/K^+-ATPase A plasma membrane ion pump (see ION CHANNELS) and antiporter for Na^+ and K^+, maintaining low internal cellular $[Na^+]$ and high external $[K^+]$. It is a P-type ATPase reversibly phosphorylated by ATP as part of the transport cycle. This is important osmotically, and consequently in cell volume homeostasis. It is the major driving force in secondary active transport (e.g., see ENTEROCYTES, PROXIMAL CONVOLUTED TUBULES) and co-transport, and in maintaining the transmembrane electrical potential of neurons and muscle fibres (see RESTING POTENTIAL). Its transcription is promoted by THYROID HORMONES.

narcolepsy A relatively rare neurological disorder, affecting 1:2,000 individuals, marked by uncontrollable urges to sleep during the daytime, especially during sedentary or non-stimulating activities, and commonly associated with sudden bouts of muscle weakness in emotional situations (cataplexy). It is the only known disorder representing abnormalities of the sleep/wake generators (see SLEEP–WAKE CYCLES) and its occurrence indicates that wakefulness and sleep are not mutually exclusive states. It is currently uncertain whether it is neurodegenerative or autoimmune in origin. Those affected lack orexin-secreting cells in the hypothalamus. It occurs most commonly in the 2nd and 3rd decades, the loss of OREXIN (hypocretin-1) neurons being very specific (not damaging the adjacent MCH-secreting neurons). The absence of orexin from the cerebrospinal fluid appears to be a unique feature of the disorder. Narcolepsy has a clear genetic component, >90% of individuals carrying the *HLA-DR2/DQ1* (aka *HLA-DR15* and *HLA-DQ6*) gene, found in <30% of the general population. However, the risk of a 1st degree relative developing narcolepsy is only 1–2%. See CIRCADIAN RHYTHM.

nasal Referring to the NOSE.

nasal septum A partition dividing the nasal cavity into left and right parts. Its anterior part is cartilage, its posterior part consists of the vomer bone and the perpendicular plate of the ethmoid bone. See PHEROMONES.

natriuretic peptides (NPs) A family of structurally related signal peptides regulating salt and water balance and dilating blood vessels. Natriuresis is excessive loss of sodium in the urine, and NPs include ATRIAL NATRIURETIC PEPTIDE, and brain natriuretic peptide. They employ RECEPTOR GUANYLYL CYCLASES on target cells.

natural killer (NK) cell A large, granular lymphoid-like cell type, deriving from a bone marrow multipotent STEM CELL and having important roles in the early stages of INNATE IMMUNITY. The immune system engages in continual 'immunosurveillance' for cells exhibiting unusual antigen profiles, which NK cells help to destroy. Unlike T and B cells, NK cells lack antigen specificity yet can attack herpes-virally infected cells. Larger than T and B lymphocytes, they have distinctive cytoplasmic granules and, through the Fc receptors on their surface, can kill antibody-coated target cells by antibody-dependent cell-mediated cytotoxicity. The $F_C\gamma$RIII (CD16) receptor of NK cells recognizes the IgG1 and IgG3 subclasses (see ANTIBODIES). They can also kill certain lymphoid tumour cell lines *in vitro* in the absence of prior immunization or activation. When activated by INTERFERONS and macrophage-derived CYTOKINES, NK cells increase their cytotoxic effects 20–100 times. Although their mechanism of cytotoxicity is similar to that of antigen-specific cytotoxic T cells, NK cells bear receptors that recognize only invariant components of infected cell surfaces, notably those infected by herpes viruses, the protozoan *Leishmania* and the bacterium *Listeria monocytogenes*. NK cells help contain VIRUS infections until the adaptive immune system can respond and clear the infection. In this respect, they produce large amounts of interferon γ in response to the combined effects of IL-12 and TNF-α. Lectin-like activating receptors ('killer receptors') on NK surfaces recognize certain carbohydrates on self cells initiate the cytotoxic response, although other immunoglobulin-like 'inhibitory receptors' (KIRs) are thought to enable NK cells to recognize cells with class 1 MHC molecules and somehow overcome the signals provided by ligand-bound killer receptors. Cells lacking surface class 1 MHC molecules do not elicit this inhibitory response.

natural selection The most widely accepted theory proposing a causal mechanism to account naturalistically for evolutionary change in populations of organisms. Proposed by Charles Darwin (see DARWINISM) and Alfred Russel Wallace in a joint communication presented in 1858 to the Linnean Society of London, it asserts that, given the diversity (both genetic and phenotypic) among the individuals making up a species population, and given that not all individuals in the population at time t_0 contribute equally to the genetic make-up of the population at time t_1, then to the extent that this non-random genetic contribution to future generations is due to the effects of inherited differences upon individuals, natural selection can be said to have occurred. *Positive selection* is the acquisition of new functions for a pre-existing gene; *negative selection* is the term used to explain the presence of functional genomic elements that, despite having undergone many random mutational events, have not changed in function. The ability to detect any such recent selection in the human genome would have profound implications for the study of human history and for medicine. The recent history of the human population (see MIGRATION) is characterized by great environmental change and emergent selective agents. Domestication of plants and animals at the start of the Neolithic gave rise to an increase in population size, and humans were confronted with the spread of new INFECTIONS, new food sources and new cultural environments.

Until the 1980s it was often taken as axiomatic that it is DNA sequences encoding proteins that have been of greatest importance in adaptive evolutionary change. Since then, several lines of evidence have indicated that coding differences between chimpanzees and humans are unexpectedly scarce. We are now in a period of active research into apparently functional DNA sequences that lie outside protein-coding sequences (see NON-CODING DNA, PSEUDOGENES).

Despite varying degrees of incompleteness, available human genomic data since 2001 (HUMAN GENOME PROJECT, HUMAN GENOME DIVERSITY PROJECT, HAPMAP project) and recent genome sequencing of chimpanzee, mouse and dog, among others, are only now enabling genome-wide studies of positive selection.

Two-way comparisons of human and chimpanzee, and three-way comparisons of these two with the mouse, are raising interesting questions. Though not entirely consistent, they suggest that positive selection in humans has been strongest for genes involved in immune response, reproduction (notably spermatogenesis) and sensory perception (notably olfaction). Many of the longest-studied and most convincing cases for selection in humans (the *LCT*, *HBB*, *FY* loci, and the MHC) also receive strong support in these empirical studies, some of them previously known to be associated with resistance to MALARIA and other infectious diseases (see SEGMENTAL DUPLICATIONS). Possible targets for selection include *NADSYN1* (NAD synthetase 1) at chromosome location 11q13 (which may be involved in prevention of PELLAGRA) and *TLR1* (TOLL-LIKE RECEPTOR 1) at 4p14, for which a role in the biology of TUBERCULOSIS and leprosy has been suggested. This reinforces the belief that infectious diseases, especially MALARIA, have been among the strongest selective pressures in recent human history.

X-linked genes (many of them sperm- and testis-specific) make up a disproportionate percentage of those that seem to be rapidly evolving. The hemizygosity of such genes in males exposes recessive alleles more strongly to selection than it does autosomal alleles, possibly promoting rapid evolution. Although the preliminary analyses used data that (i) were not specifically developed for the study of selection, (ii) may have been biased in their choice of genetic markers, and (iii) were impaired by the incomplete nature of the SINGLE NUCLEOTIDE POLYMORPHISM (SNP) database, the results have nevertheless been heuristically valuable.

Under the neutral selection theory, the great majority of DNA sequence changes within a species do not contribute to reproductive success, only a small subset of changes providing the molecular basis for phenotypic changes and adaptations. Under purifying selection, genes are expected to have lower rates of non-synonymous (K_A) relative to synonymous (K_S) codon changes, and adaptively evolving genes are predicted to have higher rates of non-synonymous relative to synonymous codon changes (the K_A/K_S ratio; see MUTATION). Strong evidence of positive selection unique to the human lineage is currently limited to a handful of genes. Initial work (2005) indicated seven categories of genes that, relative to the chimpanzee, appear to have been subjected to accelerated evolution in the human lineage. The clearest cases lie among transcription factor genes, including homeotic, forkhead and others with key roles in early development.

On the other hand, selective constraints appear to have been relaxed in hominid NON-CODING DNA sequences when compared with murids, possibly reflecting the greater involvement of stochastic processes such as GENETIC DRIFT in altering DNA sequences in hominoids, where the EFFECTIVE POPULATION SIZE is smaller than in murids. If true, this might explain why most human MENDELIAN DISEASES are due to changes in protein-coding regions where selective constraints are higher.

Sometimes, a small group of people migrate, taking with them only a subset of the population's variants of a particular gene. As a result, the frequencies of those variants will probably be inflated relative to the source population, but not because of selection. Any demographic factor, such as a population bottleneck, should affect all genes to an equal degree. But selection would affect specific parts of the genome, which is why genome-wide data are more informative than single-gene studies in assessing which genome regions have been targets of selection. It is also possible that gene variants harboured by the 'leading edge' of a human population expanding into new territory will increase in frequency because individuals there happen by chance to be more prolific than those in the main population.

Then again, the NON-CODING RNA sequence *HAR1* (see HAR1F GENE) appears to have accumulated 18 substitutions since the human-chimpanzee common ancestor, a far larger number than expected on the basis of the slow rate of change in this region in other amniotes. One of the human accelerated regions (HARs), *HAR1* is a 118-bp region expressed specifically in the Cajal–Retzius cells of the CEREBRAL CORTEX, many of the HARs being associated with transcriptional regulation and neurodevelopment. HARs are candidates for positive selection favouring enlargement of BRAIN SIZE in the human lineage. A similar situation has recently been described for *HACNS1* (see LIMBS).

When alleles increase in frequency in a population by *positive selection* (i.e., when selectively favoured), they leave distinctive patterns of genetic variation, or 'signatures'. These can be compared with the background genetic variation which, in humans, is usually considered to evolve

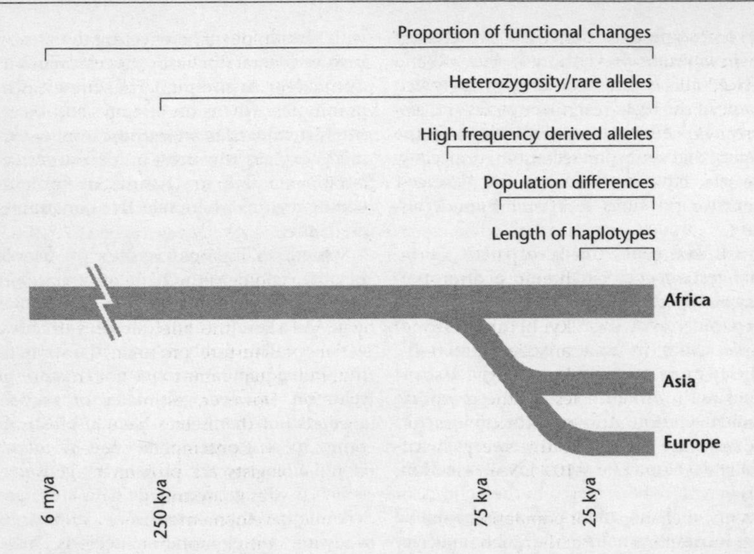

FIG 128. Diagram indicating five 'genomic signatures' that are allegedly indicative of positive selection, with their approximate timescales of persistence. See NATURAL SELECTION for further details. From *Science* Vol. 312, No. 5780 (16 June 2006), p. 1615. Reprinted with permission from AAAS.

largely in accordance with the NEUTRAL THEORY OF MOLECULAR EVOLUTION.

Despite some controversy over the issue, five such genomic signatures (see Fig. 128) are now routinely searched for in the human genome, each thought capable of revealing positive selection over different time scales. Signatures may contain: (i) a high proportion of function-altering mutations useful for comparisons between species over a timespan of millions of years; (ii) a localized reduction in genetic diversity (reduced heterozygosity) as occurs when a rare allele rapidly rises to fixation (100% frequency), carrying the HAPLOTYPE on which it occurs to high frequency due to variants chromosomally linked to the favoured allele 'hitch-hiking' (see 'SELECTIVE SWEEP'). This signature is useful over a timespan <250 Kyr, the chimpanzee genome providing the critical baseline for accurate assessment of these genomic signatures. Preliminary results using the draft chimpanzee sequence (2005) indicated six regions showing a high proportion of SNPs with significantly reduced diversity relative to divergence, combined with a high proportion of high-frequency (≥80%) derived alleles. Although not totally convincing, and despite there being no functional elements implicated, these data are highly suggestive of recent strong selective sweeps in the human lineage. In addition, a genomic region on chromosome 7q contains several regions with high diversity-divergence scores when human and chimpanzee sequences are compared. It contains the *FOXP2* and *CFTR* genes – the former a subject of much controversy concerning the selective origins of SPEECH and language, the latter as a target of selection in European populations (see CYSTIC FIBROSIS). Two genes (*ASPM* and *MCPH1*) that cause microcephaly (small cerebral cortex) are also indicative of positive evolutionary selection, manifested by a high K_A/K_S ratio. Convincing proof of past selection in the 7q region will require careful analysis of the precise pattern of genetic variation and the identification of a likely functional target of selection (see PHENOTYPE, PROTAMINES, SKIN COLOUR);

(iii) high-frequency derived alleles arising by new mutation. In a selective sweep, derived alleles can hitch-hike to high frequency but fail to reach complete fixation (frequency 100%) because of incompleteness of the sweep or recombination with the selected allele (see DUFFY BLOOD GROUP), useful for <80 Kyr; (iv) differences between species populations, as arises when gene flow is restricted by at least partial reproductive isolation (in humans, after the major migrations out of Africa) and useful for periods up to 50–75 Kyr BP; and (v) long haplotypes, in which an allele has both high-frequency and long-range associations with other alleles on the chromosome (e.g., see LACTOSE INTOLERANCE) – useful for detecting partial selective sweeps, with low allele frequency (~10%). See Y CHROMOSOME.

A major challenge for population geneticists is to determine whether such putative signatures of positive selection are indeed due to selection or to confounding effects of past population demography, including BOTTLENECKS, expansions and population subdivisions. See HAMILTON'S RULE, TUBERCULOSIS.

nature–nurture debate It has been said that when we understand the relationship between the social and biological sciences we will have solved, in general form, the nature–nurture problem. A long philosophical tradition in the West, including Plato, Descartes and Kant, has subscribed to the notion of innate ideas. The more recent empiricist tradition, whose *tabula rasa* view of the human mind favoured the supremacy of nurture, includes Bacon, Hobbes, Locke, Hume and Berkeley. The theory that provided the scientific basis for the analysis of nature and nurture was contained in Darwin's *The Origin of Species* (1859). At that time, the ease with which 'nature' could be equated with 'instinct' can be contrasted with the difficulties raised by equating 'nurture' with 'INTELLIGENCE'. By the 1930s, genetics, originally an ally of nature over nurture, was already encountering the phenomenon of variable expression of genes in the phenotype. Embryology, and developmental sciences in general (including developmental psychology), firmly supported these findings. By the 1950s, no attempt had been made to explain how learning was to be understood in the light of genetics and development, even though in social psychology a form of developmental interactionism had been prevalent since the 1920s (but only began to impinge on 1950s ETHOLOGY). Social psychologists were quick to adopt the statistical method developed in 1930 by R.A. Fisher: the analysis of variance. The nature–nurture problem, as applied to an individual, came to have no meaning whatever. However, estimates of covariance do not themselves have any causal connotations. Contemporary developmental psychologists are providing a coalescence of causes around a complex and dynamic developmental process, refusing to partition the phenotype between genetic and environmental determinants, having much in common with the approach of Waddington (see EPIGENETICS). An important factor which some developmental approaches to behaviour may ignore is the importance of historical causes which lead, via genes, to constraints on processes and structures operating in the present: the 'neural machinery' by which information is stored. See CRITICAL PERIOD, HERITABILITY, LANGUAGE, NEURAL PLASTICITY, PHENOTYPE, TWIN STUDIES, WILLIAMS–BEUREN SYNDROME.

NCAMs See NEURAL CELL ADHESION MOLECULES.

Neanderthals See HOMO NEANDERTHALENSIS.

nearsightedness (myopia, shortsightedness) A refraction abnormality, in which the plane of focus of light is in front of the retina. It is correctable by lenses. See ACCOMMODATION, CORNEA.

NEFA Non-esterified fatty acids. See FREE FATTY ACIDS.

negative feedback Neural, endocrine and neuroendocrine loops in which misalignment detectors gauge (i.e., monitor) the value of specific physiological variables and produce output signals which ultimately restore the values of the variables after displacement from their homeostatic norms. These servomechanisms effectively reduce the signs (+ve or –ve) of the displacements. Such loops are key features in HOMEOSTASIS. They can be subtle, as in receptor desensitization in SIGNAL TRANSDUCTION. Their combined effects within a single cell also can be stupifyingly complex, as in many signal transduction pathways a few of the interactions of which are illustrated in figures to, and in entries on, CREB, GLUTAMATE, RAS SIGNALLING PATHWAYS, GENE EXPRESSION, GLUCONEOGENESIS and GLYCOLYSIS. See FEEDBACK, FEEDFORWARD, POSITIVE FEEDBACK.

negative selection See NATURAL SELECTION.

neglect syndrome An ATTENTION disorder condition in which a part of the body, or a part of the external world (e.g., the entire left visual field), is ignored or suppressed and its very existence denied. It may result from lesions in the POSTERIOR PARIETAL CORTEX, most commonly in the right hemisphere, and commonly improves or disappears over time. See MICROSACCADES, SPATIAL AWARENESS.

neighbour-joining algorithm A fast algorithm employed in construction of PHYLOGENETIC TREES.

Neisseria *Neisseria gonorrhoeae,* a Gram-negative coccus, is the bacterium causing gonorrhoea. Cephalosporin ANTIBIOTICS are a useful 'post-penicillin' treatment, although β-lactam resistance since the 1980s led to ceftriaxone being favoured, new treatment modalities being recommended almost annually. Causing an extremely localized infection, it does not elicit a systemic immune response and so reinfection of a cured individual is not uncommon. In men, ~50% of infections are asymptomatic, the commonest presentations being urethritis, ascending infections causing epididymitis and rarely prostatitis. In women, ~80% of infections are asymptomatic, although urethritis and cervicitis are common. Wider invasion can lead to perihepatitis, with high fever, neutrophilia and lower abdominal back pain. Measurement of the rise in titre of antibody following vaccination is a major method of determining that the vaccination is effective. Although less common than non-specific genital infection, *N. gonorrhoeae* still causes millions of infections worldwide. In Africa, the prevalence rates for gonorrhoea and chlamydia infections are up to 66% in high-risk populations and up to 40% in low-risk populations. *Neisseria meningitidis*, a small Gram-negative diplococcus, is the causative bacterium of bacterial MENINGITIS. See CHLAMYDIA TRACHOMATIS.

Nei's standard genetic distance See GENETIC DISTANCE.

neocortex A part of the CEREBRAL CORTEX found only in mammals (*contra* the olfactory cortex, hippocampus and other parts of the temporal lobe). Developing from the dorsal PALLIUM, it is separated from the olfactory cortex by a sulcus, the rhinal fissure. It has expanded enormously during the course of hominin evolution, forms the majority of the cortex and has a characteristic six-layered organization. Comparing the neocortex of primates that depend heavily upon vision, such as humans, with mammals such as rats and cats, the number of secondary sensory areas has greatly increased, as have the association areas, particularly those in the frontal and temporal lobes. These association areas are usually linked to the emergence of 'mind'. Lesions in the frontal cortex may profoundly affect an individual's personality. See BRAIN.

Neolithic A human cultural period, beginning ~10,000 years ago, marked by the appearance in the archaeological record of industries such as polished stone and metal tools, pottery, animal domestication and agriculture.

neonatal screening See SCREENING.

neonate Infants between 1 and 5 months have a unique immunological profile which includes a decrease in maternal antibodies and the start of immunological competence. Sympathetic innervation of the heart is not complete until about 6 months of age.

neopallium The non-olfactory part of the PALLIUM.

neoplasia Term referring either (i) to the state of cancerous growth (see CANCER), or (ii) to any of a disparate collection of cancerous growths having an abnormal appearance and proliferation pattern. Some researchers include both benign and malignant tumours as neoplastic; others restrict the term to include only malignant growths. See MALIGNANT TUMOURS.

neostriatum (dorsal striatum) A major forebrain nucleus integrating cortical and thalamic afferents and forming the input nucleus of the BASAL GANGLIA (see entry) and crucial to successful execution of INTENTIONAL MOVEMENTS. It comprises the caudate nucleus and putamen, split anatomically by the internal capsule, which receive afferents from all parts of the CEREBRAL CORTEX (somatotopically), from the thalamus and the substantia nigra. Striatal neurons project to the substantia nigra pars reticulata (direct pathway) or to the lateral globus pallidus (indirect pathway), and imbalances between activities in these two pathways have been implicated in causing the profound motor deficits underlying PARKINSON'S DISEASE and HUNTINGTON'S DISEASE. Neurons in both pathways seem to express similar forms of SYNAPTIC PLASTICITY. In mouse models, the medium spiny neurons of the striatum receive convergent glutamate-mediated inputs from the cerebral cortex providing details of the context in which behaviours occur. These neurons also receive dopamine-mediated inputs from the midbrain that report on the significance of the behaviour (e.g., whether it produces an unexpected reward). When an important event occurs, the striatum stimulates circuits projecting via the thalamus to the prefrontal cortex, leading to formation of memories that can later guide planning and control of behaviour. See CORPUS STRIATUM.

nephron The functional unit of the KIDNEY, each comprising a glomerulus and an associated renal tubule (i.e., PROXIMAL CONVOLUTED TUBULE, LOOP OF HENLE, DISTAL CONVOLUTED TUBULE and COLLECTING DUCT).

nerve A bundle of axons of the PERIPHERAL NERVOUS SYSTEM, with accompanying glial cells, blood vessels, connective tissue, etc., collectively surrounded by a common connective tissue sheath, the PERINEURIUM. Only one collection of axons in the central nervous system is so-called: the optic nerve. *Mixed nerves* contain both sensory and motor axons. Mixed nerves contain both sensory and motor neurites.

nerve fibres (axons) See AXONS.

nerve growth, nerve growth factor See NEURAL REGENERATION, NGF.

nerve impulses The propagated ACTION POTENTIALS that travel along an axon membrane. The speed of conduction varies, increasing with axon diameter (varying between <1 and 25 μm in humans), MYELINATION (action potentials being conducted from one NODE OF RANVIER to the next in *saltatory conduction*) and temperature (largely through increased rate of transmitter diffusion during synaptic transmission). The velocity of conduction varies from 0.25 m sec^{-1} in very small unmyelinated axons to 100 m sec^{-1} in very large myelinated ones (see AXONS).

nerve regeneration See NEURAL REGENERATION.

nerve tissue See NERVOUS TISSUE.

nerve tracts Bundles of nerve fibres (see TRACT) which may travel long distances and link quite distant regions. Sensory and motor tracts are white matter tracts running in the spinal cord (hence *spinal tracts*), their names indicating where in the spinal cord they begin and end. Thus, the anterior spinothalamic tract is located in the anterior white column, beginning (i.e., with its neuronal cell bodies and dendrites) in the spinal cord and ending in the thalamus.

nervous system The organ system comprising the CENTRAL NERVOUS SYSTEM (the brain and spinal cord) and PERIPHERAL NERVOUS

SYSTEM (the somatic, autonomic and enteric nervous systems).

nervous tissue Tissues, ectodermal in origin, deriving from the differentiated progeny of neural stem cells, and including GLIAL CELLS and NEURONS. See Appendix VI.

nested cladistic analysis (NCA) A method, devised by Templeton in 1998, for investigating patterns of genetic diversity within different segments of a HAPLOTYPE phylogeny separately. This was thought necessary since the geographical distribution of different parts of the phylogeny may have been subject to different causal processes, e.g., historical and geographical, making analysis of the tree as a single unit impossible. The technique involves subdivision of the known phylogeny into separate parts, apportioning it into nested clades, beginning at the tips where haplotypes are one mutational step from the ancestral haplotype. After grouping the tip haplotypes into nested clades, unresolved 'interior' clades are likewise assigned to nested clades of the same level, and so on iteratively until the whole tree is represented by a single nested clade. Then explanations for those nested clades with significant geographical differentiation can be sought in terms of (i) restricted gene flow (isolation by distance), (ii) past fragmentation of populations, (iii) expansion of range, either contiguous range expansion or long distance colonization taking haplotypes into previously unoccupied regions. Statistics are then calculated for each nested clade to enable clades that exhibit significant geographical differentiation to be identified and, if sufficient data permit, evaluate which biological process is most likely to explain the present distribution of genetic diversity within the clade. Despite evidence that this method can yield correct prehistorical inferences, fully statistical methods of phylogeographic inference are still to be developed.

net filtration pressure The effective pressure causing filtration of tissue fluid from capillaries (see EXTRACELLULAR FLUID). In the KIDNEYS, it is the filtration capillary pressure, ~10 mm Hg, the result of: glomerular hydrostatic blood pressure (50 mm Hg) – hydrostatic pressure in Bowman's capsule (10 mm Hg) – blood colloid osmotic pressure (30 mm Hg).

netrins The prototypical axonal attractants, first identified as extracellular factors secreted from the floor plate, attracting commissural axons towards the midline (see AXONAL GUIDANCE). Netrins have other roles, being expressed in Schwann cells and being required for NEURAL REGENERATION, and in maintenance of the CNS. They are also implicated in development of mammary gland, lung, pancreas and blood vessels (see VASCULOGENESIS). The dual role of netrins in guiding neurons and blood vessels suggests that they may have unique therapeutic potential, possibly in diabetics.

network oscillators See CENTRAL PATTERN GENERATORS.

neural cell adhesion molecule (NCAM, CD56) A member of the immunoglobulin superfamily of glycoproteins (see IMMUNOGLOBULINS), expressed by a variety of cell types, including neurons, glial cells and skeletal muscle. At least 27 forms occur as a result of alternative splicing of mRNA (see RNA PROCESSING), and these have been variably implicated in cell–cell adhesion, neurite outgrowth, synaptic plasticity, and learning and memory.

neural circuits See NEURAL NETWORKS.

neural cliques Functional coding units of the CA1 subfield of the HIPPOCAMPUS.

neural coding The activity patterns of neurons which encode information. Neuroscientists search for correlations between patterns of firing and behavioural functions. The two leading neural coding theories are a 'rate code' and a 'temporal code'. In the rate code, all the information is conveyed in the changes of the firing of the neuron; in the temporal code, information is also conveyed in the precise interspike intervals. Rhythmic fluctuations in neuronal excitability (see ELECTROENCEPHALOGRAM) can modulate both output spike timing and sensitivity to synaptic input. In 2005, records of neural activity of rats placed in various settings showed that hippocampal neurons have independent coding schemes for spatial and non-spatial information. Changes in spatial location were represented as changes in the location of firing and firing rate of active cells, whereas non-spatial changes were represented by changes in firing rate while the location of firing remained constant. These independent encoding schemes may enable simultaneous representation of spatial and episodic memory information. Activity patterns generated by large neuronal populations (ensembles) in the brain are believed to enable it to achieve its real-time encoding, processing and execution of cognitive information. Simultaneous monitoring of cells in such ensembles is now feasible, as is population-vector analysis of ensemble firing patterns. In 2006, using mice as subjects, a large-scale ensemble recording technique, combined with novel categorical episodic memory paradigms and mathematical tools (multiple discriminant analysis and principal component analysis), was employed in work concerning network-level memory traces, their patterns, and the brain abilities mentioned (see HIPPOCAMPUS). The storms of electrical pulses that sweep through the CNS somehow translate into thoughts, emotions and sensations. Unlike earlier hypotheses of sensory perception which envisaged strictly linear signal transmission along discrete neural pathways from receptor to higher brain processing centres, work on the rodent somatosensory system suggested an alternative model (distributed representation, or population, neural coding), in which recordings taken from large populations of single neurons enable far better prediction of an animal's behaviour. One neuron's membership of such populations (cliques, or ensembles) is probably fluid, it being party to many cliques simultaneously. Such a cell's firing properties could also change continuously in accordance with the state of the sensory periphery, the animal's past perceptual experiences, internal brain dynamics, whether the animal is actively or passively sampling its environment, and even with the animal's expectations about the future. A mathematical tool, matrix inversion, has enabled conversion of activities of neural cliques into binary code, enabling prediction with 99% accuracy of the events mice have just experienced and where, in the animals' environment, it had happened. Such a binary code may have applicability in studies of COGNITION, and has potential for real-time brain-to-machine communication (see NEUROROBOTICS).

neural crest Cells arising from embryonic ectoderm along each edge (the crest) of the neural folds and extending down the length of the neural tube (see NEURULATION). They migrate laterally to form dorsal root (sensory) ganglia, sympathetic neuroblasts, Schwann cells, pigment cells, odontoblasts, meninges and mesenchyme of the PHARYNGEAL ARCHES. *PAX3* and the WNT/β-catenin signalling pathway (see WNT SIGNALLING PATHWAYS) are essential for neural crest induction. See Appendix VI.

neural degeneration (neurodegeneration) Although the complex molecular pathophysiology of neurodegeneration is largely unknown, major advances have been made in elucidating the genetic defects underlying Mendelian forms of these diseases. One of the major carriers of cholesterol in the brain, ApoE (see entry), modulates susceptibility to the most common diseases involving neural degeneration. The loss of pigmented neurons (those producing neuromelanin; see MELANINS) from specific brain nuclei is associated with a variety of neurodegenerative diseases. PGC-1 has also been implicated, while a growing body of evidence implicates dysfunction of NMDA RECEPTORS with stroke and several neurodegenerative (e.g., PARKINSON DISEASE, HUNTINGTON'S DISEASE) and psychiatric diseases (e.g., SCHIZOPHRENIA), which are often associated with impaired learning and memory, eventually leading to dementia (e.g., ALZHEIMER DISEASE). Transplantation of neural STEM CELLS, or of progenitor cells from foetal brain, has attracted much attention as a potential regenerative therapy, although neural grafts can be rejected from the normally immunologically privileged central nervous system (CNS) if there is local inflammation enabling entry of lymphocytes.

Common pathophysiological pathways include enhanced oxidative stress (see MITOCHONDRION, RETINITIS PIGMENTOSA), protein misfolding and aggregation and dysfunction of the ubiquitin-proteasome system (see UBIQUITINS). Ubiquitination of misfolded proteins that aggregate in the cytoplasm and/or nucleus of CNS neurons and forming abnormally staining deposits, or *amyloids*, are a key characteristic of neurodegenerative diseases and misfolded proteins have been identified in many of them; but those in frontotemporal lobar degeneration (FLTD-U, the most common frontotemporal dementia, or FTD) and ALS have been enigmatic (but see TAU and O-LINKED B-N-ACETYLGLUCOSAMINE). Recently, DNA-binding protein 43 (TDP-43) has been identified as the major ubiquitinated disease protein in FTLD-U and sporadic ALS. After Alzheimer's disease, FTDs are the most common cause of dementia under age 65, some patients developing parkinsonism or motor neuron disease. More than 30% of FTDs are familial.

The MOTOR NEURON diseases (MNDs), for which there is no present cure or standard treatment, are a group of progressive neurological disorders that destroy cells that control essential muscle activity such as speaking, walking, breathing and swallowing. Disruptions in signals from upper and lower motor neurons can result in gradual muscle weakening, wasting and uncontrollable twitching (fasciculations). Eventually, the ability to control voluntary movement can be lost. MNDs may be inherited or sporadic (acquired), and they occur in all age groups. In adults, symptoms often appear after age 40. In children, especially in inherited or familial forms of the disease, symptoms can be present at birth or appear before the child learns to walk. The causes of sporadic MNDs are not known, but environmental, toxic, viral or genetic factors have been implicated. Common MNDs include AMYOTROPHIC LATERAL SCLEROSIS (ALS), progressive bulbar palsy, primary lateral sclerosis and progressive muscular atrophy. Other MNDs include the many inherited forms of spinal muscular atrophy and post-polio syndrome, a condition that can strike polio survivors decades after their recovery from poliomyelitis.

A recent mouse model which involved temporally and spatially restricted induction of neuronal loss indicated that environmental enrichment reinstated learning behaviour and re-established access to long-term memories even after significant brain atrophy and neuronal loss had occurred. This enrichment correlated with histone acetylation by inhibitors of HISTONE DEACETYLASES, inducing sprouting of DENDRITES and synaptogenesis (see MEMORY). See DIABETIC NEUROPATHY, NEURAL REGENERATION, TOXINS.

neural maps (topographic maps) Although lacking the fixed analogue properties of familiar land maps, these inputs to the central nervous system are nevertheless spatially organized. Sensory pathways, other than those concerned with taste and smell, are so anatomically organized that they preserve the spatial relations of receptors in the periphery. Axons of adjacent primary receptors project to higher centres without losing their relational topography. In the somatic sensory system this is termed *somatotopy* (see PRIMARY SOMATOSENSORY CORTEX), in the visual system RETINOTOPY, and in the auditory system *tonotopy* (see COCHLEA).

neural networks Any collection of neurons potentially interacting as a system. Much of the nervous system consists of such networks (circuitry), in which the signal processing often involves DIVERGENCE, CONVERGENCE, FEEDBACK and FEEDFORWARD. CENTRAL PATTERN GENERATORS are neural networks producing autonomous cyclical patterns of activity. Identification of neural cliques as memory-coding units of the HIPPOCAMPUS (see entry) raises the possibility that the concept of these cliques as basic, self-organizing processing units could apply to many, if not all, of the brain's neural networks. Activity of CREB is required for maintenance of all neural circuitry.

neural plasticity In contrast to 'hard-wired' neural circuitry, some neural pathways are subject to continual 're-wiring', typically involving neurogenesis, synaptogenesis and BDNF signalling. *Developmental plasticity*, studied especially in the visual system, is circuitry that is conditional on early sensory experience, shaping subsequent perception. This rewiring usually involves synapse elimination, either through apoptosis of neurons, or because inappropriate axon branches are retracted. Similarly to the situation in the LATERAL GENICULATE NUCLEUS (LGN), afferents serving the two eyes are initially intermingled in layer IV of the PRIMARY VISUAL CORTEX (PVC; see Fig. 142), but segregate under the influence of activity (see MECP2 for activity-dependent gene expression). In kittens, this plasticity of the primary visual cortex is responsible for formation of OCULAR DOMINANCE columns, and hence binocular vision, for several weeks postnatally, whereas in primates it is completed by birth. However, monocular deprivation in monkeys during a 2-week postnatal CRITICAL PERIOD shows that LGN axons and their synapses in layer IV are highly dynamic even after birth, being *experience-dependent*. In fact, the critical period for anatomical plasticity of layer IV in the macaque lasts until ~6 weeks after birth, at the end of which LGN afferents seem to lose the capacity for growth and retraction (see AXONAL GUIDANCE, DENDRITES). In adult rats, the absence of the plasticity associated with establishment of ocular dominance columns has largely been attributed to maturation of intracortical inhibition, and recent (2008) work on these adult animals indicates that chronic administration of the ANTIDEPRESSANT fluoxetine reduced this inhibition, increased BDNF expression in the visual cortex and promoted visual cortical plasticity (with possible clinical significance for AMBLYOPIA). NEUROTROPHINS are involved in visual plasticity, overexpression of BDNF in mice accelerating establishment of ocular dominance columns, and GABA transmission by cortical interneurons seems to determine the critical period. Modifications to binocularity following monocular deprivation do not occur under anaesthesia, even though cortical neurons still respond to visual stimulation. Such evidence favours the view that SYNAPTIC PLASTICITY in the cortex requires extraretinal 'enabling factors', such as may be associated with the level of ATTENTION or other behavioural state. The huge neural plasticity of the infant cerebral cortex is evidenced by their ability to develop fluency of SPEECH despite excision of the left hemisphere (see LANGUAGE). LEARNING and MEMORY are dependent upon neural plasticity (see, e.g., HEBBIAN MODIFICATION), in which CREB-related gene activity appears to be involved. See NEURAL REGENERATION.

neural plate The thickened ectoderm overlying the prechordal mesoderm of the embryo, comprising neurectoderm cells that will roll up to form the NEURAL TUBE. See NEURULATION.

neural regeneration If an axon in the *peripheral nervous system* is severed, the portion connected to the cell body will probably reform a growth cone and regrow to its target (see AXONAL GUIDANCE), a process which depends on the NEUROLEMMA. Myelination seems to prevent regeneration, and regrowth of damaged neurons in *central nervous system* is very limited, especially in the adult, the neuronal environment apparently being hostile to regrowth. SEMAPHORIN 3F is a potent inhibitor of axon growth, and is upgraded on injury. Severance of an axon here usually results in neuronal death, those axons which do survive producing a disorganized mass (neuroma) near the point of the cut. When a peripheral nerve is severed, the portion distal to the cut dies. Schwann cells surrounding the axon undergo dedifferentiation, phagocytose the myelin and begin dividing mitotically. Existing RNA in the cell bodies, much of it encoding proteins concerned with synaptic functions, is destroyed and new RNA relating to axon regrowth is synthesized.

Regrowth of axons, both sensory and motor, is usually non-specific so that normal relations between a motor axon and its effector, or a sensory neuron and its sensory field, are lost. ApoE (see entry) has been implicated in regrowth of synapses.

The EXTRACELLULAR MATRIX (ECM) can be a block to growth. One key class of such inhibitory ECM molecules are the chondroitin sulphate proteoglycans (CSPGs), which bacteria such as *Proteus vulgaris* can modify using the enzyme chondroitinase ABC. This enzyme may be of future use as a 'molecular machete' in clearing the way for nerve regeneration in injured spinal cords. The neurite outgrowth inhibitory protein (Nogo) expressed by supportive glial cells also acts as a 'stop signal', especially in CNS myelin – but apparently not in PNS myelin. Transgenic expression of Nogo-A in PNS oligodendrocytes blocked axon recovery after sciatic nerve crush. Cultured retinal ganglion cells from embryonic animals grow axons more rapidly than do those from older animals, even under optimal growth conditions, suggesting this may be an intrinsic property of young neurons. A better understanding of the switch which occurs in neuronal development from axonal to dendritic growth may hold the key to regeneration in the central nervous system (see NEURAL PLASTICITY). The cannabinoid receptor CB_1 (see CANNABINOIDS) is involved in neurite outgrowth, as are the downstream activated transcription factors CREB, RAR_{α}, STAT3, SMAD3 and PAX6 (see PAX GENES), while BRCA1 may also be involved in neuronal development. See NEUROGENESIS, NEUROTROPHINS, STEM CELLS.

neural tube A hollow cylinder of neurectoderm cells forming after fusion of the neural folds (see NEURULATION, Appendix VI). Once the neural tube closes, neuroepithelial cells begin to develop into neuroblasts, characterized by a large spherical nucleus containing pale nucleoplasm and a dark-staining nucleolus. These primitive nerve cells form a layer, termed the *mantle layer*, around the neuroepithelial layer and later give rise to the grey matter of the spinal cord. Dorsal and ventral regions of the spinal cord differentiate under the influence of opposing concentration gradients of bone morphogenetic proteins and the product of the SONIC HEDGEHOG GENE, SHH. BMP4 and BMP7 are secreted by ectoderm overlying the neural tube, establishing a signalling centre in the roof plate and inducing a cascade of TGF-β proteins, including BMP5 and BMP7, activin and dorsalin in the roof plate and adjacent tissue. In the ventral region, only SHH is released. The relative concentrations of these morphogens lead to gene expression that regulates the production of dorsal and ventral roots of SPINAL NERVES. High concentrations of BMPs and low concentrations of SHH in the dorsal neural tube activate *PAX3* and *PAX7* controlling sensory neuron differentiation. High SHH and low BMP concentrations in the ventral-most neural tube lead to activation of *NKX2.2* and *NKX6.1* leading to growth of neurons forming the ventral root of the spinal nerve. Between these two poles, slightly lower SHH concentrations and slightly higher BMP concentrations initiate transcription of *NKX6.1* and *PAX6*, transcription factors influencing determination of the ventral horn cells. Neural tube defects (NTDs) variably involve the meninges, vertebrae, muscles and skin. Severe forms of disorder occur, varying with geography, but may be as high as ~1:100 births in some areas of northern China. See NEUROGENESIS, SPINA BIFIDA.

neurectoderm Cells of the NEURAL PLATE.

neurites (neuronal processes) Long projections from the cell body of a NEURON, either dendrites or axons.

neuroblast An immature NEURON, prior to cell differentiation (see NEUROGENESIS). See NEURAL TUBE.

neuroblastomas A tumour of primitive neuronal precursor cells of the peripheral nervous system and adrenal medulla. The proto-oncogene N-*MYC* is amplified in ~30% of human neuroblastomas, rising to ~50% in advanced cases in which gene amplification may be 1000-fold (see ONCOGENE).

neurocranium See SKULL.

neurodegeneration See NEURAL DEGENERATION.

neurodegenerative diseases See NEURAL DEGENERATION, references there and CLONING.

neuroepithelial cells Pseudostratified layer of cells linked by junctional complexes, forming the wall of the recently closed neural tube. Initially proliferative, they give rise to primitive nerve cells (neuroblasts) forming the MANTLE LAYER of the spinal cord (later its grey matter), and then to primitive glial cells (gliablasts). Once these have been produced, it gives rise to ependymal cells lining the central canal of the spinal cord.

neurofibromatosis (NF) Autosomal dominant disorder of the nervous system, and one of the commonest genetic disorders (about 1 in 3,500 births overall), with extremely variable phenotypic effects. It gained notoriety when it was suggested that the 'Elephant Man', Joseph Merrick, was probably affected (his disorder being subsequently identified as the much rarer Proteus syndrome). Two main types of NF occur: NF1 (incidence 1:3,000) and NF2 (incidence 1:35,000 and prevalence 1:200,000). Both conditions might be regarded as familial cancer syndromes. About 50% of cases are sporadic, due to new mutations, the estimated MUTATION rate of 1:10,000 gametes being ~100 times higher than the average human mutation rate per generation per locus. The NF1 gene spans over 350 kb and contains at least 59 exons and lies close to the centromere on the long arm of chromosome 17 (17q) close to the *TP53* gene. Its product, neurofibromin, has structural homology with the GTPase-activating proteins involved in down-regulating *Ras* gene activity and acts as a tumour suppressor, interacting with the RAS proto-oncogene product, mRNA editing of the NF1 gene transcript leading to neurofibronin truncation that inactivates its tumour suppressor function. The NF2 locus was mapped by linkage analysis in 1987 to chromosome 22q. The gene product, Merlin, is believed to be a cytoskeletal protein that acts as a tumour suppressor.

The most notable features of NF1 are café-au-lait spots (lightly pigmented areas of the skin, the colour of coffee or milk), freckles in the armpits and a few skin neurofibromata, benign but disfiguring small fleshy tumours arising from the fibrous coverings of nerves in late childhood and increasing with age. Most individuals enjoy a normal healthy life, but a third of childhood cases have mild developmental mental delay characterized by a non-verbal learning disorder and a small number develop epilepsy, a central nervous system tumour, or a scoliosis. Somatic MOSAICISM can manifest with features limited to a particular part of the body. There is no cure for NF1 at present. In NF2, the most characteristic feature is development in early adult life of tumours involving the 8th cranial nerve (vestibular Schwannomas), and cataracts are frequent but often subclinical.

neurogenesis The production of NEURONS. Initially, most cells generated by neural stem cells lining the NEURAL TUBE (see entry) become neuroblasts and proliferate mitotically. Non-dividing progeny eventually migrate away from the epithelium towards the surface of the neural tube and differentiate into neurons. Whether a local niche in the central nervous system is neurogenic or gliogenic is determined to some extent by ASTROCYTES, which secrete a variety of growth factors and cytokines. In contrast to much of the rest of the adult brain, which is gliogenic, the olfactory bulb and hippocampus remain neurogenic into adult life. One explanation lies in the secretion by the ependymal cells of noggin, which binds BMPs secreted by astrocytes and blocks their effects, redirecting local adult stem cell fates into neurogenic pathways. See ALZHEIMER'S DISEASE, DEPRESSION, MEMORY.

neurogenic Describing muscle contractions in which the signal for contraction originates in motor neurons (within the central nervous system). This is the typical situation for skeletal muscle. Cardiac muscle is fundamentally MYOGENIC, although the rate of contraction can be modified by neurogenic signals (see CARDIAC CYCLE).

neuroglia See GLIAL CELLS.

neuroglobin A monomeric protein encoded by a gene on chromosome 14 (14q24) and expressed in brain tissue, increasing oxygen availability.

neurohypophysis The posterior PITUITARY GLAND.

neuroinformatics A spectrum of tools being made available on the Internet via an electronic gateway. It ranges from brain imaging to the study of genes and proteins involved in neurological disorders.

neurolemma The outer (nucleated) cytoplasmic layer of the SCHWANN CELL covering of one or more axons in the peripheral nervous system. Unlike the axon and the myelin sheath, the neurolemma does not degenerate after a nerve has been cut or crushed; the hollow tube formed by the neurolemma is instrumental in regenerating the nerve fibre (see NEURAL REGENERATION). It is not present around axons in the CNS because the OLIGODENDROCYTE cell body and nucleus do not envelop the axons. Compare MYELIN SHEATH.

neuroleptics Antipsychotic drugs. Their effectiveness seems to lie in blockading the D_2 DOPAMINE receptors in the brain (see DARPP-32). Although side-effects include some movement disorders, some also block the serotonin $5HT_2$ receptors, which may alleviate these. See PSYCHOSIS.

neuroligins Proteins expressed on the postsynaptic density (see SYNAPSE) and, by interacting with neurexins on the presynaptic membrane tether the pre- and postsynaptic membranes of a synapse together. See AUTISM.

neurological disorders Those disorders, such as PARKINSON DISEASE, ALZHEIMER DISEASE and HUNTINGTON'S DISEASE, whose origins are clearly attributable to lesions in neural pathways and/or to NEURAL DEGENERATION. The distinction between neurological and psychological disorders is increasingly subtle, perhaps even artificial (see AFFECTIVE DISORDERS), the former lying at one end of a continuum in which causes are more obviously anatomical in origin.

neuromodulators, neuromodulation Substances, e.g. neurotransmitters, catecholamines, cholecystokinin, cortisol, growth hormone and vasoactive intestinal polypeptide, which bind METABOTROPIC RECEPTORS on a postsynaptic membrane of a SYNAPSE and set in motion complex intracellular responses with onset latency of several seconds and duration of a few minutes or more (*neuromodulation*). An example would be the effect of ACETYLCHOLINE on the muscarinic K^+ channel. See AUTORECEPTORS, DIFFUSE MODULATORY SYSTEMS, SYNAPTIC PLASTICITY.

neuromuscular junctions (NMJs) Synapses formed between a motor axon terminal and a muscle fibre membrane, several of which may be part of a MOTOR UNIT. The postsynaptic membrane, the MOTOR END PLATE, is often folded and bears receptors for the ACETYLCHOLINE (ACh), universally released from synaptic vesicles by the nerve terminal (there being no inhibitory synapses onto vertebrate skeletal muscle). An *end-plate potential* is the depolarization of a motor end-plate resulting from arrival of a single action potential at a motor neuron terminal. Quantal analysis of transmission at the NMJ indicates that a single action potential in the presynaptic terminal triggers exocytosis of ~300 synaptic vesicles, resulting in an EPSP of 40 mV or more, far greater than that at axodendritic synapses. The ACh in each of the vesicles (~4,000 molecules) produces a *miniature end-plate potential*, and an end-plate potential results from summation of the miniature end-plate potentials caused by each of ~300 synaptic vesicles released simultaneously from ~1,000 active zones (the presynaptic membrane regions specialized for transmitter release) when a single action potential arrives at the axon terminal.

The NMJ has evolved to be fail-safe, achieving muscle contraction without computation of EPSPs and IPSPs (see SYNAPSES). The presynaptic neuron is responsible for secreting ACETYLCHOLINESTERASE into the cleft, degrading ACh. But if this is inhibited (an effect of some nerve gases), the uninterrupted exposure to high concentrations of ACh after several seconds leads to *desensitization*, in which transmitter-gated channels close despite the presence of ACh, leading to paralysis. See MYASTHENIA GRAVIS.

neuronal competition See AXONAL GUIDANCE.

neuronal maps See NEURAL MAPS.

neuronal plasticity See NEURAL PLASTICITY.

neuronal processes Collective term for axons and dendrites (i.e., neurites).

neurons, neurones Nerve cells specialized for the conduction of impulses (propagated action potentials) which, when mature, are some of the longest-living cell types in the mammalian body (see NEURAL TUBE, NEUROGENESIS). Each neuron is surrounded by GLIAL CELLS, and each usually comprises a cell body (perikaryon) containing the nucleus, and neurites (long projections from the cell body) which include one or more DENDRITES, and an AXON (up to a metre in length) arising from the axon hillock. The PI3K-to-PAX6 signalling pathway is important in neuronal differentiation. Neurite outgrowth is influenced by the cannabinoid receptor CB_1 (see CANNABINOIDS), and possibly by BRCA1. Axons terminate in synaptic terminals (see SYNAPSES). There are ~10^{11} neurons in the body, their cell bodies aggregating into compact groups (nuclei and ganglia) or into sheets (laminae), as in the retina, within the grey matter of the central nervous system (CNS), or in specialized ganglia in the peripheral nervous system (PNS). The neuron-restrictive silencer factor NSRF is a mammalian transcriptional repressor that silences expression of neuronal genes in non-neuronal cell types and neuronal progenitor cells (see DNA-BINDING PROTEINS). See BIPOLAR NEURON, MULTIPOLAR NEURON, MYELIN SHEATH, NERVOUS SYSTEM, SYNAPSES, UNIPOLAR NEURON.

neuropeptides Peptide NEUROTRANSMITTERS.

neuropeptide Y (NPY) A hypothalamic peptide neurotransmitter encoded by a CREB-regulated gene and released by neurons of the ARCUATE NUCLEUS. It is orexigenic (see APPETITE and Fig. 14) and reduces thermogenesis and its secretion and action are regulated by LEPTIN and the neuropeptides melanocortin, CRH and GLP-1. The enteric hormone PYY_{3-36} (see entry) is a member of the NEUROPEPTIDE Y family. See COPY NUMBER.

neurophysins Byproducts of post-translational processing of prohormone within neurosecretory vesicles of the magnocellular neurons of the paraventricular (PVN) and supraoptic nuclei (SON) of the HYPOTHALAMUS. Neurophysin I is a 10-kDa carrier protein for OXYTOCIN during its transport through the long axon of the PVN neuron to the posterior pituitary, preventing its diffusion out of the axon. Neurophysin II is a carrier protein for VASOPRESSIN. Mutations in the encoding gene for neurophysin II, leading to vasopressin deficiency, result in familial neurogenic DIABETES INSIPIDUS.

neuropil The space between cell bodies of neurons, containing the cytoplasmic extensions of both neurons and glia.

neuropilins (NPs) Cell surface transmembrane receptors for SEMAPHORINS. NP-1 and NP-2 were identified by their ability to bind the secreted class 3 semaphorins. Sema3A signalling is mediated by a receptor complex involving NP-1, whereas Sema3F signalling occurs through NP-2, and heteromeric NP-1/NP-2 complexes are thought to serve as functional receptors for Sema3C. NP-1 is also a receptor for VEGF (see GROWTH FACTORS).

neuroplasticity See NEURAL PLASTICITY.

neurorobotics Georgopoulos, in 1982, suggested that cells in the motor cortex specify the direction of a limb's movement.

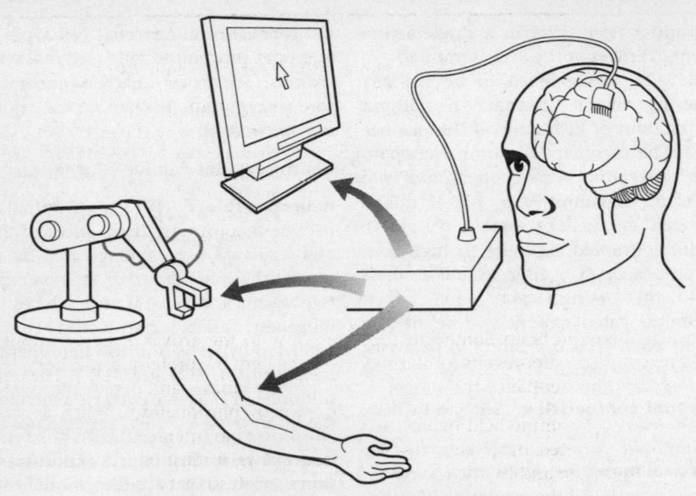

FIG 129. Diagram indicating how signals from cortical arrays implanted in the motor cortex can be applied usefully, including signalling desired movements in a paralysed subject. See NEUROROBOTICS. From R.B. Stein and V. Mushahwar, © 2005 *Trends in Neurosciences*, with permission from Elsevier.

Adding the contributions of many cells over time gives a population vector accurately predicting a limb's movement, even predicting this in real time using multielectrode arrays in the cortex (see NEURAL CODING). In 1999, Chapin and others illustrated the potential of this method by recording signals from an animal in North Carolina, processed them, and sent them over the Internet to control a robot arm in Massachusetts. By 2002, learning algorithms improved control, and work with monkeys showed that cortical signals could be dissociated from activation of spinal motor neurons and that control required just tens of cortical neurons rather than the thousands that some had supposed.

In principle, restoration of functional movements could be achieved by stimulation at various levels in the nervous system (functional electrical stimulation, FES). The most successful such device is the artificial PACEMAKER, now implanted in ~0.5 million people annually. Another is the cochlear stimulator (see COCHLEA), which has been implanted in >40,000 people in the past decade alone. Epidural stimulators are commonly used for PAIN control. In principle, cortical stimulations could be used to control (i) a cursor on a computer screen, (ii) movements of a robotic manipulator, or (iii) an FES system for the arm in which hand movements could be achieved by stimulation of intact muscles that are not receiving cortical innervation as a result of injury or diseases affecting the central nervous system (see Fig. 129). The goal of such devices would be to help those who are high quadruplegics, or 'locked-in' as a result of AMYOTROPHIC LATERAL SCLEROSIS or similar disorders, to negotiate their environments and communicate with those caring for them.

neurosecretions (neurohormones) Neurotransmitters secreted directly into the blood and acting, like hormones, over long distances. They include OXYTOCIN and VASOPRESSIN.

neurosteroids Natural metabolites of steroid hormone synthesis from cholesterol, primarily in the gonads and adrenals but

Amino acids	*Amines*	*Peptides*
Gamma-amino butyric acid (GABA)	Acetylcholine (ACh)	Cholecystokinin (CCK)
Glutamate (Glu)	Dopamine (DA)	Dynorphin
Glycine (Gly)	Adrenaline	Enkephalins (Enk)
	Histamine	*N*-Acetylaspartylglutamate (NAAG)
	Noradrenaline	Neuropeptide Y
	Serotonin (5-HT)	Somatostatin
		Substance P
		Thyrotropin-releasing hormone (TRH)
		Vasoactive intestinal polypeptide (VIP)

TABLE 7 *Major* NEUROTRANSMITTERS.

also in glial cells of the brain. Some bind the $GABA_A$ receptor and suppress its inhibitory function; but others enhance this.

neurotensin A 13 amino acid neuropeptide hormone secreted by N cells of the lower small intestine and by the hypothalamus, implicated in the regulation of luteinizing hormone and prolactin release and in interactions with the dopaminergic system. First isolated from extracts of bovine hypothalamus, where it causes a visible vasodilation in the exposed cutaneous regions of anaesthetized rats.

neurotransmitters (transmitters) Small diffusible substances released from synaptic vesicles or secretory granules at the terminal of a neuron, then diffusing across the synaptic cleft and binding to cognate receptors on the postsynaptic membrane (see SYNAPSES). At least 30 have been identified (see Table 7), with many more as candidates. Many important neurotransmitters are amines derived from amino acids in simple pathways including decarboxylation. See NEUROSECRETIONS.

Crucial to the functions of these molecules is the type of receptor to which they bind: i.e., whether it is IONOTROPIC or METABOTROPIC. Fast-acting neurotransmitters include glutamate (excitatory) and γ-aminobutyric acid (GABA, inhibitory), whereas all the effects of biogenic amine and peptide neurotransmitters, and many of the effects of glutamate and GABA, are achieved by slow synaptic transmission (see SYNAPSES and CREB, Fig. 60). The efficiency of neurotransmitter release from the presynaptic terminal of a neuron, in response to the arrival of nerve impulses, is regulated by protein phosphorylation by protein kinases and dephosphorylation by protein phosphatases. Once a neurotransmitter has interacted with postsynaptic receptors, it must be removed from the synaptic cleft so that another round of synaptic transmission can occur. Simple diffusion out of the cleft is one method; but for most amino acid and amine transmitters, diffusion is aided by reuptake into the presynaptic axon terminal through specific transporter proteins in the presynaptic membrane, after which they may be reloaded into synaptic vesicles. Uptake also occurs into glial cells. Alternatively, enzymatic degradation of transmitter may occur in the cleft (as with ACETYLCHOLINESTERASE in the neuromuscular junction). *Intrinsic neurotransmitters* (e.g., glutamate, GABA) originate in the cerebral cortex; *extrinsic neurotransmitters* (e.g., acetylcholine, noradrenaline, serotonin, dopamine) originate in subcortical regions. See AUTORECEPTORS.

neurotrophic factors, neurotrophins Substances released by glial cells and neurons and essential for neuronal growth and survival, trophic deprivation of which triggers a cellular 'suicide' programme (apoptosis). They normally activate and bind to the tyrosine receptor kinase (Trk) receptors and, via activity of phospholipase C and phosphoinositol 3-kinase, coordinate neuronal survival (see CREB, Fig. 60). They include NERVE GROWTH FACTOR (NGF) and members of the neurotrophin (NT) family (e.g., neurotrophin-3 and neurotrophin-4/5). BRAIN-DERIVED NEUROTROPHIC FACTOR

(BDNF) is required for the proliferation, maturation and survival of brain neurons and for expression of the DOPAMINE D3 receptor (see HUNTINGTON'S DISEASE). It binds Trk B receptors, enhancing cortical synaptic plasticity, probably through synergism with electrical activity of neurons to produce LONG-TERM POTENTIATION (also increasing CREB-mediated gene expression). BNDF also targets mechanoreceptor dorsal root ganglion cells, motor neurons and retinal ganglion cells (DRGs). NT3 binds Trk C more readily than it does Trk A and Trk B, targeting large diameter DRGs and γ-motor neurons. NT4/5 binds Trk B on cranial nerve afferents, sympathetic neurons and midbrain neurons.

neurulation Starting with induction of neural plate cells to form neuroectoderm, the developmental process by which the embryonic neural tube is formed (~day 20–24). In primary neurulation, the chordamesoderm induces the overlying ectoderm to thicken (forming *neural plate*), fold up (forming *neural folds*), invaginate, and pinch off from the surface to form a hollow tube. As neural folds elevate and fuse, cells at the lateral border, or crest, of the neuroectoderm begin to dissociate as NEURAL CREST cells. Up-regulation of FGF signalling combined with inhibition of BMP4 and up-regulated expression of *chordin* and *noggin* by the PRIMITIVE NODE cause induction of neural plate. In secondary neurulation, the neural tube arises from a solid cord of cells that sinks into the embryo and then hollows out (cavitates) to form the neural tube. In mice, and probably in humans, primary neurulation is the method from the anterior end of the embryo up to about the 35-somite stage, after (and posterior to) which secondary neurulation applies. Neural tube closure, like the earlier GASTRULATION, is dependent upon the coordinated cell movements of convergent extension, itself dependent upon the non-canonical WNT SIGNALLING PATHWAY. The central nervous system is then represented by a closed tubular structure with a narrow caudal portion (the SPINAL CORD) and a much broader cephalic portion characterized by BRAIN VESICLES (see BRAIN, Fig. 28). See Appendix V, PRIMITIVE STREAK.

neutral fats Triglycerides. These are 'neutral' because the polar hydroxyl groups of glycerol and the polar carboxylates of the fatty acids are bound in ester linkages, making them non-polar. See LIPIDS.

neutral theory of molecular evolution The accepted null model for the evolution of DNA sequences, postulating that at the molecular level most of the evolutionary changes and much of the variability observed within species are caused by random GENETIC DRIFT of mutant alleles that are selectively neutral, or nearly neutral. It can be proved that in a finite population of N individuals, the process of random genetic drift will not be significantly altered if the intensity of natural selection, s, on an allele is less than $1/N$. The class of evolutionarily neutral mutations includes those having absolutely no phenotypic effect, and those whose effects on fitness are less than $1/N$ – i.e., *effectively neutral*. The neutral theory is compatible with the view that many or most mutations in nucleotide sequences that are not neutral or nearly neutral are deleterious and therefore subject to negative NATURAL SELECTION.

If μ is the rate of appearance of effectively neutral mutations at a locus, per gene copy per generation, the absolute number of new mutational copies that will appear in a population of N diploid individuals is $2N\mu$, and each of these has a probability of $1/(2N)$ of reaching fixation. So, the absolute rate of replacement of old alleles by new ones at a locus, per generation, is $2N\mu \times 1/(2N) = \mu$.

neutrophils Short-lived (~10 days) phagocytic and inflammatory cells of the innate immune system, abundant in the blood but not normally in healthy tissues, and the first immune cells to arrive in large numbers at a site of INFECTION. Their cell surfaces carry the fMLP receptor and receptors for chemokines (e.g., released by MACROPHAGES) and complement fragments C5a, C3a and C4a (all G-PROTEIN-COUPLED RECEPTORS), and KININS, all of which are neutrophil chemoattractants. Like macrophages, they have surface receptors for common bacterial constituents and for complement, and they are the main host defence for engulfing and destroying invading microorganisms. During phagocytosis neutrophils, like macrophages, produce nitric oxide, the superoxide ion and hydrogen peroxide, all directly toxic to bacteria. See DIAPEDESIS; INFLAMMATION; LEUKAEMIA; STEM CELLS.

newborn screening See SCREENING.

NF-κB (nuclear factor-κB) A family of transcription factors (five in mammals), commonly heterodimeric, that are major components of signalling pathways involved in cell death (e.g., through oxidative stress; see APOPTOSIS), cell division, inflammation, immunoglobulin gene expression, and cancer. In its latent form, NF-κB is sequestered in the cytoplasm by the polypeptide IκB (inhibitor of NF-κB) and in this state the signalling system is silent. But in response to any of a wide variety of cell-surface signals, IκB becomes phosphorylated by IκB kinases (IKKs; often termed 'IKK signalling') and tagged for rapid destruction, releasing NF-κB. NF-κB then enters the nucleus and proceeds to activate at least 150 target genes (each NF-κB activating its own gene set). It is now known to enter and leave the nucleus in oscillatory waves. The kinase that tags IκB for destruction is itself activated by stimuli as diverse as the inflammatory cytokines TNF-α and IL-1β (see INFLAMMATION), bacterial LIPOPOLYSACCHARIDE, reactive oxygen species signalling cell stress (see ATHEROSCLEROSIS), anti-cancer drugs and γ-irradiation. NF-κB, a master regulator of pro-inflammatory responses, serves in gut epithelial cells to control epithelial integrity and the interaction between the mucosal immune system and the GUT FLORA (see COLON). See COLORECTAL CANCERS, TOLL-LIKE RECEPTORS.

NGF The first NEUROTROPHIN discovered, promoting the survival (preventing apoptosis) during development of sympathetic neurons, of sensory neurons in the dorsal root ganglia, and of basal forebrain cholinergic neurons. It binds the Trk A receptor. See NEURAL REGENERATION.

NHEJ proteins (non-homologous-end-joining proteins) 'Caretaker' proteins involved in some DNA REPAIR MECHANISMS.

niacin (vitamin B$_3$) See VITAMIN B COMPLEX.

niche For cell niches, STROMA.

nicotinamide adenine dinucleotide See NAD, NADH.

nicotinamide adenine dinucleotide phosphate See NADP, NADPH.

nicotine Used socially to provide a sense of well-being, an extremely toxic and addictive alkaloid present in leaves of *Nicotiana tabacum* (tobacco) and hence also in cigarette and pipe tobacco, snuff, chewing tobacco and some gums and producing dependence with continued use (see ADDICTION). It is responsible for more damage to health in the UK than all other drugs, including ALCOHOL, combined (e.g., see CORONARY HEART DISEASE). Small amounts can be lethal. A mimic of acetylcholine, it binds nicotinic ACh receptors, especially those in the autonomic ganglia, and produces stimulation raising blood pressure (through constriction of blood vessels), heart rate, release of adrenaline, and increasing the tone and activity of muscles of the gastrointestinal tract. Because it binds nicotinic receptors of the postganglionic neurons of both the sympathetic and parasympathetic divisions, much of the variability of its effects results from the opposing actions of these divisions. A general stimulant of the central nervous system, nicotine raises behavioural activity and may cause vomiting and nausea. It increases alertness, decreases irritability and reduces sensory output from muscle spindles, reducing skeletal muscle tone. It has powerful immunosuppressive and anti-inflammatory effects, binding ACh receptors on MACROPHAGES and reducing their output of pro-inflammatory cytokines – which may explain the positive effects nicotine has on such diverse illnesses as schizophrenia, Alzheimer and Parkinson diseases,

Tourette syndrome and ulcerative colitis. It can also reduce fever and protect against otherwise lethal infection by influenza A virus. Withdrawal of tobacco causes a syndrome lasting 2–3 weeks, and includes irritability, craving for tobacco, hunger and often weight gain.

The conventional view that smokers become addicted to the pleasure of smoking is difficult to equate with the decreasing pleasure over time reported by adolescents. Only those previously exposed to nicotine crave it and, to the casual observer, the temporary suppression of craving is the only effect smoking brings. In rats, one dose of nicotine can increase noradrenaline synthesis by the hippocampus for at least one month. In the early stages of addiction in humans, a single cigarette can suppress withdrawal for weeks even though the nicotine disappears from the blood in one day; but most long-term smokers can forgo smoking cigarettes for only an hour or two. The symptom-free interval between the last cigarette and the onset of withdrawal is the *latency to withdrawal* (LTW), and is longer for new smokers than for long-term smokers. With repeated cigarette use, TOLERANCE develops and the impact of each new cigarette diminishes (dependence-related tolerance). Counselling and nicotine replacement therapy (NRT) can reduce symptoms, NRT including use of chewing gum, nasal sprays and skin patches, or the drug amfebutamone (buproprion), originally used as an antidepressant.

nicotinic receptor A subtype of ACETYLCHOLINE receptor. See NICOTINE.

NIDDM Type 2, non-insulin-dependent, DIABETES MELLITUS.

Niemann–Pick disorders Lysosomal storage disorders. Type A and B disorders arise from defects in the enzymatic breakdown of lipids in late endosomes and LYSOSOMES. Type C disorder (NPC) appears to be the result of defective lipid trafficking rather than an enzyme deficiency, causing very slow efflux of cholesterol from late endosomes.

night blindness A disorder resulting from months of inadequate vitamin A intake, symptoms occurring prior to those of XEROPHTHALMIA (see entry).

nigral (Adj.) Relating to the SUBSTANTIA NIGRA.

nigrostriatal tract See SUBSTANTIA NIGRA.

Nissl stain A class of basic dyes, leaving a usually blue or violet deposit on the somas of neurons.

nitric oxide (NO) A hydrophobic local signalling molecule that is also small enough to pass readily across a target cell's plasma membrane and has a half-life of ~5–10 seconds. One of its functions is to promote smooth muscle relaxation (see ENDOTHELIUM, PENIS). It is formed by deamination of arginine to NO and citrulline, a reaction catalysed by nitric oxide synthases (NOSs, to which heat-shock protein Hsp90 binds), and stimulated by intracellular calcium increase which generates the cofactor Ca^{2+}-calmodulin (see CALCIUM). NO bioactivity is conveyed largely through the covalent modification of cysteine sulphur atoms to form more stable *S*-nitrosothiols (SNOs), notably in the haem prosthetic group of haemoglobin. One SNO, *S*-nitrosoglutathione (GSNO), represents a major source of bronchodilatory NO bioactivity (see GLUTATHIONE), directly affecting smooth muscle tone. A downstream kinase of the nitric oxide signalling pathway is PKG which phosphorylates Hsp20, one significant phosphoprotein involved in the process. NO can directly dilate bronchi and cause *S*-nitrosylation, affecting signalling molecules, enzymes, transcription factors, etc. GSNO-reductases (GSNORs) that degrade GSNO to oxidized glutathione and ammonia are elevated in ASTHMA. Many cells use NO as a signalling molecule, including activated MACROPHAGES and neutrophils. It binds to iron in the active site of cytosolic GUANYLYL CYCLASE, initiating cyclic GMP synthesis. In the heart, the resulting cGMP brings about less forceful contractions by stimulating the ion pump(s) that expel calcium (Ca^{2+}) from the cytosol. NO can also act as a RETROGRADE MESSENGER, increasing SYNAPTIC STRENGTH (see ADDICTION).

NK cells See NATURAL KILLER CELLS.

NMDA (*N*-methyl D aspartate) receptors (NRs, NMDARs) Heteromeric IONOTROPIC receptors for GLUTAMATE, and cation channels activated on binding glutamate in association with GLYCINE (although the 'unnatural' amino acid D-serine may be the main physiological NMDAR coagonist). They are fundamental to the physiology of the mammalian central nervous system (the molecule NMDA itself is a pharmaceutical product). Their functional diversity lies in their assembly as obligate heteromers of glycine-binding NR1, glutamate-binding NR2 and glycine-binding NR3 subunits, thought to be organized as a dimer of dimers (similar to AMPARS). They are unusual in being gated not only by these two amino acids but also by membrane depolarization big enough to remove Mg^{2+} ions from the channel. This is because they are blocked on the extracellular side by Mg^{2+}, which inhibits current flow even when glutamate is bound, a voltage-gated blockade that is progressively removed by depolarization. They are therefore *coincidence detectors* (see HEBBIAN MODIFICATION) capable of integrating chemical and electrical stimuli into a Ca^{2+} signal that is crucial for the activity-dependent SYNAPTIC PLASTICITY (see Fig. 169) underlying such higher functions as learning and memory (see CALCIUM SIGNALLING). In addition to these brain functions, NMDARs are found on interneurons in the spinal cord, where glutamate is also the main excitatory neurotransmitter. Some interneurons respond to activation of NMDARs with rhythmic depolarizations (see CENTRAL PATTERN GENERATORS). Controlling an ion channel of high conductance to cations – calcium (Ca^{2+}), sodium (Na^+) and potassium (K^+) – they are characterized by slow channel opening and deactivation, this slow deactivation governing the duration of the excitatory postsynaptic potential, a measure of the 'strength' of synaptic signalling. NMDAR activation is the first step in inducing LONG-TERM POTENTIATION, the entry of calcium initiating SECOND MESSENGER signal transduction cascades. A growing body of evidence implicates dysfunction of NMDARs with stroke and several neurodegenerative and psychiatric diseases, including Parkinson disease, Huntington's disease and schizophrenia. They are also implicated in ADDICTION. NMDA NRs and AMPARs coexist at many synapses in the brain, so that most glutamate-mediated EPSPs exhibit components of both. See GABA, SYNAPTIC PLASTICITY.

NNRTIs Non-nucleoside-related reverse transcriptase inhibitors. See HIV.

nociception, nociceptors Nociceptors are receptors for noxious, pain-generating stimuli. Typically, they lack background firing rates, and are classified by what excites them. Most are polymodal, responding to mechanical, thermal and chemical stimuli that can cause, or are caused by, tissue damage; but others are more selective. Mechanical nociceptors respond to intense mechanical force, generating a sharp, pricking pain, each being the ending of one of 5–20 branches of an Aβ afferent (see AXON). Thermal nociceptors are a subset of THERMORECEPTORS (see entry) and their afferents are Aδ or C fibres (see AXONS). Polymodal nociceptors respond to puncture, temperature >45°C, and to substances liberated by tissue damage (see PAIN entry). Their afferents are C fibres. Release of ATP from cells by tissue damage binds nociceptor membrane ion-gated channels and causes depolarization, as does associated release of potassium ions (K^+). Primary nociceptor afferents co-release glutamate and peptides (notably SUBSTANCE P), contributing to neurogenic INFLAMMATION, and the potent vasodilator calcitonin-gene-related peptide (CGRP). Opioids produced by the PERIAQUEDUCTAL GREY MATTER are involved in supraspinal modulation of nociception.

Nociceptors are characterized, in part, by their sensitivity to capsaicin, the natural product of capsicum peppers and an ingredient of many 'hot' and spicy foods. Capsaicin excites nociceptors by binding its receptor and increasing the permeability of plasma membranes to cations, mimicking a physiological stimulus, or endogenous ligand produced by tissue injury. The capsaicin receptor serves both as a pain and thermal sensor and is located in small-diameter neurons in sensory ganglia, a vanilloid moiety being an important component of the capsaicin receptor (more generally referred to as a vanilloid receptor). See REFERRED PAIN, Fig. 147; SOMATIC SENSORY SYSTEM.

Nod proteins (nucleotide-binding oligomerization domain proteins) Nod1 and Nod2 (aka caspase activation and recruitment domain proteins CARD4 and CARD15) are present in the cytosols of epithelial cells and immune cells. Nod2 is largely restricted to monocytes and has been shown to bind intracellular endotoxin, genetic variants causing increased susceptibility to development of CROHN'S DISEASE. Oligomerization of Nod1 or Nod2 allows them to bind the protein kinase RICK through the latter's caspases-recruitment domain, leading to activation of the transcription factor NF-κB and its pathway. Cellular NOD-like receptors (NLRs) detect bacteria (see INFECTION, INFLAMMATION).

Nodal See PRIMITIVE NODE.

node See PRIMITIVE NODE.

nodes of Ranvier Unmyelinated regions, ~0.5 μm wide, of otherwise myelinated AXONS, in direct contact with the extracellular fluid and therefore having correspondingly lower electrical impedance (see MYELIN SHEATH). Local circuit currents ahead of an ACTION POTENTIAL and arriving at the next downstream node cause the axon to depolarize beyond the threshold and initiate further propagation of the action potential. This iterative process is termed *saltatory conduction*, since it effectively 'jumps' the intervening axon membrane between nodes. These unmyelinated regions have ~100-fold greater density of voltage-gated sodium channels in their membrane than do myelinated regions, resulting in a considerably lower node threshold and reducing the risk of firing failure on account of the long distances between nodes. The distance between adjacent nodes varies with axon diameter, being greater in those with larger diameter.

***Noggin* gene** A gene whose secreted protein product is produced by the notochord and floor plate of the neural tube and, with sonic hedgehog (SHH), induces the ventromedial portions of SOMITES to become sclerotome and to express *PAX1*.

non-adaptive immune responses See INNATE IMMUNITY.

non-allelic homologous recombination (NAHR) Recombination between PARALOGOUS genomic segments, often occurring between region-specific low-copy repeats that are brought together by chromosome/chromatid misalignment. This mechanism may have played a significant role in karyotypic evolution during primate speciation.

non-coding DNA A term originally coined to refer to non-protein-coding DNA, i.e., DNA sequences that did not encode amino acid sequences (see HUMAN GENOME). Sequences encoding RNAs (tRNAs and rRNAs) involved directly in protein synthesis have been familiar since the 1960s and do not encode protein. Nor do INTRONS, *cis*-acting REGULATORY ELEMENTS or non-coding EXONS. In the last 20 years, it has become evident that a large proportion of functional DNA in the human genome does not encode protein, although distinguishing it from neutrally evolving DNA using comparative approaches has proved difficult. One estimate (in 2002) was that perhaps twice as much non-coding DNA as coding DNA is under selective constraints. If non-coding DNA is untranscribed, it will be absent from cDNA libraries, making regulatory sequences much harder to analyse than coding sequences: there are currently no algorithms from which to infer biological function from tracts of intergenic or intronic sequences, or that decipher how base-pair changes affect function. As we now appreciate, much transcribed DNA is nevertheless non-coding (see NON-CODING RNAS, 'DARK MATTER'). However, we do know that regulatory sequences are central to changes in GENE EXPRESSION and, in consequence, to an organism's morphology (see MICRORNAS).

A recent comparative analysis of selective constraint in non-coding regions in two mammalian lineages, averaged over all genes, indicated that many *cis*-regulatory elements lie in ~2 kb windows upstream and downstream of genes, that selective constraint appears to drop dramatically beyond this, and that far less constraint has been involved in the two hominoids, humans and chimpanzees, than in mice. To discover whether conserved non-coding sequences (CNSs) in the human genome bear the signature of accelerated evolution, a 2006 study located the positions of human-specific substitutions in >110,000 CNSs previously identified by whole genome sequence comparisons. The results gave 992 positively selected cases, 79% more than expected by chance, and closest neighbour analysis indicated that the cellular component most significantly enriched in accelerated CNSs was the BASAL LAMINA, most of which were associated with the dystrophin-associated glycoprotein complex (see MUSCULAR DYSTROPHIES), disruptions in which cause muscle and neuronal disease. Indeed, in this study, cell adhesion was the only biological process displaying a significant excess of CNSs accelerated in humans (see CELL ADHESION MOLECULES), many of them associated with genes involved in neuronal cell adhesion. These included genes for cadherins and protocadherins, contactins and neuroligins. Extending the work to chimpanzee and mouse genomes, the former revealed 1,050 accelerated CNSs, only 34 of which were among those found in the human genome; but, again, many of these lay near cell adhesion genes. No enrichment near neuronal cell adhesion genes was found among the 4,607 mouse CNSs. See BRAIN, BRAIN SIZE, NATURAL SELECTION.

non-coding RNAs Molecules of RNA (see entry) which are either (i) non-protein-coding RNAs (ncRNAs, or structural RNAs), namely tRNA, rRNA and small nuclear RNAs (snRNAs) and small nucleolar RNAs (snoRNAs); or (ii) the more recently discovered regulatory RNAs (e.g., microRNAs, see entry; and long RNAS) with varied roles in such processes as CHROMATIN REMODELLING, RNA PROCESSING and modification, mRNA and protein stability and GENE SILENCING (see HUMAN GENOME). Piwi-associated RNAs (piRNAs), generated from long single-stranded precursors independently of Drosha and Dicer, associate with a group of Argonaute proteins (Piwi proteins) and together with their Piwi partners are somehow essential for development of germ cells. See HUMAN GENOME PROJECT, HUMAN TRANSCRIPTOME, MICRORNAS, RNA INTERFERENCE, X CHROMOSOME INACTIVATION.

non-communicable diseases See DISEASES.

non-declarative memory (implicit memory) See MEMORY.

nondisjunction Failure of the two members of a pair of chromosomes (a bivalent) to separate at 1st meiotic anaphase, or of chromatids to separate at the 2nd meiotic anaphase or during a mitotic anaphase, leading to trisomy and monosomy in cell lines (most of which fail to survive, depending on the chromosome involved; see BIRTH DEFECTS). Notwithstanding the current state of ignorance of underlying molecular defects, the correlation between maternal age and meiotic nondisjunction in human oocytes is well known: for women in their early twenties, < 3% of all clinically recognized pregnancies are trisomic, although the risk of conceiving a trisomic foetus rises exponentially for women in their thirties and approaches 35% for women in their early forties. Such chromosome segregation errors are the leading cause of miscarriages and birth defects. See DOWN SYNDROME, KLINEFELTER SYNDROME, MOSAICISM, MUTATIONS, TURNER SYNDROME.

non-essential amino acids Those AMINO ACIDS (see entry) that can be synthesized in sufficient amounts to meet bodily requirements at the appropriate stage of development or in adult life.

non-essential fatty acids Those FATTY ACIDS (see entry) that can be synthesized in sufficient amounts to meet bodily requirements at the appropriate stage of development or adult life.

non-homologous end-joining (NHEJ) See DNA REPAIR MECHANISMS.

non-homologous recombination The independent assortment of non-homologous chromosomes into the products of the first meiotic division, forming two haploid nuclei. Even in the absence of CROSSING-OVER, a single human meiosis can produce 2^{23} genetically different nuclei by this means. It should not be termed 'recombination', which involves exchange of sequences between DNA molecules.

non-identical twins (fraternal twins) Non-monozygotic TWINS.

nonM-nonP cells See GANGLION CELLS.

non-paternity, non-paternity rates Systematic studies of non-paternity rates are very few, and rates may vary between populations. Early cystic fibrosis screening studies (1991) found 7 non-paternities in 521 European families (1.35%); but an estimate of ~12% has been given for one Mexican population. See Y CHROMOSOME.

non-penetrance See PENETRANCE.

non-REM sleep A sleep phase distinguished in part by well-defined patterns of oscillatory electrical activity of the brain (see ELECTROENCEPHALOGRAM) and divided into four stages. Stage 1 is a transitional phase between full wakefulness and sleep, characterized by mixed-frequency waves; stage 2 is typified by sleep spindles (12–14 Hz), while delta waves (1–4 Hz) distinguish NREM stages 3 and 4, collectively known as *slow-wave sleep*. Sleep spindles and delta waves rise and fall in concert with a still slower oscillatory pattern (<1Hz), the former being associated with large influxes of calcium ions into cortical neurons where they can trigger molecular cascades known to be associated with increased SYNAPTIC STRENGTH between neurons. See MEMORY, SLEEP, REM SLEEP.

nonsense-mediated (mRNA) decay (NMD) The RNA surveillance mechanism that detects and degrades mRNA transcripts harbouring premature nonsense termination ('stop') codons that occur at least 50 nucleotides upstream of the last splice junction (see RNA PROCESSING) and which would otherwise be translated into potentially lethal truncated polypeptides. The antibiotic gentamicin so interferes with ribosomes that they tend to 'read through' such stop codons and insert a normal amino acid instead, usually leading to some functional protein being produced. The drug PTC124 is being trialled with those patients with cystic fibrosis and Duchenne muscular dystrophy whose mutations are of the nonsense variety. If successful, this therapeutic approach may be effective in treating other diseases arising through this special class of mutation, e.g., β-thalassaemias (see HAEMOGLOBINOPATHIES).

non-shivering thermogenesis See TEMPERATURE REGULATION.

non-specific immunity See INNATE IMMUNITY.

non-starch polysaccharides (NSPs) NSPs present in the human diet include cellulose, and hemicelluloses (e.g., pectins, xylans, galactomannans, β-glucans, gums and mucilages). See BALANCED DIET, FIBRE, GUT FLORA.

non-steroidal anti-inflammatory drugs (NSAIDs) Drugs, including non-prescription aspirin, ibuprofen and naproxen, as well as other prescription drugs, commonly used to treat rheumatoid arthritis and the inflammatory effects of ALZHEIMER'S DISEASE. They have analgesic, antipyretic and, at higher doses, anti-inflammatory effects, but are ineffective in treatment of visceral pain, which requires opioid analgesics. They block cyclooxygenase activity and prevent conversion of arachidonate to prostaglandins and thromboxanes.

non-transmissible diseases See DISEASES.

noradrenaline (norepinephrine) A CATECHOLAMINE precursor of ADRENALINE (see Appendix I), from which it differs in lacking the N-terminal methyl group. It is released from neurons originating in the LOCUS COERULEUS, axons which are a major component of the noradrenergic diffuse modulatory system of the BRAIN (see Fig. 30a), and from postganglionic axons of the sympathetic division of the AUTONOMIC NERVOUS SYSTEM. Most circulating noradrenaline is overflow from these nerve terminals although small amounts are secreted by specialized cells of the adrenal medulla (see ADRENAL GLANDS). It acts via METABOTROPIC RECEPTORS and is important in ATTENTION. See ANTIDEPRESSANTS, Fig. 12; SYMPATHOMIMETICS.

noradrenergic Of neurons secreting or sensitive to NORADRENALINE. See ADRENERGIC; BRAIN, Fig. 30c; DIFFUSE MODULATORY SYSTEMS; LOCUS COERULEUS.

Northern blotting A variant of the SOUTHERN BLOTTING technique, in which the nucleic acid being digested is RNA rather than DNA. RNA is very unstable because of intrinsic cellular ribonucleases, so ribonuclease inhibitors are used to isolate mRNA which, if run on an electrophoretic gel, can be transferred to a filter. A gene which has been cloned can be used as a probe and hybridized against samples of RNA from different tissues, helping to determine the expression patterns of genes and the relative abundance of RNA transcripts. If the transcripts are of different sizes, they may indicate that different isoforms of a gene are being expressed. In a gene disorder in which a mutation has not been identified in the coding sequences, an alteration in the size of the mRNA transcript from normal suggests that the lesion may be in a non-coding region of the gene – e.g., the splice junction of the intron–exon border. See DNA PROBES, POSITIONAL CLONING.

nose The nose forms from five facial prominences during development of the FACE (see Fig. 74), of which it is a prominent feature. The frontal prominence gives rise to the bridge, consisting of the nasal bones and extensions of the frontal and maxillary bones; the merged medial nasal prominences provide the cartilages of the crest and tip, and the lateral nasal prominences form the sides (alae). See OLFACTION, Appendix VI.

Notch signalling pathway Possibly the most widely used signalling pathway in animal development, with a major role in determining cell fate decisions during development (e.g., the decision by a lymphocyte progenitor cell to differentiate into a T CELL and not a B CELL). The Notch receptor is a single-pass transmembrane protein with repeated EGF-like domains on its exposed portion on the cell surface and whose cytoplasmic 'tail' is successively cleaved (including proteolysis by a membrane-bound γ-secretase complex) when the receptor binds its ligand (e.g., Delta, or a Delta-like family member) on the surface of an adjacent cell. The tail domain then enters the nucleus where, upon binding appropriate transcription factors and transcription co-activators, it activates the expression of target genes. See WNT SIGNALLING PATHWAYS.

notochord Cells originating by ingression through the cranial region of the node of the PRIMITIVE STREAK.

NREM sleep See SLEEP.

NRTIs Nucleoside-related reverse transcriptase inhibitors. See HIV.

NSAIDs See NON-STEROIDAL ANTI-INFLAMMATORY DRUGS.

NST See NUCLEUS TRACTUS SOLITARIUS.

NTS See NUCLEUS TRACTUS SOLITARIUS.

nuchal Referring to the back of the neck (e.g., nuchal ligament). The nuchal lines on the occipital bone are attachment sites for the tendons of several neck muscles.

nuclear envelope The double-membranous covering of the NUCLEUS.

nuclear-hormone-receptor family See NUCLEAR RECEPTOR SUPERFAMILY.

nuclear magnetic resonance imaging (NMRI) Former term for MAGNETIC RESONANCE IMAGING.

nuclear receptors, nuclear receptor superfamily (nuclear hormone receptor superfamily) A very large superfamily of genes and their encoded receptor proteins, all structurally related, although only ~50 have so far been detected in humans. The protein products are all ligand-activated TRANSCRIPTION FACTORS, members sharing two conserved domains: a DNA-binding domain containing zinc fingers and binding its response element as a dimer, and a longer ligand-binding domain which either inactivates the DNA-binding domain in the absence of ligand or binds an inhibitory protein (as with the glucocorticoid receptor).

Some members are activated by metabolites produced by SIGNAL TRANSDUCTION pathways; others bind to steroid hormones (e.g., androgens, oestrogens, progesterone), thyroid hormones, retinoids or VITAMIN D. When steroid hormones activate their cognate receptors (ligand-activated transcription factors), the receptors dimerize, bind specific RESPONSE ELEMENTS, and engage the general transcription apparatus. Steroid hormone receptors interact with a broad array of COREGULATORY PROTEINS to alter TRANSCRIPTION, and cell-specific expression of coactivators and corepressor proteins, and their regulation by post-translational modifications, enable exquisite tissue-specific and temporal regulation of GENE EXPRESSION (which depends on the promoter context and its ligand-binding status). Some receptors, e.g. the cortisol receptor (see GLUCOCORTICOIDS), are bound to inhibitory heat-shock proteins and located mainly in the cytosol, entering the nucleus only on ligand binding. Others receptors, e.g. those for thyroid hormones and retinoids (e.g., LXR, RXR), are normally bound to inhibitory protein complexes that are themselves bound to specific nuclear DNA sequences, even in the absence of their ligand (see PPARS). But in either case, complexing with its ligand causes allosteric changes in the receptor's conformation resulting in dissociation of the inhibitory complex. Once free, the receptor can bind coregulatory proteins and alter transcription of those gene targets with the appropriate hormone response element. As indicated, responses to steroid and thyroid hormones, as with those to extracellular ligands generally, depend as much on the nature and state of the target cells as on the nature of the signal molecule. See RETINOIC ACID.

nuclear transplantation See CLONING, SOMATIC CELL NUCLEAR TRANSFER.

nuclease-hypersensitive site A chromosomal region that is highly accessible to cleavage by deoxyribonuclease 1 (DNase 1) and typically associated with open chromatin conformations and with transcriptional activity.

nucleic acid amplification test (NAAT) A DNA-BASED DIAGNOSTIC TESTING involving the polymerase chain reaction to amplify the concentration of potentially diagnostic DNA. See CHLAMYDIA TRACHOMATIS..

nucleic acid hybridization See DNA PROBES.

nucleic acids The polynucleotides DEOXYRIBONUCLEIC ACID and ribonucleic acid (see RNA).

nucleolar organizer regions The 13-kb polycistronic transcription units encoding 18S, 5.8S and 28S ribosomal RNAs, and their accompanying 27-kb non-transcribed spacer, that are tandemly repeated ~30–40 times near the tips of the short arms of the five human acrocentric CHROMOSOMES (see Fig. 52), each cluster being ~1.5 Mb in length, there being 10 clusters in a human diploid cell. See NUCLEOLUS, RNA GENES, RNA POLYMERASE.

nucleolus The most noticeable structure within an interphase nucleus formed by assembly of ribosomal subunits after transcription and processing of ribosomal RNA (rRNA). It has been described as a ribosome-producing factory and its size reflects the number of ribosomes the cell is producing. It is a large aggregate of macromolecules, including the rRNA genes themselves (see NUCLEOLAR ORGANIZER), precursor and mature rRNAs, rRNA-processing enzymes, snoRNPs (see entry), ribosomal protein subunits and partially assembled ribosomes. Other important RNA protein complexes, e.g., telomerase and the signal recognition particle (see SIGNAL RECOGNITION), are thought to be assembled in the nucleolus, and tRNA processing also takes place there. Each time a nucleus undergoes mitotic division, the nucleolus breaks up, and the ten chromosome tips coalesce as the nucleolus reforms after mitosis. Transcription of rRNA genes by RNA POLYMERASE I is required for this process.

nucleosomes The fundamental repeating units of CHROMATIN (see Fig. 51) into which eukaryotic genomes are packaged. Each comprises 160–240 DNA base pairs (varying with the species and tissue) wrapped around a core of basic proteins (see HISTONES). Assembling during DNA REPLICATION behind the replication fork, each contains a 147-base pair stretch of DNA that is sharply bent and tightly wrapped around a histone protein octamer, this bending occurring at every DNA helical repeat (~10 bp), when the major groove of the DNA faces inwards towards the histone octamer, but also occurs ~5 bp away, and with opposite direction, when the major groove faces outwards. Nucleosome particles prevent access by most DNA-binding proteins such as polymerase, regulatory, repair and recombination complexes; yet nucleosomes recruit other proteins through interactions with their histone tail domains.

Neighbouring nucleosomes are distanced from each other by 10- to 50-bp-long sequences of unbound *linker DNA*, so that ~75–90% of genomic DNA is wrapped in nucleosomes (technically, a nucleosome is one nucleosome core particle plus one of its adjacent DNA linkers).

Their distribution along the DNA is not random, nor is their stability uniform. Their static roles include being positioned so that they can expose authentic functional sites associated with the initiation of gene expression, while at the same time masking spurious recognition sites that are bound to occur in large genomes. Non-static roles include their being positioned so as to block the individual sites required for the stepwise formation of functional protein complexes until their assembly is required, and then sliding away on demand with the aid of the ATP-dependent enzymes involved in CHROMATIN REMODELLING. Nucleosome mapping has demonstrated that nucleosomes take up preferred positions with respect to DNA sequence, the sequence-dependent bending flexibility of DNA being one significant factor, but not the only one. In their association with histones, the two phosphodiester backbones of duplex DNA can either face inwards, towards the histone core, or outwards and away from it. For the ~150 base pairs of DNA that are tightly wrapped up in each nucleosome's core, the dinucleotides AA/TT/TA are favoured where the DNA faces inwards, while GC dinucleotides are favoured where the backbones face outwards.

Preliminary work on yeast chromatin indicates that the most stable nucleosomes are in genes containing the TATA SEQUENCE (signalling the start for TRANSCRIPTION during GENE EXPRESSION) and lie immediately downstream of the sequence, without engulfing it. The greatest number of high-occupancy nucleosomes occurred between genes and are probably regulatory in function, while the most stable nucleosomes were found in the most tightly packed chromosome regions, the CENTROMERES. Nucleosomes had relatively low occupancy on the functional DNA-binding sites for gene regulatory factors.

nucleus 1. The organelle housing the cell's CHROMATIN and replicated during MITOSIS. Proteins synthesized on cytoplasmic ribosomes and destined for the nucleus include histones, DNA and RNA polymerases, transcription factors, RNA-processing proteins, etc. These usually require a localization signal embedded in their structure, typically comprising 4–8 positively charged amino acids and neighbouring proline residues, although the SIGNAL SEQUENCE is often bipartite, in two blocks of 2–4 amino acids separated by 10 other amino acids. Some nuclear proteins get into the nucleus without a signal sequence by binding to others that do.

Although it has been known since the 1970s that regional compartmentalization of the X chromosome can occur in the interphase nucleus (e.g., of insects), it has become clear since the 1990s, particularly since the advent of interphase FISH, and three- and four-dimensional chromosome conformation capture (4C–3C), that dynamic physical intra- and interchromosomal interactions occur between very specific gene loci, and that these regulate their transcriptional initiation and silencing (see GENE EXPRESSION, GENE SILENCING). As cellular differentiation proceeds, specific loci appear to be physically repositioned from *chromosome territories* (the nuclear space occupied by individual chromosomes, separated by an interchromatin compartment) into distinct subnuclear regions conducive to transcriptional activity or silence. A chromosome's neighbour in the nucleus is thus far from random, varying between cell types and affecting a chromosome's ability to interact *in trans* with other components of the genome (see TRANSLOCATIONS). It is anticipated that such higher order chromatin rearrangements will be a common feature of gene organization in the nuclei of most higher organisms. For some genes, interior, exterior or surface positioning relative to the chromosome territory correlates with their activity or inactivity; however, the 'snapshots' gained from 4C analysis indicate that the main DNA sequences captured by a given locus are, on average, other regions of the *same* chromosome. It has been suggested that transcription and RNA PROCESSING occur in the space between chromosome territories, in an area termed the *interchromosome domain compartment*.

Genes can relocate some distance outside their chromosome territory. This occurs in chromatin domains of constitutively high gene expression or, sometimes, when gene expression is induced. Relocation and intermingling of DNA from different chromosomes are reduced when RNA POLYMERASE II (Pol II) is inhibited experimentally, suggesting that the process of transcription may itself be partly responsible for relocation outside chromosome territories (although not all active genes lie outside their chromosome territory). The mode by which a gene is activated may affect whether it is relocated. Pol II can be excluded from certain chromosome territories (see X CHROMOSOME INACTIVATION), while 'transcription factories' within the nucleus contain focal concentrations of the enzyme, each factory housing several expressed genes from the same and different chromosomes (see T CELLS for one example, involving naïve $CD4^+$ cells). Some active genes seem to congregate at splicing-factor-enriched 'speckles' containing specific pairs of genes, both in *cis* and *trans*. In mice, at least one imprinting control region (ICR), that for *H19*, has been shown to co-associate simultaneously with multiple gene loci in a single nucleus. The *H19* ICR is methylated on the paternal chromosome, and 75% of interactions are with the maternally derived allele. Genes involved in stability of key nuclear structures, including chromatin and the nuclear matrix, include *PALB2* (partner and localizer of *BRCA2*). See CELL CYCLE, MEIOSIS.

2. A clearly distinguishable group of neurons (e.g., the LATERAL GENICULATE NUCLEI) within the brain, generally lying deep. Compare LOCUS, SUBSTANTIA.

nucleus accumbens (NAc, ventral striatum) The ventral component of the CORPUS STRIATUM and a key location (neural substrate) for the rewarding actions of many drugs of abuse (see OPIATE RECEPTORS, REINFORCEMENT AND REWARD). It is a component of the affective basal ganglia loop (see LIMBIC SYSTEM), projecting to the substantia nigra, modulating activities of the motor and cognitive striato-thalamo-cortical circuits and providing for some of the motor and cognitive aspects of emotional states. It is the major site for D_3 dopamine receptor. Repeated exposure to opiates or stimulant drugs increases activity of the cAMP–PKA pathway within the NAc (see CREB, Fig. 60). Direct activation of PKA and consequent CREB phosphorylation activate CREB-mediated transcription in the NAc and reduce the rewarding effects of cocaine while inhibition of PKA has the opposite effect. Increased CREB function here appears to cause tolerance and dependence to drugs of abuse; and although the target genes of CREB activity contributing to this are unknown, one candidate gene within the NAc is DYNORPHIN. In rodents, disruption of CREB function within the NAc produces anxiety-like effects (contrast the AMYGDALA).

nucleus ambiguus (NA) Nucleus of the medulla oblongata receiving projections from the NUCLEUS TRACTUS SOLITARIUS and, with the dorsal vagal nucleus, giving rise to preganglionic parasympathetic fibres projecting via the vagus nerve to the heart. See CARDIOVASCULAR CENTRE.

nucleus cuneatus Medullary nucleus containing cell bodies of second-order neurons synapsing with axon termini from the dorsal column FASCICULUS CUNEATUS.

nucleus gracilis Medullary nucleus containing cell bodies of second-order neurons synapsing with axon termini from the dorsal column FASCICULUS GRACILIS.

nucleus of Meynert (nucleus basalis) A region of the primate basal rostral forebrain (the *substantia innominata*) underlying the corpus striatum and containing magnocellular nuclei whose cholinergic axons project extensively to the cortex and is equivalent to the nucleus basalis magnocellularis of the rat. It undergoes NEURAL DEGENERATION during ALZHEIMER DISEASE.

nucleus tractus solitarius (NTS, nucleus of the solitary tract, NST) Nucleus of the dorsal MEDULLA OBLONGATA in which visceral afferents of the vagus, including TASTE fibres, terminate. It projects to the hypothalamus, conveying information from the ANS through baroreceptors (see BLOOD PRESSURE) and chemoreceptors (see RESPIRATORY CENTRE, CEREBROSPINAL FLUID), as well as neural and hormonal signals (e.g., CCK) from the gastrointestinal tract. Its glucosensors detect hypoglycaemia (e.g., insulin-induced), raising its activity and sympathetic activation of the liver to raise hepatic glucose output. See APPETITE, Fig. 14; ENERGY BALANCE.

null hypothesis The starting hypothesis for most investigations, which assumes that there is no correlation between the variables being studied and enables prediction of expected results. Departures of observed data from such results can be tested for significance by an appropriate statistical test.

nulliparous Of a woman who has never given birth to a child.

numts Nuclear insertions of mitochondrial sequences. See MITOCHONDRION, MTDNA.

nurse cells Cells which synthesize many of the cytoplasmic contents of cells they support, passing them through cytoplasmic extensions, and/or serve an immunologically protective role for them. Vertebrate oocytes, unlike those of many invertebrates, do not have supporting nurse cells; but the SERTOLI CELLS surrounding developing spermatids are a human example.

nurture–nature debate See NATURE–NURTURE DEBATE.

nutrition See HEALTHY DIET, PREGNANCY.

nystagmus Involuntary eye movements, often related to faults in neural connections between eye and brain and often associated with conditions causing vertigo, and with albinism.

O

obesity The most common nutritional disorder in Western society, and a COMPLEX DISEASE, developing chronically when energy intake exceeds energy expenditure (see ENERGY BALANCE, FATTY ACIDS). Epidemiological studies indicate that obesity is an independent risk factor for CARDIOVASCULAR DISEASES and in the long run predisposes individuals to development of these and to type 2 DIABETES MELLITUS (T2D), HYPERTENSION, endometrial carcinoma, OSTEOARTHRITIS and gallstones (see BILE). There is now considerable evidence that human obesity is characterized by abnormal cardiovascular control by the sympathetic nervous system (SNS; see CARDIOVASULAR CENTRE). Although catecholamine studies have been conflicting, weight loss studies have shown that a fall in blood pressure correlates with reduction in plasma NORADRENALINE (NA) levels. High NA levels in obesity correlate with regional overactivity of the kidneys, and high renal SNS activity can lead to abnormal glomerular filtration and raised BLOOD PRESSURE. Hyperinsulinaemia of obesity may interact with the SNS to elevate blood pressure. Unlike hypertensive obesity, normotensive obesity develops in the absence of raised blood pressure.

Since obesity involves increased deposition of body fat (see ADIPOSE TISSUE), which is well vascularized, the peripheral resistance to blood flow through this added tissue may also contribute to raised blood pressure. The widespread INFLAMMATION resulting from nutrient excess also affects the ENDOTHELIUM of blood vessels, causing them to release less nitric oxide. The resulting reduction in vasodilation may also contribute to raising blood pressure. Fundamental to understanding obesity is the fact that, like body temperature, body fat stores are normally maintained within a narrow range through a process termed 'energy homeostasis' (see ENERGY BALANCE) involving brain areas controlling APPETITE and energy metabolism as well as circulating molecular signals (ADIPOKINES) conveying information about the level of body fuel reserves (see BODY MASS INDEX).

Hepatic overproduction of VLDL (see LIPOPROTEINS) seems to be the crucial defect in obesity – a result of hepatic steatosis. Until recently, it was not known why inflammation develops in obesity (see CIGARETTE SMOKING). Since the millennium, it has been clear that ADIPOCYTES (see entry) are a key factor. Some experimental evidence suggests that ENDOPLASMIC RETICULUM stress in these cells, in which proteins are not properly folded, is critical to the initiation and integration of pathways of inflammation and insulin action in obesity and T2D. Fat-laden adipocytes secrete IL-6 and TNF-α, and the tissue becomes invaded by macrophages which contribute to a downward spiral of inflammation and to the development of INSULIN RESISTANCE in peripheral tissues in obese individuals. Some evidence indicates the possible role of leptin as a regulator of fat mass, although leptin resistance in obese individuals has reduced enthusiasm for it as a therapeutic drug.

Many genes interact with the environment to produce obesity and diabetes, the most frequent mutation in the case of obesity being that in melanocortin-4 receptor (see MELANOCORTINS), accounting for up to 4% of cases of severe obesity. PGC-1 has also been implicated, as has polymorphism in the promoter region of the gene (*UCP2*) for an uncoupling protein in the mitochondrial matrix. Mutations in the *POMC* gene, which encodes the PROOPIOMELANOCORTIN from which melanocortins are derived, can also cause the disorder. The GUT FLORA appears to be an additional contributing factor to the pathophysiology of obesity. A recent UK study has found a positive relationship between childhood obesity and mothers from higher income families returning to work early after having children. Lack of breast-feeding up to the recommended 6 months (reliance on formula milk), reduction in time spent preparing food, and greater reliance on convenience food, are possible contributory factors.

object recognition It is unclear the extent to which activity in object-related areas of the human visual cortex can be modulated by shifts of visual ATTENTION. A hallmark of object recognition is its invariance despite changes in location, size, view-point, etc., and the neural machinery making this possible is poorly understood. Functional MAGNETIC RESONANCE IMAGING studies have revealed three brain regions of preferential activation for objects: area V4 (see EXTRASTRIATE CORTEX), the lateral occipital complex (LOC, located mainly laterally and ventrally in the human occipital lobe and extending into the posterior aspect of the temporal lobe), and dorsal foci comprising area V3a and a region just anterior to it. In all three, attending pictures of complex natural objects yielded stronger activation than did attending a rotating central arrow superimposed on each picture (an opposite effect was noted in the VISUAL MOTION CENTRE, MT/V5). LOC appears to be preferentially activated by objects regardless of the physical cues defining them: objects defined by relative motion, luminance differences or texture differences, all activated LOC more than did noise and texture patterns defined by the same cues. ELECTROENCEPHALOGRAM studies reveal increase of induced gamma-band (>30 Hz) responses elicited by familiar objects, a frequency-specific change in distinct brain regions believed to indicate the dynamic formation of local neuronal assemblies during activation of cortical object representations. Familiar (but not unfamiliar) objects engage widespread reciprocal information transfer during human object recognition localized to the INFEROTEMPORAL CORTEX, superior PARIETAL CORTEX and frontal cortex. Unfamiliar objects appear to entail a sparse number of merely unidirectional connections converging on parietal areas. Object recognition is often achieved when only partial or degraded views of an object are available to the observer, during which perceptual 'filling-in' (closure) is apparently achieved by brain centres which include LOC. See BINDING PROBLEM, FACIAL RECOGNITION, VISUAL PERCEPTION.

OBLA See ONSET OF BLOOD LACTATE ACCUMULATION.

oblique See ANATOMICAL PLANES OF SECTION, Fig. 10.

obsessive-compulsive disorder (OCD) See FEAR AND ANXIETY, TOURETTE SYNDROME.

occipital bone The bone forming the posterior part and most of the base of the cranium (see SKULL, Fig. 157), in the inferior part of which lies the FORAMEN MAGNUM. The *occipital condyles* are oval processes with convex surfaces, one on either side of the foramen magnum, and articulating with depressions on the 1st cervical vertebra to form the atlanto-occipital joint. The *ligamentum nuchae* extends from the external occipital protuberance (a prominent posterior projection in the midline of the bone) to the 7th cervical vertebra, helping to support the head. The *posterior cranial fossae* are bilateral depressions between the inferior nuchal lines and the foramen magnum, formed by the presence of the cerebellum. See OCCIPITAL CREST.

occipital condyles See OCCIPITAL BONE.

occipital crest Either of the two ridges on the OCCIPITAL BONE connecting the occipital protuberances and FORAMEN MAGNUM and serving for neck muscle attachment: (i) a median ridge on the outer surface of the occipital bone that with the external occipital protuberance gives attachment to the LIGAMENTUM NUCHAE; aka the *external occipital crest*, or *median nuchal line*; (ii) a median ridge also situated on the inner surface of the occipital bone, bifurcating near the foramen magnum; aka the *internal occipital crest*, or *inferior nuchal line*.

occipital lobe One of four major subdivisions of each hemisphere of the CEREBRAL CORTEX, lying caudal to the parietal lobe under the occipital bone. It includes the PRIMARY VISUAL CORTEX and has a major role in vision.

ocular cortex See VISUAL CORTEX.

ocular dominance, ocular dominance columns Ocular dominance is the tendency for BINOCULAR CELLS to show a preference for one eye. Cells with the same ocular dominance occupy ocular dominance columns situated in long stripes, ~500 μm across, within a HYPERCOLUMN in the PRIMARY VISUAL CORTEX (PVC; see Fig. 142). Those representing ipsilateral (I) and contralateral (C) input alternate regularly over the cortex. In primates, unlike kittens, ocular dominance columns are established before birth, although they remain plastic until 6 weeks after birth (see CRITICAL PERIOD, NEURAL PLASTICITY). But even in the absence of light, neuronal activity is still required, provided by spontaneous firing of retinal GANGLION CELLS that spreads in waves across the retina. The bursts of activity from the two retinae are out of phase, possibly accounting for the segregation of inputs from the left and right eyes. Blocking of NMDA RECEPTORS disrupts this segregation. *Monocular deprivation*, which replaces patterned activity in one eye with random activity, disrupts binocular connections in the PVC, neurons outside layer IV (i.e., in layers II and III), which normally have binocular fields, responding only to stimulation of the uncovered eye after even a brief period of deprivation (an ocular dominance shift). After a brief period of STRABISMUS, there is a total loss of binocular receptive fields and binocular cells in the PVC are almost completely absent, cells in the visual cortex being driven by one of the two eyes, not both (binocular competition). An ocular dominance shift after monocular deprivation leaves both macaques and humans visually impaired in the deprived eye, with elimination of depth perception. But neither effect is irreversible, if corrected soon enough in the critical period.

oculomotor nerve Cranial nerve III, containing somatic motor neurons innervating the eye's extrinsic muscles arising in the oculomotor nucleus in the periaqueductal grey matter of the midbrain, and preganglionic parasympathetic neurons arising from a nearby nucleus controlling the smooth muscle within the eye. Fibres from both sources pass ventrally through the midbrain tegmentum, many traversing the red nucleus, and exit on the medial aspect of the crus cerebri.

odds ratios The ratio of the odds of an event occurring in one group to the odds of it occurring in another group, or to a sample-based estimate of that ratio. Employed, e.g., in ascertaining the effectiveness of clinical trials and procedures.

odorant receptors (ORs) See OLFACTORY RECEPTORS.

odours Human odours are often produced by microorganisms living in moist areas of the body (e.g., mouth, axillae, genital region and feet). The type and number of microorganisms on different areas interacting with skin and other gland secretions give rise to a variety of odours from different regions. Apocrine glands are the most important odour producers. See OLFACTION, SEBACEOUS GLANDS, SWEAT.

oedema An abnormal accumulation of TISSUE FLUID. See RENAL FAILURE.

oesophagus The region of the ALIMENTARY CANAL between the inferior laryngopharynx (see PHARYNX) and the STOMACH. A collapsible muscular tube, it passes through the MEDIASTINUM and pierces the diaphragm before ending in the cardiac region of the stomach. Its mucosa consists of non-keratinized stratified squamous epithelium, lamina propria (areolar connective tissue) and muscularis mucosae, there being more mucous glands closer to the stomach than more anteriorly. The upper oesophageal sphincter regulates the entry of food from the laryngopharynx to the oesophagus, the elevation of the LARYNX causing the sphincter to relax (see SWALLOWING). Progressive passage of food along the oesophagus involves coordinated contractions and relaxations of the circular and longitudinal layers of the muscularis (see PERISTALSIS).

17β-oestradiol See OESTROGENS.

oestrogens (estrogens) Steroid hormones synthesized mainly from blood CHOLESTEROL (see Fig. 49), and to a far lesser extent from acetyl coenzyme A, in the ovaries. It promotes division and growth of cells responsible for female SECONDARY SEXUAL CHARACTERISTICS and plays a major role in the MENSTRUAL CYCLE. The three most important oestrogens are: 17 *β-oestradiol* (the principal ovarian oestrogen), *oestrone* (produced in minute quantities by the ovaries, but also by the adrenal cortices from ANDROGENS) and *oestriol* (a weak oxidative liver product of oestradiol and oestrone). The physiological potency of 17 β-oestradiol is said to be 12 times that of oestrone and 80 times that of oestriol. They tend to be transported in the blood attached to the protein albumin and specific binding globulins (e.g., SEX HORMONE-BINDING GLOBULIN).

Genomic effects of oestrogens are mediated by two different steroid NUCLEAR RECEPTORS that function as transcription factors. These regulate gene expression through binding to DNA enhancer elements and subsequently recruiting factors such as coactivators that modulate their transcriptional activity. ARNT (the aryl hydrocarbon receptor nuclear translocator; see ARH), which is the obligatory heterodimerization partner for the aryl hydrocarbon receptor and hypoxia inducible factor 1α, functions as a potent coactivator of ERα- and ERβ-dependent transcription. The coactivating effect of ARNT depends on physical interaction with the ERs and involves its C-terminal domain. Mostly located in the nucleus, but also in the cytoplasm, the ERα receptor occurs mainly in the endometrium, breast cancer cells and ovarian stroma. The ERβ receptor mainly occurs in granulosa cells of the Graafian follicle, developing spermatids, and non-reproductive tissues such as kidney, intestinal mucosa, lung parenchyma, bone marrow, bone, brain, endothelial cells and the prostate gland. Non-genomic effects of oestrogen result from direct binding of oestrogen to cell membrane receptors (largely uncharacterized) or from enzyme-mediated effects of the ERα and ERβ receptors. These include short-term vasodilation and reduction of vascular tone, and activation of growth factor-related signalling pathways: oestrogens are mitogenic agents but can also function as tumour promoters (i.e., non-mutagenic carcinogens).

Oestrogens are key inhibitors of IL-6 production (e.g., by KUPPFER CELLS), and at all ages women are less likely than men to develop colon cancer. Oestrogen replacement therapy also reduces colorectal cancer in postmenopausal women, differences which may extend to other forms of CANCER (but see BREAST CANCER).

Oestrogen's genomic effects include up-regulation of expression of apolipoprotein genes and of lipoprotein receptor expression, leading to decreased serum concentrations of total cholesterol and LDL but increases in HDL, triglycerides and serum lipoprotein A levels. Oestrogen decreases levels of circulating fibrinogen and antithrombin III, elevated oestrogen levels being associated with increased potential for fibrinolysis. Other genomic effects include promotion of bone maturation and closure of epiphyseal plates in long bones, their overall effects being antiresorptive, conserving bone mass and suppressing BONE REMODELLING, maintaining balanced rates of bone formation and bone resorption. As a result, oestrogens (both androgen-derived and therapeutic) stop bone loss in postmenopausal women by preventing bone resorption. They lead the epiphyses of long bones to unite with the shafts so that women's height increase ceases two or three years earlier than that in men (see ADOLESCENT GROWTH SPURT). In a non-pregnant woman, oestrogens are secreted in significant quantities only by the ovaries, although the adrenal cortices do produce small quantities. During pregnancy, the placenta produces large quantities of oestrogens. They influence the expression of receptors for vitamin D, growth hormone and progesterone. They also modulate responsiveness to PTH (see PARATHYROID). See GONADOTROPINS.

oestrus Although this seasonal receptivity by females to males is a feature of most primates, it is not noticeably so of modern humans.

OFF-bipolar cells (off-centre bipolar cells) See BIPOLAR CELLS.

off-centre ganglion cell Ganglion cells in the retina which are maximally excited when there is no light hitting the centre of its receptive field and light hitting the surrounds. It has a low rate of discharge in the light, but can signal rapid decreases in light intensity in its receptive field centre by increase in firing rate. See RETINA, Figs 151 and 152; contrast ON-CENTRE GANGLION CELL.

OH The two letters (for 'Olduvai hominid') prefixing a number specifying a particular fossil (e.g., OH7) discovered in the Olduvai beds, these being mostly lava flows, riverine and stream deposits surrounding an ancient lake basin, in northern Tanzania. See OLDOWAN CULTURE.

Okazaki fragments See DNA REPLICATION.

Oldowan culture A Lower Palaeolithic stone tool culture, most discovered sites being along the Rift Valley system of Eastern Africa and dating from ~2 Mya. The earliest Oldowan industry comprises simple cores made from cobbles and associated flakes, the latter often retouched by striking small chips from one or more edges. Evidence of such culture extends across Asia, among the best-known being at Zhoukoudian near Beijing, dating to ~500 kya, and associated with *Homo erectus*. At this site, there are choppers and core-scrapers and many larger flakes. See ACHEULIAN CULTURE.

olfaction The sense of smell. Dendrites emerging from OLFACTORY RECEPTOR NEURONS (ORNs) of the OLFACTORY EPITHELIUM extend to the surface of the epithelium, each forming a swelling that gives rise to a cluster of 6–12 immobile olfactory cilia projecting into a layer of mucus secreted by the supporting cells. Olfactory perception is initiated when odorous ligands bind OLFACTORY RECEPTORS (ORs) located on the surface membrane of the cilia. On binding an odorant, these G PROTEIN COUPLED RECEPTORS activate adenylyl cyclase, the transduction pathway being mediated by cyclic AMP. This opens cAMP-gated cation channels allowing Na^+ influx and generation of impulses. The axons of ORNs run in the olfactory nerve and make excitatory synapses with dendrites of mitral (or tufted) cells (M/T) and short axon inhibitory periglomerular cells in the OLFACTORY BULBS. Synapses between ORNs and M/T and periglomerular cells occur in olfactory glomeruli in the olfactory bulbs. Each glomerulus receives the terminals of ~25,000 ORNs which respond *to the same odours*, making the glomeruli odour-specific. Each glomerulus contains dendrites from ~75 M/T cells which, by CONVERGENCE, integrate weak inputs from a large number of ORNs within a glomerulus to form a strong signal. Responses from glomeruli with slightly different odour specificities are dampened by LATERAL INHIBITION involving dendrodendritic synapses between M/T cells and inhibitory interneurons (granule cells), heightening odour discrimination. M/T cell axons project via the olfactory tract to the *olfactory cortex*. This is part of the palaeocortex, connected to the HIPPOCAMPUS ventrally and laterally, composed of two cell layers, and connected to the olfactory bulb anteriorly. It is separated by a sulcus, the rhinal fissure, from the more elaborate NEOCORTEX. Its five regions have distinct connections and functions, though all receive input from the olfactory tract. Fibres of the lateral olfactory tract project to the posterior HYPOTHALAMUS as well as to the ipsilateral olfactory cortex, although they probably do not play an important role in triggering sexual behaviour in humans, unlike the situation in some other mammals. This pathway, with that to the corticomedial amygdala, is involved with affective and emotional aspects of odours, directly influencing feeding. A pathway from the olfactory tract to the entorhinal cortex, and thence to the HIPPOCAMPUS, probably encodes olfactory components of episodic memory. Higher primates have a greatly reduced sense of smell relative to other mammals. See CINGULATE CORTEX, OLFACTORY RECEPTORS; compare TASTE.

olfactory bulbs Grey matter arising by the 7th week of development and containing the cell bodies of mitral cells or tufted cells (M/T), neurons which synapse (in ~2,000 *olfactory glomeruli*) with OLFACTORY RECEPTOR NEURONS and with those neurons which form the olfactory nerves, of which the bulbs are the anterior termini. There is a topographic organization to the fibres of the olfactory nerve and their projections to the olfactory bulb. Each olfactory bulb projects to the ipsilateral olfactory cortex via the lateral olfactory tract and to the contralateral olfactory cortex via the anterior commissure. See OLFACTION.

olfactory cortex See OLFACTION.

olfactory epithelium Thin sheet of cells in the dorsal nasal cavity, comprising OLFACTORY RECEPTOR NEURONS, glia-like supporting cells and basal cells that give rise to new receptor cells. See OLFACTION.

olfactory nerves The first cranial nerves, arising in the olfactory mucosa and terminating in the OLFACTORY BULBS. See OLFACTION, OLFACTORY TRACT.

olfactory receptor neurons (ORNs) Bipolar odorant-sensitive neurons of the olfactory epithelium (see OLFACTION), replaced on a 4- to 8-week cycle by adjacent basal cells. Their unmyelinated axons pass through the bony cribriform plate and synapse with neurons in the OLFACTORY BULBS. Their exposed termini are modified cilia which project from the olfactory epithelium and bear the cell-surface receptors and signal transduction system. Each odorant receptor, unlike the G PROTEIN-COUPLED RECEPTORS for neurotransmitters, binds a range of related odour molecules with various affinities, and each ORN expresses just one subtype of receptor. They operate via the olfaction-specific G_{olf} G PROTEIN (see OLFACTORY RECEPTORS). On binding their ligand, a rise in cAMP occurs within ~50 ms, activating a cyclic nucleotide-gated channel whose non-specific cation conductance allows entry of Na^+, K^+ and Ca^{2+} ions, causing a depolarization of the dendritic knob of the ORN, which spreads passively to the axon hillock producing a graded generator potential related to the strength of the stimulus. High odour concentrations, or high Ca^{2+}, activate haem oxygenase 2, producing carbon monoxide, CO. This can activate guanylyl cyclase in adjacent ORNs to produce cGMP, which binds to and opens their cyclic-nucleotide-gated channels, spreading the odorant effect to a cluster of ORNs.

olfactory receptors, odorant receptors, ORs The gene family encoding vertebrate olfactory receptors was discovered in 1991, by Buck and Axel. The initial sensors, on OLFACTORY RECEPTOR NEURONS of the olfactory epithelium, are olfaction-specific G-PROTEIN-COUPLED RECEPTORS known as G_{olf} G PROTEINS, numbering >1,000 in some mammalian species and capable of recognizing tens, perhaps hundreds, of thousands of volatile odorants. Human OR genes are arranged in 25 large clusters scattered throughout the genome, some of which are understood to be rapidly diverging in the human and chimpanzee genomes (see GENE CLUSTERS, Table 3). Primates have a higher proportion of olfactory receptor pseudogenes than do rodents or dogs, and humans have more than do chimpanzees. This suggests that some influence has allowed the entire ape lineage, and humans in particular, to get by with a comparatively reduced sense of smell (see PSEUDOGENES). Some research indicates that the greatest loss of olfactory receptor genes (associated with the greatest increase in OR pseudogenes) occurred in ape and monkey lineages that evolved trichromatic vision, the theory being that a sensory trade-off took place when better eyesight made an acute sense of smell less critical. However, regions of the anterior cingulate cortex and INSULA with significant olfactory and gustatory capabilities in other primates appear to have been usurped in the later hominoids, in favour of higher cognitive roles. See OLFACTION.

olfactory tract (OT) See OLFACTION.

oligodendrocytes Glial cells of the central nervous system (CNS) distinguished from ASTROCYTES by their less numerous and thinner processes and lack of gap junctions. Like the SCHWANN CELLS of the peripheral nervous system, they form myelin sheaths around axons, but unlike Schwann cells usually contribute to more than one axon, enveloping them with concentric layers of plasmalemma (there being no NEUROLEMMA). Unmyelinated neurons of the CNS are not surrounded by oligodendrocytes. Their dysfunction has been linked with BIPOLAR DISORDER and SCHIZOPHRENIA. See ADENOSINE.

***O*-linked β-*N*-acetylglucosamine (*O*-GlcNAc)** A sugar which competes directly with phosphate for serine/threonine residues, glycosylation with *O*-GlcNAc modulating signalling pathways dependent on SERINE/THREONINE KINASES and influencing protein expression, degradation and trafficking. It is one of the commonest post-translational modifications of proteins. The C-terminal domain of a subpopulation of RNA POLYMERASE II is extensively *O*-GlcNAcylated, and almost all Pol II transcription factors are modified by the sugar. >500 proteins have been found to be *O*-GlcNAcylated, these being involved in almost all aspects of cellular metabolism. Mammals seems to have only one gene encoding the catalytic subunit of *O*-GlcNAc transferase (OGT), deletion of which, e.g., causes hyperphosphorylation of TAU in neurons. The genes encoding both OGT and *O*-GlcNAcase map to chromosomal regions associated with neurodegenerative disease (see NEURAL DEGENERATION).

olivary nuclei See SUPERIOR OLIVARY NUCLEI.

omega-3 fatty acids, omega-6 fatty acids Polyunsaturated fatty acids that cannot be synthesized by humans and which are therefore ESSENTIAL FATTY ACIDS (EFAs). The omega-3 class is represented by α-linolenic (LNA), eicosapentaenoic (EPA) and docosahexaenoic (DHA) acids, the omega-6 class including linoleic acid (LA). The omega (ω) carbon atom is that furthest from the carboxyl end of the molecule, so ω3 and ω6 refer to the position of the double bond closest to the methyl end of the molecule (counting of carbon atoms begins at the methyl end of fatty acid molecules). The major terrestrial source of omega-3 acid is LNA (chloroplast-containing plant tissues, linseed, rapeseed, walnut and soybean oils), the major aquatic sources (e.g., in fish oils) being EPA and DHA. It seems that humans evolved on a diet rich in these lipids, although in the past 100 years the Western diet has shifted in favour of omega-6 fatty acids, leading to an omega-3-impoverished diet. Important omega-3 fatty acids in human nutrition are: α-linolenic acid, eicosapentaenoic acid and docosahexanoic acid. A decrease in DHA may play a significant role in visual impairments during ageing, since DHA is required for normal photoreceptor functioning. It is also found in large amounts in the grey matter of the brain and in the retinal membranes, accumulating in brain neurons between weeks 26 and 40 of gestation (see AFFECTIVE DISORDERS). The omega-3 class are thought to be protective against ALZHEIMER DISEASE, and may reduce the levels of enzymes that phosphorylate the protein TAU. EFA requirements for pregnancy are estimated to lie between 600 and 650 g for both LA and LNA since rapid brain tissue growth occurs in the third trimester of development. Human cerebral cortex, retina, testis and sperm are exceptionally rich in DHA. Humans can convert dietary LNA to EPA and DHA

omentum A fold in the peritoneum. The lesser omentum arises as a fold in the serosa of the stomach and duodenum and suspends the stomach and duodenum from the LIVER; the greater omentum is larger, a double sheet hanging as a 'fatty apron' over the transverse colon and small intestine and containing more lymph nodes, contributing macrophages and plasma cells. The coeliac and mesenteric arteries run in the greater omentum.

OMIM The Online Mendelian Inheritance In Man website: www.nchi.nlm.hih.gov/omim/. See MENDELIAN INHERITANCE.

ON-bipolar cells (on-centre bipolar cells) See BIPOLAR CELLS, RETINA.

on-centre ganglion cell Ganglion cell in the RETINA (see Figs 151 and 152) which is maximally excited when there is light hitting the centre of its RECEPTIVE FIELD and no light hitting the surrounds. It has a low rate of firing under dim illumination but can signal increases in light intensity in its receptive field centre by rapid increase in firing rate. See OFF-CENTRE GANGLION CELL.

onchocerciasis (river blindness) Disease cause by the parasitic nematode worm *Oncocerca volvulus* whose vector may be any of a number of simuliid flies (black flies) which breed in moving water. Female worms produce microfilaria which swarm under the skin and cause disease and itching, but can also migrate to the eyes and gradually cause blindness. It is the endotoxin of the symbiotic bacterium *Wolbachia*, present in the worm, which causes the eye pathology. It appears that the adult worms can be killed by the treatment with the antibiotic doxycycline which kills *Wolbachia* – maybe because the bacterium may supply a critical nutrient for the worm. See BLINDNESS.

oncogenes When genes with related DNA sequences were discovered in the DNA of normal cells, it was mooted that they might play a role in normal cell growth and differentiation and that their malfunction could lead to CANCER (see entry). In the late 1960s it was shown that cells in culture could be made cancerous (see TRANSFORMATION) by several DNA viruses and retroviruses. Later, it became apparent that a single viral gene could achieve this (the first being *Src* from the Rous sarcoma virus), such genes being known as oncogenes – their encoded protein products being *oncoproteins*. Normal cellular genes with highly conserved DNA sequences homologous to viral oncogenes were termed PROTO-ONCOGENES, but could mutate into oncogenes (cellular ONCOGENES, or c-*onc*s) if one of the normally recessive homozygous alleles became dominant (gain-of-function mutation), either as a result of viral infection or through over-expression (see CANCER, Fig. 33). The signalling molecules that enable cells to sense the presence of growth factors in their environments and process this information by SIGNAL TRANSDUCTION pathways (see Fig. 155) are usurped by oncoproteins (e.g., see FUSION GENES). By taking charge of the normal growth-promoting machinery of the cell, oncoproteins trick a cell into behaving as though it had encountered growth factors in its surroundings so that it begins to proliferate (see CELL CYCLE). Several types of TUMOUR arise in this way, because oncogene DNA sequences have important roles in signal transduction pathways as regulators of cell growth, maintaining orderly progression through the cell cycle, cell division and differentiation. The gene for cyclin D1, for example, is over-expressed in many of human BREAST CANCERS. Expression of certain oncogenes accelerates de-repression of the *CDKN2a* (cyclin-dependent kinase inhibitor 2a) locus, a phenomenon termed 'oncogene-induced senescence'. Oncogene expression can also lead to over-expression of a proto-oncogene: the oncogenes *Myc*, *Wnt*-1, *Neu* and *Ras* cause oncogenic expression of some D-cyclins in breast cancer; *Ras* is mutated in 40% of lung cancers, and *Braf*, in 60% of melanomas. Inappropriate demethylation of their promoter regions is thought to switch on expression of oncogenes in some tumours (contrast TUMOUR SUPPRESSOR GENES). Therapeutic agents for some of the most frequently encountered oncogenes (e.g., members of the *Ras* and *Myc* gene families) are under development (see RAS SIGNALLING PATHWAYS).

The idea that cancer cells depend upon mutant oncogenes not just for growth, but for their viability ('oncogene dependence' or 'oncogene addiction'; see P53 GENE), is supported by the fact that mice made transgenic for oncogenes can have these genes regulated by transcriptional control so that a wide variety of tumour types regress – mainly by apoptosis – when the oncogenic proteins are de-induced. Oncogenic dependence is still poorly understood and we do not understand how a cell that was viable without an oncogene becomes ready to die (apoptosis) if deprived of it.

Oncogene amplification occurs when an oncogene undergoes GENE AMPLIFICATION, possibly increasing the copy number several hundred-fold per nucleus. Approximately 10% of mammalian tumours contain cells with small complex 'extra' chromosomes, separated from the normal chromosomes, termed *double-minute chromosomes* (or *double minutes*) resulting from such amplified chromosomal DNA sequences (e.g., see NEUROBLASTOMAS) often derived from different chromosomes.

oncoproteins Proteins encoded by ONCOGENES.

oncosuppressors See TUMOUR SUPPRESSION GENES.

onset of blood lactate accumulation (OBLA) Normally understood as an increase in blood lactate concentration to a systematic value of 4.0 mM, although the maximum sustainable lactate level varies considerably among individuals. See LACTATE THRESHOLD.

ontogeny The development of an individual organism, as opposed to PHYLOGENY (the evolutionary origin of the lineage to which it belongs).

oocytes For primary and secondary oocytes, see OOGENESIS. See EMBRYONIC STEM CELLS, FERTILIZATION, IN VITRO FERTILIZATION, MENOPAUSE.

oogenesis The ovary starts to produce germ cells early in foetal life, so that by 15 weeks of gestation PRIMORDIAL GERM CELLS proliferate and migrate to the GENITAL RIDGE, where some continue to divide mitotically and others undergo apoptosis. By the end of the 3rd month, OOGONIA and primary oocytes lie in clusters in the ovary, each surrounded by a layer of flat epithelial cells. Some oogonia begin to divide meiotically and by 6 months after birth all have become oocytes arrested in diplotene of the first meiotic prophase. These *primary oocytes* either die by apoptosis or are recruited to grow into mature ova, those remaining lying at the ovary surface. It is estimated that at birth oocyte numbers range from 6–8 $\times$ 10^5 and soon after birth, they recruit somatic FOLLICULAR CELLS and become organized into 'resting' primordial follicles. Primary oocytes remain in diplotene until puberty, through the action of oocyte maturation inhibitor secreted by follicular cells. During childhood, most of them undergo apoptosis, so that ~4 $\times$ 10^5 remain at puberty, although fewer than 500 will be ovulated (see AGE-RELATED DISORDERS). At puberty, a pool of growing follicles is established and maintained from the primordial follicle supply, and each month 15–20 follicles begin to mature, developing through the pre-antral, antral and pre-ovulatory stages (see GRAAFIAN FOLLICLES). At mid-cycle of the MENSTRUAL CYCLE an LH surge elevates levels of maturation-promoting factor, causing the pre-ovulatory growth phase during which oocytes complete the 1st meiotic division. Two daughter cells of unequal size are produced, the larger containing most of the cytoplasm becoming the secondary oocyte while the other, the first polar body, lies between the zona pellucida and the cell membrane of the secondary oocyte in the perivitelline space. The secondary oocyte then enters the 2nd meiotic division, arresting in metaphase ~3 hours prior to ovulation. Meiosis II is only completed on FERTILIZATION, when the first polar body also undergoes a second division. The LH stimulates progesterone production by follicular stromal cells, causing follicular rupture and OVULATION. In the absence of fertilization, the oocyte degenerates ~24 hours after ovulation. During oogenesis (as in SPERMATO-

GENESIS), *de novo* DNA methylation occurs in sex-specifically imprinted genes (see GENOMIC IMPRINTING) and in transposons. In mice, the protein stella (aka DPPA3), which may protect specific genome sequences against demethylation, is present in large amounts in oocytes and, after fertilization, translocates to both zygotic pronuclei. Deletion of *Stella* from the oocyte, and resulting absence of stella from the zygote, leads to early pre-implantation embryo lethality and to loss of methylation of some maternally methylated genes.

oogonia Developing from PRIMORDIAL GERM CELLS that have migrated to the gonad of a genetic female, these cells undergo several mitotic divisions so that, by the end of the 3rd month, they are arranged in clusters surrounded by a layer of flat epithelial cells (follicular cells). Some oogonia arrest their nuclear division in first meiotic prophase, forming primary oocytes. See OOGENESIS.

open reading frame (ORF) DNA sequences which begin with an initiation triplet, usually ATG, and end with a termination triplet: TAA, TAG or TGA. Because human genes have large intron sequences embedded in them, computers also search for intron/exon boundary sequences, or characteristic upstream regulatory sequences. See HUMAN GENOME.

operant conditioning See INSTRUMENTAL CONDITIONING, LEARNING.

operants See INSTRUMENTAL CONDITIONING.

opioid receptors (opiate receptors) G-PROTEIN COUPLED RECEPTORS linked to ION CHANNELS and widely distributed in the central nervous system. Activation of μ- and δ-receptors causes hyperpolarization of neurons, their activation reducing cAMP levels, opening K^+ channels and closing Ca^{2+} levels, inhibiting presynaptic transmitter release. Agonists include morphine, meperidine and fentanyl; weak agonists include methadone; antagonists include naloxone. Endogenous opioid binding to μ-opioid receptors (encoded by *Orpm* genes) is believed to mediate natural rewards, and has been proposed as the basis of infant attachment behaviour. Located on some, but not all, brain neurons, these receptors bind the natural opioid neurotransmitters: enkephalins (except κ-receptors), dynorphin (especially κ-receptors) and β-endorphin. The hedonic experience, the 'liking' of a stimulus (e.g., a palatable food), is associated with μ-opioid receptor signalling in the NUCLEUS ACCUMBENS and adjacent forebrain structures, activated in part by DOPAMINE release. The opioid system controls nociceptive and addictive behaviours (see ADDICTION).

opioids Compounds with effects that are antagonized by naloxone, with a morphine-like action in the body and used mainly for PAIN relief. These agents work by binding to OPIOID RECEPTORS, which are found principally in the central nervous system and the gastrointestinal tract. Derived from large precursor molecules, there are three families, encoded by separate genes. PROOPIOMELANOCORTIN (POMC) gives rise to β-ENDORPHIN and other non-opioid peptides. The other two groups of opioid peptide neurotransmitters are ENKEPHALINS and DYNORPHINS. They are generally co-expressed with classical transmitters such as GABA and SEROTONIN, and are usually inhibitory. They have inhibitory effects on synapses in the CNS and gut. Brain areas known to be rich in opioid receptors include the brain stem and rostral anterior cingulated cortex (ACC). The latter exchanges information with a network of brain regions, including the orbitofrontal cortex (which processes emotions). Diamorphine (heroin) and other opioids have a high misuse and dependence potential owing to the intense euphoria they produce. Substitution of oral long-acting drugs (methadone or buprenorphine) reduces the harm of heroin addiction (e.g., infection, criminality) and can be a stage in detoxification by gradual dose reduction. See ADDICTION, PROOPIOMELANOCORTIN, PSYCHOACTIVE DRUGS.

opposable digits See FOOT, HAND.

opsin Any of a family of proteins present in

RHODOPSIN (see entry) and IODOPSINS and involved in activation of G proteins during phototransduction. See COLOUR BLINDNESS, COLOUR PERCEPTION.

opsonins Substances that alter pathogen and other particle surfaces in such a manner as to make them more readily ingested by phagocytes. Some ANTIBODIES and COMPLEMENT factors are opsonizing molecules. See B CELLS, Fig. 19.

optic chiasma (optic chiasm; pl. o. chiasmata) Region at the base of the skull, just rostral to the hypothalamus, where the optic nerves of the two eyes converge, and where about one million of the fibres of each nerve cross over (decussation) to the contralateral side. It is only the fibres from the nasal hemiretina of each VISUAL FIELD (see Fig. 130) which project to the LATERAL GENICULATE NUCLEUS of the thalamus, partial decussation producing the two optic tracts. Each optic tract contains axons from the temporal hemiretina on its own side of the head and axons from the nasal hemiretina of the opposite side of the head.

optic cup See EYES.

optic disc (optic nerve head) The region of the retina where the ganglion cell axons of the optic nerve leave the eye. It is responsible for the so-called 'blind spot'.

optic nerve The axons of retinal ganglion cells leave by way of the optic disc, becoming myelinated and collectively forming the optic nerve. Although regarded as the second CRANIAL NERVE, each optic nerve is strictly a cerebral tract running out from the wall of the diencephalon of the brain, originally through the embryonic optic

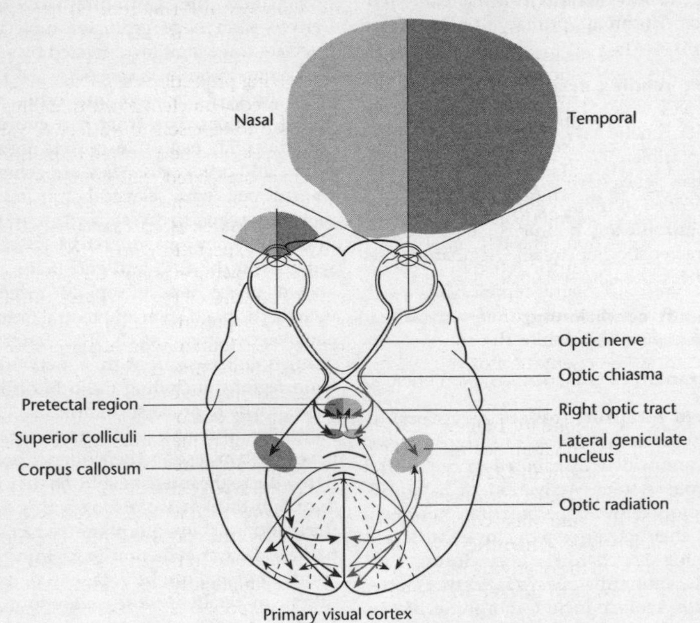

FIG 130. Diagram showing the position of the optic chiasma in the visual pathway, with some of the efferent projections from the primary visual cortex to subcortical structures indicated on the right side only. The two visual cortices are connected by axons passing through the corpus callosum. The afferent fibres from the two hemiretinas are distinguished by black (left) and grey (right) lines. From *Fundamentals of Sensory Physiology*, R.F. Schmidt © 1981, Springer. Redrawn with permission of the publisher.

stalk. Each optic nerve contains 1.2 million myelinated axons of the retinal ganglion cells and undergoes partial decussation in the OPTIC CHIASMA (see Fig. 130). They transmit visual information via the LATERAL GENICULATE NUCLEI (LGN) to the PRIMARY VISUAL CORTEX of the brain. The relatively few fibres mediating the papillary reflex leave the optic nerve before reaching the LGN and terminate in the pretectal area and the superior colliculus.

optic radiation The collection of fibres projecting from each LATERAL GENICULATE NUCLEUS to the thalamus.

optic tectum (pl. optic tecta) See SUPERIOR COLLICULUS.

optic tracts Fibres leaving the OPTIC CHIASMATA (see Fig. 130) and projecting mainly to the lateral geniculate nuclei of the THALAMUS, the PRIMARY VISUAL CORTEX (processing visual information) and to the midbrain (the pretectum, involved in pupillary reflexes; and the superior colliculus, mediating eye movements guided by visual information). In addition, a few fibres form synapses in the hypothalamus (see CIRCADIAN RHYTHMS, SLEEP–WAKE CYCLES), and a further 10% innervate the midbrain. These TRACTS run just under the pia and along the lateral surfaces of the diencephalon. The left optic tract contains axons from the left half of each retina carrying a neural representation of the right hemifield of vision (see VISUAL FIELD and Fig. 179), while the right optic tract carries a representation of the entire left visual hemifield.

optic vesicles Secondary bulges off the forebrain (prosencephalon; see BRAIN, Fig. 28), appearing in about the 22-day embryonic diencephalon. They continue to grow laterally from the neural tube, connected to it by the optic stalks. When these vesicles make contact with the overlying head ectoderm, they induce it to thicken to become the lens placodes.

oral cavity The opening of the foregut with the external environment. It develops from the stomodaeum. See DENTITION, MOUTH, SALIVARY GLANDS.

oral contraceptives See CONTRACEPTION.

oral rehydration therapy In DEHYDRATION, as following persistent DIARRHOEA, absorption of glucose across the small intestine is unimpaired, and because of the manner in which glucose and water are absorbed by ENTEROCYTES, oral (as opposed to intravenous) rehydration is achieved by drinking a solution of salt (sodium chloride) and glucose so that osmotic uptake of water follows. Sucrose or starch can substitute for glucose, being more widely available. The solution must not be significantly hypertonic, or water loss will be worse.

orbital plane The line joining the superior and inferior margins of the orbits. See SAHELANTHROPUS TCHADENSIS, Fig. 154.

orbitofrontal cortex (OFC) A region of association cortex involved in cognition and lying within the frontal lobe (see BRAIN, Fig. 28), resting above the orbits of the eyes. It is defined as the PREFRONTAL CORTEX receiving projections from the magnocellular, medial, nucleus of the thalamus. It is believed to represent the affective value of reinforcers, and be involved in planning behaviour associated with sensitivity to reward and punishment (see REINFORCEMENT AND REWARD), as well as in DECISION MAKING (see entry) and expectation, and processing of EMOTIONS. Growing evidence suggests that a neural correlate of the expected value of outcomes is present, perhaps generated, in the OFC. Human neuroimaging studies show that blood flow increases in the OFC during anticipation of expected outcomes, and when the outcome is modified or not delivered. When macaques are presented with visual cues paired with differently preferred rewards, OFC neurons fire selectively in accordance with whether an anticipated outcome is the preferred or the non-preferred reward in the trial, and that neurons fire differently depending on the reward's expected size, the anticipated time of delivery, and with possible aversive consequences associated with inappropriate behaviour. These responses can also be observed in the absence of any signalling cues and are a common feature in the OFC across many tasks where events occur sequentially and predictably. Without input from the OFC, neurons of the basolateral AMYGDALA (which receive input from the OFC) respond more slowly to associative encoding, especially when cue-outcome associations are reversed. Lacking this signal, animals behave maladaptively, being driven by antecedent cues and response habits instead of by cognitive representation of an outcome or goal. This approach to OFC function may have a bearing on drug addiction. See WILLIAMS–BEUREN SYNDROME.

orbits (eye sockets) Cone-shaped fossae, their apices directed posteriorly, whose bones provide protection for the EYES and attachment for its extrinsic muscles. The bones contributing to its structure are: the frontal, sphenoid, zygomatic, maxilla, lacrimal, ethmoid and palatine bones.

orexigenic APPETITE-stimulating.

orexinergic (hypocretinergic) Secreting, or stimulated by, OREXINS.

orexins (hypocretins) Neuropeptides produced by the posterior part of the lateral hypothalamus, stimulating APPETITE and acting on several brainstem centres involved in the regulation of SLEEP–WAKE CYCLES. They reinforce arousal systems (see RETICULAR FORMATION) and deficiency of either of them, or of their receptors, is associated with NARCOLEPSY. See NEUROPEPTIDE Y.

ORF See OPEN READING FRAME.

organizer See PRIMITIVE NODE.

organ of Corti (spiral organ) A coiled sheet of epithelial cells (hair cells and supporting cells) forming the sensory epithelium within the cochlear duct of the COCHLEA (see Fig. 58). It lies on the basilar membrane and its hair cells are in contact with the overlying tectorial membrane. Between them, they transduce the energy of vibrations in the perilymph into nerve impulses in the cochlear branch of the vestibulocochlear nerve and enable hearing. During formation of the otic vesicle, small groups of cells break away from its walls to form, with neural crest cells, the statoacoustic ganglion, the cochlear portion of which supplies the sensory cells of the organ of Corti.

organogenesis The development of organs, mostly accomplished between the 4th and 9th embryonic weeks. See entries on specific organs.

organ transplantation Transplant surgeons currently perfuse whole donor organs with a special fluid enabling them to be banked at just above 0°C for up to a few days (see CRYOBIOLOGY). At present, most people who receive an organ TRANSPLANT face a lifetime's treatment with IMMUNOSUPPRESSANT drugs, such as cyclosporin and steroids that, although possibly increasing the life of the transplant by several years, often fail to prevent its eventual rejection and put patients at risk of infections, cancers and kidney failure.

organs Functional anatomical structures consisting of two or more tissues.

orientation columns See HYPERCOLUMNS.

orientation-selective (orientation-sensitive) cells Bipolar and ganglion cells of the retina, and PRIMARY VISUAL CORTEX (see Fig. 146 and entry for some details) responding to contours, or to contrasts in either colour or luminescence, which have straight edges at particular orientations. *Simple cells* in the visual cortex respond to bars or edges with a particular orientation within a fixed portion of visual space, while *complex cells*, of the interblob regions of layer III, respond similarly but to bars or edges with a particular width placed anywhere within a larger region of visual space. Complex cells receive convergent inputs from a small number of simple cells (they mostly have binocular receptive fields, responding to stimulation of both eyes) and between them are sensitive to the outline of an object, but not to their interiors or backgrounds. They do not have distinct ON and OFF regions in their receptive fields but instead give ON and OFF responses to stimuli throughout the receptive field. It is the information carried by edges that allows us to recognize objects in a picture even if only sketched in outline. Because of their larger sensory fields and lack of clear excitatory or inhibitory regions, complex cells can respond to orientation over a range of retinal positions and may be a neuronal component in the achievement of *positional invariance*: the ability to recognize the same feature anywhere in the visual field. *End-stopped cells* are subsets of simple and complex cells which respond only if the contours of a stimulus end within their receptive field, or change in direction, responding to discontinuities in contour, as in curves, corners and ends of lines. These probably receive convergent inputs from three complex cells of similar properties but slightly staggered visual fields, the central one being excitatory and the outer two inhibitory.

origin For a muscle's origin, see MUSCLES.

origin of species See SPECIATION.

origins of replication See DNA REPLICATION.

ornithine cycle (urea cycle) Ornithine is an AMINO ACID not encoded by the genetic code but important in the UREA cycle, where its production by arginase in the cytosol is followed by uptake into the mitochondrion and its conversion to citrulline, which is exported back to the cytosol.

orphan drugs A term introduced in the USA, referring to products that treat a rare disease, affecting <200,000 individuals. The Orphan Drugs Act, which became USA law in 1983, offers incentives to induce pharmaceutical companies to develop these and other medical products for the small markets involved. See DRUGS.

orphan gene A protein-coding gene with little or no homology to genes in distantly related species.

orphan nuclear receptors Proteins of the NUCLEAR RECEPTOR SUPERFAMILY that have been identified only by DNA sequencing, their ligands being currently unknown.

Orrorin tugenensis Dating to 5–6 Myr, and described in 2001 (hence 'Millennium Man'), this genus is currently represented by 13 fossils from the Tugen Hills in Kenya. Included is a fragmentary thighbone whose structure indicated upright walking, while the thickened enamel on its small molars led its discoverers to link it to the human lineage, although other teams are more circumspect. See ARDIPITHECUS RAMIDUS, SAHELANTHROPUS TCHADENSIS.

orthognathous Straight-jawed; having the front of the head, or the SKULL, nearly perpendicular rather than retreating backwards above the jaws. As a result, the profile of the face is vertical, or nearly so. Contrast PROGNATHOUS.

orthologues Genes that have evolved by vertical descent from a common ancestral gene, in contrast to paralogues (see PARALOGOUS GENES), which originate from GENE DUPLICATIONS within a genome. See GENOME MAPPING.

osmolarity The number of osmoles in one kg of solution, where one osmole is the product of the molecular mass in grams of a solute and the number of ions into which it dissociates in solution. One milliosmole (mOsm) is 10^{-3} osmoles.

osmoreceptors (i) Receptors responding to changes in blood OSMOLARITY, lying in the vascular organ of the lamina terminalis (OVLT), one of the CIRCUMVENTRICULAR ORGANS (CVO) of the brain lying on the blood side of the BLOOD–BRAIN BARRIER. These cells increase their tonic firing rate with rising blood concentration and synapse with neurons in the paraventricular nucleus and supraoptic nucleus of the HYPOTHALAMUS. The axons of these neurons project to another CVO, the posterior lobe of the pituitary, and increase release of vasopressin into the systemic circulation. (ii) Receptors in the wall of the hepatic portal vein, responding to increased osmolarity of plasma in the absence of sufficient intake of water with a meal. Their input to the medulla inhibits feeding. See OSMOREGULATION.

osmoregulation The homeostatic control of the amount of water present in the body in relation to all solutes combined, proportions of the various ions being unimportant. Water potential of blood plasma is detected by OSMORECEPTORS, mainly in the hypothalamus, and responses to reduced water potential include the release of VASOPRESSIN by the posterior pituitary and consequent involvement of AQUAPORINS in the COLLECTING DUCTS of the kidneys. Total blood volume affects stretch receptors and a reduction in stretch of certain arterial walls brings consequent release of angiotensins from ANGIOTENSINOGENS. Ultimately, reduction in blood volume can only be reversed by fluid intake by drinking (see THIRST).

ossification (osteogenesis) The process by which bone is formed. The embryonic skeleton consists of fibrous connective tissue membranes formed by condensed mesenchyme, or pieces of hyaline cartilage resembling bones in shape (e.g., see LIMBS). Acting as a template for later ossification, starting during the 6th or 7th week of life, these structures may develop in to either MEMBRANE BONES, by intramembranous ossification or, as in most human bones, into endochondral bones by endochondral ossification.

Intramembranous ossification involves clustering of mesenchymal cells into centres of ossification and their differentiation into osteogenic cells and then into OSTEOBLASTS. These secrete the organic matrix of bone, becoming completely surrounded by it. The osteoblasts then develop into mature bone cells, OSTEOCYTES, and within a few days calcium and mineral salts are deposited and the matrix hardens (calcifies). The matrix develops into trabeculae which fuse together to form spongy bone. Blood vessels invade the spaces between the trabeculae and the mesenchyme along the surface of the new bone (see BONE, Fig. 26), connective tissue associated with these developing into red bone marrow. Mesenchyme condenses on the outside of the bone to form the periosteum and most superficial layers of the bone develop into compact bone, although spongy bone remains in the centre and the bone is slowly transformed into its adult form by BONE REMODELLING.

Endochondral ossification involves the replacement of cartilage by bone, beginning in the primary ossification centre in the middle of the bone and spreading outward, producing *cartilage bones*. Secondary ossification centres arise in the ends of the bones, but a persistent thin cartilage layer, the *epiphyseal growth plate*, enables longitudinal growth, which is accompanied by BONE REMODELLING. See ADOLESCENT GROWTH SPURT.

osteoarthritis Degenerative joint disease, apparently due to ageing and irritation and wear of the joints. See ARTHRITIS.

osteoblasts Bone-producing cells; develop from osteogenic stromal STEM CELLS in bone marrow. Synthesize and secrete collagen and promote deposition of calcium phosphate (hydroxyapatite) crystals. Growth hormone and oestrogen stimulate osteoblast production and production of type 1 collagen and alkaline phosphatase. They express parathyroid hormone receptors, and their activity is also promoted by THYROID HORMONES. The osteoblast lineage is the source of the fibroblast growth factor FGF23. See HAEMATOPOIETIC STEM CELLS, OSSIFICATION.

osteoclasts Very large multinucleated bone-resorbing cells resulting from fusion of as many as 50 monocytes in the endosteum, and rich in lysosomal enzymes. From microvilli on the surface adjacent to the bone, an osteoclast secretes enzymes and acids that digest the proteins and minerals (esp. hydroxyapatite) of the bone matrix. This activity is a part of normal bone growth and repair. The cytokines TNF-α, TNF-β, IL-1α, IL-1β and IL-6 all increase the proliferation and differentiation of osteoclast precursors and their activities and are potent promoters of bone resorption. Their activity is promoted by the THYROID HORMONES T_3 and T_4, but inhibited by CALCITONIN. See BONE REMODELLING.

osteocytes The most numerous cells in BONE, terminally differentiated from

OSTEOBLASTS, lying in lacunae in the bone matrix and extending narrow cytoplasmic connecting processes into canaliculi that radiate from them in all directions and interconnect them to one another and to osteoblasts on the bone surface. They ultimately undergo apoptosis or phagocytosis during osteoclastic resorption (see BONE REMODELLING), their contribution to this process being incompletely understood. The appearance of osteocytes among cartilage cells at ~8 weeks marks the transition between a human EMBRYO and a foetus.

osteogenesis imperfecta See BRITTLE BONE DISORDER.

osteomalacia Disorder resulting from resorption of bone exceeding bone production, generally resulting from low body CALCIUM levels. Kidneys needed to convert vitamin D to CALCITRIOL. See BONE DENSITY, RENAL FAILURE, RICKETS.

osteomyelitis Bone inflammation, often resulting from bacterial infection, which can lead to complete destruction of the bone. It may result from *Staphylococcus aureus* which has entered through wounds. Bone tuberculosis, a form of osteomyelitis, results from spread of *Mycobacterium tuberculosis* (see TUBERCULOSIS). See ANALGESIA, OSTEOSCLEROSIS.

osteon See HAVERSIAN SYSTEM.

osteoporosis Reduction in the overall quantity of bone tissue, occurring when the rate of bone resorption exceeds the rate of bone formation. Its occurrence increases with age in both sexes, but is 2.5 times more common in women; women can lose up to half their cancellous bone, men losing up to a quarter. In post-menopausal women, falling OESTROGEN levels can be responsible; but other causes of reduced oestrogen (e.g., anorexia nervosa, cigarette smoking) can also cause osteoporosis. Oestrogen is a key factor in BONE REMODELLING and its decline during MENOPAUSE reduces osteoblast activity although osteoblasts continue to remove bone mineral and break down collagen, so that women lose ~1–2% of bone mass during this period, especially from trabecular bone. Poor diet and lack of exercise, especially in youth, are contributory factors increasing the risk of hip fractures in post-menopausal women (25% lower risk in Asian than in Caucasian women). In males, reduction in testosterone levels promote osteoporosis; but their greater average bone density enables it to perform better with the same absolute bone loss, and their testosterone levels tend to decline later, at age ~65 years. Miacalcin, a calcitonin extract from salmon, is used as a drug to treat the condition. Contrast BRITTLE BONE DISEASE.

osteosclerosis Abnormal hardening or increased density of bone detected on radiographs and accompanying many disorders including hyperparathyroidism, osteoarthritis, sickle-cell anaemia, Paget's disease, leukaemias and osteomyelitis. Hodgkin's disease may be characterized by osteosclerosis alone, or by this and osteolysis. It occurs throughout the body in osteopetrosis.

otic vesicles (otocysts) See EARS.

outbreeding The converse of INBREEDING.

outgroup A molecule, species or higher monophyletic taxon employed in PHYLOGENETICS, to determine which of two alternative sequences, or homologous characters, may be inferred to be apomorphic.

Out-of-Africa models There are currently two forms of 'Out-of-Africa' models for *Homo*: 'Out-of-Africa 1' relating to the origin of HOMO ERECTUS in Africa and its later dispersal into Asia, while 'Out-of-Africa 2' relates to the origin of HOMO SAPIENS in Africa and its subsequent global expansion. It is this latter model that is referred to as the RECENT AFRICAN ORIGIN MODEL.

ovarian cycle See MENSTRUAL CYCLE.

ovarian follicle See GRAAFIAN FOLLICLES, OOGENESIS.

ovaries The paired female gonads, ~2–3.5 cm in length and 1–1.5 cm in width when mature, each developing on a urogenital ridge medial to the mesonephros. In the absence of a Y chromosome (see SEX DETERMINATION) the primitive sex cords become cell clusters and the surface epithelium continues to proliferate, so that by the 7th week it gives rise to cortical cords which penetrate the underlying mesenchyme. Remaining close to the surface, these are invaded by PRIMORDIAL GERM CELLS, which develop into oogonia as surrounding epithelial cells from the original surface form FOLLICULAR CELLS of the primordial follicles (see GRAAFIAN FOLLICLES for details) which, after puberty, are recruited during MENSTRUAL CYCLES.

Descent of the ovaries is less dramatic than in the case of the testes. They finally come to lie just below the rim of the true pelvis, attached to the broad ligament by a peritoneal fold, the mesovarium, and by the suspensory ligament extending from the mesovarium to the body wall, while the ovarian ligament and round ligament attach them to the superior margin of the uterus, the latter extending into the labia majora.

The visceral peritoneum covering the ovary surface is called the *germinal epithelium* (because these cuboidal cells were formerly thought to produce oocytes). Below this lies the thin but tough fibrous connective tissue of the tunica albuginea, beneath which lie the primordial follicles. Under this lies the denser *cortex* of the ovary within which the follicles develop, while looser connective tissue forms the inner *medulla* in which blood and lymphatic vessels enter from the mesovarium. The cortex and medulla form the stroma of the ovary. Mutations in the breast susceptibility gene *BRCA1* (see entry) greatly increase a woman's risk of developing breast and ovarian cancer.

over-expression See GENE OVER-EXPRESSION.

overloading of muscles Exercising at intensities greater than normal induces a variety of specific adaptations in physiological function enabling an athlete to perform more efficiently. It requires an appropriately planned systematic and progressive manipulation of training frequency, intensity and duration in accordance with the EXERCISE mode. See OVERTRAINING SYNDROME.

overtraining syndrome A 'staleness' experienced by 10–20% of athletes. Two clinical forms are: (i) the more common parasympathetic (addisonoid) form, characterized by high vagal activity during rest and EXERCISE, often resulting from excessive and protracted overload with insufficient recovery and rest, and involving chronic fatigue experienced during workouts and subsequent recovery periods. Besides impoverished exercise performance, associated features include altered sleep patterns and appetite, altered immune and reproductive function, acute and chronic systemic inflammatory responses, and mood disturbances; (ii) a less common sympathetic form (basedowian for hyperthyroidic patterns), characterized by impaired exercise performance and increased sympathetic nervous system activity during rest, typified by hyperexcitability and restlessness which may reflect the emotional stress involved in competitive sport.

overweight Sometimes regarded as mild OBESITY, a person's body mass may be considerably in excess of an average weight-for-height standard yet still rate as being 'underfat' for body composition. The 'extra' mass may be due to muscle mass. See BODY MASS INDEX, ENERGY BALANCE.

oviducts (uterine tubes, Fallopian tubes) Developing from the cranial and horizontal parts of the paramesonephric ducts (MÜLLERIAN DUCTS) after descent of the ovaries, ducts (~10 cm in length) conveying the ova from the ovary to the uterus. They lie between the folds of the broad ligaments of the uterus, the funnel-shaped portion (infundibulum) of each, close to the ovary, ending in a fringe of finger-like projections (*fimbriae*), attached to the lateral end of the ovary. Each has an internal mucosa containing ciliated columnar epithelial cells which assist movement of the ovum or morula along the tube, and secretory cells whose microvilli may provide it with nutrients. Circular and longitudinal muscles engage in peristaltic waves of contraction and propel the ovum/morula towards the uterus into which they open. After ovulation, currents produced by the fimbriae sweep the ovum into the oviducts. See FERTILIZATION.

ovulation The rupture of a Graafian follicle at the ovary surface (see MENSTRUAL CYCLE), close to the Fallopian tubes, and release of the contained secondary oocyte and corona radiata into the peritoneal cavity. Follicle rupture is an inflammatory process involving regulated lysis of the extracellular matrix within the thecal layers and tunica albuginea at the ovarian surface.

ovum (egg cell) The mature secondary oocyte. A large cell, ~120 μm in diameter, surrounded by a perivitelline space and an extracellular envelope, the zona pellucida, to which sperm must bind. The ovum is surrounded by a layer of cells, the cumulus cells, serving to nurture the cell prior to, and just after, ovulation. See FERTILIZATION, MENSTRUAL CYCLE (for ovulation).

β-oxidation The first step in the mitochondrial oxidation of fatty acid molecules (see FATTY ACID METABOLISM), involving the oxidative removal of successive 2-carbon units in the form of acetyl-CoA.

oxidative capacity A measure of a muscle's maximal capacity to use oxygen in microlitres (μL) of oxygen consumed per gram of muscle per hour. Factors which affect the oxidative capacity of muscles include the activity of oxidative enzymes (e.g., succinate dehydrogenase), fibre-type composition (see SKELETAL MUSCLE) and availability of oxygen.

oxidative stress See MITOCHONDRION; STRESS **2**, and references there.

oxidoreductases Members of the class of enzymes carrying out oxidative and reductive (redox) reactions.

OXM See OXYNTOMODULIN.

oxygen debt Excess oxygen consumption during recovery from exercise. Whatever the work rate and duration of EXERCISE, oxygen consumption in recovery always exists in excess of the resting value. Originating in 1922, this *recovery oxygen consumption* was believed to be due to the aerobic metabolism of stored macronutrients needed to replenish intramuscular ATP and PHOSPHOCREATINE, reconversion of muscle myoglobin and blood haemoglobin to their pre-exercise oxygenated levels (the *alactacid oxygen debt*), and reconversion of lactate to liver glycogen (the *lactacid oxygen debt*). Contrary to this view, conventional thought during the latter part of the 20th century, it is now believed that most of the elevated post-exercise $\dot{V}O_2$ is used by heart, liver, kidney and skeletal muscle fibres in oxidizing lactate as an energy substrate and in restoring high-energy phosphate levels, rather than in the re-synthesis of glycogen. See OXYGEN DEFICIT.

oxygen deficit A term quantitatively expressing the difference between the total oxygen actually consumed during exercise and the total that would have been consumed had steady-state aerobic metabolism been reached from the start of exercise. The high-energy phosphates (ATP and PHOSPHOCREATINE) are substantially depleted in exercise that generates ~3–4 dm^3 oxygen deficit, further exercise only progressing through either anaerobic glycolysis or aerobic breakdown of macronutrients. See OXYGEN DEBT.

oxygen-haemoglobin dissociation curve See HAEMOGLOBIN.

oxygen toxicity Effect of breathing pure oxygen gas. At a pressure of one atmosphere (1 atm), it results in tracheobronchial irritation and after a few days leads to thickening of the alveolar-capillary membrane and even alveolar collapse beyond an obstructed air passage. The newly born are more rapidly affected. Breathing pure oxygen at 3 atm for even short periods affects the central nervous system and causes convulsions. Although the resting oxygen consumption can be satisfied by physically dissolved oxygen in the plasma, the dissolved carbon dioxide level in the plasma rises. See HYPERVENTILATION.

oxyhaemoglobin See HAEMOGLOBINS.

oxyntic cells See PARIETAL CELLS.

oxyntomodulin (OXM) A satiety peptide (see APPETITE, Fig. 14) released by L cells in the distal ileum and colon in response to free fatty acids and carbohydrate in the ileal lumen. Anorectic, it inhibits gastric and intestinal motility and, when injected into the rat arcuate nucleus, inhibits refeeding after a fast. It causes a similar pattern of neuronal activation to GLUCAGON-LIKE-PEPTIDE-1 after peripheral administration. See CHOLECYSTOKININ.

oxytocin Nonapeptide hormone (neurosecretion) produced mainly by the magnocellular neurons of the paraventricular nuclei of the HYPOTHALAMUS (see also PITUITARY GLAND, Fig. 137), secreted along with neurophysin I to which it is bound in the neurons. It binds cell membrane $G_{q/11}$ protein-coupled oxytocin receptors expressed in the uterus, mammary glands and brain, exerting its effects through signal pathways involving PHOSPHOLIPASE C (see Fig. 135) and IP3. The raised intracellular calcium levels in myoepithelial cells surrounding ducts and alveoli of the mammary glands cause them to contract during milk ejection in response to suckling (see LACTATION); and similar responses occur in the myometrium during BIRTH. The role of oxytocin in males is unclear although it may be involved in ejaculation. Oxytocin is widely involved in the animal kingdom in promoting social attachment, including male–female bonding after copulation, mother–infant bonding after childbirth, and assorted sexual behaviours. It is known to act on the amygdala and nucleus accumbens and modulates the activity of cognitive neural networks that result in enhanced trusting behaviour. Its secretion can be disturbed by some forms of STRESS, and is stimulated by the cry of a hungry baby and during play prior to its being breast-fed. Compare VASOPRESSIN.

P

p21 A universal inhibitor of cyclin kinases. The *p21* gene is transcribed in the presence of p53 (see entry) and its over-expression inhibits the proliferation of mammalian cells. See G1.

***p53* gene (TP53) and signalling pathway** A tumour suppressor and transcriptional activator gene, *p53* (*TP53* in humans) is located in band p13 of chromosome 17, and often described as 'the guardian of the genome' providing the most important genetic defences against CANCER. These include induction on its activation of (i) cell-cycle arrest (with features of cell senescence) in sarcomas and (ii) APOPTOSIS in lymphomas. These are responses to cellular stresses such as DNA damage, and although the cascade of events signalling from DNA lesions to P53 stabilization and activation is controversial, poly(ADP-ribosylation) of different nuclear acceptors by PARP1 is an early event when a single-strand DNA lesion is produced. Another almost universal feature of *p53* activation in malignant tumours is continuous oncogenic signalling ('oncogene addiction'; see ONCOGENES).

p53 encodes a DNA-binding protein (P53) with approximate molecular mass 53,000 which acts as a transcription factor for a handful of genes, one of which encodes the cyclin-binding protein P21 which arrests the CELL CYCLE at the major G1–S-phase checkpoint. P53 protein may normally offer protection against GENOMIC INSTABILITY and chromosome rearrangements by removing cells with unrepaired, or slowly repaired, double-stranded DNA breaks, either by APOPTOSIS or by permanent blockage to the cell-division cycle. It binds to DNA as a tetramer (a pair of dimers), many faulty P53 proteins suffering from mutations in the area where the two units of the dimer join. In the great majority of human tumour cells that are mutant at the *p53* locus, the alternative wild-type allele has been cut out and replaced by a copy of the mutant allele by gene conversion. In normal, unstressed cells, P53 levels must be kept very low, since high levels would shut down cell proliferation or induce apoptosis. It is regulated by numerous post-translational modifications, including lysine methylation. In human cells, the histone lysine-specific demethylase LSD1 interacts with P53 to repress P53-mediated transcriptional activation and to inhibit its role in promoting apoptosis.

Mutations in *p53* have been detected in about one-half of all human cancers, especially those of the lung and colon/rectum (see LI–FRAUMENI SYNDROME). It is a mutagenic target of people who smoke; but even when mutated, the p53 pathway remains intact. This makes pharmacological reactivation of *p53* a promising target of drug designers.

Almost alone, P53 is entrusted with the task of receiving stress signals that indicate (i) a lack of nucleotides, (ii) UV-irradiation, (iii) ionizing radiation, (iv) oncogene signalling, (v) hypoxia, (vi) blockage of transcription, and (vii) aberrant growth signals, notably those deregulating the pRb–E2F cell cycle control pathway. In these situations, its level in the nucleus increases (by post-translational stabilization) and may cause a cell to enter any of the following: cell cycle arrest, the non-growing state termed senescence, or the apoptotic suicide programme.

Several forms of DNA damage can activate the *p53* gene, preventing cell division; but in the absence of functional p53 this check on cell cycle progression is lacking and the DNA damage can be transmitted to daughter cells, increasing the risk of cancer. Aberrant cell growth caused by deregulation of the pRb-E2F cell cycle control pathway also activates p53. P53 is allosterically stabilized by phosphorylation, some molecules then adhering to mitochondrial membranes (especially in cells directed towards apoptosis) and to centrosomes and the spindle apparatus. P53 is degraded in unstressed cells by association with Hdm2 (Mdm2 in mice) which recognizes it as a target for ubiquitylation (degradation) and for export from the nucleus, and by binding to it prevents its action. This gives the p53 protein a brief 20-minute half-life in unstressed cells. However, p53-phosphorylation protects p53 from the attentions of Hdm2, so that its concentration in the cell can increase rapidly, leading to p53-dependent transcription of such genes as $p21^{WAF1/CIP1}$ that inhibit cell-cycling, and *PUMA* and *PIG3* that control apoptosis. Because the Hdm2 gene is amplified in sarcomas and human lung tumours, by mechanisms that remain unclear, these tumour cells have a distinct advantage by avoiding apoptosis and the prevention of cell-cycling.

The *14-3-3σ* gene responds to p53 activation after DNA damage, absence of 14-3-3σ resulting in impaired division leading to binucleate and multinucleate daughter cells. 14-3-3σ binds to essential regulators of the cell cycle (e.g., cyclin B1 and cdc2), localizing them in the cytoplasm after DNA damage and preventing their entry to the nucleus where they function. The 14-3-3σ protein is lost or inactivated early in the development of several human tumours, including breast and prostate cancers, suggesting that it is a tumour suppressor. See DNA REPAIR MECHANISMS, VIRUSES.

***p57*^KIP2^** A cyclin-dependent kinase inhibitor.

$P2X_3$ receptor An ion channel in the plasmalemmas of NOCICEPTORS which opens on binding extracellular ATP released from adjacent damaged cells. Such channels desensitize in less than a second unless associated with certain other membrane receptors. Another related receptor, $P2X_4$, is a ligand-gated cation channel, activated by ATP, that seems to be required for neuropathic pain. It is found on activated microglia in the spinal cord. See PAIN.

pacemaker Most often, activity of a neuron is triggered by neurotransmitter released by other neurons. But some neurons are spontaneously active (*myogenic*), firing rhythmic action potentials at a particular endogenous frequency, which may be modified by environmental stimuli. Such *pacemaker cells* may sometimes entrain others around them to generate action potentials with the same frequency. The pacemaker of the heart, the SA NODE (see for artificial pacemaker), is modified cardiac muscle tissue rather than nerve tissue; but neurons with pacemaker properties occur in the SUBSTANTIA NIGRA, in the HYPOTHALAMUS and MEDULLA and probably more widely in the human body (see CENTRAL PATTERN GENERATORS). In most pacemaking neurons, spontaneous sub-threshold depolarization is possible because sodium enters through ION CHANNELS that can open at membrane potentials lower than the threshold. These channels may be hyperpolarization-activated cation (I_h) channels, voltage-gated sodium channels, or both (or calcium channels, in the substantia nigra). The current that enters then depolarizes the membrane to the threshold for action potentials.

Pacinian corpuscle Rapidly adapting MECHANORECEPTOR (see Fig. 120) of the deep dermal tissues of both hairy and hairless skin, extremely sensitive to pressure vibrations (detecting very small amplitudes of the order of 1 μm), particularly between 150 and 300 Hz. They comprise relatively large (≤2 mm long x 1 mm wide) neuronal end structures enclosed by a connective tissue capsule arranged in 20–70 onion-like layers, apparently making the corpuscle insensitive to low-frequency stimuli. When the terminus of the nerve fibre within the capsule is deformed by pressure, ION CHANNELS are opened creating a RECEPTOR POTENTIAL, which is depolarizing. If the depolarization is sufficient, the local current within the capsule spreads to the first node of Ranvier, within the capsule, and may then cause an action potential to be produced and propagated. With the capsule removed, the adaptation rate is slowed. Release of pressure may also generate action potentials. Also found in tendons and fascia of muscles, periosteum and joint capsules, they relay proprioceptive information about joint positions, transmitting signals along myelinated type Aβ fibres (see AXON), which can carry up to 1,000 impulses per second. Compare MEISSNER'S CORPUSCLES, MERKEL CELLS, RUFFINI ENDINGS.

pain Two modal types of pain are recognized: fast (sharp, acute, jabbing) pain ('good' pain), and slow (chronic, dull, throbbing, nauseous) pain ('bad' pain). Some believe that the former evolved to allow reflex withdrawal from a potentially damaging stimulus, while the latter evolved to permit part of the body to be reflexly immobilized while it heals (e.g., from fracture or gangrene). Others hold that chronic pain serves no defensive, or otherwise helpful, function.

Pain receptors are all non-adapting free nerve endings, eliciting stimuli normally being of three kinds: mechanical, thermal and chemical (see NOCICEPTION). Distinct receptors and pathways are associated with the pain types. Fast-sharp pain fibres are of the Aδ variety (see AXONS) and transmit at between 6 and 30 m sec^{-1}, usually in response to mechanical and thermal stimuli; slow-chronic pain fibres are type C fibres, transmitting at velocities of 0.5-2.0 m sec^{-1}, usually in response to chemical stimuli. Both fibre types enter the spinal cord via Lissauer's tract and make synapses in the dorsal horn of the spinal cord; but they do so in different layers (laminae). Second order fibres decussate (see REFERRED PAIN, Fig. 147) and pass contralaterally via the anterolateral system (particularly the spinothalamic tract, see SPINAL CORD, Fig. 167) up to the brainstem. Most slow pain fibres terminate in the RETICULAR FORMATION and thalamus. But the vast majority of fast pain fibres continue to other basal areas of the brain and to the somatosensory areas of the cortex.

Tissue damage releases bradykinin, proteolytic enzymes, potassium ions and ATP (see $P2X_3$ RECEPTOR), all of which are thought to cause intense pain of the slow variety. Serotonin, histamine, substance P, prostaglandins, acids and acetylcholine all excite pain. Tissue ISCHAEMIA is a common cause of pain, especially when associated with high metabolic rate of the tissues whose blood supply is reduced or occluded. Lactic acid build-up may be responsible; but again, local tissue damage may be the immediate cause. Bradykinin and prostaglandin E_2 reduce the threshold of nociceptors to mechanical and thermal stimuli, making the site of injury more painful (hyperalgesia). Such SENSITIZATION of pain signalling systems is a key feature of chronic inflammatory pain, in which normally non-painful stimuli elicit pain (allodynia). These sensory changes can occur at sites of injury and in surrounding tissues, suggesting that the sensitization occurs within the CNS as well as within nociceptor terminals. Similarly, in 'type 1 chronic pain', what was a minor injury develops into an excruciatingly painful experience in which INFLAMMATION results in the part of the body becoming permanently swollen and immobilized. This so-called 'learned pain' occurs where the very act of attempting to move the body part is associated with pain. It has recently been shown that, similar to some cases of referred pain, these patients benefit from MIRROR VISUAL FEEDBACK. In 2005, researchers reported using fMRI to teach people with chronic pain to monitor and control their own brain activity. The scanner showed a flickering flame that reflected activity in the patient's right anterior cingulate cortex, a region implicated in pain perception. Those who best learned to minimize the flame reported the greatest reduction in pain symptoms. As a control, those who learned to minimize the flame when this monitored the level of activity in the posterior cingulate cortex (not associated with pain processing) experienced no such pain reduction (see BIOFEEDBACK). Often termed *neuropathic pain*, it typically develops through peripheral nerve injury (PNI) during surgery, bone compression, in cancer, diabetes or infection. Such pain is often refractory to conventional analgesics. Growing evidence indicates that a loss of SYNAPTIC INHIBITION in the dorsal horn of the spinal cord is a critical factor in such chronic pain. A rational therapy might be to attempt reversal of this disinhibition, and work on 'knock-in' mice has identified specific $GABA_A$ receptor subtypes (see GABA) as keys to this spinal pain.

Increased CYTOKINE production in tissues causes pain, enabling information to pass from the immune system to the brain and from there to the efferent VAGUS NERVE. Growing evidence supports the view that PNI leads to activation of the microglia in the spinal cord with resulting up-regulated expression of several cell-surface molecules, including the ATP-binding $P2X_4$ receptor, and of the cytoplasmic p38 MAPK – both apparently key molecules associated with tactile allodynia. The first gene to be linked to neuropathic pain was *GCH1*, which controls the production of BH4 (tetrahydrobiopterin), a cofactor required for the synthesis of NITRIC OXIDE. Its administration to rodents increases excitability of neurons. Work on the rare familial pain syndrome erythromelalgia, in which a person feels excruciating pain in hands and feet when exposed to slight warmth, indicated that a mutation in a sodium channel causes sensory neurons to fire with little provocation and raises the possibility that there may be single nucleotide polymorphisms affecting people's pain thresholds. Perhaps their particular combination of genetic mutations explains why only 5–15% of those wounded in car accidents, by gunshots, or suffering from shingles, develop chronic neuropathic pain.

The enzyme catechol-*O*-methyltransferase is thought to modulate the μ-opioid system that helps control pain, and single nucleotide polymorphisms in its encoding gene have been linked to differences in the risk of developing temporomandibular joint disorder, a form of musculoskeletal pain. See ANALGESIA, EICOSANOIDS, ENDOCANNABINOIDS, ENDORPHINS, ENKEPHALINS, TRP RECEPTORS.

palaeocortex (paleocortex, archaecortex) The 'old' CEREBRAL CORTEX, which includes the OLFACTORY CORTEX, resembling the forebrain cortex of non-mammalian vertebrates in having three or fewer cell layers.

palate (adj. palatal) The *hard palate*, the bony partition between the oral and nasal cavities, and the anterior portion of the roof of the MOUTH, is formed from the maxillae and palatine bones, and covered by a mucous membrane. The *soft palate*, forming the posterior portion of the roof of the mouth, is an arch-shaped muscular partition between the oropharynx and nasopharynx and partly lined by mucous membrane. The partition between the nasal and oral cavities enables humans to chew and breathe at the same time (and enables babies to gaze at their mother while breast-feeding and breathing). See PHARYNX.

palatine bones Paired viscerocranial bones (facial bones) of the SKULL.

palindromic DNA Any stretch of DNA whose nucleotide sequence 'reads' identically on both strands when read in the same (e.g., 3′ → 5′) direction. Examples include many restriction enzyme recognition sites. See Y CHROMOSOME.

pallidal (adj) Referring to the globus pallidum (see BASAL GANGLIA).

pallidum (globus pallidum) See BASAL GANGLIA.

pallium Tissue layer giving rise to the CEREBRAL CORTEX (see Fig. 44) and lining the dorsal and medial surfaces of the lateral ventricle of a 7-week embryo. The evolutionarily more ancient olfactory *palaeopallium* (archaepallium) lies immediately lateral to the corpus striatum. The *neopallium*, non-olfactory and expanded in mammals, lies between the hippocampus and the palaeopallium. The pallium is further divided into the medial pallium (giving rise to the hippocampal formation, limbic lobe), the dorsal pallium (giving rise to the neocortex), the lateral pallium (giving rise to the olfactory/piriform cortex), and the ventral pallium (giving rise to the claustrum and parts of the amygdala).

Programmes of regional identity and morphogenesis in the pallium are influenced by signalling centres, located initially along the edges of the midline of the neural plate and then along and flanking the midline of the vesicles of the telencephalon. Sonic hedgehog, expressed in the ventral telencephalon and hypothalamus, is involved in regionalization of the subpallium and regulates morphogenesis and patterning in the pallium. Bone morphogenetic protein (Bmp) and Wnt families control patterning of the medial and dorsal pallium, including the choroid plexus. In the neopallium, waves of neuroblasts migrate to a position beneath the pia mater and then differentiate into neuroblasts. The next wave to arrive migrates through the early-formed neuroblasts to reach a subpial position, and succeeding arrivals achieve increasingly superficial positions so that at birth the cortex has a stratified appearance.

Pan The genus for CHIMPANZEES. See also PRIMATES.

pancreas A mixed gland, originating as two pancreatic buds (dorsal and ventral) from the endodermal lining of the duodenum at ~30 days of development. Rotation of the duodenum brings these together and their ducts fuse to form one main duct which, with the bile duct, enters the duodenum.

The major part of the organ is a compound acinar gland whose exocrine products are twofold: (i) an alkaline aqueous component produced mainly by columnar epithelial cells lining the smaller ducts and under the control of SECRETIN; and (ii) an enzymatic component (involved in digestion of food) produced by the acinar cells at the termini of the ducts under the control of CHOLECYSTOKININ and including the proteolytic enzymes TRYPSIN, CHYMOTRYPSIN and CARBOXYPEPTIDASES (secreted as pro-enzymes), pancreatic amylase (unlike salivary amylase, able to digest uncooked as well as cooked starch), pancreatic LIPASE, ribonuclease, deoxyribonuclease and elastases. Although, as with gastric secretion, pancreatic secretion is regulated both by the activity of the vagus (favouring zymogen granule release from the acinar cells and increase in blood flow) and by the hormones mentioned, the latter play the more important role. The endocrine component of the organ comprises pancreatic islets (islets of Langerhans), whose cells develop from parenchymatous tissue within the pancreatic buds. The two major cell types are INSULIN-secreting β-cells (insulin secretion begins at ~5 months) and GLUCAGON-secreting α-cells. δ-cells secrete SOMATOSTATIN.

β-cells function as fuel sensors, releasing insulin in response to integrated signals from nutrients in the plasma (glucose, amino acids), hormones (insulin, glucagon-like peptide 1, somatostatin, adrenaline), and neurotransmitters (noradrenaline, acetylcholine). Normally, insulin secretion by β-cells is under autonomic control, through excitation by the parasympathetic and inhibition by the sympathetic division. Lesions in the ventromedial HYPOTHALAMUS in animals cause overstimulation by the parasympathetic and reduction in sympathetic inhibition leading to oversecretion of insulin and excessive storage of triglycerides in adipose tissue. But since β-cells have their own insulin receptors, insulin can stimulate its own (autocrine) release; and glucose exerts a permissive effect for other modulators of insulin secretion. The β-cells take up glucose via GLUT-2 glucose transporters, whereupon phosphorylation by GLUCOKINASE leads to entry of the glycolytic pathway and production of acetyl-CoA and ATP within mitochondria. Raised ATP levels close the cell's ATP-sensitive K^+ channels (mediated by the sulphonylurea receptor), reducing loss of K^+ ions, depolarizing the cell membrane and opening voltage-gated Ca^{2+} channels. Raised intracellular Ca^{2+} ion levels then trigger exocytosis of insulin secretory granules and release of insulin into the tissue fluid and blood. Similarly, elevated amino acid plasma levels and direct depolarization of the β-cell membrane can raise intracellular Ca^{2+} levels. Catecholamines and somatostatin inhibit insulin secretion via G protein-coupled receptor pathways, inhibition of adenylyl cyclase, and modulation of the gating of the β-cell cation channel. When the β cell is healthy, the adaptive response to INSULIN RESISTANCE involves changes in both function and mass resulting in ~fivefold increase in insulin release than in insulin-sensitive individuals, resulting in normal glucose tolerance. But when β-cells are dysfunctional, impaired glucose tolerance, impaired fasting glucose and (in the extreme) type 2 diabetes mellitus result. Causal factors that have been suggested include inflammation and impaired IRS-PI3K signalling (see INSULIN). This extends the 'β cell exhaustion' hypothesis, which states that type 2 DIABETES MELLITUS results when pancreatic β cells can no longer meet the heightened demand for insulin secretion imposed by insulin resistance. Liver X receptor/retinoid X receptor (LXR/RXR) signalling inhibits cell proliferation, and induces apoptosis, in pancreatic β-cells. In addition to mutations in the GLUCOKINASE gene and in *G6PC2* (whose gene product IGRP is selectively expressed in pancreatic islets; see BLOOD GLUCOSE REGULATION), affecting β-cell function, mutations in at least five additional genes encoding transcription factors required for β-cell development are known. *G6PC2* may regulate fasting plasma glucose level by modulating the set-point for glucose-stimulated insulin release by the β cells. Genes for hepatocyte nuclear factors 1α (*HNF1α*) and 1β (*HNF4α*) are associated with a more severe, progressive form of DIABETES, those in *HNF1α* being the most common cause of maturity-onset diabetes of the young (MODY) in most populations.

α-cells respond to decreased plasma glucose, raised plasma amino acids, cholecystokinin, and sympathetic and parasympathetic stimulation by increased glucagon secretion. Insulin and somatostatin have negative effects on this. See PANCREATIC POLYPEPTIDE, PANCREATITIS.

pancreatic polypeptide (PP) A 36-amino acid hormone whose detailed effects are unknown. It is a satiety peptide (see APPETITE, Fig. 14) secreted, mainly by the pancreas, after a protein-rich meal, during fasting, prolonged exercise, and diabetes. Initially, it increases secretion of pancreatic enzymes; but it then decreases this (i.e., it is biphasic). It is an antagonist of CHOLECYSTOKININ and promotes gastric emptying and gut motility, relaxes the pyloric and ileocaecocolic sphincters, the colon, and gall bladder. See PYY.

pancreatitis Inflammation of the PANCREAS. Its more severe, acute, form is often associated with heavy alcohol intake or biliary tract obstruction, symptoms including increased concentrations of pancreatic digestive enzymes (e.g., amylase) in the blood and their decrease in the gut. Pancreatic cells may release trypsin instead of trypsinogen (or insufficient amounts of trypsin inhibitor) so that pancreatic cells are digested. The disorder favours promotion of tumours of the pancreas. See CYSTIC FIBROSIS.

pandemic disease An epidemic affecting all or most countries in the world at the same time. See DISEASE; compare ENDEMIC DISEASE.

Paneth cells One of the four differentiated cell types of the CRYPTS OF LIEBERKÜHN, lying at the bottom of the crypt (SMALL INTESTINE, see Fig. 159) and secreting antimicrobial agents (see INNATE IMMUNITY).

panic See FEAR AND ANXIETY.

panmictic Describing a diploid population in which each individual of a particular sex has an equal probability of producing offspring with any other member of the opposite sex in the POPULATION.

pantothenic acid See VITAMIN B COMPLEX.

Papez circuit Discovered by James Papez in the 1930s, a neural circuit formerly considered to be a possible neurological basis for an 'emotion system', in which the CINGULATE CORTEX projects to the hippocampus, closing a loop (mammillary bodies–anterior thalamic nuclei–cingulate cortex–hippocampus–hypothalamus), the CA3 axons of the hippocampus projecting to the hypothalamus via the fornix. It is now thought to be involved in explicit learning during emotional states (e.g., MEMORY of the location of a wasps' nest after being stung). See CONSCIOUSNESS, LIMBIC SYSTEM.

papillae **1**. Of hair; **2**. renal; **3**. of tongue.

papillomas (adenomas) Benign skin cancers.

papillomaviruses Human papillomaviruses (HPVs) are small non-enveloped viruses with an icosahedral capsid and a circular genome of double-stranded DNA. There are ~100 types, loosely associated with types of cutaneous, genital and laryngeal warts. 'Early' proteins involved in establishing infection include E5 (which prevents acidification of the endosome), and E6 and E7 which are oncogenic (e.g., causing cervical cancer). See VIRUSES.

PAR1, PAR2 PSEUDOAUTOSOMAL regions of the SEX CHROMOSOMES, lying at the tip of the short arm of the X chromosome and the tip of the long arm of the Y chromosome, respectively.

paracellular transport Transport of solutes and water through the modified TIGHT JUNCTIONS between epithelial cells if solute concentration is raised sufficiently to drive diffusion in the right direction (e.g., across the mucosal epithelium of the ileum after a meal). Contrast TRANSCELLULAR TRANSPORT.

paracrine cells (pl. signalling) Cells releasing signal molecules that affect nearby cells as opposed to signalling to themselves or to distant cells. Compare AUTOCRINE SIGNALLING, ENDOCRINE GLANDS.

parafollicular cells (C cells) Scarce cells of neural crest origin and located in the THYROID GLAND, mostly in its central region, secreting the hormone CALCITONIN.

parahippocampal gyrus The limbic region of cortex (see BRAIN, Fig. 29) and part of ENTORHINAL CORTEX (see HIPPOCAMPUS and Fig. 91).

parallel mutations Any mutation that independently generates a derived state already observed within the population.

parallel processing Possibly best understood in the context of visual processing, although also occurring in the SOMATIC SENSORY SYSTEM, the subdivision and abstraction of different aspects (submodalities) of complex sensory information into distinct but interdependent neural pathways (subroutines) which, somehow, are subsequently combined together (the BINDING PROBLEM) and presented for perception simultaneously, as a complex whole. It seems to depend upon circuitry in which interactions between repeating modular units is restricted to neighbours. It contrasts with serial processing, in which a task is similarly 'analysed' into several subroutines each of which must be completed in sequence, one after the other. Parallel processing has the advantage of being the faster method. Once considered to be serial, the processing involved in VISUAL PERCEPTION is now thought to involve at least three parallel processing systems, relating to depth and form (shape), motion, and colour. These pathways extend beyond the PRIMARY VISUAL CORTEX (V1; see Fig. 142) to the EXTRASTRIATE CORTEX (V2 and V4; see Fig. 72). Parallel processing has also been identified in, at least, the RETINA, CEREBELLUM, BASAL GANGLIA, THALAMUS, and pathways involved in HEARING (see ATTENTION).

paralogons Distinct chromosomal regions within a genome that share a set of paralogues. See PARALOGOUS GENES.

paralogous genes, chromosomes and chromosome segments Non-allelic genomic segments with very similar primary DNA sequences (sequence HOMOLOGY) arising from a duplication event (see GENE DUPLICATION) and subsequent descent with modification (see NEUTRAL THEORY OF MOLECULAR EVOLUTION, NATURAL SELECTION). Compare ORTHOLOGUES; see HUMAN GENOME, Fig. 97.

paramesonephric ducts See MÜLLERIAN DUCTS.

Paranthropus aethiopicus A robust AUSTRALOPITHECINE from Ethiopia and Kenya, slightly earlier in date than *P. boisei*, with a more prognathous face and with the smallest brain size (~410 mL).

Paranthropus boisei A robust AUSTRALOPITHECINE from eastern Africa, fossils dated from ~2 to 1.2 Mya. Slightly taller, at 1.3 m, than *Paranthropus robustus*, but apparently no heavier; with an even broader face, incisor-like canines and large, molar-like premolars.

Paranthropus robustus A robust AUSTRALOPITHECINE from southern Africa, fossils dated from ~2 to 1 Mya, with a flat, broad face, very large molars, rather small incisors and canines, a sagittal crest, a very tall and thick lower jaw, and a zygomatic arch far forwards on the upper jaw. Estimated to have been ~1.2 m in height, and weighing 40–80 kg, jaw sizes suggest that the size differences between males and females may have been equivalent to those between male and female gorillas today (males being up to twice the size of females).

paraphyletic A grouping of taxa that contains a subset of descendants from a common ancestor, i.e., excludes some members of the same clade. Paraphyletic groups include the common ancestor of the taxa they contain whereas POLYPHYLETIC groups do not. But only in MONOPHYLETIC groups is the ancestor unique and universal.

paraplegia Paralysis of the lower body. Compare QUADRIPLEGIA.

parasagittal A vertical plane (see ANATOMICAL PLANES OF SECTION, Fig. 9) that does not pass through the midline and divides the body, or organ, into unequal left and right portions.

parasomnias Movement, or motor behaviour, that occurs unexpectedly during NREM SLEEP when motor pattern generators become activated beyond the ability of the brain to quell motor output by inhibition. They include sleep walking, sleep talking and tooth grinding.

parasympathetic nervous system One of the two divisions of the AUTONOMIC NERVOUS SYSTEM.

parasympathomimetic A drug, also called a cholinergic, prescribed to stimulate the parasympathetic nervous system. Used to stimulate urinary bladder contraction.

parathyroid gland Any of four (usually) small endocrine glands embedded in the posterior surfaces of the lateral lobes of the THYROID GLAND. The inferior parathyroid develops from the fourth PHARYNGEAL POUCH (see Fig. 134) and the superior parathyroid develops from the fourth. It exerts its major Ca^{2+}-reabsorbing effect on the distal convoluted tubules of KIDNEY nephrons. See PARATHYROID HORMONE.

parathyroid hormone (PTH) Secreted by the PARATHYROID GLAND, PTH increases extracellular calcium (Ca^{2+}) levels and reduces extracellular PHOSPHATE ion levels. Elevated Ca^{2+} levels inhibit, and reduced Ca^{2+} levels stimulate, its release. It increases the number and activity of OSTEOCLASTS, mobilizing Ca^{2+} and phosphate ions into the extracellular fluid. PTH decreases the reabsorption of phosphate by kidney tubules so that a greater proportion of the tubular phosphate is lost in the urine. However, PTH increases the rate of Ca^{2+} reabsorption by DISTAL CONVOLUTED TUBULES. It also increases the rate of endogenous active VITAMIN D production. Its lack results in rapid decline in extracellular Ca^{2+} concentration through reduced rate of absorption from the gut, increased excretion by the kidneys and reduced bone resorption. This can lead to death through tetany of the respiratory muscles. See OESTROGENS.

paraurethral glands Mucus-secreting glands in the wall of the URETHRA in females, developing from its cranial part and homologous to the PROSTATE GLANDS of males.

paraventricular nucleus A region of the HYPOTHALAMUS (see Fig. 98) involved in regulation of the autonomic nervous system and in controlling secretion of pituitary thyrotropin and corticotropin. Its magnocellular neurosecretory neurons, with those of the supraoptic nucleus, send axons to the posterior lobe of the pituitary, secreting VASOPRESSIN and OXYTOCIN. It also receives projections from the ARCUATE NUCLEUS.

paraxial mesoderm (unsegmented mesoderm) Thick bands of mesodermal cells forming by the 17th day of development on either side of the notochord dorsal mesoderm by migration from the lateral edges of the NODE and from the cranial end of the PRIMITIVE STREAK and giving rise to SKELETAL MUSCLE. By the start of the 3rd week, it separates into segmental blocks of cells (somitomeres) as the neural streak regresses and neural folds begin to disappear. In the head region, somitomeres form in association with neural plate segmentation, forming neuromeres which contribute to head mesenchyme. From the occipital region, somitomeres develop craniocaudally and develop further into SOMITES.

parietal cells (oxyntic cells) Gastric acid-secreting cells of the STOMACH mucosa, located principally in its fundus and body, the majority secreting only hydrochloric acid only after the arrival there of food. An elaborately branched tubular system derived from the endoplasmic reticulum, and lined by microvilli, fuses to form deep invaginations of the apical membrane. Protons, derived from dissociation of water, are pumped out in exchange for potassium ions by an energetically very costly H^+, K^+-ATPase (see the ion pumps entry in ION CHANNELS), down an electrical gradient but against a huge concentration gradient. Parietal cells are therefore richly supplied with mitochondria. Chloride ions can leave the cells by two routes: diffusion through a chloride channel, and by a K^+/Cl^- symport. Potassium ions therefore move in and out across the cell's apical surface. Hydroxyl ions from the dissociation of water combine with carbonic acid (produced by CARBONIC ANHYDRASE) to generate hydrogen carbonate ions, which leave the cell in exchange for chloride ions at the basolateral surface of the cell, resulting in venous blood leaving the stomach after a meal being more alkaline than arterial blood (the 'alkaline tide').

parietal cortex With the cingulate gyrus, often functionally included in the motor cortex of each cerebral hemisphere because of its connections with the PRIMARY MOTOR CORTEX. Left-side parietal lobe lesions may impair actions requiring a sequence of movements on command (APRAXIA). The posterior parietal cortex is thought to integrate visual, sensory and motor information to determine the location of a movement target (e.g., a piece of food) and how to reach it. The superior parietal cortex appears to be involved in OBJECT RECOGNITION.

parietal lobe One of four major subdivisions of each hemisphere of the CEREBRAL CORTEX, having roles in sensory processes, ATTENTION and LANGUAGE. Lesions in the left lobe may lead to inability to name objects (anomia), inability to read, and inability to calculate (aculculia). Lesions in the right lobe may lead to APRAXIA.

Parkinson('s) disease (PD) The most common hypokinetic disorder, causing a 4–7-Hz tremor, especially of the limbs (which reduces with INTENTIONAL MOVEMENT), an increase in muscle tone, rigidity of all limb muscles, difficulty in initiating movements (akinesia), movements that are made being slow (bradykinesia). Pathologically, it is characterized by loss of pigmented neurons in the SUBSTANTIA NIGRA and the presence of *Lewy bodies* (distinctive cytoplasmic inclusions that immunostain for α-synuclein and ubiquitin). Pigmented neurons are those producing neuromelanin (see MELANINS). The phenomenon of EXCITOTOXICITY has also been implicated, and one theory is that, at least in dendrites, the unrelenting pacemaking-associated calcium entry of the dopamine-producing cells of the substantia nigra constitutes an energetic stress that makes these neurons particularly susceptible to oxidative stress and death. This would be consistent with research indicating that dysfunctional NMDA RECEPTORS are implicated.

The substantia nigra projects to the dorsal striatum. Dopamine-depletion in animal models of PD has revealed major differences in the cellular and synaptic properties of striatal medium spiny neurons (MSNs) in the direct and indirect pathways to the BASAL GANGLIA. Damage to these basal ganglia connections causes generalized increase in muscle tone producing abnormal muscle rigidity. Indirect pathway MSNs are more excitable and express dopamine-dependent and endocannabinoid-dependent long term depression (dd-LTD and eCB-LTD respectively, see SYNAPTIC PLASTICITY). Dopamine depletion in animal models of PD blocks the generation of eCB-LTD, whereas its rescue by a dopamine D2 receptor agonist or inhibitor of eCB degradation improved parkinsonian motor deficits. Dopaminergic neurons of the substantia nigra accumulate MTDNA deletions above the threshold for mitochondrial dysfunction and at the expense of normal mtDNA. Although confirmatory studies are needed, this accumulation appears to be a major factor in the appearance and progression of severe clinical symptoms of PD. These neurons seem to have a propensity to accumulate mtDNA deletions, whereas hippocampal neurons (for example) do not, and it has been suggested that reactive oxygen species (ROS) produced by dopamine metabolism could be a cause of mtDNA breaks (see MITOCHONDRION). ROS are a major cause of cell death in PD. The superoxide anion ($^{\cdot}O_2^-$) is converted by superoxide dismutase to hydrogen peroxide, subsequently reduced to water by glutathione peroxidase. But in the substantia nigra pars compacta (SNpc) of PD sufferers, glutathione concentrations are below half normal while the amount of iron in nucleomelanin is greater. This promotes the Fenton reaction, in which hydrogen peroxide is converted to the highly toxic hydroxyl radical ($^{\cdot}OH^-$).

By 2006, mutations or polymorphisms in mtDNA and at least nine nuclear genes had been identified as causing syndromic PD or affecting PD risk. Of the *nuclear* genes, those for α-synuclein, parkin and four others directly or indirectly involve mitochondria.

Levodopa (L-dopa) is effective in reducing symptoms in most patients with PD since L-dopa is converted to dopamine, which crosses the BLOOD–BRAIN BARRIER and activates the dopamine receptors whose inactivation brings on the symptoms. Sadly, many patients become refractory to L-dopa treatment, probably through down-regulation of dopamine receptors. This is a challenge for drug designers, for it should be possible to target intracellular modulatory proteins downstream of the dopamine receptor to achieve more refined treatment. Hopefully, PET scans will become routine in monitoring drug effectiveness, and suitable PET markers (e.g., radioactive iodine-containing dopamine transporters) are already undergoing testing. There has been some success in treatments involving DEEP BRAIN STIMULATION (DBS) when the targets are the subthalamic nuclei and the globus pallidus.

Hereditary parkinsonism with dementia is caused by loss-of-function mutations in a neuronal P-type ATPase gene, *ATP13A2*, underlying an autosomal recessive form of early-onset parkinsonism with pyramidal degeneration and dementia. Several early-onset disease loci lie in a 9-cM region on chromosome 1p. See NEURAL DEGENERATION, PARKINSONISM, THALAMUS.

Parkinsonism Non-hereditary parkinsonism exhibits neurological symptoms, e.g., hand tremor, muscle rigidity and slowness of movement, resembling those of PARKINSON'S DISEASE, which may be brought on by prolonged treatment with neuroleptic drugs. The insecticide rotenone, which prevents electron transfer from the Fe-S centres to ubiquinone of mitochondrial complex I, causes parkinsonian phenotype.

PARP1 (poly[ADP-ribose] polymerase) An enzyme facilitating DNA repair by binding to DNA breaks and attracting other proteins in DNA REPAIR MECHANISMS. See APOPTOSIS, CHEMOTHERAPY, *p53* GENE.

parthenogenesis Development of an unfertilized egg. It does not occur in mammals, probably because of the importance of GENETIC IMPRINTING in development. But see PARTHENOTES.

parthenotes (pseudo-embryos) Produced from an unfertilized egg cell 'tricked chemically' into becoming diploid, as though it had been fertilized. Non-viable if implanted into uterus – but easier to grow than nuclear transfer embryos (SCNT) and having only half the normal antigen combinations, making it easier to match them to patients. See CLONE, EMBRYONIC STEM CELLS.

parturition See BIRTH.

parvicellular See PARVOCELLULAR references.

parvocellular-blob/interblob pathways Two neural systems extending from the parvocellular layers of the LATERAL GENICULATE NUCLEI (see Fig. 111) to the striate cortex and the extrastriate cortex and forming two of the three parallel processing visual pathways in the VISUAL CORTEX (the other being the MAGNOCELLULAR PATHWAY). The parvocellular-blob pathway originates in PARVOCELLULAR CELLS of the retina and enters layer IVCβ of the striate cortex (V1) where its cells synapse in the BLOBS (which are non-oriented, see ORIENTATION-SELECTIVE CELLS). The parvocellular-blob/interblob pathway has a similar origin, but its cells synapse in the deeper interblobs of area V1, which are oriented. Both pathways project into area V2 (secondary visual cortex), blobs targeting the thin stripe regions while interblobs project to the pale stripe (thin stripe) regions (see EXTRASTRIATE CORTEX). Both pathways eventually project to the INFERIOR TEMPORAL CORTEX. The parvocellular-blob pathway is involved in COLOUR PERCEPTION while the parvocellular-interblob pathway is involved in high-resolution analysis of form (i.e. with *what* is seen), with depth perception and to a lesser extent with colour perception. See OCULAR DOMINANCE COLUMNS.

parvocellular cells (P cells) **1**. One class of retinal GANGLION CELL, projecting to the parvocellular layers of the LATERAL GENICULATE NUCLEUS (see Fig. 111) as the parvocellular pathway. Slow and tonic responses, responding only weakly to movement; but fine achromatic and chromatic characteristics, involved in colour and form perception. See PARVOCELLULAR-BLOB/INTERBLOB PATHWAYS. **2**. One of two classes of neurosecretory cells of the HYPOTHALAMUS, whose projections terminate in the medial eminence, brain stem and spinal cord and releasing small amounts of releasing or inhibiting neurohormones (e.g., CRH, dopamine, TRH, LHRH, GHRH, somatostatin) controlling anterior pituitary function. See MAGNOCELLULAR CELLS (2).

passive immunity An IMMUNITY acquired by transfer of ANTIBODIES across the PLACENTA, in breast MILK, or of antiserum or purified antibodies, rather than by the individual's *de novo* production of antibodies. Contrasted with ACTIVE IMMUNITY.

Patau syndrome A genetic disorder involving TRISOMY of chromosome 13, most frequently of maternal origin (85%). This trisomy is present in 2% of MISCARRIAGES. The condition shares many features in common with Edwards syndrome (caused by trisomy of chromosome 18), both having incidence figures of 1:5,000, with a very poor prognosis (most affected infants die within the first days or weeks of life). Cardiac abnormalities (an atrioventricular canal is common) occur in ~90% of individuals and learning difficulties present if long-term survival ensues. ROBERTSONIAN TRANSLOCATIONS are often responsible in cases where mosaicism or unbalanced rearrangements are involved (see TRANSLOCATION).

Patched (Ptc) The cell surface receptor for the products of the HEDGEHOG GENE family notably, during development, the mammalian *Sonic hedgehog* and *Indian hedgehog* proteins.

patella (kneecap) Small, triangular sesamoid bone developing in the tendon of the quadriceps femoris muscle anteriorly to the rest of the KNEE JOINT. A ligament attaches it to the tibial tuberosity and it serves to increase leverage in this tendon and maintain its position when the knee is bent. For the *patella tendon stretch reflex*, see MUSCLE SPINDLES.

pattern formation, patterning The events enabling a cell to determine its position in relation to other cells, crucial to its initiation of appropriate GENE EXPRESSION. In human embryos, as in those of all vertebrates, regionally-specific cell differentiation depends upon gradients of signal molecules, MORPHOGENS, which have different effects on equivalent target cells at different concentrations. The morphogen in the developing LIMB seems to be sonic hedgehog (see SONIC HEDGEHOG GENE). The source of the morphogen gradient guiding HOX GENE expression in the determination of the craniocaudal axis is the PRIMITIVE NODE (see also AXIS). See RETINOIC ACID, SOMITES.

pattern recognition receptors (PRRs) Transmembrane (e.g., TOLL-LIKE RECEPTORS) and intracellular receptors recognizing conserved molecular patterns on pathogens (e.g., viral nucleic acids, components of bacterial and fungal cell walls, flagellar proteins) which may sometimes be activated by a variety of normal host proteins and signals released by dying cells (see INNATE IMMUNITY). Two new intracellular families of PRRs have been described: NLRs (Nod-like receptors) and RLHs (RIG-like helicases, or retinoic acid-inducible gene-like helicases), comprising soluble proteins which survey the cytoplasm for signs of intracellular pathogens. NOD1 and NOD2 detect bacterial peptidoglycan and then drive activation of mitogen-activated protein kinases (see MAPK PATHWAY) and NF-κB, leading to an inflammatory response. The minimal structural requirement for activation of NOD2 is muramyl dipeptide, released by hydrolases (e.g., lysozyme) from both Gram-positive and Gram-negative bacterial walls. Muramyl peptide also activates a complex termed the NALP3 inflammasome, which includes caspase 1, and results in the processing of proIL-1β to IL-1β, also leading to an inflammatory response. RIG-like helicases are cytoplasmic sensors of virally derived dsRNA, triggering NF-κB on activation.

Pavlovian conditioning See CLASSICAL CONDITIONING, LEARNING.

***PAX* genes (Paired-box genes)** A gene family encoding tissue-specific transcription factors containing a *Paired box* containing a *paired domain* of ~128 amino acids, and usually a partial or complete paired-type homeodomain, and important in early animal development in specification of tissues. Nine *PAX* genes have been identified in association with developmental disorders. In mice, these have important roles in the developing nervous system and vertebral column; in humans, loss-of-function mutations in five *PAX* genes are associated with developmental abnormalities (e.g., *PAX3* with Type 1 WAADENBURG SYNDROME). *PAX1* is expressed by sclerotome (see SOMITES) in response to presence of Sonic hedgehog (SHH) and noggin. It is involved in control of chondrogenesis and formation of vertebrae. The *PAX2* product has been identified with kidney, retina and optic nerve development, mutations causing renal-coloboma syndrome; that of *PAX3* (a neural crest-associated transcription factor gene, important in several lineages) with ear, eye and facial development, being expressed in response to Wnt proteins in the early embryonic dermatomyotome of the paraxial mesoderm (see SOMITES). *PAX3* (hsa 2q35) is involved in MELANOCYTE development. The *PAX5* gene is the most frequent target of mutations in B-cell progenitors that cause acute lymphoblastic LEUKAEMIAS. It is essential for B-lineage commitment and maintenance. The *PAX6* gene encodes at least three protein isoforms, canonical PAX6 being involved in development of EYES (see CRYSTALLINS), other sensory organs and homologous neural and epidermal tissues, usually derived from ectoderm. Mutations in *PAX6* lead to absence of the iris (aniridia). See WAGR SYNDROME.

P cells See PRINCIPAL CELLS.

PCR See POLYMERASE CHAIN REACTION.

PdtIns Phosphatidylinositol. See PHOSPHATIDYLINOSITOL SIGNALLING SYSTEM.

pectoral girdle The shoulder girdle (see Fig. 131), forming part of the appendicular skeleton and comprising the SCAPULA, CLAVICLE and associated muscles and connective tissue. Unlike the PELVIC GIRDLE, it does not directly articulate with the vertebral column, and the sockets (the glenoid fossae) for the FORE LIMBS are shallow and maximize movement whereas those (the acetabula) of the HIND LIMB in the pelvic girdle are deeper and allow less movement. See GAIT, SKELETAL SYSTEM.

pedunculopontine nucleus (PPN) Nucleus of the brainstem tegmentum lying at the boundary of the midbrain and the pons and an important component of the ascending arousal system (see AROUSAL). Its cholinergic efferents project to the relay nuclei and reticular nuclei of the THALAMUS. In lower mammals, it is termed the MESENCEPHALIC LOCOMOTOR REGION as a result of its involvement in quadrupedal locomotion. See BASAL GANGLIA, Fig. 18.

pellagra A multiple vitamin-deficiency disorder, notably of VITAMIN B_3 (niacin), whose symptoms include mental illness, sometimes alongside digestive and skin problems (e.g., flushing of face, neck and hands). Potential liver damage is a risk. For most people, a mere 10 mg per day of vitamin B_3 will prevent the disorder. A gene showing evidence of NATURAL SELECTION in which pellagra is the selective agent is *NADSYN1* (NAD synthetase 1) at 11q13. See MALNUTRITION.

pelvic girdle (hip girdle) That part of the appendicular skeleton forming the attachment for the lower limbs (legs) to the vertebral column and supporting both the vertebral column and pelvic viscera (see Fig. 132). It comprises the two hip (coxal) bones, united anteriorly at the pubic symphysis and posteriorly with the sacrum at the sacroiliac joints. Each hip bone is formed by fusion of the ilium, ischium and pubis. The space enclosed by the rim of the pelvic girdle (termed the *pelvic inlet* superiorly and the *pelvic outlet* inferiorly) is wider and more circular in women than in men and has to accommodate the baby's head at birth.

The male's pelvis is usually more massive than the female's. The sacrum is broader in females, the inferior part being more posteriorly directed while the sacral promontory does not project as far anteriorly as in males. The subpubic angle is $<90°$ in males, and 90° or more in females. Ischial spines are further apart in females, while ischial tuberosities (where posterior thigh muscles

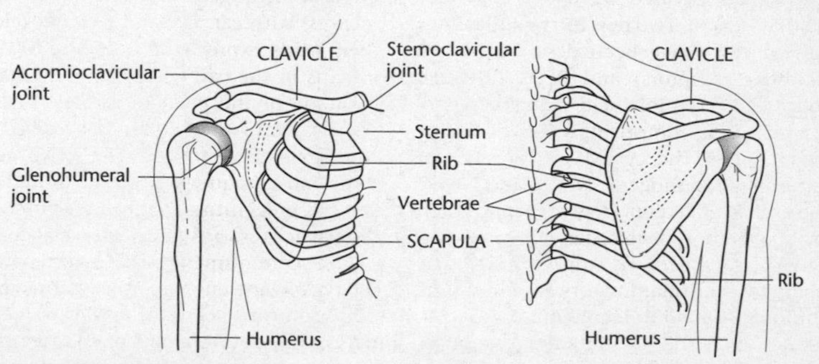

FIG 131. The skeleton of the right PECTORAL GIRDLE, in anterior view (left) and posterior view (right). From *Principles of Anatomy and Physiology*, G.J. Tortora & S.R. Grabowski, © 2000. Reproduced with permission of John Wiley & Sons, Inc.

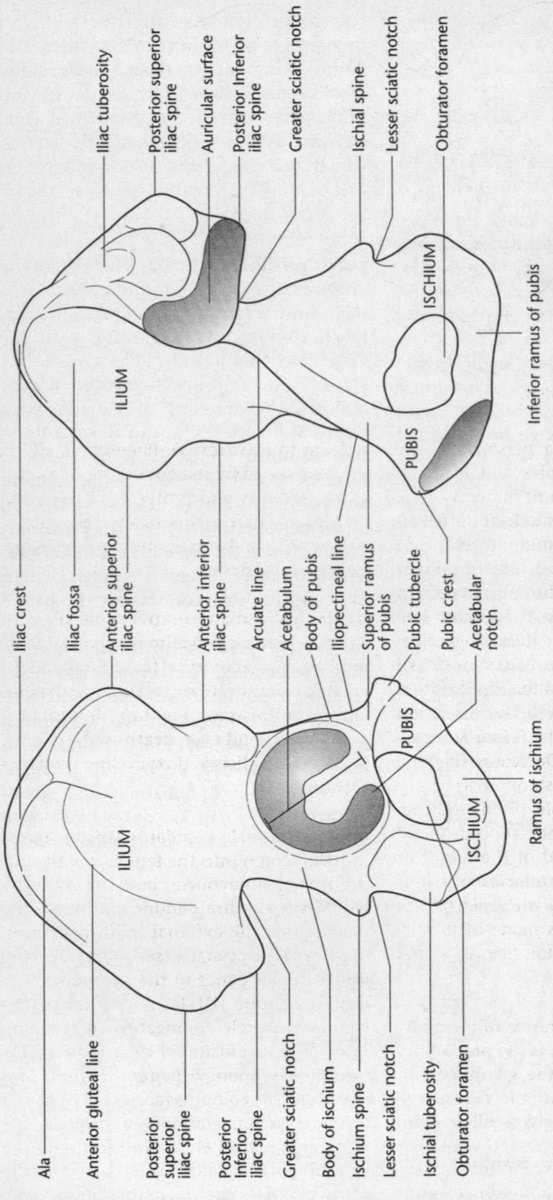

FIG 132. The female PELVIC GIRDLE in lateral view (above left), medial view (above right) and anterior view (page 649). Articular cartilage is shaded. From *Principles of Anatomy and Physiology*, G.J. Tortora & S.R. Grabowski, © 2000. Reproduced with permission of John Wiley & Sons, Inc.

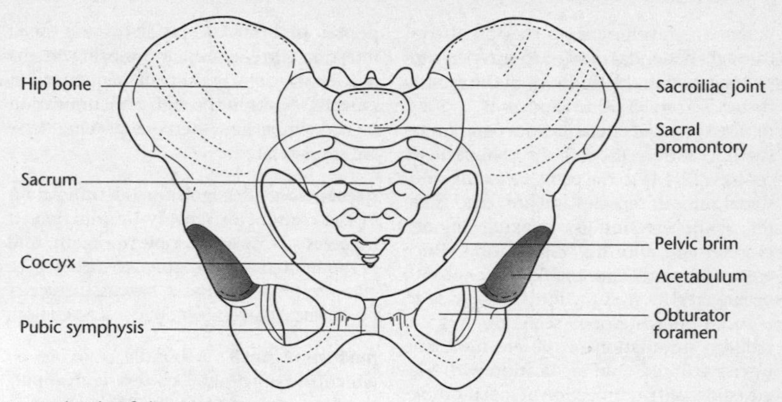

FIG 132.(*continued*) Anterior view.

attach, and on which a person sits) are turned laterally in females and medially in males. It is the origin of many muscles acting on the thigh. Thus, the gluteus maximus originates on the ilium (as well as the sacrum and coccyx) and inserts on the femur; the gluteus medius originates on the ilium and inserts on the greater trochanter of the femur; the deep thigh rotators originate, *inter alia*, on the ischial tuberosity and ischial spine. The iliopsoas originates on the iliac fossa and inserts on the lesser trochanter of the femur and capsule of the hip joint. Decreases in bone mass resulting from osteoporosis make hip fracture more probable, often requiring surgical treatment. The procedure of replacing either the head of the femur or the acetabulum is *hemiarthroplasty*; that of replacing both is *total hip arthroplasty*. The acetabulum prosthesis is made of plastic; that of the femur is metal. See Appendix VIII and Appendix X.

penetrance The penetrance of a character, for a given genotype, is the probability that a person who has the genotype will manifest the disorder; or, the percentage of individuals with a given allele who exhibit the PHENOTYPE associated with that allele. Many mutations exhibit incomplete penetrance (even non-penetrance), not every individual with the genotype expresses the corresponding phenotype (e.g., as in POLYDACTYLY). Phenotypic effects are modulated by the environment as well as by MODIFIERS and protective alleles that can affect the frequency of affected individuals sharing a disease genotype. See ENDOPHENOTYPE, EXPRESSIVITY.

penicillin A small ANTIBIOTIC molecule with a highly reactive β-lactam ring crucial to its antibacterial activity. The most frequent allergic reactions to drugs occur with penicillin, administration by injection causing anaphylaxis and even death (see ALLERGIES). In penicillin allergy, doxycycline is often a substitute.

penis The male copulatory organ, transmitting semen into the female vagina during sexual intercourse, and the route by which the urethra conducts urine to the outside (via the external urethral orifice). Male EXTERNAL GENITALIA (see entry) develop under the influence of the ANDROGEN dihydrotestosterone (DHT). A portion of the genital tubercle elongates, developing into the penis, fusion of the urethral folds forming the spongy (penile) urethra. Two of the erectile columns (corpora cavernosa) form the dorsum and sides of the penis; the third (corpus spongiosum) forming the ventral portion, expanding to form the glans penis (homologous to the CLITORIS). During sexual AROUSAL, the penis becomes enlarged with blood and rigid, action potentials travelling from the spinal cord through pudendal nerves to arteries supplying blood to the sinusoids of the erectile tissues. Acetylcholine and NITRIC OXIDE, released by vagal efferents from parasympathetic centres (S2–S4) or sympathetic centres (T2–L1) to the penis, relax the sinusoidal smooth muscles and are responsible for penile erection by relaxing smooth muscles and allowing vasodilation. Once erect, the penis can enable EJACULATION to occur. Erectile dysfunction can be due to reduced testosterone secretion, or to reduced stimulation of the erectile tissue by nerve fibres. Oral medication with Viagra (sildenafil), or injection of specific drugs to the penis base, can increase blood flow into the sinusoids. Viagra acts by blocking activity of the enzyme that converts cGMP to GMP, cGMP being the immediate cause of muscle relaxation by nitric oxide. The penis and clitoris become swollen during REM SLEEP.

pentose phosphate pathway (phosphogluconate pathway) An important route for cellular production of NADPH and D-ribose. Glucose-6-phosphate is first dehydrogenated by GLUCOSE-6-PHOSPHATE DEHYDROGENASE to 6-phosphoglucono-δ-lactone, $NADP^+$ accepting an electron in an overall equilibrium that favours NADPH synthesis. The second NADPH-generating step is the conversion of 6-phosphogluconate to D-ribulose 5-phosphate, from which D-ribose 5-phosphate can be produced. Tissues synthesizing large amounts of reduced compounds such as FATTY ACIDS recycle the pentose phosphates into glucose 6-phosphate via non-oxidative reactions.

pepsins, pepsinogens 'Pepsin' is the collective term for proteolytic enzymes secreted by the chief cells of gastric glands in the GASTRIC PITS, secreted as inactive precursors (zymogens) known as 'pepsinogens'. In the acid environment of the secreting stomach, pepsinogens are converted to active pepsins, their optimum pH being <3. Gastric pepsins are endopeptidases, liberating peptides and a few free amino acids.

peptic ulcers Ulceration of the small intestine (three-quarters occurring in the DUODENUM) as a result of reduction in mucous secretion through overstimulation by the sympathetic nervous system. Many are stress-related.

perception Recognition (identification) of the content of sensory information. It involves CATEGORIZATION by the brain, and is modulated by the GLUCOCORTICOIDS. See COGNITION, CONSCIOUSNESS, DREAMS, HALLUCINOGENIC DRUGS, TIME PERCEPTION, VISUAL PERCEPTION.

perforant path A bundle of axons by which the entorhinal cortex sends inputs to the HIPPOCAMPUS (see Fig. 91).

periaqueductal grey matter (PAG) Surrounding the cerebral aqueduct in the midbrain, PAG stimulation in rats produces a powerful suppression of PAIN responses. In animal studies, its activation is implicated in organizing defensive responses to natural and artificial predators (see FEAR AND ANXIETY). It is a region rich in endogenous opioids, and important in the supraspinal modulation of nociception, exerted via brainstem serotonergic and noradrenergic neurons (see ANALGESIA). Electrode implantation in the human PAG is used for treatment of intractable pain. It receives projections from the medial HYPOTHALAMUS, and its stimulation can evoke behavioural responses (e.g., affective aggression) via the somatic motor system. Activation of CREB-mediated gene expression here can produce detrimental effects, such as dependence.

pericardium The membrane surrounding the HEART, confining it to the MEDIASTINUM while allowing enough freedom for its contractions. The superficial *fibrous pericardium* (epicardium) is a tough, inelastic, and dense connective tissue preventing overstretching of the heart, its open end fused to the connective tissues of blood vessels entering and leaving the heart. The deeper *serous pericardium* has two layers between which, in the pericardial cavity (~2mm wide), is a thin film of serous fluid, reducing friction. The outer layer is fused to the fibrous pericardium while the inner layer is firmly attached to the heart surface.

pericentric Of DNA regions adjacent to a chromosome CENTROMERE.

pericentric duplications (pericentromeric duplications) See SEGMENTAL DUPLICATIONS.

pericytes MESENCHYME-like cells, associated with the walls of small blood vessels (see VASCULOGENESIS). As relatively undifferentiated cells, they support these vessels; but they can differentiate into fibroblasts, smooth muscle cells, or macrophages if required. In order to migrate into the vessel interstitium, a pericyte must cross the barrier formed by the BASEMENT MEMBRANE, which can be accomplished by fusion with the membrane. They are important in BLOOD–BRAIN BARRIER stability.

perilymph The fluid within the scala vestibuli and scala tympani of the COCHLEA (see Fig. 58). Secreted by arterioles of the periosteum, it has an ionic content similar to that of CEREBROSPINAL FLUID. Compare ENDOLYMPH; see HAIR CELLS.

perimysium The fibrous connective tissue sheath surrounding a bundle of skeletal muscle fibres (muscle FASCICLE).

perineum The area inferior to the pelvic diaphragm between the thighs, extending from the coccyx to the pubis. It is divided into two triangles by muscles. The anterior, *urogenital triangle*, contains the base of the penis and scrotum in the male and external genitalia and opening of the vagina and urethra in the female. The posterior, *anal triangle*, contains the anal opening in both sexes. See EXTERNAL GENITALIA.

perineurium The connective tissue sheath surrounding a nerve FASCICLE.

periosteum The thick, double-layered, connective tissue sheath covering the entire surface of a BONE other than the articular surface.

peripersonal space Perception of the space of (or filled by) objects within reaching distance. See PARIETAL CORTEX, PREMOTOR CORTEX.

peripheral nervous system (PNS) All parts of the NERVOUS SYSTEM other than the brain and spinal cord; divisible into the SOMATIC NERVOUS SYSTEM, AUTONOMIC NERVOUS SYSTEM, and the ENTERIC NERVOUS SYSTEM. The somatic PNS comprises all the spinal nerves innervating the skin, joints and voluntary muscles (apart from the cell bodies of motor neurons, which lie in the central nervous system).

peripheral neuropathy See CONNEXINS, DIABETIC NEUROPATHIES.

peristalsis The coordinated waves of longitudinal and circular muscle contraction passing along the alimentary canal, modified in the stomach by enteroendocrine control, and in the ileum, where it causes segmentation. See ENTERIC NERVOUS SYSTEM.

peritoneum The largest serous membrane (see SEROSA) of the body. The *parietal peritoneum* lines the wall of the abdominal and pelvic cavity; the *visceral peritoneum* covers some of the organs lying in this cavity, forming their serosa. See MESENTERIES.

pernicious anaemia A deficiency disease due to lack of vitamin B_{12} uptake across the colonic mucosa. See VITAMIN B COMPLEX.

peroxisome-activated receptors See PPARS.

peroxisomes Small, quite diverse, single-membrane-bound vesicles, lacking a genome but (like the ER) being self-replicating. They are major sites of molecular oxygen consumption, protecting the cell from the damaging free radicals and hydrogen peroxide produced by some of the reactions generated by the breakdown of amino acids and fatty acids according to the following:

$$RH_2 + O_2 \rightarrow R + H_2O_2$$

Containing the enzyme catalase, peroxisomes catalyse the breakdown of hydrogen peroxide to water and oxygen:

$$2H_2O_2 \rightarrow H_2O + O_2$$

Peroxisomes, like mitochondria, can carry out β-oxidation of fatty acids (see FATTY ACID METABOLISM), but via a slightly different pathway. They are the site of production of the initial precursors in plasmalogen synthesis (see MYELIN SHEATH).

persistent vegetative state See CONSCIOUSNESS.

personal space Perception of the body surface and its contact with the immediate environment. Integrated with the PERIPERSONAL SPACE.

person-to-person outbreaks A propagating outbreak of infectious disease lacking a common source, being maintained by chains of transmission between infected individuals. Contrast COMMON-SOURCE OUTBREAKS.

pertussis See WHOOPING COUGH.

PET (positron emission tomography) scanning A technique for detecting regions of increased metabolic activity in the brain, generating qualitative data rapidly and avoiding microinjection and microelectrode techniques – procedures that are not permitted in research on humans. PET has been of enormous value in research into SLEEP and dreaming, allowing us to 'see the brain in action'. Small doses of positron-emitting radionuclides of metabolites such as glucose or oxygen, injected into the blood, become concentrated in metabolically active parts, such as synapses, where the positrons react with electrons to produce γ rays detected by the PET scanner. Brain density is altered by the increasing blood flow associated with these regions (probably brought about by effects of neuromodulators), and then, as with a CT SCAN, these 'three-dimensional' signals are computerized in ways that can alter both the angle of image and the depth of focusing. The computer then produces vast numbers of pixcels, points measuring density, and plots them in two dimensions as a 'slice', colour-coding the density patterns for easy qualitative inspection. The short life of the label requires expensive cyclotrons to be on-hand for their production. Radioactive neurotransmitters can also be located by this method. The safer and cheaper functional MAGNETIC RESONANCE IMAGING (fMRI) also has greater resolution. However, PET is more useful than fMRI in distinguishing cases of ALZHEIMER'S DISEASE (which reduces metabolism in temporal lobe structures such as the hippocampus) from frontotemporal dementia (which reduces metabolism in the frontal lobes) and may sometimes detect Alzheimer's disease before symptoms appear. It can also detect some CANCERS. See LANGUAGE.

petit mal See EPILEPSY, SEIZURES.

Peyer's patches Large, isolated, clusters of LYMPH NODE-like structures, similar to tonsils, in the basal lamina of the mucosa of the ileum. During embryonic life, haematopoietic cells in the gut exhibit a random pattern of motility before aggregating into the Peyer's patch primordial. They are a major part of the GUT-ASSOCIATED LYMPHOID TISSUES, providing sites where B cells can become committed to synthesizing IgA. Specialized antigen-presenting cells of the lymphoid follicular epithelium, M (multifold) cells, lack a brush border and are specialized antigen-presenting cells, channelling antigens and pathogens from the gut lumen to Peyer's patches. These are often targeted by pathogens (e.g., SHIGELLA). See MUCOSAL-ASSOCIATED LYMPHOID TISSUES.

PFGE See PULSED FIELD GEL ELECTROPHORESIS.

PGC-1 **(PPARγ coactivator 1) genes** A gene family, conserved among chordates, encoding coactivators which are key players in integrating signalling pathways involved in the control of cellular and systemic metabolism, including oxidative metabolism in the MITOCHONDRION, and the maintenance of glucose, lipid and ENERGY BALANCE (see COREGULATORY PROTEINS), and therefore involved in diabetic phenotypes. They are highly responsive to a variety of environmental cues, such as temperature, nutritional status and physical activity, coordinately regulating metabolic pathways and biological processes in a tissue-specific manner (see GENE EXPRESSION), and recently shown to be intimately linked with the CIRCADIAN CLOCK. The first member to be discovered, PGC-1α, is a PPART (see PPARS) interacting protein from brown ADIPOCYTES, where it regulates adaptive thermogenesis. PGC-1α is regulated by the SIRTUIN SIRT1 and coactivates most members of the NUCLEAR RECEPTOR SUPERFAMILY, including nuclear respiratory factors 1 and 2 (NRF-1/2) and oestrogen-related receptor (ERR) proteins, often in a ligand-dependent way. It exerts this function by interacting with target genes and recruiting to them various proteins involved in CHROMATIN REMODELLING, especially histone acetyltransferases such as CBP, p300, P/CAF and SRC-1. It shares extensive sequence identity with PGC-1β in its N-terminal activation domain, its central regulatory domain and its C-terminal RNA-binding domain. PGC-1β has a different pattern of expression from that of PGC-1α, but is also implicated in the mitochondrial metabolism (see MITOCHONDRION, SIRTUINS), and in active thermogenesis. Both are powerful transcription activators when linked to a heterologous DNA-binding domain, or when locked onto and docking any of a number of TRANSCRIPTION FACTORS, proving to be highly versatile coactivators, activating distinct and tissue-specific biological programmes, notably in adipocytes, liver, skeletal muscle and brain. They are known to be involved in regulation of genes involved in haem biosynthesis, ion transport, and in mRNA translation in mitochondria. PGC-1α has been linked to several pathological conditions, e.g., OBESITY, DIABETES and NEURAL DEGENERATION. Both PGC-1α and PGC-1β are alternatively spliced to generate multiple ISOFORMS, although their functional significance awaits analysis.

P-glycoprotein See BLOOD–BRAIN BARRIER, MULTIDRUG RESISTANCE.

pH A measure of the acidity of a solution, defined quantitatively by the formula:

$$\mathrm{pH} = -\log_{10}\{[\mathrm{H}^{+}(\mathrm{aq})]/\mathrm{mol\ dm}^{-3}\}$$

Blood pH is homeostatically regulated (see ACID–BASE BALANCE), and two types of intercalated cells of collecting ducts of KIDNEY nephrons secrete protons (H^+).

phage display libraries A technique in which a gene segment of interest is fused to a gene encoding a coat protein of a bacteriophage, the phage then being used to infect bacteria. Resulting progeny phage particles have coats expressing the fusion protein and a collection of such recombinant phage, each displaying a different fusion protein, is a phage display library. See MONOCLONAL ANTIBODIES.

phagocytosis Cellular ingestion of large particles (≥0.5 μm in diameter) by the invaginated cell membrane. It begins with activation of cell surface receptors by ligands (e.g., the F_C DOMAIN of antibodies) present on the foreign particle, some components of COMPLEMENT, or Gram-negative antigens such as LIPOPOLYSACCHARIDE. Receptor-mediated ENDOCYTOSIS ensues, triggering local polymerization of actin and resulting in formation of cell protrusions (pseudopods) that engulf and seal the particle. It is a process requiring phosphatidylinositol-4,5-bisphosphate (see PHOSPHATIDYLINOSITOL SYSTEM) and local production of phosphatidylinositol-3,4,5-triphosphate for actin nucleation. Sealing, mediated in part by PLC, is coupled with shedding of actin and fusion of phagosomes with the endosome/LYSOSOME compartment, resulting in degradation of the ingested material, undigested material forming residual bodies. Some internalized membrane proteins do not undergo degradation because they are recycled from the phagosome to the plasma membrane in transport vesicles (see VESICLES). Three human cells are 'professional phagocytes': MACROPHAGES, NEUTROPHILS and DENDRITIC cells, each bearing on its cell surface membrane specialized receptors linked to the cell's phagocytic machinery (see Appendix XI). See INFECTION.

phagosome An intracellular membrane-bound compartment with many of the characteristics of an early ENDOSOME. It fuses with lysosomes so that any engulfed bacteria within it are normally digested (see INFECTION, Fig. 100).

phalanx (pl. phalanges) Any bone of the fingers or toes distal to the metacarpals or metatarsals, forming the distal parts of the hand and foot. The thumb (pollex) and big toe (hallux) each have two phalanges; the other fingers and toes each have three. See FOOT, FORE LIMB, LIMBS.

phantom limb syndrome Continued sensation of a missing limb, including pain from it. An 'invention of the brain', the cells normally receiving input from the missing limb raise their excitability as though 'seeking the missing input'. Eventually the severed nerves attempt to grow back into the missing limb, their cut ends becoming tangled and generating spontaneous impulses, but also responding to pressure located as permanent tender spots on the stump. Chemical differences in the new environment lead to increased excitability of the neurons and often lead to barrages of sensory input or over-reaction to sensory signals in the remaining uninjured part of the limb. One currently favoured explanation is that afferent pathways normally occupying areas of the cortex adjacent to that formerly serving the missing limb expand into the 'unused' area (*remapping of referred sensations*). Thus, touching an area on the face or upper arm may produce a tactile sensation as though it emanated from a missing hand. See MIRROR VISUAL FEEDBACK, REFERRED PAIN.

pharmacodynamics Term describing the effects or actions of a drug on the body (e.g., bronchodilation, pain relief).

pharmacogenetics The study of the genetics of DRUG effectiveness.

pharmacokinetics Term used to describe how the body deals with a DRUG; e.g., its absorption into the blood, distribution, breakdown and excretion.

pharyngeal arches (branchial arches) The most characteristic external features of the embryonic development of the head and neck, appearing in the 5th week of development (see Appendix Vb). Initially consisting of bars of mesenchymal tissue separated by deep clefts (pharyngeal clefts), they become invaded by NEURAL CREST cells and develop at the same time as outpocketings, the PHARYNGEAL POUCHES (see Fig. 133 and Appendix VI). Mammals have five branchial arches, each containing an arch artery (numbered 1–4 and 6). Although initially the arterial system forms symmetrically, in the mature mammal the left 4th and 6th arches persist and give rise to the aortic arch and pulmonary trunk while the right 4th and 6th arches regress. The result of this remodelling is the asymmetric development of the great vessels of the ventricular outflow tract, which forms part of the left and right ventricles.

The 1st pharyngeal arch consists of a dorsal portion (the maxillary process), and a ventral portion (the mandibular process, containing Meckel's cartilage). Meckel's cartilage all but disappears, but two small portions give rise to the incus and malleus (see EAR OSSICLES). The mandible is formed by membranous ossification of mesenchymal tissue surrounding Meckel's cartilage, while similar tissue of the maxillary process similarly gives rise to the premaxilla, maxilla, zygomatic bone and part of the temporal bone. Musculature of the 1st arch includes the muscles of mastication (temporalis, masseter and pterygoids). Cartilaginous components of the 4th and 6th arches give rise to the cartilages of the LARYNX. Humans, like all mammals, lack a 5th arch. Although the development of pharyngeal arches, clefts and pouches resembles the formation of gills in fish and amphibians, real gills never form in humans and the term 'pharyngeal' is therefore adopted in place of 'gill'. Indeed, the pharyngeal clefts give rise to only one structure: the external auditory meatus. Pharyngeal arches contribute to the formation of the neck and play an important role in the formation of the FACE. Among the most common congenital birth defects are abnormalities in the development of the cardiovascular system (see CONGENITAL HEART DEFECTS), especially those in the asymmetric remodelling of a transient structure termed the branchial arch artery apparatus that occurs in 6- to 7-week-old human embryos. The transcription factor Pitx2, induced by Nodal, is the product of the only gene known to be asymmetrically expressed in the embryonic tissue that generates the branchial arch. Another factor contributing to normal arch-artery remodelling is asymmetric expression of vascular growth factors such as PDGF and VEGF, inhibition of these factors leading to loss of both left and right sixth arch arteries.

pharyngeal clefts (branchial clefts) See PHARYNGEAL ARCHES.

pharyngeal pouches The human embryo has five pairs of pharyngeal pouches, developing inside the pharyngeal arches as outpockets along the lateral walls of the pharyngeal gut and penetrating surrounding mesenchyme. In contrast to the situation in fishes and amphibians, they do not establish open communication with the external pharyngeal clefts, although they arise opposite each other (see Fig. 133). Their endoderm gives rise to a number of endocrine glands and part of the middle ear (see EUSTACHIAN TUBE). Their arteries derive from the AORTIC ARCHES (see Fig. 134).

pharynx The common opening of the digestive and respiratory systems. It receives air from the nasal cavity and air, food and liquid from the oral cavity. It is connected to the respiratory system inferiorly at the LARYNX, and to the digestive system at the oesophagus. It is divisible into the nasopharynx (posterior to the choanae and superior to the soft PALATE, and into which the Eustachian tubes lead); the oropharynx (extending from the soft palate to the epiglottis); and the laryngopharynx (extending from the tip of the epiglottis to the oesophagus, passing posterior to the larynx). The nasopharynx is lined with a mucous membrane containing pseudostratified ciliated columnar epithelium with goblet cells and its posterior surface contains the pharyngeal tonsil, or ADENOIDS.

phasic muscle fibres Somewhat redundant nowadays when applied to SKELETAL MUSCLE, a term usually restricted to contraction of SMOOTH MUSCLE fibres that has a short duration, of a few seconds, followed by a period of relaxation. Contractions are initiated by spike potentials which may be single or in train, the contraction intensity being determined by the number of spikes in the initiating train. Contraction frequency in a given segment of the gastrointestinal tract is determined by the frequency of the slow wave activity in that segment, peristalsis and segmentation being the dominant phasic activities of the tract. If a skeletal muscle is composed

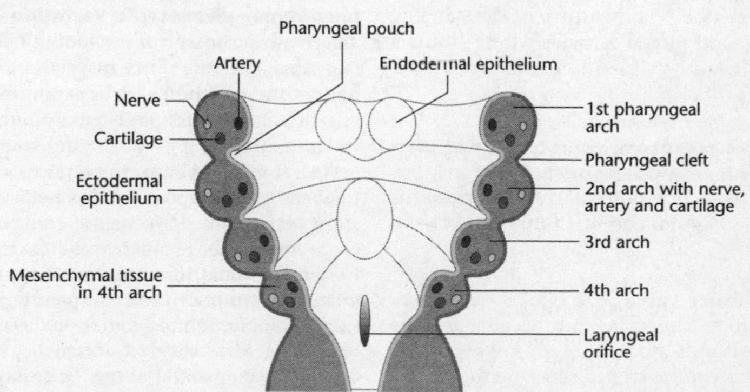

FIG 133. A four-week embryo, showing the roof of the pharynx and four pairs of pharyngeal arches. Each arch contains a cartilaginous component, a cranial nerve, an artery and a muscular component. See PHARYNGEAL POUCHES. Redrawn from *Langman's Medical Embryology*, T.W. Sadler. © 2004 Lippincott, Williams & Wilkins, with permission of Wolters Kluwer.

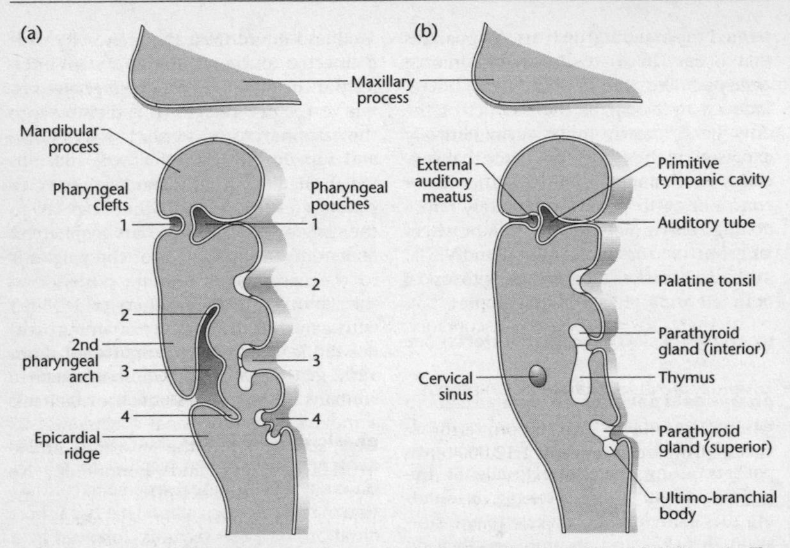

FIG 134. Development of the pharyngeal clefts and pouches between the fourth (a) and fifth (b) weeks. The 2nd pharyngeal arch grows over the 3rd and 4th, burying the 2nd, 3rd and 4th pharyngeal clefts. Structures formed by the various PHARYNGEAL POUCHES are indicated on the right. Arabic numerals indicate pharyngeal arches. Redrawn from *Langman's Medical Embryology*, T.W. Sadler. © 2004 Lippincott, Williams & Wilkins, with permission of Wolters Kluwer.

mainly of slow twitch fibres, it could be termed a tonic (endurance) muscle; if made up mostly of fast twitch fibres, it could be termed a phasic (low endurance) power muscle. Sometimes, a distinction is made between postural muscles, whose contraction is constitutive of the standing state, and phasic muscles whose contraction requires intention (see INTENTIONAL MOVEMENTS).

phasic receptors Receptors which show rapid ACCOMMODATION, generating just one or a few action potentials at the beginning of the stimulus period. Contrast TONIC RECEPTORS.

phenetics The approach to biological classification which takes into account as many observable features of organisms as possible and constructs relationships on the basis of overall similarity and difference. It tends to regard the degree of morphological divergence between two related species (assumed to be adaptive) as more important than their recency of common ancestry. It also gives each characteristic equal weight. Compare CLADISTICS, which has been in the ascendancy as a method of building phylogenies. See PHENOTYPE.

phenotype, phenotypic variation All aspects of an organism, including behaviour, available to sensory inspection. Phenotypic variance is held to be explicable by genetic, epigenetic, and environmental variance (see INHERITANCE, NATURE–NURTURE DEBATE). Adaptive changes comprise only a small minority of the total genetic variation between any two species since most evolutionary DNA sequence change is due to genetic drift (see NEUTRAL THEORY OF MOLECULAR EVOLUTION). Consequently, the extent of phenotypic variation between organisms is not strictly related to the degree of sequence variation: the gross phenotypic variation between human and chimpanzee is far greater than that between the two mice *Mus musculus* and *Mus spretus*, yet the degree of sequence difference is similar in the two cases (see GENE EXPRESSION). Domestic dogs show considerable phenotypic variation despite having little overall sequence variation (~0.15%). An alternative possibility, at the phenotypic level, is that most differences between species are adaptive and fixed by positive NATURAL SELECTION. *Phenotypic discordance*, in which a given genetic trait is expressed in only one member of a pair of monozygotic twins, is especially interesting. AUTISM and BIPOLAR DISORDER are common complex traits with high frequencies of discordance. Compare ENDOPHENOTYPE.

phenotypic discordance See PHENOTYPE.

phenylketonuria An inborn error of metabolism (occurrence in 1:12,000 births in USA), caused by inability of a liver enzyme (phenylalanine hydroxylase) to convert the amino acid phenylalanine to tyrosine, often through its deficiency, causing abnormally high levels of phenylalanine in the blood and brain, and producing severe brain growth stunting if untreated. The usual treatment is low dietary phenylalanine with tyrosine supplement. See HOMOCYSTINURIA, MENTAL RETARDATION, SCREENING.

pheromones Non-volatile chemical stimuli detected by the VOMERONASAL ORGAN (VNO), an anatomical specialization of the nose in terrestrial vertebrates that is distinct from the main OLFACTORY EPITHELIUM. Their detection is thought to be mediated by two distinct families of G-protein-coupled receptors. Despite commercial enterprise, there is no substantive evidence supporting the possibility of rapid, odour-mediated changes in human behaviour analogous to those in the insects that gave us the pheromone concept. Human social systems are also too complicated for human behaviour to be dominated by the drastic releaser effects of pheromones observed in other species, although this is not to say that pheromones did not influence the behaviour and physiology of our hominid/hominin ancestors; or that they might not have subtle influences on humans today. Those released by human females can certainly influence the MENSTRUAL CYCLES of other women. The VNO is present in early human development and the vomeronasal nerves play a vital role in guiding migration of developing LHRH neurons to the hypothalamus. But the epithelial diverticulum in the nasal septum of human adults (referred to as the VNO) is structurally and functionally different from that of the rodent VNO whose vomeronasal sensory neurons (VSNs) have a different transduction mechanism from olfactory sensory neurons, one dependent in part on the transient-receptor-potential channel 2 gene (*Trpc2*). The equivalent gene in humans, *TRPC2*, is a PSEUDOGENE which was removed from functional constraints ~23 Mya, shortly before the separation of Old World monkeys and hominoids. No vomeronasal nerves projecting to the human brain have been identified; nor has the expected target of such neurons, the accessory olfactory bulb, been identified. One of the main candidates for a source of human pheromones is the apocrine-gland secretions of the axilla. When subject to microbial action, these give rise to a mixture of odorants responsible for body odour, including androgen derivatives and volatile acids. See ODOURS, MAJOR HISTOCOMPATIBILITY COMPLEX, SEXUAL ORIENTATION.

Philadelphia (Ph^1) chromosome See LEUKAEMIAS.

phobias A disorder involving irrational, persistent fear of certain situations, objects, activities, or persons. It is an aberrant form of fear learning, the main symptom being an excessive, unreasonable desire to avoid the feared subject. When the fear is beyond one's control, or if the fear is interfering with daily life, then a diagnosis under one of the anxiety disorders can be made (see FEAR AND ANXIETY). By presenting the phobic stimulus in an environment in which fear responses are reduced (e.g., using anxiolytic drugs), phobias may sometimes be completely abolished. See AMYGDALA.

phoneme In human spoken LANGUAGE, the theoretical representation (mental abstraction) of a sound, without reference to its position in a word. In sign languages, the equivalent movements (cheiremes) were termed phonemes after it was realized that the mental abstractions involved are essentially those employed in spoken languages. Compare MORPHEME.

phosphagens See ADENOSINE TRIPHOSPHATE, Fig. 2; HIGH-ENERGY PHOSPHATES.

phosphate Roughly 80% of body phosphate is in the form of calcium phosphate in the BONE matrix. Much of the remainder is within cells, covalently bound to organic molecules such as phospholipids, proteins and carbohydrates, and as important components of DNA, RNA, ATP and NADP and in the phosphodiester bonds of other dinucleotides. Phosphates are hugely important in regulatory enzyme and SIGNAL TRANSDUCTION pathways (see KINASES) as well as in the phosphate buffer system (see ACID–BASE BALANCE). PHOSPHORYLASE and PHOSPHATASE enzymes respectively add and remove phosphates from their substrates. The ability of the kidneys to reabsorb orthophosphate ions ($H_2PO_4^-$, HPO_4^{2-}; commonly abbreviated as P_i) is limited, although decrease in phosphate levels in the extracellular fluid results in a decreased rate of their loss in the urine (see PROXIMAL CONVOLUTED TUBULES). Chronically low phosphate diets can lead to increased rate of reabsorption. Excess phosphate can bind to plasma calcium, decreasing the $[Ca^{2+}]$ of the plasma and leading to decreased production of calcitriol, the hormonally active form of vitamin D. This leads to increased output of PARATHYROID HORMONE (PTH) and demineralization of bone. By promoting bone resorption, PTH can play a major role in regulating extracellular phosphate levels, but it increases phosphate loss in the urine. There is a dynamic and complex interplay between O-LINKED B-N-ACETYLGLUCOSAMINE and orthophosphate, which compete for the same serine or threonine hydroxyl moieties on proteins. See ADENOSINE TRIPHOSPHATE.

phosphatases Enzymes which selectively remove a PHOSPHATE group from a substrate, antagonizing PHOSPHORYLASE activity. Protein phosphatases are affected adversely by reactive oxygen species (see MITOCHONDRION). Compare KINASES.

phosphatidylcholine (lecithin) The predominant phospholipid of the outer leaflet of cell-surface membranes and one of the brain's most abundant structural phospholipids, a component of which is docosahexanoic acid (DHA, see OMEGA-3 FATTY ACIDS). It is also present in the alveolar fluid, secreted mainly by type II alveolar cells, where its surfactant properties enable lungs to be inflated. In the human foetus, secretion does not occur until the 30th week, so premature babies are deficient in lung surfactant (newborn respiratory distress syndrome) and reduced production in adult life can be a factor in adult respiratory distress syndrome. See ALVEOLUS, CHOLINE.

phosphatidylethanolamine (PE) One of the brain's most abundant structural phospholipids, a component of which is docosahexanoic acid (DHA, see OMEGA-3 FATTY ACIDS). See PHOSPHATIDYLINOSITOL SYSTEM.

phosphatidylinositol (PtdIns, PI) See PHOSPHATIDYLINOSITOL SIGNALLING SYSTEM.

phosphatidylinositol-4,5-bisphosphate (PIP_2, $PtdIns(4,5)P_2$) Possibly the most important lipid in the cytosol-facing leaflet of the plasma membrane lipid bilayer. A membrane lipid acting as a docking site for intracellular proteins. See MARCKS, PHOSPHATIDYLINOSITOL 3-KINASE, PHOSPHATIDYLINOSITOL SIGNALLING SYSTEM.

phosphatidylinositol-3,4,5-triphosphate (PIP_3, $PtdIns(3,4,5)P_3$) A membrane lipid acting as a docking site for intracellular proteins. See PHOSPHATIDYLINOSITOL 3-KINASE, PHOSPHATIDYLINOSITOL SIGNALLING SYSTEM.

phosphatidylinositol 3-kinase (PtdIns-3-OH kinase, PI_3-kinase, PI_3-OH kinase, PI3K) Any of a group of kinases (PI3Ks) signalling downstream of multiple cell-surface receptor types, and phosphorylating phosphoinositides to release lipid second messengers that initiate multiple signalling cascades. They are activated in response to such extracellular signals as those activating receptor tyrosine kinases (e.g., INSULIN, INSULIN-LIKE GROWTH FACTORS; see PKB), and G PROTEIN-COUPLED RECEPTORS (e.g., CATECHOLAMINES). PI3Ks are therefore involved in signalling pathways activated by GROWTH FACTORS (e.g., the RAS SIGNALLING PATHWAY), notably signals that are MITOGENS, and in cell motility (e.g., see ANGIOGENESIS). They help to carry the signal into the cytoplasm by phosphorylating inositol phospholipids at the C3 position in the inositol ring, generating the lipids PI(3,4)P_2 and PI(3,4,5)P_3, which are attached to the membrane and serve as docking sites for intracellular signalling proteins on the cytosolic surface of the plasma membrane. It is involved in the formation of endosomes (see VESICLES) and perhaps because of this has important functions in the immune system – its impaired signalling can lead to immunodeficiency, whereas its unrestrained signalling contributes to autoimmunity and leukaemia. As part of the cell signalling pathway that mediates insulin action (the insulin receptor substrate (IRS)-PI3K pathway), the PI3K here is particularly sensitive to inactivation by molecules that promote INFLAMMATION (see ENERGY BALANCE). Besides regulating nutrient utilization in peripheral tissues, IRS-PI3K signalling is also implicated in the neuronal actions of insulin and leptin, which can also be subverted by nutrient excess. It may be that because short-term exposure to an energy-dense and palatable diet interferes with the brain's response to insulin and leptin, the brain's response to on-going increase in body fat stores is impaired. It is Class IA PI3K isoforms that couple to tyrosine kinases, each consisting of a p110 catalytic subunit constitutively bound to one of five distinct p85 regulatory subunits. Of the eight catalytic isoforms existing in mammals, p110α and p110β are ubiquitously expressed, while p110δ (enriched in leukocytes) and p110γ are considered to be promising pharmacological targets for the treatment of autoimmune, inflammatory and allergic diseases, among others. See INSULIN RESISTANCE, OBESITY, PHOSPHATIDYLINOSITOL SIGNALLING SYSTEM.

phosphatidylinositol signalling system Phosphatidylinositol (PtdIns, PI) is unique among membrane lipids in undergoing reversible phosphorylation at multiple sites to generate a variety of distinct inositol phospholipids. It typically represents <15% of the total phospholipids in eukaryotic cells, is synthesized principally in the endoplasmic reticulum and then delivered to other membranes by vesicle transport or via specific transfer proteins. It is an important attachment site for proteoglycans of the extracellular matrix. Reversible phosphorylation of its inositol ring by PHOSPHATIDYLINOSITOL 3-KINASE (PI3K) at positions 3, 4 and 5 results in the production of seven phosphoinositide species including phosphatidylinositol-4,5-bisphosphate (PIP_2, PtdIns(4,5)P_2) and phosphatidylinositol-3,4,5-triphosphate (PIP_3), which are generally less abundant in total by one order of magnitude than PtdIns. Concentrated at the cytosolic surface of membranes, these phosphoinositides recruit specific intracellular signalling proteins to the cytosolic face of the plasma membrane, mediating acute responses; but they also act as constitutive signals helping to define organelle identity (see CELL MEMBRANES, ARF PROTEINS). Each intracellular organelle has a different array of kinases and phosphatases that produce and break down phosphoinositides. This ensures their restricted distribution and prevents transfer of the lipid between cell compartments. Thus, the Golgi has a high level of PtdIns(4)P, but little PIP_2. The phosphoinositides characterizing a given organelle are intimately involved in membrane trafficking (see VESICLES). In addition to classical signal transduction at the cell surface, they mediate and regulate membrane traffic, the cytoskeleton, gene transcription, RNA editing, and membrane permeability and transport functions. They mediate the recruitment of proteins to membranes by using their phosphorylated head groups, specificity increasing when acting as coreceptors with other molecules, such as members of the small GTPase Ras family.

PIP_2 has a claim to being the most important lipid in the plasma membrane. *Inter alia*, it is the source of three second messengers, activates many ION CHANNELS, is a binding site for anchoring PLECKSTRIN HOMOLOGY proteins to the membrane and is involved in endocytosis and exocytosis. It is attracted by the numerous clusters of basic and hydrophobic residues on proteins, to which the Ca^{2+}/calmodulin also binds (see MARCKS), removing PIP_2 from the membrane and freeing it to interact with other important molecules. See INFECTION, INSULIN, INSULIN-LIKE GROWTH FACTORS, PHAGOCYTOSIS.

phosphatidylinositol-3-OH-kinase pathway See PHOSPHATIDYLINOSITOL 3-KINASE.

phosphatidylserine (PS) The predominant phospholipid of the inner leaflet of cell-surface membranes and one of the brain's most abundant structural phospholipids, a component of which is docosahexanoic acid (DHA, see OMEGA-3 FATTY ACIDS). Synthesized primarily on the cytosolic aspect of the endoplasmic reticulm, it nevertheless fails to accumulate on the cytosolic surfaces of organelles of the secretory pathway, possibly because it is translocated to their luminal leaflet. It serves as a molecular 'beacon' for proteins containing the structural C2-domain motif and contributes substantially to the negative charge of the cytoplasmic face of the plasmalemma (see CELL MEMBRANES), promoting recruitment of positively charged, polycationic proteins. Receptors for PS have been discovered and include Tim-1 and Tim-2 in mammalian cells.

phosphenes Sensations of light.

phosphocreatine (PCr) A high-energy phosphagen of skeletal muscle, generated from creatine (Cr) by a transfer of phosphate from ATP by *creatine kinase*. Muscle fibres store ~4–6 × more PCr than ATP, and use it in the regeneration of the fibre's ATP from ADP, especially during EXERCISE of short duration and high intensity. Having a more negative standard free-energy release of hydrolysis than ATP (see Fig. 2), it can drive the reversible creatine kinase-mediated reaction in the direction of ADP phosphorylation:

$$\text{PCr} + \text{ADP} \leftrightarrow \text{Cr} + \text{ATP}$$

Transient increases in ADP within the muscle during muscle contraction shift the creatine kinase reaction toward PCr hydrolysis and ATP production, a reaction which does not require oxygen and reaches a maximum energy yield in ~10 sec. This is known as the ATP:PCR system. PCr supplies ~50% of the fuel for ATP production during a 100-metre sprint, this value decreasing to 6% in a 800-metre run, and reaching zero for longer runs. If it has been resynthesized during a long race, it will be used in the sprint to the finish. CREATININE is a breakdown product of PCr. See OXYGEN DEBT.

phosphodiesterases (PDEs) Enzymes hydrolysing phosphodiester linkages, e.g., those of cyclic nucleotides (e.g., cAMP, cGMP) and of the sugar-phosphate 'backbones' of nucleic acid molecules, of which many types and subtypes exist.

phosphofructokinase-1 (PFK-1) See GLYCOLYSIS.

phosphofructokinase-2 (PFK-2) See GLUCONEOGENESIS.

phosphoinositides (PtdIns) Important membrane lipids, with easily modifiable headgroups. See PHOSPHATIDYLINOSITOL SIGNALLING SYSTEM.

phosphoinositol 3-kinases (PI3Ks) Lipid kinases regulating signalling pathways involved in cell proliferation, adhesion, survival and motility – all important for neoplasia. See PHOSPHATIDYLINOSITOL-3-OH-KINASE.

phospholipase A (PLA) An enzyme present in saliva and tears. Cytoplasmic $PLA_{2\alpha}$ uses phospholipids in the cell membrane to generate ARACHIDONIC ACID, used in prostaglandin synthesis. PLA_2 also converts phosphatidic acid to lysophosphatidic acid (see

LYSOPHOSPHOLIPIDS). It is activated by α_1 ADRENERGIC receptors.

phospholipase C (PLC) Any of a group of enzymes bound to the cytosolic surface of the plasma membrane and activated by extracellular signals to cleave a phosphoinositide in the membrane's cytosolic monolayer into two fragments (see Fig. 135). One of the fragments, DIACYLGLYCEROL, remains in the membrane and helps activate protein kinase C, while the other, IP_3 (inositol 1,4,5-triphosphate), is released into the cytosol and causes the release of CALCIUM ions, a key second messenger, from the endoplasmic reticulum. Phospholipase C-β (PLC-β) cleaves phosphatidylinositol 4,5-bisphosphate, PIP_2, and is activated by certain G PROTEINS while phospholipase C-γ (PLC-γ) catalyses the same reaction but is activated by certain RECEPTOR TYROSINE KINASES. Signal molecules acting via PLC and IP_3 include: acetylcholine (at muscarinic M_1 receptors), α_1-adrenergic agonists, angiotensin II, ATP, glutamate, gonadotropin-releasing hormone, histamine, oxytocin, platelet-derived growth factor, serotonin (at 5-HT-1c receptors), thyrotropin-releasing hormone and vasopressin (please consult these entries). Many TRP channels are activated or modulated downstream of neurotransmitter, hormone or growth factor receptors that stimulate PLC pathways. See ADRENERGIC (SYMPATHOMIMETIC), PHOSPHATIDYLINOSITOL SIGNALLING SYSTEM.

phosphorylases Enzymes (e.g., GLYCOGEN PHOSPHORYLASE) transferring a PHOSPHATE group, often from inorganic phosphate ions, to an organic compound, which is thereby phosphorylated. Phosphorylases are hugely important in both intracellular signalling (e.g., see SIGNAL TRANSDUCTION) and extracellular signalling (e.g., see BLOOD CLOTTING, COMPLEMENT). All KINASES are phosphorylases. Phosphatase enzymes remove phosphate from their substrates, antagonizing phosphorylase activity.

photophobia Not so much 'fear of light' but a condition of abnormal sensitivity to light leading to pain or discomfort. Usually, the IRIS is unable to constrict sufficiently to reduce the light level incident on the retina. It may be a symptom of corneal inflammation, or ocular ALBINISM. Some drugs (e.g., amphetamines and antihistamines) and/or poisons (e.g., strychnine) can cause photophobia by causing pupil dilation. See PUPILLARY LIGHT REFLEX.

photopia, photopic vision See LIGHT ADAPTATION.

photoreactivation See DNA REPAIR MECHANISMS, ULTRAVIOLET IRRADIATION.

phototransducion The process by which the absorption of light energy by photoreceptor cells is transduced into a change in their membrane potentials. See ROD CELLS.

phrenic nerves Nerves arising from the CERVICAL PLEXUS (cervical nerves C3–C5) and innervating the diaphragm, supplying its motor fibres. See VENTILATION.

phylogenetic footprinting The method of identifying genes from the results of GENOME MAPPING by comparing the genomes of different species and detecting conserved sequences. The theory is that such conserved sequences *are* conserved because any alteration in their sequence would be deleterious, and be selected against. Such sequences, it is held, probably encode important products and are, in consequence, gene sequences. See MOUSE MODELS, PHYLOGENETIC SHADOWING.

phylogenetics See PHYLOGENETIC TREES.

phylogenetic shadowing A variant of PHYLOGENETIC FOOTPRINTING, examining sequences of closely related species to indicate conserved coding and non-coding functional regions, while taking into account the phylogenetic relationship of the species compared. It has been used to reveal primate-specific gene regulatory elements that comparisons with distantly related non-human species miss.

phylogenetic trees Phylogenetics comprises all those methods used to construct trees and networks from genetic data. The ultimate aim is to relate groups of mol-

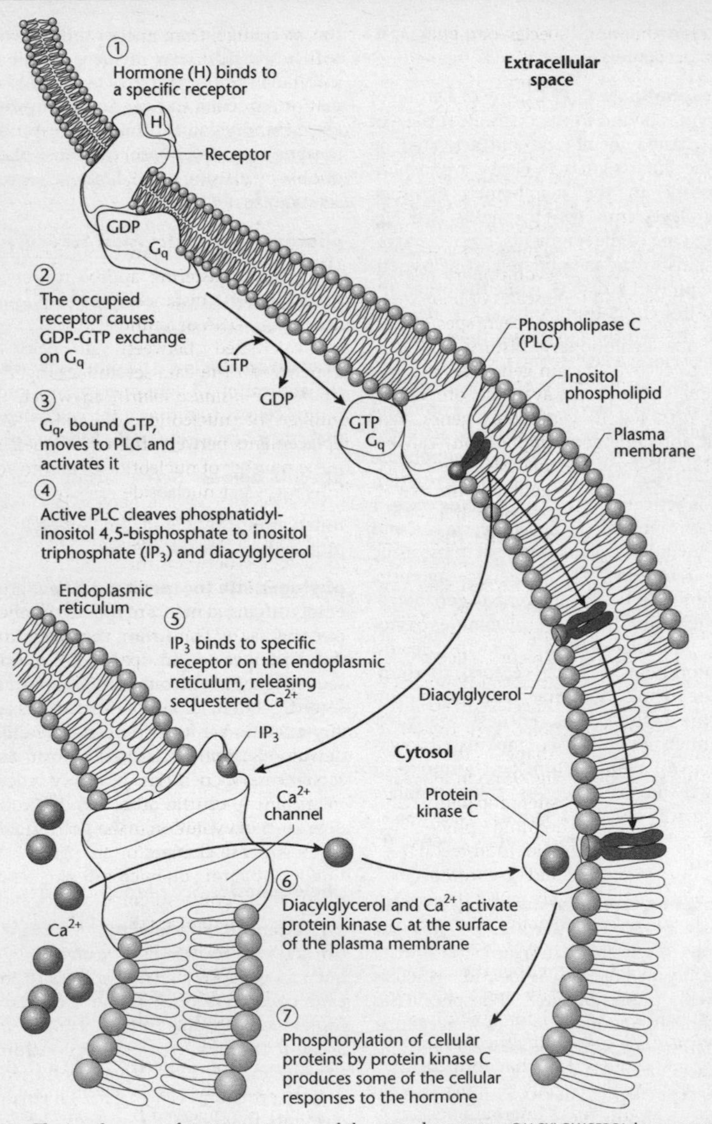

FIG 135. The involvement of PHOSPHOLIPASE C and the second messenger DIACYLGLYCEROL in response to some extracellular signals. © 2000 From *Lehninger: Principles of Biochemistry* by D.L. Nelson and M.M. Cox, published by Worth Publishers. Redrawn with permission from the publishers.

ecules, populations, species, or higher *taxa*, each occupying a special *node* (branching point) in the tree. The process of lineage branching is termed *cladogenesis* (see ANAGENESIS).

The objects involved may be composed of either hypothetical or actual ancestral taxa, and their branches never coalesce. *Species trees* and *gene trees* may look quite different. Each DNA segment in a species' genome has its own evolutionary history whose nodes and time-scales of branching may differ. By contrast, each species has just one evolutionary history and can only be recovered from gene trees if complete sets of genes are available; otherwise, only orthologous genes (i.e., genes that have not undergone duplication) can be used in phylogenetic analysis. One consequence of this is that genetic estimates of the age of a clade (i.e., branch or haplogroup) will always be older than estimates using the fossil record, since genetic data identify the earliest stages of divergence, and these predate the important morphological changes used to demarcate species in the fossil record. The EFFECTIVE POPULATION SIZE of the ancestral population and the time between species' divergences determine the proportion of gene trees that do not have the same shapes as the species tree (see HOMINOIDEA, Fig. 95). On average, therefore, there is a greater probability that, in reconstructing hominid phylogeny, gene trees from haploid (non-recombining) mitochondrial and Y-chromosomal data will more accurately reflect the species tree than will trees generated from autosomal data (see MTDNA, Y CHROMOSOME).

Trees may represent either the mechanisms by which diversity arose or, more often, express a model for how diversity arose without representing the mechanism by which it did so. In that form of tree termed a *cladogram*, only terminal nodes represent taxa whereas internal nodes represent hypothetical ancestors. A *rooted tree* includes a specified taxon termed the OUTGROUP, which has the most ancestral divergence compared to all other included taxa. This outgroup orientates the rooted tree to evolutionary time, giving the tree a direction of change, from ancestral to derived. With the publication of the chimpanzee genome in 2005, this species provides a convenient outgroup for many rooted human trees.

Phylogenetic reconstruction can employ *clustering methods* or *search-based methods*. Data employed may be either GENETIC DISTANCES between populations, molecules, etc. (calculated from raw data), or *characters*. In the clustering approach, genetic (evolutionary) distances between molecules (e.g., DNA or amino acid sequences) are calculated between all pairs of sequences in the data set and arranged in a table, or *distance matrix*, in which the number of nucleotide or amino acid replacements between the two sequences, or the number of nucleotide or amino acid differences per nucleotide (or amino acid) site, gives a value for each pair. Sequences are then linked according to the evolutionary distance between them, e.g., the two sequences with the smallest distance score being connected with a root between them (see Fig. 136). The mean distances from each member of this pair to the third node is computed for the next step in the distance matrix, and so on iteratively until all sequences are located on the tree. The critical assumption is that a constant MOLECULAR CLOCK pertains.

Characters, on the other hand, are discrete units of evolution (HAPLOTYPES, including single base changes or changes in the number of mini- or microsatellite repeat units) and permit inferences about the character content of ancestors. They permit qualitative inferences to be made concerning the place of the most recent common ancestor for a given genomic region and the MIGRATION routes of derived haplotypes (see HOMO SAPIENS).

Neighbour-Joining (NJ) is a clustering method that does not require that all lineages have diverged by equal amounts per unit time. It uses an iterative algorithm to find a tree with the minimal 'shortest sum of branch lengths', S. All possible pairs of taxa are analysed in turn and the pair with the shortest overall tree (shortest S-value) is selected and combined in a

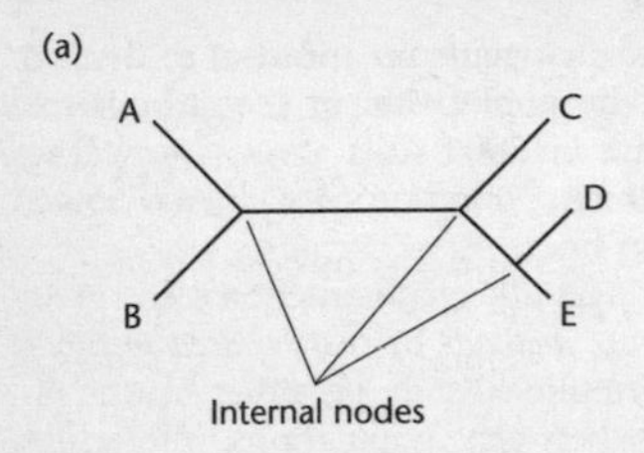

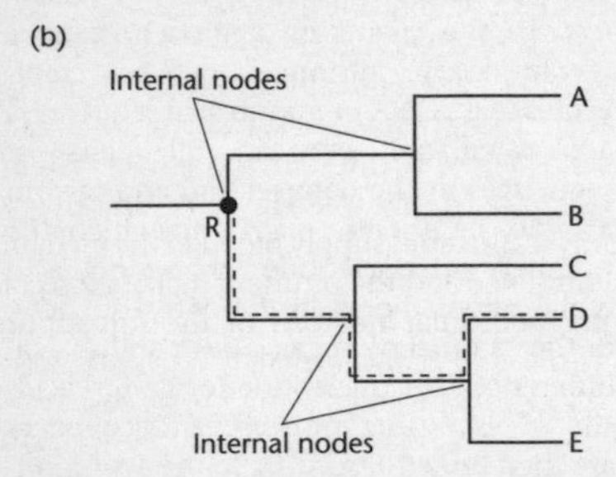

FIG 136. Evolutionary trees. In the unrooted tree (a), there are five *external nodes* (A to E), linked by lines (branches) intersecting at *internal nodes*. Such trees specify the relationships between the organisms under study but do not define the evolutionary path. In the rooted tree (b), from one particular internal node, *the root* R, there is a unique evolutionary path leading to any other node, such as that leading to D. See PHYLOGENETIC TREES. © 2004 From *Human Molecular Genetics 3* (3rd edition), by T. Strachan & A.P. Read. Reproduced by permission of Garland Science/ Taylor & Francis LLC.

single taxon. Then distances from this composite taxon to all other taxa are recalculated and again, the arrangement giving the shortest S is selected for calculation of the S-value until all interior branches have been found. Computing by this method is very rapid.

Search-based methods take account of all possible trees and select the one which best fits the data in line with certain optimality criteria. In order to cut down the amount of computation required for comparison of all possible trees from the given taxa (astronomically large, even for a few tens of taxa), only a representative subset of trees is sought. Search-based methods take longer than clustering methods but, unlike those, do not assume equal rates of evolution for all taxa. The principle of Maximum Parsimony (MP) defines the best tree as that which requires the smallest number of evolutionary changes to accord with the data, branch lengths being determined by the number of evolutionary changes along that branch. If a tree is unrooted, and a sample contains ~100 individuals with a similar number of genetically informative sites in the genome, the number of possible trees is astronomical, and the number of shortest trees runs into billions! Further, MP will give false tree topologies if mutation rates along branches differ. A faster mutation rate increases the chance that lineages will share parallel mutations, and MP will generate the wrong tree if these are more numerous than those distinguishing their common ancestors. Despite these analytical difficulties, existing human GENETIC DIVERSITY (see Fig. 82) has provided critical evidence in the evaluation of human evolution (see HOMO SAPIENS).

Phylogenetic reconstruction can also take account of the fact that some biological processes (e.g., hybridization, recombination) can cause lineages to merge, while others (e.g., parallel mutation) makes them appear to do so. One possible solution is to construct not trees, but four-sided closed structures termed *reticulations*, or cycles. Another makes use of phylogenetic trees known as *networks*, where the connecting lines between taxa (whether internal or solely terminal) are termed links. The level of confidence we can have in a phylogenetic reconstruction can be tested by BOOTSTRAPPING. Improved tree construction can be achieved by the application of PHYLOGENOMICS.

phylogeny The evolutionary history of a species, or group of species. See PHYLOGENETIC TREES; and PRIMATES for primate phylogeny.

phylogeography The analysis of the geographical distributions, or patterns, of evolutionary lineages (clades) within a phylogeny. This combines information about the ancestral-descendant relationships in the haplotype tree with the frequency and distribution of haplotypes in a sample of DNA sequences. See NESTED CLADISTIC ANALYSIS.

physical anthropology The study of humans through analysis of their physical, rather than social or cultural, characteristics.

P_i Inorganic phosphate, an important nutrient in cell metabolism, required for synthesis of DNA, RNA and numerous phosphorylated metabolic intermediates.

PI(3)K See PHOSPHATIDYLINOSITOL-3-OH-KINASE.

pia mater (pia) The innermost and most closely applied of the meninges to the brain tissue. Blood vessels supplying the embryonic brain lie originally on its surface; but their ends penetrate inwards, carrying a layer of pia mater. A perivascular space lies between the pia and each vessel.

pigmentation See MELANINS, SKIN COLOUR.

pineal gland A circumventricular glandular organ attached by the pineal stalk to the dorsal aspect of the diencephalon (see CEREBELLUM, Fig. 42), whose main secretion, during the hours of darkness, is MELATONIN, a derivative of serotonin.

pinna (auricle) The sound-collecting component of the outer EAR. See HEARING.

pinocytosis A constitutive from of ENDOCYTOSIS; compare PHAGOCYTOSIS.

PIP_2 (phosphatidylinositol 4,5-bisphosphate) See PHOSPHATIDYLINOSITOL SYSTEM.

PIP_3 (phosphatidylinositol 3,4,5-triphosphate) See PHOSPHATIDYLINOSITOL SYSTEM.

piRNAs (Piwi-interacting RNAs) Small RNAs whose biosynthesis is currently unclear but known not to require Dicer (contrast microRNAs and siRNAs). They bind to Piwi, a member of the Argonaute family that is apparently expressed only in cells of the germline (at least in *Drosophila*). It is thought that piRNAs function as master controllers of TRANSPOSABLE ELEMENTS. Discrete genomic loci give rise to ssRNA sequences that are processed by ~27 piRNAs.

P_i system See α_1 anti-trypsin.

pituitary gland (hypophysis) A composite gland, the anterior part (adenohypophysis) developing as an ectodermal outpocket of the stomodeum anterior to the buccopharyngeal membrane, termed Rathke's pouch, and a downward extension of the diencephalon, the infundibulum, linking it anatomically and physiologically with the HYPOTHALAMUS (see Fig. 97, and BRAIN, Fig. 28) through the portal system of blood supply (see Fig. 137). The hypophyseal arteries supply blood to the median eminence and the pituitary. Long axons of magnocellular neurons of the supraoptic nuclei and paraventricular nuclei terminate in the posterior pituitary, while axons of parvocellular neurons do so in the median eminence. Long portal veins drain the median eminence, transporting neuropeptides from the primary capillary plexus to the secondary capillary plexus which supplies blood to the anterior pituitary.

Hormones of the anterior pituitary comprise the glycoproteins derived from PROOPIOMELANOCORTIN (POMC), and those belonging to the growth hormone (GH) and prolactin family. Their release is controlled by stimulating and inhibiting factors from the median eminence of the hypothalamus, and mediated by G PROTEIN-COUPLED RECEPTORS on different subclasses of anterior pituitary cells: thyrotrophs, gonadotrophs, corticotrophs, somatotrophs and lactotrophs. Stimulation of these cells by an appropriate extracellular signal induces repeated oscillatory calcium spikes (see CALCIUM), each associated with a burst of hormone secretion.

PKA (protein kinase A, cAMP-dependent protein kinase) A tetrameric regulatory protein serine/threonine kinase mediating the effects of increasing levels of CYCLIC AMP within cells (e.g., during signalling by ADRENALINE, GLUCAGON, CORTICOTROPIN and CALCITONIN). On binding cAMP, inactive PKA is allosterically activated by

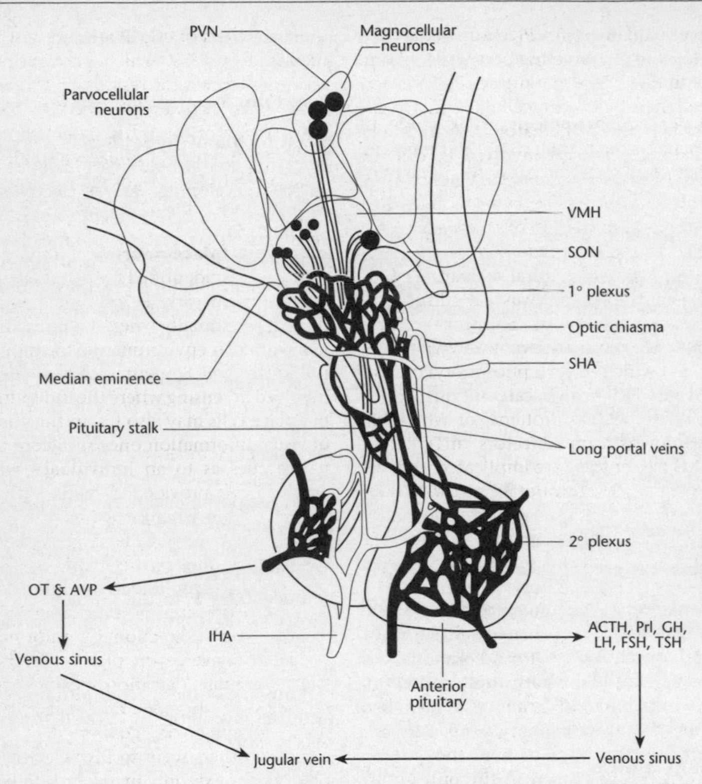

FIG 137. Anatomical and physiological relationships between the PITUITARY GLAND and hypothalamus, viewed from the right side. SHA = superior hypophysial artery; IHA = inferior hypophysial artery; PVN = paraventricular nucleus; SON = supraoptic nucleus; VMH = ventromedial hypothalamus; ACTH = corticotropin; GH = growth hormone; TSH = thyrotropin; Prl = prolactin; LH = luteinizing hormone; FSH = follicle-stimulating hormone; OT = oxytocin; AVP = vasopressin. Redrawn from *Endocrine Physiology* © 2004, P.E. Molina (Lange Medical Books/McGraw-Hill), with permission of the McGraw-Hill Companies.

release of two catalytically active subunits. The active PKA phosphorylates a number of enzymes, e.g., activating phosphorylase *b* kinase. An important effect of this is inactivation of glycogen synthase *b*, and activation of glycogen phosphorylase *a*, favouring GLYCOGEN breakdown and its release by liver cells into the plasma. The catalytically active subunits can also enter the nucleus and phosphorylate the transcription factor CRE-binding protein (see CREB) which binds to CREs in certain promoters.

PKA was first discovered during research on the cAMP-dependent effects of adrenaline and glucagon on glycogen breakdown by muscle and liver cells. It was subsequently found to be far more abundant in brain tissue than in liver, and to be concentrated in the synaptic regions of neurons. Soon afterwards, a second class of regulatory protein kinase, PKG, was found. PKA is a major component in presynaptic facilitation, and in the slow synaptic transmission pathways involved in MEMORY. See also GLUTAMATE, Fig. 84; LONG-TERM POTENTIATION.

PKB (Akt, Akt/PKB) A serine/threonine

protein kinase with a Pleckstrin homology (PH) domain directing it to the plasma membrane when PI3-kinase (PI3K) is activated there by an extracellular signal (e.g., a MITOGEN), and a mediator in intracellular signalling pathways involved in cell survival (see SURVIVAL SIGNALS). One of these pathways delivers the cytosolic effects of INSULIN (e.g., inhibition of GLUCONEOGENESIS), while INSULIN-LIKE GROWTH FACTOR-1 is one of the most potent natural activators of the AKT signalling pathway. Binding of a GROWTH FACTOR to its receptor activates PI3K (see RAS SIGNALLING PATHWAY; DISEASE, Fig. 64) which in turn phosphorylates and activates PKB which can, in turn, phosphorylate >9,000 proteins, of which the two downstream effectors mTOR and FOXO (see entries) are implicated in development of CANCERS, notably LEUKAEMIAS. In this context, another protein phosphorylated is the CDK inhibitor $p27^{Cip1}$. This normally blocks advance through the CELL CYCLE; but its phosphorylation prevents it from moving from the cytoplasm into the nucleus, where its effects would be mediated. In this way, PKB is a component of signal transduction pathways accelerating cell division. PKB can also be anti-apoptotic, by phosphorylating Bad and decreasing its ability to hold the channel for CYTOCHROME C open in the outer mitochondrial membrane, explaining the anti-apoptotic effect of Akt/PKB kinase, which phosphorylates caspase 9 (see CASPASES).

PKC A membrane-bound enzyme, docking onto and being activated by the second messenger diacylglycerol, and then phosphorylating serine or threonine residues of specific target proteins depending on the cell type (though mostly different from those targeted by PKA). It is also activated by another second messenger, calcium ions (Ca^{2+}). Inhibition of the PKC β isoform protects cells against HYDROGEN PEROXIDE challenge. See LONG-TERM POTENTIATION; PHOSPHOLIPASE C, Fig. 135.

PKG (cGMP-dependent protein kinase) A regulatory protein kinase (see PKA) activated selectively by cyclic-GMP. Indeed, most activities of cGMP are thought to be mediated by PKG, which phosphorylates serine and threonine residues in target proteins when activated by cGMP. PKG is downstream of the nitric oxide signalling pathway, and phosphorylates a small heat-shock protein Hsp20, correlating well with smooth muscle relaxation. See ROD CELLS.

place cells (place neurons) Active PYRAMIDAL CELLS, first identified in the rat HIPPOCAMPUS during the early 1970s, which respond by firing maximally when the individual is in a particular environmental location, the cell's *place field*. Sometimes, visual cues are involved in sensing where the individual is; but place cells may also fire in the absence of visual information once sufficient alternative cues as to an individual's whereabouts have provided it with sufficient evidence as to where it thinks it is. Place cells respond to a wide range of spatial inputs, including extrinsic landmarks and translational and directional movement signals. Expression of both metric positional, and directional, information seems to converge on place cells. While some exhibit location specific firing regardless of whether a rat engages in random foraging or goal-oriented food retrieval, others seem to fire selectively at their place fields only in association with a particular kind of experience. Place fields seem to share some features with receptive fields, in particular they are dynamic: place cells appear to 'learn' by altering their receptive fields in adjusting to a modified environment ('remapping'): different cells may be active in different environments or the same cell may be active at different locations in different environments, and remapping takes some cells longer than it does others. See HIPPOCAMPUS, MIRROR CELLS.

placebo A pharmacologically inert preparation (a so-called 'sugar pill') which may have a therapeutic effect (the placebo effect) by altering the symptoms that are presenting. This is usually considered by the medical profession to be due to the 'power of suggestion' of the patient (subject-expectancy effects) and symptoms may be alleviated or exacerbated. Some fMRI evidence indicates that foreknowledge leading to expectation of taste sensations (e.g., brand information) can alter the brain region in which information concerning TASTE is processed when compared with 'blind taste tests' in which information derives solely from sensation. Placebo-controlled studies (see DOUBLE-BLIND STUDY), in addition to providing evidence of a clinical treatment's effectiveness, can also reveal undesirable side effects. See SEASONAL AFFECTIVE DISORDER.

placenta A composite organ, having extraembryonic and maternal components, providing a large exchange surface between maternal and foetal blood, across which oxygen, nutrients and foetal wastes can pass during PREGNANCY (as do foetal red blood cells in small numbers). As the syncytiotrophoblast (see BLASTOCYST, Fig. 20), forming part of the CHORION, invades the endometrium (decidual layer) of the uterine wall it encounters and surrounds maternal spiral arteries, digesting their walls so that blood lacunae, still connected to the maternal vessels, surround the villi. Cords of cytotrophoblast surround the lacunae and syncytiotrophoblast as the villi sprout branches which protrude into the lacunae. Being non-antigenic, the syncytiotrophoblast does not elicit an immune response. The chorion, now the entire extraembryonic structure facing the maternal tissues, secretes HUMAN CHORIONIC GONADOTROPIN. Extraembryonic mesoderm and blood vessels from the yolk sac grow into the lacunae, as secondary and tertiary villi develop to form a cytotrophoblastic shell that will anchor the villi to the endometrium and provide a vast surface area for the diffusion and carrier-mediated transport of solutes. In the mature placenta (see Fig. 138) the cytotrophoblast disappears so that embryonic blood vessels are separated from maternal blood by the chorion alone, comprising the embryonic capillary endothelium, its basement membrane, and a thin layer of syncytiotrophoblast. This 'placental barrier' is very thin, although the syncytiotrophoblast consists of several cell layers lacking PARACELLULAR TRANSPORT pathways. IgG (see IGG) can cross the placenta. In effect, then, the placental barrier is somewhat impermeable. But the dialysis nature of placental blood flow, in which maternal blood spaces provide a large volume of well-mixed blood with moderate turnover and the foetal capillaries have a far smaller volume of blood but a faster turnover, optimizes the diffusion conditions for removal of foetal wastes and supply of its solutes. Rapid turnover of foetal blood maintains a steep diffusion gradient of wastes, while the relatively large maternal blood space (~250 ml) enables solute concentrations to remain relatively unaffected by diffusion into the foetal capillaries and keeps waste concentrations low, despite the lower blood flow on the maternal side. Respiratory gases pass by diffusion across the villi. Blood in the umbilical arteries is both highly deoxygenated (pO_2 = 24 mm Hg, or 3.2 kPa) and hypercapnic (pCO_2 = 50 mm Hg, or 6.6 kPa), the equivalent values for the maternal blood space being: pO_2 = 42 mm Hg, or 5.6 kPa and pCO_2 = 46 mm Hg, or 6.1 kPa. However, blood returning to the foetus in the umbilical vein is not in equilibrium with maternal blood. This is due in part to not all the maternal blood being in contact with the villi, and in part to the high metabolic activity of the placental tissue itself (~20% of O_2 being consumed before reaching the foetal capillaries). However, the foetal circulation makes best use of the impoverished O_2 supply because of shunts enabling the two sides of the foetal heart to operate in parallel rather than in series (see DUCTUS ARTERIOSUS, DUCTUS VENOSUS, FORAMEN OVALE).

The placenta stores nutrients including glycogen, proteins, calcium and iron which are released as the foetus grows. Glucose crosses the membranes of the syncytiotrophoblast cells by facilitated diffusion and although this is normally desirable, the foetus can become obese if the maternal blood is hyperglycaemic, as when the mother is diabetic. Amino acid levels indicate that they are actively transported across the villi, and specific transporters for the essential amino acids have been

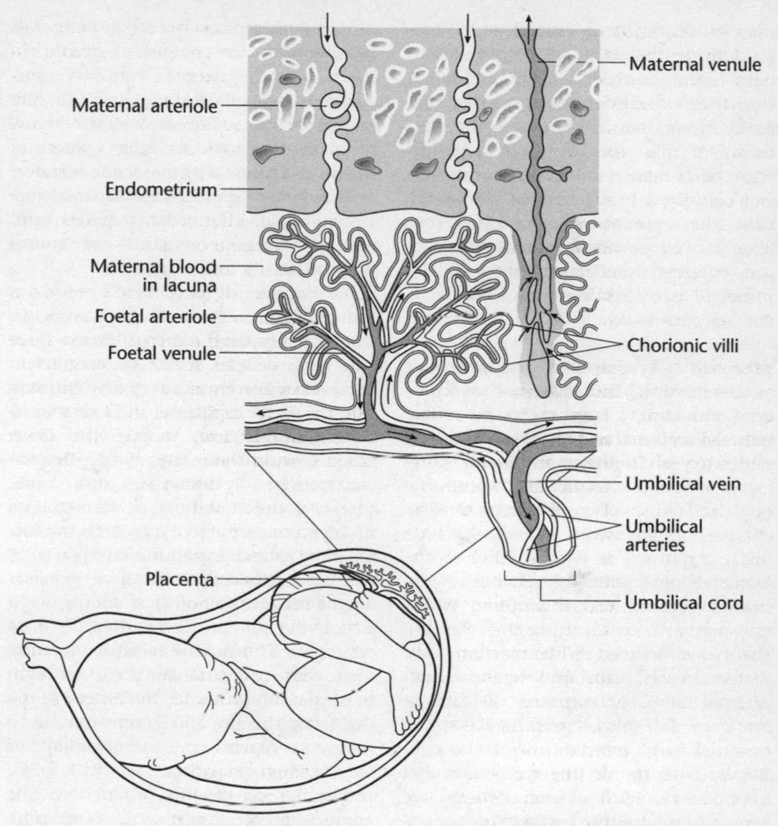

FIG 138. Foetus *in utero*, with enlargement of the tissues of the PLACENTA. Arterial blood vessels are unshaded; venous blood is shaded. Redrawn from *Anatomy & Physiology*, R.R. Seeley, T.D. Stephens and P. Tate. © 2006 McGraw-Hill, with the permission of The McGraw-Hill Companies.

found. Free fatty acids are the main lipid source for the foetus, and these diffuse down their concentration gradients to the foetal blood, passing to the foetal liver where phospholipids are synthesized. Foetal urea is generated as its tissue remodelling destroys proteins, ~40% of nitrogen in amino acids entering the foetal circulation being removed as urea. Although the foetal liver is relatively immature, it is capable of urea production; but breakdown of haemoglobin results in release of bilirubin unattached to glucuronic acid since the liver cannot produce sufficient conjugating enzymes. It crosses the placenta and is conjugated instead by the maternal liver prior to its excretion in bile. Inadequate removal from the foetal blood can result in it crossing the blood–brain barrier and damaging the brain, especially the basal ganglia, leading to *kernicterus*: permanent motor loss.

The placenta is also a protective barrier for the foetus, because most microorganisms cannot cross it. However, the AIDS virus (HIV), rubivirus (cause of rubella), varicella virus (cause of chickenpox), measles virus and the viruses causing encephalitis and poliomyelitis, can do so. After the first trimester of pregnancy, growth of the foetus outpaces the ability of the placenta to supply foetal nutritional needs, and foetal capillaries become clogged by clots as pregnancy nears term. The decreasing efficiency of this *senescent* placenta as an organ of exchange (*placental insufficiency*) may contribute to the onset of labour (see BIRTH). The foetus carries paternal MHC proteins and minor histocompatibility antigens different from the mother's, and no comprehensive explanation of why it is not rejected is available. The placenta, which is part foetal, appears to keep maternal T cells away. The trophoblast does not express classical MHC class I or class II proteins, making it resistant to attack by these cells. NK cell attack may be avoided by expression of HLA-G, a minimally polymorphic class I molecule, which binds the KIR1 and KIR2 NK cell receptors. The enzyme indoleamine 2,3-dioxygenase (IDO), expressed at a high level at the maternal–foetal interface, catabolizes the essential amino acid tryptophan which may starve T cells (nutrient depletion). There is also direct evidence from mice that maternal T cells are tolerant against paternal MHC alloantigens. Release of cytokines TGF-β, IL-4 and IL-10 at the maternal–foetal interface might also contribute. Both the uterine epithelium and trophoblast secrete these, which are known to suppress T_H1 responses (see MISCARRIAGE). It is also possible that IMMUNOSUPPRESSION by regulatory T cells is involved.

At full term, the placenta is discoid and has a diameter of 15–25 cm, is ~3 cm thick and weighs ~500–600 g. It is pulled from the uterine wall ~30 minutes after birth, when 15–20 slight bulges (the cotyledons) can be seen on its maternal side under the covering decidua basalis, between which are grooves formed by the decidual septa. On the foetal side, covered by the chorionic plate, are numerous arteries and veins (chorionic vessels) converging towards the umbilical cord.

Human growth hormone (hGH) and human placental lactogen (hPL) are important in regulating maternal and foetal metabolism, and foetal growth and development. The variant of hGH produced by the placenta is hGH-V, and becomes the predominant GH in the mother but is not released in the foetus although it seems to affect placental growth in an autocrine/paracrine manner. It stimulates IGF-I production, moderating maternal intermediary metabolism, thereby increasing glucose and amino acid availability for the foetus.

The placenta begins to secrete oestrogen and progesterone from the 8th week, so that by 3 months the corpus luteum is no longer needed to maintain pregnancy. Because the placenta has no 17α-hydroxylase (or 17,20-desmolase), it is unable to convert progesterone to oestrogen or to produce androgens. This, with placental inactivation of maternal and foetal adrenal-derived androgens through its strong aromatase activity, protects the foetus from masculinization. But placental oestrogen production therefore depends upon aromatization of these androgens to estriol, the principal syncytiotrophoblast oestrogen. The main effects of placental oestrogens are to stimulate uterine growth, prostaglandin synthesis, thickening of the vaginal epithelium and its sensitization to oxytocin, growth and development of mammary gland epithelium (with hPL), and inhibition of milk production. The production of a corticotropin-releasing hormone (CRH), similar to pituitary CRH, increases exponentially during pregnancy, mostly in the third trimester, peaking during labour – which it may help to initiate (see BIRTH). Its effects include promotion of prostaglandin production and maintenance of placental blood flow. A placental thyrotropin-like hormone increases thyroxine (T_4) output, although both the major THYROID HORMONES are deactivated by the placenta, protecting the foetus. See IMPLANTATION.

plague (**'Black Death'**) Disease caused by the flea-borne bacterium *Yersinia pestis* which has occurred in three pandemics accounting for more than 200 million human lives (including one-third of the population of Europe between 1347 and 1350).

planar cell polarity (PCP) The orientation of cells along an axis orthogonal to the apical–basal axis. It involves one form of non-canonical WNT SIGNALLING PATHWAY. See CILIUM.

plantar arch The longitudinal arch of the foot, helping to maintain mid-tarsal rigidity for powered plantar flexion during toe-off and acting as an important spring during running (see GAIT). Analysis of the Hadar and Sterkfontein specimens of *Australopithecus* suggests that these hominins may have had a partial arch only, sharing with chimpanzees a larger and more weight-bearing arch than the diminutive equivalent in modern humans.

plasma See BLOOD PLASMA.

plasma cells Mature antibody-secreting B CELLS.

plasma clearance A calculated value representing the volume of plasma that is cleared of a specific substance per minute by passage through the kidneys. If the clearance value for a substance is 100 mL min^{-1}, the substance is completely removed from 100 mL of plasma each minute. See TUBULAR MAXIMUM.

plasma proteins Proteins make up most of the suspended substances of the BLOOD plasma, contributing 7–9% of its mass and giving it its colloidal properties. They include ALBUMINS (~54%), GLOBULINS, and the BLOOD CLOTTING factors fibrinogen (~2%) and prothrombin (see appropriate entries for their roles).

plasmin (fibrinolysin) A protein-degrading complex, arising by activation of the plasma protein plasminogen. It digests fibrin and other protein blood coagulants and causes lysis of a clot (see BLOOD CLOTTING).

plasminogen activator See PLASMIN, TISSUE PLASMINOGEN ACTIVATOR.

Plasmodium The genus of apicomplexan protoctists that causes MALARIA.

plasticity See NEURAL PLASTICITY.

platelet-derived growth factor (PDGF) A potent stimulator of fibroblasts, released by PLATELETS as they aggregate during blood clot formation. It acts as a mitogenic GROWTH FACTOR by activating the RAS SIGNALLING PATHWAYS. It attracts fibroblasts to the wound site and stimulates their proliferation, exerting its effects through signal pathways involving PHOSPHOLIPASE C (see Fig. 135) and IP3.

platelets (thrombocytes) Minute round or oval discs, 1–4 μm in diameter produced in bone marrow by fragmentation from megakaryocytes (see STEM CELLS) and normally present in the blood in the concentration range $1.5 \times 10^5 – 3 \times 10^5$ per μL. They last ~10 days in an inactive form in the circulation until high local concentrations of calcium ions and adhesive proteins promote activation, after which they are destroyed by tissue macrophages, over half of them being degraded in the spleen. Low platelet count is termed thrombocytopenia (see VON WILLEBRAND FACTOR).

If endothelial cells are lost by desquamation (as in ATHEROSCLEROSIS and chronic HYPERTENSION) the underlying connective tissue collagen is exposed, itself a potent platelet activator and inducer of platelet filopodia and lamellipodia, helping to bind the endothelium and seal it. Most platelet adhesion is mediated through von Willebrand factor (vWF), a protein secreted by endothelial cells. It forms a bridge between collagen and platelets by binding to platelet surface receptors and collagen. After adhesion, platelets become activated – in which macrophages and eosinophils play a part by release of platelet-activating factor, PAF. Membranous organelles within the platelet include dense granules and alpha granules and these are released on activation by exocytosis, as receptors previously on the inside of the granule membranes now lie exposed on the cell surface. Dense granules release calcium ions (Ca^{2+}), ADP, ATP and serotonin.

Alpha granules contain growth factors (e.g., see PLATELET-DERIVED GROWTH FACTOR) stimulating fibroblast cells in blood vessels to increase protein synthesis and repair damage. Fibrinogen binds activated platelets together – *platelet aggregation* – to form a platelet plug while other adhesion proteins bind platelets to the ENDOTHELIUM of the blood vessel. Also released by activation is thromboxane A_2 (TxA_2), responsible for much of the vasoconstriction after rupture or cut of smaller blood vessels (vascular spasm). ADP and TxA_2 in turn activate nearby platelets, their stickiness causing them to adhere to the already-activated platelets, forming a platelet plug that blocks blood loss if the opening in the blood vessel is small. Activated platelets express phospholipids (platelet factor III) and coagulation factor V which are important in both the intrinsic and extrinsic pathways of BLOOD CLOTTING, during which fibrin threads form, attaching tightly to the platelets and constructing a more persistent plug. See THROMBUS.

Platyrrhini (platyrrhines) The primate infraorder (New World monkeys) characterized by forward-facing nostrils separated by a wide partition (nasal septum). Contrast CATARRHINI.

PLC See PHOSPHOLIPASE C.

Pleckstrin homology (PH) domain Domains on intracellular signalling proteins (first discovered on the platelet protein Pleckstrin) that act as docking sites, enabling them to bind to the cell membrane inositol phospholipids PIP_2 and PIP_3 (see PHOSPHATIDYLINOSITOL SYSTEM). They were first discovered on the platelet protein, Pleckstrin, but occur on at least 200 human proteins (e.g., SOS). See ARF PROTEINS, GENETIC DISORDERS.

pleiotropy The ability of alternative allelic substitutions at a gene locus to affect, or be involved in, development of more than one aspect of phenotype. A gene locus may have pleiotropic effects if its product is an enzyme involved in more than one metabolic pathway, a growth factor, or a regulatory protein (e.g., transcription factor): in fact, any of the proteins involved in signal transduction. See RNA SPLICING.

plesiomorphy In phylogenetics, describing the original pre-existing member of a pair of homologous characters, the evolutionary novelty being the apomorphic member of the pair (see APOMORPHY). A symplesiomorphy is any character that is a shared homologue of two or more taxa but which is believed to have arisen as an evolutionary novelty in an ancestor earlier than their earliest common ancestor.

pleural membrane (pleura) Two layers of SEROUS MEMBRANE enclosing and protecting each LUNG, between which lies a small space, the pleural cavity, containing lubricating fluid secreted by the membranes and reducing friction during VENTILATION. Inflammation of the membrane is known as *pleurisy*.

plexus (pl. plexuses) A network of nerves, veins, or lymphatic vessels.

pluripotent stem cells Cells which can in theory give rise to every cell type in the body other than those of embryonic tissue, and derive not from adult but rather from embryonic tissues. These include: embryonal carcinoma (EC) cells, the stem cells of testicular tumours; embryonic stem (ES) cells, derived from pre-implantation embryos; and embryonic germ (EG) cells, derived from primordial germ cells of the post-implantation embryo. See EMBRYONIC STEM CELLS, STEM CELLS. For induced pluripotent stem cells (iPS cells; see STEM CELLS). See EMBYONIC STEM CELLS, TOTIPOTENT STEM CELLS.

pneumoconiosis (pneumonoconiosis) A disease of the lungs caused by the habitual inhalation of irritant particles (usually mineral or metallic particles in dust) resulting in fibrosis (inflammatory nodule formation) and scarring of lung tissue. Silicosis, the most common occupational disease, is the form resulting from inhalation of quartz dust. This can promote lung cancer. If coal dust is inhaled for several years the deposition of dust leads to 'black lung disease'. Breathlessness is a symptom of these diseases. Asbestos crystals are subject to 'frustrated phagocytosis' and remain trapped at the engulfing macrophage's cell surface, where cytoskeletal actin filaments form. Work on mice indicates that asbestos and silica are sensed by the Nalp3 inflammasome (see INFLAMMATION), whose subsequent activation leads to interleukin-1β (IL-1β) secretion. Nalp3 appears to be a proinflammatory 'danger' receptor in particulate matter-related pulmonary diseases. The inflammasome activation is triggered by reactive oxygen species (ROS; see MITOCHONDRION) generated by a NADPH oxidase when particles undergo phagocytosis. Mice deficient in Nalp3 are resistant to asbestos-induced lung injury.

pneumonia Acute infection or inflammation of the alveoli, and one of the most common infectious causes of death. Release of toxins, especially by *Streptococcus pneumoniae*, in lungs of susceptible individuals stimulates inflammation and immune responses that damage alveoli and bronchial mucous membranes. Inflammation and oedema cause the alveoli to fill with debris, interfering with gaseous exchange and ventilation. Those most at risk are the immunocompromised (e.g., AIDS sufferers, or those taking immunosuppressive drugs), infants, the elderly, cigarette smokers and those with CHRONIC OBSTRUCTIVE PULMONARY DISEASE. Most cases are preceded by an upper respiratory infection, commonly viral. Individuals then develop fever, chills, a productive or dry cough, chest pain and malaise. Treatment may involve antibiotics, bronchodilators, oxygen therapy, increased fluid intake and physiotherapy.

pneumotaxic centre See RESPIRATORY CENTRE.

pneumothorax The result of an injury allowing air to enter the intrapleural space, from either the outside or from the alveoli, so that the pleural cavity is filled with air. The intrapleural pressure then equals that of the atmosphere and surface tension and recoil of elastic fibres cause the lung to collapse. See ALVEOLUS, RESIDUAL VOLUME.

PNS The PERIPHERAL NERVOUS SYSTEM.

point-source outbreak See DISEASE.

Pol I, Pol II, Pol III See RNA POLYMERASES.

polar bodies Haploid nuclei, surrounded by very little cytoplasm, which is produced during OOGENESIS, eventually degenerating. Their presence may be useful in preimplantation genetic diagnosis (see SCREENING).

poliomyelitis (polio) An enteroviral and notifiable disease, of which three serotypes (1, 2 and 3) occur. Poliovirus enters the tissues via the alimentary canal and the oropharynx. Viraemia follows, during which the virus enters the meninges and spinal cord, infecting and killing ventral horn cells and leading to lower motor neuron degeneration. Non-paralytic poliomyelitis is a viral MENINGITIS symptomatically indistinguishable from other enteroviral meningitides. Paralytic poliomyelitis begins with fever, often with myalgia, followed by asymmetrical paralysis 2–3 days later, developing over ~48 hours. Control is now achieved by the inactivated polio vaccine (IPV), given in three doses at monthly intervals beginning at 2 months of age, with booster doses before entry to school and at year 15. Live oral polio vaccines (OPV) strains protect against mucosal invasion and are excreted in faeces for up to six weeks. They are used to protect close contacts of those infected, but cause paralysis in ~1–2 per million vaccinated. Cases should be isolated in a hospital. In outbreaks, the virus circulates extensively and vaccination of a wide network of contacts is required. It may be possible to eradicate the disease, as has happened with smallpox. See NEURAL DEGENERATION, VIRUSES.

polyadenylation Addition of a poly-A tail to the final RNA transcript during RNA PROCESSING in the nucleus. Some mRNAs are later subject to deadenylation, the resulting short oligo(A) tail preventing their translation on ribosomes. Such inactive mRNAs, including those stored in the oocytes prior to fertilization, can be activated by cytoplasmic polyadenylation, restoring the normal tail length. The same type of poly (A) polymerase activity is employed as in the nucleus, although a uridine-rich cytoplasmic polyadenylation element needs to be attached upstream first. See ISOFORMS.

polycistronic The situation in which two or more adjacent genes are transcribed into a common (multigenic) RNA molecule. This primary transcript may then be spliced by snoRNAs into the individual functional RNA products. Most nuclear genes are transcribed individually, rather than polycistronically (RDNA being an exception); but mitochondrial DNA is often polycistronically transcribed (see MTDNA).

Polycomb group (PcG) Genes of the Polycomb family encode highly conserved but poorly understood proteins that epigenetically repress many target genes by regulating their level of HISTONE methylation. They include genes required for maintenance of homeotic gene repression in pluripotent EMBRYONIC STEM CELLS during development, their protein products contributing to the PcG-protein repressive complex (PRC). Genes so repressed include the developmental gene families homeobox (*Hox*), distal-less homeobox (*Dlx*), paired box (*Pax*) and sine-oculis-related (*Sir*). Both polycomb and trithorax proteins are chromatin modifiers (see CHROMATIN REMODELLING). PcG repressive complexes mark the histones associated with these genes by inducing trimethylation of the lysine residue at position 27 of histone H3 (H3K27), and are required for the transcriptional silencing of many genes involved in somatic processes (see HISTONE METHYLASES). The H3K27me3 mark is associated with the unique epigenetic state of STEM CELLS. Genes so targeted by the PcG proteins suppressor of zeste 12 (SUZ12) and embryonic ectoderm development (EED), which form the Polycomb repressive complex 2 (PRC2), have been identified in human embryonic stem cells. See GENE SILENCING.

polycystic kidney disease (PKD) The adult form of this disease (APCKD) is common and often lethal, in which cysts develop in the liver, pancreas, spleen and kidneys. Presently untreatable, its symptoms do not usually manifest before age 40 but lead to renal failure by the average age of 51 unless dialysis or kidney transplant is available. Use of DNA PROBES has led to the discovery of several marker loci flanking the APCKD locus (*PKD1*, chromosome site 16p) which enable presymptomatic diagnosis of the disease. 2–3% of cases are due to mutations elsewhere. Loss of function of the polycystin genes *PC1* and *PC2* results in clonal expansion of kidney epithelial cells, which form cysts that crowd out normal nephrons and cause kidney failure. Their protein products localize to the primary cilia (see CILIUM) of kidney epithelial cells and seem to cause straightening of the cilia when they should normally bend as a result of urine flowing past them. This seems to block Ca^{2+} flux into the cilium causing activation of the regulated intramembrane proteolysis pathway (RIP) which releases a portion of the PC1 protein. This enters the nucleus where, with STAT6 and p100, it activates transcription. Constitutive activation of the canonical WNT SIGNALLING PATHWAY may be associated with human PKD, and mutations in the primary CILIUM protein inversin certainly are. See DNA-BASED DIAGNOSTIC TESTING.

polydactyly The occurrence of supernumerary digits of hands or feet. Usually dominant in effect, and with incomplete PENETRANCE, it may be caused by about half a dozen different genes, including the ectopic expression of the gene *Sonic hedgehog* (*Shh*), and by the related gene *Indian hedgehog*. Expansion of a polyalanine repeat in *Hoxd13* triggers the condition when heterozygous, and SYNDACTYLY when homozygous.

polygenic inheritance The involvement of more than one gene locus in the expression of a phenotypic character, each locus usually having a slight (often incremental) additive effect on the character (see Table 8)

Number of 'dominant' alleles	Phenotype (arbitrary units)	Representative genotype	Number of possible different genotypes
0	40	g/g h/h i/i	1
1	45	G/g h/h i/i	3
2	50	G/G H/h i/i	6
3	55	G/g H/h I/i	7
4	60	G/G H/H i/i	6
5	65	G/G H/H I/i	3
6	70	G/G H/H I/I	1
			Total 27

TABLE 8 *An additive model of* POLYGENIC INHERITANCE *involving three gene loci.*

and behaving in a Mendelian manner. Such polygenically determined characters are often termed 'quantitative' characters. The phenotypes of quantitative characters tend to be subject to considerable environmental variation and to exhibit a continuous range of variation (e.g., see SKIN COLOUR).

polymerase chain reaction (PCR) A very sensitive and robust technique for selectively amplifying often minute amounts of sample DNA, even when the tissues or cells themselves are badly degraded. Since its invention in 1983, it has revolutionized genomics (e.g., by making the HUMAN GENOME PROJECT achievable in such a comparatively short time), TISSUE TYPING, and numerous other applied fields (see DNA PROBES and references). Given prior knowledge of at least some parts of a gene or DNA sequence of interest (the target sequence), two short single-stranded oligonucleotide DNA or RNA primers, ~20 nucleotides long, can be designed with base sequences complementary to the DNA either side of the sequence to be amplified. The 3′ sequence of one primer should not be complementary to any region of the other primer used. Correct base-pairing at the extreme 3′ end of bound primers has enabled methods to be developed which allow distinction between alleles differing at a single nucleotide. Human DNA to be amplified is first heated to 93–95°C to denature the DNA, breaking hydrogen bonds and separating the two strands (see Fig. 139). The temperature of the reaction mixture is then reduced to 50–70°C and the primers are added in excess. A suitably heat-stable DNA polymerase (e.g., Pfu polymerase from *Pyrococcus furiosus*, which also has proof-reading exonuclease activity) is added along with the four deoxynucleoside triphosphates, when the primers initiate synthesis of new DNA strands complementary to the DNA strands of the target DNA. Orientation of the primers is such that new strand synthesis occurs in the direction from one primer towards the other primer binding site so that the new strands can serve as templates for new DNA synthesis. The procedure is performed over and over again, with exponential increase in product. Even within a few cycles, the predominant DNA is identical to the sequence bracketed by, and including, the two primers in the original template. In Fig. 139, just three cycles produce 16 DNA strands, 8 of which are the same length as, and correspond exactly to, one or other of the original strands. Just three more cycles would produce 240 such strands out of 260.

PCR limitations include the short sizes (0–5 kb) of the amplification products compared with cell-based DNA cloning methods (~2 Mb), although *long-range PCR* methods, which involve use of two types of heat-stable polymerases and include a proof-reading activity, can raise this to ~42 kb. However, the yields tend to be poor. The method was originally time-consuming and expensive. But new microfabricated systems relying on reduction in thermal mass permit more rapid cooling and heating, with continuous flow of very small volumes through different temperature zones. PCR can thus be integrated with

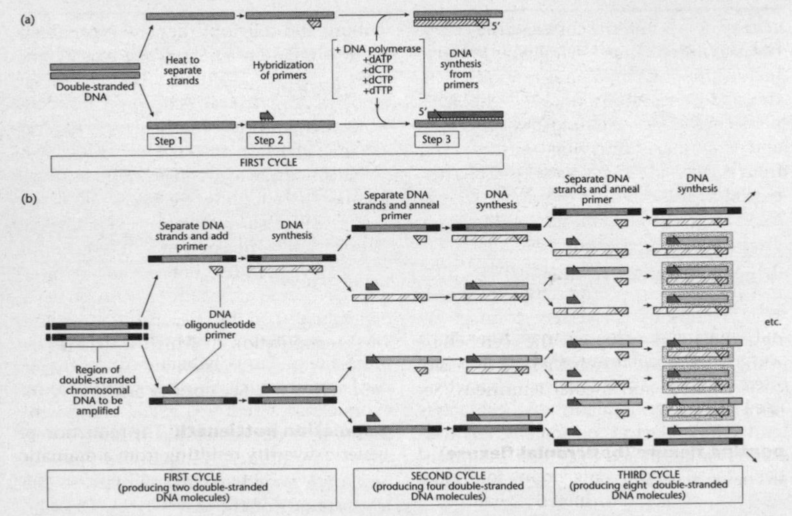

FIG 139. Diagram illustrating the principles of PCR. Three cycles of the process are shown, after which eight of the sixteen DNA chains produced (in stippled boxes) have the same length as, and correspond exactly to, one or other strand of the original sequence bracketed by the primers. The other strands contain extra DNA downstream of the original sequence, which is replicated in the first few cycles. See POLYMERASE CHAIN REACTION for explanation. © 2002 From *Molecular Biology of the Cell* (4th edition) by B. Alberts, A. Johnson, J. Lewis, M. Raff, K. Roberts and P. Walter. Reproduced by permission of Garland Science/ Taylor & Francis LLC.

reverse transcription within a single microchip, 20 cycles of DNA amplification lasting only 5 minutes. See ANCIENT DNA.

polymorphism The existence of two or more variants (of DNA sequences, proteins, chromosomes, phenotypes) at such frequencies in a population that recurrent mutation alone cannot explain it. For DNA, the convention is that any sequence variant present at ≥1% frequency is regarded as polymorphic. For trans-species polymorphism, see HOMINOIDEA. See MAJOR HISTOCOMPATIBILITY COMPLEX, SINGLE NUCLEOTIDE POLYMORPHISMS.

polymorphonuclear leukocytes (PMNs) Myeloid white blood cells with multilobed nuclei and cytoplasmic granules, usually designating older NEUTROPHILS, younger neutrophils having more rod-shaped nuclei. The nuclei of BASOPHILS and EOSINOPHILS tend to be bilobed, but these cells are not normally regarded as PMNs. See GRANULOCYTES; STEM CELLS.

polyphyletic A grouping of taxa that includes members of more than one clade. Compare MONOPHYLETIC, PARAPHYLETIC.

polyploidy (whole genome duplication) See GENE DUPLICATION. Compare ANEUPLOIDY.

polyps (adenomas, adenomatous polyps) See ADENOMAS.

POMC See PROOPIOMELANOCORTIN.

Pongo The genus which includes the orang-utan. See HOMINOIDEA.

pons (adj. pontine) That part of the rostral hindbrain (see BRAIN, Figs 28 and 29) lying ventral to the cerebellum and the 4th ventricle and anterior to the medulla oblongata, constituting a major part of the BRAINSTEM and containing both nuclei and tracts. The pontine nuclei are clusters of nuclei relaying information between the

CEREBRAL CORTEX and the cortex of the CEREBELLUM, serving as a huge 'switchboard'. Nuclei include the trigeminal motor nucleus, and the abducens motor nucleus, the facial nerve motor nucleus (see CRANIAL NERVES), and the nuclei receiving sensory input from the VESTIBULOCOCHLEAR NERVE. The LOCUS COERULEUS is located in the pons (whence input to the cerebellum via noradrenergic fibres) as are some of the RAPHE NUCLEI. The parabrachial nucleus is a somewhat enigmatic component of the RESPIRATORY CENTRE. Tracts running transversely connect the right and left sides of the cerebellum while longitudinal tracts include the corticospinal tract and medial lemniscus. See DENTATE NUCLEI.

pontine flexure (basicranial flexure) A bend in the developing brain, formed at ~32 days between the metencephalon and the myelencephalon and produced by the growth of the cerebellum at the front end of the hindbrain.

pontine nuclei See PONS.

population Although humans from any part of the world are capable of interbreeding and producing fertile offspring (the 'biological species concept'), they do not form a randomly mating population. In one sense, all human mating is non-random since it involves conscious choice, but it will be non-random in a genetical sense if it is biased by the genetic make-up of the individuals involved (i.e., is genetically assortative). All geographically widespread species, such as our own, consist of groups of individuals with differing probabilities of interbreeding. Groups whose members have the greatest probability of choosing a mate from among locally distributed individuals are populations in the biological sense. But even such populations may have an internal *structure* in terms of mating preferences and probabilities (see ISLAND MODEL OF POPULATION STRUCTURE). It is not easy to decide whether or not people belong to the same biological population and criteria to be factored in would include geographical proximity, a common language and shared ethnicity, culture and religion: they are more likely to bear children if they share history and values. Individuals tend to belong to the population with which they identify. Some take the view that when we explore genetic variation in our own species, we should in future focus on individuals rather than on populations (see GENETIC DIVERSITY).

It is a tenet of population genetics that the probability of a new mutation reaching fixation is dependent not only on its selective magnitude and sign (positive or negative) but also on the size, and effective size, of the population in which it occurs (see DEME, EFFECTIVE POPULATION SIZE, NATURAL SELECTION, PANMICTIC). See COMPLEX DISEASES.

population bottleneck The reduction of genetic diversity resulting from a dramatic reduction in population size. See FOUNDER EFFECT, GENETIC DRIFT, MTDNA.

porins Multipass transmembrane proteins with different numbers of antiparallel β-pleated strands (forming β-barrels) forming water-filled pores allowing selected hydrophilic solutes to cross the lipid bilayer. Mainly restricted to bacterial, mitochondrial and chloroplast membranes. Compare AQUAPORINS.

positional cloning Procedure in which a Mendelian disease gene is identified in the absence of any prior knowledge about it other than its approximate chromosomal location (e.g., see MUSCULAR DYSTROPHIES). Nowadays, with current sequence data and high-resolution marker maps, the location technique is very much more rapid since difficulties arise if the initial candidate region is large. First, if unambiguously affected individuals can be determined, family HAPLOTYPES are examined in the hope that LINKAGE DISEQUILIBRIUM will enable the locus to be localized to a region of about 1 Mb of DNA, which depends upon the number of informative meioses.

Then, a contig of clones, arrived at by joining together overlapping sequences, is constructed across the candidate region, after which a map of all gene and exon transcripts within the candidate region is drawn up, using as probes genomic clones from the candidate region. This would include a search for unsuspected alternative splicing products, or additional exons. However, if nothing is known about the expression pattern of the predicted gene, the cDNA library, or source tissue, used for the search may have been inappropriate – although the pathology of the disease should provide some clues as to which tissues to obtain cDNA libraries from.

Hopefully, one or more genes in the candidate region will show an expression pattern consistent with its aberrant function in disease so that, with the help of databases, these genes can be prioritized for amplification by the POLYMERASE CHAIN REACTION, or tested by NORTHERN BLOTTING or SAGE (see GENE EXPRESSION). The most detailed information on expression patterns is obtained by *in situ* hybridization against mRNA in tissue sections (e.g., see FISH, MOUSE MODELS). Again, if the candidate gene's function is already known, DNA SEQUENCING may reveal whether or not the expressed gene is a good candidate on the basis of recognizable structural motifs; and its homology to a relevant PARALOGOUS or orthologous genes (see ORTHOLOGUES) involved in a related phenotype would also provide a clue as to its candidature.

positional information Information enabling cells in a developing animal to respond as though they appreciated their spatial position within the embryo. It would typically include information concerning position along the major body axes (see AXIS). See MORPHOGEN, PATTERN FORMATION.

positional invariance The ability to recognize the same feature anywhere in the visual field. See ORIENTATION-SELECTIVE (ORIENTATION-SENSITIVE) CELLS.

positive feedback The situation in which change in the value of a variable sets in train a process that promotes further change in the same direction. Much less widespread than examples of NEGATIVE FEEDBACK, examples occur in control of the MENSTRUAL CYCLE, in BLOOD CLOTTING (where enzymes in the clot (clotting factors) activate adjacent enzymes to produce more clots); in axon depolarization when sodium entry causes more sodium channels to open and allow more sodium in (see ACTION POTENTIAL). Positive feedback also occurs in signalling pathways during development. Positive feedback is normally only transient.

positive selection See NATURAL SELECTION.

positron emission tomography See PET.

postcentral gyrus The PRIMARY SOMATOSENSORY CORTEX of the BRAIN (see Fig. 29), lying within the parietal lobe immediately posterior to the central sulcus. Receives projections from the ventral posterior nucleus.

posterior (caudal) (adj.) Directionally, towards the tail (or in humans, the coccyx). See ANATOMICAL PLANES OF SECTION, Fig. 9. Contrast ANTERIOR (rostral).

posterior parietal cortex (inferior parietal cortex, Areas 5 and 7) An association area of the BRAIN (see Fig. 29), essential for the perception and interpretation of spatial relationships, accurate body image, and learning tasks involving coordination of the body in space. It projects to the premotor area (PA) of the premotor cortex, with which it is principally involved in planning movements and providing 'instructions' for the MOTOR CORTEX to 'act' upon (see CONSCIOUSNESS), providing sensory input for targeted movements. It is one of the cortical areas in which simple, segregated streams of sensory information from various systems converge to form complex neural representations. It receives somatosensory and vestibular sensory input, and visual information from the medial temporal cortex (see EXTRASTRIATE CORTEX). Its neurons have large receptive fields, with elaborate stimulus preferences, showing selectivity for size and orientation of objects and firing when goal-oriented hand movements are made (e.g., in grasping an object). Many of its neurons fire in a manner that depends on where the gaze is directed. Damage to these areas can cause AGNOSIA, and NEGLECT SYNDROME.

posterior temporal lobe The left posterior temporal lobe (see BRAIN, Fig. 29) is usually larger than the right at birth and in 95% of people this and the left angular gyrus become dominant over the right equivalents (see CEREBRAL HEMISPHERES). In the remaining 5%, the left and right develop simultaneously and have dual function; but more rarely the right side becomes more highly developed, with full dominance. HANDEDNESS is not automatically associated with cerebral dominance, for 95% of right-handed and 50% of left-handed people have their speech centre on the left.

post-translational modification Modification of the primary structure of polypeptides after production during TRANSLATION (see ENDOPLASMIC RETICULUM; see e.g., INSULIN).

post-traumatic stress disorder See AMYGDALA, FEAR AND ANXIETY.

postural reflex pathways Neural pathways, routed via the VESTIBULOSPINAL TRACT and RETICULOSPINAL TRACT (see DESCENDING MOTOR TRACTS), which counteract destabilizing forces (e.g., gravity) which tend to shift the centre of mass, and also maintain the centre of mass during limb movements such as occur during walking, running or dancing. Feedforward motor commands from the MOTOR CORTEX and CEREBELLUM enable postural adjustments in anticipation of movement; and must be learned. But postural reflexes required to compensate for unpredicted disturbances are unlearned and depend on negative feedback, and are organized by the brain stem.

Such responses are achieved largely through antigravity muscles: extensors of the back and legs, and arm flexors (see BALANCE, GAIT, Appendix X). In destabilization resulting from displacement of the surface one is standing on, as in a moving vehicle, the resulting swaying is countered by stimulation of distal prior to proximal muscles, most movement occurring at the ankle joint. If the standing surface is rotated or tilted, bending of the hips results. Maintenance of posture here depends upon the tonic component of the muscle stretch reflex (see MUSCLE SPINDLES). In extreme situations, the reflexes are such that they either prevent falling, or coordinate limb movements so as to brace against falling; these reflexes depend more upon the phasic component of the stretch reflex.

Postural reflexes organized by the brain stem involve inputs from the otolith organs and semicircular canals of the VESTIBULAR APPARATUS, and proprioceptor (e.g., MUSCLE SPINDLES, Golgi TENDON ORGANS and joint receptors) and visual inputs routed via the SUPERIOR COLLICULUS. These inputs project to the medial and inferior vestibular nuclei of the medulla and are relayed via the medial vestibulospinal tract, controlling motor neurons to neck muscles. These medullary nuclei integrate vestibular and proprioceptor inputs. Input from the vestibular apparatus is involved in maintaining orientation of the head and body as a unit. Tilting or rotation of this *unit* reflexly activates motor neurons that maintain the head vertical with respect to gravity: vestibulocollic reflexes to the neck muscles, and vestibulospinal reflexes to the arm extensor and leg flexor muscles (reducing impact on landing). Turning of the head *relative* to the body stimulates spindles in neck muscles and afferents from the cervical vertebral joints, evoking synergistic reflexes: neck muscle contractions reorientating the head on the body (cervicocollic reflexes), and contraction of limb muscles in response to rapid head movements (cervicospinal reflexes). Vestibular afferents synapsing in the lateral vestibular nucleus project via the undecussated lateral vestibulospinal tract to all segments of the spinal cord and are implicated in some of the vestibulospinal reflexes involved in the control of limb muscles. See MEDIAL MOTOR PATHWAYS.

potassium For regulation of K^+, see DISTAL CONVOLUTED TUBULE.

potentiation Any change which makes another more probable.

poxviruses Large double-stranded DNA viruses, the genome encoding >100 viral proteins. Their complex symmetry gives a brick-like shape, ~200 × 250 × 250–300 nm in size. Viral replication takes place in the host cytoplasm and produces large inclusion bodies. The three major groups are: orthopoxviruses (monkeypox, cowpox, vaccinia and variola), variola (smallpox) being declared extinct in 1979; parapoxviruses, including orf; and unclassified viruses, including molluscum contagiosum and tanapox. Those infecting the skin are molluscum contagiosum, cowpox, orf and vaccinia. Cowpox and orf are zoonoses, cowpox having a predilection for eczematous skin. Once endemic in many human populations (~30% mortality, depending on age and other factors), smallpox was eradicated in 1979, largely as a result of mass vaccination, viral samples are officially retained in two locations. Unofficial sources cannot be ruled out, however, and contingency planning for outbreaks demands a rational assessment of the scale of casualties that a smallpox attack might cause and the adverse effects of vaccination experienced by a significant minority of individuals, were mass vaccination ever to be regarded in such planning.

PPARs (peroxisome-proliferator activated receptors) A family of nuclear hormone receptors, several of which (along with members of the liver X receptor family, LXRS) are involved in modulating the intersection between metabolic, inflammatory and innate immune responses. PPAR-γ, encoded by *PPARG*, is coactivated by PGC-1α (see PGC-1 GENES) and is the target of thiazolidinedione drugs used to treat diabetes by increasing sensitivity to insulin (see INSULIN RESISTANCE, GROWTH HORMONE) and promoting LIPOGENESIS.

PPI See PROTEIN PHOSPHATASE I.

Prader–Willi syndrome A rare disorder involving a small deletion at q11-13 of chromosome 15 (see SEGMENTAL DUPLICATIONS) and GENOMIC IMPRINTING. It involves growth and behavioural defects, newborns having poor muscle tone and feeding problems. As children, individuals become compulsive overeaters and obese, have small hands and feet, are mildly to moderately retarded and of short stature. If the deletion is inherited from the father, Prader–Willi syndrome results, but if inherited from the mother, ANGELMAN SYNDROME results. The deletion includes the large *SNURF–SNRPN* transcription unit encoding certain snoRNAs (see entry). See COPY NUMBER, EPIGENETICS.

Praeanthropus africanus The AUSTRALOPITHECINE whose well preserved postcranial skeleton presents a combination of features not found in living primates. Some argue for its having a modern humanlike posture and commitment to terrestrial BIPEDALISM; but others claim that its relatively long and curved proximal phalanges, high humerofemoral index, and mobile joints were adaptations to suspensory and climbing activities.

pRB (pRb) See RETINOBLASTOMA PROTEIN.

precentral gyrus The PRIMARY MOTOR CORTEX of the brain, lying in the frontal lobe immediately in front of the central sulcus. Receives projections from the ventral lateral nucleus and ventral anterior nucleus.

precipitin reaction An early quantitative assay for antibody concentration of increasing known amounts of soluble antigen are added to a fixed amount of serum containing the antibody. In the equivalence zone, when all the antibody complexes with antigen, the largest antigen: antibody complexes are formed, from which the antibody titre can be calculated given the valence of the antibody: the number of antigen molecules to which an antibody can bind (which depends upon its structural class).

predisposition to disease (diathesis) See GENETIC PREDISPOSITION.

prednisolone, prednisone Two anti-inflammatory steroid drugs derived from cholesterol.

preeclampsia (toxaemia of pregnancy) An unpreventable condition affecting ~3% of pregnant women with the potential to kill the mother, baby, or both. The syndrome is typified by a rapid rise in arterial blood pressure to hypertensive levels during the 3rd trimester, leakage of large amounts of protein into the urine, and often by excess salt and water retention by the mother's kidneys and her consequent weight gain through oedema. It is also associated with a systemic maternal inflammatory response of a greater magnitude than during normal pregnancy.

In placental preeclampsia, the problem arises from a placenta that is under hypoxic conditions, with oxidative stress. In maternal preeclampsia it arises from the interaction between a normal placenta and a maternal constitution susceptible to microvascular disease (as with long-term hypertension or diabetes), pregnancy amounting to a metabolic and vascular 'stress test', revealing a woman's health in later life. It is common for both to occur together.

In placental preeclampsia, invasive trophoblasts fail to get full access to maternal supplies. They meet many NK CELLS in the decidua, and recent work suggests that the contact between maternal NK cells and CYTOTROPHOBLAST cells leads to some form of immune recognition, preeclampsia having long been considered a form of maternal immune rejection of the genetically foreign foetus. It is more prevalent in women homozygous for the inhibitory HLA-A haplotypes (AA) that inhibit NK cytokine production than in women who are homozygous for stimulator HLA-B genes and it seems that preeclampsia is less common if trophoblast strongly stimulates maternal NK cells *in utero*. Eclampsia is an exaggerated form of preeclampsia, in which convulsions and even coma result. Recently found mutations in the mineralocorticoid receptor, a kidney protein involved in the handling of salt, explains some of these cases (see ALDOSTERONE). See HYPERTENSION.

prefrontal cortex (PFC) The cortex of the frontal lobe anterior to the PREMOTOR CORTEX (see BRAIN, Fig. 29) with the most protracted development of any brain area, not fully maturing until completion of ADOLESCENCE. It is greatly enlarged in primates, preferentially so in great apes and most especially in modern humans. An association area, it has rich connections with the parietal, temporal and occipital cortices through long association fibres in the subcortical white matter. Its afferents arise mainly in nuclei of the THALAMUS. Divided into lateral, ORBITOFRONTAL and medial prefrontal areas, it has been implicated in high order functions in COGNITION, including intellectual, judgemental and predictive roles, the planning of complex cognitive behaviour, in WORKING MEMORY, in integrating information from outside the body with that from within (derived from the limbic system and hypothalamus) and in expression of personality and moderating correct social behaviour. Lesions here increase distractability and decrease the period of concentration on a task. Prefrontal lobotomy (see FRONTAL LOBE), once a treatment for severe depression, leaves a patient's speech, movement, vision, hearing and intellectual skills apparently unaffected; but there tend to be mood swings and other personality changes, mainly through loss of interests. The PFC is important for actively maintaining information by sustained neural firing, which is robust in the face of potentially distracting information such as working memory. This robustness can guide a sequence of behaviours (e.g., appropriate turnings involved in the plan 'go to the hardware store before going home') in the face of possibly stronger competing actions (e.g., 'go directly home'), giving them consistency. The PFC can also rapidly update what is being maintained, essential in behavioural flexibility, and this may depend on a class of gating mechanism involving DOPAMINE D1 and D2 receptors. Dopamine is transmitted to the PFC by the midbrain ventral tegmental area and detailed computational modelling has made some sense of the manner in which this transmitter modulates PFC circuits.

Those with damage to PFC areas often exhibit 'environmental dependency syndrome' or more simply, a loss of free will, behaviour being driven more by the external environment than by internal plans or goals. The PFC is one of the primary brain areas deactivated during DREAMING.

There is an emerging picture that whereas the rest of the cortex can be characterized fundamentally as an analog system operating on graded, distributed information, the PFC has a more discrete, digital character, its robust activity being supported by a form of BISTABILITY so that neurons switch between two stable states (off or on), similar to bits in a computer: one possible factor in the origin of human INTELLIGENCE. See AMYGDALA, COGNITION, WILLIAMS–BEUREN SYNDROME.

pregnancy The interval (gestation) between IMPLANTATION of the blastocyst and parturition (see BIRTH), lasting ~38 weeks. It is subdivided into three 3-month *trimesters*, in the 1st of which the changes in the mother's body prepare her for the later stages of gestation, during which pregnancy represents a state of 'accelerated starvation' for the mother's body in which nutrients are conserved for use by the foetus. By the 2nd trimester, the foetus is making insulin and urinating. During the 3rd trimester, the foetus is moving more regularly and strongly and may survive if born prematurely.

A mother's cardiac output increases during pregnancy (and further during labour) through both raised heart rate and stroke volume, catering for increased uterine and foetal nutritional needs and foetal waste removal. The resulting raised glomerular filtration rate (GFR) results in reduced urea and creatinine levels in maternal plasma. Fibrinogen and Factors VII–X often increase, with increased risk of deep vein thrombosis. Tidal volume may increase by as much as 40% by term, and displacement and compression of the abdominal organs. Increasing OESTROGEN levels lead to increase in maternal RENIN output, eliciting a rise in ANGIOTENSIN levels. But any hypertensive effect is lessened by a reduction in sensitivity of blood vessels to these hormones, although angiotensin does increase aldosterone secretion by 6- to 8-fold, resulting in reabsorption of salt and water from the distal tubular fluid and compensating for the raised glomerular filtration rate (between 30% and 50%). Maternal tissues become more sensitive to INSULIN, and so carbohydrate loads are readily assimilated. This may lead to a small fall in plasma glucose ~6 weeks into gestation, during which time protein synthesis increases, the net effect being to stimulate growth of the uterus, breasts and essential weight-bearing musculature of the mother. In the second half of pregnancy, the mother's metabolism enters a state resembling that during STARVATION, maternal tissues showing a fall in insulin sensitivity enabling her plasma glucose level to supply the foetus with the 25 g glucose it needs per day, and to maintain the function of her central nervous system. PROGESTERONE secreted during pregnancy is both natriuretic and potassium-sparing in its effects on the renal tubules, counteracting the effects of aldosterone; but overall there is a slight retention of both sodium and potassium during gestation, leading to increase in total body fluid volume and adding to maternal weight gain. The pituitary gland enlarges during pregnancy, secreting additional corticotropin and melanocyte-stimulating hormone, adding to the effects of hormones produced by the PLACENTA, although pituitary gonadotropin and growth hormone production are reduced during pregnancy (the latter an effect of hPL). Enhanced output (8-fold) of PROLACTIN contributes to the lipid mobilization and storage required by the MAMMARY GLANDS for milk synthesis and secretion. During late gestation, a mother requires a small increase of ~6–10 g per day of protein, mainly for foetal protein synthesis. Women on a strict vegetarian diet will need additional vitamin B_{12} intake (required for foetal and maternal red blood cell production), and additional FOLATE is required before and during pregnancy. A VITAMIN A intake of ~750 mg per day is recommended for adult women, and does *not* alter during pregnancy, in contrast to most vitamins. CALCIUM, IRON and ZINC are the most important minerals in the diet during a woman's pregnancy.

Nutritional requirements of pregnancy

depend to some extent upon the number of conceptuses present; but although women with a higher BODY MASS INDEX are usually advised to gain less weight before pregnancy than those who are underweight, even overweight women will need to gain a minimum of ~6 kg in order to maximize the opportunity of delivering a baby of healthy birth weight. Excessive maternal weight gain is associated with prolonged gestation. An increased daily energy intake of ~1050–1250 kJ (250–300 kcal) is needed, amounting to ~15% increase. During the first trimester, ~3 kg of maternal fat is laid down in preparation for the energy requirement of the final trimester, when foetal GROWTH is especially rapid. See DIABETES MELLITUS (for diabetic pregnancy), MISCARRIAGE, PLACENTA.

preimplantation genetic diagnosis (PGD) See SCREENING.

preinitiation complex (PIC) See TRANSCRIPTION.

preload See END-DIASTOLIC PRESSURE, STROKE VOLUME.

premenstrual tension (premenstrual syndrome, PMS, PMT) Changes in mood suffered by some women prior to menstruation, often leading to aggressive and other socially unacceptable behaviour. It is often thought that hormonal changes associated with the MENSTRUAL CYCLE are the trigger, and some women have had success with steroid therapy, although this is not always the case. Reduction in intakes of caffeine, alcohol, sugar and animal fat also help some women. Because symptoms vary between those affected and are not easily monitored, and partly because 'premenstrual' has not been clearly defined, this is a controversial condition, and its precise causation and physiology are unknown. Some women suffer characteristics of DEPRESSION as a symptom, which can sometimes be more easily treated.

premolars Permanent teeth (see DENTITION) distal to the canines and preceded by deciduous molars. In primitive mammals there are four premolars per quadrant, although the most mesial two have been lost in New World monkeys, apes and humans. Paleontologists therefore refer to human premolars as P3 and P4 (or, P_3 and P_4). By contrast, New World anthropoids have retained one of them. Sectorial premolar teeth are typical of monkeys and (non-human) apes, and some of the earliest hominins have lower 1st premolars that closely resemble a sectorial premolar.

premotor area (PMA) See PREMOTOR CORTEX.

premotor cortex (Area 6, secondary motor cortex) The region immediately rostral to the PRIMARY MOTOR CORTEX, Area 4, lying between the precentral gyrus and the PREFRONTAL CORTEX (see Fig. 29, CEREBRAL HEMISPHERES). Electrical stimulation of Area 6 can elicit complex movements on either side of the body, supporting the view that this is a 'higher' motor area. It contains two somatopically organized motor maps, one lateral (the premotor area, or PMA) and one medial (the supplementary motor area, or SMA). These perform similar functions on different muscle groups, SMA sending axons that innervate motor units directly, PMA connecting mainly with reticulospinal neurons innervating proximal motor units. The SMAs of each hemisphere form motor loops with the basal ganglia, are closely linked via the corpus callosum, and both are required for tasks involving coordinated actions involving both sides of the body, e.g., the two hands (see APRAXIA), usually previously learnt. Neurons of both the SMA and PMA are selectively active well before movements involving the appropriate musculature are given the 'command' to be actuated (see BASAL GANGLIA), indicative of their role in planning actions (see MOTOR CORTEX). The left lateral premotor cortex is a LANGUAGE AREA whose activation is related to processes involved in analysing syntactic structures.

premotor speech area See BROCA'S AREA.

prenatal diagnosis Although AMNIOCENTESIS and CHORIONIC VILLUS SAMPLING permit collection of foetal cells, they carry a 1%

risk of miscarriage. Non-invasive foetal diagnosis currently employs foetal DNA from the ~1 per million nucleated foetal cells present in maternal blood to test for paternally inherited genes causing such diseases as cystic fibrosis and β-thalassaemia, but are currently not so successful as the invasive methods at identifying Down syndrome. As yet there is no way of distinguishing foetal DNA from the maternal DNA present in maternal blood, so that it is not possible to be sure that a foetus has inherited a mutation possessed by the mother. Epigenetic markers may resolve this issue (see GENOMIC IMPRINTING). Another approach may be to analyse foetal mRNA produced by the placenta and present in maternal blood. Trophoblasts collectable in cervical swabs make up ~1 per 10^5 cells in the swab and should be easier to distinguish than foetal blood cells, although these may be so high in the cervix as to make their collection invasive. Cell-free foetal DNA is also present in maternal blood, probably deriving from the placenta, and has been used to diagnose potentially lethal Rhesus incompatibility since 2001. Some groups have reported detection of paternally inherited genes for cystic fibrosis, β-thalassaemia, achondroplasia and Huntington disease.

One study indicates that mothers who go to term knowing that their foetus has a disorder are better able to cope psychologically once the child is born than are those who learn of their baby's disorder at birth. The field is rife with ethical concerns, not least because foetal DNA gender testing is currently an available laboratory service and may encourage marketing of gender-testing kits – not least in 'one-child' countries. The impact of widespread accurate foetal DNA testing on abortion rates is also a concern. See FISH, SCREENING.

prenatal screening See DIAGNOSTIC SCREENING.

preoptic nuclei Nuclei of the anterior HYPOTHALAMUS.

presbyopia Increase with age in the near point of vision, caused by increasing rigidity of the LENS.

presenilins Proteins (PS1, PS2) which, with associated proteins, form a series of high molecular-mass, membrane-bound complexes necessary for γ-secretase and ε-secretase cleavage of selected type 1 transmembrane proteins such as amyloid precursor protein (see ALZHEIMER DISEASE, Notch and cadherins.

presynaptic facilitation See MEMORY, SYNAPTIC PLASTICITY.

presynaptic inhibition See SYNAPTIC PLASTICITY.

presynaptic receptors Receptors (presynaptic AUTORECEPTORS) in the presynaptic membrane of a synapse which bind the neurotransmitter released by the membrane and modulate its release, usually inhibiting it.

pretectum Region of the midbrain receiving projections of the optic tracts and involved in PUPILLARY LIGHT REFLEXES. See OPTIC CHIASMA, Fig. 130.

prevalence (of diseases, genetic disorders) The proportion of people having a characteristic, such as a DISEASE or genetic disorder, at a particular time. Contrast INCIDENCE; see FAMILIAL RECURRENCE RISK.

primary afferents Neurons whose cell bodies lie in the dorsal roots of spinal nerves and whose axons bring impulses from the somatic sensory receptors of the skin and olfactory receptors (see PRIMARY RECEPTORS), to which they are attached. They synapse with first-order neurons in the dorsal column (see SPINAL CORD, Fig. 163).

primary auditory cortex (A1, area 41) A region in the superior TEMPORAL LOBE of the BRAIN (see Fig. 30) sharing structural similarities with the VISUAL CORTEX. Layer I contains few cell bodies while layers II and III contain mainly small pyramidal cells. Layer IV, in which axons from the medial geniculate nucleus terminate, comprises densely packed granule cells. Layers V and VI contain mostly pyramidal cells, somewhat larger than those in layers II and III. A1 TONOTOPY is such that low frequency sounds are represented rostrally and laterally while high frequencies are represented caudally and medially, isofrequency bands running mediolaterally across A1. Electrode penetration indicates that cells with the same frequency responses are arranged vertically in a columnar organization, although no clear division into simple or complex receptive fields has proved possible and not all cells in a column are equally frequency-specific. In addition to the frequency-specificities of A1 neurons, some are amplitude-specific and others respond to clicks, bursts of noise, frequency-modulated sounds, and animal vocalizations. See APHASIAS, HEARING, LANGUAGE AREAS, Fig. 109.

primary cilia See CILIUM.

primary follicle A preantral ovarian follicle that is developing from a primordial follicle prior to production of receptors for oestrogens or pituitary FSH (granulosa cells) or for LH (thecal cells). It contains a primary oocyte. See OOGENESIS.

primary immune response See ADAPTIVE IMMUNITY.

primary lymphoid organs Organs in which stem cells divide and mature into B cells and T cells. They include red bone marrow and the thymus gland.

primary motor cortex (Area 4, M1) Anatomically, the precentral gyrus of the frontal lobe of each cerebral hemisphere (see BRAIN, Fig. 28), but functioning with the premotor cortex and supplementary motor cortex in the control of voluntary, skilled, movements (see MOTOR CORTEX). Electrical stimulation reveals a detailed SOMATOTOPIC MAP – the area of cortex allocated to each body region being proportional to the degree of motor precision achieved (see CEREBRAL CORTEX, Fig. 44 for 'motor homunculus'). Stimulation of the primary motor cortex elicits muscle contraction on the opposite (contralateral) side of the body, with ~30% of corticospinal (PYRAMIDAL TRACT) and corticobulbar fibres arising from neurons here. The principal projections into the primary motor cortex originate in the ventral lateral nucleus of the THALAMUS (see Fig. 173). See NEUROROBOTICS.

primary oocytes Developing from OOGONIA, these cells lie within PRIMARY FOLLICLES of the ovaries, and some of them undergo further stages of OOGENESIS.

primary receptors Neuronal receptor cells in which stimulus transduction occurs in the cell itself rather than in an associated cell of non-neuronal origin. Stretch receptors, olfactory receptors and nerve endings in the dermis of the skin are examples. Their axons are PRIMARY AFFERENTS. Contrast SECONDARY RECEPTORS.

primary somatosensory cortex (SI; Brodman's areas 1, 2, 3a and 3b) Located in the postcentral gyrus of the parietal lobes of the cerebral hemispheres (see BRAIN, Fig. 28), receiving projections from neurons of the ventroposterolateral nuclei (VPL) of the THALAMUS (see Fig. 173). These in turn have received projections from the ventroposterolateral nuclei (VPL) of the THALAMUS (see Fig. 173) from receptors in the skin, joints and muscles via the DORSAL COLUMN–MEDIAL LEMNISCAL SYSTEM (DCMLS). SI is concerned with tactile discrimination and the ability to perceive shape by touch (*stereognosis*). Thalamic inputs terminate mainly in layer IV (see Fig. 140) and then project to cells in other layers. In each hemisphere, the contralateral half of the body is represented in SI as an inverted *somatotopic map* resembling that of the PRIMARY MOTOR CORTEX (see CEREBRAL CORTEX, Fig. 46, for 'somatosensory homunculus'). This was discovered by recording the activities of single neurons and determining the site of its somatosensory RECEPTIVE FIELD on the body. Similar maps were discovered by electrically stimulating the SI surface and recording the somatic sensations generated in specific parts of the body. Neurons with similar inputs and responses are stacked vertically into radially arranged columns extending through the cortical layers, resembling those in the PRIMARY VISUAL COR-

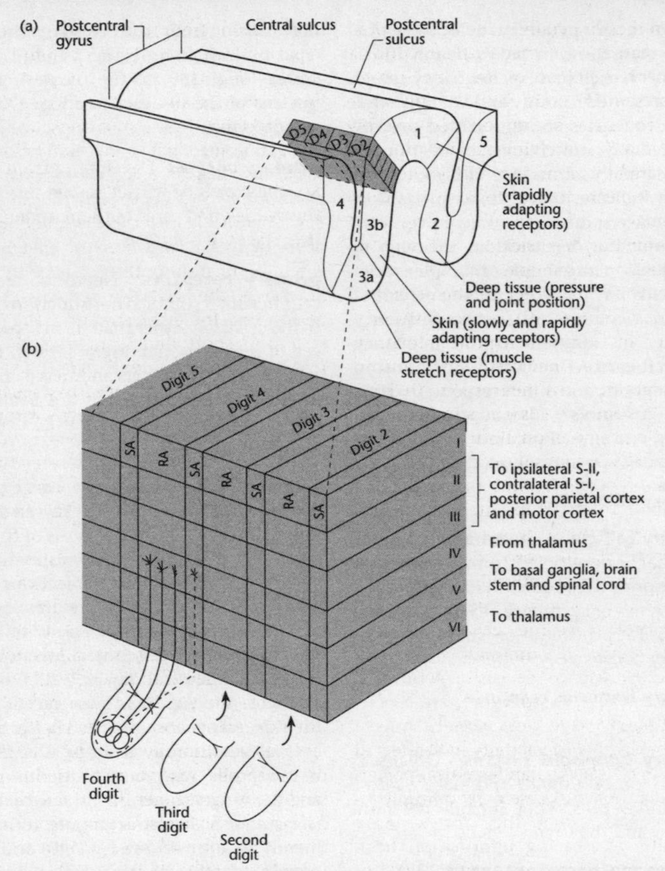

FIG 140. Diagram indicating how input to the PRIMARY SOMATOSENSORY CORTEX is organized by submodality, this example illustrating input from the digits. In (a), the somatotopic map in Brodman's area 3b is shown representing inputs from the skin of the digits. In (b), the columnar arrangement of these inputs is shown, with more detailed information concerning the six cortical layers in each vertical column. RA, rapidly adapting sensors; SA, slowly adapting sensors. From *Essentials of Neural Science and Behaviour*. E.R. Kandel, J.H. Schwartz and T.M. Jessell © 1995 by Appleton & Lange (Simon & Schuster). Reproduced with permission of The McGraw-Hill Companies.

TEX. Each receives input from a single type of receptor in a particular location, adjacent places having adjacent columns. Synaptic connections within each column are numerous, but few occur between columns (see PARALLEL PROCESSING).

The first step in somatosensory processing occurs in 3b (see Fig. 140), which receives most of the thalamic input. 3b neurons project to layer IV of areas 1 or 2. Receptive fields of neurons in 3b are simpler than those in areas 1 and 2. Each of the four regions of SI has a distinct map: cutaneous input maps to the core of the VPL and then to areas 1 (concerned with analysis of texture) and 3b (concerned largely with tactile discrimination: the size, and shape of an object). Proprioceptor input maps more peripherally in the VPL and projects to areas 2 (concerned with direction of movement of objects on the skin surface from the entire body surface, and with stereognosis, the perception of shape by touch) and 3a (receiving further proprioceptive input and projecting back to area 2). Area 2 has reciprocal connections with the primary MOTOR CORTEX (see entry), possibly providing feedback on the sensory consequences of movement.

Fibres from SI project to the *secondary somatosensory cortex* (SII), and to all the subcortical structures which project to it, to the basal ganglia and brain stem, and to the thalamus. Many neurons in SII have bilateral receptive fields (i.e., stimuli in corresponding regions on both sides of the body evoke a response), made possible by decussation of the DCMLS and by the fact that inputs from the ipsilateral body surface enter SII from the contralateral side of the brain via the CORPUS CALLOSUM. Integrating inputs from both sides of the body, SII plays a major role in formation of a unified perception of the whole body and enables tactile discriminations learned using one hand to be easily performed using the other (see SELF-AWARENESS).

In the mid-1990s, work on rats showed that blocking neuron activity in SI affected responses of their thalamic ventral posterior medial nuclei. These descending feedback signals from the cortex were modulating ascending information from the rat brain stem. In humans, ~40% of axons in the CORTICOSPINAL TRACTS originate in the primary somatosensory cortex, or in other regions of the parietal cortex. These descending fibres terminate in either the dorsal horns of the SPINAL CORD or the reticular nuclei of the brain stem where they, too, regulate sensory input. These back projections have a somatotopic mapping in register with the DCML pathway and are probably a vehicle for somatosensory input filtering during ATTENTION. See NEURAL CODING.

primary visual cortex (striate cortex, V1, Brodman's area 17) 'Visual area 1' (V1), being the first cortical area to receive information from the LATERAL GENICULATE NUCLEUS (LGN; see Fig. 111 and OPTIC CHIASMA, Fig. 130), and comprising the grey matter of the medial surface of the occipital lobes. It is ~2 mm thick, and occupies gyri immediately above and below the calcarine sulcus (see Fig. 46, CEREBRAL HEMISPHERES), in whose depths much of it is hidden. There are at least three parallel streams of input to V1, and it is the first region where input from both eyes is combined. Retaining the LGN's retinotopy, it is responsible for deconstructing the VISUAL FIELD into short line segments, the cells involved being sensitive to particular orientations of bars of light, an early stage in the PARALLEL PROCESSING of an object's form and motion (see ABSTRACTION, PARVOCELLULAR-BLOB/INTERBLOB PATHWAYS). Information relating to colour vision is processed through regions termed BLOBS, pillars of cytochrome oxidase-rich cells.

The striate cortex receives thalamic projections from the LGN and has at least nine anatomical cell layers; but in order to maintain the convention (due to Brodmann) that there are six neocortical layers, layer IV is subdivided into layers IVA, IVB and IVC. Most afferent fibres from the LGN terminate (and synapse with stellate cells) in layer IVC, retinotopy ensuring that neighbouring cells receive their inputs from neighbouring regions in the retina (compare PRIMARY SOMATOSENSORY CORTEX). Axons from the form-and-wavelength-selective parvocellular (P) cells in the LGN terminate mostly in layer IVCβ, those from movement-sensitive magnocellular (M) cells terminating principally in layer IVCα, and those from koniocellular layers, signalling average illumination, projecting to layers II and III (see Fig. 141). These streams remain quasi-independent throughout the visual system.

A subset of cells in layer III also receives direct input from the LGN. Inputs from the two eyes are combined through the *ocular dominance columns* (see Fig. 146, and HYPERCOLUMNS). The koniocellular pathway arises from a different subset of ganglion cells that are neither P cells nor M cells whose axons project into the koniocellular layers

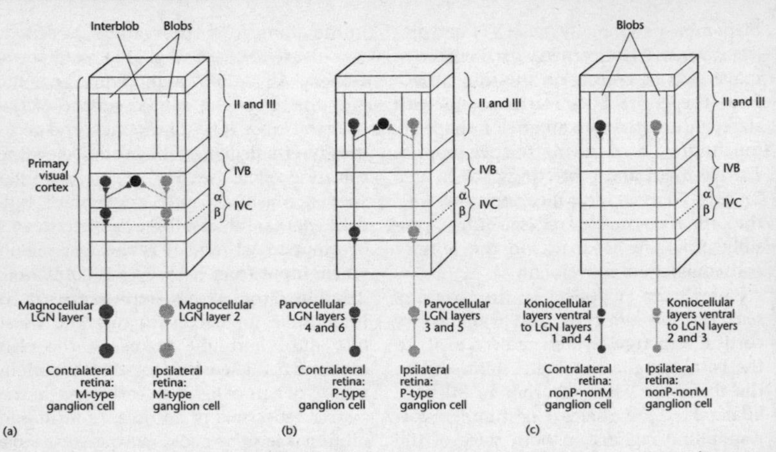

FIG 141. Diagram showing three parallel pathways from the retina to the PRIMARY VISUAL CORTEX (see entry for details). (a) Magnocellular pathway; (b) parvocellular pathway; (c) koniocellular pathway. © 2001 Redrawn from *Neuroscience* (2nd edition) by M.F. Bear, B.W. Connors and M.A. Paradiso, published by Lippincott, Williams & Wilkins. Reproduced with the kind permission of the publisher.

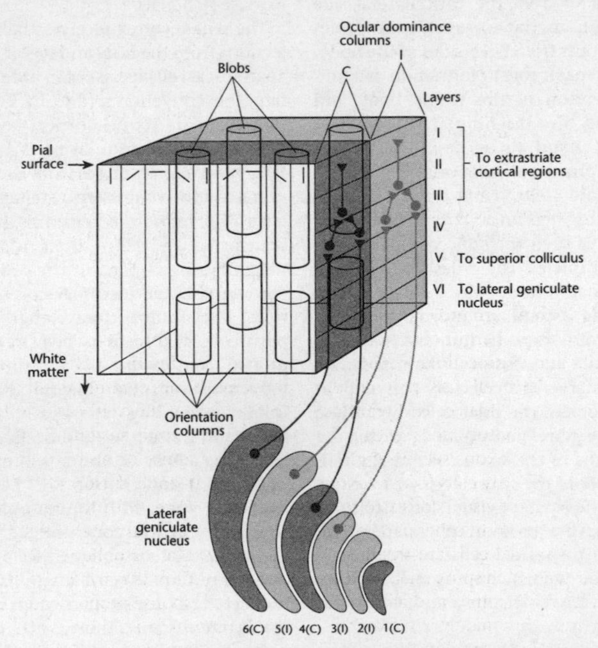

FIG 142. Diagram indicating how the visual field input from the lateral geniculate nucleus (below) is represented in the PRIMARY VISUAL CORTEX (above) by a regular pattern of hypercolumns. When visualized at the level of IVC, the columns representing ipsilateral (I) and contralateral (C) input alternate in the cortex similar in appearance to the stripes of a zebra. From *Essentials of Neural Science and Behaviour*. E.R. Kandel, J.H. Schwartz and T.M. Jessell © 1995 by Appleton & Lange (Simon & Schuster). Reproduced with permission of The McGraw-Hill Companies.

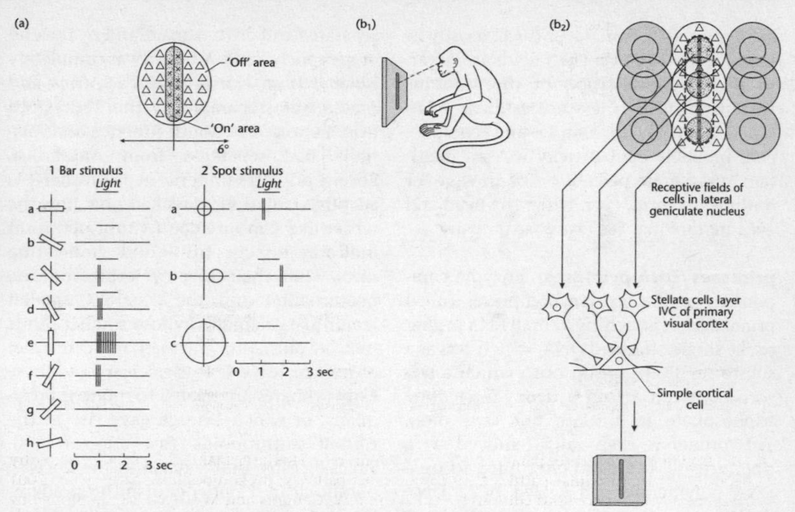

FIG 143. Results of experiments by Hubel and Wiesel (1959, 1962) indicating that when a macaque faces a target screen on which bars of light are projected (b_1) a simple cell in the PRIMARY VISUAL CORTEX responds maximally if its receptive field has a narrow rectangular central excitatory area (x) flanked by symmetrical inhibitory areas (Δ). Patterns of action potentials fired by the cell in response to two types of stimuli are shown, response to the bar of light being strongest when the bar's orientation is parallel to the axis of the excitatory zone in the cell's receptive field (orientation e). Spots of light elicit weak responses, no response, or weak inhibitory responses, depending on their position in the receptive field. b_2: The rectilinear receptive field of a simple cell may be generated by convergence of excitatory inputs from three (or more) cells in the lateral geniculate nucleus. In this model, converging cells have similar centre-surround organizations, being excited together when light falls along a straight line in the retina and strikes the receptive fields of all these cells. Convergence would create an elongated excitatory region in the simple cell's receptive field, indicated by the dashed outline linking the three stippled receptive field centres in the top diagram. © 2001 Redrawn from *Neuroscience* (2nd edition) by M.F. Bear, B.W. Connors and M.A. Paradiso, published by Lippincott, Williams & Wilkins. Reproduced with the kind permission of the publisher.

of the LGN which in turn projects into layer III blobs, which thus permit CONVERGENCE of parvocellular and koniocellular inputs.

There are two basic cell classes: pyramidal cells, and non-pyramidal cells. Pyramidal cells are large, excitatory cells with long spiny dendrites and axons secreting glutamate. They are the only cells to send axons out from the primary visual cortex to form synapses with neurons in other parts of the brain. Non-pyramidal cells are small, stellate, cells among which spiny stellates have numerous dendritic spines and are excitatory (secreting glutamate) while others, smooth stellates, have no dendritic spines and are inhibitory (secreting GABA). Cells in all regions, other than the blobs in the superficial layers, respond best to stimuli with linear properties rather than to 'spots' of light. So-called *simple cells*, located near the input layer from the LGN (layer IVCβ), are pyramidal cells with receptive fields that have either a rectilinear 'on' (excitatory) zone surrounded by an 'off' inhibitory zone, or else a rectilinear 'off' (inhibitory) zone surrounded by an 'on' excitatory zone, with the long axis of the rectangular central zone being either vertical, horizontal or oblique (see Fig. 143a, and ORIENTATION-SELECTIVE CELLS). Cells with the same response properties are organized into columns and those with the same function communicate by horizontal links. *Complex cells* are also pyramidal cells with rectilinear receptive fields and specific axes of orientation, but lie at a greater distance from layer IVCβ, mostly in layers II, III, V and VI. Their fields are larger, so the precise position of the stimulus within the field is less critical. The combined activities of simple and complex cells probably explain why we can maintain an accurate perception of an edge, or outline of a form, even when the head and eyes are moving. See EXTRASTRIATE CORTEX.

primases (DNA primases) Enzyme components of the protein complexes called primeosomes attaching a small RNA primer to the single-stranded DNA, which acts as a substitute 3´ OH group onto which a DNA POLYMERASE can attach a deoxy-nucleotide triphosphate to initiate DNA REPLICATION. This primer is eventually removed by a ribonuclease, the gap being filled using a DNA polymerase and sealed by a DNA ligase.

Primates The term 'Primate' (*'primum'*, Latin for 'first') was first coined by Linnaeus, although the order Primates is not well defined by a suite of derived morphological characters. Nevertheless, these do include: binocular vision and a shortened muzzle (snout), grasping hands and feet and considerably larger brains relative to their body size than other mammals. The plesiadapiforms, a group of New World fossils, probably resemble the earliest primates. Inhabiting wet and warm broadleaf evergreen forests in the Palaeocene (~65–54 Mya), they were small to medium-sized mammals, some terrestrial others arboreal, lacking binocular vision. Having claws rather than nails on their digits, they are not usually considered primates. The recently discovered *Carpolestes simpsoni* had hands and feet capable of grasping, a trait which probably characterized the common ancestor of modern primates. Its eyes were incapable of binocular vision, evidence that grasping hands and feet preceded movement of eyes to the front of the head in primate evolution.

With the development of wetter and warmer global conditions in the Eocene (~54–34 Mya), large tropical forests spread while Eurasian and North American plates separated and drifted apart, and by the end of the epoch South America was completely isolated from North America. Africa and Arabia were separated by the Tethys Sea from Eurasia, and South America and Australia had separated from Antarctica. Eocene primates underwent a considerable adaptive radiation, classified now into the tarsier-like Omomyidae (whose dentition indicates insect-, fruit- and gum-eating diets, while their large eye orbits indicate nocturnality) and the generally smaller lemur-like Adapidae (whose smaller orbits indicate diurnality and teeth indicate a diet of fruit or leaves). It is unclear how these early primates are related to modern prosimians, or which lineage gave rise to the earliest anthropoids (see ANTHROPOIDEA). Such ancestry may extend to the Eocene; but the clearest evidence of anthropoid fossils occurs at Fayum, in Egypt, whose deposits extend from 36 to 33 Mya and straddle the Eocene–Oligocene boundary. These diverse primate deposits indicate a complex primate genealogy, including the diverse parapithecids (possibly ancestral to both Old World and New World monkeys); propliopithecids (e.g., *Aegyptopithecus zeuxis*), which were probably arboreal quadrupeds and possibly ancestral to the modern Old World monkeys; and oligopithecids, whose dental formula was the same as that of modern Old World monkeys and apes, although may have been arrived at by evolutionary convergence (i.e., they may be homoplasious; see CLADISTICS).

During the Oligocene (34–23 Mya), Atlantic sea levels were lowest during the middle of the epoch, and temperatures generally fell throughout. Tropical broadleaf evergreen forests were replaced by broadleaf deciduous forests. It is possible that a transatlantic bridge between South America and Africa made it possible for anthropoids to colonize the New World by this route; but a northern colonization route is difficult to equate with the complete absence of North American fossil anthropoids from any time in history. Primates appear in South America for the first time in the late Oligocene, both these and Caribbean fossils resembling African forms in the Fayum. The incompleteness of the primate fossil record probably leads to our underestimating the age of phylogenetic lineages, possibly by as much as 40%. If so, the base of the anthropoid clade may lie as far back ~50 Mya, providing good reason for current uncertainty.

At its start, the Miocene (~23–5 Mya) was generally warmer than the Oligocene and tropical evergreen forests spread once more across Eurasia, although the rise of the Himalayas was probably instrumental in the cooling which took place in the Late Miocene. After Africa joined Eurasia at ~18 Mya, the East African Rift started to appear, the newly elevated mountains causing the tropical rain forests of East Africa to be replaced by drier woodlands and savannas. Although a number of Early Miocene hominoids are found in Africa, *Proconsul* is the best-known early genus. It had several derived ape-like features, e.g., lack of a tail and ischial callosities and a relatively large brain in proportion to body size. But in its arm-to-leg length ratio, and its narrow but deep thorax, it resembled other contemporary quadrupedal primates although its opposable thumb probably indicates a departure in climbing and/or feeding mode. The Middle Miocene (~15–10 Mya) arboreal hominoids include *Kenyapithecus* from East Africa, *Oreopithecus* and *Dryopithecus* from Europe, and *Sivapithecus* from southern Asia. Their thicker tooth enamel (a hominin trait), flared ZYGOMATIC ARCHES, and more robust lower jaws indicate a diet of harder and more fibrous foods than was eaten by the proconsulids, indicative of a subtropical climate that included a lengthy dry season. The Late Miocene *Ankarapithecus* has relatively small canines, once considered to be a hominin trait but now recognized in several fossil ape groups. However, *Ankarapithecus* represents a significant group of fossil apes which, although four-footed, display some measure of ground-adapted anatomy. The dryopithecine/sivapithecine and ankarapithecine groups do not appear to be closely related. But during the Late Miocene, perhaps ~7–5 Mya, putative human ancestors become recognizable, usually evidenced by signs of adaptations to BIPEDALISM (see HOMININI).

Cladistic classification divides living primates into strepsirhines (lemuroids and lorisoids) and haplorhines (tarsiers and anthropoids). The latter are further subdivided into platyrrhines (New World monkeys and marmosets) and catarrhines (Old World monkeys, Cercopithecoidea; and Hominoidea: gibbons, great apes and humans).

primitive character Term referring to a character which is a synapomorphy for the members of a particular clade but whose evolutionary origin pre-dated the origin of the clade. Contrast DERIVED CHARACTER.

primitive groove See PRIMITIVE STREAK.

primitive node (organizer, node, Hensen's node) Small slightly elevated region surrounding the small primitive pit at the cephalic end of the PRIMITIVE STREAK and largely comprising prospective notochord. It is maintained by expression of *HNF-3β*. The primary cilia (see CILIUM) of nodal cells waft fibroblast growth factor-8 (FGF-8) to the left of the embryo and establish secretion on the left side, near the node, of the TGFβ superfamily member and morphogen, *Nodal*, which maintains the primitive streak and is essential for development of the left–right AXIS (see entry). Indeed, nodal cilia seem to be at the heart of mammalian symmetry-breaking. The node is alone among cells of the embryonic disc in expressing *goosecoid*, whose transcription factor product activates *chordin*, whose secreted product, along with those of *noggin* and *follistatin*, antagonizes BMP4 activity. All mesoderm would be ventralized were it not for the blockage of BMP4 by expression of these genes by the node (termed the *organizer* on this account). The node secretes increasing quantities of RETINOIC ACID as it regresses so that posterior cells lie in an environment containing a greater concentration of this morphogen, leading to progressive activation of more HOX GENES in these posterior embryonic regions. See BILATERAL SYMMETRY, SOMITES.

primitive streak Longitudinal thickening at GASTRULATION in the EMBRYONIC DISC caused by cell migration from the posterior EPIBLAST towards the centre (see AMNIOTIC CAVITY, Fig. 8). It is initiated and maintained by expression by the primitive node of *Nodal*, a member of the *TGF-β* gene family. Once formed, a number of genes regulate formation of dorsal and ventral mesoderm, and head and tail structures. The streak narrows, thickens and extends anteriorly, marking the anterior-posterior embryonic AXIS. In humans, at 14–15 days of development, cells converge on the streak and a depression (the *primitive groove*) forms within it, through which cells migrate into the blastocoele, displacing the hypoblast to form endoderm, while mesodermal cells ingress between the epiblast and nascent endoderm at day 16. Cells remaining in the epiblast come to form ectoderm. Regions of the epiblast migrating and ingressing through the primitive streak have been mapped and their cell fates discovered. Cells ingressing through the cranial region of the node become NOTOCHORD; those migrating at the lateral edges of the node and from the cranial end of the streak become PARAXIAL MESODERM (see SOMITES); those migrating through the midstreak region become INTERMEDIATE MESODERM; those migrating though the more caudal part form LATERAL PLATE MESODERM, and cells passing through the most caudal part of the streak contribute, with the primitive yolk sac, to EXTRAEMBRYONIC MESODERM. Normally disappearing by the end of the 4th week, if it persists in development it may give rise to sacrococcygeal TERATOMAS.

primordial germ cells (PGCs) Cells of the germ line, from which the gametes will ultimately be derived. First appearing in the early post-implantation embryo among the endoderm cells in the wall of the yolk sac, close to the allantois, these amoeboid cells migrate along the dorsal mesentery of the hindgut and arrive in the primitive gonads at the start of the 5th week and invade the genital ridges in the 6th. Their failure to arrive results in a lack of gonad development and indicates their importance as inducing agents. PGCs of the embryo begin with highly methylated DNA inherited from germ cells, but progressive demethylation occurs during development (see CHROMATIN REMODELLING) so that by the time they enter the primitive gonads it is almost complete, epigenetic reprogramming which also involves methylation of histone H3K9. The somatic gene-expression programme needs to be silenced, however, and *de novo* methylation takes place as PGCs develop, leading to substantial methylation of sperm and eggs – more so during SPERMATOGENESIS than OOGENESIS. Sex-specific differences in methylation patterns are evident, notably at imprinted loci (see GENOMIC IMPRINTING), and the cytosine methyltransferase gene *Dnmt1* in mice is known to have sex-specific exons. Pluripotency-associated genes, and those with roles in later germ-cell development, are also repressed by DNA methylation, and their expression commences with its erasure. PGCs at this time have similar properties to EMBRYONIC STEM CELLS, including ability to form embryonic germ cells in culture.

primordial plexiform layer (PPL) See CEREBRAL CORTEX.

principal cells (P cells) Kidney cells lining the end of the DISTAL CONVOLUTED TUBULE, and COLLECTING DUCTS, reabsorbing Na^+ and secreting K^+, especially in response to ALDOSTERONE, and responsible for the kidney's response to VASOPRESSIN (ADH). See INTERCALATED CELLS.

prions Prion diseases (transmissible spongiform encephalopathies (TSEs)) are now defined by the transmission and involvement in neuropathogenesis of an abnormal form of prion protein (PrP). Its normal cellular form, PrP^C, is a normal glycoprotein constituent of brain tissue that is usually monomeric, protease-sensitive and linked to cell membranes through a glycophosphatidylinositol (GPI) anchor. Its function is currently unknown. Pathological forms of PrP vary, although most are commonly multimeric (aggregated) and more protease-resistant than PrP^C. Collectively, these are usually known as PrP^{Sc} (Sc for 'scrapie') although different terms have been used. It is the interaction of this with PrP^{C} that causes the latter to be converted to PrP^{Sc}. In inherited forms of prion diseases, a mutation in the gene for PrP produces a single amino acid substitution that is thought to make the conversion of the normal to the pathological form of PrP more likely. See CREUTZFELDT–JAKOB DISEASE.

PRM1, PRM2 See PROTAMINES.

proctodaeum Depression of ectoderm forming in the embryonic hindgut and developing into the anus. See Appendix VI.

***prodynorphin* gene (*PDYN*)** A gene whose protein product in neurons is a precursor for opiate compounds involved in perception, pain, social behaviour, and learning and memory. The gene is under the influence of a 68-bp regulatory sequence, of which humans have four copies compared to the single copy of other great apes and monkeys. Five base changes have occurred in the sequence since the hominin and chimpanzee lineages separated, which seems to enable human cells to make more of the protein.

progenitor cells Multipotent dividing cells, derived from STEM CELLS, that have made some of the decisions committing them to a particular differentiation pathway (e.g., see COLON, Fig. 59). Some believe that they could provide more suitable therapeutic agents than stem cells because they avoid the problem of uncontrolled differentiation that has been prevalent in MOUSE MODELS using embryonic stem cells.

It is thought that transformation of progenitor cells into cancer cells requires adoption of a stem cell-like ability for SELF-RENEWAL, and the WNT SIGNALLING PATHWAY has emerged as a critical regulator of both progenitor and stem cells. Overactivation of this pathway and dysregulation of progenitor cell differentiation in CRYPTS OF LIEBERKÜHN leads to ADENOMAS (e.g., COLORECTAL CANCERS), and in haematopoietic cells leads to LEUKAEMIAS.

progerias Rare inherited age-accelerating syndromes. There is a growing list of 'progeroid' syndromes that are linked to defects in DNA REPAIR MECHANISMS but which do not exactly equate with an accelerated process of normal AGEING, primarily because the symptoms are tissue-specific. These include Cockayne syndrome, WERNER SYNDROME, ataxia telangiectasia (see ATAXIA TELANGIECTASIA MUTATED PROTEIN), and trichothiodystrophy.

progesterone A steroid hormone, synthesized from CHOLESTEROL (see Fig. 50) by the theca and granulosa cells of the corpus luteum, when it is produced mainly in the luteal phase of the MENSTRUAL CYCLE and by the syncytiotrophoblast cells of the PLACENTA (when the cholesterol is derived from circulating LDL). By about 8 weeks of gestation, the placenta is synthesizing and releasing progesterone in large amounts, with important effects in reducing myometrial contraction during PREGNANCY, preventing endometrial shedding, inhibiting prostaglandin synthesis, and modulating the immune response in order to preserve pregnancy. Most (~80%) of plasma progesterone circulates bound to albumin, the principal targets being the reproductive tract and the hypothalamic–pituitary axis and its degradation occurs mainly in the liver. Two progesterone NUCLEAR RECEPTORS occur, A and B, encoded by a single gene, acting as ligand-inducible transcription factors and up-regulated by OESTROGEN and down-regulated by progesterone in most target tissues (see MENSTRUAL CYCLE for much of progesterone's activity). *Mifepristone* is a progesterone antagonist (see BRCA1, CONTRACEPTION).

progestins Steroid hormones produced by the ovaries. The most important is progesterone, secreted by the corpus luteum in the second half of the MENSTRUAL CYCLE, and by the PLACENTA. See OESTROGENS.

prognathous, prognathic Describing the positional relationship of the mandible and maxilla to the skeletal base where the jaws protrude beyond a predetermined imaginary line in the sagittal plane of the skull, producing a distinct muzzle. Contrast ORTHOGNATHOUS.

programmed cell death See APOPTOSIS.

projection fibres Nerve fibres passing between the cerebral cortex and such subcortical structures as the THALAMUS, STRIATUM, BRAIN STEM and SPINAL CORD. Compare ASSOCIATION FIBRES, COMMISSURAL FIBRES.

prolactin Structurally related to GROWTH HORMONE, a polypeptide synthesized by lactotrophs of the anterior PITUITARY GLAND, the percentage of these cells increasing dramatically with raised oestrogen levels (esp. during PREGNANCY). Its levels are higher in females than in males (highest during sleep), its role in males being imperfectly understood. Its release is under tonic inhibition by DOPAMINE from paraventricular and arcuate nuclei of the hypothalamus, but stimulated by serotonergic and opioidergic pathways and GnRH. The prolactin receptor is a CYTOKINE RECEPTOR, found in epithelial cells of the MAMMARY GLANDS (in very high densities), ovaries and various brain regions, whose dimerization on binding prolactin activates tyrosine kinase phosphorylations via the JAK/STAT PATHWAY. It stimulates glucose and amino acid uptake by mammary gland, synthesis by the mammary epithelial cells of milk protein (casein), α-lactalbumin, lactose and milk fats, and maintenance of milk secretion. The medium-chain fatty acids caprylic (octanoic) and capric (decanoic) acids are produced specifically by mammary gland and no other tissue.

Lipid is manufactured in the endoplasmic reticulum of the alveolar cell and leaves in the form of lipid droplets which coalesce and eventually push against the plasma membrane forming a bulge which 'pinches off' as lipid is released into the lumen of the alveolus. This is believed to be prolactin-induced. During pregnancy, prolactin prepares the breast for LACTATION. It can act as a cytokine, being released by immune cells and regulating lymphocyte responses paracrinally and autocrinally. Its oversecretion is a common cause of female INFERTILITY.

prometaphase The phase of MITOSIS when the nuclear envelope disintegrates.

promoter, promoter region A combination of short nucleotide sequence REGULATORY ELEMENTS (hence non-coding), usually just upstream (on the 5′ side) of the transcription start site of a gene, to which RNA POLYMERASE (Pol) binds just prior to the gene's TRANSCRIPTION (see entry). In the case of protein-coding genes, the minimal DNA sequence needed to specify non-regulated, or basal, transcription is known as the *core promoter*. This serves to position Pol II and the general transcription factors. *Promoter-proximal elements* are sequences located within 100–200 bps of the transcription initiation site to which other TRANSCRIPTION FACTORS (TFs) bind and increase transcription. These elements, which include the CCAAT box and GC-rich elements further upstream, are bound by TFs that are always constitutively expressed, in all cells and at all times, so that they can promote transcription in all cell types. The core promoter's function is supplemented by ENHANCERS or SILENCERS, possibly thousands of nucleotide pairs away, where regulatory proteins bind to activate or repress transcription of the gene. Several mammalian genes have two or more alternative promoters, driving transcription from alternative versions of the first EXON which is then spliced in each case to a common set of downstream exons. Some alternative promoters are positioned in more distal portions of a gene (e.g., the DYSTROPHIN gene) and drive expression of truncated products. Use of alternative promoters typically involves use of alternative exons for the start of transcription. Methylated promoter regions correlate with transcriptional silencing (e.g., see FRAGILE X SYNDROME), and some repressors involved in gene silencing recognize docking sequences present in many gene promoters. Promoters and enhancers can be identified by the modification state of the nucleosomal histones in which they reside. The ability to attach green fluorescent protein (GFP) and similar fluorescent tags to chromosomal proteins has enabled the temporal relationships of protein–protein and protein–DNA interactions (e.g., using UV-ChIP; see DNA-BINDING PROTEINS) in nuclei and on chromosomes to be analysed. An early (1986) finding in the fruit fly *Drosophila* was that heat shock genes had a Pol II complex associated with their promoter and engaged in transcription before activation, but that transcription was repressed by two further complexes, one of which included the 5-subunit negative elongation factor, NELF. Recent evidence appears to suggest that this 'promoter-pausing' may be a more widespread phenomenon for Pol II notably in vertebrate genes (e.g., human STEM CELLS). The long-established view that the major mechanisms of gene regulation in higher eukaryotes resides either at the level of recruitment of Pol II to promoters, or transcription initiation, may have to be reviewed in the light of these findings. How activator proteins interface with paused Pol II and influence its escape from elongation arrest remains to be discovered. See CPG DINUCLEOTIDE, TRANSCRIPTION FACTORS.

promoter-proximal elements (upstream promoter elements, UPEs) See PROMOTER.

pronation Rotation of the forearm so that the anterior surface is down. See ELBOW.

pronephros The most cranial of the three kidney systems to develop in humans, albeit rudimentary and non-functional. At the beginning of the 4th week, it is represented by 7–10 solid groups of cells in the cervical region. These form vestigial excretory units, nephrotomes, that regress as more caudal ones form. By the end of the 4th week, all indications that it existed have disappeared. See MESONEPHROS, METANEPHROS.

pronuclei The two haploid gamete nuclei when co-present in the mature ovum but prior to release of their chromosomes. See FERTILIZATION, OOGENESIS, SPERM.

proof-reading During transcription, errors arise when the growing mRNA incorporates a nucleotide that is not complementary to the DNA template. In bacteria at least, RNA polymerase within an erroneous complex back-steps along the DNA and RNA, the terminal non-complementary nucleotide then participating in catalytic self-removal, along with the penultimate nucleotide. This RNA cleavage provides a new, reactive RNA end and a free adjacent substrate site, allowing transcription to continue. See DNA POLYMERASES, DNA REPAIR MECHANISMS, MISMATCH REPAIR, RIBOSOMES

proopiomelanocortin (POMC) A large polypeptide pro-hormone produced by corticotrophs of the anterior PITUITARY GLAND that is cleaved progressively to give the hormones CRH, corticotropin (ACTH), α-MSH, γ-MSH, β-lipotropin, γ-lipotropin, β-endorphin and β-MSH. The melanocortin, corticotropin and β-endorphin opiate receptors are also by-products. It is now known to be produced in skin and hair follicles (e.g., see HAIR COLOUR). Expression of the *POMC* gene is promoted by LEPTIN, via the JAK/STAT pathway, and suppressed by glucocorticoid hormones. Mutations in *POMC* can cause metabolic disorders, including adrenal insufficiency and OBESITY.

prophase The first 'stage' of MITOSIS or of the first or second divisions of MEIOSIS.

propionic acid (propionate) A 3-carbon fatty acid added as a mould inhibitor to some breads and cereals and so entering the diet. Absorbed into the blood, it is oxidized in the liver and other tissues. See FATTY ACID METABOLISM.

proprioception, proprioceptors Proprioceptive information is non-visual, non-auditory sensory information relating to the body's position and acceleration in space, and thereby influencing its equilibrium. The sensory receptors involved are MECHANORECEPTORS, located in muscles and joints, and include MUSCLE SPINDLES, TENDON ORGANS and JOINT KINESTHETIC RECEPTORS. Proprioceptor pathways from the neck, arms and upper trunk (conscious upper body proprioception) are relayed in the dorsal columns to the cuneate nucleus, whence it follows that same path as touch sensation from these areas. The pathway for conscious lower body proprioception (from the lower trunk and legs) is distinct, afferents entering the dorsal nucleus in lamina VII of the medial dorsal horn and extending along the spinal cord (C8–L3). Second-order neurons then ascend ipsilaterally forming the dorsal SPINOCEREBELLAR TRACT, whence collaterals enter the medulla and then via the medial lemniscus to the cortex. See SOMATIC SENSORY SYSTEM.

prosaccade Eye movements towards a target. Contrast ANTISACCADE; see SACCADES.

prosencephalon The brain vesicle (see BRAIN, Fig. 28) giving rise to the forebrain. Comprises the telencephalon (forming the cerebral hemispheres) and the diencephalon (forming the optic cup and stalk, pituitary, thalamus and epiphysis).

Prosimii (prosimians) Primitive, nocturnal and mostly arboreal primates (Suborder Prosimii), classically including lemurs, lorises and tarsirs. The eye orbits are incompletely enclosed in bone at the rear and less frontally positioned than in anthropoids (see ANTHROPOIDEA), while the olfactory lobes are larger and the snout longer than in anthropoids, terminating in a wet muzzle (rhinarium). The incisors and canines of the prosimian lower jaw form a dental comb/scraper, the diet being more insectivorous than in anthropoids and the brain size being smaller relative to the body. Neither frontal bone nor the two halves of the lower jaw were fused. While evolutionary taxonomists include tarsirs in the Prosimii, cladistic taxonomists would move them into the Anthropoidea.

prosopagnosia In its congenital form, the deficit in face processing apparent from early childhood in the absence of any underlying neurological disorder and in the presence of intact sensory and intellectual function. Acquired prosopagnosia has usually been documented in those who have sustained brain damage as adults. See FACIAL RECOGNITION, FUSIFORM GYRUS.

prostacyclin An ANTICOAGULANT prostaglandin of endothelial origin. See BLOOD CLOTTING.

prostaglandins (PGs) Lipids derived from the unsaturated fatty acid ARACHIDONIC ACID. Their diversity of effects in different tissues is due to the occurrence of different prostaglandin receptors and the particular G proteins that mediate their signalling cascades. PGE_1 and PGE_2 relax vascular smooth muscle and are powerful vasodilators (e.g., in bronchi), but in the gut and uterus they cause smooth muscle contraction instead (but see STOMACH). Placental $PGF_{2\alpha}$ promotes uterine contraction during BIRTH.

prostate gland A single, doughnut-shaped gland lying inferior to the urinary bladder and surrounding the prostatic urethra. Increasing slowly in size from birth to puberty, it then expands rapidly to the size of a chestnut. Further enlargement may occur in later life. It secretes a milky, slightly acidic fluid released during ejaculation (see SEMEN) and containing citric acid (used by sperm for ATP production), acid phosphatase and several proteolytic enzymes (e.g., prostate-specific antigen, pepsinogen, lysozyme, amylase and hyaluronidase). It is homologous to the female paraurethral glands and derived from the Wolffian ducts. The risk of prostate cancer has been reliably associated with chromosome region 8q24.

protamines Small proteins, evolutionarily related to HISTONE H1, and unique among proteins of male reproduction in having functional analogues in somatic histones and, evolving rapidly in the human lineage, their genes are at the opposite end of the spectrum in terms of evolutionary stability (see NATURAL SELECTION). They have very low lysine content, and with >50% of their residues arginine, this accounts for their high DNA-binding affinity and possibly for their better chromatin condensing properties. Whereas most mammals have only one protamine, mice and humans have two. Protamine 1 (PRM1) is a sperm-specific protein that compacts sperm DNA (see SPERM,

SPERMATOGENESIS) and its encoding *PRM1* gene has 13 non-synonymous and 1 synonymous differences between humans and chimpanzees, suggestive of recent positive selection. The *PRM2* gene appears to have undergone less sequence divergence, its product undergoing phosphorylation by Ca^{2+}/calmodulin-dependent protein kinase IV. Protamine phosphorylation seems to lead to incorporation into DNA. The chromosomes of many sperm dispense with the DNA-binding histones of somatic cells, being packed instead with these positively charged proteins. The timing of the histone-to-protamine transition by the paternal genome during the long G1, prior to the DNA replication of the zygote, is a finely tuned one.

proteasomes Abundant ATP-dependent proteases constituting ~1% of cellular protein, present in many copies and dispersed throughout the cytosol and nucleus. A typical human cell contains ~30,000, constituting the final disposal apparatus for polypeptide chains that have been targeted for rapid degradation by ubiquitinization (see UBIQUITINS). These ribosome-sized structures comprise some 28 protein subunits, generating a 26S proteolytic complex containing a 20S degradative core proteasome, normally associated at each end with a large 19S cap complex of ~20 distinct polypeptides, at least six of which hydrolyse ATP. These are thought to unfold proteins to be digested, resulting in short peptide products. The caps act as 'gates' to the core proteasome, and also act as binding sites for ubiquitinized substrates. Proteasomes also degrade proteins whose functions depend upon their being 'short-lived', either constitutively or conditionally. For instance, the collapse of various cyclin species as a cell advances from one CELL CYCLE phase to the next is due to their rapid degradation, triggered by the actions of highly coordinated ubiquitin ligases. Two subunits of the proteasome are encoded within the major histocompatibility complex. Along with MHC Class I molecules, their expression is induced by interferons produced in response to viral infections. See CHAPERONES, IMMUNOPROTEASOMES.

protective alleles See GENETIC RESISTANCE.

protein kinases Regulatory enzymes that phosphorylate their protein substrates at very specific amino acid residues, often activating them as a result (e.g., see MAPK PATHWAY). The pleiotropic actions of a protein kinase usually stem from its ability to phosphorylate and thereby modify the functional state of a number of distinct protein substrates. Many are involved in SIGNAL TRANSDUCTION pathways and include hormone-, neurotransmitter- and calcium/calmodulin-dependent forms. Many are activated by SECOND MESSENGERS such as cAMP, cGMP, calcium or diacylglycerol. Several phosphorylate voltage-gated sodium, potassium and calcium ION CHANNELS, modulating their activities, while others have transcription factors as their substrates. Some are basophilic, preferring to phosphorylate a residue with basic neighbours; others have specificities for residues near a proline. Most protein kinases catalyse autophosphorylation, which involves covalent attachment of a phosphoryl group from ATP. This is generally intramolecular and modulated by regulatory ligands. Either serine/threonine or tyrosine can serve as the phosphoacceptor, and several sites on the same kinase subunit are usually autophosphorylated. Autophosphorylation affects the functional properties of most protein kinases. Mitogenic signalling pathways operate via RECEPTOR TYROSINE KINASES whereas other signalling pathways rely almost exclusively on SERINE/THREONINE KINASES. Often, kinases at the end of cascades (terminal kinases) trigger a cell response by directly phosphorylating transcription factors, co-regulatory proteins or histones. They are affected adversely by reactive oxygen species (see MITOCHONDRION), and their effects are opposed by PROTEIN PHOSPHATASES. See CALCIUM/CALMODULIN-DEPENDENT PROTEIN KINASE II, JAKs (in JAK/STAT PATHWAY), HUMAN KINOME, PKA, PKB, PKC, PKG, SRC-KINASES.

protein phosphatase I (PPI) A broad substrate specificity phosphatase, controlling the phosphorylation states and activities of several physiologically important substrates such as neurotransmitter receptors, voltage-gated ION CHANNELS, ion pumps and transcription factors. It is inhibited by phosphorylated DARPP-32. See PSYCHOSIS, SYNAPTIC PLASTICITY.

protein phosphatases Enzymes removing PHOSPHATE groups, often with exquisite sensitivity, from proteins, often at very specific tyrosine, threonine or serine residues. They are of critical importance in SIGNAL TRANSDUCTION pathways, acting in opposition to PROTEIN KINASES.

protein synthesis See ENDOPLASMIC RETICULUM, GENE EXPRESSION, TRANSCRIPTION, TRANSLATION.

protein trafficking See VESICLES.

proteoglycans (PGs) Covalently linked complexes of a core protein and one or more GLYCOSAMINOGLYCANS forming important components of the EXTRACELLULAR MATRIX, especially of connective tissue (e.g., aggrecan in cartilage). Some are huge molecules, up to 95% of the molecule comprising carbohydrate, including oligosaccharides. The polypeptide chain is synthesized on rough ER and the carbohydrate side-chains are attached to serine residues within the GOLGI APPARATUS. Some PGs are secreted and others remain bound to the plasma membrane, either by a transmembrane component or by attachment to the lipid bilayer by a glycophosphatidylinositol anchor. The PG decorin is widespread in connective tissues, where it binds to type I collagen. Other PGs bind secreted signal molecules, such as GROWTH FACTORS. Heparin sulphate side chains bind fibroblast growth factors (FGFs) so that these oligomerize and activate their cell surface receptors. Other GFs, such as the TGF-β family, bind the core proteins of several PGs and are inactive in the bound state. Other PGs bind, and regulate the activities of, secreted proteases and protease inhibitors (see ENDOTHELIUM). Compare GLYCOPROTEINS.

proteomics A field covering the analysis of protein expression, protein structure and protein interactions. The expression levels of all the proteins in cells (their *proteomes*) vary from tissue to tissue, between different stages of development, and between non-pathological and pathological states. Proteins can be detected and identified by specific molecular interactions, notably by binding antibodies and other ligands as probes, although unlike nucleic acids there is no available technique for amplifying or cloning rare proteins. Unlike TRANSCRIPTOME studies, proteomics samples the functional (post-translational) products of the cell, and can distinguish phosphorylated from dephosphorylated proteins – crucial in signalling pathway analysis. It complements DNA-sequence and RNA-expression analyses, offering a more direct measure of cellular responses than does measurement of nucleic acid status. It can also play a part in predicting and investigating adverse or inefficient drug responses by indicating changes in the proteome after drug treatment, whereas many drugs will not affect mRNA levels studied by transcriptome analyses.

Currently, expression proteomics uses two-dimensional gel electrophoresis (2DGE) to separate complex protein mixtures into their components, resolving up to 10,000 proteins on a single gel. Interesting features can then be detected (annotated) by using high-throughput mass spectrometry (MS), which can be implemented on these large scales. Such techniques aim at determining the current and changing protein status of cell systems, including differences between diseased and healthy samples. One approach is to label proteins from two different samples by conjugation with different fluorophores (e.g., the green Cy3 and the red Cy5), as with DNA MICROARRAYS. A problem with 2DGE is that it is not suited to automation, although disruption of large molecules such as proteins during MS has been avoided by soft-ionization methods. Compare GENOMICS, HIGH-THROUGHPUT GENOTYPING.

prothrombin (Factor II) An α_2-globulin

plasma protein (MW 68,700) produced by the liver and present in blood plasma in a normal concentration of ~15 mg dL^{-1}. It is unstable and splits into smaller products, one of which is thrombin. VITAMIN K is required for its formation, and decrease in its production can result in a bleeding tendency. See BLOOD CLOTTING.

protocadherins (PCDHs) A family of proteins playing critical roles in the development of the nervous system. See CADHERINS.

protocortex In its most extreme version, the hypothesis that thalamic afferent axons, via activity-dependent mechanisms, impose cortical areal identity on an otherwise homogenous cortex. Compare PROTOMAP and see CEREBRAL CORTEX.

protomap In its most extreme version, the hypothesis that cortical areas in the adult brain reflect an organization already present (the 'protomap') in the ventricular (mitotic) zone of the foetal telencephalon. Compare PROTOCORTEX and see CEREBRAL CORTEX.

proto-oncogenes Normal cellular genes which can be turned into ONCOGENES, e.g. by viruses or by over-expression (see CANCER, Fig. 33). They include many genes involved in control of the CELL CYCLE. A proto-oncogene can be incorporated into a retrovirus, whereupon the gene sequence may be changed or truncated so that it codes for an abnormal, but functional, protein; or it may be brought under control of strong promoters and enhancers in the viral genome so that its product is made in excess or otherwise inappropriately (e.g., ectopically). Also, a retrovirus may become inserted into the host genome within, or close to, proto-oncogenes, causing activation of the gene (insertional mutation). Most proto-oncogenes encode proteins in the SIGNAL TRANSDUCTION pathways by which cells interact socially, leading them to divide, differentiate or die. Thus although deregulated expression of the proto-oncogene c-*myc* is frequent in cancer cells, its growth-promoting activities are not normally actuated because of its corresponding growth-inhibiting properties. See APOPTOSIS.

proximal convoluted tubule (proximal tubule, PCT) The first part of the KIDNEY nephron distal to the renal corpuscle, consisting of cuboidal epithelium, whose cells have microvilli on their apical surfaces (in contact with glomerular filtrate) and whose basal surfaces rest on a basement membrane separated from peritubular blood capillaries by tissue fluid and engaged in passive diffusion, facilitated diffusion and active transport (see Table 9). Active transport of sodium (Na^+) out of the basal surface in exchange for K^+ generates a reduced intracellular Na^+ concentration and provides the energy source favouring uptake of Na^+ from the filtrate across the apical microvilli. But Na^+ can only cross the microvilli by facilitated diffusion through COTRANSPORTERS in association with other solutes (see table 9), a process described as 'secondary active transport'. Each transport protein is specific to one of the cotransported substances and the number of transport molecules limits the rate at which a substance can be carried across (see TUBULAR MAXIMUM). This cotransport raises the intracellular solute concentration and draws water across the apical microvilli osmotically (compare ENTEROCYTES). In addition to the ammonia passively transported into the PCT, its cells produce additional ammonia by deamination of the amino acid glutamine, generating both NH_3 (which binds H^+ to form NH_4^+) and new HCO_3^-. The NH_4^+ can substitute for H^+ aboard Na^+–H^+ antiporters in the apical membrane and be secreted into the tubular fluid for excretion. Some solutes (K^+, Ca^{2+}, Mg^{2+}) also diffuse between the epithelial cells, their rate of diffusion increasing as their concentration in the filtrate rises with the loss of water from the filtrate. For a normal person with a plasma phosphate level of 1.1 mmol L^{-1} ~180 mmol of PO_4^{3-} are filtered per day, of which ~80% is reabsorbed by a carrier-mediated process in the microvilli. The activity of this Na^+–PO_4^{3-} cotransporter in phosphate homeostasis is somehow modulated by the bone-derived fibroblast growth factor FGF23, whose deficiency leads to increased renal phosphate reabsorption in mouse models. The absorbed PO_4^{3-} leaves via the basolateral membrane by a poorly understood anion exchange process. Because the renal threshold for PO_4^{3-} is just below the normal plasma level, phosphate will normally be excreted.

Urate is largely reabsorbed by the PCT, being taken up across the microvilli by an anion exchanger that can utilize other organic ions, such as lactate, as counter-ions.

By the end of the PCT, the filtrate volume (water content) has been reduced by ~65% although its osmolarity remains the same as that of plasma (300 mOsm dm^{-3}) on account of the tubule's water permeability. Plasma proteins escaping into the filtrate become bound to the microvilli where they are taken in by receptor-mediated endocytosis within coated pits, absorbed into endosomes and degraded by the lysosomal compartment.

Prozac An antidepressant drug whose effects result from its selective blockage of serotonin reuptake by brain neuron terminals (noradrenaline reuptake being unaffected). See SSRIS.

PS1, PS2 See PRESENILINS.

pseudoautosomal region A small region of homology (also known as PAR1) between the X and Y chromosomes, on the short arms of both, at which one cross-over is obligatory in spermatogenesis and for which reason genes here behave as though they were autosomally inherited – despite being located on sex chromosomes. Another region of homology, PAR2, is

Reabsorption of major solutes from the PCT

Apical membrane (microvilli) of PCT cells

Cotransport with Na^+
K^+, Cl^-, Ca^{2+}, Mg^{2+}, HCO_3^-, PO_4^{3-}, amino acids, glucose, fructose, galactose, lactate, succinate, citrate
Anion exchange
A Na^+–H^+ exchanger couples Cl^- movement to efflux of an organic anion (e.g., urate), which is recycled by a Cl^--anion exchanger. Urate is exchanged by another anion exchanger using lactate as a counter-ion
Cation exchange
Creatinine is secreted
Diffusion between nephron cells
K^+, Ca^{2+}, Mg^{2+}

Basal membrane of PCT cells

Cation exchange
Na^+ exchanged for K^+ by Na^+K^+-ATPase (active transport)
Anion exchange
p-aminohippurate enters in exchange for di- and tricarboxylate ions
Facilitated diffusion
K^+, Cl^-, Ca^{2+}, HCO_3^-, PO_4^{3-}, amino acids, glucose, fructose, galactose, lactate, succinate, citrate
Symport
Na^+–di-, and tricarboxylate symport
Voltage-gated transport
Urate (uric acid)

Waste metabolites, many ionized at physiological pH, and actively secreted into the tubular fluid by anion exchange or cation exchange
H^+, hydroxybenzoates, *p*-aminohippurate, creatinine, neurotransmitters (dopamine, acetylcholine, adrenaline), prostaglandins, bile pigments, urate, drugs and toxins (salicylates, penicillin, atropine, morphine, saccharin)

Substances passively transported out of PCT
Urea

Substances passively transported into PCT
Ammonia

TABLE 9 *Some of the transport methods, and the solutes they move, in the* PROXIMAL CONVOLUTED TUBULE *of the kidney.*

located at the tips of the long arms of X and Y chromosomes, although no gene loci have been found here and cross-overs are not obligatory.

pseudogenes First discovered in the late 1970s during attempts to locate genes on the chromosomes (see GENOME MAPPING), these DNA sequences resemble genes but lack apparent function, functional parts of the gene's anatomy having been disabled by mutation. Only with the completion of the Human Genome Project has it been possible to estimate the numbers of such pseudogenes in the human genome, a figure currently standing at >19,000, just smaller than the total number of protein-coding genes believed to be present.

A small number of pseudogenes appear to be once-functional genes that for some reason are no longer employed. But most are disabled *duplicates* of currently employed genes that either suffered permanent damage during GENE DUPLICATION ('dead on arrival'), or have accumulated disabling mutations subsequent to duplication. The other process by which pseudogenes can arise is by RETROTRANSPOSITION in which, during gene expression, mRNA acts as a template for DNA synthesis by reverse transcriptase encoded by the transposable elements known as LINES. Because they contain no intron DNA, it having been removed during the production of the mRNA from which they originated, they are termed '*processed pseudogenes*', or '*retrogenes*'.

Only about one-quarter of human gene families is associated with a pseudogene, but some have generated several. Ribosomal RNA genes have given rise to ~2,000 processed pseudogenes, their abundant expression probably offering more opportunities for their creation. Analysis of primate genes for OLFACTORY RECEPTORS indicates that humans have lost many of these functional genes since their common ancestor with chimpanzees, our greater number of olfactory receptor pseudogenes hinting at what happened to some of them.

Although pseudogene recognition relies primarily on their similarity to genes, establishing their lack of function is more problematic. Compared with functional genes, many bear inactivating mutations that disrupt or delete splice sites and exons, or (e.g., 'stop' mutations) that interrupt or shift open reading frames. The NEUTRAL THEORY OF MOLECULAR EVOLUTION holds that non-functional DNA sequences can change freely, in the absence of selective constraints, and comparisons among genomes has revealed that a few pseudogenes appear to be 'better preserved' in terms of DNA sequence than would be expected were they free from selective constraints, and some of them overlap transcriptionally active intergenic regions of the genome. Several large-scale genomic studies (e.g., the ENCODE pilot project) have discovered many pseudogenes to be transcribed, and to have regulatory factors binding upstream of them. Some pseudogene transcripts have been shown to act as natural siRNAs (see RNA INTERFERENCE). Slight differences in the pseudogene complements of individual humans have been found, some olfactory receptor pseudogenes being non-functional in some people but functional in others. It is tempting that this may in part explain different people's sensitivities to odours. In yeast, at least, some pseudogenes for cell-surface proteins can be reactivated by a stressful environment, although this has not yet been demonstrated in humans. A recent study of polymorphism in the X-linked ribonucleotide reductase M2 polypeptide pseudogene (RRM2P4) yields a gene tree that roots in East Asia, and is an exception to the general rooting of X-linked loci in Africa (see HOMO SAPIENS, PHYLOGENETIC TREES).

pseudohermaphroditism The occurrence in an individual of gonads of one sex only, but EXTERNAL GENITALIA which are the opposite of the gonadal sex, ambiguous, or otherwise abnormal. Male pseudohermaphroditism arises through undersecretion of androgenic hormones and MÜLLERIAN INHIBITING SUBSTANCE, or when androgen receptors (see ANDROGEN INSENSITIVITY SYNDROME), or enzymes in the response to androgen in target cells, are defective. In others, the phenotype arises through deficiency of one form (type 2) of the enzyme 5-α-reductase, which catalyses the conversion of testosterone to dihydrotestosterone in tissues which normally become the male external genitals, and such 46,XY children are often brought up as females; but in the presence of some testosterone and the type 1 form of the enzyme, the penis enlarges and normal male voice deepening and muscle growth occur and by puberty most affected individuals feel they are males. Early diagnosis is important so that appropriate hormonal therapy can be employed and after a decision as to gender is made the genitals can be surgically corrected. Female pseudohermaphroditism is most commonly caused by CONGENITAL ADRENAL HYPERPLASIA, biochemical abnormalities in the adrenal glands resulting in decreased steroid hormone production. 21-hydroxylation is inhibited, so that 17-hydroxyprogesterone (17-OHP) is not converted to 11-deoxycortisol. Due to this defective cortisol production, an increase in corticotropin production occurs, leading to ever-increasing amounts of 17-OHP and excessive androgen production. Individuals have a 46,XX chromosome complement and ovaries, but masculinization of the external genitalia varying from enlargement of the clitoris to almost male genitalia. The labia majora may partially fuse, giving a scrotum-like appearance. See SEX DETERMINATION.

pseudohypoaldosteronism type II (PHAII) A rare autosomal dominant disorder characterized by hypertension and elevated serum potassium levels and normal renal function. See ALDOSTERONE.

Pseudomonas aeruginosa A Gram-negative bacterium frequently associated with infections of the urinary and respiratory tracts (e.g., see CYSTIC FIBROSIS) and in patients receiving treatment for severe burns. It is an opportunistic pathogen, not an obligate parasite, and is an important denitrifying soil bacterium. It is commonly found in the hospital environment (e.g., associated with catheterizations, tracheostomies and intravenous infusions).

psoriasis A chronic inflammatory AUTOIMMUNE DISEASE of the skin, marked by excessive proliferation of keratinocytes resulting in silvery-white scaly patches overlying inflamed skin. A complex disease, specific causes are unknown, although bacterial skin infections, mild trauma and stress, and genetic factors, underlie its development. Accumulation of inflammatory cells, mainly activated T CELLS and ANTIGEN-PRESENTING CELLS (notably plasmacytoid dendritic cells, pDCs), precedes other aspects of disease pathology. High levels of cytokines and antimicrobial peptides (e.g., β-defensin 2 and cathelicidin LL37) in the lesions are typical; but whereas high levels of interferon-κ (IFN- κ) are usual in bacterial infections of the skin, large amounts of IFN-α produced by pDCs occur instead. Recent studies indicate that LL37 can bind to self-DNA, the resulting complexes breaking immune tolerance and signalling IFN-α production.

psychedelic drugs See HALLUCINOGENIC DRUGS, PSYCHOACTIVE DRUGS.

psychoactive drugs Drugs altering MOOD, PERCEPTION, or behaviour, and used in the management of neuropsychological illness (see Fig. 144). They include non-selective CNS depressants (e.g., BARBITURATES, non-barbiturate hypnotics, ethyl ALCOHOL and general ANAESTHETICS); antianxiety agents (e.g., BENZODIAZEPINES); psychomotor stimulants (e.g., sympathetic nervous system activators such as AMPHETAMINES, COCAINE, CAFFEINE and NICOTINE); ANTIDEPRESSANTS; mood stabilizers (e.g., lithium and some anticonvulsants), and narcotic ANALGESICS (opioids, morphine and its analogues). Lysergic acid diethylamide (LSD) is an extremely potent synthetic HALLUCINOGENIC DRUG (25 μg produces as great a hallucinogenic effect as 560 mg of aspirin). Its chemical structure and those of the hallucinogenic components of *Psilocybe* mushrooms (used by the Maya) and peyote (a product of the peyote cactus used by the Aztec) is very similar to that of SEROTONIN (see this entry for 'ecstasy'). The α_2-ADRENERGIC receptors are implicated in several central nervous system disorders and their agonists are used in the suppression of opiate withdrawal. See ADDICTION, CANNABINOIDS, NEUROLEPTICS, SYNAPSES, and references in NEUROTRANSMITTERS.

psychopharmacology The study and application of psychoactive drugs in therapeutic medicine. See references in PSYCHOACTIVE DRUGS.

psychosis A generic term used in psychiatry for a mental state often described as involving a 'loss of contact with reality': a severe mental disorder, with or without organic damage, characterized by derangement of personality and loss of contact with reality and causing deterioration of normal social functioning. Sufferers may experience hallucinations or delusional beliefs (e.g., grandiose or paranoid delusions; see AFFECTIVE DISORDERS), and may exhibit personality changes and disorganized thinking, often accompanied by lack of insight into the unusual or bizarre nature of their behaviour, as well as difficulty with social interaction and difficulty in carrying out the activities of daily life. Some reality-distorting experiences are a feature of most people's lives, at some time, and may be one end of a spectrum of normal conscious experience. The term includes SCHIZOPHRENIA and bipolar affective disorder (BPAD; see BIPOLAR DISORDER), each affecting 1% of individuals worldwide. An abnormal sensitivity to DOPAMINE is thought to mediate psychosis and its unmodulated action in REM SLEEP may contribute to the 'madness' of dreams. Antispychotic agents exert powerful effects on the cerebral cortex, yet act mainly on targets which are very sparse there, notably the DOPAMINE D2 receptor. If transmission through the thalamus becomes excessive, the integrative capacity of the cortex may break down, and confusion or psychosis will ensue. In the striatum, dopamine is counterbalanced by a powerful corticostriatal glutamatergic system derived from all parts of the cortex. There is support for the view that in addition to dopamine, glutamate and serotonin are very much involved in schizophrenia, and it is likely that other neurotransmitters will soon be added to the list. See PROTEIN PHOSPHATASE I.

***PTEN* gene (phosphate and tensin homologue)** A TUMOUR SUPPRESSOR GENE encoding a lipid phosphatase that inhibits PI(3)K, terminating the signalling pathway of lipid second messengers in the PHOSPHATIDYLINOSITOL-3-OH-KINASE pathway. It

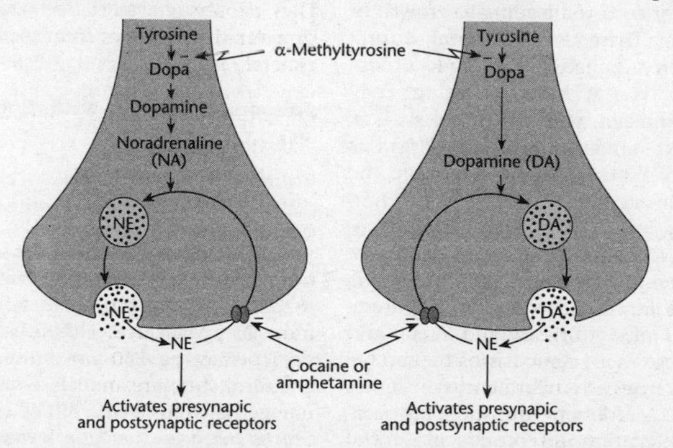

FIG 144. The PSYCHOACTIVE DRUGS cocaine and amphetamine exert their stimulant effects by blocking catecholamine reuptake, but synthetic inhibitors (e.g., α-methyltyrosine) can abolish these effects by blocking dopamine synthesis. © 2001 Redrawn from *Neuroscience* (2nd edition) by M.F. Bear, B.W. Connors and M.A. Paradiso, published by Lippincott, Williams & Wilkins. Reproduced with kind permission of the publisher.

is an upstream regulator on mTOR (see entry). PTEN function is lost in 5–10% of MELANOMAS. See CANCER.

P-type ganglion cells See GANGLION CELLS.

pubertial growth spurt See ADOLESCENT GROWTH SPURT.

puberty The physiological and mental transition between childhood and adulthood, taking place over ~4 years and involving the development of the SECONDARY SEXUAL CHARACTERISTICS and ADOLESCENT GROWTH SPURT. The trigger for puberty is not known, although a metabolic signal reflecting either growth or body mass is probably involved in girls: a critical mass of 30 kg seems to be required, which needs to reach 47 kg before menarche. The onset of puberty is accompanied by a decreased GABA inhibition of hypothalamic neurons that release GONADOTROPIN RELEASING HORMONE (GnRH) and during puberty there is a dramatic circadian rhythm in gonadotropin output. It is possible that LEPTIN levels from adipose tissue may be a contributory factor in this process.

In males, puberty is preceded by adrenarche (at ~6–8 years of age), in which adrenals release increased amounts of DHEA and androstenedione (without rising corticotropin or cortisol), leading to growth of pubic hair. These hormones peak during late puberty, although the mode of upregulation is not known. During early puberty in males, pulses of LUTEINIZING HORMONE increase (mainly during sleep) and its plasma level rises. These LH surges become regular throughout the day and lead to increasing circulating testosterone, promoting GROWTH HORMONE secretory bursts. In females, puberty onset (at ~8–13 years of age) is associated with low-amplitude nocturnal pulses of gonadotropin (LH and FSH) release. With approaching menarche, pulsatile release of gonadotropins becomes established, leading to rise in serum oestradiol concentration and consequent skeletal maturation, epiphyseal fusion and cessation on linear growth, and the onset of MENSTRUATION (MENARCHE). Adrenarche also occurs in females prior to menarche, although its involvement in its onset is not clear. The slowing of growth rate in adolescent females shows great plasticity according to environmental circumstances; there is neither a fixed height nor age at menarche. Malnourished girls reach menarche later, and with lower height. The secular trend towards earlier puberty in well-nourished populations is noted everywhere.

Poor attachment, family discord, and low investment in children are thought to affect the timing of onset of puberty, the conjunction of these stressors with early puberty contributing to conflict with parents, lower self-esteem, and to associations with problematic peers. See ADOLESCENCE; MORTALITY, Fig. 126.

pubic symphysis Fibrocartilaginous disc joining the two hip bones across the two pubes. See PUBIS.

pubis (pubic bone; pl. pubes) The most anterior and inferior of the three bones contributing on each side of the body to the hip bone (coxal bone) and joined by the pubic symphysis. See PELVIC GIRDLE.

pulmonary arteries The left and right pulmonary arteries branch off the pulmonary trunk and carry blood to the two lungs. They are elastic arteries (see ARTERY) with a complex development from the 6th AORTIC ARCH (see Fig. 13).

pulmonary minute ventilation See MINUTE VENTILATION.

pulmonary respiration The GASEOUS EXCHANGE occurring between the lungs and the blood.

pulsed field gel electrophoresis (PFGE) A technique enabling separation of DNA fragments of up to 2 million base pairs in size, contrasting with conventional agarose gel electrophoresis which resolves DNA fragments of up to ~20,000 bp (see RESTRICTION FRAGMENT LENGTH POLYMORPHISMS). DNA is digested by infrequently cutting restriction enzymes having 6–8 base pair nucleotide recognition sequences rather than the shorted sequences in the conventional method. The fragments are alternately subjected to two approximately perpendicular fields in the gel and analysis of the fragments *in situ* in the gel allows construction of physical maps of relatively large stretches of DNA.

pupillary light reflex Reflexes co-ordinated by the pretectal area of the midbrain, using inputs from the RETINA to the brainstem, dilating or constricting PUPIL diameter in response to dim and bright light, respectively. The sensory axons involved leave the optic nerve before it reaches the lateral geniculate nucleus and terminate just rostral to the superior colliculus, where they synapse with parasympathetic neurons leaving by the oculomotor nerve and controlling the sphincter pupillae muscle of the IRIS. Even if only one retina is illuminated, the pupils of both eyes constrict (i.e., the reflex is consensual), lack of a consensual pupillary light reflex often being taken as a sign of serious neurological disorder of the brain stem. See EYE-BLINK REFLEX, PHOTOPHOBIA.

pupils The apertures of the irides (see IRIS, PUPILLARY LIGHT REFLEX).

purinergic Of fibres, receptors and synapses responding to purine derivatives, e.g. adenosine.

purines The heterocyclic nitrogenous bases adenine and guanine, synthesized in a complicated pathway involving 5-phosphoribosyl 1-pyrophosphate (PRPP), glutamine, glycine, aspartate and formate, and hydrogen carbonate. After production of inosinate (IMP), the pathway then branches. Further aspartate, and GTP, are required in adenylate (AMP) synthesis, while further glutamine and ATP are required for synthesis of guanylate (GMP). With PYRIMIDINES, they are components of nucleosides, nucleotides and nucleic acids; but both are recycled by salvage pathways far simpler than that involved in *de novo* synthesis. These salvage pathways are particularly needed in brain tissue, which may account for the central nervous system damage in children with LESCH–NYHAN SYNDROME. Their degradation produces URIC ACID. See ADENOSINE DEAMINASE DEFICIENCY.

Purkinje cells (P. neurons) One form of NEURON, having a very large soma (cell body, 50–80 μm in diameter), very large dendritic trees, and very long axons extending well beyond the region of the nervous system in which their soma resides (contrast INTERNEURONS). They are particularly abundant in the cortex of the CEREBELLUM, where they exhibit LONG-TERM DEPRESSION, project into the deep cerebellar nuclei and are involved in motor learning. Not to be confused with PURKINJE FIBRES.

Purkinje fibres Large modified cardiac muscle fibres forming an extensive network beneath the endocardium and spreading excitation from the left and right bundles of His (see CARDIAC CYCLE) to the ventricular myocytes via gap junctions. Conduction through them is much more rapid than through the myocardium itself, so that all parts of the ventricles contract at much the same time, although initiated at the apex. Not to be confused with PURKINJE CELLS.

Purkinje shift The shift between scotopic and photopic vision (see DARK ADAPTATION). Depending upon whether the retina is adapted for night vision (scotopia) or day vision (photopia), the threshold sensitivity to dim light varies. In the former, the eye is most adapted to green (507 nm), while in the latter it is most adapted to yellow (555 nm).

putamen See BASAL GANGLIA.

pyramidal cells (pyramidal neurons) One form of very large NEURON, their cell bodies being pyramidal in shape, and comprising some 60% of neurons of the CEREBRAL CORTEX (see Fig. 44) and having dendrites extending to fill a pyramidal space. As with the PURKINJE CELLS of the cerebellar cortex, pyramidal cells have long axons extending some distance from the soma. These may occur at both its basal and apical ends. The CORTICOSPINAL TRACT consists of axons of ~1 million pyramidal cells in layer V of the cerebral cortex. See HIPPOCAMPUS, PRIMARY VISUAL CORTEX.

pyramidal tracts See CORTICOSPINAL TRACTS.

pyramids See MEDULLA OBLONGATA.

pyridoxine (vitamin B_6) See VITAMIN B COMPLEX for pyridoxine and pyridoxal phosphate.

pyrimidines The monocyclic nitrogenous bases cytosine, thymine and uracil, synthesized from 5-phosphoribosyl 1-pyrophosphate (PRPP), aspartate and cytosolic carbamoyl phosphate (see UREA). Their degradation generally leads to AMMONIA (NH_4^+) production, and hence to UREA synthesis.

pyrogens See ENDOGENOUS PYROGENS.

pyruvate See GLYCOLYSIS, PYRUVATE CARBOXYLASE.

pyruvate carboxylase The regulatory enzyme reversibly catalysing the carboxylation of pyruvate to oxaloacetate, which is a major anaplerotic reaction in mammalian liver and kidney. It is almost inactive in the absence of acetyl-CoA, its allosteric modulator. Whenever acetyl-CoA is in excess, pyruvate carboxylase is stimulated to produce more oxaloacetate, enabling the KREBS CYCLE to use more acetyl-CoA in synthesis of citrate.

pyruvate dehydrogenase A multienzyme complex

$PYY_{3\text{-}36}$ A 36-amino acid satiety peptide (see APPETITE, Fig. 14; ENERGY BALANCE), a member of the NEUROPEPTIDE Y family, and the most abundant form of circulating peptide YY (from which it is produced by truncation of its amino-terminal end). It is released by the same enteroendocrine L cells that produce GLUCAGON-LIKE PEPTIDE-1 in the distal small intestine and colon in response to food, secretion occurring prior to the actual arrival of food in these regions although arrival increases output. Anorectic in its actions, it decreases gastric motility and increases water and electrolyte absorption by the colon. It modulates neuronal activity within both corticolimbic and higher-cortical areas, in addition to homeostatic brain regions. In conditions of high plasma PYY concentration, changes in neural activity within the caudolateral orbital frontal cortex predict feeding behaviour independently of meal-related sensory experiences. Its plasma levels remain elevated between meals, decreasing food intake through action at the arcuate nucleus of the hypothalamus. In conditions of low PYY levels, fMRI indicates that hypothalamic activation predicts food intake. It is structurally homologous to PANCREATIC POLYPEPTIDE (PP) and neuropeptide Y (NPY) and exerts its effects via NPY receptors. See ENTEROENDOCRINE CELLS.

Q

Q fever An atypical pneumonia or feverish illness caused by the mycoplasma (Rickettsiaceae) *Coxiella burnetti* (see MYCOPLASMA PNEUMONIAE). Probably <1% of infections become chronic, with swinging fever, granulomatous hepatitis (granulomata having peripheral zones of eosinophils), moderate jaundice and, usually, endocarditis. Acute Q fever is a zoonosis, the reservoir being sheep, and it can cause placental damage and abortion. It is occasionally passed on by parturient cats, and in unpasteurized milk. Farmers' wives or women working with animals are the population most at risk during pregnancy. Tetracycline is normally the treatment of choice, although chloramphenicol is used in pregnancy.

quadriplegia Paralysis of the body from the neck down. Compare PARAPLEGIA; see NEUROROBOTICS.

R

Rab proteins Small monomeric GTPases, >60 occurring in mammals, some ubiquitous and others cell-type specific, and associated with specific organelles (see CELL MEMBRANES). They act as molecular switches and as targeting determinants for a wide variety of proteins, including MOTOR PROTEINS and tethering factors. Rab GTPases have a central role in determining when and where peripheral membrane proteins are recruited to organelles, and are recruited to specific membranes by proteins governing the GDP/GTP cycle (see GUANOSINE TRIPHOSPHATE), some of which are highly localized. The GDP-bound form is cytosolic whereas the GTP-bound form is membrane-associated. To date, all Rabs have been found associated with specific organelles, anchoring taking place by attachment of two lipid prenyl groups to the carboxy-terminus of the protein (hidden in the cytosolic form by GDP-dissociation inhibitor, GDI). Displacement factors release Rab from this complex, enabling its attachment to a membrane. Rabs often cooperate with ARF PROTEINS and bind Rab effectors, proteins enabling vesicle docking (see VESICLES).

rabies A neurotrophic lyssavirus (Rhabdoviridae) infection transmitted from infected animals to humans in saliva as a result of a bite or open skin wound. Sources include bats, dogs, cats, squirrels and occasionally horses. Rabies virus is an enzootic virus that occasionally becomes epizootic, the genome encoding five proteins of which the G protein is the antigen that induces neutralizing antibodies and is involved in binding the virus to host cell acetylcholine receptors. The virus enters peripheral nerves and migrates to the CNS, initiating encephalomyelitis. Viruses then pass down nerves to the salivary and lachrymal glands, and to other tissues. Incubation period varies from 3 to 4 weeks if the bite is to the face or head, and ≥60 days after a bite to the foot. The symptoms of 'furious rabies' include aching at the site of infection followed by basal encephalitis, fever, and periods of extreme agitation and personality alteration in which stimulation of the face or mouth during attempts to drink or from air draughts cause facial spasms and regurgitation of saliva (typical of 'hydrophobia'; see FEAR AND ANXIETY). Pre-exposure immunization is recommended for those occupationally involved and for some travellers to regions where rabies is endemic in domesticated and feral animals. The usual treatment is post-exposure immunization with high-potency cell-culture-derived vaccine and human rabies immunoglobulin (HRIG) at the site of the wound, as quickly as possible, and intramuscularly elsewhere. See VIRUSES.

races See GENETIC DIVERSITY.

RAD51 A protein which binds the BRCA2 gene product, BRCA2, during DNA repair (see DNA REPAIR MECHANISMS). See CARETAKER GENES.

radiation damage See MUTAGENS.

radiation hybrid mapping See GENOME MAPPING, SOMATIC CELL HYBRIDIZATION.

radius One of the two bones of the forearm, articulating proximally with the capitulum of the humerus and the radial notch of the ulna, and distally with three bones (carpals) of the wrist. It is connected to the ulna by an interosseous membrane.

RAG-1, RAG-2 See SOMATIC RECOMBINATION.

RAGE (receptor for advanced glycation end-products) A cell surface receptor protein of the immunoglobulin superfamily which binds potentially damaged glycated proteins. It also binds AMPHOTERIN. See AGEING.

randomized control trials (RCTs) Procedures, commonly employed as part of experimental, preclinical and clinical methodology, in which eligible subjects are randomly assigned into groups which either receive or do not receive one or more interventions being compared. The results are then compared and tested for significance. The aim is to achieve 'statistical equivalence' (i.e., lack of bias) in the groups compared. See DOUBLE-BLIND STUDY, DRUGS.

RANTES (Regulated on Activation, Normal T Expressed, and Secreted; CCL5) See CHEMOKINES.

raphe nuclei One of the DIFFUSE MODULATORY SYSTEMS of the BRAIN (see Fig. 30d) located in the midbrain, pons and medulla, whose nine paired subcortical nuclei project into different regions of the central nervous system. Those more caudal (lying in the medulla) innervate the spinal cord and modulate pain-related (nociceptor) sensory input (see REFERRED PAIN, Fig. 147). Those more rostral (in the pons and midbrain) innervate most of the brain, including the hypothalamus, in the manner of the LOCUS COERULEUS, and some of these may be involved in the expression of anxiety (see FEAR AND ANXIETY). However, unlike the latter's noradrenergic neurons, those of the raphe nuclei are serotonergic, secreting SEROTONIN. Serotonin pathways are also implicated in SLEEP, MOOD (see SSRIS), and control of cerebrospinal fluid secretion and cerebral blood flow. See RETICULAR FORMATION.

***Ras* oncogene (*ras* oncogene)** See RAS PROTEINS.

Ras proteins (RAS proteins in humans) A family of proteins (name originating from Harvey rat sarcoma virus), the three human forms (H-RAS, K-RAS, N-RAS) being encoded by three *ras* proto-oncogenes and covalently anchored to the cytoplasmic side of cell membranes coupling GROWTH FACTOR receptors to downstream signalling pathways that control cell growth, proliferation, survival and transformation (see RAS SIGNALLING PATHWAYS). Although predominantly targeted to the plasma, Ras proteins also seem to occur in the membranes of the endoplasmic reticulum and Golgi apparatus. Although not G PROTEINS, Ras proteins are monomeric GTPases and share with G proteins the property of being activated when bound to GTP in response to activation of RECEPTOR TYROSINE KINASES (RTKs) by growth factors. These receptors are often connected to *guanine-nucleotide-exchange factors* (GEFs) which replace GDP with GTP and convert inactive Ras to its active form (see GUANOSINE TRIPHOSPHATE). One of the most important of these GEFs is the Sos ('son-of-sevenless') protein. Two other proteins, Shc (pronounced 'shick') and Grb2 (pronounced 'grab two'; see GRB2), function as adaptor proteins. They form physical bridges between the growth factor receptor and Sos.

Acting as a binary switch, Ras therefore flips between active signal-emitting and inactive quiescent states. When active, Ras activates several signalling pathways, the end-point of which is usually some alteration in gene expression within the nucleus. When mutated, invariably by missense MUTATIONS, the *Ras* genes encoding the Ras proteins are known as ONCOGENES (dominant, gain-of-function mutations) and these mutated proteins pay a causal role in more than a quarter of human CANCERS. The abnormal oncoproteins with amino acid substitutions in the GTPase pockets are overactive, binding GTP but being incapable of hydrolysing it. Inability of these mutated proteins to shut off their own activity floods the cell with a relentless stream of mitogenic signals for extended periods of time, causing inappropriate signal for cell proliferation to be transmitted to the nucleus. Many types of cell also produce and release growth factors (e.g., TGF-α) inappropriately. Mutations of the *Ras* oncogene are found in 40% of human lung cancers. Some retroviruses occasionally pick up a damaged or misregulated form of a *Ras* gene and, by infection, trigger tumour production in new animal hosts – but seemingly not in humans (see RETROVIRUSES). A recent finding is that germline activating mutations can occur in genes encoding Ras proteins and other components of these signalling pathways, resulting in developmental defects. See ARF PROTEINS, RAB PROTEINS.

***Ras* signalling pathways** One of the multiple, and evolutionarily highly conserved, pathways lying downstream of GROWTH FACTOR receptors. It was already well established in the common ancestor of all extant metazoan, >600 Mya. The Ras proteins belong to a large family, the Ras superfamily of monomeric GTPases (see GUANOSINE TRIPHOSPHATE), and are G PROTEIN-like proteins divisible into two subfamilies, Rho and Rab. They contain a covalently attached lipid moiety that anchors the protein to a membrane, often the cytoplasmic surface of the cell membrane. They operate in a very similar manner and will here be termed 'Ras'. At least three major downstream signalling cascades radiate from the Ras protein.

When Ras binds GTP (see RAS PROTEINS) a domain on the Ras protein (its effector loop) can interact with several alternative downstream partners, 'Ras effectors', which have little affinity for GDP-bound Ras. Ras was first discovered as the product of a mutant *ras* gene. Its hyperactivity, resulting from loss of its GTPase activity, caused the cell to proliferate as it would have done had the cell's receptor tyrosine kinase been activated by a growth factor (see Fig. 145). As a result, Ras was unable to turn itself off, a negative feedback essential to normal Ras function. We now know that ~30% of human tumours have a hyperactive *ras* mutation. Ras functions as a 'switch' and cycles between two conformational states: an active state when bound to GTP, and an inactive state when bound to GDP. Growth factors activate Ras through RECEPTOR TYROSINE KINASES and promote, among others, the BRAF, MEK and ERK cascades. The MAP kinase cascade leads to increased levels of the regulatory protein Myc (see MYC).

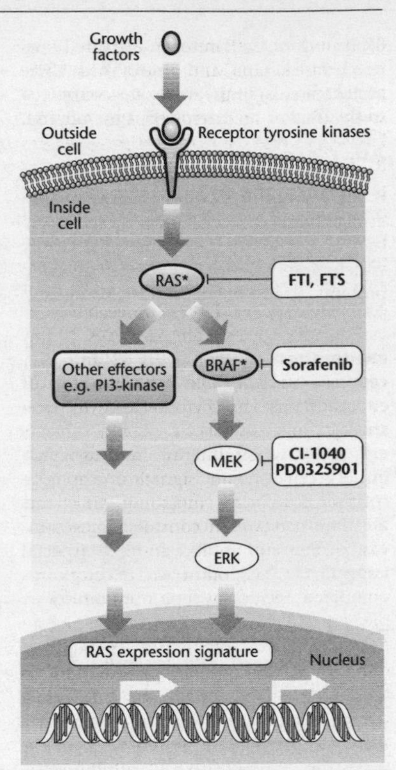

FIG 145. Diagram indicating aspects of the growth factor-activated RAS SIGNALLING PATHWAY. Overexpression or mutation of components shown in bold ovals is often present in lung tumours and melanomas (asterisks indicate activating mutations). The unique gene expression signatures of such oncogenic pathways can be used to assess their levels of activity in a cancer cell, as well as the sensitivity to drugs (shown in bold type, right) targeting those pathways, blockage being indicated by a T-bar. From J. Downward, *Nature* © 2006, with permission from Macmillan Publishers Ltd.

Signalling from RAS bifurcates to two further downstream targets in addition to

BRAF and the RAF-mitogen-activated protein kinase kinase, MEK. One of these is the pathway involving PHOSPHATIDYLINOSITOL 3-KINASE (PI3K), mediated via the Akt/PKB kinase (see PKB). This bifurcation gives Ras its mitogenic (via GSK-3β), cell growth-promoting (via mTOR) and anti-apoptotic (via Bad) abilities. A third Ras-regulated pathway involves a pair of Ras-like proteins, Ral-A and Ral-B, whose activation also involves replacement of GDP with GTP. Once activated, these proceed to activate further downstream pathways, prominently those activating Cdc42 and Rac, which in turn enable cell motility via effects on the actin cytoskeleton (important, e.g., in METASTASIS).

There is much interest in drugs that inhibit downstream signalling cascades controlled by RAS, including MEK and PI3K, both of which control CELL GROWTH. The BRAF isoform of RAF and PI3K (specifically its p110α subunit) are effectors frequently activated by point mutation in human cancers (see ONCOGENES). Phosphorylated JAK proteins (see JAK/STAT PATHWAYS) can also phosphorylate substrates in the *Ras* signalling pathway. *Ras* is mutated in 40% of lung cancers, and *Braf* in 60% of MELANOMAS. Activating mutations in BRAF make the cells sensitive to inhibitors of MEK, whereas those in RAS do not, probably because it can signal through an alternative pathway. Potent inhibitors of MEK, the direct target of RAF, have been promising in clinical trial (see Fig. 145; CHEMOTHERAPY).

rat models See CARDIOVASCULAR DISEASE, MOUSE MODELS.

Rb (pRb) See RETINOBLASTOMA PROTEIN.

rDNA See RIBOSOMAL DNA.

reactive oxygen species (ROS) See ENDOPLASMIC RETICULUM, MITOCHONDRION, PARKINSON'S DISEASE.

reading See LANGUAGE AREAS.

Recent African Origin model (RAO, Uniregional Hypothesis, Out-of-Africa 2 model) See MIGRATION (2).

receptive field (RF) The area of receptive sheet (e.g., skin or retinal surface) whose stimulation affects the activity of a cell in a sensory pathway; e.g. a receptor cell, a relay (or projection) neuron, or neurons in the sensory cortex. RFs of cells lower in the organizational hierarchy tend to be smaller than those higher up. The size of RF of a sensory receptor determines the spatial resolution of a sensory system: the smaller the RF, the greater the resolution. Although individual pathways are organized serially, several pathways of a system are organized in parallel. RFs may be excitatory or inhibitory, or mixtures of both, such as the *centre-surround receptive fields* of BIPOLAR CELLS and GANGLION CELLS of the RETINA (see Figs 150 and 151). Some RFs (as in the secondary somatosensory cortex) are bilateral, responding to stimuli from corresponding regions of both sides of the body. In its most general sense, an RF also includes the pattern of the stimulus altering the response of the cell: e.g., the shape, colour, orientation, movement, direction and binocularity of the effective stimulus. Because neurons higher up the organizational hierarchy respond with different latencies to stimulation of a receptor, the spatial domain of their RFs can alter as a function of post-stimulus time. This means that definition of the centre and boundaries of a given neuron's RF is impossible without temporal specification (see NEURAL CODING).

See ORIENTATION-SELECTIVE CELLS and PRIMARY VISUAL CORTEX for receptive fields of simple and complex cells. See also COLOUR-OPPONENT CELLS.

receptor clustering Cross-linking of antigen receptors on lymphocyte cell surface membranes by antigen-binding, resulting in activation of intracellular signal molecules (usually via receptor tyrosine kinases). See B CELLS, FAB FRAGMENTS, T CELLS..

receptor guanylyl cyclases (RGCs) Single-pass transmembrane proteins bearing an extracellular binding site for a signal ligand and an intracellular guanylyl cyclase catalytic domain. Binding of the signal molecule directly activates the cyclase domain, which produces cGMP. This in turn binds and activates PKG. Among signal molecules using RGCs are odorants (see OLFACTORY RECEPTOR NEURONS), the natriuretic peptides (e.g., ATRIAL NATRIURETIC PEPTIDE) and brain natriuretic peptide. An increasing number of RGCs are being found, most of them orphan receptors. See CYCLIC GMP.

receptor-mediated endocytosis See ENDOCYTOSIS.

receptor potential A low amplitude, graded, change in the membrane potential (voltage) of a SENSORY RECEPTOR caused by a stimulus (mechanical deformation, ligand binding, change in temperature, or electromagnetic radiation) which alters the membrane's ion permeability. In humans and all vertebrates, this is depolarizing at all receptors other than photoreceptors (ROD CELLS and CONE CELLS) and inner ear HAIR CELLS, which hyperpolarize. Receptor potentials are passively conducted, and decay with distance and time. If sufficiently large, a receptor potential will trigger an action potential, and if it does so directly is termed a GENERATOR POTENTIAL. The greater the amplitude of the receptor potential, the more frequently are action potentials initiated.

receptor trafficking Movement of receptors from one membrane domain to another via vesicular carriers and critical in the regulation of neural signalling. GLUTAMATE receptor trafficking is critical to various forms of SYNAPTIC PLASTICITY. See AMPARS.

receptor tyrosine kinases (RTKs) The most numerous of the enzyme-linked receptors, binding a large variety of secreted GROWTH FACTORS and hormones. They are characterized by a (highly variable) ligand-binding domain on the extracellular surface of the plasma membrane and a quite similar tyrosine kinase domain on the cytosolic side, the two domains linked by a single transmembrane segment (see RAS SIGNALLING PATHWAY, Fig. 145). The tyrosine kinase domain is an SH1 domain very similar to that of SRC-KINASES. They can act as docking sites for enzymes such as PLC-γ, PI(3)K (see RAS SIGNALLING PATHWAYS) and the cytoplasmic tyrosine Src-kinases. Mitogenic signalling pathways are the main employers of RTKs (see CELL GROWTH). The EGF, FGF, HGF, IGF-1, INSULIN, PDGF, VEGF and Eph receptors (the most numerous; see EPHRINS) are all RTKs, the IGF-1 receptor being the prototype for this group.

Mutations in genes encoding these receptor molecules can cause ligand-independent firing (deregulated signalling, see CANCER). Binding of their ligand is commonly a prelude to dimerization or higher oligomerization of these receptors, the ligand-binding ectodomain undergoing a stereochemical shift transmitted through the plasma membrane to the cytosolic kinase domain. Dimerization usually involves the signal protein in simultaneously binding two adjacent receptors together, and this together with the allosteric changes, cause the tyrosine kinase domains to use ATP to transphosphorylate one other on multiple tyrosines (autophosphorylation) and then to phosphorylate other intracellular proteins that in turn bind them. This is the prelude to their activating a SIGNAL TRANSDUCTION pathway (see MALIGNANT TUMOURS). Non-receptor TYROSINE KINASES are very common. See JAK/STAT PATHWAY, RECEPTOR GUANYLYL CYCLASES.

receptors A term applicable to molecules (e.g., membrane receptors such as those involved in SIGNAL TRANSDUCTION; or B-CELL and T-CELL RECEPTORS and NUCLEAR RECEPTORS), or to cells (e.g., SENSORY RECEPTORS). Many cell-surface receptors depend on cytoplasmic tyrosine kinases (see SRC-KINASES) for their activity, while others (the RECEPTOR TYROSINE KINASES) have their own kinase domain. Because molecular recognition plays such a large part in biological systems there are many senses in which the term might be applicable: enzyme active sites, antigen-binding sites of ANTIBODIES, PROMOTER regions of genes, etc. Some relevant entries include PRESYNAPTIC RECEPTORS, GLUTAMATERGIC, PURINERGIC.

reciprocal inhibition (r. innervation) The circuitry by which action potentials (e.g., along a 1a afferent) stimulate contraction of one muscle and inhibit contraction of an antagonistic muscle. See ELBOW, MUSCLE SPINDLES.

reciprocal translocations See TRANSLOCATIONS.

recombination An enzyme-mediated process, divisible into two main forms: (i) *general recombination* (homologous recombination), occurring during crossing-over in the first prophase of MEIOSIS and in double-strand break homologous end-joining DNA REPAIR MECHANISMS, which can result in GENE CONVERSION; and (ii) *site-specific recombination*, involving TRANSPOSABLE ELEMENTS which, although rare in humans, has possibly been the evolutionary origin of SOMATIC RECOMBINATION. In general recombination events, new chromosome ends produced by a double-strand break are available to degradative processes and expose a single strand with an overhanging 3´ end; and in both references mentioned this strand seeks out a region of unbroken duplex DNA with the same nucleotide sequence, undergoing a synaptic reaction with it catalysed by RAD51 or a similar protein. Compilation of the HapMap (see HAPLOTYPE) provided new evidence that meiotic recombination (CROSSING-OVER) occurs at specific 'hot spots', almost half of the 30,000 recombination sites now identified being packed into 3% of the genome. These hot spots seem to be associated with particular DNA signature sequences. Cultured cells from individuals with BLOOM SYNDROME have dramatically enhanced sister chromatid exchange during mitosis, but it is not certain how this relates to the increase in GENOMIC INSTABILITY in the disorder. See PHYLOGENETIC TREES, Y CHROMOSOME.

rectum A straight, muscular tube, beginning at the termination of the sigmoid COLON and ending at the anal canal (see ANUS). See EGESTION.

recurrent collaterals Branches of a neuron that generate FEEDBACK signals, sometimes to its own dendrites or axons.

recurrent excitation A form of FEEDBACK in neural networks, often made by recurrent collaterals, as occurs in the HIPPOCAMPUS.

recurrent inhibition A form of FEEDBACK in neural networks. See RENSHAW CELLS.

recurrent laryngeal nerves Branches of the VAGUS NERVE initially supplying the 6th pharyngeal arches but serving laryngeal muscles in the adult. When the heart descends into the thoracic cavity, they hook around the 6th aortic arches and ascend again to the larynx, accounting for their recurrent course. The right nerve moves up and hooks around the right subclavian artery; on the left, the nerve does not need to move up since the ductus arteriosus is available to position it.

red blood cells (erythrocytes) Biconcave cells ~7.5 μm in diameter, produced from myeloid progenitor cells in 'erythroblast islands' in foetal liver and adult red BONE MARROW (see also STEM CELLS), nurtured by a macrophage which supplies nutrients and growth factors and removes the nucleus extruded from the late erythroblast. Mature human erythrocytes develop from reticulocytes (immature erythrocytes). Each contains ~300 million HAEMOGLOBIN molecules, amounting to ~34% by mass, enabling oxygen loading (96% O_2 saturated in the pulmonary veins) and unloading (64% O_2 saturated in the venae cavae). The VITAMIN B COMPLEX is required in their formation, which low blood oxygen levels stimulate (see ERYTHROPOIETIN). Males may have ~5.4 million red cells per microlitre of blood, females ~4.8 million per microlitre (see HAEMATOCRIT). They are produced at a rate of ~2 million per second, surviving in the circulation for ~120 days in males and ~110 days in females, being disposed of by macrophages in the SPLEEN, liver and other lymphoid organs (see BILE PIGMENTS). Lacking mitochondria, erythrocytes generate their ATP by glycolysis and are responsible for ~one-third of blood LACTIC ACID. Much of the ATP produced is required for maintenance of the Na^+/K^+-ATPase in the membrane whose activity is required for osmotic balance and appropriate cell volume. Blood group antigens are also located in the membrane (see HAEMOLYTIC ANAEMIA). See ANAEMIA.

Heterodimers of the protein spectrin are concentrated just under the plasma membrane, forming a two-dimensional web linked by short ACTIN filaments (see Fig. 146). The junctional complexes where they meet also include tropomyosin, adducin and band 4.1 proteins. The latter are globular proteins attaching actin to spectrin and binding to the transmembrane band 3 proteins and the enigmatic glycophorin. Spectrin also attaches this cytoskeleton indirectly to the membrane via ankyrin molecules which bind the band 3 proteins (see DYNEINS). This network creates a stiff cell cortex, providing mechanical support for the plasma membrane and enabling the cell to spring back to its original shape after the deformation that occurs as it squeezes through a capillary.

Erythrocyte membranes contain an anion exchanger which is essential for CO_2 transport to the lungs from active tissues. When CO_2 enters the cell it is converted by carbonic anhydrase to HCO_3^- in exchange for chloride ions (Cl^-), exchange of these two ions being known as the *chloride shift*. The HCO_3^- enters the blood plasma for transport to the lungs, and because it is much more soluble than CO_2, this increases the blood's capacity for carrying CO_2. HCO_3^- then reenters the red blood cell and is converted to CO_2 and released across the respiratory membranes in the lung. The chloride-bicarbonate exchanger (anion exchanger) increases the permeability of the erythrocyte membrane to HCO_3^- more than a million-fold. Like the cell's uniport GLUT-1 GLUCOSE TRANSPORTER, it is an integral membrane protein that probably spans the membrane 12 times. For each HCO_3^- ion that moves in one direction, one Cl^- ion moves in the opposite direction, and is therefore an antiport system. As erythrocytes break down glucose they produce 2,3-bisphosphoglycerate (BPG), which greatly reduces the affinity of haemoglobin for oxygen.

red bone marrow See BONE MARROW.

red muscle A muscle whose fibres are rich in myoglobin and are therefore slow-contracting and aerobic. See SKELETAL MUSCLE.

red nucleus Located in the MIDBRAIN, and pinkish when freshly dissected, the paired red nuclei are organized somatotopically, and their activities precede intentional movements, correlating with such parameters as force, velocity and direction in the

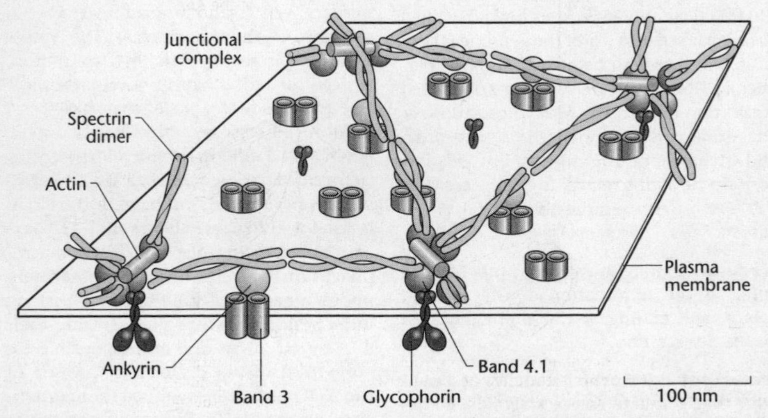

FIG 146. Some of the protein interactions on the cytoplasmic surface of a RED BLOOD CELL plasma membrane. © 2002 From *Molecular Biology of the Cell* (4th edition) by B. Alberts, A. Johnson, J. Lewis, M. Raff, K. Roberts and P. Walter. Reproduced by permission of Garland Science/Taylor & Francis LLC.

manner of the corticospinal tract neurons. It receives axons from pyramidal cells in layer V of the motor cortex, while its efferent fibres cross in the ventral tegmental decussation of the pons and descend to the SPINAL CORD in the RUBROSPINAL TRACT, an indirect pathway which originates in the lower portion of the red nucleus (the magnocellular portion) and contains neurons similar in size to the BETZ CELLS of the motor cortex. This magnocellular portion receives efferent projections from the deep nuclei of the CEREBELLUM, with which the red nucleus has similar neural connections to those between the motor cortex and the cerebellum. Work on non-human primates suggests that while the rubrospinal tract is active when previously learnt automated movements are performed, the corticospinal tract is involved when new motor patterns are being learnt. When a new movement has been learnt, its execution is switched from corticospinal to rubrospinal tract pathways, a switch which is reversed when an automated motor pattern needs modification. This switch involves the INFERIOR OLIVARY NUCLEUS.

Reelin A secreted glycoprotein expressed in the marginal zone (MZ) near the pial surface of the neocortex by CAJAL–RETZIUS CELLS during embryogenesis. Neurons arriving into the adjacent CORTICAL PLATE are devoid of Reelin but express a number of other proteins, some functioning downstream of Reelin and together forming the Reelin signalling pathway. Mutations in the *reelin* gene give rise to inversion of the layers of the CEREBRAL CORTEX, and ataxia in mice. Reelin binds extracellular lipoprotein receptors, including VLDLR and ApoER2, leading to increased tyrosine phosphorylation of Dab1. Dab1 accumulates in the absence of Reelin pathway activation. Reelin is known to bind a number of cell surface molecules present in migrating neurons.

Reference Daily Intakes (RDIs) Values for certain vitamins and minerals based on the 1968 Recommended Dietary Allowances (daily nutrient intakes sufficient to meet the needs of nearly all people in certain age and gender groups) and updated every 4–5 years. They are set for four groups

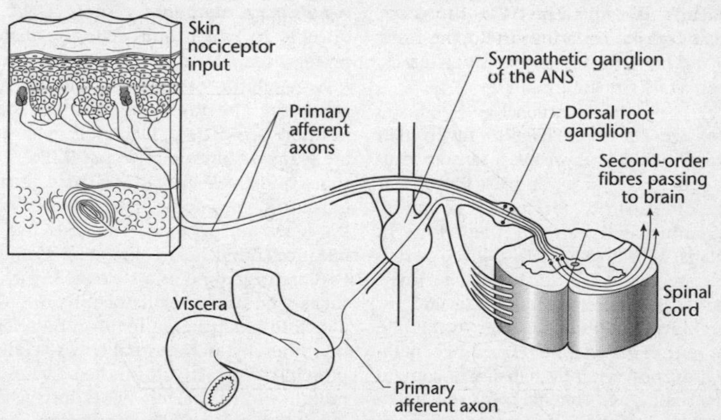

FIG 147. The convergence of nociceptor input from the viscera and skin of the abdominal wall around the navel responsible for the REFERRED PAIN experienced during a case of appendicitis. Second order fibres decussate and pass contralaterally via the anterolateral system, particularly the spinothalamic tract, up to the brainstem. ANS = autonomic nervous system. © 2001 Redrawn from *Neuroscience* (2nd edition) by M.F. Bear, B.W. Connors and M.A. Paradiso, published by Lippincott, Williams & Wilkins. Reproduced with the kind permission of the publisher.

of people: infants, toddlers, people over 4 years of age, and pregnant or lactating women, usually at the highest value in each category. See DAILY VALUES, DIETARY REFERENCE VALUES.

referred pain The feeling of pain in a part of the body remote from the tissue causing the pain. Often the cause is in one of the visceral organs, although it is experienced on the body surface. In the first stages of labour, for instance, the cause of pain is the generation of nerve impulses in A DELTA and C FIBRES in the cervix and uterus. These excite the spinal cord in the lower back and the pain is referred to distant structures. The phenomenon probably arises when sensory neurons from visceral and skin regions served by the same segment of the spinal cord share connections with the same relay neurons in the cord's dorsal horn (see Fig. 147 and see PAIN).

Thus, nociceptor axons from the viscera enter the spinal cord by the same route as the cutaneous receptors and considerable mixing of information results, a cross-talk (see SPINAL CORD, Fig. 162) in which visceral nociceptive activation is perceived as cutaneous sensation, as when heart muscle ischaemia is experienced as the acute pains of *angina pectoris* in the upper chest wall and left arm. See MIRROR VISUAL FEEDBACK, PHANTOM LIMB SYNDROME, SHINGLES.

reflex arc The neural circuitry involved in a motor reflex, comprising a sensory neuron, a motor neuron, and usually one or more interneurons (excitatory or inhibitory) interposed between these two. In humans, all but one reflex arc (the stretch reflex; see MUSCLE SPINDLES) include interneurons and therefore contain several SYNAPSES. Monosynaptic reflex arcs contain no interneuron; disynaptic reflex arcs contain one, and polysynaptic reflex arcs contain more than one. The majority of human reflex arcs involve synapses in the SPINAL CORD (see Fig. 162), and are termed spinal reflexes. The connections made by several sensory neurons or interneurons overlap and provide for possible integration, which may be non-linear (facilitation) or little affected by additional inputs (occlusion). The time between the stimulus and response is the *reflex latency*, and includes the time taken for each synaptic delay (~0.5–1 ms). See, e.g., CLASSICAL CONDITIONING, INSTRUMENTAL CONDITIONING, POSTURAL REFLEX PATHWAYS, PUPILLARY LIGHT REFLEX, and Fig. 166 for the regulation of STOMACH secretions.

refractory period The time interval, during and after passage of an ACTION POTENTIAL through a point on an excitable (e.g., muscle or axon) membrane, when the membrane is either incapable of relaying a second action potential (the ABSOLUTE REFRACTORY PERIOD) or will do so only if the stimulus is stronger than normal, and sometimes then with a smaller amplitude of depolarization (the RELATIVE REFRACTORY PERIOD). These periods prevent fusion of impulses, allowing each impulse to be discrete, and are explicable in terms of the allosteric properties of the voltage-gated sodium and potassium ION CHANNELS in the membrane.

regulated intramembrane proteolysis See RIP.

regulatory elements Those DNA sequences to which *trans*-acting regulatory proteins (capable of binding both *cis*- and *trans*-regulatory elements) bind, thereby regulating the timing of GENE EXPRESSION. Multiple *cis*-acting regulatory elements are required for RNA POLYMERASE II (Pol II) to transcribe DNA into RNA. These include PROMOTERS, promoter-proximal elements, ENHANCERS and SILENCERS. The latter two are distance-independent *cis*-acting elements that are organized as a series of sequences that are bound by regulatory proteins. One model for such action involves DNA looping, whereby architectural (DNA-bending) proteins bring activator proteins bound to promoter-proximal elements into association with activator proteins bound to distant enhancers, in this way stabilizing the Pol II initiation complex (see TRANSCRIPTION). Unlike enhancers, LOCUS CONTROL REGIONS can overcome repressive effects of condensed heterochromatin. Regulatory elements, especially promoters, are often patchworks of regulatory sequences, reflecting the evolutionary histories of each gene. New sequences, binding activators or repressors, may be added and lost during evolution as a gene becomes co-opted into a new pathway or is no longer involved in one, respectively. See also INSULATORS.

regulatory T cells See T CELLS.

reinforcement and reward Reinforcers are unconditioned stimuli whose pairing with an operant response alters the probability or strength of the response, positive reinforcers (rewards) increasing, and negative reinforcers (aversive reinforcers) decreasing, response strength or probability (see INSTRUMENTAL CONDITIONING, LEARNING). The ORBITOFRONTAL CORTEX is implicated in planning behaviour associated with sensitivity to reward and punishment. Evidence indicates that ascending dopaminergic neurons of the ventral tegmental area (VTA; see BRAIN, Fig. 30; DARPP-32, DOPAMINE) are responsible for activating motivated behaviour, generated by the somatic motor system and incited to occur by activity of the lateral HYPOTHALAMUS. The exact role of dopamine in the motivational system is debated, although animals are motivated to perform behaviours associated with the delivery of drugs that stimulate dopamine release in the NUCLEUS ACCUMBENS, and these are therefore strongly reinforced (see ADDICTION). The VTA projects to the nucleus accumbens (together known as the mesolimbic system) and to the frontal cortex (as the mesocortical system). The mesolimbic system is believed to signal the hedonic (pleasurable, positively reinforcing) qualities of a stimulus, activating the appropriate goal-seeking behaviour (see INTENTIONAL MOVEMENTS). The nucleus accumbens is part of the anterior cingulate (affective) basal ganglia circuit which is thought to translate motivation into the motor activity required for a goal-directed behaviour. But firing of the mesolimbic system in response to a natural reward or conditioned stimulus with which it has been paired depends on the reward's predictability. The predictive reward value a person puts on a particular visual cue is encoded in two distinct brain regions: amygdala and orbitofrontal cortex. One hypothesis is that CREB is a key regulator of the reactivity of the brain reward circuits, thereby regulating sensitivity to emotional stimuli. Unexpected rewards elicit a strong response initially whereas predicted rewards have little effect; but omission of an expected reward reduces mesolimbic activity. However, in instrumental conditioning at least, such omission usually leads to perseverance with the operation linked with its provision; so reduced mesolimbic firing when a predicted reward is missed might signal not to switch from the behavioural focus of ATTENTION.

Reinforcers vary in their effectiveness in such associative learning, as illustrated by *food aversion* studies. If they survive at all, animals in their natural environment tend to learn to avoid food baited with poison if the poison can be neurally associated with some environmental stimulus and causes subsequent nausea, even if this unconditioned response (nausea) occurs some hours after the food ingestion. For most species, humans included, food-aversion conditioning only occurs when a taste stimulus is associated with the subsequent aversive feeling and develops poorly or not at all if the taste is followed only by pain, and similarly with visual and auditory stimuli. It would seem likely that selection pressures have favoured some neural pathways over others in constraining what can be learnt by way of association (see NATURAL SELECTION), supported by known variability in this regard both within and between species. It may be a means by which people learn, intentionally or otherwise, to regulate their diets in such a way as to avoid unpleasant consequences of certain foods. Genetic and experiential factors can affect the effectiveness of a reinforcer, results obtained with a particular class of reinforcer varying enormously between individual humans. The malaise and nausea associated with certain forms of cancer chemotherapy may induce aversive conditioning to certain foods normally present in the patient's diet and tasted shortly before treatment. Food aversion has been employed as a therapy in treating chronic alcoholism. A response need not evoke a rewarding stimulus for instrumental conditioning to occur. It will sometimes take place if the response prevents the occurrence of an aversive stimulus. See ADDICTION, BASAL GANGLIA.

reinforcers See REINFORCEMENT AND REWARD.

relatedness, *R* Excluding incestuous matings (see INCEST), the degree of relatedness between two humans (male A and female B) is found by first locating their nearest common ancestor(s) and counting the number of generations passed by moving from B to the common ancestor(s) and then on to A (this is the *generation distance*). This number is the power to which $\frac{1}{2}$ is first raised and then multiplied by the number of nearest common ancestors. Two second cousins, e.g., have two nearest common ancestors, so their degree of relatedness, *R*, is given by $R = 2 \times (\frac{1}{2})^6 = 1/32$; meaning that two second cousins are expected to share 1 in 32 of their genes. See CONSANGUINITY, HAMILTON'S RULE.

relative refractory period The period after the passage of an impulse along an excitable membrane during which the membrane is hyperpolarized until the voltage-gated potassium channels close, and a greater-than-normal stimulus (i.e., more depolarizing current) is required to achieve the threshold for passage of a further impulse due to voltage-gated potassium channels remaining open (see REFRACTORY PERIOD). Compare ABSOLUTE REFRACTORY PERIOD.

relaxin A polypeptide of the insulin/IGF hormone family, regulating the synthesis of metalloproteinases. One of the first reproductive hormones to be described, its effects are mediated by two G-PROTEIN COUPLED RECEPTORS and involve increase in intracellular cAMP. Relaxin is secreted by the ovary and regulates growth and remodelling of reproductive tissues during late pregnancy, and in model animals promotes expansion of the birth canal and relaxation of the cervix during BIRTH (see also DECIDUA). In humans, the peak of circulating relaxin occurs in the 1st trimester, coinciding with implantation, and circumstantial evidence indicates that disruption of its circulation during early pregnancy is associated with miscarriage. It specifically induces vascular endothelial growth factor (VEGF) in the endometrium, being responsible for growth of new blood vessels. It appears to be released by other organs paracrinally, and is involved in ANGIOGENESIS during wound healing. It is also vasoactive, and its levels decrease in post-menopausal women.

relay neuron An interneuron (e.g., see MUSCLE SPINDLE).

relay nuclei Term used to indicate brain stem nuclei, and thalamic nuclei such as the ventral posterior nucleus, whose role has sometimes been thought simply to be the transference of sensory information unchanged from its arrival in the dorsal column nuclei of the spinal cord until its projection to the sensory cortex. On the contrary, synapses in both dorsal column nuclei and thalamic nuclei transform this information, in particular through inhibitory interactions between adjacent sets of input in the dorsal column–medial lemniscal pathway which enhance the response to tactile stimuli (see LATERAL INHIBITION). Some of these synapses exhibit changes of synaptic strength as a result of recent activation. Thalamic and dorsal column nuclei are also under influence from the cerebral cortex, so that cortical output can influence its input.

REM sleep See SLEEP.

remodelling Modification of pre-existing structures; e.g., remodelling of neuronal topography by sensory input during development (see NEURAL PLASTICITY); BONE REMODELLING, CHROMATIN REMODELLING, and wound-healing (e.g., see RELAXIN).

renal capsule A layer of fibrous connective tissue surrounding each KIDNEY.

renal corpuscle The enlarged proximal end of a nephron comprising a glomerulus and its surrounding Bowman's capsule. See KIDNEY.

renal cortex and medulla See KIDNEY.

renal dialysis See KIDNEY MACHINE.

renal failure Decrease, or cessation, of glomerular filtration (see KIDNEYS). It causes oedema due to salt and water retention; acidosis; increased blood urea levels; elevated potassium levels (which may lead to cardiac arrest); anaemia, due to reduced erythropoietin output; and osteomalacia due to lack of conversion of vitamin D to calcitriol. Chronic renal failure is the progressive and usually irreversible form of the disease, often resulting from glomerulonephritis, polycystic disease or traumatic loss of renal tissue. Those with *end-stage renal failure* require dialysis therapy or, possibly, kidney transplant surgery. It is often linked to type 2 diabetes and hypertension – the associated risk varying with the geographical POPULATION. For *acute renal failure*, see ACUTE DISEASE.

renal fraction That proportion of the total cardiac output that passes through the kidneys. It varies from 12% to 30% of the cardiac output in healthy individuals, but averages 21%.

renin Enzyme produced by the juxtaglomerular cells of the kidneys when blood pressure/volume decreases. It converts angiotensinogen to angiotensin I (see ANGIOTENSINS).

renin-angiotensin-aldosterone system The system responsible for preserving circulatory homeostasis in response to a loss of salt and water. See ANGIOTENSINS.

Renshaw cells One kind of small INTERNEURON, located in the ventral horn of the SPINAL CORD in close association with the motor neurons. They receive collateral fibres from these motor neurons soon after the motor axons leave the body of the anterior motor neuron. Renshaw cells then transmit inhibitory signals back to the surrounding motor neurons (see FEEDBACK), so that stimulation of each motor neuron tends to inhibit the surrounding motor neurons, a phenomenon termed *recurrent inhibition*. This probably allows unabated transmission of the primary motor signal in the right direction, suppressing the tendency for the signal to spread laterally. A similar principle is known to 'sharpen up' signals in the sensory system (see LATERAL INHIBITION).

repairosome See DNA REPAIR MECHANISMS.

repetitive DNA Repeat sequences account for at least 50%, and probably much more, of human DNA. Often described as 'junk' DNA, they provide valuable insights into the history of the HUMAN GENOME, which they have helped to reshape. They include LINES, SINES, LTR TRANSPOSONS and DNA transposons, MINISATELLITE DNA, MICROSATELLITES and SATELLITE DNA. About 10% of human genes show evidence of *intragenic* DNA duplication (e.g., those for COLLAGENS, serum albumin, the C region of IgE and dystrophin), often originating by exon duplication (see GENE DUPLICATION). Repeating domains are often advantageous in proteins with major structural roles. Some moderately repetitive portions of DNA include sections repeated tens to thousands of times and include some genes with well known functions such as those for ribosomal and transfer RNAs, while other functional genes with repetitive elements include the gene families for HISTONES and ANTIBODIES. In some classifications, *repetitive DNA* involves repeat sequences present in a few to ~10^5 copies per genome, with *highly repetitive DNA* sequences present in ~10^5–10^7 copies per genome (some classes encoding their own RNA-directed DNA POLYMERASE). Nearly 30 human hereditary disorders result from an increase in the copy number of simple repeats in genomic DNA, often in non-coding regions of their resident genes, leading to altered GENE EXPRESSION. Most of these are due to trinucleotide repeats; but disorders due to tetranucleotide, pentanucleotide and even dodecanucleotide repeats are known. GC-rich CAGG repeat elements appear to mediate recent (10–20 Myr) hominoid genomic duplications from diverse chromosomal locations to pericentric locations. Only a handful of all DNA repeats expand progressively between generations in the dynamic manner of some TRIPLET REPEATS (for replication slippage, see MUTATION). Expandable repeats can be located in (i) coding regions of the resident gene, as in numerous diseases mediated by polyglutamine or polyalanine runs in proteins; (ii) 5´ untranslated regions (5´ UTRs), as in FRAGILE-X SYNDROME, fragile X mental retardation associated with the *FRAXE* site, fragile X tremor and ataxia syndrome, and spinocerebellar ataxia 12; (iii) 3´ UTRs, as in myotonic dystrophy 1, spinocerebellar ataxia 8, and Huntington-disease-like 2 (see RNA PROCESSING); (iv) introns, as with myotonic dystrophy 2, Friedreich's ataxia and spinocerebellar ataxia 10; and (v) promoter regions, as in progressive myoclonic epilepsy. Most diseases associated with repeat expansions show dominant inheritance, possibly through gain of function at the RNA level ('toxic' RNA). See CPG DINUCLEOTIDE, HUNTINGTON'S DISEASE, TELOMERES.

replacement theory For discussion of this model of human evolution and of the rival multiregional theory, see HOMO.

replication fork See DNA REPLICATION.

replication slippage See MUTATION, Fig. 127.

reproductive variance See EFFECTIVE POPULATION SIZE.

reserpine A drug used in the 1960s to control hypertension until the discovery that it caused psychotic depression in one-fifth of patients. See AFFECTIVE DISORDERS.

reservoir of infection The human or animal population or environment in which a pathogen exists and from which it can be transmitted.

residual volume (RV, residual lung volume) The volume of gas that cannot be expelled from the lungs (see SPIROMETER, Fig. 165). Its value increases with age, probably on account of decreasing lung elasticity. It can only leave the lungs if they collapse away from the wall of the chest (e.g., in pneumothorax), and even then ~200 cm^3 remains (the minimal volume). In the helium dilution technique, RV can be measured after maximum expiration by breathing a mixture of oxygen containing a known concentration of helium. After a number of rapid, deep breaths, in a closed system, the helium will mix with and add to the gases in the RV and become diluted. An equilibrium is reached between the gases in the spirometer and those in the subject's lung volume, carbon dioxide being removed and oxygen being continually added to replace that consumed. RV computes from the equation:

$$\begin{aligned}&\text{Initial helium volume } (V_1)\\&\quad\times \text{initial helium concentration } (C_1)\\&= \text{final gas volume}(V_2)\\&\quad\times \text{final helium concentration}(C_2)\end{aligned}$$

See FUNCTIONAL RESIDUAL CAPACITY.

resistin Peptide adipokine induced during adipogenesis, and antagonizing INSULIN action.

respiration An enzymatic release of energy from substrates involving electron transport and generation of a proton gradient in which the energy is temporarily stored prior to its use in ATP synthesis. In animals, humans included, the aerobic phase of respiration (oxidative phosphorylation) occurs in mitochondria, molecular oxygen being the final electron acceptor. GLYCOLYSIS, which does not involve electron transport and occurs independently of respiration, is a form of fermentation rather than respiration.

respiratory burst See MACROPHAGES.

respiratory centre The respiratory centre of the MEDULLA OBLONGATA comprises two dorsal and two ventral groups of intercommunicating cells. The dorsal groups lie in the nucleus of the solitary tract (see NUCLEUS TRACTUS SOLITARIUS, NTS) and are most active during inspiration, the two ventral groups (in the caudal and rostral ventrolateral medulla) being active during both inspiration and expiration. The respiratory group of neurons in the pons (formerly the 'pneumotaxic centre') contains neurons some of which are active in either inspiration or expiration. The functions of this pontine group (the parabrachial nucleus) are unclear, although its connections with the medullary centres may be involved in the switching between inspiration and expiration.

Dorsal neurons in the medullary 'inspiratory' centre continually receive inputs via the vagus from receptors monitoring blood respiratory gas levels, pulmonary stretching, blood pressure, temperature, and muscle and joint movements, as well as from higher centres of the brain. Baroreceptor input inhibits inspiration, so that if blood pressure falls, depth of inspiration increases. The spinal motor neurons in the phrenic, neck, and intercostal nerves are driven by premotor neurons in the ventrolateral medullary groups, where the ventilatory rhythm is generated by a network of cells (the *pre-Botzinger complex*) acting as a CENTRAL PATTERN GENERATOR. Inspiratory and expiratory neurons here make excitatory (glutamatergic) connections with the spinal neurons of the phrenic and intercostal nerves; and there are several groups of ventral respiratory group neurons, some with intrinsic pacemaker properties, each firing a burst of action potentials at a specific phase of the ventilatory cycle (see VENTILATION). Inspiration starts when the combined input from sources to the NTS initiates volleys of efferent impulses along spinal motor neurons in the phrenic and intercostal nerves, firing rate increasing for ~2 seconds as more and more of the neurons are recruited. Inspiration is stopped by certain ventral medullary neurons (the 'expiratory' centre) as these receive inputs from the pontine respiratory group, stretch receptors in the lungs (the Hering–Breuer reflex) and probably from elsewhere. When these neurons are activated during quiet breathing, they inhibit the dorsal neurons causing inspiration, terminating each inspiratory volley. The resulting relaxation of the inspiratory muscles results in expiration, lasting ~3 seconds. The expiratory centre also projects to expiratory muscles involved in more vigorous breathing. At these times, in response to increasing pCO_2 or falling pH (the NTS is sensitive to both), the frequency of impulses per volley from the inspiratory centre increases, leading to greater contractility of the inspiratory muscles, the interval between each volley decreasing, so reducing the time between breaths. As a result, pulmonary stretch receptors increase their firing, having normally been fairly inactive during quiet breathing. During deep inspiration these send inhibitory impulses via the vagus to the inspiratory centre and discharge impulses to the expiratory centre causing expiratory muscle contraction. Exhalation reduces the input from the pulmonary stretch receptors so that the inspiratory centre is freed from inhibition and fires another succession of volleys to the inspiratory muscles, starting another ventilation cycle.

During normal resting conditions and during exercise blood carbon dioxide levels (pCO_2) are a major regulator of ventilation, both rate and depth of ventilation increasing with even a small rise in its level: an increased pCO_2 of 5 mm Hg causing a 100% increase in ventilation, thereby helping to eliminate CO_2 from the blood and restoring its acid–base balance. The chemosensory regions of the medulla respond indirectly to a rise in pCO_2 through its effect on blood pH, while chemosensory neurons in the carotid and aortic bodies respond directly but are responsible for only 5–10% of the total ventilatory response to changes in pCO_2 or pH (see HYPERVENTILATION). Changes in blood pO_2 also affect ventilation, but within the normal pO_2 range these effects are small compared to those of CO_2 and pH. Only after pO_2 levels have fallen to ~50% of their normal value do they begin to influence breathing rate. This is because of the high percentage saturation of haemoglobin at any pO_2 level higher than 80 mm Hg. Carotid and aortic body chemoreceptors respond to reduced pO_2 levels by increased stimulation of the medullary respiratory centre; but if the pO_2 level falls sufficiently, the respiratory centre can fail, resulting in death. See ALTITUDE, CARDIOVASCULAR CENTRE, CHEYNE–STOKES BREATHING, SPIROMETER.

respiratory compensation The increase in MINUTE VENTILATION which compensates for metabolic acidosis during heavy exercise.

respiratory exchange ratio (RER) The ratio between the percentages of carbon dioxide and oxygen in inhaled and exhaled air. The measurement of this ratio can be used in estimating the RESPIRATORY QUOTIENT, and is often used in conjunction with the VO_2 max test.

respiratory pump See VEINS.

respiratory quotient (RQ) The ratio of carbon dioxide production to oxygen consumption during metabolism of respiratory substrates (see RESPIRATORY EXCHANGE RATIO). In 1920, remarkably accurate work by Krogh and Lindhard determined the respiratory gas exchange of subjects who had consumed varied diets, and enabled them to determine the macronutrient combustion during rest and exercise. Subjects who had previously consumed exclusively either lipid or carbohydrate, with protein held constant, were put within a closed chamber containing a cycle ergometer and two fans. A gas collection apparatus outside the chamber was connected to it via a small-bore tube. They discovered that energy expended to perform a standard physical effort differed and was inversely proportional to the RQ, calculated by the equation

$$RQ = CO_2 \text{ produced} \div O_2 \text{ consumed}$$

The RQ values did not indicate combustion of fat only or carbohydrate only, however. When they quantified the relationship between RQ and the relative amounts of fat and carbohydrate oxidized, they discovered that the percentage of total energy derived from fat oxidation approximated a straight-line function of the RQ. Fat released less energy than carbohydrate per dm^3 oxygen consumed during exercise, and further work revealed that neither fat not carbohydrate was the sole source of energy during EXERCISE and that a blend of the macronutrients served as the fuel source.

Because of the differences in the state of reduction in carbohydrate, fat and protein, different amounts of oxygen are required to complete their oxidation to carbon dioxide and water. RQ provides a convenient rule of thumb for approximating the nutrient mixture catabolized for energy during rest and aerobic exercise. The values are given by the following equations:

For carbohydrate:

$$C_6H_12O_6 + 6\ O_2 \rightarrow 6\ CO_2 + 6\ H_2O$$
$$RQ = 6\ CO_2 \div 6\ O_2 = 1.00$$

For fat:

$$C_{16}H_{32}O_2 + 23\ O_2 \rightarrow 16\ CO_2 + 16\ H_2O$$
$$RQ = 16\ CO_2 \div 23\ O_2 = 0.696$$

(a value of 0.70 is usually taken as representative)

For protein:

$$C_{72}H_{112}N_{18}O_{22}S + 77\ O_2 \rightarrow 63\ CO_2 + 38\ H_2O + SO_3 + 9\ CO(NH_2)_2$$
$$RQ = 63\ CO_2 \div 77\ O_2 = 0.818$$

(a value of 0.82 is usually taken as representative)

Their results indicated that: (i) the efficiency of constant-load exercise is greater with carbohydrate as the fuel source than with fat; (ii) performance deteriorates in high-intensity exercise when fat rather than carbohydrate serves as the preferential energy source; (iii) pre-exercise nutrition influences the metabolic mixture during both rest and exercise; (iv) the RQ changes during the transition from rest to moderate exercise, increasing with high-intensity exercise and indicating a greater reliance then on carbohydrate oxidation; (v) fat oxidation predominates during the latter stages of a 1-hour constant-intensity exercise. Measurement by indirect calorimetry of heat release by the body requires measurement of both RQ and oxygen consumption. See ENERGY BALANCE.

respiratory zones The subdivision of the ventilatory system comprising the alveoli (see ALVEOLUS), representing the location of gas exchange, occupying ~2.5–3.0 dm^3 and constituting the largest portion of the lung volume.

response element Specific regulatory sequences, often within a promoter, binding a specific repressor, activator or other regulatory molecule and either enhancing or suppressing specific gene expression. Hormone response elements (HREs) bind NUCLEAR RECEPTOR–hormone complexes. See CREB.

resting metabolic rate See BASAL METABOLIC RATE.

resting potential (resting membrane potential) The equilibrium position across a plasma membrane, in which there is no net ion flow. Its value varies in humans between −35 mV and −90 mV, depending upon the cell type. It decays only relatively slowly (within minutes) when the Na^+K^+-ATPase (see ION CHANNELS) is inactivated. A small drop occurs immediately on inactivation since the Na^+-K^+ pump makes a small direct contribution to the membrane potential; but the abolition of the resting potential is due mainly to the leakage of K^+ through its leak channels and slow entry of Na^+ (the membrane is not totally impermeable to Na^+, even though no Na^+-leak channels are involved). As Na^+ enters, the cell's osmotic balance is upset and water enters. If it does not burst, the cell eventually arrives at a new state in which Na^+, K^+ and Cl^- are all at equilibrium across the plasma membrane. The more permeable the membrane is for a given ion, the more strongly the membrane potential tends to be driven towards the equilibrium value for that ion.

restriction enzymes (restriction endonucleases) Any of a large number of nucleases that can cleave a DNA molecule at specific recognition sites (short nucleotide sequences). They are employed in recombinant DNA technology, restriction mapping and in DNA PROFILING.

restriction fragment length polymorphisms (RFLPs) Fragments of DNA of varying length produced by cutting DNA with restriction enzymes. These were the first DNA MARKERS to be extensively investigated, arising from base substitutions that create alternative alleles whose existence is detected by the effect they have on the patterns of migration of the fragments (restriction fragments) of DNA in gel electrophoresis after digestion by restriction enzymes (see SOUTHERN BLOTTING). The fragments move different distances on the gel in a given time depending on their relative sizes, and their location can be detected by using single-stranded DNA PROBES. The heterozygous state is very common and can be easily detected. See DNA PROFILING, PULSED FIELD GEL ELECTROPHORESIS, SINGLE NUCLEOTIDE POLYMORPHISMS.

restriction mapping Techniques employed to produce a diagrammatic representation of a DNA molecule indicating the sites of cleavage by various RESTRICTION ENZYMES. See MTDNA.

restriction site polymorphisms (RSPs) Those SINGLE NUCLEOTIDE POLYMORPHISMS which cause loss or gain of a restriction site for a restriction enzyme. See POLYMERASE CHAIN REACTION, RESTRICTION FRAGMENT LENGTH POLYMORPHISMS.

restriction site variation The basis of the production of polymorphisms revealed by restriction enzymes. These enzymes cleave DNA at very specific recognized sequences, so that DNA sequences caused by mutations within restriction sites will no longer be cut by an enzyme that did so previously.

***RET* gene** A proto-oncogene (hsa10q11.2) whose protein product has three major domains: an extracellular domain binding to glial cell line-derived neurotrophic factor (GDNF); a transmembrane domain; and an internal tyrosine kinase domain involved in signal transduction. GDNF-binding induces dimerization. Loss-of-function mutations are found in 50% of familial cases of HIRSCHSPRUNG DISEASE, while very specific missense gain-of-function mutations are found in familial medullary thyroid carcinoma and the related more extensive multiple endocrine neoplasia type 2.

reticular formation (reticular activating system, reticular system) A complex, net-like (reticular), meshwork of neurons and fibres occupying a considerable part of the BRAINSTEM from the midbrain to the medulla, just under the cerebral aqueduct and fourth ventricle. Involved in a great variety of afferent and efferent pathways (e.g., its dorsal portion controls the VOMITING reflex), it is essential to the sensory regulation of *arousal* from sleep, receiving inputs from the spinal trigeminal nucleus and from all other afferent cranial nerves. As von Economo discovered in the early 20th century, lesions at the junction of the midbrain and forebrain produce profound and long-lasting impairment of arousal. This, we now appreciate, is because they block both ascending pathways of the reticular activating system.

The main components of the *ascending reticular activating system* (or ascending arousal system) involved in arousal from sleep and increasing arousal and responsiveness of the sensory cortex to its input (see Fig. 148), are parts of the noradrenergic and cholinergic diffuse modulatory systems of the BRAIN (see Fig. 30). It splits into two branches in the diencephalon. One branch bypasses the thalamus and projects through the lateral hypothalamus to the cerebral cortex from several sources, including noradrenergic neurons of the LOCUS COERULEUS, serotonergic cells of the RAPHE NUCLEI, dopaminergic ventral periaqueductal grey matter and histaminergic neurons of the tuberomammillary nuclei (TMN). Input to the cortex is augmented by lateral hypothalamic peptidergic neurons (LHA) containing melanin-concentrating hormone (MCH) or orexin/hypocretin. Lesions in the LHA and rostral midbrain produce profound and long-lasting sleep, or even coma and their monoaminergic nuclei fire fastest during wakefulness, slowing their rate of firing during NREM sleep, and stop firing during REM sleep. Orexin neurons in the LHA are also most active during wakefulness, while MCH neurons are active during REM sleep. Most cholinergic neurons of the basal forebrain are active during both waking and REM sleep.

The other branch of the ascending arousal system comprises histaminergic neurons in the TMN and cholinergic neurons in the pedunculopontine and lateral dorsal tegmental nuclei (PPT/LDT) in the pons and midbrain, respectively. These send a profusion of impulses to the thalamus and thence to the sensory cortex and subcortical areas, firing most rapidly during wakefulness and REM sleep (see SLEEP) and crucial in activating the thalamic relay neurons that transmit information to the cerebral cortex. They are augmented by cholinergic neurons in the basal forebrain. Other small diameter slow-conducting neurons synapse mainly in the intralaminar nuclei of the thalamus and in the RETICULAR NUCLEI covering its surface, from where numerous small fibres project throughout the cerebral cortex. These excitatory inputs are important in controlling the long-term background excitability of the cerebrum. One cell group in the hypothalamus, the ventrolateral preoptic nucleus (VLPO), sends inhibitory projections to all the other major cell groups in the hypothalamus and brain stem that participate in arousal. Its neurons are principally active during sleep and contain the inhibitory neurotransmitters galanin and GABA. Both noradrenaline and serotonin (5-HT) inhibit VLPO neurons, so that these can be inhibited by the very arousal system that it inhibits during sleep. Another inhibitory reticular area is located more ventrally in the medulla, exciting inhibitory serotonergic neurons. Its input and output links with the hypothalamus and limbic system mediate affective-emotional effects of sensory stimuli, especially of pain stimuli carried by the anterolateral funiculus (see RETICULOSPINAL TRACTS). See STRESS RESPONSE.

reticular nuclei Thin layers of cells of the lateral thalamus, receiving collaterals of

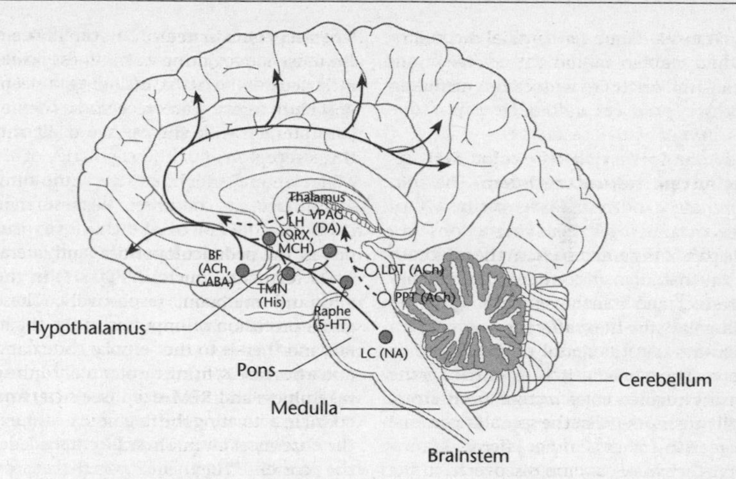

FIG 148. A schematic representation of some key components of the ascending arousal system, a major component of the RETICULAR FORMATION (see entry for details). Its major input to the relay and reticular nuclei of the thalamus (shown as dotted lines) originates in the cholinergic (ACh) cell groups in the upper pons, the pedunculopontine (PPT) and laterodorsal tegmental nuclei (LDT). A second pathway (unbroken lines) activates the cerebral cortex, facilitating processing of thalamic inputs, and this arises from neurons of the monoaminergic cell groups, such as the tuberomammillary nucleus (TMN) containing histamine (His), the A10 cell group containing dopamine (DA), dorsomedial raphe nuclei containing serotonin (%-HT), and the noradrenergic (NA) locus coeruleus (LC). Input also comes from orexigenic (ORX) or MCH-secreting neurons in the lateral hypothalamus (LH), as well as from GABAergic and cholinergic basal forebrain (BF) neurons. From C.B. Saper, T.E. Scammell and J. Lu, © *Nature* 2005, with permission from Macmillan Publishing Ltd.

both thalamocortical and corticothalamic fibres passing between other thalamic nuclei and the cerebral cortex. They receive cholinergic fibres from the upper pons which facilitate thalamocortical transmission as part of the ascending arousal system of the RETICULAR FORMATION. In the medulla and pons, reticular nuclei are important in modulating ventilation, heart rate and blood pressure (see CARDIOVASCULAR CENTRE), as well as sourcing axons of the RETICULOSPINAL TRACTS (see CENTRAL PATTERN GENERATORS). The reticular formation of the rostral pons and midbrain is critical for the maintenance of CONSCIOUSNESS (lesions in this area result in coma), and reticular nuclei (such as the paramedian pontine reticular formation, PPRF) are important for mediating movement of the EYES.

reticulospinal tracts Ventromedial DESCENDING MOTOR TRACTS of the SPINAL CORD, arising mainly in the RETICULAR NUCLEI of the pons and medulla and concerned with maintenance of posture. Axons arising in the pons, with inputs from the semicircular canals of the vestibular apparatus, descend ipsilaterally as the medial (or pontine) reticulospinal tract and stimulate antigravity reflexes of the spinal cord, facilitating EXTENSOR MUSCLES of the lower limbs and inhibiting proximal limb FLEXOR MUSCLES, reflexly helping to maintain equilibrium during a standing posture (see POSTURAL REFLEX PATHWAYS). Axons arising in the medullary reticular nuclei, whose input connections originate in the saccule and utricle, descend bilaterally in the lateral (or medullary) reticulospinal tracts and produce polysynaptic inhibition of axial and proximal limb extensors, having the opposite effect from the medial reticulospinal tract fibres, exciting proximal limb flexors, freeing antigravity muscles from reflex control and damping activity within the spinal cord, controlling static movement. Some of them also produce monosynaptic inhibition of neck and axial motor neurons. Loss of this medullary pathway produces profound extensor tone. See CENTRAL PATTERN GENERATORS.

retina (pl. retinas, retinae) The thin, laminated, innermost layer of the eyeball (see EYES), lining its most posterior three-quarters and containing the photoreceptor cells that transduce light energy into altered membrane potentials that are ultimately the trigger for the action potentials that travel along the optic nerve to the brain. The retina does far more than generate information concerning the patterns of light and dark falling upon it. In conjunction with the rest of the visual pathway (e.g., the LATERAL GENICULATE NUCLEI), it extracts and abstracts information concerning numerous aspects of what is perceived to be the external world, including the shapes, relative sizes, positions in space, and movements, of objects (see PARALLEL PROCESSING), as well as generating colour from a spectrum of visible wavelengths.

Developing from the optic cup as a two-layered structure, the retina of a 6-week embryo consists of an outer melanin-containing pigmented layer (sometimes regarded as a part of the choroid) and a more complex inner neural layer whose fibres are already converging on the optic stalk which develops into the optic nerve, the whole completely embedded in undifferentiated mesenchyme. The intraretinal space between the two layers has all but disappeared by the 7th week. The neural portion of the retina is an outgrowth from the lateral wall of the brain and has a similar laminar organization. The cell layers seem functionally 'inside out' since light must pass through the ganglion cell layer and bipolar neuron layer before it can reach the photoreceptors (see Fig. 149), a distance of several hundred micrometres. But these intervening cells are relatively transparent, with relatively few mitochondria (cytochromes absorb light) and distort the image very little. The ganglion cell layer and nuclear layer are laterally displaced in the fovea (see below), so that light can strike its cones directly. The capillaries of the choroid provide nutrients to the posterior surface of the retina, including the outer segments of the photoreceptors. They do not interfere with the passage of light to these receptors; but the capillaries of the central retinal artery that supply the anterior surface do. However, because these capillaries do not overlie the fovea they do not compromise its visual acuity.

The retina's innermost layer, the ganglion cell layer, contains the cell bodies of ganglion cells; the inner nuclear layer contains the cell bodies of bipolar, horizontal and amacrine cells, the outer nuclear layer contains the cell bodies of the ROD CELLS and CONE CELLS whose outer segments comprise the outermost layer. These lie embedded in the pigmented epithelial layer that absorbs any light passing through the retina, preventing any loss of acuity through reflection (see ALBINISM). Synaptic contacts between these receptors and neurons lie in the outer and inner plexiform layers, respectively.

Retinal structure is not uniform, the peripheral retina having a higher ratio of rods to cones and of photoreceptors to ganglion cells, so making it more sensitive to dim light. This enables it to detect faint stars at night, but makes it relatively poor at resolving fine details in daylight (see VISUAL ACUITY).

The retina is said to be *duplex* because under nighttime lighting (see DARK ADAPTATION), only rods contribute to vision whereas under daytime lighting (see LIGHT ADAPTATION) cones are far more involved than rods. Only ganglion cells produce action potentials, all other neurons in the retina (with the exception of some amacrine cells) respond to stimulation by graded changes in membrane potential. Information flows most directly from a cone photoreceptor to a bipolar cell to a ganglion cell; but at each synapse the response is modified by lateral connections with horizontal cells and amacrine cells. Rods and cones are depolarized in the dark and hyperpolarized in the light, so that they release more of the neurotransmitter glutamate in the dark than in the light. It is therefore useful to consider themas 'preferring' dark to light as a stimulus, so that when a shadow passes across them they respond by depolarizing and releasing more glutamate. Each bipolar cell receives direct synaptic contact from a cluster of photoreceptors by convergence (see Fig. 150), the number varying from one in the foveola to thousands in the retinal periphery. Bipolar cells also receive synaptic input from HORIZONTAL CELLS (see entry for details) that synapse with a ring of photoreceptors surrounding those photoreceptors with which it synapses directly, the whole forming the cell's RECEPTIVE FIELD (the area of retina which, when stimulated by light, alters the cell's membrane potential). One millimetre on the retina corresponds to ~3.5 degrees of visual angle and bipolar cell receptive field diameters vary from a fraction of a degree in the central retina to several degrees at its periphery. BIPOLAR CELLS (see entry for details) have either on-centre or off-centre receptive fields (see Fig. 151), a specificity maintained by the synaptic connections between bipolar and ganglion cells in the inner plexiform layer. These different types of bipolar neuron respond differently to the same transmitter (glutamate) since they have

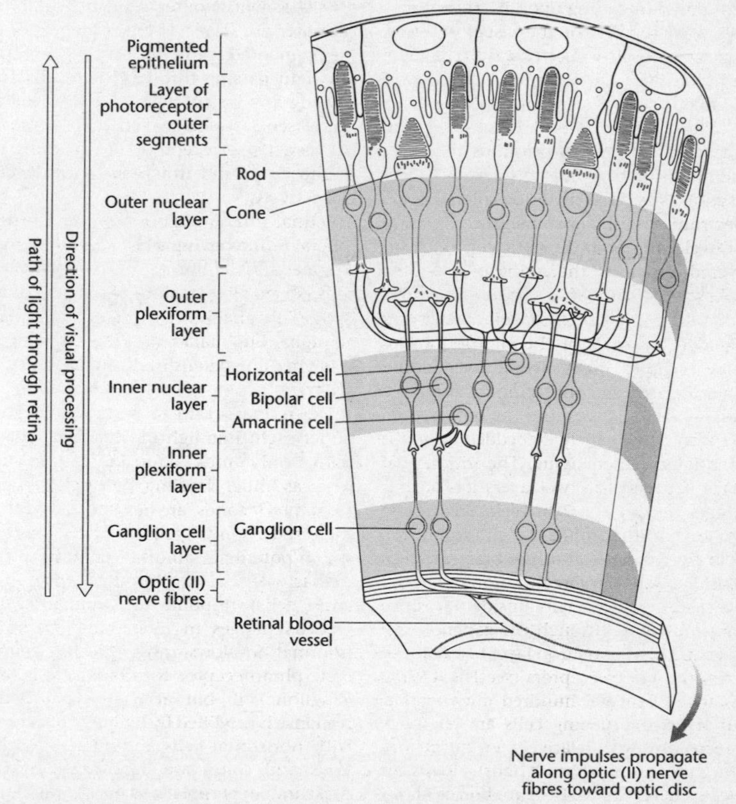

FIG 149. The cell layers and anatomy of the RETINA. From *Principles of Anatomy and Physiology*, G.J. Tortora & S.R. Grabowski, © 2000. Reproduced with permission of John Wiley & Sons, Inc.

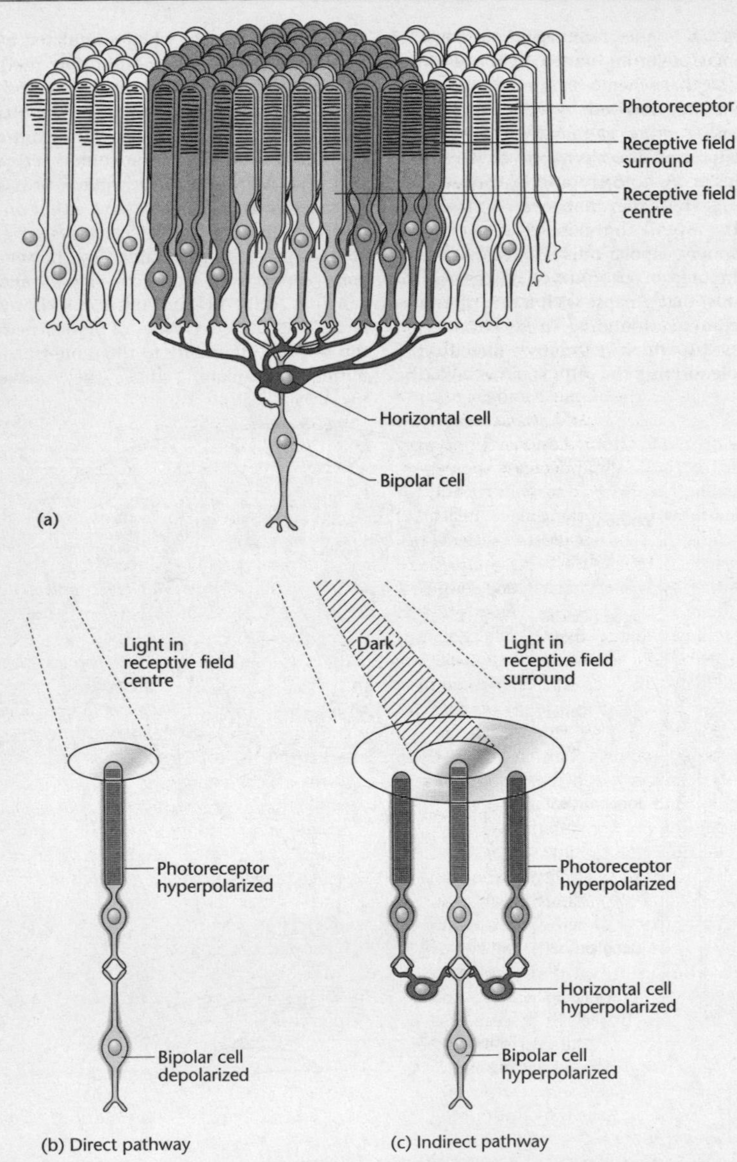

FIG 150. The direct and indirect pathways from photoreceptors to bipolar cells in the RETINA. (a) Bipolar cells receive direct synaptic input from a cluster of photoreceptors, forming the *receptive field centre*, and from surrounding photoreceptors forming the *field surround*. (b) An ON-centre bipolar cell is depolarized when light strikes its receptive field centre via the direct pathway. (c) As a result of HORIZONTAL CELL surround inhibition, light striking the field surround hyperpolarizes an ON-centre bipolar cell via the indirect pathway. Whatever the type of bipolar cell, the effect of a horizontal cell is to affect the surround photoreceptors in a manner opposite to the effect of light on the centre photoreceptors. © 2001 Redrawn from *Neuroscience* (2nd edition) by M.F. Bear, B.W. Connors and M.A. Paradiso, published by Lippincott, Williams & Wilkins. Reproduced with the kind permission of the publisher.

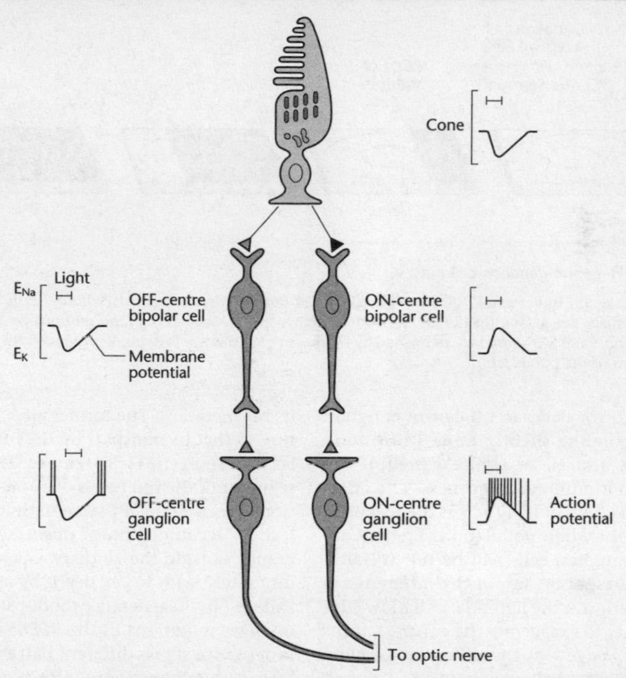

FIG 151. Diagram showing how light's effect on a cone cell has different effects on OFF-centre and ON-centre bipolar cells (see RETINA). Light hyperpolarizes the cone, causing decrease in glutamate release. The transmitter has different effects (shown by the membrane potentials and action potentials) because the two types of bipolar cell have different postsynaptic receptors gating different types of ion channel. From *Essentials of Neural Science and Behaviour*. E.R. Kandel, J.H. Schwartz and T.M. Jessell © 1995 by Appleton & Lange (Simon & Schuster). Reproduced with permission of The McGraw-Hill Companies.

different postsynaptic receptors gating different types of ION CHANNELS. AMACRINE CELLS form lateral connections with ganglion cells in this layer and help to specify their receptive fields in ways as yet imperfectly understood.

The specificity of synaptic connections between bipolar cells and ganglion cells maintains continuity between on-centre and off-centre receptive fields. An on-centre ganglion cell depolarizes and responds with a barrage of action potentials when a small spot of *light* stimulates the centre of its receptive field, whereas an off-centre ganglion cell will respond in just this way to a small *dark* spot in the centre of its receptive field. However, in both cases the response to stimulation of the centre is negated by the response to stimulation of the surround, resulting in most retinal ganglion cells being fairly unresponsive to changes in illumination that include both the receptive field centre and surround (see Fig. 152) and responding mainly to differences in illumination *within* their receptive fields. So, when a shadow enters the surround region of the receptive field of an off-centre ganglion cell, it hyperpolarizes the neuron and decreases its firing rate; but as it begins to spread into the centre of the receptive field the partial inhibition of the surround is overcome and the cell responds by firing impulses. When the shadow fills the entire surround the response is inhibited again. And the response is little different in uni-

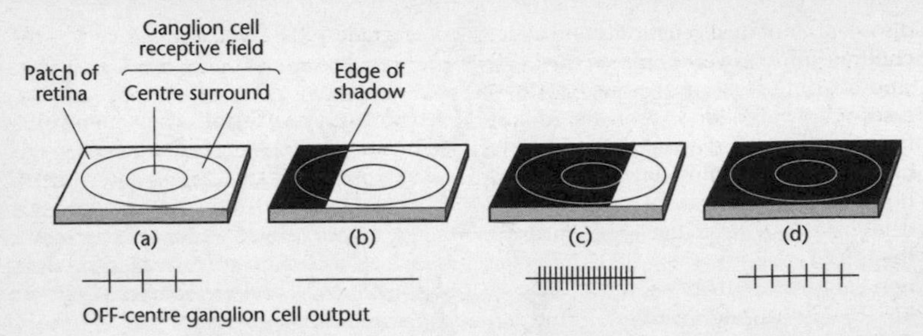

FIG 152. Diagram indicating OFF-centre ganglion cell output when a light–dark border falls within its receptive field. See RETINA for details. © 2001 Redrawn from *Neuroscience* (2nd edition) by M.F. Bear, B.W. Connors and M.A. Paradiso, published by Lippincott, Williams & Wilkins. Reproduced with the kind permission of the publisher.

form light or dark, an off-centre ganglion cell responding slightly more in uniform darkness and an on-centre ganglion cell doing so in uniform brightness. The effect of the shadows moving across the receptive fields of the whole population of retinal off-centre ganglion cells will be not to faithfully represent by output the difference in illumination at the light–dark shadow edge but rather to exaggerate the contrast at the shadow's edge – as the different ganglion cell outputs below indicate (*contrast enhancement*; see LATERAL INHIBITION). Our perception of light must therefore be relative, not absolute. A grey rectangle surrounded by a paler one appears darker than the same rectangle surrounded by a black one; and the print on this page appears as it does in both a dimly lit room and outdoors in bright sunlight: differences in light intensities appear to remain constant despite the change in illumination.

retinal convergence Retinal pathways in which several photoreceptor cells converge on, and synapse with, each BIPOLAR CELL. It is a feature of the rod, but far less of the cone, system. Signals from many ROD CELLS are thus pooled by the bipolar cell, reinforcing one another and increasing the ability of the brain to detect dim light, but reducing VISUAL ACUITY. See ABSTRACTION, CONVERGENCE, RETINA.

retinitis pigmentosa A common form of retinal degeneration caused by mutations in the genes for RHODOPSIN production or its regeneration. The former alter its structure so that its transport by the intraflagellar transport (IFT) system to the outer segment of the rod cell is disrupted. Mutations affecting the IFT system itself can also lead to accumulation of opsin and membranes outside the primary CILIUM, which ultimately leads to cell death by apoptosis. Failure to regenerate rhodopsin often involves mutations in the *RPE65* gene. Its inheritance shows different patterns: autosomal dominance or recessiveness, X-linked recessiveness, and mitochondrial. In the latter instance, it is usually due to a single substitution (T>G, position 8993) in the coding region of subunit 6 of ATPase, which is also associated with NEURAL DEGENERATION and ataxia. Clinical abnormalities used in carrier detection include mosaic retinal pigmentation and abnormal electroretinogram (see SCREENING). See GENE THERAPY.

retinoblastoma Rare CANCER of the eyes arising from cells in the retina, probably cone cells, that have two mutated sites (deletions) at q14 on the long arm of chromosome 13 including the *RB1* locus. Unlike the case in many human tumours, retinoblastoma does not involve mutation in the *p53* tumour suppressor gene, but expresses the wild-type form; however, recent work has discovered a gene amplification that suppresses the Arf-MDM2/MDMX-p53 pathway in human retinoblastoma. In 40% of cases (~1:20,000 births) the disorder is inherited genetically and affects children under 3 years, since retinal cells stop dividing early in life; but in 60% of cases the distribution is sporadic, indicating an INHERITED PREDISPOSITION which later became cancerous through a second mutation in a cell already heterozygous for the deletion (see Fig. 34 in the CANCER entry). In the non-hereditary form of the disease, non-cancerous cells of the body have no *RB1* defects while cancerous cells are defective on both chromosomes. See RETINOBLASTOMA PROTEIN.

retinoblastoma protein (pRB, pRb) Protein encoded by the TUMOUR SUPPRESSOR GENE *RB1*, regulating the late-G1 CELL CYCLE checkpoint and normally expressed in most cells of the body. In normal cells, this 110-kDa nuclear protein is inactivated by phosphorylation and activated by dephosphorylation, being required for differentiation of several tissue types through interactions with other proteins (e.g., Mood and Id2). When active, it binds and inactivates the transcription factor E2F, blocking transcription of S-phase genes and preventing the cell from entering S-phase (the G1 blockade). This acts as a main brake on progress around the cell cycle, which is withdrawn when pRB loses its affinity for E2F on being phosphorylated by a cascade of cyclins, cyclin-dependent kinases (CDKs; see INTERFERONS) and cyclin-dependent kinase inhibitors 2–4 hours before a cell enters S-phase. E2F can then activate S-phase gene expression. Mutations in *RB1* initiate RETINOBLASTOMAS and may play a key role in susceptibility to other forms of cancer, although other mutations would need to occur as well. The *MDM2* gene product binds pRB and targets it for ubiquitinization, any such loss of pRB function contributing to tumour development by preventing cell differentiation (see Myc). Aberrant growth signals resulting from deregulation of the pRB-E2F cell cycle control pathway activate p53 (see P53 GENE). See CANCER (and Fig. 34 in that entry). Several viral oncoproteins (adenovirus E1A, SV40T antigen, human papillomas E7 protein) bind and sequester or degrade pRB, favouring cell cycle progression. See ARF GENE.

retinofugal projection The neural pathway leaving the eye, including the optic nerve, optic chiasma, optic tract, lateral geniculate nucleus (dorsal thalamus), primary visual cortex (striate cortex) and over two dozen extrastriate cortical areas in the temporal and parietal lobes.

retinohypothalamic tract A population of small glutamatergic retinal ganglion cells, driven by cone cells over a wide area of the retina, that synapse directly with neurons in the core of the SUPRACHIASMATIC NUCLEUS. They encode total luminescence information. See CIRCADIAN CLOCKS.

retinoic acid (RA) A vitamin A-derived signalling molecule and MORPHOGEN, produced by the PRIMITIVE NODE and controlling transcription of several genes (RIGs, or retinoic acid-inducible genes) by acting as a ligand for nuclear RA receptors (see NUCLEAR RECEPTORS). During embryonic development, it regulates key genes involved in PATTERN FORMATION, notably several *Hox* genes involved in patterning the hindbrain and spinal cord as neurectoderm emerges from the PRIMITIVE STREAK, and in generating bilateral symmetry of left and right SOMITES as presomitic mesoderm emerges from the primitive streak. It is known to antagonize FGF8 generated at the posterior end of the embryo during somitogenesis, without which segmentation cannot proceed. It plays a role in organizing the cranio-caudal axis since it can cause respecification of the cranial segments into more caudal ones by regulating expression of homeobox genes, but normally limits *FGF8* expression to the most posterior ectoderm (epiblast), preventing ectopic *FGF8* expression more anteriorly. These findings suggest that human vertebral BIRTH DEFECTS, such as congenital scoliosis, may be caused by a defect in RA signalling during somitogenesis. Retinoic acid-inducible genes (RIGs) include SONIC HEDGEHOG, and some that are critical in detecting viral double-stranded RNA and controlling infection. See HAEMA-

TOPOIETIC STEM CELLS, INNATE IMMUNITY, INTERFERONS, LIMBS.

retinoids, retinoid receptors Retinoids are analogues or metabolites of RETINOIC ACID.

retinol binding protein-4 (RBP4) A blood transporter for vitamin A (retinol) and an ADIPOKINE thought to link GLUT4 glucose transporter suppression in adipose tissue and INSULIN RESISTANCE, and to be a central mediator in INSULIN action in general (see ADIPOCYTES). Its effects favour retention of glucose in blood plasma, triggering problems associated with DIABETES MELLITUS.

retinotopy (adj. retinotopic) The manner in which sensory pathways leaving the two retinas are anatomically organized to preserve the spatial relations of the ganglion cells, each lateral geniculate nucleus containing a NEURAL MAP of the retina, although this is not a point-for-point correspondence (the fovea having a much larger representation than the periphery of the retina). Likewise, the PRIMARY VISUAL CORTEX is retinotopically organized (see Fig. 143). See AXONAL GUIDANCE; compare SOMATOTOPIC MAPS.

retrogenes See PSEUDOGENES.

retrograde amnesia Memory loss of events prior to trauma.

retrograde inhibition One form of retrograde signalling, first discovered through the effects of ENDOCANNABINOIDS on GABA-secreting neurons in rat cerebellum and hippocampus. Depolarization-induced suppression of inhibition (DSI) is a form of short-term SYNAPTIC PLASTICITY of GABAergic synaptic transmission occurring in cerebellar Purkinje cells and hippocampal CA1 pyramidal cells. It involves the release of a calcium-dependent RETROGRADE MESSENGER by the somatodendritic compartment of the postsynaptic cell. Both glutamate and endogenous cannabinoids have been proposed as retrograde messengers and such retrograde inhibition can be triggered by activation of either postsynaptic metabotropic or ionotropic glutamate receptors, being restricted to synapses activated with high-frequency bursts. Endocannabinoids therefore allow neurons to inhibit specific synaptic inputs in response to a burst, thereby dynamically fine-tuning the properties of synaptic integration.

retrograde messengers Molecules that travel in the 'wrong' direction across a synapse, from the postsynaptic to the presynaptic neuron. Nitric oxide seems to be one such, diffusing out of the postsynaptic cell after its Ca^{2+}-dependent synthesis by nitric oxide synthase. In the CA3–CA1 synapse of the hippocampus, the NMDA RECEPTOR Ca^{2+} signal causes NO to cross the synaptic cleft and stimulate guanylyl cyclase, enhancing the probability of glutamate release during LONG-TERM POTENTIATION. See RETROGRADE INHIBITION.

retrograde propagation (back-propagation) See LONG-TERM POTENTIATION.

retrograde transport Movement of material within vesicles and along microtubules from the terminal of an axon towards the cell body. Employed, as with ANTEROGRADE TRANSPORT, in tracing neural connections in the brain.

retroposons See RETROTRANSPOSONS.

retrotransposition The process by which cellular reverse transcriptases use RNA (e.g., mRNA) as a template for cDNA production. The cDNA then inserts into a new chromosomal location, many interspersed repeats having arisen from such cDNAs that have integrated in the genome at sites other than their production. The retrotransposition machinery has had a major impact on gene evolution by making copies of exons and moving them about (see EXON SHUFFLING), providing one mechanism for gene duplication. See GENE DUPLICATION, PSEUDOGENES, SPERMATOGENESIS, TRANSPOSABLE ELEMENTS.

retrotransposons (retroposons) Those TRANSPOSABLE ELEMENTS that encode and use reverse transcriptase to transpose via an RNA intermediate. Those having long terminal repeats at their ends are the LTR-retrotransposons; those that do not are non-LTR retrotransposons. They also encode the integrase enzyme (INT), and ribonuclease H (RH) that degrades the DNA-RNA hybrids obtained during transposition. See HERVS, RETROTRANSPOSITION.

retroviruses RNA VIRUSES (see entry), such as HIV, whose single-stranded RNA genomes cannot be used as mRNA but are used as templates for DNA synthesis by REVERSE TRANSCRIPTASE (some endogenous retroviruses encode their own) before this DNA is inserted into the host cell's genome. The virally encoded reverse transcriptase is packaged within the virion, such viruses getting their name from the resulting 'reversing' of the normal flow of genetic information. Specific DNA sequences near the ends of the resulting double-stranded DNA are held together by a virally encoded integrase enzyme that creates activated 3′-OH viral DNA ends that can directly break a target DNA molecule in a manner resembling that of the cut-and-paste DNA-only TRANSPOSONS. A retrovirus may pick up a damaged proto-oncogene from its current location in one host genome and, by transposition, integrate it in the genome of a second host. While infection with such exogenous retroviruses can cause cancers in some animal species (chickens, mice, rats, cats and the woolly monkey), none has been found to transform human cells (see caveats concerning use of retroviruses to reprogramme skin cells in the EMBRYONIC STEM CELLS entry). Nevertheless, human tumours may contain activated oncogenes related to those picked up by retroviruses from their host genomes (e.g., the *ras* oncogene; see RAS PROTEINS). Endogenous retroviruses (ERVs), those that are not transmissible from one individual to another by infection, became integrated as proviruses into the human germline genome long before the emergence of the human species. However, they have become all but extinct in the human and chimpanzee lineages: only HERV-K is still active, and in both lineages. A few other ERV classes persisted in the human genome after the human–chimpanzee split, leaving ~nine human-specific insertions before dying out. The chimpanzee has two active retroviral elements (PtERV1 and PtERV2) that are unlike older elements and must have infected the germ line. Recombinant DNA technology often employs retroviruses as vectors for transgenes, transducing cultured cells *in vitro* and cells of living tissues *in vivo* (see GENE THERAPY).

Rett syndrome A monogenic epigenetic neurodevelopmental disorder involving mutations in the methyl CpG-binding protein 2 (*MeCP2*) gene, whose product encodes a protein binding to methylated DNA sequences. DNA methylation proceeds normally, but epigenetic silencing is faulty because of a failure to recognize this mark. Prenatal and early infant development is normal, erosion of the neurodevelopmental milestones being undetected until later childhood. See AUTISM.

Rev-erb A family of orphan nuclear receptors involved in the control of CIRCADIAN CLOCK circuitry.

reverberatory circuit (oscillatory circuit) A neural circuit in which positive feedback results in re-excitation of the original input so that the circuit continues to discharge for a long time.

reverse genetics Efforts which attempt to determine the function of a gene from a knowledge of its sequence. This often involves deliberately mutating the gene to alter its sequence *in vitro*, after which it is transferred with the appropriate regulatory region into a cell, where it can integrate into the cell's genome so that all the cell's descendants contain the mutant gene. The use of GENE KNOCKOUTS is also a form of reverse genetics, as is GENE SILENCING.

reverse transcriptase, r. transcription See DNA POLYMERASES for RNA-directed DNA polymerases. See also RETROVIRUSES, such as HIV.

reward See REINFORCEMENT AND REWARD.

reward systems See REINFORCEMENT AND REWARD.

RFLPs See RESTRICTION FRAGMENT LENGTH POLYMORPHISMS.

Rhesus blood group system Discovered by LANDSTEINER and Wiener in 1937, the Rhesus antigen is inherited like the AB antigens of the ABO BLOOD GROUP SYSTEM. The system involves two closely linked loci for the allele pairs, *C/c*, *D/d* and *E/e*. One locus determines D and d antigens, the other determining both C/c and E/e by alternative splicing (see EXON SHUFFLING). Rhesus-negative individuals are homozygous for a deletion of the D locus. Analysis of reticulocyte cDNA from Rh-negative individuals homozygous for dCe, dcE and dce enabled identification of the genomic DNA sequences responsible for the different antigenic determinants. The D product is very strongly antigenic and individuals are, for practical purposes, either Rh-positive (possessing the D antigen) or Rh-negative (lacking the D antigen). In fact, there appear to be two types of Rh red cell membrane polypeptide: one corresponding to the D antigen, and the other to the C and E series of antigens. All human populations are polymorphic for the two alleles, and a simple theoretical prediction would be that the polymorphism is unstable and should have disappeared. This is because of the selective disadvantage of the Rh^- allele since Rh-positive children (Rh^+/Rh^-) born to Rh-negative mothers (Rh^-/Rh^-) often suffer a HAEMOLYTIC ANAEMIA (haemolytic disease of the newborn) as newborns. If fatal, this is termed ERYTHROBLASTOSIS FOETALIS. The condition results because their mothers produce antibodies against the foetal blood cells, red cells of foetal origin being capable of entering the maternal circulation either during pregnancy, on the birth of a previous Rh-positive child, or as a result of previous miscarriage. It is routine to screen all Rh-negative women during pregnancy for the development of Rh antibodies. In order to avoid sensitizing a mother, Rh-negative blood must always be used in a blood transfusion to a woman, and sensitization of a Rh-negative mother by her baby's Rh antigens can be prevented by giving her an injection of Rh antibodies (so-called anti-D) to agglutinate any foetal cells in her circulation.

rheumatic fever An acute systemic inflammatory disease, usually occurring after streptococcal infection of the throat and commonly affecting many of the body's connective tissues, including the heart valves. The whole heart may be weakened, but most commonly the bicuspid (mitral) and aortic semilunar valves are damaged to the extent that they fail to open and close properly (stenosis), or may constantly leak.

rheumatoid arthritis (RA) An AUTOIMMUNE DISEASE manifesting itself after puberty, with onset between the years 30 and 60. Characterized by chronic inflammation and destruction of the synovial joints, it is a COMPLEX DISEASE, with susceptibility and severity determined by both genetic and environmental factors. Sibling recurrence risk estimates range from 5 to 10, and an association with alleles of the *HLA-DRB1* (see MAJOR HISTOCOMPATIBILITY COMPLEX) locus has been long established. In common with type 1 diabetes, RA is also associated with the T allele of *PTPN22* (protein tyrosine phosphatase, non-receptor type 22). With *HLA-DRB1*, these associations explain ~50% of the familial aggregation of RA. See ARTHRITIS.

rhinal fissure (rhinal sulcus) The sulcus separating the olfactory cortex from the NEOCORTEX.

Rho See RAS SIGNALLING PATHWAYS.

rhodopsin (visual purple) A conjugated protein densely packed in the membranes of the stacked discs of ROD CELL outer segments, comprising a G PROTEIN RECEPTOR, opsin, and a lipid prosthetic group, retinal, derived from vitamin A by retinol dehydrogenase. Isomerization of retinal from 11-*cis* retinal to *trans* retinal occurs in a few picoseconds on its absorption of light (absorption maximum ~500 nm). This photoexcited rhodopsin (R*) functions as an enzyme and activates many molecules of the G PROTEIN transducin, also in the disc membrane. Transducin then exchanges its bound GDP for GTP, activating phosphodiesterase (PDE), which breaks down the cGMP present in the rod cell's cytoplasm in the dark to 5′-GMP. cGMP normally keeps open the cation channels in the outer segment, and its reduced concentration terminates the dark current of sodium and calcium ions, hyperpolarizing the rod cell. Within a few seconds the bond between retinal and opsin in photoexcited rhodopsin hydrolyses and the all-*trans* retinal diffuses away from the opsin. In high light, most of the rhodopsin is in this dissociated ('bleached') state, the rod being 'saturated'. Within about a second another enzyme, rhodopsin kinase, phosphorylates and inactivates the photoexcited rhodopsin, which then binds arrestin, blocking binding by transducin. The entire cascade then reverses back to the normal state with open cation channels. Rhodopsin is regenerated in the dark, retinal isomerase catalysing conversion of the all-*trans* isomer to the 11-*cis* isomer, which reassociates with the opsin. It is known that the product of the *RPE65* gene, an isomerohydrolase, is critical in this regeneration, mutations in *RPE65* being associated with inherited retinal dystrophies (see GENE THERAPY). Resynthesis of cGMP is also catalysed by guanylate cyclase, which underlies DARK ADAPTATION.

Rhodopsin and the discs in which it occurs are continually degraded and replaced by a pathway of synthesis in the Golgi apparatus of the inner segment of the rod cell. Newly formed discs, formed at a rate of 3–4 per hour, join those at the base of the outer segment, whence they are gradually displaced towards the tip and phagocytosed by cells of the retinal pigmented epithelium. This does not occur in cone cells. See LIGHT ADAPTATION; compare IODOPSINS.

rhombencephalon The primary brain vesicle (see BRAIN, Fig. 28) giving rise to the HINDBRAIN. It comes to consist of the myelencephalon, the most caudal secondary brain vesicle, which gives rise to the medulla oblongata, and the metencephalon, which extends from the pontine flexure to the rhombencephalic isthmus, and gives rise to the CEREBELLUM and PONS. See ALAR PLATES, BASAL PLATES, RHOMBOMERES.

rhombomeres Seven swellings dividing the rhombencephalon rostrocaudally after neural tube closure into smaller, metameric, compartments, demarcated by grooves. The 1st gives rise to the cerebellum, each subsequent pair containing a reiterated set of motor nuclei contributing fibres to one CRANIAL NERVE. See HOX GENES.

ribosomal DNA (rDNA) There are probably 700–800 human rRNA genes, mostly organized in five clusters (i.e., 10 in the diploid nucleus), one located on the short arm of each of the five acrocentric chromosomes at the nucleolar organizer regions. Owing to the selection of restriction enzymes used in the human genome project and the delay in sequencing low complexity DNA, the precise number could not be inferred from the draft sequence. The rRNA genes, encoding 18S, 5.8S and 28S rRNAs, are consecutively arranged on a common 13-kb polycistronic transcription unit which, together with an adjacent 27-kb non-transcribed 'spacer', is tandemly repeated 30–40 times. These are transcribed by RNA polymerase I (see RNA POLYMERASES), the primary transcript, a 45S precursor rRNA, being cleaved by various snoRNAs into the three rRNA classes. Such polycistronic transcription is unusual for nuclear DNA, resembling the mitochondrial situation (see mtDNA). The ~200–300 5S rDNA genes also occur in tandem arrays, the largest being located on chromosome 1q41-42. 5S rDNA is transcribed by RNA polymerase III. See CHROMOSOME, Fig. 52, for rDNA chromosomal locations.

ribosomal RNA (rRNA) The most abundant cellular RNAs (see RNA), comprising ~80% of the RNA in rapidly dividing cells. Encoded by RIBOSOMAL DNA, three of the four eukaryotic rRNA types (18S, 5.8S and 28S) are each present in one copy per ribosome and are formed by modification (notably methylations, and isomerizations of uridines to pseudouridines) and cleavage of a single 13-kb precursor rRNA. Small nuclear RNAs (see SNRNAS) and small nucleolar RNAs (see SNORNAS) are involved in these modifications. The other rRNA type, 5S rRNA, is produced from a separate gene cluster and by a different RNA POLYMERASE and is unmodified. The 16S and 23S rRNAs are components of the small ribosomal subunit of the cytosol while the 28S, 5.8S and 5S rRNAs are components of the large subunit of the cytosol. The 16S subunit is useful in identifying bacteria in the mouth.

ribosomes Non-membranous cytoplasmic organelles composed of two-thirds RNA and one-third protein, forming the sites of the TRANSLATION phase of protein synthesis. Structural revelations in 2000 made clear the importance of the rRNA, rather than protein, in the overall structure and catalytic activities of the organelle. The proteins tend to lie on the surface and to fill in gaps between the crevices of the folded rRNA, although some extend into the ribosome core. Prokaryotic ribosomes (70S) are present in mitochondria, the rest of a cell's ribosomes being of the eukaryotic variety, with a diameter ~23 nm and a sedimentation coefficient of ~80S, and are either free in the cytosol or attached to the ENDOPLASMIC RETICULUM and outer nuclear membrane. The functional ribosome only forms in the presence of mRNA and Mg^{2+}, the 40S subunit comprising 18S rRNA and ~33 proteins, the 60S subunit comprising 5S rRNA, 28S rRNA, 5.8S rRNA and ~49 proteins (see RIBOSOMAL RNA). A typical eukaryotic cell contains millions of ribosomes, and the protein variety in ribosomes is huge. These proteins are synthesized in the cytoplasm and enter the nucleus along with proteins involved in processing rRNA to form a complex, the large ribonucleoprotein particle, from which the large ribosomal subunit (60S subunit) and small subunit (40S subunit) are made. These leave the nucleus and assemble on mRNA, usually near its 5′ end, to form the functional cytoplasmic ribosome. Each ribosome has four RNA-binding sites: one for mRNA and three (A, P and E) for tRNAs. See ELONGATION, PROOF-READING.

riboswitches Sequences of nucleotides within some mRNA molecules which contain structural domains, APTAMERS, which undergo a conformational change on binding their metabolite and alter the mRNA's access to machinery required either for its own transcription from a gene, or for its translation on a ribosome. Riboswitches regulate intracellular levels of metabolites in bacteria; but a riboswitch has been discovered in the mould *Neurospora crassa* that regulates gene expression involved in vitamin B_1 biosynthesis.

ribs (costae) Twelve pairs of curved bones of the thorax, each articulating with its corresponding thoracic vertebra. The 1st to the 7th pairs ('true' ribs) also attach via their costal cartilages directly to the sternum anteriorly, while the remaining five pairs ('false' ribs) either attach to it indirectly (three pairs of vertebrosternal ribs) or not at all (two pairs of 'floating' ribs, terminating in the abdominal muscles). Spaces between ribs are occupied by intercostal muscles, blood vessels and nerves. The middle ribs are those most commonly fractured. The ribs form a 'cage' protecting the heart and lungs. The diaphragm muscle has its origin, in part, on the costal cartilages of the six most inferior ribs. In a typical rib (3rd–9th), the *head* is the projection at its posterior end, bearing superior and inferior facets articulating with facets on the bodies of the thoracic vertebra to form a vertebrocostal joint; the *neck* is a narrow portion between the head and the knob-like *tubercle* bearing a facet that articulates with that of a transverse process of the more inferior vertebra of the two to which the head of the rib is connected. An abrupt change in curvature (the *costal angle*) of the rib then occurs, leading into the *body* (*shaft*) which terminates in a costal cartilage uniting the rib to the sternum. See INTERCOSTALS; SKELETAL SYSTEM, Fig. 156.

ricin A protein toxin from the castor bean (*Ricinus communis*) which can be extracted from castor beans. Its average lethal dose in humans is 0.2 mg. It consists of two distinct protein chains (almost 30 kDa each) linked to each other by a disulfide bond. *Ricin A* is an N-glycoside hydrolase that specifically removes an adenine base from ribosomal RNA, resulting in an inhibition of PROTEIN SYNTHESIS. *Ricin B* is a LECTIN that binds galactosyl residues, assisting ricin A's entry into a cell by binding with a cell-surface component. Many plants, including barley, produce the A chain but not the B chain. Ricin A is of extremely low toxicity in the absence of the B chain.

rickets Disorder in which young growing bones fail to calcify, resulting in 'soft' and easily deformable bones as new BONE formed at the epiphyseal plates fails to ossify. Osteomalacia ('adult rickets') develops when there is inadequate calcium salt deposition in the matrix of bone growing after fracture (remodelled bone). Inadequate levels of circulating CALCIUM and VITAMIN D are often responsible for these disorders. See MALNUTRITION, OSTEOPOROSIS.

rima glottidis The space between the two membranous folds of the GLOTTIS.

RIP (regulated intramembrane proteolysis) In cell signalling, the cleavage by proteases of transmembrane proteins within the plane of the membrane to liberate cytosolic fragments that enter the nucleus and control GENE EXPRESSION. Processes affected are diverse, including cell differentiation, lipid metabolism and the response to unfolded proteins. See ENDOPLASMIC RETICULUM.

RISC (RNA-induced silencing complex) See RNA INTERFERENCE.

risk factors Determinants of disease associated with increased or decreased risk of disease, including inherited and acquired predispositions to disease (see individual diseases). Their effects are usually identified by analysis of case-control or cohort studies (see EPIDEMIOLOGY).

An allele might increase disease risk in carriers (a susceptibility allele), such as the *APOE4* variant in ALZHEIMER DISEASE. Sometimes, as with high blood pressure (see HYPERTENSION) or high blood cholesterol levels, both inherited and acquired factors are involved in compounding a risk factor (see COMPLEX DISEASES). Ageing is the greatest risk factor for neurodegenerative diseases. See http://www.who.int/whr/2002/en/; BREAST CANCER, GENETIC RISK, LIFESTYLE.

risk genotype See GENETIC RISK.

river blindness See ONCHOCERCIASIS.

RNA A polynucleotide in which ribose forms the pentose and uracil replaces the thymine of DNA. All are produced by TRANSCRIPTION by RNA POLYMERASES within the nucleus, but only mRNA encodes polypeptides (see TRANSLATION). There are many non-protein-coding genes in the HUMAN GENOME, including those for rRNAs (see RIBOSOMAL RNA), tRNAs (see TRANSFER RNAS), small nucleolar RNAs (SNORNAS) and MICRORNAS (see entries). Within mitochondria, in addition to the 16S and 23S mitochondrial rRNA molecules, there are three associated with the larger ribosomal subunit (28S, 5.8S and 5S rRNAs), transcribed as a single transcription unit, and one (18S RNA) with the smaller ribosomal subunit.

Significant fractions of the human genome give rise to RNA, much of it unannotated and with reduced protein-coding potential. Recently, the genomic origins of human nuclear and cytosolic polyadenylated RNAs >200 nucleotides (long RNAs, lRNAs), and whole-cell RNAs <200 nucleotides (short RNAs, sRNAs), were investigated for their potential as primary transcript precursors of short RNAs.

7SL RNA is part of the complex by which newly synthesized polypeptides are secreted through the endoplasmic reticulum (see ALU REPEAT FAMILY). See NON-CODING RNAS, RNA EDITING, RNA PROCESSING.

RNA aptamers Small RNA molecules binding specific proteins.

RNA capping Processing of pre-mRNA within the nucleus involves attachment of a *cap* of 7-methylguanosyl triphosphate to their 5′-end. Ribosomes recognize this cap (see INITIATION FACTORS) and commence TRANSLATION. Several viral RNAs do not require a 5′ cap for translation in eukaryotic cells (see VIRUSES). See RNA PROCESSING.

RNA editing A form of post-transcriptional processing which can involve the insertion or deletion of nucleotides, or substitution of single nucleotides, at the RNA level. Its significance remains unclear, and there is no evidence of insertion or deletion RNA editing in mammals. But substitution editing has been detected in a restricted number of genes and usually involves deamination of very particular cytosine or adenine residues, so that cytosine is converted to uracil and adenine to inosine. These changes are catalysed by a family of RNA-dependent deaminases (e.g., adenosine deaminase acting on RNA, ADAR); but transamination can also occur (see WILMS TUMOUR). It can provide an additional way of creating protein diversity (see ISOFORMS). See ADENOSINE DEAMINASE DEFICIENCY.

RNA genes There are probably ~3,000 human RNA genes (>10% of the total gene number). Although most of these encode rRNA (RIBOSOMAL RNA) and tRNA (TRANSFER RNA) and are involved in translation of mRNA, many other RNA gene products are involved in RNA maturation (both cleavage and base-specific modification) of these and other RNAs. There are ~700–800 nuclear rRNA genes, most of them arranged in tandemly repeated gene clusters (see NUCLEOLAR ORGANIZER REGIONS). Genes for tRNA are unusual in having internal promoters, and those for tRNA and rRNA often contain Group I and II INTRONS. See RNA, RNA POLYMERASES, SINES.

RNA-induced silencing complex (RISC) See RNA INTERFERENCE.

RNA interference (RNAi; post-transcriptional gene silencing, PTGS) Discovered by Fire and Mello in 1998, this involves various types of small 'guide' RNA sequences in the regulation of protein

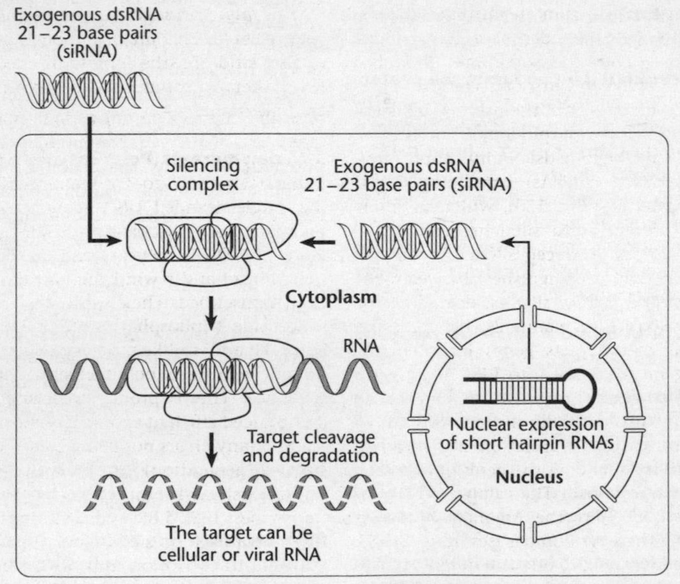

FIG 153. Diagram illustrating how RNA INTERFERENCE is believed to occur in mammalian cells. The original source of the double-stranded RNA (dsRNA) may lie outside the cell, or it may be produced within the cell. In either case, the enzyme Dicer generates short interfering RNAs (siRNAs) which are exported to the cytoplasm and target complementary mRNA molecules.

levels by targeting mRNA for degradation. Initially, it was the process whereby double-stranded RNA molecules (dsRNA) could mediate GENE SILENCING (see also GENE EXPRESSION). Recent work implicates RNAi mechanisms in targeting and maintaining HETEROCHROMATIN. RNA transcripts from one DNA sequence can activate or repress another gene, or interfere with translation of that gene's mRNA transcript into a functional protein. For example, to repress translation or facilitate degradation of mRNAs, microRNAs (see entry) bind to the latter's untranslated regions (see UTRS). An antisense RNA strand can block gene expression, but double-stranded RNA comprising both the sense and antisense strands of a target gene can inhibit that gene's activity even more effectively. RNA viruses replicate through dsRNA intermediates and multi-copy transgenes can produce low levels of dsRNAs, too.

RNAi is triggered when a cell encounters a long dsRNA molecule that might have been produced from an introduced transgene, a viral intruder or a transposable element. They are sometimes produced from longer precursors, typically RNAs that fold back on themselves by base-pairing to form hairpin loops. The enzyme Dicer cuts the long dsRNA into short interfering RNAs (siRNAs) containing 21–23 base pairs (see Fig. 153), which assemble with proteins into silencing complexes. Such an RNA-induced silencing complex (RISC) then distinguishes between the two strands of the siRNAs, degrading the sense strand. In mammals, the antisense strand then targets genes for silencing, being incorporated into RISC to target a complementary mRNA for destruction so that it can repeatedly degrade the mRNA and prevent its translation into protein, effectively silencing the gene from which it was transcribed. The catalytic components of RISC are the Argonaute proteins (Ago). These endonucleases bind siRNA fragments and degrade the mRNA strands bound by complementarity to their siRNA. Links have also been found between RNAi and transcriptional silencing (see TRANSCRIPTION). It is thought that RNAi evolved as an ancient mechanism for silencing viruses and rogue TRANSPOSABLE ELEMENTS that make dsRNA intermediates.

Many companies are seeking to apply RNAi to medical conditions as diverse as hepatitis and cancer using the method to turn off oncogenes or shut down viral replication, and it is now routine for laboratories to test the function of a protein, or generate animal models of diseases, by using RNAi to silence genes. Indeed, at least two PSEUDOGENES have been described that behave in this manner. Apolipoprotein B (ApoB) is involved in the metabolism of cholesterol and its concentrations in human plasma correlate with those of cholesterol. In 2004, synthesized siRNAs targeting apolipoprotein B mRNA and complexed to cholesterol were injected intravenously in mice and were taken up by several tissues. The levels of ApoB were reduced by more than 50% in the liver and by 70% in the jejunum, resulting in lowered blood cholesterol comparable to that in mice in which the ApoB gene had been deleted. Encouraging though results of this kind are, the long-term effects of RNAi therapy based on such techniques are uncertain.

RNA polymerases (Pols) Enzymes synthesizing RNA in the 5′-to-3′ direction, employing single-stranded DNA sequences as a template (see TRANSCRIPTION). As they move along the DNA, they unwind the double helix ahead and rewind the part that has been transcribed. Their substrates are the nucleoside triphosphates ATP, UTP, CTP and GTP and, unlike DNA POLYMERASES, they do not require a polynucleotide primer sequence. This is probably because they do not need efficient exonucleolytic proofreading (any errors not being passed on to the next generation). In eukaryotes, three classes exist: *RNA polymerase 1* (Pol I) which transcribes 18S, 5.8S and 28S ribosomal RNAs organized on a common 13-kb polycistronic transcription unit; *RNA polymerase II* (Pol II) which transcribes both introns and exons of structural (polypeptide-encoding) genes, and certain species of snRNA (see NUCLEUS for exclusion compartment); and *RNA polymerase III* (Pol III) which transcribes the 'housekeeping genes' encoding small RNAs such as tRNAs and 5S rRNAs. Each RNA polymerase must interact with PROMOTERS and TRANSCRIPTION FACTORS before transcription occurs, each Pol having its own set of general transcription factors. The Pol II TRANSCRIPTION machinery is the most complex, with a total of nearly 60 polypeptides of which only a few are required for transcription by the other Pols (see also TRANSCRIPTION FACTORS). RNA viruses have relatively unstable genomes, largely owing to the lack of proof-reading during their replication. See ALU REPEAT SEQUENCES, RNA PROCESSING; compare DNA POLYMERASES.

RNA probes See DNA PROBES.

RNA processing Those processes resulting in modifications to the primary mRNA (pre-mRNA) transcript before it leaves the nucleus, including removal of unwanted internal segments followed by rejoining of the remaining segments (*RNA splicing*). In the case of RNA polymerase II transcripts (see GENE), a specialized cap is added to the 5′ end of the primary transcript (see RNA CAPPING), the transcript is cleaved to form the 3′ end and ~200 adenylate (AMP) residues are added as a tail to the 3′ end (polyadenylation) by poly-A polymerase. Alternative polyadenylation is quite common in human mRNA, the polyadenylated transcripts showing tissue specificity (see ISOFORMS). In many genes (e.g., the GROWTH HORMONE receptor gene), two or more polyadenylation signals are found in the 3′ UTR sequence (see UTRS), but sometimes the signals come into effect after alternative splicing. Special poly-A binding proteins bind to the tail, somehow determining its final length, and guide it through the nuclear pore complexes to the cytoplasm. It has been suggested that transcription and RNA processing occur in the space between chromosome TERRITORIES, in an area termed the *interchromosome domain compartment*.

Both spliceosomal INTRON and exon sequences are transcribed into RNA, the former being removed by RNA splicing. Each splicing event removes one intron RNA sequence and involves two transesterifications in which the two exons bordering the intron are joined together while the intron is excised. Accuracy is believed to be achieved by the *spliceosomes*, large nuclear assemblages of RNA and protein molecules with complexities approaching those of ribosomes. They assemble on pre-mRNA from separate components, their RNA molecules (small nuclear RNAs) being catalytically active. Each of these five snRNA species (U1, U2, U4, U5, U6) is complexed to seven protein subunits to form an snRNP component of the spliceosome core. Introns almost always start with the dinucleotide GT (GU at the RNA level) and end with AG. Alone, these are insufficient to signal the presence of an intron, and sequences immediately adjacent to GT and AG dinucleotides also show considerable sequence conservation, while a third conserved intronic sequence (the *branch site*), at most 40 nucleotides from the terminal AG, and other exonic and intronic sequences, also positively affect splicing. It is thought that a group of special serine-arginine dipeptide-rich spliceosome proteins (the SR RNA-binding proteins) and certain HnRNPs somehow distinguish intron sequences from exons, assemble on exon sequences and mark out their 3′ and 5′ splice sites, starting at the 5′ end of the RNA. The major (GU–AG) spliceosome and the minor (AU–AC) spliceosome differ in their snRNA components, the latter processing the rare AT–AC introns and having U11 and U12 snRNAs instead of U1 and U2.

In *alternative splicing* the primary transcript, or pre-mRNA, is spliced in different ways in the nucleus to produce different mRNAs (only after the 5′ and 3′ end processing and splicing have taken place is the transcript termed mRNA). This occurs when the boundaries between exons and introns (splice sites) are differently selected by the splicing machinery to generate different mRNAs from the same pre-mRNAs. This enables different polypeptides to be encoded by the same gene in different cells, different combinations of coding exons and non-coding exons, and exon length variants, being possible. In almost every case, the regulation of splicing together of exons to form splice junctions is achieved through interplay between the pre-mRNA and protein factors (see NONSENSE-MEDIATED DECAY). At least 50%, possibly far more, of human genes undergo alternative splicing, giving rise to different protein ISOFORMS (see HUMAN GENOME). It may help to explain the small number of genes in the HUMAN GENOME. The HIV genome is also alternatively spliced. Some bacterial mRNA molecules, and those in some eukaryotes (e.g., some fungi), contain RIBOSWITCHES that enable the choice of splice sites to be dictated entirely by a metabolite-induced change in RNA folding brought on by binding of an APTAMER. The possibility that splicing patterns of genes involved in the same biochemical pathway can be altered by feedback from the same aptamer in higher eukaryotes, as in some fungi, is intriguing. See RNA EDITING, SYSTEMIC LUPUS ERYTHEMATOSUS.

RNA splicing See RNA PROCESSING.

RNPs (ribonucleoproteins) Formed in the NUCLEOLUS, these complexes of RIBOSOMAL RNA and proteins (the latter originating in the cytoplasm) include small nuclear RNPs (snRNPs; see entry), such as the U6 snRNP involved in pre-mRNA splicing (see RNA PROCESSING), and small nucleolar RNPs (SNORNPS; see entry).

Robertsonian fusion A chromosomal rearrangement converting two acrocentric chromosomes into one metacentric or submetacentric chromosome, sometimes called centric fusion, although the point of exchange is in the proximal short arm rather than the centromere.

Robertsonian translocations Those TRANSLOCATIONS involving exchanges between the proximal short arms of the acrocentric chromosomes (chromosomes 13, 14, 15, 21 and 22) in which the centromeres of both chromosomes are present but function as one so that the resulting fusion chromosome is stable. Although a small acentric fragment is lost, it has no pathological significance since it will contain only rDNA sequences repeated elsewhere. See DOWN SYNDROME; GENETIC DISEASES AND DISORDERS, Fig. 81.

ROBO proteins Proteins involved in AXONAL GUIDANCE, ROBO1 and ROBO2 acting as receptors for Slit proteins, while ROBO3 blocks their actions.

rod cells Bipolar photoreceptor cells of the retina (120×10^6 per eye; see RETINA, Fig. 150) occurring over most of its surface except the fovea. Responsible for scotopic vision (see DARK ADAPTATION), each has a cylindrical outer segment, which is the expanded tip of a primary CILIUM. As in CONE CELLS, it requires the continuous intraflagellar transport-mediated movement of large amounts of lipids and proteins into the cilium for maintenance of its photoreceptor machinery (see RETINITIS PIGMENTOSA). Rods are 1,000-fold more sensitive to light energy than are cones, and more sensitive to scattered light. Their high signal amplification is capable of detecting a single photon, but their sensitivity is also due in part to their ability to integrate responses to incoming photons over a long period (~100 ms). However, this makes for poor temporal resolution, and they are unable to resolve flickering faster than ~12 Hz. The outer segment contains more discs and more photopigment than that of a cone cell and, unlike cones, rods are achromatic (not colour-sensitive). Also unlike cone cells, rod cell discs are produced at the base, pass up the outer segment, and are removed at its tip (see RHODOPSIN). The rod system is highly convergent in terms of its retinal pathways (see RETINA), making for poor spatial resolution (see VISUAL ACUITY), but making the peripheral retina very sensitive to motion ('seeing out of the corner of one's eye'). Signalling by rod cells depends upon light intensity. During high illumination, rod cells are saturated, and cones alone operate (photopic vision). In the partially dark-adapted eye (see DARK ADAPTATION), rod cells are functional, signalling through gap junctions to adjacent cones and maintaining visual acuity.

The process of phototransduction by

photoreceptors (rod cells and cone cells) shares several features with some forms of synaptic transmission. In the dark, the membrane potential of a rod cell's outer segment is ~–40 mV as a result of the steady influx of sodium (Na^+) and calcium (Ca^{2+}) through cGMP-gated cation channels in its membrane (the 'dark current'). Na^+ is actively extruded by the Na^+–K^+ ATPase in the inner segment and Ca^{2+} leaves the photoreceptor via a Na^+–K^+–Ca^{2+} transporter. The voltage-gated calcium channels in rod cell synaptic terminals are therefore open and calcium enters the terminals triggering the release of the neurotransmitter GLUTAMATE. The cGMP which keeps the cation channels open in the dark is produced from GTP by the enzyme guanylate cyclase. But in light of ~500 nm, bleaching of the rod's photopigment rhodopsin leads to a reduction in cytoplasmic cGMP levels (see RHODOPSIN). The cation channels in the outer segment close, causing the membrane potential to hyperpolarize to ~–80 mV (the receptor potential), reaching a peak in ~3 seconds and lasting more than 1 second, far more slowly than in cone cells. In both rods and cones, conduction of this receptor potential from the outer segment to the synapse with a bipolar cell is by electrotonic conduction and not by action potentials. This enables a graded conduction of signal strength, the strength of the hyperpolarizing output signal being directly related to the (logarithm of the) light intensity. The eye can therefore discriminate light intensities through a range many thousand times as great as would be possible if the response to light signals were to be carried by action potentials. Under optimal conditions, a single photon of light can cause a measurable receptor potential of ~1 mV, while 30 photons cause half-saturation of the rod. See the RETINA entry and its references for further information on visual pathways. See AMACRINE CELLS, HORIZONTAL CELLS.

roof plates The dorsal portion of the developing NEURAL TUBE, lacking neuroblasts but serving primarily as a pathway for nerve fibres crossing from one side to the other, as do FLOOR PLATES.

ROR A family of orphan nuclear receptors oscillating in a circadian manner and involved in regulating the machinery of the CIRCADIAN CLOCK. See NUCLEAR RECEPTOR SUPERFAMILY.

ROS Reactive oxygen species. See MITOCHONDRION.

rostral Anatomically ANTERIOR. Contrast CAUDAL.

RPA (replication protein A) A single-stranded DNA- (ssDNA-)binding protein involved in DNA REPLICATION.

RSKs (ribosomal S6 kinases) MAPK-activated kinases that enter the nucleus and phosphorylate CREB at serine 133, enhancing recruitment of CRE and leading to CREB-related gene expression.

R-spondins A family of orphan ligands that activate Wnt and β-catenin signalling pathways. Their expression overlaps with Wnt expression in many tissues. See WNT SIGNALLING PATHWAYS.

rubella A systemic viral infection with many symptoms, including a rash. Infection commonly produces subclinical or trivial disease; but even this subclinical viraemia can infect the developing foetus, resulting in severe tissue damage and developmental defects. In the absence of mass vaccination (e.g., MMR vaccine; see VACCINES), rubella epidemics occur roughly every 6 years, children being those predominantly affected (average age ~8 years), and up to 5% of susceptible women may be infected, which results in subsequent epidemics of congenital rubella syndrome and rubella-associated terminations of pregnancy (nearly doubling the risk of foetal death). The virus, rubivirus, is an alphavirus belonging to the Togaviridae (see VIRUSES) and is spread by droplet infection and direct physical contact. Its pathological course is similar to that of MEASLES, with primary and secondary viraemia. After incubation (17–18 days), a mild sore throat and conjunctivitis follow, and a fine macular rash appears on the 2nd or 3rd day, coalescing into a generalized 'blush'; but fever is rarely high. Arthralgia in the small joints is common in young adults. Infected persons should avoid proximity to pregnant women, and pregnant women with suspected rubella infection must be isolated from other antenatal patients. The risk of congenital rubella varies with the stage of pregnancy at which infection occurs. Early infection (e.g., in the 1st month of pregnancy) gives rise to severe, multisystem damage (e.g., patent ductus arteriosus with or without pulmonary stenosis; mild to profound deafness, and/or large cataracts). Most such infants have a severe brain syndrome which inhibits motor, sensory and intellectual development. Later infection causes isolated defects, particularly hearing defects, low birth weight, or a seemingly normal infected neonate. Rubella vaccine is live and attenuated, being produced from the RA 27/3 strain of the virus. A single dose results in protective antibodies in 95% of recipients, and the duration of protection afforded by the vaccine is not known since it was only produced in the 1960s.

rubrospinal tracts Lateral DESCENDING MOTOR TRACTS of the SPINAL CORD, originating in the magnocellular part of the RED NUCLEUS of the midbrain tegmentum and exerting excitatory effects on limb flexor muscle tone. It provides an indirect, non-pyramidal route by which the motor cortex and cerebellum can influence spinal motor activity. In non-human primates, rubrospinal axons have the same distribution to proximal and distal limb motor neurons as the CORTICOSPINAL TRACT, their activities moving individual digits. The distinction between these two lateral pathways is less clear in humans, and its functions seem to have been subsumed by the corticospinal tract where it is involved in coarse rather than fine movements. See DESCENDING MOTOR TRACTS.

Ruffini endings (Ruffini's end organs) Located in the DERMIS of the skin, mainly in the fingers, these MECHANORECEPTORS (see Fig. 120) respond to continuous touch or pressure directly superficial to the receptor, and to stretching of adjacent skin. Compare MERKEL CELLS, PACINIAN CORPUSCLES.

running See GAIT.

RXR A partner protein which binds the THYROID HORMONE, T_3, in association with its nuclear receptor (T_3R), the complex then binding promoters of T_3 target genes. It forms heterodimers with LXRS.

S

saccades Very fast conjugate eye movements, horizontal or vertical, that move the fovea to target a different point in visual space (see PROSACCADE). Visual, auditory and somatosensory inputs are involved in determining the eye rotation required to adjust the gaze. Head rotation detected by the semicircular canals of the VESTIBULAR APPARATUS trigger equal and opposite rotation of both eyes. These reflexes are coordinated by the SUPERIOR COLLICULUS. See MICROSACCADES.

saccule One of two sacs of the vestibule lying between, and connected to, the utricle and cochlea and forming part of the membranous labyrinth of the inner EAR. Like the utricle, it contains a MACULA in its wall and contains receptors for static and dynamic equilibrium. See VESTIBULAR APPARATUS.

sacral plexus Nerve plexus formed from the anterior (ventral) rami of spinal nerves L4–L5 and S1–S4, supplying the buttocks, perineum and lower limbs. The sciatic nerve (the combined tibial and common peroneal nerves) is the largest nerve in the body and arises between L4 and S3. Injury to the sciatic nerve and its branches results in pain (sciatica) down the posterior and lateral aspect of the leg and lateral aspect of the foot. Damage to the peroneal nerve causes plantar flexion of the foot, i.e., movement of the foot towards the coronal plane of the body, described as 'footdrop'; and inversion, or supination (a condition termed equinovarus), in which the ankle turns so that the plantar surface of the foot faces medially.

sacrum (adj. sacral) A triangular bone formed by the fusion of five sacral vertebrae, fusion usually commencing by 18 years and reaching completion by 30 years of age. A female's sacrum is shorter, wider and more curved than is the male's. It articulates with the ilium of each hip bone at the sacroiliac joints and with the fifth lumbar vertebra at the lumbosacral joints. It serves for attachment of ligaments.

SAGE (serial analysis of gene expression) This was the first high-throughput sequence sampling technique to be developed. Poly(A)+ RNA is extracted from the source and used to generate cDNA. A short sequence tag of just 9 base pairs from this cDNA can in principle distinguish 4^9 (>262,000) transcripts. Sequence tags 12–15 bp long can be even more discriminatory. Tags from different transcripts are covalently linked together (concatemerized) in a serial manner within a single clone and sequenced to identify the tags in the clone, so that up to 100 different transcripts can be assayed in one sequencing reaction. Relative transcript levels are deduced from the frequency with which the tag occurs. See DNA MICROARRAYS, GENE EXPRESSION.

sagittal Of a plane parallel to the midsagittal plane. See ANATOMICAL PLANES OF SECTION, Fig. 9; MIDLINE.

sagittal crest A raised bony crest running from front to back along the top of the cranium. Found in 'robust' australopithecines (see PARANTHROPUS entries) and in gorillas, supporting the large temporalis muscles.

sagittal fissure Deep FISSURE separating the two cerebral hemispheres of the brain.

Sahelanthropus tchadensis Currently, the earliest recorded hominin (taxa phylogenetically closer to humans than to chimpanzees; see HOMININI). Discovered in 2002 from Toumaï, Chad, Central Africa, it dates from the late Upper Miocene, ~6.5–7.4 Myr BP, its faunal associations suggest that it lived not far from a lake and quite close to sandy desert – perhaps similar conditions to those of the Okavango basin of Botswana today. It is distinguished from living forms as follows: from *Pongo* by its non-concave lateral facial profile, a wider interorbital pillar, an anteroposteriorly short face, robust supraorbital morphology and numerous dental characters; from *Gorilla*, it is distinguished by smaller size, a narrow and less prognathic lower face, no supraorbital sulcus, smaller canines and lower-cusped cheek teeth; from *Pan*, by its anteroposteriorly shorter face, thicker and more continuous supraorbital torus with no supratoral sulcus, relatively longer braincase and narrower basicranium with a flat nuchal plane and large external occipital crest and cheek teeth with thicker enamel. Its estimated endocranial volume, at 360–370 mL, is the smallest documented for an adult hominin, but is within the range of chimpanzees. Several lines of evidence suggest that *S. tchadensis* may have been bipedal. Humans and non-human primates tend to move with their orbital planes (the line joining the superior and inferior margins of the orbits; see Fig. 154) approximately perpendicular to the ground. In addition, they orient the upper cervical ver-

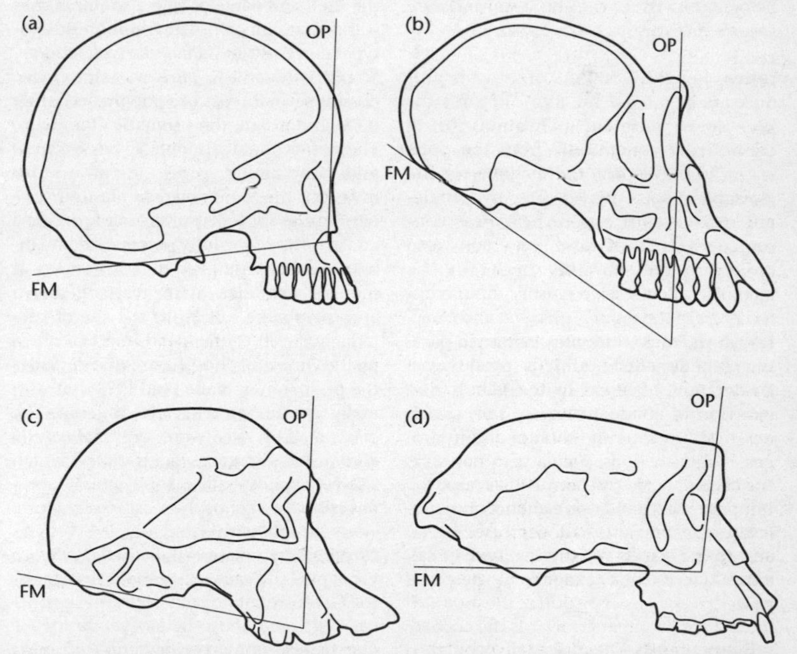

FIG 154. Diagrams showing the relationship between the foramen magnum (FM) and orbital planes (OP) of (a) *Homo sapiens*, (b) *Pan troglodytes*, (c) *Australopithecus africanus*, and (d) specimen TM266 of *Sahelanthropus tchadensis* (see entry). From C.P.E. Zollikofer, M.S. Ponce de León, D.E. Lieberman, F. Guy, D. Pilbeam, A. Likius, H.T. Mackaye, P. Vignaud and M. Brunet. © *Nature* 2005, with permission from Macmillan Publishing Ltd.

tebrae roughly perpendicular to the plane of the foramen magnum, with a limited range (~10°) of flexion and extension possible at the cranio-cervical joint. As a result, the angle between the foramen magnum and the orbital plane in *Homo sapiens* is nearly perpendicular, but more acutely angled in *Pan* and other species with more pronograde postures. The foramen magnum angle relative to the orbital plane in specimen TM 266 of *S. tchadensis* is very similar to that of later bipedal hominins (e.g., *Australopithecus afarensis*, *A. africanus* and humans), and favours bipedalism rather than quadrupedalism. The flat nuchal plane orientation, within the range of *Australopithecus* and *Homo*, but not of *Pan*, and the downward-lipping nuchal crest, also favour bipedalism. But much more post-cranial material is required before this issue can be resolved. See ORRORIN TUGENENSIS, X CHROMOSOME.

saliva Roughly 1500 mL of saliva is produced daily (~0.5–7 mL min^{-1}) from SALIVARY GLANDS, secretion (salivation) being controlled autonomically from the pons via facial nerves, and the medulla via the glossopharyngeal nerves. The hypothalamic FEEDING CENTRE and cortical areas concerned with TASTE and OLFACTION also project to the salivatory nuclei in the medulla which receive sensory input from receptors in the mouth, pharynx and olfactory areas, most responses being via parasympathetic efferents. This produces a greater flow of blood to the glands and secretion of abundant watery saliva, rich in AMYLASE, mucins and sodium and hydrogen carbonate ions. Saliva also contains antibacterial LYSOZYME and PHOSPHOLIPASE A. Sympathetic stimulation enhances the production of amylase, but decreases blood flow to the glands and total volume of saliva produced (the dry mouth in stress and fear).

salivary glands The major salivary glands are compound alveolar glands, a considerable number of them are scattered throughout the oral cavity. Three pairs of large glands exist: parotids, serous glands located just anterior to the ears and producing watery saliva released on to the surface of the 2nd upper premolar (the viral infection mumps is the most common type of parotiditis); submandibulars, mixed glands with more serous than mucous alveoli lying just under the mandible; and sublinguals, mixed glands with more mucous than serous alveoli and lying just beneath the mucous membrane in the floor of the mouth. See SALIVA, SWALLOWING.

Salmonella A genus of facultatively aerobic Gram-negative rod-shaped enteric bacteria, closely related to *Escherichia*. FOOD-POISONING serotypes of *Salmonella* infect both humans and domestic animals, most being capable of causing salmonellosis and most human infections are food-borne. Poultry meat is the commonest source, the shell and white of eggs also sometimes being contaminated. The commonest serotypes currently are *Salmonella typhimurium*, *S. enteritidis* and *S. virchow*. Salmonellae possess several types of lipopolysaccharide (LPS) endotoxin, their somatic O antigen. The non-typhoid salmonellae possess plasmids containing genes important for invasion, the *S. typhimurium* plasmid containing the *rck* gene preventing formation and activity of the fully polymerized membrane attack complex of COMPLEMENT. It induces extensive actin rearrangements and membrane ruffling at the site of bacteria–host cell contact, and injects a phosphatidylinositol phosphatase that degrades the plasma membrane lipid PIP_2 that normally recruits proteins that organize the cortical actin (see INFECTION). *Salmonella* also arrests endosome maturation through a secreted phosphatase, SopB, which modifies endosomal phosphoinositides and prevents fusion with Golgi-derived vesicles carrying hydrolases. *Salmonella enterica* var typhi, the cause of typhoid fever, uses the CTFR protein to enter intestinal epithelial cells, and those heterozygous for CF may be resistant to typhoid and to cholera toxin (see CYSTIC FIBROSIS). *S. enterica* var paratyphi A and B causes paratyphoid. The organism replicates in the reticuloendothelial cells, including within PEYER'S PATCHES (which often ulcerate). Typhoid is the typical enteric fever, symptoms being insidious, including fever, abdominal discomfort and coughing. Early typhoid is often mistaken for influenza. The incubation period averages 2 weeks, the fever reaching its peak after 7–10 days. Relapse is common. Almost all virulent *S. typhi* possess a Vi antigen inhibiting phagocytosis and masking the O antigen. Antibodies to Vi develop during recovery from the disease. Widespread use of antibiotics has led to resistance, when the treatment of choice is ceftriaoxone injected intravenously. Avoidance of high-risk foods and untreated drinking water can greatly reduce chance exposure, and vaccination with Vi-polysaccharide vaccine (often combined with hepatitis A vaccine) is 70–75% effective, a booster being recommended after 3 years. See SHIGELLA.

SA node (sinoatrial node, SAN, pacemaker) Originating in caudal part of the left cardiac tube (see HEART), with the tube's function being later performed by the sinus venosus, this 3-mm wide and 1- to 2-cm-long mass of specialized cardiac muscle tissue eventually resides within the posterior wall of the right atrium, near the opening of the right atrium. Its spontaneous rhythm of depolarization and repolarization sets the rhythm of the CARDIAC CYCLE, and hence HEART RATE, because its cells have the greatest ability to initiate spontaneous action potentials (see PACEMAKER). Special non-gated Na^+ channels allow Na^+ entry and cause depolarization, favoured by continued closure of voltage-gated K^+ channels from the previous action potential. This depolarization causes voltage-gated Ca^{2+} channels to open, entry of Ca^2+ into the pacemaker cells causing further depolarization. When the resulting prepotential voltage reaches threshold, many more voltage-gated Ca^{2+} channels open, causing the 'pacemaker current' which is largely responsible for the depolarization phase of the action potential – a situation in which pacemaker cells differ from normal cardiac muscle fibres. Repolarization follows as the voltage-gated Ca^{2+} channels close and voltage-gated K^+ channels open, as in normal cardiac muscle fibres. After the resting membrane potential is re-established, production of the prepotential begins again owing to the inherent tendency of its cells to depolarize. The cell with the fastest rate of depolarization leads the rest to fire with it simultaneously.

On their own, fibres of the SAN initiate action potentials ~90–100 times per minute, but hormones, neurotransmitters and ionic inbalances can alter the rate. In a person at rest, acetylcholine from the vagus slows SAN pacing to ~75 action potentials per minute by hyperpolarization of the SAN. Stretching of the SAN fibres can directly increase heart rate by 10–15%.

Artificial pacemakers are either temporary (external) or permanent (implantable) devices, artificial pacing being used in treatment of ARRHYTHMIAS. Battery-powered, their microprocessors receive feedback about such input parameters as carbon dioxide and oxygen concentrations in the arteries and veins, body temperature, adrenaline and ATP levels. The pulse generator then dynamically regulates either the ventricles alone (single-chamber pacemakers) or both atria and ventricles (dual-chamber pacemakers). The latter enable better timing of atrial contraction prior to that of the ventricles, normalizing the efficiency of the cardiac cycle; but debate surrounds their advantage over single-chamber pacemakers, especially in the elderly. See EXERCISE, STROKE VOLUME.

sarcomas A general term for CANCERS arising from mesenchymal (i.e., non-epithelial) tissues, e.g., muscle or connective tissue. Cell types include fibroblasts, adipocytes, osteoblasts and myocytes. Contrast CARCINOMAS.

SARS (sudden acute respiratory syndrome) A pneumonia-like respiratory disease caused by a not-very contagious coronavirus (see VIRUSES), which is primarily a virus of bats, striking <9,000 people in China in 2002 and leaving hundreds dead before being brought under control. Its genome, published in 2003, differs from those of three previously known coronaviruses, but is remarkably invariant. This close brush with a pandemic disaster was a 'wake-up call' for those planning responses to the next INFLUENZA pandemic.

satellite DNA Multiple tandem repeats of short nucleotide sequences, each repeat unit usually fewer than 10 base-pairs in length but the satellite sequence often being hundreds of kilobases in length. Not to be confused with 'satellites' on karyograms (see Appendix III), their existence was first indicated by their separating out apart from the main DNA band in a caesium chloride ultracentrifugation tube on account of its different proportion of G+C base-pairs. *Alpha satellite repeats* have sequences with a high (A+T) ratio. Probes can be prepared for these simple-sequence DNA, which bind complementarily to partially denatured chromosomes. The bulk of satellite DNA is located in the heterochromatic regions around centromeres (see HETEROCHROMATIN). See CHROMOSOME BANDING, MICROSATELLITES, REPETITIVE DNA.

satellites See the legend to Appendix III.

satiety centre See FEEDING CENTRE.

savant syndrome (savantism) A predominantly male condition, apparently associated with left cerebral hemisphere abnormalities, in which extraordinary powers of memory develop – usually, but not always, in a specific domain such as music, art or mathematics, commonly independently of any particular interest in the savant field. It is more commonly linked with AUTISM than is any other single disorder.

scabies An infestation of the skin by the microscopic mite *Sarcoptes scabei*. Infestation is common and worldwide, affecting people indiscriminately and spreading rapidly under crowded conditions where there is frequent skin-to-skin contact between people, such as in hospitals, institutions, child-care facilities, and nursing homes.

scaphoid Bone of the WRIST.

scapula (shoulder blade) A large, flat, triangular bone of the pectoral girdle, situated bilaterally in the superior part of the posterior thorax between the levels of the 2nd and 7th ribs and forming a major component of the shoulder joint. Across its posterior surface runs a sharp ridge, or spine, whose lateral end projects as a process, the acromion, which articulates with the clavicle. Inferior to the acromion is a shallow depression, the glenoid cavity, which receives the head of the humerus. The coracoid process is a projection of the superior and lateral border of the scapula which serves for attachment of muscle tendons. The fossae above and below the spine serve as attachment sites for the tendons of shoulder muscles.

Schaffer collateral See HIPPOCAMPUS, Fig. 91; LONG-TERM POTENTIATION.

schistosomiasis (bilharzia) A parasitic disease caused by several species of trematode flatworm (blood flukes) which inhabit the visceral veins. Peak prevalence of infection is in children aged 5–15 years. Despite its low mortality rate, it can be very debilitating, the chronic illness resulting from infection of the blood in which the larva travels to the liver, where the sexes pair and produce thousands of eggs daily. The spiny eggs work their way into adjacent organs, causing debilitation, liver and intestinal damage. Eventually they are released in faeces and urine, but not before producing an intense inflammatory response, especially if they lodge. The disease is found in tropical countries in Africa, the Caribbean, eastern South America, east Asia and in the Middle East. *Schistosoma mansoni* (locates in veins of the small intestine) is found in parts of South America and the Caribbean, Africa and the Middle East; *S. haematobium* (locates in veins of vesical plexus) in Africa and the Middle East; and *S. japonicum* in the Far East. *S. mekongi* and *S. intercalatum* are found focally in Southeast Asia and central West Africa, respectively. An estimated 207 million people have the disease, 120 million symptomatic. Infection is associated with a powerful T_H2

response, characterized by high IgE levels, circulating in tissue EOSINOPHILS, and a harmful fibrotic response to the eggs, leading to hepatic fibrosis. Freshwater snails are the parasite's species-specific secondary host. Molluscicides are used to kill them, but in some countries urbanization, pollution, and/or consequent destruction of snail habitat has reduced exposure and new infections. The most common way of becoming infected in developing countries is by penetration of the exposed skin by the cercarial larval stage of the worm when in water harbouring the snails (the acute cutaneous response is called 'swimmer's itch'). Praziquantel is the treatment of choice.

schizophrenia A devastating neurodevelopmental disorder, and PSYCHOSIS, characterized by such positive manifestations as auditory hallucinations, delusions, thought disorders and behavioural disturbances. It typically arises in late adolescence or early adulthood, with a worldwide prevalence of ~1%. Negative symptoms include social withdrawal and emotional apathy. Evidence indicates origins in developmental abnormalities involving the medial temporal lobe, temporal and frontal lobe cortices. The selective deactivation of the frontal lobe which occurs may account for the difficulties patients have in organizing their thoughts, in integrating them with EMOTION, and translating them into appropriate actions. The genetic basis for the disorder is still far from understood, the condition having many similarities with BIPOLAR DISORDER, with which there is evidence of an overlap in genetic susceptibility: the gene 'disrupted in schizophrenia 1' (*DISC1*) has been implicated in susceptibility to both. OLIGODENDROCYTE dysfunction has also been implicated in the pathophysiology of both disorders (see HERVS). Genomic microduplications and microdeletions (aka structural variants, or COPY NUMBER variants) have been implicated, as have karyotypic abnormalities that led to the discovery of disease-causing mutations in *DISC1*, *PDE4B* and *NPAS3*. Microarray-based methods now enable much smaller events to be detected. Phenotypic discordance between identical twins is 55% (see TWIN STUDIES). Intrauterine and obstetric events may also be involved in its aetiology, and dysfunctional NMDA RECEPTORS have been implicated. Neuroleptic drugs relieve some of the positive symptoms, such as hallucinations and delusions, and these are all antagonists at dopamine receptors, suggesting that schizophrenia is linked to increased activity of the dopaminergic mesolimbic and/or mesocortical pathway (see BRAIN, Fig. 30). There is evidence linking heavy cannabis use (see CANNABINOIDS) with symptoms of schizophrenia in predisposed individuals, and low levels of circulating HISTAMINE are found in many patients suffering with its symptoms. Structural mutations involving genes in neurodevelopmental pathways, such as the neuregulin and glutamate pathways, include a deletion disrupting *NRXN1* in affected siblings, a *de novo* duplication involving *APBA2*, and deletions containing *CNTNAP2*. Deletions at 22q11.2 are also associated with schizophrenia.

Schwann cells Versatile myelin-producing GLIAL CELLS, originating from the neural crest and myelinating peripheral nerves. 'General purpose' Schwann cells migrate peripherally to wrap themselves around axons 8–10 times and form the NEUROLEMMA sheath (see MYELIN SHEATH), a white covering to peripheral neurons appearing especially from about the fourth month of foetal life, when they begin to function (compare OLIGODENDROCYTES). Except for the innermost turn, most of the Schwann cell cytoplasm gets excluded from this, so most of the layers simply comprise a double layer of membrane. Terminal Schwann cells are specialized glia that ensheath the synaptic junction between motor nerve endings and muscles. Each Schwann cell myelinates ~0.15–1.5 mm of axon, and the thicker the axon the longer the region of it that a single Schwann cell myelinates. Between adjacent ensheathed regions lies a tiny gap (0.5 μm) of naked axon, the *node of Ranvier*, at which the axon membrane is exposed to the extracellular fluid. Unmyelinated axons still have a Schwann cell covering, but the pseudopodium-like mesaxon which initiates the wrapping process does not proceed further. A cut axon can regenerate and find its way back to the abandoned neurotransmitter receptors on the neuromuscular junction by following the Schwann cells that remain; but if this is not possible, adjacent Schwann cells develop branches that extend to the damaged synapse and form a bridge along which the damaged axon can grow a new projection to the receptors. Influx of calcium (Ca^{2+}) by terminal Schwann cells occurs during neurotransmission at synapses.

sciatic nerve See SACRAL PLEXUS.

sciatic notch (ischiadic notch) The greater sciatic notch, through which the sciatic nerve passes, lies on the posterior side of the ilium, just inferior to the inferior posterior iliac spine. The lesser sciatic notch lies just inferior to the ischial spine (see PELVIC GIRDLE, Fig. 132).

SCID Severe combined immunodeficiency disease. About half of those with ADENOSINE DEAMINASE DEFICIENCY develop SCID, but it is also a fatal X-linked disorder, for which trials of a treatment using RNA interference (RNAi) were given the go-ahead in 2004. An approach employed in 2000 involved *ex vivo* infection of bone marrow cells expressing CD34 (a marker for haematopoietic stem cells) with a retroviral preparation encoding the γc chain of the cytokine receptor gene *IL2R*. Cultured cells were then returned to the patients, of whom nine out of eleven were apparently cured and enabled to lead a normal life. Two of these patients subsequently developed a leukaemia, almost certainly as a result of treatment-induced insertional activation of the LMO2 oncogene, and clearly we do not have the technology to ensure non-pathological insertion of transgenes. All trials involving retroviral transduction of large pools of lymphocytes were quickly suspended worldwide. See CYSTIC FIBROSIS, GENE THERAPY.

sclera The firm, opaque, white outer layer of the posterior five-sixths of the EYE (see Fig. 73), continuous with the cornea and comprising dense collagenous connective tissue with elastic fibres. It helps provide the shape of the eye and is an attachment for the extrinsic eye muscles moving it. The 'white 'of the eye is the small portion of the sclera visible from the front. It is white in humans, in contrast to great apes – where it might be an unsettling signal in face-to-face contact.

sclerotome Mesodermal cells which originally form the ventral and medial walls of the SOMITES (see entry), but become polymorphous and form a loosely arranged tissue, the mesenchyme. Some of the cells form tendons, but the remaining cells surround the spinal cord and notochord and form the cartilage and bone of the vertebral column (see VERTEBRAE).

SCNT See SOMATIC CELL NUCLEAR TRANSFER.

scoliosis Lateral curvature of the spine, involving asymmetric fusion of two successive vertebrae, or one-half of a vertebra missing. It may be congenital (see RETINOIC ACID), idiopathic, or a result of some other condition, e.g., cerebral palsy. See MARFAN SYNDROME, NEUROFIBROMATOSIS.

scotoma An area of reduced or lost vision in the VISUAL FIELD.

scotopia See DARK ADAPTATION.

screening *'Targeted'*, or *family*, screening involves screening of individuals and couples known to be at significant or high risk of passing a 'disease gene' to their offspring because they have a positive family history. This includes heterozygote, or carrier, screening as well as presymptomatic testing. Carrier testing for autosomal recessive and X-linked disorders can often involve biochemical or haematological techniques, as is the case for such inborn errors of metabolism as TAY–SACHS DISORDER and such haemoglobinopathies as SICKLE-CELL ANAEMIA. Carriers can be detected here with a high degree of certainty; but with other single-gene (Mendelian) disorders, such as haemophilia, these procedures only confirm carrier status in a proportion of individuals. Mild clinical manifestations of carrier status are common in some X-linked disorders, but require careful clinical examination. The mosaic pattern of retinal pigmentation in female carriers of X-linked ocular ALBINISM is such an instance. Not so female carriers of HAEMOPHILIA, where a tendency to bruise easily overlaps variation in the general population. In other single gene disorders, e.g., Duchenne MUSCULAR DYSTROPHY (DMD), the creatine kinase abnormality used in the diagnosis is not itself the defective gene product but a consequence of a secondary (downstream) process. The biochemical identification of carrier females of X-linked recessive abnormalities is complicated by X CHROMOSOME INACTIVATION, but the advent of recombinant DNA technology has made carrier detection in DMD simpler by demonstrating linkage between a polymorphic GENETIC MARKER and the disease locus. Non-penetrance is a major pitfall in genetic counselling, and counsellors need to know the usual degree of penetrance of each dominant syndrome. There is a full spectrum from fully penetrant dominant Mendelian to multifactorial characters (see COMPLEX DISEASES), with increasing influence of other loci and/or the environment, with no logical break between the extremes. A complicating factor in diagnosis is that many single-gene autosomal dominant disorders either have delayed age of onset or exhibit reduced PENETRANCE OF EXPRESSIVITY. Age-of-onset curves are important tools for the counsellor, enabling estimation of the probability that an at-risk but asymptomatic person will later develop the disorder.

Biochemical and microscopical examination of stained or fluorescent chromosomes has been supplemented by *DNA-based diagnostic testing*, the most recent technique used in testing for genetic disorders, involving examination of the DNA itself. It can be applied in carrier screening, preimplantation genetic diagnosis, prenatal diagnostic testing, newborn screening, presymptomatic testing (e.g., in predicting adult-onset disorders such as HUNTINGTON DISEASE and risk estimation of adult-onset cancers and ALZHEIMER DISEASE), conformational diagnosis of a symptomatic individual, and in forensic/identity testing. The methods commonly involve DNA hybridization (see DNA PROBES), DNA MICROARRAYS, DNA SEQUENCING and DNA PROFILING, in which comparisons can be made between the individual's DNA and normal sequences (e.g., SINGLE NUCLEOTIDE POLYMORPHISMS) and those of relatives.

Neonatal screening programmes for some disorders are widespread. They are techniques aimed at detecting in the newborn Mendelian *inborn errors of metabolism* and other diseases. A few drops of blood are taken from the heel and put onto several diagnostic circles on a piece of filter paper. Since the 1960s, every child born in the USA has been tested this way for PHENYLKETONURIA, usually by paper chromatography or by the bacterial inhibition (Guthrie) assay, in which the level of an amino acid in the blood – in this case phenylalanine – is assayed by growing strains of bacteria which require that amino acid for growth (auxotrophic mutants) and measuring resulting colony sizes. Soon afterwards, testing for GALACTOSAEMIA became routine. Immunoassay measurement of the thyroid hormone T_4 (thyroxine) in neonatal blood screens for CONGENITAL HYPOTHYROIDISM is one of the triumphs of modern medicine, preventing CRETINISM. A similar test for trypsin in dried blood screens for CYSTIC FIBROSIS has been introduced in several countries. Separation of the haemoglobin polypeptides from dried cord blood screens for sickle-cell disease. Tandem mass spectroscopy can now be used to detect abnormal signatures of fragmented biomolecules in these last two, and other, diseases. Screening for other HAEMOGLOBINOPATHIES is routine where a high Afro-Caribbean community exists. More sophisticated DNA-BASED DIAGNOSTIC TESTING is likely to be increasingly demanded as routine in developed countries. The occurrence of false-positive and false-negative results should be kept as low as possible by adopting the highest standards of quality control. See MUSCULAR DYSTROPHIES.

Population screening (community genetics) is the offer of screening on an equitable basis to members of a defined, but general, population, who are at low risk. Its aims are to enhance autonomy through enabling individuals to be better informed about genetic risks and reproductive options and to prevent morbidity and suffering due to genetic disease.

Preimplantation screening is a method used to detect genetic defects in embryos created through IN VITRO FERTILIZATION before their implantation into the uterus. The most common technique involves removing a single blastomere from an 8-cell (3-day) MORULA and analysing it to detect genetic abnormalities. The 7-cell morulae that have no genetic abnormalities are then transferred to the mother's uterus for implantation and pregnancy. An alternative is to analyse the first and second polar bodies, whose genotype can predict that of the oocytes. There is a limited time window of 24–48 hours for blastomere removal, genetic analysis and transfer, there being concerns about possible perturbations in genomic imprinting while *in vitro* if the delay is longer. Preimplantation genetic diagnosis (PGD) does not, apparently, increase the birth defect frequency in babies. See PRENATAL DIAGNOSIS.

scrotum The musculocutaneous sac containing the TESTES. See EXTERNAL GENITALIA.

scurvy A disorder arising from deficiency of dietary VITAMIN C.

seasonal affective disorder (SAD, winter depression) A condition diagnosed by the occurrence in autumn and/or winter of at least two episodes of serious depression, which disappear in the spring and summer, and in which there are no clear-cut, seasonal, psychosocial precipitating factors. Light therapy is the treatment of choice, the PLACEBO component of which is apparently very large. Pharmacological treatment involves drugs selectively inhibiting reuptake of serotonin (see MELATONIN), and dual-action antidepressants acting on both serotoninergic and chatecholaminergic systems. Alterations in the secretion levels of melatonin, or related neurochemicals, may underlie the condition.

sebaceous gland Accessory SKIN structure of the squamous interfollicular epidermis but actually lying mainly in the dermis. Simple or compound alveolar glands, they secrete sebum, an oily white substance rich in lipids, when the cells die (hence holocrine glands). Most are connected by a duct to the upper part of a HAIR FOLLICLE (see Fig. 86) from which the sebum gains access to the skin surface, helping to prevent drying and bacterial colonization. A few are located in the lips, eyelids and EXTERNAL GENITALIA, but are not associated with hair and secrete directly onto the surface.

second messengers Organic molecules and certain metal ions which act as intracellular signal molecules, their production amplifying an extracellular signal received at the cell surface. The cyclic nucleotides CYCLIC AMP and CYCLIC GMP are such regulatory molecules, as are DIACYLGLYCEROL, IP_3 (see PHOSPHATIDYLINOSITOL SIGNALLING SYSTEM) and CALCIUM ions (Ca^{2+}).

secondary active transport See ACTIVE TRANSPORT.

secondary auditory cortex (A2) Immediately inferior to the primary auditory cortex (A1) and less well tonotopically organized than it, a region apparently required for neural coding of word sounds (see LANGUAGE AREAS).

secondary immune response See ADAPTIVE IMMUNITY.

secondary lymphoid organs Lymph nodes, spleen and lymphatic nodules, in which T cells await activation.

secondary receptors Receptors comprising two linked elements, a nerve ending and another cell of non-neuronal origin, in which the sensory transduction process takes place in the non-neuronal cell. Merkel's cell in the dermis and sensory cells in taste buds are examples. Contrast PRIMARY RECEPTORS.

secondary sexual characteristics Those characteristics, other than the external genitals, their ducts and associated glands, which distinguish the two sexes and which develop during PUBERTY. In women, these include development of the stromal tissue and deposition of fat in the breasts and growth of the associated ducts. The softer skin texture of women and greater deposition of subcutaneous fat in the buttocks and thighs are attributable to the effects of OESTROGENS, producing the characteristic female figure. In men, they include narrowing and lengthening of the pelvis compared with women; hypertrophy of the laryngeal mucosa and enlargement of the larynx, deepening of the voice; broadening of the shoulders and chest; thicker bone growth; heavier muscle development (on average 50% more muscle than women) and growth of facial hair. Axillary and pubic hair growth occurs in both sexes and is attributable to the effects of ANDROGENS, notably testosterone. Compare EXTERNAL GENITALIA. See KLINEFELTER SYNDROME, SEX DETERMINATION.

secondary somatosensory cortex (SII) See PRIMARY SOMATOSENSORY CORTEX.

secondary visual cortex Area V2 of the EXTRASTRIATE CORTEX.

second-order neurons Afferent neurons in the dorsal medulla making the first synapse in a sensory pathway with fibres that have passed uninterrupted from a dorsal column of the SPINAL CORD (see Fig. 163). They in turn project to the thalamus, where they synapse with third order neurons in the pathway.

β-secretase An enzyme required for the formation of myelin, but also involved in the formation of β-amyloid in ALZHEIMER DISEASE.

secretin A peptide hormone secreted by the S cells of the duodenal mucosa in response to acid chyme and digestion products of fats. It reaches the stomach via the blood, where it inhibits GASTRIN release and reduces sensitivity of the PARIETAL CELLS to gastrin, decreasing gastric motility. It also stimulates bile secretion, mainly by increasing its water and hydrogen carbonate ion content.

secretor The majority of persons secrete their ABO BLOOD GROUP antigens in their body fluids, including saliva, and are *secretor positive*, a condition which is genetically dominant to the inability to do so (*secretor negative*). Two alleles, on chromosome 19q, determine a person's secretor status, which is sometimes regarded as an 'honorary' blood group.

sectorial Of a tooth having its front surface long and blade-like, for shearing. Its single cusp (protoconid) creates a more acute axis of the tooth in relation to the tooth row than does the bicuspid first premolar of later hominins, which has a protoconid and a metaconid. See DENTITION.

segmental duplications (region-specific low-copy repeats) Mutations in which chromosome regions, often 1–200 kb segments, are duplicated – either tandemly (when the duplicated segments are located adjacently) or insertionally (producing non-adjacent duplications). The mechanism of occurrence is only partially understood; but initial analyses of the HUMAN GENOME revealed a large amount of such material, which may have played a significant role in gene and genomic evolution. The human genome is notable for its high proportion of recent segmental duplications, which are of great medical interest because their unusual structure tends to predispose them to deletion or rearrangement, with resulting phenotypic effects (e.g., see WILLAMS–BEUREN SYNDROME, Y CHROMOSOME). Segmental duplications are selectively enriched for genes involved in immune responses (see HIV) and some such regions (hotspots) are 'evolutionary nurseries' in which coding sequences are undergoing strong positive selection (see NATURAL SELECTION). Duplications of sequence contribute more to the genetic differences between humans and chimpanzees (70 megabases of material) than do single base pair substitutions (see DNA MICROARRAYS). Whole-genome shotgun sequence detection (WSSD) and fluorescence *in situ* hybridization techniques (see FISH) have made possible the identification and quantification of duplications. Representing ≥5% of the genome, they are frequently found in pericentromeric and subtelomeric locations, the former seeming readily to accept copies of sequences from other regions and to exchange them in turn, while the latter have tended to foster evolutionary novelty in other organisms by allowing one copy of a gene to adapt to a new function without disrupting the original. GC-rich CAGG repeat elements appear to mediate recent (10–20 Myr) hominoid genomic duplications from diverse chromosomal locations to pericentric locations (see REPETITIVE DNA). In 2005, Eichler and colleagues found that although segmental duplications in the chimpanzee outnumber those in humans, humans have a greater variety of them. Base-for-base, large segmental duplication events have had a greater impact (2.7%) in altering the genomic landscapes of humans and chimpanzees than have single base-pair substitutions (1.2%). Comparison between human and chimpanzee genomes indicates a genomic duplication rate of 4–5 megabases per Myr since divergence – events that have resulted in gene expression differences between the species. There is a particular bias for human-specific duplications on chromosomes 5 and 15, predisposing humans to PRADER–WILLI SYNDROME. See COPY NUMBER, PSEUDOGENES, SUBTELOMERES.

segmentation **1**. For body segmentation, see SOMITES. **2**. The movements of the small intestine controlled by the ENTERIC NERVOUS SYSTEM such that migrating motility complexes strengthen when most nutrients and water have been absorbed (when the intestine walls are less stretched). With more vigorous peristalsis, chyme moves along to the colon at up to ~10 cm sec^{-1}, the first meal remnants reaching the colon in ~4 hours.

seizures (‘**fits**’) The most extreme form of synchronized brain activity, never occurring during normal behaviour and usually accompanied by very large ELECTROENCEPHALOGRAM patterns. A *generalized seizure* involves the entire cerebral cortex, of both hemispheres. They include tonic-clonic seizures (grand mal), involving loss of consciousness and convulsions; and petit mal, in which there is loss of consciousness for a few seconds, accompanied by a 3-Hz EEG originating in the thalamus. A *partial seizure* is initiated in a more restricted cortical area, typically the MOTOR CORTEX or the TEMPORAL LOBE. In simple partial seizures there is no loss of consciousness; in complex partial seizures there is. See EPILEPSY, TAY–SACHS DISORDER, TOXINS.

selectins A family of LECTINS (carbohydrate-binding proteins) located in plasma membranes and mediating calcium-dependent cell–cell recognition and adhesion in a wide range of cellular processes within the blood. L-selectin occurs on leukocytes; P-selectin on platelets and activated endothelial cells during inflammation; and E-selectin, also on endothelial cells. IL-8 induces neutrophils to express selectins, and while this is a normal protective feature enabling these cells to cross the ENDOTHELIUM, in severe inflammation the endothelial cells are attacked by the activated neutrophils (the respiratory burst; see MACROPHAGES). Endothelial cells in lymphoid organs express oligosaccharides which bind selectins on lymphocytes, making them loiter; but during inflammation, endothelia express selectins which recognize oligosaccharides on white blood cells and platelets. See DIAPEDESIS.

selective sweep The ‘hitch-hiking’ by which selectively neutral derived variation which happens to be genetically linked to a positively selected allele (see LINKAGE DISEQUILIBRIUM, NATURAL SELECTION) is driven to fixation along with it – or in adjacent regions after fixation. The physical (i.e. chromosomal) limits of a sweep will depend upon the strength of the linkage between the selected and adjacent sites. Subsequently, there may be a relatively high frequency of rare polymorphisms as variation again builds up relatively slowly (mutations being rare) – and there is a statistic (*D*) that tests for selective neutrality. If population polymorphism frequencies are skewed in favour of rare types, *D* has a negative value, which is thought to reflect relative recency of a selective sweep. This is of particular value as a test because it can be applied to data from a single species, regardless of whether the sequences are coding or not. The comparatively large negative *D* value of the human *FOXP2* locus, for example, is indicative of recent positive selection in the *Homo sapiens* lineage. Effects similar to selective sweeps are produced if population BOTTLENECKS occur, followed by expansion, and information from multiple unlinked loci is required to distinguish them as selective sweeps affect variation at single loci while bottlenecks involve numerous loci simultaneously.

selenocysteine Modified form of the AMINO ACID cysteine in which a selenium atom replaces the sulphur atom. It is produced from a serine attached to a special tRNA molecule that base-pairs with the 'stop' codon UGA. The mRNAs for proteins (e.g., deiodinase enzymes) in which selenocysteine occurs carry an additional nucleotide sequence causing this recoding event.

self-awareness The secondary somatosensory cortex, SII (see PRIMARY SOMATOSENSORY CORTEX), seems to play a major role in formation of a unified perception of the whole body. See COGNITION, MIRROR SELF-RECOGNITION.

selfish gene theory The recognition that genes, not individuals, are the units of selection, view prepounded most potently by W.D. Hamilton (see HAMILTON'S RULE). As Richard Dawkins has eloquently reiterated, it is perfectly compatible with selfish gene theory that such genes should, *inter alia*, help to construct altruistic bodies: they can sometimes improve their biological fitness by building social, cooperative and altruistic bearers. See ALTRUISM, SOCIOBIOLOGY.

self-renewal The ability of STEM CELLS to divide asymmetrically so that either one of the two cells produced by the cell cycle is another stem cell (asymmetrical division), or the 'parent' stem cell forms two further parent stem cells (symmetrical division). Thus, a neural stem cell can remain a neural stem cell or can differentiate into neurons. In asymmetrical division, the other daughter cell is often a 'transit-amplifying cell' that undergoes repeated rounds of growth and division, expanding exponentially. Stem and PROGENITOR CELLS are maintained in a state favouring their self-renewal by the Wnt and Hedgehog signalling pathways. In some CANCERS, transformation of a progenitor cell can lead to activation of self-renewal, a property of 'cancer stem cells' (see CANCER CELLS). Continuous reactivation of the self-renewal programme can lead to stem cell exhaustion, resulting in either their depletion or quiescence. See HAEMATOPOIETIC STEM CELLS, WNT SIGNALLING PATHWAYS.

self-tolerance An individual's tolerance of their own potentially antigenic substances, best understood for $CD4^+$ T cells. In *central tolerance*, immature lymphocytes that happen to recognize self antigens in generative lymphoid organs (bone for B cells and the thymus for T cells) die by apoptosis. In *peripheral tolerance*, mature self-reactive lymphocytes encounter self antigens in peripheral tissues and are killed or shut off. See AUTOIMMUNE DISEASE, T CELLS.

semaphorins A large family of secreted and transmembrane signalling proteins regulating AXONAL GUIDANCE in the developing nervous system and regrowth during NEURAL REGENERATION. Semaphorin receptors (see NEUROPILINS) also act in such diverse processes as lymphocyte activation, control of vascular endothelial cell motility, and lung-branching morphogenesis. Thus, the cell-surface semaphorin receptor Neuropilin-1 is also a receptor for VEGF (see GROWTH FACTORS). Plex is another semaphorin receptor. See CEREBRAL CORTEX, EPHRINS.

semen A composite of SPERM cells and secretions from the male reproductive glands. SEMINAL VESICLES produce ~60% of the fluid (containing fructose, citric acid, and other sperm nutrients, androgens, and the SEMENOGELINS and fibrinogen involved in a weak coagulation of semen after EJACULATION), the PROSTATE GLAND produces ~30% (thin, milky secretion with a high pH, containing clotting factors and fibrinolysin), the TESTES ~5% (acidic secretions) and the BULBOURETHRAL GLANDS ~5%. It contains a high concentration of zinc (Zn^{2+}), known to reduce the activity of prostate-specific antigen. Semen is discharged from the epididymis into the prostatic urethra during emission. See INFERTILITY.

semenogelins Gel proteins (SgI and SgII) secreted by the seminal vesicles into SEMEN digested by prostate-specific antigen (PSA), resulting in liquefaction and release of motile spermatozoa. They are rapidly degraded on EJACULATION. They can serve as substrates for transglutaminase and activators of sperm hyaluronidase, suggesting they are involved in degradation of the zona pellucida (see FERTILIZATION). They have also been found in kidney, trachea and the gastro-intestinal tract. Their genes show positive selection in chimpanzees.

semicircular canals See VESTIBULAR APPARATUS.

seminal vesicles Sac-shaped glands, derived from the Wolffian ducts and lying adjacent to the ampullae of the vasa deferentia. Each is ~5 cm in length, encapsulated in fibrous connective tissue and smooth muscle fibres and tapering into a short duct joining the ampulla to form the ejaculatory duct. Their cells bear androgen-binding protein (see SEX HORMONE-BINDING RECEPTOR), concentrating androgens in the seminal fluid. Their secretions are alkaline, contain fibrinogen and SEMENOGELINS, and make up ~60% of the SEMEN volume.

senescence **1**. Cell senescence is a non-growing state of cells during which they exhibit distinctive phenotypes and remain viable for extended periods but are unable to replicate, probably because their chromosomes have reached a critical TELOMERE length. Cells that have been *in vitro* for extended periods become senescent, but fibroblasts deficient in p53 and pRB, or those harbouring viral oncoproteins, continue beyond senescence and hit *crisis*, at which time telomeres are probably so short that chromosomes can no longer be protected from rearrangement. Senescence can be regarded as a stress-induced barrier limiting the proliferative potential of damaged cells. There are abundant senescent cells within tumours (see CANCER CELLS, GENOMIC INSTABILITY), a situation believed to be triggered mainly by oncogenic signals that function in part by de-repressing the *CDKN2a* locus (see ONCOGENES). There is rapid *in vivo* clearance of tumour cells that have undergone P53-triggered senescence. **2**. See AGEING.

sensations The common properties of sensations include modality, intensity, duration and location in space. See SENSORY MODALITIES, SOMATOSENSORY SYSTEM.

sensitization **1**. At the cellular level, the situation in which the same intensity of signal brings about an increased level of response, when repeated, than when applied in a single dose. Thus serotonin (5HT) can set in motion a cascade that sensitizes a sensory axon terminal so that it lets in more calcium ions (Ca^{2+}) per action potential (see MEMORY, SYNAPTIC PLASTICITY). See PAIN. **2**. Behavioural sensitization. A more complex form of non-associative learning than HABITUATION in which, after a harmful stimulus, an individual typically learns to respond more vigorously to a variety of *otherwise neutral* stimuli. We might experience this as 'jumping out of our skin' at a tap on the shoulder after hearing a loud explosion, becoming sensitized to it and startled by an otherwise innocuous stimulus. Defensive reflexes thereby become 'sharpened' for withdrawal and escape. For *some* addictive drugs, sensitization (reverse TOLERANCE) to their rewarding effects is observed. See FEAR AND ANXIETY.

sensory adaptation The decrease in intensity of a sensation with persistence of the stimulus (compare HABITUATION), brought about by a decrease in the rate of firing of a receptor despite being stimulated at the same intensity (see ACTION POTENTIALS). Each action potential allows entry of Ca^{2+} through voltage-gated Ca^{2+} channels, and the intracellular Ca^{2+} concentration may build up to levels that open Ca^{2+}-activated K^+ channels. The resulting increased permeability of the membrane to K^+ makes it more difficult to be depolarized, increasing the delay between one action potential and the next. A continuously activated neuron thereby becomes less responsive to a constant stimulus, so helping us to detect a light touch on the skin while ignoring the constant pressure of our clothing. It appears that in the nerve terminal of the sensory neuron, calcium (Ca^{2+}) channels become progressively and persistently less effective when they are repeatedly opened.

sensory modalities Different qualities (kinds) of SENSATION. Five such modalities have been recognized since ancient times: vision, hearing, touch, taste and smell. Each of these has many submodalities – elemental senses from which more complex sensations are constructed. Thus flavours are the results of combining the four submodalities of taste: sweet, sour, salty and bitter. The cerebral cortex has a map of the body for each submodality of sensation.

sensory receptors Cells responding to a stimulus with a change in their membrane potential, the RECEPTOR POTENTIAL. They may be classified according to stimulus into MECHANORECEPTORS, THERMORECEPTORS, CHEMORECEPTORS and NOCICEPTORS. See RECEPTIVE FIELDS; see RECEPTORS for membrane receptors.

sensory threshold The lowest stimulus intensity a subject can detect, for a given sensory modality. They can be modified by contextual cues.

septic shock Usual term for the consequences of 'blood poisoning' (sepsis). Any bacterial infection so widely disseminated that it passes from the blood to tissue after tissue, causing extensive damage. Common causes include peritonitis from infection or rupture of the gut; generalized body infection, e.g. of streptococcal or staphylococcal origin; generalized gangrenous infection; infection (e.g., by *E. coli*) spreading from kidney or urinary tract to the blood. Involves 'sludging' of blood caused by red cell agglutination in response to degenerating tissues, and micro blood clots (disseminated intravascular coagulation) resulting in haemorrhages, especially in the gut wall. Such massive consumption of clotting proteins results in the person's blood no longer clotting normally. This condition frequently leads to the failure of organs such as kidneys, liver, heart and lungs. The cytokine TNF-α (see TUMOUR NECROSIS FACTORS) plays a major role in these processes, and endothelin-1 (see ENDOTHELINS) is often released.

septum **1**. A wall dividing two cavities (e.g., left and right sides of the HEART; nasal septum). **2**. The septal region, or septal area, of the LIMBIC SYSTEM (see Fig. 114) comprises nuclei lying beneath the rostral part of the corpus callosum, receiving input from the AMYGDALA and projecting to the hypothalamus via the precommissural fornix. Some of its nuclei also receive input from monoaminergic nuclei in the brain stem (see DIFFUSE MODULATORY SYSTEMS) via fibres which project to HABENULAR NUCLEI of the epithalamus, posterior to the thalamus. It also receives input from these nuclei, linking the hippocampus to the hypothalamus. A component of the LIMBIC SYSTEM (see Fig. 114) receiving input from the habenular nuclei and linking the hippocampus to the hypothalamus via the precommisural fornix.

sequence-tagged site mapping (STS mapping) The most powerful method for physically mapping a genome (compare EXPRESSED SEQUENCE TAGS). An STS is a unique identifiable region, 1–20 nucleotides long, deliberately embedded in the genome. Here it acts as a special 'barcode' for subsequent rapid identification by hybridization techniques or amplification by PCR, serving as a physical, as opposed to a genetic, marker. STS mapping can be followed up by restriction mapping to provide more detail. See GENOME MAPPING, HAPLOTYPE.

serial processing The process(es) by which one, or very few, of available objects in the (e.g.) visual field are selected for specialized analysis. See ATTENTION; compare PARALLEL PROCESSING.

serine An AMINO ACID. Its 'unnatural' D-form (D-serine) may be the main NMDA RECEPTOR coagonist. D-serine released from glia and from neurons may play distinct roles in regulation of NMDAR signalling.

serine/threonine kinases PROTEIN KINASES involved in SIGNAL TRANSDUCTION pathways other than those involving mitogens. PKA, PKB, PKC (see PHOSPHOLIPASE C, Fig. 135), PKG, CAM-KINASES and MTOR are all examples. O-linked β-*N*-acetylglucosamine (*O*-GlcNAc) competes directly with phosphate for serine/threonine residues, and glycosylation with *O*-GlcNAc modulates signalling, influencing protein expression, degradation and trafficking. See CELL GROWTH, HUMAN KINOME.

serosa (serous membrane) The lining of any body cavity not opening directly to the exterior, also covering organs lying in that cavity (e.g., portions of the gastrointestinal tract suspended in the abdominopelvic cavity). It consists of areolar connective tissue covered by simple squamous epithelium (here termed *mesothelium*) and is arranged in two layers: the portion attached to the cavity wall is the *parietal layer*, while the portion covering the organs inside the cavity and continuous with the parietal layer is the *visceral layer*. The mesothelium secretes a watery serous fluid lubricating the organs and enabling them to glide over each other and slide against the wall of the cavity. That lining the thoracic cavity and covering the lungs is the pleura; that lining the abdominal cavity and covering its organs is the PERITONEUM.

serotonergic Of neurons and synapses that produce and release SEROTONIN.

serotonin (5-hydroxytryptamine, 5-HT) Synthesized from the AMINO ACID tryptophan, serotonin is an amine NEUROTRANSMITTER found in a series of nuclei, the RAPHE NUCLEI (see entry) scattered along the ventral region of the brain stem (see BRAIN, Fig. 30d; DIFFUSE MODULATORY SYSTEMS), and in enterochromaffin cells of the mucosal epithelium of the small intestine. Animals in which these nuclei are depleted of their serotonin are unable to SLEEP, and numerous studies have found that aggressive and impulsive animals, including humans, have on average lower levels of the serotonin metabolite tryptophan in their cerebrospinal fluid (5-hydroxytryptophan, 5-HTP, is the intermediate product). Serotonin transmission is important in MOOD determination, and its deficiency is associated with DEPRESSION and increased risk of suicide. Serotonin exerts its effects through signal pathways coupling G PROTEINS to PHOSPHOLIPASE C (see Fig. 135), IP_3 (see PHOSPHATIDYLINOSITOL SYSTEM) and CYCLIC AMP. It is critical for LEARNING and MEMORY, especially fear learning by the amygdala and hippocampus (see FEAR AND ANXIETY), since only those synapses that are being marked by serotonin make use of proteins produced by CREB-mediated gene expression involved in synaptic growth and persistent changes in synaptic strength (see SYNAPTIC PLASTICITY).

At least 14 subtypes of serotonin receptor occur in the brain, all except the 5-HT_3 receptor (a ligand-gated ION CHANNEL) being METABOTROPIC (see SYNAPSES). Most 5-HT_1, 5-HT_2 and 5-HT_3 receptors are located postsynaptically, although the 5-HT_{1B} receptor is a presynaptic AUTORECEPTOR (5-HT_{1A} is a postsynaptic autoreceptor). Agents reducing serotonin transmission, notably 5-HT_2 and 5-HT_3 antagonists and 5-HT_{1A} agonists, are potent anxiolytics. The psychostimulant *ecstasy* (3,4-methylene-dioxymethamphetamine) competes with serotonin for the serotonin re-uptake system involving a saturable Na^+/Cl^--dependent transporter in the synapse, by which serotonin's action is terminated, and animal studies suggest that serotonergic neurons may be killed by the drug. This transporter is inhibited by SSRIS, and by tricyclic ANTIDEPRESSANTS. Oxidative deamination of serotonin by MONOAMINE OXIDASES yields its main metabolite, 5-hydroxyindoleacetic acid (5-HIAA), an auxin.

serpentine receptors See SEVEN-MEMBRANE-SPANNING RECEPTORS.

Sertoli cells (sustentacular cells) Cells originating during the 4th month of development from the surface epithelium of the developing TESTES and lying among spermatogonia in the wall of the seminiferous tubules, where they act as nurse cells (see BLOOD–TESTIS BARRIER) during SPERMATOGENESIS. Their induction is by *SRY* and *SOX9* expression in the foetus and their proliferation is under the control of FOLLICLE STIMULATING HORMONE. *SRY* triggers their differentiation from supporting cell precursors, which would otherwise give follicle cells. In addition to producing androgen-binding protein (see SEX HORMONE-BINDING PROTEIN), they produce inhibin-B (see INHIBINS) and MÜLLERIAN INHIBITING SUBSTANCE. See SEX DETERMINATION.

serum The clear fluid remaining after the cellular components of the blood have adhered to one another and contracted as a 'blood clot'. It contains growth factors.

sesamoid bones Small bones embedded in tendons where tension or friction develop; e.g. in thumb and great toe, palms and soles. Individuals differ in their number; but the patellae (kneecaps) are normally present in everyone.

seven-membrane-spanning receptors Serpentine receptors, comprising a protein with seven membrane-spanning α-helices, commonly involved in activation of G PROTEINS (they are therefore termed G-PROTEIN-COUPLED RECEPTORS). They include the photopigments RHODOPSIN and IODOPSINS. Their signalling activity is terminated by ARRESTINS.

sex chromosomes Chromosomes with a strong causal role in SEX DETERMINATION and present in humans as a homologous pair in nuclei of females (the homogametic sex) but occurring as a very unequal pair in nuclei of males (the heterogametic sex). Human sex chromosomes, like those of other animals, evolved perhaps 300 Mya from a pair of autosomes and then diverged in sequence so that today there are relatively short regions at either end of the Y CHROMOSOME that remain identical to corresponding (homologous) regions of the X CHROMOSOME (see entry for further details) and between which recombination occurs during sperm production. Recombination between the chromosomes is believed to have been regionally suppressed, probably by inversions, starting soon after the divergence of the lineages leading to mammals and birds and subsequently in discrete steps.

Conventional theory reasons that when a primitive Y chromosome stops recombining, the effectiveness of natural selection decreases because selection cannot act independently on different Y-linked mutations, slowing accumulation of advantageous mutations but speeding that of mildly deleterious ones. The fitness of a Y-linked allele wanes relative to its X-linked homologue, and being redundant to its X-linked homologue it may ultimately be silenced by persistent accumulation of deleterious mutations. It is usually held that because most of the Y chromosome does not engage in sexual recombination, it has degenerated substantially, both in size and gene content, in comparison with the X chromosome. Humans vary least in their sex chromosomes, the SNP (single nucleotide polymorphism) variation between different X chromosomes being ~4.69 differences per 10 kb and very much lower for the Y chromosome, at 1.51 differences per 10 kb. This is because fewer ancestors have contributed to the sex chromosomes, and because sex chromosomes have patterns of mutation and crossing-over that differ both from each other and from the non-sex chromosomes.

Recent work on the fly *Drosophila* has challenged the view that the Y chromosome is locked into continual decay and indicates that it influences the expression of genes that are more strongly expressed in males and those that are more strongly influenced by environmental stress (e.g., those encoding heat shock proteins) and many associated with sperm development. Gene loci influencing mitochondria are also over-represented in the set of genes influenced by the Y chromosome. Whereas the Y chromosome spends every generation in males, the X chromosome and autosomes alternate between the sexes across generations. A mutation favouring males that is positioned on the X chromosome or autosomes can only increase its frequency when selection in females is concordant, absent, or not very strongly discordant. However, for Y-linked mutations there can be no counter-selection in females so that the Y chromosome presents a favourable platform for mutations improving male gene expression. Males lacking the Y chromosome in *Drosophila afinis* (a species in which the Y chromosome is not required for fertility) sire many fewer offspring when competing with XY males. The Y chromosome in fruit flies and humans may have evolved to modulate gene expression rather than turn certain genes on or off, so that its effects may be continuous rather than discrete, making its effects much more difficult to detect. This might explain why so few Y-linked traits have been discovered in these species. See DOSAGE COMPENSATION, SEX-LINKED INHERITANCE, X-CHROMOSOME INACTIVATION.

sex determination It is thought that of about 50 known 'sex-determining loci' about 20 are X-linked, 2 are Y-linked and the rest are autosomal. In presumptive males, by about the 7th week of development, the DNA-binding testis-determining factor (TDF) is produced as a result of expression of the *SRY* gene on the Y chromosome. This causes the inner parts of the genital ridges to develop as TESTES, although the products of autosomal genes are also involved. Somatic SERTOLI CELLS of the testes then produce MÜLLERIAN-INHIBITING SUBSTANCE, causing apoptosis and atrophy of the Müllerian ducts; while LEYDIG CELLS produce TESTOSTERONE (some of which is converted to the more potent dihydrotestosterone), probably under the influence of HUMAN CHORIONIC GONADOTROPIN (see role of SF1, later). The Wolffian ducts then develop to form the prostate glands, seminal vesicles and vasa deferentia. Neural pathways in the brain involved in the female menstrual cycle are modified while dihydrotestosterone alters the genital tubercle to become a penis as adjacent swellings develop to form the scrotum.

In presumptive females, coordinated action of several genes by about the 12th week of development causes the outer parts of the genital ridges to begin to develop as OVARIES. The 'anti-testis' gene *DAX1* on chromosome X is important here in opposing the effects of *SRY*. Several autosomal genes affect ovary development as the Wolffian tubes (mesonephric ducts) regress in the absence of testosterone. Instead, Müllerian tubes (paramesonephric ducts) develop superiorly into oviducts and fuse inferiorly to form the uterus and vagina, while the external genitals develop as the clitoris and the labia majora and minora. The female state is therefore not, as was once thought, the 'default' embryonic condition, from which maleness is derived by expression of the *SRY* gene and downstream targeting by SRY product. Rather, *positive* expression of genes on the X chromosome and autosomes are required in the regulation of ovarian development. Over-expression of such genes (e.g., *DAX* and *WNT4A*) can feminize XY individuals even if they possess a functional *SRY* gene.

It was originally thought that SRY would be found to bind many downstream targets that affect various aspects of the testis development programme. But subsequent research indicated that the closely related autosomal gene, *SOX9*, a transcription regulator, is up-regulated immediately downstream of *SRY* and can also activate the testis pathway. Its product, SOX9, is known to bind the promoter region of the gene for MÜLLERIAN INHIBITING HORMONE (MIH). SRY and SOX9 appear to act in conjunction in testis differentiation, one or both of them inducing the testis to secrete FGF9, a chemotactic factor causing tubules from the mesonephric duct to penetrate the genital ridge and form primitive vasa efferentia. *SRY*, either directly or indirectly through *SOX9*, up-regulates production of steroidogenesis factor 1 (SF1), which stimulates the differentiation of Sertoli and Leydig cells. With *SOX9*, SF1 elevates the concentration of MIH, leading to regression of the Müllerian (paramesonephric) ducts, while SF1 also up-regulates genes for testosterone synthesis. Testosterone receptor complexes mediate virilization of the mesonephric ducts to form the vas deferens, seminal vesicles, efferent ductules and epididymis. Dihydrotestosterone receptor complexes modulate the differentiation of the male EXTERNAL GENITALIA.

An antagonistic relationship seems to exist between the FGF and WNT SIGNALLING PATHWAYS, translated into two opposite outcomes: activation or repression of *SOX9*. *WNT4* is the ovary-determining gene, up-regulating expression of *DAX1* (a nuclear hormone receptor gene) that inhibits *SOX9* in precursors of supporting cells. Interestingly, a single case of *WNT4* duplication in humans did lead to XY SEX REVERSAL to female. This suggests a new model of sex determination in which the fate of somatic cells in the gonad hinges on the predominance of SOX9 versus Wnt downstream signals. In humans and other mammals, SRY normally acts as the testis determinant through promotion of *SOX9* expression; but this outcome can also be favoured by loss of the R-SPONDIN gene *RSPO1*, whose product is a ligand activating Wnt and β-catenin signalling pathways. Down-regulation of Wnt and β-catenin pathways by loss of *RSPO1* activity would favour predominance of SOX9. *RSPO1* is required to block the testis pathway in humans and is up-regulated in XX gonadal cells during the critical bipotential gonad stage of sex determination, and one theory is that it is the critical gene that cooperates with *Wnt4* to repress the male pathway. However, it has not yet been shown that β-catenin is activated during XX gonad development or that R-spondin 1 regulates this pathway in the gonad. Oestrogens are also involved in female sexual differentiation, and under their influence the Müllerian (paramesonephric) ducts are triggered to form the uterine tubes (oviducts), uterus, cervix and upper vagina. They also act on the EXTERNAL GENITALIA at the indifferent stage to form the labia majora, labia minora, clitoris and lower vagina.

There is much evidence that XX cells can switch to male cell fate under certain conditions in adult life, and the critical balance between Fgf and Wnt/β-catenin signalling pathways helps explain the underlying property of 'bipotentiality' in cells of the gonad in molecular terms.

Hermaphrodites arise when neither testes nor ovaries develop, when both occur, or when gonadal sex does not accord with chromosomal sex. *True hermaphrodites* (usual chromosome complement 46,XX) are extremely rare and roughly two-thirds are raised as males. In many individuals, the paternally derived X chromosome carries Y-specific DNA sequences as a result of crossing-over between X and Y chromosomes during meiosis I of spermatogenesis. Both testicular and ovarian tissues occur, either on different sides of the body or both together in the gonad (termed an ovotestis). They may produce eggs, but rarely sperm, and are usually sterile. Most have a uterus and some menstruate. A few individuals are found to be CHIMAERAS, with both 46, XX and 46,XY cell lines. In PSEUDOHERMAPHRODITISM, individuals have either testes or ovaries, but the external genitals are either ambiguous or of the opposite from what would be expected from the gonadal sex. One such case occurs in TESTICULAR FEMINIZATION, an X-linked recessive trait occurring in males who are chromosomally male, i.e. 46,XY. In female pseudohermaphroditism, the karyotype is female and the external genitalia are either virilized and resemble a normal male's, or are ambiguous. CONGENITAL ADRENAL HYPERPLASIA is the most important cause and is a common cause of incorrect sex assignment in the newborn.

sex differences It is important to avoid sex-sterotypical remarks in what is an objective biological field of enquiry. Sex differences other than those involved in SEX DETERMINATION and SECONDARY SEXUAL CHARACTERISTICS tend to be differences of degree rather than of kind, some originating during ADOLESCENCE and PUBERTY. Despite evidence of a 'huge number of differences' between men's and women's brains, it is hard to know which are functionally relevant, and researchers are cautious about their conclusions. Androgens (e.g., TESTOSTERONE) produced by the testes in foetal and neonatal life act on the brain to produce sex differences in neural structure and function. In the cerebral cortex of males, androgen receptor (AR) binding by testosterone is greater in the right frontal lobe and left temporal lobe – an asymmetry absent from females. There is some evidence, hotly debated, suggesting that the male brain is more strongly lateralized than is the female brain. Brain-imaging studies are generating data on sex differences, revealed as different brain activity, and in different parts of the brain, for many tests. Females exhibit more widespread activation of the amygdala (see FEAR AND ANXIETY), corresponding with increased heart rate and sweating, than do males in response to pictures of people with fearful expressions, and normal females show more reaction to stress (anticipating a pain on the wrist) in the subgenual prefrontal cortex, a region linked to anxiety and depression. Men appear to be more prone to 'externalizing' unhappiness through physical behaviour that includes drinking alcohol, drug abuse and violence, whereas women are more likely to 'internalize', leading to depression and such disorders as anorexia. Among those with schizophrenia, brain scans show that men tend to have greater deficits than do women in attention, language, visual-spatial perception and other areas (e.g., olfaction and motor skills) ruled by the cerebral cortex – areas that are sexually dimorphic in normal subjects.

Although males and females do not differ in general INTELLIGENCE, performance in specific cognitive tasks does reveal sex differences, for which experiments in non-human subjects suggest a biological foundation. There is a substantial degree of overlap, but considered at the population level – not, it should be emphasized, at the level of the individual – there is compelling evidence for SEXUAL DIMORPHISM in the brain, cognition and behaviour. Brain imaging suggests that LANGUAGE AREAS are less lateralized in women than in men. Males tend to be favoured in intelligence tests involving mental rotation and spatial navigation (including map reading) and tend to play with mechanical toys as children. Males also tend to be favoured as adults in engineering and physics problems. Females tend to be favoured in tests of emotion recognition, social sensitivity and verbal fluency, tend to start speaking earlier than boys, and are more likely to play with dolls as children. When one-day-old human babies are presented with either a live face or a mechanical mobile, girls allocate more time looking at the face, whereas boys prefer the mechanical object. Young male vervet monkeys prefer to play with trucks, whereas young female vervets prefer dolls.

Because another person's emotional states depend upon subtle differences in their current and past emotional experience, they are all but impossible to parametize formally (see AUTISM). Empathizing can be thought of as the capacity to predict and to respond to the behaviour of agents (usually people) by inferring their mental states and by responding to these with an appropriate emotion. Systemizing, on the other hand, can be regarded as the capacity to predict and to respond to the behaviour of non-agentive, deterministic, systems by analysing input-operation-output relations and inferring the rules governing such systems. According to the empathizing-systemizing (E-S) theory of psychological sex differences, innate biological differences lead to stronger systemizing in males and stronger empathizing in females.

The caveat made above, that psychometric data should be considered at the population level and not at the individual level, applies equally to brain morphometric data, in which there is considerable variance. That said, the cerebrum as a whole is ~9% larger in men than in women, due mainly to a larger volume of white matter, three-dimensional studies indicating that the ratio of corpus callosum to total cerebral volume is smaller in men, in line with studies suggesting that increased BRAIN SIZE is predictive of decreased interhemispheric connectivity.

Larger muscle fibres and larger total muscle mass in male athletes are the principle sex differences in muscle physiology. The near-linear relation between strength and flexor muscle size indicates little difference in strength for the same size of muscle in men and women. In weight-lifting, there is not much disparity between the sexes in the maximum weights lifted in the lighter-weight categories, but women of 75–82.5 kg body mass lift only about 60% of the maximal weight lifted by their male counterparts which reflects differences in body composition (see FAT-FREE BODY MASS). For some sex differences in death rates, see MORTALITY (see Fig. 126). See SEXUAL DIMORPHISM.

sex hormone-binding globulin (SHBG, androgen-binding protein, ABP) A serum β-globulin binding the ANDROGENS testosterone and dihydrotestosterone, and to a lesser extent 17-β oestradiol. Its cell surface receptor, SEX HORMONE-BINDING GLOBULIN RECEPTOR (SHBGR), has been identified functionally in several tissues, e.g., prostate, testis, breast and liver. SHBG appears to be incapable of binding SHBGR if already bound to the steroid, and must bind SHBGR first before the steroid can bind the complex. SHBG is secreted by SERTOLI CELLS, binding specifically to the two androgens testosterone, dihydrotestosterone (DHT), and to 17-β oestradiol. By attaching to the two androgens, it probably concentrates them within the lumenal fluid of the seminal vesicles enabling SPERM maturation, and its retention by Sertoli cells permits SPERMATOGENESIS in the seminiferous tubules.

sex hormone-binding globulin receptor (SHBGR) A membrane β-globulin glycoprotein whose post-translational modification, by addition of different sugar moieties, produces the isoform SEX HORMONE-BINDING GLOBULIN.

sex hormones See ANDROGENS, OESTROGENS, and references there.

sex-linked inheritance (sex-linkage) Primate X and Y chromosomes differ considerably in size, the Y bearing very few coding loci while the X carries many (see SEX CHROMOSOMES, X CHROMOSOME, Y CHROMOSOME). Normal females carry two X chromosomes; but since normal males carry one X and one Y, any locus that is present on the X but not on the Y, or present on the Y and not on the X, will occur in the hemizygous condition in males (i.e., the second allele is absent). Alleles at such loci will therefore tend automatically to be expressed in males, whether or not the phenotype they produce in the female genome is dominant or recessive. Such loci are said to be *sex-linked*. Sex-linked loci are involved in several inherited disorders, including haemophilia and MUSCULAR DYSTROPHIES. Some forms of COLOUR BLINDNESS are also sex-linked. For Y-linkage, see SRY GENE and SEX REVERSAL. For other X-linked examples, see entries prefaced by X-LINKED-, FRAGILE-X SYNDROME.

sex reversal Roughly 1:20,000 male births is a phenotypic male with a 46,XX karyotype, and phenotypic females with a 46,XY karyotype are born with roughly the same frequency. Mutation in, or rearrangement of, genes associated with sex determination are one cause, but gain or loss of the *SRY* gene (encoding the DNA-binding testis-determining factor and located close to the pseudoautosomal region on the short arm of the Y chromosome) is often responsible. Such sex-reversed males are sterile and, as adults, sometimes show clinical features associated with KLINEFELTER'S SYNDROME. Intelligence is unaffected and mean height is shorter than in 46,XY males but taller than in 46,XX females. Phenotypic sex-reversed females have Fallopian tubes, an underdeveloped uterus and poorly developed gonads that render them sterile and in which tumours may develop. The absence of menstruation and secondary sexual characteristics can be overcome by hormonal treatment. Both types of individual may lead normal lives, apart from their sterility. See INTERSEX.

The puzzling existence of XX individuals who develop testes and complete female-to-male sex reversal, but who carry no *SRY* gene, may have been resolved. A closely related autosomal gene, *SOX9*, is up-regulated immediately downstream of *SRY* and can also activate the testis pathway. and findings in mice suggest that this may be the only requisite target of the *Sry* gene product in these mammals – in which case, gain or loss of function mutations in *SOX9* in humans would mimic the effects of *SRY* (see SEX DETERMINATION).

sexual arousal See AROUSAL, SOMATIC SENSORY SYSTEM.

sexual dimorphism Systematic difference in form between individuals of different sex but of the same species (see SEX DIFFERENCES). In monogamous species, males and females are typically of the same size. The notion that body-size dimorphism represents the outcome of competition among males for access to females is not without evidence; but it may also reflect the anti-predator protective role of males in social groups (this is not without circularity). Canine teeth in primates also tend to be larger in males than in females of the same species. Smaller females may breed earlier, and selection for earlier breeding might explain sexual dimorphism in some mammals. The possible occurrence of sexual dimorphism poses a difficulty in interpreting fossil hominoid material, creating controversy over whether remains belong to one species or two. Notwithstanding this factor, evidence for a marked reduction in sexual dimorphism in the hominid fossil record, possibly reflecting a change in sexual practices, appears at ~1.8 Myr (see BRAIN SIZE). See BODY MASS, SIZE AND SHAPE.

sexual orientation The tendency to become erotically aroused by the opposite, or the same, sex. Although differences in the size of one interstitial nucleus (INAH-3) of the anterior HYPOTHALAMUS have been detected between adult heterosexual and homosexual men, this tells us little about whether these differences are present at birth. There is a tendency for male homosexuality to have a familial disposition associated with the man's maternal lineage, and some research suggests that one portion of DNA at the tip of the X chromosome shows a tendency to be the same in (non-twin) gay brothers. However, most studies find that considerably fewer than 100% of identical twins share the same sexual orientation. Examples of genetic males with 5-α-reductase deficiency (see TESTOSTERONE) suggest a role for hormones in sexual orientation; but it presently remains unclear whether or not there is a neuroendocrine and/or genetic basis to sexual orientation. Multifactorial causation is possible.

SF1 Gene encoding steroidogenesis factor-1. See SRY.

SH2 and SH3 domains Src homology regions, first discovered in the Src protein. The numerous SH2 domains are components of adaptor proteins that link receptor activation to downstream members of signalling pathways. They interact very specifically with a short oligopeptide sequence containing one or more phosphotyrosine residues generated by receptor-associated tyrosine kinases. The human genome is estimated to encode at least 117 distinct SH2 groups, each constituting a domain of a larger protein, and each seemingly having affinity for binding a particular phosphotyrosine together with a flanking oligopeptide sequence. SH3 domains allow an activated receptor to bind to proline-rich regions of other signalling molecules. Some proteins (e.g., GRB2; see entry) consist almost entirely of SH2 and SH3 domains and act as 'adaptor proteins' involved in coupling tyrosine-phosphorylated proteins to others lacking SH2 domains, such as the signalling protein Ras (see RAS PROTEINS, RAS SIGNALLING PATHWAY). See JAK/STAT PATHWAY, SRC-KINASES.

Shigella Bacillary bacteria causing dysentery in humans. Until recently, *Shigella* was considered to be a genus; but it is now considered to be eight separate strains of ESCHERICHIA COLI. In Western countries, the endemic *Shigella* spp. generally cause mild self-limiting illnesses; but tropical shigelloses tend to be more severe and persistent, especially in children. They mainly cause mucosal infection of the distal ileum, antigen presentation leading to release of cytokines which recruit neutrophils. These break down the tight junctions between the epithelial cells, enabling invasion of the submucosa and uptake by M cells of PEYER'S PATCHES. The bacteria exploit the host Arp2/3 complex (see ACTIN) to become pushed against the host cell membrane, creating a protrusion in the adjacent epithelial cell membrane and migrating into the cell via the outer membrane protein. DNA analyses suggest that two of the three main strains evolved 50–270 Kyr BP while the third did so 35–170 Kyr BP. See DISEASES, INFECTION, LISTERIA MONOCYTOGENES, SALMONELLA.

shingles (herpes zoster) Increasingly common with age, an inflammatory disease caused by CHICKENPOX viruses (varicella) that have inhabited the dorsal root ganglia since the patient had that disease as a child. These suddenly multiply in huge numbers, migrate along sensory neurons to the skin supplied by the ganglion, which is where the inflammation occurs, leaving the skin scarred. In some cases, sensory neurons are destroyed and experiences similar to REFERRED PAIN result for the rest of the patient's life. The reactivated virus increases the excitability of sensory neurons, leading to low thresholds of firing and spontaneous activity. Normally the virus is reactivated in only one dorsal root ganglion, so that symptoms are restricted to the skin innervated by its axons, marking the skin territory of that DERMATOME. See CAPSAICIN.

shivering See TEMPERATURE REGULATION.

shock See STRESS.

short bones Bones which are nearly equal in length and breadth and somewhat cube-shaped. Consist mainly of spongy bone, with a thin layer of compact bone at the surface. Include carpal (wrist) and tarsal (ankle) bones. Contrast LONG BONES.

shortsightedness (myopia) See NEAR-SIGHTEDNESS.

short-term memory See MEMORY.

shotgun sequencing See DIVERGENCE, Fig. 65; GENOME MAPPING.

shoulder See PECTORAL GIRDLE.

sialic acid (*N*-acetylneuraminic acid) An acidic sugar, and derivative of *N*-acetylmannosamine, which is a component of many glycoproteins and glycolipids, added in the GOLGI APPARATUS.

sibling recurrence risk, l_s See FAMILIAL RECURRENCE RISK.

sickle-cell anaemia, sickle-cell disease, sickle-cell trait See HAEMOGLOBINOPATHIES.

SIDS See SUDDEN INFANT DEATH SYNDROME.

sigh A reflexly generated single, deep, breath occurring after a period of quiet breathing. See ALVEOLUS, YAWNING.

signal amplification One of the features of SIGNAL TRANSDUCTION pathways, in which a very specific but low level signal (be it molecular, photon or mechanical stimulus) leads rapidly to a far larger number of molecular target molecules being produced or altered.

signal recognition Polypeptides directed to the endoplasmic reticulum for inclusion in a cell membrane, or for secretion, have a signal sequence comprising the first 20 or so amino acids. The signal sequence is guided to the ER by a signal recognition particle (SRP), a complex of 7SL RNA and six specific proteins, probably assembled in the NUCLEOLUS. The SRP complex binds both the growing polypeptide chain and the ribosome and directs them to an SRP receptor protein on the cytosolic surface of the ENDOPLASMIC RETICULUM (ER), after which the polypeptide can enter the ER lumen. SRP function involves hydrolysis of GTP, one of the six protein subunits containing an N-terminal GTPase domain required for dissociation of the particle from the receptor complex on the ER.

signal sequence A localization signal, usually a short peptide sequence, embedded in the structure of a polypeptide, directing it to the correct address in the cell. This is usually then removed by a signal peptidase. See SIGNAL RECOGNITION, and LYSOSOME, MITOCHONDRION and NUCLEUS, for other localization signals.

signal transduction Mechanisms by which a receptor molecule, typically in the cell surface membrane, is able to detect, often with extreme sensitivity, either an energy source (e.g., photons), mechanical distortion, or a diffusible or membrane-bound signal molecule (its ligand), and activate a response by transducing it across the membrane into intracellular changes (see Fig. 155), often resulting in altered GENE EXPRESSION. Neurotransmitters, peptide hormones, steroid hormones (see NUCLEAR RECEPTOR SUPERFAMILY) and growth factors all operate by signal transduction pathways. Specificity (see Fig. 155a) is achieved by subtle molecular affinities of the RECEPTORS for their signals, and by the fact that, through fate-determining developmental events, receptors and the transduction pathway components for a given signal

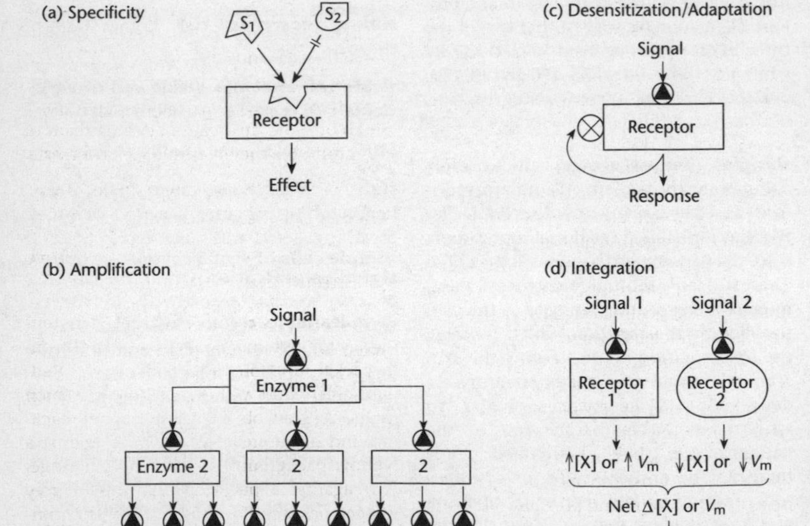

FIG 155. Four features of SIGNAL TRANSDUCTION. (a) Specificity of the receptor, in this case a ligand-binding receptor, is achieved by complementarity with the signal (S), although signals need not be particles. (b) Signal amplification is achieved by enzyme cascades in which the number of affected molecules increases geometrically. (c) Desensitization (adaptation) of the receptor, such as occurs when its activation triggers a feedback circuit shutting off the receptor (cross within a circle) or removing it from the surface. (d) Integration of the regulatory outcome of two signals, as when each has an opposite effect on some key component in a pathway (e.g., the concentration [X] of a second messenger, or the membrane potential (V_m) of the cell). Activation of a pathway component is indicated by a triangle within a circle. © 2000 From *Lehninger Principles of Biochemistry* by D.L. Nelson and M.M. Cox, published by Worth Publishers. Redrawn with permission from the publisher.

are present only in cells of certain types. Sensitivity of transduction pathways is achieved through cooperative interaction between the receptor and its ligand and by amplification of the signal by cascades involving protein kinases (see Fig. 155b). These cascades begin by activation of an enzyme, often a kinase, associated with the receptor. This activates many molecules of a second enzyme (often another kinase), each of these activating many molecules of a third, and so on. This enables a cell's response to a single signal to be amplified by several orders of magnitude within milliseconds. Desensitization of the receptor (see Fig. 155c) can be achieved by *receptor inactivation*, temporary removal to the cell interior (*receptor sequestration*), or destruction by lysosomes after removal (*receptor down-regulation*). This is often achieved through negative feedback. Pathways often bifurcate, the route taken usually depending on the phosphorylation state of key components. Alternatively, the effects of different signals can be integrated (see Fig. 155d).

Receptors involved include RECEPTOR TYROSINE KINASES and those linked to RECEPTOR GUANYLYL CYCLASES (see also PROTEIN KINASES). But activation by agonist binding of G PROTEIN-COUPLED RECEPTORS controls most signal transduction processes, many of which transduce extracellular signals via the cytosol to the nucleus, where they alter gene transcription (e.g., see PHOSPHOLIPASE C, Fig. 135; CREB, Fig. 60). The JAK/STAT PATHWAY exemplifies one of the most direct. Pathways involving the TRANSFORMING GROWTH FACTOR-β SUPERFAMILY (including TGF-β itself, BMPs and activins) also operate speedily via receptors that bind and phosphorylate any of five cytosolic gene regulatory proteins of the Smad family. The Smad protein then dissociates, binds Smad 4, and enters the nucleus as a protein complex that associates differentially with other regulatory proteins to activate a particular gene set (a different set for each Smad complex).

Signals that are continuously present can desensitize the receptor system until they fall below a certain threshold (e.g., see SCOTOPIA), while some signal transducing systems receive multiple signals and integrate them into a unified response, different pathways conversing at different levels to generate adaptive (e.g., homeostatic) responses far beyond the cells immediately involved. Microarrays are increasingly used to analyse the gene expression profiles ('expression signatures') resulting from activation of different signalling pathways. Recently, signatures of ~100 genes have been found to correlate with mutational activation of the Myc, Ras, Src or β-catenin proteins, or loss of the Rb tumour suppressor gene (see CANCER), and signalling pathways are implicated in other forms of DISEASE (see Fig. 64 and PKB).

signing, sign language See LANGUAGE.

silencers Those *cis*-acting REGULATORY ELEMENTS that bind repressors, thereby inhibiting activators and decreasing expression of a gene. Unlike promoter-proximal elements, they can act at a distance, sometimes 50 kb or more upstream or downstream of the promoter they control. See GENE EXPRESSION, GLOBINS.

silicosis See PNEUMOCONIOSIS.

simple cells of visual cortex See ORIENTATION-SELECTIVE CELLS, PRIMARY VISUAL CORTEX.

simultaneous colour contrast The tendency for certain colours to stand out from a background of a different colour. Red-and-green and yellow-and-blue are such pairs. A grey object against a red background has a green tinge, while against a yellow background it has a purple tinge; but against a purple background a grey object has a yellow tinge. This phenomenon may be explicable in terms of the properties of double-opponent cells of the primary visual cortex.

SINEs (short interspersed nuclear elements) Class 1 TRANSPOSABLE ELEMENTS that are short (100–400 bp) DNA sequences that have successfully colonized mammalian genomes by inserting into INTRONS, positions unlikely to cause mutations subject to negative selection. Some, such as the ALU

REPEAT FAMILY, are primate-specific. SINES are best described as non-autonomous LINES, because they have the same structural features at their 3′ end as LINEs but do not encode their own reverse transcriptase. In fact, they do not encode any proteins and therefore are not autonomous (cannot copy and integrate on their own), but by parasitizing the LINE element transposition machinery SINES have attained very high copy numbers. Those that originated from copies of tRNA are transcribed by RNA polymerase III and are unusual in having internal promoters. Older SINE elements tend to lie in gene-rich (G + C)-rich regions, whereas younger SINE elements tend to lie in gene-poor (A + T)-rich regions where (LINE)-1 elements also accumulate, consistent with the fact that Alu retrotransposition is mediated by L1. See DNA REPEATS.

single-locus probe See GENE PROBE.

single nucleotide polymorphisms (SNPs, pronounced 'snips') Any polymorphic variation at a single nucleotide, SNPs are often used as markers for studying susceptibility to common disease, and several high-throughput genotyping methods can now detect them (see DNA MICROARRAYS, DNA PROBES, DNA SEQUENCING, DHPLC, SSCP). About 90% of human genetic variation has been ascribed to SNPs, single base pair differences between individuals in a population's DNA sequence occurring at a frequency of greater than 1%. The HUMAN GENOME PROJECT has revealed that there may be as much as one nucleotide substitution every 1,000–2,000 base pairs in human DNA, and perhaps 1.6–3.2 million such snips in the whole genome, resulting in an estimated 10 million common human SNPs (see GENOMICS). Non-synonymous SNPs are those that cause substitution of one amino acid for another in an encoded protein during its synthesis, usually thought (perhaps erroneously) to be selectively neutral or of minor phenotypic effect. Many alleged SNPs, and 12% in public and private databases, remain unconfirmed; many are paralogous sequence variants, and other 'real' SNPs do not get into the appropriate databases. Validation of an SNP is therefore very important. In 2007, a suite of studies associated increased risk of developing heart disease on some hitherto unsuspected genetic 'culprits' closely linked on a long genomic region on chromosome 9 (hsa 9p21.3). Despite the strength of this association, these SNPs are not necessarily the causative mutations and may not lie in gene-coding or regulatory elements and may not contribute to risk, even indirectly (see COMPLEX DISEASES). As larger populations are scanned for SNPs, more will be associated more reproducibly with more diseases. But such 'SNP-spotting' alone does not detect environmental contributions to disease, or the contributions of duplicate genes to disease susceptibility (see COPY NUMBER). At present, family history is a better predictor of personal disease risk than are these SNP association studies, and there is no evidence that results of genetic tests encourage healthier LIFESTYLES. The 60-year-old Framingham Heart Study, whose database contains thousands of clinical variables from blood analyses to vascular imagery and lifestyle surveys, provides a rich vein of data in which 550,000 SNPs in more than 9,000 individuals from the Massachusetts area have been genotyped (now termed the SNP Health Association Resource, or SHARe). In all such large-scale epidemiological studies, poor analyses of data can result in missed associations, or even false positives. It is widely recognized that large-scale studies that have been made possible by the Human Genome Project work best when data are shared (as on the Data of Genotype and Phenotype, dbGaP). Many loci being found today have no link to traditional risk factors of disease, indicating that we probably know very little about the underlying biology of complex diseases.

Estimates of SNP frequency in gene-containing regions appear to converge on 8.00 per 10 kb, the genome-wide frequency being 7.51 per 10 kb. The variation between individual non-sex chromosomes is small, lying between 5.19 (chromosome 21) and 8.79 (chromosome 15) differences per 10 kb, humans varying least in their SEX CHROMOSOMES. In the vast bulk of cases, these SNPs will not be polymorphic in non-human primates. Comparison with the chimpanzee genome makes it possible to determine which variants in human SNPs represent the 'original form', an important factor in genetic association studies aimed at mapping the locations of gene variants in complex diseases such as diabetes and high blood pressure. *Restriction site polymorphisms (RSPs)* are those SNPs which cause loss or gain of a restriction site, formerly typed by Southern blotting methods – the assay detecting RESTRICTION FRAGMENT LENGTH POLYMORPHISMS (RFLPs), although RSPs are now better typed using a PCR-based method. See GENETIC DIVERSITY, HAPLOTYPE, HUMAN GENOME, NATURAL SELECTION.

single opponent cells See COLOUR-OPPONENT CELLS.

single-stranded binding proteins (SSBPs) Proteins that are important in maintaining stability of the replication fork during DNA REPLICATION. Single-stranded DNA is very labile (unstable) and these proteins bind to it while it remains single-stranded and prevent it from being degraded. See TRANSCRIPTION.

sinoatrial node (pacemaker) See SA NODE.

sinuses A hollow in a bone, or other tissue; or, any cavity having a narrow opening. **1**. Mucous membrane-lined air spaces within certain SKULL bones (e.g., the ETHMOID BONE) and facial bones (see CONCHAE). **2**. The marginal and medullary sinuses of LYMPH NODES (see Fig. 118). **3**. Venous sinuses, similar in structure to sinusoidal capillaries but even larger in diameter, e.g., in red BONE MARROW. **4**. See SINUS VENOSUS.

sinusoidal capillaries See CAPILLARIES.

sinusoids Large-diameter sinusoidal CAPILLARIES, occurring between LIVER hepatocytes and in BONE MARROW.

sinus venosus Originally a sinus connecting the left and right sinus horns (each receiving blood from embryonic umbilical, vitelline and cardinal veins) to the atrium (see HEART, Fig. 90), the entrance of the sinus shifts to the right during the 5th week of development as the left sinus horn disappears and the right sinus horn becomes part of the right atrium.

siRNAs (short interfering RNAs) Small-interfering RNAs of ~21 nucleotides produced through cleavage of double-stranded RNA (dsRNA). These bind to the Argonaut-family effector Ago2 and function in defence against external nucleic acids. See RNA INTERFERENCE.

sirtuins (SIRTs) An evolutionarily conserved group of NAD^+-dependent protein deacetylases (the Sir2, or silent information regulator 2, family) with, in lower organisms, a wide range of cellular activities affecting LIFE EXPECTANCY, including modulating how tightly DNA is packaged within nuclei. In mammals, where several forms (>7) occur, they act as regulators of apoptosis and cell maturation. Some act on non-histone proteins (e.g., the human SIRT1 promotes cell survival by negatively regulating the p53 tumour suppressor) while others decacetylate specific HISTONES. Yeast Sir2 is a NAD-dependent histone deacetylase, the oxidized form of NAD enhancing its activity, SIRT1 being the closest mammalian homologue (where it also deacetylates other proteins, including MyoD). Extra copies of the Sir2 gene are sufficient to increase lifespan in yeast and the worm *Caenorhabditis elegans*. The plant polyphenol resveratrol, present in red wine, is an activator of Sir2 and stimulates SIRT1 activity. Sirtuins may also affect MITOCHONDRIA, beyond the expression of FOXO-regulated antioxidant and pro-apoptotic proteins. PGC-1α (see PGC-1) is regulated by SIRT1, which also has direct links to autophagy and possible links to life expectancy. Other sirtuins also localize to mitochondria.

situs inversus Condition in which transposition of the viscera occurs in the thorax (including the HEART) and abdomen. Cilia of the PRIMITIVE NODE usually occur on the ventral surface and possibly result in defective patterning by wafting morphogens in the wrong direction and, in ~20% of cases, abnormal cilia also occur in the respiratory tract. See CILIUM.

size constancy See VISUAL PERCEPTION.

skeletal muscle (striated muscle, voluntary muscle) Deriving from PARAXIAL MESODERM, MUSCLES of the axial skeleton, body wall and limbs develop from SOMITES (see entry for genes involved) from the occipital to sacral regions, while muscles of the head develop from head somitomeres. They are innervated by the pathways originating in primary MOTOR CORTEX. During differentiation, precursor myoblasts fuse to form long, syncytial, multinucleated muscle fibres. Myofibrils appear in the cytoplasm, and by the 3rd month the cross-striations typical of skeletal muscle appear. Tendons attaching muscles to bones derive from sclerotome cells adjacent to myotomes. For the positions and names of the superficial skeletal musculature, see Appendices Xa and Xb.

Skeletal muscle contains a variety of fibre types, with different metabolic and contractile properties (see Table 10), but they all express the GLUT-4 GLUCOSE TRANSPORTER. A common technique for determining the specific fibre type is to assess the myosin molecule's heavy chain, which exists in three isoforms differing in the sensitivities of the ATPase to altered pH. This affects the rapidity of ATP hydrolysis and hence sarcomere shortening. Fast twitch fibres often rely on a well-developed, short-term glycolytic system for energy transfer, their activation predominating in anaerobic sprint activities and other forceful muscle actions depending on anaerobic energy metabolism (see EXERCISE). Slow twitch fibres generate ATP mainly through aerobic energy transfer, making them fatigue-resistant and suited for prolonged aerobic exercise. Both slow and fast twitch fibre types contribute during near-maximum aerobic and anaerobic exercise (e.g., middle-distance running, football), which combine high levels of aerobic and anaerobic energy transfer. If a skeletal muscle is composed mainly of slow twitch fibres, it could be termed a tonic (endurance) muscle; if made up mostly of fast twitch fibres, it could be termed a phasic (low endurance) power muscle.

Skeletal muscle accounts for ~50% of a resting person's oxygen consumption (see OXIDATIVE CAPACITY) and up to ~90% during heavy EXERCISE. It can use free fatty acids, ketone bodies or glucose as fuels, depending on the level of activity. Triglyceride provides more than half the energy requirement of resting skeletal muscle fibres; moderately active muscle uses blood glucose, in addition to fatty acids and ketone bodies. Glucose uptake by skeletal muscle is mediated through GLUCOSE TRANSPORTERS in the sarcolemma, the GLUT-4 transporter being unique in being recruited to the cell surface from Golgi-derived cytosolic vesicular compartments when INSULIN binds its membrane receptor (see INSULIN RESISTANCE).

But in maximally active muscles, the ATP demand outstrips blood oxygen and fuel supply and, especially under ADRENALINE stimulation, stored GLYCOGEN is broken down to lactic acid by fermentation. Although striated muscle stores relatively little glycogen, only ~1% of its total mass, the accumulating LACTATE and fall in pH reduce their efficiency of contraction. Much of the OXYGEN DEBT is used in oxidation of lactate by liver, an organ which provides blood glucose for skeletal muscles to replenish their glycogen reserves. Skeletal muscle uses the glycerol 3-phosphate shuttle (see GLYCEROL) to pass reducing equivalents from cytosolic NADH into the mitochondria for oxidative phosphorylation.

PGC-1α (see PGC-1 GENES) functions as a sensitive 'rheostat' in skeletal muscle, responding to neuromuscular input and the ensuing contractile activity. Its expression is readily inducible by EXERCISE training in rodents and humans. Subsequent activation of calcium signalling pathways seems to play a major role in PGC-1α transcription via calcineurin and Ca^{2+}-dependent protein kinases, culminating in activation of such transcription factors

Structural characteristics	*Slow-oxidative (S-O) (slow twitch) fibres. Type 1*	*Fast-oxidative-glycolytic (FOG) (fast twitch) fibres. Type 2a (IIa)*	*Fast glycolytic (FG) (fast twitch) fibres. Type 2b (IIb)*
Fibre diameter	Smallest	Intermediate	Largest
Myoglobin content	High	High	Low
Mitochondria	Many and large	Many	Few
Capillaries	Many	Many	Few
Colour	Red	Red-pink	White (pale)
Development	Favoured by PGC-1α expression		
Functional characteristics			
Capacity for ATP generation, and method used	High; aerobic (mitochondrial); high carnitine level reduces inhibition of pyruvate dehydrogenase activity, favouring conversion of pyruvate and lactate to acetyl coenzyme A	Intermediate; both aerobic (mitochondrial) and anaerobic (glycolysis). High level of aerobic succinate dehydrogenase and anaerobic phosphofructokinase	Low; anaerobic (glycolysis)
Rate of ATP hydrolysis by myosin ATPase	Low	High	High
Contraction velocity	Low (slow twitch)	High (fast twitch)	Very high (fast twitch)
Force production	Low	Moderate	High
Fatigue resistance	High	Intermediate	Low
Glycogen stores	Low	Intermediate	High
Order of recruitment	First	Second	Third
Where fibres are abundant	Postural muscles, e.g. neck. Arm and leg muscles (~50% of fibres)	Leg muscles	Arm muscles
Main function of fibres	Maintaining posture; suited for prolonged aerobic exercise, e.g. endurance activities	Walking, sprinting	Rapid, intense movements of short duration

TABLE 10 *Some of the major* SKELETAL MUSCLE *fibre types, their characteristics, roles and distributions.*

as CREB and MEF2. Transgenic expression of PGC-1α in FAST TWITCH FIBRES leads to activation of genetic programmes characteristic of slow twitch fibres, apparently coupling the metabolic and contractile programmes of muscle-fibre specification, functionally transforming fast twitch fibres into more oxidative SLOW TWITCH FIBRES. See MOTOR CORTEX, MOTOR UNIT, MUSCLE SPINDLES.

skeletal muscle pump See VEINS.

skeletal musculature For the positions and names of the superficial skeletal musculature, see Appendices Xa and Xb.

skeletal system Originating from SCLEROTOME, cartilages and the 206 bones in an adult human, grouped into two systems (see Fig. 156 and Table 11): the APPENDICULAR SKELETON and the AXIAL SKELETON (see SKULL, HYOID, EAR OSSICLES). See GAIT, LEVERAGE. See Appendix VI.

skeleton See SKELETAL SYSTEM.

skin The organ forming the entire external

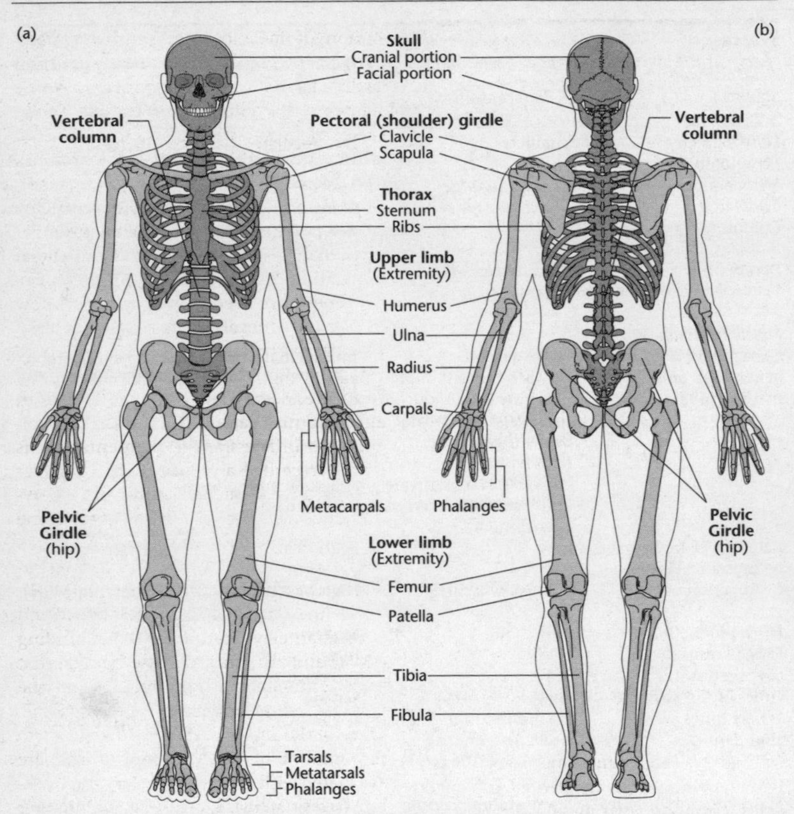

FIG 156. (a) Anterior view, (b) posterior view, of the human SKELETAL SYSTEM, the axial skeleton shown dark grey, the appendicular skeleton light grey. From *Principles of Anatomy and Physiology*, G.J. Tortora & S.R. Grabowski, © 2000. Reproduced with permission of John Wiley & Sons, Inc.

body surface, except for the eyes. It has two major tissue layers: the EPIDERMIS and the DERMIS (see entries for details and references). These two layers vary in thickness from ~0.5 mm in the eyelids, to 5.0 mm for the back and shoulders, most of this attributable to variation in thickness of the dermis: not to be confused with *thick skin* and *thin skin*, the skin of the back actually being thin skin (see EPIDERMIS), despite its overall thickness. Accessory skin structures include HAIR, SEBACEOUS GLANDS, SWEAT GLANDS, MAMMARY GLANDS and ceruminous glands (see CERUMEN). Water-proofing is provided by lipids (see KERATINOCYTES), its sensory role being catered for by MECHANORECEPTORS (see Fig. 120), thermoreceptors (see TEMPERATURE REGULATION) and NOCICEPTORS, some of which occur in the subcutis (hypodermis). Immunologically, the skin is served by a dense network of dendritic cells, termed LANGERHANS CELLS. See SKIN CANCER, SKIN COLOUR.

skin cancers MALIGNANT TUMOURS of the skin, of which three forms occur: basal cell carcinoma (BCC, the most common form of cancer, with about a million new cases estimated in the USA each year) arising in keratinocytes of the basal layer of the EPIDERMIS; squamous cell carcinoma; and

Structure	Number of bones
Axial skeleton	
Skull	8 (cranium)
	14 (facial bones)
Hyoid	1
Auditory ossicles	6
Vertebral column	26
Thorax	1 (sternum)
	24 (ribs)
Subtotal	80
Appendicular skeleton	
Pectoral (shoulder) girdles	2 (clavicles)
	2 (scapulae)
Upper limbs	2 (humeri)
	2 (ulnae)
	2 (radii)
	16 (carpals)
	10 (metacarpals)
	28 (phalanges)
Pelvic (hip) girdles	2 (coxal bones)
Lower limbs	2 (femurs)
	2 (fibulae)
	2 (tibiae)
	2 (patellae)
	14 (tarsals)
	10 (metatarsals)
	28 (phalanges)
Subtotal	126
TOTAL	206

TABLE 11 *The distribution of bones of the* SKELETAL SYSTEM.

MELANOMA (occurring in MELANOCYTES), being the rarest but most regularly metastatic (see METASTASIS). Basal cell and squamous cell carcinomas are usually fairly easy to treat if detected early by biopsy, although 5–10% can be resistant to treatment, or locally aggressive; but they are not usually life-threatening and have very low rates of metastasis. Skin cancers are most closely associated with chronic inflammation of the skin, including SUNBURN (see ULTRAVIOLET IRRADIATION), especially in early life (UVA and UVB have both been implicated). Thirty times more UV-irradiation is required to produce sunburn – the prime cause of skin cancer – in black people than white people.

skin colour Variation in skin pigmentation is a continuously varying (quantitative) character and results partly from differences in the number, size and distribution of melanosomes within KERATINOCYTES of the epidermis (dark skin contains many large, very dark, melanosomes; lighter skin contains smaller, less dense, melanosomes); but the types of melanin within the melanosomes is a major factor). The degree of pigment production manifests as skin 'phototype' (colour and ease of TANNING) and is the most useful predictor of human SKIN CANCER risk in the general population. Although many human ALBINISM genes have been cloned, and we know of over 100 genes affecting melanin synthesis and coat colour in the MOUSE, many of which have corresponding human phenotypic effects, very few genes are known to affect 'normal' variation in human skin colour. Nevertheless, skin pigmentation is usually regarded as a polygenic character (see POLYGENIC INHERITANCE). The highly polymorphic melanocortin 1 receptor gene *MC1R* (see MELANOCORTINS) is one such, whose gene product, the receptor for α-melanocyte stimulating factor (α-MSH), lies in the MELANOCYTE cell membrane and triggers an elevation of cAMP on binding α-MSH and initiating a kinase cascade that leads ultimately to production of the mature eumelanogenic melanosome. Three of the alleles in the *MC1R* gene affect the amount of cAMP produced and are associated with red hair, fair skin and freckling (see MELANINS). Increase in the number of non-synonymous MUTATIONS an individual carries in the *MC1R* gene compared with the wild-type consensus sequence is associated with increasingly lighter skin, although sharing of identical *MC1R* haplotypes among individuals with very different skin colours indicates that other loci must be involved: there are many fair-skinned but dark-haired individuals in whom MC1R is unlikely to limit skin pigmentation on its own. At least two other loci, *OCA2* and *OCA4*, responsible for different forms of oculocutaneous albinism, act dominantly in modern African populations and argue for the view that black hair, brown eyes and dark skin colour are to be considered the original state for humans prior to transcontinental MIGRATION. Epidemiological studies suggest that pigmentation is under functional constraint in the African continent, although it is not clear whether this is necessitated by UV-induced vitamin D production over protection from UV-induced DNA damage or as a result of an as yet undiscovered critical pathway. *MC1R* variant alleles have been linked to κ-opioid analgesia in both mice and humans.

One such is the gene *SLC24A5*, discovered during a search for cancer genes in the Zebrafish (*Danio rerio*), where mutation in the fish orthologue of the human gene (see GENOME MAPPING) causes the *golden* phenotype and encodes a putative cation exchanger, probably in the melanosome membrane. Its function in humans has yet to be determined. SINGLE NUCLEOTIDE POLYMORPHISM (SNP) within *SLC24A5* revealed alternative alleles encoding either alanine or threonine at amino acid 111 in the third exon, the only coding SNP then current in the International Haplotype Map (see HAPLOTYPE). The presence of ala^{111} in all other known members of the gene family to which *SLC24A5* belongs clearly pointed to its being the ancestral situation. The allele frequency for the Thr^{111} variant ranged from 98.7% to 100% among several European-American population samples, while that for the ancestral Ala^{111} ranged from 93% to 100% in African, Indigenous American and East Asian population samples, a difference in allele frequencies that ranks within the top 0.01% of SNP markers in the HapMap database, consistent with the possibility that this SNP has been a target of natural or sexual selection, further evidenced by the striking reduction in heterozygosity near the *SLC24A5* gene (see MUTATION, NATURAL SELECTION). See ALBINISM, SUNBURN.

skinfold calipers See ADIPOSE TISSUE.

skull The 29 bones of the skull (see Fig. 157) are part of the axial skeleton (see SKELETAL SYSTEM) and can be analysed in two groups: the neurocranium, forming a protective case around the brain, and the viscerocranium, forming the skeleton of the face. Bones from both groups contribute to the ORBITS of the eyes.

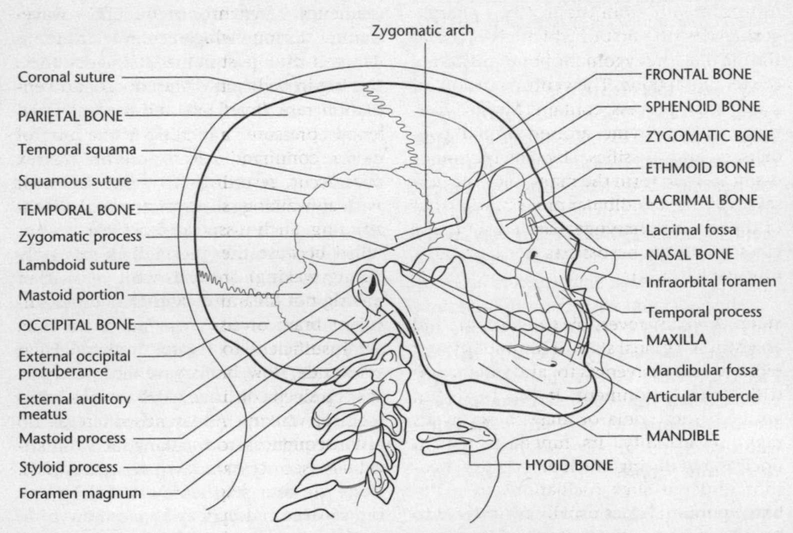

FIG 157. Right lateral view of the adult human SKULL. The hyoid bone, although not a part of the skull, is included for reference. From *Principles of Anatomy and Physiology*, G.J. Tortora & S.R. Grabowski, © 2000. Reproduced with permission of John Wiley & Sons, Inc.

The neurocranial (or *cranial bone*) group, serving primarily to surround and protect the brain, comprises the paired parietals and temporals, and the solitary frontal, occipital, ethmoid and sphenoid bones. The neurocranium, or braincase, comprises the flat MEMBRANE BONES surrounding the brain as a vault (cranial vault), and the cartilaginous components (chondrocranium) which develop into bones at the base of the skull. They contain spongy BONE in their interiors. The chondrocranium consists initially of a number of separate cartilages, the most anterior deriving from neural crest cells while those posterior to the rostral limit of the notochord derive from paraxial mesoderm, fusing to form the base of the skull by endochondral ossification.

The viscerocranial (*facial bone*) group, mainly involved in protection and functioning of the eyes, nose and mouth, comprises the two nasal bones, two maxillae, two lacrimals, two inferior nasal CONCHAE, the vomer, two palatines, two zygomatic bones (forming the zygomatic arches) and the mandible. The viscerocranium is formed mainly from the first two pharyngeal arches, the first of which gives rise to the maxilla, the zygomatic bone and part of the temporal bone. The ventral portion, or mandibular process, contains Meckel's cartilage, mesenchyme around which condenses and ossifies by membranous ossification to form the mandible. The dorsal tip of the mandibular process, with that of the second pharyngeal arch, eventually gives rise to the EAR OSSICLES (see PHARYNGEAL POUCHES, Fig. 134).

sleep A readily reversible (*contra* coma and anaesthesia) global state of immobility and reduced responsiveness to, and interaction with, the environment. It can be distinguished from coma or anaesthesia by its rapid reversibility. Its functions remain unclear, but during much of it cortical neurons undergo slow oscillations in membrane potential. It is usually considered to be induced by active processes, the SLEEP-WAKE CYCLE in humans being the most obvious circadian rhythm (see CIRCADIAN CLOCKS). There is considerable evidence that sleep enhances MEMORY (see later) (see CIRCADIAN CLOCKS). One line of research is to determine the critical minimum period of sleep for such memory enhancement. Another possibility is that during sleep the brain is repaired and detoxified. Sleep deprivation is followed by extra recovery sleep proportional to the sleep loss.

ELECTROENCEPHALOGRAM (EEG) and other physiological measures point to the existence of two sleep states (see Fig. 158): *non-rapid eye movement (NREM) sleep* and *rapid eye movement (REM) sleep*. The REM/non-REM cycle is ultradian. At the start of sleep, the EEG changes from fully awake to NREM sleep state by progressive decrease in EEG frequency, but increase in amplitude. In NREM sleep, the body is active and the brain inactive, and there is relatively little dreaming. There are often limb movements, and parasomnia (sleep-walking), during NREM sleep. REM sleep, by contrast, is characterized by rapid eye movements and relative absence of muscle tone.

1. *NREM sleep* has high amplitude, low frequency (synchronized) EEG waveforms, during which muscle tone is retained and postural adjustments (e.g., turning in bed) sometimes occur and ventilation rate, heart rate and mean arterial blood pressure all fall. Activation of motor command centres in the cortex may occur, resulting in PARASOMNIAS such as sleep-walking, sleep-talking and tooth-grinding. Such dissociated behaviours (so-called because they normally occur only during waking) are the result of neither waking nor dreaming. Rather, there is sufficient brain arousal to support walking, but insufficient to support waking. High amplitude, slow, brain waves characteristic of deep sleep continue to be recorded during sleep walking. Apparently, there are no ill consequences from waking such an individual (see CENTRAL PATTERN GENERATORS). There are four stages of NREM sleep (see Fig. 158), each deeper and with a lower frequency than the last. Stages 1 and 2 are considered light sleep, stage 2 being interspersed with higher-frequency bursts

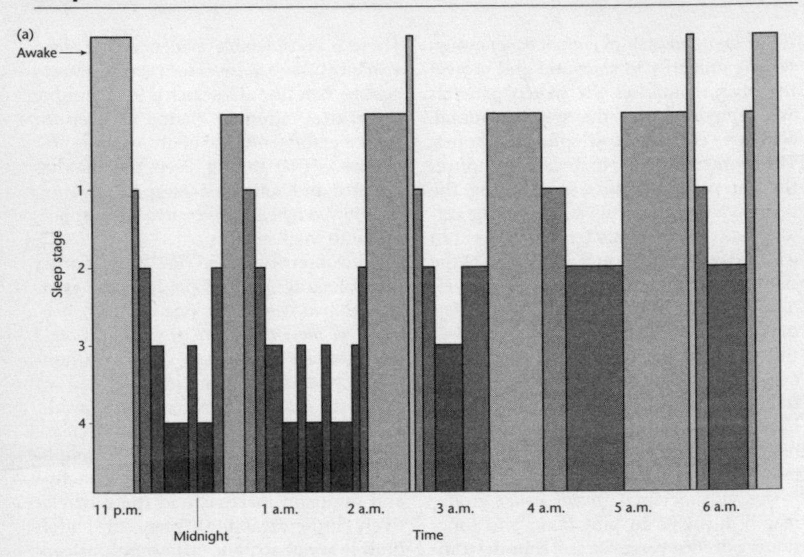

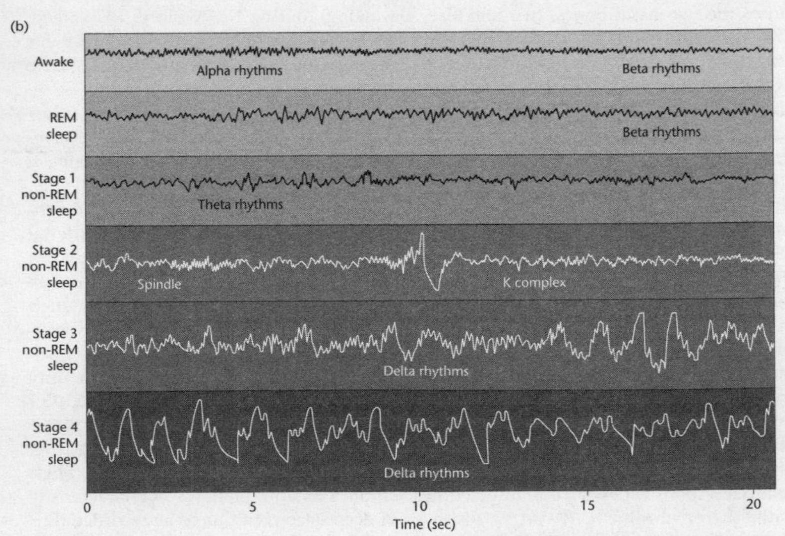

FIG 158. Diagrams indicating the stages of SLEEP (see entry) during one night. (a) Histogram representing falling asleep at 11 p.m. and entering NREM sleep stage 1, progressively deeper stages of NREM sleep, and then REM sleep. This cycle is repeated several times, each cycle tending to have shorter and shallower NREM periods and longer REM periods. Vertical height, and numbers, represents the stage of REM sleep, as shown in (b), which shows the EEG rhythms occurring during each of the sleep stages. Shadings in (a) and (b) match. © 2001 Redrawn from *Neuroscience* (2nd edition) by M.F. Bear, B.W. Connors and M.A. Paradiso, published by Lippincott, Williams & Wilkins. Reproduced with the mind permission of the publisher.

termed *sleep spindles*. Stages 3 and 4, deep sleep ('slow-wave sleep'), are characterized by slow-wave activity: delta (Δ) waves. After ~90 minutes of NREM sleep, the subject enters REM sleep. Burst firing of thalamic relay neurons occurs during NREM sleep when these cells are hyperpolarized by loss of input from the ascending arousal system, and it is this that drives synchronized bursts of cortical cells. It is usually held that this limits information flow between the thalamus and the cortex. During the deepest NREM sleep, thalamocortical neurons are so hyperpolarized that they are silent, so that cortical neurons fire with their intrinsic Δ rhythm.

NREM sleep phenomena can be generated by the isolated forebrain, where groups of sleep-active neurons ('sleep on' neurons) are maximally active at these times. Their axons inhibit aminergic, cholinergic and orexinergic neurons in the forebrain and brain stem that maintain waking, and lie in the preoptic and anterior hypothalamic regions, regions which are also involved in control of body and brain temperature. Many sleep-active neurons are thermosensitive, increasing activity at higher temperatures. Damage to these regions greatly reduces sleep, while heating preoptic regions increases it. Sleep is reduced at temperatures outside the thermoneutral zone and REM sleep lengths are maximal at the upper levels within this zone.

Since the mid-1990s a wealth of evidence has accumulated for at least one function of sleep: the consolidation of memories. Diverse approaches indicate that during sleep memories are replayed, modified, stabilized and even enhanced. The proposed mechanisms involved rely on two brain effects that are peculiar to sleep: alterations in the levels of neurotransmitters, and distinctive oscillations in electrical activity. It is during the occurrence of sleep spindles in NREM sleep that consolidation for memorizing word-pairs seems to occur, but apparently only if the levels of circulating acetylcholine and cortisol are low. However, different memory systems consolidate in different stages of sleep, and possibly in different components of those stages. Inducing slow oscillation-like potential fields by application of oscillating transcranial electrode potentials (0.75 Hz) during early nocturnal NREM sleep enhances retention of HIPPOCAMPUS-dependent declarative memory (e.g., memorizing of word-pairs) in healthy human subjects (see LONG-TERM POTENTIATION).

2. *REM (rapid eye movement) sleep* is 'dreaming sleep' (see DREAMING) in which the EEG closely resembles that of an active, waking brain (hence aka 'paradoxical' sleep). Typified by fast, low-voltage fluctuations, these sleep periods are the best predictors of vivid dreaming, and 90–95% of those awakened during or after it report dreaming. The brain's oxygen consumption is even higher than during waking, although there is an almost complete loss of muscle tone (atonia) and ventilatory activity is much reduced. The muscles controlling eye and ear ossicle movement are the only notably active muscles, eyes darting back and forth under closed lids. Compared with waking, PET studies reveal blockade of external sensory inputs in REM sleep, selective activation of the cortical region responsible for visuospatial integration, selective activation of the AMYGDALA and other parts of the LIMBIC SYSTEM, and selective deactivation of the dorsolateral prefrontal cortex, thought to be essential to working memory, directed attention and volition and normally activated in the waking state. Similarly, the brain's noradrenergic, serotonergic and histaminergic diffuse modulatory systems, normally firing at their highest rate in alert awake humans, are all deactivated. These neurons are termed 'wake-on/REM-off' cells. The dopaminergic neuronal system, by contrast, is active at waking levels throughout all states of sleep. REM sleep phenomena can be generated by the isolated brain stem, in particular by the pons and adjacent midbrain, where there is a subgroup of cholinergic neurons ('wake-on/REM-on' cells) that are maximally active during REM sleep, causing a complete loss of muscle tone in postural muscles by triggering both inhibition and withdrawal of excitation to motor neurons. These pedunculopontine and laterodorsal tegmental nuclei, cells which are active during wakefulness *and* during REM sleep, are responsible for the desynchronization of the EEG in these states. They are much less active during NREM sleep, when cortical activity is low. Damage to these neurons reduces or prevents REM sleep for long periods, while prolonged REM sleep can be achieved by microinjection of acetylcholine agonists into specific regions of the pons. During REM sleep, one of the commonest experiences is of imagined motion, and to ensure that we do not actually move there must be a blockade of motor commands preventing movement, although these are hallucinated as dream movement. The penis and clitoris become swollen with blood, irrespective of dream content. Sympathetic system activity is dominant during REM sleep and body temperature drifts lower as the body's TEMPERATURE REGULATION system is inexplicably shut down. In humans, the duration of REM sleep episodes increases progressively throughout the sleep period and is maximal at the expected time of awakening.

Most theories of sleep, in addition to its role in memory consolidation, associate NREM sleep with energy conservation and nervous system recuperation, while REM sleep might play a role in periodic brain activation during sleep, in localized recuperative processes and in emotional regulation. After a period of sleep deprivation, NREM sleep is typically repleted first. Extracellular ADENOSINE, acting on A_1 receptors, has been proposed as a homeostatic accumulator of the need to sleep, its levels rising in some brain regions, including the basal forebrain, during prolonged wakefulness (the intracellular ATP:AMP ratio then allegedly being low). This, or some other somnogen, may lead to the activation of the ventrolateral preoptic nuclei of the HYPOTHALAMUS that is required to trigger a sleep episode (see RETICULAR FORMATION).

REM sleep is apparently designed for rest, as muscle tension throughout the body is reduced and movement is minimal. Bodily temperature and energy consumption are lowered. Parasympathetic activity is increased, so heart rate, ventilation and kidney function are reduced, while digestive activity increases. Brain neurons fire at their lowest rate and EEG rhythms indicate they are oscillating in relatively high synchrony, and most sensory input is blocked from reaching the cortex. However, although the body is capable of movement, only rarely does the brain command it to do so – and then usually only to adjust the body's position. Other peripheral expression of the intense central motor activation during REM sleep is blocked by inhibition of motor neurons.

One theory of the biological significance of REM sleep in neonatal life is that the activity of the visual system that is known to occur during it compensates for any asymmetrical, abnormal or absent visual input (e.g., strabismus) that would otherwise 'prune away' unused connections in the LATERAL GENICULATE NUCLEUS during a CRITICAL PERIOD. As for theories of REM sleep function in adults, it is known that animals awakened from NREM sleep have poor sensory-motor function compared with those awakened from REM sleep. It is possible that REM sleep stimulates the adult brain to reverse the effects of NREM sleep on immediate post-waking behaviour: waking in an alert state would presumably carry a selective advantage. See SLEEP–WAKE CYCLE.

sleeping sickness (African trypanosomiasis) See TRYPANOSOMIASIS.

sleep–wake cycles Early studies on cats made clear that the brain stem regulates the sleep–wake cycle. Cutting the spinal cord at C1 does not prevent electroencephalograms characteristic of the normal cycle; but midbrain transection, isolating the forebrain, gives permanently synchronized slow-wave EEGs characteristic of sleep. Inability to wake follows severance of thalamic projections of the midbrain RETICULAR FORMATION (see Fig. 148) and while the midbrain certainly contains neurons required for the awake state, the caudal brain stem contains neurons *actively* involved in generating sleep (indicated by the EEG desynchronization caused by injecting barbiturates, normally an anaesthetic, into this region of sleeping cats).

The start of NREM sleep is under control of hypothalamic ventrolateral preoptic area (VLPO) and SUPRACHIASMATIC NUCLEI, and involves suppression of both branches of the ascending RETICULAR FORMATION (see Fig. 148). Activity of monoaminergic neurons of the ascending reticular system is required for the awake state, and VLPO cells suppress the cortical branch by inhibiting all monoaminergic and orexigenic neurons. GABAergic neurons of the VLPO inhibit histaminergic neurons in the tuberomammillary nucleus, the resulting loss of excitation on thalamic relay neurons causing them to hyperpolarize and go into burst firing, characteristic of NREM sleep. The lack of orexigenic firing shuts off the cholinergic basal forebrain cortical arousal system and reduces noradrenergic and serotonergic (wake-on/REM-off) cell activity and pontine (wake-on/REM-on) cholinergic cell activity.

The ultimate trigger(s) for NREM sleep are not well understood, but the SUPRACHIASMATIC NUCLEUS is certainly involved, elevated core temperature brought about by CIRCADIAN CLOCK activity being a likely factor. GABAergic neurons of the dorsomedial nuclei (DMH) of the hypothalamus project to the VLPO, promoting wakefulness by inhibiting sleep. Other glutamatergic and THYROTROPIN RELEASING HORMONE-containing DMH axons project to the lateral hypothalamus (LHA) and are thought to be excitatory and to promote wakefulness. The DMH has little input to the brain stem components of the ascending arousal system, although the orexigenic neurons have. Orexigenic neurons probably stabilize whichever state (awake or REM) the brain is in. In animals, cell-specific lesions in the VLPO reduce both NREM and REM sleep by >50% and cause *insomnia*. Optic tract fibres are also involved in hypothalamic synchronization of sleep–wake cycles to light–dark cycle. See AGEING, ALZHEIMER'S DISEASE, NARCOLEPSY, SLEEP.

slipped-mispairing mechanism See MUTATION, see Fig. 127.

Slit proteins A family of proteins involved in the regulation of axon branching (see AXONAL GUIDANCE).

slow synaptic transmission See SYNAPSE.

slow twitch fibres (slow phasic fibres, slow oxidative fibres) The smallest diameter SKELETAL MUSCLE fibres, appearing dark red because of their large quantity of myoglobin and rich blood capillary supply. Their numerous large mitochondria generate ATP aerobically, making them resistant to fatigue and capable of prolonged sustained contractions over many hours. The ATPase in the myosin heads hydrolyses ATP relatively slowly, giving the fibres a low contraction velocity (a twitch lasts 100–200 msec). They take longer than FAST TWITCH FIBRES to reach maximum tension despite being better at utilizing oxygen. See FAST OXIDATIVE-GLYCOLYTIC FIBRES.

Smad family proteins See SIGNAL TRANSDUCTION.

small intestine That part of the ALIMENTARY CANAL (see Fig. 4) comprising the DUODENUM (~25 cm), JEJUNUM (~2.5 m) and ILEUM (~3.5 m), ranging in all from 4.6 to 9 m in length. Its folded surface is covered with finger-like projections, the villi, between 0.5 and 1.5 mm in height, responsible for most of the digestion and absorption of food – the bulk occurring in the villi of the duodenum and jejunum (see ENTEROCYTES). Each villus is covered by columnar absorptive epithelium, their apical surfaces bound by tight junctions and covered by microvilli. Uptake of calcium ions occurs both transcellularly and paracellularly, the former predominating in a vitamin D-dependent manner when Ca^{2+} is scarce. Other epithelial or nerve cells of the villus are endocrine, paracrine or neurocrine in function, variously producing cholecystokinin (neurocrine), gastric-inhibitory peptide, glucagon-like peptides, neurotensin (neurocrine), secretin, serotonin and somatostatin. The intestinal epithelium is critically involved in maintaining intestinal immune homeostasis (see also COLON), acting as a physical barrier separating luminal bacteria and immune cells and expressing ANTIMICROBIAL PEPTIDES, while the basal lamina of the ileum contains PEYER'S PATCHES (see also GUT-ASSOCIATED LYMPHOID TISSUES). The epithelial tight junctions are dynamically regulated in response to cytokines, TNF-α, IL-17, IFN-γ, chemokines and the underlying innate and adaptive immune cell network. The epithelium is also in constant communication with the luminal GUT FLORA, and its cells express toll-like receptors, NOD 1 and 2, and receptors for different chemokines and antibody-specific F_C. Goblet cells and Paneth cells in the epithelium of the crypts are also antimicrobial in their secretions. There is considerable evidence for the role of T_{reg} cells in maintaining intestinal homeostasis (see IMMUNOSUPPRESSION).

Motility of the small intestine is controlled by the ENTERIC NERVOUS SYSTEM, segmentations (a localized mixing kind of contraction where the intestine is distended by large volumes of chyme) being of great importance in mixing the chyme with the digestive enzymes and facilitating absorption. Peristaltic waves rarely travel >10 cm (short-range contractions) after a meal, although during FASTING individual waves travel up to 70 cm before dying out, taking 1–2 hours to travel the length of the small intestine. The electrical activity underlying this wave, repeated every 70–90 minutes, is termed the *migrating motility complex*, probably serving to remove debris and bacteria from the lumen. After a meal, 'housekeeper' contractions, linked with those of the stomach and under the control of extrinsic nerves, respond to distension of the ileal wall. The gastro-ileal reflex involves increased motility of the terminal ileum when increased secretory and/or motility of the stomach occurs.

Villi (see Fig. 159) differ in appearance along the small intestine, being broad in the duodenum, leaf-like in the jejunum and more finger-like in the ileum, and they can change their length by smooth muscle contraction. Each contains a modified lymph vessel (lacteal) and blood vessels. Goblet cells in the villous epithelium secrete mucus (supplemented in the duodenum by an alkaline mucus Brunner's gland in the submucosa), while differentiated PANETH CELLS secrete lysozyme, the CRYPTS OF LIEBERKÜHN secreting an alkaline fluid. The villi show a piston-like contraction and relaxation, thought to improve removal of the digestion products of fats from the lacteals (see CHYLOMICRONS). The epithelium is self-renewing, a rapid cell turnover resulting in complete renewal every 6 days from a population of STEM CELLS (now thought to be base columnar cells, as in the COLON) and proliferative PROGENITOR CELLS within the crypts. As cells destined for the gut surface leave this population and reach the crypt–villus junction they enter cell cycle arrest and stop dividing, then mature and differentiate, migrating into the crypts and up the side of the villus (see Fig. 159). Loss of cells at the tips of the villi releases enzymes from the brush borders of the enterocytes into the intestinal lumen. Evidence indicates

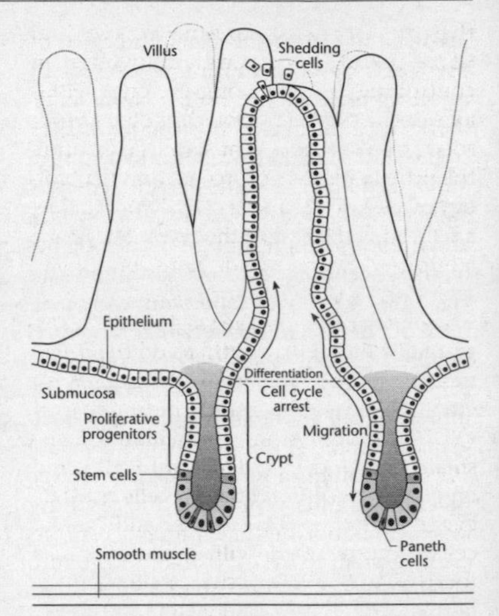

FIG 159. Diagram showing some of the tissue anatomy of the SMALL INTESTINE, the location of stem cells (stippled), and the directions of cell migration. From T. Reya and H. Clevers, © *Nature* 2005, with permission from Macmillan Publishers Ltd.

that the WNT SIGNALLING PATHWAY, BMPS and NOTCH SIGNALLING PATHWAYS are involved in controlling cell fate along the crypt–villus axis (see ADENOMATOUS POLYPOSIS COLI, JUVENILE POLYPOSIS SYNDROME). Wnt signals constitute the principal driving force behind the biology of the crypt. Paneth cells do not divide, and occupy the base of the crypt. They have a finite lifetime and are continually replaced from progenitor cells.

The diameter of the region decreases gradually along its length, as do the thickness of the intestinal wall, the number of circular folds (rugae) and the number of villi. The absorptive epithelium of the small intestine is ordered into villi and crypts, with differentiated cells (ENTEROCYTES, ENTEROENDOCRINE CELLS and goblet cells) occupying the villi while cells of a fourth type, PANETH CELLS, reside in the bottom of the crypts and secrete antimicrobial agents. The remainder of the crypts comprises the stem/progenitor compartment.

smallpox See POXVIRUSES.

SMC1 Structural maintenance of chromosomes 1-protein. See DNA REPAIR MECHANISMS.

smell For the sense of smell. See OLFACTION.

smoking See CIGARETTE SMOKING.

Smoothened (Smo) A transmembrane protein in the hedgehog signalling pathways. See HEDGEHOG GENE, SONIC HEDGEHOG GENE.

smooth muscle (visceral muscle) All smooth MUSCLE develops from LATERAL PLATE MESODERM, comprising numbers of elongated spindle-shaped cells, sometimes syncytial, lacking transverse striations or other obvious ultrastructure, the position of the single nucleus varying widely. Responsible, *inter alia*, for peristalsis and often arranged in muscle coats within connective tissue as bundles of cells averaging ~100 μm in diameter. It occurs principally in visceral, vascular and other locations where hollow organs occur (e.g., the uterus), and is almost always under autonomic control, but often has intrinsic pacemakers.

SNAREs (soluble NSF-attachment protein (SNAP) receptors) A set of conserved proteins mediating the fusion of a vesicle with its target membrane (NSF is *N*-ethylmaleimide-sensitive fusion protein). See VESICLES.

snRNAs (small nuclear RNAs) A major RNA class (see RNA) involved in assisting general GENE EXPRESSION, and encoded by ~100 genes which are dispersed but show some degree of clustering. These RNAs tend to be uridine-rich (hence the 'U' in their names; and the numbering indicates order of discovery, so that the U3 snRNA was the third to be classified). Many pseudogenes for snRNAs exist (there were 1,135 U6 snRNA-related sequences discovered in the draft sequence of the Human Genome Project).

snoRNAs (small nucleolar RNAs) A large class of RNAs (see RNA) encoded by >100 genes, and so named because they perform their functions in the nucleolar subcompartment of the nucleus. As with snRNAs, these functions include assisting gene expression or modification (including base modification) of precursor RIBOSOMAL RNA or of stable RNAs such as U6 SNRNA. Many are encoded by INTRONS of other genes, especially introns of genes encoding ribosomal proteins. They are synthesized by RNA POLYMERASE II and processed from excised intron sequences. The genes are dispersed, although some clustering does occur (see GENE CLUSTERS). The C/D box snoRNAs are mainly involved in guiding site-specific ribose methylations, while the H/ACA snoRNAs are mainly involved in guiding site-specific pseudouridylations. See NON-CODING RNAS, PRADER–WILLI SYNDROME, RNPS.

snoRNPs (small nucleolar RNPs) Complexes of snoRNAs (see entry) and ribosomal proteins. See RIBOSOMAL RNA.

SNPs See HAPLOTYPES, SINGLE NUCLEOTIDE POLYMORPHISMS.

snRNAs (small nuclear RNAs) The RNA molecules of spliceosomes. They recognize intron–exon boundaries in pre-mRNA and are involved in the major form of RNA splicing. See RNA PROCESSING.

snRNPs Small nuclear ribonucleoproteins, such as U6 snRNP, forming subunits of the spliceosome and involved in RNA splicing. See RNA PROCESSING.

sociobiology A field of inquiry which, during the 1970s, effectively replaced ethology as the dominant methodology in the scientific study of animal behaviour. Its central tenets belong to SELFISH GENE THEORY, the implication being that organisms are predicted to behave in ways maximizing the INCLUSIVE FITNESS of genes contributing to the genetic variance of behaviour. As with other observational behavioural sciences, it proves extremely challenging to control out alternative explanations of behaviour which may have little or nothing to do with the relatedness of participants. Attempts to interpret human behaviour sociobiologically have been decried by some as 'pop' sociobiology. See ALTRUISM, NATURE–NURTURE DEBATE.

sodium Required mineral ion for water potential regulation and nerve and muscle function. Deficiency causes nausea, vomiting, exhaustion and dizziness (see SWEATING). Its estimated REFERENCE DAILY INTAKE is 500 mg for people over 10 years of age, its maximum DAILY VALUE being 2,400 mg. See ALDOSTERONE, PROXIMAL CONVOLUTED TUBULE.

sodium pumps See the *ion pumps* section of ION CHANNELS.

solar plexus (coeliac plexus) A major autonomic plexus at the level of the last thoracic (T12) and first lumbar (L1) vertebrae, serving the abdomen and pelvis. The largest autonomic plexus, it surrounds the coeliac and anterior mesenteric arteries and contains two large coeliac ganglia (which receive input from the greater SPLANCHNIC NERVE) and a dense autonomic fibre network. Through this plexus, preganglionic vagal efferents supply terminal ganglia in the walls of the gall bladder, biliary ducts, pancreas and small intestine, ascending colon and transverse colon. Its sympathetic axons project to the diaphragm, stomach, spleen, liver, adrenal glands, kidneys, testes and ovaries.

solitary tract (tractus solitarius) See NUCLEUS TRACTUS SOLITARIUS.

soma All cells (somatic cells) other than those of the GERM LINE.

somatic cell hybridization A procedure involving fusion of somatic cells taken from human and from a non-human species (made easier by the presence of Sendai virus or use of polyethylene glycol). The chromosome constitution of the resulting hybrid cell shows a preferential loss of human chromosomes, depending on the interspecific hybrid used. A panel of somatic cell hybrids retaining different human chromosomes can be selected enabling assignment of any DNA sequence, or expressed enzyme or other protein, to a particular human chromosome. Cells from individuals with a chromosome rearrangement (e.g., deletion or translocation) can be used to map a gene to a particular chromosome more precisely. As a refinement, cells exposed to radiation that fragments their chromosomes (*radiation hybrids*) enables assignment of a gene or DNA sequence to a specific chromosome region.

somatic cell nuclear transfer (SCNT) Conventional SCNT involves replacement of the nuclear genome of an unfertilized egg (oocyte) with that of a somatic (i.e., non-germ) cell. Simulated fertilization, induced by electrical or chemical triggers, is followed by several rounds of cell division and implantation into a surrogate mother – with possible development to term. At the time of writing, 12 species have been cloned by this method, which in 2004 was used to generate human EMBRYONIC STEM CELLS (see entry) and has recently (2007) been used to generate from a 3.5-day-old mouse BLASTOCYST using embryonic stem cells (ESCs) or tail-tip cell nuclei (see CLONING, Fig. 55). Although mitotic, embryonic and somatic cells have previously been used in donor nuclear transfer work, this is the first time in which the donor nuclei have been introduced into mitotic cytoplasm. Arresting the mitotic process with different inhibitors enabled the zygote mitotic spindle and chromosomes to be optically removed and replaced with condensed donor chromosomes. The failure of SCNT in humans and monkeys has been attributed to possible fundamental differences in the method of mitotic spindle formation from those of non-primate unfertilized eggs. This work provides hope that it may now be possible to circumvent the need for freshly ovulated human oocytes in the production of human ESCs. Animal oocytes are a more readily available source than are human oocytes. Interspecies SCNT, in which a human cell is transferred into an animal egg and cultured for up to 14 days, is planned to advance understanding of genetic reprogramming and nuclear–mitochondrial interaction (see MTDNA). The success of induced pluripotent stem cell (iPS) lines is highly dependent on an understanding of embryo-derived stem cells, and study of reprogramming through SCNT may improve induced pluripotency and help produce therapeutically useful cells (see EMBRYONIC STEM CELLS).

somatic hypermutation Modification by mutation of the rearranged V regions of assembled immunoglobulin genes creating additional immunoglobulin diversity within an expanding clone of B cells that are responding to antigen in the germinal centre of a LYMPH NODE (see ANTIBODY DIVERSITY). This allows the affinity of the antibody response to increase (AFFINITY MATURATION) and those who cannot mutate their antibodies suffer recurring bacterial and viral infections, and do not respond to vaccination. Hypermutation only occurs in B cells that have been activated by antigen. All four nucleotides are mutated with similar frequency in the V genes, and transitions are favoured over transversions. Activation-induced cytosine deaminase (AID) produces uracils in the DNA, which is removed either by uracil glycosylase or another glycosylase, SMUGI, whereupon three alternative pathways are possible. Low-fidelity DNA POLYMERASES can bypass the gap during replication, filling it more or less randomly. During error-prone base-excision repair, the gap may be filled incorrectly by DNA polymerase ι, inserting thymine opposite guanine. And during short-patch repair synthesis, polymerase η could fill the gap and synthesize several bases by strand displacement opposite adenines and thymines. The mismatch repair proteins MSH2 and MSH6, and polymerase ζ, may also be involved. Somatic hypermutation only occurs in somatic cells, not in germ line cells, and does not occur in the generation of T-cell receptor variation. If a hypermutated self-reactive B cell in a germinal centre experiences strong cross-linking of its B-cell receptor to self antigens, it undergoes apoptosis rather than further proliferation. Compare SOMATIC RECOMBINATION.

somatic mesoderm A derivative of the LATERAL PLATE MESODERM.

somatic motor system The combined motor output from the brain innervating and commanding skeletal muscle fibres via CORTICOSPINAL TRACTS. Its upper motor neurons leave the brain and synapse with lower motor neurons that innervate the peripheral target organs (see SPINAL CORD, Fig. 162). With the autonomic nervous system (ANS), it constitutes the entire neural output of the CNS; but unlike the ANS, all cell bodies of the lower motor neurons lie within the CNS, either in the ventral horn of the spinal cord or the brain stem. See INTENTIONAL MOVEMENTS, REINFORCEMENT AND REWARD; compare SOMATIC SENSORY SYSTEM.

somatic nervous system The component of the PERIPHERAL NERVOUS SYSTEM comprising the 31 pairs of SPINAL NERVES and 12 pairs of CRANIAL NERVES. See SOMATIC MOTOR SYSTEM, SOMATIC SENSORY SYSTEM.

somatic peripheral nervous system See PERIPHERAL NERVOUS SYSTEM.

somatic recombination A rearrangement of genomic DNA segments occurring in the B-CELL lineage, but not in other cells, by which segments encoding V- and C-regions are brought into closer contact during generation of B-cell receptor and ANTIBODY DIVERSITY. It involves the looping-out and excision of intervening DNA sequences performed by an enzyme complex, the V(D)J recombinase. The recombination process is imprecise, with random addition of nucleotides at the joins between gene segments, so that it is a matter of chance whether the juxtaposed J sequence and the μ constant (C)-region sequence downstream result in production of a correct reading frame. A large number of B cells are lost because they produce non-productive rearrangements. Those B cells that are self-reactive and cross-link their receptors on binding multivalent self-antigen develop no further but continue to undergo light-chain rearrangements that can rescue the cell by deleting or displacing the rearranged light chain that was part of the self-reactive receptor. Two of the enzymes involved, recombination-activating genes, are lymphoid-specific and encoded by the *RAG-1* and *RAG-2* genes which may have originated as a transposase that was co-opted by vertebrates in the evolution of ADAPTIVE IMMUNITY. No genes homologous to these have been found in invertebrates. Similar events are involved in ISOTYPE SWITCHING. Compare SOMATIC HYPERMUTATION.

somatic sensory system The receptors and neural pathways involved in information relating to touch, proprioception, nociception and temperature senses. Only three synaptic relays are involved in conveying impulses to the CEREBRAL CORTEX: the dorsal root ganglia in the medulla; axons which project from there to the posterior nucleus of the thalamus; and finally the synapses thalamic axons make with the PRIMARY SOMATOSENSORY CORTEX (see Fig. 140) of the anterior parietal lobe of the cerebrum. There are two main routes conveying this information along the SPINAL CORD (see Fig. 163): the DORSAL COLUMN–MEDIAL LEMNISCAL SYSTEM composed of large myelinated fibres conducting precise tactile and proprioceptive details rapidly, with a high degree of spatial orientation of fibres with respect to their origins (see TOUCH); and the *anterolateral (spinothalamic) system* (see SPINOTHALAMIC TRACT), conducting more slowly, with less precision of sensory detail and with poorer spatial localization of signals (e.g., see PAIN). Two further tracts, the SPINOCEREBELLAR TRACTS, convey proprioceptive information to the CEREBELLUM. Although the anterolateral pathway is not as fine-tuned as is the dorsal column–medial lemniscal system, it relays some sensory modalities not dealt with by the latter: pain, temperature, tickle, itch, and sexual sensations, as well as rude touch and pressure. Initially, sensory information is processed serially in discrete pathways; but in each sensory system these pathways operate in parallel with each other. Such PARALLEL PROCESSING by different components of each sensory system, and by all, together, is crucial to the manner in which the brain forms our perceptions of the external world. A common feature in information processing in the somatic sensory system is CONTRAST ENHANCEMENT. Compare SOMATIC MOTOR SYSTEM.

somatomedins See INSULIN-LIKE GROWTH FACTORS.

somatosensory cortex See PRIMARY SOMATOSENSORY CORTEX.

somatosensory system See SOMATIC SENSORY SYSTEM.

somatostatin (growth hormone-inhibiting hormone) Produced as a 116-amino-acid precursor by the D cells of the epithelium of the small intestine, it is released either as a 14- or 28-residue peptide in response to the presence of fats, glucose and bile salts in the intestinal lumen. Acting via G protein-linked receptors (whose expression is modulated by the nutritional state of the individual), it inhibits gastric acid and pepsin secretion, gastrin release, pancreatic enzyme secretion, acetylcholine release from enteric neurons (downgrading intestinal responses to cholinergic stimulation), insulin and glucagon release. It is also a neurotransmitter synthesized in most brain regions, but released especially by parvicellular neuron terminals in the median eminence of the hypothalamus, inhibiting release of GROWTH HORMONE by somatotrophs. A small number of pancreatic δ-cells in the islets also release somatostatin.

somatotopic maps Diagrams of brain regions (e.g., the cerebral cortex in coronal section), over the surface of which are drawn the corresponding affecting or affected body parts in proportion to the density of receptors (if a somatosensory region) or to the accuracy with which movements are made (if a motor region). The neostriatum, for example, receives information from the whole cerebral cortex in such a somatotopic fashion. See PRIMARY MOTOR CORTEX, PRIMARY SOMATOSENSORY CORTEX.

somatotrophs Cells in the anterior PITUITARY GLAND releasing GROWTH HORMONE.

somatotropin See GROWTH HORMONE.

somatotypes See BODY MASS, SIZE AND SHAPE.

somites Appearing first in the occipital region by the 20th day of development, blocks of segmented mesoderm originating from unsegmented PARAXIAL MESODERM on both sides of the neural tube that will give rise to many of the connective tissues of the back (bone, muscle, cartilage, dermis). Somites must remain bilaterally symmetrical, a process driven by symmetrical waves of gene expression, only made possible because somites are actively masked from left–right cues present in their physical territory and required for patterning somites along the antero-posterior AXIS that would otherwise cause them to develop asymmetrically (see ASYMMETRY). It is dependent on a 'segmentation clock' established by *cyclic genes*, including members of the Notch and Wnt signalling pathways (determining left–right symmetry), expressed in an oscillating pattern in presomitic mesoderm. These signals periodically activate other segment-patterning genes, the antero-posterior boundaries for their expression being regulated by RETINOIC ACID (RA) from the PRIMITIVE NODE and *FGF8*. RA up-regulates somite-patterning genes, while *FGF8* represses RA activity and inhibits maturation of presomitic mesoderm into somites. RA is expressed in a rostrocaudal gradient, while *FGF8* is expressed in a caudorostral gradient. Approximately three pairs are added per day in a craniocaudal direction until, at the end of the 5th week, 42–44 pairs are present: 4 occipital, 8 cervical, 12 thoracic, 5 lumbar, 5 sacral and 8–10 coccygeal pairs. By the start of the 4th week, cells forming the ventral and medial walls of each somite lose their compactness and shift to the surrounding notochord. These cells, forming the *sclerotome*, form a loosely woven tissue, the *mesenchyme*, some cells of which form tendons while those remaining surround the spinal cord and form the vertebral column. Cells at the dorsolateral portion of somites migrate to give rise to precursors of limb and body wall musculature (see LIMBS). Subsequently, cells of the dorsomedial portion of the somite proliferate and migrate down the ventral surface of the somite's dorsal epithelium to form the medially positioned *myotome* layer: groups of cells giving rise to the vertebral muscles spanning the vertebrae and allowing the back to bend. What remains of the dorsal epithelium forms the *dermatome*, giving rise to mesenchymal connective tissue of the dorsal skin: the DERMIS. The dermatome and myotome together form the *dermatomyotome*.

Genes involved in somite differentiation include *Noggin*, *Sonic hedgehog*, *PAX1* and *PAX3*, *Wnt* (see WNT SIGNALLING PATHWAYS), the muscle-specific genes *MYF5* and *MYOD* (see MyoD entry), *BMP4* and probably genes for FIBROBLAST GROWTH FACTORS, and *NT3*, which encodes neurotrophin 3 secreted by the dorsal neural tube and inducing dermis formation.

The antero-posterior patterning of somites is specified by the combined action of Notch, Wnt, FGF8 and retinoic acid (RA) signalling pathways, although the Notch, FGF8 and RA pathways are also required for left–right determination. Antagonism between FGF8 and RA is essential for segmentation to proceed.

somitomeres Segmentally arranged concentric whorls of cells of PARAXIAL MESODERM, forming in association with the neural plate, appearing in the cephalic region by the 3rd week of development and progressing cephalocaudally. The first seven, cranial, somitomeres go on to form the striated muscles of the face, jaw and throat; but the remainder develop into discrete blocks of segmental mesoderm, the SOMITES.

***Sonic hedgehog* (*SHH*) gene** A RETINOIC ACID-inducible gene (RIG) and member of the HEDGEHOG family, whose product, the protein MORPHOGEN SHH, is secreted by the notochord and floor plate of the NEURAL TUBE (see entry), establishing the ventral midline of the CNS (see PATTERN FORMATION). Regulated movements of key proteins into and out of a cell's primary CILIUM provide a switch by which this fundamental and conserved signalling system is turned on and off. SHH binds to the transmembrane protein Patched (Ptc) at the cell surface, whereupon Ptc loses its inhibitory effect on another transmembrane protein, Smoothened (Smo). Smo then transduces a signal to the nucleus via glioma (Gli) transcription factors, activating selected gene expression. Several members of the signalling pathway localize to the tip or base of the primary cilium, including Gli and proteins regulating late Gli activity at the tip. In the absence of SHH, Ptc translocates to the primary cilium, blocking ciliary localization of Smo, while Gli is either degraded endosomally or processed to repressors (GliR). But on binding of SHH to Ptc on the cilium, Ptc leaves the cilium and Smo enters the cilium from internal vesicles, moving up from its base.

Secreted by the prechordal plate, SHH induces expression of *NKX2*.1, a homeodomain-containing gene regulating development of the HYPOTHALAMUS. With the *noggin* gene product, it causes the ventral part of each SOMITE to form sclerotome and to express *PAX1*, which then controls chondrogenesis and formation of vertebrae. It is now known to be involved in AXONAL GUIDANCE and to have a role in FOETAL ALCOHOL SYNDROME. It acts indirectly, since it is unable to diffuse more than a few cell widths from its source; it may function as a repressor for left-sided gene expression on the right of the embryo and, with *Lefty-1*, act as a barrier to prevent left-sided signals from crossing over the embryonic midline (see AXIS, LIMBS). See HOLOPROSENCEPHALY.

sonography See ULTRASOUND.

Sos (son-of-sevenless) A major guanine nucleotide exchange factor (GEF, see GUANOSINE TRIPHOSPHATE), linking a growth factor receptor to Ras (see RAS PROTEINS). It has Pleckstrin homology domains.

Southern blotting A method in which target DNA is digested with one or more restriction endonucleases, generating fragments that are size-fractionated by agarose gel electrophoresis (see Fig. 164), denatured and transferred to a nitrocellulose or nylon membrane for hybridization (see DNA PROBES). DNA fragments that are smaller move faster on the gel than those that are larger (all move towards the positive electrode, since DNA is negatively charged in the buffer conditions used). An important application in human genetics is to use a probe to identify related sequences in the same genome, or in other genomes (including those from other species). See DNA PROFILING, NORTHERN BLOTTING, VNTRS.

***SOX* genes ((SRY)-type HMG box genes)** A family of 18 genes, each containing the *SRY*-like HMG (high mobility group) box which encodes a domain of ~79 amino acids. The HMG domain activates transcription by bending DNA so that other regulatory factors can bind to the promoter regions of genes encoding important structural proteins. SOX genes are important transcription regulators, expressed in specific tissues during embryogenesis, *SOX2* being expressed in embryonic stem cells and forming a part of an autoregulatory pluripotency network. Loss-of-function mutations in *SOX9*, on the hsa17 chromosome, causes the rare condition campomelic DYSPLASIA involving bowing of the long bones, sex reversal in 46,XY males and poor long-term survival. Mouse models have shown that *SOX9* is expressed in embryonic skeletal primordia, regulating type II collagen expression, as well as in the genital

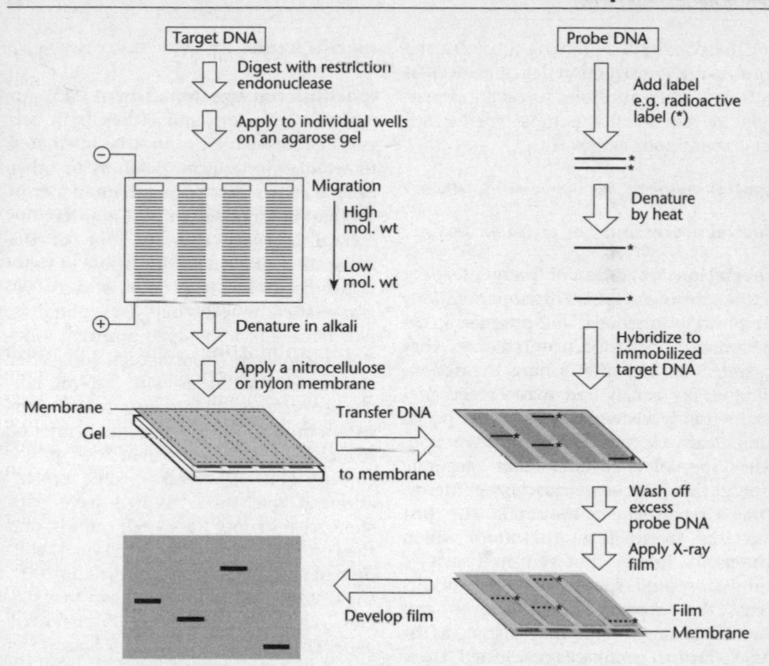

FIG 160. A pictorial outline of the method of SOUTHERN BLOTTING. Asterisks and bold lines indicate radioactively labelled material; but other methods of fragment identification may be used. From *Human Molecular Genetics 3* (3rd edition), by T. Strachan & A.P. Read. Reproduced by permission of Garland Science/Taylor & Francis LLC.

ridges and early gonads (see SEX DETERMINATION for *SOX9*). Mutations in *SOX10* on hsa22 cause a rare form of WAARDENBURG SYNDROME in which affected individuals have a high incidence of HIRSCHSPRUNG DISEASE.

spatial awareness The posterior PARIETAL CORTEX (PP) is an important component of the 'where' (dorsal stream) of visual processing (see EXTRASTRIATE CORTEX, and Fig. 72), as are the synapses between the entorhinal cortex and dentate gyrus of the HIPPOCAMPUS. PP cells have large receptive fields, show selectivity for size and orientation of objects, and fire when a macaque makes hand movements in grasping an object (see ATTENTION). The other major projection of the dorsal stream is to the medial superior temporal cortex. Lesions to either of these regions in primates, including humans, leads to optic ataxia, in which visuospatial competence, such as reaching for an object, is profoundly affected but leaves object recognition (the ventral processing stream) unaffected. In *hemispatial neglect,* which often accompanies blindness in parts of the visual field after a stroke, a person loses awareness of the half of the space around him lying opposite a brain lesion in the right posterior parietal cortex. Recent work with people who have hemispatial neglect without blindness indicates that damage to the superior temporal lobe is responsible for the condition and that the parietal lobe is less involved in spatial awareness than in matching sensory input to specific patterns of motor output. Damage to parts of the frontal lobe sometimes cause symptoms of spatial neglect. Studies of individuals with WILLIAMS–BEUREN SYNDROME (WS) implicate half-a-dozen genes within the chromosome 7 deletion responsible for the condition as having a role in the visuospatial construction deficit associated with WS, while autopsies reveal abnormalities in the superior parietal cortex. See BLINDSIGHT, NEGLECT SYNDROME.

spatial memory See HIPPOCAMPUS, MEMORY.

spatial processing See SPATIAL AWARENESS.

speciation Separation of lineages leading to the subsequent failure of the populations involved to interbreed and produce fertile offspring. Fossil evidence indicates that within 5–10 Myr after a mass extinction, biodiversity equals and may exceed pre-extinction levels while the rate of speciation greatly exceeds the rate of extinction. The speciation events that separate lineages may occur so closely together in time that genetic variation in the first ancestral species (e.g., that from which the gorilla lineage diverged) may survive until a second speciation event occurs (e.g., that separating chimpanzee and human lineages). This means that, as far as the human genome is concerned, there is not just one history to describe its relationship to those of the African apes: different segments of the genome have different histories. Human–chimpanzee genetic DIVERGENCE varies from <84% to >147% of the average, giving a range of more than 4 Myr for their speciation. A recent (2006) analysis indicates that human–chimpanzee speciation occurred <6.3 Mya and probably more recently, conflicting with some interpretations of ancient fossils. Some features of the X CHROMOSOME indicate a very recent divergence time, close to the genome minimum along most of its length, which would be explained if the human and chimpanzee lineages initially diverged, but later exchanged genes prior to a final permanent separation. See HABITAT HYPOTHESIS.

specific cognitive abilities See INTELLIGENCE.

specific dynamic action (s.d. effect, obligatory thermogenesis) See TEMPERATURE REGULATION.

specific immunity See ACQUIRED IMMUNITY.

specific language impairment (SLI) An umbrella term for familial deficits in language ability which cannot be attributed to retardation, autism, deafness or other general causes. More concordant in identical than in fraternal twins, it is a developmental language disorder, one of the hallmarks being a delay in the use of functional morphemes (e.g., *the*, *a*, *is*) and other grammatical morphology (e.g., plural *-s*, past tense *-ed*), individuals omitting function morphemes from speech long after age-matched children with typical language development demonstrate their consistent production. The disorder cannot be reduced solely to a problem with motor control. One of several possible explanations is that such children have poor short-term MEMORY for speech sounds, and they often perform worse than typical children in repeating nonsense words. In 2001, the disorder was narrowed down to a specific gene on chromosome 7, the FOXP2 gene. The gene encodes a transcription factor whose expression in early development seems to be linked to normal cerebral cortex development and is required in the homozygous condition. Mutation in one of the two alleles appears to result in difficulties in articulating speech sounds and in controlled movements of the tongue (via the hypoglossal nerve) and mouth. Disorders of spoken language are termed APHASIAS. See LANGUAGE, SPEECH.

speech Speech consists of sounds generated by coordinated interactions of the LARYNX (see entry) in association with the lungs, diaphragm, abdominal muscles, throat, neck muscles and mouth. The lips and tongue are very important in providing sensory information during speech-making, which is probably reflected in the large area of the somatosensory cortex devoted to their inputs (see CEREBRAL CORTEX, Fig. 45). By contrast, primate calls are controlled not by the cortex but by neural circuits in the forebrain and midbrain that are also responsible for emotional and physiological arousal.

The INSULA is involved in planning the articulatory movements involved in speech, which is often analysed into: (i) *phonation* – expired air generating vibrations of the free edges of the vocal folds of the glottis, resulting in alternate compression and rarefaction of air producing variations in pitch, loudness or quality of the voice; (ii) *resonance*, in which the chest, neck and head – and the thyroid cartilage – participate. Resonance gives the voice its characteristic acoustic pattern, and can amplify the sound; (iii) *articulation*, much of which is controlled by the lips, tongue, teeth and pharynx to form the various consonants and vowels. The larynx also plays a role, coordinating the starts and ends of phonation in accordance with the upper articulators mentioned, producing voiced and unvoiced sounds. See PHONEME.

Speech in infants correlates well with increase in brain volume (see BRAIN), developing from babbling at ~6–8 months, to the one-word stage at 10–12 months, to the two-word (phrase) stage at ~2 years. Sign systems and other LANGUAGE outputs are spontaneously acquired by both deaf and hearing infants in a similar developmental course, but speech perception and even grammatical knowledge develop much earlier, within the first months after birth (see SPECIFIC LANGUAGE IMPAIRMENT). Repeated efforts to train primates to mimic even simple speech sounds have had little success, although signing has been more successful. Speech perception is characterized by loss of ability with age, with 4- to 6-month-olds discriminating phonetic differences that distinguish syllables in both their native and unfamiliar languages, whereas 10- to 12-month-olds can only discriminate the phonetic variations used in their native language (see FACIAL RECOGNITION). Two brain regions, usually associated with the left hemisphere, are especially active during speech: BROCA'S AREA (the premotor area for speech) and WERNICKE'S AREA (a sensory area concerned with understanding language). See COCHLEA, HYOID BONE, SPHENOID.

sperm, spermatozoa When fully formed by SPERMATOGENESIS, sperm cells (which contribute the paternal genome to the zygote) enter the lumen of the seminiferous tubule whence they are pushed to the epididymis by smooth muscle fibres in the tubule walls. Each consists of a head (4–5 μm long × 2.5–3.5 μm wide), comprising the nucleus, and the ACROSOME (which covers two-thirds of the nucleus), and a tail. The tail may be subdivided into a neck, middle piece (5–7 μm long × 1 μm wide), main piece (~45 μm long) and end piece (~5 μm long). The main tail components are the axoneme complex (see CILIUM) and the surrounding plasma membrane; but in the middle piece there is a mitochondrial sheath with >50 helically arranged mitochondria. Final maturation of sperm takes place in the EPIDIDYMIS. After EJACULATION during coitus sperm may remain alive in the female reproductive tract for several days. Up to 10% of sperm have visible defects, especially with the head or tail. Some are giants or dwarfs; others are joined. Abnormal sperm are usually immotile and probably do not fertilize oocytes. Sperm CAPACITATION begins in the vagina, and movement of sperm from the cervix to the oviduct takes 2–7 hours and is mainly due to their own motility, possibly assisted by ciliary currents from the uterus; but they become less motile on reaching the isthmus. On ovulation, however, they resume motility and swim to the AMPULLA where FERTILIZATION (see entry) usually occurs. Soon after fertilization, many paternally derived DNA sequences become demethylated. This occurs after the removal of PROTAMINES (basic proteins associated with DNA in sperm) and the acquisition of histones by the paternal genome during the long G1 prior to DNA replication in the zygote. See INFERTILITY, SPERM COMPETITION.

sperm competition This is said to occur when sperm from more than one male have the opportunity to fertilize eggs of a single female. Larger testes size (leading to greater sperm production capacity), social dominance, age, body weight and behavioural differences are all features that have been shown to influence relative numbers of sperm contributed by different males in multimale mating systems. Paternity identification work has revealed previously unsuspected levels of extra-pair paternity across many species. Observations strongly suggest that some aspects of sperm quality *per se* determine fertilization success since paternity is often still skewed when the factors above are taken into account. See Y CHROMOSOME.

spermatic cord A supporting structure of the male reproductive system, ascending out of the scrotum and comprising the VAS DEFERENS, the testicular artery, autonomic nerves, veins draining the testes and carrying testosterone, lymphatic vessels and the cremaster muscle. With the ilioinguinal nerve, it passes through the inguinal canal.

spermatogenesis The process of male gamete production, taking place after puberty in the seminiferous tubules of the TESTES under the protection of SERTOLI CELLS. There are many more cell divisions required to make a sperm cell than are required to make an oocyte (see OOGENESIS) and, unlike that process, the number increases with paternal age. The number needed for spermatogenesis at puberty is 35, increasing by 23 per annum thereafter. Since base substitutions are associated with DNA REPLICATION, this may explain much of the excess of mutations in men compared to women, and for the paternal age effect. Spermatogonia, deriving from PRIMORDIAL GERM CELLS, give rise (one division every 16 days giving the annual increase in numbers of stem cell divisions quoted) to primary spermatocytes which undergo the first meiotic division, each of the two secondary spermatocytes giving rise to spermatids which develop into spermatozoa. The primary spermatocytes leave the basement membrane of the tubule and enter the adluminal tubular compartment, apparently by transient disruption of the tight junctions between the Sertoli cells. Spermatids, connected by cytoplasmic bridges, are originally spherical ('round spermatids'), but in close association with Sertoli cells reorganize their nucleus and cytoplasm and acquire a flagellum, comprising an axoneme originating from a basal body. The process from spermatocytes to motile sperm takes ~70 days in humans. Connected *residual bodies* of cytoplasm from the round spermatids are left behind prior to release of the sperm, which complete their maturation in the EPIDIDYMIS. During spermatogenesis, the X and Y chromosomes condense during first meiotic prophase to form a partially synapsed transcriptionally inactive XY body (the 'sex body') – a process termed meiotic sex-chromosome inactivation (MSCI). Likewise, autosomal chromatin is also silenced during pachytene (meiotic silencing of unsynapsed chromatin, MSUC). Both MSCI and MSUC are dependent on histone H2A.X phosphorylation, mediated by the ATR kinase, after which there is considerable replacement of nucleosomes within unsynapsed chromatin (see PROTAMINES). Nucleosomal eviction results in exclusive incorporation of the H3.3 histone variant, formerly associated exclusively with transcriptional activity (see HISTONES). This epigenetic reprogramming of sex chromatin is probably required for GENE SILENCING in the male germ line. Autosomal copies of X-linked genes (derived by retrotransposition) are the source of proteins that are functional at this time. Recent (2007) work on mice indicates the importance of the histone H3K9me2/1-specific demethylase JHDM2A (JmjC-domain-containing histone demethylase 2A; aka JMJD1A) in spermatogenesis. Mice deficient in the enzyme exhibit post-meiotic chromatin condensation defects, JHDM2A binding directly to and controlling expression of transition nuclear protein 1 (*Tnp1*) and protamine 1 (*Prm1*) genes, whose products are required for packaging and condensation of sperm chromatin. See SPERM.

S-phase The period in the CELL CYCLE during which DNA REPLICATION and CENTROSOME replication take place, lasting 6–8 hours. This is the phase during which most HISTONE synthesis occurs, the level of histone mRNA increasing ~50-fold. S-phase gene expression is initiated by the regulatory protein E2F, in response to which a protein kinase is activated that assembles the remaining components of the replication machinery onto the pre-replicative protein complex assembled on each origin recognition complex during G1.

sphenoid The central bone of the base of the CRANIUM, reduction in which during hominid evolution has been linked to the origin of SPEECH.

sphingomyelins See MYELIN SHEATH.

spike The depolarization and repolarization back to the resting potential which occurs during one ACTION POTENTIAL in an axon. Immediately after the spike, the membrane hyperpolarizes, an after-hyperpolarization which decays over a few milliseconds (see RELATIVE REFRACTORY PERIOD). For calcium spikes, see CALCIUM.

spina bifida One of the most serious vertebral defects resulting from imperfect fusion or non-union of the laminae of the vertebral arches (cleft vertebra). It may involve only the bony vertebral arches, leaving the spinal cord intact, when the bony defect is covered by skin (*spina bifida occulta*). More serious cases (*spina bifida cystica*) also involve failure of the NEURAL TUBE to close, leading to protrusion of the meninges and leaving neural tissue exposed. Amniocentesis can then detect elevated levels of α-foetoprotein in the amniotic fluid. It has an incidence of 1:1,000 births (sex ratio 2:3 male:female) and may often be prevented by a prospective mother taking FOLATE prior to, and then during, pregnancy. It exhibits multifactorial inheritance, unaffected parents having a 4–5% risk of having a 2nd affected child. As with ANENCEPHALY (see entry), it can be detected by prenatal ULTRASOUND. See VERTEBRAE.

spinal cord The portion of the closed tubular structure developing from the NEURAL TUBE and lying outside the cranium,

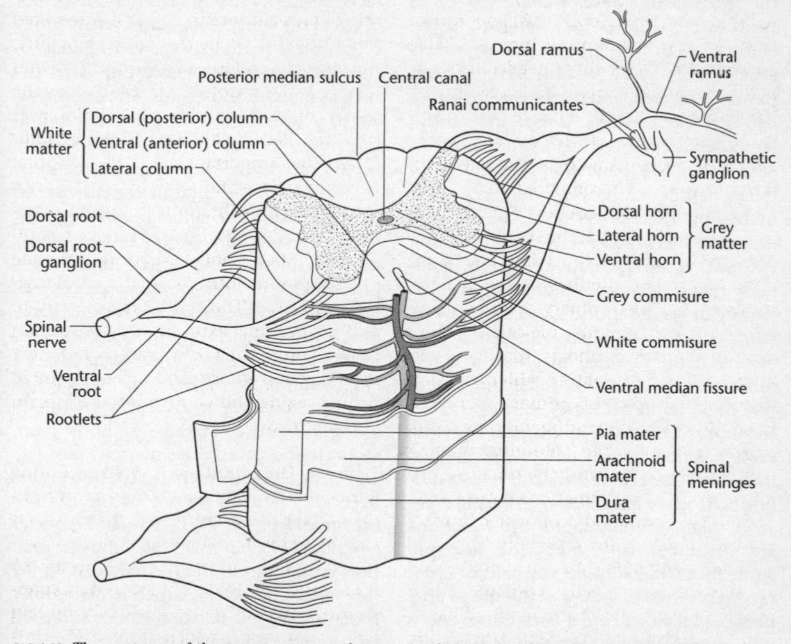

FIG 161. The anatomy of the SPINAL CORD, and one pair of SPINAL NERVES, in anterolateral view.

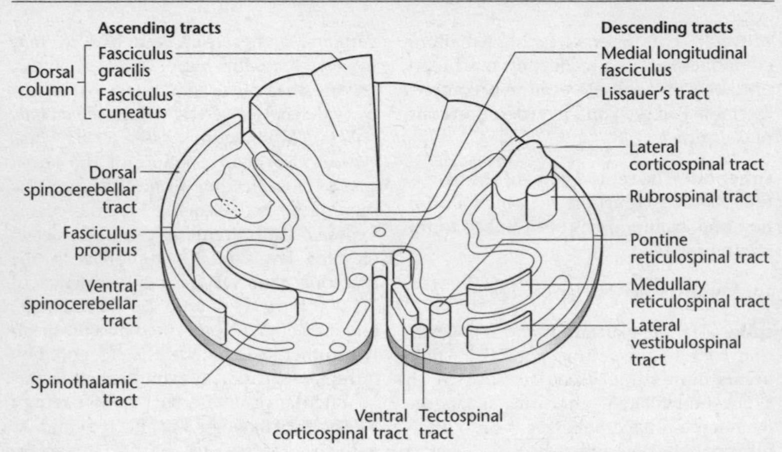

FIG 162. Direct motor pathways from the cerebral cortex to the spinal nerves via the corticospinal (pyramidal) tracts. From *Principles of Anatomy and Physiology*, G.J. Tortora & S.R. Grabowski, © 2000. Reproduced with permission of John Wiley & Sons, Inc.

i.e. caudal to the broader cephalic portion, the BRAIN, with which it is continuous. In the 3rd month of development, the spinal cord extends the entire length of the embryo, all of the 31 pairs of SPINAL NERVES except the 1st pair and those in the sacrum passing through intervertebral foramina between the vertebrae. Dorsal and ventral are dependent on the concentration gradients of the TGFβ (bone morphogenetic proteins) family of growth factors. With increasing age, the vertebral column and dura lengthen more rapidly than the neural tube so that the end of the spinal cord gradually moves relatively anteriorly, at birth ending at the level of the 3rd lumbar vertebra (L3). In the adult, the spinal cord begins at the foramen magnum and terminates at the level of L2 to L3 and is composed of an inner core of grey matter surrounded by a thick covering of white matter, often considered to be divided into dorsal, lateral and ventral *columns*, or funiculi. Nerve fibres (axons) sharing common origins, terminations and functions are organized into *tracts*, or fasciculi. Some fibres interconnect adjacent or distal cord segments. Each half of the spinal grey matter is divided into a *dorsal horn* and a *ventral horn*, between which lies the *intermediate zone* (lateral horn) of grey matter (see Fig. 161). In general, dorsal horn cells receive sensory input from the dorsal root fibres of the spinal nerves and ventral horn cells project axons into the ventral roots of the spinal nerves, innervating effectors such as muscles and glands. The intermediate zone cells are interneurons that modulate motor outputs in response to sensory inputs and descending brain commands.

The large dorsal column contains axons carrying ascending somatic sensory information ipsilaterally (see ASCENDING SENSORY PATHWAYS). Ascending spinal tracts carry impulses to the brain from proprioceptive, pain, temperature and tactile sensors, some of it impinging on consciousness and projecting to the cerebral cortex, some of it not (e.g., terminating in the cerebellum). Information reaching consciousness travels along a sequence of three neurons: first-, second- and third-order neurons (see Fig. 162). The level at which decussation occurs depends upon the source of the sensory information (see Fig. 163, and REFERRED PAIN, Fig. 147); but in all cases the information terminates in the contralateral cortex (see AXONAL GUIDANCE), which explains why touching the left side of the body is sensed by the right side of the brain.

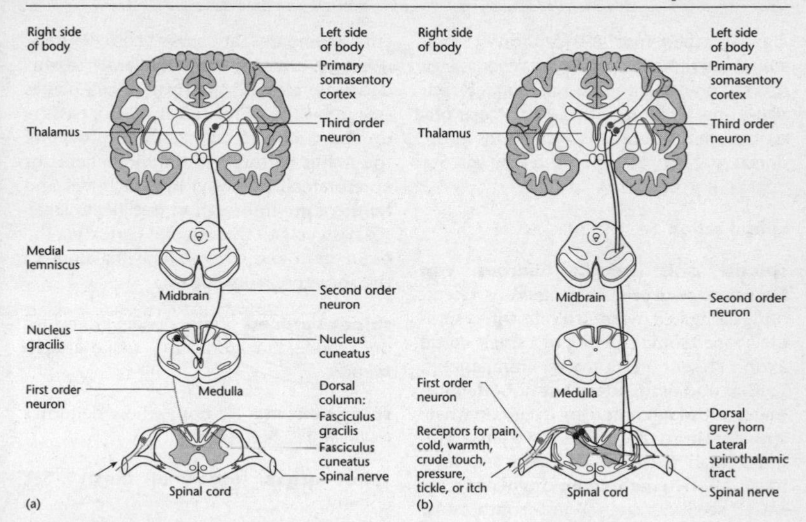

FIG 163. The two major pathways of the SOMATIC SENSORY SYSTEM which project to the cerebral cortex. (a) The dorsal-column medial-lemniscus pathway; (b) the spinothalamic (anterolateral) pathway (see SPINAL CORD). From *Principles of Anatomy and Physiology*, G.J. Tortora & S.R. Grabowski, © 2000. Reproduced with permission of John Wiley & Sons, Inc.

The lateral and ventral columns contain both ascending and descending tracts. The main lateral ascending tract is the dorsal SPINOCEREBELLAR TRACT. Lateral descending tracts include the lateral CORTICOSPINAL TRACT (axons of which innervate neurons of the intermediate zone and ventral horn, communicating signals controlling voluntary movement), the RUBROSPINAL TRACT, the medullary RETICULOSPINAL TRACT and the lateral VESTIBULOSPINAL TRACT (see POSTURAL REFLEX PATHWAYS). Most of the lateral tracts contain either sensory or motor axons, not both.

The ventral ascending tracts include the ventral SPINOCEREBELLAR TRACT and spinothalamic tract, while ventral descending tracts include the pontine reticulospinal tract, ventral corticospinal tract and TECTOSPINAL TRACT.

spinal nerves Thirty-one pairs of nerves arising from each region of the SPINAL CORD, supplying specific anatomical regions of the body. In the 4th week of development, motor axons grow out from the ventral regions of the developing spinal cord, collecting in bundles: ventral nerve roots. Under the influence of morphogens secreted in the dorsal and ventral regions of the NEURAL TUBE (see entry), dorsal nerve roots form as a collection of fibres whose cell bodies lie in the *dorsal root ganglia*, while central processes from these cell bodies form bundles that grow into the spinal cord opposite the dorsal horns. Distal processes from these dorsal root cell bodies grow out and fuse with the ventral nerve roots, and almost immediately the resulting spinal nerves divide into dorsal and ventral rami (see SPINAL CORD, see Fig. 162; and Appendix VI). Dorsal rami innervate most of the deep muscles of the dorsal trunk involved in movement of the vertebral column. In the thoracic region, ventral rami (as intercostal nerves) innervate intercostal muscles; ventral rami of the remaining spinal nerves form five major plexuses (cervical, brachial, lumbar, sacral and coccygeal). Additional rami (rami communicantes) from the thoracic and upper lumbar spinal cord regions carry axons forming part of the sympathetic division of the autonomic nervous system (see

Fig. 17). Apart from C1 (cervical 1), each spinal nerve has a specific cutaneous sensory distribution, the dorsal roots of each pair of nerves innervating a particular DERMATOME. Survival of sensory neurons in the dorsal root ganglia is promoted by NGF. See CEREBRAL CORTEX.

spinal reflex See REFLEX ARC.

spindle cells (spindle neurons, von Economo neurons) Large neurons, resembling elongated pyramidal cells with a spindle-shaped soma tapering to a single apical axon. Present in humans, chimpanzees, gorillas and orang-utans but not other primates, although occurring in certain whale species. Those of humans have larger volumes than those in chimpanzees, their localization in the anterior cingulate cortex (ACC; see PAPEZ CIRCUIT) and frontoinsular cortex having implicated them in communication, language comprehension and autonomic functions. Signals from the ACC are received in Brodman's area 10 in the frontal polar cortex, where disambiguation between alternatives is thought to occur.

spinocerebellar tracts (SCTs) The major pathways by which conscious and subconscious proprioceptive information (see PROPRIOCEPTION), and on-going activity in spinal cord interneurons, ascends to the cerebellum for the control of posture and coordination of movement (see SPINAL CORD, Fig. 163). They comprise: (i) the dorsal SCT, relaying ipsilateral information from muscle spindles, Golgi tendon organs and tactile receptors via the inferior cerebellar peduncle, and (ii) the ventral SCT conveying similar contralateral information from spinal interneurons involved in locomotion (see CENTRAL PATTERN GENERATORS) via the superior cerebellar peduncle. It arises from the ventral horn of the spinal cord.

spinothalamic tract The tract of the SOMATIC SENSORY SYSTEM containing second-order neurons whose cell bodies lie in the contralateral dorsal horn and receive inputs from small diameter primary afferents (compare the DORSAL COLUMN–MEDIAL LEMNISCAL SYSTEM). Spinothalamic axons decussate close to their parent cell bodies (see SPINAL CORD, Figs 162 and 163), passing up the contralateral tract, and providing the major pathway (the spinothalamic or anterolateral pathway) by which PAIN and temperature information (see THERMORECEPTORS) ascend to the cerebral cortex via the brain stem and contralateral thalamus to the somatosensory cortex.

spinous process Sharp, slender projection on a bone; for connective tissue attachment.

spiny neurons Neurons whose dendrites have spines. See DENDRITES.

spiral organ (organ of Corti) See COCHLEA.

spirometer An apparatus (see Fig. 164) for recording breathing movements, volumes (see Fig. 165), and oxygen uptake in which a lid, attached to a pen recording kymograph, is pivoted at one end and sealed from the atmosphere by water while the chamber under the lid is filled with air or, more usually, with medical grade oxygen. When a subject breathes into the apparatus, one of two hoses connected to the mouthpiece contains soda lime, absorbing carbon dioxide, and a two-way valve ensures that the same air is not immediately rebreathed but flows in one direction through the circuit. The lid is counterweighted so that it exerts no pressure and the subject can breathe normally. A noseclip is worn to prevent air entering or leaving the apparatus except via the subject, and although the amount of oxygen in the apparatus falls as breathing occurs, the percentage of oxygen remains the same. A tap at the end of the hose can either isolate the subject from the apparatus or connect him/her to it, and after the subject has been quietly breathing atmospheric air, the tap is opened at the end of an expiration so that from then the system is closed and the subject draws gas from the spirometer. If air is used instead of oxygen, the soda lime must be taken out of the circuit

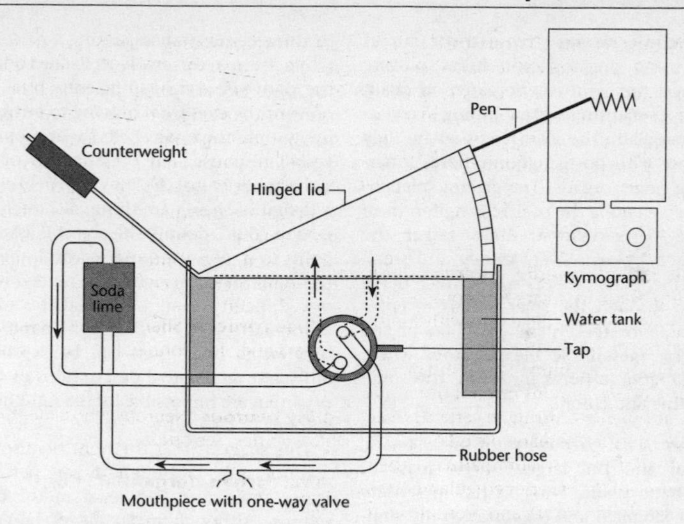

FIG 164. A simplified sectional view of a SPIROMETER. © 1990 Hodder Headline. Redrawn from *New Perspectives in Advanced Biology*, M. Hanson, with the permission of the publisher.

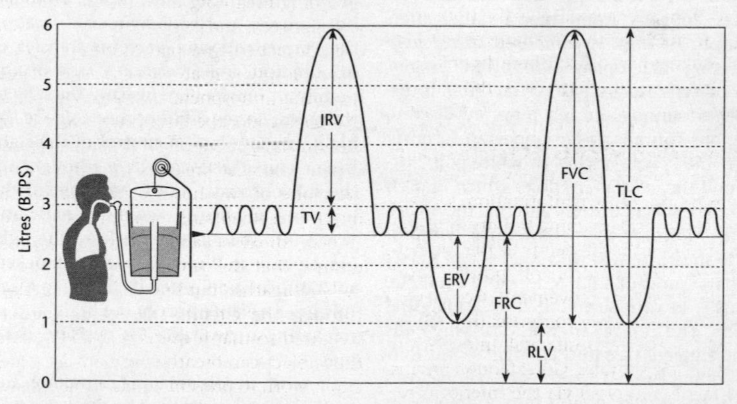

FIG 165. Pen recordings of static lung volumes, obtained by using a SPIROMETER. TV = tidal volume; IRV = inspiratory reserve volume; ERV = expiratory reserve volume; TLC = total lung capacity; RLV = residual volume; FVC = forced vital capacity; FRC = functional residual capacity (see entries). BTPS refers to the gas volume expressed at body temperature (37°C), ambient pressure and saturated with water vapour. © 2001 Redrawn from *Exercise Physiology* (5th edition) by W.D. McArdle, F.I. Katch and V.L. Katch, published by Lippincott, Williams & Wilkins. Reproduced with the kind permission of the publisher.

because as the percentage of oxygen falls, unless there is a rise in pCO_2 of air inhaled there will be insufficient pCO_2 in the plasma to stimulate inspiration until the pO_2 has fallen low enough in the carotid artery for the subject to have lost consciousness. See LUNGS, VENTILATION.

splanchnic mesoderm A derivative of the LATERAL PLATE MESODERM.

splanchnic nerves Sympathetic nerves containing preganglionic fibres passing through the sympathetic ganglion chain without synapsing and extending to collateral ganglia in the SOLAR PLEXUS where they synapse with postganglionic fibres innervating target organs. The greater splanchnic nerve enters the coeliac ganglion from where postganglionic fibres target the stomach, spleen, liver, kidney and small intestine. The lesser splanchnic nerve passes through the solar plexus to enter the superior mesenteric ganglion of the superior mesenteric plexus, from where postganglionic fibres innervate the small intestine and colon.

spleen A fist-sized organ of the LYMPHATIC SYSTEM, and the largest single mass of lymphatic tissue, lying in the mesentery of the stomach and organizationally similar to a LYMPH NODE. But, unlike lymph nodes, the spleen is perfused with (and collects antigens from) blood rather than lymph. Its bulk is composed of *red pulp* housing macrophages involved in red blood cell disposal (see KUPFFER CELLS). Interspersed among the red pulp are discrete regions containing lymphocytes, forming the *white pulp*. Patches of white pulp surround the central arterioles which branch off the trabecular artery entering the organ, T CELLS forming the periarteriolar lymphocyte sheath (PALS) with a corona of B CELLS around most of a PALS, forming a follicle. This is in turn surrounded by a marginal zone. The central arterioles carrying cells and antigens pass into a marginal sinus surrounding the PALS and B-cell corona, and this eventually drains into a trabecular vein.

spliceosome See RNA PROCESSING.

splicing For RNA splicing, see RNA PROCESSING.

'split-brain' phenomenon The left CEREBRAL HEMISPHERE (see entry) is usually dominant for language (see LANGUAGE AREAS), and if the CORPUS CALLOSUM is severed, information from the right and left visual fields project only to the opposite cerebral hemisphere. In these circumstances, asking a person to name the image of an object flashed only in the right visual field sufficiently briefly to prevent the eyes from moving to bring the image onto the fovea elicits the appropriate verbal response. But presentation of the same image to just the left visual field elicits a denial of seeing anything, although the person could identify the object if asked to point to it. Numbers and words flashed to the right visual field are also named without difficulty, and objects that can be manipulated only by the right hand (out of view to both eyes) can be described, although such verbal descriptions of sensory input are impossible for the right hemisphere.

This suggests that the right hemisphere 'cannot talk', although it can perceive, learn, remember and issue motor commands. Although incapable of language output, the right hemisphere can process rather simple linguistic input, although not complex words or sentences. Cutting the corpus callosum prevents transfer of information from WERNICKE'S AREA of the dominant hemisphere (usually the left) to the motor cortex of the opposite side of the brain, causing loss of control of the right motor cortex and voluntary (though not the subconscious) motor functions of the left hand and arm. Likewise, somatic and visual information transfer from the right hemisphere to Wernicke's area is blocked, such information generally failing to reach this interpretative area of the brain and so remaining unavailable for decision making.

In most people, reading, speaking and writing are all controlled by the left hemisphere. But in one case, words seen by the left hemisphere could be spoken but not written while words shown to the right hemisphere could be written but not spoken. This indicates that there is not necessarily a single brain system for all aspects of language in one hemisphere.

Results of split-brain studies indicate that the two hemispheres can function as independent brains, and have different LANGUAGE abilities. One patient, asked to arrange blocks with his right hand (left hemisphere) to match a pattern on a small card, struggled to do so until the left hand (right hemisphere), which knew how to do it, reached in and pushed the right hand out of the way. When intact, the corpus callosum seems to permit synergistic interactions between the two hemispheres for language and other functions. See ANTERIOR COMMISSURE.

spongy bone (cancellous bone) See BONE.

spontaneous abortion See MISCARRIAGE.

sporadic Of DISEASE occurrence that is nonfamilial.

***src* gene** The first viral ONCOGENE to be discovered (1975–76). The normal chicken cellular oncogene c-*src* has been captured and exploited by the retrovirus Rous sarcoma virus (RSV), becoming integrated into the viral genome as a viral oncogene: v-*src*. This gene has the ability to transform infected cells into CANCER CELLS. Its product, Src, is an src-kinase (see entry), and is constitutively active in many melanomas, where its oncogenic function phosphorylates Stat3 (see JAK/STAT PATHWAY). The tyrosine kinase activity of c-Src is autoinhibited by interaction between this domain and the Src-homology domains SH2 and SH3. Disruption of these interactions by proteins binding these domains, or by dephosphorylation of tyrosine 527 of the SH2 domain, activates the c-Src kinase. See MAPK PATHWAY.

Src oncoprotein See SRC GENE.

Src-kinases Pronounced 'sark'-kinases. The largest family of mammalian cytoplasmic tyrosine kinases, all of which phosphorylate cell-surface receptors (i.e., they are receptor-linked kinases) when these bind their ligand. As with many subsequently analysed protein kinases, the catalytic SH1 domain is itself composed of two domains (N- and C-lobe kinase domains), between which lies the ATP-binding site where catalysis takes place. The SH2 and SH3 domains of the protein are involved in substrate recognition and the regulation of catalytic activity. Different family members associate with different receptors and phosphorylate overlapping but distinct sets of target proteins (see INTEGRINS). They include the first cellular oncoprotein ever studied (Src itself) and all are located on the cytoplasmic side of the plasma membrane. Their covalently attached lipid chains (attached by myristilation and palmitoylation of the N-terminus) also stabilize their location on the cytosolic surface of the plasma membrane (see CHOLESTEROL, Fig. 48). Src and some other members can also dock onto and interact with other activated RECEPTOR TYROSINE KINASES, reinforcing each other's influence by strengthening and prolonging the signal.

SREBPs (sterol regulatory element-binding proteins) A family of transcription factors binding SRE elements within promoters, and directly activating the expression of >30 genes dedicated to the synthesis and uptake of CHOLESTEROL, FATTY ACIDS, triglycerides and phospholipids, as well as the NADPH required for their synthesis (see also ENERGY BALANCE). Three isoforms are encoded by mammalian genomes, alternative transcription of one gene giving rise to two of them, while a second gene encodes the third. Synthesized as inactive precursors bound to the endoplasmic reticulum, each SREBP must be released proteolytically in order to reach the nucleus. One of the three proteins involved in this is the escort protein and sterol sensor SCAP, the two others being proteases. SREBP-1a is a potent activator of all SREBP-responsive genes, including those mediating synthesis of cholesterol, fatty acids and triglycerides. SREBP-1c preferentially enhances transcription of genes required for fatty acid, but not cholesterol, synthesis. SREBP-2 preferentially activates cholesterol synthesis. In liver, PGC-1α docks and coactivates the SREBP and LXR families of transcription factors, coupling triglyceride and cholesterol synthesis, lipoprotein transport (as micelles in the bile) and VLDL secretion (see COREGULATORY PROTEINS, PGC-1 GENES). At least one SREBP increases the expression of low density lipoprotein receptor mRNA, promoting hepatocyte LDL receptor density. In mice, at least, SREBP also increases production of the secreted enzyme PCSK9, one of whose effects is to destroy LDL receptors. See STATINS.

***SRY* gene (sex-determining region of the Y chromosome)** The critical male-determining genetic locus, encoding a transcription factor that activates downstream genes required for testis development. It is Y-linked, and appears to act directly on the gonadal ridge and indirectly on the mesonephric ducts, inducing testes to secrete a chemotactic factor causing tubules from the mesonephric duct to penetrate the gonadal ridge and stimulate further testicular development. It also up-regulates steroidogenesis factor-1 (*SF1*), which acts through another transcription factor, *SOX9*, to induce differentiation of Sertoli and Leydig cells (see SOX GENES). Work on mice indicates that SRY binds to multiple elements within a *Sox9* gonad-specific enhancer, and that it does so in conjunction with SF1, an orphan nuclear receptor. Overexpression of such genes as *DAX* and *WNT4A* can feminize XY individuals even if they possess a functional *SRY* gene (see SEX REVERSAL).

SSCP (single-strand conformation polymorphism) A popular method of mutation detection that does not rely on heteroduplex formation (see DHPLC), in which altered mobilities of single-stranded mutant DNA are detected on non-denaturing gels. PCR products of double-stranded DNA, if heated to make them single-stranded, fold up to make a three-dimensional structure, and an alteration in the DNA sequence can result in a different conformation which, under the correct gel conditions, results in different electrophoretic mobility. Putative SNPs discovered must be checked by direct DNA SEQUENCING. See DNA MICROARRAYS.

SSRIs (selective serotonin reuptake inhibitors) Antidepressant and anxiolytic DRUGS, such as fluoxetine, Prozac, developed in the early 1970s, which block serotonin reuptake at synapses without also blocking noradrenaline reuptake. See ANTIDEPRESSANTS, Fig. 12; SEROTONIN.

standard neutral model A population genetics model that assumes that all individuals in a population are replaced by their offspring each generation, so that the population size remains constant, that mating occurs randomly (i.e., non-assortatively) and that each parent produces a Poisson-distributed number of offspring. Under such conditions, the model predicts the fate of mutations that are not affected by natural selection.

stapes See EAR OSSICLES.

Staphylococcus A genus of catalase-positive and facultatively aerobic Gram-positive cocci, commonly parasites of humans, whose members can divide in more than one plane. *S. epidermidis* is a non-pigmented and non-pathogenic form found on skin and mucous membranes. *S. aureus*, a yellow pigmented form, is a major cause of FOOD POISONING and is able to produce up to 15 enterotoxins (see TOXINS), one of which is coagulase, an enzyme-like factor that causes fibrin to coagulate and form a blood clot. The major toxins of *S. aureus* are the pyrogenic SUPERANTIGENS, also secreted by *S. pyogenes*, one of which (SPE A) binds tightly to endotoxin and is then toxic to immune cells. *S. aureus* causes impetigo and furunculosis in children suffering frequent mild skin trauma and with little immunity to the bacterium, and adults can also be affected by 'secondary impetiginization' of skin lesions, pus being highly infectious. Sweating or dampness of the skin macerates the epidermal surface, increasing susceptibility. Staphylococcal infections therefore spread very easily, and dead skin (dust) carries the bacterium. Most clinical isolates of *S. aureus* express 1 of 11 polysaccharide capsular types, over 75% of which belong to serotypes 5 or 11, immunization against which protects from lethal infection. One of its surface proteins, protein A, has anti-phagocytic properties related to its ability to bind the F_C domain (see entry) of immunoglobulins (e.g., see ANTIBODIES). The cell wall components peptidoglycan and lipoteichoic acid stimulate release of TNF-α, IL-1, IL-6 and IL-8, leading to SEPTIC SHOCK and coagulopathy (see INNATE IMMUNITY). See MRSAS.

stargazin One of several transmembrane AMPA receptor regulatory proteins, or TARPs. See AMPARS.

Starling's law of the heart See STROKE VOLUME.

starvation conditions When the fasted state is pushed into a state of starvation, or under conditions of increased energy substrate demands (e.g., by prolonged strenuous EXERCISE), an individual's net metabolism and ENERGY BALANCE become catabolic (see FASTING). This is a STRESS RESPONSE, in which up-regulation of HEAT-SHOCK PROTEINS occurs, autophagy (see AUTOPHAGOSOMES) is stimulated, and inactivation of mTOR (see entry) occurs, associated in many organisms with increased LIFE EXPECTANCY (see also FORKHEAD PROTEINS). Both starvation and exercise also up-regulate the number of INSULIN receptors on cell surfaces. Triglycerides stored in ADIPOSE TISSUE could provide sufficient fuel to maintain a basal rate of metabolism for about 3 months, and a very obese individual has sufficient energy to endure a fast for more than a year. During these conditions, the effects of the counterregulatory hormones cortisol, adrenaline, noradrenaline and glucagon are significant and those of insulin are all but abolished. Ketone bodies, generated by the liver, will be oxidized to release much of the energy required by such tissues as skeletal and cardiac muscle and the renal cortex. The brain, which preferentially uses glucose as fuel, can adapt to use ACETOACETATE under starvation conditions, but this could result in death. When fat reserves are gone, degradation of essential proteins begins, and UREA production increases. Eventually, the resulting loss of heart and liver function result in death.

statins Drugs inhibiting HMG-COA REDUCTASE, the rate-controlling enzyme in CHOLESTEROL synthesis, and thereby reducing liver cholesterol and increasing the number of hepatic LOW-DENSITY LIPOPROTEIN receptors through activation of the transcription factor SREBP which increases expression of LDL receptor mRNA, lowering plasma LDL concentration.

STATs (signal transducers and activators of transcription) Proteins which, when phosphorylated (by receptor-associated JAKs, see JAK/STAT PATHWAY), enter the nucleus and bring about selective gene expression.

steatosis (fatty change) In cellular pathology, the process describing the abnormal retention of lipids within a cell. It reflects an impairment of the normal processes of synthesis and breakdown of triglyceride fat. For hepatic steatosis, see OBESITY. See also VITAMIN A.

stellate cells See PRIMARY VISUAL CORTEX.

stem cells Cells with the unique ability for both SELF-RENEWAL and production of differentiated cell-types. Pluripotent stem cells can give rise to all cell types in a body, the most primitive being the EMBRYONIC STEM CELL (ESC), although others include germline stem cells and their derivatives, epiblast stem cells derived from post-implantation epiblast-stage embryos, and induced pluripotent stem (iPS) cells derived from somatic cells. Apart from the last-mentioned, stem cells always develop in association with cells of the STROMA, niches that are important in determining whether symmetrical or asymmetrical stem cell division occurs during self-renewal (see CANCER, CANCER CELLS). Stem cells from a single tissue, but at different developmental stages, can have distinct differences in self-renewal abilities. In most tissues, stem cells form PROGENITOR CELLS with a more limited potential for differentiation, and once these start to differentiate they usually lose their ability for self-renewal. However, developmental regulators can act directly on more mature stem cell populations, as when addition of the growth factor sonic hedgehog (SHH; see SONIC HEDGEHOG GENE) increases self-renewal of adult neural progenitor cells. But most progenitor cells and further-differentiated cells cannot self-renew, although some types of differentiated cell (e.g., lymphocytes) can. At the time of writing, the prevailing view is that the switch between self-renewal and differentiation is regulated by competition between transcription factors. It is possible that the signalling pathways controlling self-renewal are conserved in all an organism's tissues and that competition between transcription factors provides the functional basis of 'stemness'.

Although self-renewal is not yet fully understood at the molecular level, it was quite recently believed that stem cell identity could be summed up as a gene-expression signature differing from those of non-stem cell populations, and that it is also hierarchical. The idea that stem cell populations might have such gene expression signatures in common has lost ground as other stem cell populations indicated little overlap of the genes in 'stemness'. Stem cells rely on POLYCOMB GROUP (PcG) proteins to reversibly repress genes encoding transcription factors required for differentiation (see SENESCENCE, CHROMATIN REMODELLING, GENE SILENCING, CANCER CELLS) and Polycomb-deficient animals seem to have severe defects in stem cell self-renewal.

PcG repressive complexes mark the histones associated with genes required for maintenance of homeotic gene repression by inducing trimethylation of the lysine residue at position 27 of histone H3 (H3K27), involved in transcriptional silencing involved in somatic processes. The H3K27me3 mark is associated with the unique epigenetic state of stem cells. Stem cell division and stem cell expansion are not the same: the result of stem cell division may be two stem cells; or one stem cell and one differentiating cell; or two differentiating cells. Evidence is growing that the proliferative programme is distinct from the molecular programme of self-renewal. Unfettered potential for division is restricted to a small number of slowly growing stem cells which typically undergo occasional asymmetric divisions to generate one daughter to replace the original and one which founds a clonal population committed to a terminal differentiation programme. In the wake of pioneering work in the past five decades on HAEMATOPOIETIC STEM CELLS, recent work indicates that most other adult tissues also harbour stem cells. These *adult stem cells* are normally required for homeostatic self-renewal processes but can be recruited to repair tissues on injury. Recent evidence suggests that the expression pattern of the *Lgr5* gene (leucine-rich-repeat-containing G-protein-coupled receptor 5 gene (aka *Gpr49*)) marks stem cells in multiple adult tissues and cancers. Accumulation of DNA damage leading to adult stem cell exhaustion has been suggested as a principal contributor to AGEING.

In grafting, autologous (self) adult stem cells (ASCs) are not rejected by the patient whereas ESCs may bear foetal antigens and suffer direct immune attack. Nor do ASCs form teratomas when transplanted, unlike ESCs. But a major practical issue with ASCs is that they lack multipotency (ability to form many cell types) and therefore need to be sourced from the relevant target tissue. But not all adult tissues harbour stem cells (e.g., those of the pancreas); or the patient's own tissue may be diseased. Allogeneic transplants of adult stem cells would require strategies to overcome rejection. Transplantation of one type of ASC, neural stem cells, has received much attention on account of the poor regenerative powers of the CNS (see NEURAL DEGENERATION).

In the absence of the protein PTEN (see AKT SIGNALLING PATHWAY), not only is there a gradual loss of haematopoietic stem cells, but their differentiation characteristics allow LEUKAEMIAS to develop in which large numbers of cells behave as CANCER stem cells. Age-induced expression of p16INK4a in adult stem cells seems to be associated with impairment of tissue regeneration.

Stem cells tend to have little resistance to injury (e.g. DNA damage), undergoing apoptosis readily – indicative of the need to prevent progenitor cells with a faulty genome from surviving.

See COLON, Fig. 59; HAIR FOLLICLE, Fig. 86; MELANOCYTE.

stenosis (pl. stenoses) A narrowed opening, as in mitral stenosis (narrowing of the aperture of the mitral valve; see RHEUMATIC FEVER), or arterial stenosis (see ATHEROSCLEROSIS). See CONGENITAL HEART DEFECTS.

stereocilia (sing. stereocilium) Giant MICROVILLI, ~4 μm long. See CILIUM, HAIR CELLS.

stereognosis The ability to perceive shape by TOUCH. See also PRIMARY SOMATOSENSORY CORTEX.

stereopsis See BINOCULAR VISION.

sternum (breastbone) A flat, narrow bone forming a part of the axial skeleton, ~15 cm long and located on the anterior midline of the thoracic wall. Its three portions are: the *manubrium* (the superior portion, articulating with the costal cartilages of the 1st and 2nd ribs, forming the sternocostal joints); the *body* (the middle and largest portion, articulating with the costal cartilages; see RIBS); and the *xiphoid process* (the inferior and smallest portion, providing attachment for some abdominal muscles).

steroids Lipids produced from CHOLESTEROL (see Fig. 49) by reactions involving hydroxylation requiring the presence of NADPH and molecular oxygen. Steroid hormones contain 21 or fewer carbon atoms, the sex hormones OESTROGENS, PROGESTERONE and TESTOSTERONE being mitogenic agents, but also acting as tumour promoters (i.e., non-mutagenic carcinogens). The MINERALOCORTICOIDS and GLUCOCORTICOIDS are also steroid hormones.

stomach Appearing in the 4th week of development, and lying originally in the midline, that region of the foregut which rotates around an anteroposterior axis so that its caudal (pyloric) part moves right and upwards while its cephalic (cardiac) portion moves to the left and slightly downward, assuming a final position with its axis running from above left to below right. Connecting the oesophagus to the duodenum, it lies in the left superior part of the abdomen, attached superiorly to the liver by the lesser omentum and inferiorly to the small intestine by the greater omentum. The cardiac and pyloric sphincter muscles surround its rostral and caudal openings, respectively. Its wall is a modification of that representing the general ALIMENTARY CANAL (see Fig. 4), the mucosa and submucosa being thrown into rugae (wrinkles) when the stomach is empty, enabling the wall to be stretched when full (see ENTERIC NERVOUS SYSTEM, ENS). Gastric motility is under the control of the ENS, and of the hormones GASTRIN (increasing motility) and SECRETIN (decreasing it). The epithelium of the gastric mucosa bounding the stomach's lumen is composed almost entirely of mucus-secreting mucous cells (goblet cells). But this lumenal surface is characterized by millions of gastric pits (~100 per mm^2), deep mucosal invaginations into which open the gastric glands. A variety of secretory cells is located in these glands (see GASTRIC PITS). The stomach's main digestive role is the hydrolysis of dietary proteins, achieved by the combined effects of chief cells (through PEPSINS) and PARIETAL CELLS (producing gastric acid). Roughly 2–3 L of gastric juice is produced daily, ~700 mL being produced per meal. Protection from self-digestion derives from the tight junctions of the surface mucous cells of the epithelium lumenal to the gastric pits which produce an alkaline mucus on their surfaces which adheres to the mucosa, forming a 5–200 μm thick protective layer, and neutralizes the hydrochloric acid. E-type prostaglandins increase the thickness of this layer, stimulate HCO_3^- production and cause vasodilation of the mucosal microvasculature, favouring repair of damaged tissues (see HELICOBACTER PYLORI). An estimated 9% of all cancer deaths worldwide arise because chronic infections by this bacterium result in gastric carcinomas.

Hydrochloric acid in the stomach is a barrier to bacteria ingested by mouth, or distributed from the lungs by coughing and swallowing. Individuals with low rates of gastric acid secretion are most prone to tuberculosis, cholera and typhoid. Scotland has the highest rates of hypersecretion-linked ulcers and (once) of the possible selective agent, *Mycobacterium tuberculosis*.

Urban overcrowding during the industrial revolution may have raised the frequency of gastric ulcers as the cost for surviving tuberculosis in pre-reproductive years.

The regulation of stomach secretion is divisible into cephalic, gastric and gastro-intestinal phases (see Fig. 166).

stomodaeum (stomodeum) Intucking, or depression, of the ectoderm where it meets the endoderm of the embryonic gut. It develops into the ORAL CAVITY. See FACE, Fig. 74.

strabismus A condition in which the eyes are not perfectly aligned (being 'cross-eyed'), a common human disorder which, if uncorrected within the CRITICAL PERIOD, can result in permanent loss of stereoscopic vision through loss of binocular receptive fields. See OCULAR DOMINANCE.

strepsirhine See HAPLORHINE, PRIMATES.

***Streptococcus pneumoniae* ('pneumococcus')** A Gram-positive coccus belonging to the *S. oralis* group and the commonest bacterial cause of community-acquired pneumonia (although MYCOPLASMA PNEUMONIAE competes with it for this honour). More than 90 serotypes exist, pathogenicity being determined by the type of capsule polysaccharide, the presence of a capsule reducing the lethal infective dose for a mouse by a factor of eight. Pneumococcal pneumonia is an infection of the alveoli, typified by high fever and such non-specific symptoms as pleuritic chest pain, headache, vomiting and loose stools. The alternative COMPLEMENT pathway is important in limiting early infection. A vaccine to the capsule polysaccharides from each of the 23 commonest serotypes is 60–70% effective, although less so in the immunosuppressed and ineffective for children <2 years old. See MENINGITIS.

streptokinase A fibrinolytic agent used in the dispersal of venous thromboses. See ASPIRIN, THROMBOSIS.

stress 1. Mental or physical 'tension'. Stress tends to be felt, as when we are exposed to 'stressful situations' such as physical danger, emotional conflict in our private lives, or an impending exam. It can be real or imagined, chronic, acute or of some intermediate duration, but whatever its source (the *stressor*) the response is the same: raised heart rate, altered blood flow to the organs and dilation of the pupils (see STRESS RESPONSE). Cardiac hypertrophy, in which CARDIAC MUSCLE fibres increase in size, is an essential adaptive physiological response to stress (see HYPERTENSION, MICRO-RNAS). Those under stress are twice as likely

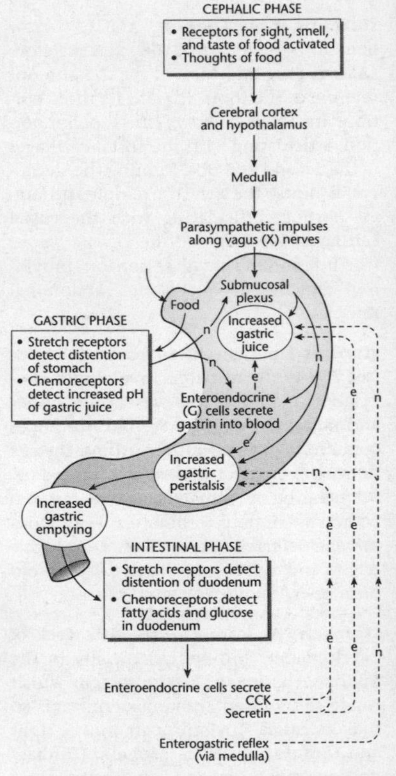

FIG 166. The cephalic, gastric and intestinal phases of digestion in the STOMACH (see entry). Pathways are either endocrinal (e) or neural (n). Stimulatory pathways are shown by unbroken arrows, inhibitory pathways by broken arrows. From *Principles of Anatomy and Physiology*, G.J. Tortora & S.R. Grabowski, © 2000. Reproduced with permission of John Wiley & Sons, Inc.

as those who are not to suffer from the common cold and influenza. If chronic, stress can itself prevent recovery and can increase the risk of disease, commonly cardiovascular disease, ulcers (e.g., PEPTIC ULCERS), and suppression of the immune system. It can often be alleviated by EXERCISE, relaxation training, massage, acupuncture, BIOFEEDBACK, realistic expectations and improved management of time. But severe chronic stress can cause great physical harm, interfering with performance at home, work or school. In response to severe stress or circulatory shock, increased sympathetic stimulation of the kidneys can cause renal blood flow to fall to such low levels that it is inadequate to maintain normal functioning if blood flow is not re-established: one reason why shock should be quickly treated. Healthy individuals often channel lightly stressful situations into productive outcomes: they handle stress well. But those with low self-esteem may view themselves as victims, or as expendable within a corporate setting, when they encounter stressful situations. It is therefore important, if at all possible, to select an environment and create a lifestyle that are as stress-free as achievable. See AMYGDALA, FEAR AND ANXIETY.

2. Cellular stress includes hydrogen peroxide entry, internal oxidative stress (see MITOCHONDRION), caloric restriction (see ENERGY BALANCE), ENDOPLASMIC RETICULUM stress and GENOTOXIC STRESS. ATR has a critical role in virtually all cellular stress responses that share inhibition of replication-fork progression as a common mechanism, and the protein p66^{Shc} may protect an organism against such stress.

stress hormones The hormones ADRENALINE and CORTISOL. See STRESS RESPONSE.

stress response The coordinated response to STRESSORS, characterized at the organismal level by avoidance behaviour, heightened ATTENTION and AROUSAL, physiologically by activation of the sympathetic division of the ANS and release of CORTISOL from the adrenals, and at the cellular level by up-regulation of HEAT-SHOCK PROTEINS. In its acute setting, referred to as the 'fight or flight' response, it is activated through the sympathetic nervous system and the adrenal medulla, with characteristically immediate effects. These include direct nervous stimulation of heart and ventilation rate, constriction of arterioles in the skin and gut and their dilation in the heart and limb muscles – all of which serve to route oxygenated blood faster to the muscles involved in exercise and away from organs performing less urgent functions. The spleen muscles also contract, squeezing their red blood cell reserve into the circulation to counter blood loss (haemorrhage) through wounding. Almost as quickly, nervous stimulation of the adrenal medulla releases adrenaline and noradrenaline into the blood, supplementing and prolonging many of the above effects, in addition to activating liver enzymes causing release of glucose from stored glycogen.

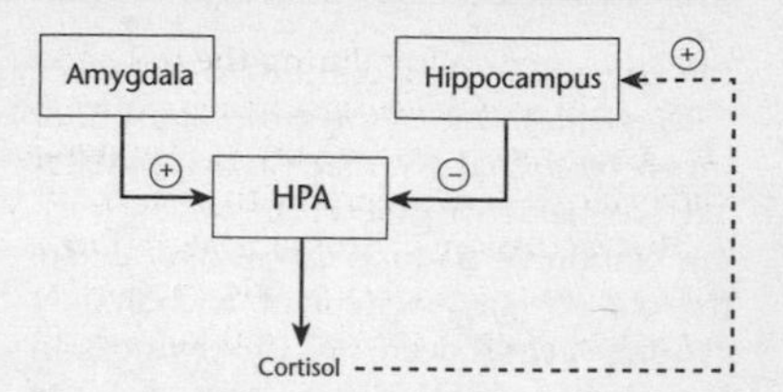

FIG 167. Simple circuit diagram of the control of the hypothalamic–pituitary–adrenal gland axis during the STRESS RESPONSE.

Few other stressors to which the body is exposed approach those of STARVATION CONDITIONS and heavy EXERCISE. Stress responses involve activation of the hypothalamic-pituitary-thyroid (HPT) and hypothalamic-pituitary-adrenal (HPA) axes (see Fig. 167). Exposure to cold, especially chronic exposure to severe cold, can increase output of THYROID HORMONES by 100% of normal values and increase basal metabolic rate by as much as 50%. Activation of the HPA axis by the amygdala increases output of GLUCOCORTICOIDS leading to mobilization of amino acids by protein breakdown, an increase in blood glucose concentration by GLUCONEOGENESIS and promotion of lipolysis. Cortisol activates a hippocampal circuit which suppresses hypothalamic release of CRH by negative feedback (but in ANOREXIA NERVOSA, the stress of feeding is thought to induce CRH release). One of the most robust findings of biological psychiatry is that the HPA axis is hyperactive in severe DEPRESSION. Various emotional traumas can also modulate thyroid output, excitement and anxiety causing acute decrease in thyrotropin secretion. Cytokines can activate hypothalamic-pituitary release of glucocorticoids and, in turn, glucocorticoids suppress further cytokine synthesis. The interaction between the nervous and immune system signalling has been implicated in the generation of intractable 'pathological' PAIN. See JNK SIGNALLING PATHWAY, STRESS.

In modern life, the normal physiological stress response can become chronic. The pathway is initiated by activation of the central nucleus of the amygdala whose projections to the stria terminalis enter the hypothalamus and activate the HPA axis, which is negatively regulated by the HIPPOCAMPUS, whose activation suppresses CORTICOTROPIN-RELEASING HORMONE (CRH) and consequently the release of cortisol. Chronic activation of this pathway, as occurs during chronic stress, can cause hippocampal neurons to wither and die in experimental animals, setting off a vicious circle in which the stress response is more pronounced still. Brain imaging studies have shown a decrease in human brain volume in some people with post-traumatic stress disorder (see FEAR AND ANXIETY).

stressors These include exercise, fasting and starvation, fright, temperature extremes, high altitude, bleeding, infection, surgery, trauma and disease. Many of these situations initiate up-regulation of HEAT-SHOCK PROTEINS. See CREB, FEAR AND ANXIETY.

stretch receptors Nerve endings responding to proprioceptive stretch information, as in the stomach and intestinal walls, TENDON ORGANS and MUSCLE SPINDLES. Stretch receptors more specifically responding to stretching of an organ by fluid pressure are termed BARORECEPTORS.

stretch reflex (myotatic reflex) See ELBOW, MUSCLE SPINDLES.

stria terminalis The principal efferent projection of the AMYGDALA, running in the wall of the lateral ventricle and terminating in the HYPOTHALAMUS. Its lateral basal nucleus is implicated in cortisol-mediated FEAR responses.

striate cortex (V1) Alternative term for the PRIMARY VISUAL CORTEX.

striated muscle See CARDIAC MUSCLE, SKELETAL MUSCLE.

striatum (adj. striatal) Term which refers collectively to (i) the NEOSTRIATUM, or dorsal striatum (caudate nucleus and putamen, collectively termed the lentiform nucleus) and the palaeostriatum (globus pallidus); and (ii) the ventral striatum, or NUCLEUS ACCUMBENS. See BASAL GANGLIA.

strokes Brain damage caused by haemorrhage or occlusion of blood vessels, often the result of thrombus formation (see BLOOD CLOTTING) and/or HYPERTENSION. Strokes damaging the motor cortex or corticospinal tract are common, their immediate consequence being paralysis of the contralateral side, although considerable recovery of voluntary movement may occur in time. Neuronal death in strokes is triggered by a shortage of oxygen supply to the brain, and APOPTOSIS is implicated, as are dysfunctional NMDA RECEPTORS. But the synthetic CANNABINOID HU-211 is neuroprotective during such brain trauma. See APHASIAS.

stroke volume The volume of blood leaving each ventricle during one heart cycle, and a major factor in CARDIAC OUTPUT. It is equal to end-diastolic volume (normally ~125 cm^3) minus end-systolic volume (~55 cm^3), although three factors affect its value: (i) a greater stretch (*preload*) on the heart walls prior to contraction, as during EXERCISE, provided by an increased volume of blood filling the ventricles at the end of diastole (END-DIASTOLIC VOLUME) and resulting in greater force of ventricular contraction: the Frank–Starling mechanism (see CARDIAC

MUSCLE), a property of all striated muscle; (ii) greater contractility (forcefulness of muscle fibre contraction) arising from sympathetic stimulation by NORADRENALINE, and by ADRENALINE, and from decreased parasympathetic release of ACETYLCHOLINE; (iii) the value of the pressure in the aorta (afterload) which needs to be overcome before the left ventricle can force blood through the aortic semilunar valve varies, increase in which (e.g., by hypertension and atherosclerosis) therefore decreases stroke volume. The forcefulness of contraction is also decreased by NITRIC OXIDE release, the resulting cGMP achieving this by stimulating the ion pump(s) that expel calcium (Ca^{2+}) from the cytosol of cardiac muscle fibres.

Starling's law of the heart (often termed the Frank–Starling relationship) states that 'the energy of contraction of the ventricle is a function of the initial length of the muscle fibres comprising its walls'. In accordance, the ventricle will eject the volume of blood that entered it during diastole, so that the output from the two ventricles is closely matched and the heart automatically adjusts its cardiac output to match its venous return. End-diastolic volume, and therefore the degree of stretch of the ventricles, is an important factor in determining cardiac output and heart work rate. In HYPERTENSION, increase in the left ventricular stroke work results in raised left ventricular END-DIASTOLIC PRESSURE. Cardiac muscle hypertrophies in compensation for the increased work load so that the ventricle is able to achieve greater stroke volume at the normal end-diastolic pressure, minimizing the increase in ventricular end-diastolic pressure which would otherwise arise from the operation of the Starling mechanism. See CARDIAC CYCLE, CARDIOVASCULAR CENTRES, EXERCISE, STRESS.

stroma (adj. stromal) The mesenchymal components (connective tissue) underlying epithelial and haematopoietic tissues and tumours (see HEDGEHOG GENE). It includes fibroblasts, adipocytes, endothelial cells, and different immunocytes and the associated extracellular matrix. Early B-CELL and T-CELL development is dependent on adhesive contacts made with non-lymphoid stromal cells of bone marrow (see STEM CELLS), which also provide soluble factors controlling B-cell and T-cell differentiation and proliferation. Other such factors are membrane bound, and include stem cell factor (SCF). Stromal cells (a network of epithelia) in the thymus provide a further unique environment for the development of T cells. See GASTROINTESTINAL STROMAL TUMOURS.

structural genomics Research aimed at providing a complete picture of the number and variety of existing protein structures encoded by human and other genomes.

STS mapping (sequence-tagged site mapping) See GENOME MAPPING.

subarachnoid space The space between the arachnoid and the pia mater and in communication with the ventricles, containing cerebrospinal fluid. This fluid is absorbed by blood vessels at arachnoid villi in the subarachnoid space. See MENINGES.

subclavian arteries A pair of arteries, supplying blood to the left and right forearms. That on the right side develops from the 4th AORTIC ARCH (see Fig. 13), the right dorsal aorta and the 7th intersegmental artery, and in the adult is a continuation of the brachiocephalic artery. On the left side, it arises from an originally quite distant branch of the AORTA, but then is brought much closer to the left common carotid artery.

subclavian veins A pair of veins, draining the left and right forearm and formed by fusion of the brachial, axillary and cephalic veins. Each joins its appropriate internal jugular vein to form the left and right brachiocephalic veins which fuse to form the superior vena cava. Although both subclavian veins receive lymph vessels through the jugular and subclavian trunks, the left subclavian vein receives the bulk of lymph drainage from regions below the heart via the thoracic duct. See LYMPHATIC SYSTEM.

subcortical Referring to BRAIN structures 'below' the CEREBRAL CORTEX in terms of anatomy. They include: forebrain structures (BASAL GANGLIA, LIMBIC SYSTEM, the THALAMUS, including the LATERAL GENICULATE NUCLEI, and HYPOTHALAMUS); midbrain structures (INFERIOR COLLICULUS, SUPERIOR COLLICULUS and TEGMENTUM); and hindbrain structures (CEREBELLUM, RETICULAR FORMATION, PONS and MEDULLA OBLONGATA).

subcutis (hypodermis, superficial fascia) The region underlying the DERMIS of the skin. It is not a part of the skin. Roughly half the body's fat occurs in the subcutaneous ADIPOSE TISSUE.

subgenual Lying immediately inferior to the LATERAL GENICULATE NUCLEI.

substance P A parasympathetic neuropeptide transmitter stored within granules in axon terminals of PAIN afferents (see SUBSTANTIA GELATINOSA) and released in response to high frequency action potentials. It causes smooth muscle contraction in the gastrointestinal tract and enhances sensitivity of pain endings in the dorsal horn of the spinal cord by prolonging the effects of glutamate (see NOCICEPTORS). Enkephalins inhibit its release. See ACUPUNCTURE, CAPSAICIN.

substantia A group of related neurons (e.g., the SUBSTANTIA NIGRA) lying deep within the brain, generally with less distinct borders than those of nuclei (see NUCLEUS).

substantia gelatinosa The tip of the dorsal horn of the SPINAL CORD's grey matter, and a region receiving collaterals of the smallest diameter myelinated and unmyelinated afferent axons. These neurons are glutamatergic and excitatory, also producing SUBSTANCE P as a neurotransmitter and ENKEPHALINS.

substantia nigra (SN) One of the component structures of the BASAL GANGLIA (see also BRAIN, Fig. 28). Roughly 80% of dopaminergic neurons are found in the SN pars compacta (SNpc), whose axons project in the nigrostriatal tract through the ventrolateral hypothalamus to the dorsal striatum (the *nigrostriatal pathway*) and are involved in basal ganglia regulation of INTENTIONAL MOVEMENTS. Lesions in these result in PARKINSON DISEASE. The pars reticulata (SNpr), anatomically and functionally similar to the globus pallidus pars interna, receives projections from the dorsal striatum and sends GABAergic inhibitory axons to the thalamus. Neurons of the SNpc are spontaneously active, even without synaptic input, firing action potentials at ~1–2 Hz. Unusually, the PACEMAKER current is via CALCIUM channels.

substitutions Point MUTATIONS in which, instead of insertion or deletion of a base pair (which would shift the reading frame during translation), the error in DNA REPLICATION involves either one pyrimidine nucleotide or one purine nucleotide replacing an existing one (transition) or one purine or pyrimidine nucleotide replacing one pyrimidine or one purine nucleotide (transversion). Transitions are the commoner of the two, C to T transitions occurring with relatively high frequency (see CPG DINUCLEOTIDE). If substitution occurs in a protein-coding sequence, because the genetic code is degenerate, it may result in no alteration in amino acid sequence (synonymous mutation), a change in amino acid sequence (non-synonymous mutation), or give rise to a stop codon, prematurely terminating amino acid chain elongation. The effect of a substitution in non-coding DNA sequences depends upon whether or not it alters the recognition of a REGULATORY ELEMENT by its ligand.

substrate-level phosphorylation The phosphorylation of a compound by means of an inorganic, as opposed to an organic, phosphate source.

subtelomeres Polymorphic patchworks of interchromosomal SEGMENTAL DUPLICATIONS at chromosome ends, arising recently in humans through repeated translocations between chromosome ends. Pieces of subtelomeric sequence have changed location and copy number with unprecedented frequency since the divergence of human and great African ape lineages, a phenomenon which may have both advantageous and pathological consequences. Much of the asymmetrical increase of duplicated DNA in the chimpanzee lineage compared with the human since their separation is to be found in the chimpanzee subtelomeric regions of chromosomes 2, 4 and 9. Translocations can result from aberrant double-strand break repair by either non-homologous end-joining or homologous recombination (see DNA REPAIR MECHANISMS). Similarly to human pericentromeric DNA, these regions have served as sinks for duplicative transposition and expansion of particular euchromatic segments. One conservative estimate (2005) is that 49% of the known subtelomeric DNA sequence was generated after humans and chimpanzees diverged. See HUMAN GENOME.

subthalamic nuclei Small paired structures (nuclei), each resembling a biconvex lens in coronal sections, lying beneath the thalamus and against the medial surface of the internal capsule. Fibres pass from it through the internal capsule to the globus pallidus, most densely so in its medial segment. Glutamatergic, they are excitatory of pallidal neurons. Similar fibres pass to the pars reticulate of the substantia nigra. See BASAL GANGLIA.

subventricular zone The region, adjacent to the walls of the lateral VENTRICLES, where in the adult brain neuroblasts are produced and migrate to the olfactory bulb where they differentiate into local interneurons. These neuroblasts follow the direction of flow of cerebrospinal fluid generated by ependymal cilia.

sucrase See DISACCHARIDASES.

sudden infant death syndrome (SIDS) 'Crib death'. Death of an infant under 1 year of age which remains unexplained after the performance of a complete postmortem investigation, including an autopsy, the examination of the death scene and the review of the case history. Since the discovery that 50% of SIDS victims had higher than normal levels of astroglial cells in the respiratory centres of the brain stem, some believe that one contributory factor in these cases may be impaired neural control of breathing. Impaired breathing (the apnoea hypothesis) might take any of several forms, including upper airway obstruction during sleep coupled with defective arousal mechanisms. Alternatively, rapid hypoxaemia brought on by shunting of poorly oxygenated blood into the systemic circulation has been suggested as a cause. Other studies suggest a link between SIDS and impaired sympathetic innervation of the heart, especially since there is evidence of previous bradycardia in some SIDS infants. Although defects in fatty acid metabolism and liver glucose-6-phosphatase have been implicated in some SIDS studies, they probably represent a small proportion of cases.

sulcus **1**. Furrow along a bone surface accommodating a blood vessel, nerve or tendon. **2**. Furrow, or groove, between gyri of the CEREBRAL CORTEX. Some are consistently located and mark the locations of functional areas. Particularly large sulci are termed fissures. See GYRUS.

summation Because synaptic vesicles (see SYNAPSES) fuse with the presynaptic membrane in a quantal fashion (a certain number fusing per action potential arriving at the presynaptic membrane), the amount of transmitter in the cleft at any one time depends upon the rate of arrival of impulses at the presynaptic membrane. If these impulses arrive in quick succession, the amount of transmitter released into the cleft will build up, or summate, and increase the probability of action potentials being generated in the postsynaptic neuron. Summation at a single synapse is termed *temporal summation*, because of its dependence on time. But cell bodies and axon membranes make hundreds or thousands of synaptic contacts with terminals of other axons. If there is inadequate transmitter release to initiate a postsynaptic action potential from any of these individually, two or more synapses may release sufficient transmitter to do so together. This kind of summation is *spatial summa-*

tion, because it results from adjacent synapses 'pooling' their transmitters.

sunburn Skin damage resulting from short-term exposure to ULTRAVIOLET RADIATION (UVB), experienced by those with both light and dark SKIN COLOUR. 'Sunburn cells' are keratinocytes in which apoptosis is triggered by UV-irradiation, and require p53 for their formation. Apart from discomfort and lowered pain threshold, there is risk of SKIN CANCER, damage to sweat glands and suppression of sweating, disrupting TEMPERATURE REGULATION. See TANNING.

superantigens Antigens, e.g., produced by STAPHYLOCOCCUS, that activate large numbers (2–20%) of T cells by binding to MHC class II molecules and V_β domains of T-CELL RECEPTORS of those cells that have already bound a peptide antigen. They are not processed into peptides, so do not initiate apoptosis. The result is an intense and destructive immune reaction that is not limited in the usual way.

superior Dorsally positioned. Contrast INFERIOR.

superior colliculus (SC) Paired elevations of the tectum of the midbrain (see BRAIN, Fig. 28; CEREBELLUM, Fig. 42), the mammalian homologue of the amphibian optic tectum, a key role being to turn sensory coordinates into motor coordinates, each point in the SC representing a location in visual and auditory space which specifies the reflex SACCADES required to point the gaze towards it (see HEARING). Each SC is divided into superficial, intermediate and deep layers. The superficial layers have a retinotopic map, receiving retinotopic projections from the retina and from the visual cortex. Representation of the visual field magnifies the central few degrees of arc because of the larger number of ganglion cells with receptive fields in or near the fovea compared with the periphery. The deep layers of each SC receive somatosensory and auditory information, and so have two maps: an *auditory map* depicting the location of sounds in space, with inputs projecting from the INFERIOR COLLICULUS; and a *somatotopic map*, in which body parts closest to the eyes receive greatest representation. The auditory map and the retinotopic map are in register. The intermediate layers are the site of a motor map, the collicular saccade-related burst neurons located here firing with high frequency 20 msec prior to a saccade. Besides generating saccades, the SC coordinates neck motor muscles involved in head rotation during orientation responses to a stimulus, neurons of the TECTOSPINAL TRACT originating here. See AXONAL GUIDANCE, EPHRINS, LATERAL GENICULATE NUCLEUS.

superior olivary nuclei (superior olive) A complex of nuclei in the medulla receiving inputs from the ventral cochlear nucleus and projecting ipsilaterally and contralaterally via the lateral lemnisci to the lateral lemniscal nuclei in the pons. See HEARING.

superior temporal gyri See TEMPORAL LOBE.

superior temporal sulcus (STS) See MIRROR NEURONS, TEMPORAL LOBE.

superoxide dismutases (SODs) Enzymes involved in neutralizing the superoxide free radical (see MITOCHONDRION), but believed by many to be involved also in development of neuron architecture (pleiotropy). Three genes encode these enzymes in humans, that for SOD1 occurring on chromosome 21. See AMYOTROPHIC LATERAL SCLEROSIS

superoxides The anion $^{\bullet}O_2^{-}$. See FREE RADICALS, MACROPHAGES, MITOCHONDRION, PEROXISOMES.

supination Rotation of the forearm, when parallel to the ground, so that the anterior surface is facing up (see ELBOW).

supplementary motor area (SMA) See PREMOTOR CORTEX.

suppression In genetics, the process by which a mutant allele at one gene locus reverses the effect of a mutation at another locus. Suppressor mutations sometimes have no effect in the absence of the other mutation; in other cases it produces its own abnormal phenotype. It is sometimes confused with EPISTASIS, the key difference being that a suppressor cancels the expression of a mutant allele and restores the wild-type phenotype. At the molecular level, a suppressor gene product may restore the functional shape of the product of the mutant allele at the other locus by binding to it, if both are proteins.

suppressor T cells (T_{reg} cells) See IMMUNOSUPPRESSION, T CELLS.

suprachiasmatic nuclei (SCN) Paired nuclei of the anterior HYPOTHALAMUS able to express *in vitro* sustained circadian cycles of electrical firing, cytosolic calcium ion concentrations and gene expression, and concerned with control of diurnal rhythms and the sleep–wake cycle. They contain the *circadian pacemaker* (see CIRCADIAN CLOCKS) and receive inputs from the retinas via the RETINOHYPOTHALAMIC TRACT and are involved in readjusting the clock to the changing light intensity during the daily light–dark cycle (photoperiod). The γ-aminobutyric acid-containing neurons of the SCN (see GABA) lie in the dorsal 'shell' of each nucleus, characterized by VASOPRESSIN expression, while neurons containing vasoactive intestinal polypeptide (VIP) are found in the more ventral nuclear 'core', both divisions innervating a range of targets. Neurons in the core of the SCN inhibit the autonomic division of the paraventricular nucleus of the hypothalamus. The SCN has relatively modest projections to the ventrolateral preoptic nucleus (VLPO) or the orexin neurons of the hypothalamus, but the bulk of its output is directed to the adjacent subparaventricular zone (SPZ) and the dorsomedial nucleus (DMH). Cell-specific lesions in the ventral SPZ disrupt circadian SLEEP–WAKE CYCLES, and locomotor activity, corticosteroid secretion and feeding; those of the DMH are apparently mainly activating and are a major source of input to the VLPO, conveying SCN influence to the sleep–wake regulatory system. Signals from the SCN trigger release of MELATONIN, but suppress its release during daylight. These signals decrease with age, possibly through calcification in the SCN or the pineal, so that by the seventies only a small fraction of the melatonin synthesized at age 35 is produced.

supraoptic nucleus See HYPOTHALAMUS.

supraorbital Above the ORBITS of the eyes, as in the *supraorbital torus* (brow ridge).

surfactant A substance with detergent-like properties. Alveolar fluid contains a mixture of phospholipids (largely desaturated phosphatidylcholine) secreted by type II alveolar cells, lowering overall alveolar surface tension and conferring alveolar stability and helping to prevent the air spaces in the alveoli from filling up with liquid. See ALVEOLUS, Fig. 5.

surround inhibition A feedforward information processing mechanism in sensory systems, underlying contrast enhancement in the RETINA (see also AMACRINE CELLS, HORIZONTAL CELLS). It is generated by signals representing average illumination. It may also contribute to selective ATTENTION. See DARK ADAPTATION.

survival factors, s. signals Signal molecules, or receptor-mediated contacts, required by one cell from other to enable it to survive and avoid CELL DEATH by apoptosis – hence, anti-apoptotic. Like mitogens and growth factors, any molecular signals (e.g., INTEGRINS) involved usually bind to cell-surface receptors but activate pathways that keep the 'death programme' leading to APOPTOSIS suppressed, often by regulating members of the Bcl-2 protein family. The PHOSPHATIDYLINOSITOL-3-OH-KINASE pathway is involved in cell growth and transmits a strong antiapoptotic signal, promoting cell survival.

susceptibility genes See RISK FACTORS.

suture A fibrous joint comprising a thin layer of dense fibrous connective tissue uniting the bones of the skull.

swallowing (deglutition) A complicated voluntary and reflex process during which the pharynx is diverted briefly from its normal function of conducting respiratory gases to the propulsion of food. In the voluntary stage, food is squeezed or rolled by the tongue posteriorly into the pharynx and backward against the palate. The food bolus or drink stimulates receptors in the oropharynx, afferent impulses passing along fibres projecting to the medulla and lower pons, initiating an involuntary (reflex) pharyngeal stage. The soft palate is pulled upward to close the posterior nares, preventing food from refluxing into the nasal cavities. The pharynx rises and widens to receive food or drink while elevation of the larynx causes the free edge of the epiglottis to descend and form a lid over the glottis, closing it off. The bolus passes through the laryngopharynx and enters the oesophagus in 1–2 seconds, the respiratory passages reopen, and breathing resumes. Damage to the 5th, 9th, 10th and 12th cranial nerves can cause paralysis of portions of the swallowing mechanism.

sweat glands See SWEAT.

sweat, s. glands (sudoriferous glands) *Eccrine sweat glands* are tubular structures distributed throughout the skin except for the lip margins, nail beds, glans penis, clitoris, labia minora and eardrums. Each comprises a deep subdermal coiled portion that secretes the sweat, and a duct portion passing outward through the dermis and epidermis of the skin. Cholinergic sympathetic fibres ending on or near the glandular cells cause production and release in the adult of ~600 mL of sweat per day (sensible perspiration) by exocytosis from gland cells, and these glands start to function soon after birth. Normally, only a small quantity of sodium ions is excreted per day in sweat, although this increases during heavy exercise in a warm environment. Mechanisms regulating sweating control the Na^+ lost through this route. Because sweat remains hypotonic to body fluids, water replacement is the immediate requirement during exercise rather than mineral replacement. For a fluid loss of <2.7 kg in adults, addition of a small amount of salt to food readily replenishes the sodium lost in sweat, while addition of sodium and potassium chloride to drinking water provides little benefit to either men or women who become sweat-loss dehydrated by 3% of body mass but who received food and water freely during each daily recovery period. During prolonged exercise, the kidneys' Na^+-conserving mechanisms generally balance sodium losses. As the rate of sweat production increases, the quantity of Na^+ lost in the urine decreases, maintaining its concentration in the EXTRACELLULAR FLUID (see also ALDOSTERONE). Eccrine sweat contains water, ions (mostly sodium and chloride), urea, uric acid, ammonia (playing a small role in the elimination of these three), amino acids, glucose and lactic acid. Sunburn reduces the ability to sweat. We do not know when the elaboration and multiplication of human eccrine sweat glands for evaporative heat loss evolved (see SODIUM, TEMPERATURE REGULATION).

Apocrine sweat glands are similarly simple, coiled tubular glands, but located mainly in the skin of the axilla (armpit), groin, areolae of the breasts and bearded regions of the face in adult males. Their secretions are slightly more viscous than those of eccrine glands. They do not start to function until PUBERTY. Despite releasing their secretions by exocytosis, they are still referred to as 'apocrine' glands.

sweating See TEMPERATURE REGULATION.

Sylvian fissure (lateral fissure, lateral sulcus) The most noticeable of all surface brain (and endocast) landmarks, the fissure demarcating the upper boundary of the temporal lobe (see BRAIN, Fig. 29). It is lower and longer on the left than the right side in most modern human brains, correlated with the larger size of WERNICKE'S AREA on the left. But corresponding asymmetries are also found in living ape and monkey brains, so a simple association with LANGUAGE function is not possible.

symmetry See ASYMMETRY, AXIS.

sympathetic nervous system One of the two divisions of the AUTONOMIC NERVOUS SYSTEM.

sympathomimetics Extraneous substances whose effects resemble in some respects those of the sympathetic neurotransmitter noradrenaline (e.g., hypertension, tachycardia, insomnia). Indirectly acting sympathomimetics (e.g., AMPHETAMINES, COCAINE) resembling noradrenaline structurally are transported by reuptake pathways into nerve terminals and displace vesicles carrying noradrenaline. This leads to MONOAMINE OXIDASE degradation of noradrenaline, but also to its release by carrier-mediated transport and activation of adrenoceptors. Directly acting sympathomimetics include ADRENALINE and NORADRENALINE.

symphysis (pl. symphyses) A cartilaginous, slightly movable joint in which the ends of the articulating bones are covered in hyaline cartilage although the bones are connected by a broad, flat disc of fibrocartilage; e.g. intervertebral joints, pubic symphysis.

symporters Transport proteins linking carriage of one particle to that of another, but not directly involving ATP hydrolysis. See ANTIPORTERS, ENTEROCYTES, PROXIMAL CONVOLUTED TUBULE.

synaesthesia The accompaniment of a sensation due to stimulation of a receptor associated with one sensory modality by a sensation characteristic of another. It is the involuntary physical experience of a cross-modal linkage (e.g., as when hearing a tone evokes an additional sensation of seeing a colour). The ability appears to be genetically transmitted (affecting ~1 in 200), resulting in a 'mingling of the senses'. It has been demonstrated that in such people there is activation of the region in the fusiform gyrus dealing with colour when they are shown numbers in black and white (normally, the colour area is activated only when coloured numbers are shown).

synapses Specialized junctions where an axon terminal makes close contact with either another neuron, muscle fibre or gland cell, forming the basic units of neural circuitry (e.g., see CONVERGENCE (1), DIVERGENCE (2)). Two basic forms of synapse occur: (i) *electrical synapses*, involving GAP JUNCTIONS, ~3 nm wide, between presynaptic and postsynaptic membranes, transmission of current across which is very rapid. If the synapse is large, this is fail-safe – an action potential in the presynaptic neuron produces an almost instantaneous postsynaptic one. Abundant between glia, epithelial cells, smooth and cardiac muscle fibres, liver cells and some glandular cells, electrical synapses are relatively rare in the CNS, occurring where highly synchronized activity between neighbouring cells is required. (ii) *Chemical synapses*, in which the communicating cells are separated by a synaptic cleft 20–50 nm wide, containing fibrous CELL ADHESION MOLECULES enabling cell adherence (see the crucial roles of ASTROCYTES and terminal SCHWANN CELLS). The presynaptic terminals harbour synaptic vesicles containing an amino acid or amine NEUROTRANSMITTER, and secretory granules containing a peptide neurotransmitter and carried from the Golgi apparatus by axoplasmic transport. Although both are frequently found in the same synapse, they are released under different conditions. Synaptic vesicles (see VESICLES), carried along microtubules in a KINESIN-dependent manner, fuse with the presynaptic membrane after Ca^{2+} entry through voltage-gated calcium channels in response to action potentials arriving at the nerve terminal. After release of the transmitter (in quantal fashion, depending upon the level of intracellular Ca^{2+}; see SUMMATION), the vesicle membrane is recovered by endocytosis from the presynaptic membrane and refilled with transmitter. Secretory granules also release their peptides by exocytosis in a Ca^{2+}-dependent manner, usually requiring high-frequency trains of presynaptic action potentials. Unlike the fast release of amino acid or amine transmitters, peptide release often takes 50 msec or longer.

Although it has become almost axiomatic that vertebrate neurons carry information in the form of a digitalized 'all-or-nothing' action potential 'spike' whose arrival at a synaptic bouton relates to quantized neurotransmitter release, it has

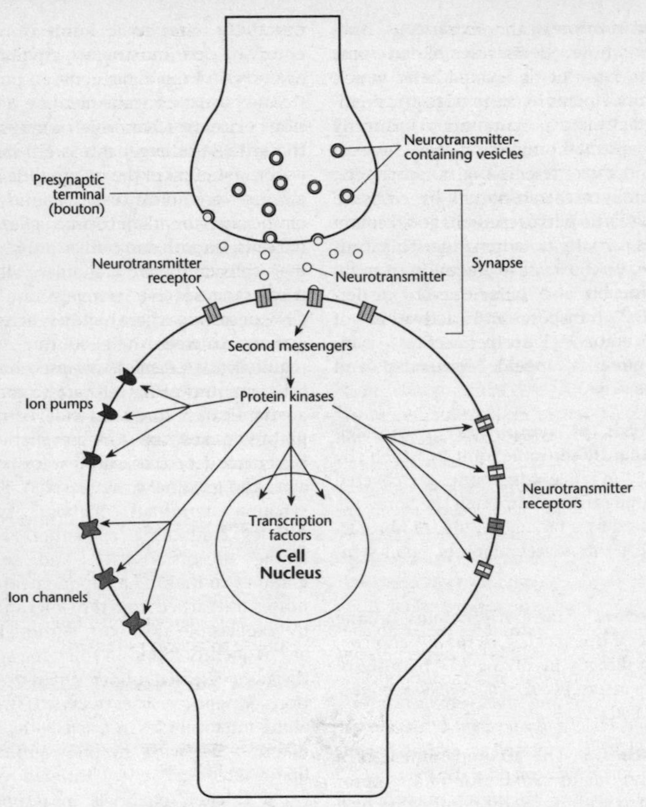

FIG 168. Some of the signalling pathways involved in *slow synaptic transmission* (see SYNAPSES). From *Science* Vol. 294, No. 5544 (2 Nov. 2001), p. 1026. Reprinted with permission from AAAS.

recently become clear that an 'analogue' process can also occur, in which graded subthreshold depolarizations of the somatic (cell body) membrane can influence release of transmitter by its terminals (boutons).

Chemical synapses can occur between an axon terminal and (i) a dendritic membrane of a cell body (an *axodendritic synapse*), (ii) another axon (an *axo-axonic synapse*), (iii) a muscle fibre membrane, the MOTOR END-PLATE of a NEUROMUSCULAR JUNCTION being functionally analogous to (i) and (ii). In certain specialized neurons, dendrites form synapses with one another (*dendrodendritic synapses*). Dense protein accumulations adjacent to, and within, the pre- and postsynaptic membranes are termed *membrane differentiations*. On the presynaptic side, pyramid-like protein complexes may intrude into the cytosol forming the *active zones* of the membrane, involved in transmitter release from synaptic vesicles and secretory granules. Proteins may accumulate thickly in, and just under, the postsynaptic membrane. This *postsynaptic density* (PSD) contains receptors for transmitters, converting this intercellular signal into an intracellular signal, in the form of a voltage change across the postsynaptic membrane and/or chemical modification of membrane receptors (see

SYNAPTIC PLASTICITY). Abnormal PSDs have been implicated in disorders of COGNITION, and the neuroligins located in it which bind to neurexins in the presynaptic membrane play an important part in spanning the synapse and tethering it (see AUTISM).

Ligand-gated ION CHANNELS that bind transmitter on the postsynaptic membrane are less selective than the voltage-gated ion channels involved in propagation of an action potential; e.g., ACh-gated ion channels at the neuromuscular junction are permeable to both Na^+ and K^+. But generally, if the open channels are permeable to Na^+, the net effect will be a depolarization of the postsynaptic cell, bringing the membrane closer to the threshold for generating action potentials (see AXON HILLOCK) and therefore excitatory. Any such transient postsynaptic membrane depolarization is termed an *excitatory postsynaptic membrane potential* (EPSP). Perhaps 50–100 excitatory presynaptic neurons need to fire simultaneously to produce a synaptic potential large enough to trigger an action potential. If the ligand-gated ion channel is permeable to Cl^-, the net effect will be to hyperpolarize the postsynaptic cell, making depolarization less likely and therefore inhibitory. Such a transient postsynaptic membrane depolarization is termed an *inhibitory postsynaptic membrane potential* (IPSP).

There are two categories of chemical transmission between neurons: fast and slow synaptic transmission, ~50% of the fast brain synapses being excitatory and involving glutamate as their transmitter, the remainder being inhibitory, involving GABA.

Fast synaptic transmission occurs in <1 msec, attributable to the ability of fast-acting transmitters (glutamate, GABA and glycine) to bind their receptor, change its conformation, and open ligand-gated ion channels in the postsynaptic membrane. In fast *excitatory* transmission (see AMPARS), the changed receptor conformation allows positively charged sodium ions to rush into the cell and depolarize the membrane. The amine acetylcholine mediates fast synaptic transmission at all neuromuscular junctions. In fast *inhibitory* transmission, it is negatively charged chloride ions that enter the cell, causing hyperpolarization of the postsynaptic membrane.

Slow synaptic transmission (see Fig. 168) takes place over hundreds of milliseconds to minutes and is much more complex in its operation, involving numerous biogenic amines (e.g., DOPAMINE), peptides and amino acids serving rather like neurotransmitters (see NEUROMODULATORS) but operating instead through non-ion channel-linked, METABOTROPIC, receptors (G PROTEIN-COUPLED RECEPTORS) via second messengers and protein kinases.

Many of the effects of the fast-acting neurotransmitters such as glutamate and GABA are the result of slow synaptic transmission pathways and are fundamental to the molecular details of memory storage (see MEMORY, Fig. 122).

synapsins A class of key protein in the presynaptic membrane of synapses, whose regulation by phosphorylation and dephosphorylation modulates the efficiency of neurotransmitter release in response to action potentials.

synaptic depression See SYNAPTIC STRENGTH.

synaptic gain See DOPAMINE; LATERAL INHIBITION, Fig. 112.

synaptic inhibition An event making an action potential at a postsynaptic membrane (see SYNAPSE) less probable. See GABA, GLYCINE, INTERNEURONS, LATERAL INHIBITION.

synaptic plasticity Although synapses form in the absence of any electrical activity, the awakening of synaptic transmission during development is very important in the refinement of connections (see NEURAL PLASTICITY). Modification of the strength of the signal passing from one neuron to the next at a synapse by changes in the SYNAPSE itself is termed *synaptic plasticity*, and is a fundamental mechanism for information storage by the nervous system built into the very molecular architecture of chemical synapses (see LEARNING, MEMORY). A classic example is HEBBIAN MODIFICATION, where there is an increase in the strength of a synapse as a result of simultaneous activation of pre- and postsynaptic neurons ('neurons that fire together wire together'). A synapse activated briefly at high frequency will produce larger responses in the postsynaptic (receiving) neuron for some while after its activity levels return to normal, a phenomenon referred to as LONG-TERM POTENTIATION (LTP). Deeper understanding of activity-dependent GENE EXPRESSION (e.g., see MECP2) should help clarify the molecular nature of the neuronal dysfunction underlying such neurodevelopmental disorders as RETT SYNDROME and AUTISM spectrum disorders.

One relatively simple form of synaptic plasticity is *presynaptic facilitation* (see SYNAPTIC STRENGTH), which occurs when an interneuron makes axo-axonal synapses with the terminal of another neuron and (typically by releasing serotonin) causes increased transmitter release from that terminal. In *short-term presynaptic facilitation*, a transient release of SEROTONIN resulting from a single stimulus leads to covalent modification of pre-existing postsynaptic proteins via a G PROTEIN-COUPLED RECEPTOR pathway. This leads to increased transmitter availability and release at the synapse. But in *long-term presynaptic facilitation*, repeated serotonergic stimulation by the interneuron causes the presynaptic sensory neuron's cAMP levels to rise and persist for several minutes. Catalytic subunits of PKA then translocate to the nucleus and recruit mitogen-activated protein kinase (see MAP KINASES), where PKA and MAPK phosphorylate and activate the cAMP response element-binding protein CREB-1. CREB-1 in turn activates several immediate-response genes whose products result in persistent activity of PKA, maintaining persistent phosphorylation of its own substrate proteins. Activation of other immediate-response genes activated by CREB-1 leads in turn to the production of elongation factors and machinery involved in the growth of new synaptic connections (*synaptogenesis*). The phenomenon termed *synaptic capture* applies in cases where gene products released from previous activity at one synapse are 'captured' by other synapses on the same postsynaptic neurons (heterosynaptic facilitation). Whereas late-LTP-associated gene products may be distributed throughout the cell, they appear to be captured and used only at synapses that have been previously 'tagged' (synaptic tagging) by their own activity.

In *presynaptic inhibition*, e.g. by GABAergic interneurons in the dorsal column nucleus of the spinal cord (see LATERAL INHIBITION), interneurons make axo-axonal synapses on the afferent terminals. The activated $GABA_A$ receptors bring about depolarization because in these sensory neurons the membrane potential is more negative than the reversal potential of the chloride current through the $GABA_A$ receptor channels. This primary afferent depolarization is inhibitory because as an action potential sweeps into the terminal its amplitude decreases, less Ca^{2+} occurs, and transmitter release from the terminal is reduced. OPIOID receptor activation also inhibits presynaptic transmitter release.

The threshold for synaptic plasticity and memory storage in the hippocampus has been shown to be determined by the relative degrees of PKA-dependent protein phosphorylation and dephosphorylation. Inhibition of the Ca^{2+}-dependent phosphatase CALCINEURIN promotes learning and strengthens short- and long-term memory during several spatial and non-spatial tasks.

Another form of long-term synaptic plasticity is *long-term depression* (LTD), or synaptic weakening. This decrease in excitatory postsynaptic potential (EPSP) was first noticed during coincident parallel fibre and climbing fibre stimulation of Purkinje fibres of the CEREBELLUM. This resembles classical conditioning in *Aplysia* (see LEARNING). However, the stimulus convergence here takes place in the postsynaptic Purkinje dendrite rather than in the presynaptic axon membrane; and again unlike classical conditioning the sensory motor synapse becomes *less* responsive to transmitter (GLUTAMATE), not more so. The glutamate receptor mediating LTD in the cerebellum is the AMPA receptor (AMPAR), synaptic weakening being brought about by AMPAR internalization. The critical

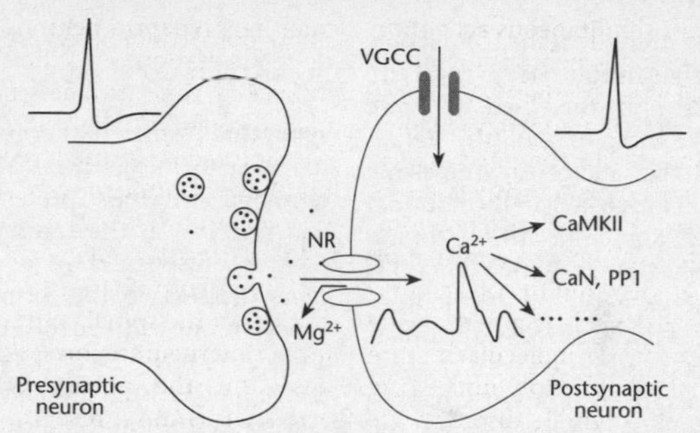

FIG 169. Diagram indicating the central role of postsynaptic calcium ions (Ca^{2+}) in the induction of activity-dependent long-term SYNAPTIC PLASTICITY (see entry). Ca^{2+} can enter the postsynaptic neuron via several sources, including NMDA receptors (NR) and voltage-gated Ca^{2+} channels (VGCCs). On arrival of a presynaptic action potential, glutamate (small dots) is released from synaptic vesicles, which bind to NRs and open them, allowing Ca^{2+} influx if the magnesium ion (Mg^{2+}) block is removed by postsynaptic depolarization. Such depolarization opens VGCCs, allowing Ca^{2+} entry. Depending on its spatiotemporal pattern, this Ca^{2+} signal (the wavy line inside the postsynaptic neuron) can activate enzymes leading to synaptic modification. These include, CaMKII, calcineurin (CaN) and protein phosphatase 1 (see PP1). Action potentials are shown outside the two neurons. From Guo-Qiang Bi and J. Rubin © 2005 *Trends in Neurosciences*, with permission from Elsevier Ltd.

step in this induction seems to be the surge of calcium ions (Ca^{2+}) into the Purkinje cell (PC) dendrite following the Na^+ entry when glutamate from parallel fibres (pfs) binds the AMPAR causing a large EPSP. This Ca^{2+} signal (see Fig. 169) is a result of activation of a second *metabotropic* glutamate receptor (the NMDA RECEPTOR, NR). Activation of this NR component (mGluR1) and of voltage-gated Ca^{2+} channels (VGCCs) in response to the large EPSP is coupled via a G protein-coupled receptor to PHOSPHOLIPASE C, activation of which generates diacylglycerol, activation of PKC, and phosphorylation of the carboxy-terminal tail of the AMPA receptor (GluR2). The AMPAR is then internalized by clathrin-mediated endocytosis, leading to synaptic depression (weakening). So, three simultaneous intracellular signals seem to be required for cerebellar LTD: (i) a rise in intracellular [Ca^{2+}] due to climbing fibre activation; (ii) a rise in intracellular [Na^+] due to AMPAR activation; and (iii) activation of PKC. Activation of mGluRs is required for translation of crucial mRNA molecules near synapses into proteins. In this model, LEARNING occurs when rises in internal [Ca^{2+}] and [Na^+] coincide with the activation of PKC, while some forms of MEMORY seem to occur when the AMPAR channels are internalized. Enhanced LTD is currently being linked to the developmental and cognitive problems associated with FRAGILE X SYNDROME. In animal models, LTD has also been shown to be inducible in the neostriatum by ENDOCANNABINOIDS, a process involving D2 dopamine receptor activation.

synaptic strength Synaptic strengthening is the increase in transmitter release from a presynaptic cell brought about by release onto the presynaptic membrane of neurotransmitter (e.g., serotonin) by a modulatory neuron (see MEMORY and LEARNING for some details). It can also result from RETROGRADE MESSENGER activity; e.g., of nitric oxide (see LONG-TERM POTENTIATION).

Synaptic depression is the decrease in synaptic strength which normally accompanies prolonged trains of action potentials and can be brought about by release of nitric oxide. In the peripheral nervous system, terminal SCHWANN CELLS (those at the neuromuscular junction) monitor synaptic activity by detecting neuron-glial signalling molecules (ATP, adenosine) released from the neuron along with neurotransmitter, integrate the activity of the synapse and balance the strength of the connection by regulating transmitter release from the presynaptic neuron. Increased synaptic strength may be brought about by signalling pathways that increase cytoplasmic levels of calcium ions (Ca^{2+}) in the terminal Schwann cells. Similarly, in the brain, ASTROCYTES ensheath synaptic junctions and use purinergic receptors for neuron-glial signalling. A rise in astrocytic cytoplasmic Ca^{2+} levels is associated with changes in synaptic strength in adjacent synapses, both in culture and in intact retinas and associated in turn with rise in extracellular ATP concentrations (see NON-REM SLEEP).

The chaperone Stargazin and its family members play crucial roles in delivering the AMPAR glutamate receptors to their synaptic sites and enhance the efficiency with which glutamate released presynaptically can depolarize a postsynaptic neuron. It is now thought that the highly coordinated spatiotemporal changes in the synaptic content of AMPARs are major factors in the way brain cells regulate their synaptic strength and a prime mechanism for the biological mechanism underlying memory (see SYNAPTIC PLASTICITY). See ADDICTION.

synaptic transmission See SYNAPSE.

synaptic vesicles See SYNAPSES, VESICLES.

synaptogenesis The formation of synaptic connections. See SYNAPTIC PLASTICITY.

synarthrosis (pl. synarthroses) An immovable joint; e.g. between the bones of the skull. See SUTURE.

synchondrosis (pl. synchondroses) A cartilaginous joint in which the connecting material is hyaline cartilage; e.g., the epiphyseal plate connecting the epiphysis and diaphysis of a growing bone.

syncope Fainting; a sudden loss of consciousness. It may be due to sudden emotional stress (vasodepressor syncope), pressure associated with urination (situational syncope), drug-induced (e.g., through use of anti-hypertensives or diuretics), or to loss of blood pressure on standing up. See CAROTID SINUSES.

syncytins Genes that apparently entered the primate lineage 25–40 Mya and encode retroviral envelope proteins (syncytin-1 and syncytin-2) that promote cell fusion (i.e., are fusogenic) in a variety of cell types, leading to the formation of giant syncytia. The virus involved appears to be a defective retrovirus, HERV-W (see HERVS), and exhibits tissue- (notably placenta-)specific expression and a distinctive receptor interaction during trophoblast cell differentiation and syncytium formation. It is possible that syncytin may mediate CYTOTROPHOBLAST fusion during human PLACENTA formation.

syncytiotrophoblast The multinucleate outer layer of the early trophoblast (see BLASTOCYST, Fig. 20), lacking distinct cell boundaries and eventually becoming part of the CHORION. Its enzymes erode maternal tissues, including endothelial walls of sinusoids, so that by day 9 of the blastocyst stage a large number of lacunae (trophoblastic lacunae) have developed within it which will be filled by blood from maternal sinusoids to establish the uteroplacental circulation. See CYTOTROPHOBLAST, IMPLANTATION, SYNCYTINS.

syndactyly The condition in which the digits fail to separate in development. Compare POLYDACTYLY.

syndesmosis (pl. syndesmoses) A fibrous joint, with a greater gap between the articulating bones than in a suture and with more fibrous connective tissue; permits slight movement; e.g., the interosseous membrane between the tibia and fibula.

syndrome X See METABOLIC SYNDROME.

syndromes Consistent and recognizable patterns of abnormalities, for which the underlying cause is often known. Chromosomal abnormalities are often responsible. See BIRTH DEFECTS.

synesthesia See SYNAESTHESIA.

syngeneic Genetically identical (see GRAFTS). Contrast ALLOGENEIC, XENOGENEIC.

synovial fluid Clear viscous fluid secreted into the synovial cavity by capillaries in synovial membrane within the articular capsule of SYNOVIAL JOINTS. It comprises HYALURONIC ACID and interstitial fluid and lubricates the joint while supplying the nutrients and removing the wastes of the chondrocytes in the hyaline articular cartilage. Pre-exercise 'warm-ups' promote secretion of synovial fluid.

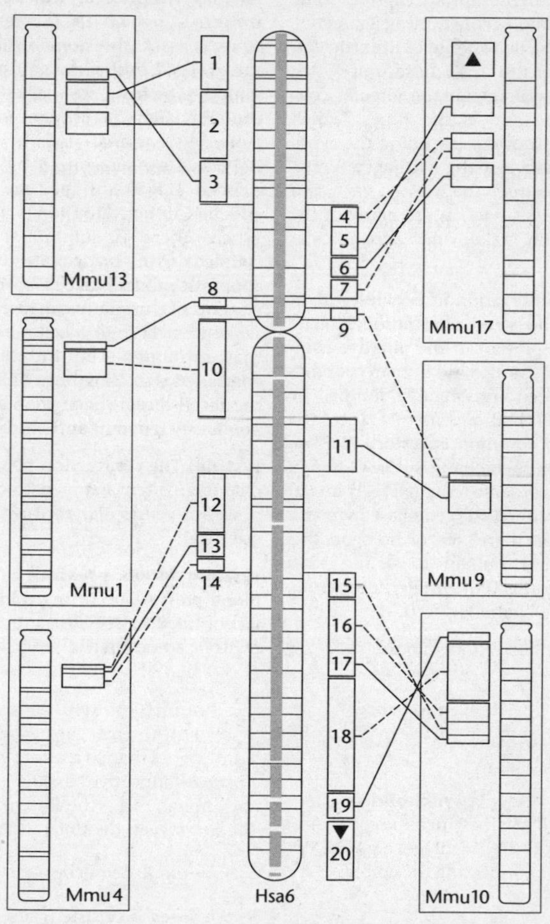

FIG 170. Diagram showing the segments of conserved chromosomal SYNTENY revealed by comparisons of human chromosome 6 (Hsa6) and six mouse chromosomes (Mmu). Numbered blocks represent regions of conserved synteny. Dashed lines joining blocks between two species indicate blocks that are inverted relative to their orientations in the mouse chromosomes. Undashed lines indicate relative repositioning without inversion. Segments are numbered alongside Hsa6 and the vertical bars in the centre of the Hsa6 indicate the contigs included in the 2002 analysis. From D.R. Bentley et al. © *Nature* 2002, with permission from Macmillan Publishing Ltd.

synovial joint Any JOINT (see Fig. 106) formed by bones between which is a synovial cavity and joined by the dense connective tissue of the articular capsule, often with associated ligaments. All freely movable joints (diarthroses) are synovial joints. The articular capsule comprises two layers: a dense fibrous capsule, some of the fibre bundles constituting ligaments, attached to the periosteum of the articulating bones, on the inner surface of which lies the synovial membrane (areolar connective tissue and elastic fibres) which secretes the SYNOVIAL FLUID filling the synovial cavity between the cartilages. Vertebrocostal joints, between RIBS and VERTEBRAE, are synovial, as are many of the joints of the axial and appendicular SKELETON.

synteny Conservation in two descendant species of the same chromosomal gene order as was present in the putative common ancestor. Mouse and human genomes have each been considerably shuffled by chromosomal rearrangements since the time of their common ancestor ~75 Mya; yet the rate of such changes is low enough for local gene order to remain largely intact. Thus one recent (2001) gene-based synteny map used more than 3,600 orthologous loci of mouse and human to define ~200 regions of conserved synteny. See HUMAN GENOME (Fig. 97) and Fig. 170.

syphilis See TREPONEMA PALLIDUM.

systemic lupus erythematosus (SLE) A systemic AUTOIMMUNE DISEASE characterized by the presence of antibodies against antigens that are found ubiquitously in the body. It has an EPIGENETIC component, with aberrant hypomethylation found in T cells of SLE patients, including in genes such as lymphocyte function-associated antigen-1, which is overexpressed in lupus cells. Anti-histone antibodies (e.g., anti-H1) and antibodies against the spliceosome proteins are often involved. Unmethylated CpG sequences are far commoner in bacterial than in mammalian DNA and are recognized by the TOLL-LIKE RECEPTOR TLR-9. But they are particularly enriched in activated lupus lymphocytes. Where there is substantial cell death coupled with inadequate clearance of apoptotic products, B cells specific for components of chromatin can internalize these sequences via their B-cell receptor so that they encounter TLR-9 intracellularly. The combined signalling from TLR-9 and from the B-cell receptor can then activate such previously ignorant anti-chromatin B cells.

systole The contraction phase of a heart chamber; often assumed to be the more powerful ventricular contraction. See CARDIAC CYCLE.

systolic blood pressure The highest blood pressure occurring during ventricular contraction (see CARDIAC CYCLE). Compare DIASTOLIC BLOOD PRESSURE.

T

θ- For these entries, see THETA-.

T_3 and T_4 Triiodothyronine and tetraiodothyronine (THYROXINE). See THYROID GLAND.

TAARs Receptors to TRACE AMINES.

tachycardia Increase in heart rate. See ARRHYTHMIAS, CARDIOVASCULAR CENTRE; compare BRADYCARDIA.

tactile discs See MERKEL CELLS, TOUCH.

talin An intermediary protein linking INTEGRINS to actin in the CYTOSKELETON.

tanning The most common example of acquired pigmentation. It is well known that α-MSH increases skin darkening, traditionally observed in patients with adrenal insufficiency whose pituitary glands overproduce it. The precursor for α-MSH is proopiomelanocortin, now known to be produced in hair follicles and skin. When produced, α-MSH is secreted by keratinocytes and MELANOCYTES, and UV-induced DNA damage may itself be important in pigment production (thymidine dimers have been shown to up-regulate tyrosinase *in vitro*). Tyrosinase is involved in melanin production, and is activated on binding of α-MSH to its MC1R receptor on melanocytes. Inability to tan reveals several genetic factors that may be instructive. Variants of the *MC1R* gene (see MELANOCORTINS) cause relative inability to tan (red-haired people cannot tan and are prone to SUNBURN). Using transgenic mice, fair-skinned mice containing epidermal melanocytes were found to be acutely UV-sensitive, and cultured human keratinocytes have been shown to increase their POMC/α-MSH production 30-fold in UV light, indicating they sense it and then secrete α-MSH in response. See MELANOMAS.

taphonomy The study of the processes occurring from the time an organism dies, to its burial and potential fossilization. These may include the manner of death, decomposition, movement, burial and chemical alteration.

tarsus (ankle) Collective term for the seven bones forming the proximal region of the foot. These tarsal bones include the talus (the only ankle bone articulating with the tibia and fibula, and the most superior) and calcaneus (the largest and strongest of the seven, forming the heel), situated in the posterior part of the foot. The talus transmits about half the body weight to the calcaneus, the remainder being transmitted to the others.

taste The sense of taste (gustation) enables avoidance of potentially noxious foodstuffs, and selection of foods with high energy content. Gustatory receptor cells lie within TASTE BUDS on the dorsum of the tongue and are sensitive to more than one taste modality, five well-defined modalities being salty, sour, sweet, bitter and umami (due to amino acids, as in monosodium glutamate) since there is no cross-adaptation between them. The excitable generator potential-producing receptor cells have microvilli on their apical surfaces carrying out sensory transduction via Na^+, K^+ and Ca^{2+} channels, each receptor projecting through taste pores in the enclosing epithelium. The sensory experience of having food in the mouth (flavour perception) is important in triggering and modifying autonomic responses to feeding, such as salivation, gastric secretion and gastrointestinal motility.

Sour solutions are always of low pH, their high H^+ concentration opening specific Na^+ channels and closing specific K^+ channels, depolarizing the receptor cell and opening voltage-gated Ca^{2+} channels and leading to exocytosis of neurotransmitter onto the appropriate taste afferent fibres exciting them. Salty solutions also open Na^+ channels, with the same excitatory result. Sweet and bitter substances bind G PROTEIN-COUPLED RECEPTORS. Sweet substances activate adenylyl cyclase via G_s, raising cAMP levels and closing certain K^+ channels, depolarization similarly leading to excitation of appropriate afferents. Bitter substances (e.g., alkaloids) activate PHOSPHOLIPASE C via G_t, the resulting raised Ca^{2+} causing depolarization and release of transmitter onto appropriate afferents. The umami taste sensation, generated by amino acids such as L-glutamate, appears to involve metabotropic GLUTAMATE receptors of the mGluR4 subtype, coupled via G_i proteins to the inhibition of adenylyl cyclase.

Afferents from taste buds leave via cranial nerves VII, IX and X and terminate in the NUCLEUS TRACTUS SOLITARIUS (NTS) of the medulla oblongata, whence information is conveyed to multiple sites in the hindbrain (e.g., the parabrachial nucleus), midbrain (ventral tegmental area) and forebrain (e.g., the nucleus accumbens, striatum, thalamus and the cerebral cortex). These collectively sense and discriminate among different tastes and textures, and assign reward value to them (see REINFORCEMENT AND REWARD). In primates, higher processing of taste information involves 'primary taste neurons' in the insular cortex, supplied by neurons of the NTS via the thalamus, which project to 'secondary taste neurons' that integrate relevant olfactory, visual and cognitive inputs with the taste information, decreasing their response as food is eaten (*hunger-dependency*). Functional MRI tends to support the view that this reduction in firing diminishes the reward value of foods, contributing to meal termination through projections to the striatum and amygdala. Compare OLFACTION.

taste buds Small clusters of 50–100 neuron-like epithelial gustatory cells and supporting cells just beneath the covering epithelium on the lateral surfaces of the folate, fungiform and vallate papillae of the TONGUE. Fungiform papillae lie anterolaterally; folate papillae occur laterally and halfway along the tongue, and vallate papillae are centrally placed towards the rear of the taste-sensitive region. The receptor cells are innervated by axons of the facial (CNVII) and glossopharyngeal (CNIX) cranial nerves, forming SECONDARY RECEPTORS. A few are also scattered around the oral cavity and on the soft palate. All modalities of taste may be detected by the taste buds, although the tongue is most sensitive to bitterness towards its base, to sour and salt along its sides, while the tip is most sweet-sensitive. At present there is no evidence that any kind of spatial segregation of the sensitivities contributes to the neural representation of taste quality. See TASTE.

TATA-binding protein, TATA factor See CLONE (1), TATA SEQUENCE.

TATA sequence (TATA box) A short, double-helical DNA consensus sequence, typically 25 nucleotides upstream of the TRANSCRIPTION start site composed mainly of thymidine and adenosyl nucleotides to which the general transcription factor TFIID binds by its TATA-binding protein (TBP). The sequence is TATAA/TAA/T (a forward slash indicating that either of the nucleotides on either side is equally probable at the site). Binding of TFIID distorts the TATA box, and this is believed to provide the physical landmark of an active PROMOTER so that DNA sequences either side of the distortion can be brought into proximity and enable subsequent protein assembly to produce the transcription initiation complex. The TATA box sequence is usually present in the vicinity of most RNA POLYMERASE II transcription start points, along with one or two other consensus sequences, e.g., BRE (G/CG/CG/ACGCC) or DPE (A/GGA/TCGTG). Those RNA polymerase II binding sites lacking a TATA box sequence typically have a 'strong' INR sequence (C/TC/TANT/AC/TC/T, where N indicates any nucleotide). See NUCLEOSOMES, TRANSCRIPTION FACTORS.

tau A cytoskeletal protein encoded by the *tau* gene, associated with and stabilizing microtubules. It is involved in the pathology of a number of diseases involving neural degeneration, including ALZHEIMER'S DISEASE (AD), where it exhibits hyperphosphorylation and oxidative modifications so that it no longer binds microtubules. The microtubules deteriorate, leading to neuron death. In the pathology of affected cells, tau is found within protein meshworks termed neurofibrillary tangles (NFTs). No mutations in *tau* have been linked to AD, and its involvement is probably downstream of Aβ protein and plaque formation. However, mutations in *tau* cause some cases of a less common dementia, FTLD (frontotemporal lobar degeneration). Abnormal forms of tau protein can cause neural degeneration, memory loss and other neurological deficits. Mice carrying a mutant human *APP* gene, and in addition two copies of *tau*, found it difficult to learn the Morris water maize as they aged; those with one *tau* gene were less impaired, and those with no *tau* gene were able to learn as well as did controls. Other research indicates that tau reduction can alleviate memory loss, and that dietary intake of omega-3 fatty acids may reduce the levels of enzymes that phosphorylate tau (see O-LINKED B-N-ACETYLGLUCOSAMINE).

taurodont Of a tooth in which the roots merge for much of their length, as in some modern humans but in fossils only common in Neanderthal teeth, where ~50% of molars exhibit taurodontism. The roots can fill with secondary dentine, which may confer an advantage where teeth suffer heavy wear since normal molars cease to function when the crown has worn down to root level. Taurodont teeth are functional for longer, as the extra dentine delays the fragmentation which eventually follows heavy wear.

Tay–Sachs disorder Inherited recessive disorder caused by mutation in the *HEXA* gene (chromosome site 15q), causing a defect in the enzyme, hexosaminidase A, which catalyses a step in lysosomal breakdown of the glycolipids termed gangliosides, abundant in neurons (see LYSOSOMES). Although the phenotype at birth is normal, symptoms, including developmental delay and seizures, become recognizable in the first year of life. Blindness develops by second year; death usually occurs before third birthday. It is more common in Jewish populations of eastern European ancestry (Ashkenazi), where heterozygote frequencies may reach 11%. There is evidence that the advantage of the gene accrues from resistance to TUBERCULOSIS. There is no treatment, but a test for carriers exists.

TB See TUBERCULOSIS.

T-cell receptors (TCRs) Each T CELL bears ~30,000 identical antigen-receptor (TCR) molecules on its surface (see IMMUNOGLOBULINS, Fig. 99), which only recognize a composite ligand consisting of a foreign peptide antigen fragment bound to a self MHC molecule (see INFECTION). Although a TCR resembles the membrane-bound Fab fragments of secreted ANTIBODIES (see Fig. 10) and although the encoding gene segments are arranged in a similar pattern to those encoding immunoglobulins, immunoglobulins recognize peptide fragments unattached to MHC molecules. Most TCRs (see Fig. 171) consist of a heterodimeric combination of two transmembrane glycoprotein chains, TCRα and TCRβ chains, although a minority of T cells bears alternative but structurally similar TCRs composed of a different pair of chains, γ and δ, and it is not clear how these differ in their antigen-recognition properties. In early development, γ:δ T cells predominate, but from birth onwards more than 90% of thymocytes express T-cell receptors of the α:β type (see MUCOSAL-

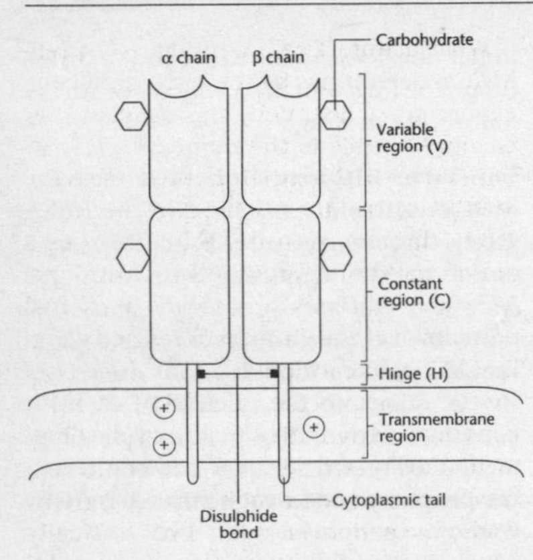

FIG 171. The structure of a T-CELL RECEPTOR. The transmembrane helices of the glycoprotein contain basic (positive) residues within the hydrophobic transmembrane segments. © 2005 From *Immunobiology* (6th edition), by C.A. Janeway Jr, P. Travers, M. Walpart and M.J. Shlomchik. Reproduced by permission of Garland Science/Taylor & Francis LLC.

ASSOCIATED LYMPHOID TISSUES). Like immunoglobulin light and heavy chains, T-cell receptor α and β chains both consist of a variable (V) amino-terminal region and a constant (C) region. A short segment, analogous to an immunoglobulin hinge region, connects the immunoglobulin domains to the membrane and contains the cysteine residue that forms the interchain disulphide bond. Unusually, the transmembrane helices of both contain positively charged (basic) residues within the hydrophobic transmembrane segment.

The TCRα locus (on chromosome 14) comprises ~80 V_α gene segments (perhaps not all functional) while 61 alternative J_α gene segments are located some way downstream of these. Still further downstream is a single C gene containing separate exons for the constant and hinge domains and transmembrane and cytoplasmic regions. The TCRδ locus interrupts the J and V gene segments on the TCRα locus. The TCRβ locus (on chromosome 7) is differently organized, with a cluster of 52 functional V_β gene segments distantly positioned from two separate clusters each of which contains one D gene segment, six or seven J gene segments and a single C gene – each C gene with similar exon arrangements to the single C gene in the TCRα locus.

The gene rearrangements which form complete V-domain exons are similar for B cells (see ANTIBODY DIVERSITY) and T cells, although those for T cells occur in the thymus.

Each individual TCR will recognize a particular combination of MHC molecule and peptide antigen and is thus MHC-restricted in its antigen recognition. Only 10–30% of T-cell receptors generated by gene rearrangement are able to recognize self-MHC:self-peptide combinations on stromal cells of the thymic epithelium and become involved in self-MHC-restricted responses to foreign antigens, and these will be positively selected for survival within the thymus. Effector T cells become activated when their TCR binds a MHC–peptide complex on an ANTIGEN-PRESENTING CELL surface. The TCR is associated with a complex of invariant Ig-like domain-bearing transmembrane proteins, termed the CD3 complex, which transduce the binding of a MHC–peptide complex into intracellular signals.

T cells (T lymphocytes) Lymphocytes developing from undifferentiated common lymphoid precursors (stem cells) in the BONE MARROW (see STEM CELLS) but which fail to develop in the absence of a functioning THYMUS. It is in the thymus that those thymocytes that are T-cell progenitors encounter crucial fate-determining cues analogous to those provided for B CELLS by bone marrow stromal cells. The immune system must discriminate between self- and non-self-antigens, and when this fails it destroys cells and tissues of the body and as a result causes AUTOIMMUNE DISEASES. T cells, which are crucial in monitoring cell-surface antigens, are able to make this discrimination as a result of their formative experiences in the thymus.

A T-cell progenitor receives a signal (Dll-4) in the thymus at its Notch1 receptor, transmitted through the NOTCH SIGNALLING PATHWAY, committing it to the T-cell lineage rather than the B-cell lineage (in which Notch signalling is blocked). Assembly of its T-CELL RECEPTOR (TCR) while in the thymus is the most significant event in the progenitor cell's life, determining which of the two major T-cell classes a cell will belong to: helper T cells (T_H) or cytotoxic T cells (T_C).

As in B-cell development, a T cell assembles its receptor in stages, the products of gene rearrangement being assembled into a receptor complex at the cell surface. This assembly is instrumental in promoting further T-cell development, giving rise either to the commoner $\alpha:\beta$ type receptor or to the rarer $\gamma:\delta$ type receptor. The two major classes of T cell, $CD4^+$ and $CD8^+$ cells, differ in the class of MHC molecule that their membrane receptor recognizes, recognition that will later be stabilized by co-stimulatory molecules from ANTIGEN-PRESENTING CELLS. The rearrangement of the $V\alpha$ gene segments continues until a receptor is positively selected by binding a peptide:self-MHC complex – one of the ways in which T-cell development differs from that of B cells (the final assembly of the immunoglobulin in B cells normally leads to cessation of gene rearrangement).

At this stage, thymocytes express both the CD4 and CD8 co-receptor molecules and are double-positive $CD4^+CD8^+$ cells. If naïve T cells were activated by self-antigens on leaving the thymus, no body cell would survive T-cell attack. The process of *central tolerance* to self-antigens ensures that this is averted (see AUTOIMMUNE DISEASES). It is interaction of the TCR with peptide self-MHC complexes that provides the survival cue to developing T cells and constitutes positive selection. Cells that fail to form functional TCRs or fail to recognize the MHC molecules present on the thymic stromal epithelium never progress further than the double-positive $CD4^+CD8^+$ stage and undergo APOPTOSIS by neglect. Thymocytes with a high affinity for peptide:self-MHC complexes could damage host cells if allowed to mature and are also deleted by apoptosis (negative selection). But those with low affinity for these complexes are positively selected into the $CD4^+$ and $CD8^+$ lineages.

What mature T cells consider to be self-MHC is determined by the MHC molecules expressed by the thymic stromal cells they encounter while in the thymus and determines the MHC restriction (MHC class I or MHC class II) of the mature TCR repertoire. T cells that receive an MHC-peptide signal but do not simultaneously receive a signal from CO-STIMULATORY MOLECULES (e.g., CD80 and CD86) on the antigen-presenting cell may also undergo apoptosis. At the end of the selection process, mature $CD4^+$ MHC class II-restricted (T_H) T cells and $CD8^+$ MHC class I-restricted (T_C) T cells move to the periphery and leave the thymus, giving rise to the peripheral pool of naïve T cells. An alternative fate for a developing $CD4^+$ thymocyte with an intermediate avidity for self-peptide ligands (i.e., not strong enough to be deleted by apoptosis), and in addition expressing the CD25 component of the IL-2 receptor, is that it will up-regulate expression of the *FoxP3* gene that encodes a transcription factor required for CD25 expression and for production of the important effector cytokine, IL-10. Such cells develop into regulatory T cells (T_{reg}; aka suppressor T cells), the most potent of which have the characteristic $CD4^+CD25^+$ phenotype (see later).

While central tolerance is taking place, in mice at least, it is possible to introduce precursors of T cells and dendritic cells from foreign bone marrow in such a way that peptides they then present to maturing T cells in the host thymus will later be treated as self antigens. Such animals later accept grafts of skin from these same donor sources.

The crucial first step in adaptive immunity is activation of naïve antigen-specific T cells by antigen-presenting cells in the lymphoid tissues (see LYMPH NODES, Fig. 118). When activated, a T cell oxidizes fatty acids in preference to glucose to fuel movement and protein synthesis and build its extensive lamellipodia. Pathogens and their products are carried to lymphoid tissue by the lymph that drains tissues, or more rarely by blood (see SPLEEN). As they migrate through the cortical region of a lymph node, naïve T cells bind briefly to each encountered antigen-presenting cell (APC), especially DENDRITIC CELLS, through interactions between their own LFA-1, ICAM-3 and CD2 proteins and ICAM-1, ICAM-2, DC-SIGN and CD58 on the dendritic cell. This provides time for naïve T cells to sample MHC molecules on each APC for the presence of any peptide:MHC combinations recognized by its own TCR. If such a rare event does occur, the signalling through the TCR alters the LFA-1 confirmation, greatly increasing its affinity for ICAM-1 and ICAM-2 and stabilizing the link between the T cell and APC for days, during which time the T cell proliferates, its progeny also binding to the APC and differentiating into activated effector T cells. If no peptide:MHC recognition takes place, the T cell delivers a survival signal to the APC and dissociates from it. The CD40 ligand on the T-cell surface binds to CD40 on the APC, transmitting activating signals to the T cell and activating the APC to present B7 CO-STIMULATORY MOLECULE. Only when an APC presents both a specific antigen to the T-cell receptor and B7 to the CD28 receptor on the T cell is the T cell activated to release IL-2, an autocrine signal which binds the secreting T cell's receptor and causes it to proliferate. Its progeny cells will differentiate into armed effector T cells; but T cells that bind antigen in the absence of co-stimulation by the same APC die, a dual requirement that helps prevent naïve T cells from responding to self antigens on tissues cells.

A T cell falls into one of two major classes on account of the type of co-receptor CD protein it bears on its cell membrane. During antigen recognition, these co-receptors associate in the plasma membrane with the T-CELL RECEPTOR (TCR). CD4-bearing T_H cells ($CD4^+$) recognize the invariant site of MHC class II molecules while CD8-bearing T_C cells ($CD8^+$) recognize the invariant site of MHC class I molecules and each stabilizes the link between the TCR and the peptide:MHC complex.

Differentiation of naïve $CD4^+$ T cells into the two major classes of CD4 effector T cell (T_H1 cells and T_H2 cells) takes place when they respond to antigen in the lymph nodes, influenced by the cytokines present during early T-cell activation. This engagement determines not only the relative numbers of the two $CD4^+$ cell types produced but also the extent of macrophage activation and the isotypes of ANTIBODIES that will predominate in a B-cell response. This in turn influences whether the type of adaptive response is directed against intracellular or extracellular pathogens. There is general agreement that IL-12 activation of STAT4 (see STATS) is necessary for optimal differentiation of naïve T cells into IFN-γ-producing T_H1 cells. IFN-γ then activates the STAT1 transcription factor and subsequent expression of the T-bet transcription factor, a step which initiates T_H1 lineage development from naïve cells both by activating T_H1 genetic programmes and by repressing the opposing T_H2 programmes. T_H1 cells are important for cell-mediated immunity and T_H2 cells are associated with humoral responses. Transient interactions between genes and their regulatory elements, e.g., the IFN-γ (*Ifng*) gene and the locus control region of the T_H2 cytokine locus in naïve $CD4^+$ cells (see NUCLEUS, GENOME), are thought to hold the *Ifng* and cytokine gene loci in a state poised for rapid response to T-cell activation by expression of both loci at low levels. After the decision to differentiate into a T_H1 or T_H2 cell has been made, and expression of *Ifng* or T_H2 cytokines is high, these interchromosomal associations are relinquished in favour of intrachromosomal ones.

T_H1 cells are pivotal in the activation of MACROPHAGES through expression of effector molecules such as IFN-γ and TNF-α, especially important when the macrophage has been infected by parasites that proliferate within its vesicles. When a T_H1 cell TCR binds a macrophage (or other APC) presenting the appropriate peptide, its CD40 ligand binds the CD40 TUMOUR NECROSIS FACTOR receptor (TNFR) on the macrophage and delivers the sensitizing signal that, with secreted IFN-γ, activates macrophage to digest those intracellular microbes (e.g., *Mycobacterium tuberculosis* and *M. leprae*) located within their phagosomes and whose antigen peptide fragments they presented (see TUBERCULOSIS entry for pathogen evasion). CD40 also activates B cells. Effector T_H1 cells also induce cytotoxic T cells to kill (e.g., virally) infected cells. They may also stimulate B cells to secrete certain subclasses of IgG antibodies that coat extracellular microbes and activate COMPLEMENT (see INFECTION).

An effector T_H2 cell, by contrast, is a relatively poor macrophage activator since it does not produce IFN-γ and produces IL-10, a macrophage-deactivating cytokine. Further, activated T_H2 cells secrete TGF-β, which with IL-10 inhibits activation and growth of T_H1 cells. Instead, T_H2 cells secrete IL-4, IL-5 and IL-13. By so doing, they will stimulate B cells to make most classes of antibody, including IgE and some classes of IgG that bind MAST CELLS, BASOPHILS and EOSINOPHILS, so that these in turn release local mediators causing coughing, sneezing or diarrhoea and help to expel extracellular microbes and larger parasites from epithelial surfaces. T_H2 cells secrete IL-3, IL-4, IL-5, IL-9 and IL-13 (all of which activate B cells, IL-4 inducing ISOTYPE SWITCHING and IL-5 promoting eosinophil production; see also ASTHMA) and GM-CSF. When activated, T_H2 cells express the CD40 ligand on their cell surfaces, like T_H1 cells.

In order for an APC to activate a cytotoxic or helper T cell, it must provide a costimulatory signal in the form of B7 proteins (e.g., CD80 and CD86) that are recognized by the co-receptor protein CD28 on the T-cell surface. B7 production is induced by pathogens during an innate immune response to the infection and is then increased by T-cell contact (compare INNATE IMMUNITY). Cytotoxic T cells (T_C cells; cytotoxic CD8 T cells) are predominantly killer T cells that recognize internally produced pathogen- (e.g., virally) derived peptides bound to MHC class I molecules on infected cells and kill these target cells by APOPTOSIS, during which process intracellular pathogens are destroyed by the same nucleases that destroy the target cell's DNA, preventing release of infectious agents. Death receptor ligands, notably FasL, are often used by T_C cells to kill CANCER CELLS, which display the Fas receptor. Alternatively, on aggregation of the T-cell receptor, proteases ('granzymes') stored in an active form within the T_C cell's lytic granules are released by calcium-dependent exocytosis. These include perforins (which polymerize on contacting target cell membranes and form transmembrane pores), serine proteases and the pro-apoptotic protein granulysin. These latter proteins appear to enter the target cell through the perforin pores. Granzyme B cleaves and activates the cellular enzyme caspase-3, activating a CASPASE proteolytic cascade. A second, perforin-independent, route to cytotoxicity by apoptosis involves the binding of Fas receptors (CD95) in the target-cell membrane to the Fas ligand (FasL) present in membranes of activated T_C cells, T_H1 cells and some T_H2 cells.

After immunization, the numbers of armed reactive T cells to a given antigen increase dramatically, falling back to persist at a level 100–1,000-fold above their initial frequency. Termed *memory T cells*, they carry cell-surface proteins more like those of armed effector cells than of naïve T cells. Changes in cell-surface proteins favour their migration from the blood into the tissues rather than directly into lymphoid tissues. CD8 memory T cells require reactivation to become cytotoxic, do not require new DNA synthesis and have a gene expression pattern (e.g., *Bcl-2* expression) favouring cell survival. Some CD4 memory T cells (*effector memory cells*) undergo CD45 isoform changes through alternative splicing, enabling better antigen recognition in preparation for activation to effector status. But both CD4 and CD8 memory cells may express CCR7 and recirculate as *central memory cells* to T-cell zones of peripheral lymphoid tissues. Being very sensitive to T-cell receptor cross-linking, they rapidly produce CD40 ligand in response, but take longer to differentiate and do not secrete as much cytokine as effector memory cells on restimulation. T_C cells make TNF-α in both soluble and membrane-associated forms, but when activated their most important membrane-associated TNF-related molecule is the Fas ligand (CD178).

All three major classes of effector T cells (T_H1, T_H2 and T_C cells) bear the integrin LFA-1 which binds ICAMs on antigen-presenting cells (APCs) and also express one or more members of the TNF family when they recognize their specific antigen on the target cell.

A fourth T-cell class, regulatory (suppressor) T_{reg} cells (T_R cells), is involved in recognition, and 'toleration', of the body's own cells and molecules as 'self', and therefore as harmless. They act on other effector cells, particularly self-reactive cells, and can suppress their activity, including their cytokine secretion. They are attracted to antigen-presenting cells and B cells by CCL4, which may play a central role in the normal initiation of their humoral responses, and actively suppress immune responses of these and other cells. They are involved in closing down immune responses after they have successfully tackled invading organisms, but they also prevent pathological self-reactivity (AUTOIMMUNE DISEASE), their genetic deficiency leading to *severe autoimmune syndrome*. Since their activity would be a hindrance during a normal immune response, it appears that this is downregulated on encountering an infectious microorganism (a response manipulated to their advantages by retroviruses, *Leishmania* and *Plasmodium*). The suppressive mechanism is a subject of active research, but appears to require cell-to-cell contact with the cell undergoing suppression. A recently discovered third class of T_H cell, the pro-inflammatory T_H17 cell, is characterized by production of IL-17A (see CYTOKINES) and appears to defend the host against specific pathogens and to be a potent inducer of autoimmunity and tissue inflammation. T_{reg} cell development is reciprocally related to that of the T_H17 cells, and the dioxin receptor AHR is a regulator of both cells' differentiation. TGF-β1 mimics dioxin's effects and induces T_{reg} cell differentiation, whereas TGF-β1 in association with IL-6 results in T_H17 differentiation. T_{reg} cells expressing the transcription factor Foxp3 control the autoreactive components of the immune system. By modulating TGF-β signalling within the nucleus, AHR is likely to shift the balance between the two T-cell populations with opposing effects. T_{reg} cell secretion of TGF-β and IL-10 has also been implicated in inhibiting T-cell proliferation, while IL-10 can also affect differentiation of dendritic cells, inhibiting their secretion of IL-12 and impairing their ability to promote T-cell activation and T_H1 differentiation. See IMMUNE TOLERANCE, IMMUNOSUPPRESSION.

tears Secretions of the LACRIMAL GLANDS.

tectospinal tract Originating in the SUPERIOR COLLICULUS of the midbrain, a descending tract of the SPINAL CORD operating in conjunction with the vestibulospinal tract in maintaining coordinated head and eye orientation so that an appropriate point in external space is imaged on the fovea. See TEGMENTUM, VESTIBULOSPINAL TRACT.

tectum The roof of the midbrain (see BRAIN, Fig. 28), differentiating into the SUPERIOR COLLICULUS and the INFERIOR COLLICULUS.

teeth See DENTITION.

tegmentum The floor of the MIDBRAIN (see BRAIN, Fig. 28). A subcortical structure, the most ventral part of which is occupied by the SUBSTANTIA NIGRA. Dopamine manufactured in these ventral tegmental area (VTA) neurons forms part of the motivation system. It also contains the more dorsal RED NUCLEUS from which axons of the RUBROSPINAL TRACT leave and pass ventromedially, crossing in the ventral tegmental decussation. Axons of the TECTOSPINAL TRACT arise in the superior colliculus of the midbrain and cross in the dorsal tegmental decussation.

telencephalon The secondary brain vesicle (see BRAIN, Fig. 28) that eventually forms the CEREBRAL HEMISPHERES.

telomerase A reverse-transcriptase-like enzyme (see DNA POLYMERASES) that replicates the ends of chromosomes. It carries an RNA component containing a short sequence near its 5′-end which is the antisense sequence of the TELOMERE repeat sequence. Recognizing the tip of a G-rich strand of an existing telomere DNA repeat sequence, it elongates it in the 5′- to 3′ direction, synthesizing a new copy of the repeat by using its own RNA template. The other parental DNA strand is then replicated using the newly synthesized strand as a template for DNA polymerase α. As a result, the 3′-end of the telomere DNA is always slightly longer than the 5′-end. With designated assistance from proteins such as TRF2, this protruding end tucks itself back into the duplex DNA so that the end of a chromosome normally has a unique structure protecting it from degradation and attempts at DNA repair. Although telomerase is present in the human germline, its absence from somatic cells other than some stem cells results in telomere length shortening by 50–100 bp per cell generation. Eventually, a signal is generated similar to the double-strand break arresting the CELL CYCLE (see Fig. 39). Cells in prolonged culture reach a point of SENESCENCE, and stop dividing. Engineering telomerase into cells that normally lack the enzyme preserves the telomeres through many cell cycles. Possible telomerase inhibition by 'type 1 interferon' is believed to lead to increased telomere erosion in T CELLS.

telomeres Repetitive DNA sequences (see DNA REPEATS), in humans GGGTTA, extending for ~10,000 nucleotides at the ends of chromosomes and replicated in the germ line and some stem cells by TELOMERASE. Because lagging strand DNA synthesis requires primers ahead of the replication machinery, when the last primer is removed a single-stranded tip remains in one of the daughter DNA molecules. When the daughter chromosome is replicated again, the strand missing the end sequences would become a shortened double-stranded molecule, becoming shorter after each replication cycle. The telomere repeat sequences permit this inevitable erosion of terminal DNA without information loss: it can be sacrificed to the replication process. Telomeres are protected by densely compact chromatin (see HETEROCHROMATIN) whose protein components include some which are involved in DNA REPAIR MECHANISMS and prepared for telomere maintenance. After many cell generations, descendant cells will inherit defective chromosomes because their tips cannot be properly replicated, signalling the cell to enter replicative cell SENESCENCE. Recent evidence indicates that telomere DNA may be transcribed into RNA, and that this may be involved in regulation of telomere length.

As telomeres shorten with age, subsequent exposure of chromosome ends can trigger their end-to-end ligation, with disastrous results for the cell and its progeny. A checkpoint enabling such a cell to decide whether to senesce or undergo apoptosis when telomeres become critically short is the checkpoint activator ataxia telangiectasia mutated (see ATM), which can signal to checkpoint arrests throughout the CELL CYCLE. In human somatic cells, the telomere repeats have been proposed to provide a cell with a 'counting mechanism' preventing unlimited proliferation of damaged cells (e.g., CANCER CELLS). It has been proposed that the telomerase system is important in the maintenance of tissue architecture, and has a bearing on AGEING. Mice have very long telomeres, so that they must go through three or more generations before telomeres have shrunk to the normal human length. Unlike humans, they retain telomerase activity in somatic cells. But transgenic mice lacking telomerase develop progressively more defects with generation number, especially in proliferative tissues, older animals showing a marked tendency to produce tumours. These animals do not age prematurely, however.

Although shortened telomeres were seen in the first cloned sheep, they were not observed in cloned calves using almost senescent embryonic cells as donors. Telomere length is adjusted postzygotically in embryos, including cloned embryos, apparently through restored telomerase activity after nuclear transfer.

telophase The stage of MITOSIS and of each division of MEIOSIS when chromosomes arrive at the poles of the spindle. In a mitotic cycle, the chromosomes then decondense and a new nuclear envelope reforms around each set. These processes also take place at the ends of both meiotic telophases during oogenesis and spermatogenesis.

temperature regulation Humans are endotherms, generating body heat from internal organs. Heat is released during food consumption, digestion, absorption and assimilation (all aspects of *obligatory thermogenesis*). The other major component of thermogenesis is *facultative thermogenesis* and relates to activation of the sympathetic nervous system and its stimulating effect on metabolic rate.

The reasons for the remarkable stability of body temperature achieved by homeothermic regulation among mammals, and why 37°C has been evolutionarily selected in humans, remain obscure. Changes in reproductive physiology that might result from a lower value might select against it, although homeotherms with a restricted calorific intake develop a low body temperature and have a prolonged lifespan (not that this would directly increase reproductive output). Although few people would choose a lifestyle that limited their caloric intake, this raises the question of whether 37°C is indeed the optimal human body temperature. At temperatures above ~37.1°C, the 'set-point' of the temperature control mechanism, the rate of heat loss > the rate of heat production while the reverse is true below this value. This is the value, set by integrated activity of non-temperature-sensitive interneurons in the HYPOTHALAMUS, above which the posterior hypothalamus initiates sweating and below which it initiates shivering. Catecholamines from the pontine reticular formation regulate these interneurons and alter the set point, lowering it by ~0.5°C during sleep and increasing it by progesterone during the luteal phase of the MENSTRUAL CYCLE. FEVER is similarly induced by ENDOGENOUS PYROGENS, while chronic cold exposure also results in long-term shifts of the set point.

The *preoptic area*, the most rostral (anterior) hypothalamus, is both thermosensitive and required for maintaining normal body temperature. Neurons in the medial preoptic region have an intense inhibitory effect on thermogenic responses (see NEUROPEPTIDE Y) while neurons in the medial hypothalamus, the paraventricular and dorsomedial nuclei, have an excitatory effect on thermogenesis (see LEPTIN) and it is the interplay between these two sets of neurons that is critical in control of body temperature over a wide range of conditions. The preoptic thermosensitive neurons, with similar receptors in the cervical spinal cord, transduce small changes in blood temperature into changes in firing rate, and project to the posterior hypothalamus. Humoral responses, and visceromotor responses such as constriction of arterioles and piloerection, are then mediated by neurons in the medial preoptic region while behavioural responses such as shivering and warmth-seeking are initiated by neurons in the lateral hypothalamic regions. *Vasodilation* of cutaneous arterioles is caused by inhibition of the sympathetic centres in the posterior hypothalamus, while *vasoconstriction* in response to external cold is caused by increased sympathetic output and stimulation of α-adrenoceptors. SWEATING is brought about by sympathetic outflow to the skin everywhere, and to some extent by adrenaline. An additional 1°C in body temperature causes both sufficient sweating to remove ~10 times the basal rate of heat production through latent heat of vaporization (sunburn reduces the ability to sweat), and a decrease in thermogenesis through inhibition of shivering and chemical thermogenesis. *Shivering*, which is thermogenic, is initiated by the stimulation of brainstem reticular neurons by the preoptic area. The reticular neurons synapse with γ-fusimotor neurons, so that increasing muscle tone in one muscle (the agonist) causes intrafusal fibres of MUSCLE SPINDLES in its antagonist to generate a stretch reflex, a repetitive cycle of the somatic motor system which begins in the masseter muscles of the jaw and spreads to the trunk and proximal limb muscles. A further heat-gain mechanism, *non-shivering thermogenesis*, is very important in human babies since they cannot shiver. It is caused by increased sympathetic activity to brown ADIPOSE TISSUE (BAT), when noradrenaline acts on β3 adrenoceptors, raising internal cAMP. This activates lipolysis, liberating free fatty acids for oxidation, and uncoupling the collapse of the mitochondrial proton gradient from ATP synthesis, liberating heat. THERMORECEPTORS in peripheral areas of the body also contribute slightly to body temperature control by altering the hypothalamic set-point temperature (see BASAL METABOLIC RATE, LOWER CRITICAL TEMPERATURE, THERMONEUTRAL ZONE). When the skin becomes cold these afferents, which run in the spinothalamic tract, drive the hypothalamic centres to the shivering threshold even though the brain temperature itself is on the hot side of normal: a low skin temperature would soon lead to a very depressed body temperature unless heat production increased. When skin temperature is high, sweating will begin at a lower temperature than when it is low (the combined effect of low skin temperature and sweating would result in too much heat loss).

Cold-sensitive neurons in the anterior hypothalamus release thyrotropin releasing hormone, initiating THYROTROPIN release by the anterior pituitary and, via THYROID HORMONES, promoting chemical thermogenesis, notably by brown ADIPOCYTES (see STRESS RESPONSE). The increased thyroxine output by the thyroid gland is thermogenic, but usually after several weeks' exposure of the body to cold: time enough for thyroid gland hypertrophy. A rise in core body temperature detected by anterior hypothalamic neurons results in slowing of metabolism through reduced thyrotropin release (shunting of blood towards the periphery to promote heat dissipation), and shade-seeking behaviour.

Most enzyme pathways are temperature sensitive and having a constant and high body temperature (T_b) enables rapid responses of cells to changing needs. Although the central nervous system can perform at low temperatures, it functions more rapidly at higher ones: for example, neurotransmitters diffuse faster across synapses. A high T_b therefore enables faster information processing and rapid response times. Muscle viscosity also declines at higher temperatures, reducing internal friction during contraction, again speeding response times. There is also an improvement in stamina with high T_b, endotherms being much more capable of sustaining aerobic respiration and continuous muscle contraction than ectotherms, in which bursts of muscle activity are usually anaerobic and sustainable for only brief periods. Endotherms have far greater independence of the environmental temperature than ectotherms, in particular allowing them to be alert and active at night when solar energy is absent. However, endotherms must consume food faster, and with a higher calorific value, than ectotherms.

Low ambient temperatures are dangerous: blood becomes viscous, veins contract and blood pressure rises making the heart work harder. Deaths due to heart disease and stroke increase by one-tenth for each three-degree drop in temperature and winter death rate for poor people is twice what it is for the well-off; Australian aborigines can sleep naked outside on frosty nights when the relatively recent colonists shiver in sleeping bags! Several derived mechanisms for dissipating metabolic heat in *Homo* include a narrow, elongated body form; an elaborated cranial venous circulation enabling venous blood that has been cooled by sweating in the face and scalp to cool hot arterial blood in the internal carotid artery prior to its reaching the brain (by a countercurrent heat exchanger in the cavernous sinus). Likewise, given the comparatively small human nasopharynx, mouth breathing, rather than panting – or nasal breathing as in apes – allows higher ventilatory demands to be met, and excess heat to be unloaded, in strenuous activities such as endurance running (see GAIT). See HEAT.

temporal bones Paired neurocranial bones of the SKULL, each joined superiorly to a parietal bone by a squamous suture and anteriorly to a sphenoid bone. A prominent feature of the temporal bone is the EXTERNAL AUDITORY MEATUS, just posterior and inferior to which is the mastoid process, seen and felt as a lump just posterior to the ear. The temporal bone contributes a zygomatic process to the zygomatic arch, which is the origin of the masseter muscle, and the mandibular (glenoid) fossa also lies in the temporal bone.

temporal lobe One of four major subdivisions of each hemisphere of the BRAIN (see Fig. 28), separated from the frontal lobe, and in part from the parietal lobe, by the lateral fissure beneath which it lies. Its lateral surface is divided into three principal gyri running parallel to the lateral fissure: the superior, middle and inferior temporal gyri, the first two of which are implicated in CONSCIOUSNESS. It functions in auditory perception, speech and complex visual perceptions. The superior temporal gyrus of the dominant hemisphere includes the PRIMARY AUDITORY CORTEX, including WERNICKE'S AREA beneath and parallel to the lateral fissure. The superior temporal sulcus harbours MIRROR CELLS. On the medial surface of the hemisphere, portions of the frontal, parietal and temporal lobes constitute components of the limbic system, the cingulate gyrus becoming continuous with the parahippocampal gyrus of the temporal lobe, the anterior temporal lobe containing the AMYGDALA. Deep to the parahippocampal gyrus, within the temporal lobe, lies the HIPPOCAMPUS, formed by an in-curling of the inferomedial part of the lobe. The INFEROTEMPORAL CORTEX is an association centre involved in the perception of form. See CEREBRAL HEMISPHERES, SEIZURES, SPATIAL AWARENESS.

temporalis muscle Originating on the temporal bone and inserting on the coronoid process of the mandibular bone, a muscle elevating and retracting the mandible and assisting in its side-to-side movements. Compare MASSETER MUSCLE; see MASTICATION.

tendon organs (Golgi tendon organs, GTOs) Proprioceptor organs, enclosed in a delicate connective tissue capsule and surrounding a bundle of tendon fascicles, each with a small swelling containing nerve terminals sensitive to *tension* of the tendon. In series with muscle fibres, they help to control muscle tension and, with MUSCLE SPINDLES (with which they work in opposition), are involved in the control of motor function by the spinal cord and in relaying the state of muscle contraction via the spinocerebellar tracts to the cerebellum. Increases in muscle tension stretch the collagen fibres in the GTO and distort terminals of the Ib afferents interwoven among them, activating a negative feedback reflex, the *inverse myotatic reflex*, opposing increase in tension. The Ib afferents synapse with Ib inhibitory interneurons in the intermediate zone of the spinal cord, which synapse in turn with α motor neurons supplying the homonymous and synergistic muscles. This inhibition is augmented by Ia muscle spindle afferents, afferents in joints, and skin mechanoreceptors onto Ib interneurons, the significance of which is unclear. Descending motor pathways may either excite or inhibit the Ib interneurons. See ELBOW.

tendons Any of the white fibrous cord of dense, regularly arranged, collagenous connective tissue continuous with the epimysium, perimysium and endomysium of skeletal muscle, attaching it to the periosteum of bone. See TENDON ORGANS; compare LIGAMENTS.

teratogens Agents that can cause malformations of an embryo or foetus.

teratomas A distinctive tumour type, although of disputed origin, containing cell types derived from all three embryonic germ layers (STEM CELLS), including bone, hair, muscle, gut epithelia. It is believed that these tumours originate from a pluripotent stem cell, able to differentiate into the above tissue types. There is some evidence that errant primordial germ cells, having deviated from their normal migration paths, may be responsible for some teratomas. Alternatively, some may arise from epiblast cells that have migrated through the primitive streak during gastrulation. If remnants of the primitive streak persist in the sacrococcygeal region, clusters of pluripotent cells there may form sacrococcygeal teratomas.

terpenoids Phenolic lipid substances containing multiple units of the 5-carbon hydrocarbon isoprene, and present especially in fruit and vegetables. They include monoterpenes (e.g., limonene, carvone or carveol), diterpenes (e.g., the RETINOIDS) and tetraterpenes (e.g., all different carotenoids, such as α- and β-carotene, lutein, lycopene, zeaxanthine and cryptoxanthine). They have been associated with reductions of oxidative stress, carcinogenesis and cardiovascular disease. See HEALTHY DIET.

territories For chromosome territories. See NUCLEUS.

testes (sing. testis) Although the male and female gonads appear during the 3rd week of development as a pair of longitudinal *genital ridges* adjacent to the mesonephric ridges (see SEX DETERMINATION), they do not acquire distinctly male or female characteristics until the 7th week. The ridges are invaded by PRIMORDIAL GERM CELLS during the 5th week by migration along the dorsal mesentery of the hindgut and failure of these amoeboid cells to arrive results in lack of gonad development, indicative of their importance in both testes and OVARIES in inducing adjacent tissues to differentiate. Around the time of arrival of the primordial germ cells, the epithelia of the genital ridges of gonads in both sexes proliferate and epithelial cells penetrate the underlying mesenchyme to form a number of irregularly shaped primitive sex cords, connected to the surface epithelium. The primitive sex cords of this *indifferent gonad* then continue to proliferate, under the influence of testis-determining factor (see SEX DETERMINATION), and penetrate the medulla of the mesonephros to form the testis (or medullary) cords, breaking up into a network of tiny cell strands that later give rise to tubules of the rete testis separated from the epithelium by tough connective tissue, the tunica albuginea. Probably under the influence of HUMAN CHORIONIC GONADOTROPIN, the testis cords become horseshoe-shaped in the 4th month and contain both primitive germ cells and sustentacular SERTOLI CELLS, while interstitial LEYDIG CELLS come to lie between the testis cords. The testis cords remain solid until PUBERTY when they acquire a lumen, become seminiferous tubules and join the rete testis tubules, which in turn enter the vasa efferentia (see later).

A serous membrane, the tunica vaginalis, derived from the peritoneum and partially covering each testis, forms during their descent from the posterior abdominal wall into the scrotum. It is not entirely clear what controls descent, normally completed by the 33rd week of development, but androgens and MÜLLERIAN INHIBITING HORMONE are involved. Outgrowth of the extra-abdominal portion of the gubernaculum leads to intra-abdominal migration and a build-up of intra-abdominal pressure due to organ growth assists passage through the inguinal canal, while regression of the extra-abdominal part of the gubernaculum completes the process. The testes descend over the rim of the pubic bone and are normally present in the scrotum at birth (see CRYPTORCHIDISM).

The outer part of each testis, the connective tissue of the tunica albuginea, enters the organ as incomplete septa, dividing it into ~300–400 conical lobules containing the seminiferous tubules (~350 m in length) which are surrounded by groups of LEYDIG CELLS. The epithelium of the seminiferous tubule overlies spermatogonia, self-renewing progenitor cells which divide mitotically and produce primary spermatocytes. Lying among the spermatogonia are SERTOLI CELLS by means of whose tight junctions the tubules are divided into two compartments. The basal compartment below the tight junctions has connections with the circulatory system and surrounds the developing spermatogonia and primary spermatocytes. The adluminal compartment, in which meiosis is completed, protects the spermatocytes from immune responses, the tight junctions forming a '*blood–testis barrier*'. The tubules empty into a set of short tubules, the tubuli recti, which merge into the tubules of the rete testis. These ducts leave the testis as

~15 vasa efferentia which pierce the tunica albuginea, their ciliated epithelium helping to move the immobile sperm into the epididymis for storage.

The androgens produced by the testes during foetal and neonatal life are important in generating SEX DIFFERENCES (see also SEX DETERMINATION). After puberty, production of TESTOSTERONE increases again under the influence of GONADOTROPIN RELEASING HORMONE from the hypothalamus, the negative feedback involved being illustrated in Fig. 172.

In a recent study of GENE EXPRESSION diversity in humans and common chimpanzees, the ratio of gene expression divergence

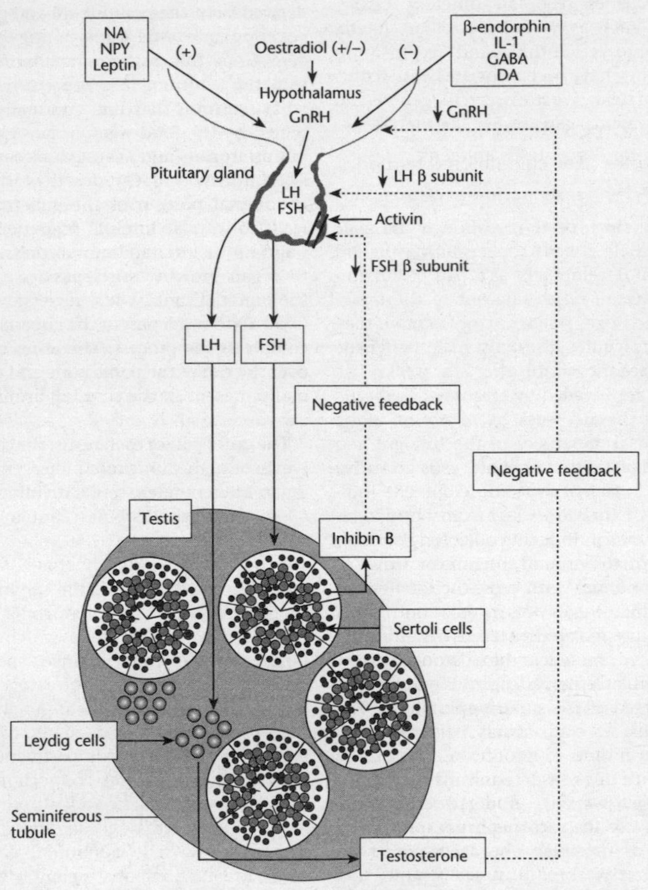

FIG 172. Schematic diagram showing the involvement of hormone production by the TESTES in the negative feedback regulation of gonadotropin release by the pituitary. The involvement of the ovaries is discussed in the MENSTRUAL CYCLE entry. DA = dopamine; GABA = γ-amino butyric acid; IL-1 = interleukin-1; NPY = neuropeptide Y; NA = noradrenaline; FSH = follicle stimulating hormone; LH = luteinizing hormone. See GONADOTROPIN-RELEASING HORMONE for more details. Redrawn from *Endocrine Physiology* © 2004, P.E. Molina (Lange Medical Books/McGraw-Hill), with permission of The McGraw-Hill Companies.

between species to gene expression diversity within species is higher for the testes than for other organs studied (brain, heart, kidney, liver), an observation that might indicate positive natural selection favouring gene expression differences. The difference might be explained were the cellular composition of the species' testes to differ more than it does for the other tissues; but although human and chimpanzee testes differ in size, there is no evidence that their cellular compositions differ. It has been suggested that strong selective constraints on genes account for the low extent of expression diversity in the testis. See CANCER for cancer-testis antigens.

testicular feminization See ANDROGEN INSENSITIVITY SYNDROME

testis-determining factor (TDF) Protein produced by the *SRY* gene on the short arm of the Y chromosome in male human foetuses, and by the *Sry* gene in mice. Triggers the gonadal ridges to begin developing into testes. Other autosomal and X-linked gene products are required for complete testis development. See SEX DETERMINATION, TESTIS.

testosterone The principal male androgen, produced by the interstitial cells of the TESTES and to a lesser extent by adrenal glands (in both sexes), controlling development of male sex organs, SECONDARY SEXUAL CHARACTERISTICS, sperm production (OESTROGENS are probably a minor contributor) and body growth. A derivative of cholesterol (see Fig. 48), its blood concentration in adult males varies (e.g., it is reduced during stress and raised during certain kinds of excitement, e.g. sexual). Its biological activity is thought to occur predominantly after entering cells by passive diffusion and binding a cytosolic androgen receptor and freeing heat-shock proteins complexed to it (see ANDROGENS). The receptor then dimerizes, enters the nucleus, and binds regulatory elements of target genes. It is one component of a negative feedback loop involving LUTEINIZING HORMONE (LH) produced by the anterior pituitary gland: low testosterone levels promote LH release and vice versa; raised LH levels stimulate testosterone release. Testosterone levels are also raised by increased FOLLICLE-STIMULATING HORMONE levels in the blood. In male embryos, some testosterone is converted by 5-α-reductase into dihydrotestosterone (DHT), an even more potent androgen, which promotes embryonic sexual differentiation of the urethra, prostate gland and EXTERNAL GENITALIA (scrotum, penis); but in the absence of HUMAN CHORIONIC GONADOTROPIN, DHT level declines after birth. Genetic males with 5-α-reductase deficiency have ambiguous external genitalia, and more than half are reared unquestioningly as girls. At puberty, a gonadal testosterone surge causes deepening of the voice, muscularity, descent of the testes and penis enlargement. Most of these girls develop erotic interest in females and change their gender identity from female to male (see SEXUAL ORIENTATION).

Testosterone is converted to oestradiol in adipose tissue by AROMATASE, whose expression is directly related to the degree of male adiposity. Testosterone is an immunosuppressant (see MORTALITY) and is inactivated in the liver. High testosterone levels *in utero* have been linked to possession of a ring-(2nd) finger as long as or longer than the index finger (see AUTISM). See SEX DETERMINATION, SEX DIFFERENCES.

tetanus Spastic, twitching paralysis caused by binding of tetanus TOXIN (TeNT), an exotoxin produced by the obligately anaerobic soil bacterium *Clostridium tetani*. A neurotoxin, the toxin is a specific protease that cleaves SNARE complexes at neuron terminals and blocks Ca^{2+}-dependent neurotransmitter release, often fatally (if ventilatory muscles are involved). If the toxin binds inhibitory interneurons it results in contraction of both members of an antagonistic pair of muscles. If muscles of the mouth are involved, lockjaw results. The tetanus TOXOID is used in tetanus vaccines. Compare BOTULISM.

tetrahydrofolate See FOLATE.

tetrodotoxin (TTX) A natural TOXIN that inactivates voltage-gated sodium channels, thereby inhibiting action potentials. Originally isolated from the ovaries of the Japanese puffer fish, it is usually fatal if ingested. It is commonly used in experiments requiring blocking of action potentials of neurons and muscle.

TGF-α, TGF-β See TRANSFORMING GROWTH FACTOR SUPERFAMILY.

T_H cells T-helper cells. See T CELLS.

thalamus Major component of the diencephalon of the forebrain and the principal bilateral relay and integration station for optic, auditory, and other sensory impulses reaching the cerebral cortex from the spinal cord, brain stem, cerebellum and other regions of the cerebrum. Lying beneath the brain's lateral ventricles (see BRAIN, Fig. 29), it is composed largely of grey matter organized into over 30 paired nuclei, organized into five groups, between which run tracts of white matter (see Fig. 173). Only 5–10% of thalamic synapses are made by afferent terminals, most of the thalamus being concerned with controlling which sensory inputs are sent to the CEREBRAL CORTEX. Except for olfaction, all sensory information reaching the cerebral cortex is first processed in the ventral or posterior groups of the thalamus, other nuclear groups being massively interconnected with cortical regions concerned with EMOTION (the anterior group) and MEMORY (the medial group).

The *intermediate mass* is a bridge of grey matter crossing the third ventricle and linking the left and right sides. The *medial geniculate nuclei* relay auditory impulses via the auditory radiation to the auditory cortex (see HEARING); the LATERAL GENICULATE NUCLEI form small eminences (geniculate bodies) on the surface, and relay visual impulses to the primary visual cortex; the *ventral posterior nuclei* (VPN) relay impulses for taste, touch, pressure, heat, cold and pain from the contralateral body surface to the PRIMARY SOMATOSENSORY CORTEX, and are highly organized somatotopically. Nuclei relaying impulses to the cerebrum include: the *ventral lateral nuclei* (from the cerebellum to the primary motor cortex) and the *ventral anterior nuclei* (from the BASAL GANGLIA via the

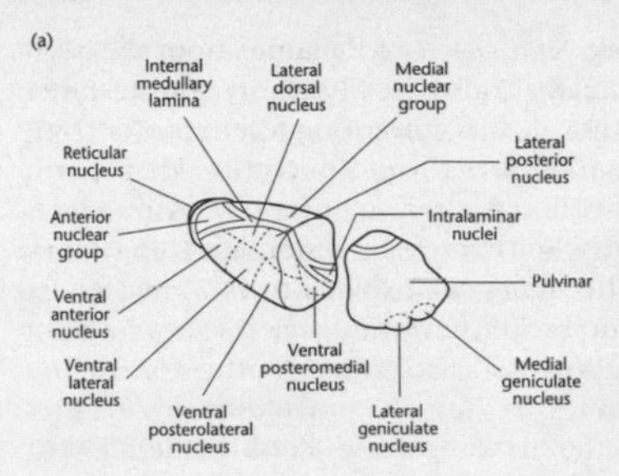

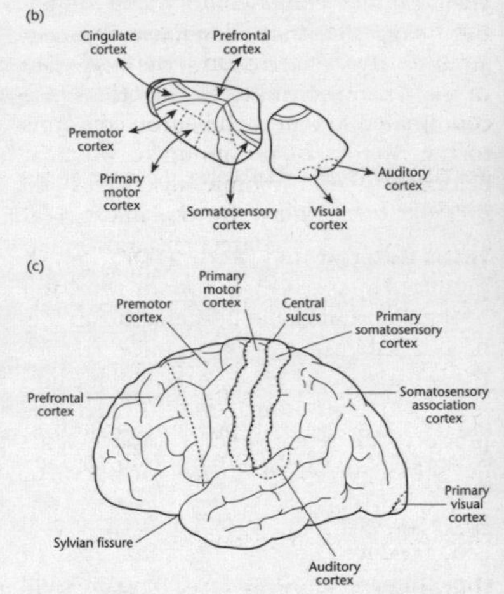

FIG 173. Diagram showing the anatomical and functional relationships between the CEREBRAL HEMISPHERES and nuclei of the THALAMUS. (a) The thalamus viewed posterolaterally, indicating the main nuclei. The posterior part is sectioned off to reveal the internal structure. (b) and (c) indicate the efferent projections from the thalamic nuclei to regions of the cerebral cortex, although all thalamic nuclei receive fibres from the cortex in a basically reciprocal fashion (indicated by two-headed arrows). The cingulate cortex is hidden. From A.R. Crossman and D. Neary © 1995 *Neuroanatomy, An Illustrated Colour Text*; redrawn with permission from Elsevier Ltd.

globus pallidus and substantia nigra to the premotor cortex). A major output of the hippocampus, via the fornix and mammillary bodies of the hypothalamus, reaches the anterior thalamus. The *anterior nuclei* project mainly to the cingulate gyrus, form part of the LIMBIC SYSTEM, and are involved in emotional responses and memory. Signals to the thalamus from the basal ganglia and cerebellum can be integrated with those representing recent motor commands from the motor cortex. Tremor and involuntary movements in PARKINSON DISEASE may be relieved by lesions in this region. The thalamus also has roles in awareness and knowledge acquisition (see COGNITION).

With its huge input to the cortex, the thalamus can act as a pacemaker. Under certain conditions, as a result of oscillatory feedback between excitatory and inhibitory neuron loops, thalamic neurons can generate rhythmic action potentials even in the absence of input to the cells. Once coordinated so that each neuron conforms to the rhythm of the group to which it belongs, these rhythmic discharges are passed to the cortex by thalamocortical axons and can stimulate a far larger group of cortical cells (see EPILEPSY). When we are awake, the thalamus allows sensory information to pass through it and be projected up to the cortex. When we are asleep, thalamic neurons enter a self-generated rhythmicity of firing that prevents organized sensory information from reaching the cortex (see SLEEP).

thalassaemias See HAEMOGLOBINOPATHIES.

therapeutic cloning See SOMATIC CELL NUCLEAR TRANSFER.

thermogenesis See TEMPERATURE REGULATION.

thermoneutral zone That temperature range over which an endotherm utilizes energy at a constant rate, its basal rate – far higher than that of an ectotherm of equivalent size. But when environmental temperatures extend further and further above and below this range the endotherm consumes increasing amounts of energy maintaining a constant body temperature. See THERMORECEPTORS.

thermoreceptors Neurons whose membranes are temperature-sensitive are clustered in the hypothalamus and spinal cord, and occur unevenly in the skin (see TEMPERATURE REGULATION). Some areas of skin 1 mm across are especially sensitive to either hot or cold, but not to both, while regions between such 'hot' and 'cold' spots may be relatively insensitive to temperature. Responses of different thermoreceptive neurons to temperature, in which HEAT-SHOCK PROTEINS are believed to be involved, indicates that they are of two classes exhibiting a static discharge frequency of action potentials: warm receptors (sensitive in the 30–48°C range, peaking at 42°C, with a rapid loss of sensitivity above 45°C) and cold receptors (sensitive in the range 10–35°C, peaking at ~25°C). Noxious cold receptors also exist, having a lower peak temperature threshold. Thermal NOCICEPTORS start firing at ~45°C, and although some cold receptors also fire at this temperature the brain interprets their impulses as a feeling of cold. Classic work in the early 1950s revealed that cold receptors are coupled to Aδ and C fibres (see AXON), while warm receptors are coupled to C fibres alone. Based on inputs from both cold and warm fibres, the CNS somehow identifies temperatures below the thermoneutral skin temperature of 33°C as cold, and temperatures above this as warm. The instantaneous responses of warm and cold receptors to temperature are mirror images of each other, implying that cooling evokes a dual response to the brain. This may explain Weber's illusion, whereby fingers coming from a bowl of cold water elicit a warm sensation when placed in water of neutral temperature while those coming from a bowl of warm water feel cold. Studies on mice implicate several TRP RECEPTOR channels in these responses. Mice lacking the menthol receptor TRPM8 fail to generate a burst in spike frequency in cutaneous C and Aδ fibres from cold receptors when the temperature is gradually lowered from 35°C to 2°C. Whether the off-response of these TRPM8-expressing cold fibres contributes to warm perception, or whether closing of heat-activated thermo-TRPs contributes to a cold response is not clear. But another warm receptor channel TRPV3, expressed in KERATINOCYTES, passes signals to sensory neurons via an unidentified messenger system. The capsaicin receptor serves both as a pain and thermal sensor, and is located in small-diameter neurons in sensory ganglia. The temperature pathway is virtually identical to that for PAIN (see SPINAL CORD, Fig. 163) and is routed through the SPINOTHALAMIC TRACT.

thermoregulation See TEMPERATURE REGULATION.

thermosensor See THERMORECEPTORS.

theta rhythms See ELECTROENCEPHALOGRAM.

(thick/thin stripe) pathways See PARVOCELLULAR BLOB/INTERBLOB PATHWAYS.

third-order neurons Afferent neurons whose cell bodies lie in thalamic nuclei and form the second synaptic connection in an ascending sensory pathway, projecting from second-order neurons to the cerebral cortex. The information carried usually reaches consciousness. See SPINAL CORD and Fig. 163.

thirst The sensation of thirst results mainly from increase in the osmolarity of the EXTRACELLULAR FLUID, and reduced blood volume and BLOOD PRESSURE, registered by cells of the supraoptic nuclei of the HYPOTHALAMUS and initiating activity in neural circuits that result in feeling thirsty. ~1–2% of body mass can be evaporated in sweat before thirst is triggered, so it is important to take proactive measures to prevent dehydration without relying on thirst to inform one of water loss. Raised angiotensin levels (e.g., during pregnancy) also increase thirst. A reduced sensation of thirst follows wetting of the oral mucosa after it has been dry, and stretching of the walls of the gastrointestinal tract by fluid. See OSMORECEPTORS.

thoracic cavity See THORAX.

thorax (chest, rib cage) Formed by the sternum and 12 pairs of RIBS (see SKELETAL SYSTEM, Fig. 156) and costal cartilages. See VEINS for the *respiratory pump*.

threshold The value of some stimulus parameter (or membrane voltage) below which a receptor (or ion channel) does not respond, and above which it does. The response may be an all-or-nothing response, or a graded response, depending on the system. See BISTABILITY.

thrombin A protein (MW 33,700) produced by splitting of the unstable PROTHROMBIN, removing four low molecular-mass peptides from each prothrombin molecule to form a fibrin monomer. Initially held together by weak hydrogen bonds, the monomers polymerize to form long fibrin fibres forming the reticulum of a blood clot. See BLOOD CLOTTING.

thrombomodulin See ANTICOAGULANTS.

thromboplastin See BLOOD CLOTTING.

thrombopoietin (TPO) A glycoprotein GROWTH FACTOR, sharing much of its amino acid sequence with ERYTHROPOIETIN, and produced mainly by the liver and the kidney. It stimulates the production and differentiation of MEGAKARYOCYTES in the bone marrow. The thrombopoietin gene is located at chromosome 3q26.3-27, mutations in which occur in some hereditary forms of thrombocytosis (high platelet count) and in some cases of leukaemia. Also produced by striated muscle and stromal cells in the red BONE MARROW. In the liver, its production is augmented by IL-6. Both STAT- and NF-κB-binding sites are required for TPO-induced Bcl-xL promoter activity. Its effects are mediated by JAK-STAT and PI(3)K (see PHOSPHATIDYLINOSITOL-3-OH-KINASE) pathways.

thrombosis The presence of a blood clot (THROMBUS) in an unbroken blood vessel, commonly a vein. If the thrombus dislodges and then settles in the pulmonary artery, pulmonary embolism results; if in the coronary artery, the result is *coronary thrombosis* (see CARDIOVASCULAR DISEASE, CORONARY HEART DISEASE). Treatment to dissolve the clot in the coronary circulation or in *deep vein thrombosis* might involve injection into a coronary artery of streptokinase or TISSUE PLASMINOGEN ACTIVATOR. Anticoagulants such as HEPARIN would also normally be infused. See STROKES.

thromboxanes A group of pro-inflammatory eicosanoids released from damaged cells, TxA_2 being released on activation of

PLATELETS and causing local vasoconstriction. The NSAID aspirin inhibits COX-1, involved in thromboxane synthesis from arachidonic acid. See CYCLOOXYGENASES.

thrombus A blood clot in the cardiovascular system formed from blood constituents. It may occlude a blood vessel, or be attached to the vessel or heart wall without occluding it. See BLOOD CLOTTING, PLATELETS, THROMBOSIS.

thymine A pyrimidine base, and a metabolite of dUMP and of methylene tetrahydrofolate, being synthesized from these by thymidylate synthase (see FOLATE). Deamination of cytosine in DNA to uracil is probably the reason why DNA contains thymine and not uracil. Thymine is also formed by the deamination of 5′-methylcytosine (see CPG DINUCLEOTIDE, DNA REPAIR MECHANISMS). Degradation of pyrimidines produces UREA.

thymocytes Lymphoid cells found in the thymus. Most of them are developing T CELLS, although some are functionally mature.

thymus A bilobed glandular LYMPHOID ORGAN, developing in the mediastinum from the 3rd PHARYNGEAL POUCH (see Fig. 134). Each lobe is surrounded by thin connective tissue, from which trabeculae extend into the substance of the gland dividing it into lobules. The framework of thymic tissue consists of epithelial cells joined by desmosomes, forming compartments in which lymphocytes reside. Most lymphocytes occupy the denser cortex, with fewer in the lighter medulla. The thymus expresses an abundance of ligands that activate Notch (see NOTCH SIGNALLING PATHWAY), notably the Delta-like family member Dll-4, creating an environment conducive to high level signalling by Notch in T-cell progenitors arriving from the bone marrow (see T CELLS). After PUBERTY, the thymus becomes very involuted, leading to a lack of distinction between cortical and medullary areas, with a corresponding reduction in thymic function. After 40–50 years, it usually has less than 10% of its maximal cellular volume, and much of the thymic lymphatic tissue has been replaced by adipose tissue. This correlates with a reduction in the export of newly derived naïve T cells. A rational method of ensuring that a patient accepts over a long term any stem cell transplant from a donor would seem to centre around employment of the same immune tolerance mechanisms the patient uses to induce tolerance to their own 'self' antigens. So it is disappointing that the main organ of the immune system that establishes and maintains this tolerance, the thymus, undergoes age-related atrophy, imposing limitations on any tolerance-induction strategy for possible use in adult patients. Reversal of this age-related atrophy, leading to rejuvenation of the thymus, would allow uptake of donor cells (e.g., HAEMATOPOIETIC STEM CELLS or their immediate PROGENITORS from BONE MARROW), subsequent development of IMMUNE TOLERANCE, and coexistence of donor and recipient cells (mixed chimaerism). Antigen-presenting cells (APCs) of donor origin migrate to or develop within the thymus, contributing to the negative selection of donor-reactive T cells and resulting in tolerance to that donor. T_{reg} cells of donor origin can be produced within the thymus and then populate the periphery, contributing to peripheral tolerance of an allograft (see GRAFT).

Several approaches to restoration of thymic function are currently in preclinical studies or clinical trials. In one, the endocrine–immune axis is targeted by analogues of GONADOTROPIN-RELEASING HORMONE, inhibiting pituitary release of adult gonadotropin and sex-steroid production. In another approach, thymic GROWTH FACTORS such as GROWTH HORMONE, IL-7 and keratinocyte growth factor (aka FGF7, which stimulates thymic epithelial cells) are all added since they are known to decline in an age-related manner. Sex steroid removal rejuvenates the thymus, increases IL-7-responsive progenitor cells in bone marrow and export of B cells, and increases the numbers of thymus-derived naïve T cells in the blood. The thymic stromal environment also improves, as does the bone marrow microenvironment. See EMBRYONIC STEM CELLS.

thyroglobulin A glycoprotein containing multiple tyrosine residues, synthesized in the follicular epithelial cells of the THYROID GLAND and secreted as a colloid into the follicular lumen through the apical membranes. It undergoes post-translational modification at the apical surfaces, multiple tyrosine residues becoming iodinated after which coupling of some of the iodotyrosine residues generates T_3 and T_4, ~70% of which is bound to it. Levels are elevated in thyroiditis and Graves disease.

thyroid gland Gland appearing as a proliferation of epithelial tissue in the floor of the pharynx behind the tongue and descending in front of the pharyngeal gut as a bilobed diverticulum, reaching its final position in front of the trachea in the 7th week of development. It begins to function by the end of the 3rd month, when follicular cells produce the THYROGLOBULIN colloid serving as the source of the thyroid hormone T_4 (and, to a much lesser extent, T_3). Of the two, T_3 has much greater physiological importance. If the thyroid malfunctions and too little thyroid hormone is produced (hypothyroidism), symptoms of slow heart rate, weight gain, lethargy and mental dullness may be attributable to low basal metabolic rate. If it occurs during pregnancy, or in newborn children, the condition can give rise to cretinism. See GRAVES DISEASE.

thyroid hormones The hormones triiodothyronine (T_3) and tetraiodothyronine (T_4) are derived from tyrosine by peroxidase activity in which THYROGLOBULIN is a protein scaffold. It folds in such a way that the ring structures of two iodinated tyrosine residues come close together and are then covalently bonded together enzymatically. Thyroid hormone production and release are under negative feedback control by TRH and THYROTROPIN (TSH), and their nuclear receptors (see NUCLEAR RECEPTOR SUPERFAMILY) are expressed in virtually all tissues. Being lipophilic, they circulate bound to specific transport proteins synthesized by the liver. They affect many cellular events, both genomic and non-genomic, entering cells through plasma membrane channels. T_3 is at least five times more potent than T_4 (the 'prohormone'), being produced from it by peripheral deiodination, principally in the liver and kidneys. Both T_4 and T_3 are converted to the biologically inactive hormone T_2 in the placenta, protecting the foetus, although thyroid hormones are essential for foetal development.

Non-genomic effects include stimulation of Ca^{2+}-ATPase activity in the plasma membrane and sarcoplasmic reticulum, and stimulation of the Na^+/H^+ antiporter. The most characteristic genomic effects include increase in O_2 uptake (by promoting transcription of cell membrane Na^+/K^+-ATPase), transcription of mitochondrial uncoupling protein and other mitochondrial enzymes, all favouring heat release and TEMPERATURE REGULATION; and regulation of protein synthesis and degradation (contributing to growth and differentiation), promotion of adrenaline-induced glycogenolysis, gluconeogenesis, cholesterol synthesis, and LDL receptor regulation.

Tissue-specific effects of thyroid hormones include promotion of osteoblast and osteoclast activity; promotion of cardiac output and increase in blood volume; promotion of adipocytes; proliferation and white ADIPOSE TISSUE differentiation; induction of liver and adipocyte lipogenic enzymes (see LIPOGENESIS); stimulation of synthesis of pituitary growth hormone and inhibition of thyrotropin; and promotion of axonal growth and development in the brain. See CONGENITAL HYPOTHYROIDISM.

thyrotrophs Cells constituting ~5% of anterior pituitary gland cells, synthesizing THYROTROPIN in response to THYROTROPIN RELEASING HORMONE.

thyrotropin (TSH) Although not a gonadotropin, thyrotropin is a member of the same group of glycoprotein hormones as them and is a dimeric protein with the same α-subunit as FSH and LH but with a unique β-subunit conferring its specificity. Secreted by the anterior pituitary THYRO-

TROPHS, under the influence of hypothalamic THYROTROPIN RELEASING HORMONE, it binds to a G_S protein-coupled receptor in the thyroid gland, activating adenylyl cyclase, the resulting increased cAMP stimulating a protein kinase A signalling pathway. This activates all events involved in THYROID HORMONE synthesis and release. Its release is under negative feedback control, especially from triiodothyronine.

thyrotropin releasing hormone (TRH) Synthesized and released by the paraventricular nuclei of the hypothalamus (mainly by parvicellular neurons) and released from nerve terminals in the median eminence. It binds a specific G protein-coupled receptor on THYROTROPH membranes and activates PHOSPHOLIPASE C, the resulting increase in cAMP leading to release of THYROTROPIN into the blood.

thyroxine (tetraiodothyronine, T_4) See THYROID HORMONES.

tibia (shin bone) The larger and more medial of the two weight-bearing bones of the leg (see FIBULA). Articulates proximally with the femur and fibula and distally with the fibula and talus bone of the ankle. It is connected to the fibula by an interosseous membrane.

tidal volume (TV) The volume of air inspired (and expired) per breath (see SPIROMETER, VENTILATION). During quiet breathing it averages 600 mL in men and 500 mL in women, although TVs of 2 L and above commonly occur during EXERCISE, increasing exercise MINUTE VOLUME to 100 L or more.

tight junction (zonula occludens) An occluding junction (see INTERCELLULAR JUNCTIONS) between epithelial cells serving as a barrier to the diffusion and mixing of certain proteins and lipids between the apical and basolateral regions of the CELL MEMBRANE, and at the same time sealing neighbouring cells together so that macromolecules and some smaller molecules, depending on the proteins forming the junction, do not pass across it (e.g., see ENTEROCYTES of the ileum, and PROXIMAL CONVOLUTED TUBULE). Some tight junctional barriers can be relaxed at times, allowing increased flow of solutes and water across the epithelial surface (see PARACELLULAR TRANSPORT). A tight junction is formed by numerous *sealing strands* that encircle the apical end of each epithelial cell in the cell sheet, the outer monolayers of the two interacting cell membranes being held together by the extracellular domains of transmembrane adhesion proteins of the sealing strands, notably the *claudins*. The effectiveness of the tight junction as a barrier increases logarithmically with the number of sealing strands present. Also present are *occludins* which, together with claudins, associate with *ZO* proteins, intracellular proteins that anchor the sealing strands to the underlying actin of the cytoskeleton. Other proteins are also present, perhaps involved in the regulating epithelial cell polarity, and delivery of components to the correct membrane domain. Compare ADHERENS JUNCTION.

tiling microarrays See DNA MICROARRAYS.

time perception Experiments using fMRI in which subjects are invited to estimate time duration reveal activity in the prefrontal and parietal cortices and basal ganglia. Many researchers, perhaps the majority, are now convinced that time perception involves a distributed neural network over multiple brain regions, and until recently the prevailing model had been that 'pacemaker' neurons (thought to reside in the dopamine-secreting neurons of the substantia nigra) release pulses of one or more neurotransmitters at periodic intervals while other 'accumulator' neurons (thought to lie in the striatum) accumulate them and linearly sum up the temporal pulses over the course of seconds to minutes. This 'pacemaker-accumulator' model, proposed in the 1970s and receiving some support into the 1990s, is now being challenged as being either too simplistic or even fatally flawed. In an alternative model, the striatum reads out intervals from a time slice of activity across a network of cortical neurons, different ensembles of which have firing rates oscillating at different frequencies but all of which are connected to the striatum. When an event is to be timed, on the 'striatal beat frequency' (SBQ) model, a cortical clock is reset through a synchronization of neuronal firing in the network. Striatal neurons track the changing pattern of activity in the network until a certain kind of reinforcement arrives, such as a pellet of food for a rat. Each reinforcement causes the substantia nigra to release a wave of dopamine onto striatal cells, helping to establish an association between their discharge and the pattern of network activity at that moment. After a number of experiences with a given duration, striatal neurons start to recognize this pattern of coincident activity across a subset of neuronal populations (a 'snapshot' which is filed in the long-term MEMORY), and step up their firing rate when the pattern occurs, indicating that the time interval is over. Opponents of the SBQ model have difficulty believing that patterns of synchronous firing among ensembles of neurons could ever be picked out among the synchronous spikes which are a ubiquitous feature of the cortex so as to invest them with significance. Some still work with the pacemaker approach, holding that time is somehow wrapped up inextricably with expectation and represented as part of an anticipatory build-up signal in each of the cortical areas involved in carrying out a task.

time to the most recent common ancestor (TMRCA) See COALESCENCE, MOST RECENT COMMON ANCESTOR.

tiredness See CIRCADIAN CLOCKS, FATIGUE.

tissue engineering Use of biomaterials (powders, solutions, doped microparticles) to stimulate local tissue repair. Cells in contact with the stimuli produce growth factors or ionic dissolution products that stimulate multiple generations of growing cells to self-assemble along the biochemical and biomechanical gradients and eventually repair the damage.

tissue fluid The major compartment of EXTRACELLULAR FLUID lying between body cells, other than those in the blood and lymph vessels. Bathing most of the body's cells, and being the medium of exchange between these and the blood, it comprises ~11 dm^3 in a normal 70-kg adult. It is formed by filtration, and then mostly reabsorbed, through the walls of blood capillaries as a result of the combined and opposing hydrostatic pressures of the blood and tissue fluid, and of their respective water potentials (osmotic pressures). The resultant net filtration pressures (see Fig. 174) are positive in an outward direction at the arteriole ends of capillaries and positive in an inward direction at the venule ends of capillaries. It has an ionic composition similar to blood plasma, but has a lower protein content since blood capillaries have a low permeability to proteins (see BASEMENT MEMBRANE). Its volume and composition are regulated in large measure by the KIDNEYS, abnormal accumulation resulting in oedema (e.g., see KWASHIORKOR).

tissue plasminogen activator (tPA) A serine protease involved in proteolysis of the extracellular matrix. After a blood clot has prevented bleeding (see BLOOD CLOTTING), tPA converts plasminogen to PLASMIN, which removes the remaining clot and opens many small previously blocked blood vessels. It may be administered following cardiac BYPASS SURGERY, transplant, or other open heart surgeries to prevent development of pulmonary embolisms. Its production by microbial gene manipulation is in high demand. See MYOCARDIAL INFARCTION.

tissue type, t. typing A person's tissue type is their particular combination of transplantation antigens, which are IMMUNOGLOBULINS encoded by the MAJOR HISTOCOMPATIBILITY COMPLEX (MHC). An important consequence of tissue types is the requirement to match them as closely as possible between the tissues of organ transplants and those of the recipient (tissue typing). Before development of the polymerase chain reaction, this involved using, as the typing reagent, serum from people who had had a transplant or had received mul-

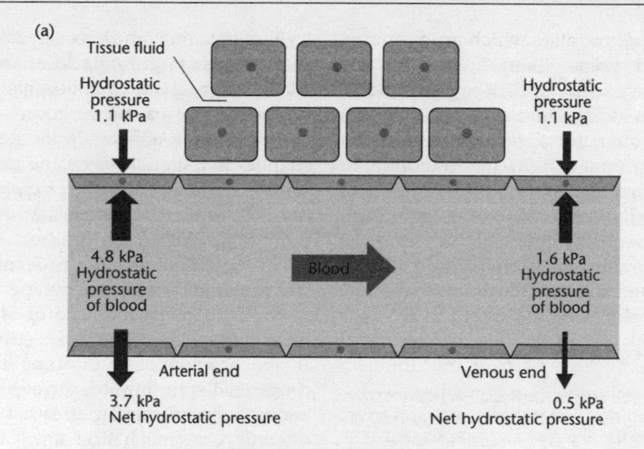

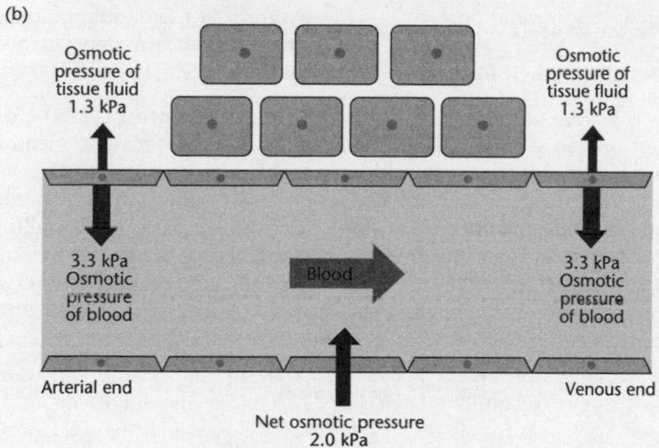

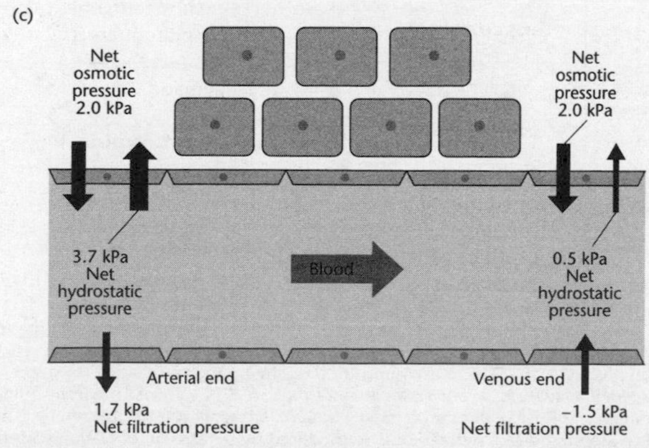

FIG 174. Factors influencing the production of TISSUE FLUID at arterial ends of capillaries and fluid return at the venous ends: (a) hydrostatic influences; (b) osmotic influences; (c) the resultant net filtration pressures. kPa = kilopascal. From *Biological Sciences Review* © 2000 Hachette Livre UK (Philip Allen Updates). Redrawn with the kind permission of the publisher.

tiple blood transfusions, or from women who had borne several children. The immune systems of such people would have been induced into making antibodies against various tissue antigens: those of the transplant donor, those of the donors of blood samples, or those of the father(s) of the numerous children. Today, a person's tissue type can be tested by DNA MICROARRAY technology, the position of the patient's PCR-amplified DNA being read directly from the chip.

TLRs See T-CELL RECEPTORS.

TMRCA See MOST RECENT COMMON ANCESTOR.

TNFs, TNFRs See TUMOUR NECROSIS FACTORS.

tocopherols See VITAMIN E.

toes See FOOT, LIMBS.

tolerance **1**. Failure of immune cells to respond to an antigen. An essential feature of the immune system is that it should be tolerant to self-antigens, without which auto-immune responses would occur. This is normally acquired during T-CELL development in the thymus. **2**. See ACCLIMATIZATION. **3**. To drugs: progressively reduced effectiveness of a drug on its repeated administration, so that increasing doses need to be taken to maintain the original action. Chronic drug administration induces poorly understood homeostatic adaptive changes in the brain, opposing the drug's action. These may involve enzyme induction, increase in membrane calcium channels, depletion in transmitter, receptor down-regulation, changes in second messenger or synthesis of an inverse agonist. Acquiring tolerance to a drug does not necessarily lead to ADDICTION, although it is one component of it. Some addictive drugs lead to SENSITIZATION (reverse tolerance) to their rewarding effects. Tolerance often develops much more slowly than the withdrawal-related adaptations; but once it emerges it becomes entrenched. See NICOTINE.

Toll-like receptors (TLRs) Usually associated with the plasma membrane, these highly conserved cell surface PATTERN RECOGNITION RECEPTORS and their homologues are part of an ancient signalling pathway employed in the innate defence mechanisms in most multicellular organisms (see INNATE IMMUNITY). The ten distinct forms

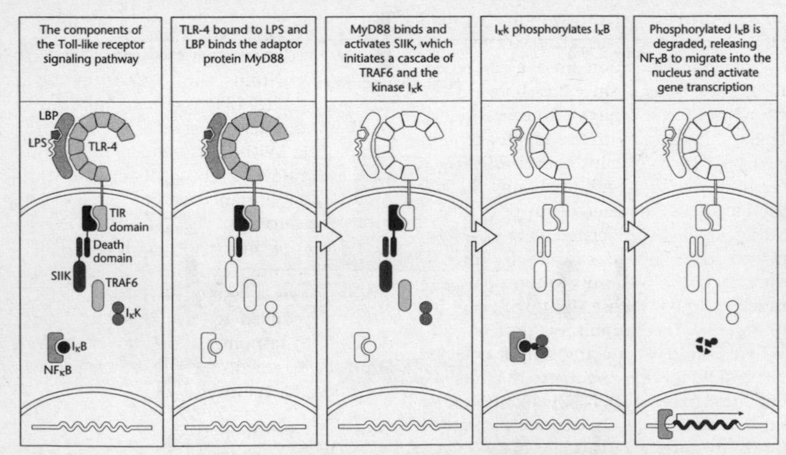

FIG 175. The involvement of TLR-4 in activation of NFκB by lipopolysaccharide (LPS). See the TOLL-LIKE RECEPTOR entry for explanation. MyD88 = an adaptor protein; SIIK = serine/threonine innate immunity kinase; TIR domain = Toll/IL-1R domain. © 2005 From *Immunobiology* (6th edition), by C.A. Janeway Jr, P. Travers, M. Walport and M.J. Schlomchik. Reproduced by permission of Garland Science/Taylor & Francis LLC.

occurring in mammals, encoded by *TLR* genes, provide a far more restricted specificity than the antigen receptors of the adaptive immune system. Different forms recognize bacteria, fungi and protozoa. In humans and other mammals TLR-4 binds to the LIPOPOLYSACCHARIDE (LPS) component of Gram-negative bacteria as well as to many other microbial surface components that can activate the Toll pathway, leading in turn to activation of the transcription factor NFκB in lymphocytes and other cells. The view that TLR3 can detect dsDNA has recently been challenged. *TLR1* is believed to have been the target of NATURAL SELECTION as a result of its probable involvement in the biology of TB and leprosy. The peptidoglycans of Gram-positive bacteria interact with TLR-2. The virus-detecting TLRs operate mainly in dendritic cells and respond to viral nucleic acids that are within endosomal compartments, leading to production of type 1 INTERFERON. Double-stranded RNA, forming the genomes of some viruses, is recognized as a distinct pattern by TLR-3. The unmethylated CpG-containing DNA and their flanking regions in bacteria are recognized by TLR9. These sequences are much more abundant in bacterial than in mammalian DNA (see SYSTEMIC LUPUS ERYTHEMATOSUS).

The cytoplasmic domain of a Toll receptor is known as TIR since it is shared by the cytoplasmic tail of the cytokine receptor for IL-1 and interacts with the adaptor protein MyD88, which contains a DEATH DOMAIN (see Fig. 175). Through this, the bound MyD88 interacts with another death domain on SIIK (aka IL-1R-associated kinase, IRAK), activating a kinase cascade leading to dimerization of subunits to form part of a multimeric IκB kinase (Iκk), which phosphorylates the inhibitory protein IκB. When bound to the transcription factor NFκB, IκB prevents its entry to the nucleus; but after IκB removal, NFκB migrates to the nucleus and binds its cognate promoters (see NFκB entry), activating genes involved in host defense against INFECTION.

tongue Formed from skeletal muscle which inserts on the hyoid bone and is attached to the anterior floor of the MOUTH by a fold of its mucous membrane covering (the frenulum), behind the lower incisor teeth. Its surface is covered with small projections (papillae) of four kinds: filiform, folate, fungiform and vallate. The organs of taste are the TASTE BUDS located on the folate, fungiform and vallate papillae. It also plays important roles in SPEECH and SWALLOWING.

tonic muscle fibres See PHASIC MUSCLE FIBRES for discussion of use.

tonic receptors Receptors showing slow ACCOMMODATION, firing repetitively and gradually decreasing frequency (i.e. longer interspike interval) in response to a prolonged constant-current stimulus (see ADAPTATION). Contrast PHASIC RECEPTORS.

tonotopy Systematic organization of characteristic frequencies of sound within an auditory structure. Tonotopic maps exist on the basilar membrane (see COCHLEA), within each of the auditory-vestibular nerve relay nuclei (cochlear nuclei, superior olivary nuclei, inferior colliculi), the median geniculate nuclei and the auditory cortex. See NEURAL MAPS, PRIMARY AUDITORY CORTEX.

tonsils GUT-ASSOCIATED LYMPHOID TISSUES developing from the 2nd PHARYNGEAL POUCHES (see Fig. 134) and being infiltrated by lymphatic tissue between the 3rd and 5th month. With the adenoids, they form a ring (Waldeyer's ring) at the back of the mouth and entrance to the alimentary canal and airways. They consist of large aggregates of mucosal lymphoid tissue, often becoming extremely enlarged in childhood as a result of recurrent infections. A reduced IgA response has been detected in response to oral polio vaccination in those who have had their tonsils and adenoids removed.

tooth See DENTITION.

topographic maps See NEURAL MAPS.

topoisomerases See DNA TOPOISOMERASES.

TOR See MTOR.

total lung capacity The volume in the lungs at maximum inspiration (~6 dm^3), being a combination of the inspiratory reserve volume, tidal volume, expiratory reserve volume and residual volume. See SPIROMETER, VENTILATION.

total peripheral resistance (TPR) Combined resistance to blood flow in all the parallel vascular beds of the systemic circuit. About 25% due to aorta and arteries, 40% due to arterioles, 20% due to capillaries and 15% due to the venous system. Controlled by intrinsic autoregulation, by sympathetic neurons and by ADRENALINE. See CARDIAC OUTPUT.

touch In its broadest sense, the sensory modality discriminating bodily (somatic) sensations, including mechanical, thermal and noxious sensations, the textures of objects and their movements across the skin (the latter most effectively in the fingertips: see MEISSNER'S CORPUSCLES). Aristotle argued for a total of five senses, including the senses of hot and cold within touch. Recent work suggests that detection of heat and of impact may be served by a polymodal transient receptor potential (TRP) membrane channel protein in which a heat-induced bilayer rearrangement alters the membrane tension gating the channel. If so, it could be that these channels have been selected to sum heat and force (e.g., into PAIN).

Touch receptors are rapidly adapting MECHANORECEPTORS (see Fig. 120), and information about touch, vibration and proprioception passes along the dorsal column–medial lemniscal pathway, a different pathway from that for pain and temperature. This touch pathway is swift and uses thick myelinated Aβ AXONS (see for details) while the pain pathway is slow and employs thin, lightly myelinated Aδ fibres and unmyelinated C fibres. These axons also differ in their connections in the spinal cord, and in the manner they transmit information to the brain (see SPINAL CORD, Fig. 30; SOMATIC SENSORY SYSTEM). Aβ fibres terminate deep in the dorsal horn and run in the dorsal column–medial lemniscal pathway, whereas the Aδ and C fibre axons immediately decussate and ascend through the spinothalamic tract to the ventral posterior nucleus of the thalamus and then to the PRIMARY SOMATOSENSORY CORTEX and ORBITOFRONTAL CORTEX, where the elaborate circuitry involved in spatial awareness and the sense of touch are located. Information is modified, particularly by lateral inhibition, at each synaptic connection in the brain; and the output of the cortex can also influence its input from the medullary dorsal column nuclei and the thalamus. Tactile information from the face is supplied via the TRIGEMINAL NERVES (cranial nerve V), the facial nerve (CNVII), glossopharyngeal nerve (CNIX) and vagus nerve (CNX). See KINESTHESIA.

Tourette syndrome (Tourette, TS) A COMPLEX DISORDER characterized by repetitive movements known as motor tics, and at least one vocal (phonic) tic, all of which characteristically wax and wane, onset usually occurring between 4 and 8 years. For most, the severity of the tics typically increases from 8 to 12 years, decreasing during adolescence. Between 1 and 11 children per 1,000 have TS, but adult TS is rare. The tics are believed to result from dysfunction in the frontal cortex and the thalamus and basal ganglia among subcortical regions. They are often temporarily suppressible, because there is usually a premonitional urge similar to that preceding an itch or sneeze. These urges are often experienced as an irresistible build-up of tension which the person chooses to release through the tic. Obsessive-compulsive disorder (see FEAR AND ANXIETY) and ATTENTION-DEFICIT HYPERACTIVITY DISORDER are often associated with the condition in referrable cases. A person with TS has ~50% probability of passing the condition to one of her/his children, although because the condition has variable EXPRESSIVITY and incomplete PENETRANCE only a minority of children will have symptoms needing referral. Both phonic and motor tics respond to antipsychotic drugs that are DOPAMINE-receptor blockers.

toxins Environmental toxins include heavy metals (which can disrupt DNA METHY-

LATION and CHROMATIN), while oestrogenic toxins (e.g., DDT, dioxin, PCBs, 2,4-D, diethylstilbestrol and chemical sunscreens; see OESTROGENS) and anti-androgenic toxins (e.g., 2,3,7,8 tetrachlorodibenzo-*p*-dioxin, TCDD) decrease male fertility by altering DNA methylation.

A toxin (β-oxalylaminoalanine), an agonist of glutamate receptors, is found in certain chickpeas. If ingested in large amounts, this can cause lathyrism, a degeneration of motor neurons. Another glutamate agonist, domoic acid, is found in contaminated mussels and can cause seizures and brain damage in small amounts. The plant toxin β-methylaminoalanine can cause symptoms resembling those of ALZHEIMER DISEASE and PARKINSON DISEASE in inhabitants of the island of Guam. Several ANTIBIOTICS are toxins; e.g., nigericin and gramicidin, maitotoxin and α-toxin form potassium channels in the plasma membrane. For genotoxins, see GENOTOXIC STRESS. See RICIN.

Exotoxins are released extracellularly by microorganisms, may travel to distant parts of the host, and act at the host cell surface. Although IgG and IgA antibodies protect against exotoxins, the initial dose may be highly toxic, highly immunogenic and often fatal. They are produced by the notable infectious agents: *Streptococcus pyogenes* (scarlet fever), *Staphylococcus aureus* (FOOD POISONING), *Corynebacterium diphtheriae* (DIPHTHERIA), *Clostridium tetani* (TETANUS), *Clostridium perfringens* (gas gangrene) and *Vibrio cholerae* (CHOLERA). Exotoxin A belongs to the group of toxins that includes diphtheria toxin, pertussis toxin, cholera toxin, C3 exoenzyme and others. They have four major targets: α-subunit of heterotrimeric G PROTEINS, ACTIN, Rho/Rac and eEF2 (see ELONGATION) in the eukaryotic cell, modifying one protein type or a very closely related protein. *Enterotoxins* are exotoxins acting on the small intestine, generally causing massive secretion of fluid into the gut lumen, leading to symptoms of diarrhoea. Exotoxin A (ETA), produced by *Pseudomonas aeruginosa*, a common problem in hospital infections, enters the host cell by receptor-mediated endocytosis and binds to elongation factor EF2 (eEF2). By mimicking a motif of 18S rRNA, it binds to the 80S ribosome, apparently at the 40S ribosomal subunit. By targeting such a central housekeeping molecule, this minimizes the risk of the host evolving resistance to the invading toxin. For the exotoxins produced by *Clostridium botulinum*, see BOTULISM.

Endotoxins are produced by Gram-negative bacteria and include LIPOPOLYSACCHARIDE and other intrinsic components of the microbe. Pyrogenic, relatively poorly immunogenic, weakly toxic and rarely fatal (except in large doses), they are released on lysis of the cell, triggering phagocytosis and the release of cytokines producing local or systemic symptoms. The host response to these is more complex than it is to exotoxins because the innate immune system has receptors to some of them (e.g., MANNOSE-BINDING LECTIN). They are produced by the infectious agents *Escherichia coli, Haemophilus influenzae, Salmonella typhi* (typhoid fever), *Shigella* and *Yersinia pestis* (BUBONIC PLAGUE). Many of these genera/species have their own entries. See FOOD POISONING, TOXOIDS.

toxoids A toxin, so modified as to retain its antigenicity (its epitopic domain) but to lose its toxicity. They are used in vaccines against human DIPHTHERIA, and TETANUS.

toxoplasmosis A congenital (non-hereditary) apicomplexan (formerly sporozoan) infection, usually contracted by the mother from raw meat and cat litter. The parasite, *Toxoplasma*, can cross the placenta and may cause severe eye injury, growth retardation, hydrocephalus, and mental disturbances linked to convulsions of the newborn. Investigations into NEWBORN SCREENING and incidence are taking place.

***TP53* gene** See P53 GENE.

T_R (T regulatory) cells See IMMUNOSUPPRESSION, T CELLS.

trabeculae Irregular lattice of thin bars of light lamellated bone, forming spongy BONE (e.g., in epiphyses). Orientated along lines of stress, they support the red bone marrow tissue. See OSSIFICATION.

trace amines Low concentration amines found in the central nervous system, including β-phenylethylamine, tyramine, tryptamine and octopamine. They have long been suspected of involvement in psychiatric disorders, and their receptors (TAARS) were discovered during a search for receptors to neurotransmitters resembling those for serotonin and dopamine. TAARs have been considered to play a role in depression and schizophrenia.

trachea (windpipe) A tubular passageway (~12 cm × 2.5 cm) located anterior to the oesophagus and extending from the LARYNX to the superior border of the 5th thoracic vertebra. Here it divides into the left and right primary bronchi (see BRONCHUS, LUNGS). Its layers, from interior to exterior, are: mucosa (pseudostratified ciliated columnar epithelium, goblet cells and elastic lamina propria), submucosa (areolar connective tissue containing seromucous glands and their dusts), hyaline cartilage (16–20 incomplete horizontal rings providing semi-rigid support, with the open part of each 'C' facing the oesophagus allowing slight expansion during SWALLOWING), and an adventitia composed of areolar connective tissue. The cartilaginous rings keep the trachea open during inspiration. There may be tracheal swelling and suffocation in ANAPHYLACTIC SHOCK.

tract A collection of central nervous system axons with a common origin and common destination (e.g., the corticospinal and pyramidal tracts). Contrast BUNDLE; see NERVE TRACT.

training See EXERCISE.

***trans*-acting regulatory elements** In the context of gene regulation, 'diffusible' regulatory elements; i.e., TRANSCRIPTION FACTORS.

transamination See DEAMINATION.

transcellular transport Movement of substances across epithelial surfaces through the cells, the tight junctions preventing passage between the cells. Facilitated diffusion, active transport and pinocytosis are employed. See ENTEROCYTES.

transchromosome A chromosome, or part of one, that has been taken from one organism and inserted into cells of another (compare TRANSGENE), a technique employed in MOUSE MODELS of DOWN SYNDROME.

transcription The processes, occurring in the NUCLEUS, by which all a cell's genomic RNA is produced, whether *de novo* or in the parent cell from which it is inherited cytoplasmically (see RNA). It is the first step in GENE EXPRESSION, and its stages are *initiation*, *elongation* and *termination*. It has been suggested that transcription and RNA PROCESSING occur in the space within the nucleus between chromosome TERRITORIES, in an area termed the *interchromosome domain compartment*. Nucleosomes within genes are obstacles to eukaryotic transcription, which is therefore accompanied by disruption of histone-DNA contacts as RNA polymerase moves along the DNA, nucleosome reformation occurring in the enzyme's wake. Positive post-translational modifications (PTMs) of HISTONES occur across transcribed regions and are linked to different stages of transcription. Early transcription is associated with H3/H4 acetylation, H2BK123 ubiquitylation and H3K4/K79 methylation. During late transcription, they include H3/H4 deacetylation, H2B de-ubiquitylation, H3K36 methylation and DNA methylation.

In addition to the nucleosomes, the large gene number and low gene density make transcription, especially its initiation step, a more complicated process in humans than in prokaryotes. A eukaryotic protein-coding gene requires the assembly of many proteins at its promoter before Pol II (see RNA POLYMERASE) can begin RNA production. Some of these, general TRANSCRIPTION FACTORS (GTFs), are themselves multiprotein complexes and must bind to promoter DNA before Pol II can do so (see PROMOTER), while others bind after it. The DNA sequence TATA (see TATA SEQUENCE) is often located ~30 bp upstream of the transcription start site, and forms part of the core promoter to which TATA-binding protein (TBP) binds. Six GTFs (TFIIA, TFIIB, etc.) serve to attract the Pol II core and to position it correctly. TBP is a part of TFIID, and once this GTF has bound others and Pol II do so, forming the preinitiation complex (PIC). Once RNA synthesis has started, Pol II dissociates and engages in RNA elongation while some of the GTFs remain at the promoter to attract the next Pol II core enzyme. For many years, the paradigm has been that gene regulation in higher eukaryotes resided at the level of either recruitment of Pol II to promoters or transcription initiation. It now appears likely that a post-recruitment and early elongation mechanism play a major role (see PROMOTER).

The enzymatic processes of elongation begin with the opening and unwinding of a small portion of the DNA double helix by the Pol, exposing the bases on both strands. Within this 'transcription bubble' the enzyme catalyses the formation, in a 5′- to 3′-direction, of phosphodiester bonds between ribonucleoside triphosphates. These align with one of the DNA strands, in accordance with base-pairing rules. The free energy release involved derives from hydrolysis of the triphosphates (see ADENOSINE TRIPHOSPHATE). Elongation continues until the Pol recognizes chain termination nucleotide sequences, usually ~40 bp in length ending in a GC-rich sequence followed by a poly-A region. RNA produced along the GC-rich DNA regions forms HAIRPIN LOOPS which cause release of the RNA from the Pol, and release the Pol from the DNA. The primary transcript, if generated by Pol II, must undergo cotranscriptional RNA PROCESSING (see entry for details) before it can be exported from the nucleus.

Transcription is the level at which many complex biological programmes are controlled. Not only are DNA-binding TRANSCRIPTION FACTORS (both activator and repressor TFs) crucial in determining whether a promoter region is accessed by RNA polymerase, but since the 1990s it has been clear that coregulatory multiprotein complexes in the nucleus (well illustrated by the PGC-1 family of coactivators) can be recruited to TFs in response to cellular signals. It is estimated that transcription of at least 10% of cellular genes oscillates in a circadian manner (see CIRCADIAN CLOCKS).

One mode of action of siRNAs (see RNA INTERFERENCE) involves silencing of a gene's transcription. Long dsRNAs might be produced by symmetrical transcription from opposing promoters, the antisense strand recruiting proteins that inhibit transcription either by base pairing with DNA or newly forming pre-mRNA transcripts.

One of the more surprising discoveries of the ENCODE pilot project (see HUMAN GENOME) is the large fraction of experimentally identified functional (i.e., transcribed) genomic elements that show no evidence of evolutionary constraint.

transcription factors (TFs) Broadly defined, any protein required to initiate or regulate eukaryotic TRANSCRIPTION, tissue-specific TFs triggering differential gene expression. They are 'diffusible' gene regulatory elements, encoded by genes which themselves require transcription. They have been identified using ChIP-chip technology (see DNA-BINDING PROTEINS). *General transcription factors* are similar for all Pol II-transcribed genes (see RNA POLYMERASES, TATA SEQUENCE). They may be recruited in a stepwise fashion on some promoters before Pol II can bind, but often form part of a large pre-assembled *RNA polymerase II holoenzyme* complex containing a 20-subunit *mediator* complex required for additional activator proteins to stimulate transcription initiation (see GENE EXPRESSION). Some TFs require to be changed from being repressors of transcription to activators (see HEDGEHOG GENE). Even before this can happen, additional general transcription factors, such as TFIID and TFIIA, may have to be assembled on the promoter separately. A subunit of TFIID contains two 120-amino acid domains termed *bromodomains*, each of which forms a binding pocket for an acetylated lysine side-chain. Some TFs are repressors of specific gene activity: the RE1-silencing transcription factor (REST), which is constitutively expressed in all cell types except neurons, recognizes the promoter region docking sequence called the neural restrictive silencer element (NRSE) so that widely dispersed but functionally similar genes are stably inactivated in all non-neuronal cells (see GENE SILENCING). Many have to be phosphorylated by protein kinases before binding their regulatory element or COACTIVATOR.

transcriptional gene silencing See TRANSCRIPTION.

transcriptome The total set of different RNA transcripts in a cell or tissue, representing the functional component of the genome (see GENOMICS). Gene-by-gene hybridization techniques have been replaced by DNA MICROARRAYS and DNA sequence sampling methods that enable global expression profiling to be carried out. See HUMAN TRANSCRIPTOME.

transducin A G PROTEIN present in the outer segment membranes of rod cells and cone cells and involved in phototransduction. See RHODOPSIN, TASTE.

transfection Because the term TRANSFORMATION is often used to describe conversion of mammalian cells into a malignant state, the introduction of DNA into these cells has been called transfection (which has another meaning in bacterial systems). See CANCER.

transfer RNAs (tRNAs) Once called adaptor-RNAs, these molecules (each ~80 nucleotides in length) bring a specific amino acid to the ribosome during the TRANSLATION phase of protein synthesis. Complementary base-pairing enables each molecule to generate four short double-helical segments, three of which fold up into 'clover-leaf' loops (D loop, T loop and anticodon loop). The conformations generated are specific. The remaining segment carries an unpaired ACC tip at its 3′ end, to which a specific amino acid becomes attached in the cytosol by activating enzymes (amino-acyl-tRNA synthetases), is an essential requirement if the genetic code is to be a possibility. Mg^{2+} is required for this amino-acyl activation (see ELONGATION). One of the loops carries a triplet anticodon sequence, which recognizes a complementary mRNA codon during translation. The genetic code is a redundant code, since most amino acids are encoded by more than one mRNA codon (i.e., some amino acids have more than one tRNA) and because some tRNAs require accurate base-pairing only at the first two positions of the codon. There are ~500 nuclear genes that encode cytoplasmic tRNA molecules, grouped into 49 families according to their anticodon sequences, and there are also ~325 tRNA-derived putative PSEUDOGENES. The tRNA genes are scattered throughout the chromosomes, although more than half reside on either chromosome 6 or on chromosome 1. The exact number of tRNA species differs from one species to the next; but their genes are always transcribed by RNA POLYMERASE III and are usually produced as large precursor tRNAs that are trimmed, have any introns removed during a special form of RNA PROCESSING, and may also have some bases altered, before leaving the nucleus. They must be successfully folded in order to leave the nucleus. See RNA.

transferrin A soluble protein, produced by the liver, carrying IRON in the plasma. On binding transferrin, cell-surface transferrin receptors deliver the transferrin/iron complex to early ENDOSOMES by receptor-mediated ENDOCYTOSIS, where the low pH causes iron to dissociate while the transferrin/receptor complex enters the tubular extensions of the early endosome and is recycled back to the plasma membrane.

transformation Any heritable alteration in the properties of a eukaryotic cell. In tissue culture, usually refers to the conversion of a non-cancerous cell to a cancerous one ('neoplastic transformation'), as by retroviruses or carcinogens. Common properties of transformed cells include increased blebbing of the plasma membrane and increased mobility of proteins within it; reduced adhesion of the cell to surfaces ('contact inhibition' in the context of adjacent cells), tending to make them rounder in appearance; increased extracellular proteolysis on account of increased release of plasminogen activator; growth by proliferation to a high cell density, with consequent 'immortality'; reduced growth factor requirement, and tendency to cause tumours when injected into susceptible host animals. Compare TRANSFECTION.

transforming growth factor-α (TGF-α) An EGF-like GROWTH FACTOR (sharing 33% homology), believed to play a role in ANGIOGENESIS during wound-healing and tissue repair and released by a variety of oncogene-transformed cells, notably by those transformed by activated *ras* oncogenes. Like EGF, it binds to, and activates, the EGF receptor.

transforming growth factor-β (TGF-β) superfamily A superfamily of dimeric secreted glycoprotein GROWTH FACTORS (also cytokines) which includes TGF-α (see entry), TGF-β, ACTIVINS, INHIBINS, Nodal (see PRIMITIVE NODE) and, the largest family, bone morphogenetic proteins (see BMPS). During development, some act as graded MORPHOGENS and help regulate PATTERN FORMATION.

TGF-β itself is a growth-inhibitory protein produced by activated $CD4^+$ T CELLS and controlling cell cycle machinery by modulating levels of CDK inhibitors (CdkIs). It induces expression of the CDK inhibitors that block both the formation and actions of cyclin D–CDK4/6 complexes in G_1, (i) preventing a cell from reaching the R point (see G_1) and (ii) antagonizing cell proliferation (overriding the signal of a mitogen). Some are bound and inactivated for a time by the PROTEOGLYCANS of the extracellular matrix. See DISEASE (Fig. 64) for signalling pathway.

transgenics The processes by which gene(s) from one species are inserted into cells of another. While it is illegal to transfer genes by these methods into human cells, transfer of human genes into mice has been fundamental to the analysis of human gene function, and comparative analysis of human and chimpanzee orthologous genes in transgenic mice is now almost certain to be pursued. While cell-based applications of transgenic procedures are an important part of gene manipulation ('genetic engineering') in animals, the term 'transgenic' is usually reserved for whole organisms. The techniques are useful in basic research, as in the study of embryological development (compare GENE KNOCKOUTS), and in biotechnological applications such as the production of high-value therapeutic proteins by transgenic farm animals. See GENE THERAPY.

transitions A form of nucleotide-pair substitution in which a purine replaces another purine or in which a pyrimidine replaces another pyrimidine. Compare TRANSVERSIONS.

translation The phase of protein synthesis that takes place on ribosomes, during which amino acids brought in by activated amino-acyl-tRNA molecules are joined by peptide bonds to form polypeptides. Control of translation occurs mainly at the initiation step, in which the 40S ribosomal subunit is recruited to mRNA and positioned at the initiation codon specifying the first amino acid of the eventual polypeptide. In general, mRNA translation is *cap-dependent* (see RNA CAPPING), the mRNA being bound by the cap-binding protein complex eIF4F comprising three subunits (eIF4A, eIF4E and eIF4G; see INITIATION FACTORS), without which ribosomes do not bind. This binding step is a major target for translation regulation by repressor proteins (4E-binding proteins, or 4E-BPs) which compete with the scaffolding protein eIF4FG for binding to eIF4E and inhibit translation. Phosphorylation of serine and threonine residues of 4E-BPs reduces their interaction with eIF4E, a process brought about by activation of the MTOR signalling pathway (see CELL GROWTH). *Cap-independent translation* also occurs in the normal eukaryotic cell cycle, during and immediately after the metaphase stage of mitosis, when cap-dependent translation is suppressed (as during infection by some

VIRUSES). The mRNAs for some cell cycle proteins contain domains termed internal ribosome entry sites (IRESs) that permit ribosome recruitment in the absence of the eIF4F complex.

Mutations in amino-acyl-tRNA synthetases (aaRSs) sometimes result in incorporation of the wrong amino acid to the growing polypeptide on the ribosome because the mutated enzyme is less able to recognize and correct ('proof-read') a mistake it makes in binding an amino acid to a tRNA molecule (see ELONGATION). Structurally similar amino acids may then be confused. Mutations leading to a sequence impairment in the 'editing site' of the aaRS, which normally ensures that incorrectly activated amino acids are removed before they can be incorporated into a polypeptide, lead to failure to discriminate between pairs of similar amino acids.

translocations Transfer of genetic material from one chromosome to another (contrast INVERSIONS). Reciprocal translocations involve breakage of at least two chromosomes followed by exchange of fragments. Known to be influenced by the relative positioning of chromosomes within the interphase NUCLEUS, it is believed that the incidence of reciprocal translocations in the general population is ~1:500 (see GENETIC DISEASES AND DISORDERS, Fig. 84). Balanced translocations do not alter the total amount of genetic material present while unbalanced translocations do, and are much more likely to cause significant disabilities, both physical and mental. Thus a translocation of the long arms of chromosomes 14 and 21 at the centromere, t(14;21), although resulting in loss of both short arms, does not produce clinical symptoms. But these individuals are at risk of producing offspring with unbalanced chromosomes since balanced reciprocal translocations cause meiotic problems. Chromosomes cannot pair normally since homologous regions now lie on more than just one pair of chromosomes, often leading to 4-chromosome clusters at first prophase: *pachytene quadrivalents*. The long arms of chromosomes 11 and 22 have a tendency to do this. This leads to considerable chromosome imbalance in the gametes, the resulting embryo having some regions of a chromosome trisomic and parts of another chromosome monosomic. Either miscarriage results, or birth of a child with multiple abnormalities. Alternatively, three chromosomes may segregate into one gamete (tertiary trisomy), leaving only one chromosome of the quadrivalent in the other. Stable reciprocal translocations occur when the exchanged fragments are both acentric (see Philadelphia chromosome in chronic myeloid LEUKAEMIA). Unstable reciprocal translocations result when centric and acentric fragments are exchanged, resulting in dicentric and acentric chromosomes. See BURKITT'S LYMPHOMA, Fig. 32; DNA REPAIR MECHANISMS; GENOTOXIC STRESS; ROBERTSONIAN TRANSLOCATION.

transmitters See NEUROTRANSMITTERS, SYNAPSE.

transplants Surgical replacement of defective tissues and organs by functioning ones, often limited by the supply and short-term viability of fresh donor organs. At present, most people who have had organ transplants face a lifetime of treatment with immunosuppressant drugs, such as cyclosporin and steroids. These prevent rejection for several years, but often fail to prevent it indefinitely while putting patients at risk of infections, cancers and kidney failures. In humans, each *HLA* gene has up to 100 allelic forms, and immune cells will kill cells that have different forms of the HLA proteins from their own, which is why *HLA* genes must be carefully matched between bone marrow donors and recipients (see MAJOR HISTOCOMPATIBILITY COMPLEX, TISSUE TYPE). Successful transplantation in 2002 of ovaries, Fallopian tubes and the upper segment of the uterus in rats after their storage in liquid nitrogen was followed by re-initiation of oestrus, which is a promising result. See CYTOMEGALOVIRUS, IMMUNOSUPPRESSION, THYMUS.

transposable elements (TEs) There is no officially agreed system for classifying transposable elements. In essence, they are any genetic unit that can insert into a chromosome, exit and relocate elsewhere in the genome. Almost half the human genome is derived from transposable elements, the vast majority of which belong to two types of retrotransposon: LINEs and SINEs. TEs are found mainly in telomeres, centromeres and intergenic DNA. Those that do insert into genes are probably capable of inserting into both exons and introns although only the latter will survive in the population because of the negative selection against the mutagenic effects of the former. Most transposable elements in humans (as in other multicellular organisms) are inactive and can neither move nor increase their copy number. Those that can move are usually subject to epigenetic regulation involving DNA METHYLATION (~90% of methylcytosines are thought to be located in retrotransposond families); but some escape these surveillance systems (see HAEMOPHILIA, NEUROFIBROMATOSIS, Duchenne MUSCULAR DYSTROPHY and BREAST CANCER). Because TEs can transpose at high frequency, with a rate of between 10^{-3} and 10^{-5} per element per generation, depending on the element, they are more powerful generators of raw genetic variation than is the classical nucleotide-base substitution rate (~10^{-8}–10^{-9} per nucleotide per generation). Certain small RNAs mediate transposon silencing (see piRNAs).

They are usually placed in two groups based on their method of transposition:

Class 1 elements ('RNA elements', retroelements, retrotransposons, retroposons). Once inserted, these remain in the genome permanently and are mobile because their transcript RNA 'copies' undergo reverse transcription into DNA that is inserted into new chromosomal locations (RETROTRANSPOSITION), similar to the manner in which pseudogenes and retrogenes are generated. Class 1 elements include the autonomous LINES, and non-autonomous SINE *Alu* elements and retrovirus-like elements with long terminal repeat sequences (LTRs). Autonomous LTRs are endogenous retroviral sequences (ERVs) that contain *gag* and *pol* genes encoding a protease (PR, cutting up precursor proteins and involved in particle assembly), reverse transcriptase, RNAse and integrase (INT, splicing the double-stranded DNA into a new place in the host genome). Most of these (accounting for 4.6% of the human genome) are defective and have rarely engaged in active transposition during the past several million years of hominid evolution (see RNA INTERFERENCE). However, some members of the very small HERV-K group (see also HERVS) have undergone comparatively recent transposition (see HUMAN GENOME). The retrotransposon SVA is the third most active transposable element since the human–chimpanzee lineage separation, creating ~1,000 dispersed copies in each lineage. A composite element (~1.5–2.5 kb), it consists of two Alu fragments, a tandem repeat and a region seemingly derived from the 3′ end of a HERV-K transcript and is probably mobilized by (LINE)-1. SVA is of particular interest because each copy carries a CpG island sequence and potential transcription factor binding sites, so that these 1,000 dispersed elements could be a source of regulatory differences between chimpanzees and humans.

Class 2 elements ('DNA elements') move by excision and reinsertion of the excised DNA into a new location by a direct 'cut-and-paste' process. If, by inserting, they cause a gene to mutate, this will be reversed once they have excised themselves.

There is a third family of transposable elements that includes active elements in vertebrates but which have only recently been studied at the molecular level. These *hAT* elements share a number of amino-acid motifs different from those shared by the elements above (the retroviral integrase superfamily). Significant similarities exist in the catalytic amino acids of one of these, the *Hermes* element, the V(D)J recombinase RAG1 and retroviral integrase superfamily transposases, so linking the movement of transposable elements and V(D)J recombination (see SOMATIC RECOMBINATION).

transposases Enzymes encoded by Class 2 TRANSPOSABLE ELEMENTS and responsible for their cut-and-paste transposition. See INTEGRASES, SOMATIC RECOMBINATION.

transposons Discrete mobile DNA segments that have been found in virtually every genome examined (see TRANSPOSABLE ELEMENTS). They can be arranged in two groups on the basis of their methods of transposition: retrotransposons (retroposons) and DNA transposons. Many transposon families are marked by DNA methylation and repressive HISTONE modifications (e.g., H3K9 methylation). Transposon silencing involves piRNAs and certain other small RNAs.

trans-species polymorphism See HOMINOIDEA.

transverse plane (horizontal plane) A plane dividing the body or organs into superior and inferior portions. See ANATOMICAL PLANES OF SECTION, Fig. 9.

transversions A form of nucleotide-pair substitution in which a purine replaces a pyrimidine or vice versa. Compare TRANSITIONS.

trapezium (trapezoid) A carpal bone of the WRIST (Fig. 182), situated in primates directly proximal to the index finger. In modern humans, it is shaped like a boot, the palmar half of the bone being radio-ulnarly and proximo-distally wide, whereas in other primates it is more wedge-shaped with a narrow palmar tip and wide dorsal base.

trapezius muscle A posterior thoracic muscle stabilizing the SCAPULA. Its origins are on the superior nuchal line of the occipital bone, ligamentum nuchae, and spines of the 7th cervical and all thoracic vertebrae, and it inserts on the clavicle, acromion and spine of the scapula. Its superior fibres elevate the scapula and help extend its head; its middle fibres adduct the scapula; while its superior and inferior fibres rotate the scapula upward. See Appendix Xb.

trapezoid body See HEARING.

Treponema pallidum The spirochaete cause of syphilis, nowadays a relatively uncommon sexually transmitted disease in industrialized countries, but important to travellers to countries where it is more prevalent. However, incidence of the disease has increased in recent years in inner cities in the USA and Russia, and in the UK. It remains more common in homosexual men than in other groups, and its early diagnosis prevents the later congenital infection and associated complications. A very infective organism, with fewer than 20 organisms being required. Primary syphilis appears ~10–70 days after infection, the organisms entering even through intact skin, their survival in the host being due in part to their poor antigenicity. Commonly infected sites are the foreskin, vulva, uterine cervix, urethra or penile shaft, a painless enlargement of local lymph nodes and a painless primary lesion (chancre, an indurated ulcer) occurs. In secondary syphilis, developing 6–8 weeks after this, fever, lesions of the skin and mucosae occur, notably 'snail-track' mouth ulcers and warty lesions around the anus. Systemic symptoms include meningitis and arthritis. Latent syphilis, an asymptomatic state, may persist for years after untreated infection. Late syphilis is characterized by the gumma, an indolent granulomatous lesion that may become mucoid and ulcerate, producing a sticky discharge in any body tissue. In cardiovascular syphilis, these lesions occur 10 or more years after the primary infection; in neurosyphilis (meningovascular syphilis), the base of the brain and upper spinal cord are affected. Long-acting penicillin (e.g., procaine penicillin) remains the treatment of choice, the length of treatment increasing (from 10 days to 17 days) with the stage of syphilis involved. Non-sexually transmitted (endemic) treponemal diseases are nowadays rare.

TRH (thyrotropin releasing hormone, TRF) Peptide hormone produced by the parvicellular neurons of the median eminence of the hypothalamus and released into the long portal veins leading to the anterior pituitary, where it causes release of THYROTROPIN.

triacylglycerols (triglycerides) Esters of glycerol and three fatty acid molecules. See LIPIDS, LIPOGENESIS.

trial-and-error learning See INSTRUMENTAL CONDITIONING.

trichotomy For the possible gorilla–chimpanzee–human trichotomy. See HOMINOIDEA.

trichromatic vision See COLOUR BLINDNESS, COLOUR PERCEPTION.

trigeminal nerve (cranial nerve V) A mixed nerve with three sensory branches, conveying sensations of TOUCH, pain, temperature and proprioception, and all terminating in the PONS. These are: the *ophthalmic branch* (from the nasal cavity, lacrimal glands, forehead and skin of upper eyelid); the *maxillary branch* (from nasal mucosa, palate, parts of the pharynx, upper teeth and lower eyelid) and the *mandibular branch* (from the anterior two-thirds of the tongue, excluding those of taste; the lower teeth, the skin over the mandible, cheek, and side of the head in front of the ear). Sensory connections of the trigeminal nerve are analogous to those of the dorsal roots of the SPINAL CORD. Large diameter sensory axons synapse onto second-order neurons in the ipsilateral trigeminal nucleus, which decussate and project to the medial part of the ventral posterior nucleus of the thalamus, and then on to the primary somatosensory cortex (see TOUCH). The motor portion of the trigeminal nerve is involved in the control of chewing. Sensory neurons projecting to the pons from the head and face initiate motor impulses that leave the pons via the mandibular branch.

triglycerides (triacylglycerols) Esters of GLYCEROL and three FATTY ACID molecules. Roughly 95% of their biologically available energy resides in their three long-chain fatty acids, the remainder contributed by the glycerol moiety. See FATTY ACID METABOLISM, LIPIDS, LIPOGENESIS.

triiodothyronine (T_3) See THYROID HORMONES.

trimesters The three trimesters divide the interval of normal gestation into three approximately 3-month periods. See PREGNANCY.

trinucleotide repeat expansion See TRIPLET REPEATS.

triplet (trinucleotide) repeats Repetition in a genome of three nucleotides in tandem, involved in several genetic disorders (e.g., HUNTINGTON DISEASE, FRAGILE-X SYNDROME), several ataxias and myotonic dystrophy (see MUSCULAR DYSTROPHIES). The progressive character of repeat expansion across generations (ANTICIPATION) is due to the fact that the longer a repeat is, the more probable it is that it expands between generations (see DYNAMIC MUTATIONS). In some non-pathogenic cases, *trinucleotide repeat expansion* gives rise to REPETITIVE DNA forming some MICROSATELLITES. See DNA REPEATS, HIGH THROUGHPUT GENOTYPING.

triploidy See CHROMOSOMAL MUTATIONS.

trisomy The presence of an extra chromosome, most of which lead to early pregnancy loss (see MISCARRIAGE). Most autosomal trisomies result from NONDISJUNCTION during the 1st meiotic anaphase, less often from nondisjunction during the 2nd meiotic anaphase. There is a preponderance of nondisjunction in maternal meioses. But nondisjunction can also occur during early mitotic division of the developing zygote (see MOSAICISM). Most cases of DOWN SYNDROME are due to an additional chromosome 21. Other autosomal trisomies compatible with survival include PATAU SYNDROME. See TRANSLOCATION.

tritanopia See COLOUR BLINDNESS.

trithorax proteins Chromatin-modifying proteins that, like POLYCOMB GROUP proteins, play a key role in epigenetic regulation of development, differentiation and maintenance of cell fates.

trivariancy That form of COLOUR PERCEPTION in which there are three primary colours, i.e. the minimum number of colours by whose varied proportional combinations we can match any colour of the visible spectrum.

tRNAs See TRANSFER RNAS.

trochanter Very large projection on bone, for connective tissue attachment. Each femur has a greater and lesser trochanter for attachment (through tendons) of some of the muscles of the buttock and thigh (see GLUTEUS MAXIMUS). Compare TUBERCLE, TUBEROSITY.

trochlear nerve (cranial nerve IV) The cranial nerve whose motor neurons originate in midbrain nuclei and innervate the inferior rectus muscle, an extrinsic muscle of the EYE (see entry). High-frequency spikes in these oculomotor neurons lead to intorsion (rotation) and depression of the eyeball.

trophectoderm See TROPHOBLAST.

trophoblast (trophectoderm) The layer of cells covering the BLASTOCYST (see Figs 20 and 21). Where it comes into contact with the endometrium after implantation, it develops into two cell layers, the CYTOTROPHOBLAST and the SYNCYTIOTROPHOBLAST. It is one of the two groups of cells produced by early cleavage divisions of the zygote (see MORULA), the other being the INNER CELL MASS, which adheres to it. The trophoblast gives rise to extraembryonic structures such as the CHORION, and through this to the extraembryonic portion of the PLACENTA.

tropocollagen See COLLAGENS.

TRP receptors (transient receptor potential receptors) Transient receptor potential (TRP) membrane ionotrophic channel proteins that detect painful stimuli, vibration, touch (see MECHANORECEPTORS), temperature (see THERMORECEPTORS) and osmotic membrane stretch. They are ionotropic channels, many of them key components of signal transduction pathways in a variety of sensory systems, and implicated in the detection of noxious cold and 'cool' substances such as menthol, eucalyptol and icilin. Pungent compounds in mustard oil and garlic also depolarize a subpopulation of primary sensory neurons activated by capsaicin and by Δ^9-tetrahydrocannabinol (THC), the active component of marijuana (see CANNABINOIDS). A heat-induced bilayer rearrangement may alter the membrane tension that gates the polymodal TRP channels. See PAIN, PHOSPHOLIPASE C.

trypanosomes Flagellated protoctist blood parasites. *T. brucei* causes sleeping sickness (see TRYPANOSOMIASIS), killing an estimated 300,000 people in Africa annually (one subspecies, *T. b. brucei* is responsible for the cattle disease nagana and does not infect humans). In South America, the related *T. cruzi* causes Chagas' disease which affects 20 million people annually. The four stages in the life cycle use different energy sources (histidine in the insect and fatty acids in humans). Humans and several African primates (e.g., baboons) carry an anti-trypanosome factor in their blood. The third member of the 'Tritryps' group is *Leishmania major* (see LEISHMANIASIS). The genomes of all three species were sequenced in 2005, from which major research it is hoped that new therapeutic drugs will ultimately emerge – particular targets being those metabolic pathways which are atypical for eukaryotes: initiation of DNA replication, nucleotide excision/repair, and bacteria-like DNA polymerases used in mitochondrial DNA replication. Recently, human serum apolipoprotein L-I (apoL-I) was found to lyze African trypanosomes. Apo LI contains a membrane pore-forming domain that targets the lysosomal membrane of incoming trypanosomes, forming an ionic pore that triggers osmotic swelling of the lysosome and leading to death of the trypanosome. This provides an innate immunity against *T. brucei*. The loss of the *Apol1* gene from the chimpanzee genome may explain the observation that humans, gorillas and baboons possess the trypanosome lytic factor, whereas chimpanzees do not.

trypanosomiasis (African trypanosomiasis, sleeping sickness) Frequently epidemic protozoan disease of 36 sub-Saharan African countries, where many people are not treated sufficiently early. The incidence may approach 300,000–500,000 cases per year. Current treatments are inadequate: current drugs are toxic, there is no prophylactic chemotherapy and little prospect of a vaccine. Caused by parasites of the genus *Trypanosoma*, which enter the blood via the bite of blood-feeding tsetse flies (*Glossina spp.*) and causing progressive illness – ultimately invading the central nervous system where it invariably leads to death if untreated. *Trypanosoma brucei rhodesiense* occurs mainly in east and southern Africa; *T. b. gambiense* occurs mainly in west and central Africa. Antigenic variation enables evasion of acquired immunity (unlike the African trypanosomes, the South American *T. cruzei* does not undergo antigenic variation). The APOL1 protein is associated with the high-density lipoprotein fraction in serum and has been proposed as the lytic factor responsible for certain subspecies of *Trypanosoma brucei*. The loss of *APOL1* in chimpanzees could explain why these animals lack trypanosome lytic factor while humans, gorillas and baboons have it. See TRYPANOMOSMES.

trypsin, trypsinogen Trypsin is a serine protease, digesting proteins to peptides. Released in the pancreatic juice as the pro-enzyme trypsinogen, this is converted to trypsin by enteropeptidase in the brush borders of intestinal epithelial cells. Trypsin then converts more trypsinogen to trypsin, and converts chymotrypsinogen and procarboxypeptidase and proelastase to their active enzymes. See α_1 ANTI-TRYPSIN.

TSH See THYROTROPIN.

tubercle Small, rounded, projection on bone (e.g., greater tubercle of humerus) for connective tissue attachment. Compare TROCHANTER.

tuberculin test Clinical test in which purified protein derivative (PPD) from the bacterium *Mycobacterium tuberculosis* (see TUBERCULOSIS) is subcutaneously injected. PPD elicits as delayed-type hypersensitivity in individuals who have had TB or have been immunized against it.

tuberculosis (TB) One of the commonest human infections, one-third of the global population being infected, and responsible for two million deaths annually. Numbers infected are expected to increase in the next few years, particularly in low-income countries with a high incidence of malnutrition, although its incidence in many high-income countries has started to rise. Symptoms of the disease may include fever, night sweating, cough, blood-spitting and loss of weight. But none of these is unique to TB. Continued ill-health is probably the most constant symptom and X-rays are often the first evidence of the disease. An extremely virulent and drug-resistant strain (XDR TB) of the bacterium in South Africans with HIV is virtually untreatable and very worrying. The symptoms of TB are more difficult to diagnose in those with AIDS.

Primarily a DISEASE of lungs, TB is believed to predate civilization. The TOLL-LIKE RECEPTOR gene *TLR1* is believed to have been subject to NATURAL SELECTION resulting from its probable involvement in the biology of TB and leprosy.

Most infections in developed countries are caused by *Mycobacterium tuberculosis* and although there are >85 species of *Mycobacterium* only *M. tuberculosis*, *M. africanum*, *M. bovis* and bacille Calmette–Guérin (BCG) are obligate pathogens. This genus of rod-shaped bacteria was first identified by Robert Koch in 1882. Mycobacteria are not stained by the Gram method because of their thick hydrophobic surface lipid content; but they are Gram-positive when this is removed, being related to *Listeria*. *Mycobacterium tuberculosis* is an obligate aerobe, causing lesions in the upper lobe of a lung where oxygen concentration is highest.

Two kinds of human tuberculosis can be distinguished: *primary* and *post-primary* (re-infection) infection. Primary infection usually results from inhalation of droplets or dust particles bearing living bacteria from an individual with an active pulmonary infection. Individuals will respond to the tuberculin test by a delayed reaction (within 8 hours) after inoculation with the bacterial surface protein, tuberculin. A person who has already been infected will become reddened at the site of inoculation and after two days a raised and pus-filled swelling will develop. Initial infection provokes a cell-mediated immune response, with local inflammation and abscess formation (collections of pus), tissue destruction and healing. These lesions dwindle to small scars – the tubercles – revealed by X-rays, and in healthy well-nourished individuals the infection is restricted to these, lesions becoming inactive.

Once ingested by antigen-presenting cells, especially lung macrophages, mycobacterial antigens are processed and presented by the phagosomal/endosomal pathway in association with class II MHC molecules. T cells bearing appropriate receptors bind these macrophages, which then secrete IL-2 and stimulate growth not only of T_H cells but also of NK CELLS. However, *M. tuberculosis* prevents fusion of macrophage phagosomes with clathrin-coated vesicles from the Golgi apparatus, so protecting itself from their bactericidal contents. Compartments containing the bacterium retain markers of early endosomal organelles (e.g., Rab 5) and exclude the late endosomal marker Rab 7 (see RAB PROTEINS). They also secrete an enzyme, Sap M, which dephosphorylates the PIP_3 on its own compartment, inhibiting progression down the endocytic route towards LYSOSOME development (see INFECTION, Fig. 100). Such bacteria may be eliminated when the macrophage is activated by a T_H1 cell, but when they resist the effects of macrophage activation, chronic infection with inflammation can develop. In such post-primary tuberculosis, re-infection or reactivation of the primary lesions occurs, possibly after weeks or years. The bacteria then grow slowly in the lungs, whereupon aggregates of activated macrophages form, termed *granulomas*. This characteristic localized inflammatory response consists of a central core of infected macrophages (including fused macrophages, multinucleated giant cells) which commonly forms a cheesy central necrosis (caseation). This core is surrounded by T cells, many of which are CD4-positive, the effect being to 'wall-off' pathogens resisting destruction, although the death of necrotic cells can cause significant pathology.

In individuals with low resistance, through malnutrition or an otherwise compromised immune system (as in AIDS), the bacteria are unopposed and an acute pulmonary infection results leading to destruction of lung tissue. The bacteria may then spread to other organs (see OSTEOMYELITIS), leading to death. But such acute infection is relatively uncommon in otherwise healthy individuals, the bacteria generally remaining localized and eventually becoming quiescent (dormant). In older people, the TB lesions may reactivate, causing a second phase of infectious disease.

Before the Second World War, TB was a major European problem, but after the war a major drive to control the disease took place. Vaccination using live, weakened bacteria (the BCG vaccine) was so successful that the disease went into decline. This led to complacency and in recent years vaccination programmes have become lax. A fierce debate also raged over the value of vaccination. We are now seeing the consequences of this inaction as TB is a newly emerging disease in economically developed countries. Grandparents infected many years ago can act as a potential pool of infection for their grandchildren. With the changing epidemiology of TB, BCG is now targeted to infants (aged 0–12 months) living in areas where the incidence of the disease is 40/100,000 or greater, or children with a parent or grandparent who was born in a country with this incidence. BCG vaccine should not be used in patients with immunosuppression.

The polymerase chain reaction is now used to identify mycobacterial DNA within a day rather than waiting 2 months for the bacteria to grow in culture. The slow growth of most mycobacteria makes treatment very prolonged. First-line mycobacterial drugs include a 6- to 9-month course of orally taken rifampicin, which kills slow-growing mycobacteria and penetrates moderately well into the cerebrospinal fluid. The nicotinamide derivative isoniazid (see N-ACETYLTRANSFERASES) is highly effective in killing rapidly dividing mycobacteria and is taken with rifampicin throughout the duration of treatment. The third first-line drug is pyrazinamide. The intensive phase of treatment lasts 2 months and involves prescriptions for all three drugs. The continuation phase of treatment lasts from 4 to 7 months, during which rifampicin and isoniazid only are taken. Rifabutin, related to rifampicin, is often used when combined with retroviral therapy, although cross-resistance between it and rifampicin is common. Because of the problem of emerging antibiotic resistance, antituberculosis drugs are always given in combination, the pattern depending on local resistance patterns. Because of the lengthy treatment and unpleasant side effects, patients often fail to complete their courses of drugs. This has led to the evolution of *Mycobacterium tuberculosis* strains resistant to rifampicin and isoniazid, leading to multidrug-resistant tuberculosis (MDRTB). As a consequence, some American cities now pay patients to have supervised delivery of drugs. In England in 2002, just $<1\%$ of cases of TB were MDRTB. Resistance to tuberculosis may be the selective advantage of the hexosaminidase A mutant producing TAY–SACHS DISORDER, and possibly of genes predisposing to the ulcers resulting from high pepsinogen I levels and high rates of gastric acid secretion. See MULTIDRUG RESISTANCE.

tuberomammillary nucleus (TMN) Monoaminergic cell groups of the HYPOTHALAMUS, containing histamine, and making excitatory synapses with thalamic relay neurons projecting to the cortex. It is activated by noradrenergic stimuli originating in the locus coeruleus of the RETICULAR FORMATION (see Fig. 148).

tuberosity Large, rounded, usually roughened projection on bone (e.g., ischial tuberosity of hip bone); for connective tissue attachment.

tuberous sclerosis An autosomal dominant condition characterized by epilepsy, impaired mental development, skin tumours and depigmented spots and bone cysts.

tubular maximum The maximum rate at which a substance can be actively reabsorbed from the nephrons, which depends upon the concentration of the substance in the glomerular filtrate and the density of transport proteins present in the nephron epithelia. See PLASMA CLEARANCE.

tubulins Two closely related and highly conserved eukaryotic globular proteins, α-tubulin and β-tubulin, occur as heterodimer subunits of MICROTUBULES, tightly bound by non-covalent bonds. Each subunit has a binding site for one GTP molecule (exchangeable for GDP on the β-subunit). As with ACTIN, tubulin molecules occur in multiple isoforms. Compare INTERMEDIATE FILAMENTS.

tumor See TUMOUR.

tumorigenesis The production of a cancerous mass of cells (tumour). See CANCER CELLS.

tumour (tumor) Any cluster of adhering cancerous cells. It may be benign (noninvasive) or malignant (invasive, metastatic). They can be broadly classified as *liquid*, containing neoplastic cells whose precursors are normally mobile (e.g., leukaemias and lymphomas) or *solid*, composed of epithelial or mesenchymal cells that are normally immobile. Solid tumours are composed of two compartments: the neoplastic epithelial cells (which always have aneuploid karyotypes) and the stromal cells (see ANGIOGENESIS). See CANCER CELLS.

tumour necrosis factors (TNFs) A CYTOKINE protein family, whose members are either secreted or are cell surface receptors (TNFRs). Most are functional as homotrimers (e.g., TNF-α), although TNF-β acts as a heterotrimer with the related lymphotoxin LT-β. Most effector T cells express TNF members on their cell surfaces, notably TNF-α, TNF-β, Fas ligand and CD40 ligand, all of which bind receptors that are TNFR family members. TNF-α and TNF-β, both secreted by activated T_H1 cells, alter the endothelial cell membrane at sites of infection so that macrophages adhere to them, a crucial first step in local INFLAMMATION (see this entry for further details).

TNF-α, also secreted by T_H2 cells, some cytotoxic T cells, activated macrophages and adipocytes, is an important cytokine and helps to contain local infections by activating macrophage NITRIC OXIDE production. Once an infection reaches the bloodstream, however, systemic release of TNF-α by macrophages in the spleen, liver and other sites causes vasodilation, leads to loss of blood pressure, increased vascular permeability, loss of plasma volume and eventual SEPTIC SHOCK. This seems to be brought about by autocrine binding of TNF-α to its receptors on the macrophage's own surface, up-regulating its own antimicrobial activity. Mice with mutant TNF-α receptors are resistant to septic shock, although they are also unable to control local infection. Monoclonal antibodies against TNF, and TNF-binding proteins, have sometimes been used successfully in treatment of rheumatoid arthritis and Crohn's disease, although not in treating bacterial invasion or sepsis. TNF-β produced by CD4 T cells inhibits B-cell growth and macrophage activation, activates neutrophils among haematopoietic cells, and is the main factor in ISOTYPE SWITCHING to IgA by B cells.

tumour suppressor genes (oncosuppressor genes) and proteins So-called 'gatekeeper genes' (contrast CARETAKER GENES), whose protein products operate to constrain or suppress cell proliferation (and invasiveness). They normally occur in the homozygous dominant condition. If both alleles mutate to recessive forms the inhibition is lost (loss-of-function mutation), resulting in CANCER (e.g., see VHL GENE). In many tumours, function of the p16 (*CDKN2A*) tumour suppressor gene is abrogated by methylation of the promoter rather than by mutation of its DNA sequence (contrast inappropriate methylation of ONCOGENES). A half dosage of just one oncosuppressor gene, *APC* (see ADENOMATOUS POLYPOSIS COLI PROTEIN), makes the intestinal epithelium susceptible to development of cells with inaccurate grasp of polarity and position and deregulated proliferation – founder cells of adenomas (e.g., see FAMILIAL ADENOMATOUS POLYPOSIS). Tumour suppressor GENE SILENCING has been linked to promoter hypermethylation, first described for *RB*, the gene encoding RETINOBLASTOMA PROTEIN. The significance of proper checkpoint signalling for the prevention of cancer is given by the fact that most checkpoint kinases and mediators are either established or emerging tumour suppressors. See CHEMOTHERAPY, GENE EXPRESSION.

tunica albuginea A dense white fibrous capsule of connective tissue lying (i) between the testis cords and overlying epithelium of the TESTES and extending inwards, dividing the testis into a series of internal compartments, or *lobules*; (ii) immediately under the epithelium of the ovary.

Turner('s) syndrome A condition resulting from NONDISJUNCTION which, although common at conception and in MISCARRIAGES, occurs in liveborn female infants with low incidence: estimates range from 1:5,000 to 1:10,000. Clinical features can present at any time from pregnancy to adulthood, but are increasingly being detected during the 3rd trimester through ULTRASOUND as a result of generalized oedema, or swelling localized to the neck. But many Turner babies appear normal at birth; others show neck webbing and remains of intrauterine oedema, a low posterior hair-line, increasing carrying angles at the elbows, short 4th metacarpals, widely spaced nipples and coarctation (narrowing) of the aorta (present in 15% of individuals). INTELLIGENCE in Turner females is normal, although some differences in social cognition and higher-order cognitive function skills have been related to the parent-of-origin of the single X CHROMOSOME. Individuals are heterosexually oriented. The main problems are short stature (apparent by mid-childhood) and ovarian failure. Haploinsufficiency of the *SHOX* gene, located on the pseudoautosomal region, is at least partly responsible for the short stature. Ovarian failure begins during the 2nd half of intrauterine life, leading almost always to amenorrhoea and INFERTILITY. Oestrogen replacement therapy, initiated at adolescence, is effective for development of SECONDARY SEXUAL CHARACTERISTICS, and long-term prevention of OSTEOPOROSIS. Chromosomally, individuals are 45,X; but, as with KLINEFELTER SYNDROME, there is chromosome mosaicism in a significant proportion of individuals, and those with a normal cell line (46,XX) have a chance of fertility. Otherwise, IN VITRO FERTILIZATION using donor eggs offers the hope of pregnancy for Turner women. See GENETIC COUNSELLING, GENOMIC IMPRINTING.

turnover-pulse hypothesis The suggestion that, if speciation and extinction are driven largely by climatic fluctuations, this pattern should be observable synchronously in all species' groups. Evidence seems to refute such an extreme (even simplistic) version of the effect of climate change.

twin concordance Since monozygotic twins are genetically identical clones, they will be concordant (see CONCORDANCE) for any genetically inherited trait, regardless of the number of genes involved (see PHENOTYPE). Exceptions to this rule are instances in which individual variability of gene expression (including epigenetic expression) are to be expected (e.g., the pattern of X-CHROMOSOME INACTIVATION in females, and the range of functional IMMUNOGLOBULINS, T-CELL RECEPTOR genes). See FAMILIAL RECURRENCE RISK, TWIN STUDIES.

twin studies Francis Galton stressed the significance of TWINS in studies of human genetics. From the higher level of TWIN CONCORDANCE among identical twins than among fraternal twins, we need not conclude that such concordant traits are genetically inherited (see FAMILIAL RECURRENCE RISK). Half of dizygotic (DZ) twins are of different sexes, whereas all monozygotic (MZ) twins are same-sex twins. Even when restricting comparisons to same-sex DZ twins, especially for behavioural traits, the case can be made that such twins are likely to be reared more similarly than DZ twins. Studies of MZ twins separated at birth who subsequently discover that they share similar tastes in clothing and make similar choices in career and partners have tended to be more anecdotal (and newsworthy) than adequately controlled, and the small sample sizes involved inevitably impact on their statistical significance. Adoption studies offer the hope of disentangling genetic and environmental factors in the aetiologies of complex traits (e.g., COMPLEX DISEASES). Relevant questions to be addressed among adopted people who suffer from a particular disease known to have high FAMILIALITY include: whether the disease runs in their biological or adoptive family; and among affected parents whose children have been adopted out of the biological family, whether their experience saved their children from the same familial disease. Twin studies have revealed the importance of the timing of menarche in BREAST CANCER risk in women: in identical twins who developed this cancer, the twin whose menarche began earlier had a 5.4 greater risk of being the first to be diagnosed.

All such studies should be genuinely 'blind', in that researchers should not be able to bias results through inappropriate knowledge of case histories. A major obstacle in adoption studies is a lack of knowledge concerning biological parentage, often hampered by the delicacy of asking such questions. Another problem is the tendency of adoption agencies to place adopted children in family environments resembling those from which they originate. Adoption studies are therefore especially difficult to control adequately, and have tended to be carried out for psychological conditions, where the NATURE-NURTURE DEBATE has perhaps been most volatile (e.g., see SCHIZOPHRENIA). Twin and adoption studies strongly suggest that violent behaviour has at least some genetic component. One 1993 study linked male AGGRESSIVE BEHAVIOUR in one family to a very rare defect in the gene encoding monoamine oxidase A (see MONOAMINE OXIDASE, MOUSE MODELS, AFFECTIVE DISORDERS). See GENETIC DETERMINISM.

twins Siblings born at approximately the same time include identical (monozygotic, MZ) twins, and fraternal (dizygotic, DZ) twins (see FERTILIZATION). A single embryo may become two monozygotic twins up to the primitive streak stage of development, after which time developmental potential is restricted by loss of cell pluripotency. If only one primitive streak appears, regional specification involving PATTERN FORMATION and spatial organization of symmetry become actualized in biologically irreversible ways. In very rare cases, two primitive streaks may form and give rise to identical twins or to conjoined (Siamese) twins. If two heads are formed by conjoined twins, it is usually accepted that two human individuals have also been formed. The frequency of MZ twinning is remarkably constant at ~4 twin pairs per 1,000 pregnancies; but the frequency of DZ twinning varies widely, being greatest among Africans and least among Asians. Nearly constant MZ rates but quite variable DZ rates are also observed as a function of maternal age. In order to produce DZ twins, two ova must be released in a given ovarian cycle, and the level of pituitary gonadotropins are known to increase from adolescence throughout the entire reproductive period, paralleled by an increase in DZ twin rates up to ~37 years of maternal age. MZ twinning shows no tendency to recur in the same mother, or to exhibit FAMILIALITY; but the incidence of DZ twinning indicates that a woman who has conceived one DZ twin-pair is slightly more likely to have a second such pair than is a non-twinning mother, and this raised incidence is shared among close female relatives of a DZ-twinning mother. Although the genetic basis of this tendency is unknown, genes involved may be inherited and transmitted by either sex. See TWIN STUDIES.

tympanic membrane See EARDRUM.

typhoid fever See SALMONELLA.

typhus A rickettsial feverish disease, one species of rickettsia (*Rickettsia prowazekii*) being spread by infected human body or head lice (causing epidemic typhus) while another species (*R. typhi*) is spread by the rat flea or rat louse (causing the less common endemic typhus). Contrast typhoid fever (see SALMONELLA).

tyrosine kinases Enzymes adding PHOSPHATE groups to specific tyrosine residues in target proteins. Many cell-surface receptors (e.g., cytokine receptors) depend for their activity on tyrosine phosphorylation but are without a tyrosine kinase domain. Such receptors act through cytoplasmic tyrosine kinases, such as SRC-KINASES, while the RECEPTOR TYROSINE KINASES have their own tyrosine kinase domain. The Jaks (see JAK/STAT PATHWAY) are soluble tyrosine kinases which bind certain activated receptors. Compare SERINE/THREONINE KINASES. See HUMAN KINOME, SH2 DOMAIN.

U

ubiquinone (coenzyme Q) An isoprenoid lipophilic electron carrier in the inner mitochondrial membrane (see MITOCHONDRION), receiving electrons from complex I and transferring them to the cytochrome *b*-c_1 complex.

ubiquitins A family of small and highly conserved proteins involved in the degradation of most cellular proteins and targeting them for rapid hydrolysis by PROTEASOMES. They are prepared for conjugation to other proteins by the ATP-dependent ubiquitin-activating enzyme (E1), activated ubiquitin being transferred to one of several ubiquitin-conjugating enzymes (E2), which act as a team with other accessory proteins (E3) that recognize degradation signals on the target proteins. Although CHAPERONES are very effective in ensuring that newly synthesized polypeptides and proteins are correctly folded, misfolding, denaturation and other degradation signals are recognized by the enzyme complex ubiquitin ligase (E2–E3) which helps form a *multiubiquitin chain* attached to a lysine residue of the protein substrate, many ubiquitin molecules being linked successively to one another at a specific lysine of the preceding ubiquitin molecule and targeting the complex for the cap complex of a proteasome. The ubiquitin–proteasome system comprises many distinct proteolytic pathways; but all have E1 at the top and the proteasome underneath and differ in their E2–E3 components.

It is now apparent that ubiquitinization is a major factor in the control of the TRANSLATION of mRNA on ribosomes, and consequently in the control of CELL GROWTH. CANCER CELLS appear more susceptible to effects of proteasome inhibitors than do normal cells.

UCPs (uncoupling proteins) A subfamily of mitochondrial carriers, including UCP1, UCP2 and UCP3, of which UCP1 occurs in brown ADIPOSE TISSUE and is essential to temperature regulation in small mammals (and infant humans). UCP1 enables protons pumped into the intermitochondrial space to re-enter the matrix without ATP synthesis so that respiration is dramatically increased and heat is released. Discovery in other tissues of genes similar to that for UCP1 suggested that their encoded proteins participated in the proton leak observed in mitochondria, and therefore were involved in the regulation of energy expenditure, including basal metabolic rate, and the regulation of body weight. They may also be involved in adaptation of cellular metabolism to an excessive supply of substrates, regulating the ATP level, the NAD:NADH ratio, metabolic pathways and limitation of superoxide production. See MITOCHONDRION, OBESITY.

ulcerative colitis See ALLERGY.

ulna The longer of the two bones of the forearm, articulating proximally with the humerus at the olecranon fossa to form a major part of the elbow joint; but its distal end is separated from the wrist by a fibrocartilaginous disc. It is bound to the radius by an interosseous membrane.

ultradian rhythms Biological cycles having faster periods than CIRCADIAN RHYTHMS.

The non-REM/REM/non-REM sleep cycle repeats about every 90 minutes. See SLEEP.

ultrasound The second oldest imaging technique (the first being radiography), first developed in the 1950s, in which high frequency sound waves strike internal organs and are reflected onto a receiver placed on the skin. Computerized analysis of the sound waves is followed by imaging on a screen (the sonogram), enabling real time movements to be observed. It can detect TURNER SYNDROME and ANENCEPHALY in pregnancy. See CT SCAN, MAGNETIC RESONANCE IMAGING.

ultraviolet (ir)radiation (UVR) Ionizing radiation (see MUTAGENS), mostly of solar origin. UV photons occur with wavelengths 315–400 nm (UV-A) and wavelengths 290–315 nm (UV-B). UV-B is the most energetic form of UVR normally reaching Earth's surface. Notwithstanding the presence of the ozone layer, UVR causes nutrient photodegradation in the skin since important nutrients such as flavins, carotenoids, tocopherol and folate degrade in its presence (see SKIN COLOUR). It is involved in the formation of previtamin D_3 (cholecalciferol) in the skin, while short-term UVR exposure causes SUNBURN. Long-term UVR exposure has DNA-damaging effects (pyrimidine dimers) which activate the ATR kinase pathway (see DNA REPAIR MECHANISMS) and causes cancers if this pathway is ineffective – notably basal cell and squamous cell carcinomas, and the often-fatal MELANOMAS (usually presenting after reproductive age and so probably not a strong selective force). See SKIN CANCER.

umbilical cord Developing from EXTRAEMBRYONIC MESODERM that becomes invaded by blood vessels (see CARDIOVASCULAR SYSTEM, Fig. 38) the umbilical cord derives from the connecting stalk and yolk sac stalk (vitelline duct) that pass through the primitive umbilical ring, the oval line of reflection marking the border of the amnion and the embryonic ectoderm. The connecting stalk contains the allantois and the umbilical vessels, two arteries and one vein. As the amnion envelops the connecting and yolk sac stalks, these are pushed together to form the primitive umbilical cord. The umbilical veins develop from chorionic villi and, passing on either side of the liver, carry oxygenated blood to the embryo. The umbilical arteries, initially paired ventral branches of the dorsal aorta, run to the PLACENTA (see Fig. 138) in close association with the allantois, acquiring during the 4th week a secondary connection with the dorsal branch of the aorta, the common iliac artery, thereby losing their original origin.

uncoupling proteins See UCP1, UCP2.

Uniregional Hypothesis See RECENT AFRICAN ORIGIN MODEL.

upstream activating sequences Those *cis*-acting regulatory sequences in yeast that are distinct from the promoter and increase expression of a neighbouring gene.

uracil A pyrimidine base, synthesized *de novo* from aspartate and occurring in uridine nucleotides and in ribosenucleic acids (RNAs). Uridylylation is the transfer of a UMP moiety to a substrate (e.g., by uridylyltransferase). Uridine 5´-triphosphate (in the ribosyl form) is the substrate used by RNA POLYMERASES, and is also converted by cytidylate synthetase to cytidine 5´-triphosphate. UTP is also required by UDP-glucose pyrophosphorylase in the synthesis of GLYCOGEN. Deamination of cytosine in DNA to uracil is probably the reason why DNA contains thymine and not uracil (see CPG DINUCLEOTIDE, DNA REPAIR MECHANISMS). Many of the uridine residues of RIBOSOMAL RNA are isomerized to pseudouridine. Degradation of pyrimidines produces UREA.

urea A small, uncharged, polar molecule ($CO(NH_2)_2$) synthesized within liver hepatocytes using the AMMONIA produced by transamination (see DEAMINATION) of excess amino acids. In common with water and glycerol, urea does not readily diffuse across cell membranes. Its production involves hepatocyte enzymes in both the mitochondria and cytosol. Carbon dioxide produced by mitochondria is used (as HCO_3^-) in an ATP-dependent reaction to form carbamoyl phosphate in the mitochondrial matrix, catalysed by a mitochondrial-specific isoform of carbamoyl phosphate synthetase I. The carbamoyl group is transferred to ornithine by ornithine transcarbamoylase, forming citrulline and releasing P_i. The citrulline enters the cytosol, where a second amino group is provided by transamination from aspartate in an ATP-dependent reaction, forming argininosuccinate. Argininosuccinate lyase then cleaves this (reversibly) to fumarate and arginine, the arginine being converted by arginase in the cytosol to produce urea and ornithine, an amino acid which is recycled back into the mitochondrion. The urea cycle was discovered by Hans Krebs and a colleague. The KREBS CYCLE and urea cycle are interconnected: fumarate produced by the urea cycle can enter the mitochondrion, and can also be converted by cytosolic isoforms of fumarase and malate dehydrogenase to produce fumarate and oxaloacetate which can also enter the mitochondrion and be substrates of the Krebs cycle (see Appendix IId). Although not toxic, urea does denature proteins in high concentrations by interfering with their hydrophobic interactions. During prolonged STARVATION, when muscle protein is a major source of metabolic energy, urea production increases substantially. For urea excretion, see KIDNEY.

ureter A small diameter tube developing from the ureteric bud, conveying urine from the KIDNEY (see Fig. 107) to the urinary bladder.

ureteric bud See KIDNEY.

urethra The duct leading from the urinary bladder to the exterior. Its epithelium is endodermal in origin, becoming surrounded by mesenchyme and smooth muscle derived from splanchnic mesoderm, so that at the end of the 3rd month of development the epithelium of the prostatic urethra proliferates to form outgrowths penetrating the surrounding mesenchyme. These buds form the prostate gland in the male, while in the female the cranial part of the urethra gives rise to the urethral and paraurethral glands. Also at this time in males, the penile urethra forms by closure of the urethral folds to form a urethral groove as the developing phallus elongates, the canal extending into the tip of the phallus during the 4th month. In males, the urethra is the shared terminal duct of the reproductive and urinary systems, serving for the passage of urine and semen.

uric acid A product of PURINE degradation and present at 0.7–8.7 mmol l^{-1} in urine, between 7 and 25 times greater than in blood plasma. See GOUT, LESCH–NYHAN SYNDROME.

uridine The URACIL nucleoside.

urinary bladder See BLADDER.

urine Produced by the KIDNEYS, the composition of urine is subject to considerable variation. Compared with blood plasma, normal urine contains little if any protein, glucose or hydrogen carbonate ions (see ACID–BASE BALANCE), but contains far higher concentrations of CREATININE, ammonium ions (NH_4^+; see AMMONIA), PHOSPHATE and UREA. See CALCIUM, POTASSIUM, SODIUM, URIC ACID, VASOPRESSIN.

urobilin, urobilinogen See BILE PIGMENTS.

urogenital ridges See MESONEPHROS.

uterine artery A branch of the internal iliac artery supplying the uterus, vagina, oviduct and ovary.

uterine tubes See OVIDUCTS.

uterus Arising from the fused paramesonephric ducts (MÜLLERIAN DUCTS), the wall of the uterus has three layers: the ENDOMETRIUM (mucosa) lining the inside wall; a thick smooth muscle layer, the myometrium; and the perimetrium, a peritoneal covering lining the outer surface of the myometrium. The oviducts open into it superiorly, while its inferior cylindrical portion is the cervix.

UTP (uridine triphosphate) See URACIL.

UTRs Untranslated regions of mRNA, typically lying between the stop codon and the start of the polyadenylated tail (see RNA PROCESSING) which are often used to direct the mRNA to specific intracellular locations prior to translation. See CYTOKINES, ISOFORMS, REPETITIVE DNA, RNA INTERFERENCE.

utricle With the saccule, one of the two macula-containing sacs of the VESTIBULAR APPARATUS.

V

V1 Area V1. See PRIMARY VISUAL CORTEX.

V2, V4, V5 See AREA V4, MT (AREA V5), EXTRA-STRIATE CORTEX.

vaccines Preparations containing foreign (non-self) antigens that on introduction to a host (orally or by injection) activate the adaptive immune system (see ADAPTIVE IMMUNITY) and result in host production of memory B cell and T cells so that any subsequent secondary immune response will be faster, and produce a higher titre of appropriate ANTIBODIES, than would have occurred in the vaccine's absence. The development of childhood vaccines has been the main factor responsible for almost doubling the life expectancy in industrialized nations during the 20th century, and to the names of pioneers such as Edward Jenner and Louis Pasteur should be added that of Maurice Hilleman (1919–2005) who, with his co-workers over 30 years, produced >40 viral and bacterial vaccines, culminating in the measles–mumps–rubella (MMR) vaccine.

Vaccines are derived from whole viruses and bacteria, or from their antigenic components. *Live vaccines* derive from attenuated strains with minimal pathogenicity but the capacity to induce durable protective immunity, multiplying in the human and providing rather prolonged antigenic stimulation although they very occasionally cause full-blown infection (notably in immunosuppressed persons). Only one dose is usually required, but their use during pregnancy should be avoided on account of potential risk to the foetus. These include the BCG VACCINE, MMR (see MEASLES), oral polio (see POLIOMYELITIS) and YELLOW FEVER vaccines. *Non-replicating vaccines* contain either inactivated whole pathogens or their antigenic components and often contain adjuvants (commonly aluminium salts) which enhance the immunogenic effect or keep the antigen in a localized or particulate form. These vaccines are safe for use during pregnancy and in the immunosuppressed, although more than one dose is usually required for protection and local reactions to them are relatively common on account of the quantity of antigen introduced. They include pertussis, influenza, *Haemophilus influenzae* type b, rabies and typhoid vaccines.

Some vaccines may be impermissible (contraindicated) in certain circumstances, e.g., in those with compromised immune systems (but see MEASLES). Pathogens vary in their ability to undergo antigenic variation, such ability being a major obstacle to commercial production of an adequate vaccine. Genes that evolve quickly are probably least suitable as targets for (e.g.) antimalarial drugs because they will probably continue to evolve quickly. Genes that have been relatively constant for millions of years and produce a good immune response are probably more rewarding. Those who cannot mutate their B-cell antibody-producing genes after an initial secondary (adaptive) immune response, suffer recurring bacterial and viral infections and do not respond to vaccination through failure in AFFINITY MATURATION. Work on DNA VACCINES has shown promise against some viral infections. See HIV, INFLUENZA.

vagus nerve (cranial nerve X, CN X) A large, mixed, cranial nerve innervating much of the alimentary canal, ventilatory system and heart (see AUTONOMIC NERVOUS SYSTEM). Its nuclei in the MEDULLA OBLONGATA receive its afferent impulses from, and provide motor impulses to, many thoracic and abdominal viscera. Afferent signals include those from stretch receptors in bronchioles and the alimentary canal, its visceral afferents terminating in the NUCLEUS TRACTUS SOLITARIUS. Stimulation of efferent vagal parasympathetic activity has been associated classically with slowing of heart rate (see CARDIOVASCULAR CENTRE), induction of gastric motility (see STOMACH) and dilation of arterioles (see ENDOTHELIUM). Inhibition of the process of INFLAMMATION can now be added to this list. Its somatic motor efferents include those to muscles of the PHARYNX and LARYNX. Preganglionic fibres of the vagus extend to the thorax to contribute to the cardiac plexus supplying the heart, and the pulmonary plexus supplying the lungs. Some preganglionic vagal fibres extend through the diaphragm to supply the wall of the stomach; others contribute to the coeliac and superior mesenteric plexuses (see SOLAR PLEXUS).

valgus angle The angle subtended by the femur at the knee, critical to the evolution of BIPEDALISM. In humans, the inward sloping angle of the thigh to the knee enables the two feet to be placed very close to the midline at rest so that the body's centre of gravity need not be shifted far laterally back and forth during each phase of walking. In chimpanzees, the femur does not slope inward to the knee as much, the feet normally being placed well apart (nor are the gluteal muscles as well developed as in humans). During bipedal walking, a chimpanzee must rotate its upper body through a considerable angle with each step, an exaggerated GAIT (see Fig. 79) made more inefficient by the chimpanzee's higher centre of gravity relative to the hip joint than it is in humans.

Valium See BENZODIAZEPINES.

variable number tandem repeats See VNTRS.

variation Phenotypic variation in a population arises from two sources: genotypic and environmental. Even identical genotypes exhibit phenotypic variation because of environmental variation. The phenotypic variance of the population ($s_p{}^2$) may be divided between the variance between genotypic means, or *genetic variance* ($s_g{}^2$), and the remaining *environmental variance* ($s_e{}^2$). This simple analysis ignores possible covariance between genotype and environment – i.e. that these parameters are themselves correlated. Were 'ability genes' to exist in humans such that (and this is hypothetical) proclivity at mathematics had a heritable component, parents of children with such genes might themselves be mathematicians and create a home environment more than usually favourable to nurturing a mathematical interest in their already genetically predisposed children (see HERITABILITY).

If phenotype is the sum of genetic and environmental contributions ($p = g + e$), then the phenotypic variance is given by the sum of the genetic variance, environmental variance and twice the covariance between genotypic and environmental contributions:

$$s_p^2 = s_g^2 + s_e^2 + 2 \text{ cov } ge$$

Unless covariance between genotypic and environmental values is included in the calculation of phenotypic variance, this covariance will remain hidden in the genotypic and environmental variances. See COPY NUMBER, GENETIC DIVERSITY.

varicella zoster virus (VZV) See CHICKENPOX.

vasa efferentia (ductuli efferentes) The remaining parts of the mesonephric excretory tubules, linking the rete testes and mesonephric (Wolffian) ducts of the developing TESTES.

vasa recta Derived from the efferent arterioles of the juxtamedullary nephrons of the

KIDNEY, vasa recta pass deep into the medulla and have sluggish blood flow. Blood entering them is isotonic with normal plasma, and their walls are permeable to salts and water so that the blood they contain becomes progressively more concentrated as it enters the deep medulla through uptake of salts and loss of water, while the reverse occurs as the blood returns via an interlobular vein in the cortex to the arcuate vein and thence to the renal vein. By means of this passive exchange with the interstitial fluid, blood leaving the vasa recta is only slightly hyperosmotic to normal plasma. The countercurrent arrangement of the vasa recta, with their low blood velocity, helps maintain the osmolarity in the renal medulla. Excess salt and water provided by the processes of the LOOP OF HENLE will have been removed.

vascular compliance The tendency for blood vessels to increase in volume as blood pressure increases. The more easily the vessel wall stretches, the greater its compliance.

vascular endothelial growth factor (VEGF) See ANGIOGENESIS, GROWTH FACTORS.

vascular patterning See ANGIOGENESIS, VASCULOGENESIS.

vasculogenesis The formation of new blood vessels (compare ANGIOGENESIS). During embryogenesis, new blood vessels arise from blood islands. Fibroblast growth factor 2 (FGF-2) binds its receptor on subpopulations of mesoderm cells, inducing them to form haemangioblasts. Vascular endothelial growth factor (VEGF), acting through two different receptors, influences these cells to become endothelial cells and to coalesce to form tubular vessels (see ENDOTHELIUM). Arteries and veins develop from these, where necessary, by addition of PERICYTES, connective tissue and smooth muscle, under endothelial influence. Different EPHRINS are involved on the arterial and venous ends of the capillary: Eph-B2 in the former and Eph-B4+ in the latter. Pericyte recruitment, in particular, depends on endothelial PDGF-B release, and even in a mature vessel signals from the endothelium to the surrounding connective tissue and smooth muscle are crucial in regulating its structure and function. Final modelling and stabilization of the vasculature are achieved by PDGF and transforming growth factor β (TGF-β). NETRINS are endothelial mitogens and chemoattractants for human microvascular endothelial cells (HMVECs), stimulating proliferation, migration and tube formation of human endothelial cells *in vitro*, apparently independently of known netrin receptors.

vas deferens (ductus deferens; pl. vasa deferentia) The persisting Wolffian (mesonephric) duct in the male, eventually attaining ~45 cm in length and ascending along the posterior border of the EPIDIDYMIS, passing through the inguinal canal, entering the pelvic cavity, looping over the ureter and passing down the posterior surface of the urinary bladder, dilating at its terminal portion (the ampulla) and joining the seminal vesicle duct to form the ejaculatory duct. It stores sperm in a viable condition for several months and conveys them from the epididymis towards the urethra by peristaltic contractions of its muscular coat during EJACULATION and forms a part of the SPERMATIC CORD. See MESONEPHROS.

vasoactive intestinal polypeptide (VIP) A modulatory neuropeptide released by preganglionic fibres of the autonomic nervous system, interacting with G-PROTEIN-COUPLED RECEPTORS, and triggering small EPSPs lasting for several minutes and making postsynaptic neurons more responsive to the fast nicotinic effects of ACh. See SUPRACHIASMATIC NUCLEUS.

vasodilation and vasoconstriction Changes in the diameter of a blood vessel lumen. Sympathetic axons innervate vascular smooth muscle, stimulation typically causing contraction of the fibres and constriction of the vessel. When stimulation ceases, or in the presence of NITRIC OXIDE, K^+, H^+ and lactic acid, smooth muscles relax and the lumen diameter increases, causing vasodilation. Vasoconstriction also occurs during haemorrhage. Vasomotor tone is under the control of the vasomotor centre (see CARDIOVASCULAR CENTRE). See ARTERIOLES, EXERCISE, TEMPERATURE REGULATION.

vasomotor centre, v. tone See CARDIOVASCULAR CENTRE.

vasopressin (arginine-vasopressin, AVP, antidiuretic hormone, ADH) Nonapeptide hormone (neurosecretion) secreted, in association with neurophysin II, mainly by magnocellular neurons of the supraoptic nuclei of the HYPOTHALAMUS in response to raised blood concentration (reduced water potential) or reduced blood volume. OSMORECEPTORS in the vascular organ of the lamina terminalis near the hypothalamus activate these neurons, which secrete the hormone from the termini of their long axons which extend into the posterior PITUITARY GLAND (see Fig. 137). The rate of impulse transmission along these axons increases as blood concentration rises and as the osmoreceptors shrink in consequence. This causes the AVP stored in secretory granules at axon termini to be released into blood capillaries, resulting in increased uptake of water from the COLLECTING DUCTS in the kidney medulla through increasing AQUAPORIN concentration and resulting water permeability of cells in the tubule walls. As a result, less copious and more concentrated urine is produced (*osmoregulation*). Return of blood concentration to a normal level reduces AVP release. AVP receptors are METABOTROPIC and of at least three subtypes, differing in the G proteins to which they are coupled and therefore in the SECOND MESSENGER system they activate. V_1 receptors increase phosphoinositide turnover in muscle membranes and elevate intracellular CALCIUM (see PHOSPHOLIPASE C). AVP is also released in response to fall in effective circulating volume of blood, as during haemorrhage. It is a potent vasoconstrictor, acting particularly on smooth muscle of the arterioles of the skin and splanchnic circulation. V_2 receptors are located in the collecting ducts of the renal medulla and involve cAMP as second messenger. See DIABETES INSIPIDUS; compare OXYTOCIN.

VCAM-1 (vascular cell adhesion molecule-1) See ADHESION MOLECULES, ATHEROSCLEROSIS.

V(D)J recombination See ANTIBODY DIVERSITY.

vector 1. An animal carrier (and transmitter) of infectious pathogens. **2**. A vehicle (e.g., a drug-resistant plasmid, or plasmid-based DNA VACCINE) having its own replication origin and capable of being designed to insert and clone other pieces of DNA (cloning vectors) or to express one or more of the included genes (expression vectors). See HUMAN ARTIFICIAL CHROMOSOMES.

VEGFs (vascular endothelial growth factors) See GROWTH FACTORS.

veins Major blood vessels carrying blood away from the organs and towards the heart. Their walls comprise a *tunica interna* (thin folds of which form the flap-like cusps of valves projecting into lumen towards the heart), *tunica media* and *tunica externa* (usually the thickest layer; see and compare ARTERIES). Small veins (venules) drain blood from capillary beds and unite to form larger veins, which in turn unite to form larger veins. About two-thirds of the body's total blood volume is located within the highly distensible large veins (the capacitance vessels). Because their walls are relatively thin and possess little elastic tissue, blood returning to the heart can pool in the veins simply by distending them. The degree of venous pooling is determined by the tone of the smooth muscle in the major veins (vasomotor tone), governed by the sympathetic nerves supplying them (see CARDIOVASCULAR CENTRE). *Venous return*, the volume of blood returning to the heart from the systemic veins, is a function of the BLOOD PRESSURE difference between venules (~16 mm Hg) and the right ventricle (0 mm Hg) and keeps pace with output from the left ventricle because the resistance of veins is so low. Two mechanisms act as pumps to boost venous return: the *skeletal muscle pump*, notably in the lower limbs, and the *respiratory pump*. The former operates as follows: when standing at rest, the distal and proximal valves in a given section of vein in the calf region are open and blood flows upward towards the heart. Contraction of leg muscles compresses the vein, pushing blood through the proximal valve while closing the distal valve in the uncompressed vein segment just below. Following muscle contraction, pressure falls in the previously compressed section of vein, causing the proximal valve to close. The distal valve now opens since blood pressure in the foot exceeds that in the leg, filling the vein with blood from the foot. The respiratory pump depends on the relative changes in pressure in the thoracic and abdominal cavities during VENTILATION. During inspiration, movement of the diaphragm causes a decrease in thoracic pressure and an increase in abdominal pressure, compressing abdominal veins and moving blood from them into the decompressed thoracic veins and then into the right atrium. As pressure reverses during expiration, valves in the veins prevent backflow of blood.

venous return The rate of blood flow into the heart from the VEINS.

ventilation 1. Pulmonary ventilation (breathing). The normal pattern of quiet breathing (*eupnoea*) usually consists of shallow, deep, or combined shallow and deep breathing. Shallow breathing (*costal breathing*) involves upward and outward movement of the chest through external intercostal muscle contraction; while abdominal breathing (*diaphragmatic breathing*) involves outward movement of the abdomen through contraction and descent of the diaphragm. The lungs can be inflated and deflated by: (i) Downward and upward movement of the diaphragm, lengthening and shortening the thoracic cavity, the movements largely responsible for normal quiet breathing (see SPIROMETER). Premotor neurons in the ventrolateral medulla (see RESPIRATORY CENTRE) drive motor neurons in spinal segments C3–C5, whose axons run in the phrenic nerves. Impulses travelling along these axons cause contraction of the diaphragm during inspiration and their absence causes relaxation. (ii) Elevation and depression of the ribs and sternum. Motor neurons in C4–L3 supply neck and external intercostal muscles during inspiration, increasing the anteroposterior diameter of the thoracic cavity during inspiration by ~20% compared with expiration. Laboured breathing involves additional impulses from the intercostal nerves to the external intercostal muscles and to the sternocleidomastoid and pectoralis minor muscles. Expiration during quiet breathing is achieved by relaxation of the diaphragm, elastic recoil of the lungs, chest wall and abdominal structures, all effecting compression of the lungs. But during heavy breathing such elasticity is insufficient for the rapid expiration required and abdominal muscles are needed to push the abdominal viscera against the underside of the diaphragm while the abdominal recti and internal intercostals pull down on the lower ribs.

Alveolar pressure is the pressure of air inside the alveoli. With the glottis open, and no air flow, the pressures throughout the respiratory tree down to the alveoli are equal to atmospheric pressure. Inward flow of air during inspiration requires alveolar pressure to fall slightly below atmospheric pressure, achieved by increased volume of the thoracic cavity during inspiration. During expiration, an equivalent positive pressure is exerted on the alveoli, forcing air out. This tidal airflow maintains the mean concentration gradients of O_2 and CO_2 across the respiratory membrane sufficient to enable adequate gaseous exchange (see ALVEOLUS, Fig. 5).

Dissolved carbon dioxide is the main stimulus to inspiration. When air containing 15% O_2 but normal CO_2 is inhaled, there is no observable effect on ventilation: only when the O_2 level falls to 12% is breathing stimulated. By contrast, even a small increase in the inhaled CO_2 concentration strongly stimulates breathing. See CHEYNE–STOKES BREATHING.

2. Alveolar ventilation. See ALVEOLUS.

ventilatory equivalent (V_E/VO_2) The ratio of minute ventilation to oxygen consumption. Several weeks of aerobic training reduce its value during submaximal EXERCISE. See LACTIC ACID.

ventilatory system The passages through which air is moved tidally during ventilation, commonly subdivided into CONDUCTION ZONES and RESPIRATORY ZONES.

ventilatory threshold, V_T The point at which pulmonary ventilation increases disproportionately with oxygen consumption during graded EXERCISE. In healthy young adults, this is ~25 (i.e., 25 dm^3 of air breathed per dm^3 of O_2 consumed) during submaximal exercise, rising to ~55% of the VO_{2max}. Higher values occur in children, but during exercise where oxygen supply is more restricted, as in swimming, the value is significantly lower.

ventral Relating to the anterior, or front side of the body. Contrast DORSAL.

ventral anterior nucleus An important component of the pathway by which the BASAL GANGLIA exert their influence on movement, and through which abnormalities of movement are mediated in basal ganglia disorders. These paired thalamic nuclei occupy the rostral part of its lateral nuclear mass. Its subcortical afferents project from the ipsilateral basal ganglia. See VENTRAL LATERAL NUCLEUS.

ventral lateral nucleus Paired nuclei, lying immediately caudal to the ventral anterior nucleus of the thalamus, in the ventral lateral nuclear complex, their inputs originate mainly in the ipsilateral globus pallidus and substantia nigra, and from the contralateral dentate nuclei of the cerebellum. A part of the motor system, this nucleus and the related VENTRAL ANTERIOR NUCLEUS have reciprocal connections with motor areas of the primary motor cortex of the precentral gyrus.

ventral pallidum The ventral globus pallidus (see BASAL GANGLIA), forming a major component of the LIMBIC SYSTEM and receiving striatal input.

ventral tegmental area Region adjacent to the substantia nigra of the midbrain (see BRAIN, Fig. 30) whose dopaminergic axons project to the BASAL GANGLIA and PREFRONTAL CORTEX. See DOPAMINE, MESOLIMBIC SYSTEM.

ventral visual pathway See EXTRASTRIATE CORTEX.

ventricle **1**. A posterior chamber of the HEART (see Fig. 90). **2**. Enlargement of the central canal of the spinal cord within the BRAIN (see Fig. 28). The ventricular system comprises: (i) The two *lateral ventricles*, which are extensive chambers within the cerebral hemispheres. (ii) The *third ventricle*, which is a laterally compressed chamber within the diencephalon, the lateral walls of which form the thalamus. (iii) The cerebral aqueduct links the third ventricle to (iv) the *fourth ventricle*, a shallow rhomboid-shaped depression on the dorsal surface of the medulla oblongata and pons, which in turn becomes the spinal canal posteriorly.

Early in development the walls of the ventricles comprise just two cell layers: an inner *ventricular zone*, and an outer *marginal zone*, adjacent to the pia.

ventricular diastole Ventricular relaxation. See CARDIAC CYCLE.

ventricular system of the brain See VENTRICLE (2).

ventricular systole Ventricular contraction (see CARDIAC CYCLE).

ventricular zone The innermost of the two cell layers forming the walls of the ventricles of the embryonic brain (see VENTRICLE (2)). Cells in the ventricular zone divide by mitosis to produce either more ventricular zone cells (by cleavage in the vertical plane) or, in sequential waves, neuroblasts of cortical layers 6-2 (by cleavage in the horizontal plane). See CEREBRAL CORTEX.

ventrolateral preoptic nucleus (VLPO) See HYPOTHALAMUS, RETICULAR FORMATION.

ventromedial pathways Four descending motor tracts of the SPINAL CORD: the VESTIBULOSPINAL TRACTS, TECTOSPINAL TRACTS and the pontine and medullary RETICULOSPINAL

TRACTS, originating in the brainstem and terminating among spinal interneurons controlling proximal and axial musculature. They use sensory information relating to balance, body position and the visual environment in maintaining balance and body posture.

ventroposterolateral nucleus (VPN) See THALAMUS.

venules Small diameter VEINS.

vertebrae There are 33 vertebrae, developing by endochondral OSSIFICATION and forming the *vertebral column* (see SKELETAL SYSTEM, Fig. 156): 7 cervical (the 1st, or *atlas*, supporting the head), 12 thoracic (supporting the RIBS), and 5 lumbar vertebrae, 5 fused sacral vertebrae and (usually) 4 fused coccygeal vertebrae. Each forms during the 4th week from the sclerotome components of the SOMITES. Cells from the bilateral sclerotome blocks migrate around the spinal cord and notochord and merge. Resegmentation then follows, as the caudal half of each sclerotome fuses with the cephalic half of each subjacent sclerotome. Under the patterning influence of HOX GENES, each vertebra therefore develops from the caudal and cranial halves of two successive sclerotomes. Mesenchymal cells filling the space between two such precartilaginous vertebral bodies contribute to the intervertebral disc. These are bridged by myotome so that they can move the vertebral column.

There are regional variations, but each vertebra typically comprises a thick, disc-shaped anterior *body* (or centrum), a *vertebral arch* extending posteriorly from the body, its fused dorsal laminae surrounding a foramen through which runs the spinal cord, and seven processes arising from the arch: a *transverse process* extending laterally on each side; usually a single *spinous process* projecting posteriorly from the laminae; and two superior articular processes articulating with the two inferior articular processes of the vertebra behind (inferior). Failure of the dorsal laminae to fuse results in SPINA BIFIDA. Articulating surfaces are *facets*, and the articulations formed between the bodies and articular facets of successive vertebrae are intervertebral JOINTS. From the 2nd cervical vertebra to the sacrum, INTERVERTEBRAL DISCS form strong joints between the bodies of the vertebrae, allowing movement of the vertebral column and absorbing shock. The main functions of the vertebral column are: (i) to support the weight of the head and trunk; (ii) to protect the spinal cord; (iii) to enable spinal nerves to exit the spinal cord; (iv) to provide sites for muscle attachment; and (v) to enable movement of the head and trunk. See SCOLIOSIS.

vertebral column See VERTEBRAE.

vesicles, vesicle Eukaryotic cells have evolved an internal communication system of membranous organelles. This enables them to regulate the delivery of newly synthesized proteins, carbohydrates and lipids to specific targets, including the plasma membrane, and to take up external macromolecules and deliver them to digestive enzymes stored in lysosomes. Compartments enclosed by the membranes of organelles are topologically equivalent and in constant communication by a system of membranous transport packages, *transport vesicles*, which include small spherical vesicles and larger, irregular, vesicles or fragments of the donor compartment. Proteins and lipids move from one compartment to another using transport vesicles. Many of the proteins mediating traffic between organelles are 'peripheral', lacking transmembrane domains and recruited to the cytosolic surface of the membrane. These proteins include the various coat proteins that generate transport vesicles, MOTOR PROTEINS that move vesicles and organelles along the CYTOSKELETON, and 'tethering proteins' attaching vesicles to their target membrane prior to fusion with it.

Most peripheral proteins involved in membrane trafficking recognize the correct organelle by binding either to specific lipids, such as phosphoinositides, or to activated forms of GTPases (see GUANOSINE TRIPHOSPHATE). Because these are usually present only on a subset of internal membranes they uniquely identify an organelle to these proteins which act on their cytosolic surface, and because they are short-lived they enable vesicles to lose the identity of the organelle from which they have budded. This plasticity may have left eukaryotic cells more susceptible to INFECTION by pathogens.

Clathrin-coated vesicles were the first coated vesicles to be discovered. The major coat protein is clathrin itself, each subunit comprising three large and three small polypeptides forming a three-legged *triskelion*, which assemble into a basket-like convex framework of hexagons and pentagons which bend the membrane on the cytosolic side to form *coated pits*. A second multisubunit complex, *adaptin*, is required for binding of clathrin to the membrane while trapping various transmembrane proteins that in turn capture soluble cargos within the vesicle. In this manner, the proteins interact to form clathrin-coated transport vesicles, different adaptins determining the donor membrane specificity, as clathrin-coated vesicles can bud from different membranes (e.g., from the *trans* Golgi network; see LYSOSOMES). Clathrin coat assembly at the plasma membrane is involved in receptor-mediated ENDOCYTOSIS.

Target membranes display complementary receptors that recognize specific vesicle surface marker proteins that indicate both their origin and their type of cargo. A critical step in the secretory pathway involves inclusion of the correct complementary SNARE proteins, of which at least 20 are known, in the membranes of the vesicle (v-SNAREs) and of the target (t-SNAREs) which determine vesicle docking and membrane fusion, as in fusion of synaptic vesicles with the presynaptic membrane (see SYNAPSES). A guanine nucleotide exchange factor (GEF; see GUANOSINE TRIPHOSPHATE) in the donor compartment membrane recognizes a specific cytosolic GDP-bound Rab protein (see entry), a small monomeric GTPase, which becomes active on binding the membrane through its lipid

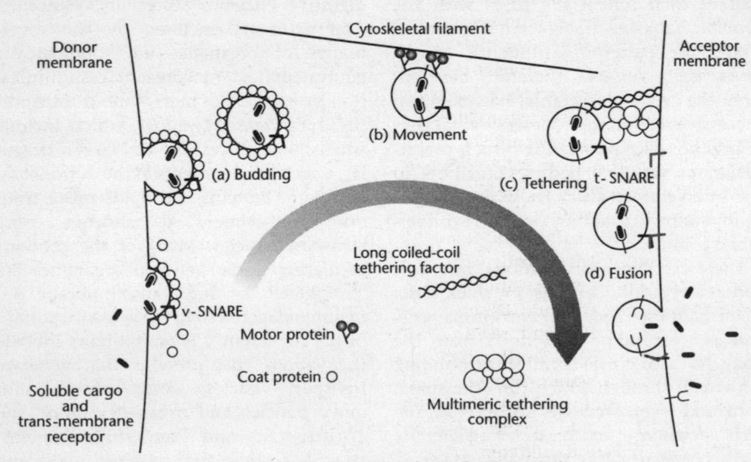

FIG 176. Some of the steps involved in vesicle transport between cell compartments. Coat proteins are recruited to the cytosolic face of the donor membrane, inducing formation of a VESICLE (see entry). After uncoating, motor proteins can be recruited, enabling the vesicle to travel along microtubules or actin filaments. Once at its destination, the vesicle becomes tethered to the acceptor (target) membrane, probably by long, coiled, proteins or multimeric tethering complexes. Complementary SNAREs bind and drive membrane fusion and delivery of the cargo within the vesicle. From R. Behnia and S. Munro, © *Nature* 2005, with permission from Macmillan Publishing Lrd.

anchor and exchanges GDP for GTP. A vesicle formed at the donor membrane carrying this Rab-GTP and an appropriate v-SNARE will then bind a specific Rab effector on the target compartment, helping the vesicle dock, and the v-SNARE will bind the appropriate t-SNARE, driving membrane fusion (see Fig. 176).

The Arf proteins (see entry) are also localized to specific organelles and recruit a wide range of effectors: Sar1 recruits the COPII vesicle coat found on vesicles and tubules originating from the ENDOPLASMIC RETICULUM, while Arf1 recruits COPI and clathrin/adaptin vesicle coats from the plasma membrane, to the GOLGI APPARATUS, COPI-coated packages also budding from pre-Golgi compartments and from cisternae of the Golgi complex.

The phosphoinositide PIP_2 is essential for the budding of clathrin-coated vesicles, because the budding machinery binds to it. Similarly, in the invagination of vesicles into late endosomes, PHOSPHATIDYLINOSITOL 3-KINASE is required, along with PtdIns(3)P-5-OH kinase.

vestibular apparatus Collective term for the paired organs of equilibrium lying within the temporal bone, each adjacent to the cochlea of the inner EAR (see Fig. 68) and developing from the otic vesicles. Each comprises a system of fine membranous and fluid-filled enclosures (the membranous labyrinth) and includes a saccule, a utricle and three semicircular canals. Its receptors, HAIR CELLS, are phylogenetically related to those of the cochlea, the stimulus for impulse generation being mechanical stretching of the membranes of the stereocilia.

In the utricles and saccules, hair cells lie in a small thickened region of their walls, the MACULA, and their stereocilia lie embedded in glycoprotein of the gelatinous otolithic membrane, over the entire surface of which extends a layer of dense calcium carbonate crystals, the *otoliths*. The main effective stimulus for the maculae is gravity (a special case of linear acceleration), enabling them to serve as receptors involved in static equilibrium. However, the maculae also respond to other forms of translational acceleration, e.g., that associated with the start and end of a run, when the otolithic membrane shifts over the underlying sensory epithelium and tilts the stereocilia. Maculae are fixed with respect to the skull, that in a utricle being approximately horizontal during upright posture while that in the saccule is approximately vertical. When the head is tilted, these organs are displaced to angles between horizontal and vertical and for every position of the skull the effect of gravity is to move the otolith membranes into different positions with respect to the sensory epithelium, resulting in a very specific constellation of excitation in the associated afferent neurons of the vestibular branch of the VESTIBULOCOCHLEAR NERVE, encoding information about the position of the skull in space (see POSTURAL REFLEX PATHWAYS).

Receptors of the *semicircular canals* respond to rotational acceleration rather than linear. At its union with the utricle, each canal enlarges to form an ampulla, within which a gelatinous membrane, the cupula, projects into the endolymph along an elevated ridge, the *crista*. Their specific gravities being equal, the cupula (unlike the otolith membrane) does not move with respect to the canal during linear acceleration. But when subjected to rotational forces of acceleration, the cupula is deformed because of the inertia of the endolymph. If the skull, initially at rest, is turned, the endolymph takes time to move, and since the cupula is attached to the wall of the canal it is pulled in the opposite direction to that of rotation. The stereocilia of the hair cells are deflected by the bending cupula and, similarly to the maculae, results in generator potentials that may lead to impulses along the afferent fibres of the vestibular branch. Bending of the cupula in one direction causes an increase in discharge rate while bending in the other causes a decrease. In the horizontal semicircular canal, the fibres are excited by displacement of the cupula towards the utricle (e.g., in the left canal when the head rotates to the left).

Reflexes elicited by the maculae of the

vestibular apparatus are of two kinds: (i) *Static reflexes*, maintaining equilibrium while standing quietly, sitting and lying down. These may involve compensatory rolling of the eyes, ensuring that horizontal and vertical lines are imaged on the retinas in the same way. (ii) *Statokinetic reflexes* occur during movements, and are movements themselves. They include the reflex turning elicited during a free fall, both the maculae and ampullae being involved. During the *lifting response*, an appropriate reaction to free fall, extensor tone is increased when the body is moved downwards and decreased when it is moved upwards. In *vestibular nystagmus*, as when the body is rotated in complete darkness, vestibular stimulation causes eye movement in the opposite direction to that of the body so that the direction of gaze is preserved (see SACCADES, HGPPS). This is most pronounced if the individual's head is tilted forward 30° while spinning so that the lateral semicircular canals are brought into the horizontal plane. The horizontal semicircular canals are, in effect, 'wired to' the medial and lateral rectus muscles that reset the eyes and counter head rotation. During large head rotations, before the eyes reach the limit of their range there is a rapid eye movement in the direction of body rotation, the eyes 'snapping ahead' and fixing on a new spot, stabilizing the retinal image. This is followed by another period of slow movement, again compensating for the rotation. If asked to walk in a straight line, the individual deviates in the direction of rotation; and if asked to point to an object, the finger used deviates in the direction of rotation. See CEREBELLUM, GAIT.

vestibular nuclei Paired groups of nuclei located at the junction of the medulla and pons, each group functioning almost as a deep cerebellar nucleus because of its direct connections with the cerebellar cortex and the flocculomodular lobe (see CEREBELLUM). Most of the vestibular nerves from the VESTIBULAR APPARATUS terminate in the medial and inferior vestibular nuclei, which give rise to the bilateral medial VESTIBULOSPINAL TRACT. Vestibular afferents to the lateral vestibular nucleus are involved in control of limb movements, its efferents projecting via the undecussated lateral vestibulospinal tract to all segments of the SPINAL CORD and facilitating extensor and inhibiting flexor motor neurons.

vestibular nystagmus See VESTIBULAR APPARATUS.

vestibule The oval central portion of the bony labyrinth of the inner EAR (see see Fig. 68).

vestibulocochlear nerve Cranial nerve VIII. A mixed (mainly sensory) nerve whose sensory and motor *vestibular branch* serves the semicircular canals, saccule and utricle (projecting to the vestibular nuclei of the medulla and thence along the VESTIBULOSPINAL TRACTS and RETICULOSPINAL TRACTS) and whose sensory and motor cochlear branch serves the organ of Corti. See COCHLEA, VESTIBULAR APPARATUS.

vestibulospinal tract A DESCENDING MOTOR TRACT (see SPINAL CORD), originating in the VESTIBULAR NUCLEI of the medulla and, with fibres of the RETICULOSPINAL TRACT, terminating in the intermediate and ventral grey matter of the SPINAL CORD. Vestibular nuclei relay sensory input from the vestibular apparatus which accompanies movement of the head. One component of the vestibulospinal tract, the *medial vestibulospinal tract*, innervates the cervical spinal circuits controlling some head and neck muscles (vestibulocollic reflexes), guiding head movements and ensuring head stability during body movement. A second component, the *lateral vestibulospinal tract*, projects ipsilaterally to the lumbar spinal cord and helps maintain muscle tone in the axial muscles and leg EXTENSOR MUSCLES (see entry), important in the reflex control of balance and posture (see POSTURAL REFLEX PATHWAYS). See MEDIAL MOTOR PATHWAYS, TECTOSPINAL TRACT.

***VHL* gene (von Hippel–Lindau gene)** A TUMOUR SUPPRESSOR GENE encoding part of a ubiquitin ligase complex which, in the presence of oxygen, degrades the transcription factor HIF-1α (see HYPOXIA) and therefore inhibits expression of cytokines such as VEGF and, in consequence, inhibits ANGIOGENESIS. VHL disease is an autosomally dominant inherited cancer syndrome, individuals heterozygous for one inactivating mutation in *VHL* being predisposed to developing a variety of TUMOUR types, such as the massively hypervascular renal clear cell carcinomas, while homozygous individuals exhibit highly elevated levels of VEGF expression caused by constitutive HIF1 activity resulting from inactivation of the second *VHL* allele.

vibration sensors See MECHANORECEPTORS, TRP RECEPTORS.

villus (pl. villi) Finger-like projections of the surface of the SMALL INTESTINE (see entry).

vimentin-like proteins A family of INTERMEDIATE FILAMENT including vimentin (in many cells of mesenchymal origin), desmin (in muscle fibres), glial fibrillary acidic protein (in astrocytes and some Schwann cells) and peripherin (in some neurons).

vinblastine, vincristine Alkaloids produced by the Madagascar periwinkle, *Catharanthus roseus*. Used in treatment of HODGKIN'S DISEASE, LEUKAEMIA and other forms of CANCER.

vinculin An intermediary protein linking actin to α-actinin in the cytoskeleton.

violence See AGGRESSIVE BEHAVIOUR.

viral haemorrhagic fevers (VHFs) A term that includes several severe viral infections in which haemorrhage is a part of the clinical presentation. The disease list includes those caused by arboviruses: flaviviral dengue haemorrhagic fever (see DENGUE FEVERS) and YELLOW FEVER, Crimean-Congo haemorrhagic fever (CCHF; nairovirus), Rift Valley fever, and others; hantaviruses (the reservoir being in mice and voles): haemorrhagic fevers with renal syndrome and haemorrhagic fevers with pulmonary syndrome; arenaviruses: lassa fever (the reservoir being the multimammate rat, *Mastomys natalensis*), Argentine haemorrhagic fever, Bolivian haemorrhagic fever, and others; and filoviruses: Marburg disease and ebola virus haemorrhagic fever (whose epidemiologies and reservoirs are not understood, although transmission of Marburg infections via laboratory vervet monkeys has occurred). Ebola infection has caused large community outbreaks, with fatality as high as 75–80%. Haemorrhage is usually due to platelet deficiency or dysfunction, and intravascular coagulation is usually a late feature, except in CCHF. Over 90% of VHF suspects have a final diagnosis of falciparum MALARIA. See VIRUSES.

viruses Usually minute infectious agents, 20–300 nm long and/or wide, that are unable to multiply except within a living cell of a host, of which they are obligate parasites and outside which they are inert (i.e., there is no enzyme activity outside the host cell). Unlike host cells, they almost never contain both RNA and DNA (human cytomegaloviruses were the first to be identified as containing both nucleic acid types). Viral DNA is structurally different in a way that marks it as foreign to a host cell (see INTERFERONS). The fully formed virus particle (virion) contains its nucleic acid or nucleoprotein core either within a naked coat of protein (the capsid, consisting of protein subunits, the capsomeres), or within a capsid enveloped by one or more membranes, of host nuclear or plasma membrane origin, acquired during exit from the cell and often modified, as by addition of specific glycoproteins involved in passage of the virus to another cell. Such enveloped viruses are sensitive to ether and other lipid-dissolving solvents and are readily inactivated by these (see Table 12).

Viral attachment to a host cell membrane receptor is crucial to INFECTION and in determining cell specificity (e.g., see HIV, INFLUENZA and Fig. 177). When inside a cell, virus particles may pack together in a crystalline condition through symmetry and regularity of their capsids. Modern virus classification is based on genetically based features, notably the type of nucleic acid and mode of transcription within the host cell; the structure and symmetry of

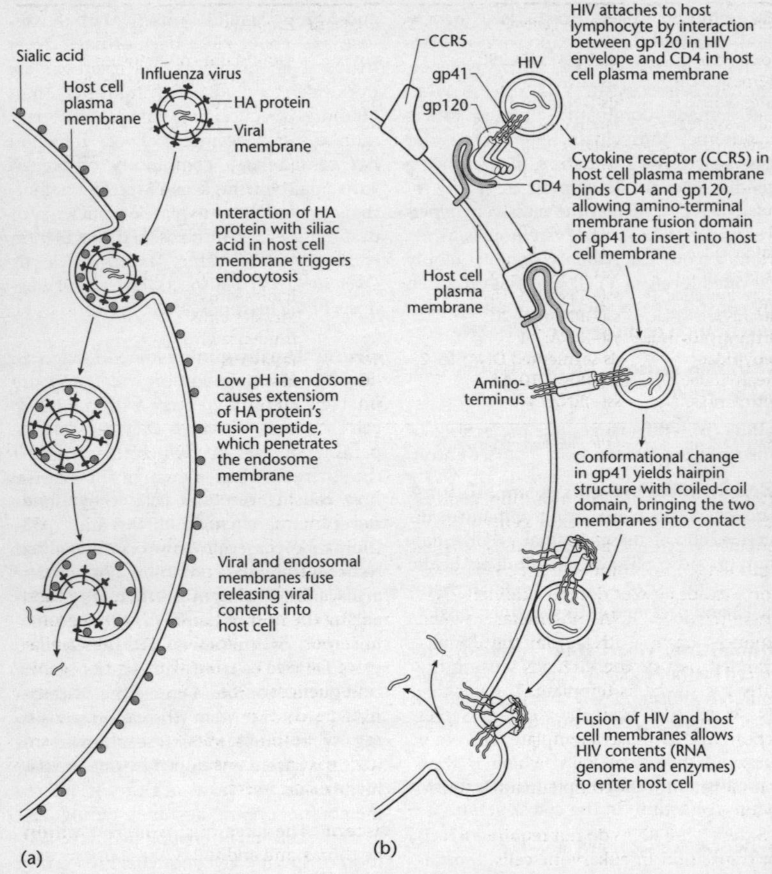

FIG 177. Two different methods by which enveloped viruses fuse with the host plasma membrane during entry to the cell: (a) influenza virus; (b) HIV (the involved glycoproteins are shown greatly enlarged). See VIRUSES. © 2000 From *Lehninger Principles of Biochemistry* by D.L. Nelson and M.M. Cox, published by Worth Publishers. Redrawn with permission from the publishers.

the capsid proteins, and the presence or absence of a lipid envelope. Viral nucleic acid is both transcribed and replicated within the host cell.

The RNA viruses adopt one of three reproductive strategies. If the RNA is positive (the coding, 'sense', strand) it can act directly as mRNA and be translated into viral proteins that include structural proteins and an RNA-dependent RNA POLYMERASE needed for viral replication. If the RNA is negative (the non-coding, 'antisense' strand), then an encoded viral RNA polymerase transcribes a positive complementary strand which serves either as a template for further viral negative strand synthesis, or as mRNA for translation into viral proteins. The third strategy is employed by RETROVIRUSES, whose positive single-strand ('sense') RNA cannot serve as mRNA. Retroviruses carry a reverse transcriptase, and their mRNA is only produced

Family of viruses	Type of genome, and size (kb)	Enveloped	Examples
Poxviridae	ds-DNA; 130–280	No	Molluscum contagiosum
Herpesviridae	ds-DNA; 120–220	Yes	Herpes simplex, Epstein–Barr
Adenovirus	ds-DNA; 36–38	No	Adenovirus
Papovaviridae	circular ds-DNA; 8	No	Human wart virus
Parvoviridae	ss-DNA; 5	No	Parvovirus
Hepadnaviridae	ds-DNA, ss portions; 3	Yes	Hepatitis B virus
Picornaviridae	ss(+)RNA; 7.2–8.4	No	Poliovirus, common cold
Togaviridae	ss(+)RNA; 12	Yes	Rubella virus
Flaviviridae	ss(+)RNA; 10	Yes	Yellow fever and West Nile viruses
Calciviridae	ss(+)RNA; 8	No	Norwalk agent
Coronaviridae	ss(+)RNA; 27–35	No	Common cold, SARS
Rhabdoviridae	ss(–)RNA; 13–16	Yes	Rabies virus
Paramyxoviridae	ss(–)RNA; 16–20	Yes	Measles virus
Orthomyxoviridae	ss(–)RNA; 14	Yes	Influenza virus
Reoviridae	ds segmented DNA; 16–22	No	Rotavirus
Areanviridae	ss(–)RNA; 10–14	Yes	Lassa
Retroviridae	ss(+)RNA; 3–9	Yes	HIV-1

TABLE 12 *Some details of the more important* VIRUS *families.*

after integration of the resulting double-stranded DNA into the host genome. The reoviruses possess a segmented double-stranded RNA genome and their reproduction includes use of double-stranded RNA-single-stranded RNA polymerase which produces positive RNA from the double-stranded part of the viral RNA using the antisense strand as template. The positive RNA is extruded from the virus and serves as both mRNA and as a template for further antisense (negative) RNA, which is then complexed with the complementary strand to form ds-RNA.

Several viral RNAs do not require a 5′ cap for translation in eukaryotic cells, translation being mediated by internal ribosome entry sites contained in their non-capped RNAs. This favours translation of viral mRNAs during infection once cap-dependent TRANSLATION of the host mRNAs has been suppressed.

Viral infection generates, initially, host production of IFN-α, IFN-β, TNF-α and IL-12. This is followed by a wave of NATURAL KILLER CELLS, which together control virus replication without viral elimination. This occurs when virus-specific CD8 T CELLS (cytotoxic T cells) are produced. Viruses have evolved ways of hijacking the process of ubiquitinization and protein degradation. Human papillomaviruses (Herpesviridae), causing genital warts, cervical and anal cancers, circumvent the host p53 defence system (see *p53* GENE) by encoding a protein that binds p53 and E3 (see UBIQUITINS), leading to ubiquitinization of p53 and making the host cell more likely to become cancerous. HIV does something similar with CD4 (see HIV), but without oncogenic consequences. See ADENOVIRUS, COMMON COLD, EPSTEIN–BARR VIRUS, HEPATITIS, HERPES VIRUSES, HIV, INFLUENZA, MEASLES, MUMPS, POLIOMYELITIS, POXVIRUSES, RABIES, RUBELLA, SARS, VIRAL HAEMORRHAGIC FEVERS.

viscera The large organs housed within the thorax and abdomen (see BODY CAVITIES). The adjective *visceral* is often contrasted with *somatic*, where the latter in this context relate to structures lying external to the coelomic cavities (see COELOM).

viscerocranium See SKULL.

vision See VISUAL PERCEPTION and other 'VISUAL' entries

visual acuity The clarity and detail of visual perception, amounting to the ability to distinguish (resolve) two nearby points. A discrete point of light falling on the retina can activate many cells, and many more cells in the target structure (e.g., the superior colliculus) of the retinofugal projection, because of the overlap of receptive fields. Good visual acuity requires a high level of illumination (photopic vision) and a low ratio of photoreceptors to ganglion cells in the RETINA, the region most specialized for high-resolution vision being the FOVEA. This region appears pit-like in cross-section on account of a lateral displacement of ganglion cells, allowing light to strike its photoreceptors (all cone cells) without any light scattering that would blur the image. The absence of short-wavelength (S) cones from the FOVEOLA prevents the chromatic aberration which would also compromise high-acuity vision, although COLOUR PERCEPTION is relatively coarse-grained. Scotopic vision is achromatic, since all rods have the same spectral sensitivity curve (see DARK ADAPTATION). In the partially dark-adapted eye, gap junctions between rod cells and cone cells remain open, augmenting cone cell function and maintaining acuity. But scotopic vision is characterized by low acuity since (i) the image formed on the peripheral retina is considerably distorted, (ii) CONVERGENCE of many rods onto a single bipolar cell leads to token information about localization of the light signal being less precise than for cone cells, and (iii) gap junctions between rods and cones are closed, so that cone cell signalling fails completely. See MELANINS, RECEPTIVE FIELD.

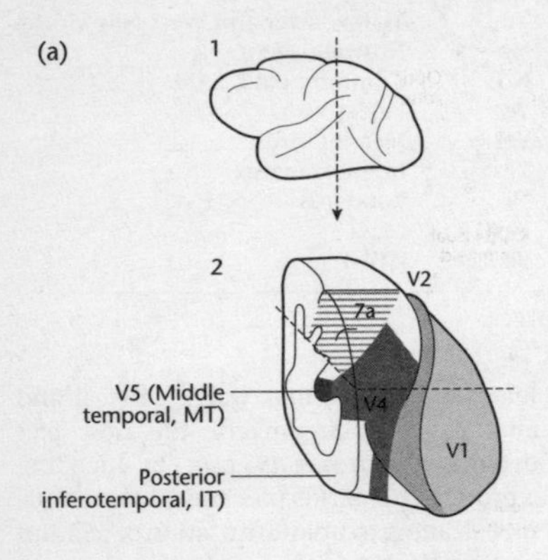

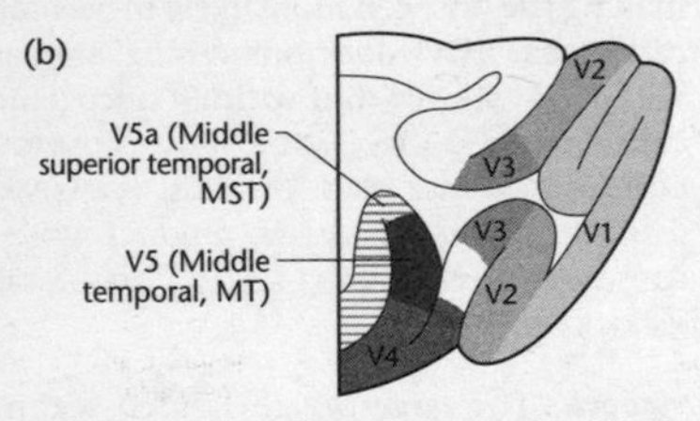

FIG 178. Diagrams of the macaque monkey's brain, indicating areas dedicated to vision that correspond to those of the human brain (see VISUAL CORTEX). (a.1) Lateral view of the right hemisphere showing exposed areas of the striate cortex (area V1) and extrastriate cortex (V2 and V4). The vertical line shows the position of the coronal section in (a.2). (a.2) Expanded anterolateral view of the occipital lobe showing areas V1, V2 and V4, and area V5. The latter lies in the buried MT area at the rostral border of the occipital lobe (see Fig. 32). (b) A horizontal section through the occipital lobe at the location indicated by the dotted line in (a.2), illustrating the approximate positions of known areas involved in visual processing at this level. From *Essentials of Neural Science and Behaviour*. E.R. Kandel, J.H. Schwartz and T.M. Jessell. © 1995 by Appleton & Lange (Simon & Schuster). Reproduced with permission of The McGraw-Hill Companies.

visual association cortex (v. a. area) See EXTRASTRIATE CORTEX.

visual cortex Those regions of the cerebral cortex (see CEREBRAL HEMISPHERES, Fig. 46) receiving projections from the LATERAL GENICULATE NUCLEI and concerned with visual experience (see Fig. 178). Organized anatomically into six vertical layers of neurons and arranged functionally in columnar systems (see HYPERCOLUMNS). See EXTRASTRIATE CORTEX, PRIMARY VISUAL CORTEX, VISUAL PROCESSING.

visual field The entire region of space, measured in degrees of visual angle, seen by the two eyes looking straight ahead and without movement of the head. This includes the left and right monocular zones and the binocular zone (see Fig. 179). The left and right OPTIC TRACTS contain fibres whose information represents the entire contralateral *visual hemifield*. When the foveae of both eyes are fixed on a single spot in space, the left half of the visual field (the left hemifield) projects on the nasal hemiretina of the left eye and on

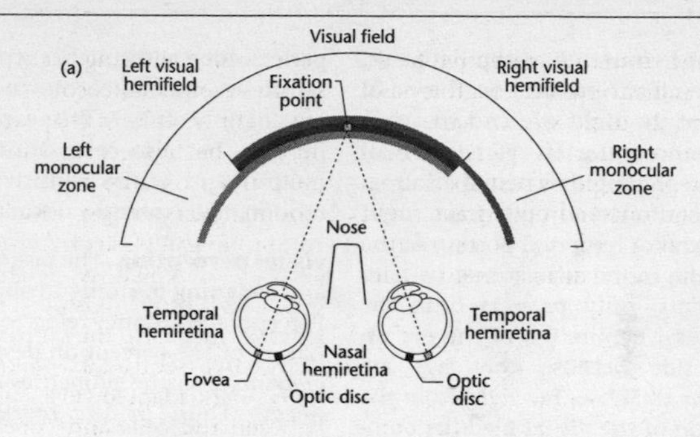

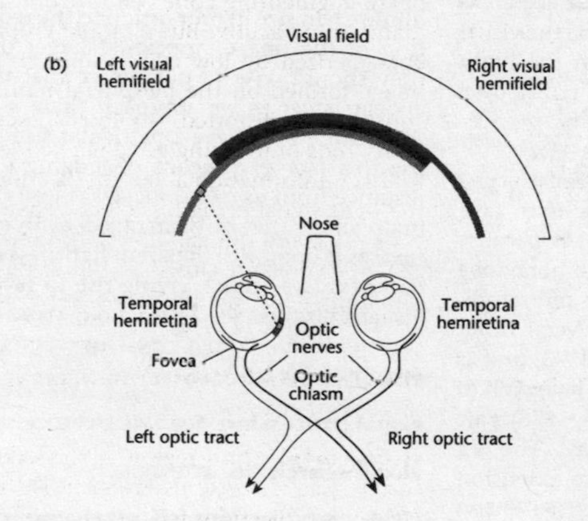

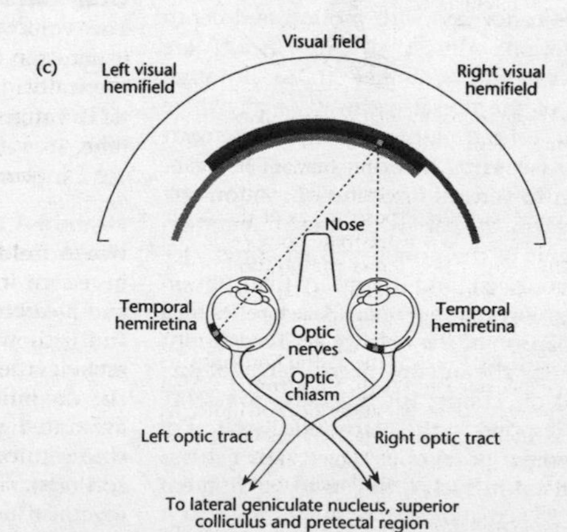

FIG 179. Schematic representation of the binocular and monocular components of the VISUAL FIELD (see entry for details). In (a), light from the binocular zone strikes both eyes, and light in the centre of the field impinges on the fovea. Light activates receptors in the temporal hemiretina and part of the nasal hemiretina of each eye. Ganglion cells leave the retina at the latter's true centre (the optic disc), which is free of receptors (see RETINA). In (b), light from a monocular zone strikes only one eye, falling on the ipsilateral nasal hemiretina but not projecting to the contralateral retina (blocked by the nose). In (c), each optic tract carries a complete representation of one half of the visual field. From *Essentials of Neural Science and Behaviour*. E.R. Kandel, J.H. Schwartz and T.M. Jessell. © 1995 by Appleton & Lange (Simon & Schuster). Reproduced with permission of The McGraw-Hill Companies.

the temporal hemiretina of the right eye. The right hemifield projects on the nasal hemiretina of the right eye and the temporal hemiretina of the left eye. Each visual hemifield is represented by neural information in the contralateral optic tract. Light from the lateral or temporal portion of the hemifield (the monocular zones, or temporal crescents) only projects onto the medial or nasal hemiretina of the eye on the same side because the rays are obstructed by the nose. But light from the central region of the visual filed (its binocular zone) enters both eyes (see BINOCULAR VISION). The lens of the eye inverts the visual image on the retina, although the brain adjusts this inversion. See also LATERAL GENICULATE NUCLEUS, Fig. 110.

visual motion centre Specialized processing of object motion is thought to take place in cortical area V5 (aka MT because of its location in the medial temporal lobe in some monkeys). V5 receives retinotopically organized input from several other cortical areas, notably V2 and V3, and is directly innervated by cells in layer IVB of the primary visual cortex (see Fig. 178), part of the magnocellular pathway. The V2 thick stripe–V3–V5 route is the extension of the MAGNOCELLULAR PATHWAY (see entry) and is concerned with motion and depth perception. Almost all cells in MT are *direction-sensitive*, unlike those in other parts of the dorsal pathway or anywhere in the ventral pathway of the EXTRASTRIATE CORTEX (see Fig. 72). Moreover, MT is organized into vertical direction-of-motion columns on a par with the orientation columns of the primary visual cortex (see HYPERCOLUMNS), and it appears that motion perception at any point in space relies on a comparison of the activity across columns spanning the full 360-degree range of preferred directions. Interestingly, some MT cells respond to the perceived direction of movement of an object even when this is contrived in such a way as to be different from its actual direction, suggesting that it makes a transformation critical to our perception of motion, the loss of which is *akinetopsia*. Beyond MT are areas in the parietal lobe with further types of motion sensitivity. In an area known as MST there are not only (as in MT) linear *motion-selective cells*, but also cells sensitive to radial motion and others sensitive to circular motion. See ATTENTION, OBJECT RECOGNITION.

visual perception The task of identifying and assigning meaning to objects in space. Perceptual constancy refers to the invariant nature of the perception despite wide discrepancies in the properties of the retinal image. Thus, in *size constancy* familiar objects (see OBJECT RECOGNITION) do not diminish in size in proportion to the reduction of the image, appearing larger than they should. We do not notice that they appear larger as we approach them since we assume they have a particular size and use this fact as a means of judging their distance from us (see BINOCULAR VISION). The brain judges the size of an object by its context, as it does with apparent lightness and darkness (see RETINA), giving rise to several visual illusions. See ABSTRACTION, ATTENTION, CATEGORIZATION, COLOUR CONSTANCY, LANGUAGE AREAS, PARALLEL PROCESSING.

visual processing See PARALLEL PROCESSING.

visual search See ATTENTION.

vital capacity (forced vital capacity) The volume in the lungs at maximum inspiration comprising the sum of the inspiratory reserve volume, inspiratory tidal volume and expiratory reserve volume. Its value averages 4.8 L in men and 3.2 L in women. See SPIROMETER.

vitamin A A retinoid present in the diet in two forms: retinol (the animal form found in red meat, fish, eggs and dairy produce); and β-carotene (found in red and orange fruit and vegetables). It is a precursor of RETINOIC ACID, and a maternal shortage of the vitamin during pregnancy might be associated with increased rates of foetal skeletal defects (e.g., hemivertebrae and scoliosis). However, it is known to be teratogenic (the cause of foetal malformations) if taken in excess during the first trimester of pregnancy and may lead to SPONTANEOUS ABORTION. Vitamin-A deficiency is common in south Asia and Africa, and millions of pregnancies here may be carried by women with vitamin-A shortage. The visual pigments rhodopsin and iodopsin contain the aldehyde of vitamin A, 11-*cis*-retinal (see NIGHT BLINDNESS).

vitamin B complex A group of water-soluble vitamins.

Vitamin B_1 (thiamine, or thiamin) is the precursor of thiamine pyrophosphate (TPP), a coenzyme in reactions involving cleavage of bonds adjacent to a carbonyl group (e.g., the decarboxylation of α-keto acids: see vitamin B_6, below). Thiamin deficiency commonly accompanies alcohol abuse, especially in those with liver disease, chronic dietary deficiency leading to beriberi. Dermatitis, insomnia, retarded growth and slow *responses* are also reported. Milk and cheese, leafy green vegetables, liver and legumes are good sources. Vitamin B_1 is also a bacterial product of DXP (see GUT FLORA).

Vitamin B_2 (riboflavin) is the prosthetic group of the coenzymes FAD and FMN and therefore required by all flavoproteins (dehydrogenase enzymes involved in oxido-reductase reactions, including those involved in the KREBS CYCLE). Deficiency may manifest itself as cracks and sores in the corners of the mouth, eye disorders, inflammation of the mouth and tongue, and skin lesions.

Vitamin B_3 (niacin) is a component of the coenzyme NAD, involved in GLYCOLYSIS and the KREBS CYCLE; deficiency causes pellagra: diarrhoea, dermatitis and nervous system disorders. It is sourced in fish, liver, red meat, grains and peas.

Pantothenic acid (vitamin B_5) is a precursor of COENZYME A (see ACETYL COENZYME A, for roles) and stored primarily in the liver and kidneys. Other food sources include yeast and green vegetables.

Biotin is a specialized carrier of 1-carbon groups in their most oxidized form, i.e., CO_2, and is the prosthetic group of mitochondrial pyruvate carboxylase, involved in GLUCONEOGENESIS, and a key vitamin in many carboxylation reactions, including those in fatty acid and nucleic acid synthesis (see FATTY ACIDS, see Fig. 75). Its dietary sources include liver, yeast and eggs, while it is also a product of the gut flora. Deficiency causes mental and muscle dysfunction, fatigue and nausea.

Vitamin B_6 (pyridoxine) is the precursor of the coenzyme pyridoxal phosphate, which functions as an intermediate carrier of amino groups at the active site of aminotransferases (i.e., it is the enzyme's prosthetic group; see DEAMINATION). It is involved in other reactions involving the α, β and γ carbons (C2–C4) of amino acids, including decarboxylations (see vitamin B_1, above).

Folate is a B vitamin derivative of folic acid (pteroylglutamic acid). The pyridoxal

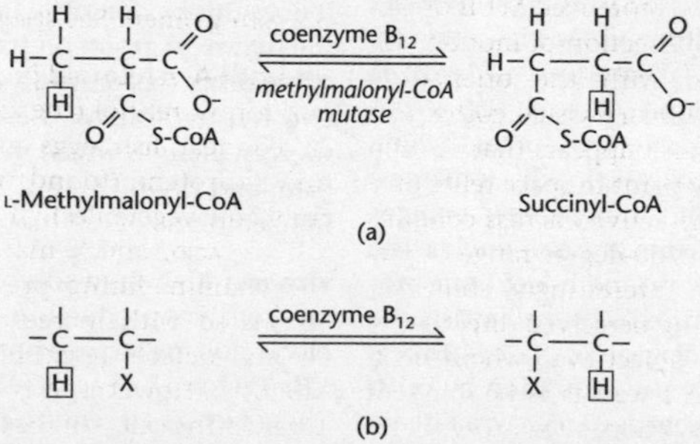

FIG 180. (a) (b) Involvement of coenzyme B_{12} in the β-oxidation of odd-number fatty acids (see VITAMIN B COMPLEX). © 2000 From *Lehninger Principles of Biochemistry* by D.L. Nelson and M.M. Cox, published by Worth Publishers. Redrawn with permission from the publisher.

form of vitamin B_6 is also a bacterial product of DXP (see GUT FLORA). See FOLATE.

Vitamin B_{12} (cobalamins) usually isolated as cyanocobalamin (with a cyano group attached to the trace element cobalt) is the source of coenzyme B_{12}, the cofactor for enzymes catalysing reactions of the general type shown in Fig. 180b, an example of which is illustrated in Fig. 180a, the methylmalonyl-CoA mutase reaction involved in the β-oxidation of odd-number fatty acids (see FATTY ACID METABOLISM). Since B_{12} is required for production of mature red blood cells, the severe disorder pernicious anaemia results from inability to absorb it adequately. It is stored in the liver, but can only be synthesized by a few microorganism species and is required at ~3 μg per day. It can only be absorbed across the mucosa in the absence of the glycoprotein INTRINSIC FACTOR, symptoms of pernicious anaemia including reduced erythrocyte production, reduced haemoglobin production and severe impairment of the central nervous system. Treatment involves lifelong intramuscular injection of large doses of vitamin B_{12}, successful even after gastrectomy. Women on strict vegetarian or vegan diets need additional B_{12} intake during PREGNANCY since it only occurs in animal products. CHOLINE is also often included as a B vitamin.

vitamin C (ascorbic acid) A water-soluble vitamin (hence readily excreted by the kidneys) that can be synthesized from D-glucose or D-galactose in most organisms other than primates, guinea pigs, and certain other mammals and certain birds. It promotes absorption of IRON across the intestinal mucosa. Dietary sources include vegetables, especially spinach, tomatoes, potatoes, broccoli, strawberries, oranges and other citrus fruits. It is required in the synthesis of COLLAGEN (and hence bone production), is an antioxidant, and assists the liver in its detoxification of plasma. Excess interferes with copper utilization by cells. Dietary deficiency leads to scurvy, a disease characterized by swollen and bleeding gums and poor wound-healing through reduced collagen production. Its deficiency also raises plasma cholesterol, triglycerides, LDLs, apoprotein and lipoprotein A, and reduces HDL levels (all reversed by the RDA intake of ~60 mg). The protective role of vitamin C against various cancers, cardiovascular disease and the common cold apparently becomes significant only when intake is >400–1,000 mg per day.

vitamin D Lipid-soluble vitamin, whose hormonally active form (*calcitriol*) is produced by two hydroxylation steps (first in the liver and then in the kidney), either from plant-derived precursors (vitamin D_2, *ergocalciferol*) or from cholesterol-derived 7-dehydrocholesterol in the skin through the action of ultraviolet sunlight (producing the previtamin D_3, *cholecalciferol*). Previtamin D_3 is isomerized to the hormonally active vitamin D_3, 1,2,5-dihydroxycholecalciferol (stimulated by PARATHYROID HORMONE), and transported in the plasma bound to vitamin D-binding proteins. It is necessary for the normal absorption of CALCIUM across the intestinal mucosa, and because of the dependency on calcium of BONE growth and BONE REMODELLING, vitamin D deficiency is a cause of rickets in children and osteomalacia in adults.

vitamin E (tocopherols, tocotrienoids) 'Vitamin E' refers to any mixture of biologically active tocopherols, lipid antioxidants located mainly in the phospholipid bilayer of plasma membranes, preventing not only oxidation of polyunsaturated fatty acids but also the oxidation of DNA. It does this by donating the hydrogen from the hydroxyl group on its ring structure to FREE RADICALS (becoming a relatively unreactive free radical in the process). Dietary sources include wheat germ, cotton seed, palm and rice oil grains. Deficiency results in haemolysis of red blood cells.

vitamin K A lipid-soluble vitamin necessary for liver production of five important blood clotting factors: prothrombin, Factor VII, Factor IX, Factor X and protein C. These factors all require production of the amino acid γ-carboxyglutamic acid, for which vitamin K is required. Vitamin K1 (phylloquinone) is found in green plant leaves (spinach, cabbage), and the related form K2 (menaquinone) is synthesized continuously by colon bacteria (except in neonates). Vitamin K deficiency often occurs during gastrointestinal disease as a result of poor lipid absorption (see BILE). It is not transferred very efficiently across the placenta. Its dietary sources include alfalfa, liver, spinach, vegetable oils and cabbage.

vitamins Organic substances not normally synthesized (or not synthesized in sufficient quantity) by the organism and therefore required in the diet, in minute amounts (contrast essential AMINO ACIDS, ESSENTIAL FATTY ACIDS). Divisible into lipid-soluble (A, D, E, K and CHOLINE) and water-soluble forms (B-vitamins, C), most of the former are stored in liver hepatocytes and are not excretable by the kidney, whereas the latter are excretable by the kidney and are usually precursors of coenzymes. Deficiency diseases result when their availability falls below that required. See individual vitamin entries, MALNUTRITION.

vitreous humour Jelly-like material filling the vitreous bodies (posterior chambers) of the EYES, playing a subordinate role to the AQUEOUS HUMOUR in maintaining intraocular pressure but holding the retina flush against the choroid. It does not undergo constant replacement, being formed during embryonic life and not replaced thereafter. Contains phagocytic cells clearing debris.

vitrification See CRYOBIOLOGY.

VLDL (very low density lipoprotein) See LIPOPROTEIN.

VNTRs (variable number tandem repeats) One form of simple-sequence length polymorphism, aka hypervariable MICROSATELLITE sequences, in which a short DNA sequence is repeated in tandem along a chromosome. Such 'loci' in humans vary from 1 to 5 kb in length and comprise variable numbers of a repeating unit, usually from 15 to 100 nucleotides in length. VNTRs with the same repeating, but different numbers of repeats, are scattered through the genome. They can be located by first digesting the whole genome with a particular restriction enzyme lacking a target site within the VNTR but with target sites in flanking sequences. In this way, the length of the fragment containing a VNTR will correspond to the number of repeats. An appropriate probe can then provide a Southern blot and will reveal a large number of differently-sized fragments – patterns often called *DNA fingerprints* (see DNA PROFILING, SOUTHERN BLOTTING). A *microsatellite VNTR* (aka short tandem repeat polymorphism) is a form of VNTR in which the array of tandem repeats is usually small (<100 bp) and the repeat unit is small (~1–4 nucleotides long). A *minisatellite VNTR* (often confusingly termed 'VNTR') has a moderately lengthy array size, the repeat unit often being 9–65 bp long. See DNA MICROARRAYS.

$\dot{V}O_2$, $\dot{V}O_2$ max $\dot{V}O_2$ is the oxygen consumption rate of an individual, oscillating in accordance with the CIRCADIAN CLOCK. $\dot{V}O_2$ max is the maximum volume of oxygen that can be consumed per kilogram of body mass per minute. See EXERCISE, OXYGEN DEBT.

voltage-gated ion channels See ION CHANNELS.

voluntary movement See INTENTIONAL MOVEMENT.

vomeronasal organ A sensor for scent signalling, functional loss of which in Old World monkeys appears to have coincided with the advent of COLOUR PERCEPTION, prompting the idea that chemical signalling gave way to visual signalling in higher primate mating systems, colourful signals being favoured in such gregarious species. See PHEROMONES.

vomiting (emesis) Frequently preceded by loss of appetite (anorexia), the forceful reflex oral expulsion of the stomach contents. Coordinated by the dorsal portion of the reticular formation of the medulla, afferent signals reach it from the pharynx, other regions of the alimentary canal, viscera (liver, urinary bladder, semicircular canals) and a variety of chemicals (e.g., general anaesthetics, opiates and digitalis) also trigger it by stimulating receptors in the floor of the fourth brain ventricle. Parasympathetic efferents leave via several cranial nerves (V, VII, IX, X and XII). Vagal (CNX) and sympathetic stimulation of the stomach, phrenic stimulation of the diaphragm and intercostal nerves stimulating the abdominal muscles from T6–T12 and L1 all bring about vomiting. It is generally protective in preventing potentially toxic substances from being absorbed, but repeated occurrence can lead to alkalosis through loss of stomach acid. It can also cause dehydration and hypokalaemia.

von Willebrand factor (VWF) A large protein circulating in the blood as multimers of varying size, a main function of which is to mediate interactions between PLATELETS and the ENDOTHELIUM of blood vessel walls, which is crucial to maintaining the balance between bleeding and BLOOD CLOTTING. The enzyme that cleaves the large multimers into smaller fragments is markedly reduced in those with thrombotic thrombocytopenic purpura (TTP), a potentially fatal human disease which results from widespread, abnormal aggregation of platelets with VWF in small blood vessels of many organs including the brain and kidneys. The accumulated clots of platelets and proteins obstruct blood flow and cause red blood cells to fragment, leading to serious neurological and renal malfunction, anaemia, fever and low platelet count.

vulva See EXTERNAL GENITALIA.

Waardenburg syndrome (WS) The core WS phenotype is seen in WS Type 2, characterized by congenital white forelock, asymmetric iris colour and sensorineural deafness, all resulting from disrupted migration of MELANOCYTES (see entry). Mutations in *PAX3* are responsible for WS, Type 1 showing autosomal dominance and characterized by sensorineural hearing loss, craniofacial deformities, areas of depigmentation of hair and skin, abnormal iris pigmentation and widely spaced inner canthi (the angles where the inner eyelids join at the medial margin of the eye). The commoner Type 2 form exhibits less widely separated inner canthi, some cases being due to mutations in the microphthalmia gene *MITF* (see EYES), a small subset of cases (WS Type 2a) having germline heterozygous mutations (see MELANOCYTES). Type 3 WS may include limb deformities. Type 1 and Type 3 differences are due to different mutations in *PAX3* (see PAX GENES).

WAGR syndrome A syndrome combining WILMS TUMOUR, aniridia, genitourinary malformations and retardation of growth and development.

Wahlmund effect See GENETIC DIVERSITY.

wake–sleep cycles See SLEEP–WAKE CYCLES.

walking See CENTRAL PATTERN GENERATORS, GAIT.

warfarin A synthetic compound inhibiting formation of active prothrombin and an invaluable anticoagulant in treating those at risk of excessive BLOOD CLOTTING.

warts Growths (adenomas) of skin epithelial tissue. See PAPILLOMAVIRUS.

waterproofing See SKIN.

Weber's law A quantitative and linear relationship between stimulus strength and discrimination proposed by E.H. Weber (1834) such that $\Delta S = K \times S$ where ΔS is the 'just noticeable difference' in strength between a reference stimulus, S, and a second stimulus and K is a constant.

Weibel–Palade bodies Granules within cells of the vascular ENDOTHELIUM containing P-selectin.

Werner syndrome A PROGERIA characterized by early onset of grey and thinning hair, skin wrinkling (see EPIDERMIS), osteoporosis, atherosclerosis and cancer. The altered *WRN* gene encodes a DNA helicase, so that cells lacking it have high mutation rates. WRN is a part of the BRCA1-associated genome surveillance complex, BASC (see DNA REPAIR MECHANISMS). See AGEING.

Wernicke's area (area 22) Region of the AUDITORY ASSOCIATION CORTEX (the posterior superior temporal gyrus in the dominant hemisphere), lying between the primary auditory cortex and the angular gyrus (from which it receives input), which is crucial for understanding the spoken word (see LANGUAGE, SPEECH, SYLVIAN FISSURE). It has connections with other LANGUAGE AREAS of the BRAIN (see Fig. 29), and is reciprocally interconnected by the arcuate fasciculus to BROCA'S AREA. See CEREBRAL HEMISPHERES, 'SPLIT-BRAIN' PHENOMENON.

West Nile fever Caused by a member of

the Japanese encephalitis antigenic complex of flavivirusus (see YELLOW FEVER), and the most widespread flavivirus, transmissible by several species of mainly bird-feeding species of culicine mosquito, and certain ticks. 80% of infected people are asymptomatic; but symptoms include febrile, influenza-like illness (incubation 3–6 days), headache (often frontal), sore throat, backache, anorexia, diarrhoea and rash (in 50% of symptomatic cases). In <15% of cases, acute septic meningitis or encephalitis, neck stiffness, vomiting, confusion and disturbed consciousness, convulsions and coma occur. Most fatal cases have occurred in patients >50 years. Geographic distribution includes Africa and Eurasia, and heavy rains followed by floods, irrigation and higher than usual temperatures are resulting in its re-emergence in Europe. The principal cycle is rural; but urban cycles also occur.

white blood cells (leukocytes) Lymphoid and myeloid cells (see STEM CELLS) produced by BONE MARROW and including lymphocytes (B CELLS, T CELLS, NATURAL KILLER CELLS), DENDRITIC CELLS, NEUTROPHILS, MONOCYTES (and MACROPHAGES), BASOPHILS and EOSINOPHILS.

white matter Generic term for a collection of central nervous system axons, appearing white when freshly cut on account of the myelin sheaths. Contrast GREY MATTER.

white muscle Muscle whose fibres contain little myoglobin and therefore tend to be fast-contracting and anaerobic.

whole-genome shotgun sequence detection (WSSD) See GENOME MAPPING.

whooping cough (pertussis) A prolonged, severe, distressing and contagious disease caused by the small Gram-negative coccobacillus *Bordetella pertussis*, typed by its cell wall proteins (agglutinogens) into 1, 2 and 3, antibodies to which are protective. Readily transmitted by the airborne route, it is now quite rare since immunization is widespread. Large epidemics did return in the 1980s after doubts about vaccine safety reduced its acceptance. Adults form the reservoir from which infants are infected. Morbidity is high in these infants, and in older children with respiratory or cardiac disease. After 14–20 days' incubation, symptoms are a simple cough and mild fever. Coughs begin to group into paroxysms over the next 4 days in which 20–30 coughs occur without inspirations in between. Thick mucus is expectorated, with difficulty, in strands. A characteristic stridulous inspiratory cry ('whoop') ends the paroxysm, often with a vomit. The coughs, or sneezing paroxysms, continue for up to 3 weeks, gradually improving over 10 days. For pertussis toxin, see TOXIN.

Williams–Beuren syndrome (WS) A disease having a prevalence of ~1:7,500 people, typical individuals share a deletion of 28 known genes in the long arm (q) of one of their two chromosome 7s (see DELETIONS, Fig. 61). At least six of these genes are implicated in pathways of neural development in the brain. Phenotypic effects include tendency to excessive sociality, non-social anxiety (e.g. fear of heights; see FEAR AND ANXIETY), mental retardation, cardiac anomalies (see CONGENITAL HEART DEFECTS), upturned noses, wide mouths and small chins (hence 'elfin face syndrome'), difficulty visualizing an object as a set of parts (visuospatial construction deficit). Although WS individuals experience no difficulty in facial recognition, a structural abnormality in the orbitofrontal cortices of WS individuals may explain their low fear response to threatening faces (see AMYGDALA). Children with WS also tend to approach strangers fearlessly and indiscriminately. In WS, genes affect social behaviour very early, and such a child's unusual social behaviour is in turn likely to construct for him/her an abnormal social environment in which others' interactions with him/her will differ from their interactions with a child lacking the syndrome. WS children who have a parent who is good at drawing tend to be better at drawing than are other individuals with the same chromosomal DELETION, which may reflect transactions with genes outside the deletion, better opportunities for drawing, and/or better adult models of how to draw. Comparing the results of functional MAGNETIC RESONANCE IMAGING (fMRI) of a WS group with normal overall intelligence with those of healthy controls, WS individuals showed lower neuronal activity in part of the brain involved in the spatial processing of visual information: the fold separating the parietal and occipital lobes (the parieto-occipital sulcus). This turned out to have an abnormally low density of nerve tissue. See NATURE–NURTURE DEBATE, SEGMENTAL DUPLICATION, SPATIAL AWARENESS.

Williams syndrome See WILLIAMS–BEUREN SYNDROME.

Wilms tumour A paediatric cancer of the kidney, believed to be derived from undifferentiated nephrogenic precursors. The *WT1* gene, an autosomal Wilms tumour suppressor expressed during development of the renal blastema, is inactivated in 5–10% of cases, the genetic underpinning of the majority of cases being largely unknown. Recent discoveries revealed another tumour suppressor gene, *FAM123B*, or *WTX*, that is X-linked and inactivated in ~30% of sporadic cases in both sexes by somatic deletions or point mutations. These two genes may give rise to distinct subclasses of Wilms tumour. The WTX protein is part of the proteasome complex normally promoting destruction of the signalling protein β-catenin (see WNT SIGNALLING PATHWAY), and mutations in *WTX* lead to increased β-catenin levels, promoting cell proliferation. See BECKWITH–WIEDEMANN SYNDROME, LI–FRAUMENI SYNDROME.

wisdom teeth The third permanent molars; their appearance marks the end of the juvenile period.

withdrawal, withdrawal syndrome Withholding a drug after dependence to it is established results in an extremely unpleasant syndrome which lasts until the long-term biological changes that brought about the dependence have reversed. As a result, it is believed that ADDICTION to a drug (of which withdrawal syndrome is a component) can be motivated as much by aversion to withdrawal as by positive reinforcing properties of the drug. See REINFORCEMENT AND REWARD.

Wnt signalling pathways Activation of the WNT (Wingless-int) signalling pathway stimulates growth and mediates developmental signals between cells. There are almost 20 human *Wnt* gene family members, whose products are secreted lipoprotein morphogens and GROWTH FACTORS that bind serpentine transmembrane receptors resembling G PROTEIN-COUPLED RECEPTORS and regulate both cell proliferation and differentiation. Other intracellular signalling pathways with roles in tissue development and differentiation are activated by NOTCH ligands (see NOTCH SIGNALLING PATHWAY) and BONE MORPHOGENETIC PROTEINS. Some ligands are produced only in particular tissues, although possibly activating common downstream signalling mechanisms. Thus, WNT2 is crucial for generation of a liver, while WNT1 is needed for production of specific neuron types in the brain. Some of them signal through G PROTEINS and the PHOSPHATIDYLINOSITOL SYSTEM. Unlike those of the HEDGEHOG GENE family, they can activate several distinct signalling pathways, although common downstream signalling mechanisms may also be involved. Such pathways are classified as either β-catenin-dependent (the canonical pathway, generally more common) or β-catenin-independent (non-canonical pathways, which often target the cytoskeleton; see METASTASIS). Signalling is initiated when Wnt class growth factors bind their cognate receptor complex (a receptor of the Frizzled family and a member of the LDL receptor family, Lrp5/6). The key player is then a cytoplasmic protein, β-catenin, whose stability is regulated by the proteasome destruction complex. WNT proteins function to stabilize β-catenin, which relocates to the nucleus and interacts with lymphoid-enhancer-binding factor (LEF) and members of the T-cell factor (TCF) family, driving expression of genes encoding such proteins as Myc (see MYC) and cyclin D (see CYCLINS). The canonical Wnt cascade (see

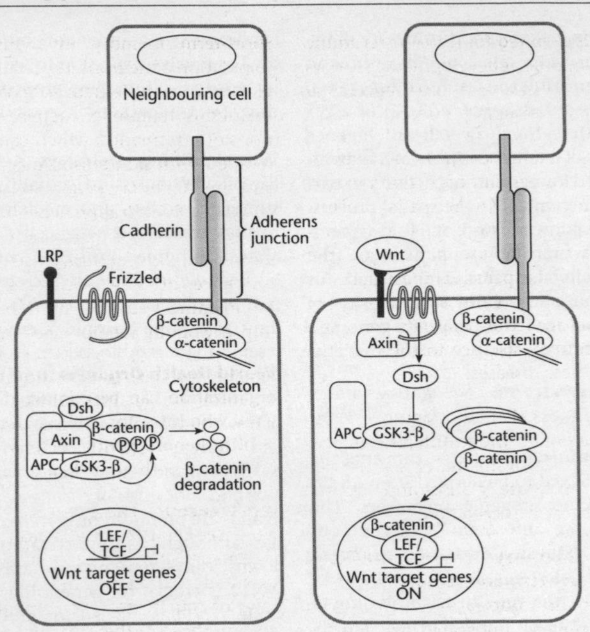

FIG 181. Diagram illustrating the canonical WNT SIGNALLING PATHWAY. In the absence of this signalling (left half of diagram), β-catenin is complexed with axin, APC and GSK3-β, becomes phosphorylated, and is degraded. In its cadherin-bound form, β-catenin regulated cell–cell adhesion. But during Wnt signalling, (right half of diagram), β-catenin is uncoupled from the degradation complex and enters the nucleus, binding Lef/Tcf transcription factors and activating target genes. APC, Axin and GSK3-β are components of the β-catenin degradation complex. Malfunctioning of this pathway is a feature of many CANCER CELLS (see Fig. 35). Dsh, dishevelled; LEF, lumphoid-enhance-binding factor; TCF, T-cell factor. From T. Reya and H. Clevers, © *Nature* 2005, with permission from Macmillan Publishers Ltd.

Fig. 181) has emerged as a critical regulator of STEM CELLS and PROGENITOR CELLS, dysregulation of which has been associated with cancer in intestinal, epidermal and haematopoietic tissues (see CANCER CELLS, Fig. 35; DISEASE, Fig. 64). Tightly regulated self-renewal mediated by Wnt signalling in stem and progenitor cells is subverted to enable malignant proliferation. Recent studies in mice indicate that the regenerative potential of skeletal muscle (which declines with age) may be reduced by activation of the Wnt pathway, converting progenitor cells from myogenic to fibrogenic fates.

The protein Dishevelled (Dsh) acts as a switch between canonical and non-canonical pathways, being present as a cytosolic pool in the former but plasma membrane-bound in the latter. In the canonical pathway, in the absence of Wnt signalling, β-catenin is bound in the destruction complex with the scaffold protein axin, APC (see ADENOMATOUS POLYPOSIS COLI PROTEIN) and the kinase GSK3-β, where it is phosphorylated and targeted rapidly for proteasome degradation (see WILMS TUMOUR). Wnt signalling releases β-catenin from the proteasome complex, upon which it enters the nucleus, binds Lef/Tcf transcriptional co-repressors, switches them to transcription activators, and activates target genes, among which is *c-myc* (see MYC for details of cell proliferation). In its cadherin-bound form, however, β-catenin regulates cell–cell adhesion (see ADHERENS JUNCTIONS). Some Wnt proteins are known to be involved in AXONAL GUIDANCE, and the pathway is required for NEURAL CREST induction and MELANOCYTE development. See BMPS, HAEMATOPOIETIC STEM CELLS, HAIR FOLLICLE, SEX DETERMINATION, SOMITES.

Wolbachia A *Rickettsia*-like bacterial genus occurring as a parasite of a wide range of invertebrate animals, where it is transmissible in egg cytoplasm but not via sperm. *Wolbachia* may be useful as a weapon against pests and parasites that cause diseases such as MALARIA and RIVER BLINDNESS, by introducing into the parasite's genome a gene encoding resistance to the causative agents of these diseases.

Wolffian ducts Paired GENITAL DUCTS, originally present in the indifferent embryo but only persisting in males, as the vasa deferentia, prostate glands and seminal vesicles. See SEX DETERMINATION.

working memory (representational memory, scratchpad memory) The ability to maintain information so that it can be manipulated, integrated with other information and then used to guide behaviour during DECISION-MAKING. A form of short-term memory involving its own population of neurons, especially cholinergic ones in the PREFRONTAL CORTEX, which maintains temporary, active, representations of information which can be rapidly recalled. Diffusion tensor imaging (see DTI) and functional MAGNETIC RESONANCE IMAGING (fMRI) analytical approaches have shown the importance of maturation of prefrontal-parietal connectivity in the performance of a working memory task. Monoamine neurotransmitters appear to impair working memory. See MEMORY.

World Health Organization (WHO) The organization can be contacted on http://www.who.int/ for the current World Health Report giving details of disease prevalence both globally and regionally.

wrist (carpus) The proximal region of the hand (see Fig. 182 and FORE LIMB) comprising eight small bones, the *carpals*, joined by ligaments, arranged in two transverse rows of four bones each. From lateral to medial, those in the proximal row are: the scaphoid, lunate, triquetrum and pisiform. Similarly, those in the distal row are: the

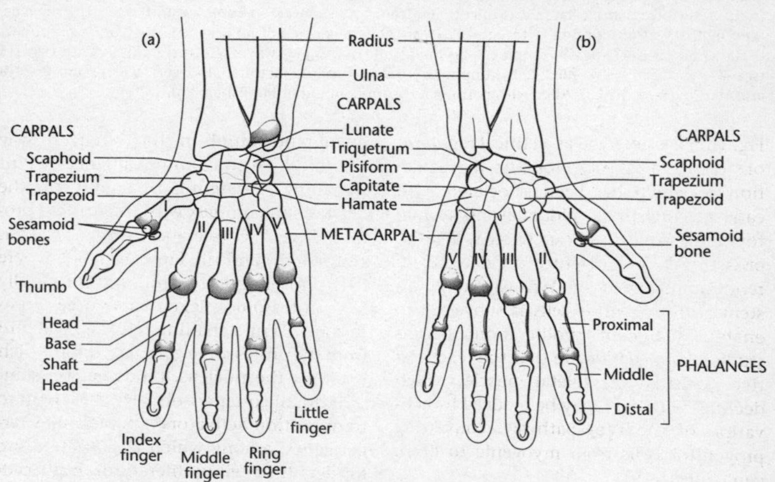

FIG 182. The relationship of the right WRIST and hand to the ulna and radius. The skeleton of the har comprises proximal carpals, intermediate metacarpals and distal phalanges. Digits are numbered. Anterior view; (b) posterior view. From *Principles of Anatomy and Physiology*, G.J. Tortora S.R. Grabowski, © 2000. Reproduced with permission of John Wiley & Sons, Inc.

trapezium, trapezoid, capitate and hamate. The capitate is the largest carpal bone, its rounded projection articulating with the lunate. Most fractures involve the scaphoid since the force of a fall on an outstretched hand is transmitted via it from the capitate to the radius.

writing See LANGUAGE AREAS, SPLIT-BRAIN PHENOMENON.

X

X chromosome The human X chromosome appears to have been remarkably stable in organization since the radiation of eutherian mammals and, some think, may even be identical to the putative original ~300-Myr-old autosomal 'PROTO-SEX CHROMOSOME' which pre-dated mammalian origins. The X chromosome seems to comprise an X-conserved region (XCR), including the long arm of the chromosome, and a separate X-added region (XAR) brought in by translocation from another autosome prior to eutherian radiation ~150 Myr ago. Homologous recombination between human X and Y chromosomes occurs at two pseudoautosomal regions, one at the tip of the short arm (PAR1) and one at the tip of the long arm (PAR2). No recombination occurs at the other homologous region between X and Y – the XAR (X-added) region. In females, however, recombination is possible along the entire lengths of the X chromosomes. Interspersed repeats account for 56% of the euchromatic X chromosome sequence compared with a genome average of 45%; its level of heterozygosity is approximately 57% of that observed on autosomes. Only 54 of the 1,098 genes located on the human X have functional homologues on the Y, 24 of them occurring in the PAR1 region and 5 in the PAR2 region. A further 15 are in the XAR while three more are shared by the X and Y copies of the 'X-transposed region', XTR. The remaining 7 XY gene pairs are thought to have descended from the proto-sex chromosomes. Few genes on the X chromosome have an active equivalent on the Y and most of these are in regions of relatively recent XY homology.

Females inherit one X chromosome from each parent, while males inherit their single X chromosome from their mother, a situation which gives rise to SEX-LINKED INHERITANCE.

It is the chromosome with the most extreme mean K_A/K_I ratio (= 0.32) among hominines (see MUTATIONS), restricted largely to genes expressed in the testes (see GENE EXPRESSION). One recent (2006) analysis indicates that DIVERGENCE from chimpanzees is reduced along nearly the entire length of the human X chromosome, the data pointing to an enormous decrease in genetic divergence for this chromosome compared with the autosomes, amounting to ~1.2 Myr. This has led to the provocative explanation that the hominin and chimpanzee lineages initially separated but then exchanged genes prior to finally separating at ~6.3 Mya (see SPECIATION). This could explain how the Toumaï specimen of SAHELANTHROPUS TCHADENSIS could have dates older than hominin speciation yet still have hominin features, the wide range of divergence times (>4 Myr) across the human–chimpanzee genomes, and the low divergence of human and chimpanzee on chromosome X. See FRAGILE X SYNDROME, KLINEFELTER SYNDROME, TURNER SYNDROME, X CHROMOSOME INACTIVATION.

X chromosome inactivation (XCI) A process occurring in all female mammals, resulting in selective inactivation of alleles on one of the two X CHROMOSOMES in the somatic cells. It provides a mechanism of

DOSAGE COMPENSATION, overcoming sex differences in the expected ratio of autosomal gene dosage to X chromosome gene dosage. It is a form of epigenetic reprogramming which, as with telomere length adjustment but unlike DNA methylation, takes place soon after the zygote has been formed. Both X chromosomes are active during preimplantation cleavage of the female embryo, imprinting and other inactivating mechanisms somehow being reversed in the inner cell mass, and it is usually a matter of chance which X chromosome is inactivated in the embryonic (epiblast) lineage of the embryo although an unidentified genetic imprint causes the paternal X chromosome to be inactivated in the TROPHECTODERM, which contributes to the placenta. The X inactivation pattern needs to be erased from generation to generation and, possibly connected with this, the X chromosome is known to be inactivated transiently during gametogenesis in both sexes.

X inactivation involves modification of chromatin structure (see CHROMATIN REMODELLING), resulting in a condensed heterochromatinized structure, the *Barr body*, which can be seen along the inside of the nuclear envelope in female cells. Not all the genes of an inactivated X chromosome are silenced, the few that are transcriptionally active include those where there is a functional homologue on the Y chromosome, and some other genes where dosage of product does not seem to be important. The HISTONE variant of H2A, MacroH2A, is enriched on the inactive (X_i) chromosome in mammalian female cells and, although a hallmark of X-inactivation, is apparently not essential for maintenance of the inactivated state.

Two important *cis*-acting elements have been implicated in the process: (i) an X-inactivation centre (*Xic*), mapping to Xq13, and controlling initiation and propagation of inactivation, only X chromosomes carrying *Xic* being capable of inactivation. Its function is dependent on *XIST* (*Xist* in the mouse), a gene specifying a long unctional NON-CODING RNA essential for initiating inactivation on its own chromosome but not for maintaining it; (ii) the X-controlling element (*Xce*) which affects the choice of which X chromosome remains active and which inactive.

XIST in uniquely expressed from the inactive X chromosome and its primary transcript undergoes splicing and polyadenylation (see RNA PROCESSING) to generate a 17-kb functional non-coding RNA. If this A-repeat unit, required for silencing, is deleted from the *Xist* RNA, the pol II exclusion compartment within the NUCLEUS still forms, but genes usually scheduled for inactivation are not silenced, or epigenetically modified. Nor do they relocate to the inside of the *Xist* chromosome compartment, as they normally do as they become silenced. This indicates a link between physical relocation of X-linked genes with the epigenetic silencing mechanism, suggesting that silencing depends on the architecture of the nucleus and chromosome.

The *TSIX* gene has a transcription unit overlapping the entire *XIST* gene, but on the antisense strand. Expressed in undifferentiated embryonic cells, it is thought to control *XIST* expression in *cis* at the onset of X inactivation. It is probable that some form of long-distance CHROMATIN REMODELLING is also involved, since a diffusible *XIST*-mediated substance would not affect just the X chromosome on which *XIST* is expressed. See GENOME, GENOMIC IMPRINTING, SCREENING.

xenobiotics Substances, including many drugs, that are foreign to the body. The term is also used in the context of pollutants, such as dioxins (see TOXINS) and polychlorinated biphenyls. Products of one organism may become xenobiotic if assimilated by another species, as in uptake of human hormones (e.g., oestrogens) by fish, crustaceans and molluscs downstream of sewage treatment outfalls. Some can induce APOPTOSIS at lower doses, and necrosis at higher doses. See AHR for the *xenobiotic response element*, XRE.

xenogeneic Referring to animals of a different species. See ALLOGENEIC, XENOGENEIC.

xenografts, xenotransplants Grafts (xenografts) and organ transplants (xenotransplants) from animals of non-human species to humans are routinely rejected by a hyperacute response. Were they usable, it would overcome the severe shortage of donor organs. Most humans and other primates have antibodies reacting to α-1,3-galactosyl (α-Gal), a ubiquitous cell-surface antigen of other mammals, including pigs. A recent development of transgenic pigs lacking the enzyme α-1,3-galactosyl transferase that adds α-Gal but expressing human decay-accelerating factor (DAF), a COMPLEMENT-regulatory protein, may be a step towards eventual success with xenotransplants. Although the T-cell-mediated graft rejection mechanisms might be very difficult to overcome with current immunosuppressive drugs (see GRAFTS, IMMUNOSUPPRESSION), some workers are trying to 'trick' the recipient into recognizing the donor organ as 'self'. One approach is to engraft cells from the donor pig thymus into the recipient while its immune system is temporarily suppressed so that as the immune system recovers it will be tolerant to the donor organs and tissues. Even if such work eventually succeeds and immune hurdles are overcome, the possibility of transplanting viruses or other pathogens along with the xenografts would have to be discounted.

xeroderma pigmentosum (XP) A group of rare diseases (frequency ~1:250,000) causing freckling, pigment splotches, open facial sores and increasing by 1,000-fold the risk of developing cancer in UV-exposed skin, and by 10-fold the risk of developing other tumours. Some forms of XP involve neurological disorders leading to microcephaly and progressive mental deterioration. XP is caused by defective global genome nucleotide excision repair of UV-induced DNA damage and involves mild mutations in the *XPF* gene. It can occur with COCKAYNE SYNDROME. See AGEING, DNA REPAIR MECHANISMS.

xerophthalmia A disorder caused by deficiency of VITAMIN A, and a major cause of blindness in the tropics. Loss of sensitivity to green light is followed by a loss of acuity in dim light then, after a matter of months, total *night-blindness* through reduced production of retinal, a component of rhodopsin required for all non-colour vision. This is because large quantities of vitamin A are normally stored in the liver and can be mobilized for use by the eyes. Once apparent, night blindness can often be reversed in less than 1 hour by intravenous injection of vitamin A. In xerophthalmia, conjunctival membranes of the eyes become dry, developing oval or triangular spots (Bitot's spots); corneas become cloudy or soft, and may become keratinized, with ulceration leading to total blindness.

X-irradiation Ionizing electromagnetic radiation of wavelength <100 nm (frequency $>3 \times 10^{15}$ Hz). As much as 80% of the energy deposited in cells by X-rays is believed to be expended in stripping electrons from water molecules, the resulting free radicals proceeding to generate reactive oxygen species (see mitochondria). Free radicals generated create single- and double-strand breaks in DNA, genotoxic stress that increases the cell's level of P53 (see entry), which transfers signals to ATM kinase (see ATAXIA TELANGIECTASIA MUTATED PROTEIN) which then phosphorylates p53 (see DNA REPAIR MECHANISMS). Unrepaired double-strand breaks may lead to chromosome breaks visible in metaphase. See MUTAGENS for more details.

X-linkage See SEX-LINKED INHERITANCE.

X-linked agammaglobulinaemia (XLA, Bruton's sex-linked agamaglobulinaemia) A genetic disorder occurring in ~1:250,000 males, in which pre-B cells fail to develop in the bone marrow. As a result, the reticuloendothelial system and LYMPHOID ORGANS are poorly developed, and peripheral lymph nodes may all be reduced in size. As a result, infections begin once transferred maternal IgG antibodies have been catabolized, typically at ~6 months of age. The locus concerned is expressed in B cells and is a member of the src family of proto-oncogenes encoding protein-tyrosine kinases (see SRC-KINASES). The kinase involved has been named Bruton's tyrosine kinase (BTK). Female carriers have no clinical manifestations.

X-linked hyper IgM syndrome See ISOTYPE SWITCHING.

XY body A transcriptionally inactive combination of condensed X and Y chromosomes formed during SPERMATOGENESIS.

Y

yawning A deep inspiration through a widely opened mouth, depressing the mandible exaggeratedly. Its precise causation is uncertain, but it is possibly brought on by drowsiness, fatigue, and often by witnessing another's yawning. See ALVEOLUS.

Y chromosome The sex chromosome unique to males. Although mitochondrial DNA is inherited by both sexes matrilinearly, only females both inherit and pass it on. By contrast, the Y chromosome is inherited only by males, and only they pass it on – and then, only to their sons (see EFFECTIVE POPULATION SIZE, SEX-LINKED INHERITANCE). Because most of the Y chromosome does not engage in sexual RECOMBINATION with a chromosome homologue, natural selection acts on most of the chromosome as a unit – effectively, a locus, or HAPLOTYPE. It is particularly vulnerable to gene deletions on the long arm and cannot retrieve lost genetic information by homologous recombination. It has been predicted that an essentially non-recombining chromosome should degenerate by accumulation of harmful mutations that cannot be corrected by gene conversion or recombinational exchange with its homologue. Such mutant alleles would either drift to fixation (as Y chromosomes with fewer mutations are lost by chance), or they would 'hitchhike' by fortuitous linkage to a favourable allele in a region protected from recombination. Some models suggest that there will be no selective pressure to retain such protected DNA sections, which will eventually be lost by DNA turnover mechanisms that include iterative deletion. As a result, these models predict that the human Y chromosome will be lost entirely, although maleness would be retained by adoption of an alternative sex determination system (but see SEX CHROMOSOME for further theoretical considerations).

Although there are relatively short regions at either end of the Y chromosome that are identical to the corresponding regions on the X chromosome, and engage in recombinational crossing-over with it, more than 95% of the present-day Y chromosome is male-specific (MSY, formerly thought to be non-recombining, NRY, but now known to undergo GENE CONVERSION), comprising some 23 million base pairs (Mb) of euchromatin and a variable amount of heterochromatin (highly repetitive DNA) which is often dismissed as non-functional. At least 15 of its protein-coding genes have no detectable X chromosome homologue.

Most of its euchromatic sequences fall into one of three classes: (i) X-transposed sequences that are 99% identical to sequences on the X chromosome (Xq21), result from a massive X/Y transposition that took place ~3–4 Mya, and are dominated by a high proportion of interspersed repetitive sequences and contain just two protein-coding genes; (ii) X-degenerate sequences, more distantly related to the X chromosome, and dotted with 16 single-copy gene and 11 pseudogene homologues of 27 X-linked genes (see below for chimpanzee comparisons); and (iii) ampliconic sequences comprising at least nine multiple-copy protein-coding genes (~35 copies of the testis-specific protein Y

(*TSPY*) gene), and long blocks of DNA tens or hundreds of kilobases in length making a series of eight palindromes (reading the same on both DNA strands) each occurring as two mirror-symmetrical arms about a central point and exhibiting up to 99.9% identity in their two arms. Such high levels of identity are believed to result from frequent gene conversion events between the two arms of the palindromic sequences, which may forestall gene decay. The male-specific H-Y ANTIGENS are encoded by the MSY region.

While many of the X-degenerate genes are expressed ubiquitously in body tissues (although the SRY GENE is expressed predominantly in the testes), the ampliconic genes are expressed in a testis-specific manner.

Comparison of the X-degenerate portions of the human and chimpanzee Y chromosomes revealed that both the gross structures and nucleotide sequences of hominoid Y chromosomes have evolved rapidly (in marked contrast to impressive level of colinearity between human chromosome 21 and its chimpanzee orthologue, chromosome 22). Whereas the human X-degenerate sequences are distributed along both arms of the Y chromosome, interrupted at several points by ampliconic, heterochromatic, or other sequences, the chimpanzee's X-degenerate sequences form a single, almost contiguous block on the long arm. Other structural differences include two large inversions, one of which occurred in the chimpanzee lineage and one in the human lineage. All 16 human X-degenerate genes and 11 pseudogenes have been found as orthologues on the chimpanzee Y chromosome. The great majority of inactivating mutations are found in these pseudogenes, indicating that none of the human X-degenerate pseudogenes has lost its functionality since the human and chimpanzee lineages split. This is compatible with models that predict a slowing in the rate of gene decay as Y chromosomes evolve.

Comparison of divergence in the human and chimpanzee X-degenerate genes' exons with their introns (acting as controls), in accordance with the prediction that genes for which the protein products are subject to purifying selection should exhibit less interspecies divergence in exons than in introns, indicates that this is the case ($P < 0.0001$) and, as expected, this disparity was not found in pseudogenes. This supports the view that purifying selection has been a potent force in maintaining X-degenerate gene function during recent human evolution. By contrast, there is evidence of significant X-degenerate gene decay in the chimpanzee lineage. This may be due to 'hitch-hiking' of deleterious genes to strongly beneficial mutations in other Y-linked chimpanzee genes, such as those in the ampliconic regions, whose testis-restricted expression patterns are favoured in such a species as the chimpanzee where females usually mate with multiple males whose sperm compete for a limited number of oocytes (see SPERM COMPETITION). Such selective forces may have been less intense in humans, a less promiscuous species. Future work on sex chromosome variability in humans and the two chimpanzee species should provide more data with which to test this explanation.

Since the Y chromosome is inherited patrilinearly (*contra* matrilineal inheritance of MTDNA), Y-chromosome polymorphisms are especially useful in tracing relationships to dead persons (and suspected NONPATERNITIES) since an individual male inherits the complete genotype from a single definable ancestor. The technique, termed 'deficiency paternity testing' relies on matching Y-chromosomal markers of the child with those of a potential grandfather or paternal uncle, evidence in favour of true paternity strengthening as the frequency of the marker becomes rarer.

The Y chromosome is also useful in tracing the descendants of historical diasporas, when a high frequency of a particular Y-chromosomal (or mtDNA) HAPLOTYPE in a certain population can sometimes be used as a 'population marker'. See INFERTILITY, SEGMENTAL DUPLICATION, X CHROMOSOME.

Y chromosome Adam The COALESCENCE of all current human Y chromosomes would terminate in a single Y chromosome of a single past individual, dubbed 'Y chromosome Adam'. This individual would almost certainly have lived at a different time from MITOCHONDRIAL EVE.

yellow fever A flavivirus infection, transmitted by mosquitoes of the genus *Aedes* and (in South America) *Haemagogus*. After an incubation period of 6–7 days, fever and myalgia occur with remission after 4–5 days; severe cases progress to clinical jaundice, bleeding, coughing up of blood (haematemesis), sepsis and multiorgan failure. It is endemic in tropical parts of Africa and South America, prevention being achieved by vaccination with the attenuated 17D strain of the virus (one dose can remain effective for 10 years). In endemic regions, mass immunization and vector control programmes are effective, although ~30,000 deaths occur annually. It has re-emerged in urban America since the 1990s (see DENGUE FEVERS), transmitted by *Aedes aegypti*.

Yersinia pestis See BUBONIC PLAGUE.

yolk sac The structure located ventral to the bilaminar germ disc (see AMNIOTIC CAVITY, Fig. 8; EMBRYONIC DISC), derived from the hypoblast and being the site of origin of the first blood cells and of the germ cells. It remains attached to the midgut via the vitelline (yolk sac) duct until late in development.

Z

zinc a constituent of a large number of enzymes, including CARBONIC ANHYDRASE, carboxypeptidases A and B, reverse transcriptase, METALLOPROTEINASES, and RNA and DNA polymerases. It is also required for insulin synthesis and activity. Many transcription factors, regulatory proteins and other DNA-binding proteins (e.g., see NUCLEAR RECEPTOR SUPERFAMILY) contain zinc finger domains which bind a zinc ion and typically interact with the major groove of the DNA double helix. Zinc deficiency impairs the immune system, and recent research suggests that zinc plays a part in signalling pathways by which dendritic cells mature and develop. An iron-to-zinc ratio greater than 3:1 interferes with absorption of zinc across the intestine, which should be considered when iron supplements are taken.

zona pellucida Clear glycoprotein layer between the primary oocyte and the granulosa cells of the Graafian follicle. Facilitates and maintains binding of a conspecific spermatozoan during the ACROSOME reaction but prevents fertilization by sperm from a different species. Contains the ligand protein ZP3 required for sperm binding and initiation of the acrosome reaction; but alters its glycoprotein composition on sperm entry as a result of release of material from the cortical granules. This blocks further sperm adhesion (the 'zona reaction') and prevents POLYSPERMY. See CUMULUS CELLS, FERTILIZATION, Fig. 76.

zone of polarizing activity (ZPA) See LIMBS.

zonula adherens (adhesion belt) Belt-like ADHERENS JUNCTION encircling the apical end of an epithelial cell and attaching it to the adjoining cell.

zoonosis Any human disease which is transmitted from non-human animals to humans.

zygomatic arch An arch of bone consisting of the fused ZYGOMATIC BONE and zygomatic process of the temporal bone, and the origin of the MASSETER MUSCLE.

zygomatic bone (cheekbone) One of the paired viscerocranial bones of the skull, anterior to the sphenoid. It unites with the zygomatic process of the temporal bone to form the ZYGOMATIC ARCH, forming the origins of the zygomaticus major and minor (muscles involved in elevation and abduction of the upper lip) and the inferolateral surface of the ORBITS.

zygote The fertilized oocyte. See FERTILIZATION.

zymogens (proenzymes) Inactive enzyme precursors that become activated by proteolytic cleavage. They include zymogens of CASPASES, several COMPLEMENT proteins and clotting factors in the blood plasma, proteins in the gut lumen involved in digestion.

APPENDICES

APPENDIX I Some common functional groups of biomolecules. The amino group of adrenaline is a secondary amino group (*s*), one of its hydrogen atoms being replaced by another group. R represents any substituent group, possibly a single hydrogen atom but more commonly a carbon-containing moiety. When more than one substituent group is present, these are indicated by upper case numerals. © 2000 From *Lehninger Principles of Biochemistry* by D.L. Nelson and M.M. Cox, published by Worth Publishers. Redrawn with permission from the publisher.

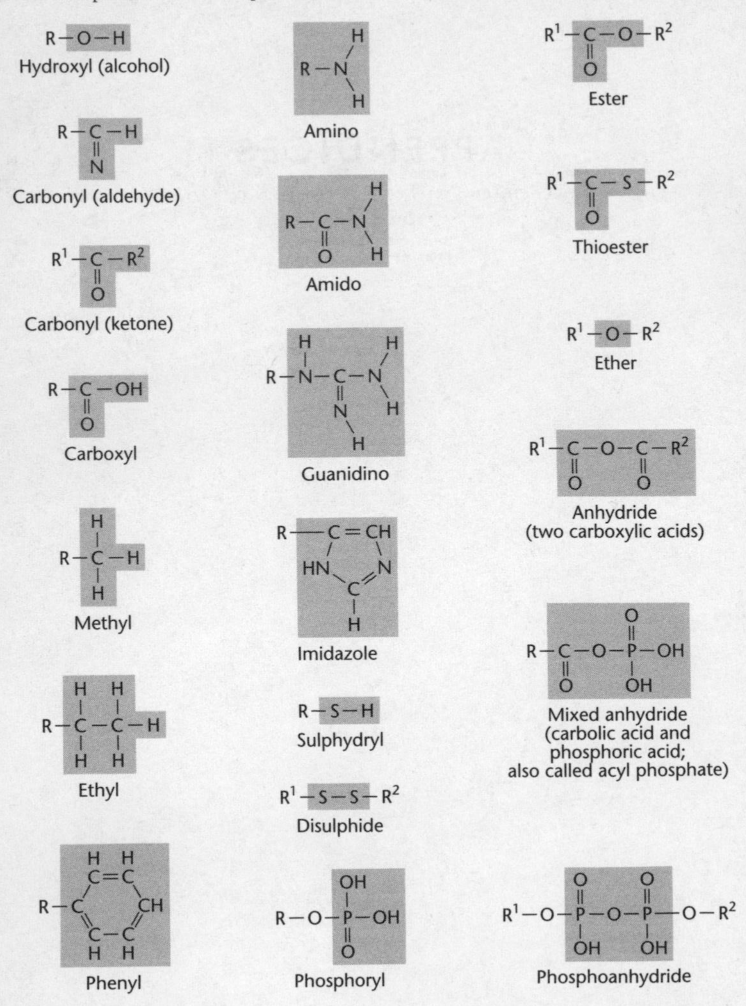

(Continued on p.901)

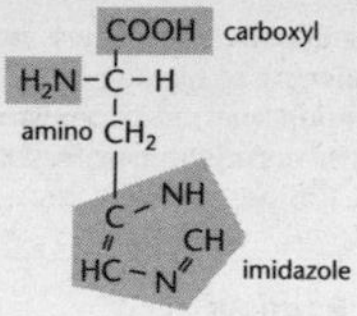

COOH carboxyl
H_2N-C-H
amino CH_2
C NH
CH
HC N
imidazole

Histidine

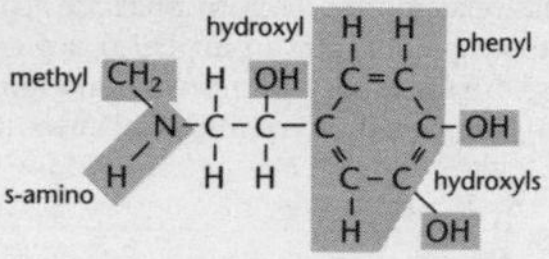

hydroxyl H H phenyl
methyl CH_3 H OH C=C
N-C-C-C C-OH
s-amino H H H C-C hydroxyls
H OH

Adrenaline

amino
imidazole NH_2
methyl phosphoanhydride
thioester amido amido H CH_3 OH OH
$CH_3-C-S-CH_2-CH_2-NH-C-CH_2-CH_2-NH-C-C-C-CH_2-O-P-O-P-O-CH_2$
O O O OH CH_3 O O
methyl
hydroxyl
HC O CH
HC CH
O OH
HO-P=O phosphoryl
OH

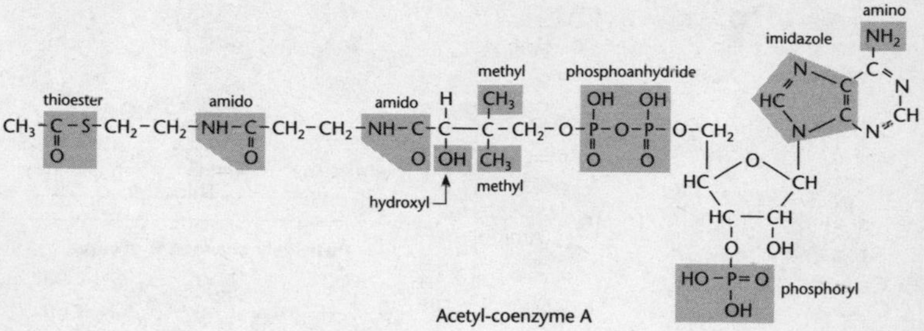

Acetyl-coenzyme A

APPENDIX IIA The 20 amino acids, excluding selenium, that are encoded by the genetic code. They are shown as having the ionization state predominating at pH 7.0, although histidine tends to have a positive charge at this pH. R-groups lie within the shaded portions. © 2000 From *Lehninger Principles of Biochemistry* by D.L. Nelson and M.M. Cox, published by Worth Publishers. Redrawn with permission from the publisher.

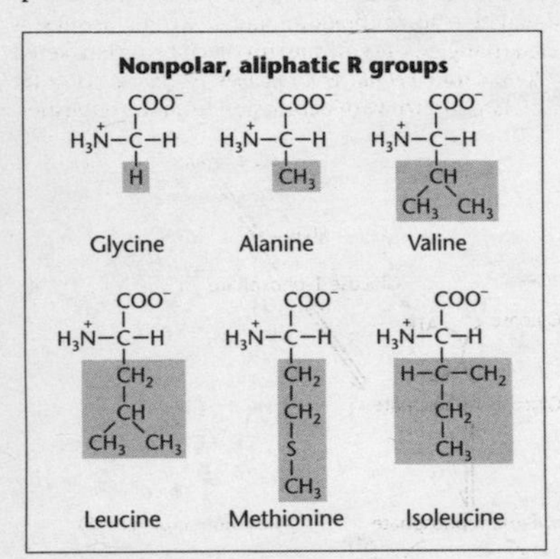

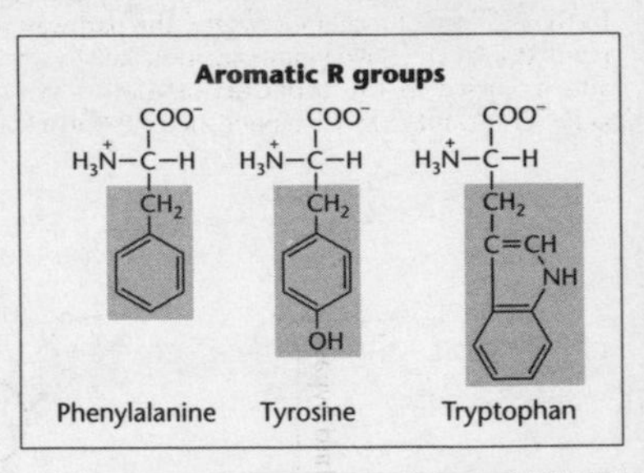

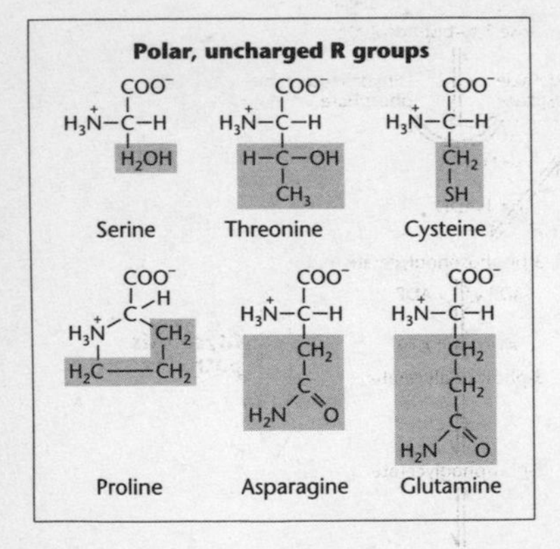

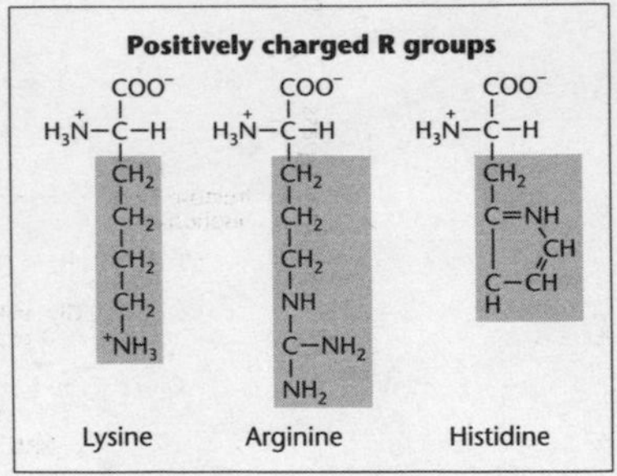

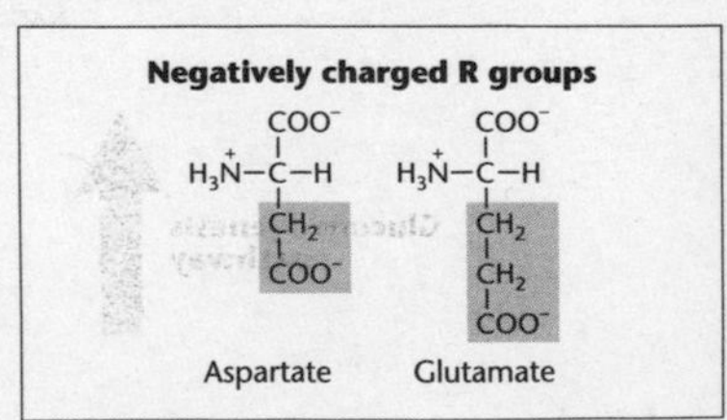

APPENDIX IIB One of the two alternative metabolic pathways leading from pyruvate to phosphoenolpyruvate. If lactate predominates as the precursor, cytosolic NADH is produced by lactate dehydrogenase in the cytosol and so does not need to be shuttled out of the mitochondrion disguised as malate. Instead, pyruvate entering the mitochondrion is converted there to oxaloacetate and then via mitochondrial PEP carboxykinase to phosphoenolpyruvate (PEP), which is then free to enter the cytosol. In the absence of available lactate, the pathway via malate (shown) predominates. Two major sites of regulation of gluconeogenesis are indicated by encircled triangles, the enzymes involved being bracketed and discussed in the GLUCONEOGENESIS entry. © 2000 From *Lehninger Principles of Biochemistry* by D.L. Nelson and M.M. Cox, published by Worth Publishers. Redrawn with permission from the publisher.

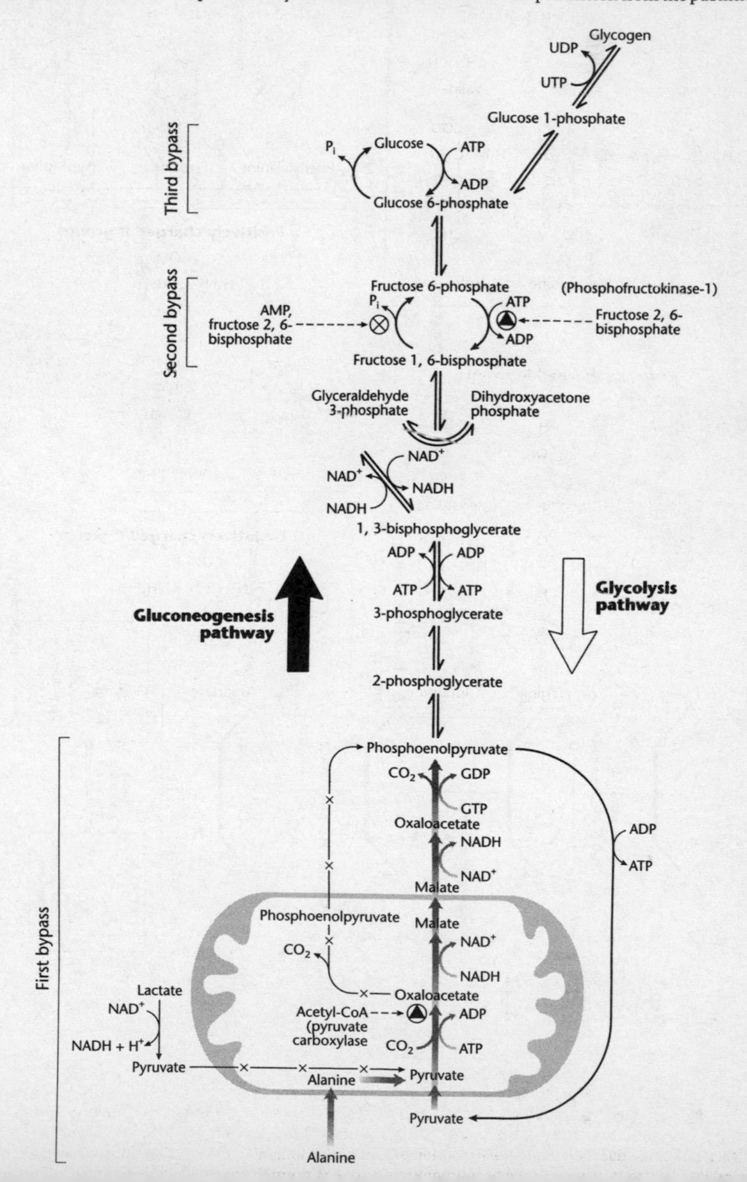

APPENDIX IIC The enzymatic steps in GLYCOLYSIS, the almost universal metabolic pathway leading from glucose to pyruvate. See also GLUCONEOGENESIS, and Appendix IIb. © 1990 John Wiley & Sons. *Principles of Anatomy and Physiology*, G.J. Tortora & N.P. Anagnostakos. Reprinted with permission of the publisher.

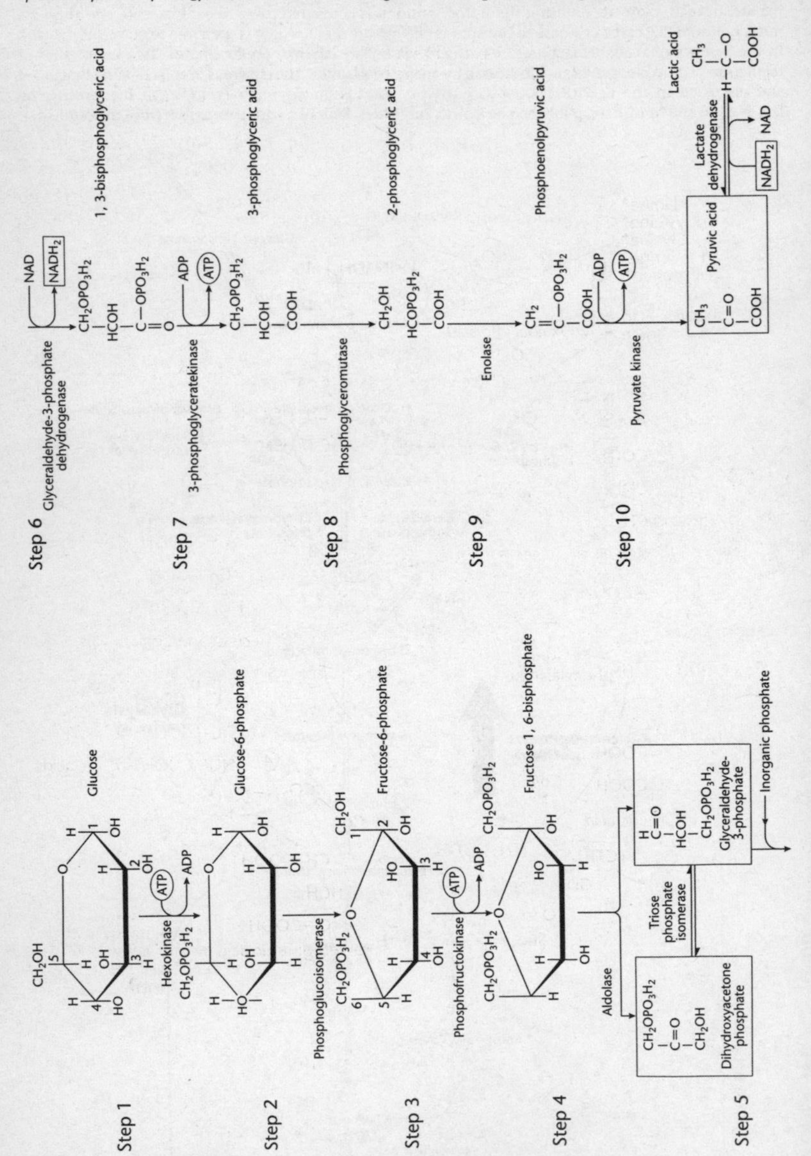

APPENDIX IID The KREBS CYCLE and its relationships: carbohydrate, fatty acyl and amino acyl inputs. Glucogenic amino acids are shown entering at their appropriate places, ketogenic amino acids being indicated by *.

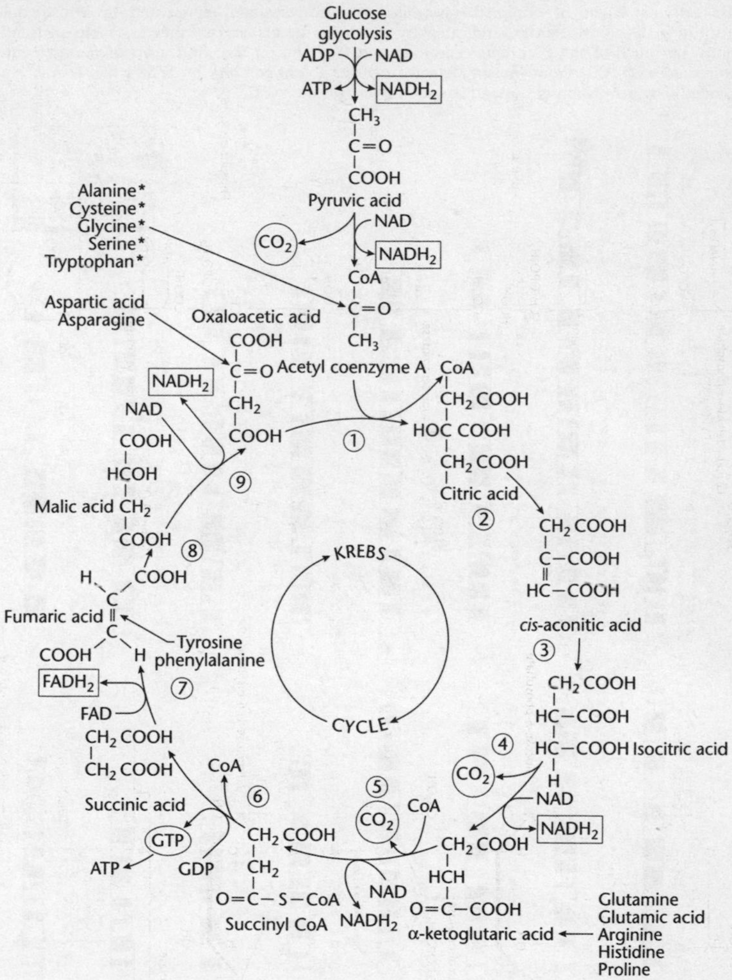

Glucose
glycolysis
ADP → ATP
NAD → $NADH_2$
CH_3
C=O
COOH
Pyruvic acid
CO_2
NAD → $NADH_2$
CoA
C=O
CH_3
Acetyl coenzyme A

Alanine*
Cysteine*
Glycine*
Serine*
Tryptophan*

Aspartic acid
Asparagine

Oxaloacetic acid
COOH
C=O
CH_2
COOH

① CoA
CH_2 COOH
HOC COOH
CH_2 COOH
Citric acid

②
CH_2 COOH
C−COOH
HC−COOH
cis-aconitic acid

③
CH_2 COOH
HC−COOH
HC−COOH Isocitric acid
H

④
CO_2
NAD → $NADH_2$
CH_2 COOH
HCH
O=C−COOH
α-ketoglutaric acid ← Glutamine, Glutamic acid, Arginine, Histidine, Proline

⑤
CO_2
CoA
NAD → $NADH_2$
CH_2 COOH
CH_2
O=C−S−CoA
Succinyl CoA

⑥
CoA
GDP → GTP → ATP
CH_2 COOH
CH_2 COOH
Succinic acid

⑦
FAD → $FADH_2$
H, COOH
C
Fumaric acid ‖ ← Tyrosine phenylalanine
C
COOH, H

⑧
COOH
HCOH
Malic acid CH_2
COOH

⑨
NAD → $NADH_2$

KREBS CYCLE

APPENDIX III A human karyogram (see KARYOTYPE) in which the CHROMOSOMES have been G-banded and the best patterns used in the compilation. The chromosome bands are dark horizontal bars. Chromosomes are numbered in order or size, except that 21 is smaller than 22. Arrays of repeated ribosomal DNA genes on the short arms (p) of the acrocentric chromosomes often appear as thin stalks carrying knobs of chromatin (*satellites*). Centromeres are represented by constrictions. Constitutive HETEROCHROMATIN, indicated by hatching, occurs at centromeres, some pericentric regions, on much of the Y chromosome long arm (Yq) and on the short arms of the acrocentric chromosomes.

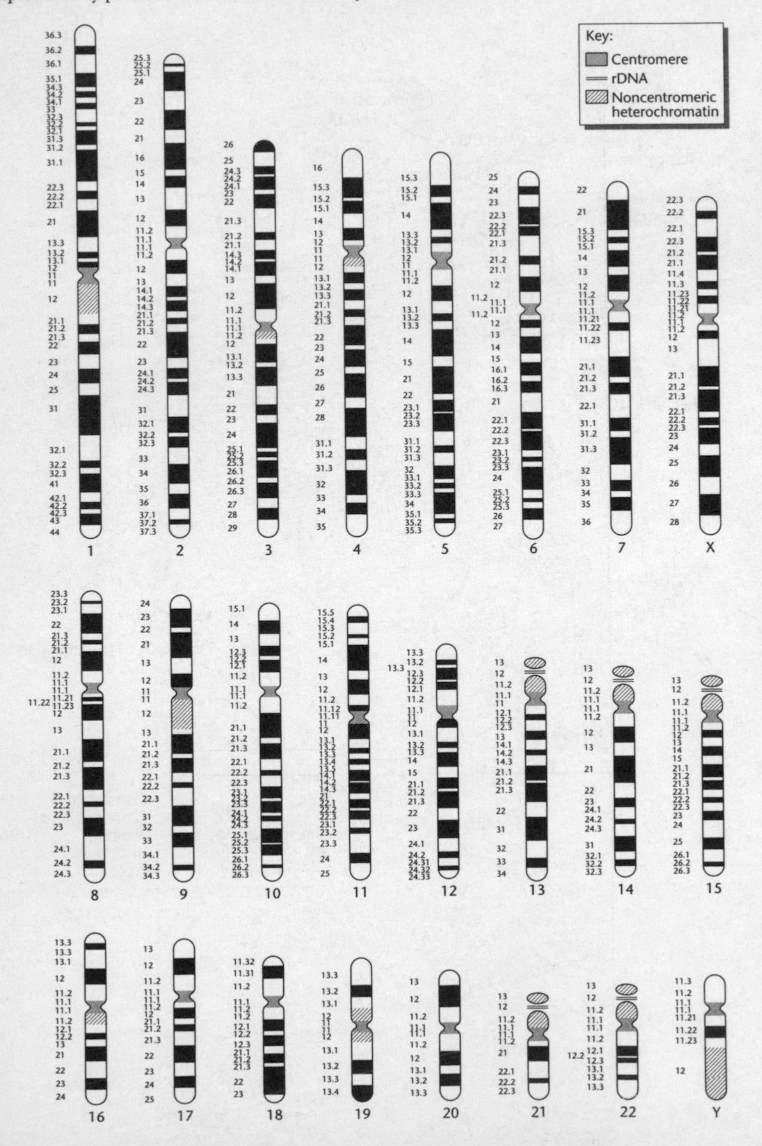

APPENDIX IV An approximation of the relative times (ka = thousand years ago) during which modern human mitochondrial DNA types are believed to have originated and spread globally (see MIGRATION, mtDNA). With permission from The Royal Society, © 2004, *Phil Trans R Soc Land* (B) 359; 258–260, and with gratitude to the author, Peter Foster.

APPENDIX IV (i) (a) 200–100 ka; (b) 80–60 ka; (c) 60–30 ka.

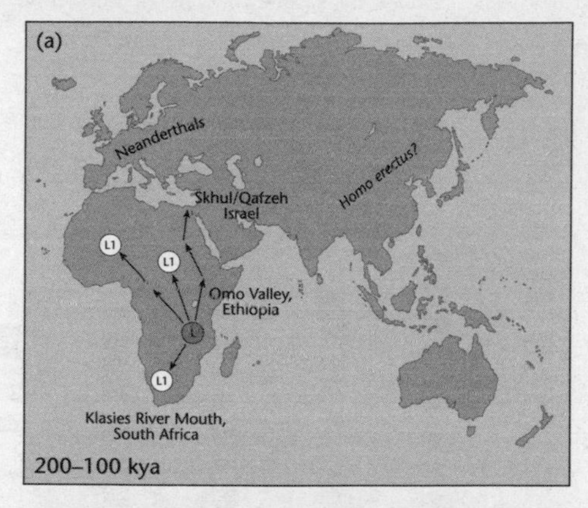

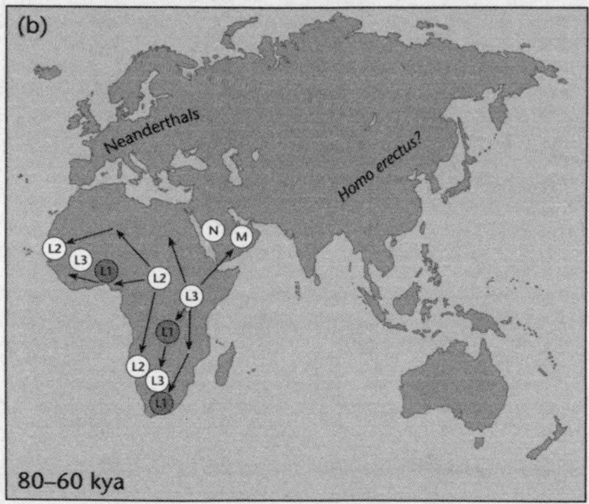

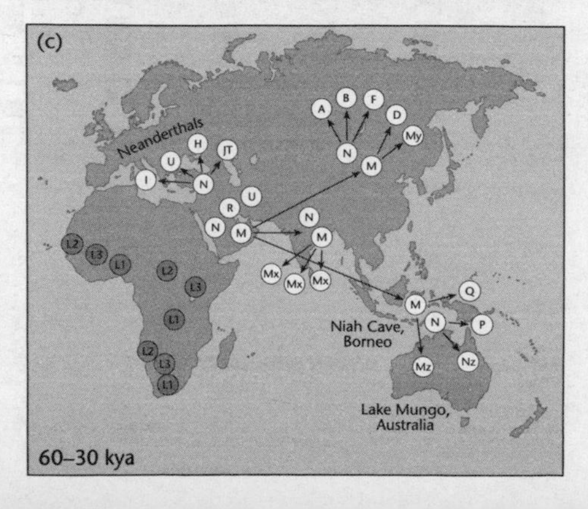

APPENDIX IV (II) (d) 30–20 ka; (e) 20–15 ka; (f) 15–2 ka.

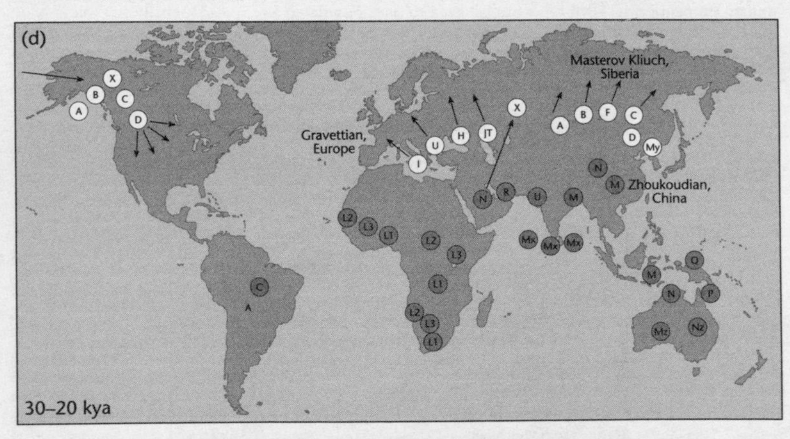

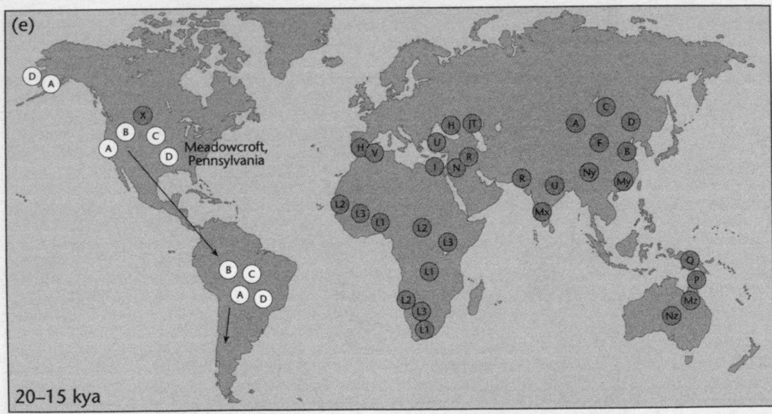

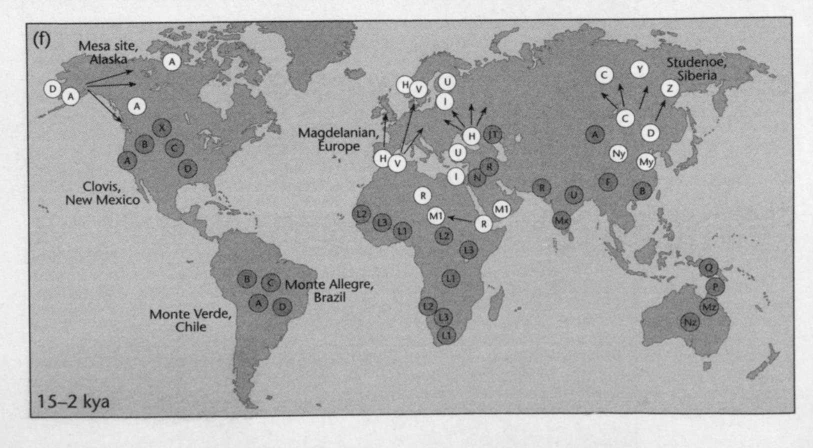

APPENDIX IV (III) (g) Less than 2 ka.

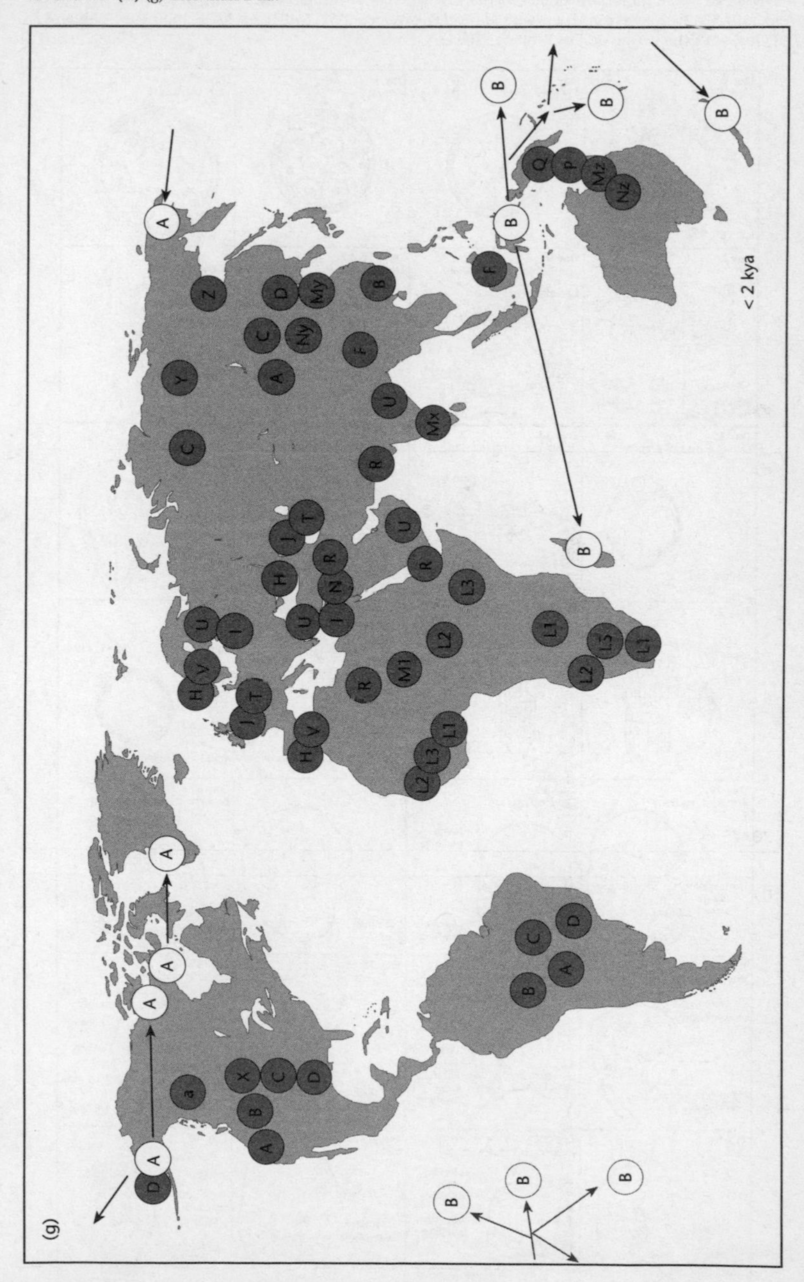

APPENDIX VA Some important events in the first 48 days of human embryonic DEVELOPMENT. See also Appendix VB. Redrawn from *Langman's Medical Embryology*, T.W. Sadler, © 2004 Lippincott, Williams & Wilkins, with the permission of Wolters Kluwer.

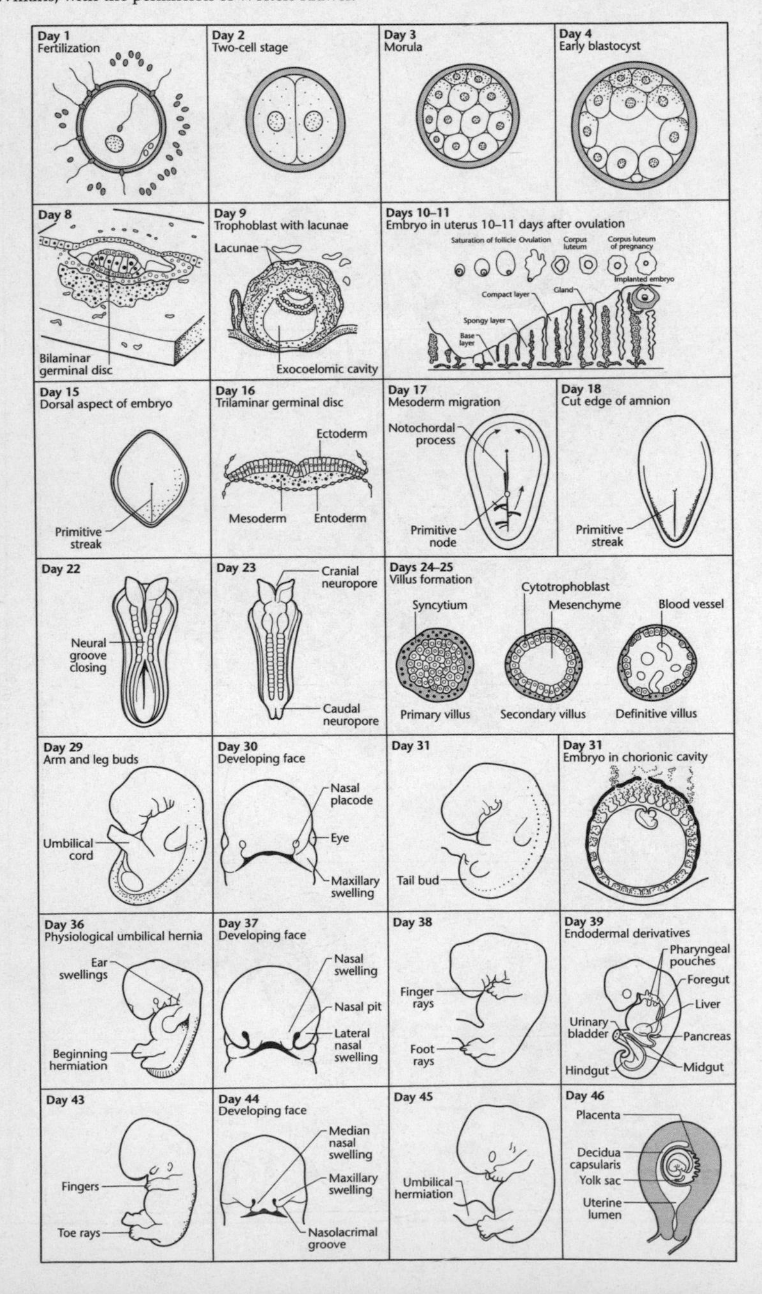

APPENDIX VB A continuation of Appendix Va.

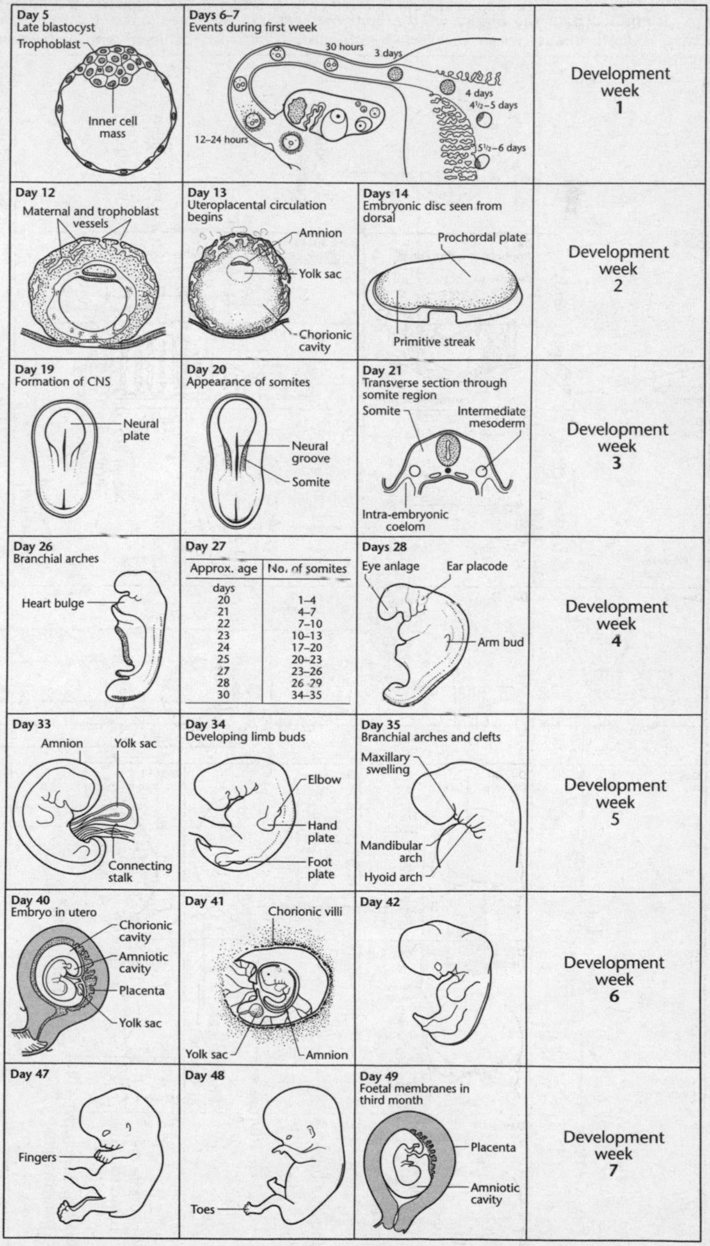

Day 5 Late blastocyst Trophoblast Inner cell mass	**Days 6–7** Events during first week 30 hours 3 days 4 days 4½–5 days 12–24 hours 5½–6 days		Development week 1
Day 12 Maternal and trophoblast vessels	**Day 13** Uteroplacental circulation begins Amnion Yolk sac Chorionic cavity	**Days 14** Embryonic disc seen from dorsal Prochordal plate Primitive streak	Development week 2
Day 19 Formation of CNS Neural plate	**Day 20** Appearance of somites Neural groove Somite	**Day 21** Transverse section through somite region Somite Intermediate mesoderm Intra-embryonic coelom	Development week 3
Day 26 Branchial arches Heart bulge	**Day 27** (see table below)	**Days 28** Eye anlage Ear placode Arm bud	Development week 4
Day 33 Amnion Yolk sac Connecting stalk	**Day 34** Developing limb buds Elbow Hand plate Foot plate	**Day 35** Branchial arches and clefts Maxillary swelling Mandibular arch Hyoid arch	Development week 5
Day 40 Embryo in utero Chorionic cavity Amniotic cavity Placenta Yolk sac	**Day 41** Chorionic villi Yolk sac Amnion	**Day 42**	Development week 6
Day 47 Fingers	**Day 48** Toes	**Day 49** Foetal membranes in third month Placenta Amniotic cavity	Development week 7

Day 27

Approx. age (days)	No. of somites
20	1–4
21	4–7
22	7–10
23	10–13
24	17–20
25	20–23
27	23–26
28	26–29
30	34–35

APPENDIX VI An indication of the cellular lineages of the human body from the three primary GERM LAYERS (grey boxes) and their subsequent subdivisions (open boxes). Mesoderm is indicated by the broken outline. Asterisks denote the origins of the epithelial parts of an organ only. © 1997 Sinauer Associates. Redrawn from *Developmental Biology* (5th edition) by Scott F. Gilbert, with permission of the publisher.

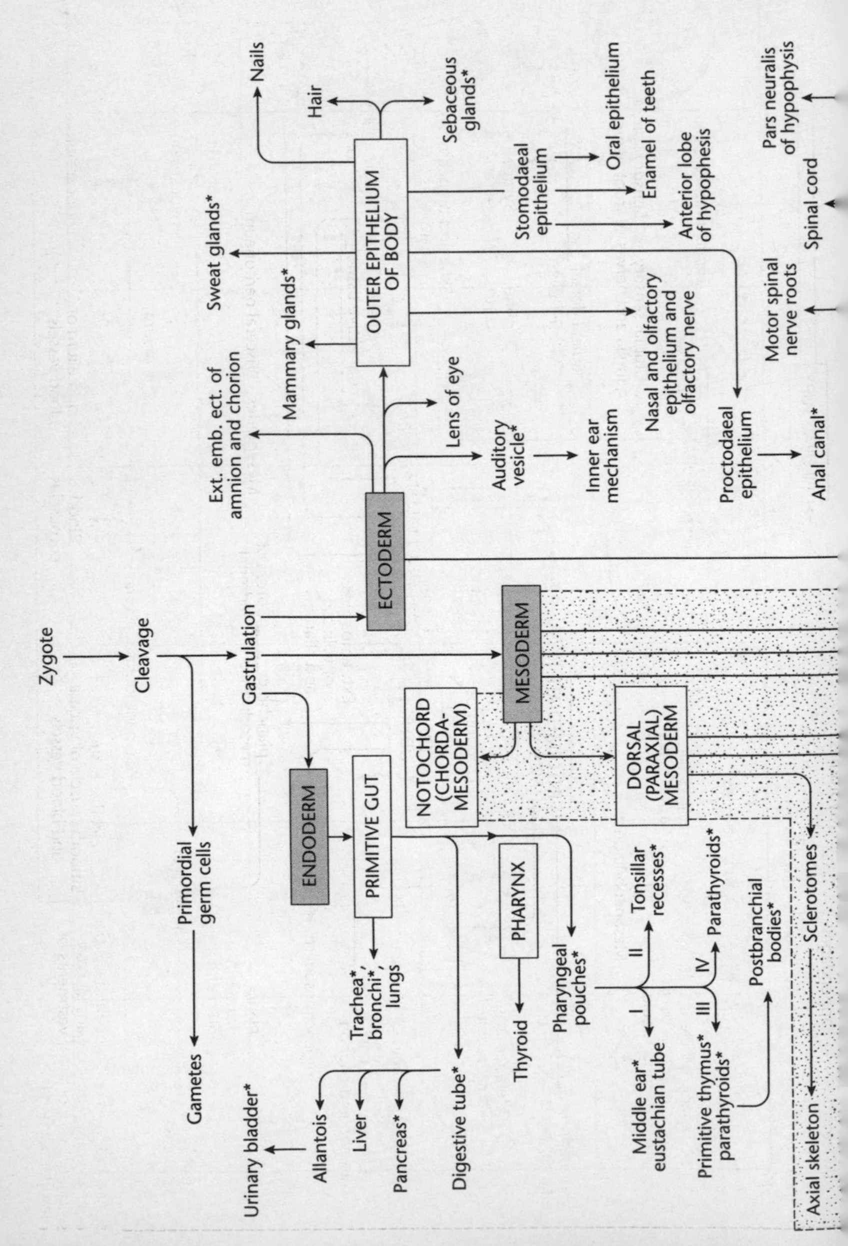

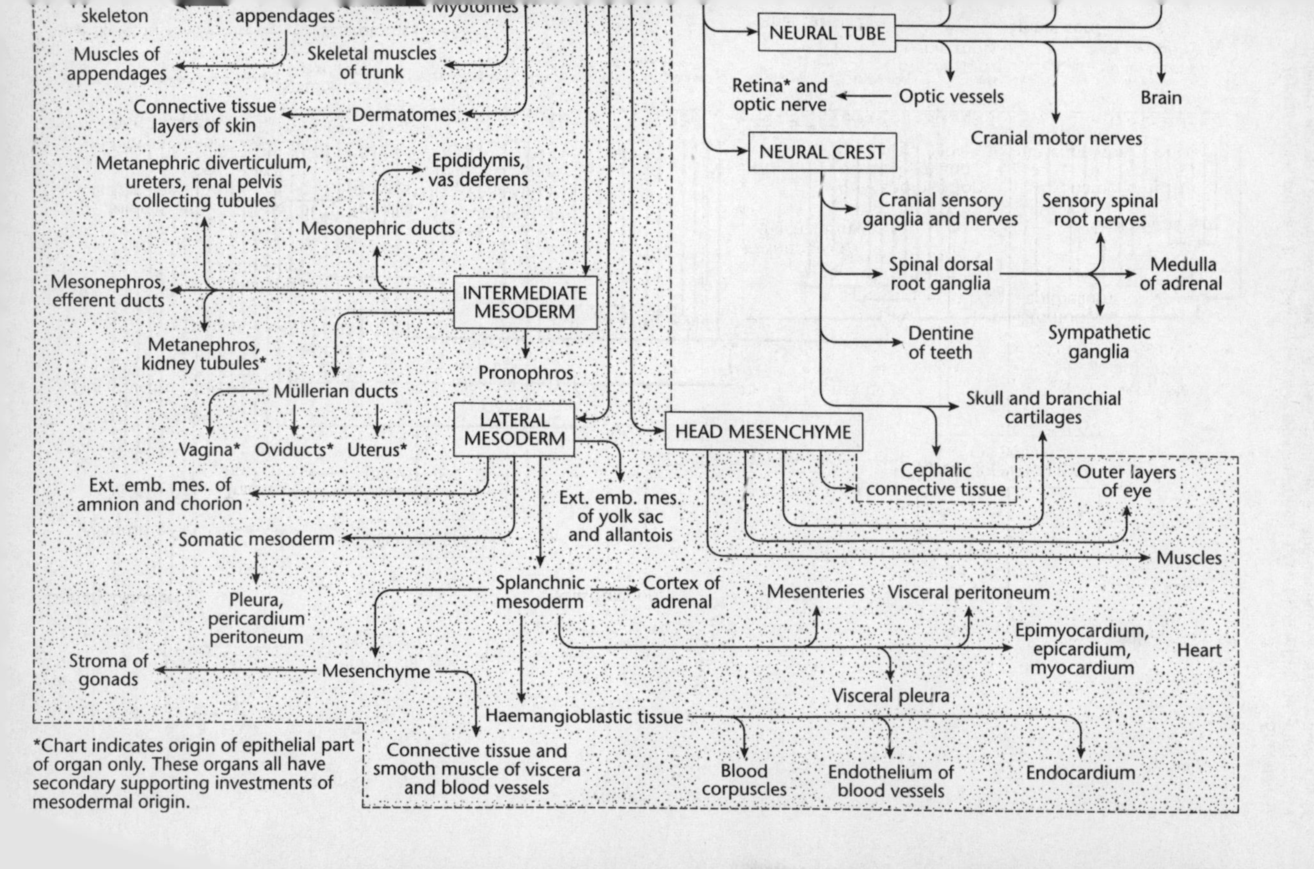

skeleton
appendages
Myotomes
Muscles of appendages
Skeletal muscles of trunk
Connective tissue layers of skin
Dermatomes
Metanephric diverticulum, ureters, renal pelvis collecting tubules
Epididymis, vas deferens
Mesonephric ducts
Mesonephros, efferent ducts
INTERMEDIATE MESODERM
Metanephros, kidney tubules*
Pronophros
Müllerian ducts
Vagina*
Oviducts*
Uterus*
LATERAL MESODERM
HEAD MESENCHYME
Ext. emb. mes. of amnion and chorion
Ext. emb. mes. of yolk sac and allantois
Somatic mesoderm
Pleura, pericardium peritoneum
Splanchnic mesoderm
Cortex of adrenal
Stroma of gonads
Mesenchyme
Haemangioblastic tissue
Connective tissue and smooth muscle of viscera and blood vessels
Blood corpuscles
Endothelium of blood vessels
Endocardium
Mesenteries
Visceral peritoneum
Epimyocardium, epicardium, myocardium
Heart
Visceral pleura
NEURAL TUBE
Retina* and optic nerve
Optic vessels
Brain
Cranial motor nerves
NEURAL CREST
Cranial sensory ganglia and nerves
Sensory spinal root nerves
Spinal dorsal root ganglia
Medulla of adrenal
Dentine of teeth
Sympathetic ganglia
Skull and branchial cartilages
Cephalic connective tissue
Outer layers of eye
Muscles
*Chart indicates origin of epithelial part of organ only. These organs all have secondary supporting investments of mesodermal origin.

APPENDIX VIIA Schematic worldwide representation of human mitochondrial DNA (mtDNA) phylogeny, the African haplogroups being in shaded boxes. See MIGRATION entry for a general explanation, and the HOMO SAPIENS entry for the importance of the asterisk. © 2004 From *Human Molecular Genetics*, M.A. Jobling, M. Hurles and C. Tyler Smith. Reproduced by permission of Garland Science/Taylor & Francis LLC.

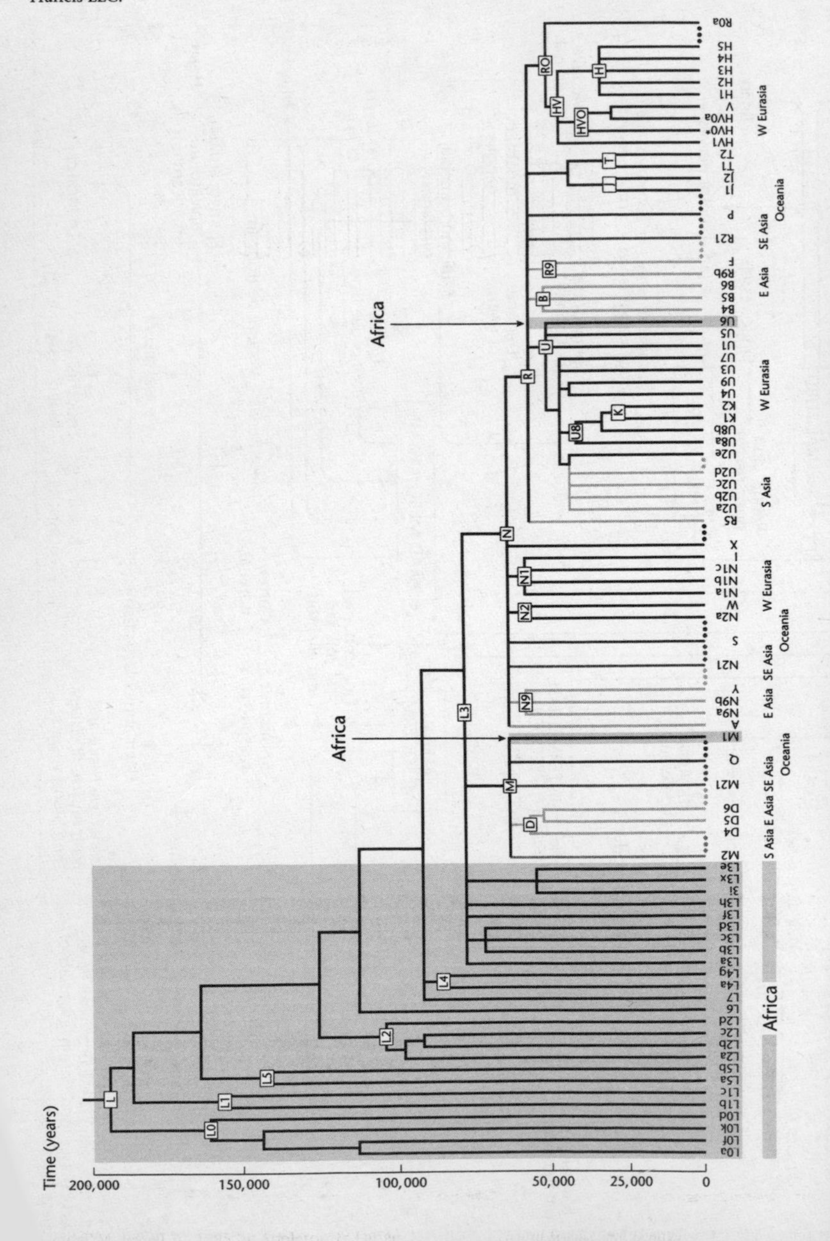

APPENDIX VIIB The human Y-chromosomal phylogeny, as interpreted in 2000. The shaded box indicates African lineages. See the HOMO SAPIENS entry for explanation. © 2004 From *Human Molecular Genetics*, M.A. Jobling, M. Hurles and C. Tyler Smith. Reproduced by permission of Garland Science/Taylor & Francis LLC.

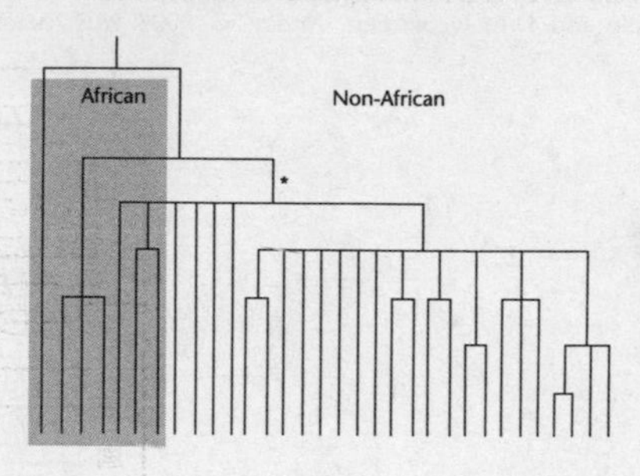

APPENDIX VIII Anatomical comparisons of the following: *Homo sapiens* (a, c); *H. erectus* (e), based largely on specimen KNM-WT 15000; chimpanzee, anterior and posterior views (b, d), the labelled muscles connecting the head and neck to the pectoral girdle being reduced or absent in humans; and *Australopithecus afarensis* (f), based largely on specimen AL-288. The numerals refer to the entry on GAIT. From B.M. Bramble and D.E. Lieberman. *Nature* © 2004 with permission from Macmillan Publishers Ltd.

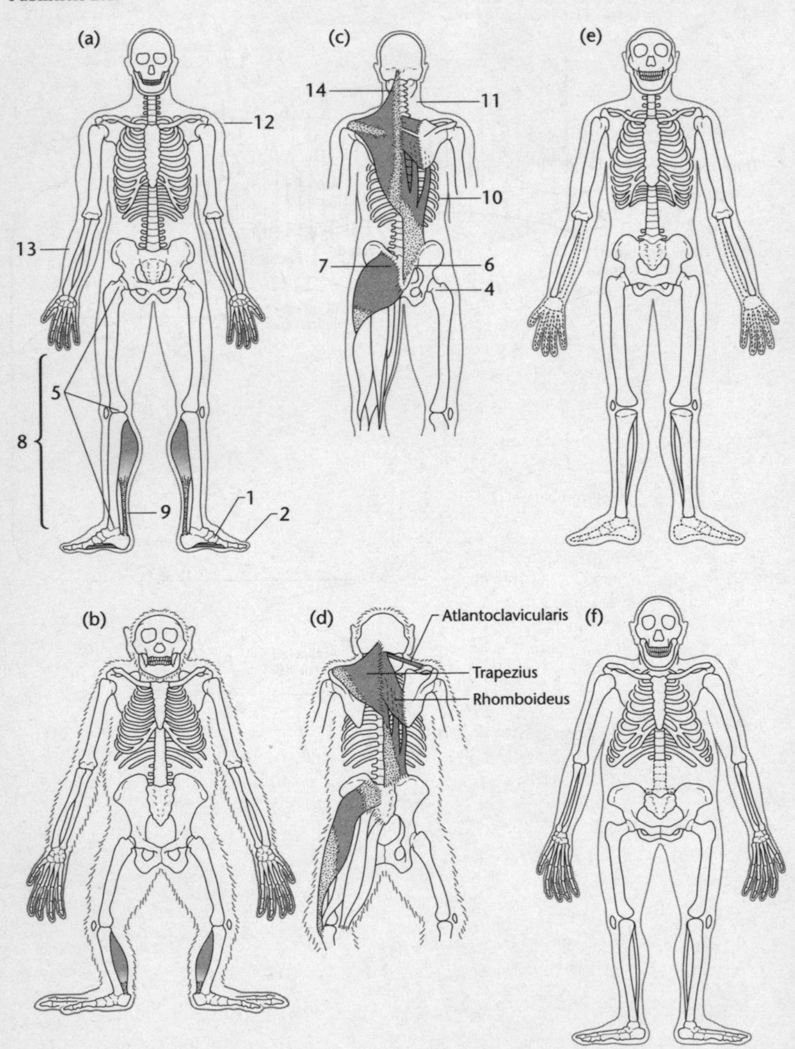

APPENDIX IX The BLOOD CLOTTING cascades. Both involve the activation of inactive clotting factors (hatched outlines) into active ones (entire outlines). Stage 1 can occur in both the extrinsic and intrinsic pathways. Redrawn from *Anatomy & Physiology*, R.R. Sealey, T.D. Stephens and P. Tate. © 2006 McGraw-Hill, with the permission of The McGraw-Hill Companies.

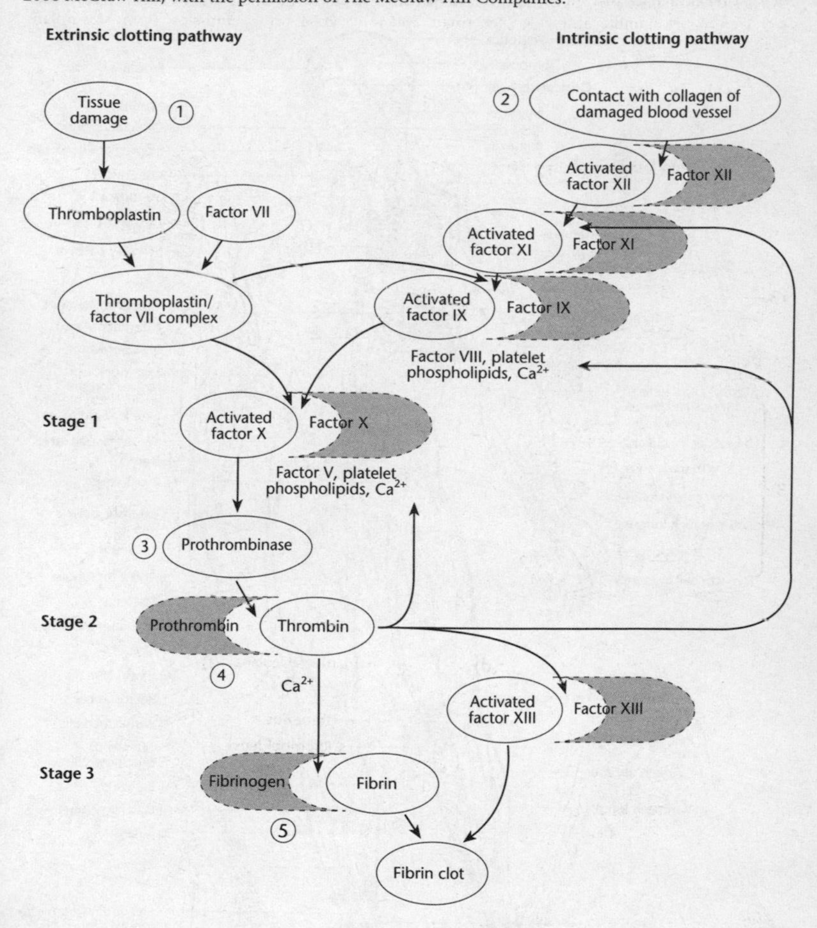

APPENDIX XA The principal superficial skeletal muscles, viewed anteriorly. Tendons are shown stippled, those of the external oblique and serratus anterior muscles overlying the rectus abdominis muscles. From *Principles of Anatomy and Physiology*, G.J. Tortora & S.R. Grabowski, © 2000. Reproduced with permission of John Wiley & Sons, Inc.

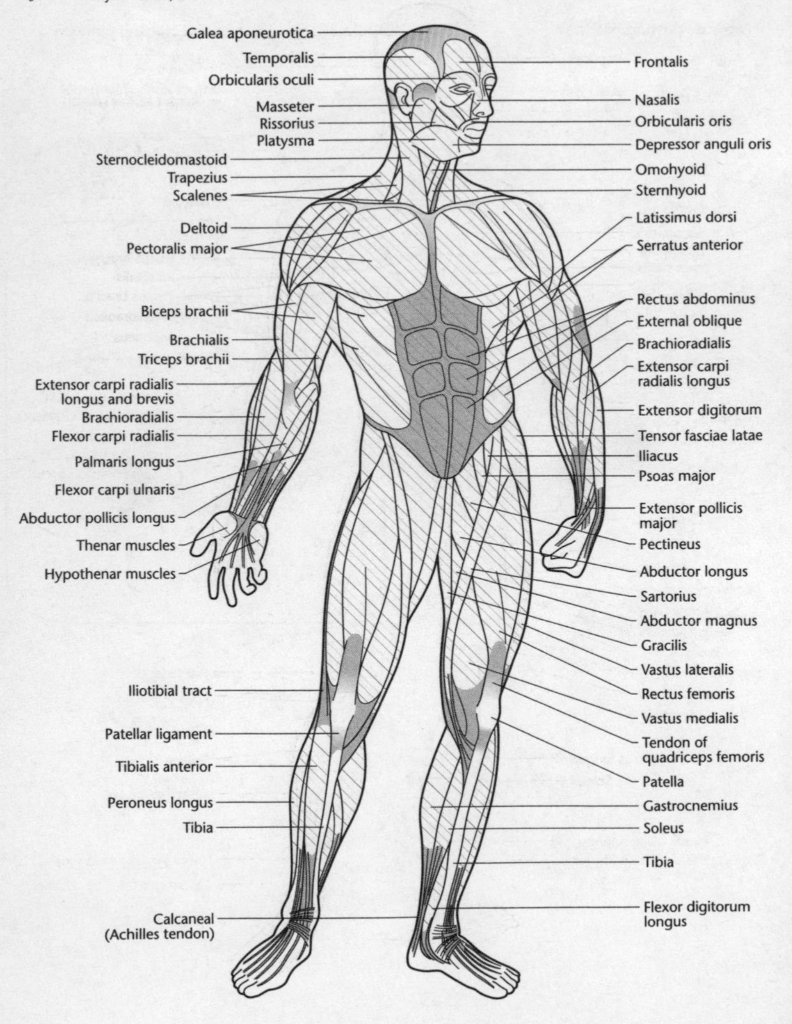

APPENDIX XB The principal superficial skeletal muscles, viewed posteriorly. Tendons are shown stippled. From *Principles of Anatomy and Physiology*, G.J. Tortora & S.R. Grabowski, © 2000. Reproduced with permission of John Wiley & Sons, Inc.

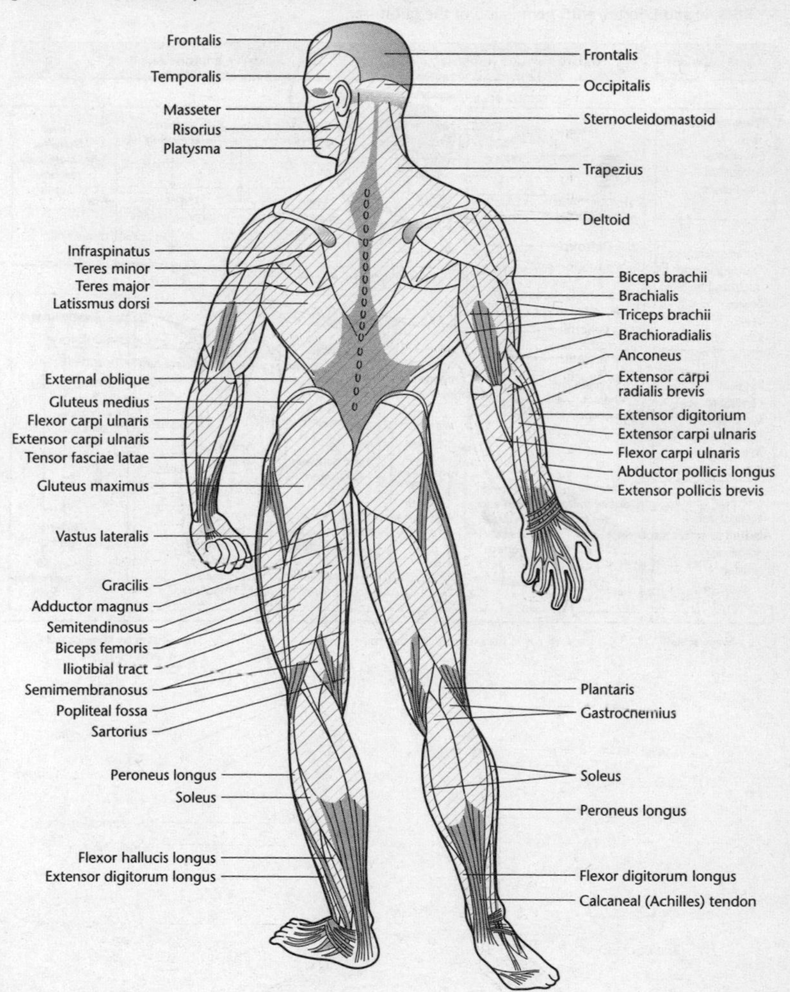

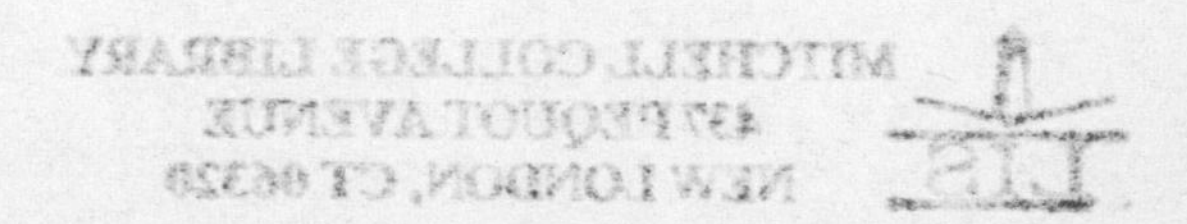

APPENDIX XI An overview of the human body's defences against INFECTION. Those termed 'local defences' are often subsumed within the category of INNATE IMMUNITY. See also ADAPTIVE IMMUNITY. © Blackwell Publishing. Redrawn from *Infection (Microbiology and Management)*, 3rd edition, by B. Bannister, S. Gillespie and J. Jones, with permission of the publisher.

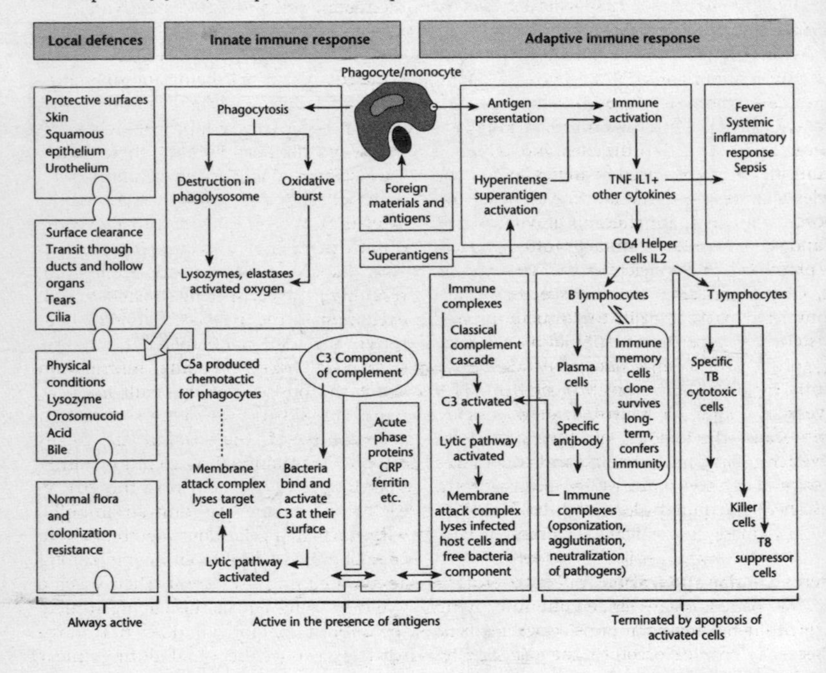

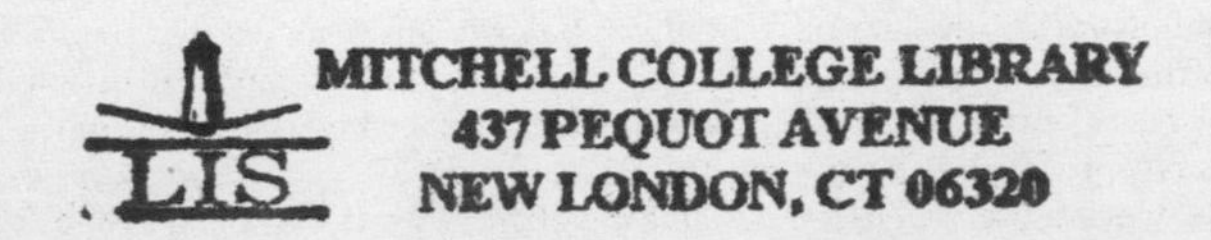